1638528
KT-539-619

Nutrition

Eighth Edition

Concepts and Controversies

Frances Sienkiewicz Sizer
Eleanor Noss Whitney

Short Loan

UNIVERSITY OF GREENWICH LIBRARY
613.
2
SIZ

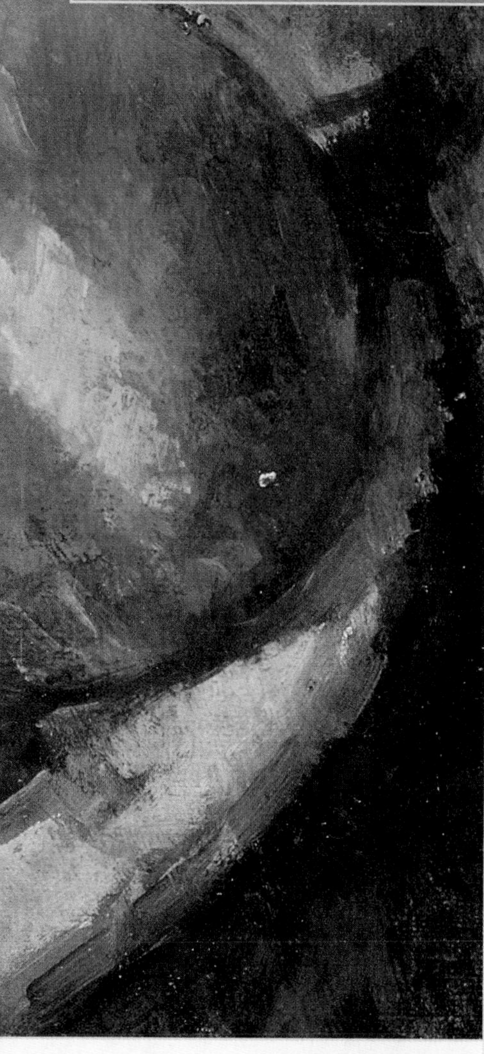

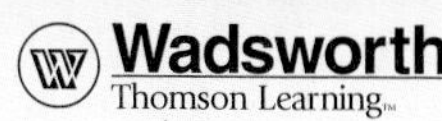
Wadsworth
Thomson Learning

Australia • Canada • Denmark • Japan • Mexico • New Zealand
Philippines • Puerto Rico • Singapore • Spain • United Kingdom
United States

Publisher: Peter Marshall
Development Editor: Laura Graham
Editorial Assistant: Keynia Johnson
Marketing Manager: Becky Tollerson
Project Editor: Sandra Craig
Print Buyer: Barbara Britton
Permissions Editor: Susan Walters
Production Service: The Book Company
Text and Cover Designer: Baugher Design Inc.
Photo Researcher: Myrna Engler
Copy Editor: Patricia Lewis
Illustrations: Impact Publications, McMahon Medical Art, EmSpace Artwork, Dorothy Reinhardt
Photographers and Food Stylists: Michael Shay, Steve Cherry, Lucy Radys, Carol Ladd
Cover Image: SuperStock
Compositor: Parkwood Composition Services
Printer: Von Hoffmann Press

COPYRIGHT © 2000 by Wadsworth, a division of Thomson Learning. Thomson Learning is a trademark used herein under license.

ALL RIGHTS RESERVED. No part of this work may be reproduced, transcribed, or used in any form or by any means—graphic, electronic, or mechanical, including photocopying, recording, taping, Web distribution, or information storage and retrieval systems—without the written permission of the publisher.

Printed in the United States of America
1 2 3 4 5 6 7 03 02 01 00 99

For permission to use material from this text, contact us by
Web: www.thomsonrights.com
Fax: 1-800-730-2215
Phone: 1-800-730-2214

COPYRIGHT © 2000 Thomson Learning. ALL RIGHTS RIGHTS RESERVED. Thomson Learning Testing Tools is a trademark of Thomson Learning.

Library of Congress Cataloging-in-Publication Data

Sizer, Frances Sienkiewicz.
Nutrition : concepts and controversies / Frances Sienkiewicz Sizer, Eleanor Noss Whitney. — 8th ed.
p. cm.
Includes bibliographical references and index.
ISBN 0–534–56466–6
1. Nutrition. 2. Food. I. Whitney, Eleanor Noss. II. Title.
QP141.S5365 2000
613.2—dc21 99–36374

Annotated Instructor's Edition ISBN 0-534-56467-4

This book is printed on acid-free recycled paper.

For more information, contact

Wadsworth/Thomson Learning
10 Davis Drive
Belmont, CA 94002-3098
USA
www.wadsworth.com

International Headquarters
Thomson Learning
290 Harbor Drive, 2nd Floor
Stamford, CT 06902-7477
USA

UK/Europe/Middle East
Thomson Learning
Berkshire House
168-173 High Holborn
London WC1V 7AA
United Kingdom

Asia
Thomson Learning
60 Albert Street #15-01
Albert Complex
Singapore 189969

Canada
Nelson/Thomson Learning
1120 Birchmount Road
Scarborough, Ontario M1K 5G4
Canada

About the Authors

FRANCES SIENKIEWICZ SIZER, M.S., R.D., F.A.D.A., attended Florida State University where, in 1980, she received her B.S., and in 1982, her M.S. in nutrition. She is certified as a charter Fellow of the American Dietetic Association. She is a founding member and vice president of Nutrition and Health Associates, an information and resource center in Tallahassee, Florida, that maintains an ongoing bibliographic database tracking research in more than 1,000 topic areas of nutrition. Her textbooks include *Life Choices: Health Concepts and Strategies; Making Life Choices; The Fitness Triad: Motivation, Training, and Nutrition;* and others. She is a primary author of *Nutrition Interactive,* an instructional college-level nutrition CD-ROM. In addition to writing, she lectures at universities and at national and regional conferences, and co-coordinates a local hunger relief effort in her community.

ELEANOR NOSS WHITNEY, Ph.D., RECEIVED HER B.A. in Biology from Radcliffe College in 1960 and her Ph.D. in Biology from Washington University, St. Louis, in 1970. Formerly on the faculty at the Florida State University, and a dietitian registered with the American Dietetic Association, she now devotes full time to research, writing, and consulting in nutrition, health, and environmental issues. Her earlier publications include articles in *Science, Genetics,* and other journals. Her textbooks include *Understanding Nutrition, Understanding Normal and Clinical Nutrition, Nutrition and Diet Therapy,* and *Essential Life Choices* for college students and *Making Life Choices* for high-school students. Her most intense interests presently include energy conservation, solar energy uses, alternatively fueled vehicles, and ecosystem restoration.

To my husband Philip Webb and our four "little fish," Caroline, Amanda, David and Spencer, and the smallest fry, David Warren.

Fran

To all of my grandchildren and stepgrandchildren: Max, Zoey, Emily, Rebecca, Sarah, Will, TootToot, and Jacob.

Ellie

Contents in Brief

Table of Contents

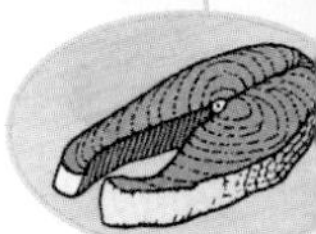

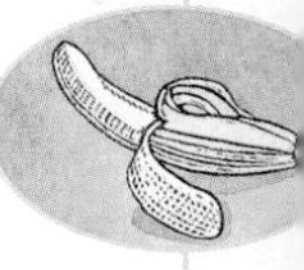

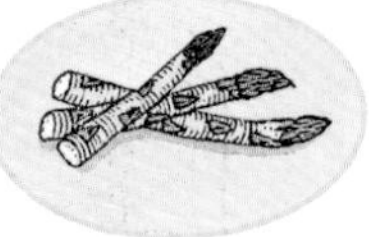

Chapter 4
THE CARBOHYDRATES: SUGAR, STARCH, GLYCOGEN, AND FIBER 93

Chapter 5
THE LIPIDS: FATS, OILS, PHOSPHOLIPIDS, AND STEROLS 133

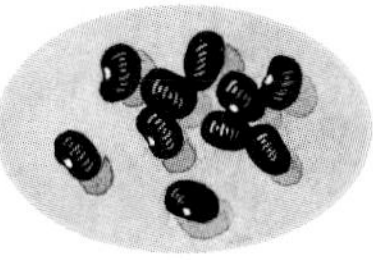

Preface

FOR MORE than 20 years and in seven previous editions, *Nutrition: Concepts and Controversies* has been serving the needs of students and professors in classrooms across North America. In this, our millennium edition, we apply what we have learned from our past readers, changing the book to meet changing times. As the 21st century is born, the World Health Organization predicts a healthier future for people living in developed nations, with longer life spans, fewer deaths in infancy and childhood, and declining disabilities due to heart and artery disease in old age. Such improvements spring from advances in the science of nutrition, from progress in medical treatment of diseases, and from improved education concerning the effects of daily life choices on people's well being. This book provides the key understandings in the science of nutrition and guides its readers toward making wise choices amidst almost unlimited options.

In this edition, we have explored many of the new frontiers in nutrition, confronting new mysteries while acknowledging their grounding in science. We have heightened our sense of personal connection with instructors and learners alike, writing to them in the informal, clear style that has become our trademark. For both verbal and visual learners, our logical presentation and our clear, colorful figures keep interest high and understanding at a peak. New photos also adorn many of our pages, adding pleasure to reading.

In this edition, you will encounter two new features. The first of these tickles the reader's interest by posing some "Frequently Asked Questions" on the opening page of each chapter. The answers are found sprinkled throughout the chapter text, in sections marked with the FAQ symbol: The second new feature, called *Fitness for Life,* draws connections among physical activity, nutrition, and health and is intended to motivate readers by addressing fitness topics as they arise in each chapter. The *Fitness for Life* section of Chapter 1 identifies the health benefits that one can expect from regular physical activity, echoing the chapter's emphasis on nutrition and disease risks. That of Chapter 2 provides guidelines for physical activity; those for Chapters 4 through 8, the nutrient chapters, emphasize the roles of the nutrients in supporting physical performance. The remaining sections spell out advantages of physical activity for weight-conscious readers, athletes, healthy adults, pregnant women, and older people.

The practical interactive activities, called *Do It!*, have been retained in this edition to invite students to apply the chapter contents in their everyday encounters with nutrition. The popular *Self Check* features provide review questions at the end of each chapter. The answers to the *Self Check* questions are in Appendix G to provide immediate feedback to the learner.

By popular demand, we have retained our *Snapshots* of vitamins and minerals, but have given them an updated look. These capsules of information depict food sources and reinforce concepts concerning the Food Guide Pyramid and the Daily Values found on food labels. In this edition, the food label format appears often in figures to familiarize readers with the process of comparing the nutrients in foods that bear labels.

We hope that you will enjoy the eighth edition of our text. Chapter 1 begins with a personal challenge to nutrition students. It asks the question so many people ask of nutrition educators: "What kinds of nutritious foods can help to prevent diseases?" We answer with a lesson in diseases on a continuum, from those almost totally preventable by nutrition to those caused by inherited genes. We then introduce the nutrients and explore their roles in the body. Finally, a discussion of the importance of scientific research lends perspective on the context in which study results may be rightly viewed. Chapter 2 brings together the concepts of diet planning through nutrient allowances, such as the new *Dietary Reference Intakes*, and food grouping systems featuring the *Daily Food Guide* with its pyramid of food choices. Chapter 3 presents a thorough, but brief, introduction to the workings of the human body with major emphasis on the digestive system. Chapters 4 through 6 are devoted to the energy-yielding nutrients—carbohydrates, lipids, and proteins. Chapters 7 and 8 present the vitamins, minerals, and water, with special emphasis on the emerging importance of the antioxidant nutrients. Chapter 9 relates energy balance to body composition, obesity, and underweight and presents weight maintenance as a lifelong effort. Chapter 10 presents details about relationships between fitness, physical activity, and nutrition to follow up on concepts introduced in the *Fitness for Life* features of other chapters. Chapter 11 applies the essence of the first ten chapters to two broad and rapidly changing areas within nutrition: immunity and disease prevention. Chapters 12 and 13 point out the importance of nutrition throughout the life span, from gestation through old age. Chapter 14 considers the problems and advantages of food technology, with emphasis on food safety. Chapter 15 touches on the vast problems of the global food supply—world hunger, pollution, overpopulation—and shows how everyday food choices link each person with the meaningful whole.

The *Controversies* of this book's title invite you to explore beyond the safe boundaries of established nutrition knowledge. These optional readings, which appear at the end of each chapter and are printed with colored borders, delve into current scientific topics and emerging controversies. Some are new to this edition and the others have been updated. Controversy 7 sets up a lively competition between food and supplements as vitamin sources, exploring the research to date on the antioxidant vitamins. Controversy 9 tackles some pressing questions surrounding the safety and effectiveness of weight-loss diets, diet profiteers, and attitudes toward overweight people in this country. Controversy 10 presents current thinking about eating disorders. Of special current interest is Controversy 11, which presents the struggle between scientific exploration of the phytochemicals in foods and consumers' need for guidance concerning their consumption of phytochemicals. Controversy 14 evaluates new food technologies and invites the reader to look forward to and evaluate future innovations. Controversy 15 explores ways in which agriculture can ensure a high-quality food supply throughout this century and beyond.

The *Food Feature* sections that appear in most chapters act as bridges between theory and practice; they are practical applications of the chapter concepts that help readers to choose foods according to nutrition principles. *Consumer Corners* present information on olestra and other fat replacers, amino acid supplements, vitamin C and the common cold, bottled water, marketing of infant formula, organic foods, and other nutrition-related marketplace issues to empower students to make informed decisions.

New or major terms in chapters are defined in the margins of the pages where they are introduced and also in the Glossary at the end of the book. Terms in Controversy sections are grouped together and defined in tables within the sections and in the Glossary. The reader who wishes to locate any term can do so by consulting the index, which lists the page numbers of definitions in boldface type.

The appendixes have been updated. Appendix A now presents the most complete and accurate listings ever of the nutrient contents of more than 2,200 foods. Appendix B, *Canadiana,* supplies the RNI, the Guidelines, the Food Guide, Food Labels, and the Choice System for our Canadian readers. Appendix C demonstrates nutrition calculations, with special emphasis on finding the percentage of calories from fat in a diet and percentages of the Daily Values. Appendix D provides full coverage with applications of the U.S. Exchange System. Appendix E offers an invaluable list of current addresses, telephone numbers, and Internet websites for those interested in additional information. We have collected all chapter and Controversy references in Appendix F. Older source notes have been removed but are easily available by consulting older editions of this book or by contacting the publisher.

As always, our purpose in writing this text is to enhance our readers' understanding of nutrition science and motivation to apply it. We hope the information on this book's pages will reach beyond the classroom into our readers' lives. Take the information you find inside this book home with you. Use it in your life: nourish yourself, educate your loved ones, and nurture others to be a part of our healthy new future. Stay up with the news, too. For despite all the conflicting messages, inflated claims, and even quackery that abound in news reports, true nutrition knowledge progresses with a genuine scientific spirit, and important new truths are constantly unfolding.

Acknowledgments

OUR THANKS to Linda Kelly DeBruyne for the newly updated Chapter 10 and Chapter 12 of this edition. Thanks to Kellie Harder Hatcher for her careful revision of Chapter 14 and her competent and willing assistance at each stage of this writing. Thanks also to Sharon Rady Rolfes for her helpful ideas throughout this work. Thanks to Sally Lorch Mayo for office tasks and for extra help when time ran short (and congratulations on graduating). Thanks, too, to our associate Lori Turner for much of the *Instructor's Manual;* thanks to Margaret Hedley, University of Guelph, who prepared the Canadian material for the manual and reviewed the Canadian resources listed in the text. For the special Instructor's Edition of the text, we thank Lori Turner, University of Arkansas–Fayetteville (lecture outlines); and Judy Kaufman, Monroe Community College (margin references to overheads, *Nutrition Interactive* CD-ROM, and *NutriLink* CD-ROM). Thanks also to Judy Kaufman who developed the *Self Check* questions for the seventh edition, most of which are still serving students in the current text, and who reviewed the acetate package for the eighth edition. Thank you, Myrna Engler, Dorothy Reinhardt, Polara Studios, David Ruppe and Sandra McMahon and Associates for bringing our figures and photos to their full potential. For his work in creating this edition's lively new design, we thank Norman Baugher.

Special thanks to our publisher, Peter Marshall, and to our editors Laura Graham, Dusty Friedman, and Sandra Craig, and to their staff, for their unflagging efforts to ensure the highest quality of all facets of this book. Our applause and gratitude goes to Becky Tollerson for her efforts in marketing our book. Special thanks to our librarians: Linda Patton, Jackie Hannick, and Lonnie Walsh, for their competent and exacting bibliographic help. Thanks also to Jana Kicklighter of Georgia State University for preparing the *Student Study Guide* and the *Test Bank.* As always, we are grateful to Bob Geltz and Betty Hands and their staff at ESHA research for Appendix A and for the computerized diet analysis program that accompanies this book. To our reviewers, many heartfelt thanks for your many thoughtful ideas and suggestions:

Jenna Anding	University of Houston
Raga M. Bakhit	Virginia Polytechnic Institute
Linda C. Barnes	Tidewater Community College
Ethan A. Bergman	Central Washington University
Debra Boardley	University of Toledo
Mallory Boylan	Texas Technical University
Pat Brown	Cuesta College
N. Joanne Caid	California State University–Fresno
Marjorie Caldwell	University of Rhode Island
Leah Carter	Bakersfield College
Thomas W. Castonguay	University of Maryland
Stacy Coseio	University of Houston
Leslie Edwards Cummings	University of Nevada
Earlene Davis	Bakersfield College
Allan Davison	Simon Fraser University
Robert DiSilvestro	Ohio State University
Carol Friesen	Ball State University
Sherrie Frye	University of Northern Colorado
Cindy J. Fuller	University of North Carolina
Leonard Gerber	University of Rhode Island
Jan Goodwin	University of North Dakota
Cynthia Gossage	Prince George's Community College
Deborah Gustafson	Utah State University
Deloy G. Hendricks	Utah State University
Ann A. Hertzler	Virginia Polytechnic Institute
Wendy T. Hunt	American River College
Kendra K. Kattelmann	South Dakota State University
Judy Kaufman	Monroe Community College
Margaret Kessel	Ohio State University
Jeanne W. Lawless	Bloomsburg University
Dalia Lima	Houston Community College
Cleo Long	Shelby State Community College
Teresa T. Marcus	Chattanooga State Technical Community College
Sharon L. McWhinney	Prairie View A & M University
Janis Mena	Santa Fe Community College (Univ of FL)
Stella Miller	Mt. San Antonio College
Ryna Levy Milne	University of British Columbia
Cherie L. Moore	Cuesta College
Mary Etta Moorachian	Johnson & Wales University
Sharon K. Morcos	Kansas State University
Leonard A. Piche	Brescia College
Clay Robinson	Lewis Clark State College
Samuel C. Smith	University of New Hampshire
Wendy Stuhldreher	Slippery Rock University
Joan Thompson	Weber State University
Dellman Walker	Middle Tennessee State University
Hope Weiler	University of Manitoba
Julian H. Williford, Jr.	Bowling Green State University
Fred H. Wolfe	University of Arizona
Cynthia Wright	Southern Utah University
Lisa Young	New York University

1 Food Choices and Human Health

Henry Church, *Still Life*
Date Unknown

Contents

Frequently Asked Questions

food medically, any substance that the body can take in and assimilate that will enable it to stay alive and to grow; the carrier of nourishment; socially, a more limited number of such substances defined as acceptable by each culture.

nutrition the study of the nutrients in foods and in the body; sometimes also the study of human behaviors related to food.

diet the foods (including beverages) a person usually eats and drinks.

nutrients components of food that are indispensable to the body's functioning. They provide energy, serve as building material, help maintain or repair body parts, and support growth. The nutrients include water, carbohydrate, fat, protein, vitamins, and minerals.

malnutrition any condition caused by excess or deficient food energy or nutrient intake or by an imbalance of nutrients. Nutrient or energy deficiencies are classed as forms of undernutrition; nutrient or energy excesses are classed as forms of overnutrition.

IF YOU CARE about your body, and if you have strong feelings about **food,** then you have much to gain from learning about **nutrition**—the study of how food nourishes the body. Nutrition is a fascinating, much talked-about subject. Everyone wants to know how food affects health—naturally, because we love food and we care about our health. Each day, newspapers, radio, and television present stories of new findings on nutrition and heart health or nutrition and cancer prevention. Magazine advertisements and television commercials constantly bombard us with multicolored pictures of tempting foods—pizza, burgers, cakes, chips, sweet drinks, alcoholic beverages, and many more.[1] Several times a day, you get hungry and turn from your other activities to eat a meal. And if you are like most people, you wonder, "Is this food good for me?" or you berate yourself, "I probably shouldn't be eating this."

When you study nutrition, you learn which foods serve you best, and you can work out ways of choosing foods, planning meals, and designing your **diet** wisely. Knowing the facts can enhance your health and your enjoyment of eating while relieving you of feeling guilty or worried that you aren't eating well.

This book devotes many chapters to the science of nutrition. This chapter provides a starting point by offering answers to the following questions:

1. What connections exist between the diets people consume and physical health?
2. What does food do for the body and its owner?
3. How can consumers judge whether to believe nutrition news from the media?
4. What constitutes a nutritious diet?

A Lifetime of Nourishment

If you live for 65 years or longer, you will have consumed more than 70,000 meals, and your remarkable body will have disposed of 50 tons of food. The foods you choose have cumulative effects on your body. At 65 years of age, you will see and feel those effects, if you know what to look for.

Your body renews its structures continuously, and each day it builds a little muscle, bone, skin, and blood, replacing old tissues with new. It may also add a little fat, if you consume excess food energy (calories), or subtract a little, should you consume less than you require. In this way some of the food you eat today becomes part of "you" tomorrow. The best food for you, then, is the kind that supports the growth and maintenance of strong muscles, sound bones, healthy skin, and sufficient blood to cleanse and nourish all parts of your body. This means you need food that provides not only energy but also sufficient '**nutrients**,' that is, enough water, carbohydrates, fat, protein, vitamins, and minerals. If the foods you eat provide too little or too much of any nutrient today, your health may suffer just a little. If the foods you eat provide too little or too much of one or more nutrients every day for years, then, by the time you are old, you may well suffer severe disease effects.

The point is that a well-chosen array of foods supplies enough energy and enough of each nutrient to prevent **malnutrition.** Malnutrition includes deficiencies of nutrients, imbalances, and excesses, any of which can take a toll on health over time.

When you choose foods with nutrition in mind, you can enhance your own well-being.

KEY POINT *The nutrients in food support growth, maintenance, and repair of the body. Deficiences, excesses, and imbalances of nutrients bring on the diseases of malnutrition.*

How Powerful Is a Nutritious Diet in Preventing Diseases?

Your choice of diet profoundly influences your long-term health prospects. Only two common lifestyle habits are more influential: smoking and other tobacco use, and excessive drinking of alcohol.[2] Many older people suffer from debilitating conditions that could have been largely prevented had they known the nutrition principles that we know today and applied those principles throughout their lives.

The poor health conveyed by a poor diet involves not only the various forms of malnutrition just described, but also other diseases, especially the **chronic diseases:** heart disease, diabetes, some kinds of cancer, dental disease, adult bone loss, and others. We should hasten to say that although diet can powerfully influence these diseases, they cannot be prevented by a good diet alone; they are to some extent determined by a person's genetic constitution, activities, and lifestyle.[3] Within the range set by your inheritance, however, the likelihood that you will develop these diseases is strongly influenced by your food choices.

chronic diseases long-duration degenerative diseases characterized by deterioration of the body organs; examples include heart disease, cancer, and diabetes.

Some people overestimate and some underestimate the influence of diet in preventing diseases and poor health. Putting diet's exact role in perspective is difficult not only for ordinary people, but also for research scientists who spend their working lives trying to figure out precisely how diet relates to health and various diseases. Some different aspects of the relationship, described next, may help to show the connections.

KEY POINT *Nutrition profoundly affects health.*

Genetics and Individuality

Consider the role of genetics. Different diseases are differently influenced by genetics and nutrition, as shown in Figure 1-1. The **anemia** caused by sickle cell disease, for example, is purely hereditary. Nothing a person eats affects the person's chances of contracting this anemia, although nutrition therapy may help ease its course. Sickle-cell anemia is shown at the left in the figure as a nutrition-unrelated, genetic disease. In contrast, a condition such as iron-deficiency anemia, listed at the right in the figure, is most often a nutrition-related condition. A person's low iron status can easily result from undernutrition, which can lead to the deficiency form of anemia. Diseases and conditions of poor health appear all along the continuum from purely genetic to purely nutritional; the more nutrition-related a disease or health condition is, the more successfully sound nutrition can prevent it.

Anemia is a blood condition in which red blood cells are inadequate or impaired, and so cannot meet the oxyen demands of the body. More about the anemia of sickle-cell disease in Chapter 6; iron-deficiency anemia is described in Chapter 8.

Furthermore, some diseases, such as heart disease and cancer, are not one disease but many. Two people may both have heart disease, but not the same form. People differ genetically from each other in thousands of ways. One person's heart disease or cancer may be nutrition-related, but another's may not be. The concept presented in Figure 1-1 is based on the experience of millions of people; in contrast, no simple statement can be made about the extent to which diet can help any one person avoid a disease or slow its progress.

KEY POINT *Choice of diet influences long-term health, within the range set by genetic inheritance. Nutrition has no influence on some diseases but is closely linked to others.*

Figure 1-1

Nutrition and Disease

Not all diseases are equally influenced by diet. Some are purely genetic, like the anemia of sickle-cell disease. Some may be inherited (or the tendency to develop them may be inherited) but may be influenced by diet, like some forms of diabetes. Some are purely dietary, like the vitamin and mineral deficiency diseases.

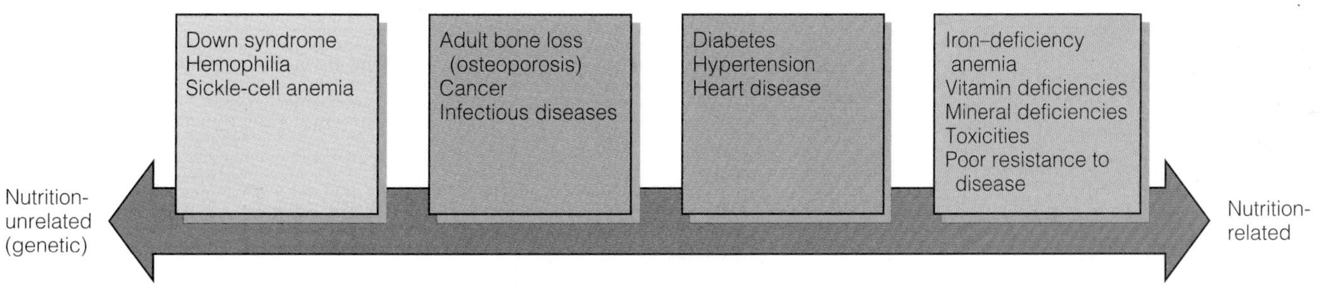
Down syndrome
Hemophilia
Sickle-cell anemia
Adult bone loss
(osteoporosis)
Cancer
Infectious diseases
Diabetes
Hypertension
Heart disease
Iron-deficiency
anemia
Vitamin deficiencies
Mineral deficiencies
Toxicities
Poor resistance to
disease
Nutrition-
unrelated
(genetic)
Nutrition-
related

energy the capacity to do work. The energy in food is chemical energy; it can be converted to mechanical, electrical, heat, or other forms of energy in the body. Food energy is measured in calories, defined on p. 7.

Why Bother to Be Active?

This chapter makes clear that a person's daily food choices can powerfully influence long-term health, but, in truth, the combination of nutrition and physical activity is more powerful still. This section and others like it in chapters to come serve a double purpose of drawing connections among physical activity, nutrition and health, and of motivating you, the reader, to take action. Perhaps you feel that the pressures of your school, job, or social life leave you no time to change clothes and work out at the gym or on the track or tennis court. Or maybe just the thought of being physically active makes you lose heart. If so, you are not alone. Sadly, only about a third of our population manages to get enough physical activity each week to support health. The rest of us are missing out on one of the most powerful modulators of disease risk known. People who are active may receive these benefits:

- Reduced risk of cardiovascular diseases.
- Reduced risk of some types of cancer (especially of the colon or breast).
- Improved mental outlook and lessened likelihood of depression.
- Improved mental functioning.
- Feeling of vigor.
- Feeling of belonging—the fun and companionship of sports.
- Strong self-image and self-confidence.
- Reduced body fatness and increased lean tissue.
- Greater bone density and, thus, lessened risk or reduced severity of bone disease later in life.
- Sound, beneficial sleep.
- A more youthful appearance, healthy skin, and improved muscle tone.
- Faster wound healing.
- Improvement or elimination of menstrual cramping.
- Improved resistance to colds and infection.

Science cannot promise that you will receive all of these benefits if you exercise, but almost everyone who is both well nourished and physically active reaps at least some of them. If even half of these rewards were yours for the asking, wouldn't you step up to claim them? In reality, they are yours to claim. Claiming them, however, requires learning about the foods and activities that support wellness and then applying that knowledge in your daily life. Taking this class is an excellent start. After that, it's up to you.

Alcohol use and abuse and their effects on body tissues are topics of Controversy 5.

Lifestyle Choices

Other lifestyle choices, besides people's choices of what foods to eat, also affect their health. Tobacco and alcohol use and other substance abuse can all be destructive of health. Other major health determinants include sleep, stress, and home and job conditions, including environmental quality. Physical activity is so important to robust health that many chapters of this book feature the connections between nutrition and fitness in Fitness for Life sections like the one above. In all of these contexts, healthful nutrition can help prevent or reduce the severity of some diseases. Table 1-1 lists some other relationships, and later chapters will examine them in detail.

KEY POINT *Personal life choices, such as use of tobacco or alcohol or staying physically active, also affect health for the better or worse.*

The Human Body and Its Food

As your body lives each day, it moves and works, and to move or work, it must use **energy.** The energy that fuels the body's work comes indirectly from the sun by way of plants. Plants capture and store the sun's energy in their tissues as they grow. When

Table 1-1

Nutrition Measures to Prevent Diseases

Adequate Intake of Essential Nutrients, Especially *Protein*, and *Energy* from Food Helps Prevent

In Pregnancy

Low birthweight
Poor resistance to disease
Some forms of birth defects
Some forms of mental/physical retardation

In Infancy and Childhood

Growth deficits
Poor resistance to disease

In Adulthood and Old Age

Poor resistance to infectious diseases
Susceptibility to some forms of cancer

Moderation in Intake of *Energy* from Food Helps Prevent

Obesity and related diseases, such as diabetes and hypertension

Moderation in *Fat* Intake Helps Prevent

Susceptibility to obesity, some cancers, and atherosclerosis

Adequate *Fiber* Intake Helps Prevent

Digestive malfunctions such as constipation and diverticulosis and possibly some cancers
Susceptibility to heart disease

Moderation in *Sugar* Intake Helps Prevent

Dental caries

Moderation in *Alcohol* Intake Helps Prevent

Liver disease
Malnutrition
Sudden death from heart failure

Adequate Intake of *Any Essential Nutrient* Prevents

Deficiency diseases such as cretinism, scurvy, and folate-deficiency anemia

Moderation in Intake of *Essential Nutrients* Prevents

Toxicity states

Adequate *Calcium* Intake Helps Prevent

Adult bone loss
Possibly colon cancer and hypertension

Adequate *Iron* Intake Helps Prevent

Iron-deficiency anemia

Adequate *Fluoride* Intake Helps Prevent

Dental caries

Moderation in *Sodium* Intake Helps Prevent

Hypertension and related diseases of the heart and kidney

Adequate *Vitamin* Intake Helps Prevent

Susceptibility to certain cancers and possibly heart disease
Certain birth defects

you eat plant-derived foods such as fruits, grains, or vegetables, you obtain and use the solar energy they have stored. Plant-eating animals obtain their energy in the same way, so when you eat animal tissues, you are eating compounds containing energy that came originally from the sun.

The body also requires six kinds of nutrients—families of molecules indispensable to its functioning—and foods deliver these. Table 1-2 lists the six classes of nutrients. Four of these six are **organic**; that is, the nutrients contain the element carbon derived from living things. The human body and foods are made of the same materials, arranged in different ways (see Figure 1-2 on the next page).

organic carbon containing. Four of the six classes of nutrients are organic: carbohydrate, fat, protein, and vitamins. Strictly speaking, organic compounds include only those made by living things and do not include carbon dioxide and a few carbon salts.

energy-yielding nutrients the nutrients the body can use for energy. They may also supply building blocks for body structures.

The Nutrients in Foods

Foremost among the six classes of nutrients in foods is water, which is constantly lost from the body and must constantly be replaced. Among the four organic nutrients, three are **energy-yielding nutrients,** meaning that the body can use the energy they

Table 1-2

Elements in the Six Classes of Nutrients

The nutrients that contain carbon are organic.

	Carbon	Oxygen	Hydrogen	Nitrogen	Minerals
Water		✔	✔		
Carbohydrate	✔	✔	✔		
Fat	✔	✔	✔		
Protein	✔	✔	✔	✔	b
Vitamins	✔	✔	✔	✔[a]	b
Minerals					✔

[a]All of the B vitamins contain nitrogen; *amine* means nitrogen.
[b]Protein and some vitamins contain the mineral sulfur; vitamin B_{12} contains the mineral cobalt.

Figure 1-2

Materials of Food and the Human Body

Foods and the human body are made of the same materials.

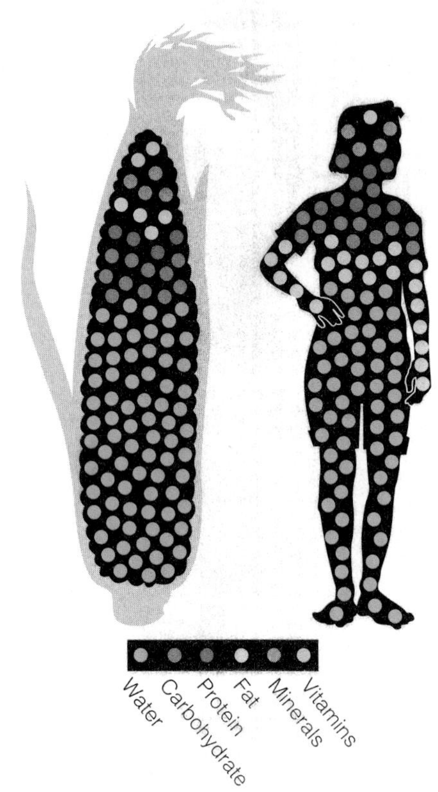

Water
Carbohydrate
Protein
Fat
Minerals
Vitamins

contain. The *carbohydrates* and *fats* (fats are properly called *lipids*) are especially important energy-yielding nutrients. As for *protein*, it does double duty: it can yield energy, but it also provides materials that form structures and working parts of body tissues. Alcohol yields energy, too, but it is a toxin, not a nutrient (see the note to Table 1-3).

The fifth and sixth classes of nutrients are the *vitamins* and the *minerals*. These provide no energy to the body. A few minerals serve as parts of body structures (calcium and phosphorus, for example, are major constituents of bone), but all vitamins and minerals act as regulators. As regulators, the vitamins and minerals assist in all body processes: digesting food; moving muscles; disposing of wastes; growing new tissues; healing wounds; obtaining energy from carbohydrate, fat, and protein; and participating in every other process necessary to maintain life. Later, a chapter is devoted to each of these six classes of nutrients in the order just named, except water, which is addressed with the minerals.

When you eat food, then, you are not just engaging in a pleasurable activity; you are providing your body with energy and nutrients. Furthermore, some of the nutrients are **essential nutrients,** meaning that if you do not receive them from food, you will develop deficiencies; the body cannot make these nutrients for itself. Essential nutrients are found in all six classes of nutrients. Water is essential; so is a form of carbohydrate; so are some lipids, some parts of protein, all of the vitamins, and the minerals important in human nutrition, too.

To support understanding of many of the discussions that follow, two definitions and a set of numbers are needed. Food scientists measure food energy in **calories,** units of heat. Food and nutrient quantities are often measured in **grams,** units of weight. The most energy-rich of the nutrients is fat, which contains 9 calories in each gram. Carbohydrate and protein each contain only 4 calories in a gram (see Table 1-3).

Scientists have worked out ways to measure the energy and nutrient contents of foods. They have also calculated the amounts of energy and nutrients various types of people need—people of both sexes, of different ages, and of different walks of life. Thus, after studying human nutrient requirements (the subject of Chapter 2 of this book), you can state with some accuracy just what your own body needs—this much water, that much carbohydrate and fat, so much protein, and so forth. Might it be possible, then, to simply take pills or **supplements** in place of food? No, because, as it turns out, food offers more than just the six basic nutrients.

KEY POINT *Food supplies energy and nutrients. The most vital nutrient is water. The energy-yielding nutrients are carbohydrates, fats (lipids), and protein. The helper nutrients are vitamins and minerals. Food energy is measured in calories; food and nutrient quantities are often measured in grams.*

essential nutrients the nutrients the body cannot make for itself (or cannot make fast enough) from other raw materials; nutrients that must be obtained from food to prevent deficiencies.

Can I Forgo Food and Live Just on Supplements?

Nutrition science has achieved the ability to state what nutrients human beings need to survive—at least for a time. Scientists are becoming skilled at making **elemental diets**—diets that have a precise chemical composition and are life-saving for people in the hospital who cannot eat ordinary food. These formulas, administered to severely ill people for days and weeks at a time, support not only continued life but also recovery from nutrient deficiencies and infections and the healing of wounds.

Lately, marketers have taken these formulas out of the medical setting for which they were developed and have advertised them heavily to healthy people as "insurance" against malnutrition. As a result, sales of the formulas have skyrocketed. The truth is that such products are not superior to a sound diet of real foods. In fact, experience with hospitalized people suggests that the opposite is true.

Formula diets are essential to help sick people to survive, but they are not sufficient to enable people to thrive over long periods. Elemental diet formulas support life but not optimal growth and health; and they often lead to medical complications.[4]

Table 1-3

Calorie Values of Energy Nutrients

Energy Nutrient	Energy
Carbohydrate	4 cal/g
Fat (lipid)	9 cal/g
Protein	4 cal/g

NOTE: Alcohol contributes 7 calories/gram that the human body can use for energy. Alcohol is not classed as a nutrient, however, because it interferes with growth, maintenance, and repair of body tissues.

Although these problems are rare and can be detected and corrected, they show that the composition of these diets is not yet perfect for all people in all settings. Healthy people who eat a regular diet especially do not need such formulas.

Even if a person's basic nutrient needs are perfectly understood and met, concoctions of nutrients still lack something that foods provide. Many times, hospitalized clients who are fed nutrient mixtures through a vein improve dramatically when they can finally eat food. Something in real food is of importance to health, but whether this effect is physical or psychological remains unknown.

Science demands explanations, though: What is this mysterious "something" food offers that cannot be provided through a needle? Part of the answer may lie in the reaction of the digestive tract to food. The stomach and intestine are dynamic living organs, and they change constantly in reaction to the foods they receive, and even to just the sight, aroma, and taste of food.[5] When a person is fed through a vein, the digestive organs, like unused muscles, weaken and grow smaller. Experiments have shown that when the intestine receives food, it releases hormones, chemical messengers that maintain the body's integrity; when not fed, the body receives too little stimulation and so deteriorates. In light of this knowledge, medical wisdom now dictates that when a hospital client has to be fed through a vein, the duration should be as short as possible, and real food, taken into the intestine, should be reintroduced as early as possible.[6]

The hormones that the intestine releases in response to food also affect the brain. The messages they deliver bring about the sense of satisfaction that makes a person feel, "There, that was good. Now I'm full." Eating offers both physical and emotional comfort: after a good meal, you can relax, enjoy entertainment, rest, or sleep. One writer says, "One cannot think well, love well, sleep well, if one has not dined well."[7]

Food does still more than maintain the intestine and convey messages of comfort to the brain. Foods are chemically complex. Even an ordinary potato contains hundreds of different compounds. People are complex, too, and the relationship between people and food is ancient. In view of all this, the fact that food gives us more than just nutrients is not surprising. If it were otherwise, that would be surprising.

KEY POINT *In addition to nutrients, food conveys emotional satisfaction and hormonal stimuli that contribute to health.*

calories units of energy. Strictly speaking, the unit used to measure the energy in foods is a kilocalorie (*kcalorie,* or *Calorie*): it is the amount of heat energy necessary to raise the temperature of a kilogram (a liter) of water 1 degree Celsius. This book follows the common practice of using the lowercase term *calorie* (abbreviated *cal*) to mean the same thing.

grams units of weight. A gram (g) is the weight of a cubic centimeter (cc) or milliliter (ml) of water under defined conditions of temperature and pressure. About 28 grams equal an ounce.

supplements pills, liquids, or powders that contain purified nutrients or other ingredients (see Chapter 7).

elemental diets diets composed of purified ingredients of known chemical composition; intended to supply all essential nutrients to people who cannot eat foods.

When you eat foods, you are receiving more than just nutrients.

nonnutrients a term used in this book to mean compounds other than the six nutrients that are present in foods.

phytochemicals nonnutrient compounds in plant-derived foods that have biological activity in the body.

What Are the Nonnutrients and Phytochemicals That I Hear about in the News?

In addition to their many nutrients, foods also contain many other compounds that act on the body in some way. One name for these compounds is **nonnutrients.**

Among the nonnutrients of great interest today are the **phytochemicals** (*phyto* means plant), and Controversy 11 provides details about them. Phytochemicals include the compound that gives hot peppers their burning taste, the compound that gives garlic its pungent flavor, the pigments that give spinach and tomatoes their dark green and dark red colors, the products from yeast cells that make bread rise, and thousands upon thousands of others. Some foods contain nonnutrients that have drug effects—one in cranberries, for example, may prevent some bacteria from clinging to the urinary tract and so prevents infection.[8] Some substances are toxic: one in cabbage, for example, can damage the thyroid gland when consumed in excess. That doesn't mean that cabbage is a harmful food, of course; it also contains phytochemicals thought to have an anticancer effect.

Many medical drugs also come from plants. The drug effects of plants, including foods, have been known to human beings for more than 20,000 years—since the Stone Age when people first began to collect plants for their medicinal properties. For all that time, and no doubt for thousands of years before that, people were developing a relationship with plants that involved much more than merely depending on them for nutrients. Humans learned to turn to food for comfort, for relief of pain, for pleasure, and even for the cure of some ailments. Chapter 11 discusses herbal and other alternative medicines.

Today, people are rediscovering the medicinal effects of foods, and they are also finding that food conveys meanings beyond the physical health of the body. Cultural and social traditions are attached to the preparing, serving, eating, and sharing of food.

KEY POINT *Foods contain compounds other than nutrients, the phytochemicals, that give them their tastes, aromas, colors, and other characteristics. Some phytochemicals are believed to play roels in disease prevention.*

The Science of Nutrition

Nutrition is a science—a field of knowledge composed of organized facts. Unlike some other areas of science, such as astronomy and physics, nutrition is a relatively young science. Most nutrition research has been conducted since 1900. The first vitamin was identified in 1897, and the first protein structure was not fully described until 1945. Much remains to be learned about the effects of foods, nutrients, and nonnutrients on the body. Because nutrition science is an active, changing, growing body of knowledge, reports of scientific findings often seem to contradict one another, and those findings may be subject to several, seemingly conflicting interpretations.

For this reason, people who don't understand how science operates may despair as they try to decipher current reports to learn what is really going on. They may even become distrustful: "When the scientists themselves can't agree on what is true, how am I supposed to know?" Yet, beyond the theories now being investigated, many facts in nutrition are known with great certainty—enough to fill this book and many more. And where there are conflicts and contradictions, researchers are energetically attempting to resolve them. The next section can help consumers understand why apparent contradictions sometimes arise in nutrition science.

If the Scientists Don't Know, How Can I?

Everyone stampedes for oat bran, red wine, or fish oil based on today's news that these products are good for health. Then tomorrow's news reports, "It isn't true after all," and everyone drops oat bran, red wine, or fish oil and takes up the next craze. Meanwhile, bewildered consumers complain in frustration, "Those scientists don't know anything."

Some foods offer beneficial nonnutrients called phytochemicals.

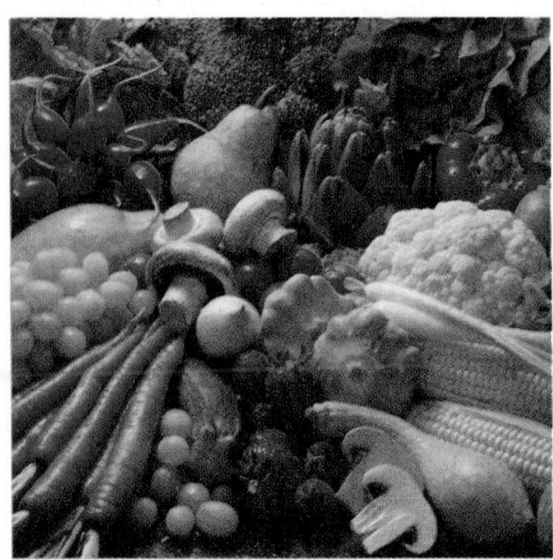

In truth, though, it is a scientist's business not to know. Scientists obtain facts by systematically asking questions—that's their job. Then they conduct experiments designed to test for various possible answers (see Figure 1-3 and Table 1-4 on the next page). When they have ruled out some possibilities and found evidence for others, they submit their findings, not to the news media, but to boards of reviewers composed of other scientists who try to pick the findings apart. If the reviewers consider the conclusions to be well supported by the evidence, they endorse the work for publication, not in the news media, but in scientific journals where still other scientists can read it.

Scientific Challenge

Once a new finding is published, it is still only preliminary. One experiment does not "prove" or "disprove" anything. The next step is for other scientists to attempt to

Figure 1-3

Research Design

The source of valid nutrition information is scientific research. Nutrition is a science, an organized body of knowledge composed of facts that we believe to be true because they have been supported, time and again, in experiments designed to rule out all other possibilities. Each fact has been established by many different kinds of experiments. For example, we know that eyesight depends partly on vitamin A because animals deprived of that vitamin and only that vitamin begin to go blind; when it is restored soon enough to their diet, they regain their sight. The same fact holds true in observations of human beings.

The type of study chosen for research depends upon what sort of information the researchers require. Studies of individuals ***(case studies)*** *yield observations that may lead to possible avenues of research. A study of a man who ate gumdrops and became a famous dancer might suggest that an experiment be done to see if gumdrops contain dance-enhancing power.*

Studies of whole populations ***(epidemiological studies)*** *provide another sort of information. Such a study can reveal a* ***correlation.*** *For example, an epidemiological study might find no worldwide correlation of gumdrop eating with fancy footwork but, unexpectedly, might reveal a correlation with tooth decay.*

Studies in which researchers actively intervene to alter people's eating habits ***(intervention studies)*** *go a step further. In such a study, one set of subjects (the* ***experimental group****) receive a treatment, and another set (the* ***control group****) go untreated or receive a* ***placebo*** *or sham treatment. If the study is a* ***blind experiment,*** *the subjects do not know who among the members receives the treatment or who receives the sham. If the two groups experience different effects, then the treatment's effect can be pinpointed. For example, an intervention study might show that withholding gumdrops, together with other candies and confections, reduced the incidence of tooth decay in an experimental population compared to that in a control population.*

Finally, ***laboratory studies*** *can pinpoint the mechanisms by which nutrition acts. What is it about gumdrops that contributes to tooth decay: their size, shape, temperature, color, ingredients? Feeding various forms of gumdrops to rats might yield the information that sugar, in a gummy carrier, promotes tooth decay. In the laboratory, using animals or plants or cells, scientists can inoculate with diseases, induce deficiencies, and experiment with variations on treatments to obtain in-depth knowledge of the process under study. Intervention studies and laboratory experiments are among the most powerful tools in nutrition research because they show the effects of treatments.*

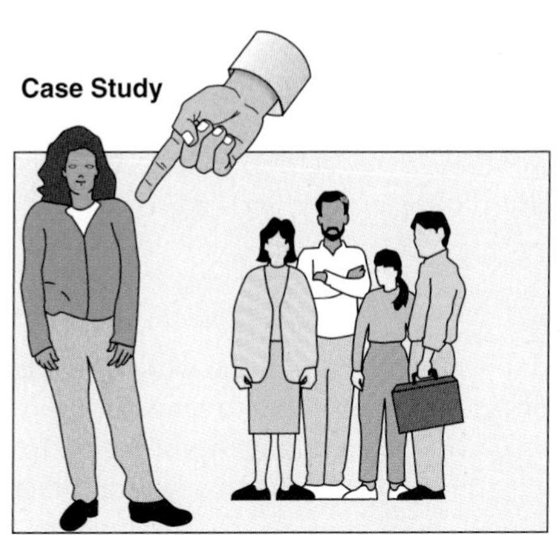
Case Study

"This person eats too little of nutrient X and has illness Y."

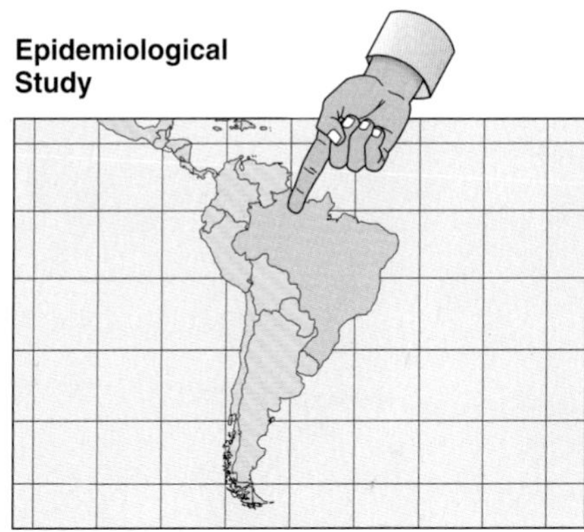
Epidemiological Study

"This country's food supply contains more nutrient X, and these people suffer less illness Y."

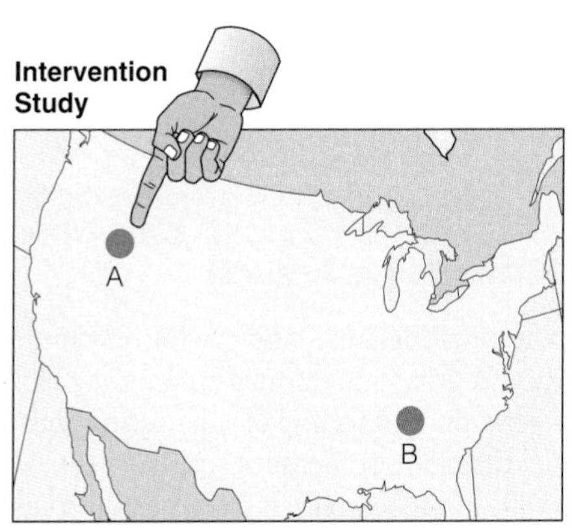
Intervention Study
A
B

"Let's add foods containing nutrient X to city A's food supply and compare illness Y rates of city A with those of city B."

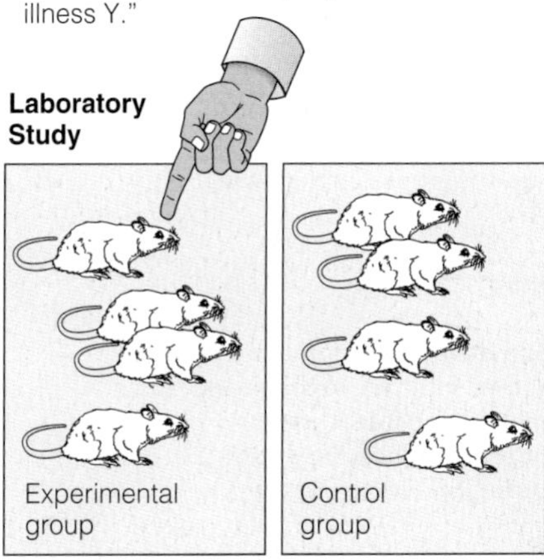
Laboratory Study
Experimental group
Control group

"Now let's prove that a nutrient X deficiency causes illness Y by inducing a deficiency in these rats."

Table 1-4

Research Design Terms

- **blind experiment** an experiment in which the subjects do not know whether they are members of the experimental group or the control group. In a *double-blind experiment,* neither the subjects nor the researchers know to which group the members belong until the end of the experiment.
- **case studies** studies of individuals. In clinical settings, researchers can observe treatments and their apparent effects. To prove that a treatment has produced an effect requires simultaneous observation of an untreated similar subject (a *case control*).
- **control group** a group of individuals who are similar in all possible respects to the group being treated in an experiment but who receive a sham treatment instead of the real one. Also called *control subjects.* See also *experimental group* and *intervention studies.*
- **correlation** the simultaneous change of two factors, such as the increase of weight with increasing height (a *direct* or *positive* correlation) or the decrease of cancer incidence with increasing fiber intake (an *inverse* or *negative* correlation). A correlation between two factors suggests that one may cause the other, but does not rule out the possibility that both may be caused by chance or by a third factor.
- **epidemiological studies** studies of populations; often used in nutrition to search for correlations between dietary habits and disease incidence; a first step in seeking nutrition-related causes of diseases.
- **experimental group** the people or animals participating in an experiment who receive the treatment under investigation. Also called *experimental subjects.* See also *control group* and *intervention studies.*
- **intervention studies** studies of populations in which observation is accompanied by experimental manipulation of some population members—for example, a study in which half of the subjects (the *experimental subjects*) follow diet advice to reduce fat intakes while the other half (the *control subjects*) do not, and both groups' heart health is monitored.
- **laboratory studies** studies that are performed under tightly controlled conditions and are designed to pinpoint causes and effects. Such studies often use animals as subjects.
- **placebo** a sham treatment often used in scientific studies; an inert harmless medication. The *placebo effect* is the healing effect that the act of treatment, rather than the treatment itself, often has.

duplicate and support the work of the first researchers or to challenge the finding by designing experiments to refute it.

Only when a finding has stood up to rigorous, repeated testing in several kinds of experiments performed by several different researchers is it finally considered confirmed. Even then, strictly speaking, science consists not of facts that are set in stone but of *hypotheses* that can always be challenged and revised. Some, though, like the hypothesis that the earth revolves about the sun, are so well supported by so many observations and experimental findings that they are generally accepted as facts. What we "know" in nutrition is confirmed in the same way: it results from years of replicating study findings. This attitude stands in sharp contrast to those reflected in most popular media reports, as described next.

Can I Trust the Media to Deliver Nutrition News?

Some newspapers, magazines, talk shows, Internet websites, and other media specialize in sensationalism that borders on quackery—see this chapter's Controversy for details.

The news media are hungry for new findings, and reporters often latch onto ideas from the scientific laboratories before they have been fully tested. Or a reporter who lacks a strong understanding of science may misunderstand complex scientific principles.[9] To tell the truth, sometimes scientists get excited about their findings, too, and leak them to the press before they have been through a rigorous review by the scientists' peers. As a result, the public is often exposed to late-breaking nutrition news stories before the findings are fully confirmed. Then, when the hypothesis being tested fails to hold up to a later challenge, consumers feel betrayed by what is simply the normal course of

UNIVERSITY OF GREENWICH LIBRARY

science at work. The Consumer Corner offers some tips for evaluating news stories about nutrition.

It also follows that people who take action based on single studies are almost always acting impulsively, not scientifically. The real scientists are trend watchers. They evaluate the methods used in each study, assess each study in light of all the evidence gleaned from other studies, and, little by little, modify their picture of what is true. As evidence accumulates, the scientists become more and more confident about their ability to make recommendations that apply to people's health and lives. Single studies are interesting, perhaps even exciting, but experienced observers learn to withhold judgment about the application of a study's findings until they have been repeated and confirmed.

Sometimes media sensationalism overrates the importance of even true, replicated findings. For example, a few years ago the media eagerly reported that oat bran lowers blood cholesterol, a lipid indicative of heart disease risk. Although the reports were true, oat bran is only one of several hundred factors that affect blood cholesterol. News reports on oat bran often failed to mention that cutting saturated fat intake is still the major step to take to lower blood cholesterol.

Also, new findings need refinements. Oat bran is truly a cholesterol reducer, but how much bran must a person eat to produce the desired effects? Do little oat bran pills or powders meet the need? Do oat bran cookies? If so, how many cookies must be eaten? Does everyone respond the same way to oat bran? How much of a dietary indiscretion can a person commit and still rectify it with oat bran? One reviewer of a report on this very topic observed, "To get the equivalent of the oat fiber in one bowl of oatmeal, it would [be] necessary to eat 90 cookies."[10] As for oatmeal, it takes a bowl-and-a-half daily to affect blood lipids. A few cookies cannot provide nearly so much and certainly cannot undo all the damage from a high-fat meal.

Today, oat bran's cholesterol-lowering effect is considered to be established, and labels on food packages can proclaim that a diet high in oats may reduce the risk of heart disease. The whole process of discovery, challenge, and vindication took almost ten years of research; establishing some other effects has taken many years longer. In science, a single finding almost never makes a crucial difference to our knowledge as a whole, but like each individual frame in a movie, it contributes a little to the big picture. Many such frames are needed to tell the whole story.

Besides conveying nutrients and nonnutrients, foods represent our cultures, our philosophies, and our beliefs. The next section explores how the special meanings that people assign to foods affect the diets they choose.

KEY POINT *Scientists uncover nutrition facts by experimenting. Single studies must be replicated before their findings can be considered valid. New nutrition news is not always to be believed; established nutrition news has stood up to the test of time.*

Cultural and Social Meanings Attached to Food

foodways the sum of a culture's habits, customs, beliefs, and preferences concerning food.

People from every country enjoy special **foodways** that reflect their own histories. As a result, the sharing of food can be symbolic: people offering foods that reflect their heritages are expressing a willingness to share cherished values with others. People accepting those foods are symbolically accepting not only the person doing the offering but the person's culture. This is why meetings of heads of state worldwide most often include a meal and why couples entering into cross-cultural marriages invite each other to share traditional holiday meals.

The same is true in a nation of mixed cultures such as the United States. Years ago, sociologists believed the United States was a "melting pot," creating a single "American" culture. In reality, though, our nation resembles a mosaic more than a melting pot because people of similar heritages tend to coalesce into distinct cultural

CONSUMER CORNER

Nutrition in the News

A NEWS READER, who had sworn off butter years ago for his heart's sake, bemoaned this headline: *Margarine Fat as Bad as Butter for Heart Health.* "Do you mean to say that I could have been eating butter all these years? That's it. I quit. No more diet changes for me." His response is understandable—diet changes, after all, take effort to make and commitment to sustain. Those who do make changes may feel betrayed when, years later, science appears to have turned its advice upside down. This reader might have avoided information burnout had he known some facts about the way science is reported in the news. He isn't alone in lacking knowledge. Most people are confused and frustrated by today's news.

It bears repeating that the findings of a single study never prove or disprove anything. Study results may constitute strong supporting evidence for one view or another, but they rarely merit the sort of finality implied by journalistic phrases such as "Now we know . . ." or "The answer has been found." Misinformed readers who look for simple answers to complex nutrition problems often take such phrases literally.

To read news stories with an educated eye, keep these points in mind:

- The study being described should be published in a peer-reviewed journal such as the *American Journal of Clinical Nutrition* and others. An unpublished study or one from a less credible source may or may not be valid; the reader has no way of knowing because the study has not been challenged or reviewed by other experts in the field.
- The report should describe the purpose of the study and the research methods used to obtain the data; it should also note their limitations. For example, it matters whether the study participants numbered eight or eight thousand, or whether the researchers personally observed participants' behaviors or relied on self-reports collected over the telephone.
- The subjects of the study may have been single cells, animals, or human beings, and the report should clearly make this distinction. If the study subjects were human beings, the more you have in common with them (age and gender, for example), the more applicable the findings may be for you.
- Valid reports also describe previous research and put the current research in proper context. Some reporters regularly follow developments in a research area and thus acquire the background knowledge they need to report meaningfully in that area; others simply "drop in" for a story.
- Useful for their broad perspective on a single topic are review articles appearing in journals such as *Nutrition Reviews.* Such articles allow judgement about a single study within the context of many other studies on the same topic.

Finally, ask yourself if the study makes common sense. Even if it turns out that the fat of margarine is damaging to the heart, do you eat enough margarine to worry about its effects? Before making a decision, learn more about the effects of fats on the arteries in Chapters 5 and 11. Then weigh new evidence against what you already know about fat and about yourself.

When a headline touts a shocking new "answer" to a nutrition question, read the story with a critical eye. It may indeed be a carefully researched report, but often it is a sensational story intended to catch the attention of newspaper and magazine buyers, not to offer useful nutrition information.

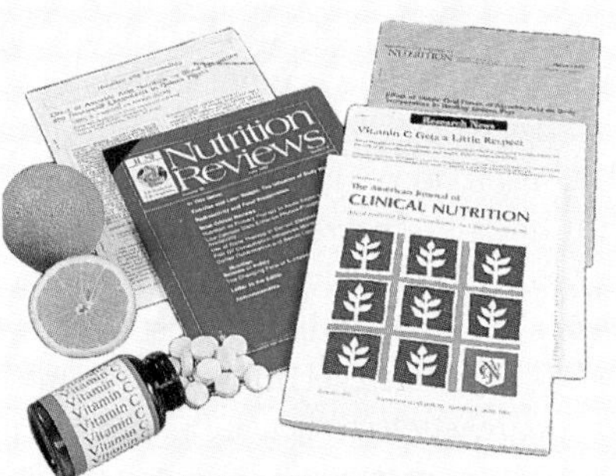
Nutrition Reviews
Vitamin C Gets a Little Respect
The American Journal of
CLINICAL NUTRITION
Vitamin C

A person wanting the whole story on a nutrition topic is wise to seek articles from peer-reviewed journals such as these. A review journal examines all available evidence on major topics. Other journals report details of the methods, results, and conclusions of single studies.

OUR GLOBE
ETHIOPIA
ETHIOPIA
LALIBELA

Sharing ethnic food is a way of sharing culture.

ethnic foods foods associated with particular cultural subgroups within a population.

cuisine a style of cooking.

omnivores people who eat foods of both plant and animal origin, including animal flesh.

vegetarians people who exclude from their diets animal flesh and possibly other animal products such as milk, cheese, and eggs.

communities. This arrangement offers a wealth of different cultural experiences for those who choose to seek them.

One of the most enjoyable ways to sample other cultures is to try some of the **ethnic foods** they have to offer—that is, to sample their **cuisine**.[11] Luckily, most people living in the United States can easily do this. A menu in an "American-style" restaurant might list spaghetti (Italian), tacos (Mexican), hot dogs with sauerkraut (German), croissants (French), stewed okra (African), baked squash (Native American), and egg rolls (Chinese). These traditional everyday foods, now adapted to locally available ingredients, have all become an integral part of the "American diet."

Cultural traditions regarding food are not inflexible; they keep evolving as people move about, learn about new foods, and teach each other.[12] Today, some people are ceasing to be **omnivores** and are becoming **vegetarians.** Vegetarians often choose this lifestyle because they honor the lives of animals or because they have discovered the health and other advantages associated with diets rich in beans, whole grains, fruits, and vegetables.[13] The Controversy of Chapter 6 explores the pros and the cons of both the vegetarian's and the meat eater's diets.

All of these and other considerations—physical, psychological, cultural, social, and philosophical—make up the framework within which people choose the foods they generally eat. Still other factors bear more immediately on a person's day-to-day food choices:

- *Advertising.* The media have persuaded you to eat them.
- *Availability.* There are no others to choose from.
- *Convenience.* They are quick and easy to prepare.
- *Economy.* They are within your means.
- *Emotional comfort.* They can make you feel better for a while.
- *Ethnic heritage.* They are the foods of your ethnic group.[14]
- *Habit.* They are familiar; you always eat them.
- *Personal preference.* You like the way they taste.[15]
- *Positive associations.* They are eaten by people you admire, or they indicate status, or they remind you of fun.
- *Region of the country.* They are foods favored in your area.
- *Social pressure.* They are offered; you feel you can't refuse them.
- *Values or beliefs.* They fit your religious tradition, square with your political views, or honor the environmental ethic.
- *Weight.* You think they will help to control body weight.
- *Nutritional value.* You think they are good for you.

Just the last two of these reasons for choosing foods assign a high priority to nutritional health. Similarly, the choice of where, as well as what, to eat is often based more on social considerations than on nutrition judgments. College students often choose to eat at fast-food and other restaurants to socialize, to get out, to save time, or to date; they are not always conscious of the need to obtain healthful food.[16] The next sections aim to set the table with some goals for the nutrition of the nation and some guidelines for choosing an adequate diet.

KEY POINT *Cultural traditions and social values revolve around food. Some values are expressed through foodways.*

Table 1-5

Proposed Nutrition-Related Health Objectives for the Nation, Year 2010[a]

Disease-Related Objectives

1. Reduce *coronary heart disease* deaths.
2. Reduce *cancer* deaths.
3. Decrease the incidence of *diabetes* (type 2), increase diagnosis of existing diabetes, and reduce rates of diabetes-related illness and death.
4. Reduce the prevalence of *osteoporosis*.
5. Reduce *dental caries*.

Nutrition Objectives

6. Increase the prevalence of *healthy weight* and decrease the prevalence of *obesity*.
7. Reduce *growth retardation* among low-income children.
8. Increase the proportion of people aged 2 and older who meet the *Dietary Guidelines* for *fat* and *saturated fat* in the diet.[b]
9. Increase intakes of *fruit and vegetables* to at least five servings a day.
10. Increase intakes of *grain products* to at least six servings a day.
11. Increase the proportion of people who meet the recommendation for *calcium*.
12. Increase the proportion of people who meet the Daily Value of 2,400 milligrams or less of *sodium* a day.
13. Reduce *iron deficiency* in children, adolescents, women of childbearing age, and low-income pregnant women.
14. Increase the proportion of mothers who *breastfeed* immediately after birth, for the first six months, and preferably, through the infant's first year of life. Increase the proportion of mothers who *breastfeed exclusively*.
15. Increase the proportion of children and adolescents whose intakes of *meals and snacks at school* contribute to overall dietary quality.
16. Increase the proportion of schools teaching *essential nutrition topics*.

Food Safety Objectives

17. Reduce the proportion of *infections* caused by foodborne pathogens and *antibiotic-resistant* pathogens.[c]
18. Reduce deaths from *food allergy* (anaphylaxisis).
19. Increase the proportion of *consumers* who practice four critical food safety behaviors when handling foods: washing hands, preventing cross contamination, cooking meats thoroughly, and chilling foods promptly.[d]
20. Reduce occurrences of improper food safety techniques in *retail food establishments*.

[a]Adapted. *Italic type* has been added to emphasize main areas of concern. Some wording may change upon final publication of the objectives. The *Healthy People 2010 Objectives* may be viewed at their website: http://web.health.gov/healthypeople/

[b]The objective specifies an average of 30 percent of calories or less from fat, and 10 percent or less from saturated fat.

[c]A pathogen is a disease-causing organism, such as a bacterium or virus.

[d]See Chapter 14 for details on these and other food safety concepts.

SOURCE: *Healthy People 2010: National Health Promotion and Disease Prevention Objectives* (Washington, D.C.: U.S. Department of Health and Human Services, 2000).

Nutrition Goals for the Nation

The U.S. Department of Health and Human Services (HHS) sets ten-year health objectives for the nation in its document *Healthy People*. The objectives for the year 2010 listed in Table 1-5 under "Nutrition" provide a quick scan of the nutrition-related objectives set for this decade. The inclusion of disease-related objectives, nutrition objectives, and those pertaining to food safety reveal that public health officials consider these areas to be top national priorities.

In 1999 researchers issued a report card evaluating the nation's progress toward achieving the objectives of the past decade.[17] On the positive side, objectives such as making food labels more informative and making low-fat foods more abundant had been fully achieved. Infant mortality had declined, and overall cancer death rates were below the year 2000 target. Deaths from heart disease and stroke had also declined substantially, but, on the negative side, heart disease was still the leading cause of death among adults. Also, the number of overweight people had jumped significantly. The objectives for the year 2010 represent a redoubling of effort to achieve some of the goals left unmet from the previous edition, and a chance to address some additional concerns at the start of the new century.

Nutrition monitoring makes it possible to assess the nutrient status and dietary intakes of the U.S. population. Among the agencies involved with these efforts are the:

- DHHS, already mentioned.
- U.S. Department of Agriculture (USDA).
- Centers for Disease Control and Prevention (CDC).

You are also likely to hear of two research projects. The National Health and Nutrition Examination Surveys (NHANES) involve:

- Asking people what they have eaten.
- Recording measures of their health status.

The Continuing Survey of Food Intakes by Individuals (CSFII) involves:

- Recording what people have actually eaten for two days.
- Comparing the foods they have chosen with recommended food selections.

All of these government efforts reflect ambitious goals for our health and food choices. Now, how should we go about achieving these goals? The next section offers some guidelines for choosing foods.

KEY POINT *The U.S. Department of Health and Human Services sets health objectives for the nation. Government agencies also monitor the nutrition status of the population.*

Agencies active in nutrition policy and monitoring:

- Department of Health and Human Services (DHHS).
- United States Department of Agriculture (USDA).
- Centers for Disease Control and Prevention (CDC).

Ongoing national nutrition research projects:

- National Health and Nutrition Examination Surveys (NHANES).
- Continuing Survey of Food Intakes by Individuals (CSFII).

The Challenge of Choosing Foods

The foods you choose should fit your tastes, personality, family and cultural traditions, lifestyle, and budget. At their best, well-planned meals convey pleasure, too, and they should also be nutritious, or at least the diet you build from them should be. Foods today come in astounding numbers and varieties. Consumers can lose track of what foods contain and how they can best be put together into health-promoting diets. A few guidelines can help.

The Variety of Foods to Choose From

A list of the variety of foods available a hundred years ago would be relatively short. It would consist of basic foods—foods that have been around for a long time such as vegetables, fruits, meats, milk, and grains. These foods have been variously called unprocessed, natural, whole, or farm foods. An easy way to obtain a nutritious diet is to consume a variety of selections from among these foods each day. Nevertheless, data from a recent CSFII report show that on a given day as many as 48 percent of our population consume no fruits or fruit juices. Also, although people generally consume a few servings of vegetables, the vegetable they most often choose is potatoes, usually prepared as french fries.[18]

In 1900, Americans chose from among 500 or so different foods; today, they choose from more than 50,000.

Ironically, the variety of foods available to us today may make it more difficult, rather than easier, to plan nutritious diets. The food industry offers thousands of foods; many are mixtures of the basic ones, and some are even constructed mostly from artificial ingredients.

Table 1-6 presents a glossary of terms related to foods. A reading of the terms will reveal that all types of food—including fast foods and processed foods—offer various constituents to the eater. You may also hear about *functional foods* or *nutraceuticals*. The term *functional foods* reflects an attempt to identify those foods that might lend protection against chronic diseases by way of the nutrients or nonnutrients they contain.[19] The trouble is, scientists trying to single out the most health-promoting foods find that almost every naturally occurring food is functional in some way with regard to human health. The term *nutraceuticals* may be used to refer to extracts or concentrates of nutrients and nonnutrients that normally occur in foods. Scientists are trying to develop universally accepted meanings for these terms, but so far there is no scientific agreement.

Controversy 11 presents details about the phytochemicals and Chapter 11 puts functional foods into perspective.

The extent to which foods support good health depends on the calories, nutrients, and nonnutrients they contain. In short, to select well among foods, as among people, you need to know more than their names; you need to know the foods' inner qualities.

Even more importantly, you need to know how to combine foods into nutritious diets. Foods are not nutritious by themselves; each is of value only in so far as it

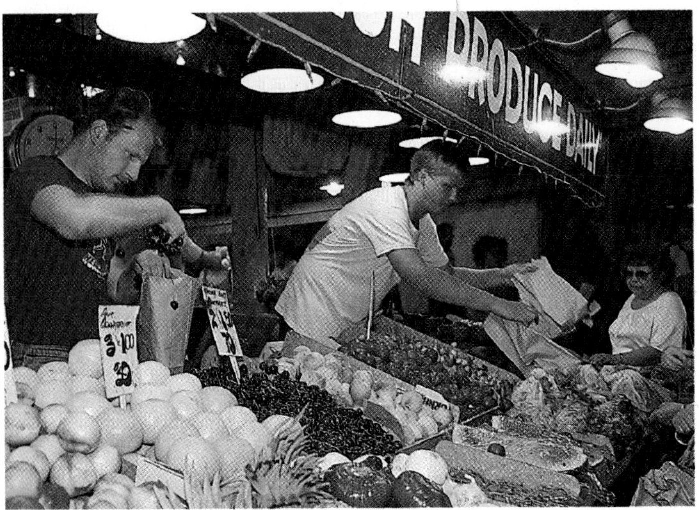
PRODUCE DAILY

Foods once looked like this . . .

Nutty Bars
Swiss Cake Rolls
Tim's CASCADE STYLE
POTATO CHIPS
INSTANT
White Rice
DELUXE
Fred Meyer
INSTANT
Mashed Potatoes
SPAM

. . . but now foods often look like this.

contributes to a nutritious diet. A key to wise diet planning is to make sure that the foods you eat daily, your staple foods, are especially nutritious.

KEY POINT *Foods come in a bewildering variety in the marketplace, but the foods that form the basis of a nutritious diet are ordinary milk and milk products; meats, fish, and poultry; vegetables and dried peas and beans; fruits; and grains.*

Table 1-6

Glossary of Food Types

The purpose of this little glossary is to show that good-sounding food names don't necessarily signify that foods are nutritious. Read the comment at the end of each definition.

- **basic foods** milk and milk products; meats and similar foods such as fish and poultry; vegetables, including dried beans and peas; fruits; and grains. These foods are generally considered to form the basis of a nutritious diet. Also called *whole foods.*
- **enriched foods** and **fortified foods** foods to which nutrients have been added. If the starting material is a whole, basic food such as milk or whole grain, the result may be highly nutritious. If the starting material is a concentrated form of sugar or fat, the result may be less nutritious.
- **fast foods** restaurant foods that are available within minutes after customers order them—traditionally, hamburgers, french fries, and milkshakes; more recently, salads and other vegetable dishes as well. These foods may or may not meet people's nutrient needs well, depending on the selections made and on the energy allowances and nutrient needs of the eaters.
- **functional foods** a term that reflects an attempt to define as a group the foods known to possess nutrients or nonnutrients that might lend protection against diseases. However, all nutritious foods can support health in some ways..
- **natural foods** a term that has no legal definition.
- **nutraceutical** a term with no legal or scientific meaning, but sometimes used to refer to foods, nutrients, or dietary supplements believed to have medicinal effects (see Chapter 11). Often used to sell unnecessary or unproved supplements.
- **organic foods** understood to mean foods grown without synthetic pesticides or fertilizers. In chemistry, however, all foods are made mostly of organic (carbon-containing) compounds. (See Chapter 14's Consumer Corner for details.)
- **partitioned foods** foods composed of parts of whole foods, such as butter (from milk), sugar (from beets or cane), or corn oil (from corn). Partitioned foods are generally overused and provide few nutrients with many calories.
- **processed foods** foods subjected to any process, such as milling, alteration of texture, addition of additives, cooking, or others. Depending on the starting material and the process, a processed food may or may not be nutritious.
- **staple foods** foods used frequently or daily, for example, rice (in the Far East) or potatoes (in Ireland). If well chosen, these foods are nutritious; certainly, they should be.

What Exactly Is a Nutritious Diet?

A nutritious diet has five characteristics. One is **adequacy:** the foods provide enough of each essential nutrient, fiber, and energy. Another is **balance:** the choices do not overemphasize one nutrient or food type at the expense of another. The third is **calorie control:** the foods provide the amount of energy you need to maintain appropriate weight—not more, not less. The fourth is **moderation:** the foods do not provide excess fat, salt, sugar, or other unwanted constituents. The fifth is **variety:** the foods chosen differ from one day to the next.

adequacy the dietary characteristic of providing all of the essential nutrients, fiber, and energy in amounts sufficient to maintain health and body weight.

balance the dietary characteristic of providing foods of a number of types in proportion to each other, such that foods rich in some nutrients do not crowd out of the diet foods that are rich in other nutrients. Also called *proportionality.*

calorie control control of energy intake; a feature of a sound diet plan.

moderation the dietary characteristic of providing constituents within set limits, not to excess.

variety the dietary characteristic of providing a wide selection of foods—the opposite of monotony.

Adequacy Any nutrient could be used to demonstrate the importance of dietary *adequacy.* Iron provides a familiar example. It is an essential nutrient: you lose some every day, so you have to keep replacing it; and you can get it into your body only by eating foods that contain it.* If you eat too few of the iron-containing foods, you can develop iron-deficiency anemia: with anemia you can feel weak, tired, cold, sad, and unenthusiastic; you may have frequent headaches; and you can do very little muscular work without disabling fatigue. If you add iron-rich foods to your diet, you soon feel more energetic. Some foods are rich in iron; others are notoriously poor. Meat, fish, poultry, and legumes are in the iron-rich category, and an easy way to obtain the needed iron is to include these foods in your diet regularly.

Balance To appreciate the importance of dietary *balance,* consider a second essential nutrient, calcium. Most foods that are rich in iron are poor in calcium. Calcium's best food sources are milk and milk products, which happen to be extraordinarily poor iron sources. A diet lacking calcium causes poor bone development during the growing years and increases a person's susceptibility to disabling bone loss in adult life. Children and adults are advised to consume enough milk, milk products, or other calcium-rich foods each day to meet their calcium needs—but not so much as to crowd iron-rich foods out of the diet.

A nutritious diet follows the a, b, c, m, v principles:

- adequacy
- balance
- calorie control
- moderation
- variety

Clearly, to obtain enough of both iron and calcium, which seldom appear together in the same foods, people have to balance their food choices. Balancing the whole diet to provide enough but not too much of every one of the 40-odd nutrients the body needs for health is a juggling act that requires considerable skill. As you will see in Chapter 2, food group plans can help you achieve dietary adequacy and balance because they recommend specific amounts of foods of each type.

Calorie Control Energy intakes should not exceed energy needs. Nicknamed *calorie control,* this diet characteristic ensures that energy intakes from food balance energy expenditures in activity. The eater of such a diet achieves control of body fat content and weight. The many strategies that promote this goal appear in Chapter 9.

Moderation Intakes of certain food constituents such as fat, cholesterol, sugar, and salt should be limited for health's sake (more on health effects in later chapters). A major guideline already mentioned is to keep fat intake below 30 percent of total calories. Some people take this to mean that they must never indulge in a delicious beefsteak or hot-fudge sundae, but they are misinformed: *moderation,* not total abstinence, is the key. A steady diet of steak and ice cream might be harmful, but once a week as part of an otherwise moderate diet plan, these foods may have little impact; as once-a-month treats, these foods would have practically no effect at all. Moderation also means that limits are necessary, even for desirable food constituents. For example, a certain amount of fiber in foods contributes to the health of the digestive system, but too much fiber leads to nutrient losses.

*A person can also take supplements of iron, but as later discussions demonstrate, this is not as effective as eating iron-rich foods.

nutrient density a measure of nutrients provided per calorie of food.

All of these factors help to build a nutritious diet.

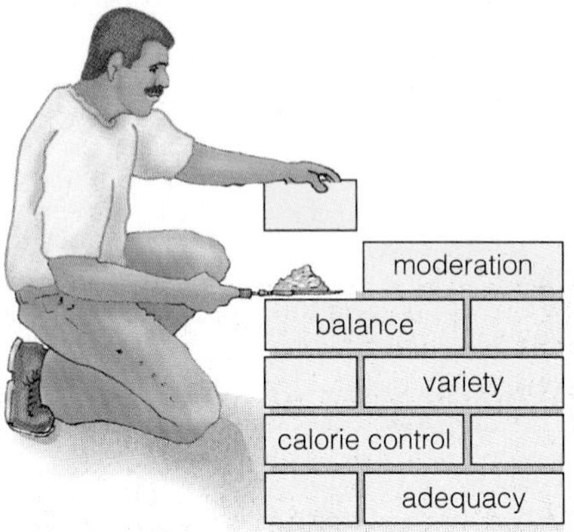

moderation
balance
variety
calorie control
adequacy

Variety As for *variety,* it is generally agreed that people should not eat the same foods, even highly nutritious ones, day after day, for several reasons. One reason is that some less well-known nutrients and some nonnutrient food components could be important to health; some foods may be better sources of these than others. Another reason is that a monotonous diet may deliver large amounts of toxins or contaminants. Each such undesirable item in a food is diluted by all the other foods eaten with it and is even further diluted if several days are skipped before the food is eaten again. Last, variety adds interest—trying new foods can be a source of pleasure.

Conclusion According to the experts, adults in the United States are not very successful at meeting these objectives.[20] In particular, people seem to consume diets that are either adequate or moderate, but not both. They behave as though they had to choose between two alternatives—getting all their nutrients but overconsuming fat calories in the process, or keeping their fat in line but running short on nutrients. Only 1 percent of several thousand adults who were surveyed managed to achieve both adequacy and moderation.[21] That finding defines the challenge for the health-conscious eater: try to achieve adequacy and moderation at the same time. Because this challenge is the key to good nutrition, this chapter's Food Feature is devoted to the skill required to meet it. The Food Feature offers a tool to help make it easier—the concept of **nutrient density.**

KEY POINT *A well-planned diet is adequate in nutrients, is balanced with regard to food types, offers food energy that matches energy expended in activity, is moderate in unwanted constituents, and offers variety. Foods of high nutrient density form the foundation of such a diet.*

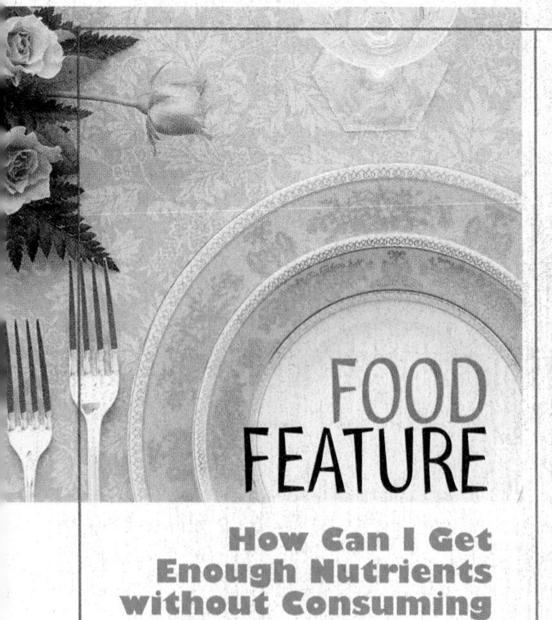

FOOD FEATURE

How Can I Get Enough Nutrients without Consuming Too Many Calories?

THE PLANNER WHO is trying to control calories while balancing the diet and making it adequate is bound to find certain foods especially useful. These foods are rich in nutrients relative to their energy contents: that is, they are foods with high nutrient density. Consider calcium sources, for example. Ice cream and fat-free milk both supply calcium, but the milk is "denser" in calcium per calorie. A cup of rich ice cream contributes more than 350 calories, a cup of fat-free milk only 85—and with almost double the calcium. Or consider iron. A 3-ounce serving of a high-fat cut of beef pot roast offers about the same amount of iron as a 4-ounce serving of baked fish, but the beef contains over 300 calories, the fish about 150. Most people cannot, for their health's sake, afford to choose foods without regard to their energy contents. Those who do very often fill up their calorie allowances while leaving nutrient needs unmet.

For a busy person who plans and prepares a meal, being conscious of nutrient density is especially important. To save both time and money, the food preparer is well advised to *center* the meal on foods of high nutrient density. The foods that offer the most nutrients per calorie are the vegetables, especially the nonstarchy vegetables such as broccoli, carrots, mushrooms, peppers, and tomatoes. Interestingly, these foods are also rich in phytochemicals thought to protect against diseases. These inexpensive foods take time to prepare, but time invested this way pays off in nutritional health. Twenty minutes spent peeling and slicing vegetables for a salad is a better investment in nutrition than 20 minutes spent fixing a fancy, high-fat, high-sugar dessert. Besides, the dessert ingredients cost more money, and they strain the calorie budget, too.

Investing resources in nutritious foods is especially important when time is limited. In today's households, although both men and women spend some 71 hours a week sleeping and taking care of personal needs, women still do most of the cooking and food shopping. Few households can afford a stay-at-home spouse, so families have very little time for food preparation—most spend less than 30 minutes preparing the evening meal, and 20 percent spend less than 15 minutes.[22] Busy chefs should seek out convenience foods, such as bags of ready-to-serve salads and frozen vegetables, that are nutrient dense. Other selections, such as most potpies, are less so because they contain too little of the vegetables and too much fat and salt.

Nutrient density is such a useful concept in diet planning that this book encourages you to think in those terms. Watch for the tables and figures in later chapters that show the best buys among foods, not necessarily in nutrients per dollar (although nutrient-dense foods are often among the least expensive), but in nutrients per calorie. Figure 1-4 offers a preview of the way this viewpoint can help you distinguish between more and less nutritious foods.

Would it take more time to prepare this dish than to prepare a batch of cookies? No, less time and less cleanup, too.

All of this discussion leads to the principle that is central to achieving nutritional health. It is not the individual foods you choose, but the way you combine them into meals and the way you arrange meals to follow one another over days and weeks that determine how well you are nourishing yourself. Nutrition is a science, not an art, but it can be used artfully to create a pleasing, nourishing diet. The remainder of this book is dedicated to helping you make informed choices and combine them artfully to meet all the body's needs.

Figure 1-4

How the Experts Judge Which Foods Are Most Nutritious

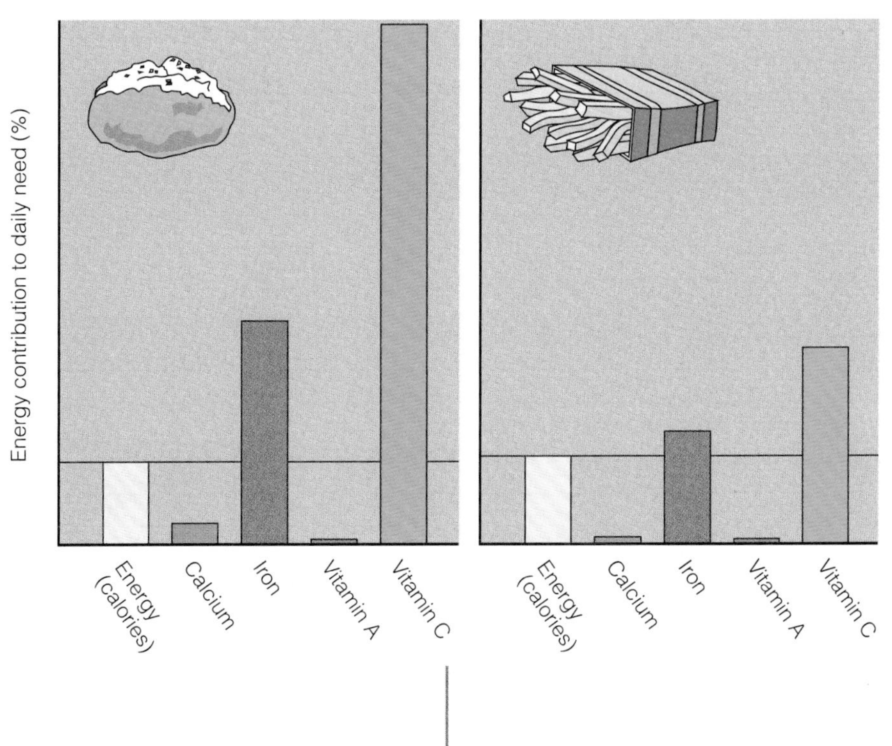
Energy contribution to daily need (%)
Energy (calories)
Calcium
Iron
Vitamin A
Vitamin C
Energy (calories)
Calcium
Iron
Vitamin A
Vitamin C

ANALYZE NUTRITION NEWS

This chapter has pointed out the importance of reading nutrition news with an educated eye. Here is a chance to practice your skills. Find an article in your local newspaper or any publication that carries nutrition news. On a copy of Form 1-1, answer these questions:

1. What sort of language does the writer use? Do the words imply sensationalism or conclusive findings? Phrases such as "startling revelation" or "now we know" or "the study proved" are clues to whether the report is a sensational one. Does the author take a tentative approach, using words such as *may, might*, or *could*? What do these words imply?
2. Is the finding placed in the context of previous nutrition findings? Does the article imply that the current finding wipes out all that has gone before it? Can you detect a broad understanding of nutrition on the writer's part? From what clues? For example, an article about folate and heart disease should say that saturated fat probably plays the major nutrition role in heart disease development.
3. Does the article mention whether the research results under discussion are published in a medical or nutrition journal? Where? The following Controversy section has more information about which types of journals publish valid scientific findings and which do not.
4. How were the results obtained? Can you tell from the article whether this was a case study, an epidemiological study, an intervention study, or a laboratory study? How does that information affect your understanding of what the results have contributed to nutrition science?
5. Does the finding apply to you? Should you change your eating patterns because of it? In what ways did the subjects resemble or differ from you? Were there enough subjects to make the study seem valid? (In a serious evaluation, a statistical analysis would be used to answer this question.)
6. Does the finding make sense to you in light of what you know about nutrition? You may not know enough to make this judgment yet, but by the end of this course, you should have developed a "feel" for identifying information that fits with reality.

This sort of assessment can guide you through the numerous nutrition articles appearing in newspapers and magazines. That way, you can avoid making nutrition decisions based on passing fads.

Form 1-1

Critiquing Nutrition News

The news report I am critiquing comes from ________________ (publication), dated ____________ (attach the report to this form).

1. I evaluate the language used in the publication as follows:

____

____

____

2. I believe the author's understanding of previously reported findings to be:

____

____

____

3. I judge the credibility of the item to be:

____

____

____

4. The methods used to obtain these results were:

____

____

____

5. The results of the study apply to the following populations:

____

____

____

6. To a reader without extensive nutritional background, the results of a study may be misleading. This report might mislead by:

____

____

____

Self Check

Answers to these Self Check questions are in Appendix G.

1. Studies of whole populations that reveal correlations between dietary habits and disease incidence are referred to as:
   a. case studies
   b. intervention studies
   c. laboratory studies
   d. epidemiological studies
2. Energy-yielding nutrients include all of the following *except:*
   a. vitamins
   b. carbohydrates
   c. fat
   d. protein
3. Organic nutrients include all of the following *except:*
   a. minerals
   b. fat
   c. carbohydrates
   d. protein
4. One of the characteristics of a nutritious diet is that it provides no constituent in excess. This principle of diet planning is called:
   a. adequacy
   b. balance
   c. moderation
   d. variety
5. A slice of peach pie supplies 357 calories with 48 units of vitamin A; one large peach provides 42 calories and 53 units of vitamin A. This is an example of:
   a. calorie control
   b. nutrient density
   c. variety
   d. essential nutrients
6. Heart disease and cancer are due to genetic causes, and diet cannot influence whether they occur. T F
7. Both carbohydrates and protein provide 4 calories per gram. T F
8. Once a new finding about nutrition is published, you can feel confident about changing your diet accordingly. T F
9. A registered dietitian (RD) has the educational background necessary to deliver reliable nutrition advice. (Read about this in the upcoming Controversy.) T F

NUTRITION ON THE NET

For further study of the topics of this chapter, access these websites and search for the phrases or words in quotation marks:

1. For the latest changes in chapter material, to find supplemental learning tools, or to access links to related websites, go to the "Nutrition: Concepts and Controversies" site at:
   www.wadsworth.com/nutrition/prod/allprod.html
2. Search the vast U.S. government health information banks for "nutrition":
   www.healthfinder.gov
3. To view "Healthy People 2010":
   web.health.gov/healthypeople
4. For the Canadian "National Plan of Action" for nutrition:
   www.hc-sc.gc.ca/datahpsb/npu
5. Visit the Mayo Clinic Nutrition Center:
   www.mayohealth.org/mayo/common/htm/dietpage.htm
6. For more on "nutrition monitoring" in the United States:
   www.cdc.gov
7. The International Food Information Council provides information on "functional foods":
   www.ificinfo.health.org
8. To find a registered dietitian in your area, select "find a dietitian" at:
   www.eatright.org
9. Search among thousands of current scientific and medical abstracts for any nutrition topic at:
   http://www.ncbi.nlm.nih.gov/PubMed/

WHO SPEAKS ON NUTRITION?

CONTROVERSY
1

TODAY MORE THAN ever before, people want to know what nutrition news they can believe and safely use. They want to know how best to take care of themselves. Some people seek miracles, too: supplements for weight loss without effort and nutrients to forestall aging or prevent baldness or increase sexual potency. People's heightened interest in nutrition translates into a deluge of dollars spent on services and products peddled by both legitimate and fraudulent businesses. Consumers who obtain legitimate products can improve their health. Those enticed into scams, however, may lose their health, their savings, or both.[1] Unfortunately, nutrition and other **fraud** (or **quackery,** defined in Table C1-1) rings cash registers to the tune of $27 billion annually.[2] Ironically, quacks spread useless or even dangerous advice, sham products, and unproven procedures that not only rob people of the very health they are seeking, but also delay their use of legitimate strategies that could truly improve health.[3] Consumers with questions about fraud or suspicions about a product or individual can contact one of the consumer fraud organizations listed in Appendix E.

The makers of fraudulent claims usually are not credentialed professionals. More often than not, their qualifications are nothing more than words on paper. Occasionally, though, a person with all the earmarks of the real thing turns out to be just plain dishonest.

How can people distinguish valid nutrition information from misinformation? One excellent approach is to notice who is purveying the information: quacks or qualified sources. At the extreme, quackery may be easy to identify, as Figure C1-1 points out.

Between the extremes of quackery and scientific thought lies an abundance of less easily recognized nutrition information and misinformation. An instructor at a gym, a physician, a health-store clerk, an author of a book (and seller of juice machines) all recommend nutrition regimens. Can you believe these people? A famous talk show host speaks out to advise viewers about the safety of hamburgers. A creator of a website on the Internet advises dieters to eat pineapples morning, noon, and night. What qualifies these people to give advice? Would following the advice be helpful or harmful? The person who would sift the meaningful nutrition information from the rubble must first learn to recognize quackery wherever it presents itself.

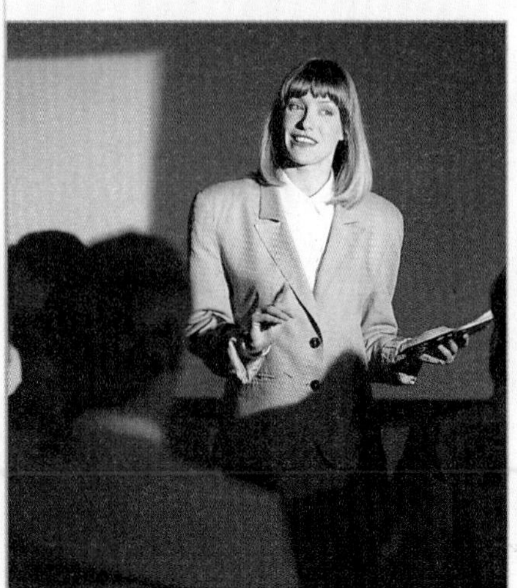

Table C1-1

Misinformation Terms

- **anecdotal evidence** information based on interesting and entertaining, but not scientific, personal accounts of events.
- **fraud** or **quackery** the promotion, for financial gain, of devices, treatments, services, plans, or products (including diets and supplements) that alter or claim to alter a human condition without proof of safety or effectiveness. (The word *quackery* comes from the term *quacksalver,* meaning a person who quacks loudly about a miracle product—a lotion or a salve.)

IDENTIFYING VALID NUTRITION INFORMATION

As Chapter 1 explained, nutrition is a science; that is, it derives information from scientific research. Scientists must systematically conduct research studies and cautiously interpret the findings before they can provide practical nutrition information. The following are characteristics of scientific research:

- Scientists test their ideas by conducting properly designed scientific experiments. They report their methods and procedures in detail so that other scientists can verify the findings through replication.
- Scientists recognize the inadequacy of **anecdotal evidence** or testimonials.
- Scientists who use animals in their research do not apply their findings directly to human beings.
- Scientists may use specific segments of the population in their research. When they do, they are careful not to generalize the findings to all people.
- Scientists report their findings in respected scientific journals. Their work must survive a screening review by their peers before it is accepted for publication.

With each report from scientists, the field of nutrition changes a little—each finding contributes another piece to the whole body of knowledge. Table C1-2 on page 24 lists some sources of credible nutrition information.

NUTRITION ON THE NET

Hundreds of millions of websites await users of the Internet. Searching for nutrition information by computer can be much like walking into an enormous media store, with millions of articles, video clips, newspaper headlines, and magazines on display. Also like a media store, the Internet makes no guarantees of the accuracy of its wares.[4] Be forewarned: much of the nutrition "information" found on the Internet is pure fiction, aimed at separating consumers from their money. Sales of unproven and dangerous products over the Internet have reached unbelievable proportions, partly because the Internet is difficult to regulate. The Internet also offers access to high-quality information, however, so it pays to learn to use it wisely. Table C1-3 provides some clues to reliable nutrition information websites.

When using the Internet, pay attention to the last few letters in a website's name. Generally, an extension of three letters following a dot gives clues about a site's affiliations. For example, a site bearing the letters "gov" is a government site and generally posts valid information. Universities use the extension "edu," standing for *educational institution;* these sites may include everything from interesting opinion chat rooms to reports of ongoing research projects. The extension "com" stands for *commercial,* a business website that typically has products or service to sell; and "org" stands for *organization,* websites that can include just about anything. Of course, these hints alone are insufficient for judging material on the net. The user must also scrutinize those posting materials, as described next.

WHO ARE THE TRUE NUTRITION EXPERTS?

Most people turn to their physicians for dietary advice. Physicians are expected to know all about health-related matters.

Figure C1-1

Earmarks of Nutrition Quackery

The more of these claims you hear about nutrition information, the less likely it is to be valid.

STEP RIGHT UP! **VIT-O-MITE** WILL MAKE YOU AS **HEALTHY** AS **A HORSE**—GUARANTEED! FEEL **STRONGER, LOSE** WEIGHT, **IMPROVE** YOUR MEMORY ALL WITH THE HELP OF **VITE-O-MITE!** OH SURE, YOU MAY HAVE HEARD THAT **VITE-O-MITE** IS NOT ALL THAT WE SAY IT IS, BUT THAT'S WHAT THE FDA WANTS YOU TO THINK! **OUR DOCTORS** AND SCIENTISTS SAY IT'S THE ULTIMATE VITAMIN SUPPLEMENT. SAY NO! TO THE WEAKENED VITAMINS IN TODAY'S FOODS. **VITE-O-MITE** INCLUDES **POTENT SECRET INGREDIENTS** THAT YOU CANNOT GET WITH ANY OTHER PRODUCT! YESSIREE FOLKS, STEP RIGHT UP!

BEFORE
AFTER
280lbs.
160lbs.
STUDY
VITE O MITE

Too good to be true
The claim presents enticingly simple answers to complex problems. It says what most people want to hear. It sounds magical.

Suspicions about food supply
The person or institution pushing the product or service urges distrust of the current methods of medicine or suspicion of the regular food supply, with "alternatives" for sale (providing profit to the seller) under the guise that people should have freedom of choice.

Testimonials
The evidence presented to support the claim is in the form of praise by people who have been "healed," "made younger," and the like by the product or treatment.

Fake credentials
The person or institution making the claim is titled "doctor," "university," or the like, but has simply created or bought the title and is not legitimate.

Unpublished studies
Scientific studies are cited, but are nowhere published and so cannot be critically examined.

Persecution claims
The person or institution pushing the product or service claims to be persecuted by the medical establishment or tries to convince you that physicians "want to keep you ill so that you will continue to pay for office visits."

Authority not cited
The studies cited sound valid, but are not referenced, so that it is impossible to check and see if they were conducted scientifically.

Motive: personal gain
The person or institution making the claim stands to make a profit if it is believed.

Advertisement
The claim is being made by an advertiser who is paid to make claims for the product or procedure. (Look for the word "Advertisement," probably in tiny print somewhere on the page.)

Unreliable publication
The studies cited are published, but in a newsletter, magazine, or journal that publishes misinformation.

Logic without proof
The claim seems to be based on sound reasoning but hasn't been scientifically tested and shown to hold up.

But are physicians the best sources of accurate and current information on nutrition? Only about a quarter of all medical schools in the United States require students to take even one nutrition course. Students attending these classes receive an average of 20 hours of nutrition instruction—an amount most graduates consider inadequate. Although many experts call for a greatly expanded role for nutrition in the medical curriculum, they acknowledge that the curriculum carries a heavy burden already. Many see it as a challenge to integrate adequate, meaningful nutrition information into already existing courses.

A decade ago, Congress passed a law mandating that "students enrolled in United States medical schools and physicians practicing in the United States [must] have access to adequate training in the field of nutrition and its relationship to human health."[5] Plans are now in the works to make nutrition education a standard course in medical schools. Enlarging on this idea, the **American Dietetic Association (ADA)** asserts that nutrition education should be part of the curriculum for all sorts of health-care professionals: physician's assistants, dental hygienists, physical and occupational therapists, social workers, and all others who provide services directly to clients.[6] This way more people would have access to reliable nutrition information.

Even though most physicians are not schooled adequately in nutrition, some are superbly qualified to speak on nutrition. All physicians appreciate the connections between health and nutrition because of their course work in biochemistry and physiology. Those who have specialized in clinical nutrition in medical schools that offer that specialty are especially well qualified. Membership in the American Society for Clinical Nutrition, whose journal is cited many times throughout this text, is another sign of nutrition knowledge. Still, few physicians have the knowledge, time, or experience to develop diet plans and provide detailed diet instruction for clients. Often physicians wisely refer their clients to nutrition specialists for diet advice. Table C1-4 lists the best specialists to choose.

Table C1-2

Credible Sources of Nutrition Information

Professional health organizations, government health agencies, volunteer health agencies, and consumer groups provide consumers with reliable health and nutrition information. Credible sources of nutrition information include:

- Professional health organizations, especially the American Dietetic Association's National Center for Nutrition and Dietetics (NCND); also the Society for Nutrition Education and the American Medical Association.
- Government health agencies such as the Federal Trade Commission (FTC), the U.S. Department of Health and Human Services (DHHS), the Food and Drug Administration (FDA), and the U.S. Department of Agriculture (USDA).
- Volunteer health agencies such as the American Cancer Society, the American Diabetes Association, and the American Heart Association.
- Reputable consumer groups such as the Better Business Bureau, the Consumers Union, the American Council on Science and Health, and the National Council Against Health Fraud.

Appendix E provides addresses for these and other organizations.

SOURCE: Data from J. M. Ashley and W. T. Jarvis, Position of the American Dietetic Association: Food and nutrition misinformation, *Journal of the American Dietetic Association* 95 (1995): 705–707.

Table C1-3

Is This Site Reliable?

To judge whether an Internet site offers reliable nutrition information, answer the following questions?

Who is responsible for the site?
Clues can be found in the three-letter "tag" that follows the dot in the site's name. For example, "gov" and "edu" indicate government and university sites, usually reliable sources of information.

Do the names and credentials of information providers appear? Is an editorial board identified?
Many legitimate sources provide e-mail addresses or other ways to obtain more information about the site and the information providers behind it.

Are links with other reliable information sites provided?
Reputable organizations almost always provide links with other similar sites because they want you to know of other experts in their area of knowledge. Caution is needed when you evaluate a site by its links, however. Anyone, even a quack, can link a Web page to a reputable site without the organization's permission. Doing so may give the quack's site the appearance of legitimacy, just the effect the quack is hoping for.

Is the site updated regularly?
Nutrition information changes rapidly, and sites should be updated often.

Is the site selling a product or service?
Commercial sites may provide accurate information, but they also may not, and their profit motive increases the risk of bias.

Does the site charge a fee to gain access to it?
Many academic and government sites offer the best information, usually for free. Some legitimate sites do charge fees, but before paying up, check the free sites. Chances are good you'll find what you are looking for without paying.

Some credible websites include:

National Council Against Health Fraud
www.ncahf.org

Stephen Barrett's Quackwatch
www.quackwatch.com

Tufts University
navigator.tufts.edu

U.S. Government
www.healthfinder.gov
Search for Quackery

SOURCE: Adapted from M. Larkin, Health information on-line, FDA Consumer, June 1996. Available from *http://vm.cfsan.fda.gov/list.html*.

Fortunately, the credential that indicates a qualified nutrition expert is easy to spot—you can confidently call on a **registered dietitian (RD).** Additionally, some states require that **nutritionists,** as well as **dietitians,** receive a **license to practice.** Meeting these established criteria certifies that an expert is the genuine article.

Dietitians are easy to find in most communities because they perform a multitude of duties in a variety of settings. They work in foodservice operations, pharmaceutical companies, the food industry, home health agencies, long-term care institutions, private clinics, public health departments, cooperative extension offices,* research centers, education settings, some fitness centers, and hospitals.

*Cooperative extension agencies are associated with land grant colleges and universities and may be found in the phone book's government listings.

Dietitians can assume a number of different responsibilities depending on their work settings and positions.[7] Dietitians in hospitals have many subspecialties. Administrative dietitians manage the foodservice system; clinical dietitians provide client care (see Table C1-5); and nutrition support team dietitians coordinate nutrition care with the efforts of other health-care professionals. In the food industry, dietitians conduct research, develop products, and market services. Dietitians who specialize in public health nutrition work in government-funded agencies to provide nutrition services to populations. Among their many roles, **public health nutritionists** help plan, coordinate, and evaluate programs; act as consultants to other agencies; manage finances; and much more.[8] Nutrition graduates with advanced degrees or course work in public health are well placed for employment in this vast field.

DETECTING FAKE CREDENTIALS

In contrast to RDs, thousands of people possess fake nutrition degrees and claim to be nutrition counselors, nutritionists, or "dietists." These and other such titles may sound meaningful, but most of these people lack the established credentials of the ADA-sanctioned dietitian. If you look closely, you can see signs that their expertise is fake.

Take, for example, a nutrition expert's educational background. The minimum standards of education for a dietitian specify a bachelor of science (BS) degree in food science and human nutrition (or related fields) from an **accredited** college or university (Table C1-6 defines this and related terms). Such a degree generally requires four to five years of study. In contrast, a fake nutrition expert may display a degree from a six-month correspondence course; such a degree is simply not the same.† In some cases, schools

†To find out whether a correspondence school is accredited, write the National Home Study Council, Accrediting Commission, 1601 Eighteenth Street NW, Washington, DC 20009, or call (202) 234-5100.

Table C1-4

Terms Associated with Nutrition Advice

- **American Dietetic Association (ADA)** the professional organization of dietitians in the United States. The Canadian equivalent is the Dietitians of Canada (DC),[a] which operates similarly.
- **dietitian** a person trained in nutrition, food science, and diet planning. See also *registered dietitian.*
- **license to practice** permission under state or federal law, granted on meeting specified criteria, to use a certain title (such as *dietitian*) and to offer certain services. Licensed dietitians may use the initials LD after their names.
- **medical nutrition therapy** nutrition services used in the treatment of injury, illness, or other conditions; includes assessment of nutrition status and dietary intake, and corrective applications of diet, counseling, and other nutrition services.
- **nutritionist** someone who engages in the study of nutrition. Some nutritionists are RDs, whereas others are self-described experts whose training is questionable and who are not qualified to give advice. In states with responsible legislation, the term applies only to people who have masters of science (MS) or doctor of philosophy (PhD) degrees from properly accredited institutions.
- **public health nutritionist** a dietitian or other person with an advanced degree in nutrition who specializes in public health nutrition.
- **registered dietitian (RD)** a dietitian who has graduated from a university or college after completing a program of dietetics. The program must be approved or accredited by the American Dietetic Association (or Dietitians of Canada). The dietitian must serve in an approved internship, coordinated program, or preprofessional practice program to practice the necessary skills; pass the five parts of the association's *registration* examination; and maintain competency through continuing education.[b] Many states also require licensing for practicing dietitians.
- **registration** listing with a professional organization that requires specific course work, experience, and passing of an examination.

[a]A new organization comprised of the former Canadian Dietetic Association (CDA) and ten provincial dietetic associations.

[b]The five content areas included on the registration examination for dietitians are nutrition services, foodservice systems, management, education and communication, and evaluation and standards. L. C. Webb and J. O. Maillet, The development of test specifications for the registration examinations, *Journal of the American Dietetic Association* 90 (1990): 1134–1135.

Table C1-5

Responsibilities of a Clinical Dietitian

The first six items on this list play essential roles in **medical nutrition therapy** as part of a medical treatment plan.

- Assesses clients' nutrition status.
- Determines clients' nutrient requirements.
- Monitors clients' nutrient intakes.
- Develops, implements, and evaluates clients' medical nutrition therapy.
- Counsels clients to cope with unique diet plans.
- Teaches clients and their families about nutrition and diet plans.
- Provides training for other dietitians, nurses, interns, and dietetics students.
- Serves as liaison between clients and the foodservice department.
- Communicates with physicians, nurses, pharmacists, and other health-care professionals about clients' progress, needs, and treatments.
- Participates in professional activities to enhance knowledge and skill.

posing as legitimate **correspondence schools** offer even less. They are actually **diploma mills**—fraudulent businesses that sell certificates of competency to anyone who pays the fees, from under a thousand dollars for a bachelor's degree to several thousand for a doctorate. Buyers ordering multiple degrees are given discounts. To obtain these "degrees," a candidate need not read any books or pass any examinations.

Lack of proper accreditation is the identifying sign of a fake educational institution. To guard educational quality, an accrediting agency recognized by the U.S. Department of Education certifies that certain schools meet the criteria defining a complete and accurate schooling, but in the case of nutrition, quack accrediting agencies cloud the picture. Fake nutrition degrees are available from schools "accredited" by more than 30 phony accrediting agencies.[†]

[†]To find out whether a correspondence school is accredited, write the Distance Education and Training Council, Accrediting Commission, 1601 Eighteenth Street, N.W., Washington, D.C. 20009, call (202) 234-5100, or visit their website (www.detc.org).

To find out whether a school is properly accredited for a dietetics degree, write the American Dietetic Association, Division of Education and Research, 216 West Jackson Boulevard, Chicago, IL 60606, call (312) 899-4870, or visit their website (www.eatright.org/caade).

The American Council on Education publishes a directory of accredited institutions, professionally accredited programs, and candidates for accreditation in *Accredited Institutions of Postsecondary Education Programs Candidates* (available at many libraries). For additional information, write the American Council on Education, One Dupont Circle NW, Suite 800, Washington, D.C. 20036, call (202) 939-9382, or visit their website (www.acenet.edu).

Table C1-6

Terms Describing Institutions of Higher Learning, Legitimate and Fraudulent

- **accredited** approved; in the case of medical centers or universities, certified by an agency recognized by the U.S. Department of Education.
- **correspondence school** a school that offers courses and degrees by mail. Some correspondence schools are accredited; others are *diploma mills.*
- **diploma mill** an organization that awards meaningless degrees without requiring its students to meet educational standards.

American Association of
Nutrition & Dietary
Consultants
Professional Member
Sassafras Herbert

Sassafras and Charlie display their professional credentials.

To dramatize the ease with which anyone can obtain a fake nutrition degree, one writer enrolled for $82 in a nutrition diploma mill that billed itself as a correspondence school. She made every attempt to fail. She intentionally answered all the examination questions incorrectly. Even so, she received a "nutritionist" certificate at the end of the course, togther with a letter from the "school" officials explaining that they were sure she must have misread the test.

In a similar stunt, Ms. Sassafras Herbert was named a "professional member" of a nutrition association. For her efforts, Sassafras has received a wallet card and is listed in a sort of fake Who's Who in nutrition that is distributed at health fairs and trade shows nationwide. Sassafras is a poodle. Her master, Victor Herbert, MD, paid $50 to prove that she could be awarded these honors merely by sending in her name. Mr. Charlie Herbert is also a professional member of such an organization; Charlie is a cat.

State laws do not necessarily help consumers distinguish experts from fakes; some states allow anyone to use the title *dietitian* or *nutritionist.* But some states are beginning to respond to the need by allowing only RDs or people with certain graduate degrees to call themselves dietitians, and many have licensing requirements. Licensing provides a way to identify people who have met minimum standards of education and experience.

By knowing who is qualified to speak on nutrition, consumers can stay one step ahead of the nutrition quacks. Does the instructor at the spa have a degree in nutrition from an accredited university? No? Better check the instructor's advice with someone who does. Is the author of the magazine article an RD? If not, you cannot know whether to believe what you've read. Have you seen the health-store clerk's license to practice as a dietitian? If not, seek a qualified source—an RD or a person with an advanced degree in nutrition.

In summary, to check a provider's qualifications, first look for the degrees and credentials listed by the person's name (such as MD, RD, MS, PhD, or LD). Next find out what you can about the reputations of the institutions that awarded the degrees. Then call your state's health-licensing agency and ask if dietitians are licensed in your state. If they are, find out whether the person giving you dietary advice has a license—and if not, find someone better qualified. Your health is your most precious asset, and protecting it is well worth the time and effort it takes to do so.

NOTES

Notes are in Appendix F.

2 Nutrition Tools—Standards and Guidelines

Paul Gauguin 1848–1903, *Nave Nave Moe (Sacred Spring-Sweet Dreams)* (detail), 1894

NAVE NAVE MOE

Contents

Frequently Asked Questions

FAQ: Why can't nutrient advice be kept simple? p. 29

FAQ: How much of a nutrient is too much? p. 32

FAQ: How can the Food Guide Pyramid help me to eat well? p. 38

FAQ: How many servings do I need each day? p. 44

FAQ: Which packaged foods and restaurant choices are best for my health? p. 49

EATING WELL is easy in theory. All you have to do is choose a selection of foods that supplies appropriate amounts of the essential nutrients, fiber, and energy without excess intakes of fat, sugar, and salt, and be sure to get enough exercise to balance the foods you eat. A few people do these things automatically, but most do not.[1] Many people are overweight, or undernourished, or suffer from nutrient excesses or deficiencies that impair their health—that is, they are malnourished. You may not think that this statement applies to you, but you may already have less-than-optimal nutrient intakes and activity without knowing it. Accumulated over years, the effects of your habits can seriously impair the quality of your life. Putting it positively, you can enjoy the best possible vim, vigor, and vitality if you learn now to nourish yourself optimally.

To master the task of meeting your nutrition needs, you may find it useful to learn the answers to several questions. How much energy and how much of each nutrient do you need? How much physical activity do you need to balance the energy you take in from foods? Which types of foods supply which nutrients? How much of each type of food do you have to eat to get enough? And how can you eat all these foods without gaining weight and without getting too much fat or sugar? This chapter begins by identifying some ideals for nutrient intakes and ends by showing how to achieve them.

Recommended Dietary Allowances (RDA) formerly, the name of the nutrient intake standards of the United States. Currently, the RDA constitute a part of the Dietary Reference Intakes (DRI). RDA are average daily amounts of nutrients considered adequate to meet the known nutrient needs of practically all healthy people.

Dietary Reference Intakes a set of four lists of values for the dietary nutrient intakes of healthy people in the United States and Canada. The values include:

- Estimated Average Requirement (EAR)
- Recommended Dietary Allowances (RDA)
- Adequate Intakes (AI).
- Tolerable Upper Intake Levels (UL)

Descriptions of the DRI values and other nutrient standards are found in Table 2-1 on page 30.

Nutrient Recommendations

Nutrient recommendations are sets of yardsticks used as standards for measuring healthy people's energy and nutrient intakes. Nutrition experts use them to make nutrient recommendations, to assess nutrient intakes, and to perform other nutrient-related tasks. For 50 years, the **Recommended Dietary Allowances (RDA)** have been the U.S. nutrient intake standards; in Canada their equivalents were the Recommended Nutrient Intakes for Canadians (RNI), listed in Appendix B. Today, both of these standards are being replaced by the **Dietary Reference Intakes (DRI).**

Making the change to the DRI requires some effort. The DRI represent a whole new way of thinking about nutrient values. For example, for each nutrient, the DRI establish two or three values where there used to be only one. This change represents a tremendous accomplishment, for each DRI value serves a different purpose (more details about these purposes later). Be assured, however, that although all the DRI values are important for their purposes, most people need to focus on only two kinds of DRI values: those that set nutrient intake goals for individuals and those that define an upper limit of safety for nutrient intakes.

People in Canada and the United States share a new set of nutrient intake recommendations: the Dietary Reference Intakes (DRI).

Directory of recommendations:

- DRI lists, inside front cover page A.
- 1989 RDA for nutrients, inside front cover page B.
- Daily Values, inside front cover page C
- 1989 RDA Recommended Energy Intakes, inside back cover page Y.

One other set of nutrient standards can assist the person striving to make wise nutrition choices. These are the **Daily Values**, familiar to anyone who has stopped to read a food label. (Read about the Daily Values and other nutrient standards in Table 2-1, page 30.) All of these nutrient standards—the RDA, DRI, and Daily Values—are used and referred to so often that they are printed on pages of the inside front and back covers of this book.

The Canadiana appendix (Appendix B) presents the 1990 RNI.

KEY POINT *Nutrient recommendations are currently undergoing change in the United States and Canada. The Dietary Reference Intakes are replacing traditional recommendations in both countries. The Daily Values are U.S. standards used on food labels.*

The DRI Nutrient Intake Recommendations

A committee of qualified nutrition experts from the United States and Canada is currently developing and publishing the DRI.* So far, the DRI committee has published recommendations for ten vitamins and four minerals. Until the committee finishes its

*This is a committee of the Food and Nutrition Board, Institute of Medicine of the National Academy of Sciences, working in association with Health Canada.

work, some of the older standards are still being used. Currently, the 1989 RDA values for protein, energy, four vitamins, and four minerals are still in effect.

 Why Can't Nutrient Advice Be Kept Simple?

One can rightly ask, "Why do we need new nutrient intake standards to replace the old ones?" The answers to this question can be found in the purposes of the new values. The next four sections spell out the goals the DRI committee had in mind when setting the DRI values.

Goal #1. Setting Intake Recommendations for Individuals One of the great advantages of the DRI values recommended intakes lies in their applicability to individuals. In the past, nutrient standards were appropriate for planning and assessing the diets of populations; their developers discouraged their use for individuals. In contrast, individuals are a prime concern of the DRI committee. The committee offers two sets of values for judging the nutrient intakes of individuals: an updated set of Recommended Dietary Allowances (RDA) and a set referred to as **Adequate Intakes (AI).**

The RDA are indisputably the bedrock of the DRI recommended intakes, for they are based on solid experimental evidence and other reliable observations. The AI values are also as scientifically based as possible, but setting them requires some educated guesswork. The committee establishes an AI value whenever scientific evidence is insufficient to generate an RDA.[2] Both the RDA and AI values are intended to be used as goals for planning nutritious diets for individuals, so there is no need to distinguish between them for the practical purposes of diet planning.[3] This book refers to them collectively as the DRI recommended intakes.[†]

Goal #2. Preventing Chronic Diseases Another advantage of the DRI is that they set the recommended intakes to take into account disease prevention, where appropriate, as well as nutrient adequacy. In the last decade, abundant new research has linked nutrients in the diet with the promotion of health and the prevention of chronic diseases, and the DRI Committee used this research to advantage. For example, the recommendation for calcium for each life stage now reflects calcium intakes thought to lessen the likelihood of osteoporosis-related fractures later in life.

Goal #3. Facilitating Nutrition Research and Policy Another set of values established by the DRI committee accomplishes other important tasks. This set, the **Estimated Average Requirements (EAR),** establishes population-wide average requirements that researchers and nutrition policymakers use in their work. Nutrition scientists may use the EAR as standards in research. Public health officials may also use them to assess nutrient intakes of populations and make recommendations.

Goal #4. Establishing Safety Guidelines A final goal of the DRI committee is to establish upper limits of intake for nutrients that can pose a hazard when they are overconsumed. These values, the **Tolerable Upper Intake Limits (UL),** are indispensable to consumers who take supplements or eat foods to which vitamins or minerals are added. People need to know how much of a nutrient is too much. (A later section returns to the idea that nutrients can cause harm when taken in excess.) The UL are also of value to public health officials who set the allowances for nutrients added to our food and water supplies.

Making Good Tools Better All in all, the new DRI values are well designed to meet the diverse needs of individuals, the scientific and medical communities, and

In 1999, DRI recommended intake values exist for:

Vitamins
Vitamin D.
Thiamin.
Riboflavin.
Niacin.
Vitamin B_6.
Folate.
Vitamin B_{12}.
Pantothenic acid.
Biotin.
Choline.

Minerals
Calcium.
Fluoride.
Magnesium.
Phosphorus.

1989 RDA still stand for:
Energy (calories).
Protein.

Vitamins
Vitamin A.
Vitamin E.
Vitamin K.
Vitamin C.

Minerals
Iron.
Zinc.
Iodine.
Selenium.

The DRI table on the inside front cover distinguishes the RDA and AI values, but both kinds of values are intended as nutrient intake goals for individuals.

See Table 2-1 on page 30 for definitions of terms on this page.

[†] As suggested by DRI committee members, personal communication, national convention of the American Dietetic Association, October 1998.

Don't let the "alphabet soup" of nutrient intake standards confuse you. Their names make sense when you learn their purposes.

others.[+] Table 2-1 sums up the names and purposes of the nutrient intake standards just introduced.

KEY POINT *The DRI provide nutrient intake goals for individuals, take into account new research on disease prevention, provide a set of standards for researchers and makers of public policy, and establish tolerable upper limits for nutrients that can be toxic in excess.*

Understanding the DRI Recommended Intakes for Individuals

Nutrient recommendations have been much misunderstood. One young woman, on learning of the DRI recommended intakes, was outraged: "You mean that some bureaucrat tells me that I must eat exactly 5.0 micrograms of vitamin D every day?" This is not the DRI committee's intention. The DRI are recommendations, not commandments. The following facts will help put the DRI recommended intakes into perspective:

- Their ongoing creation is funded by the government, but the committee that determines the values is composed of scientists representing a variety of specialties.
- The values are based on available scientific research to the greatest extent possible and are updated periodically in light of new knowledge.
- The values are recommendations for optimal and safe intakes, not minimum requirements. They include a generous margin of safety and meet the needs of virtually all healthy people in a specific age and gender group.

Table 2-1

Nutrient Standards

Standards from the DRI Committee

Dietary Reference Intakes (DRI) a set of nutrient intake values for the dietary nutrient intakes of healthy people in the United States and Canada. These values are used for planning and assessing diets and include these four lists of values:

1. **Recommended Dietary Allowances (RDA)**
   Nutrient intake goals for individuals.[a] Derived from the Estimated Average Requirements (see below).
2. **Adequate Intakes (AI)**
   Nutrient intake goals for individuals.[a] Set whenever scientific data are insufficient to allow establishment of an RDA value.
3. **Tolerable Upper Intake Levels (UL)**
   Suggested upper limits of intake for potentially toxic nutrients. Intakes above the UL are likely to cause illness from toxicity.
4. **Estimated Average Requirements (EAR)**
   Population-wide average nutrient requirements used in nutrition research and policymaking. The basis upon which RDA values are set.

Daily Values

Daily Values (DV)
Nutrient standards used on food labels, in grocery stores, and on some restaurant menus. The DV allow comparisons among foods with regard to their nutrient content.

[a]For simplicity, this book combines the two sets of nutrient goals for individuals (AI and RDA) and refers to them as *DRI recommended intakes.*

The Daily Values, discussed later, that appear in figures of this book compare the nutrient contents of foods.

- The values are chosen in reference to specific indicators of nutrient adequacy, such as blood nutrient concentrations, normal growth, and reduction of risk of certain chronic diseases or other disorders.
- The values are recommended daily intakes to be achieved on average, over time. They assume that intakes will vary from day to day, and they are set high enough to ensure that body nutrient stores will meet nutrient needs during periods of inadequate intakes lasting a day or two for some nutrients and up to a month or two for others.
- The recommendations apply to healthy persons only.

Separate recommendations are made for specific sets of people: Men, women, pregnant women, children, and other life-stage groups vary in their needs. Children aged four to eight years, for example, have their own DRI recommended intakes. Each individual can look up the recommendations for his or her own age and gender group.

The DRI recommended intakes are generous allowances. Even so, they do not necessarily cover every individual for every nutrient. On average, one should probably try to get 100 percent or more of the DRI recommended intake for every nutrient to ensure an adequate intake over time.

The DRI are designed for health maintenance and disease prevention in healthy people, not for the restoration of health. Under the stress of serious illness or malnutrition, a person may require a much higher intake of certain nutrients or may not be able to handle even the DRI amount. Therapeutic diets adjust the recommended intakes to account for the needs of medical conditions, such as recovery from surgery, burns, fractures, illnesses, or addictions.

KEY POINT *The DRI used in the United States and Canada represent suggested daily intakes of energy and selected nutrients for healthy people in the population. The DRI are comprised of the RDA, AI, UL, and EAR lists of values. The Daily Values are nutrient intake standards used on food labels.*

How the Committee Establishes DRI Values—An RDA Example

At this point you may be wondering how the DRI committee goes about its work of setting the DRI values. A theoretical discussion will help to explain the process. Suppose we are the DRI committee members with the task of setting an RDA for nutrient X (an essential nutrient).[‡] Ideally, our first step will be to find out how much of that nutrient various healthy individuals need. To do so, we review studies of deficiency states, nutrient stores and their depletion, and the factors influencing them. We then select the most valid data for use in our work. Of the DRI family of nutrient standards, the setting of an RDA value demands the most rigorous science and tolerates the least guesswork.

One experiment we might review or conduct is a **balance study.** In this type of study, scientists measure the body's intake and excretion of a nutrient to find out how much intake is required to balance excretion. For each individual subject, we can determine a **requirement** to achieve balance for nutrient X. With an intake below the requirement, a person will slip into negative balance or experience declining stores that could, over time, lead to deficiency of the nutrient.

With additional study, we find that different individuals, even of the same age and gender, have different requirements. Mr. A needs 40 units of the nutrient each day to maintain balance; Mr. B needs 35; Mr. C, 57. If we look at enough individuals, we find that their requirements are distributed as shown in Figure 2-1—with most requirements near the midpoint (here, 45), and only a few at the extremes.

balance study a laboratory study in which a person is fed a controlled diet and the intake and excretion of a nutrient are measured. Balance studies are valid only for nutrients like calcium (chemical elements) that do not change while they are in the body.

requirement the amount of a nutrient that will just prevent the development of specific deficiency signs; distinguished from the DRI recommended intake value, which is a generous allowance with a margin of safety.

[‡]This discussion describes how an RDA value is set; to set an AI value, the committee would use some educated guesswork as well as scientific research results to determine an approximate amount of the nutrient most likely to support health.

Figure 2-1

Individuality of Nutrient Requirements

Each square represents a person. A, B, and C are Mr. A, Mr. B, and Mr. C. Each has a different requirement.

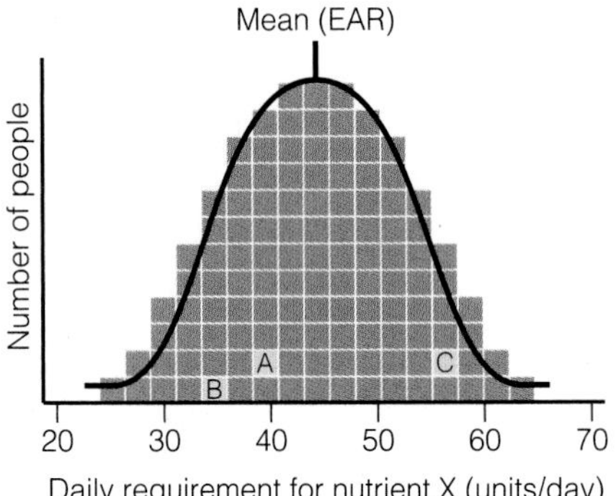

Mean (EAR)
Number of people
B
A
C
20
30
40
50
60
70
Daily requirement for nutrient X (units/day)

To set the value, we have to decide what intake to recommend for everybody. Should we set it at the mean (45 units in Figure 2-1)? This is the Estimated Average Requirement (EAR) for nutrient X, mentioned earlier as valuable to scientists, but not appropriate as an individual's nutrient goal. The EAR value is probably close to everyone's minimum need, assuming the distribution shown in Figure 2-1. (Actually, the data for most nutrients indicate a distribution that is much less symmetrical.) But if people took us literally and consumed exactly this amount of nutrient X each day, half the population would begin to develop internal deficiencies and possibly even observable symptoms of deficiency diseases. Mr. C (at 57) would be one of those people.

Perhaps we should set the recommendation for nutrient X at or above the extreme, say, at 70 units a day, so that everyone will be covered. (Actually, we didn't study everyone, so some individual we didn't happen to test might have an even higher requirement.) This might be a good idea in theory, but what about a person like Mr. B, who requires only 35 units a day? The recommendation would be twice his requirement, and to follow it, he might spend money needlessly on foods containing nutrient X to the exclusion of foods containing other nutrients he needs.

The decision we finally make is to set the value high enough so that the bulk of the population will be covered but not so high as to be excessive (the graph on the left of Figure 2-2 illustrates such a value). In this example, a reasonable choice might be 63 units a day. Moving the DRI further toward the extreme would pick up a few additional people, but it would inflate the recommendation for most people, including Mr. A and Mr. B. The committee makes judgments of this kind when setting the DRI recommended intakes for nutrients. In theory, relatively few healthy people have requirements that are not covered by the DRI recommended intakes.

In contrast to the recommendations for nutrients, the value set for energy intake is not generous; instead it is set at the average of the population's estimated energy requirements (see the graph on the right in Figure 2-2). Too much energy is as bad for health as too little because excess energy leads to obesity, whereas too little energy may cause undernutrition. The 1989 RDA for energy intakes are found on page y of the inside back cover.

KEY POINT *The DRI are based on scientific data and are designed to cover the needs of virtually all healthy people in the United States and Canada.*

How Much of a Nutrient Is Too Much?

Beyond a certain point, it is unwise to consume large amounts of any nutrient. Nutrient needs fall within a range, and there is a danger zone both below and above that range. Figure 2-3 illustrates this point. As supplements become more and more

Figure 2-2

The Differences Between Recommended Intakes of Nutrients and Energy

The nutrient intake recommendations are set so that they will meet the requirements of nearly all people (boxes represent people). The recommended intake of energy is set at the average, or mean, so that half the population's requirements will fall below and half above the recommended level.

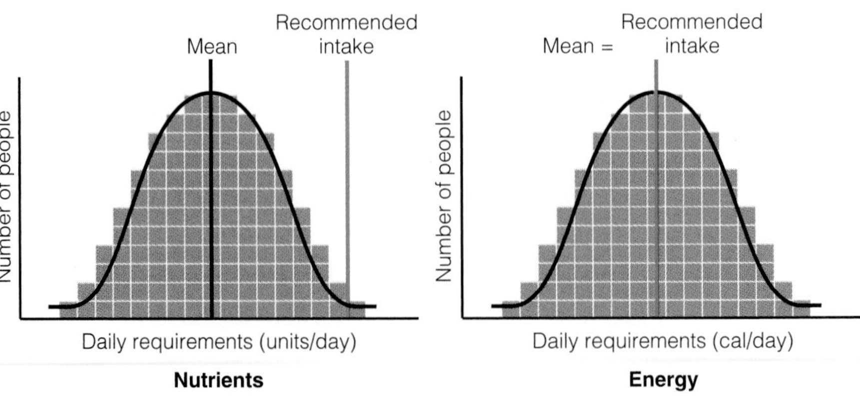

Mean
Recommended intake
Number of people
Daily requirements (units/day)
Nutrients
Mean =
Recommended intake
Number of people
Daily requirements (cal/day)
Energy

popular, and as manufacturers add more and more nutrients to fortified foods, people need to know how much of a nutrient is too much. As discussed earlier, the DRI committee sets Tolerable Upper Intake Levels (UL) to suggest upper limits for nutrient intakes that are likely to be safe for most healthy people. The UL appear on page A of the inside front cover and also as part of Table 7-3 in Chapter 7.

People's tolerances for high doses of nutrients vary. The scientists who developed the UL values urge individuals to use caution when applying them. UL values are still under development, so some nutrients are missing from the list. The absence of a nutrient does not imply that it is safe to consume in any amount. It means only that the committee has yet to complete its work for that nutrient, or that insufficient data exist to establish a value.

KEY POINT *The DRI committee also sets Tolerable Upper Intake Levels as suggested upper limits for intakes of nutrients believed to be tolerable by most healthy people.*

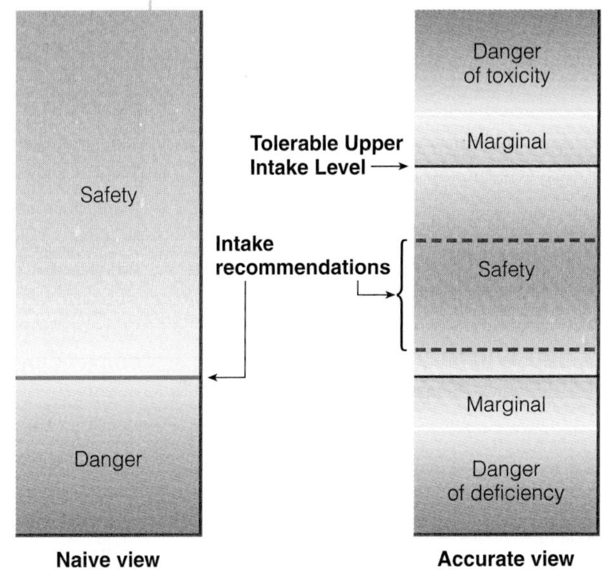

Safety
Danger
Naive view
Tolerable Upper Intake Level
Intake recommendations
Danger of toxicity
Marginal
Safety
Marginal
Danger of deficiency
Accurate view

Figure 2-3

The Naive View versus the Accurate View of Optimal Nutrient Intakes

Consuming too much of a nutrient endangers health, just as consuming too little does. The DRI recommended intakes fall within a range of safe intake levels with the UL marking the tolerable upper limit.

Daily Values

Most careful diet planners are already familiar with the Daily Values as a set of standards already mentioned as those used on U.S. food labels. Those who learn about the DRI ask why yet another set of standards is needed for food labels. Why not just use the DRI? One answer is that the DRI values vary from group to group, whereas on a label, one set of values must apply to everyone. The Daily Values reflect the needs of an "average" person—someone eating 2,000 to 2,500 calories a day. Another answer is that a label must specify daily amounts of food constituents not yet covered by the DRI, such as carbohydrate, fat, and fiber and Daily Values are set for those constituents.

Tolerable Upper Intake Levels (UL) are listed on p. A inside the front cover.

The Daily Values are ideal for their intended purpose of allowing comparisons among *foods*. This strength is also their limitation, however. Because the Daily Values apply to all people, from children of age four through aging adults, they are much less useful as nutrient intake goals for individuals. Details about how to use the Daily Values appropriately in making comparisons among foods are offered in this chapter's Consumer Corner.

KEY POINT *The Daily Values are standards used on food labels to enable consumers to compare the nutrient values among foods.*

Other Nutrient Standards

Many nations and international groups have published sets of standards similar to the DRI. They differ from the DRI in some respects, though, partly because of different interpretations of the data from which they are derived and partly because people in different parts of the world have somewhat different food intakes and daily lives.

Many countries use recommendations developed by two international groups: the World Health Organization (WHO) and the Food and Agriculture Organization (FAO). The WHO/FAO recommendations are considered sufficient for the maintenance of health in nearly all healthy people worldwide.

Appendix E, Nutrition Resources, provides addresses for WHO, FAO, and other agencies.

KEY POINT *Many nations and groups issue recommendations for nutrient and energy intakes appropriate for specific groups of people.*

Table 2-2

The Dietary Guidelines for Americans

- Eat a variety of foods.
- Balance the food you eat with physical activity—maintain or improve your weight.
- Choose a diet with plenty of grain products, vegetables, and fruits.
- Choose a diet low in fat, saturated fat, and cholesterol.
- Choose a diet moderate in sugars.
- Choose a diet moderate in salt and sodium.
- If you drink alcoholic beverages, do so in moderation.

SOURCE: *Nutrition and Your Health: Dietary Guidelines for Americans,* United States Department of Agriculture, Home and Garden Bulletin Number 232 (Washington, D.C.: Government Printing Office).

The *Dietary Guidelines for Americans* suggest these percentages of energy-yielding nutrients in a diet that supports health:

- <30% calories from fat.
- 15–20% calories from protein.
- 55–60% calories from carbohydrate.

Chapter 11 considers the issue of evaluating individual risks of chronic diseases, and Chapter 13 discusses the nutrient needs of growing children.

Dietary Guidelines

As you have learned, nutrient intake recommendations provide standards for people's nutrient and energy consumption. Why, then, are "guidelines" needed as well? One reason is that although nutrient intake recommendations do much to ensure nutrient adequacy, they do little for moderation. Nutrient recommendations address intakes of protein, vitamins, and minerals. They also make some general statements about energy intakes, but they do little to protect people from excess intakes of fat, sugar, salt, and other food constituents believed to be related to chronic diseases. Guidelines take up where nutrient recommendations leave off, and they go a step further in recommending physical activity to improve or maintain body weight. Also, the DRI refer to nutrients, not foods. Guidelines specify healthful uses of foods and nutrients because many people need guidance in selecting the foods they consume each day.

Some people find fault with the concept of dietary guidelines, however, because, like the Daily Values, guidelines address the entire population as if "one-size-fits-all," without regard for people's differing ages, genetic inheritances, and lifestyles.[5] For example, a middle-aged man at risk for heart disease may benefit from advice to eat a diet low in fat; the same advice applied stringently to a growing child by overzealous adults may do more harm than good because some children have trouble obtaining enough energy from bulky, low-fat foods to grow normally.[6] Many experts are calling for a separate set of official dietary guidelines for children.[7]

To ensure dietary moderation where needed, the governments of many countries have published sets of recommendations that best apply to the genetic makeup of their populations and to the environment in which they live.[8] These recommendations include one set from the United States, two from Canada, and one intended for all the people of the world. They are, respectively, the *Dietary Guidelines for Americans,* which are updated every five years (Table 2-2), the *Nutrition Recommendations for Canadians* (Table 2-3), *Canada's Guidelines for Healthy Eating* (Table 2-4), and the World Health Organization's (WHO) *Population Nutrient Goals* (Table 2-5). Only the WHO Goals set both upper and lower limits for nutrients, and they have been proposed as an international set of guidelines. Many other sets of recommendations have been published, and

Table 2-3

The Nutrition Recommendations for Canadians

- The Canadian diet should provide energy consistent with the maintenance of *body weight* within the recommended range.
- The Canadian diet should include *essential nutrients* in amounts recommended.
- The Canadian diet should include no more than 30% of energy as *fat* (33 grams/1,000 calories or 39 grams/5,000 kilojoules) and no more than 10% as saturated fat (11 grams/1,000 calories or 13 grams/5,000 kilojoules).
- The Canadian diet should provide 55% of energy as *carbohydrate* (138 grams/1,000 calories or 165 grams/5,000 kilojoules) from a variety of sources.
- The *sodium* content of the Canadian diet should be reduced.
- The Canadian diet should include no more than 5% of total energy as *alcohol,* or two drinks daily, whichever is less.
- The Canadian diet should contain no more *caffeine* than the equivalent of four regular cups of coffee per day.
- Community water supplies containing less than 1 milligram per liter should be *fluoridated* to that level.

NOTE: Italics added to highlight areas of concern.
SOURCE: Health and Welfare Canada, *Nutrition Recommendations: The Report of the Scientific Review Committee* (Ottawa: Canadian Government Publishing Centre, 1990).

Table 2-4

Canada's Guidelines for Healthy Eating

- Enjoy a variety of foods.
- Emphasize cereals, breads, other grain products, vegetables, and fruits.
- Choose lower-fat dairy products, leaner meats, and foods prepared with little or no fat.
- Achieve and maintain a healthy body weight by enjoying regular physical activity and healthy eating.
- Limit salt, alcohol, and caffeine.

SOURCE: These guidelines derive from *Action Towards Healthy Eating: The Report of the Communications/Implementation Committee and Nutrition Recommendations. . . . A Call for Action: Summary Report of the Scientific Review Committee and the Communications/Implementation Committee,* which are available from Branch Publications Unit, Health Services and Promotion Branch, Department of Health and Welfare, 5th Floor, Jeanne Manice Building, Ottawa, Ontario K1A 1B4.

Dietary recommendations encourage habits that support the health of individuals and are also best for the earth. Chapter and Controversy 15 explore the relationships among people, their food choices, and the planet's well-being.

all offer similar advice on which nutrients people should emphasize and which they most often need to control for health's sake.[9]

Notice that these guidelines do not require that you give up your favorite foods or eat strange, unappealing foods. Almost anyone's diet, with some adjustments, can fit most of these recommendations. The secret for most people seems to be to modify the diet in four ways. First, learn to watch portion sizes, especially of fat-rich foods such as high-fat meats, dairy products, and desserts. Second, strictly limit a few foods, especially pure fats and sugar, such as margarine and sugary soft drinks. Third, make substitutions, such as fat-free or low-fat for high-fat dairy products. Finally, eat more whole grains, fruits, and vegetables. These four tactics, together with physical activity to balance energy intake, can change a potentially harmful diet into a nutrient-dense one that supports nutrition and health superbly. The Fitness for Life feature near here

Table 2-5

The WHO Population Nutrient Goals

	LIMITS FOR POPULATION AVERAGE INTAKES	
	Lower Limit	Upper Limit
Total fat	15% of energy	30% of energy[a]
Saturated fatty acids	0% of energy	10% of energy
Polyunsaturated fatty acids	3% of energy	7% of energy
Dietary cholesterol	0 mg/day	300 mg/day
Total carbohydrate	55% of energy	75% of energy
Complex carbohydrate[b]	50% of energy	75% of energy
Dietary fiber[c]	27 g/day	40 g/day
Sugars[d]	0% of energy	10% of energy
Protein	10% of energy	15% of energy
Salt	0 g/day	6 g/day[c]

NOTE: The lower limit defines the minimum intake needed to prevent deficiency diseases, and the upper limit expresses the maximum intake compatible with the prevention of chronic diseases. This set of guidelines is proposed by WHO for acceptance as a set of international standards.

[a]An interim goal for nations with high fat intakes; further benefits would be expected by reducing fat intake toward 15% of total energy.

[b]A daily minimum intake of about 2 cups vegetables and fruits, including about a half-cup of legumes, nuts, and seeds.

[c] From mixed food sources.

[d]Added refined sugars, not the sugars found naturally in fruits, vegetables, and milk.

SOURCE: Reproduced, by permission, from *Diet, Nutrition and the Prevention of Chronic Diseases. Report of a WHO Study Group.* Geneva, World Health Organization, 1990, p. 108 (WHO Technical Report Series, No. 797).

Dietary guidelines allow you to choose foods that you enjoy.

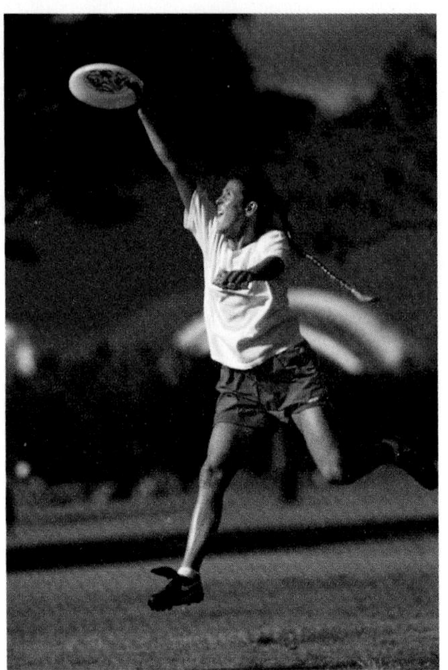

***The Dietary Guidelines* suggest that physical activity should balance food intake for a healthy body weight.**

provides some guidelines on how much physical activity is needed to help achieve these benefits.

The changes just described sound simple, but do not be deceived. Even nutrition experts struggle to design diets of appealing, nutrient-dense foods that meet both nutrient recommendations for adequacy and the *Dietary Guidelines* for moderation.[10] The achievement of such a diet is a worthy goal, however, and details to help accomplish it are provided throughout this book. You will see these recommendations again wherever diet and the health of the body are discussed.

If the experts who develop such guidelines were to ask us, we would add one more recommendation to their lists: choose foods that you enjoy. Though choosing foods that meet nutrient needs is of prime importance, seeking out delicious foods that meet the needs for pleasure and fun is also important. The joys of eating are physically beneficial to the body because they trigger health-promoting changes in the nervous, hormonal, and immune systems. Pleasure from food ensures that people will eat and thus obtain the nutrients needed for healthy body systems, as well as for the healthy skin, glossy hair, and natural good looks that accompany health. People tend to repeat what brings them pleasure, so they are most likely to stay with foods they like. Remember to enjoy your foods.

KEY POINT *The Dietary Guidelines for Americans, Nutrition Recommendations for Canadians, and other recommendations address the problems of overnutrition and undernutrition. To implement the recommendations requires controlling portions, limiting fat intake, substituting nutrient-dense for fat-rich foods, and amplifying servings of low-fat grains, fruits, and vegetables.*

FITNESS FOR LIFE

How Much Activity Is Enough?

Research on health and physical activity leads to a single conclusion: people need exercise as well as nutrition to stay healthy and live long.[1] Many groups make recommendations about how much physical activity is enough, but the answer for any one person depends on personal goals.

The *Dietary Guidelines* suggest balancing calorie intake and output to achieve a healthy, trim body weight. A 1996 Surgeon General's report observes that 30 minutes of cumulative physical activity each day brings benefits and that the activity need not involve sports. Any kind of physical activity counts. A few minutes spent climbing stairs, another few spent pulling weeds, and several more spent walking the dog all contribute to the day's total. The American College of Sports Medicine (ACSM), an authority in exercise, makes these recommendations:

- Obtain daily physical activity.
- Exercise at a comfortable effort level (can be moderate, such as brisk walking).
- Exercise for a duration of at least 30 minutes total daily (can be intermittent).

The ACSM also makes detailed recommendations for those who wish to gain greater fitness; these are presented in Chapter 10.

Just for fun, Figure 2-4 presents a pyramid of activities to remind you to exercise each day. If you are the forgetful type, post a copy of the pyramid where you'll be sure to see it daily. Of course, safety dictates that medical advice should be sought before beginning any exercise program. No one yet knows how much activity is too much, but excessive physical activity can compromise health in some people. For the vast majority of people, however, the problem is not too much activity, but rather too little—so get up and go.

Chapter 10 provides more details about the kinds of physical activities best suited to promoting health and fitness.

Diet Planning with the Daily Food Guide and the Food Guide Pyramid

Diet planning connects nutrition theory with the food on the table. To help people plan menus, **food group plans** describe food groups and dictate numbers and sizes of servings to choose each day. The Daily Food Guide provides an example. Another planning tool, the **exchange system** (see Appendix D), can help people estimate the amounts of carbohydrate, fat, protein, and energy (calories) that each type of food provides. The Canadian food group plan, *Food Guide to Healthy Eating,* is presented in full in Appendix B, and a brief description appears in Table 2-6 in the margin of the next page.

food group plans diet planning tools that sort foods into groups based on origin and nutrient content and then specify that people should eat certain minimum numbers of servings of foods from each group.

exchange system a diet planning tool that organizes foods with respect to their nutrient contents and calorie amounts. Foods on any single exchange list can be used interchangeably. See the U.S. Exchange System, Appendix D, for details.

The Daily Food Guide

In past decades, schoolchildren learned about the Four Food Group Plan, which taught generations of people to recognize key nutrients provided by certain related groups of foods. Today the Daily Food Guide (see Figure 2-5 on pp. 40 and 41) is based on five groups instead of four, but many of the original concepts still apply. The Food Guide Pyramid is a visual representation of the Daily Food Guide. The Food Guide Pyramid applies to adults only; children have their own pyramid, presented and explained in Chapter 13.

The Food Guide Pyramid for Young Children is in Chapter 13.

The foods in each group are well-known contributors of certain key nutrients, but you can count on them to supply many other nutrients as well. If you design

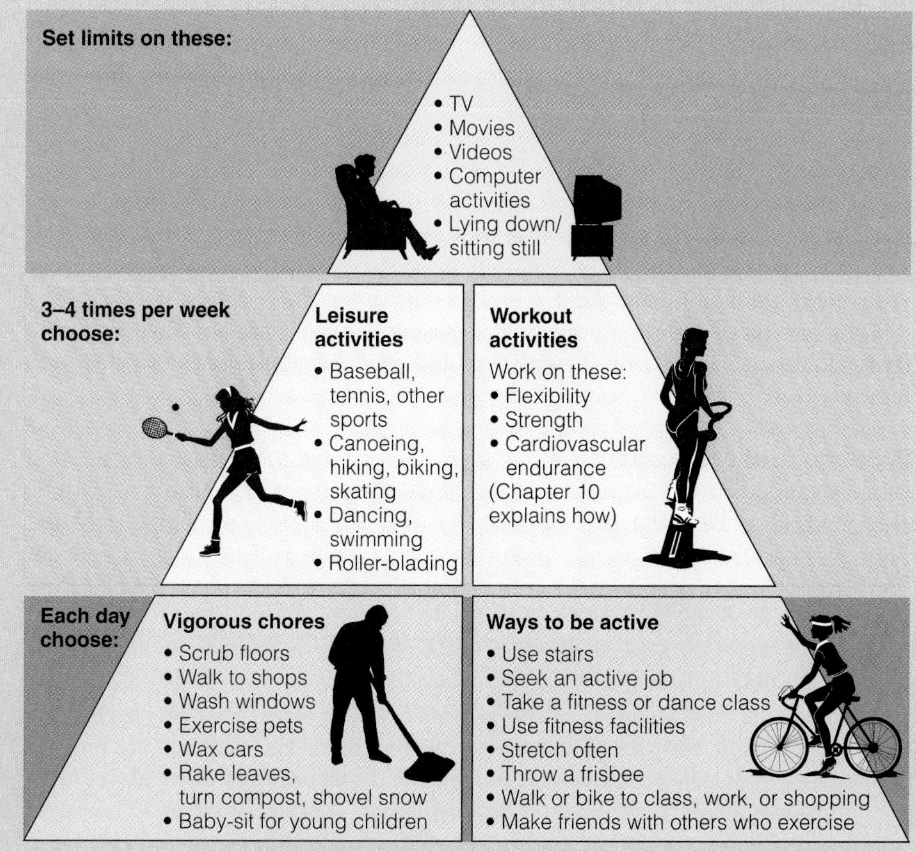

Set limits on these:
• TV
• Movies
• Videos
• Computer activities
• Lying down/ sitting still
3–4 times per week choose:
Leisure activities
• Baseball, tennis, other sports
• Canoeing, hiking, biking, skating
• Dancing, swimming
• Roller-blading
Workout activities
Work on these:
• Flexibility
• Strength
• Cardiovascular endurance
(Chapter 10 explains how)
Each day choose:
Vigorous chores
• Scrub floors
• Walk to shops
• Wash windows
• Exercise pets
• Wax cars
• Rake leaves, turn compost, shovel snow
• Baby-sit for young children
Ways to be active
• Use stairs
• Seek an active job
• Take a fitness or dance class
• Use fitness facilities
• Stretch often
• Throw a frisbee
• Walk or bike to class, work, or shopping
• Make friends with others who exercise

SOURCE: Adapted from F. Sizer-Webb, E. Whitney, and L. DeBruyne, *Health: Making Life Choices* (Cincinnati: West, 1999), p. 242.

Figure 2-4

Physical Activity Pyramid

The daily activities shown in this pyramid's bottom third and middle third serve as examples of ways to be active. The sedentary activities at the very top of the pyramid are not intrinsically harmful, of course, but must be balanced with more physically demanding ones if the body is to remain healthy.

Table 2-6

Canada's Food Guide to Healthy Eating[a]

Food Group	Servings/Day
Grain products	5–12
Vegetables and fruits	5–10
Milk products	
Children aged 4–9 years	2–3
Youth aged 10–16 years	3–4
Adults	2–4
Pregnant and breast-feeding women	3–4
Meat and alternatives	2–3

[a]Canada's *Food Guide* and other Canadian guidelines are presented in full in the Canadiana appendix (Appendix B).

your diet around this plan, it is assumed that you will obtain adequate amounts of not only the nutrients named in the figure but also the other two dozen or so essential nutrients because they are distributed among the same groups of foods. This is true in theory. In practice, however, diet planners must be sure to choose mostly nutrient-dense foods in each group because some processes strip foods of some nutrients and add calories from fat. Figure 2-5 identifies a few foods in each group as having high, moderate, or low nutrient density to give you an idea of which are which. With this caution, the Daily Food Guide can provide a reasonable road map for diet planning.

KEY POINT *The Daily Food Guide and other food group plans sort foods into groups based on their nutrients and origins. Then the plans suggest patterns of intake by group that will cover nutrient needs.*

How Can the Food Guide Pyramid Help Me to Eat Well?

The Food Guide Pyramid makes applying the Daily Food Guide easier to do. The Pyramid graphic assists you in planning a day's meals with the needed number of servings from each group of nutritious foods. By using it wisely and by learning about the energy nutrients in various foods (as you will do in coming chapters), you can achieve the goals of a nutritious diet first mentioned in Chapter 1: adequacy, balance, calorie control, moderation, and variety.

Chapters to come give details about the energy nutrients in foods within the Pyramid.

By design, the Pyramid provides guidance as to adequacy, balance,* moderation, and variety. By specifying serving sizes and suggesting numbers of servings for individuals, it implies calorie control, too. However, the makers of the Pyramid leave it up to you to regulate your calorie intake by choosing wisely among various foods. Figure 2-5 eases food selection by dividing foods into three color-coded nutrient density categories. A later section provides guidance in choosing an appropriate number of servings and points out the importance of holding serving sizes in line with recommendations.

Adequacy To achieve adequacy, adults using the Food Guide Pyramid must choose at least 6 servings from the bread, cereal, rice, and pasta group; 3 from the vegetable group; 2 from the fruit group; 2 from the meat, poultry, fish, dry beans, eggs, and nuts group; and 2 from the milk, yogurt, and cheese group. To help remember this pattern, think of the numbers: 6, 3, 2, 2, and 2.

Remember: 6, 3, 2, 2, and 2.

These are the minimum numbers of servings. The plan's makers suggest that to meet additional energy needs, a person should choose more servings of foods from these very same groups.

Balance and Moderation The broad base of grains at the bottom of the Food Guide Pyramid conveys the idea that you should eat more grain foods than anything else; grains form the foundation of a healthful diet. Next in volume are the fruits and vegetables. Meats and milks are dense in nutrients such as protein and are important sources of vitamins and minerals, but the number of servings must be limited because these foods can also be high in fat and calories.

Fats, oils, and sweets occupy only a tiny triangle at the top of the Food Guide Pyramid, an indication that they should be used sparingly. These foods do not comprise a food group, because servings of them are optional; that is, they are not required to promote health. They do provide some essential lipids and vitamin E along with abundant energy. Alcoholic beverages provide scant nutrients and are excluded from the Food Guide Pyramid altogether. They are high in calories, however, and must be

*Also called "proportionality."

counted in a day's tally, so a bottle of wine appears in Figure 2-5 as a reminder. Spices, coffee, tea, and diet soft drinks, also excluded from the Pyramid, provide few, if any, nutrients, but can add flavor and pleasure to meals as well as some potentially beneficial phytochemicals, such as those in tea or certain spices.

The five groups are:

1. Bread, cereal, rice, and pasta.
2. Vegetables.
3. Fruits.
4. Meat, poultry, fish, dry beans, eggs, and nuts.
5. Milk, yogurt, and cheese.

The fats, oils, and sweets are extra and are not counted among the groups.

Variety The beauty of the Food Guide Pyramid lies in its simplicity. Also, although it may appear rigid, it can actually be very flexible once its intent is understood. For example, the user can substitute cheese for milk because both supply the key nutrients for the milk, yogurt, and cheese group. The user can choose legumes (beans) and nuts as alternatives to meats. One can adapt the plan to mixed dishes such as casseroles and to national and cultural cuisines as well, as Figure 2-6 on pp. 42 and 43 shows.

As mentioned, the Food Guide Pyramid tends to de-emphasize meats and animal products such as milk, cheese, and eggs and to emphasize grains, fruits, and vegetables. This scheme can assist vegetarians in their food choices, while encouraging others to choose foods from plants most often. The food group that includes the meats also includes *meat alternates*—foods such as legumes, nuts, and tofu. As for the food group that includes milk and milk products, people who choose not to use dairy foods can substitute soy "milk"—a product made from soybeans that fills the same nutrient needs, provided that it is fortified with calcium and vitamin B_{12}. In short, people who choose to eat no meats or products taken from animals can still use the Food Guide Pyramid to make their diets adequate.

More on phytochemicals in foods in Controversy 11.

Vegetarians will find more tips for choosing the right foods to supply the nutrients they need in the chapters to come.

Drawbacks to the Food Guide Pyramid The Food Guide Pyramid does have drawbacks, however. As mentioned, it does not limit food choices to foods low in calories. People who select the minimum number of servings from among the most nutrient-dense foods in each group and who strictly limit their use of fats, sweets, and alcoholic beverages can keep their energy intakes low. Even then, zinc and vitamin E are often lacking. However, people who use the higher-calorie foods in each group and who eat large servings, even without extra fats, sweets, or alcohol added in, can easily obtain too many calories.

In the cheese-for-milk substitution mentioned earlier, sufficient cheddar cheese to meet a day's calcium requirement would also provide almost 600 calories, over 70 percent of them from fat. A day's calcium from fat-free milk comes with about 300 calories and with hardly any fat. Less obvious but significant over time are energy differences between sliced bread and biscuits, fish and hot dogs, nuts and legumes, or even green beans and sweet potatoes—all proper substitutions according to the Food Guide Pyramid. High-calorie choices may be just what some people, such as athletes, need to meet their large energy requirements, but for others such choices can dramatically boost energy intakes and, over time, add too much to body fat stores.

The Food Guide Pyramid attempts to caution consumers about fats and added sugars in food groups by sprinkling symbols for those two constituents across the pictures of the groups most likely to contain them. This warning may be difficult to put into practical use, however. The bread group, for example, is sprinkled with the symbols for both fat and added sugar because baked goods can be high in those constituents. From the symbols alone the user cannot point to individual high-fat or high-sugar foods within the groups. Most people in the United States consume far more fat and sugar than dietary wisdom would allow—and fall short in the areas of nutritious foods. They need more guidance in selecting foods.

Another criticism of the Food Guide Pyramid is that a person may choose the right number of servings from each group, yet make consistently nutrient-poor choices, and so fail to meet the day's needs for some nutrients. A diet can easily lack vitamin E or certain essential fatty acids, for example, because these nutrients are easily destroyed in processing or refined out of foods.[11]

Figure 2-5

The Daily Food Guide and the Food Guide Pyramid

KEY: Nutrient Density

- *Foods generally highest in nutrient density (preferable first choice).*
- *Foods moderate in nutrient density (reasonable second choice).*
- *Foods lowest in nutrient density (limit selections).*

BREAD, CEREAL, RICE, AND PASTA

These foods contribute complex carbohydrates and fiber, plus riboflavin, thiamin, niacin, iron, protein, magnesium, and other nutrients.

6 to 11 servings per day.

Serving = 1 slice bread: ½ c cooked cereal, rice, or pasta: 1 oz ready-to-eat cereal; ½ bun, bagel, or English muffin; 1 small roll, biscuit, or muffin; 3 to 4 small or 2 large crackers.

- Whole grains (wheat, oats, barley, millet, rye, bulgur), enriched breads, rolls, tortillas, cereals, bagels, rice, pastas (macaroni, spaghetti), air-popped corn.
- Pancakes, muffins, cornbread, crackers, low-fat cookies, biscuits, presweetened cereals, granola.
- Croissants, fried rice, doughnuts, pastries, sweet rolls.

VEGETABLES

These foods contribute fiber, vitamin A, vitamin C, folate, potassium, and magnesium.

3 to 5 servings per day (use dark green, leafy vegetables and legumes [dried beans] several times a week).

Serving = ½ c cooked or raw vegetables; 1 c leafy raw vegetables; ½ c cooked legumes;[a] ¾ c vegetable juice.

- Bean sprouts, broccoli, brussels sprouts, cabbage, carrots, cauliflower, cucumbers, eggplant, green beans, green peas, bell peppers, leafy greens (spinach, mustard, and collard greens), legumes, lettuce, mushrooms, summer and winter squash, tomatoes.
- Cassava, corn, potatoes, sweet potatoes, yams.
- French fries, olives, tempura vegetables.

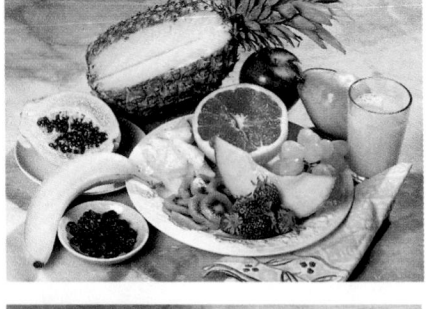

FRUITS

These foods contribute fiber, vitamin A, vitamin C, and potassium.

2 to 4 servings per day.

Serving = typical portion (such as 1 medium apple, banana, or orange, ½ grapefruit, 1 melon wedge; ¾ c juice; ½ c berries; ½ c diced, cooked, or canned fruit; ¼ c dried fruit.

- Apples, apricots, bananas, cantaloupe, grapefruit, kiwi, oranges, orange juice, papaya, peaches, pears, pineapple, strawberries.
- Canned or frozen fruit.
- Avocados, dried fruit.

MEAT, POULTRY, FISH, DRY BEANS, EGGS, AND NUTS

These foods contribute protein, phosphorus, vitamin B_6, vitamin B_{12}, zinc, magnesium, iron, niacin, and thiamin.

2 to 3 servings per day.

Serving = 2 to 3 oz lean, cooked meat, poultry, or fish (total 5 to 7 oz per day); count 1 egg, ½ c cooked legumes,[a] or 2 tbs peanut butter as 1 oz meat (or about ⅓ serving).

- Poultry, fish, lean meat (beef, lamb, pork, veal), legumes, egg whites.
- Fat-trimmed beef, lamb, pork; refried beans; egg yolks, tofu, tempeh.
- Hot dogs, luncheon meats, peanut butter, nuts (including coconut), sausage, bacon, fried fish or poultry, duck.

MILK, YOGURT, AND CHEESE

These foods contribute calcium, riboflavin, protein, vitamin B_{12}, and, when fortified, vitamin D and vitamin A.
2 servings per day.
3 servings per day for teenagers and young adults, pregnant/lactating women, women past menopause.
4 servings per day for pregnant/lactating teenagers.
Serving = 1 c milk or yogurt; 2 oz process cheese food; 1½ oz cheese.

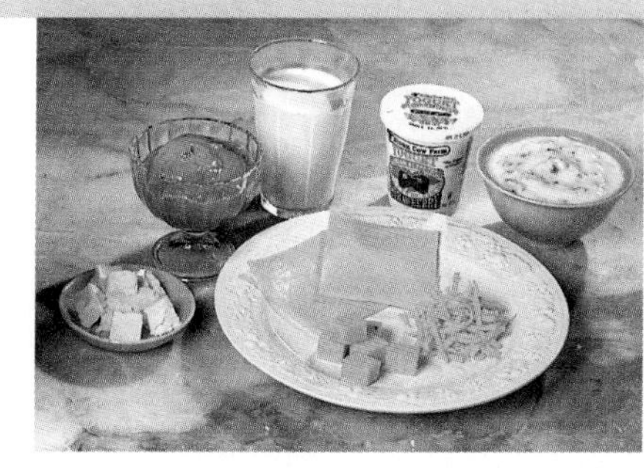

- Fat-free and 1% low-fat milk (and fat-free products such as buttermilk, cottage cheese, cheese, yogurt); fortified soy milk.
- 2% reduced-fat milk (and low-fat products such as yogurt, cheese, cottage cheese); sherbet; ice milk.
- Whole milk (and whole-milk products such as cheese, yogurt, cottage cheese[b]); custard; milkshakes; pudding; ice cream.

FATS, OILS, AND SWEETS

These foods contribute sugar, fat, alcohol, vitamin E, and food energy (calories). Their consumption should be limited because these foods provide few nutrients. Alcoholic beverages are not classed as foods on the pyramid; they contribute few nutrients, but they do contribute calories, and so are mentioned here.

- Foods high in fat include butter, margarine, lard, salad dressings, oils, mayonnaise, cream, sour cream, cream cheese, gravy, and sauces.
- Foods high in sugar include candy fruit rolls, other candies, soft drinks, fruit drinks, jelly, syrup, gelatin, desserts, sugar, and honey.
- Alcoholic beverages include wine, beer, and liquor.

---

[a]The Daily Food Guide lists legumes (dried beans, lentils, and peas) both under vegetables, for their starch, fiber, and vitamins, and under meats, for their protein and minerals.
[b]Cottage cheese is lower in calcium than most cheeses; 1 cup cottage cheese counts as ½ serving from the Milk, Yogurt, and Cheese group.
NOTE: Pregnant women may require additional servings of fruits, vegetables, meats, and breads to meet their higher needs for energy, vitamins, and minerals.

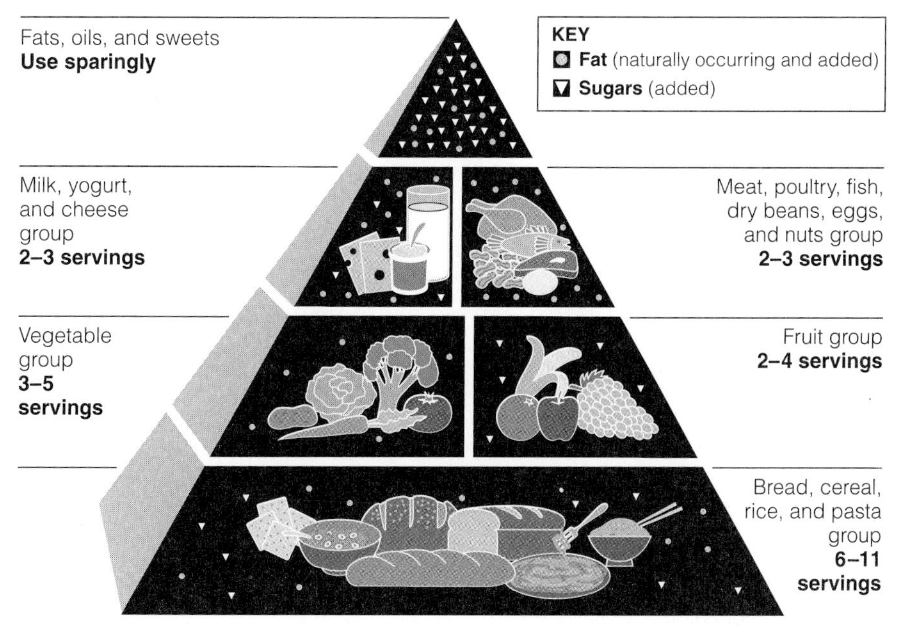
Fats, oils, and sweets
Use sparingly
KEY
Fat (naturally occurring and added)
Sugars (added)
Milk, yogurt, and cheese group
2–3 servings
Meat, poultry, fish, dry beans, eggs, and nuts group
2–3 servings
Vegetable group
3–5 servings
Fruit group
2–4 servings
Bread, cereal, rice, and pasta group
6–11 servings

Food Guide Pyramid: A Guide to Daily Food Choices

The breadth of the base shows that grains (breads, cereals, rice, and pasta) deserve the most emphasis in the diet. The tip is smallest: use fats, oils, and sweets sparingly.

Figure 2-6

Adding Variety with Ethnic and Regional Foods

CHINESE[a]

Bread, Cereal, Rice, and Pasta

- Millet; rice; rice noodles.
- Fried rice.

Vegetables

- Bamboo shoots; bean sprouts; cabbages; scallions; seaweed; snow peas; soybeans; water chestnuts.
- Fried vegetables.

Fruits

- Oranges, pears, plums, and other fresh fruit.

Meat, Poultry, Fish, Dry Beans, Eggs, and Nuts

- Broiled or stir-fried fish and seafood; egg whites.
- Broiled or stir-fried beef or pork; egg yolks; tofu.
- Deep-fried meats and seafood; egg foo young; pine nuts and cashews.

Fats, Oils, and Sweets

- Lard or oil for deep-frying.

Seasonings and Sauces[b]

- Bean sauce; garlic; ginger root; hoisin sauce;[c] oyster sauce;[c] plum sauce;[c] rice wine; scallions; soy sauce.[c]
- Sesame oil; other oils.

GREEK

Bread, Cereal, Rice, and Pasta

- Greek breads.

Vegetables

- Eggplant; lentils and beans; onions; peppers; tomatoes.
- Olives.

Fruits

- Dates; figs; grapes; lemons; raisins.

Milk, Yogurt, and Cheese

- Low-fat yogurt.
- Feta cheese; goat cheese.

Meat, Poultry, Fish, Dry Beans, Eggs, and Nuts

- Egg whites; fish and seafood; lentils and beans.
- Egg yolks; lamb; poultry; beef.
- Ground lamb; ground beef; gyros; walnuts; almonds.

Fats, Oils, and Sweets

- Olive oil; baklava (honey-soaked nut pastry); honey; cakes.

Seasonings and Sauces

- Garlic; herbs; lemons; egg and lemon sauce.
- Olive oil.

KEY: Nutrient Density

- *Foods generally highest in nutrient density (preferable first choice).*
- *Foods moderate in nutrient density (reasonable second choice).*
- *Foods lowest in nutrient density (limit selections).*

[a]Traditional cuisines of China and of West African influence exclude fluid milk as a beverage for adults and use few or no milk products in cooking. Calcium and certain other nutrients of milk are supplied by other foods, such as small fish eaten with the bones or large servings of leafy green vegetables.

[b]Most Chinese sauces are fat-free.

[c]May be high in sodium.

MEXICAN

Bread, Cereal, Rice, and Pasta

- Cereal; corn or flour tortillas; macaroni; rice.
- Graham crackers.
- Fried tortilla shells; tortilla chips.

Vegetables

- Cabbage; cactus; iceberg lettuce; legumes; squash; tomatoes.
- Corn; potatoes.
- Olives.

Fruits

- Bananas; guava; oranges; papaya; pineapple.
- Avocados.

Milk, Yogurt, and Cheese

- Evaporated low-fat milk; powdered fat-free milk.
- Cheddar or jack cheese; custard.

Meat, Poultry, Fish, Dry Beans, Eggs, and Nuts

- Fish; lean beef, poultry, lamb, and pork; many bean varieties.
- Egg yolks; refried beans.
- Bacon; fried fish, pork, or poultry; nuts; sausages.

Fats, Oils, and Sweets

- Butter; candy; cream cheese; lard; margarine; pastries; soft drinks; vegetable oil; sour cream.

Seasonings and Sauces[d]

- Herbs; hot peppers; garlic; pico de gallo; salsas; spices.
- Guacamole; lard.

DEEP SOUTH (WEST AFRICAN INFLUENCE)[a]

Bread, Cereal, Rice, and Pasta

- Rice.
- Biscuits; cornbread; pastries.

Vegetables

- Beans; black-eyed peas; collards (other leafy greens); okra; tomatoes.
- Corn; sweet potatoes; hominy.
- Fried green tomatoes; fried okra.

Fruits

- Apples; berries; melons; peaches; pears.
- Fried pies; fruit pastries.

Meat, Poultry, Fish, Dry Beans, Eggs, and Nuts

- Beans and peas; grilled or smoked poultry and fish.
- Braised or roasted meats (beef, poultry).
- Bacon; boiled peanuts;[e] fried chicken or pork; ham hocks; peanut butter; salted pork; sausage; spareribs.

Fats, Oils, and Sweets

- Butter; lard; shortening; gravy.

---

[d]Many Mexican sauces are fat-free.

[e]The peanut, a native of South America, was carried to West Africa by Portuguese explorers.

KEY POINT *The Daily Food Guide and its visual image, the Food Guide Pyramid, convey the basics of planning a diet adequate in nutrients. The Food Guide Pyramid fails to show how to use nutrient-dense foods to form the bulk of food selections from each food group.*

A Note about Exchange Systems

Exchange systems can be useful to careful diet planners, especially those wishing to control calories (weight watchers), those who must control carbohydrate intakes (people with diabetes), and those who should control their intakes of fat and saturated fat (almost everyone). The system presented in Appendix D (Appendix B for Canada) lists the estimated carbohydrate, fat, saturated fat, and protein contents of food portions, as well as their calorie values. The values in the exchange lists differ from the exacting values given for individual foods in Appendix A because exchange lists estimate values for whole groups of foods. With these estimates, exchange system users can make an educated approximation of the nutrients and calories in almost any food they might encounter.

The exchange system also highlights a fact that the Food Guide Pyramid overlooks: most foods provide more than just one energy nutrient. Meat, for example, is famous for protein, but meats like bacon and sausage deliver many more calories from fat than from protein. Pasta and bread contain significant protein with their carbohydrates. Milk products provide carbohydrates and protein, but their fat values vary, and so on. This focus on nutrients in foods leads to some unexpected food groupings in the exchange lists. The high-fat meats mentioned above and also many cheeses are listed together as "high-fat meats" because fat constitutes the predominant form of energy in these foods, followed by protein. Potatoes and other vegetables high in starch are listed with the breads because one serving of bread and one serving of a starchy vegetable contain about the same amount of carbohydrate. To explore this powerful aid to diet planning, spend some time studying Appendix D or B.

KEY POINT *Exchange lists facilitate calorie control by providing an understanding of how much carbohydrate, fat, and protein are in each food group.*

Table 2-7

How Many Servings?

	Sedentary Women, Some Older Adults	Children, Teenage Girls, Active Women, Sedentary Men	Teenage Boys, Active Men
Calories[a]	About 1,600	About 2,000	About 2,800
Breads, cereals, rice, and pasta group	6	9	11
Vegetable group	3	4	5
Fruit group	2	3	4
Milk, yogurt, and cheese group[b]	2–3	2–3	2–3
Meat, poultry, fish, dry beans, eggs, and nuts group	2 (5 oz total)	2 (6 oz total)	3 (7 oz total)
Total fat (g)	53	73	93
Added sugar (tsp)	6	12	18

[a]Assumes mostly low-fat and low-calorie food choices.

[b]Women who are pregnant or lactating, teenagers, and young adults to age 24 need three servings. In fact, given the 1997 DRI, which raised the calcium recommendation, all individuals may need an additional milk serving to meet their calcium need.

SOURCE: U.S. Department of Agriculture, *Home and Garden Bulletin* 252 (1992): 9.

How Many Servings Do I Need Each Day?

Note that for each food group the Food Guide Pyramid presents a range of numbers of servings. Find yourself among the people described at the top of Table 2-7; then look at the column of numbers below for the approximate number of servings to take from each food group to meet your calorie goal.

Clearly, a sedentary person can meet the plan's requirements and still eat only about 1,600 calories (see Table 2-8). If you are only moderately active, you can probably eat an additional 600 to 1,200 calories without gaining weight. The more active you are, the higher the energy allowance you "earn." A wise choice is to invest many of these additional calories in additional nutrient-dense vegetables, legumes, fruits, and whole-grain foods and only a few in luxury items such as sweet desserts, butter, margarine, oil, or alcohol. If you make

Table 2-8

Sample Diet Planned with the Food Guide Pyramid

Breakfast: Cornflakes with milk and sugar; toast; coffee; orange juice.
Lunch: Small cheeseburger, macaroni salad; banana; diet cola.
Supper: Chili with beans, beef, and rice; spinach salad with dressing; corn on the cob with margarine; water.

Pattern from the Pyramid	Example	Energy (cal)[a]
Breads, cereals, rice, and pasta group—6 servings	½ c cooked white rice	103
	½ c low-fat macaroni salad	126
	1 oz cornflakes	100
	1 slice toast	119
	1 bun	123
Meat, poultry, fish, dry beans, eggs, and nuts group—2 servings (2 to 3 oz each)	½ c chili beans	126
	3 oz extra lean ground beef	225
Fruit group—2 servings	1 banana	109
	¾ c orange juice	84
Vegetable group—3 servings	½ c tomato sauce	37
	1 c spinach leaves	12
	1 medium corn on cob	72
Milk, yogurt, and cheese group—2 servings	1 c fat-free milk	85
	2 oz processed cheese	210
Added fat	2 pats reduced-fat margarine	32
	1 tbs low-calorie salad dressing	21
Added sugar—1 tsp	12 oz diet cola	0
	1 tsp sugar	16
		Total: 1,600

[a]Values from the Table of Food Composition, Appendix A.

WITHDRAWN FROM UNIVERSITIES AT MEDWAY LIBRARY

additions from the latter group, make them by conscious choice rather than through the unintentional use of high-calorie foods. With judicious selections, the diet can supply all the necessary nutrients and provide some luxury items as well.

KEY POINT *The Daily Food Guide specifies how many servings of foods from each group people need to consume to meet their nutrient requirements.*

3.5 SERVINGS
1.3 SERV
2.2 SERVINGS
2 SERVINGS
1 SERVING
5.1 SERVINGS

Here's how a typical U.S. diet stacks up.

SOURCE: National Livestock and Meat Board, courtesy of the National Cattlemen's Beef Association.

Serving Sizes versus Helpings

To use the Food Guide Pyramid meaningfully, a person still must clarify what is meant by the word *serving*, for it can mean different things to different people.[12] Owners of restaurants, whose patrons want their money's worth, often deliver colossal helpings to ensure repeat business; a server on a cafeteria line may be instructed to deliver "about a spoonful" of the foods offered; makers of frozen entrées fill up partitioned trays with whatever amounts of foods fit within the sections; fast-food burgers range from a one-ounce miniburger to a half-pound double deluxe; and so on. The trend in the United States has been toward consuming larger food portions, especially of foods rich in fat and sugar (see Figure 2-7) and, as the drawing in the margin shows, the result is an unbalanced diet. At the same time, body weights have been creeping upward, suggesting an increasing need to control portion sizes.

Figure 2-7

U.S. Trend Toward Colossal Cuisine

Food	Food Guide Pyramid	Typical 1977	Colossal 2000
Cola	—	10 oz bottle, 120 cal	40–60 oz fountain, 580 cal
Bagel	½ bagel, 90 cal	2–3 oz, 230 cal	5–7 oz, 550 cal
French fries	10, 160 cal	about 30, 475 cal	about 50, 790 cal
Hamburger	2–3 oz meat, 240 cal	3–4 oz meat, 330 cal	6–8 oz meat, 650 cal
Steak	2–3 oz, 170 cal	8–12 oz, 690 cal	16–22 oz, 1,260 cal
Pasta	½ cup, 100 cal	1 cup, 200 cal	2–3 cups, 600 cal
Baked potato	3–4 oz, 110 cal	5–7 oz, 180 cal	one pound, 420 cal
Candy bar	—	1½ oz, 220 cal	3–4 oz, 580 cal
Popcorn	—	1½ cups, 80 cal	8–16 cup tub, 880 cal

NOTE: Calories are rounded values for the largest portions in a given range.
SOURCE: Data for most entries from L. R. Young and M. Nestle, Portion sizes in dietary assessment: Issues and policy implications, *Nutrition Reviews* 53 (1995): 149–158.

1977 2000

1977 2000

1977 2000

In contrast to the random-sized helpings found elsewhere, most serving sizes in the Food Guide Pyramid are specific and precise and can be relied upon to deliver certain amounts of key nutrients in foods. Among volumetric measures, 1 "cup" refers to an 8-ounce measuring cup (not a teacup or drinking glass), filled to level (not heaped up, or shaken, or pressed down). Tablespoons and teaspoons refer to measuring spoons (not flatware) filled to level (not rounded). Ounces signify weight, not volume. Two ounces of meat, for example, refers to ⅛ pound of cooked meat. One ounce (weight) of granola cereal measures ¼ cup (volume), but take care: 1 ounce of crispy rice cereal measures a full cup. Also some foods are specified as "medium," as in "one medium apple," but the word *medium* means different things to different people. When college students are asked to bring medium-sized foods to show the class, they bring bagels weighing anywhere from 2 to 5 ounces; muffins from about 2 to 8 ounces; baked potatoes from 4 to 9 ounces; and so forth.[13] The Food Guide Pyramid provides these standards for "medium": a bagel weighs 2 ounces; a muffin, 1.5 ounces; and a potato, 3.9 ounces. The Table of Food Composition, Appendix A, lists both weights and volumes of a wide variety of foods.

Wise diners also read labels of packaged foods to help them determine the foods' nutrient and energy contents and to decide how the foods may fit into their total eating plan. This chapter's Consumer Corner explains how to gain insight from the information on food labels.

KEY POINT *A person wishing to avoid overconsuming calories must pay attention to serving sizes.*

CONSUMER CORNER

Checking Out Food Labels

A PACKAGE OF potato chips must tell you on its label's ingredient list that it contains potato, fat, and salt; and it must also reveal details about its nutrient composition on its "**Nutrition Facts**" panel (see Table 2-9). These requirements allow you to use packaged foods artfully in diet planning—if you can interpret their labels. The Nutrition Education and Labeling Act of 1990 set the requirements for label information. According to the law, every packaged food must state the following somewhere on its label:

- The common or usual name of the product.
- The name and address of the manufacturer, packer, or distributor.
- The net contents in terms of weight, measure, or count.
- The nutrient contents of the product, presented in accordance with the rules governing the Nutrition Facts panel.

Then, the label must list the following in ordinary language:

- The ingredients, in descending order of predominance by weight.

If the product meets strict criteria, the label may also display certain claims:

- **Nutrient claims** concerning the product's nutritive value.
- **Health claims,** concerning its health value.

Most food labels must conform with all these requirements.[1] The size of a package makes a difference, however. A large package, such as the box of cereal in Figure 2-8, must provide all of the information listed above. A smaller label, such as the label on a can of tuna, provides some of the same information in abbreviated form. A label on a roll of candy rings provides only a phone number, as is allowed for the tiniest labels. Much nutrition information is located on a label's Nutrition Facts panel, described next. The Canadian version of a food label can be found in Appendix B.

The Nutrition Facts Panel

Somewhere on the label of most packaged foods is a Nutrition Facts panel, like the one shown in Figure 2-8, to inform consumers of the nutrient contents of the food. Many grocers also voluntarily post placards or offer handouts in fresh-food departments to provide consumers with similar sorts of nutrition information for the most popular types of fresh fruits, vegetables, meats, poultry, and seafoods.

When you read a Nutrition Facts panel, be aware that only the top portion

Table 2-9

Nutrition Facts and Claims on Food Labels

health claims claims linking food constituents with disease states; allowable on labels within the criteria established by the Food and Drug Administration.

nutrient claims claims using approved wording to make statements about the nutrient value of foods, such as a claim that a food is "high" in a desirable constituent or "low" in an undersirable one.

Nutrition Facts on a food label, the panel of nutrition information required to appear on almost every packaged food. Grocers may also provide the information for fresh produce, meats, poultry, and seafoods.

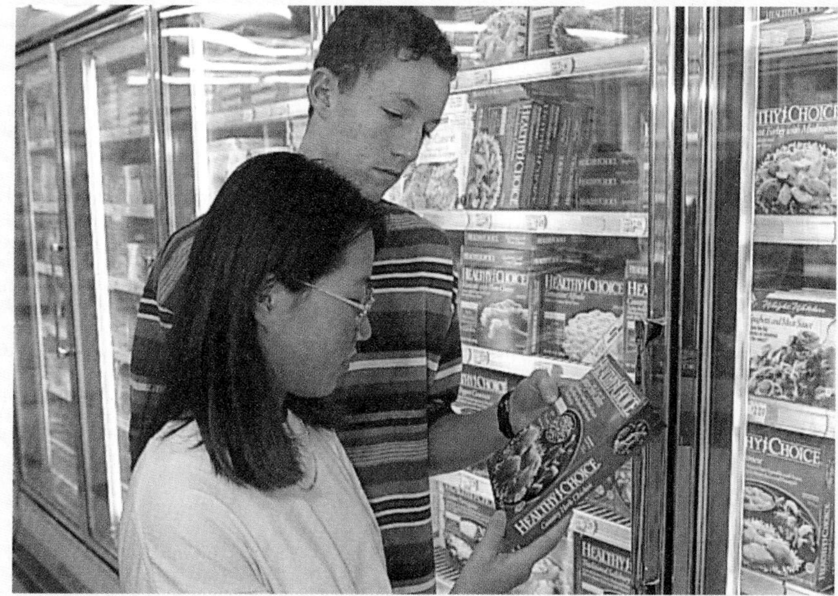
HEALTHY CHOICE

Modern-day entertainment: Reading food labels.

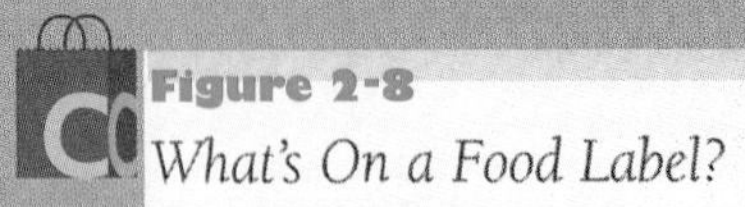

Figure 2-8

What's On a Food Label?

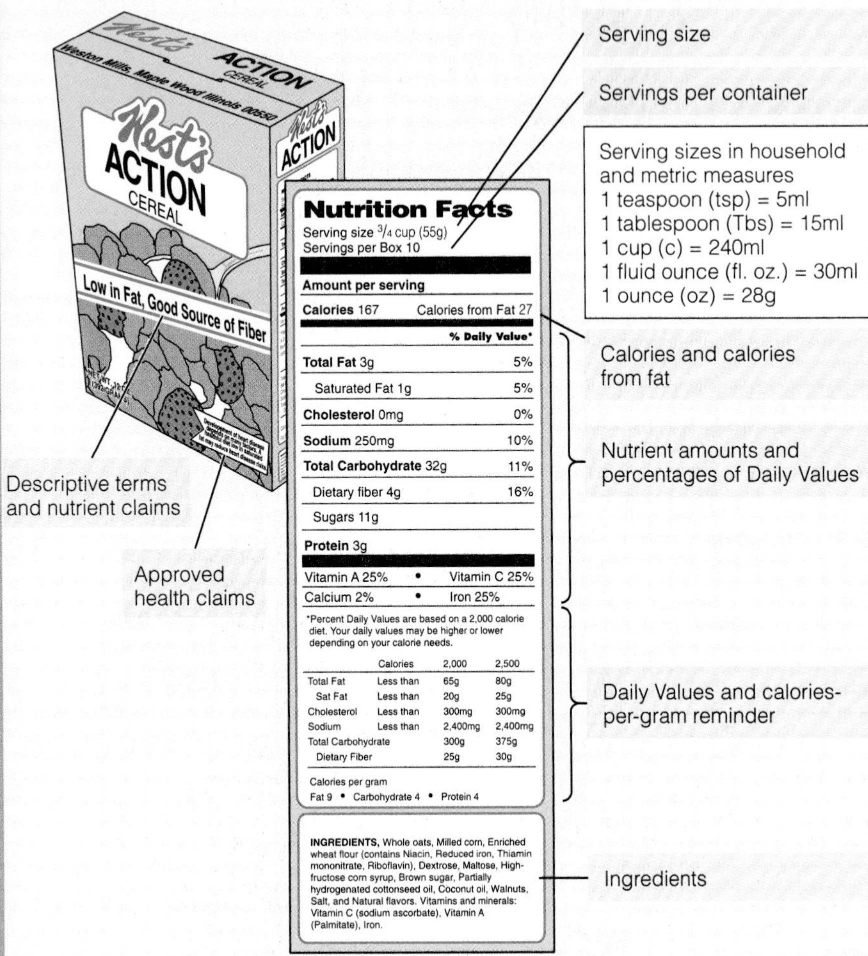

Hest's ACTION CEREAL
Weston Mills, Maple Wood Illinois 00550
Low in Fat, Good Source of Fiber
Serving size
Servings per container
Serving sizes in household and metric measures
1 teaspoon (tsp) = 5ml
1 tablespoon (Tbs) = 15ml
1 cup (c) = 240ml
1 fluid ounce (fl. oz.) = 30ml
1 ounce (oz) = 28g
Nutrition Facts
Serving size 3/4 cup (55g)
Servings per Box 10
Amount per serving
Calories 167 Calories from Fat 27
% Daily Value*
Total Fat 3g 5%
Saturated Fat 1g 5%
Cholesterol 0mg 0%
Sodium 250mg 10%
Total Carbohydrate 32g 11%
Dietary fiber 4g 16%
Sugars 11g
Protein 3g
Vitamin A 25% • Vitamin C 25%
Calcium 2% • Iron 25%
*Percent Daily Values are based on a 2,000 calorie diet. Your daily values may be higher or lower depending on your calorie needs.
Calories 2,000 2,500
Total Fat Less than 65g 80g
Sat Fat Less than 20g 25g
Cholesterol Less than 300mg 300mg
Sodium Less than 2,400mg 2,400mg
Total Carbohydrate 300g 375g
Dietary Fiber 25g 30g
Calories per gram
Fat 9 • Carbohydrate 4 • Protein 4
INGREDIENTS, Whole oats, Milled corn, Enriched wheat flour (contains Niacin, Reduced iron, Thiamin mononitrate, Riboflavin), Dextrose, Maltose, High-fructose corn syrup, Brown sugar, Partially hydrogenated cottonseed oil, Coconut oil, Walnuts, Salt, and Natural flavors. Vitamins and minerals: Vitamin C (sodium ascorbate), Vitamin A (Palmitate), Iron.
Calories and calories from fat
Nutrient amounts and percentages of Daily Values
Daily Values and calories-per-gram reminder
Ingredients
Descriptive terms and nutrient claims
Approved health claims

of the panel conveys information specific to the food inside the package. The bottom portion of the label lists the Daily Values. These values are identical on every label.

The highlighted items in this section correspond with those of Figure 2-8, which shows the location of the items listed below.

- Serving size. Common household and metric measures to allow comparison of foods within a food category. This is the amount of the food that constitutes a single serving and the portion that contains the nutrient amounts listed. A serving of chips may be 10 chips, but if you eat 50 chips, you will have consumed five times the nutrient amounts listed on the label.
- Servings per container. Number of servings or portions per box, can, package, or other unit.
- Calories/calories from fat. Total food energy per serving, and energy from fat per serving.
- Nutrient amounts and percentages of Daily Values. This section provides the core of information concerning these nutrients:
  - *Total fat.* Grams of fat per serving with a breakdown showing grams of *saturated fat* per serving.
  - *Cholesterol.* Milligrams of cholesterol per serving.
  - *Sodium.* Milligrams of sodium per serving.
  - *Total carbohydrate.* Grams of carbohydrate per serving, including starch, fiber, and sugars, with a breakdown showing grams of dietary *fiber* and *sugars.* The sugars include those that occur naturally in the food plus any added during processing.
  - *Protein.* Grams of protein per serving.

In addition, percentages of the Daily Values must be stated for these nutrients:

- *Vitamin A.*
- *Vitamin C.*
- *Calcium.*
- *Iron.*

Other nutrients present in significant amounts in the food may also be listed on the label. The percentages of the Daily Values are given in terms of a person requiring 2,000 calories each day.

- Daily Values and calories-per-gram reminder. This portion lists the Daily Values for a person needing 2,000 or 2,500 calories a day, and a calories-per-gram reminder as a handy reference for label readers.

An often neglected but highly valuable body of information comes next. Be sure to note the list of:

- Ingredients. The product's ingredients must be listed in descending order of predominance by weight.

Knowing how to read an ingredient list puts you many steps ahead of the naive buyer. Whatever is listed first is the ingredient that predominates by weight. Consider the ingredient list on an orange drink powder whose first three ingredients are "sugar, citric acid, orange flavor." You can tell that sugar is the chief ingredient. Now consider a canned juice whose ingredient list begins with "water, orange juice concentrate, pineapple juice concentrate." This product is clearly made of reconstituted juice. Water is first on the label because it is the main constituent of juice. Sugar is nowhere to be found among the ingredients because sugar has not been added to the product. Sugar occurs naturally in juice, though, so the label does specify sugar grams; details are in Chapter 4.

Now consider a cereal whose entire list contains just one item: "100 percent shredded wheat." No question, this is a whole-grain food with nothing added. Finally, consider a cereal whose first three ingredients are "puffed milled corn, sweeteners (sugars: corn syrup,

sucrose, honey, dextrose), salt." If you recognize that sugar, corn syrup, honey, and dextrose are all different versions of sugar (and you will, after Chapter 4), you might guess that this product contains close to half its weight as sugar.

More about Percentages of Daily Values As mentioned, some of the Daily Values are printed on each label in the Nutrition Facts panel. The entire list can be found on the inside cover of this text, p. C. The calculations that determine the "% Daily Value" figures that are printed on the upper right side of the Nutrition Facts panel are based on a 2,000-calorie diet. For example, if a food contributes 13 milligrams of cholesterol per serving, and the Daily Value is 300 milligrams, then a serving of that food provides about 4 percent of the Daily Value for cholesterol.

The Daily Values are of two types. Some, such as those for fiber, protein, vitamins, and most minerals, are akin to other nutrient intake recommendations. They suggest an intake goal to strive to reach; below that level, some people's needs may go unmet. Other Daily Values, such as those for cholesterol, total fat, saturated fat, and sodium, constitute healthy daily maximums.

Of course, people's actual calorie intakes vary widely; some people need fewer than 2,000 calories, and some need many more. Therefore, the Daily Values are most useful for comparing one food with another. Still, by examining a food's general nutrient profile, a consumer can determine whether the food contributes "a little" or "a lot" of a nutrient, whether it contributes "more" or "less" than another food, and how well it fits into the person's overall diet. Chapter 5's Do It! provides some practice in comparing the percentages of the Daily Values for fat, saturated fat, and cholesterol on food labels. Other figures in later chapters use the Daily Values for comparing the nutrients of one food with those of another.

Descriptive Terms and Nutrient Claims The Daily Values are intended to be easy to interpret. They are the basis upon which foods may claim to be "low" in cholesterol or a "good source" of vitamin A. Table 2-10 lists terms used on food labels along with their definitions. These definitions can help consumers in choosing foods. For example, any food providing 10 percent or more of the Daily Value for a nutrient is considered to be a good source of the nutrient; a food providing 20 percent is considered "high in" the nutrient. Additionally, as a rule of thumb, any food containing less than 5 percent of a Daily Value provides just a small amount of the nutrient per serving. For nutrients that must be limited, such as fat or sodium, foods providing less than 5 percent may be desirable. For hard-to-get nutrients such as iron or calcium, a reasonable goal might be to choose foods that are "good sources" or "high" in those nutrients several times each day. The vitamin and mineral Snapshot features in Chapters 7 and 8 point out "good sources" of the vitamins and minerals.

WHICH PACKAGED FOODS AND RESTAURANT CHOICES ARE BEST FOR MY HEALTH?

Increasingly, people have little time to cook their foods from scratch—they rely on packages of convenience foods and restaurant meals. When chosen wisely, such foods can support nutritional health but more often they carry too much fat, saturated fat, and salt and too little fiber, calcium, iron, and other nutrients to qualify as nutritious staple foods in the diet.[2] Luckily, health and nutrient claims on labels and menus can act as a sort of short cut to identifying the packaged or restaurant foods that provide the nutrients the body needs but limit the constituents that are generally oversupplied. The trick lies in learning the meanings of the claims.

Approved Health Claims

Claims linking nutrients and food constituents to disease states are allowed on food labels in the United States under FDA guidelines. The same claims made on a restaurant menu are held to the standards set for labels of packaged foods.[3] The following list describes the claims that can be made, and Table 2-10 defines the terms in bold type. Health claims are permitted on food labels when the claims are well supported by the available scientific evidence.

Labels may make statements concerning the following relationships:

1. *Calcium and osteoporosis.* A food making this claim must be **high** in calcium.
2. *Sodium and hypertension (high blood pressure).* The food must be **low sodium.**
3. *Dietary fat and cancer.* The food must be **low fat.**
4. *Dietary saturated fat and cholesterol and coronary heart disease.* The food must be **low saturated fat, low cholesterol,** and **low fat.**
5. *Fiber-containing grain products, fruits, vegetables, and cancer.* The food must be low in fat and have no added fiber. It must be a **good source** of dietary fiber.
6. *Fruits, vegetables, and grain products that contain fiber, especially soluble fiber, and the risk of coronary heart disease.* The food must be **low saturated fat, low fat,** and **low cholesterol.** It must also contain at least 0.6 gram of soluble fiber (explained in Chapter 4) per serving.
7. *Fruits and vegetables and cancer.* The food must be **low fat,** and without added nutrients, it must be a **good source** of fiber, vitamin A, or vitamin C. (Fresh fruits and vegetables may make this claim whether or not they meet the criteria for vitamin A or vitamin C.)
8. *Diets high in oatmeal or oat bran and the risk of coronary heart disease.* The food must be **low saturated fat, low cholesterol**, and **low fat**; provide at least 13 grams of oat bran or 20 grams of oatmeal; and, without fortification, provide at least 1 gram of soluble fiber per serving.
9. *The vitamin folate and birth defects of the brain and spinal cord (neural tube defects).* The food must be a **good source** of the vitamin folate and contain no more than 100

percent of the Daily Value for vitamin A or vitamin D.

10. *Sugar alcohols and tooth decay.* Sugar alcohols do not promote tooth decay; frequent between-meal snacks that are high in sugar and starch promote tooth decay.

The claims may say only that a substance "may" or "might" reduce disease risks. This tentative wording reflects the fact that science is still accumulating evidence concerning the roles of diet in diseases. Also, claims must state that the development of a disease rests on many factors. A permissible health claim might look like this:

> Development of heart disease depends on many factors. A healthful diet low in saturated fat and cholesterol may lower blood cholesterol levels and may reduce the risk of heart disease.

When you choose a food with the word **healthy** as part of its name, you can rely on it to live up to its claim. To qualify, a serving of the food must contain at least 10 percent of at least one of these:

- Vitamin A.
- Vitamin C.
- Iron.
- Calcium.
- Protein.
- Fiber.

Some exceptions to these requirements exist. Fresh fruits and vegetables, and some canned and frozen varieties, can be labeled *healthy,* even if a serving falls short of the 10 percent mark for any of these nutrients. These foods support health, regardless.

Foods labeled as *healthy* or those making any kind of a health claim cannot contain any nutrient or food constituent in an amount known to increase disease risk. Specifically, a serving of the product may contain no more than 20 percent of the Daily Value for the following:

- Total fat.
- Saturated fat.
- Cholesterol.
- Sodium.

Table 2-10

Descriptive Terms Used on Food Labels

Energy Terms

- **low calorie** 40 calories or fewer per serving.
- **reduced calorie** at least 25% lower in calories than a "regular," or reference, food.
- **calorie free** fewer than 5 calories per serving.

Fat Terms (Meat and Poultry Products)

- **extra lean**
  - less than 5 g of fat *and*
  - less than 2 g of saturated fat *and*
  - less than 95 mg of cholesterol per serving.
- **lean**[a]
  - less than 10 g of fat *and*
  - less than 4 g of saturated fat *and*
  - less than 95 mg cholesterol per serving.

Fat and Cholesterol Terms (All Products)

- **cholesterol free**
  - less than 2 mg cholesterol *and*
  - 2 g or less saturated fat per serving.
- **fat free** less than 0.5 g of fat per serving.
- **low cholesterol**
  - 20 mg or less of cholesterol *and*
  - 2 g or less saturated fat per serving.
- **low fat** 3 g or less fat per serving.
- **low saturated fat** 1 g or less saturated fat per serving.
- **percent fat free** may be used only if the product meets the definition of *low fat* or *fat free.* Requires disclosure of g fat per 100 g food.
- **reduced** or **less cholesterol**
  - at least 25% less cholesterol than a reference food *and*
  - 2 g or less saturated fat per serving.
- **reduced saturated fat**
  - 25% or less of saturated fat *and*
  - reduced by more than 1 g saturated fat per serving compared with a reference food.
- **saturated fat free**
  - less than 0.5 g of saturated fat *and*
  - less than 0.5 g of *trans*-fatty acids.

This means that whole milk, even though it is a rich source of calcium, may not make a claim about osteoporosis because whole milk contains too much saturated fat to qualify. Lowfat and fat-free milks, however, do qualify to bear the calcium and osteoporosis claim (milk's names and fat contents are clarified in the Food Feature of Chapter 5).

Conclusion

Health claims on labels and menus are so carefully controlled that consumers can rely on them instead of worrying about grams, percentages, and other mathematical speed bumps that would slow them down in grocery store aisles and at restaurants. Food manufacturers have now performed much of the math

Table 2-10

Terms Used on Food Labels (Continued)

Fiber Terms

- **high fiber** 5 g or more per serving. (Foods making high-fiber claims must fit the definition of low fat, or the level of total fat must appear next to the high-fiber claim.)
- **good source of fiber** 2.5 g to 4.9 g per serving.
- **more** or **added fiber** at least 2.5 g more per serving than a reference food.

Other Terms

- **free, without, no, zero** none or a trivial amount. *Calorie free* means containing fewer than 5 calories per serving; *sugar free* or *fat free* means containing less than half a gram per serving.
- **fresh** raw, unprocessed, or minimally processed with no added preservatives.
- **good source** 10 to 19% of the Daily Value per serving.
- **healthy** low in fat, saturated fat, cholesterol, and sodium and containing at least 10% of the Daily Value for vitamin A, vitamin C, iron, calcium, protein, or fiber.
- **high in** 20% or more of the Daily Value for a given nutrient per serving; synonyms include "rich in" or "excellent source."
- **less, fewer, reduced** containing at least 25% less of a nutrient or calories than a reference food. This may occur naturally or as a result of altering the food. For example, pretzels, which are usually low in fat, can claim to provide less fat than potato chips, a comparable food.
- **light** this descriptor has three meanings on labels:
  1. a serving provides one-third fewer calories or half the fat of the regular product.
  2. a serving of a low-calorie, low-fat food provides half the sodium normally present.
  3. the product is light in color and texture, so long as the label makes this intent clear, as in "light brown sugar."
- **more, extra** at least 10% more of the Daily Value than in a reference food. The nutrient may be added or may occur naturally.

Sodium Terms

- **low sodium** 140 mg or less sodium per serving.
- **sodium free** less than 5 mg per serving.
- **very low sodium** 35 mg or less sodium per serving.

[a]The word *lean* as part of the brand name (as in "Lean Supreme") indicates that the product contains fewer than 10 grams of fat per serving.

previously required of nutrition-conscious people. Between the honest and accurate numbers and the carefully defined words, consumers who take time to learn and use the lingo can choose among foods with confidence.

FOOD
FEATURE

Getting a Feel for the Nutrients in Foods

FIGURE 2-9 ILLUSTRATES a playful contrast between two day's meals. "Monday's Meals" were selected following the recommendations of this chapter. "Tuesday's Meals" were chosen more for convenience and familiarity than out of concern for nutrition. The two sets of meals are similar in energy (calories) so that other differences will stand out.

Now, how can a person compare the nutrition that these sets of meals provide? One way is to look up each food in a table of food composition, write down the food's nutrient values, and compare each one to a standard such as the Daily Values, as we've done in Figure 2-9. The computer is a time saver—it performs nutrient calculations with lightning speed. This convenience may make working with paper, pencils, and erasers seem a bit old-fashioned, but computers are rarely available on line for diners in cafeterias or fast-food counters where real-life decisions must be made. Those who can "see" the nutrients in their foods can make informed choices before eating meals, while others must wait until they visit their computers to find out how well they did in retrospect. By the time you reach Chapter 11 of this text, you will be ready to test your skill at "seeing" the nutrients in foods in that chapter's Do It! section. This chapter's Do It! provides an alternative method for judging the adequacy of diets that employs the Food Guide Pyramid as its standard.

Monday's Meals
Nutrient Dense

Tuesday's Meals
Convenient

Peanut
Butter
Grape
Jelly

DIET
COLA

DIET
COLA

How Do these Two Day's Meals Differ?

Figure 2-9

Two Days' Meals Compared with Five Daily Values

Monday's Meals

Monday's meals reflect nutrient-dense choices from the Food Guide Pyramid.

Foods	Energy (cal)	Fiber (g)	Total Fat (g)	Saturated Fat (g)	Vitamin C (mg)
Before heading off to class, a student eats breakfast:					
1 c sweetened cold cereal	166	1	—	—	—
1 c 1% low-fat milk	102	—	3	2	2
½ banana (sliced)	52	1	—	—	5
Then goes home for a quick lunch:					
1 turkey sandwich on roll with mayonnaise and mustard	294	1	12	2	—
1 c vegetable juice	46	2	—	—	67
While studying in the afternoon, the student eats a snack:					
4 whole-wheat crackers	80	2	3	—	—
1 oz low-fat cheddar cheese	49	—	2	1	—
1 apple	82	3	—	—	8
That night, the student makes dinner:					
A salad:					
1 c raw spinach leaves, shredded carrots, and sliced mushrooms	29	3	—	—	19
⅓ c garbanzo beans	135	4	2	—	1
5 lg olives and 1 tbs ranch salad dressing	80	1	8	1	—
A main course:					
1 c spaghetti with meat sauce	332	6	12	3	22
½ c green beans	18	2	—	—	6
2 tsp butter	68	—	8	5	—
And for dessert:					
Strawberry shortcake made with:	293	3	8	4	102
1¼ c strawberries					
1 piece spongecake					
2 tbs whipped cream					
Later that evening, the student enjoys a bedtime snack:					
3 graham crackers	89	—	2	—	—
1 c 1% low-fat milk	102	—	3	2	2
Totals:	2,017	29	63	20	234
Daily Values:[a]	2,000	25	65	20	60
Percentage of Daily Values:	101%	116%	97%	100%	390%
Percentage of calories from fat: 30%					

[a]Daily Values based on a 2,000-calorie diet.

Peanut Butter
Grape Jelly

DIET COLA

DIET COLA

Figure 2-9

Two Days' Meals Compared with Five Daily Values (Continued)

Tuesday's Meals

Tuesday's meals are lower on the nutrient density scale.

Foods	Energy (cal)	Fiber (g)	Total Fat (g)	Saturated Fat (g)	Vitamin C (mg)
Today, the student starts the day with a fast-food breakfast:					
1 c coffee	5	—	—	—	—
1 English muffin with egg, cheese, and bacon	383	2	20	9	1
Between classes, the student returns home for a quick lunch:					
1 peanut butter and jelly sandwich on white bread	350	3	14	3	—
1 c whole milk	150	—	8	5	2
While studying, the student has:					
12 oz diet cola	—	—	—	—	—
Bag of chips (14 chips)	228	2	15	5	13
That night for dinner, the student eats:					
A salad: 1 c lettuce 1 tbs blue cheese dressing	84	1	8	1	3
A main course:					
6 oz steak	343	—	14	5	—
½ baked potato (large)	110	2	—	—	13
1 tbs butter	102	—	12	7	—
1 tbs sour cream	31	—	3	2	—
12 oz diet cola	—	—	—	—	—
And for dessert:					
4 sandwich-type cookies	189	1	8	2	—
Later on, a bedtime snack:					
2 creme-filled snack cakes	214	—	8	2	—
1 c herbal tea	—	—	—	—	—
Totals:	2,189	11	110	41	32
Daily Values:[a]	2,000	25	65	20	60
Percentage of Daily Values:	109%	44%	169%	205%	53%
Percentage of calories from fat: 45%					

[a]Daily Values based on a 2,000-calorie diet.

do it!

SCORE YOUR DIET WITH THE FOOD GUIDE PYRAMID

This activity examines your diet for adherence to the Food Guide Pyramid and for variety. How can you tell if your diet meets the ideals of the Food Guide Pyramid? Part of the story is told by the number of servings the diet provides from each food group. More is revealed by the variety of foods *within* each group. The rest of the tale unfolds as you learn, in the chapters to come, about the carbohydrates, lipids, protein, vitamins, and minerals in foods.

In this activity, your diet will receive six scores (Form 2-2). When added together, these six scores yield a single number. The first five scores measure how well your diet meets the ideals of the Food Guide Pyramid. The sixth score is awarded for variety among your food choices.

In this exercise, 60, not 100, is the highest score attainable, as a reminder that the Food Guide Pyramid gives only a partial accounting of a diet's attributes.* The *Dietary Guidelines* concerning moderation in fat, salt, sugar, and alcohol are important, too. Try to keep them in mind as you review your food choices.

Your study begins with the next section. Make several copies of Form 2-1, one to use now and others in case you want to repeat this activity later.

Preparing Your Food Record

Step 1. Record all of the foods you ate and all of the beverages you drank in one typical 24-hour period. Write them on the left-hand side of Form 2-1.

Step 2. As accurately as possible, record the numbers of servings or fractions of servings you obtained from each food group (see Figure 2-5 for serving sizes, pp. 40–41). List the numbers in the appropriate columns and total each column. To do this accurately, as you eat the food, make careful note of the amount. Estimate the amount to the nearest ounce, quarter-cup, tablespoon, or other common measure. The Aids to Calculation Appendix (Appendix C) can help with conversion factors. Foods that belong in the tip of the Pyramid, the fats, oils, and sweets are ignored in this exercise.

In guessing at serving sizes, use these rules of thumb:

- A 3-ounce serving of meat is about the size of the palm of a woman's hand or a deck of cards.
- A standard piece of fruit or potato is the size of a regular (60-watt) lightbulb.
- A 1½-ounce piece of cheese is the size of a nine-volt battery.
- A standard slice of lunch meat or American-type cheese weighs 1 ounce.
- A pat (1 tsp) of a quarter-pound stick of butter or margarine is about as thick as 250 pages of this book (pressed together).

You may have to break down mixed dishes into their ingredients to decide how many servings a food represents. For example, one cup of tuna noodle casserole may provide:

- An ounce of tuna (½ serving of meat, poultry, fish, dry beans, eggs, and nuts)
- A half-cup of noodles (1 serving of breads, cereals, rice, and pasta)
- A quarter-cup of a combination of peas, carrots, and onions (½ serving of vegetables).

Errors of up to 20 or 30 percent are expected and tolerated.

Step 3. On Form 2-1 write the suggested total number of servings from each food group for someone whose energy need is similar to yours. Table 2-7 on p. 44 lists energy needs for some groups of people.

Scoring against the Food Guide Pyramid

Step 4. Compare your own totals on Form 2-1 with the suggested total number of servings for each food group and score yourself. Then transfer your scores to the pyramid of Form 2-2.

Score as follows:

- 10 points for each food group in which you ate all or more of the recommended number of servings.
- 0 points for no servings from a group.
- For values between these extremes, find your score by dividing the number of servings you ate by the recommended number and multiplying by 10.

To help clarify the process, we use a student's diet as an example. Joe is an active young adult who needs 2,800 calories.

	Joe Needed This Many Servings	Joe Consumed
▪ Breads, cereals, rice, or pasta:	11	6
▪ Vegetables:	5	5
▪ Fruit:	4	0
▪ Milk, yogurt, or cheese:	2	3
▪ Meat, poultry, fish, dry beans, eggs, and nuts:	3	5

*This exercise represents six of the ten test categories of the *Healthy Eating Index (HEI)*, a diet assessment tool developed by the United States Department of Agriculture's (USDA's) Center for Nutrition Policy and Promotion in 1995. The full HEI also tests the diet for adherence to the Dietary Guidelines for Americans. For a copy of the entire HEI, contact the USDA (see the Nutrition Resources Appendix).

Form 2-1

Food Record

NUMBER OF SERVINGS FROM EACH FOOD GROUP

FOOD/AMOUNT	Breads, Cereals, Rice, and Pasta	Vegetables	Fruit	Milk, Yogurt, and Cheese	Meat, Poultry, Fish, Dry Beans, Eggs, and Nuts
Breakfast:					
Snack:					
Lunch:					
Supper:					
Snack:					
Your totals:					
Suggested totals (From Table 2-7, p. 44)					

Your *Score* (Transfer to Form 2-2)					

do it!

Form 2-2

Do It Scoreboard

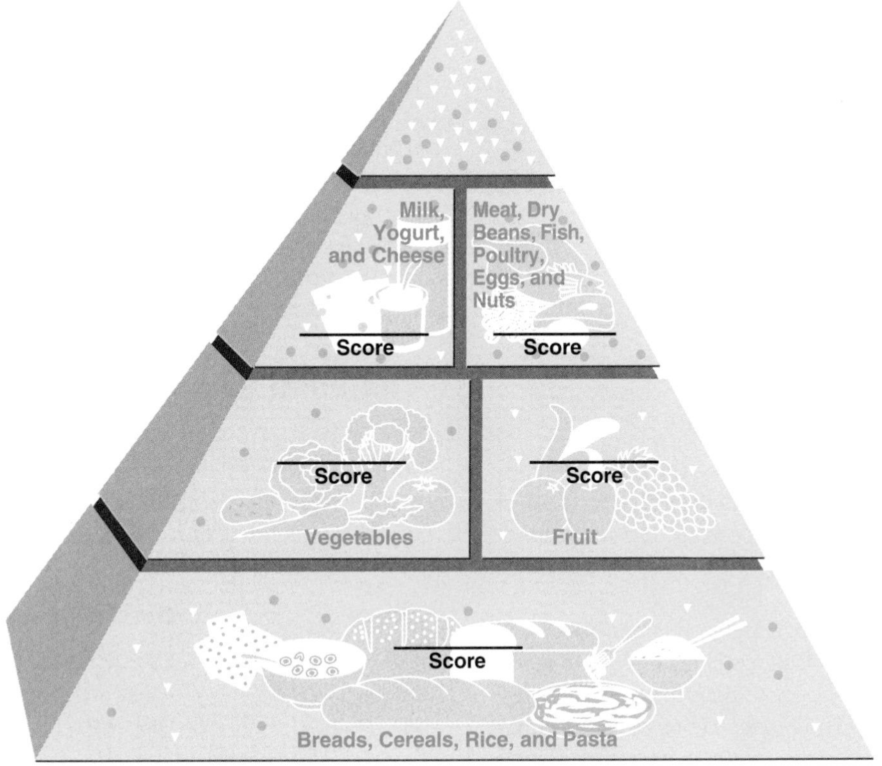
Milk, Yogurt, and Cheese
Score
Meat, Dry Beans, Fish, Poultry, Eggs, and Nuts
Score
Score
Vegetables
Score
Fruit
Score
Breads, Cereals, Rice, and Pasta

A. Transfer your scores from Form 2-1 to the boxes of the pyramid above. Add the 5 scores to obtain a single number as your pyramid total.

Pyramid total: ____________

B. Determine your score for variety by the method described in the text. Add your variety score and your pyramid total to obtain your total score:

Variety score: ____________

  Pyramid total: ___________
+ Variety score: ___________

Final score: ___________

Table 2-11

Scoring Variety (see text instructions)

If You Ate This Many Servings of Different Types of Foods:	You Receive This Score:
9.5 or more	10
9.0	9
8.5	8
8.0	7
7.5	6
7.0	5
6.5	4
6.0	3
5.5	2
5.0	1
4.5 or fewer	0

After analyzing his diet on Form 2-1, Joe finds that he consumed 6 bread, cereal, rice, or pasta foods. Thus, he ate 6 of 11 recommended servings of breads, cereals, rice, or pasta.

$$(6 \div 11) \times 10 = 5.5 \text{ points}$$

He consumed 5 vegetable servings, right on target, so he gets 10 points for these. He ate no fruit, so here he scores a zero.

From the milk group, he ate more than the recommended 2 servings, but he gets no extra points. He receives the maximum 10 points. The same is true for the meat group: a maximum of 10. After adding the points earned for each of the five food groups, Joe finds that his total points were 35.5 out of a possible 50.

Add up your scores and write the total on the pyramid total line of Form 2-2, part A.

Scoring Variety

The final scoring component, variety, is subjective. It asks you to use your own judgment in your assessment.

Step 5. Look over your food intake record and jot down the approximate number of different types of foods listed. Note that the food types referred to are not the *groups* of foods specified in the Food Guide Pyramid. Here food types mean any of the individual foods that make up those groups. For example, a person choosing milk and cheese in a day counts them as two types of foods because milk and cheese differ significantly from each other. A person choosing milk and pudding, on the other hand, counts them both as the same food because pudding is made from fluid milk and so is counted as the same type of food as milk. Other examples:

- Beef and lamb differ from each other enough to count as two types of food, but hamburger and beefsteak are counted as the same type.
- Different fruits are counted as different foods, but apples, applesauce, and apple juice are all of one type.
- Whole-wheat bread and enriched white bread are different foods, but enriched white hamburger rolls, enriched white biscuits, and enriched white bread are counted as one type of food, and so on.

Divisions among food types are not perfectly described. Give yourself half a credit for half-servings of foods in mixtures, such as ¼ cup tomato sauce (a whole serving is ½ cup) in lasagna, or ¼ cup chopped tomato or onion that dresses a hot dog.

Scoring: The top honor of 10 points is awarded to the diet that includes 9.5 or more servings a day of different types of foods. A day with fewer than four types earns a zero. Use the table on Form 2-2 to obtain a score; enter your score on Form 2-2, part B.

The Final Score

Step 6. Total your score by adding all six scores (five pyramid scores plus variety) on the Do It Scoreboard (Form 2-2). You may interpret the score as follows: a score of 60 means that the diet is excellent with regard to both number of servings in each group and variety in each group.

Score	Diet characteristics
60	Excellent variety and choices
50–59	Fairly adequate and varied diet
Less than 50	Diet needs study and improvement

Analysis

Answer these questions:

1. How well did your diet's overall score compare with the perfect score of 60 points?
2. Did you consume the recommended minimum number of servings for each of the five groups of the Food Guide Pyramid?
3. Which groups of foods, if any, are underrepresented in your diet? Reasons for failing to consume all of the servings from all of the food groups include allergies and intolerances, but people often simply choose their diets impulsively with inadequate planning for nutrition. If you did not consume the minimum number of servings from each group, list some reasons why.
4. Did your diet provide an adequate variety of foods, or were your choices monotonous? How can you expand your field of choices?

Keep your extra copies of Form 2-1 to help you track changes in your score over time. If you perform this analysis every week or so throughout the course, you will have a fairly accurate representation of your food habits and the adequacy and variety of your diet. You may also see changes in your eating habits as you learn more about the effects of nutrition on health.

Self Check

Answers to these Self Check questions are in Appendix G.

1. The nutrient standards used include all of the following *except:*
   a. Adequate Intakes (AI)
   b. Daily Minimum Requirements (DMR)
   c. Daily Values (DV)
   d. a and c
2. The Dietary Reference Intakes were devised for which of the following purposes:
   a. to set nutrient goals for individuals
   b. to suggest upper limits of intakes, above which toxicity is likely
   c. to set average nutrient requirements for use in research
   d. all of the above
3. According to the Food Guide Pyramid, which foods should form the foundation of a healthy diet?
   a. vegetables
   b. breads, cereals, rice, and pasta
   c. fruits
   d. milk, yogurt, and cheese
4. Which of the following adjustments in one's diet would agree with the *Dietary Guidelines for Americans*?
   a. eating baked potatoes rather than french fries
   b. drinking fat-free rather than 2% milk
   c. eating fruits rather than cakes and pies
   d. all of the above
5. According to the World Health Organization's (WHO) *Population Nutrient Goals,* the upper limit of total fat in one's diet should be:
   a. 5 percent
   b. 10 percent
   c. 20 percent
   d. 30 percent
6. The energy intake recommendation is centered around the average requirements for each age and sex group. T F
7. The Dietary Reference Intakes (DRI) are for all people, regardless of their medical history. T F
8. People who choose not to eat animals or their products need to find an alternative food guide to use instead of the Food Guide Pyramid when planning their diets. T F
9. By law, food labels must state as a percentage of the Daily Values the amounts of vitamin C, vitamin A, niacin, and thiamin present in a food. T F
10. To be labeled "low fat," a food must contain 3 grams of fat or less per serving. T F
11. Overall, the diets of the Mediterranean are high in animal protein and low in carbohydrates and fiber. (Read about this in the upcoming Controversy.) T F

NUTRITION ON THE NET

For further study of the topics of this chapter, access these websites and search for the phrases or words in quotation marks:

1. For the latest changes in chapter material, to find supplemental learning tools, or to access links to related websites, go to the "Nutrition: Concepts and Controversies" site at: **www.wadsworth.com/nutrition/prod/allprod.html**
2. Search the National Academy Press site for more information on "Dietary Reference Intakes (DRI)": **www.nap.edu/readingroom**
3. For more on "Food Labels": **www.healthfinder.gov/searchoptions/topicsaz.htm**
4. Canadian information on "nutrition guidelines" can be found at: **www.hc-sc.gc.ca**
5. The Food and Agriculture Organization and the World Health Organization (FAO/WHO) provide information on "other nutrient standards": **www.fao.org and www.who.org**
6. To learn more about "Dietary Guidelines" for Americans: **www.nal.usda.gov/fnic/dga/dguide95.html**
7. To view the "Food Guide Pyramid": **www.nal.usda.gov/fnic/Fpyr/pyramid.html**
8. Visit the American Diabetes Association site for "exchange lists": **www.diabetes.org**

CONTROVERSY

2

THE MEDITERRANEAN DIET: DOES IT HOLD THE SECRET FOR A HEALTHY HEART?

MUCH HAS BEEN said lately in favor of eating the Mediterranean way. People who eat traditional Mediterranean diets die much less frequently from heart disease and certain cancers than do people who eat diets typical of northern Europe and North America. Would people in the United States be healthier if they abandoned traditional American foods and adopted Mediterranean ones? Does this diet of ample grains, vegetables, fruits, olive oil, and cheese, accompanied by wine, hold special benefits? Some experts think so, but the issues have sparked some lively debates.[1] This Controversy explores this idea from the scientific point of view.

Anyone beginning to explore these ideas runs into some paradoxes. For example, the diet of one Mediterranean country, Greece, provides up to 42 percent of its calories as fat, mostly from olive oil and olives.[2] Many Greeks carry more body fat than is considered prudent in the United States. According to U.S. guidelines, then, the Greek population would be urged to eat less fat and to lose weight to protect their hearts. Yet Greeks living in Greece enjoy one of the longest life expectancies worldwide and die from cardiovascular disease far less often than do Americans.[3]

Should we be eating the Mediterranean way?

SHOULD AMERICANS EAT LIKE GREEKS?

Some have suggested that perhaps a diet similar to that of the Greeks might achieve heart disease rates here as low as those in Greece. Such a diet might also be easier to follow than the very-low-fat diet traditionally prescribed for heart health in the United States.[4]

Greece is not alone in having heart disease risks lower than those of the United States; many nations of the Mediterranean region do, and nobody knows why. It was once thought that genetics might bestow resistance to heart disease upon Mediterranean people. This idea has been all but abandoned, however, because the risks seem to follow regions, not individual people. Mediterranean immigrants to the United States who adopt an American diet and lifestyle suffer heart disease and cancer at the same rates as native-born U.S. citizens. Further, as "American-style" foods advance across the Mediterranean region, disease rates shift toward those more typical of northern Europe and North America.[5] This evidence supports the idea that diet, not genetics, is the primary variable that confers disease risk or protection. Indeed, many experts are concerned that countries with low rates of chronic diseases are abandoning their traditional diets in favor of heavily marketed American-style fatty foods.[6]

DEFINING THE MEDITERRANEAN DIET

Scientists who try to define a single "Mediterranean" diet run into a problem.[7] The vast Mediterranean region includes the diets of Spain, Portugal, France, Syria, Israel, and others, all different from one another. As an example, Greeks eat abundant fat as olive oil, but southern Italians keep total fat intakes low.

In days of old, say, 3000 B.C. or so, the diet of the region was founded on grain foods such as crusty breads, honey-sweetened cakes, rice, seeds including lentils and beans, fish, other seafood, goat cheese, olives and olive oil, vegetables, fruits (especially grapes and figs), and wine mixed half-and-half with water. Eventually, oranges and lemons arrived from the Far East, and tomatoes, eggplant, rice, beans, and potatoes from other lands became central to Mediterranean meals. Today, traditional diets of those countries follow much the same patterns. Meats are for special occasions only; the favored cooking fat is oil pressed from olives; butter is shunned. Overall, then, the diets of the Mediterranean are:

- Low in saturated fat.
- High in monounsaturated fats (such as olive oil).
- Low in animal protein.
- High in carbohydrates and fiber.
- High in antioxidant nutrients and phytochemicals.[8]

Based on these observations, a group of researchers have proposed a Mediterranean Pyramid, similar to the U.S. Food Guide Pyramid. Figure C2-1 shows the pyramids side by side, and the rest of this Controversy compares them. Later chapters revisit many of the issues discussed here because they are among the top concerns of nutrition scientists today.

THE WEIGHTED BASE—CARBOHYDRATES AND FIBER

Right away, some similarities between the pyramids of Figure C2-1 are apparent. Both have wide bases of breads, cereals, pasta, rice, and other grains, signifying that the two diets are built on the same foundation. This wise choice supplies abundant carbohydrate-rich foods to the eater, foods that in their unrefined state provide needed energy, vitamins, minerals, and fiber with almost no fat. Such foods meet the body's needs for these nutrients without elevating the person's risk of disease. In fact, diets high in fiber are thought to reduce risks of heart disease and cancer substantially. So far, both plans agree.

VEGETABLES AND FRUITS

In both pyramids vegetables and fruits appear next in predominance but with one important difference: the Mediterranean plan calls for servings of legumes (dried beans) or nuts every day. In Spain, a plateful of raw or cooked vegetables commonly precedes the main meal; melon or figs serve as dessert. So much is said in later chapters about the virtues of vegetables and fruits in the diet, let it suffice here to say that the case for their role in disease prevention is virtually ironclad. Vegetables provide fiber, phytochemicals, antioxidants, and other nutrients (see Controversies 7 and 11), all of which correlate with reduced incidence of chronic diseases. Further, with few exceptions, these foods are extraordinarily low in fat. Without a doubt, vegetables and fruits are allies to those who seek a healthful diet.

THE CLASSIFICATION OF LEGUMES

As noted, a difference between the U.S. and Mediterranean Pyramids is their recommendations for legumes (dried beans and peas). Legumes are difficult to classify in any food grouping plan because they have desirable qualities in common with so many other foods. For example, the U.S. Pyramid classes legumes with meats because, like meats, legumes are high in protein and also with the vegetables by virtue of their fiber, carbohydrate, and vitamins. Legume protein is not quite identical to that of meat. Still, a serving or two of either meat or legumes in a balanced diet amply meets the body's need for protein.

Legumes also supply iron and other minerals typically associated with meats. The similarities end there, however. Many meats are notoriously high in fat, much of it the heart-clogging saturated kind, and they provide no fiber or carbohydrate. Legumes, on the other hand, supply abundant fiber and carbohydrate with little or no fat, a combination the heart prefers. In some ways, certain legumes even resemble dairy products: they combine ample calcium and protein in one food.

In light of legumes' high fiber and mineral contents, the Mediterranean system lists legumes separately from meats and on a plane with the vegetables. Mediterranean main dishes often include some form of legumes, sometimes as part of the dish, sometimes as garnish. In Spain, cooked legumes are deep-fried in olive oil and served as snack foods, similar to potato chips in this country. Although legumes are hard to categorize, the following evidence concerning their health effects leaves no doubt about their importance.

ARGUMENTS FOR REQUIRING LEGUMES

"Why should I eat a bowl of beans every day when I can afford to eat meat?" asks a student who observes

Figure C2-1
Two Pyramid Plans

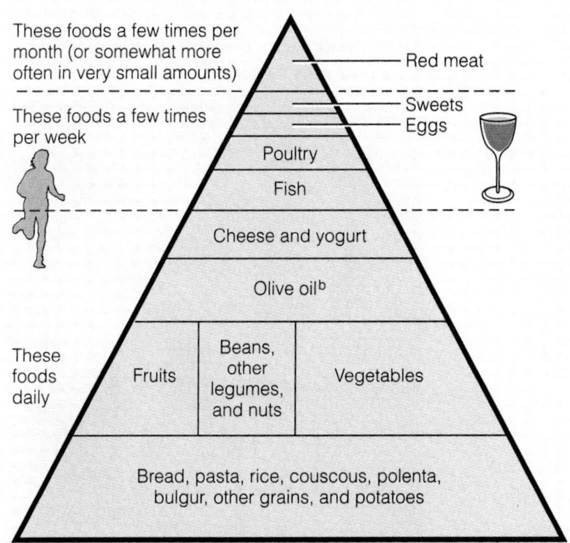
MEDITERRANEAN DIET PYRAMID[a]
These foods a few times per month (or somewhat more often in very small amounts)
Red meat
These foods a few times per week
Sweets
Eggs
Poultry
Fish
Cheese and yogurt
Olive oil[b]
These foods daily
Fruits
Beans, other legumes, and nuts
Vegetables
Bread, pasta, rice, couscous, polenta, bulgur, other grains, and potatoes

[a]The authors of this pyramid also recommend regular physical exercise and moderate consumption of wine.
[b]Other oils rich in monounsaturated fats, such as canola or peanut oil, can be substituted for olive oil. People who are watching their weight should limit their oil consumption.

SOURCE: 1994 Oldways Preservation & Exchange Trust, by permission.

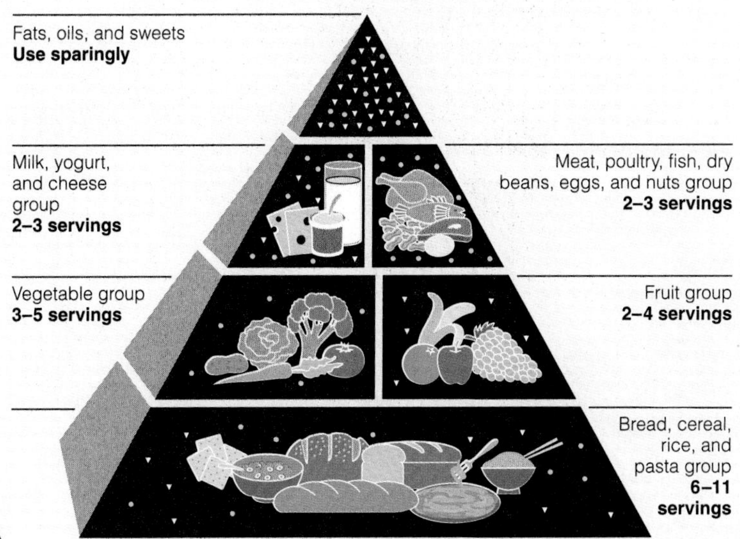
USDA's FOOD GUIDE PYRAMID
Fats, oils, and sweets
Use sparingly
Milk, yogurt, and cheese group
2–3 servings
Meat, poultry, fish, dry beans, eggs, and nuts group
2–3 servings
Vegetable group
3–5 servings
Fruit group
2–4 servings
Bread, cereal, rice, and pasta group
6–11 servings

SOURCE: U.S. Department of Agriculture/U.S. Department of Health and Human Services

KEY TO SYMBOLS
Fat (naturally occurring and added) **Sugars** (added)
These symbols show that fat and added sugars come mostly from fats, oils, and sweets, but can be part of or added to foods from the other food groups as well.

that legumes and meats are interchangeable in the U.S. Pyramid. The answer is found in evidence on legumes' fiber and their health effects.

The fiber of legumes is a constituent often lacking in U.S. diets. A reviewer reported recently that "an overwhelming consensus among health organizations advises increased consumption of fruits and vegetables, dried peas and beans, and whole grains" to help people meet current dietary recommendations.[9] Dry beans have more dietary fiber per serving than almost any other unprocessed food, and they provide a balance of the types of fibers that exert various beneficial effects on health.

Fiber alone cannot explain all of the findings about legumes and health, however. Consider the evidence:

- Endurance athletes who eat legumes in the hours before competition may extend their glucose fuel availability during competition.
- Legumes contain a form of starch believed to help control blood glucose in people with diabetes.[10]
- Legumes are rich in the type of fiber most strongly associated with low blood cholesterol and low rates of heart disease.[11]
- Replacing meat *protein* in the diet with soybean protein may lower blood lipids, thereby lowering heart disease risk, even when dietary *fats* are held constant.[12]
- Legumes contain substances believed to inhibit the growth of certain cancers (more about these substances in Controversy 11).[13]
- A meal that includes legumes satisfies hunger longer than other meals.[14]

The last of these attributes may have importance for weight control. A person who achieves a feeling of fullness sooner during a meal and stays full longer has less trouble resisting unneeded snacks.

As wonderful as legumes are, however, they are not "perfect foods" nor do they comprise an adequate diet when eaten alone. They lack the vitamins A and C of vegetables and fruits and the vitamin B_{12} of meats and dairy products. To overconsume legumes to the exclusion of other foods would be as poor a choice as never to include a bean in the diet.

Legumes: remarkable foods

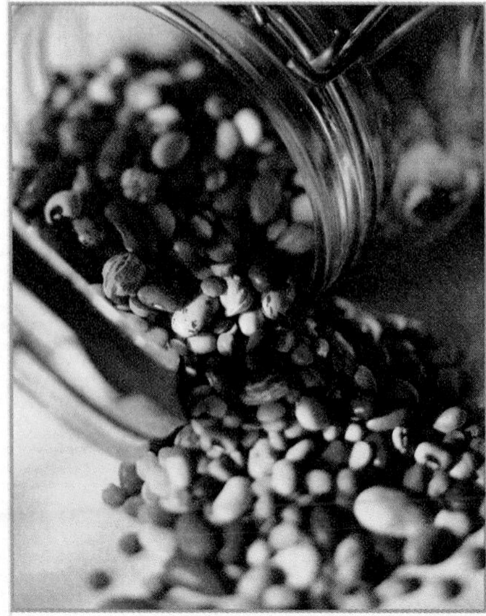

ARGUMENT CONCERNING OLIVE OIL

One of the most startling differences between the two pyramids, and the one that draws the most controversy, concerns fat. The Mediterranean plan suggests that the next most predominant daily dietary constituent after fruits and vegetables should be olive oil—a source of pure fat. In the U.S. Pyramid, olive oil joins other fats at the tip-top, flagged with a warning to use all fats sparingly.

As noted, the Mediterranean plan delivers much more fat than would be thought prudent under the U.S. guidelines. People following the Mediterranean plan receive some 40 percent of their calories from fat, mostly from olive oil chosen instead of butter, margarine, cooking oils, or other fats. Advocates of this high fat intake believe that such a diet may reduce the risk of heart disease and help control diabetes while providing more variety and greater satisfaction to the eaters.

Some scientific evidence to support this view exists. For example, researchers have found that olive oil is metabolized differently in the body than the oil of soybeans.[15] As a result, olive oil may conserve a beneficial blood lipid (HDL) associated with low risk of heart disease, while lowering the levels of riskier types of blood lipids.[16] Olives and the dark oils made from them contain potent antioxidants that may be partly responsible for low disease risks in those who consume them regularly.[17] Furthermore, olive oil stands out among unsaturated fats in being resistant to oxidation (see Chapter 5), which is a possible cause of heart disease.[18]

Not everyone agrees that research to date has made a sufficient case for a protective effect from olive oil. Not enough data exist, some say, to make recommendations for its use.[19] In addition, critics of the Mediterranean Pyramid note that foods such as avocados, nuts, and canola oil have oils similar to the oil of olives, but these foods are ignored by the plan. Also, critics point out that olive oil is fattening, providing 9 calories per gram, as do all fats. Though some very active people of the Mediterranean may burn off the excess calories from their high-fat diets, many people from that region are also overweight. Obesity and its many associated ills remain major threats to health in the United States. Critics therefore conclude that the recommendation to consume a high-fat diet is unwise, even if the fat itself is easy on the heart.

Olives and their oil contain phytochemicals that may benefit health.

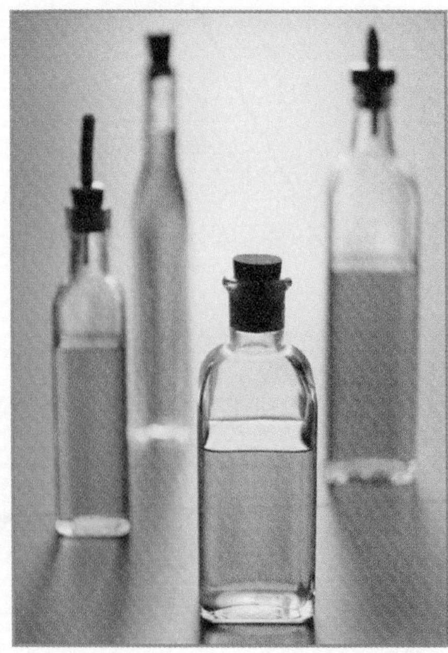

A MEDITERRANEAN APPROACH TO MEATS AND SWEETS

Returning to the differences between the two pyramids, the Mediterranean plan does not restrict total fat, but does limit *animal* fat (and this is mostly saturated fat). The U.S. Pyramid makes no such distinction. The Pyramid allows high-fat red meats, with their associated saturated fat, at every meal if a diner so chooses. Yet the diner who eats this much meat cannot help exceeding the recommended limit on fat intake.

People following the Mediterranean plan typically consume less than 10 percent of their calories from saturated fat—a goal of both plans. People following the U.S. plan find meeting this goal difficult, however. Notice that fish dominates poultry in the Mediterranean plan, and that eggs and sweets dominate red meats. Notice, too, that recommendations for consuming poultry and red meats are in terms of servings *per week or month,* not two to three servings *per day* as suggested by the U.S. plan. The average daily consumption of meat in the United States is more than half a pound per person per day; in the Mediterranean region, it is about half a pound per person per *week.* The difference in meat intake, and therefore in saturated fat intake, is significant.

As for the treatment of sweets in the plans, look back at Figure C2-1 and note the position of sweets in the two pyramids. The Mediterranean plan suggests that only one food—red meats—be eaten less frequently than sweets. The U.S. plan gives no specific guidance concerning intake of sweets but cautions us to "use sparingly."

EVIDENCE ABOUT YOGURT

The Mediterranean plan suggests daily cheese or yogurt but fails to mention fat-free milk, the staple milk source in the U.S. plan. Mediterranean people eat yogurt or cheese each day, and they may use milk to lighten their coffee, but milk by itself is not used as a beverage by adults. Cheese contains the nutrients of milk but is typically very high in saturated fat. Yogurt is really just fermented milk. Besides all of the nutrients in milk, yogurt contains the fermenting agent, *Lactobacillus,* a type of bacterium, and the products formed during fermentation. These attributes have generated interest in yogurt's effects on the body.

Interest in yogurt often stems from a century-old study of some healthy, aged people of another culture who credited their long lives to daily meals of yogurt. This study was discredited when researchers discovered that this culture highly valued the elderly, so people often made false claims about their age. Still, researchers are currently studying yogurt and its *Lactobacillus* to find out if they have special health-promoting qualities.

Early research on yogurt seemed to point to a connection between yogurt consumption and cancer prevention. Occasionally, studies have suggested a correlation between tumor suppression and yogurt or *Lactobacillus* consumption. No serious conclusions are possible about implications for human cancer at this time, however.

Another claim made for yogurt is that it prevents infections by promoting an immune response in tissues of the digestive tract.[20] Researchers found a correlation between live-culture fermented milk (yogurt) consumption and a fourfold increase in an indicator of immune system activity.[21] Another possible effect is the interference by *Lactobacillus* with the activities of a bacterium thought to cause ulcers, and a reduction in its ability to adhere to the stomach wall and cause damage.[22] The same characteristic of reducing adhesion of disease-causing organisims is also under study for the prevention of some vaginal and urinary tract infections in women who consume yogurt with live *Lactobacillus* cultures.[23] In fact, the research on *Lactobacillus* and disease prevention is part of a new field of study called *probiotics,* in which beneficial microorganisms are pitted against diseases to reduce reliance on antibiotic drugs. We can only wait for further research to support or refute these ideas.

The Mediterranean preference for yogurt over milk may stem from the likelihood that adults of the region may have problems digesting the milk sugar lactose and therefore experience digestive distress when they consume milk. Yogurt contains somewhat less lactose than does fresh milk (Chapter 4 provides details). People of northern European descent, however, retain milk-digesting abilities throughout adult life. In whatever form, milk and its products provide calcium in abundance, and most adults need more calcium in their diets.

EVIDENCE ABOUT WINE

The most controversial suggestion of the Mediterranean Pyramid is the advice to partake moderately of wine. The major concern, of course, is that encouraging wine drinking may invite development of alcoholism or an increase in alcohol-related traffic accidents. These consequences would be too severe to risk even if wine's supposed action against heart disease was a certainty—and it is not.

Does red wine protect the French and other Mediterranean people from heart disease? The idea probably has some validity. Studies show low rates of heart disease in populations that use moderate amounts of wine (red or white) as a beverage with meals. An interesting observation throws doubt on the idea that wine protects the heart, however. Wine-drinking peoples have been observed to consume a diet that is more supportive of health than the diets of other people, so it may well be that some other dietary factor is responsible for the protective effect.[24]

Alcohol itself (one or two drinks a day) may reduce the risk of heart disease by raising levels of beneficial blood lipids (HDL) and preventing blood clot formation.[25] These benefits are most apparent in people over age 50 and in those most likely to develop heart disease.[26] Studies also report, however, that heavy alcohol consumption (three or more drinks a day) *increases* the risk of death from other causes.[27] Regular alcohol consumption probably also increases the risk of breast cancer for some women and has many negative effects on body systems; the Controversy section of Chapter 5 describes these effects.

SHOULD AMERICANS EAT THE MEDITERRANEAN WAY?

Many questions surrounding the Mediterranean Pyramid remain unanswered. For one, though heart disease rates are low in Mediterranean countries, no one can explain why the incidence of stroke is almost double that of the United States. If the diet can receive the credit for heart health, should it also be blamed for the increased incidence of stroke?

Also, though the link between diet and heart disease overall is strong, no one knows for sure what parts of the diet form that link. Perhaps the olive oil is a key factor, but both nutrients and phytochemicals found in vegetables, legumes, seafood, and seasonings such as garlic or herbs offer protective effects, too. Active lifestyles may play a role, as may differences in tobacco use; exercise reduces heart disease risk, and smoking greatly increases it. The vitamin-conserving Mediterranean cooking techniques and preferences for raw fruits and vegetables may also matter.

Should everyone abandon the U.S. Pyramid, then, and take up eating Mediterranean style? Critics of the Mediterranean plan have expressed concerns that it may be inadequate in calcium and iron—two problem nutrients for many people, especially women and children.[28] Because these nutrients are typically lacking in many people's diets, it seems unwise to restrict selections of calcium-rich cheeses and yogurt and iron-rich meats to a few times a month.

Certainly, though, benefits are likely to follow if a few suggestions from the Greeks and other Mediterranean peoples are adopted. Including legumes as part of a balanced daily diet seems like a good idea, as does *replacing* saturated fats such as butter and meat fat with unsaturated phytochemical-rich fats like olive oil. The authors of this book would not stop there, however. They would urge you to reduce fats from all sources; choose small portions of the leanest meats, fish, and poultry; and include fresh foods from all the groups each day. Also, exercise daily, as the Mediterraneans do. As for wine, this is a personal choice, but one to approach with extreme caution, for no other dietary choice has such destructive potential.

CONCLUSION

Recommendations will continue to evolve as nutrition science unfolds. Meanwhile, you must choose foods every day. Be assured that you can confidently build your diet according to the unchanging foundation principles of adequacy, balance, calorie control, moderation, and variety.

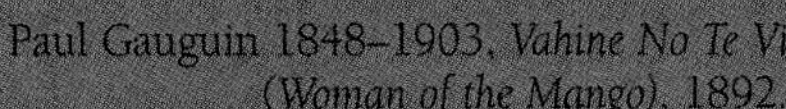

Paul Gauguin 1848–1903, *Vahine No Te Vi* (*Woman of the Mango*), 1892.

Contents

Frequently Asked Questions

genes units of a cell's inheritance, made of the chemical DNA (deoxyribonucleic acid). Each gene directs the making of a protein to do the body's work. (Proteins are described fully in Chapter 6.)

cells the smallest units in which independent life can exist. All living things are single cells or organisms made of cells.

AT THE MOMENT of conception, you received from your mother and father the **genes** that determine how your body works. Many of these genes are thousands of centuries old and have not changed since the Stone Age, when your ancestors walked the earth clad in skins and carrying clubs. Your body has changed very little since then, but you are living with the food, the luxuries, the smog, the additives, and all the other pleasures and problems of the twenty-first century. Faced with a variety of cultures and traditions, you may not have learned any time-tested and proven way of patterning your food intake. There is no guarantee that your diet, haphazardly chosen, will meet the needs of your Stone Age body. Unlike your ancestors, you must learn how your body works and what it needs from food to serve it best. Hence the study of nutrition and the workings of the body, so that you can put your mind to the task of nourishing your body.

The Body's Cells

The human body is composed of trillions of **cells,** and none of them knows anything about food. *You* may get hungry for fruit, milk, or bread, but each cell of your body needs nutrients—the vital components of foods. The ways in which the body's cells cooperate to obtain and use nutrients are the subjects of this chapter.

Each of the body's cells is a self-contained, living entity (see Figure 3-1), although each depends on the rest of the body to supply its needs. Among the cells' most basic needs are energy and the oxygen with which to burn it. Cells also need water to maintain the environment in which they live. They need building blocks and control systems. They especially need the nutrients they cannot make for themselves, the essential nutrients first described in Chapter 1, which must be supplied from food. The first principle of diet planning is that the foods we choose must provide energy and the essential nutrients, including water.

As living things, cells also die off, although at varying rates. Some skin cells and red blood cells must replenish themselves every 10 to 120 days. Cells lining the digestive

Figure 3-1

A Typical Cell (Simplified Diagram)

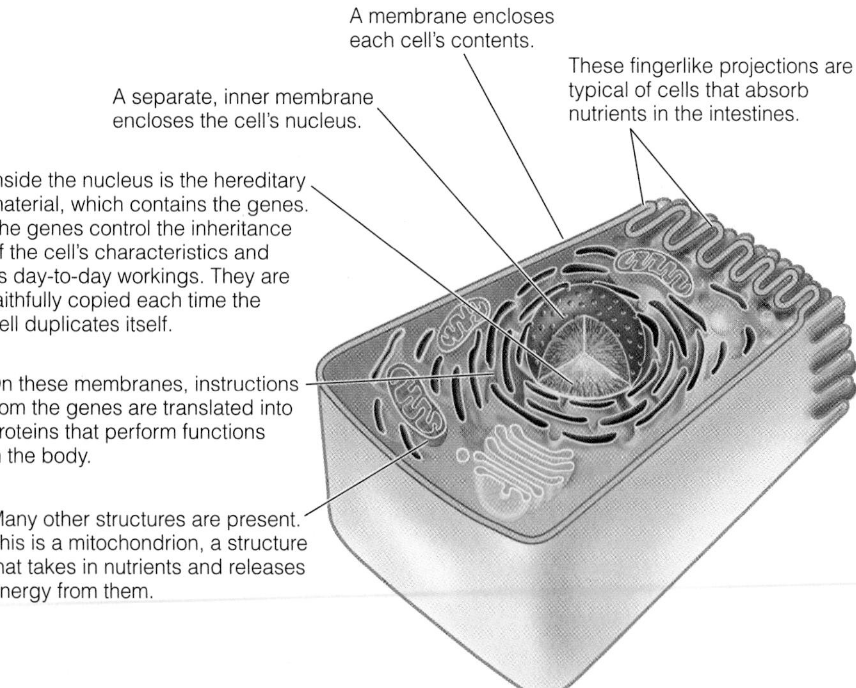
A membrane encloses each cell's contents.
These fingerlike projections are typical of cells that absorb nutrients in the intestines.
A separate, inner membrane encloses the cell's nucleus.
Inside the nucleus is the hereditary material, which contains the genes. The genes control the inheritance of the cell's characteristics and its day-to-day workings. They are faithfully copied each time the cell duplicates itself.
On these membranes, instructions from the genes are translated into proteins that perform functions in the body.
Many other structures are present. This is a mitochondrion, a structure that takes in nutrients and releases energy from them.

tract replace themselves every three days. Under ordinary conditions, many muscle cells reproduce themselves only once every few years. Liver cells have the ability to reproduce quickly and do so whenever repairs to the organ are needed. Certain brain cells do not reproduce at all; if damaged by injury or disease, they are lost forever.

In the human body, every cell works in cooperation with every other cell to support the whole. A cell's genes determine the nature of that work. Each gene is a blueprint that directs the production of a piece of protein machinery, often an **enzyme,** that helps to do the cell's work. Each cell contains a complete set of genes, but different ones are active in different types of cells. For example, in some intestinal cells, the genes for making digestive enzymes are active; in some of the body's **fat cells,** the genes for making enzymes that metabolize fat are active.

Nutrients affect the genes' activities within the cells. For example, several vitamins are known to enter the cell nucleus where they help to direct the making of various proteins. Because proteins often function as working machinery within the body, nutrient and gene interactions have sweeping effects on body functioning, affecting everything from fetal development to the strength of a person's bones or the risk of developing a chronic disease.[1]

Cells are organized into **tissues** that perform specialized tasks. For example, individual muscle cells are joined together to form muscle tissue, which can contract. Tissues, in turn, are grouped together to form whole **organs.** In the organ we call the heart, for example, muscle tissues, nerve tissues, connective tissues, and other types all work together to pump blood. Some body functions are performed by several related organs working together as part of a **body system.** For example, the heart, lungs, and blood vessels cooperate as parts of the cardiovascular system to deliver oxygen to all the body cells. The next few sections present the body systems with special significance to nutrition.

KEY POINT *The body's cells need energy, oxygen, and nutrients, including water, to remain healthy and do their work. Genes direct the making of each cell's machinery, including enzymes. Specialized cells are grouped together to form tissues and organs; organs work together in body systems.*

enzyme a protein that promotes a chemical reaction, described in Chapter 6.

fat cells cells that specialize in the storage of fat and that form the fat tissue.

tissues systems of cells working together to perform specialized tasks. Examples are muscles, nerves, blood, and bone.

organs discrete structural units made of tissues that perform specific jobs. Examples are the heart, liver, and brain.

body system a group of related organs that work together to perform a function. Examples are the circulatory system, respiratory system, and nervous system.

blood the fluid of the cardiovascular system; composed of water, red and white blood cells, other formed particles, nutrients, oxygen, and other constituents.

lymph (LIMF) the fluid that moves from the bloodstream into tissue spaces and then travels in its own vessels, which eventually drain back into the bloodstream.

arteries blood vessels that carry blood containing fresh oxygen supplies from the heart to the tissues (see Figure 3-2).

veins blood vessels that carry blood, with the carbon dioxide it has collected, from the tissues back to the heart (see Figure 3-2).

capillaries minute, weblike blood vessels that connect arteries to veins and permit transfer of materials between blood and tissues (see Figures 3-2 and 3-3).

plasma the cell-free fluid part of blood and lymph.

extracellular fluid fluid residing outside the cells.

lungs the body's organs of gas exchange. Blood circulating through the lungs releases its carbon dioxide and picks up fresh oxygen to carry to the tissues.

The Body Fluids and the Cardiovascular System

Body fluids supply the tissues continuously with energy, oxygen, and nutrients, including water. The fluids constantly circulate to pick up fresh supplies and deliver wastes to points of disposal. Every cell continuously draws oxygen and nutrients from those fluids and releases carbon dioxide and other waste products into them.

The body's main fluids are the **blood** and **lymph.** Blood travels within the **arteries, veins,** and **capillaries,** as well as within the heart's chambers (see Figure 3-2 on the next page). Lymph travels in separate vessels of its own. Circulating around the cells are other fluids such as the **plasma** of the blood and the fluid surrounding muscle cells (see Figure 3-3). Fluid surrounding cells **(extracellular fluid)** is derived from the blood in the capillaries; it squeezes out through the capillary walls and flows around the outsides of cells, permitting exchange of materials. Some of the fluid outside the cells returns to the blood by reentering the capillaries. The fluid remaining outside the capillaries forms lymph, which travels around the body by way of lymph vessels. The lymph eventually returns to the bloodstream near the heart where large lymph and blood vessels join. In this way, all cells are served by the cardiovascular system.

The fluid inside cells provides a medium in which all cell reactions take place. Its pressure also helps the cells to hold their shape. The fluid inside cells is drawn from the fluid on the outside that bathes the cells.

As the blood travels through the cardiovascular system, it delivers materials cells need and picks up their wastes. The blood picks up oxygen in the **lungs** and also

Interactions between vitamins and minerals and the genes are addressed in Chapters 7 and 8; other nutrition and gene interactions are addressed in the chapters on pregnancy and disease prevention.

Figure 3-2

Blood Flow in the Cardiovascular System

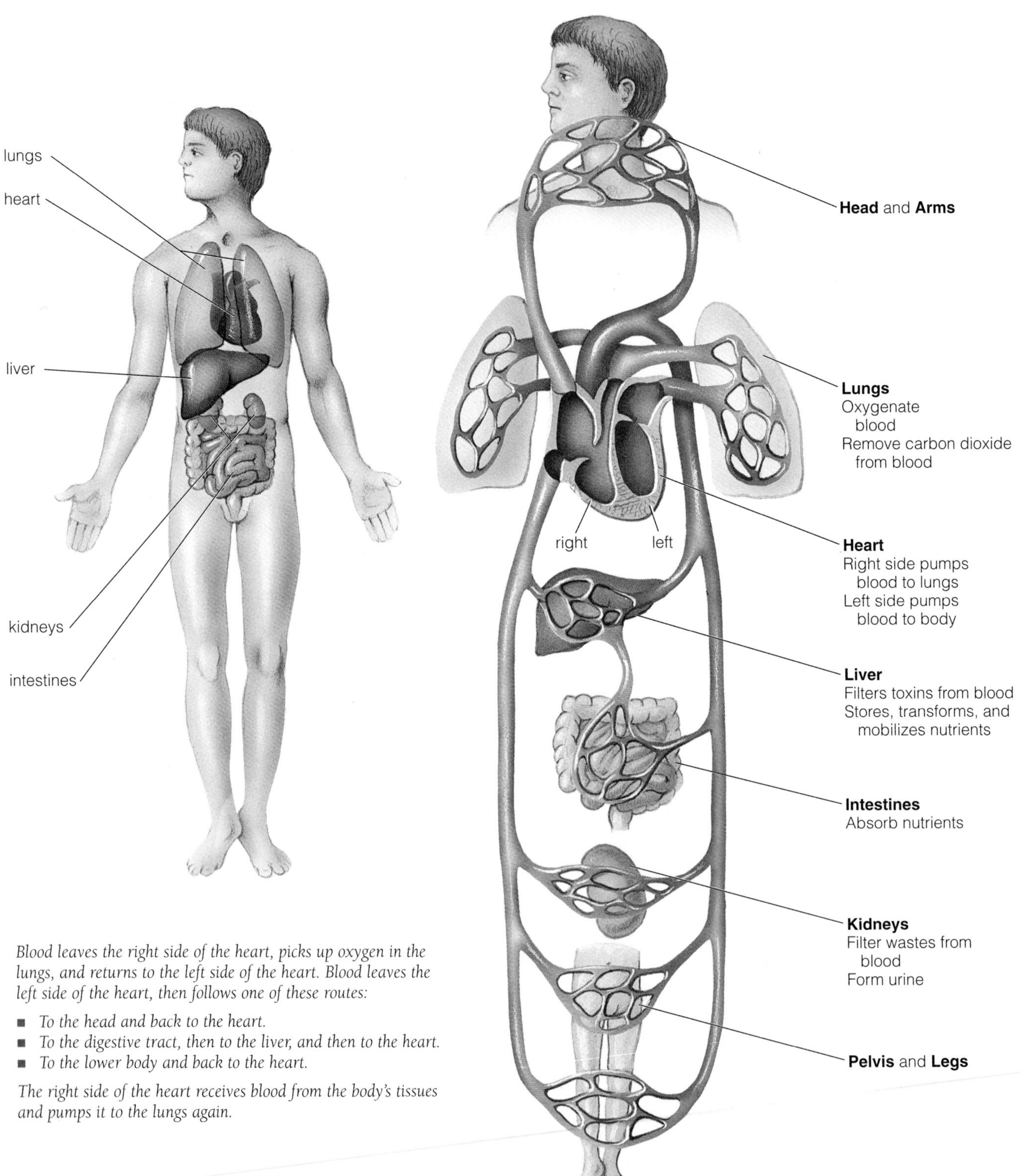
lungs
heart
liver
kidneys
intestines
Head and Arms
Lungs
Oxygenate blood
Remove carbon dioxide from blood
Heart
Right side pumps blood to lungs
Left side pumps blood to body
right
left
Liver
Filters toxins from blood
Stores, transforms, and mobilizes nutrients
Intestines
Absorb nutrients
Kidneys
Filter wastes from blood
Form urine
Pelvis and Legs

Blood leaves the right side of the heart, picks up oxygen in the lungs, and returns to the left side of the heart. Blood leaves the left side of the heart, then follows one of these routes:

- *To the head and back to the heart.*
- *To the digestive tract, then to the liver, and then to the heart.*
- *To the lower body and back to the heart.*

The right side of the heart receives blood from the body's tissues and pumps it to the lungs again.

Figure 3-3

How the Body Fluids Circulate Around Cells

The upper box shows a tiny portion of tissue with blood flowing through its network of capillaries (greatly enlarged). The lower box illustrates the movement of the extracellular fluid.

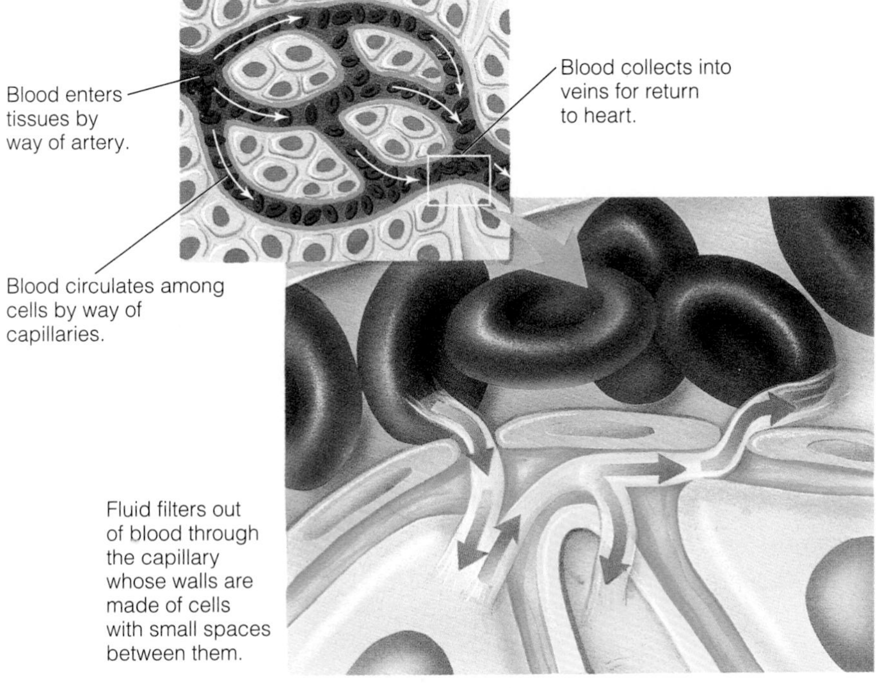

Blood enters tissues by way of artery.
Blood collects into veins for return to heart.
Blood circulates among cells by way of capillaries.
Fluid filters out of blood through the capillary whose walls are made of cells with small spaces between them.
Exchange of materials takes place between cell fluid and extracellular fluid.
Fluid may flow back into capillary or into lymph vessel. Lymph enters the bloodstream later through a large lymphatic vessel that empties into a large vein.

intestine the body's long, tubular organ of digestion and the site of nutrient absorption.

liver a large, lobed organ that lies just under the ribs. It filters the blood, removes and processes nutrients, manufactures materials for export to other parts of the body, and destroys toxins or stores them to keep them out of the circulation.

kidneys a pair of organs that filter wastes from the blood, make urine, and release it to the bladder for excretion from the body.

Chapter 8 offers guidelines for water intake.

releases carbon dioxide there, as Figure 3-4 shows. All the blood circulates to the lungs, then returns to the heart, where it receives powerful impetus from the pumping heartbeats that push it out to all body tissues. Thus, all tissues receive oxygenated blood fresh from the lungs.

As it passes through the digestive system, the blood delivers oxygen to the cells there and picks up most nutrients other than fats from the **intestine** for distribution elsewhere. Lymphatic vessels pick up most fats from the intestine and then transport them to the blood. All blood leaving the digestive system is routed directly to the **liver**, which has the special task of chemically altering the absorbed materials to make them better suited for use by other tissues. Later, in passing through the **kidneys,** the blood is cleansed of wastes. In summary, the blood is routed as follows (look again at Figure 3-2):

- Heart to tissues to heart to lungs to heart (repeat).

The portion of the blood that flows by the intestine travels from:

- Heart to intestine to liver to heart.

To ensure efficient circulation of fluid to all your cells, you need an ample fluid intake. This means drinking sufficient water to replace the water lost each day. Cardiovascular fitness is essential, too, and constitutes an ongoing project that requires attention to both nutrition and physical activity. Healthy red blood cells also play a role, for they carry oxygen to all the other cells, enabling them to use fuels for energy. Since red blood cells arise, live, and die within about four months, your body replaces them constantly, a manufacturing process that requires many essential nutrients from food.

hormones chemicals that are secreted by glands into the blood in response to conditions in the body that require regulation. These chemicals serve as messengers, acting on other organs to maintain constant conditions.

pancreas an organ with two main functions. One is an endocrine function—the making of hormones such as insulin, which it releases directly into the blood (*endo* means "into" the blood). The other is an exocrine function—the making of digestive enzymes, which it releases through a duct into the small intestine to assist in digestion (*exo* means "out" into a body cavity or onto the skin surface).

insulin a hormone from the pancreas that helps glucose enter cells from the blood (details in Chapter 4).

Figure 3-4

Oxygen–Carbon Dioxide Exchange in the Lungs

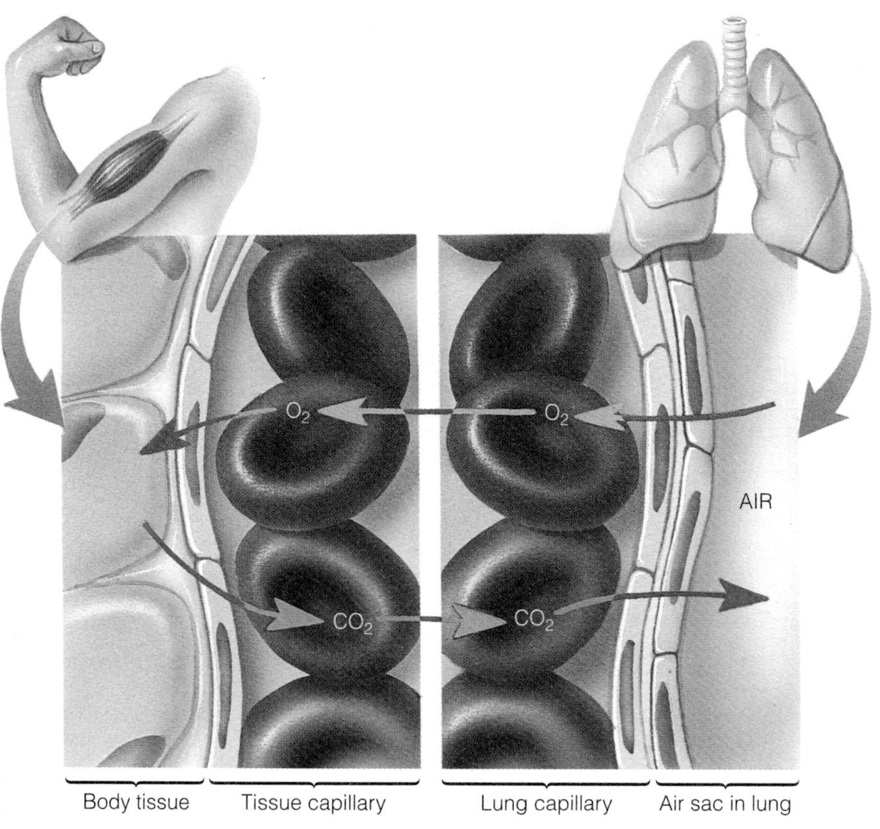

O_2
O_2
AIR
CO_2
CO_2
Body tissue
Tissue capillary
Lung capillary
Air sac in lung

In body tissues, red blood cells give up their oxygen (O_2) and absorb carbon dioxide (CO_2).

In the air sacs of the lungs, the red blood cells give up their load of carbon dioxide (CO_2) and absorb oxygen (O_2) from air to supply to body tissues.

The blood often serves as an indicator of disorders caused by dietary deficiencies or imbalances of vitamins or minerals; the blood is very sensitive to malnutrition.

KEY POINT *Blood and lymph deliver nutrients to all the body's cells and carry waste materials away from them. Blood also delivers oxygen to cells. The cardiovascular system ensures that these fluids circulate properly among all organs.*

The Hormonal and Nervous Systems

In addition to nutrients, oxygen, and wastes, the blood also carries chemical messengers, **hormones,** from one system of cells to another. Hormones communicate changing conditions that demand responses from the body organs.

What Do Hormones Have to Do with Nutrition?

Hormones are secreted and released directly into the blood by organs known as glands. Glands and hormones abound in the body. Each gland monitors a condition and produces one or more hormones to regulate it. Each hormone acts as a messenger that stimulates various organs to take appropriate actions.

For example, when the **pancreas** (a gland) detects a high concentration of the blood's sugar, glucose, it releases **insulin,** a hormone. Insulin stimulates muscle and other cells to remove glucose from the blood and to store it. When the blood glucose level falls, the pancreas secretes another hormone, **glucagon,** to which the liver responds by releasing into the blood some of the glucose it stored earlier. Thus, normal blood glucose levels are maintained.

Nutrition affects the hormonal system. Fasting, feeding, and exercise alter hormonal balances. People who become very thin have an altered hormonal balance that may make them unable to maintain their bones. People who eat high-fat diets have hormone levels that may make them susceptible to certain cancers.

Hormones also affect nutrition. Along with the nervous system, they regulate hunger and affect appetite. They carry messages to regulate the digestive system, telling the digestive organs what kinds of foods have been eaten and how much of each digestive juice to secrete in response. Hormones also regulate the menstrual cycle in women, and they affect the appetite changes many women experience during the cycle and in pregnancy. An altered hormonal state is thought to be at least partly responsible, too, for the loss of appetite that sick people experience. Hormones also regulate the body's reaction to stress, suppressing hunger and the digestion and absorption of nutrients. When questions about a person's nutrition are asked, the state of that person's hormonal system is often part of the answer.

KEY POINT *Glands secrete hormones that act as messengers to help regulate body processes.*

How Does the Nervous System Interact with Nutrition?

The body's other major communication system is, of course, the nervous system. With the brain and spinal cord as central controllers, the nervous system receives and integrates information from sensory receptors all over the body—sight, hearing, touch, smell, taste, and others—which communicate to the brain the state of both the outer and inner worlds, including the availability of food and the need to eat. The nervous system also sends instructions to the muscles and glands, telling them what to do.

The nervous system's role in hunger regulation is coordinated by the brain. The sensations of hunger and appetite are perceived by the brain's **cortex,** the thinking, outer layer. Much of the brain's regulatory work, however, goes on in the deep brain centers without the person's (or the cortex's) awareness. Deep inside the brain, the **hypothalamus** (see Figure 3-5) monitors many body conditions, including the availability of nutrients and water. To signal hunger, the physiological need for food, the digestive tract sends messages to the hypothalamus by way of hormones and nerves. The signals also stimulate the stomach to intensify its contractions and secretions, causing hunger pangs (and gurgling sounds). When your cerebral cortex becomes conscious of hunger, you eat. The conscious mind of the cortex, however, can override such signals and allow a person to choose to delay eating despite hunger or to eat when hunger is absent.

A marvelous adaptation of the human body, the ability to respond to physical danger, involves the workings of both the hormonal and nervous systems. Known as the **fight-or-flight reaction** or the *stress response,* this adaptation is present with only minor variations in all animals, showing how universally important it is to survival. It is a magnificently well-coordinated response. When danger is detected, nerves fire releasing **neurotransmitters** and glands supply the compounds **epinephrine** and **norepinephrine.*** Every organ of the body responds and **metabolism** speeds up. The pupils of the eyes widen so that you can see

*Strictly speaking, norephinephrine is a neurotransmitter; see Controversy 13.

glucagon a hormone from the pancreas that stimulates the liver to release glucose into the bloodstream.

cortex the outermost layer of something. The brain's cortex is the part of the brain where conscious thought takes place.

hypothalamus (high-poh-THAL-uh-mus) a part of the brain that senses a variety of conditions in the blood, such as temperature, glucose content, salt content, and others. It signals other parts of the brain or body to adjust those conditions when necessary.

fight-or-flight reaction the body's instinctive hormone- and nerve-mediated reaction to danger. Also known as the *stress response.*

neurotransmitters chemicals that are released at the end of a nerve cell when a nerve impulse arrives there; they diffuse across the gap to the next cell and alter the membrane of that second cell to either inhibit or excite it.

epinephrine the major hormone that elicits the stress response.

norepinephrine a compound related to epinephrine that helps to elicit the stress response.

metabolism the sum of all physical and chemical changes taking place in living cells; including all reactions by which the body obtains and spends the energy from food.

Details about hormones, menstruation, and the bones appear in Controversy 8 and Controversy 10.

All the body's cells live in water.

Figure 3-5

Cutaway Side View of the Brain Showing the Hypothalamus and Cortex

The hypothalamus monitors the body's conditions and sends signals to the brain's thinking portion, the cortex, which decides on actions. The pituitary gland is called the body's master gland, referring to its roles in regulating the activities of other glands and organs of the body.

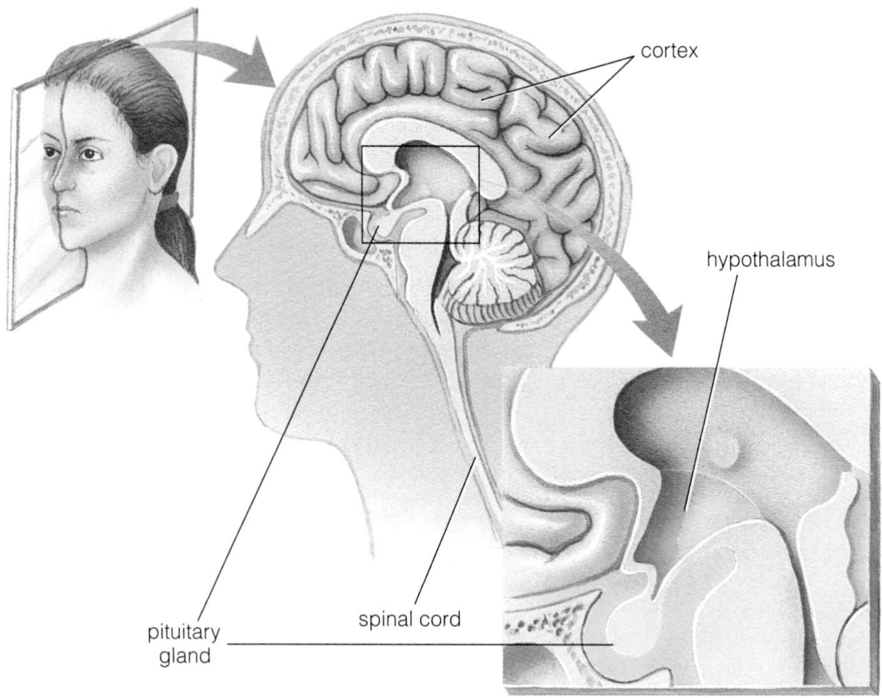
cortex
hypothalamus
pituitary gland
spinal cord

better; the muscles tense up so that you can jump, run, or struggle with maximum strength; breathing quickens and deepens to provide more oxygen. The heart races to rush the oxygen to the muscles, and the blood pressure rises to deliver efficiently the fuel the muscles need for energy. The liver pours forth glucose from its stores, and the fat cells release fat. The digestive system shuts down to permit all the body's systems to serve the muscles and nerves. With all action systems at peak efficiency, the body can respond with amazing speed and strength to whatever threatens it.

In ancient times, stress usually involved physical danger, and the response to it was violent physical exertion. In the modern world, stress is seldom physical, but the body's reaction to it is still the same. What stresses you today may be a checkbook out of control or a teacher who suddenly announces a pop quiz. Under these stresses, you are not supposed to fight or run as your Stone Age ancestor did. You smile at the "enemy" and suppress your fear. But your heart races, you feel it pounding, and hormones still flood your bloodstream with glucose and fat.

Your number-one enemy today is not a saber-toothed tiger prowling outside your cave, but a disease of modern civilization: atherosclerosis. Years of fat and other constituents accumulating in the arteries and stresses that strain the heart often lead to heart attacks, especially when chronic underexertion pairs with sudden high blood pressure. Daily exercise as part of a healthy lifestyle releases pent-up stress and helps to protect the heart against atherosclerosis.

KEY POINT *The nervous system joins the hormonal system to regulate body processes through communication among all the organs. Together, the hormonal and nervous systems respond to the need for food, govern the act of eating, regulate digestion, and call for the stress response.*

The Immune System

Many of the body's tissues cooperate to maintain defenses against infection. The skin presents a physical barrier, while the body's cavities (lungs, digestive tract, and others) are lined with membranes that resist penetration by invading **microbes** or unwanted substances. Vitamin and other nutrient deficiencies easily damage these linings, and health-care providers inspect both the skin and the inside of the mouth to detect signs of malnutrition. (Later chapters present details of the signs of deficiencies.) If an **antigen** invader penetrates the body's barriers, the **immune system** rushes in to defend the body against harm.

Of the 100 trillion cells that make up the human body, one in every hundred is a white blood cell. The actions of two types of white blood cells, the phagocytes and the **lymphocytes** known as T-cells and B-cells, are briefly described below:[2]

- **Phagocytes.** These scavenger cells travel throughout the body and are the first to defend body tissues against invaders. When a phagocyte recognizes a foreign particle, such as a bacterium, the phagocyte engulfs the invader, by forming a pocket in its outer membrane, surrounding the particle. Then, the phagocyte digests and destroys the particle. Phagocytes also leave a chemical trail that helps other immune cells to join the defense against infection.
- **T-cells.** T-cells recognize chemical messages from phagocytes and "read" the identity of an invader from the messages. The T-cells then seek out and destroy all foreign particles having the same identity. T-cells defend against fungi, viruses, parasites, some bacteria, and some cancer cells. They also pose a formidable obstacle to a successful organ transplant—the physician must prescribe immunosuppressive drugs following surgery to hold down the T-cells' attack against the "foreign" organ. People suffering from the disease AIDS are rendered defenseless against diseases because the human immunodeficiency virus (HIV) selectively attacks and destroys their T-cells.
- **B-cells.** B-cells respond rapidly to infection by dividing and releasing invader-fighting proteins, **antibodies,** into the bloodstream. Antibodies travel to the site of the infection and stick to the surface of the foreign particles, killing or inactivating them. The B-cells retain a chemical memory of each invader, and should the encounter recur, the response is swift. This is how immunizations work—a disabled or harmless form of a disease-causing organism is injected into the body so that the B-cells can learn to recognize it. Afterward, should the real, live infectious organism invade, the B-cells quickly release antibodies to destroy it.

Many other categories of white blood cells exist. Chapter 11 describes the roles of nutrition in supporting the body's defense system.

KEY POINT *The immune system enables the body to resist disease.*

microbes bacteria, viruses, or other organisms invisible to the naked eye, some of which cause diseases. Also called *microorganisms.*

antigen a microbe or substance that is foreign to the body.

immune system a system of tissues and organs that defend the body against antigens, foreign materials that have penetrated the skin or body linings.

lymphocytes (LIM-foe-sites) white blood cells that participate in the immune response; B-cells and T-cells.

phagocytes (FAG-oh-sites) white blood cells that can ingest and destroy antigens. The process by which phagocytes engulf materials is called *phagocytosis.* The Latin word *phagein* means "to eat."

T-cells lymphocytes that attack antigens. *T* stands for the thymus gland of the neck, where the T-cells are stored and matured.

B-cells lymphocytes that produce antibodies. *B* stands for bursa, an organ in the chicken in which B-cells were first identified.

antibodies proteins, made by cells of the immune system, that are expressly designed to combine with and to inactivate specific antigens.

Chapter 11 discusses the roles of nutrition support in AIDS and HIV infection.

The Digestive System

When your body needs food, your brain and hormones alert your conscious mind to the sensation of hunger. Then, when you eat, your taste buds guide you in judging whether foods are acceptable.

The taste buds contain on their surfaces structures that detect four basic chemical tastes: sweet, sour, bitter, and salty. A fifth taste may sometimes be included on this list: the taste of monosodium glutamate, sometimes called *savory* or by its Asian name, *umami* (ooh-MOM-ee). Other factors that affect a food's flavor include aroma, texture, temperature, and other flavor elements present. In fact, the ability to detect a food's aroma is thousands of times more sensitive than the sense of taste. The nose can detect just a few molecules responsible for the aroma of frying bacon, for example, even when they are diluted in several rooms full of air.

digestive system the body system composed of organs that break down complex food particles into smaller, absorbable products. The *digestive tract* and *alimentary canal* are names for the tubular organs that extend from the mouth to the anus. The whole system, including the pancreas, liver, and gallbladder, is sometimes called the *gastrointestinal,* or *GI,* system.

digest to break molecules into smaller molecules; a main function of the digestive tract with respect to food.

absorb to take in, as nutrients are taken into the intestinal cells after digestion; the main function of the digestive tract with respect to nutrients.

Why Do People Like Sugar, Fat, and Salt?

Sweet, salty, and fatty foods seem to be universally desired, but most people have aversions to bitter and sour tastes in isolation (see Figure 3-6).[3] The enjoyment of sugars and fat encourages people to consume ample energy, especially in the form of foods containing sugars, which provide the energy fuel for the brain. Likewise, foods containing fats provide energy and essential nutrients needed by all body tissues. The pleasure of a salty taste prompts eaters to consume sufficient amounts of two very important minerals—sodium and chloride. The aversion to bitterness discourages consumption of foods containing bitter toxins and also affects people's liking for foods in general. People born with great sensitivity to bitter tastes are apt to avoid foods with slightly bitter flavors, such as turnips and broccoli.[4]

The instinctive liking for sugar, fat, and salt can lead to drastic overeating of these substances. Sugar has become available in pure form only in the last hundred years, so it is relatively new to the human diet. Although fat and salt are much older, today all three substances are being added liberally to foods by manufacturers to tempt us to eat their products.

KEY POINT *The preference for sweet, salty, and fatty tastes seems to be inborn and can lead to overconsumption of foods that offer them.*

The Digestive Tract

Once you have eaten, your brain and hormones direct the many organs of the **digestive system** to **digest** and **absorb** the complex mixture of chewed and swallowed food. A diagram showing the digestive tract and its associated organs appears in Figure 3-7. The tract itself is a flexible, muscular tube extending from the mouth through the throat, esophagus, stomach, small intestine, large intestine, and rectum to the anus, for a total length of about 26 feet. In a sense, the human body is itself a tube surrounding this digestive canal. When you have swallowed something, it still is not inside the body; it is only inside the inner bore of this tube. Only when a nutrient or

Figure 3-6

The Innate Preference for Sweet Taste

This newborn baby is (a) resting, (b) tasting distilled water, (c) tasting sugar, (d) tasting something sour, and (e) tasting something bitter.

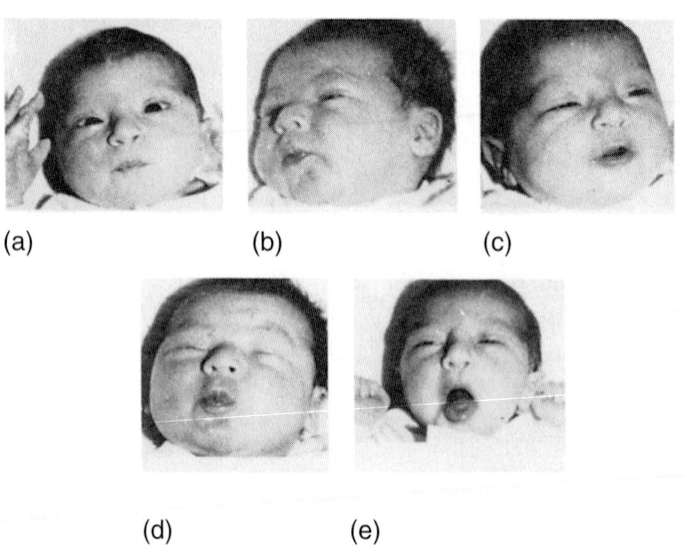
(a)
(b)
(c)
(d)
(e)

SOURCE: Taste-induced facial expressions of neonate infants from the classic studies of J. E. Steiner, in *Taste and Development: The Genesis of Sweet Preference,* ed. J. M. Weiffenbach, HHS publication no. NIH 77–1068 (Bethesda, Md.: U.S. Department of Health and Human Services, 1977), pp. 173–189, with permission of the author.

Figure 3-7

The Digestive System

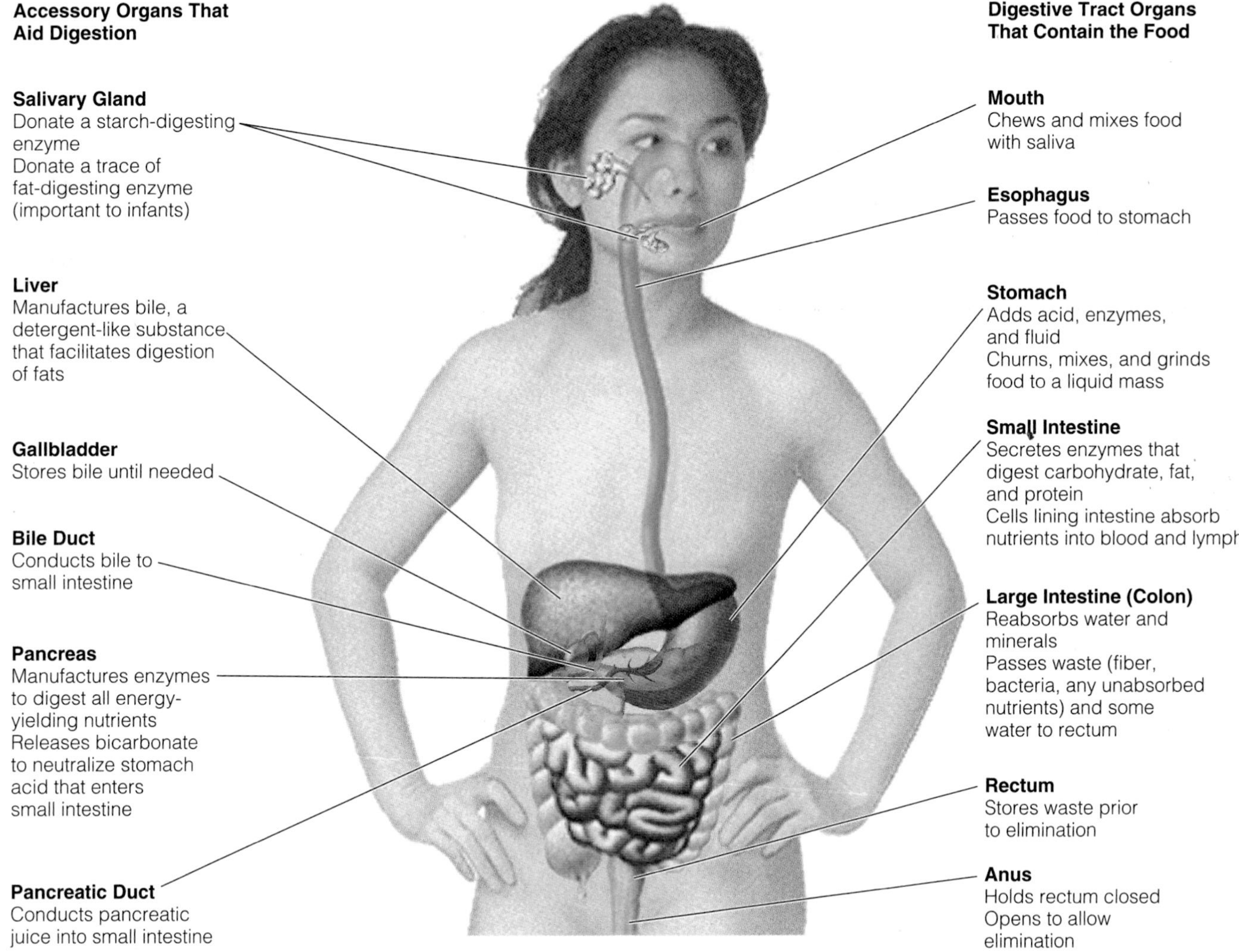

Accessory Organs That Aid Digestion
Salivary Gland
Donate a starch-digesting enzyme
Donate a trace of fat-digesting enzyme (important to infants)
Liver
Manufactures bile, a detergent-like substance that facilitates digestion of fats
Gallbladder
Stores bile until needed
Bile Duct
Conducts bile to small intestine
Pancreas
Manufactures enzymes to digest all energy-yielding nutrients
Releases bicarbonate to neutralize stomach acid that enters small intestine
Pancreatic Duct
Conducts pancreatic juice into small intestine
Digestive Tract Organs That Contain the Food
Mouth
Chews and mixes food with saliva
Esophagus
Passes food to stomach
Stomach
Adds acid, enzymes, and fluid
Churns, mixes, and grinds food to a liquid mass
Small Intestine
Secretes enzymes that digest carbohydrate, fat, and protein
Cells lining intestine absorb nutrients into blood and lymph
Large Intestine (Colon)
Reabsorbs water and minerals
Passes waste (fiber, bacteria, any unabsorbed nutrients) and some water to rectum
Rectum
Stores waste prior to elimination
Anus
Holds rectum closed
Opens to allow elimination

other substance passes through the wall of the digestive tract does it actually enter the body's tissues. Many things pass into the digestive tract and out again, unabsorbed. A baby playing with beads may swallow one, but the bead will not really enter the body. It will emerge from the digestive tract within a day or two.

The digestive system's job is to digest food to its components and then to absorb the nutrients and some nonnutrients, leaving behind the substances, such as fiber, that are appropriate to excrete. To do this, the system works at two levels: one, mechanical; the other, chemical.

KEY POINT *The digestive tract is a flexible, muscular tube that digests food and absorbs its nutrients and some nonnutrients.*

The Mechanical Aspect of Digestion

The job of mechanical digestion begins in the mouth, where large, solid food pieces such as bites of meat are torn into shreds that can be swallowed without choking. Chewing also adds water in the form of saliva to soften rough or sharp foods, such as fried tortilla chips, to prevent them from tearing the esophagus. Saliva also moistens and coats each bite of food, making it slippery and able to pass easily down the esophagus.

peristalsis (perri-STALL-sis) the wavelike muscular squeezing of the esophagus, stomach, and small intestine that pushes their contents along.

stomach a muscular, elastic, pouchlike organ of the digestive tract that grinds and churns swallowed food and mixes it with acid and enzymes, forming chyme.

sphincter (SFINK-ter) a circular muscle surrounding, and able to close, a body opening.

chyme (KIME) the fluid resulting from the actions of the stomach upon a meal.

small intestine the 20-foot length of small-diameter intestine, below the stomach and above the large intestine, that is the major site of digestion of food and absorption of nutrients.

pyloric (pye-LORE-ick) **valve** the circular muscle of the lower stomach that regulates the flow of partly digested food into the small intestine. Also called *pyloric sphincter.*

large intestine the portion of the intestine that completes the absorption process.

colon the large intestine.

feces waste material remaining after digestion and absorption are complete; eventually discharged from the body.

Nutrients trapped inside indigestible skins, such as seeds, must be liberated by breaking these skins before they can be digested. Chewing bursts open kernels of corn, for example, which would otherwise traverse the tract and exit undigested. Once food has been mashed and moistened for comfortable swallowing, longer chewing times provide no additional advantages to digestion. In fact, for digestion's sake, a relaxed, peaceful attitude during a meal aids digestion much more than chewing for an extended time.

Other organs take up the task of liquefying foods through the various mashing and squeezing actions of the stomach and intestines. The best known of these actions is **peristalsis,** a series of squeezing waves that start with the tongue's movement during a swallow and pass all the way down the esophagus (see Figure 3-8). The stomach and the intestines also push food through the tract by waves of peristalsis. Besides these actions, the **stomach** holds swallowed food for a while and mashes it into a fine paste; the stomach and intestines also add water so that the paste becomes more fluid as it moves along.

Figure 3-9 shows the muscular stomach. Notice the circular **sphincter** muscle at the base of the esophagus. It squeezes the opening at the entrance to the stomach to narrow it and prevent the stomach's contents from creeping back up the esophagus as the stomach contracts. The stomach stores swallowed food in a lump in its upper portion and squeezes the food little by little to its lower portion. There the food is ground and mixed thoroughly, ensuring that digestive chemicals mix with the entire thick liquid mass, now called **chyme.** Chyme bears no resemblance to the original food. The starches have been partly split, proteins have been uncoiled and clipped, and fat has separated from the mass.

The process of digestion is complicated, and the stomach acts as a holding tank, releasing only small amounts of food into the **small intestine** at one time. The muscular **pyloric valve** at the stomach's lower end (look again at Figure 3-9) controls the exit of the chyme, allowing only a little at a time to be squirted forcefully into the small intestine. Within a few hours after a meal, the stomach empties itself by means of these powerful squirts. The small intestine contracts rhythmically to move the contents along its length.

By the time the intestinal contents have arrived in the **large intestine** (also called the **colon**), digestion and absorption are nearly complete. The colon's task is mostly to reabsorb the water donated earlier by digestive organs and to absorb minerals, leaving a paste of fiber and other undigested materials, the **feces,** suitable for excretion. The fiber provides bulk against which the muscles of the colon can work. The rectum

Figure 3-8

Peristaltic Wave Passing Down the Esophagus

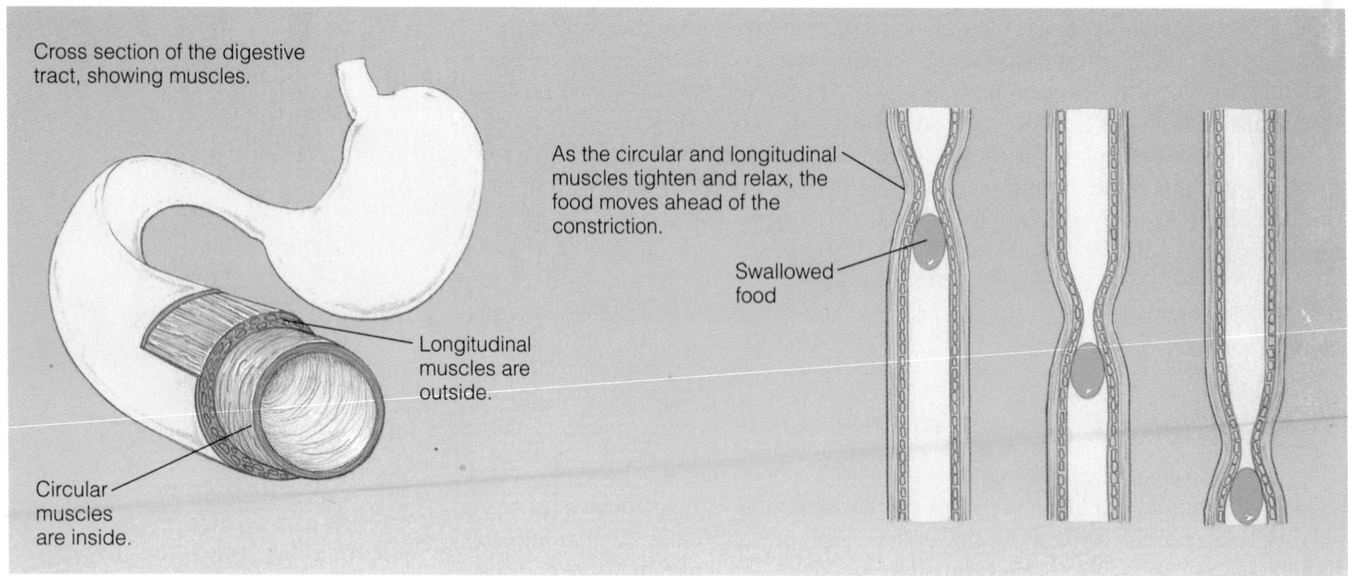
Cross section of the digestive tract, showing muscles.
As the circular and longitudinal muscles tighten and relax, the food moves ahead of the constriction.
Swallowed food
Longitudinal muscles are outside.
Circular muscles are inside.

Figure 3-9
The Muscular Stomach

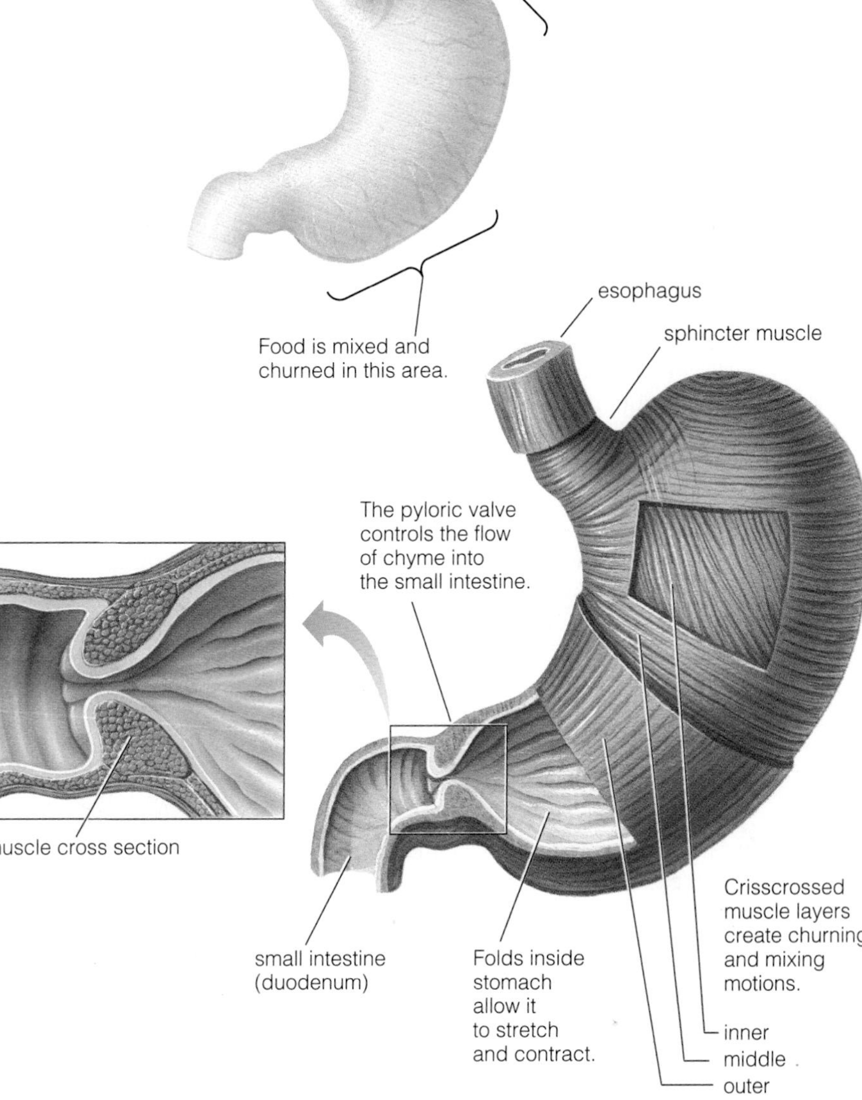
Food is stored in this area.
Food is mixed and churned in this area.
esophagus
sphincter muscle
The pyloric valve controls the flow of chyme into the small intestine.
muscle cross section
small intestine (duodenum)
Folds inside stomach allow it to stretch and contract.
Crisscrossed muscle layers create churning and mixing motions.
inner
middle
outer

stores this fecal material to be excreted at intervals. From mouth to rectum, the transit of a meal is accomplished in as short a time as a single day or as long as three days.

Some people wonder whether the digestive tract works best at some hours in the day and whether the timing of meals can affect how a person feels. Timing of meals is important to feeling well, not because the digestive tract is unable to digest food at certain times, but because the body requires nutrients to be replenished every few hours. Digestion is virtually continuous, being limited only during sleep and exercise. For some people, eating late may interfere with normal sleep. As for exercise, it is best pursued a few hours after eating because digestion can inhibit physical work (see Chapter 10 for details).

KEY POINT *The digestive tract moves food through its various processing chambers by mechanical means. The mechanical actions include chewing, mixing by the stomach, adding fluid, and moving the tract's contents by peristalsis. After digestion and absorption, wastes are excreted.*

gastric juice the digestive secretion of the stomach.

pH a measure of acidity on a point scale. A solution with a pH of 1 is a strong acid; a solution with a pH of 7 is neutral; a solution with a pH of 14 is a strong base.

mucus (MYOO-cus) a slippery coating of the digestive tract lining (and other body linings) that protects the cells from exposure to digestive juices (and other destructive agents). The adjective form is *mucous* (same pronunciation). The digestive tract lining is a *mucous membrane.*

emulsifier (ee-MULL-sih-fire) a compound with both water-soluble and fat-soluble portions that can attract fats and oils into water to form an emulsion.

bile a compound made by the liver, stored in the gallbladder, and released into the small intestine when needed. It emulsifies fats and oils to ready them for enzymatic digestion (described in Chapter 5).

pancreatic juice fluid secreted by the pancreas that contains both enzymes to digest carbohydrate, fat, and protein and sodium bicarbonate, a neutralizing agent.

bicarbonate a common alkaline chemical; a secretion of the pancreas; also, the active ingredient of baking soda.

The Chemical Aspect of Digestion

Several organs of the digestive system secrete special digestive juices that perform the complex chemical processes of digestion. Digestive juices contain enzymes that break nutrients down into their component parts. The digestive organs that release digestive juices are the salivary glands, the stomach, the pancreas, the liver, and the small intestine. Their secretions are listed in Figure 3-7 (on p. 75).

How Do "Digestive Juices" Work? Digestion begins in the mouth. An enzyme in saliva starts rapidly breaking down starch, and another enzyme initiates a little digestion of fat, especially the digestion of milk fat, important in infants. Saliva also helps maintain the health of the teeth in two ways: by washing away food particles that would otherwise foster decay and by neutralizing decay-promoting acids produced by bacteria in the mouth.

In the stomach, protein digestion begins. Cells in the stomach release **gastric juice,** a mixture of water, enzymes, and hydrochloric acid. A strong acid is needed to activate a protein-digesting enzyme and to initiate digestion of protein. As you might guess from the presence of acid and enzymes, protein digestion is the stomach's main function. The strength of an acid solution is expressed as its **pH.** As Figure 3-10 demonstrates, saliva is only weakly acidic, while the stomach's gastric juice is much more strongly acidic.

Upon learning of the powerful digestive juices and enzymes within the digestive tract, students often wonder how the tract's own cellular lining escapes being digested along with the food. Indeed, if it were not for specialized cells that secrete a thick, viscous substance known as **mucus,** the structures of the tract lining would be exposed to chemical attack. Mucus coats and protects the digestive tract lining.

In the small intestine, the digestive process gets under way in earnest. The small intestine is *the* organ of digestion and absorption, and it finishes what the mouth and stomach have started. The small intestine works with the precision of a laboratory chemist. As the thoroughly liquefied and partially digested nutrient mixture arrives there, hormonal messengers signal the gallbladder to contract and to squirt the right amount of the **emulsifier, bile,** into the intestine. Other hormones notify the pancreas to release **pancreatic juice** containing the alkaline compound **bicarbonate** in amounts precisely adjusted to neutralize the stomach acid that has reached the small intestine. All of the actions just described alter the intestinal environment to perfectly support the work of the digestive enzymes.

Meanwhile, as the pancreatic and intestinal enzymes act on the chemical bonds that hold the large nutrients together, smaller and smaller pieces are released into the intestinal fluids. The cells of the intestinal wall also hold some digestive enzymes on their surfaces; these enzymes perform last-minute breakdown reactions required before nutrients can be absorbed. Finally, the digestive process releases pieces small enough for the cells to absorb and use. Digestion and absorption of carbohydrate, fat, and protein are essentially complete by the time the intestinal contents enter the colon. Water, fiber, and some minerals, however, remain in the tract. Table 3-1 on page 80 provides a summary of all the processes involved.

KEY POINT *Chemical digestion begins in the mouth, where food is mixed with an enzyme in saliva that acts on carbohydrates. Digestion continues in the stomach, where stomach enzymes and acid break down protein. Digestion then continues in the small intestine; there the liver and gallbladder contribute bile that emulsifies fat, and the pancreas and small intestine donate enzymes that continue digestion so that absorption can occur.*

Are Some Food Combinations More Easily Digested Than Others? The digestive system can adjust to whatever mixture of foods is presented to it. People sometimes wonder if the digestive tract has trouble digesting certain foods in combination—for example, fruit and meat. Proponents of the fad of "food combining" claim that the digestive tract cannot perform certain digestive tasks at the same time, but this is a gross underestimation of the tract's capabilities.

The truth is that all foods, regardless of identity, are broken down by enzymes into the basic molecules that make them up. In fact, scientists who study digestion suggest that the tract analyzes the diet's nutrient contents and delivers juice and enzymes appropriate for digesting those nutrients. The pancreas is especially sensitive in this regard and has been observed to adjust its output of enzymes to digest carbohydrate, fat, or protein to an amazing degree. The pancreas of a person who suddenly consumes a meal unusually high in carbohydrate, for example, would begin increasing its output of carbohydrate-digesting enzymes within 24 hours, while reducing outputs of other types. This sensitive mechanism ensures that foods of all types are used fully by the body.

The next section reviews the major processes of digestion by showing how the nutrients in a mixture of foods are handled. The foods used to illustrate the digestive process are those in a peanut butter and banana sandwich. But whether the nutrients in the digestive tract occurred originally in this meal or in a chili dog makes little difference to the digestive tract, which can promptly polish off either or both.

KEY POINT *The healthy digestive system is capable of adjusting to almost any diet and can handle any combination of foods with ease.*

If "I Am What I Eat," Then How Does a Sandwich Become "Me"?

The process of rendering foods into nutrients and absorbing them into the body fluids is remarkably efficient. Within about 24 to 48 hours of eating, a healthy body digests and absorbs about 90 percent of the carbohydrate, fat, and protein in a meal. Here, we follow a peanut butter and banana sandwich on whole-wheat, sesame seed bread through the tract for the purpose of reviewing digestive processes in order of their occurrence in the body.

In the Mouth In each bite, food components are crushed, mashed, and mixed with saliva by the teeth and the tongue. The sesame seeds are crushed and torn open by the teeth, which break through the indigestible fiber coating so that digestive enzymes can reach the nutrients inside the seeds. The peanut butter is the "extra crunchy" type, but the teeth grind the chunks to a paste before the bite is swallowed. The carbohydrate-digesting enzyme of saliva begins to break down the starches of the bread, banana, and peanut butter to sugars. Each swallow triggers a peristaltic wave that travels the length of the esophagus and carries one bite of food to the stomach.

In the Stomach The stomach collects bite after swallowed bite in its upper storage area, where starch continues to be digested until the gastric juice mixes with the salivary enzymes and halts their action. Small portions of the mashed sandwich are pushed into the digesting area of the stomach where gastric juice mixes with the mass. Acid in gastric juice unwinds proteins; an enzyme clips into pieces the protein strands from the bread, seeds, and peanut butter. The sandwich has now become chyme. The watery carbohydrate- and protein-rich part of the chyme enters the small intestine first; a layer of fat follows closely behind.

In the Small Intestine Some of the sweet sugars in the banana require so little digesting that they begin to cross the linings of the small intestine immediately on contact. Nearby, the liver donates bile through a duct into the small intestine. The bile blends the fat from the peanut butter and seeds with the watery enzyme-containing digestive fluids. The nearby pancreas squirts enzymes into the small intestine to break

Figure 3-10

pH Values of Digestive Juices and Other Common Fluids

A substance's acidity or alkalinity is measured in pH units. Each step down the scale indicates a tenfold increase in concentration of hydrogen particles, which determine acidity. For example, a pH of 2 is 1,000 times stronger than a pH of 5.

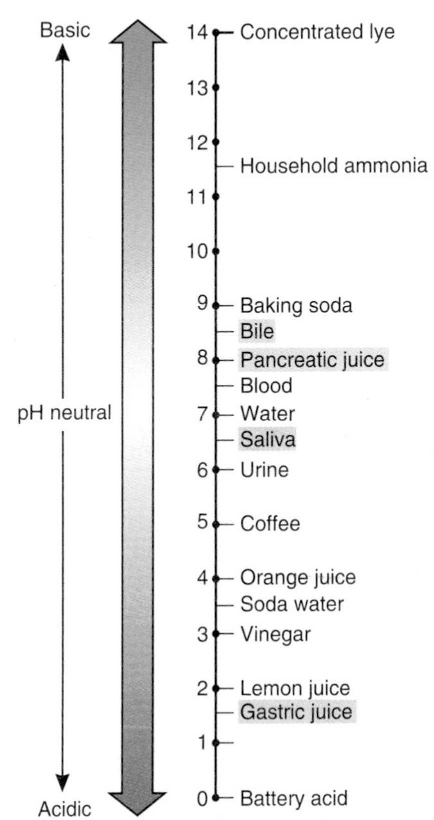
Basic
pH neutral
Acidic
14 — Concentrated lye
13
12
Household ammonia
11
10
9 — Baking soda
Bile
8 — Pancreatic juice
Blood
7 — Water
Saliva
6 — Urine
5 — Coffee
4 — Orange juice
Soda water
3 — Vinegar
2 — Lemon juice
Gastric juice
1
0 — Battery acid

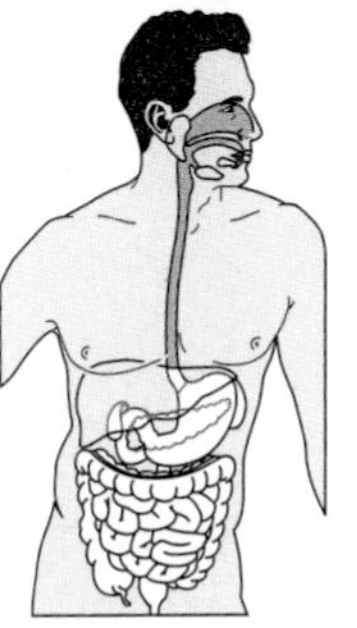

Time in mouth, less than a minute.

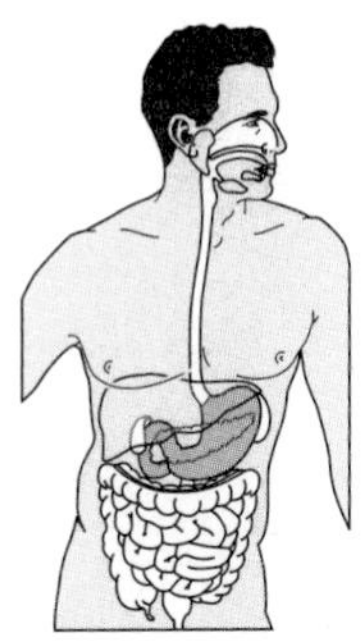

Time in stomach, about 1–2 hours.

Table 3-1

Summary of Chemical Digestion

	Mouth	Stomach	Small Intestine, Pancreas, Liver, and Gallbladder	Large Intestine (Colon)
Sugar and Starch	The salivary glands secrete saliva to moisten and lubricate food; chewing crushes and mixes it with a salivary enzyme that initiates starch digestion.	Digestion of starch continues while food remains in the upper storage area of the stomach. In the lower digesting area of the stomach, hydrochloric acid and an enzyme of the stomach's juices halt starch digestion.	The pancreas produces a starch-digesting enzyme and releases it into the small intestine. Cells in the intestinal lining possess enzymes on their surfaces that break sugars and starch fragments into simple sugars, which then are absorbed.	Undigested carbohydrates reach the colon and are partly broken down by intestinal bacteria.
Fiber	The teeth crush fiber and mix it with saliva to moisten it for swallowing.	No action.	Fiber binds cholesterol and some minerals.	Most fiber excreted with feces; some fiber digested by bacteria in colon.
Fat	Fat-rich foods are mixed with saliva. The tongue produces traces of a fat-digesting enzyme that accomplishes some breakdown, especially of milk fats. The enzyme is stable at low pH and is important to digestion in nursing infants.	Fat tends to rise from the watery stomach fluid and foods and float on top of the mixture. Only a small amount of fat is digested. Fat is last to leave the stomach.	The liver secretes bile; the gallbladder stores it and releases it into the small intestine. Bile emulsifies the fat and readies it for enzyme action. The pancreas produces fat-digesting enzymes and releases them into the small intestine to split fats into their component parts (primarily fatty acids), which then are absorbed.	Some fatty materials escape absorption and are carried out of the body with other wastes.
Protein	Chewing crushes and softens protein-rich foods and mixes them with saliva.	Stomach acid works to uncoil protein strands and to activate the stomach's protein-digesting enzyme. Then the enzyme breaks the protein strands into smaller fragments.	Enzymes of the small intestine and pancreas split protein fragments into smaller fragments or free amino acids. Enzymes on the cells of the intestinal lining break some protein fragments into free amino acids, which then are absorbed. Some protein fragments are also absorbed.	The large intestine carries undigested protein residue out of the body. Normally, almost all food protein is digested and absorbed.
Water	The mouth donates watery, enzyme-containing saliva.	The stomach donates acidic, watery, enzyme-containing gastric juice.	The liver donates a watery juice containing bile. The pancreas and small intestine add watery, enzyme-containing juices; pancreatic juice is also alkaline.	The large intestine reabsorbs water and some minerals.

down the fat, protein, and starch in the chemical soup that just an hour ago was a sandwich. The cells of the small intestine itself produce enzymes to complete these processes. As the enzymes do their work, smaller and smaller chemical fragments are liberated from the chemical soup and are absorbed into the blood and lymph through the cells of the small intestine's wall. Vitamins and minerals are absorbed here, too. They all eventually enter the bloodstream to nourish the tissues.

In the Large Intestine (Colon) Only fiber fragments, fluid, and some minerals are absorbed in the large intestine. The fibers from the seeds, whole-wheat bread, peanut butter, and banana are partly digested by the bacteria living in the colon, and some of the products are absorbed.[5] Most fiber is not absorbed, however, and, along with some other components, passes out of the colon, excreted as feces.

KEY POINT *The mechanical and chemical actions of the digestive tract break foods down to nutrients, and large nutrients to their smaller building blocks, with remarkable efficiency.*

villi (VILL-ee, VILL-eye) fingerlike projections of the sheets of cells that line the intestinal tract. The villi make the surface area much greater than it would otherwise be (singular: *villus*).

microvilli (MY-croh-VILL-ee, MY-croh-VILL-eye) tiny, hairlike projections on each cell of every villus that can trap nutrient particles and transport them into the cells (singular: *microvillus*).

Absorption and Transportation of Nutrients

Once the digestive system has broken food down to its nutrient components, the rest of the body awaits their delivery. First, though, every molecule of nutrient must traverse one of the cells of the intestinal lining. These cells absorb nutrients from the mixture within the intestine and deposit them in the blood and lymph. The cells are selective: they recognize some of the nutrients that may be in short supply in the body. The mineral calcium is an example. The less calcium in the body, the more calcium the intestinal cells absorb. The cells are also extraordinarily efficient: they absorb enough nutrients to nourish all the body's other cells.

The cells of the intestinal tract lining are arranged in sheets that poke out into millions of finger-shaped projections **(villi)**. Every cell on every villus has a brushlike covering of tiny hairs **(microvilli)** that can trap the nutrient particles. Each villus (projection) has its own capillary network and a lymph vessel so that as nutrients move across the cells, they can immediately mingle with the body fluids. Figure 3-11 provides a close look at these details.

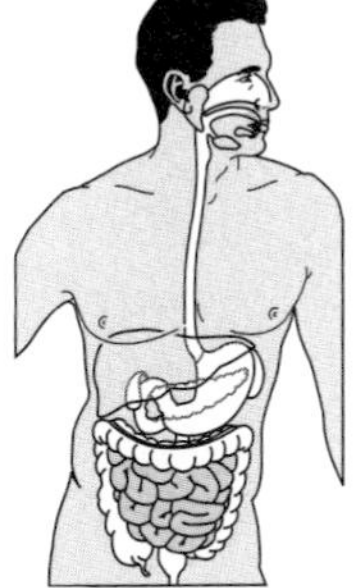

Time in small intestine, about 7–8 hours.*

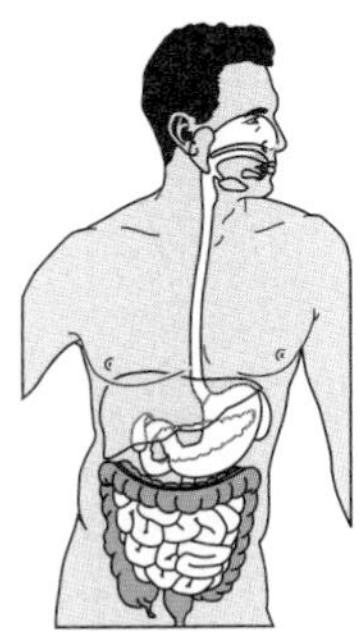

Time in colon, about 12–14 hours.*

*Based on a 24-hour transit time. Actual times vary widely.

The small intestine's lining, villi and all, is wrinkled into thousands of folds, so that its absorbing surface is enormous. If the folds, and the villi that poke out from them, were spread out flat, they would cover a third of a football field. The billions of cells of that surface weigh only 4 to 5 pounds, yet they absorb enough nutrients to nourish the other 150 or so pounds of body tissues.

After the nutrients pass through the cells of the villi, the blood and lymph take over the job of transporting the nutrients to their ultimate consumers, the body's cells. The lymphatic vessels initially transport most of the products of fat digestion and a few vitamins, later delivering them to the bloodstream. The blood vessels carry the products of carbohydrate and protein digestion, most vitamins, and the minerals from the digestive tract to the liver. Thanks to these two transportation systems, every nutrient soon arrives at the place where it is needed.

The digestive system's millions of specialized cells are themselves sensitive to an under supply of energy, nutrients, or dietary fiber. In cases of severe undernutrition of energy and nutrients, the absorptive surface of the small intestine shrinks. The surface may be reduced to a tenth of its normal area, preventing it from absorbing what few nutrients a limited food supply may provide. Without sufficient fiber to provide an undigested bulk for the tract's muscles to push against, the muscles become weak from lack of exercise. Malnutrition that impairs digestion is self-perpetuating because impaired digestion makes malnutrition worse. In fact, the digestive system's needs are few, but important. The body has much to say to the attentive listener, stated in a language of symptoms and feelings that you would be wise to study. The next section takes a lighthearted look at what your digestive tract might be trying to tell you.

What is your digestive tract trying to tell you?

KEY POINT *The digestive system feeds the rest of the body and is itself sensitive to malnutrition. The folds and villi of the small intestine enlarge its surface area to facilitate nutrient absorption through uncountable cells to the blood and lymph. These transport systems then deliver the nutrients to all the body cells.*

Figure 3-11

Details of the Small Intestinal Lining

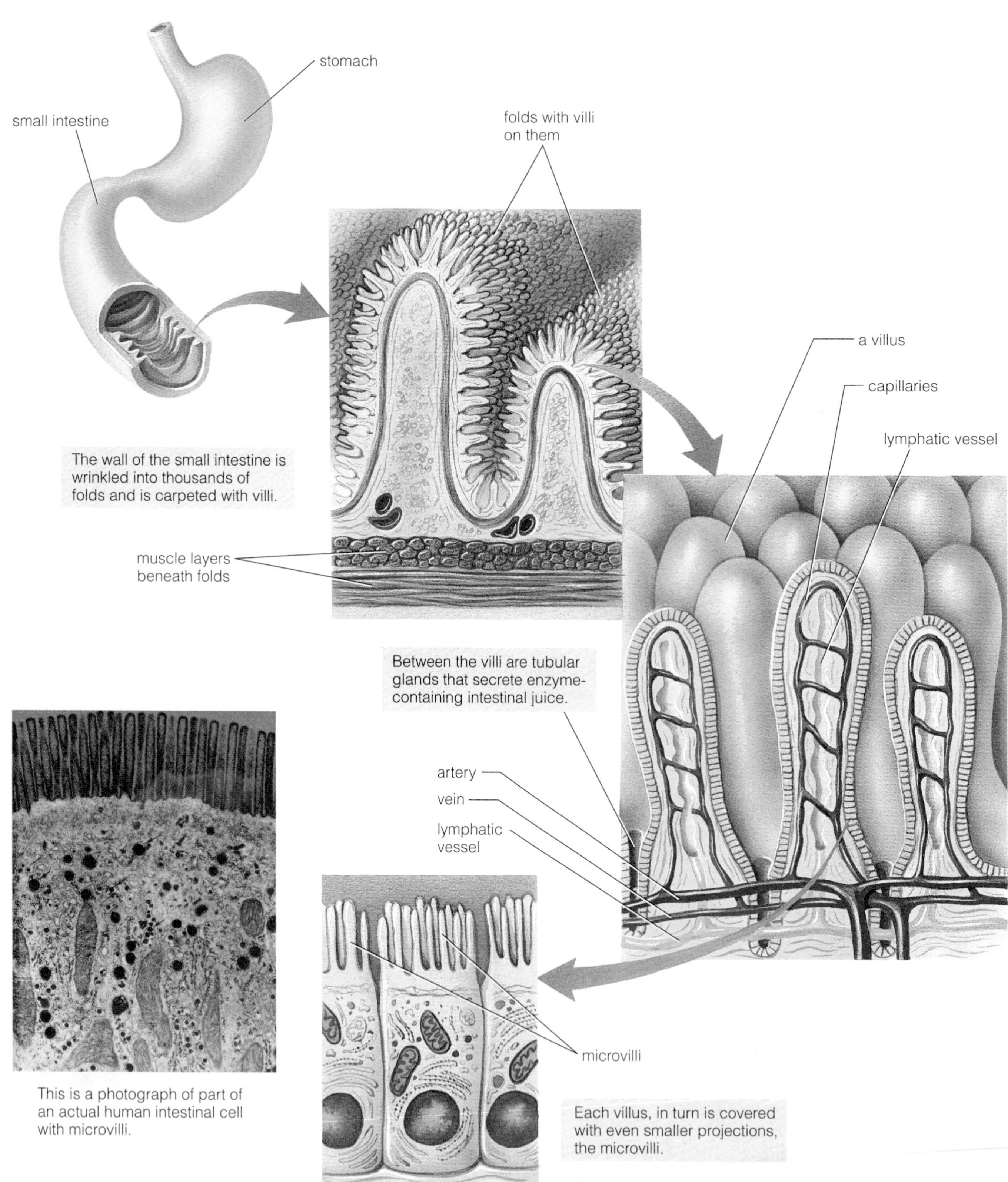
stomach
small intestine
folds with villi on them
a villus
capillaries
lymphatic vessel
The wall of the small intestine is wrinkled into thousands of folds and is carpeted with villi.
muscle layers beneath folds
Between the villi are tubular glands that secrete enzyme-containing intestinal juice.
artery
vein
lymphatic vessel
microvilli
Each villus, in turn is covered with even smaller projections, the microvilli.

This is a photograph of part of an actual human intestinal cell with microvilli.

A Letter from Your Digestive Tract

To My Owner,

You and I are so close; I hope that I can speak frankly without offending you. I know that sometimes I *do* offend with my gurgling noises and belching at quiet times and, oh yes, the gas. But please understand that when you chew gum, drink carbonated beverages, or eat hastily, you gulp air with each swallow. I can't help making some noise as I move the air along my length or release it upward in a noisy belch. And if you eat or drink too fast, I can't help getting **hiccups.** Please sit and relax while you dine. You will ease my task and we'll both be happier.

Also, when someone offers you a new food, you gobble away, trusting me to do my job. I try. It would make my life easier, and yours less gassy, if you would start with small amounts of new foods, especially those high in fiber. The bacteria that break down fiber produce gas in the process. I can handle just about anything if you introduce it slowly. But please: if you do notice more gas than normal from a specific food, avoid it. If the gas becomes excessive, check with a physician—the problem could be something simple, or it could be serious.

When you eat or drink too much, it just burns me up. Overeating causes **heartburn** because the acidic juice from my stomach backs up into my esophagus. Acid poses no problem to my healthy stomach, whose walls are coated with thick mucus to protect them. But when my too-full stomach squeezes some of its contents back up into the esophagus, the acid burns its unprotected surface. Also, those tight jeans you wear constrict my stomach, squeezing the contents upward into the esophagus. Just leaning over or lying down after a meal may allow the acid to escape up the esophagus because the muscular closure separating the two spaces is much looser than other such muscles. And if we need to a lose a few pounds, let's get at it—excess body fat can also squeeze my stomach, causing acid to back up. When heartburn is a problem, do me a favor: try to eat smaller meals; drink liquids an hour before or after, but not during, meals; wear reasonably loose clothing; and relax after eating, but sit up (don't lie down).

Sometimes your food choices irritate me. Specifically, chemical irritants in foods, such as the "hot" component of chili peppers, chemicals in coffee, fat, chocolate, soda pop, and alcohol, may worsen heartburn in some people. Avoid the ones that cause trouble. Above all, do not smoke. Smoking makes my heartburn worse—and you should hear your lungs bellyache about it.

By the way, I can tell you've been taking heartburn medicines again. You must have been watching those TV commercials and letting them mislead you. You need to know that **antacids** are designed only to temporarily relieve pain caused by heartburn by neutralizing stomach acid for a while. But the antacids trigger my stomach, which is normal and healthy, to produce *more* acid. That's because when my normal stomach acidity is reduced, I respond by producing more acid to restore the normal acid condition. Also, the ingredients in antacids can interfere with my ability to absorb nutrients. And don't decide that you need to take the heavily advertised **acid reducers** and **acid controllers** for my sake; these restrict my ability to produce acid so much that my job of digesting food becomes harder. In fact, the drugs often *cause* indigestion and diarrhea. Also, any heartburn medicine can mask the symptoms of **ulcer, hernia,** or the severe destructive form of chronic heartburn known as **gastro-esophageal reflux disease (GERD).**[6] This is serious business, because if not treated with antibiotic drugs, the bacterium that causes stomach ulcer may progress to cause a condition that can sometimes lead to stomach cancer.[7] A hernia can cause food to back up into the esophagus, and so it can feel like heartburn, but many times hernias require corrective treatment by a physician. GERD can feel like heartburn, too, but may require surgery or drug therapy to prevent serious damage to tissues.

When you eat too quickly, I worry about choking (see Figure 3-12). Please take time to cut your food into small pieces, and chew it until it is crushed and moistened with saliva. Also, refrain from talking or laughing before swallowing, and never

hiccups spasms of both the vocal cords and the diaphragm, causing periodic, audible, short, inhaled coughs. Can be caused by irritation of the diaphragm, indigestion, or other causes. Hiccups usually resolve in a few minutes, but can have serious effects if prolonged. Breathing into a paper bag (inhaling carbon dioxide) or dissolving a teaspoon of sugar in the mouth may stop them.

heartburn a burning sensation in the chest (in the area of the heart) area caused by backflow of stomach acid into the esophagus.

antacids medications that react directly and immediately with the acid of the stomach, neutralizing it. Antacids are most suitable for treating occasional heartburn. More about antacids appears in Controversy 8.

acid reducers and **acid controllers** drugs that reduce the acid output of the stomach. They are most suitable for treating severe, persistent forms of heartburn, but are useless for neutralizing acid already present in the stomach. Previously sold as prescription ulcer medications, the drugs are now sold freely, but the packages bear warnings of side effects; some types interfere with the stomach's ability to destroy alcohol, so more of the alcohol in a drink enters the bloodstream.

ulcer an erosion in the topmost, and sometimes underlying, layers of cells that form a lining. Ulcers of the digestive tract commonly form in the esophagus, stomach, or upper small intestine.

hernia a protrusion of an organ or part of an organ through the wall of the body chamber that normally contains the organ. An example is a *hiatal* (high-AY-tal) *hernia,* in which part of the stomach protrudes up through the diaphragm into the chest cavity, which contains the esophagus, heart, and lungs.

gastro-esophageal reflux disease (GERD) a severe and chronic splashing of stomach acid and enzymes into the esophagus, throat, mouth, or airway that causes inflammation and injury to those organs. Untreated GERD may increase the risk of esophageal cancer; treatment may require surgery or management with medication.

constipation infrequent, difficult bowel movements often caused by diet, inactivity, dehydration, or medication. (Also defined in Chapter 4.)

diarrhea frequent, watery bowel movements usually caused by diet, stress, or irritation of the colon. Severe, prolonged diarrhea robs the body of fluid and certain minerals, causing dehydration and imbalances that can be dangerous if left untreated.

irritable bowel syndrome intermittent disturbance of bowel function, especially diarrhea or alternating diarrhea and constipation; associated with diet, lack of physical activity, or psychological stress.

For more information concerning ulcers and medication, call the Centers for Disease Control and Prevention at 1-888-MY-ULCER. The call is toll-free.

Figure 3-12

Swallowing and Choking

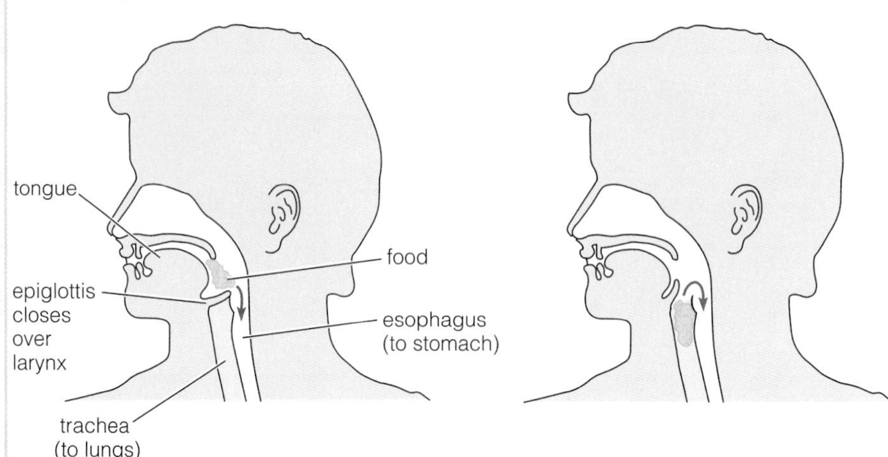
tongue
food
epiglottis closes over larynx
esophagus (to stomach)
trachea (to lungs)

A normal swallow. The epiglottis acts as a flap to seal the entrance to the lungs (trachea) and direct food to the stomach via the esophagus.

Choking. A choking person cannot speak or gasp because food lodged in the trachea blocks the passage of air. The red arrow points to where the food should have gone to prevent choking.

attempt to eat when you are breathing hard. Also, for our sake and the sake of others, learn the Heimlich maneuver as shown in Figure 3-13.

When I'm suffering, you suffer, too. When **constipation** or **diarrhea** strikes, neither of us is having fun. Slow, hard, dry bowel movements can be painful, and failing to have a movement for too long brings on headaches and ill feelings. Don't rely on laxatives, though. They often contain stimulants that can cause side effects.† Instead of using laxatives, listen carefully for my signal that it is time to defecate, and make time for it even if you are busy. The longer you ignore my signal, the more time the colon has to extract water from the feces, hardening them. Also, please choose foods that provide enough fiber (Chapter 4 lists some of these foods). Fiber attracts water, creating softer, bulkier stools that stimulate my muscles to contract, pushing the contents along. Fiber helps my muscles to stay fit, too, making elimination easier. Be sure to drink enough water because dehydration causes the colon to absorb all the water it can get from the feces. And please make time to be physically active; exercise strengthens not just the muscles of your arms, legs, and torso, but those of the colon, too.

When I have the opposite problem, diarrhea, my system will rob you of water and salts. In diarrhea my intestinal contents have moved too quickly, drawing water and minerals from your tissues into the contents. When this happens, please rest a while and drink fluids. To avoid diarrhea, try not to change my diet too drastically or quickly. I'm willing to work with you and learn to digest new foods, but if you suddenly change your diet, we're both in for it. I hate even to think of it, but one likely cause of diarrhea is dangerous food poisoning (*please* read, and use, the tips in Chapter 14 to keep us safe). Also, if diarrhea lasts longer than a day or two, or if it alternates with constipation, the cause could be **irritable bowel syndrome,** and you should go see a physician.

Thank you for listening. I know we'll both benefit from communicating like this, because you and I are in this together for the long haul.

Affectionately,

Your Digestive Tract

Figure 3-13

The Heimlich Maneuver

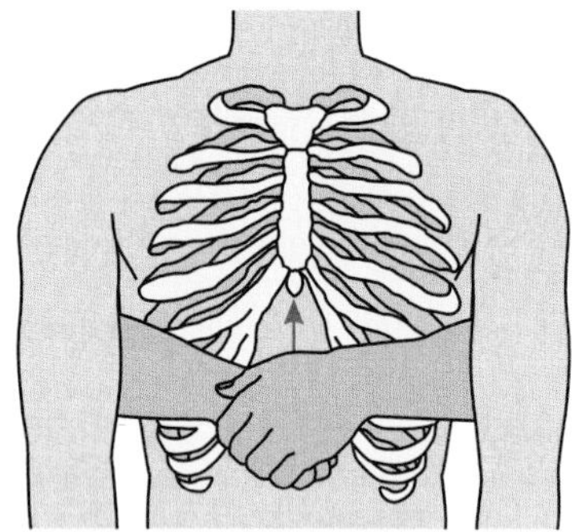

Rescuer positions fist directly against victim's abdomen as shown.

1. Rescuer stands behind victim and wraps her or his arms around victim's waist.
2. Rescuer makes a fist with one hand and places the thumb side of the fist against the victim's abdomen, slightly above the navel and below the rib cage.
3. Rescuer grasps fist with other hand and rapidly squeezes it inward and upward three or four times in rapid succession.
4. Rescuer repeats the process if necessary.

If the victim is alone, the victim positions himself or herself over edge of fixed horizontal object, such as a chair back, railing, or table edge, and presses abdomen into edge with quick movement.

†One such stimulant, phenolphthalein, was recently banned because of an association with colon cancer.

So much for digestion of foods, absorption of nutrients, and the workings of the digestive tract. The next section introduces the body's primary organs of waste removal, the kidneys, and describes how these marvelous organs monitor and adjust the amounts of dissolved substances in the blood.

KEY POINT *The digestive tract has many ways to communicate its needs. By taking the time to listen, you will obtain a complete understanding of the mechanics of the digestive tract and its signals.*

The Excretory System

Cells generate a number of wastes, and all of them must be eliminated. Many of the body's organs play roles in removing wastes. Carbon dioxide waste from the cells travels in the blood to the lungs, where it is exchanged for oxygen, as already mentioned. Other wastes are pulled out of the bloodstream by the liver. The liver processes these wastes and either tosses them out into the digestive tract with bile, to leave the body with the feces, or prepares them to be sent to the kidneys for disposal in the urine. Organ systems work together to dispose of the body's wastes, but the kidneys are waste- and water-removal specialists.

The kidneys straddle the cardiovascular system and filter the passing blood. Waste materials, dissolved in water, are collected by the kidneys' working units, the **nephrons.** These wastes become concentrated as urine, which travels through tubes to the urinary **bladder.** The bladder empties periodically, removing the wastes from the body. Thus the blood is purified continuously throughout the day, and dissolved materials are excreted as necessary. One dissolved mineral, sodium, helps to regulate blood pressure, and its excretion or retention by the kidneys is a vital part of the body's blood pressure–controlling mechanism. As you might expect, the kidneys' work is regulated by hormones secreted by glands that respond to conditions in the blood (such as the sodium concentration).

Because the kidneys remove toxins that could otherwise damage body tissues, whatever supports the health of the kidneys supports the health of the whole body. A strong cardiovascular system and an abundant supply of water are important to keep blood flushing swiftly through the kidneys. In addition, the kidneys need sufficient energy to do their complex sifting and sorting job, and many vitamins and minerals serve as the cogs in their machinery. Exercise and nutrition are vital to healthy kidney function.

KEY POINT *The kidneys adjust the blood's composition in response to the body's needs, disposing of everyday wastes and helping remove toxins. Nutrients, including water, and exercise help keep the kidneys healthy.*

nephrons the working units in the kidneys, consisting of intermeshed blood vessels and tubules.

bladder the sac that holds urine until time for elimination.

Storage Systems

The human body is designed to eat at intervals of about four to six hours, but cells need nutrients around the clock. Providing the cells with a constant flow of the needed nutrients requires the cooperation of many body systems. These systems store and release nutrients to meet the cells' needs between meals. Among the major storage sites are the liver and muscles, which store carbohydrate, and the fat cells, which store fat as well as other things.

When I Eat More Than My Body Needs, What Happens to the Extra Nutrients?

Nutrients collected from the digestive system sooner or later all move through a vast network of capillaries that weave among the liver cells. This arrangement ensures that liver cells have access to the newly arriving nutrients for processing. Later chapters provide the details, but it is important to know now that the liver converts excess

glycogen a storage form of carbohydrate energy (glucose), described more fully in Chapter 4.

energy-containing nutrients into two forms. It makes some into **glycogen** (a carbohydrate) and some into fat. The liver stores the glycogen to meet the body's ongoing glucose needs. Liver glycogen can sustain cell activities when the intervals between meals become long. Without glucose absorbed from food, the cells (including the muscle cells) draw on liver glycogen. Should no food be available, the liver's glycogen supply dwindles; it can be effectively depleted within as few as three to six hours. Muscle cells make and store glycogen, too, but selfishly reserve it for their own use.

Whereas the liver stores glycogen, it ships out fat in packages (see Chapter 5) to be picked up by cells that need it. All body cells may withdraw the fat they need from these packages, and the fat cells pick up the remainder and store it to meet long-term energy needs. Unlike the liver, fat tissue has virtually infinite storage capacity. It can continue to supply the body's cells with fat for days, weeks, or possibly even months when no food is eaten.

These storage systems for glucose and fat ensure that the body's cells will not go without energy even if the body is hungry for food. Body stores also exist for many other nutrients, each with a characteristic capacity. For example, liver and fat cells store many vitamins, and bones provide reserves of calcium, sodium, and other minerals. Stores of nutrients are available to keep the blood levels constant and to meet cellular demands.

Variations in Nutrient Stores

Some nutrients are stored in the body in much larger quantities than others are. For example, certain vitamins are stored without limit, even if they reach toxic levels within the body. Other nutrients are stored in only small amounts, regardless of the amount taken in, and these can readily be depleted. As you learn how the body handles various nutrients, pay particular attention to their storage so that you can know your tolerance limits. For example, you needn't eat fat at every meal, because fat is stored abundantly. On the other hand, you normally do need to have a source of carbohydrate at intervals throughout the day because the liver stores less than one day's supply of glycogen.

KEY POINT *The body's energy stores are of two principal kinds: fat in fat cells (in potentially large quantities) and glycogen in muscle and liver cells (in smaller quantities). Other tissues store other nutrients.*

Other Systems

In addition to the systems just described, the body has many more: bones, an immune system, muscles, reproductive organs, and others. All of these cooperate, enabling each cell to carry on its own life. Each system ensures, through hormonal or nerve-mediated messengers, that its needs will be met by the others, and each contributes to the welfare of the whole by doing its own specialized work. For example, the skin and body linings defend other tissues against microbial invaders, while being nourished and cleansed by tissues specializing in these tasks. Each system needs a continuous supply of many specific nutrients to maintain itself and carry out its work. Calcium is particularly important for bones, for example; iron for muscles; glucose for the brain. But all systems need all nutrients, and every system is impaired by an undersupply or oversupply of them.

While external events clamor and vie for attention, the body quietly continues its life-sustaining work. Of the billions of cells in the body, only a small percentage make up the cortex of the brain, where the conscious mind resides. When the cells of the cortex receive messages from other cells, the person "becomes conscious" of a need for decision and action. In modern life, the need may be complex, as when a person feels anxious and decides to consult an adviser, or simple, as when a person feels hungry and decides, "I'd better eat."

Most of the body's work is directed automatically by the unconscious portions of the brain and nervous system, and this work is finely regulated to achieve a state of well-being. But you need to involve your cerebral cortex, your consciousness, so as to cultivate an understanding and appreciation of your body's needs. In doing so, attend to nutrition first. The rewards are liberating—ample energy to tackle life's tasks, a robust attitude, and the glowing appearance that comes from the best of health. Indulge in the foods that best meet your body's needs and support its health. Read on, and learn to let nutrition principles guide your choices.

KEY POINT *To achieve optimal function, the body's systems require nutrients from outside. These have to be supplied through a human being's conscious food choices.*

Self Check

Answers to these Self Check questions are in Appendix G.

1. All blood leaving the digestive system is routed directly to the:
   a. heart
   b. kidney
   c. liver
   d. lungs
2. Which of the following can affect the hormonal system?
   a. fasting
   b. eating
   c. exercise
   d. all of the above
3. Chemical digestion of which nutrient begins in the mouth?
   a. starch
   b. vitamins
   c. protein
   d. all of the above
4. Which chemical substance released by the pancreas neutralizes stomach acid that has reached the small intestine?
   a. mucus
   b. enzymes
   c. bicarbonate
   d. bile
5. Which nutrient passes through the large intestine mostly unabsorbed?
   a. starch
   b. vitamins
   c. minerals
   d. fiber
6. Organs that work together to perform a function are parts of a body system. T F
7. The immune system is so important that its tissues cannot be damaged by malnutrition. T F
8. The process of digestion occurs mainly in the stomach. T F
9. To digest food efficiently, people should not combine certain foods, such as meat and fruit, at the same meal. T F
10. The gallbladder stores bile until it is needed to emulsify fat. T F
11. Absorption of the majority of nutrients takes place across the mucus-coated lining of the stomach. T F
12. Stone Age people achieved their abundant nutrient intakes using only two of the four food groups: meats and milk and milk products. (Read about this in the upcoming Controversy.) T F

NUTRITION ON THE NET

For further study of the topics of this chapter, access these websites and search for the phrases or words in quotation marks:

1. For the latest changes in chapter material, to find supplemental learning tools, or to access links to related websites, go to the "Nutrition: Concepts and Controversies" site at: **www.wadsworth.com/nutrition/prod/allprod.html**
2. For research abstracts on many topics in "medicine," choose libraries, Pub Med at: **www.healthfinder.gov**
3. To learn more about digestive disorders such as "constipation", "diarrhea", and "ulcers": **www.niddk.nih.gov/health/health.htm**

CONTROVERSY 3

SHOULD WE BE EATING THE "NATURAL" FOODS OF ANCIENT DIETS?

Some people are unwilling to follow the traditional nutrition teaching. They'd rather have simpler rules to follow. One popular notion is that we can "just eat the way our ancestors did—they were healthy." This Controversy looks at that proposition and applies some scientific criticism to it.

Though you may be clothed in the latest fashions and drive the latest model car, your body is of prehistoric design—an ancient body in the modern world. It is a caveperson body. (Actually, our earliest ancestors were not cave dwellers and are properly called the people of the Stone Age.) You have come a long way from the Stone Age in many ways: in language skills, in the arts, in medicine, and especially in the use of machinery. But your body handles food and physical activity in virtually the same way as your ancient ancestors' bodies did. The muscles, heart, and lungs have hardly changed at all, and the brain, too, responds in the same ways to what the body sees, smells, and tastes..

The world, however, has changed vastly, especially in the last hundred years. The earth is far more crowded than it used to be. Our ancestors roamed the wilderness; today, more than half of the world's people live in cities. Extremes contrast with each other: poverty and death from malnutrition and related causes are common in the developing world; health problems caused by overabundance are common in the developed countries. The processed foods of developed countries are often much higher in fat, salt, and sugar and lower in fiber, vitamins, and minerals than the wild foods eaten by our ancestors. Too, most people's lifestyles offer much less physical activity than did the lives of their ancestors. Many people also face new technological by-products in their environments, such as smog and water pollution. Your Stone Age body is stressed by contending with these new problems. Naturally, people are inclined to wonder whether we should live and eat more nearly as our ancestors did.

THE TIME AND WORLD OF OUR ANCESTORS

Before arriving at any conclusions, researchers need to know what our ancestors ate and how healthy they were. First, though, the researchers must decide which ancestors to study—our farmer forebears and the early agricultural people who preceded them, or the hunter-gatherer people who lived still longer ago.

People have farmed the land for 10,000 years or more, 50 times longer than the period since the beginning of the industrial era, about A.D. 1800. Ten thousand years may seem long, but try a second comparison. Imagine all of the three-million-year history of human existence on earth compressed into the last 24 hours. Then the agricultural era would have begun only 3½ minutes ago, and the industrial era would have begun 4 *seconds* ago. People who grow their own food have thus occupied the earth for only a few *hundred* generations, whereas people who hunted for their food roamed the earth for *thousands* of generations. Figure C3-1 depicts the magnitude of the differences between the earlier people's times on earth and our own.

One group of people to study therefore is the earliest one, the hunter-gatherers before agriculture. They were the people of the Stone Age, the Paleolithic period, which lasted from about 500,000 years ago to about 10,000 years ago (*paleo* means "ancient"; *lith* means "stone," referring to the stone tools they used). Thus, Stone Age people's way of life and diet persisted for close to half a million years. Enough time elapsed to permit natural selection of genes for traits that favored survival of generation after generation under the conditions of that period including the diet. Many of the genes for those traits persist in our inheritance

Figure C3-1

A Perspective on Modern Human Beings' Time on Earth

If all human existence on earth is collapsed into 1 year's time, then agriculture began yesterday, and the current A.D. *calendar began 4 hours ago. In the pyramid below, the space occupied is in proportion to the time spent on earth.*

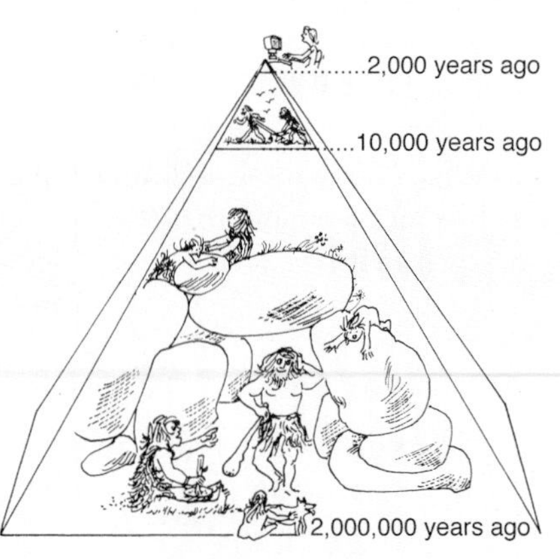
.......2,000 years ago
....10,000 years ago
2,000,000 years ago

today. Thousands of years old, those genes still support physical characteristics in us that favored survival in those early times. The more we know about the Stone Age people, then, the better we can understand our bodies' needs today.

Modern-day hunting-gathering peoples are also worth studying in this regard. A few such groups remain on earth today, and from them we can learn of the effects on health of a changing diet.

THE ANCIENT BODY IN THE MODERN WORLD

Food was not available all the time for Stone Age people. Times of plenty alternated with times of famine. The human body adapted well to this state of affairs. It was omnivorous—able to digest and to use the nutrients from both plants and animals. This made wide food choices possible and reduced the likelihood of starvation in the face of a food supply that depended on place and season. The body could also store excess energy in its fat tissues when food was plentiful. Then people could draw on that fuel supply during famine or illness.

Our bodies can do the same things today. We too can eat many kinds of food. We too efficiently store surplus energy in body fat. Today, though, these abilities confer less of an advantage. Not all the foods available to us today benefit our health, and in food-abundant societies, times of famine never come. Our overeating and storage of excess fat often produce conditions that shorten life. In susceptible people, excess body fatness precipitates diabetes, aggravates high blood pressure, renders certain cancers more likely, and worsens arthritis, among other things.

The body feels hungry at approximately four- to six-hour intervals, even though it may have sufficient fat stores to last for many days. This adaptation served Stone Age people well, for it drove them to continue stocking fuel within their bodies as often as their digestive systems could perform the task. They ate whenever food was available, even when they had sufficient body fat and nutrient stores for temporary needs. As a result, they were able to maintain ample stores and then live on them for long times when the food supply ran out.

Furthermore, Stone Age people's appetites were especially stimulated whenever they encountered foods that were rich in food energy—those with the taste of oils or fats or the sweetness of concentrated sugar. Foods that tasted of salt also appealed to their taste buds, for pure salt was rarely available in very early times and the essential nutrient sodium was hard to come by. Novel foods also appealed to them—for good reasons. For long periods their diets were monotonous; their eagerness to try new foods probably helped to ensure that they would obtain the nutrients their regular diet might have lacked. Today, people still respond this way to foods. We prefer tasty, high-fat foods and will eat even when full if new delicious foods present themselves. This is why the dessert cart can entice you to stuff yourself, even after you've eaten a large meal.

The sense of taste is also the front line of the body's defense against poisons: people refuse foods that don't taste "right." The second line of defense is the stomach's rejection response: the body vomits up or washes out via diarrhea many toxins that enter the digestive system. The third line is the liver's filtering and detoxifying systems. Toxins that get into the bloodstream are removed from it by the liver cells, which then render the toxins harmless and put them away in permanent storage or release them for excretion in the urine.

For example, protection against the harmful effects of one ancient and familiar substance—alcohol—is built into the body's genes. One of those genes, expressed in the liver, codes for an enzyme that converts alcohol into substances the body can use or excrete. So long as the liver is not overwhelmed with alcohol, the system works efficiently.

The same is not true of all poisons. Alcohol has been around since the first fruit ripened and fermented, so there have been millions of years for natural selection to mold a detoxifying system for it. On the other hand, most of the contaminants, pollutants, and toxins that get into foods by mistake are new to the body. If it can't excrete them, it may accumulate harmful quantities or convert them to odd, unfamiliar substances that can interfere with metabolism or cause cancer or birth defects. An important new area of study in nutrition is concerned with the body's handling of these substances.

In other ways, too, the body is adapted to the conditions of earlier times. Heredity has given each human being a body that can *develop* itself to run after prey, fight enemies, or carry heavy burdens long distances; it responds to physical exertion by becoming stronger and swifter. Among the muscles that become stronger in response to exercise are the heart and muscles that work the lungs, and they also (like all muscles) become weaker without exercise. In ancient times, vigorous activity and hard physical work were part of everyone's life, but today people may sit around for months at a time. We have to make special efforts to plan exercise into our daily routines if our muscles are not to become weak.

These and other differences add up to a set of circumstances that challenge your body and mind to maintain health against many odds. You are living with the food, the labor-saving devices, the medical miracles, the contaminants, and all the other pleasures and problems of the twenty-first century. However, you are housed in a body adapted to a world in which strong men and women survived on simple foods obtained through hard physical labor. There is no guarantee that your diet and exercise routine, haphazardly chosen, will meet the needs of your Stone Age body.

Only with your brain can you compensate for these disadvantages of modern life. Stone Age people used their brains to discover ways to obtain food; you must use yours, sometimes, to refuse delicious food and to battle the ancient instincts that cry out for you to eat. Stone Age people used their

ingenuity to save their energy when they could; you may have to use yours to find ways to spend energy so that you can maintain appropriate weight and keep your heart and muscles fit. You have an advantage, though. Unlike your ancestors, you can learn more about how your body works and what it needs from food.

THE STONE AGE DIET

Researchers have studied what Stone Age people actually ate. Analysis of their probable diet from fossilized remains of their meals and excretions has produced a picture of what foods they ate and how much of each nutrient they received.[1] The figures are undoubtedly not exact, but it is interesting to compare them with those of today. Early people probably consumed 3,000 calories per day and apparently were never obese. (Today, we consume about 2,000 calories a day, yet many of us are obese because we get so little exercise.)

Given their large energy allowances and the exclusively nutrient-dense food options from which they were forced to choose, Stone Age people probably met all of their nutrient needs well. For example, although they drank no milk and made no cheese, their intakes of calcium were probably close to 1,500 milligrams a day, thanks to the fruits and vegetables and to the softened ends of bones that they consumed. Today, we fall short of 800 milligrams, although recommendations state that many people (and especially young people) need more than 1,000 milligrams to preserve the integrity of their bones. Stone Age people probably consumed close to 400 milligrams of vitamin C a day, whereas we take in less than 100 milligrams. They ingested much more fiber than we do today, 45 grams or so, compared with our 20 grams or less. Counterbalancing these pluses, they also ate more dirt, and even gravel. Figure C3-2 provides other comparisons of the diets of hunter-gatherers, agriculturist peoples, and the U.S. population.

Figure C3-2

Selected Differences Between Ancient and Modern Diets

The percentages referred to here are the percent of total daily caloric intake from fat, starch, and refined sugar. Daily fiber is shown in grams.

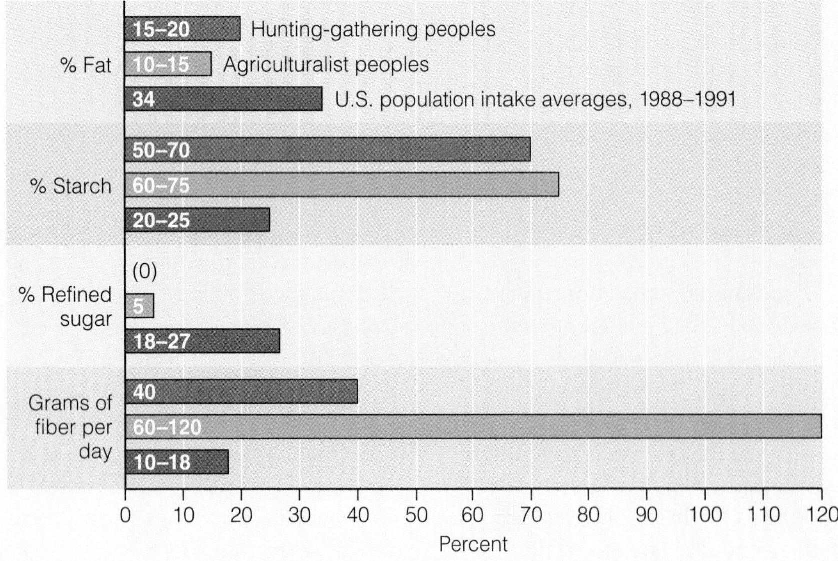

% Fat
15–20
Hunting-gathering peoples
10–15
Agriculturalist peoples
34
U.S. population intake averages, 1988–1991
% Starch
50–70
60–75
20–25
% Refined sugar
(0)
5
18–27
Grams of fiber per day
40
60–120
10–18
0
10
20
30
40
50
60
70
80
90
100
110
120
Percent

SOURCE: Data for hunter-gatherers and agriculturists from WHO Study Group, Diet, nutrition, and the prevention of chronic diseases. World Health Organization Technical Report Series 797 (Geneva: WHO, 1990). 43: Data for U.S. population averages from *Diet and Health: Implications for Reducing Chronic Disease Risk* (Washington, D.C.: National Academy Press, 1989), pp. 41–84 and Interagency, Board for Nutrition Monitoring, *Third Report on Nutrition Monitoring in the United States* (Washington, D.C.: Government Printing Office, 1995), pp. ES-3, 49.

Stone Age people achieved their abundant nutrient intakes using only three of the five groups of foods we think of as important: meat, fruits, and vegetables. Their intakes of meat, and therefore of protein, were two to five times higher than ours are today, but most of their meats were lean, whereas many of ours are high in fat.[2] The fats in their meats included a type thought to be preventive against heart disease and cancer and known to be lacking from our meats today, the so-called omega-3 fatty acids described in Chapter 5. They apparently consumed cereal grains rarely, if at all, and they had no dairy foods whatsoever. (Remember, they lived before the dawn of agriculture, and kept no cattle.)

Although their total energy intakes were higher than ours, the people of the Stone Age had lower intakes of two no-no's that plague modern eaters: fat and sodium. Their cholesterol intakes were similar to ours, however, because even lean meat contains cholesterol. Also, their diet seldom contained concentrated sweets such as honey, and there was no such thing as table sugar.

Were earlier people healthier, then, than we are today? Not necessarily. Many died of starvation during times of climatic extremes. Many must have suffered vitamin deficiencies and food-borne diseases. Stone Age people died younger than we do, and this is one reason why the so-called degenerative diseases of old age were less common then than they are today.

Our longer lives are not the only reason for the modern prevalence of cancer and heart disease, though; our diets share the blame. Researchers have learned this from studying primitive people living today: tribes in Africa and other places whose diets and ways of life resemble those of Stone Age people. These people's lifestyles enable them to attain the age of 60 years relatively free of degenerative diseases. Others living today within a generation or two of ancient foodways are Native Americans. The traditional diets of the tribes most often studied are a blend of agricultural and hunting-gathering styles and are of interest because of their rapid rates of change.

NATIVE AMERICAN DIETS

Within two hundred years, the foodways of Native Americans have undergone unprecedented changes. The hunter-gatherer and agricultural lifestyles of history have given way to a modern food culture relying on refined grains and purified sugars, fast foods, abundant high-fat meats, dairy products, and alcohol.[3] Researchers conclude that for Native Americans, the effects of the changes on health have been overwhelmingly negative.

A well-studied example of a group that suffers enormously from the effects of a "modernized" diet are the Pima Indian tribe of central Arizona. For thousands of years up to 1930, the Pima diet consisted of a wide variety of wild and cultivated desert legumes, cactus leaves and fruit, fish, venison (deer meat), small seeds, mesquite pods, acorns, and corn. Then in 1930, a change began. The tribe largely replaced their traditional wild foods, which had become scarce, with modern ones such as refined wheat flour, lard, sugar, coffee, and ready-to-eat cereals that were limited in variety but easily obtained. The result has been tragic—the Pima now suffer the highest per capita rate of diabetes (roughly 50 percent) known among any people of the world. Genetics may make diabetes especially likely to develop in Pimas who adopt modern foodways.[4] Likewise, the Sioux Indians rarely ever suffered heart disease when consuming their traditional diets, but now suffer one of the highest known rates of heart and artery disease. Navajos also now suffer high rates of diabetes and cardiovascular disease.[5] And Alaskan natives, among whom obesity and diabetes were rare before 1960, now suffer greatly from these twin threats.[6]

The Pima, Navajo, and Sioux tribes changed not only their diets but also their highly active lifestyles. Modern-day Pima and Sioux no longer hunt game on the windswept plains, toil in the fields, or gather wood to cook over stone fireplaces as their ancestors once did. They now drive to supermarkets to purchase convenience foods to cook in microwave ovens. The Sioux also traded their occasional ceremonial pipes for daily cigarette smoking, a new habit that is especially damaging to the heart.

While changed diets and lifestyles almost certainly have contributed to the changes in Native American health, their native diets were not perfect, either.[7] They did not provide adequate amounts of some nutrients, and availability of foods depended on such unpredictable factors as weather conditions and herd movements. Whereas modern foods are usually safe and sanitary, Native Alaskans eating traditional foods suffer more botulism (a deadly food poisoning) than any other group worldwide. Modern descendants changed the ancient methods of preserving meat, fish, and blubber (fat) in slight but critical ways that encourage the growth of the bacterium that causes botulism.[8] The point is that all diets, even those that have supported human beings through many centuries, have drawbacks. Still, by studying them, scientists are learning that when Native Americans consumed their ancient diets of high-fiber, low-fat native foods, their hearts and bodies benefited.

Does this mean we should abandon today's foods and eat only meat and wild fruits and vegetables? No—even if we did, we would not be eating like the Stone Age people. Our foods are different. Our meats differ in the amounts and types of fat they contain. Our fruits and vegetables are totally different. Ancient people's environments, with their vast, cityless open spaces, no longer exist for us. And, sadly, much cultural knowledge, including traditional foodways, has been all but lost to modern generations.[9]

Still, even if we cannot eat the same foods they did, we can attempt to duplicate their activity and nutrient intake levels using the foods available to us. Clearly, we should emulate them in incorporating more physical activity into our days. If we exercise more, we can eat more without getting fat; if we eat more, we can obtain more nutrients; and if we obtain more nutrients, we are better protected against deficiencies.

In conclusion, no one can go back in time to live as the ancient people did. We can learn from them, though, the importance of getting more exercise and of eating wholesome foods that will support our health as well as, or even better than, theirs did.

4 The Carbohydrates: Sugar, Starch, Glycogen, and Fiber

Shiva Dayal Lal, *Women Selling Grains and Vegetables,* mid 19th Century

Contents

Frequently Asked Questions

FAQ: If I want to lose weight, should I avoid carbohydrates? p. 98

FAQ: How does fiber in food affect my health, and how much do I need to stay healthy? p. 98

FAQ: Can my diet have too much fiber? p. 101

FAQ: Why do some people have trouble digesting milk? p. 107

FAQ: What is diabetes? p.112

FAQ: If I feel dizzy between meals, do I have hypoglycemia? p. 114

carbohydrates compounds composed of single or multiple sugars. The name means "carbon and water," and a chemical shorthand for carbohydrate is CHO, signifying carbon (C), hydrogen (H), and oxygen (O).

complex carbohydrates long chains of sugar units arranged to form starch or fiber; also called *polysaccharides.*

simple carbohydrates sugars, including both single sugar units and linked pairs of sugar units. The basic sugar unit is a molecule containing six carbon atoms, together with oxygen and hydrogen atoms.

IT IS IMPOSSIBLE to single out the most important nutrient. Nutrients work together in harmony, each affecting the functions of many others. **Carbohydrates,** however, are ideal to meet your body's energy needs, keep your digestive system fit, feed your brain and nervous system, and, within calorie limits, help keep your body lean. False propaganda about carbohydrate's supposed "fattening power" misleads millions of weight-conscious people to avoid carbohydrate-rich foods, a counterproductive tactic. In truth, people who wish to lose fat and to maintain lean tissue can do no better than to control calories and design their diets around low-fat foods that supply carbohydrates in balance with other energy nutrients. Digestible carbohydrates, together with fats and protein, add bulk to foods and provide energy and other benefits for the body.[1] Indigestible carbohydrates, which include most of the fibers in foods, yield little or no energy but provide other important benefits.

All carbohydrates are not equal as far as nutrition is concerned. This chapter invites you to learn to distinguish between foods containing the **complex carbohydrates** (starch and fiber), which are put to good use in the body, and those made of the **simple carbohydrates** (sugars), which can be less valuable to people's health. The Controversy then asks whether the sugar added to foods harms health and whether the alternative sweeteners designed to replace sugar are preferable.

This chapter on the carbohydrates is the first of three on the energy-yielding nutrients. Chapter 5 deals with the fats and Chapter 6 with protein. The Controversy of Chapter 5 addresses one other contributor of energy, alcohol.

Figure 4-1

Carbohydrate—Mainly Glucose—Is Made by Photosynthesis

The sun's energy becomes part of the glucose molecule—its calories, in a sense. In the molecule of glucose on the leaf here, dots represent the carbon atoms; bars represent the chemical bonds that contain energy.

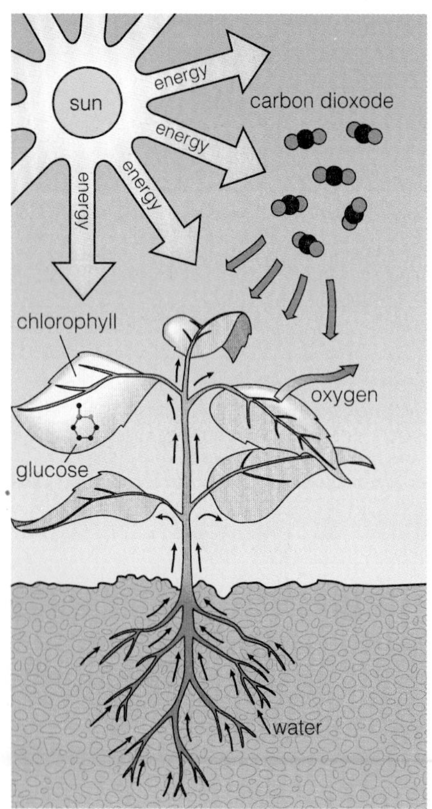
sun
energy
energy
energy
energy
carbon dioxode
chlorophyll
oxygen
glucose
water

A Close Look at Carbohydrates

Carbohydrates contain the sun's radiant energy, captured in a form that living things can use to drive the processes of life. Thus they form the first link in the food chain that supports all life on earth. Carbohydrate-rich foods come almost exclusively from plants; milk is the only animal-derived food that contains significant amounts of carbohydrate.

Green plants make carbohydrate through **photosynthesis** in the presence of **chlorophyll** and sunlight. In this process, water (H_2O), absorbed by the plant's roots, donates hydrogen and oxygen, and carbon dioxide gas (CO_2), absorbed into its leaves, donates carbon and oxygen. Water and carbon dioxide combine to yield the most common of the **sugars,** the single sugar **glucose.** Scientists know the reaction in the minutest detail, but have never been able to reproduce it from scratch; green plants are required to make it happen (see Figure 4-1).

Light energy from the sun drives the photosynthesis reaction. The light energy becomes the chemical energy of the bonds that hold six atoms of carbon together in the sugar glucose. Glucose provides energy for the work of all cells of the stem, roots, flowers, and fruits of the plant. For example, in the roots, far from the energy-giving rays of the sun, each cell draws upon some of the glucose made in the leaves, breaks it down (to carbon dioxide and water), and uses the energy thus released to fuel its own growth and water-gathering activities.

Plants do not use all of the energy stored in their sugars, so it remains available for use by the animal or human being that consumes the plant. For this reason, the energy captured by photosynthesis is important to our survival. The next few sections describe the forms assumed by carbohydrates with their treasures of stored energy awaiting use in the human body.

KEY POINT *Through photosynthesis, plants combine carbon dioxide, water, and the sun's energy to form glucose. Carbohydrates are made of carbon, hydrogen, and oxygen held together by energy-containing bonds: carbo means "carbon"; hydrate means "water."*

Sugars

Altogether, six sugar molecules are important in nutrition. Three are single sugars, or **monosaccharides.** The other three are double sugars, or **disaccharides.** All of their chemical names end in *ose,* which means *sugar,* and although they all sound alike to the newcomer, they exhibit distinct characteristics to the nutrition enthusiast who quickly gets to know each individually. Figure 4-2 on the next page shows the relationships among the sugars.

The three monosaccharides are glucose, already described, **fructose,** and **galactose.** Fructose or fruit sugar, the intensely sweet sugar of fruit, is made by rearranging the atoms in glucose molecules. Fructose occurs mostly in fruits, in honey, and as part of table sugar. Glucose and fructose are the most common monosaccharides in nature.

The other monosaccharide, galactose, has the same number and kind of atoms as glucose and fructose but in yet another arrangement. Galactose is one of two single sugars that are bound together to make up the sugar of milk. It does not occur free in nature, but instead is tied up in milk sugar until it is freed during digestion.

The three other sugars important in nutrition are disaccharides, which are linked pairs of single sugars. All three contain glucose. In **lactose,** the sugar of milk just mentioned, glucose is linked to galactose.

In malt sugar, or **maltose,** there are two glucose units. Maltose appears wherever starch is being broken down. It occurs in germinating seeds and arises during the digestion of starch in the human body.

The last of the six sugars, **sucrose,** is the most familiar. It is table sugar, the product most people think of when they refer to *sugar.* In sucrose, fructose and glucose are bonded together. Table sugar is obtained by refining the juice from sugar beets or sugar cane, but sucrose also occurs naturally in many vegetables and fruits. It tastes sweet, as does fruit sugar, because it, too, contains the sweet monosaccharide fructose. Sucrose is of major importance in human nutrition, and research about its effects on the human body is a topic of Controversy 4.

When you eat a food containing single sugars, you can absorb them directly into your blood. When you eat disaccharides, though, you must digest them first. Enzymes in your intestine must split the disaccharides into separate monosaccharides so that they can enter the bloodstream. The blood then delivers all products of digestion first to the liver, which possesses enzymes to modify nutrients, making them useful to the body. Glucose is the most-used monosaccharide inside the body, so the liver quickly converts fructose or galactose to glucose or to smaller pieces that can serve as building blocks for either glucose or fat.

When people learn that the energy of fruits and many vegetables comes from sugars, they may think that eating sweet-tasting fruit is the same as eating concentrated sweets such as candy or cola beverages. Not so. Like vegetables, fruits differ from concentrated sweets in nutrient density. The sugars of fruits arrive in the body diluted in large volumes of water, packaged with fiber, and mixed with many needed vitamins and minerals. In contrast, all types of refined sugars, even honey, arrive in the body in concentrated form, practically devoid of nutrients. From the body's point of view, fruits are vastly different from purified sugars, except that both provide glucose in abundance.

KEY POINT *Glucose is the most important monosaccharide in the human body. Most other monosaccharides and disaccharides become glucose in the body.*

photosynthesis the process by which green plants make carbohydrates from carbon dioxide and water using the green pigment chlorophyll to capture the sun's energy (*photo* means "light"; *synthesis* means "making").

chlorophyll the green pigment of plants that captures energy from sunlight for use in photosynthesis.

sugars simple carbohydrates, that is, molecules of either single sugar units or pairs of those sugar units bonded together.

glucose (GLOO-cose) a single sugar used in both plant and animal tissues for quick energy; sometimes known as blood sugar or *dextrose.*

monosaccharides single sugar units (*mono* means "one"; *saccharide* means "sugar unit").

disaccharides pairs of single sugars linked together (*di* means "two").

fructose (FROOK-tose) a monosaccharide; sometimes known as fruit sugar (*fruct* means "fruit"; *ose* means "sugar").

galactose (ga-LACK-tose) a monosaccharide; part of the disaccharide lactose (milk sugar).

lactose a disaccharide composed of glucose and galactose; sometimes known as milk sugar (*lact* means "milk"; *ose* means "sugar").

maltose a disaccharide composed of two glucose units; sometimes known as malt sugar.

sucrose (SOO-crose) a disaccharide composed of glucose and fructose; sometimes known as table, beet, or cane sugar.

polysaccharides another term for complex carbohydrates; compounds composed of long strands of glucose units linked together (*poly* means "many"). Also called *complex carbohydrates.*

Single sugars are monosaccharides.

Pairs of sugars are disaccharides.

Strands of many sugar units are polysaccharides.

Starch

Glucose occurs in foods not only in sugars but also in long strands of thousands of glucose units strung together. These are the **polysaccharides** (see Figure 4-3, page 97). Starch is a polysaccharide as are glycogen and most of the fibers.

Figure 4-2

How Monosaccharides Join to Form Disaccharides

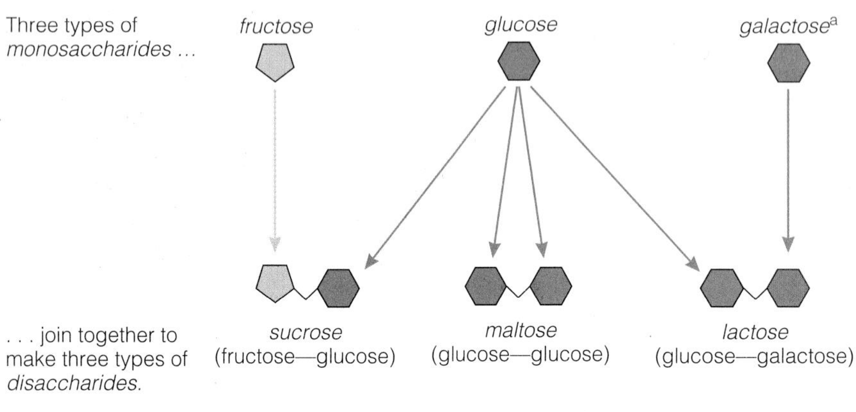
Three types of
monosaccharides . . .
fructose
glucose
galactose[a]
. . . join together to
make three types of
disaccharides.
sucrose
(fructose—glucose)
maltose
(glucose—glucose)
lactose
(glucose—galactose)

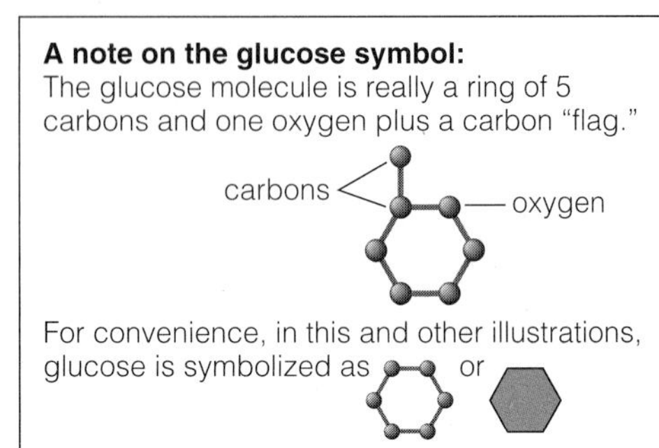
A note on the glucose symbol:
The glucose molecule is really a ring of 5 carbons and one oxygen plus a carbon "flag."
carbons
oxygen
For convenience, in this and other illustrations, glucose is symbolized as or

[a]Galactose does not occur in foods singly but only as part of lactose.

Starch is a plant's storage form of glucose. As a plant matures, it not only provides energy for its own needs but also stores energy in its seeds for the next generation to use. For example, after a corn plant reaches its full growth and has many leaves manufacturing glucose, it stores packed clusters of **starch** molecules in **granules** and packs the granules into its seeds to provide energy for the growth of new plants next season. Glucose is soluble in water and would be washed away by rains while the seed lay in the soil. Starch is an insoluble substance that will stay with the seed and nourish it until it forms shoots with leaves that can catch the sun's rays. A kernel of corn, then, is really a seed packed with this nutritive material. Most of the starch of corn and other foods is nutritive for human beings, too, because they can digest the starch to glucose and extract the sun's energy stored in its chemical bonds. A later section describes starch digestion in greater detail.

KEY POINT *Starch is the storage form of glucose in plants that is also nutritive for human beings.*

Glycogen

Just as plants store glucose in long chains of starch, animal bodies store glucose in long chains of **glycogen.** Glycogen resembles starch in that it consists of glucose molecules linked together to form chains, but its chains are longer and more highly branched (see Figure 4-3). Unlike starch, which is abundant in grains, potatoes, and other foods from plants, glycogen is virtually undetectable in meats because glycogen breaks down rapidly upon the slaughter of the animal. A later section describes how the human body handles its own packages of stored glucose.

KEY POINT *Glycogen is the storage form of glucose in animals, including human beings.*

Fiber

The **fibers** of a plant form the supporting structures of its leaves, stems, and seeds. Most fibers are polysaccharides—chains of sugars—just as starch is, but in fibers the sugar units are held together by bonds that human digestive enzymes cannot break. Most fibers therefore pass through the human body without providing energy for its use. The best-known fibers are *cellulose* (shown in Figure 4-3), *hemicellulose,* and *pectin.* (Other fibers are *gums, mucilages,* and *lignins.*) Cellulose and hemicellulose are found in the familiar strings of celery, the skins of corn kernels, and the membranes surrounding kernels of wheat. In the body, these fibers provide what a grandparent

starch a plant polysaccharide composed of glucose. After cooking, starch is highly digestible by human beings; raw starch often resists digestion.

granules small grains. Starch granules are packages of starch molecules. Various plant species make starch granules of varying shapes.

glycogen (GLY-co-gen) a polysaccharide composed of glucose that is made and stored by liver and muscle tissues of human beings and animals as a storage form of glucose. Glycogen is not a significant food source of carbohydrate and is not counted as one of the complex carbohydrates in foods.

fibers the indigestible polysaccharides in food, consisting mostly of cellulose, hemicellulose, and pectin. Also called *nonstarch polysaccharides.*

Figure 4-3

How Glucose Molecules Join to Form Polysaccharides

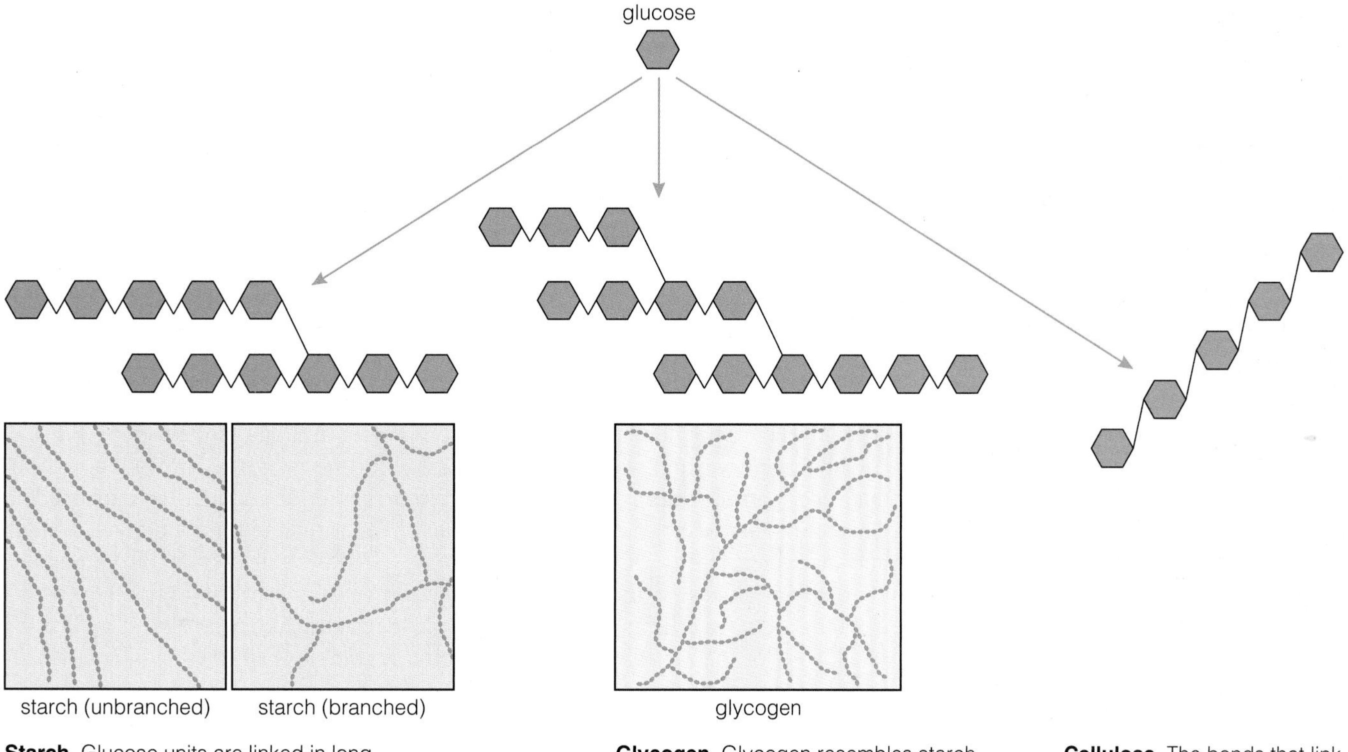
glucose
starch (unbranched)
starch (branched)
glycogen

Starch Glucose units are linked in long, occasionally branched chains to make starch. Human digestive enzymes can digest these bonds, retrieving glucose. Real glucose units are so tiny that you can't see them, even with the highest-power light microscope.

Glycogen Glycogen resembles starch in that the bonds between its glucose units can be broken by human enzymes, but the chains of glycogen are more highly branched.

Cellulose The bonds that link glucose units together are different from the bonds in starch or glycogen. Human enzymes cannot digest them.

might have called **roughage**—fiber that aids in digestion and elimination. Pectin, isolated from plants such as apples or citrus fruits, may be used as a food additive to thicken jelly, keep salad dressing from separating, and otherwise alter the texture and consistency of processed foods.

The term *dietary fiber* refers to substances that cannot be broken down by human digestive enzymes. They are, however, somewhat vulnerable to breakdown by the enzymes of bacteria that reside in the digestive tracts of human beings. Intestinal bacteria change some fibers and some other undigested substances into products that are absorbed and contribute a few calories' worth of energy and may provide health benefits or have other effects on the body. The amount of fiber that is broken down varies greatly depending upon the nature of both the fiber and the bacteria in the tract.[2]

Some animals, such as cattle, depend heavily on their intestinal bacteria to make the energy of glucose available from the abundant fiber cellulose in their fodder. When we eat beef, we receive indirectly some of the sun's energy that was originally stored in the fiber of the plants the cattle ate and converted into the energy contained in meat. Beef, of course, contains no fiber itself; no meats or dairy products contain fiber.

One way to classify fibers is according to how readily they dissolve in water.* Some fibers are **insoluble fibers;** others are **soluble fibers.** Each type of fiber exerts important effects on people's health; these effects will be described later.

roughage (RUFF-idge) the rough parts of food; an imprecise term that has largely been replaced by the term *fiber.*

insoluble fibers the tough, fibrous structures of fruits, vegetables, and grains; indigestible food components that do not dissolve in water.

soluble fibers food components that readily dissolve in water and often impart gummy or gel-like characteristics to foods. An example is pectin from fruit, which is used to thicken jellies. Soluble fibers are indigestible by human enzymes but may be broken down to absorbable products by bacteria in the digestive tract.

*Another way to classify fibers is by chemical name without reference to solubility. See Joint FAO/WHO Expert Consultation, *Carbohydrates in Human Nutrition* (Geneva: Food and Agriculture Organization, World Health Organization, 1998).

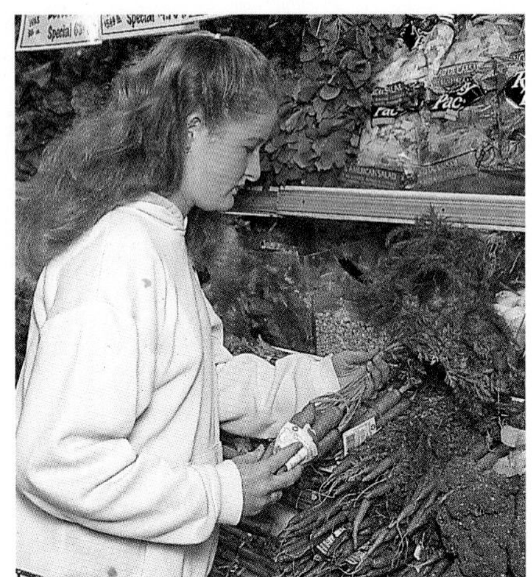

The sugars in these carrots are diluted with water and packaged with vitamins, minerals, and fiber.

Chapter 15 revisits humankind's relationship with the earth's food chain.

In summary, carbohydrate plays a prominent role in the global carbon cycle. Carbon dioxide, water, and energy are combined in plants to form glucose; the plants may store the glucose in the polysaccharide starch. Then animals or people eat the plants and retrieve the glucose. In the body, the liver and muscles may store the glucose as the polysaccharide glycogen, but ultimately it becomes glucose again. The glucose delivers the sun's energy to fuel the body's activities. In the process, glucose breaks down to waste products, carbon dioxide and water, which are excreted. Later, these compounds are used again by plants as raw materials to make carbohydrate.

KEY POINT *Little fiber is digested by the enzymes in the human digestive tract. Much of the fiber passes through the digestive tract unchanged.*

The Need for Carbohydrates

Glucose from carbohydrate is the preferred fuel for most body functions. Only two other nutrients provide energy to the body: protein and fats. Protein-rich foods are usually expensive, and when used to make fuel for the body, they provide no advantage over carbohydrates. In fact, their overuse has disadvantages, as explained in Chapter 6. Fats are not normally used as fuel by the brain and central nervous system, and diets high in fats are associated with many disease states. Thus, of the possible alternatives, glucose is the preferred energy source. It is especially important as the chief fuel of nerve cells, including those of the brain, which depend almost exclusively on glucose for their energy. And starchy foods, or complex carbohydrates, are the preferred source of glucose.

If I Want to Lose Weight, Should I Avoid Carbohydrates?

Complex carbohydrates are often wrongly accused of being the "fattening" ingredients of foods. Some people are startled to hear that most people need to consume more starchy foods rather than less. Yet much evidence supports this assertion. Gram for gram, carbohydrates donate fewer calories than do dietary fats, so a moderate, balanced diet based mostly on high-carbohydrate foods is likely to be lower in calories than a diet of high-fat foods. Also, to convert glucose to fat in the body requires chemical conversions that cost many of the glucose's original calories, making glucose even less fattening. Government agencies in many countries, recognizing the value of complex carbohydrates, urge their citizens to consume abundant foods that contain them. Table 4-1 reviews the U.S. recommendations and goals first presented in Chapters 1 and 2, as well as the World Health Organization's recommended upper and lower limits for carbohydrate intakes.

Unlike complex carbohydrates, pure sugars displace nutrient-dense foods from the diet. Purified, refined sugar (sucrose) contains no other nutrients—protein, vitamins, minerals, or fiber—and so can be termed an empty-calorie food. If you choose 400 calories of sugar in place of 400 calories of starchy food such as whole-grain bread, you lose not only the starch but also the vitamins, minerals, and fiber of the bread. You can afford to do this only if you have already met your nutrient needs for the day and still have calories to spend. This chapter's Food Feature offers more about the sugars in foods.

KEY POINT *Complex carbohydrates are the preferred energy source for the body.*

How Does Fiber in Food Affect My Health, and How Much Do I Need to Stay Healthy?

Foods containing starch offer additional benefits if fibers come with the starch, as they do naturally. Fibers may benefit health in all these ways:

Table 4-1

Recommendations Concerning Intakes of Carbohydrates

1. Recommendations for Complex Carbohydrates

Dietary Guidelines

- Every day eat 5 to 9 servings[a] of a combination of vegetables and fruits. Also, increase intake of starches and other complex carbohydrates by eating 6 to 11 daily servings of a combination of breads, cereals, and legumes.

Daily Values[b]

- 300 grams of complex carbohydrate, or 60% of total calories.

Proposed Healthy People 2010

- Increase intakes of fruits and vegetables, including legumes, to at least five servings a day.
- Increase intakes of grain products to at least six servings a day.

World Health Organization

- Lower limit: 50% of total calories from complex carbohydrates.
- Upper limit: 75% of total calories from complex carbohydrates.

2. Recommendations for Refined Sugars

Dietary Guidelines

- Use sugars only in moderation.

World Health Organization

- Lower limit: 0% of total calories from refined sugars.
- Upper limit: 10% of total calories from refined sugars.

3. Recommendations for Dietary Fiber

Dietary Guidelines

- Increase your fiber intake by eating more of a variety of foods that contain fiber naturally.

Daily Values[b]

- 25 grams of fiber per day, or 11.5 grams per 1,000 calories.

World Health Organization

- Lower limit: 27 grams of dietary fiber a day.
- Upper limit: 40 grams of dietary fiber a day.

[a]Serving sizes were presented in Figure 2-5 of Chapter 2.
[b] Daily Values are for a 2,000-calorie diet.

- Improve the body's handling of glucose and the hormone insulin, perhaps by slowing the digestion or absorption of carbohydrate.[3] A habitual lack of fiber in the diet along with an abundance of highly refined carbohydrate foods is likely to increase the risk of developing diabetes.[4]
- Possibly reduce the risk of colon cancer.[5] Insoluble fibers of whole grains, fruits, and vegetables may bind or dilute cancer-causing materials and speed their transit through the colon. (Data collected over many years support this idea, although one recent survey reported no effect. See Chapter 11 for details.)
- Reduce energy consumption by displacing calorie-dense concentrated fats and sweets from the diet while donating little energy. Fibers therefore can help in weight control.
- Possibly reduce the risk of heart and artery disease by several mechanisms. Soluble fibers may lower blood cholesterol by delaying absorption in the digestive tract (see Figure 4-4).[6] Some soluble fibers are digested by intestinal bacteria to yield small, fatlike products that, when absorbed, may lower LDL cholesterol (the harmful kind).[7] Finally, fibers displace fatty, cholesterol-raising foods from the diet.
- Promote feelings of fullness because they absorb water and swell. Soluble fibers in a meal also slow the movement of food through the upper digestive tract, so you feel full longer.

See Figure 4-12 on page 122 for lists of the fiber contents of foods.

Figure 4-4

One Way Fiber in Food May Lower Cholesterol in the Blood

In some ways, the liver is like a vacuum cleaner, sucking up cholesterol from the blood, converting the cholesterol to bile, and discharging the bile into its storage bag, the gallbladder. The gallbladder empties its bile into the intestine, where bile performs necessary digestive tasks. In the intestine, some of the bile links up with fiber and is carried out of the body in feces.

A. When the diet is rich in fiber, much of the cholesterol (as bile) is carried out of the body.

B. When the diet is low in fiber, most of the cholesterol is reabsorbed and returned to the bloodstream.

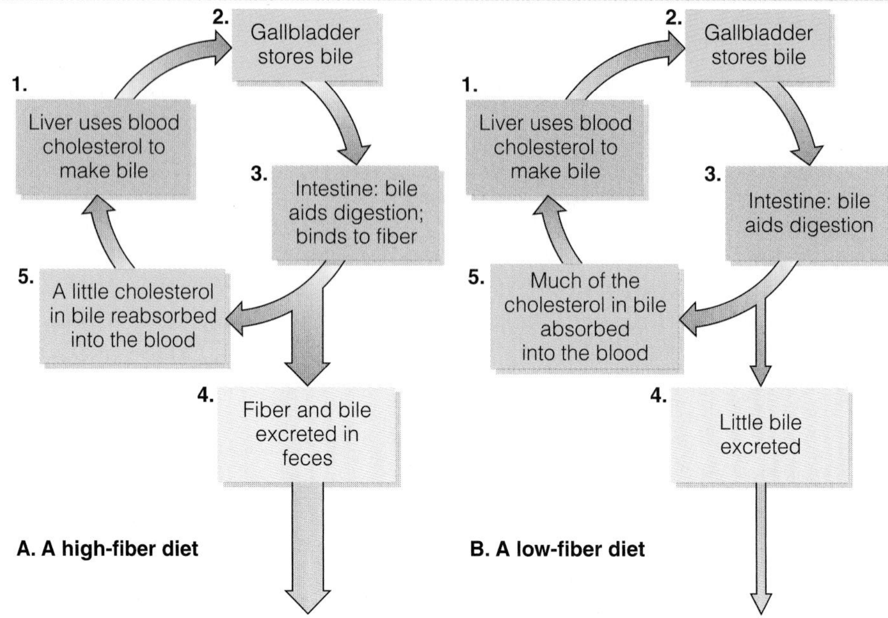
1.
Liver uses blood cholesterol to make bile
2.
Gallbladder stores bile
3.
Intestine: bile aids digestion; binds to fiber
4.
Fiber and bile excreted in feces
5.
A little cholesterol in bile reabsorbed into the blood
A. A high-fiber diet
1.
Liver uses blood cholesterol to make bile
2.
Gallbladder stores bile
3.
Intestine: bile aids digestion
4.
Little bile excreted
5.
Much of the cholesterol in bile absorbed into the blood
B. A low-fiber diet

- Help prevent **constipation, hemorrhoids,** and other intestinal problems by keeping the contents of the intestine moist and easy to eliminate.
- Help prevent bacterial infection of the appendix **(appendicitis)** by the same mechanism.
- Stimulate the muscles of the digestive tract so that they retain their health and tone. This prevents **diverticulosis,** in which the intestinal walls become weak and bulge out in places (see Figure 4-5 in the margin).

In general, then, fibers:

- *Moderate rates of absorption of nutrients* and other molecules by entrapping molecules and preventing their contact with absorptive surfaces.
- *Delay cholesterol absorption,* probably by the same mechanism.
- *Stimulate bacterial fermentation* in the colon (described below).
- *Increase stool weight* by holding water within the feces.

These four actions underlie the many health benefits attributed to dietary fibers.[8]

People choosing high-fiber foods in hopes of receiving some of these benefits are wise to seek out a variety of fiber sources and to drink extra fluids to help the fiber do its job. Wheat bran, which is composed mostly of insoluble fibers, is one of the most effective stool-softening fibers; oat bran and other more soluble fibers have a greater cholesterol-lowering effect.[9] The fibers of legumes, apples, and carrots may also lower blood cholesterol.

Table 4-2 shows the diverse effects of different fibers; it also shows that most unrefined plant foods contain a mix of fiber types. To consumers, this means that although a food may play a starring role in providing one type of fiber, to receive the whole range of fiber benefits, one must choose a variety of whole foods each day.

In the United States, people report an average intake of about 14 to 15 grams of fiber per day—too low to meet recommendations.[10] The Daily Value for fiber is 25 or 30 grams per day for people eating a 2,000-calorie or 2,500-calorie diet, respectively. Most experts agree that healthy people should meet their fiber need by eating unprocessed, fiber-containing foods and not by eating refined fiber sources, because many of the benefits attributed to fiber may come from other constituents of fiber-containing foods, and not from fiber alone.[11] As Table 4-1 showed, the World Health Organization recommends a daily intake of 27 to 40 grams of fiber.[12] This chapter's Consumer Corner provides detailed information about choosing wisely among grain foods, and the Do It! feature guides you in analyzing the fiber in your own diet. You can get a quick approx-

Figure 4-5

Diverticulosis

Diverticula are abnormal bulging pockets formed in the colon wall. These pockets can entrap feces and become painfully infected and inflamed, requiring hospitalization, antibiotic therapy, or surgery.

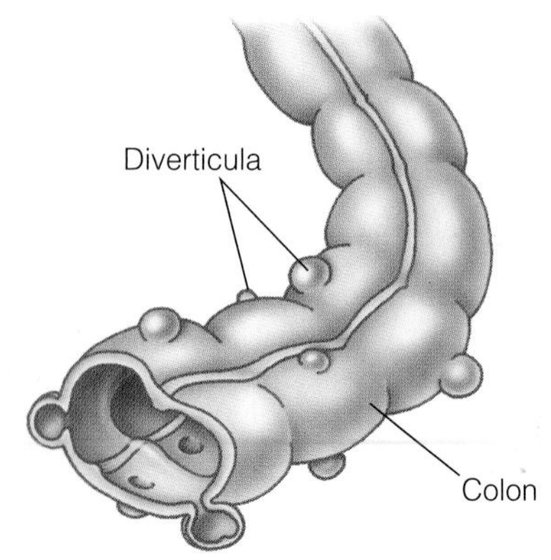
Diverticula
Colon

imation of the fiber in your diet right now by following the instructions of Table 4-3 on page 102. Use your food record from the Do It! section of Chapter 2.

KEY POINT *Fibers aid in maintaining the health of the digestive tract and help to prevent or control certain diseases. Most people probably need between 20 and 40 grams of fiber each day.*

Can My Diet Have Too Much Fiber?

Like any other substance, fibers can cause harm if taken in excess. The addition of purified fibers, such as oat or wheat bran, to foods is easily taken to extreme. One enthusiastic eater of oat bran muffins required emergency surgery for a blocked intestine; he had eaten so much bran that his digestive system had been unable to function. This doesn't mean that people should avoid bran-containing foods, but that they should use bran, separated from its original food product, with moderation. A purified fiber such as cellulose may not affect the body the same way as the cellulose in whole grains.

Less extreme concerns are that purified fiber might displace nutrients from the diet or cause them to be lost from the digestive tract. Purified fibers are like refined sugars in one way: the nutrients that originally accompanied the fibers have been lost. Binders in some fibers act as **chelating agents** and link chemically with nutrient minerals (iron, zinc, calcium, and others) and then carry them out of the body. The mineral iron is mostly absorbed at the beginning of the intestinal tract, and excess insoluble fibers may limit absorption by speeding foods through the upper part of the digestive tract. Too much bulk in the diet can also limit the total amount of food consumed and cause deficiencies of both nutrients and energy. The malnourished, the elderly, and children who consume no animal products are especially vulnerable to this chain of events. Fibers also carry water out of the body and can cause dehydration. The addition of fiber to the diet calls for an extra glass or two of water to go along with it.

KEY POINT *Fiber needs are best met with whole foods. Purified fiber in large doses can have undesirable effects.*

constipation hardness and dryness of bowel movements, associated with discomfort in passing them from the body.

hemorrhoids (HEM-or-oids) swollen, hardened (varicose) veins in the rectum, usually caused by the pressure resulting from constipation.

appendicitis inflammation and/or infection of the appendix, a sac protruding from the intestine.

diverticulosis (dye-ver-tic-you-LOH-sis) outpocketing or ballooning out of areas of the intestinal wall, caused by weakening of the muscle layers that encase the intestine.

chelating (KEE-late-ing) **agents** molecules that surround other molecules and are therefore useful in either preventing or promoting movement of substances from place to place.

Chelating agents are often sold by supplement vendors to "remove poisons" from the body. Some valid medical uses such as treatment of lead poisoning exist, but most of the chelating agents sold over-the-counter are promoted based on unproven claims.

Table 4-2

Water Solubilities, Sources, and Health Effects of Fiber

Fiber Type	Major Food Sources	Possible Health Effects
Soluble Gums, mucilages, pectins, psyllium,[a] some hemicellulose	Barley, fruits, legumes, oats, oat bran, rye, seeds, vegetables	These fibers lower blood cholesterol; slow glucose absorption; slow transit of food through upper digestive tract; hold moisture in stools, softening them; are partly fermentable into fragments the body can use.
Insoluble Cellulose, lignin, some hemicellulose	Brown rice, fruits, legumes, seeds, vegetables, wheat bran, whole grains	These fibers soften stools; regulate bowel movements; speed transit of material through small intestine; increase fecal weight and speed fecal passage through colon; reduce colon cancer risk; reduce risks of diverticulosis, hemorrhoids, and appendicitis.

[a]Psyllium, a fiber laxative and a cereal additive, has both soluble and insoluble properties.

CONSUMER CORNER

Refined, Enriched, and Whole-Grain Bread

Figure 4-6

A Wheat Plant and a Single Kernel of Wheat

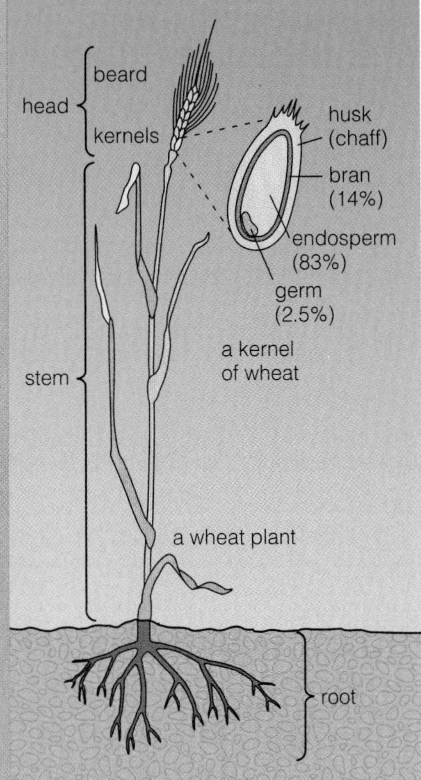

head
beard
kernels
husk
(chaff)
bran
(14%)
endosperm
(83%)
germ
(2.5%)
a kernel
of wheat
stem
a wheat plant
root

For many people, bread supplies much of the carbohydrate, or at least most of the starch, in a day's meals. Any food used in such abundance in the diet should be scrutinized closely, and if it doesn't measure up to high nutrition standards, it should be replaced with a food that does. For people who eat bread, the meanings of the words associated with the wheat flour that makes up the bread—**refined, enriched, fortified,** and **whole grain**—hold the key to understanding this product, in which they invest many calories per day (see Table 4-4).

The part of the wheat plant that is made into flour and then into bread and other baked goods is the seed or kernel. The wheat kernel (a whole grain) has four main parts: the **germ,** the **endosperm,** the **bran,** and the **husk,** as shown in Figure 4-6. The germ is the part that grows into a wheat plant and therefore carries with it concentrated food to support the new life. It is especially rich in vitamins and minerals. The endosperm is the soft, white, inside portion of the kernel, containing starch and proteins that help nourish the seed as it sprouts. The kernel is encased in the bran, a protective coating that is similar in function to the shell of a nut; the bran is also rich in nutrients and fiber. The husk, commonly called chaff, is the dry outermost layer and is inedible for human beings but can be used in animal feed.

In earlier times people milled wheat by grinding it between two stones,

Table 4-4

Terms That Describe Grain Foods

- **bran** the protective fibrous coating around a grain; the chief fiber donator of a grain.
- **brown bread** bread containing ingredients such as molasses that lend a brown color; may be made with any kind of flour, including white flour.
- **endosperm** the bulk of the edible part of a grain, the starchy part.
- **enriched, fortified** refers to the addition of nutrients to a refined food product. As defined by U.S. law, these terms mean that specified levels of thiamin, riboflavin, niacin, folate, and iron have been added to refined grains and grain products. The terms *enriched* and *fortified* can refer to the addition of more nutrients than just these five; read the label.[a]
- **germ** the nutrient-rich inner part of a grain.
- **husk** the outer, inedible part of a grain.
- **refined** refers to the process by which the coarse parts of food products are removed. For example, the refining of wheat into flour involves removing three of the four parts of the kernel—the chaff, the bran, and the germ—leaving only the endosperm, composed mainly of starch and a little protein.
- **stone ground** refers to a milling process using limestone to grind any grain, including refined grains, into flour.
- **unbleached flour** a beige-colored endosperm flour with texture and nutritive qualities that approximate those of regular white flour.
- **wheat flour** any flour made from wheat, including white flour.
- **white flour** an endosperm flour that has been refined and bleached for maximum softness and whiteness.
- **whole grain** refers to a grain milled in its entirety (all but the husk), not refined.
- **whole-wheat flour** flour made from whole-wheat kernels; a whole-grain flour.

[a]Formerly, *enriched* and *fortified* carried distinct meanings with regard to the nutrient amounts added to foods, but a change in the law has made these terms virtually synonymous.

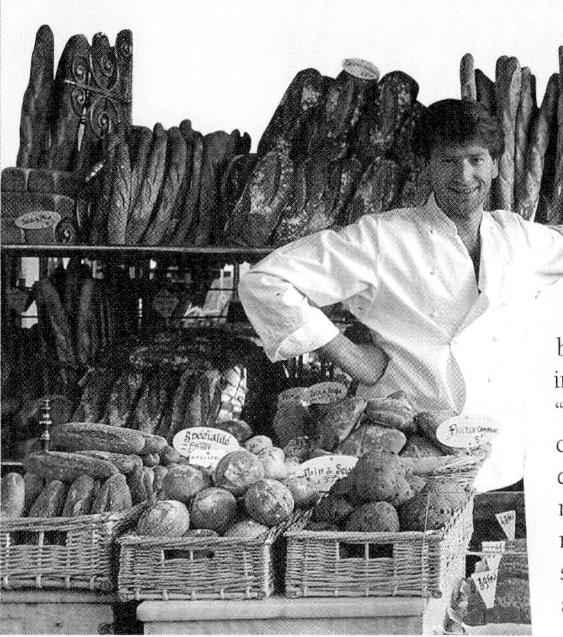

In Western societies, bread is the staff of life.

blowing or sifting out the chaff, and retaining the nutrient-rich bran and germ as well as the endosperm. Then milling machinery was "improved," or so the makers thought, and it became possible to remove the dark, heavy germ and bran as well, leaving a whiter, smoother-textured flour. People came to look on this flour as more desirable than the crunchy, dark brown, "old-fashioned" flour.

In turning to white bread, bread eaters suffered a tragic loss of needed nutrients. Many people developed deficiencies of iron, thiamin, riboflavin, and niacin—nutrients that they had formerly received from whole-grain bread. Finally, the problem was recognized, and Congress passed the Enrichment Act requiring that iron, niacin, thiamin, and riboflavin be added to refined grain products before they were sold. The Enrichment Act of 1942 is still in effect in the United States today but was amended in 1996 to include the vitamin folate, or "folic acid," as it is sometimes called on food labels. This doesn't make a single slice of refined bread "rich" in these nutrients, but people who eat several or many slices of bread a day obtain significantly more of the nutrients than they would from unenriched white bread, as Figure 4-7 on the next page shows. Today you can almost take for granted that all breads, grain products such as rice, macaroni, and spaghetti, and all types of cereals have been enriched with at least the nutrients just mentioned.

To a great extent, the enrichment of grain products eliminated known deficiency problems, but many other deficiencies went undetected for many more years. The trouble with enriched flour is that it is comparable to whole grain only with respect to the added nutrients and not with respect to others. Enriched products still contain less magnesium, zinc, vitamin B_6, vitamin E, and chromium than whole-grain products do. When a grain is refined, fiber is lost, too (see Table 4-5). Bread sold for weight-reduction dieting may be fortified with pure cellulose, but adding cellulose alone is not enough; the bread still lacks other fibers.

Table 4-5

Grams of Fiber in One Cup of Flour

Dark rye, 18 g.
Whole-grain cornmeal, 9 g.
Whole wheat, 15 g.
Light rye, 14 g.
Buckwheat, 12 g.
Enriched white, 3 g.

Only *whole-grain* flour contains all nutritive portions of the grain. Notice the distinctions between **wheat flour** and **whole-wheat flour** and **white flour** and **unbleached flour** among the terms that describe grain foods; also notice that neither **brown bread** nor **stone ground** on a label is a guarantee that the bread has been made with whole-grain flour. If bread is a staple food in your diet—that is, if you eat it every day—you would be well advised to learn to like the hearty flavor of whole-grain bread.

Figure 4-7

Nutrients in Whole-Grain, Enriched White, and Unenriched White Breads

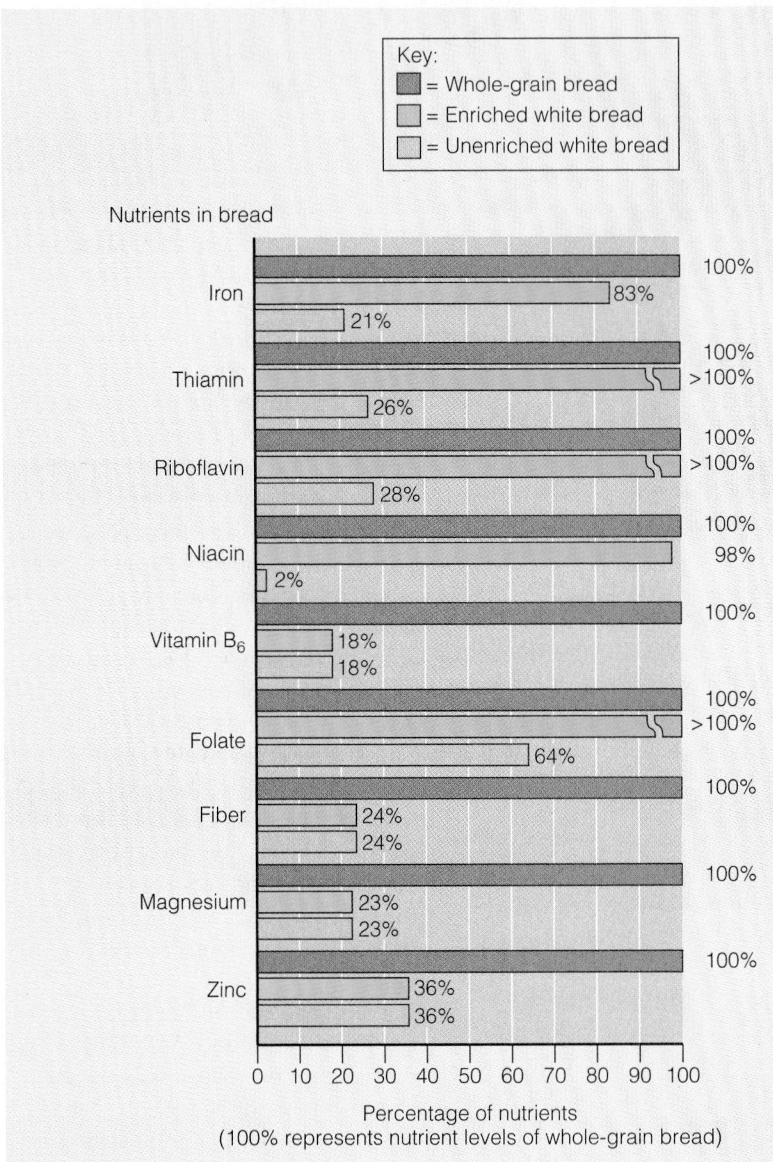

Key:
= Whole-grain bread
= Enriched white bread
= Unenriched white bread
Nutrients in bread
Iron
100%
83%
21%
Thiamin
100%
>100%
26%
Riboflavin
100%
>100%
28%
Niacin
100%
98%
2%
Vitamin B_6
100%
18%
18%
Folate
100%
>100%
64%
Fiber
100%
24%
24%
Magnesium
100%
23%
23%
Zinc
100%
36%
36%
0 10 20 30 40 50 60 70 80 90 100
Percentage of nutrients
(100% represents nutrient levels of whole-grain bread)

From Carbohydrates to Glucose

The body's cells cannot use foods such as bread or even whole molecules of lactose, sucrose, or starch for energy, but they need the glucose in those molecules, and they need it continuously. An important task of the various body systems, then, is to make glucose available to the cells, not all at once when it is eaten, but at a steady rate all day.

Digestion and Absorption of Carbohydrate

To obtain glucose from newly eaten food, the digestive system must first render the starch and disaccharides from the food into monosaccharides that can be absorbed through the cells that line the small intestine. The largest of the digestible carbohydrate molecules, starch, requires the most extensive breakdown. Disaccharides, in contrast, must be split only once before they can be absorbed.

Digestion of most starch begins in the mouth, where an enzyme in saliva mixes with food and begins to split starch into maltose. While chewing a bite of bread, you may have noticed that a slightly sweet taste develops. This is because maltose is being liberated from starch by the enzyme. The salivary enzyme continues to act on the starch in the swallowed bite of bread while it remains tucked in the stomach's storage area together with other swallowed bites. Slowly, each chewed lump is pushed downward, to be thoroughly mixed with the stomach's acid and other juices. Enzyme molecules are made of protein, and as such, they eventually succumb to deactivation by the stomach's protein-digesting acid. (The only exception is the protein-digesting enzyme that works in the stomach; its structure protects it from the stomach's acid.) Starch digestion therefore ceases in the stomach, but it resumes at full speed in the small intestine, where another starch-splitting enzyme is delivered by the pancreas. This enzyme breaks starch down entirely into disaccharides and small polysaccharides.[13]

Some forms of starch are easily digested. The starch in bread made of refined white flour, for example, breaks down rapidly to glucose that is absorbed high up in the small intestine.[14] Some starch, such as that of cooked beans, digests more slowly and releases its glucose later on in the digestion process. Other starch, called **resistant starch** is found in raw foods such as raw corn starch, inside the impenetrable hulls of seeds or in foods subjected to intense heating that renders the starch indigestible. Some resistant starch may be digested, but slowly, or it may remain intact until the bacteria of the colon eventually break it down.[15] The rate of starch digestion may affect the body's handling of its glucose, as a later section explains.

Sucrose and lactose from food, along with maltose and small polysaccharides freed from starch, undergo one more split to yield free monosaccharides before they are absorbed. This split is accomplished by enzymes that are attached to the cells of the lining of the small intestine. The conversion of a bite of bread to nutrients for the body is completed when monosaccharides cross these cells and are washed away in a rush of circulating blood that carries them to the waiting liver. Figure 4–8 on page 106 presents a quick review of carbohydrate digestion.

Once in the body, the absorbed carbohydrates (glucose, galactose, and fructose) travel to the liver, which converts fructose and galactose to glucose or products of glucose metabolism (such as fats). The circulatory system transports the glucose and fats to the cells. Liver and muscle cells may store circulating glucose as glycogen; all cells may split glucose for energy.

As mentioned, molecules of most fiber and of some resistant starch are not changed by human digestive enzymes. Many of these molecules can, however, be readily digested by the billions of living inhabitants of the human digestive tract, the resident bacteria. So active are these inhabitants in breaking down substances from food that one expert claims they constitute "an organ of intense metabolic activity that is involved in nutrient salvage."[16] Digestion of soluble fibers and resistant starch by resident bacteria yields waste products, mainly small fat fragments that the body absorbs and can use to provide a tiny bit of energy.[17] Some of these fragments may play roles in lowering the blood lipids associated with heart disease risk.[18]

resistant starch the fraction of starch in a food that is slowly digested, or not digested, by human enzymes.

Table 4-3

A Quick Method for Estimating Fiber Intake

To quickly estimate fiber in a day's meals:

1. Multiply servings[a] of fruits and vegetables by 1.5 g.[b]
   Example: 5 servings of fruits and vegetables × 1.5 = 7.5 g. fiber
2. Multiply servings of refined grains by 1.0 g.
   Example: 4 servings of refined grains × 1.0 = 4.0 g fiber
3. Multiply servings of whole grains by 2.5 g.
   Example: 3 servings of whole grains × 2.5 = 7.5 g fiber
4. Add fiber values for servings of legumes, nuts, seeds, and high-fiber cereals and breads, look these up in Appendix A.
   Example: ½ c black beans = 8.0 g fiber
5. Add up the grams of fiber from the previous lines.
   Example: 7.5 + 4.0 + 7.5 + 8.0 = 27 g fiber

Day's total fiber = 27 g fiber

[a]Use standard serving sizes presented in Chapter 2.
[b]Juices do not count toward this total.
SOURCE: Adapted from J. A. Marlett and T.-F. Cheung, Database and quick methods of assessing typical dietary fiber intakes using data for 228 commonly consumed foods, *Journal of the American Dietetic Association* 97 (1997): 1139–1148, 1151.

Figure 4-8

How Carbohydrate in Food Becomes Glucose in the Body

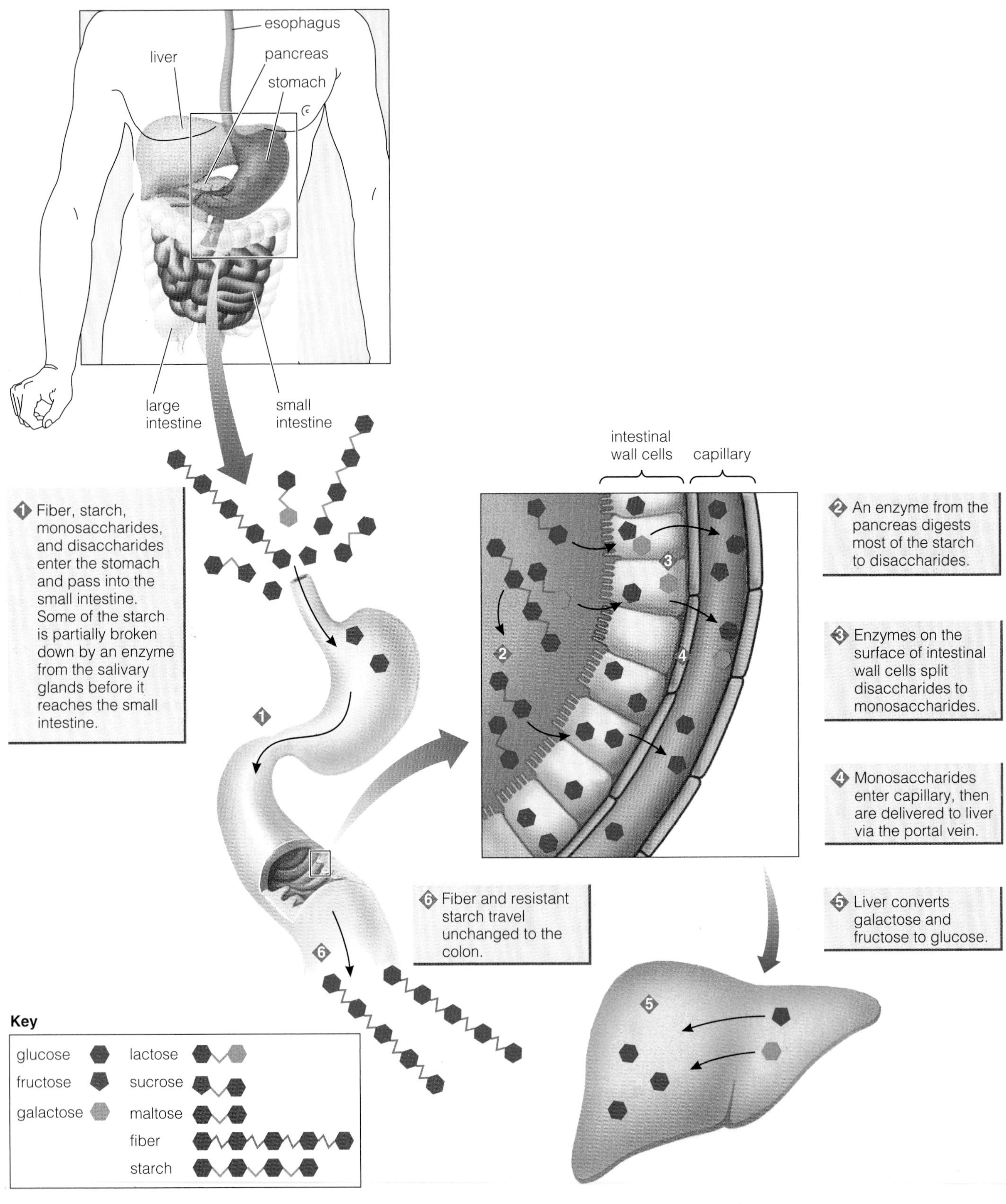

esophagus
liver
pancreas
stomach
large intestine
small intestine
intestinal wall cells
capillary
1 Fiber, starch, monosaccharides, and disaccharides enter the stomach and pass into the small intestine. Some of the starch is partially broken down by an enzyme from the salivary glands before it reaches the small intestine.
2 An enzyme from the pancreas digests most of the starch to disaccharides.
3 Enzymes on the surface of intestinal wall cells split disaccharides to monosaccharides.
4 Monosaccharides enter capillary, then are delivered to liver via the portal vein.
5 Liver converts galactose and fructose to glucose.
6 Fiber and resistant starch travel unchanged to the colon.
Key
glucose
fructose
galactose
lactose
sucrose
maltose
fiber
starch

A by-product of metabolism by the bacteria in the colon can be any of several odorous gases, which may make people want to avoid fiber-containing foods altogether. Don't give up on high-fiber foods if they cause gas. Instead, start with small servings and gradually increase the amount consumed over several weeks; chew foods thoroughly to break up hard-to-digest lumps that can ferment in the intestine; and try a variety of fiber-rich foods until you find some that do not cause the problem. Some people also find relief from excess gas when they use commercial enzyme preparations sold for use with beans. Such products contain enzymes that may help to break down some of the indigestible fibers in foods before they reach the colon. In other people, persistent painful gas may indicate that the digestive tract has undergone a change in its ability to digest the sugar in milk, a condition known as lactose intolerance.

lactase the intestinal enzyme that splits the disaccharide lactose to monosaccharides during digestion.

lactose intolerance inability to digest lactose due to a lack of the enzyme lactase.

KEY POINT *With respect to starch and sugars, the main task of the various body systems is to convert them to glucose to fuel the cells' work. Fibers help regulate digestion and contribute a little energy.*

Why Do Some People Have Trouble Digesting Milk?

About 75 percent of the world's people, as they age, lose most of their ability to produce enough of the enzyme **lactase** to digest the milk sugar lactose.[19] Lactase, which is made by the small intestine, splits the disaccharide lactose into its component monosaccharides glucose and galactose, which are then absorbed.

People with **lactose intolerance** experience nausea, pain, diarrhea, and excessive gas on drinking milk or eating lactose-containing products. The undigested lactose remaining in the intestine demands dilution with fluid from surrounding tissue and ultimately from the bloodstream. Intestinal bacteria use the undigested lactose for their own energy, a process that produces gas and intestinal irritants. Infants produce abundant lactase, which helps them absorb the sugar of breast milk and milk-based formulas; a few who suffer inborn lactose intolerance must be fed solely on lactose-free formulas.[20]

Approximate percentages of people with lactose intolerance:

90% Asian Americans.
80% Native Americans.
80% African Americans.
70% Mediterranean peoples.
60% Inuits (Native Alaskans).
50% Hispanics.
25% U.S. population.
<15% Northern Europeans.

The failure to digest lactose affects people to differing degrees. Many can tolerate as much as a cup or two of milk a day; some can tolerate lactose-reduced milk; only a rare few cannot tolerate lactose in any amount.[21] Often people overestimate the severity of their lactose intolerance, blaming it for all sorts of symptoms that are most probably caused by something else.[22] Tragically, disadvantaged young children of the developing world sustain the most severe consequences of lactose intolerance when it combines with disease, malnutrition, or parasites to produce a loss of nutrients that greatly reduces the children's chances of survival.

Because milk is an almost indispensable source of the calcium every child needs for growth, a milk substitute must be found for any child who becomes lactose intolerant. Women who fail to consume enough calcium during youth may later develop weak bones, so it is urgent that young women, too, search for substitutes if they become unable to tolerate milk. Sometimes yogurt or aged cheese makes an acceptable substitute: the bacteria or molds that help create these products digest lactose as they convert milk to a fermented product.[23] Yogurts that contain added milk solids also contain added lactose, milk solids are listed among the ingredients on the label.

Alternatively, people may choose milk products that have undergone treatment with lactose-digesting enzymes, or they may treat the products themselves. Enzyme pills and drops can be purchased over-the-counter. The pills are taken with milk-containing meals, and the drops are added to milk-based foods; both products help to digest lactose by replacing the missing natural enzymes. In all cases, the trick is to find ways of splitting lactose to glucose and galactose so that the body can absorb the products, rather than leaving the lactose undigested to feed the bacteria of the colon.

Sometimes sensitivity to milk is due not to lactose intolerance but to an allergic reaction to the protein in milk. Milk allergy arises the same way other allergies do—from sensitization of the immune system to a substance. In this case, the immune system overreacts when it encounters the protein of milk.

Food allergies are a topic of Chapter 13.

protein-sparing action the action of carbohydrate and fat in providing energy that allows protein to be used for purposes it alone can serve.

ketosis (kee-TOE-sis) an undesirable high concentration of ketone bodies, such as acetone, in the blood or urine.

ketone (kee-tone) **bodies** acidic, fat-related compounds that can arise from the incomplete breakdown of fat when carbohydrate is not available.

Children and adults with milk allergy often cannot tolerate cheese or yogurt either, and they have to find nondairy calcium sources. Good choices are calcium-fortified orange juice or soy milk, where available, or canned sardines or salmon with the bones. Controversy 8 examines the topic of milk in adult diets in relation to the adult bone disease osteoporosis.

KEY POINT *Lactose intolerance is a common condition in which the body fails to produce sufficient amounts of the enzyme needed to digest the sugar of milk. Uncomfortable symptoms result and can lead to milk avoidance. Lactose-intolerant people and those allergic to milk need milk alternatives that contain calcium.*

The Body's Use of Glucose

Carbohydrates serve structural roles in the body, such as forming part of the mucus that provides a protective coating for the internal organs, but their main role is to serve as an energy source. Glucose is not only the main original unit from which carbohydrate-rich foods are made, but it is also the basic carbohydrate unit that each cell of the body uses for energy. The body handles its glucose judiciously. It maintains an internal supply for use in case of need, and it tightly controls its blood glucose concentration to ensure that glucose remains available for ongoing use.

Splitting Glucose for Energy

Glucose fuels the work of most of the body's cells. When a cell splits glucose for energy, it performs an intricate sequence of maneuvers that are of great interest to the biochemist—and of no interest whatever to most people who eat bread and potatoes. One fact that everybody needs to understand, though, is that there is no good substitute for carbohydrate. That is why carbohydrate was described as *essential* in Chapter 1. The following details illustrate this point.

The Point of No Return At a certain point, glucose is forever lost to the body, and this can have serious consequences. Inside a cell, glucose is broken in half, releasing some energy. Two pathways are then open to these glucose halves. They can be put back together to make glucose again, or they can be broken into smaller fragments. If they are broken further, they can never again be reassembled to form glucose. The smaller fragments can yield still more energy and in the process break down completely to carbon dioxide and water; or they can be hitched together into units of body fat. Figure 4-9 shows how glucose is broken down to yield energy and carbon dioxide.

Figure 4-9

The Breakdown of Glucose Yields Energy and Carbon Dioxide

Cell enzymes split the bonds between the carbon atoms in glucose, liberating the energy stored there for the cell's use. The first split yields two 3-carbon fragments. The two-way arrows mean that these fragments can also be rejoined to make glucose again. Once they are broken down further into 2-carbon fragments, however, they cannot rejoin to make glucose. The carbon atoms liberated when the bonds split are combined with oxygen and released into the air, via the lungs, as carbon dioxide. Although not shown here, water is also produced at each split.

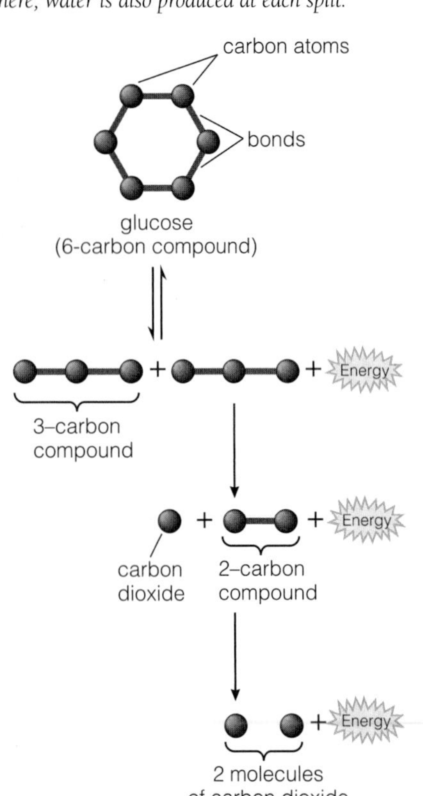

carbon atoms
bonds
glucose
(6-carbon compound)
Energy
3–carbon
compound
Energy
carbon
dioxide
2–carbon
compound
Energy
2 molecules
of carbon dioxide

Below a Healthy Minimum Although glucose can be converted into body fat, body fat can never be converted into glucose to feed the brain adequately. This is one reason why fasting and low-carbohydrate diets are dangerous. When it faces a severe carbohydrate deficit, the body has two problems. Having no glucose, it must turn to protein to make some (the body has this ability), thus diverting protein from vitally important functions of its own such as maintaining the body's immune defenses. Protein's functions in the body are so indispensable that carbohydrate should be kept available precisely to prevent the use of protein for energy. This is called the **protein-sparing action** of carbohydrate.

Without sufficient carbohydrate, the body cannot use its fat in the normal way. (Carbohydrate has to combine with fat fragments before they can be used for energy.) Using fat without the help of carbohydrate causes the body to go into **ketosis,** a condition in which unusual products of fat breakdown **(ketone bodies)** accumulate in the blood, disturbing the normal acid-base balance. Ketosis during pregnancy can cause brain damage to the fetus resulting in irreversible mental retardation after birth.

The minimum amount of carbohydrate needed to ensure complete sparing of body protein and avoidance of ketosis is around 100 grams of digestible carbohydrate a day for an average-sized person. Considerably more (three or four times more) than this minimum is recommended. The servings of vegetables, fruits, and grains recommended in Table 4-1 (p. 99) would deliver 125 grams at a minimum and 200 to 400 grams on average.

KEY POINT *Without glucose the body is forced to alter its uses of protein and fats. The body breaks down its own muscles and other protein tissues to make glucose and converts its fats into ketone bodies, incurring ketosis.*

Storing Glucose as Glycogen

After a meal, as blood glucose rises, the pancreas is the first organ to respond. It releases the hormone **insulin,** which signals the body's tissues to take up surplus glucose. Muscle and liver cells use some of this excess glucose to build the polysaccharide glycogen. The muscles hoard two-thirds of the body's total glycogen and use it just for themselves. The liver stores the other one-third and is more generous with its glycogen, making it available as blood glucose for the brain or other organs when the supply runs low.

Glycogen is wondrously designed for its task of releasing glucose on demand. Unlike starch, which has long chains with occasional branches that are cleaved linearly during digestion, glycogen is many branched with hundreds of ends extending from each molecule's surface (refer back to Figure 4-3 on page 97). When the blood glucose concentration drops and cells need energy, a pancreatic hormone, **glucagon,** floods the bloodstream. Thousands of enzymes within the liver cells respond by attacking a multitude of glycogen ends simultaneously to release a surge of glucose into the blood for use by all the other body cells. Another hormone, epinephrine, does the same thing as part of the body's defense mechanism in times of danger.

To a person living in the Stone Age, this internal source of quick energy was indispensable. Life was fraught with physical peril. The person who stopped and ate before running from a saber-toothed tiger did not survive to produce our ancestors. The quick-energy response in a stress situation works to our advantage today as well. For example, it accounts for the energy you suddenly have to clean up your room when you learn that a special person is coming to visit. To meet such emergencies, we are well advised to eat and to store carbohydrate every four to six waking hours.

You may rightly ask, "What kind of carbohydrate?" Candy bars and sugary beverages supply sugar energy quickly, but they are not the best choices. Balanced meals, eaten on a regular schedule, help the body to maintain its blood glucose. Meals containing starch and fiber along with some protein and a little fat slow digestion so that glucose enters the blood gradually in an ongoing steady supply. Such meals also provide an assortment of other nutrients, not found in candy and soft drinks, that help cells to use their glucose.

KEY POINT *Glycogen is the body's form of stored glucose. The liver stores glycogen for use by the whole body. Muscles have their own private glycogen stock for their exclusive use. The hormone glucagon acts to liberate stored glucose from liver glycogen.*

insulin a hormone secreted by the pancreas in response to a high blood glucose concentration. It assists cells in drawing glucose from the blood.

glucagon a hormone secreted by the pancreas that stimulates the liver to release glucose into the blood when blood glucose concentration dips.

Returning Glucose to the Blood

Should your glucose supplies ever fall too low, you would feel dizzy and weak. Should your blood glucose ever climb abnormally high, you might become confused or have difficulty breathing. Both conditions could be dangerous, but luckily the healthy body guards against them.

Regulation of Blood Glucose The maintenance of a normal blood glucose concentration depends on the two safeguards already mentioned. When blood glucose starts to fall too low, it is replenished by drawing on liver glycogen stores. When it starts to rise too high, the body siphons off the excess into the liver, to be converted to glycogen or fat, and into the muscle, to be converted to glycogen.

glycemic (gligh-SEEM-ic) **effect** a measure of the extent to which a food raises the blood glucose concentration and elicits an insulin response as compared with pure glucose.

To replenish blood glucose, the hormone glucagon triggers the breakdown of liver glycogen to free glucose. Other hormones also act in this manner, including epinephrine (a stress hormone) and some that promote the conversion of protein to glucose. The liver's glycogen stores can be depleted within half a waking day, however. As for protein, only a little can be spared. When body protein is used, it is taken from blood, organ, or muscle proteins; no surplus of protein is stored specifically for emergencies. As for fat, it cannot regenerate enough glucose to make a difference.

Obviously, when blood glucose falls and stores are depleted, a meal or a snack can replenish the supply. The meal or snack you choose may flood the blood with glucose, however, requiring the body to protect itself against too *high* a blood glucose concentration.

The Glycemic Effect of Foods Within limits, some foods elevate blood glucose and insulin concentrations higher than others do. The effect, called the **glycemic effect,** is worth a moment's attention. Scientists measure the glycemic effect by administering a food or a meal and then observing how fast and how high the blood glucose rises and how quickly the body responds by bringing it back to normal. Foods providing carbohydrate that is rapidly digested and absorbed, such as simple sugars or highly digestible starches, usually rank high in their glycemic effect.

Most people can quickly adjust to changes in blood glucose, but people with abnormal carbohydrate metabolism may experience extremes in blood glucose levels. These people do well to choose most often foods with a low glycemic effect such as dried beans, pasta, barley, bulgur (wheat), pumpernickel bread, and any food with the soluble fiber psyllium added.[24] These foods produce a slow, sustained rise in blood glucose.[25] They may also modulate appetite, providing a lasting feeling of fullness. (Chapter 9 provides details.) In addition, eating small, frequent meals spreads glucose absorption across the day and prevents a large influx.

Many factors work together to determine a food's glycemic effect, and the result is not always what a person might expect. Ice cream, for example, produces less of a response than potatoes; baked potatoes produce less of a response than mashed; a sweet, juicy apple produces a low response (probably due to the apple's soluble fiber); and dried beans and legumes of all kinds are notable for keeping blood glucose remarkably steady (probably because of their digestion-resistant starch). Importantly, a food's glycemic effect differs depending on whether it is eaten alone or as part of a mixed meal. The glycemic effects of foods in mixed meals tend to balance each other. Most people eat a variety of foods in a meal and so need not worry much about the glycemic effect of the foods they choose. This chapter's Controversy discusses research concerning a possible relationship between a *diet* with a high glycemic effect and the development of diabetes.[26]

KEY POINT *Blood glucose regulation depends mainly on the hormones insulin and glucagon. Certain carbohydrate foods produce a greater rise and fall in blood glucose than others do. Most people have no problem regulating their blood glucose, especially when they consume regular mixed meals.*

Converting Glucose to Fat

When food is tempting, people may continue to eat beyond the amount they need. After meeting the body cells' immediate energy needs and filling glycogen stores to capacity, the body takes a third path for handling incoming carbohydrates. Suppose you have eaten dinner and are now sitting on the couch, munching pretzels and drinking cola as you watch a ball game on television. Your digestive tract is delivering molecules of glucose to your bloodstream, and your blood is carrying these molecules to your liver and other body cells. The body cells use what glucose they can for their energy needs of the moment. Excess glucose is linked together and stored as glycogen until the muscles and liver are full to capacity with glycogen. Still the glucose keeps coming, and the liver has no choice but to handle the excess. The liver breaks the extra glucose into small fragments and puts them together into more permanent energy-storage compounds—fats. (This would happen with excess protein or fat, too.) The fats are then released into the blood, carried to the fatty tissues of the body, and deposited there. Unlike the liver cells, which can store only about four to six hours' worth of glycogen, the fat cells can store practically unlimited quantities of fats. Moral: you had better play the game if you are going to eat the food.

Even though excess carbohydrate is converted to fat and stored, a balanced diet that is high in complex carbohydrates helps control body weight and maintain lean tissue. Chapter 5 presents a few more details, but the main point is that, calorie for calorie, carbohydrate-rich foods contribute less to body fatness than do fat-rich foods.[27] Had you chosen fatty potato chips instead of low-fat pretzels for your ball game snack, your body would have stored even greater amounts of fat for the calories taken in. Thus, if you want to stay within your calorie limits, eat until full, never skip a meal, and remain lean, you should make every effort to choose foods that, together, provide a diet with 55 percent or more of its calories from mostly unrefined sources of complex carbohydrates and 30 percent or less from fats. This chapter's Food Feature provides the first set of tools required for the job of designing such a diet. Once you have learned to identify the carbohydrates in foods, you must then learn where the fats come in (Chapter 5's Food Feature) and how to obtain adequate protein without overdoing it (Chapter 6).

KEY POINT *The liver converts extra energy compounds into fat, a more permanent and unlimited energy-storage compound than glycogen.*

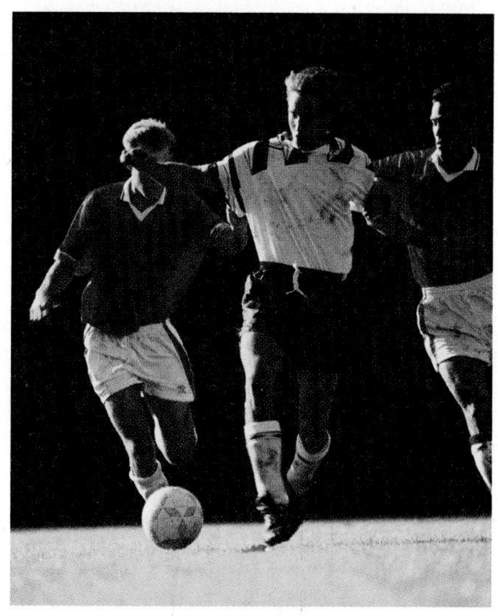

You had better play the game, if you are going to eat the food.

What Can I Eat to Make Workouts Easier?

FITNESS FOR LIFE

A working body needs carbohydrate fuel, and when it runs low, physical activity can seem more difficult. The body's glycogen stores are sufficient to comfortably fuel almost two hours of activity, but a diet that chronically provides too little carbohydrate can deplete muscles of their glycogen and make workouts seem harder.

If your workouts seem to drag and never get easier, take a look at your diet. Are your meals regularly timed? Do they provide abundant carbohydrate to fill up glycogen stores to last through a workout? Or do you try to work out after skipping meals or eating foods that lack the complex carbohydrates your muscles and brain need to prevent physical and mental fatigue? If so, change your choices to include more whole grains, fruits, milk, legumes, vegetables, and other foods that help to hold blood glucose steady.

Here's a trick: about two hours before your workout, eat a small snack of about 300 calories of foods rich in complex carbohydrates, and drink some extra fluid. The snack provides glucose at a steady rate to spare glycogen, and the fluid helps to maintain hydration. Half of a peanut butter sandwich on whole grain bread and a glass of fat-free milk would serve the purpose. If you can tolerate caffeine, a cup of tea or coffee, or one serving of another mild caffeine source might be a good choice of beverage with your snack: a little caffeine may lift your mood, stimulate the release of stress hormones, and ready your body for the physical exertion ahead.

Once you have begun exercising, be sure to drink enough water to replace lost fluids (Chapter 10 provides fluid intake guidelines). Sports drinks are probably not needed for glucose during ordinary workouts lasting less than two hours, and they contain many calories of sugar—counterproductive to the weight-conscious exerciser. For those whose workouts last longer, though, the glucose in the drinks can often help to supplement the exerciser's carbohydrate stores and to sustain the activity longer than possible otherwise.

Diabetes and Hypoglycemia

Some people have physical conditions that render their bodies unable to handle carbohydrates normally. One of these, **diabetes,** is common in developed nations and can be detected by a timed blood test. Another, hypoglycemia, is rare as a true disease condition, but many people believe they experience its symptoms at times.

diabetes (dye-uh-BEET-eez) a disease (technically termed *diabetes mellitus*) characterized by elevated blood glucose and inadequate or ineffective insulin, which renders a person unable to regulate blood glucose normally.

type 1 diabetes the type of diabetes in which the person produces no or very little insulin; also known as *juvenile-onset* or *insulin-dependent diabetes,* although some cases arise in adulthood.

type 2 diabetes the type of diabetes in which the person makes plenty of insulin, but the body cells resist insulin's action; also called *adult-onset* or *noninsulin-dependent diabetes.*

What Is Diabetes?

Diabetes is a chronic disease characterized by elevated blood glucose concentrations. Diabetes is among the top ten killers of adults, and it can lead to or contribute to a number of other serious diseases. Diabetes causes more new cases of blindness in the United States than any other cause, and its many other possible complications include amputations, heart and kidney disease, and premature death.[28] An estimated half of those suffering from diabetes are unaware of their condition and so fail to take action to prevent its damaging effects on the body. The early stage of the most common form of diabetes often presents few or no warning signs typically associated with diabetes (see Table 4-6). Therefore, the American Diabetes Association is calling for everyone over age 45 years and younger people with risk factors such as overweight to be tested regularly for diabetes.[29]

Several diseases have been called diabetes, but by far the most common ones are the two main forms of diabetes mellitus described here. Both types are disorders of blood glucose regulation. The characteristics of both are summarized in Table 4-7.

Type 1 Diabetes **Type 1 diabetes** is less common overall (about 10 to 20 percent of cases), but it is the leading chronic disease among children and young adults.[30] In this type of diabetes, the person's own immune system attacks the cells of the pancreas that normally synthesize the hormone insulin. Soon the pancreas can no longer produce insulin, and after each meal, blood glucose remains elevated, even though body tissues are simultaneously starving for glucose. The person must receive insulin from an external source to assist the cells in taking up the needed glucose from the blood; therefore, this type of diabetes has also been called *insulin-dependent diabetes mellitus (IDDM).*

Researchers concur that genetics, viral infection, other diseases, toxins or other substances from the environment, and a disordered immune system are probable culprits in provoking an immune system attack on the pancreas.[31] Some people may hear of a possibility that feeding formula based on cow's milk to infants younger than six months of age may sometimes trigger an immune response associated with the development of type 1 diabetes.[32] Limited research supports this idea, but much more evidence is needed before milk can be implicated in diabetes causation. Under most circumstances, breastfeeding is best for young babies.

More details about the benefits of breastfeeding are found in Chapter 12.

Insulin is a protein, and if it were taken orally, the digestive system would digest it. Insulin must therefore be injected, either by daily shots or by an insulin pump about the size of a personal pager worn next to the abdomen that delivers insulin through an implanted needle or tube. New fast-acting and long-lasting forms of insulin and other drugs are allowing more flexibility in managing meals and treatments, but users must still plan ahead to balance blood insulin and glucose concentrations.[33] Medical advances may soon eliminate the need for insulin shots—an insulin nasal spray and an inhaler that delivers insulin to the lungs are proving useful in clinical studies. Medical researchers are also coming closer to their ultimate goal of developing a vaccine or other therapy to prevent type 1 diabetes from occurring.[34]

Table 4-6

Warning Signs of Diabetes[a]

- Excessive urination and thirst.
- Glucose in the urine.
- Weight loss with nausea, easy tiring, weakness, or irritability.
- Cravings for food, especially for sweets.
- Frequent infections of the skin, gums, vagina, or urinary tract.
- Vision disturbances; blurred vision.
- Pain in the legs, feet, or fingers.
- Slow healing of cuts and bruises.
- Itching.
- Drowsiness.
- Abnormally high glucose tolerance test results.

[a]These signs appear reliably in type 1 diabetes and, often, in the later stages of type 2 diabetes.

Type 2 Diabetes The predominant type of diabetes mellitus, **type 2 diabetes** (about 90 percent of cases), is generally characterized by insulin resistance of the body's cells, including fat cells. Insulin may be present, often in abnormally large amounts, and it may stimulate cells to take up glucose, but they do so more slowly than normal. Blood glucose rises too high, as in type 1, but in this case blood insulin also rises. Another name for this type of diabetes, therefore, is *noninsulin-dependent diabetes mellitus (NIDDM)*. Eventually, the pancreas becomes less able to make insulin. At this point, some people with type 2 diabetes must take insulin to supplement their own supply. If drugs are necessary, a preferred therapy is to take a drug that stimulates the person's own pancreas to secrete insulin or one that improves the uptake of glucose by the tissues.[35]

Type 2 diabetes tends to occur late in life and to run in families. People with the disease often become obese because they overeat due to their cells' resistance to insulin—while they are waiting for their cells to be fed, so to speak. Figure 4-10 depicts one theory on how this may become a cycle—the larger the fat cells become, the more insulin resistant they become, and the more obese the person becomes. Obesity then worsens insulin resistance, which, in turn, worsens the obesity. Weight loss alone in overweight people with diabetes often helps control the disease. Even moderate weight gain in adults has been observed to predict diabetes.[36]

The incidence of this type of diabetes also increases with age, for in all people, the pancreatic cells that produce insulin progressively lose their function with time.[37] In some people, this age-related decline in cell function is more rapid or more severe than in others, and these people especially need to beware of excess weight gain and also to watch their alcohol intakes. Heavy use of alcohol makes development of type 2 diabetes more likely.

Diagnosing and Controlling Diabetes Diabetes can be diagnosed by means of a fasting blood glucose test or a **glucose tolerance** test. In the former, a person has blood drawn after a night of fasting so that a clinician may measure an indicator of blood glucose to determine whether it falls within the normal range for fasting. In the latter test, the body is challenged to handle a sudden, large amount of glucose. After fasting overnight, the subject is fed a sugary drink. Four or six hours later, when blood glucose should be normal, the diabetic person's blood glucose will still be elevated **(hyperglycemia)**, and often the blood insulin will be, too.

Although the symptoms of diabetes are controllable for the most part, its effects can be severe and may progress even when drugs control blood glucose.[38] Problems may include impaired circulation leading to disease of the feet and legs, often necessitating amputation; kidney disease, sometimes requiring hospital care or kidney transplant; impaired vision or blindness due to cataracts and damaged retinas; nerve damage; skin damage; and strokes and heart attacks.[39] The root cause of all these conditions is

glucose tolerance the ability of the body to respond to dietary carbohydrate by promptly regulating its blood glucose concentration to a normal level.

hyperglycemia (HIGH-per-gligh-SEEM-ee-uh) an abnormally high blood glucose concentration (hyper means "too much"; *glyce* means "glucose"; *emia* means "in the blood").

Figure 4-10

The Obesity-Diabetes Cycle

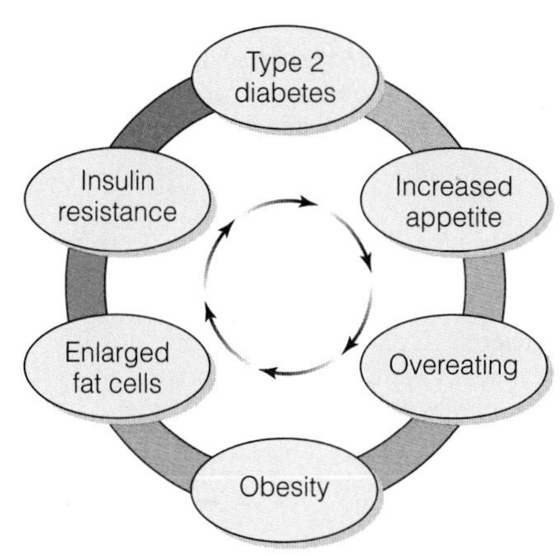
Type 2
diabetes
Increased
appetite
Overeating
Obesity
Enlarged
fat cells
Insulin
resistance

Table 4-7

Diabetes Types 1 and 2 Compared

	Type 1 Diabetes	Type 2 Diabetes
Age of onset	Childhood or mid-life	Adulthood
Body cells	Responsive to insulin action	Resistant to insulin action
Body fatness	Generally low to average	Generally high
Insulin shots required	Yes	Possibly[a]
Insulin-stimulating drugs or other drugs may be effective	No	Yes
Natural insulin	Pancreas makes too little or none.	Pancreas makes enough or too much.
Pancreatic function	Insulin-producing cells impaired or nonfunctional	Insulin-producing cells normal
Severity of symptoms	Relatively severe; many are apparent on diagnosis.	Relatively mild; few or none may be present on diagnosis.

[a]People past age 40 who suffer from type 2 diabetes may lose pancreatic function and become dependent on insulin.

hypoglycemia a blood glucose concentration below normal, a symptom that may indicate any of several diseases, including impending diabetes.

postprandial hypoglycemia a drop in blood glucose that follows a meal and is accompanied by symptoms of the stress response; also called *reactive hypoglycemia.*

fasting hypoglycemia hypoglycemia that occurs after 8 to 14 hours of fasting.

The exchange system introduced in Chapter 2 and presented in full in Appendix D was developed to help people with diabetes control calorie, carbohydrate, sugar, and fat intakes.

Chapter 12 discusses a form of diabetes seen only in pregnancy—gestational diabetes.

probably the same. Diabetes causes blockage or destruction of capillaries that feed the body organs, and tissues die from lack of nourishment. A new line of thinking holds that some damage may result from oxidation that accompanies elevated blood glucose.[40] The hope is that a diet high in vegetables and fruits that contain antioxidants may be protective, but much more research is needed to confirm the oxidation theory. Meanwhile the person with diabetes is advised to control not only weight but also all possible risk factors that might contribute to heart and blood vessel disease (atherosclerosis and hypertension, discussed in Chapter 11).

A diet constructed of a balanced pattern of foods is best for controlling diabetes and also for controlling weight and supporting physical activity. Such a diet also has all the characteristics important to prevention of chronic diseases and meets most of the *Dietary Guidelines for Americans.* The diet can vary, depending on personal tastes and on how much restriction is required to control an individual's blood glucose and lipid values.[41] In general, the diet:

- Is adequate (deficiencies in trace minerals, especially chromium, may hasten diabetes onset).
- Provides the recommended amount of fiber (fiber helps regulate blood glucose concentration).
- Is moderate in concentrated sugar (the amount allowed varies with an individual's blood glucose response).[42]
- Is high in complex carbohydrates (thought to assist in blood glucose regulation).
- Is low in saturated fat (thought to worsen insulin response and cardiovascular disease) and may provide some monounsaturated fat (thought to be harmless to the heart).[43]
- Is not too high in protein (to protect the kidneys).[44]

A person at risk for diabetes can do no better than to adopt such a diet long before any symptoms appear.

Also, the benefits from regular physical activity in preventing and controlling diabetes can hardly be overstated.[45] Exercise not only helps to maintain a desirable body weight, but it also heightens tissue sensitivity to insulin and may help prevent or forestall type 2 diabetes.[46] Like a juggler who keeps three circus balls constantly in motion, the person with diabetes must constantly balance three lifestyle factors—diet, exercise, and medication—to control the blood glucose level.

KEY POINT *Diabetes is an example of the body's abnormal handling of glucose. Inadequate or ineffective insulin leaves blood glucose high and cells undersupplied with glucose energy. This causes blood vessel and tissue damage. Weight control and exercise may be effective in preventing the predominant form of diabetes (type 2) and the illnesses that accompany it. A person diagnosed with diabetes must establish patterns of eating, exercise, and medication to control blood glucose.*

If I Feel Dizzy Between Meals, Do I Have Hypoglycemia?

The term **hypoglycemia** refers to a *symptom,* low blood glucose, and to a variety of conditions, including *disease conditions* that cause that symptom. One such condition is **postprandial hypoglycemia,** literally, "low blood glucose after (or caused by) a meal." The symptoms are fatigue, weakness, confusion, dizziness, irritability, a rapid heartbeat, anxiety, sweating, trembling, hunger, and headaches. These symptoms are so general and common to so many conditions that people can easily misdiagnose themselves as having postprandial hypoglycemia.[47]

It takes more than guesswork to diagnose postprandial hypoglycemia, though. A diagnosis requires a test to detect low blood glucose while the symptoms are present to confirm that both occur simultaneously. When they do, the diagnosis is confirmed.

A person who has symptoms while well advanced into the fasting state (for example, overnight) has a different kind of hypoglycemia. The symptoms of **fasting hypo-**

glycemia are different from those of postprandial hypoglycemia: headache, mental dullness, fatigue, confusion, amnesia, and even seizures and unconsciousness.

Only a few people suffer from truly abnormal conditions that cause hypoglycemia. Many of the cases result from serious disease that endangers health and life. Conditions such as cancer, pancreatic damage, infection of the liver with accompanying damage (hepatitis), or advanced alcohol-induced liver disease can all produce hypoglycemia.[48]

A very few other people experience symptoms together with a drop in blood glucose that cannot be explained by a disease state and that can be relieved by eating. In a study of people with postprandial hypoglycemia, researchers determined that a portion of hard-to-digest starch (raw cornstarch in this case) eaten with sugar effectively prevented a drop in blood glucose together with the associated symptoms. The authors of the study suggest what some people have suspected for a long time—that a sugary meal promotes symptoms in those few people who have postprandial hypoglycemia, but that certain starches and other nutrients taken with the sugar can prevent the problem. It may be that people with postprandial hypoglycemia respond too vigorously to the insulin released after a meal and so store too much of their blood glucose, or perhaps their response to the glucose-releasing hormone glucagon is too weak to oppose insulin's effects.[49]

Regularly timed, balanced meals help to hold blood glucose steady.

Hypoglycemia can be experimentally produced in just about everyone but only through extreme measures. To produce even mild hypoglycemia and its symptoms in normal, healthy people requires administering drugs that overwhelm the body's glucose-controlling hormones, insulin and glucagon. Without such intervention, those hormones rarely fail to keep blood glucose within normal limits. Symptoms that people ascribe to hypoglycemia rarely correlate to low blood glucose in blood tests. Perhaps these people experience symptoms when their blood glucose drops slightly, but still remains within the range considered normal.

For people who seem to experience postprandial hypoglycemia after meals, it may help to avoid oscillating between low-carbohydrate dieting and sudden large sugar doses and to eat regularly timed, balanced meals to hold blood glucose steady.[50] It may also help to avoid alcoholic beverages because alcohol impairs normal regulation of blood glucose in otherwise healthy people.[51]

Part of eating right is choosing wisely among the many foods available. Two features that follow can help with choices of carbohydrate-containing foods. First, the Food Feature explains how to integrate foods into a diet that meets the body's needs for carbohydrates. The Do It! section then asks you to critique your diet with regard to the fiber it contains.

KEY POINT *Postprandial hypoglycemia is a rare medical condition in which blood glucose falls too low. It can be a warning of organ damage or disease. Many people believe they experience symptoms of hypoglycemia, but their symptoms normally do not accompany below-normal blood glucose.*

So far, this chapter has explored the body's responses to carbohydrate, processes that occur largely without your awareness. Now it asks that you take the controls by learning which foods supply the carbohydrates your body needs.

FOOD FEATURE

Meeting Carbohydrate Needs

The exchange system appendix (Appendix D) lists carbohydrate values for a variety of foods.

To best support health, a diet must supply enough carbohydrate-rich foods to meet the body's needs. The Daily Value suggests a goal of 300 grams of mostly complex carbohydrate each day for a person who eats a 2,000-calorie diet. This Food Feature illustrates how you can obtain the carbohydrate-rich foods you need while using the Food Guide Pyramid as a guide. Breads, cereals, vegetables, fruits, and milk are noted for their contributions of starches and dilute sugars, both valuable energy-yielding carbohydrates. Many of these foods are also rich sources of fiber, and this chapter's Do It! section helps to point them out.

Bread, Cereal, Rice, and Pasta

A serving of most foods in this group—a slice of whole-wheat bread, half an English muffin or bagel, a 6-inch tortilla, or a half-cup of rice, pasta, or cooked cereal—provides about 15 grams of carbohydrate, mostly as starch.* People who like breads and other starchy foods are happy to learn that nutrition authorities encourage people to use them in abundance. If calories are a problem, people should first limit foods with added sugar or fat and limit total calories. Some foods in this group, especially baked goods such as biscuits, croissants, muffins, and snack crackers, do contain added sugar and/or fat, however.

Vegetables

Some vegetables are major contributors of starch in the diet. Just one small white or sweet potato or a half-cup of cooked dry beans, corn, peas, plantain, or winter squash provides 15 grams of carbohydrate, as much as in a slice of bread, though as a mixture of sugars and starch. A half-cup of carrots, okra, onions, tomatoes, cooked greens, or most other nonstarchy vegetables or a cup of salad greens provides about 5 grams as a mixture of starch and sugars.

Fruits

Different forms of fruit are assigned different serving sizes. A typical fruit serving—three-quarters of a cup of juice; a small banana, apple, or orange; a half-cup of most canned or fresh fruit; or a quarter-cup of dried fruit—contains an average of about 15 grams of carbohydrate, mostly as sugars, including the fruit sugar fructose. Fruits vary greatly in their water and fiber contents, and therefore their sugar concentrations vary also. With the exception of avocado, which is high in fat, fruits contain insignificant amounts of fat and protein.

Milk, Cheese, and Yogurt

A serving (a cup) of milk or yogurt is a generous contributor of carbohydrate, donating about 12 grams. Among cheeses, cottage cheese provides about 6 grams of carbohydrate per cup, but most other types contain little if any carbohydrate. These foods also contribute high-quality protein (a point in their favor), as well as several important vitamins and minerals. All milk products vary in fat content, an important consideration in choosing among them; Chapter 5 provides the details.

Cream and butter, although dairy products, are not equivalent with milk because they contain little or no carbohydrate and insignificant amounts of the other nutrients important in milk. They are appropriately placed with the fats at the top of the Pyramid.

Meat, Poultry, Fish, Dry Beans, Eggs, and Nuts

With two exceptions, foods of this group provide almost no carbohydrate to the diet. The exceptions are nuts, which provide a little starch and fiber along with their abundant fat, and dry beans, revered by diet watchers as low-fat sources of both

*Gram values in this section are adapted from the 1995 exchange system.

starch and fiber. Just a half-cup serving of beans provides 15 grams of carbohydrate, an amount equaling the richest sources in the Food Guide Pyramid. Among providers of fiber, beans are peerless, totaling 8 grams in a half cup.

Fats, Oils, and Sweets

Fats are devoid of carbohydrate, of course, but sweets supply carbohydrate, so it is useful to account for them in the diet. Most people enjoy sweets and frequently include them in their diets, so it is useful, too, to learn something of the nature of these foods.

In the last decade, scientists have been arguing about what constitutes "sugar" and how to measure it in the diet. (Table 4-8 defines sugar terms.) Some experts wish to abandon the idea of measuring the sugars in foods. They accurately point out that a sugar molecule arising in an orange by way of photosynthesis is indistinguishable in laboratory tests from one added at the jam factory to sweeten orange marmalade. How can we measure the **added sugars,** they ask, when we cannot separate them from the **naturally occurring sugars** in foods? Besides, they say, the body handles all the sugars in the same ways, whatever the source.

Nutritionists argue back that chemical structures of sugars are not at issue; the addition of a concentrated energy source reduces the nutrient density of the foods, and this can affect the body adversely. Manufacturers measure sugar

added sugars sugars added to a food for any purpose, such as to add sweetness or bulk or to aid in browning (baked goods).

naturally occurring sugars sugars that are not added to a food but are present as its original constituents, such as the sugars of fruit or milk.

carbohydrate sweeteners ingredients composed of carbohydrates that contain sugars used for sweetening food products, including glucose, fructose, corn syrup, concentrated grape juice, and other sweet carbohydrates.

Table 4-8

Terms That Describe Sugar

Note: The term *sugars* here refers to all of the monosaccharides and disaccharides. On a label's ingredient list, the term *sugar* means sucrose. See Controversy 4 for terms concerning *artificial sweeteners* and *sugar alcohols.*

- **brown sugar** white sugar with molasses added, 95% pure sucrose.
- **concentrated fruit juice sweetener** a concentrated sugar syrup made from dehydrated, deflavored fruit juice, commonly grape juice; used to sweeten products that can then claim to be "all fruit."
- **confectioner's sugar** finely powdered sucrose, 99.9% pure.
- **corn sweeteners** corn syrup and sugar solutions derived from corn.
- **corn syrup** a syrup, mostly glucose, partly maltose, produced by the action of enzymes on cornstarch. *High-fructose corn syrup (HFCS)* is mostly fructose; glucose (dextrose) and maltose make up the balance.
- **dextrose** an older name for glucose.
- **fructose, galactose, glucose** the monosaccharides.
- **granulated sugar** common table sugar, crystalline sucrose, 99.9% pure.
- **honey** a concentrated solution primarily composed of glucose and fructose produced by enzymatic digestion of the sucrose in nectar by bees.
- **invert sugar** a mixture of glucose and fructose formed by the splitting of sucrose in an industrial process. Sold only in liquid form and sweeter than sucrose, invert sugar forms during certain cooking procedures and works to prevent crystallization of sucrose in soft candies and sweets.
- **lactose, maltose, sucrose** the disaccharides.
- **levulose** an older name for fructose.
- **maple sugar** a concentrated solution of sucrose derived from the sap of the sugar maple tree, mostly sucrose. This sugar was once common but is now usually replaced by sucrose and artificial maple flavoring.
- **molasses** a syrup left over from the refining of sucrose from sugar cane; a thick, brown syrup. The major nutrient in molasses is iron, a contaminant from the machinery used in processing it.
- **raw sugar** the first crop of crystals harvested during sugar processing. Raw sugar cannot be sold in the United States because it contains too much filth (dirt, insect fragments, and the like). Sugar sold as "raw sugar" domestically is not actually raw but has gone through more than half of the refining steps.
- **turbinado** (ter-bih-NOD-oh) **sugar** raw sugar from which the filth has been washed; legal to sell in the United States.
- **white sugar** pure sucrose, produced by dissolving, concentrating, and recrystallizing raw sugar.

Figure 4-11

Sugar in Processed Foods

½ c canned corn = 3 tsp sugar[a]
12 oz cola = 8 tsp sugar
1 tbs ketchup = 1 tsp sugar
1 tbs creamer = 2 tsp sugar
8 oz sweetened yogurt = 7 tsp sugar
2 oz chocolate = 8 tsp sugar

[a]Values based on 1 tsp = 4 g.

Whole
Kernel
CORN
Non-dairy
Creamer
Tomato
Ketchup
Sweetened
Yogurt
COLA

Sugars on the Nutrition Facts panel of a food label reflect both added and naturally occurring sugars in foods. Sugars listed among the ingredients are all added. Products listing sugars among the first few ingredients contain substantial amounts per serving.

Sugar alcohols, discussed in this chapter's Controversy, help protect against tooth decay.

when they add it to food products, so laboratory tests are not needed to determine the amounts in foods.

Both sides of the argument have validity, and as the spat continues, some useful distinctions between sugars are emerging. A new term, **carbohydrate sweeteners,** shifts the focus from separating sugars by their chemical structures to including sweet sugars from all sources. It is clear that the carbohydrate sweeteners from beets, corn, grapes, honey, and sugarcane are alike. All arise naturally and, through processing, are purified of most or all of the original plant material—bees process honey, and machines process the other types. Nutrition authorities discourage the liberal use of added sugars, but we consume dramatically more added sugar than people did a century ago when most forms of purified sugar were unknown. Today, our intakes of added sugars exceed recommendations, and nutrition authorities recommend that we use them more sparingly (see Controversy 4).

The Nature of Sugar

Each teaspoonful of any sweet can be assumed to supply about 20 calories and 4 grams of carbohydrate. You may not think of candy or molasses in terms of *teaspoons,* but we've done this to emphasize that all sugary items are like white sugar, in spite of many people's belief that some are different or "better." (Look again at Table 4-8.) For a person who uses ketchup liberally, it may help to remember that a tablespoon of it contains a teaspoon of sugar. And for the soft drink user, a 12-ounce can of sugar-sweetened cola contains about 8 or more teaspoons of sugar. Figure 4-11 shows that processed foods contain surprisingly large amounts of sugar.

What about the nutritional value of a product such as molasses or honey compared to white sugar? Molasses contains more than 3 milligrams of iron per tablespoon, so if used frequently it can contribute some of this important nutrient. Molasses is less sweet than the other sweeteners, however, so more molasses is needed to provide the same sweetness as sugar. Also, the iron comes from the iron machinery in which the molasses is made and is in the form of an iron salt not easily absorbed by the body. And honey is no better for health than other sugars by virtue of being "natural." As a matter of fact, honey is chemically almost indistinguishable from sucrose. Honey contains the two monosaccharides glucose and fructose in approximately equal amounts. Sucrose contains the same monosaccharides but joined together in the disaccharide form. Spoon for spoon, however, sugar contains fewer calories than honey because the dry crystals of sugar take up more space than the sugars of honey dissolved in its water. No form of sugar is "more healthy" than white sugar, as Table 4-9 shows.

It would be absurd to rely on any sugar for nutrient contributions. A tablespoon of honey (64 calories) does offer 0.1 milligram of iron, but it would take 180 tablespoons of honey—11,500 calories—to provide 100% of the Daily Value of 18 milligrams of iron. The nutrients of honey just don't add up as fast as its calories. Thus if you choose molasses, brown sugar, or honey, choose them not for their nutrient contributions but for the pleasure they give. These tricks can help magnify the sweetness of foods without boosting their calories:

- Serve sweet food warm (heat enhances sweet tastes).
- Add sweet spices such as cinnamon, nutmeg, allspice, or clove.
- Add a tiny pinch of salt; it will make food taste sweeter.
- Try reducing the sugar added to recipes by one-third.
- Select fresh fruits or fruit juice, or those prepared without added sugar.
- Use small amounts of sugar substitutes in place of sucrose.
- Read food labels for clues on sugar content.

Finally, enjoy whatever sugar you do eat. Sweetness is one of life's great sensations, and you need not forgo it completely. The person who cares about nutrition and loves sweets can artfully combine the two by using moderate amounts of sugar with creative imagination to enhance the flavors of nutritious foods.

Table 4-9

The Empty Calories of Sugar

At first glance, honey, jelly, and brown sugar look more nutritious than plain sugar, but when compared with a person's nutrient needs, none contributes anything to speak of. The cola beverage is clearly an empty-calorie item, too.

Food	Energy (cal)	Protein (g)	Fiber (g)	Calcium (mg)	Iron (mg)	Magnesium (mg)	Potassium (mg)	Zinc (mg)	Vitamin A (RE)	Thiamin (mg)	Riboflavin (mg)	Niacin (mg)	Vitamin B_6 (mg)	Folate (µg)	Vitamin C (mg)
Sugar (1 tbs)	46	0	0	0	0.0	0	0	0.0	0	0	0	0.0	0	0	0
Honey (1 tbs)	64	0	0	1	0.1	0	11	0.0	0	0	0	0.0	0	<1	0
Molasses (1 tbs)	48	0	0	34	3.6	50	188	0.1	0	0	0	0.4	1.3	0	0
Jelly (1 tbs)	49	0	0	1	0.0	1	12	0.0	0	0	0	0.0	0	0	<1
Brown sugar (1 tbs)	34	0	0	8	0.2	3	31	0.0	0	0	0	0.0	0	0	0
Cola beverage (12 fl oz)	153	0	0	11	0.1	4	4	0	0	0	0	0.0	0	0	0
Daily Values	2,000	56	25	1,000	18.0	400	3,500	15.0	1,000	1.5	1.7	20.0	2.0	400	60

do it!

INVESTIGATE YOUR FIBER INTAKE

This activity guides you in estimating the grams of the fiber in your diet and helps you to recognize the fiber in the foods of the Food Guide Pyramid.

Preparing Your Food and Fiber Record

Step 1. List on a copy of Form 4-1 all the foods you ate and beverages you drank in one day. Take care to record portion sizes accurately. It makes a big difference whether you ate a half-cup or a quarter-cup of beans, berries, or any other foods. If you need guidance in doing this, reread "Preparing Your Food Report" in the Do It! section of Chapter 2 (p. 55).

Step 2. List the fiber grams in each of the foods. Fiber values for many common foods are listed in Figure 4-12 (p. 122); look there first for convenience. The fiber values of other foods can be found in Appendix A.

Step 3. Add the numbers in the vertical columns to obtain subtotals of fiber; then add the subtotals across to obtain your grand total fiber intake for the day. Find the percentage of the Daily Value for fiber that your diet provided. Keep in mind that a person eating more than 2,000 calories a day needs proportionally more fiber.

Analysis

Answer the following questions:

1. Did your fiber total meet 100 percent of the Daily Value for a 2,000-calorie diet (25 grams)? Did it approach or exceed the maximum value of 40 grams?
2. Do you think your intake was too low, just right, or too high? Why do you think so?
3. Did your diet meet the minimum number of servings of foods from each fiber-containing group? One of the reasons the Food Guide Pyramid recommends a minimum of 3 servings of vegetables, 2 of fruit, and 6 of grains is to meet fiber needs.
4. Which fiber-containing groups fell short of the recommended intake?
5. Using the foods you listed on Form 4-1, analyze the fiber in this day's meals using the quick method described in Table 4-3 on p. 102. How do the results of the two methods of estimating fiber compare? Which method would you prefer to use, and why?
6. Which specific foods provided the most fiber to the day's meals? Which provided the least? Identify trends in your food choices that would affect your fiber intakes. For example, if you consistently choose whole grains, your fiber intake benefits.
7. What alterations might you make among your vegetable, fruit, meat and alternates, or grain choices to increase the fiber in your meals? (Remember to increase your fluid intake when you increase fiber.)
8. Looking at the foods listed in Figure 4-12, how do juices differ from whole foods with regard to fiber? Turn to Appendix A and look up some of the energy values for whole vegetables and fruits and their juices. If a person consistently chooses juices over whole foods, how does this choice affect the diet's fiber and calorie values?
9. Among the juices listed in Figure 4-12, is any relatively rich in fiber? What type of fiber do you think juices might contain? (Hint: The main constituent of juice is water.)
10. What contributions do meats or milk products make to the day's fiber total? What advice about fiber would you give to someone who emphasizes meat and milk products at each meal? (Hint: A quick way to direct someone to the fiber in foods is to guide them to the *bottom half* of the Food Guide Pyramid, which accounts for 11 of the 15 food servings recommended for a day.)
11. Did your meals include fiber-rich bean dishes, such as chili, bean burritos, beans in a salad, or split pea soup? Anyone interested in obtaining fiber should find ways to eat some legumes each day.

Constructed mostly of refined foods and meats and lacking in plant-derived foods, the average U.S. diet does a disservice to the health of the eater. Does this mean that you should never consume low-fiber foods such as white rolls, white rice, meats, potato chips, or apple juice? No, but such foods should be included in moderation among abundant fruit, vegetables, legumes, and whole grains. Constructed from a variety of whole foods, the diet can easily meet the recommended 25 or more grams of dietary fiber daily.

Form 4-1

Food and Fiber Record

Instructions: In the left-hand column, list the foods and amounts that you ate. Then list the grams of fiber in the food in the appropriate column to the right. For example, for a piece of wheat toast eaten at breakfast, list 2 grams of fiber in the Breads, Cereals, Rice, and Pasta column. For a glass of milk, list 0 grams in the Milk, Yogurt, and Cheese column.

FOOD/AMOUNT	Breads, Cereals, Rice, and Pasta	Vegetables	Fruit	Milk, Yogurt, and Cheese	Meat, Poultry, Fish, Dry Beans, Eggs, and Nuts
Breakfast:					
______	______	______	______	______	______
______	______	______	______	______	______
______	______	______	______	______	______
______	______	______	______	______	______
______	______	______	______	______	______
______	______	______	______	______	______
Snack:					
______	______	______	______	______	______
______	______	______	______	______	______
______	______	______	______	______	______
______	______	______	______	______	______
______	______	______	______	______	______
______	______	______	______	______	______
Lunch:					
______	______	______	______	______	______
______	______	______	______	______	______
______	______	______	______	______	______
______	______	______	______	______	______
______	______	______	______	______	______
______	______	______	______	______	______
Supper:					
______	______	______	______	______	______
______	______	______	______	______	______
______	______	______	______	______	______
______	______	______	______	______	______
______	______	______	______	______	______
______	______	______	______	______	______
Snack:					
______	______	______	______	______	______
______	______	______	______	______	______
______	______	______	______	______	______
______	______	______	______	______	______
______	______	______	______	______	______
______	______	______	______	______	______
Your subtotals:					

Daily Value (2,000-calorie diet): 25 g

Your Grand Total: ______ g Your Percent of Daily Value ______ % Calculation: (Your intake ÷ Daily Value) × 100 = % Daily Value

Figure 4-12
Finding the Fiber in Foods

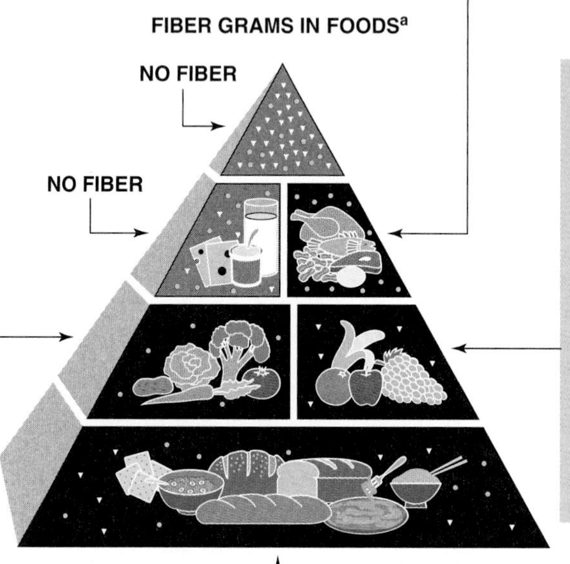

FIBER GRAMS IN FOODS[a]
NO FIBER
NO FIBER

Meat, Poultry, Fish, Dry Beans, Eggs, and Nuts Group

Food	Fiber g
Dried beans, 1/2 c	8
Lentils or peas, 1/2 c	5
Nuts, 1/4 c	2
Peanut butter, 2 tbs	2

Vegetable Group

Food	Fiber g
Baked potato with skin, 1	5
Brussels sprouts, 1/2 c	3
Carrot juice, 3/4 c	2
Broccoli, 1/2 c	2
Asparagus, 1/2 c	2
Corn, 1/2 c	2
Celery, 1/2 c	2
Green beans, 1/2 c	2
Spinach, 1/2 c	2
Baked potato, no skin, 1	2
Cauliflower, 1/2 c	2
Carrots, 1/2 c	2
Cabbage, 1/2 c	2
Onions, 1/2 c	1
Tomato, raw, 1 medium	1
Eggplant, 1/2 c	1
Lettuce, raw, 1 c	1
Bell peppers, 1/2 c	1
Dill pickle, 1 whole	1
Tomato juice, canned, 3/4 c	1

Fruit Group

Food	Fiber g
Prunes, cooked, 1/4 c	4
Pear, raw, 1 medium	4
Blackberries/raspberries, raw, 1/2 c	4
Apple/orange, raw, 1 medium	3
Apricots, raw, 3 each	3
Banana, raw, 1	2
Other berries, raw, 1/2 c	2
Peach, raw, 1 medium	2
Fruit cocktail, canned, 1/2 c	1
Raisins, dry, 1/4 c	1
Cantaloupe, raw, 1/2 c	1
Cherries, raw, 1/2 c	1
Apple juice, 3/4 c	<1
Orange juice, 3/4 c	<1

Breads, Cereals, Rice, and Pasta Group

Food	Fiber g
100% bran cereal, 1 oz	8
Barley, 1/2 c	7
Muffin, bran, 1	4
Wheat flakes, 1 oz	3
Shredded wheat, 1 large biscuit	2
Oatmeal, 1/2 c	2
Puffed wheat, 1 1/2 c	2
Whole-wheat bread, 1 slice	2
Light rye bread, 1 slice	2
Pumpernickel bread, 1 slice	2
Popcorn, 2 c	2
Brown rice, 1/2 c	2
Cheerios, 1 oz	2
Corn flakes, 1 oz	1
Pasta,[b] 1/2 c	1
Muffin, blueberry, 1	1
White rice, 1/2 c	<1
White bread, 1 slice	<1

[a]All values are for ready-to-eat or cooked foods, unless otherwise noted. Fruit values include edible skins. All values are rounded values.
[b]Pasta includes spaghetti noodles, lasagna noodles, and other noodles made from enriched white flour. Whole-wheat pastas have significantly more fiber.

Self Check

Answers to these Self Check questions are in Appendix G.

1. The dietary monosaccharides include:
   a. sucrose, fructose, and glucose
   b. glucose, fructose, and galactose
   c. lactose, maltose, and glucose
   d. glycogen, starch, and fiber
2. The primary form of stored glucose in plants is:
   a. galactose
   b. glycogen
   c. cellulose
   d. starch
3. The polysaccharide that helps form the supporting structures of plants is:
   a. cellulose
   b. maltose
   c. glycogen
   d. sucrose
4. Digestible carbohydrates are absorbed as ________ through the small intestinal wall and are delivered to the liver where they are converted to ________ .
   a. disaccharides; sucrose
   b. glucose; glycogen
   c. monosaccharides; glucose
   d. galactose; cellulose
5. When the body uses fat for fuel without the help of carbohydrate, this results in the production of:
   a. ketone bodies
   b. glucose
   c. starch
   d. galactose
6. When blood glucose concentration rises, the pancreas secretes ________ , and when blood glucose level falls, the pancreas secretes ________ .
   a. glycogen; insulin
   b. insulin; glucagon
   c. glucagon; glycogen
   d. insulin; fructose
7. In the body, the liver and muscles store glucose as the polysaccharide glycogen. T F
8. The body converts excesses of glucose into glycogen or fat. T F
9. Type 2 diabetes is characterized by insulin resistance of the body's cells. T F
10. Type 1 diabetes is most often controlled by successful weight-loss and maintenance. T F
11. Around the world, most people are lactose intolerant. T F
12. By law, enriched white bread must equal whole-grain bread in nutrient content. T F
13. The fiber-rich portion of the wheat kernel is the bran layer. T F
14. Using artificial sweeteners has been proven to help people lose weight. (Read about this in the upcoming Controversy.) T F

NUTRITION ON THE NET

For further study of the topics of this chapter, access these websites and search for the phrases or words in quotation marks:

1. For the latest changes in chapter material, to find supplemental learning tools, or to access links to related websites, go to the "Nutrition: Concepts and Controversies" site at:
   www.wadsworth.com/nutrition/prod/allprod.html
2. Search the International Food Information Council site for information on "fiber":
   ificinfo.health.org
3. Visit the U.S. Government site for "lactose intolerance":
   www.healthfinder.gov/searchoptions/topicsaz.htm
4. To learn more about "diabetes" and related disorders:
   www.diabetes.org
5. For more on "sugars":
   ificinfo.health.org
6. The main website for the Food, Nutrition, and Consumer Services can be located at:
   www.fns.usda.gov/fncs/
7. Search among thousands of current scientific and medical abstracts for any topic related to carbohydrates at:
   http://www.ncbi.nlm.nih.gov/PubMed/

CONTROVERSY 4

SUGAR AND ALTERNATIVE SWEETENERS: ARE THEY "BAD" FOR YOU?

ALMOST EVERYONE finds sweet tastes pleasing—after all, the preference for sweets is inborn. Children, especially, love sweets, and for them the sweeter food tastes, the better.[1] In adults, the preference for sweets is somewhat diminished, but if their consumption of sugars and sweeteners is any indication, sweets still hold enormous appeal.

Imagine pouring almost three quarters of a measuring cup of sugar onto your foods and into your beverages before consuming them today.*[2] This represents the average U.S. per capita consumption of 32 teaspoons a day of added sugars, enough to provide every man, woman, and child with over 100 pounds per year, and an all time high by some ways of reckoning.[3] A steady upward trend (see Figure C4-1) toward increasing sugar consumption is largely the result of a dramatic increase in consumption of commercially prepared foods and beverages.[4] Food manufacturers are adding the sugars before people purchase the foods. In contrast, people are adding less sugar from the sugar bowl to foods at home.

*The sugar referred to here includes cane sugar, high-fructose corn syrup, and all other added sugars.

EQUAL
NUTRASWEET
SWEET'N LOW

Higher figures for sugar consumption are often reported in other sources (including older editions of this text). Up to now, sugar data was available only as "sugars sold," but not all of the sugar (or, more correctly, carbohydrate sweeteners) available on the market is consumed by people. Some is purchased but discarded: the sugary syrup of canned fruit, the brine of sweet pickles, or jam that spoils and is thrown away; some is used in fermentation; still more sugar is used up in pet and livestock feed. Today's reporting of sugars consumption takes these and other uses into account, and reflects only the sugars actually eaten by consumers. By any method of accounting, sugars intakes have increased substantially and appear to be rising higher.

In addition to consuming more sugar, people are also consuming more artificial sweeteners, such as aspartame and saccharine.[5] Instead of substituting artificial sweeteners for sugar, however, people seem to be choosing more sweet foods and beverages in all their forms.

Does all this sugar harm people's health? And if it does, are sugar substitutes a better choice? This Controversy addresses these questions and, in the process, demonstrates how nutrition researchers pursue their answers, step by step, via scientific inquiry.

EVIDENCE CONCERNING SUGAR

Sugar is accused by some of causing nutrition problems. It is said to (1) promote and maintain obesity, (2) cause and aggravate diabetes, (3) increase the risk of heart disease, (4) disrupt behavior in children and adults, and (5) cause dental decay and gum disease. Is sugar guilty or innocent of these charges?

Does Sugar Cause Obesity?

The evidence suggests that sugar is unlikely to cause obesity. For example, rats fed a sucrose-rich diet do not become fatter, but their fat distribution changes. Rats on the high-sucrose diet seem to develop more belly fat than do rats fed a diet of regular rat chow. This change might be significant if it held true for people because central obesity is associated with human heart disease. So far, evidence for the effect in people is lacking.

More direct evidence on obesity in people comes from population studies. In many developing countries, incidence of obesity increases as sugar consumption rises. But this evidence does not all point to sugar as the sole cause. In general, when sugar intake increases, fat and total calorie intakes also rise. Simultaneously, physical activity declines. Furthermore, obesity also occurs where sugar intakes are low, and obese people in many instances eat less sugar than thin people do.[6] Also, obese people report preferring sweet foods no more often than normal-weight people.[7] Fat is more calorie dense than sugar and often occurs together with sugar in sweet treats and snacks. Studies of populations by themselves cannot separate the effects of eating sugar from those of eating too much fat or of exercising too little.

Concentrated sweets do make it easy for people to consume large amounts of calories quickly, however, and that is why most diet plans recommend avoiding such foods. Some people believe that eating even small amounts of sugar triggers eating binges; for them, conscientious sugar avoidance is an important part of weight management. For others, the inclusion of small amounts of sugar in a weight-loss plan makes the plan easier to follow. In short, the effects of sugar on a person's eating style and body weight depend on the user.

Does Sugar Cause or Contribute to Type 2 Diabetes?

Recall from the chapter that in diabetes, insulin secretion or

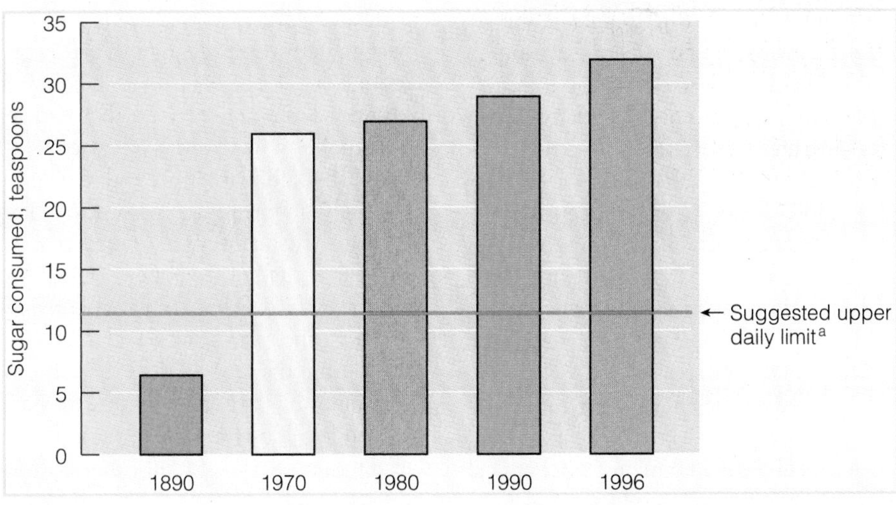
Sugar consumed, teaspoons
35
30
25
20
15
10
5
0
1890
1970
1980
1990
1996
Suggested upper daily limit[a]

Figure C4-1

Sugars Consumption, United States 1980–1996

[a]Recommended upper limit for a 2,200-calorie diet.
SOURCE: U.S. Department of Agriculture, Economic Research Sevice, *A Dietary Assessment of the U.S. Food Supply* (AER #772) Washington, D.C.: Government Printing Office 1998, p. 25.

tissue responsiveness to it becomes abnormal. This, of course, affects the body's ability to manage sugar. At one time people thought that eating sugar caused diabetes by "overstraining the pancreas," but now we know that this is not the case. Body fatness is more closely related to diabetes than diet is. High rates of diabetes have not been reported in any society where obesity is rare. Still, evidence on whether people with the genetic tendency to develop type 2 diabetes should avoid eating sugar is conflicting and interesting.

In populations around the world, a profound increase, by as much as ten-fold, in the incidence of diabetes has occurred simultaneously with an increase in sugar consumption. This has been true for the Japanese, Israelis, Africans, Native Americans, Eskimos, Polynesians, and Micronesians. Yet, in some other populations, no relationship has been found between sugar intake and diabetes. Wherever starch and the fiber that accompanies it, rather than sugar, are the major carbohydrates in the diet, diabetes is rare. But this does not prove that sugar causes diabetes or that starch prevents it. The apparent protective effect of starch might be due, for example, to the chromium or fiber that comes with it.

Supporting evidence comes from researchers who tracked the dietary habits of over 65,000 women and almost 43,000 men for six years.[8] The researchers were trying to discover whether diabetes development is related to a diet that contains many servings of foods with a high glycemic effect. As Chapter 4 explained, some carbohydrate-rich foods are so easily digested and rapidly absorbed that they evoke an especially large insulin response. Examples include highly refined grains, such as white bread, English muffins, and white rice; sources of concentrated sugars such as regular colas and jams; mashed white potatoes; and some highly refined cold breakfast cereals.

The researchers discovered that the women who most often chose foods with a high glycemic effect developed diabetes more often than those who consistently ate foods with a low glycemic effect, such as fiber-rich whole-grain breads and cereals, legumes, and fruits and vegetables. The fairest conclusion that can be drawn from studies so far is not that sugar alone is culpable in type 2 diabetes causation, but that it may contribute both to obesity, a major causal factor, and to a total diet profile that may increase diabetes risk.

Can a person who already has diabetes use moderate amounts of sugar? Most authorities agree that, an amount of sucrose equaling 5 to 10 percent of total calories consumed is acceptable as part of the carbohydrate in a controlled diet. *Other* body responses must also be considered, however, among them raised blood lipids, which suggest a high risk of heart disease (discussed next).

Does Eating Sugar Increase the Risk of Heart Disease?

Again, a research study using rats provides some clues to the relationship between sugar and heart disease. When researchers fed rats a diet with sucrose as the only carbohydrate source, the rats sustained microscopic damage to their arteries, and their blood tested high for both saturated fat and cholesterol.[9] Rats fed starch instead of sugar did not develop the damage or the elevated blood lipids. When researchers tried a similar study on human beings, they confirmed that people, too, respond to diets with extremely high sugar contents by releasing saturated fat, probably made from the sugar, into the blood; when fed diets high in starch instead of sugar, the same subjects dramatically reduced their synthesis and release of saturated fats.[10]

Keep in mind that both the rats and the people in the studies just described consumed sucrose amounting to 75 percent or more of their daily calories; even people with a highly unusual craving for sweets wouldn't choose to live on such a diet. Other studies among human beings also show that blood lipids associated with heart disease rise in response to diets high in sucrose and fructose, and those believed to be protective fall.[11] The diets evoking this response contain almost twice the nation's average intake of added sugars, though; current intake levels are thought to be below the amount that may adversely affect blood lipids.[12] Still, the American Heart Association's *Dietary Guidelines* suggest choosing a diet moderate in sugar because high-sugar diets are often high in calories and low in fiber and nutrients.[13]

Saturated fat is clearly the major *dietary* culprit in the heart disease susceptibility of most people, but there is a *hereditary* culprit, too, and some people may have inherited the tendency to develop raised blood lipid levels in response to carbohydrate and alcohol. If their heart disease risk is assessed as high, they are told to restrict their intakes of carbohydrate and alcohol. Throughout many years of research, no one has shown conclusively that moderate amounts of sugar (10 percent of total calories) affect the disease process in healthy human beings.

What about Sugar and Behavior? Twenty years ago, claims appeared that eating sugary foods caused children to become unruly and adolescents and adults to exhibit antisocial and even criminal behavior. Research since then has yielded only mixed or negative results. Though most experts agree that the "sugar-behavior" theory has been put to rest, many teachers, parents, grandparents, and others still believe that the children they know react behaviorally to sugar.

Meanwhile, research continues. Sugar might influence behavior in many ways: by altering the levels of chemicals in the brain that affect mood, by inducing nutrient deficiencies, by the series of hormones the body releases after consuming sugar, and others. It is known that blood glucose is regulated not by diet but by hormones, and one of those is the stress hormone, norepinephrine.

To test for any relationship between sugar intake and norepinephrine, researchers fed a syrupy beverage to 9 adults and 14 children and then tested their blood norepinephrine levels three hours later, after insulin had had time to store the sugar. The blood *sugar* levels of both groups had dropped only slightly, but the blood *norepinephrine* was elevated. In the children, it had shot up to double the level seen in the adults. The children also complained of symptoms such as weakness and nervousness during the test period. Though it is tempting to declare that sugar elevates blood norepinephrine levels in children and that this effect leads to behavior changes, many more studies are needed before the theory can be confirmed as fact.

Another way sugar has been theorized to affect behavior is by providing energy (the so-called Halloween effect). Studies do not suggest, however, that sugar, specifically, has a negative effect on behavior. Indeed, several well-controlled studies have shown that sugar calms normal children, a finding consistent with convincing biochemical evidence. One such study showed no differences in activity, social interactions, learning performance, or mood in children given artificial sweeteners, but found that sugar made the children less active. In other studies, sugar has calmed juvenile delinquents with pronounced behavioral problems.

Overwhelmingly, studies have failed to demonstrate any consistent effects of sucrose on behavior in either normal or hyperactive children.[14] In conclusion, occasional behavioral reactions to sugar may be possible, but until research proves otherwise, the idea that sugar alone directly affects behavior adversely in most healthy children or adults has been all but ruled out.

Does Sugar Cause Dental Caries? **Dental caries** are a serious public health problem. They afflict the majority of people in the country, half by the age of two. (A very lucky few *never* get caries because they have an inherited resistance; others have had sealants applied to teeth that stop caries before they can begin.) One of the most successful measures taken to reduce the incidence of dental decay is fluoridation of community water. But sugar has something to do with dental caries, too.

Caries develop as acids produced by bacterial growth in the mouth eat into tooth enamel. Bacteria establish colonies known as **plaque** whenever they can get a foothold on tooth surfaces. Once established, they multiply and affix themselves more and more firmly unless they are brushed, flossed, or scraped away. Eventually, the acid of plaque creates pits that deepen into cavities. Below the gum line, plaque works its way down until the acid erodes the roots of teeth and the jawbone in which they are embedded, loosening the teeth and leading to infections of the gums. Gum disease severe enough to threaten tooth loss afflicts the majority of our population by their later years. Flossing every 24 hours greatly reduces this risk. Table C4-1 defines some terms related to caries.

Bacteria thrive on carbohydrate. Carbohydrate as sugar has been named as the main causative factor in the formation of caries. However, starch also supports bacterial growth if the bacteria are allowed sufficient time to work on it. Of prime importance is the length of time the food stays in the mouth, and this depends on the food's composition, how sticky it is, how often you eat it, and especially on whether you brush your teeth afterward.[15]

Bacteria produce acid for 20 to 30 minutes after exposure to sugar. Thus, when you eat three pieces of candy, one right after the other, your teeth are exposed to approximately 30 minutes of acid demineralization. When you eat the candy pieces at half-hour intervals, the acid exposure time is 90 minutes. Likewise, slowly sipping a sugary soft drink may be more harmful than drinking quickly and emptying the mouth of sugar.

Some forms of candy, such as milk chocolate and caramels, may be less

Table C4-1

Dental Terms

- **dental caries** decay of the teeth (*caries* means "rottenness").
- **plaque** (PLACK) a mass of microorganisms and their deposits on the crowns and roots of the teeth, a forerunner of dental caries and gum disease. (The term *plaque* is also used in another connection—arterial plaque in atherosclerosis. See Chapter 11.)

harmful than once believed because the sugar dissolves completely and is washed away in saliva. Breads, granola bars, sugary cereals, oatmeal cookies, raisins, salted crackers, and chips, however, may be worse than once thought because particles get stuck in the teeth and do not dissolve. These particles may remain in contact with tooth surfaces for hours, providing a feast for bacteria and greatly increasing the likelihood of caries. A table in Chapter 13 lists foods of both high and low caries potential. The punchline seems to be: Brush your teeth after eating.

Total sugar intakes still play a major role in caries incidence, though, and populations whose diets provide more than 10 percent of calories from sugar have an unacceptably high incidence of dental caries. Worldwide, many governing agencies urge their citizens to consume no more than 10 percent of calories from sugar because of sugar's link with dental caries. It is clear that sugar is an energy source for the bacteria that cause tooth decay and that when exposure is sufficient in susceptible people, sugar is guilty as charged.

PERSONAL STRATEGY FOR USING SUGAR

Some of sugar's health effects still await scientific clarification, but other concerns about sugar have been all but resolved in the minds of scientists.[16] Meanwhile, consumers wonder whether they should avoid sugar or reduce their intakes. The *Dietary Guidelines* suggest only that people "use moderation" concerning sugar, not that they avoid it altogether. The World Health Organization recommends that sugar contribute no more than 10 percent of a person's total calorie intake. A person who eats 2,000 calories of energy a day, then, is allowed 200 calories from sugar. Those 200 calories of sugar, 13 teaspoons or so, sound like quite a lot. But when you add up all the sugar teaspoons present in common foods, 200 calories may seem restrictive.

People who wish to avoid sugar may choose from two sets of alternative sweeteners: the sugar alcohols, which are energy-yielding sweeteners sometimes referred to as nutritive sweeteners, and the artificial sweeteners, which provide virtually no energy and are thus sometimes referred to as nonnutritive sweeteners. These options do provide sweetness without sucrose, but people are curious about their safety.

EVIDENCE CONCERNING SUGAR ALCOHOLS

The sugar alcohols are familiar to people who use special dietary products. Many new low-calorie food products appearing on grocery shelves depend on sugar alcohols for their bulking and sweetening powers.[17] Among the sugar alcohols frequently encountered in the marketplace are **mannitol, sorbitol, isomalt,** and **xylitol.**[18] These and other sugar alcohols can be metabolized by human beings, and generally speaking, most provide about as much energy as sucrose or just a little less (2 to 4 calories per gram).

A proven benefit of sugar alcohols is that ordinary mouth bacteria metabolize them less rapidly than other carbohydrates. As a result, sugar alcohols do not contribute to dental caries, and foods that contain them can say so on their labels.

Mannitol is the least satisfactory of the sugar alcohols just named. It is less sweet than sucrose, so large amounts have to be used to obtain the same sweetness (see Table C4-2). It lingers unabsorbed in the intestine for a long time, available to intestinal bacteria for their energy. As they consume the mannitol, the bacteria multiply, attract water, produce irritating waste, and cause diarrhea.

More commonly used than mannitol, sorbitol sweetens sugar-free gums and candies, but it, too, has drawbacks. At least two teaspoons as much sorbitol (with twice the calories) must be used to deliver the sweetness of one teaspoon of sucrose. Also, like mannitol, it reliably causes diarrhea when consumed in large amounts. One woman, after seven years of unsuccessful treatments for her severe chronic diarrhea, finally received a correct diagnosis: her daily chewing gum, sweetened with sorbitol, was causing her problem.[19] When she cut back on the gum, her diarrhea vanished.

A relative newcomer to the sugar alcohol list is isomalt. Chemically,

Table C4-2

Sweetness of Sugar Substitutes

Sugar Substitute	Relative Sweetness[a]
Sugars	
Sucrose	1.0
Fructose	1.7
Sugar Alcohols	
Sorbitol	0.5
Mannitol	0.7
Isomalt	0.6
Xylitol	1.0
Noncaloric Sweeteners	
Acesulfame-K	200.0
Cyclamate	45.0
Aspartame	200.0
Saccharin	300.0
Sucralose	600.0
Alitame	2,000.0

[a]The relative sweetness depends on the temperature, acidity, and other flavors of the foods in which the substance occurs. The sweetness of pure sucrose is the standard with which the approximate sweetness of sugar substitutes is compared.

isomalt combines molecules of the sugar alcohols mannitol and sorbitol with molecules of glucose, yielding a product that provides about half the sweetness and half the calories of sucrose. Because it is heat stable, isomalt can be used in cooked or baked products. Isomalt's weak sweetening power necessitates adding artificial sweeteners to products to make them taste acceptably sweet to consumers. Like other sugar alcohols, isomalt reduces the risks of tooth decay. Isomalt may be less likely than the others to cause diarrhea because its larger chemical structure attracts less water into the colon.[20]

Xylitol is popular, especially in chewing gums. It not only does not support caries-producing bacteria, but it may actually inhibit their production of acid and prevent them from adhering to the teeth. Xylitol occurs naturally in many fruits and also arises in the body during normal metabolic processes. Most people can tolerate small amounts, and it has been suggested as a useful sugar replacer in diets for treating diabetes.[21] In large amounts, xylitol slows down the emptying of the stomach, but also stimulates release of a hormone (motilin) that speeds up intestinal activity and so causes diarrhea.

The person who wishes to reduce energy intake should be aware that the sugar alcohols *do* provide energy. The body handles them differently from sugar, but unlike artificial sweeteners, they are not calorie-free.

EVIDENCE CONCERNING ARTIFICIAL SWEETENERS

Like the sugar alcohols, artificial sweeteners make foods taste sweet without promoting dental decay. Unlike sugar alcohols, they have the added attraction of being calorie-free. Also unlike sugar alcohols, the human taste buds perceive the artificial sweeteners as supersweet. But are they safe?

All substances are toxic if high enough doses are consumed. Artificial sweeteners, their components, and metabolic by-products are not exceptions. The questions to ask are whether artificial sweeteners are harmful to human beings at the levels normally used, and how much is too much. The Food and Drug Administration (FDA) has proposed answers by setting **acceptable daily intake (ADI)** levels for some of the artificial sweeteners used in the United States. Table C4-3 defines some sugar substitute terms. The major synthetic sweeteners today are **saccharin, acesulfame-K, sucralose,** and **aspartame.**

Saccharin Saccharin has had a rocky history of acceptance, although it is now consumed by millions of Americans, primarily in prepared foods and beverages and secondarily as a tabletop sweetener. Questions about its safety surfaced in the late 1970s, when experiments suggested that it caused bladder tumors in rats. As a result, the FDA proposed banning it. The public outcry in favor of retaining it was so loud, however, that Congress placed a moratorium on any action, and the ban proposal was eventually withdrawn. At this writing, products containing saccharin still must carry the warning label familiar to all consumers of diet products: "Use of this product may be hazardous to your health. This product contains saccharin, which has been determined to cause cancer in laboratory animals."

Does saccharin cause cancer? Some worry that it might.[22] The evidence that it does so in animals is as follows. Rats that had been fed diets containing saccharin from the time of weaning to adulthood were mated. The offspring of those rats were then fed saccharin throughout their lives and were found to have a higher incidence of bladder tumors than comparable animals not fed saccharin. In Canada, on the basis of these findings, saccharin was banned except for use as a tabletop sweetener to be sold in pharmacies with a warning label.

In human beings, a large-scale population study involving 9,000 people seemed to show a slightly elevated risk of cancers in women who drank two or more saccharin-sweetened diet sodas a day and in men and women who both smoked heavily and used artificial sweeteners. Other studies involving more than 5,000 people showed no excess risk of bladder cancers.[23]

A solid clue from the laboratory is based on some physiological differences between the urinary systems of rats and human beings.[24] Rats excrete far less water in their urine than people do. As a result, rats can make highly concentrated solutions of substances in just small amounts of water in their urine. Dissolved substances in such high concentrations are likely to crystallize. In safety tests, saccharin overdoses caused

Table C4-3

Sugar Substitute Terms

- **acceptable daily intake (ADI)** the estimated amount of sweetener that can be consumed daily over a person's lifetime without any adverse effects.
- **acesulfame** (AY-sul-fame) **potassium,** also called **acesulfame-K** a zero-calorie sweetener approved by the FDA and Health Canada.
- **alitame** a noncaloric sweetener formed from the amino acids L-aspartic acid and L-alanine. In the United States, the FDA is considering its approval.
- **aspartame** a compound of phenylalanine and aspartic acid that tastes like the sugar sucrose but is much sweeter. It is used in both the United States and Canada.
- **cyclamate** a zero-calorie sweetener under consideration for use in the United States and used with restrictions in Canada.
- **isomalt, mannitol, sorbitol, xylitol** sugar alcohols that can be derived from fruits or commercially produced from dextrose; absorbed more slowly and metabolized differently than other sugars in the human body and not readily used by ordinary mouth bacteria.
- **saccharin** a zero-calorie sweetener used freely in the United States but restricted in Canada.
- **sucralose** a noncaloric sweetener derived from a chlorinated form of sugar that travels through the digestive tract unabsorbed. Recently approved by the FDA for use in the United States.

crystals to form in the rats' bladders, and the crystals probably caused the tumors. Human beings cannot concentrate urinary substances to such a degree, so they would never form saccharin crystals, even if they consumed larger-than-normal doses of saccharin. They would, however, lose large amounts of water as the kidneys struggled to free the blood of the overload.

It goes without saying that overloading on huge saccharin doses is probably not safe, but consuming moderate amounts almost certainly does not cause bladder cancer in human beings. Although no ADI has been set for saccharin, the amount of saccharin that can be commercially added to foods or drinks is limited to about 30 milligrams per serving. The FDA also suggests that adults not exceed total daily saccharin intakes of 1,000 milligrams and that children not exceed 500 milligrams.[25]

Acesulfame-K For 15 years of testing and use, the artificial sweetener acesulfame potassium (or acesulfame-K) has been used without reported health problems. An ADI of 15 milligrams per kilogram of body weight was set for acesulfame-K on its approval. Marketed under the trade names Sunette and Sweet One, this sweetener is about as sweet as aspartame and is used in chewing gum, beverages, instant coffee and tea, gelatins, and puddings, as well as for table use. Acesulfame-K holds up well during cooking.

Acesulfame-K is 200 times as sweet as sucrose but leaves a slight aftertaste. Blending it with other sweeteners solves the problem. Acesulfame-K is not recognized by the body's metabolic equipment and therefore is excreted unchanged by the kidneys.

Sucralose Recently approved for use as a sweetener in the United States, sucralose is the only artificial sweetener made from sucrose. Three chlorine atoms substitute for three hydrogen and oxygen groups on the structure of sucrose, making a product that provides 600 times the sweetness of sugar. Many years of testing have deemed sucralose safe to use and, specifically, not a cause of cancer. Sucralose is not recognized by the body as sugar and therefore passes through unchanged. The ADI for sucralose has been set at 5 milligrams per kilogram of body weight per day for all ages, including pregnant and lactating women. Sucralose is heat stable and so is useful for cooking and baking; it will soon be showing up in more commercially prepared products and as a tabletop sweetener. Time will tell whether consumers will embrace sucralose or pass it by.

Aspartame Aspartame is one of the most thoroughly studied substances ever to be approved for use in foods.[26] Manufacturers use aspartame under the name *NutraSweet* to sweeten foods that require sweetness but are not exposed to cooking temperatures, as aspartame is not heat stable.[27] A gram of aspartame provides 4 calories, as does a gram of protein, but because so little is needed, it is virtually calorie-free as used in foods and beverages. Under various brand names, such as *Equal* or *NutraSweet,* aspartame is also available as a powder to use at home in place of sugar. In powdered form it is mixed with lactose, so a 1-gram packet contains 4 calories.

The amazing popularity of aspartame is mostly due to its flavor, which is almost identical to that of sugar. Furthermore, aspartame is touted as safe for children, so families wishing to limit their children's sugar intakes are offering them NutraSweet products instead.

Aspartame is a simple chemical compound: two protein fragments (the amino acids phenylalanine and aspartic acid) joined together. In the digestive tract, the two fragments are split apart, absorbed, and metabolized just as they would be if they had come from protein in food. The flavors of the components give no clue to the combined effect; one of them tastes bitter, and the other is tasteless. Yet aspartame is 200 times sweeter than sucrose.

With its phenylalanine base, aspartame poses problems for people with an inherited metabolic disease known as phenylketonuria (PKU). People with PKU have the hereditary inability to dispose of phenylalanine eaten in excess of need. Unusual products made from phenylalanine build up and damage the tissues. PKU causes irreversible, progressive brain damage if left untreated in early life. Newborns in the United States are tested for PKU; if they have it, the treatment is to limit dietary intake of phenylalanine.

For a compelling reason, children with PKU should not get their phenylalanine from aspartame. Phenylalanine occurs in such protein-rich and nutrient-rich foods as milk and meat, and the PKU child is allowed only a limited amount of these foods. The child has difficulty obtaining the many essential nutrients, such as calcium, iron, and the B vitamins, found along with phenylalanine in these foods. To suggest that such a child squander any of the limited phenylalanine allowance on the purified phenylalanine of aspartame, with none of the associated nutrients to support normal growth, would be to invite nutritional disaster. Product labels carry special warnings for people with PKU.

Other concerns about aspartame's safety have had to do with compounds that arise briefly during its metabolism. These compounds (methyl alcohol, formaldehyde, and diketopiperazine) are not toxic at the levels generated from the ADI amount of aspartame, and concerns about them have been laid to rest.

A recent uproar about aspartame's safety began when a scientist wrote an article in a scientific journal noting a parallel between an increasing rate of brain tumors beginning in the 1980s and the approval of aspartame for public use in 1981.† The article put forth an observation, but presented no data to support a scientifically based relationship between aspartame and brain tumors; in fact, no documentation exists to show whether any of the brain cancer patients cited in the article had ever consumed aspartame.[28] Still, newspapers, magazines, and other

†The article appeared in the *Journal of Neuropathology and Experimental Neurology.*

media reported the story with gusto and "taught" the whole nation, wrongly, that a scientist had proved that aspartame causes brain cancer. Soon, stories circled the globe on the Internet accusing aspartame of causing everything from Alzheimer's disease and brain cancer to nerve disorders and skin warts. Meanwhile, a year after the release of the original published opinion, other researchers had finished their orderly scientific investigations into the theory and found no relationship between aspartame intake and brain tumors, behavior, mood, or brain chemistry.[29] They did, in fact, find a disturbing unexplained nationwide acceleration in the incidence of brain tumors, but the greatest increase was recorded several years before aspartame's approval. Of course, the media had lost interest by this time—the finding that aspartame is safe will not sell newspapers.

Another concern about aspartame use is headaches. Although no experimental evidence has shown a connection, the Centers for Disease Control (CDC) has received many thousands of individual complaints.

Every day, millions of people use aspartame. Every day, millions of people have headaches. Anyone who claims, on this basis, that aspartame causes headaches is using personal experience to jump to conclusions. Only casual reports of a link, and no true connections, have been shown. Some of the headache sufferers might indeed be reacting to the artificial sweetener, but they might also be reacting to another substance such as caffeine or to factors in their lives unrelated to foods. Obviously, people who believe aspartame gives them headaches should use a different sweetener.

On approving aspartame for U.S. consumers, the FDA set the ADI of 50 milligrams per kilogram of body weight in a day. In Canada, the acceptable level is set at 40 milligrams per kilogram. These seem to be reasonable numbers. Still, the ADI amount is not impossible to exceed. For a 132-pound person, it adds up to 80 packets of aspartame sweetener or 15 soft drinks sweetened only with aspartame. A child who drinks a quart of Kool-Aid on a hot day and who also has pudding, chewing gum, cereal, and other products sweetened with aspartame can pack in more than the daily ADI limit. Infants or toddlers under two years old should probably not be fed artificially sweetened foods and drinks.

Other Artificial Sweeteners

Two other artificial sweeteners are awaiting FDA approval—**cyclamate** and **alitame.** Cyclamate was once approved in the United States, but later was banned when it was suspected, but never proved, to cause cancer in rats. In Canada, cyclamate is restricted to use as a tabletop sweetener on the advice of a physician and as a sweetening additive in medicines. Alitame resembles aspartame in being composed of two amino acids, but unlike aspartame it remains stable when heated.

Do Artificial Sweeteners Help with Weight Control? Many people eat and drink products sweetened with artificial sweeteners in the belief that the products help control weight. Do they work? Ironically, studies of rats report that intense sweeteners, such as saccharin, stimulate appetite and lead to weight *gain* instead of loss. Many studies on *people,* however, find either no change or a decline in feelings of hunger. Researchers conclude that most people's food intakes do not change much when they use artificial sweeteners.

In studying the effects of artificial sweeteners on food intake and body weight, different researchers ask different questions and take different approaches in searching for the answers.[30] It matters, for example, whether the subjects of a study are of a healthy weight or obese and whether they are on weight-loss diets. Motivations for using sweeteners differ, too, and this influences a person's actions. For example, a person might drink a low-calorie beverage now so as to be able to eat a high-calorie food later. This person's energy intake might stay the same or increase. In contrast, a person trying to control food energy intake might use the artificial sweetener and then choose low-calorie foods consistently. This person might achieve a lower-than-normal energy intake in this way.

Researchers must also distinguish between the effects of the experience of tasting something sweet and the physiological effects of a particular substance on the body. If a person experiences hunger or feels full shortly after eating an artificially sweetened snack, is that because tasting something sweet stimulates or depresses the appetite? Or is it because the artificial sweetener itself somehow affects the appetite through nervous, hormonal, or other means? Furthermore, if appetite is stimulated, does that actually lead to increased food intake?

One recent study was designed to answer questions concerning appetite and artificial sweeteners. Researchers fed normal-weight people one of four breakfasts and then measured their food intakes at later meals throughout the day.[31] Two of the breakfasts provided 700 calories: one contained sucrose, and the other aspartame with enough starch to equalize the calories. The other two breakfasts provided 300 calories: one was plain, and the other contained aspartame. Subjects who ate either lower-calorie breakfast, regardless of sweetness, were hungrier later. Those who ate either higher-calorie breakfast stayed fuller longer. Sweet taste and the presence of aspartame seemed to have no effect on hunger or subsequent food energy intakes. At the end of the day, both groups who had eaten the 700-calorie breakfasts had higher total energy intakes because, although they were less hungry at lunch, they still consumed ample food energy at lunch and supper.

Overall it seems that artificial sweeteners alone do not stimulate or depress appetite. If the sweeteners have the effect of lowering calorie intakes, however, this may leave people hungry, so they compensate at later meals. Whether a person overcompensates or partially or fully compensates for the reduction in energy intake depends on several factors, including the person's characteristics. Using artificial sweeteners will not automatically lower energy intake; to control energy intake successfully, a person will need to make informed diet and activity decisions throughout the day (as Chapter 9 explains).

Another sweetener, however, has shown a clear, strong ability to cut the appetite and to reduce food intake. That sweetener is sugar.[32] The common belief that sugar "spoils the appetite" has proved true.

PERSONAL STRATEGIES FOR USING ARTIFICIAL SWEETENERS

Current evidence indicates that moderate intakes of artificial sweeteners pose no health risks.[33] For those who choose to include artificial sweeteners in their diets, moderation is the key. When used in the context of a nutritious diet, artificial sweeteners are generally safe for consumption by healthy people.[34] Though they are clearly not magic bullets in fighting overweight, they probably do not hinder weight-loss efforts either, and they are safer for the teeth than carbohydrate sweeteners.

5 The Lipids: Fats, Oils, Phospholipids, and Sterols

Tall Sandwich KERRY Jean Damianakes

Kerry Damianakes 1996, *Tall Sandwich*

Contents

Frequently Asked Questions

lipid (LIP-id) a family of compounds soluble in organic solvents but not in water. Lipids include triglycerides (fats and oils), phospholipids, and sterols.

cholesterol (koh-LESS-ter-all) a member of the group of lipids known as sterols; a soft, waxy substance made in the body for a variety of purposes and also found in animal-derived foods.

fats lipids that are solid at room temperature (70° F or 25° C).

oils lipids that are liquid at room temperature (70° F or 25° C).

cardiovascular disease (CVD) disease of the heart and blood vessels; also called *coronary heart disease (CHD)*. The two most common forms of CVD are atherosclerosis and hypertension.

triglycerides (try-GLISS-er-ides) one of the three main classes of dietary lipids and the chief form of fat in foods. A triglyceride is made up of three units of fatty acids and one unit of glycerol (fatty acids and glycerol are defined later). Triglycerides are also called *triacylglycerols*.

phospholipids (FOSS-foh-LIP-ids) one of the three main classes of dietary lipids. These lipids are similar to triglycerides, but each has a phosphorus-containing acid in place of one of the fatty acids. Phospholipids are present in all cell membranes.

lecithin (LESS-ih-thin) a phospholipid manufactured by the liver and also found in many foods; a major constituent of cell membranes.

sterols (STEER-alls) one of the three main classes of dietary lipids. Sterols have a structure similar to that of cholesterol.

essential fatty acids fatty acids that the body needs but cannot make in amounts sufficient to meet physiological needs.

YOUR BILL FROM a medical laboratory reads, "Blood **lipid** profile—$125." A health-care provider reports, "Your blood **cholesterol** is high." Your physician advises, "You must cut down on the saturated **fats** in your diet and replace them with **oils** to lower your **cardiovascular disease (CVD)** risk." Blood lipids, cholesterol, saturated fats, and oils—all contribute to health and detract from it.

Introducing the Lipids

The lipids in foods and in the human body fall into three classes. About 95 percent are **triglycerides.*** The other classes of the lipid family are the **phospholipids** (of which **lecithin** is one) and the **sterols** (cholesterol is the best known of these).

No doubt you are expecting to hear that these fat-related compounds have the potential to harm your health. It may come as a surprise to hear that lipids are also valuable. In fact, lipids are absolutely necessary, and some lipids must be present in your foods if you are to maintain good health. The low-fat diet recommended for health doesn't mean a "no-fat" diet.[1] Luckily, traces of fats and oils are present in almost all foods, so you needn't make an effort to eat any extra.

Usefulness of Fats in the Body

When people speak of fat, they are usually talking about triglycerides. The term *fat* is more familiar, though, and we will use it here. Fat is the body's chief storage form for the energy from food eaten in excess of need. The storage of fat is a valuable survival mechanism for people who must live a feast-or-famine existence: stored during times of plenty, fat enables them to remain alive during times of famine. In addition, fats provide most of the energy needed to perform much of the body's work, especially muscular work.

Most body cells can store only limited fat, but some cells are specialized for storing fat. These fat cells seem able to expand almost indefinitely. The more fat they store, the larger they grow. An obese person's fat cells may be many times the size of a thin person's. A fat cell is shown in Figure 5-1.

You may be wondering why the carbohydrate glucose is not the body's major form of stored energy. As mentioned in Chapter 4, glucose is stored in the form of glycogen. One characteristic of glycogen is that it holds a great deal of water, and as a result, it is quite bulky and heavy. The body cannot store enough glycogen to provide energy for very long. Fats, however, pack tightly together without water and can store much more energy in a small space.[2] The body fat found on a normal-weight, healthy person contains sufficient energy to fuel a marathon run to the finish or to give a sick person who cannot eat the energy to battle disease.

Fat serves many other purposes in the body, too. Pads of fat surrounding the vital organs serve as shock absorbers. Thanks to the fat pads cushioning your internal organs, you can ride a horse or a motorcycle for many hours with no serious internal injuries. The fat blanket under the skin also insulates the body from extremes of temperature, thus assisting with internal climate control.

Some essential nutrients are soluble in fat and therefore are found mainly in foods that contain fat and are absorbed most efficiently from them. These nutrients are the fat-soluble vitamins: A, D, E, and K. Other essential nutrients, the **essential fatty acids,** constitute parts of the fats themselves. As a later section explains, the essential fatty acids serve as raw materials from which the body makes molecules it needs. Lipids are also important to all the body's cells as part of their surrounding envelopes, the cell membranes.

KEY POINT *Lipids not only serve as energy reserves but also cushion the vital organs, protect the body from temperature extremes, carry the fat-soluble nutrients, serve as raw materials, and provide the major material of which cell membranes are made.*

*Another name for triglyceride is *triacylglycerol*.

Figure 5-1

A Fat Cell

Within the fat cell, lipid is stored in a droplet. This droplet can greatly enlarge, and the fat cell membrane will grow to accommodate its swollen contents. More about fat cells and obesity in Chapter 9.

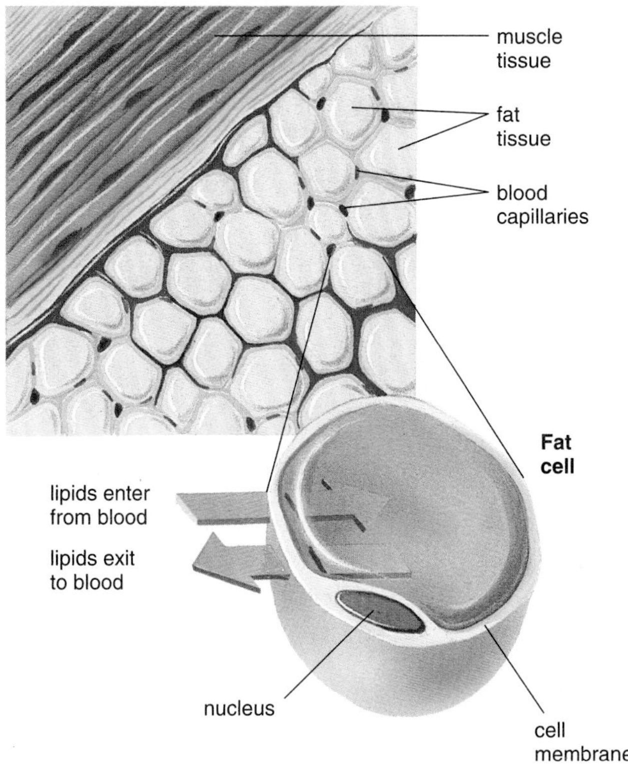
muscle tissue
fat tissue
blood capillaries
Fat cell
lipids enter from blood
lipids exit to blood
nucleus
cell membrane

satiety (sat-EYE-uh-tee) the feeling of fullness or satisfaction that people experience after meals.

A reminder from Chapter 1:
1 g carbohydrate = 4 calories.
1 g fat = 9 calories.
1 g protein = 4 calories.

Usefulness of Fats in Food

The energy density of fats makes foods rich in fat valuable in many situations. A gram of fat or oil delivers more than twice as many calories as a gram of carbohydrate. A hunter or hiker needs to consume a large amount of food energy to travel long distances or to survive in intensely cold weather. As Figure 5-2 shows, such a person can carry more energy in fat-rich foods than in carbohydrate-rich foods. But for a person who is not expending much energy in physical work, those same high-fat foods may deliver many unneeded calories in only a few bites.

People naturally like high-fat foods.[3] In fact, around the world, as fat becomes less expensive and more available in a given food supply, people seem to choose diets providing greatly increased amounts of fat.[4] Fat carries with it many dissolved compounds that give foods enticing aromas and flavors, such as the aroma of frying bacon or french fries. In fact, when a person refuses food, foods flavored with some fat may tempt that person to eat again. Fat also lends tenderness to foods such as meats and baked goods.

Fat also provides **satiety**, the lasting satisfaction of feeling full after a meal. Fat in food slows digestion and helps sustain feelings of fullness until the next mealtime. Recent research has revealed that carbohydrate and protein also help to regulate the appetite, but their timing is different from that of fat. During a meal, fat consumption sends no or only weak signals of fullness, so fat-rich foods can easily be overconsumed.[5] Carbohydrate and protein act faster and send stronger signals during the meal to stop the diner from continuing to eat. Chapter 9 revisits the topic of appetite and its control. Table 5-1 sums up the usefulness of fats, both in foods and in the body.

KEY POINT *Lipids provide more energy per gram than carbohydrate and protein, enhance food's aroma and flavor, and contribute to satiety, or a feeling of fullness, after a meal.*

Figure 5-2

Two Lunches

Both lunches contain the same number of calories, but the fat-rich lunch takes up less space and weighs less.

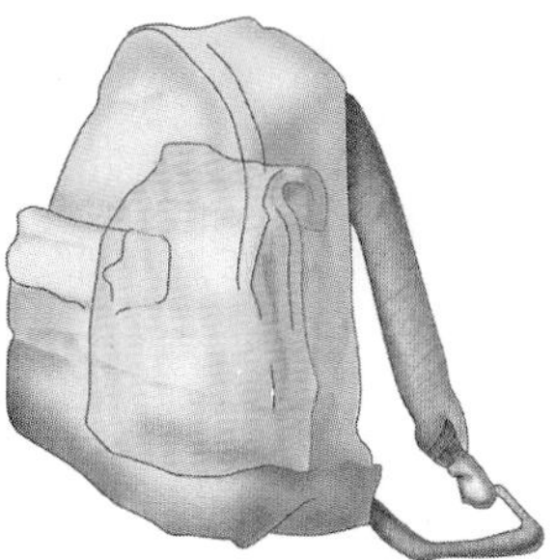

carbohydrate-rich lunch
1 low-fat muffin
1 banana
2 oz carrot sticks
8 oz fruit yogurt

calories = 550
weight (g) = 500

fat-rich lunch
6 butter-style crackers
1 1/2 oz American cheese
2 oz trail mix with candy

calories = 550
weight (g) = 115

Table 5-1

The Usefulness of Fats

Fats in Food	Fats in the Body
■ *nutrient* fats provide essential fatty acids. ■ *energy* fats provide a concentrated energy source in foods. ■ *transport* fats carry fat-soluble vitamins A, D, E, and K, and assist in their absorption. ■ *raw materials* fats provide raw material for making needed products. ■ *sensory* fats contribute to taste and smell of foods, and ■ *appetite* fats stimulate the appetite. ■ *satiety* fats contribute to feelings of fullness. ■ *texture* fats help make foods tender.	■ *energy stores* fats are the body's chief form of stored energy. ■ *muscle fuel* fats provide most of the energy to fuel muscular work. ■ *emergency reserve* fats serve as an emergency fuel supply in times of illness and diminished food intake. ■ *padding* fats protect the internal organs from shock through fat pads inside the body cavity. ■ *insulation* fats insulate against temperature extremes through a fat layer under the skin. ■ *cell membranes* fats form the major material of cell membranes. ■ *raw materials* fats are converted to other compounds, such as hormones, bile, and vitamin D, as needed.

A Close Look at Fats

As mentioned, the term *fat* refers to triglycerides, the major form of lipid found in foods. Triglycerides, in turn, are made of fatty acids and glycerol.

Triglycerides: Fatty Acids and Glycerol

Very few **fatty acids** are found free in the body or in foods. Usually, the fatty acids are incorporated into large, complex compounds: triglycerides. The name almost explains itself: three fatty acids *(tri)* are attached to a molecule of **glycerol.** Figure 5-3 shows how glycerol and three fatty acids combine to make a triglyceride molecule. Tissues all over the body can easily assemble triglycerides or disassemble them as needed. Many triglycerides eaten in foods are transported to the fat depots—muscles, breasts, the insulating fat layer under the skin, and others—where they are stored.

Figure 5-3

Triglyceride Formation

Glycerol, a small, water-soluble carbohydrate derivative, plus three fatty acids, equals a triglyceride.

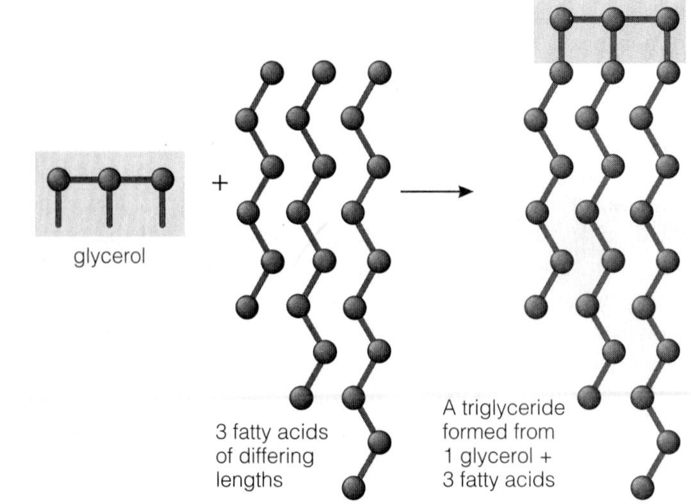

glycerol
+
3 fatty acids of differing lengths
A triglyceride formed from 1 glycerol + 3 fatty acids

fatty acids organic acids composed of carbon chains of various lengths. Each fatty acid has an acid end and hydrogens attached to all of the carbon atoms of the chain.

glycerol (GLISS-er-all) an organic compound, three carbons long, of interest here because it serves as the backbone for triglycerides.

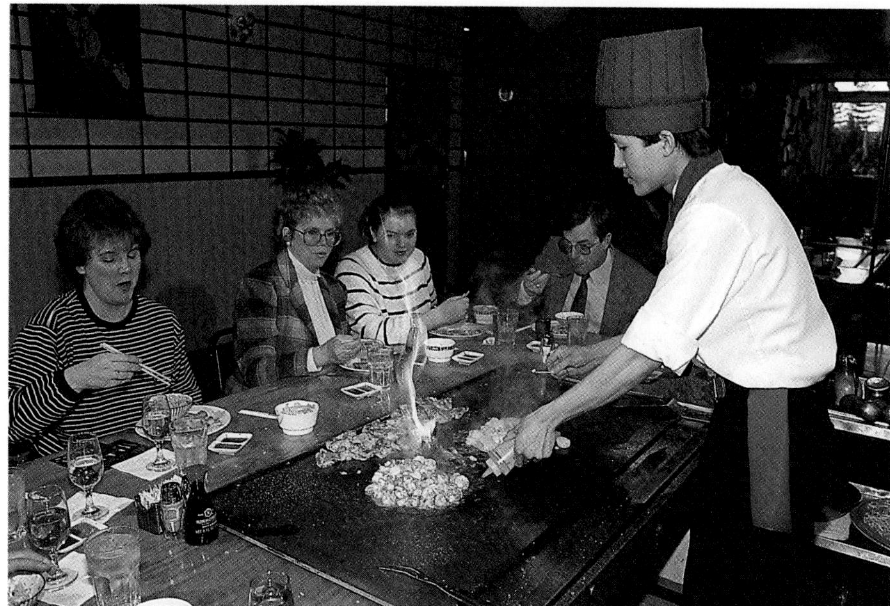

Small amounts of fat offer eaters both pleasure and needed nutrients.

Fatty acids may differ from one another in two ways: in chain length and in degree of saturation (explained next). Triglycerides usually include mixtures of various fatty acids. Depending on which fatty acids are incorporated into a triglyceride, the resulting fat will be soft or hard. Triglycerides containing mostly the shorter-chain fatty acids or the more unsaturated ones are softer and melt more readily. Each species of animal (including people) makes its own characteristic kinds of triglycerides, a function governed by genetics. Fats in the diet, though, can affect the types of triglycerides made. For example, animals raised for food can be fed diets containing softer or harder triglycerides to give the animals softer or harder fat, whichever consumers demand.

KEY POINT *The body combines three fatty acids with one glycerol to make a triglyceride, its storage form of fat. Fatty acids in food influence the composition of fats in the body.*

Saturated versus Unsaturated Fatty Acids

Saturation refers to the number of hydrogens a fatty acid chain is holding. If every available bond from the carbons is holding a hydrogen, the chain forms a **saturated fatty acid;** it is filled to capacity with hydrogen. The zigzag structure on the left in Figure 5-4 represents a saturated fatty acid.

Sometimes, especially in the fatty acids of plants and fish, the chain has a place where hydrogens are missing, an "empty spot," or **point of unsaturation.** A fatty acid carbon chain that possesses one or more points of unsaturation is an **unsaturated fatty acid.** If there is one point of unsaturation, then the fatty acid is a **monounsaturated fatty acid** (see the second structure in Figure 5-4). If there are two or more points of unsaturation, then it is a **polyunsaturated fatty acid,** sometimes abbreviated as **PUFA** (see the third structure in Figure 5-4; other examples are given later in the chapter).

The degree of saturation of fatty acids in a fat affects the temperature at which the fat melts. Generally, the more unsaturated the fatty acids, the more liquid the fat is at room temperature. In contrast, the more saturated the fatty acids, the firmer the fat is. Thus, of three fats—lard (which comes from pork), chicken fat, and safflower oil—lard is the most saturated and the hardest; chicken fat is less saturated and somewhat soft; and safflower oil, which is the most unsaturated, is a liquid at room temperature. If a health-care provider recommends limiting **saturated fats** and using **monounsaturated fats** or **polyunsaturated fats** instead, you can generally judge by the hardness of the

saturated fatty acid a fatty acid carrying the maximum possible number of hydrogen atoms (having no points of unsaturation). A saturated fat is a triglyceride that contains three saturated fatty acids.

point of unsaturation a site in a molecule where the bonding is such that additional hydrogen atoms can easily be attached.

unsaturated fatty acid a fatty acid that lacks some hydrogen atoms and has one or more points of unsaturation. An unsaturated fat is a triglyceride that contains one or more unsaturated fatty acids.

monounsaturated fatty acid a fatty acid containing one point of unsaturation.

polyunsaturated fatty acid (PUFA) a fatty acid with two or more points of unsaturation.

saturated fats triglycerides in which most of the fatty acids are saturated.

monounsaturated fats triglycerides in which most of the fatty acids have one point of unsaturation (are monounsaturated).

polyunsaturated fats triglycerides in which most of the fatty acids have two or more points of unsaturation (are polyunsaturated).

The more unsaturated a fat, the more liquid it is at room temperature. The more saturated a fat, the higher the temperature at which it melts.

The FATTY ACIDS
SAT-
URATED
MONO-
UNSATURATED
POLY-
UNSATURATED

A trio of fatty acids—saturated, monounsaturated, and polyunsaturated.

Figure 5-4

Three Fatty Acids

The more carbon atoms in a fatty acid, the longer it is. The more hydrogen atoms attached to those carbons, the more saturated the fatty acid is.

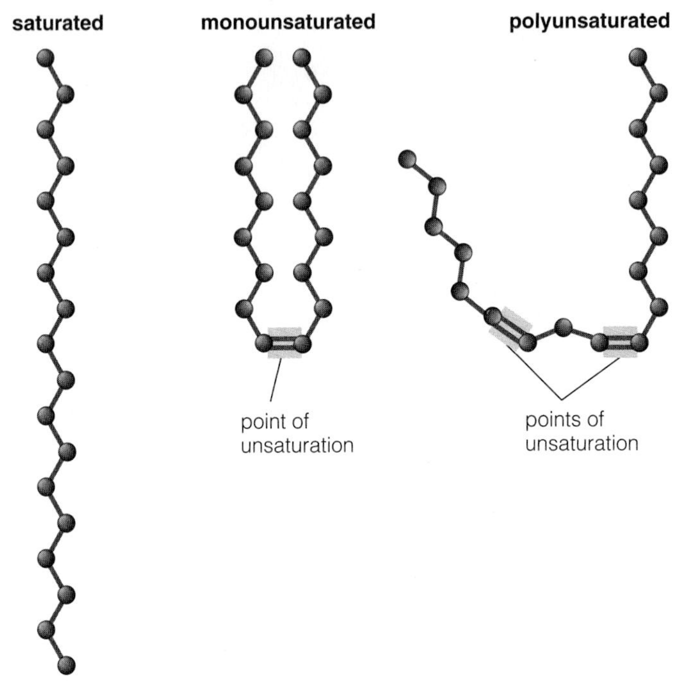
saturated
monounsaturated
polyunsaturated
point of unsaturation
points of unsaturation

fats which ones to choose. To determine whether an oil you use contains saturated fats, place the oil in a clear container in the refrigerator and watch for cloudiness. The least saturated oils remain the clearest.

Generally speaking, vegetable and fish oils are rich in polyunsaturates. Some vegetable oils, especially olive oil, are also rich in monounsaturates, and animal fats are generally the most saturated. But you have to know your oils. To obtain polyunsaturated oils, it is not enough to choose foods with labels claiming plant oils over those containing animal fats. Some nondairy whipped dessert toppings use coconut oil, one of the so-called tropical oils, in place of cream (butterfat). Coconut oil does come from a plant, but it disobeys the rule that plant oils are less saturated than animal fats; the fatty acids of coconut oil are actually more saturated than those of cream and are of a type that seems to add to heart disease risk. Similarly, palm oil, used frequently in food processing, is also highly saturated and has been added to the list of fats that elevate blood cholesterol.[6]

A benefit to health is seen when monounsaturated fat is used in place of saturated fat in the diet.[7] As Controversy 2 in Chapter 2 made clear, olive oil does not harm the health of the heart. When it replaces other fats in the diet, olive oil may even benefit health, if the low rates of heart disease and breast cancer among the people of Mediterranean regions serve as an indicator.[8] Canola oil, another rich source of monounsaturated fatty acids, also stocks U.S. grocery shelves. Figure 5-5 compares the percentages of saturated, monounsaturated, and polyunsaturated fatty acids in various fats and oils, and Figure 5-6 identifies the sources of saturated fats in the U.S. diet. Note that the bread, cereal, rice, and pasta group is not represented in the figure, but as a later section points out, manufacturers often add fats to these foods.

KEY POINT *Fatty acids are energy-rich carbon chains that can be saturated (filled with hydrogens) or monounsaturated (with one point of unsaturation) or polyunsaturated (with more than one point of unsaturation). The degree of saturation of the fatty acids in a fat determines the fat's softness or hardness.*

Figure 5-5

Fatty Acid Composition of Common Food Fats

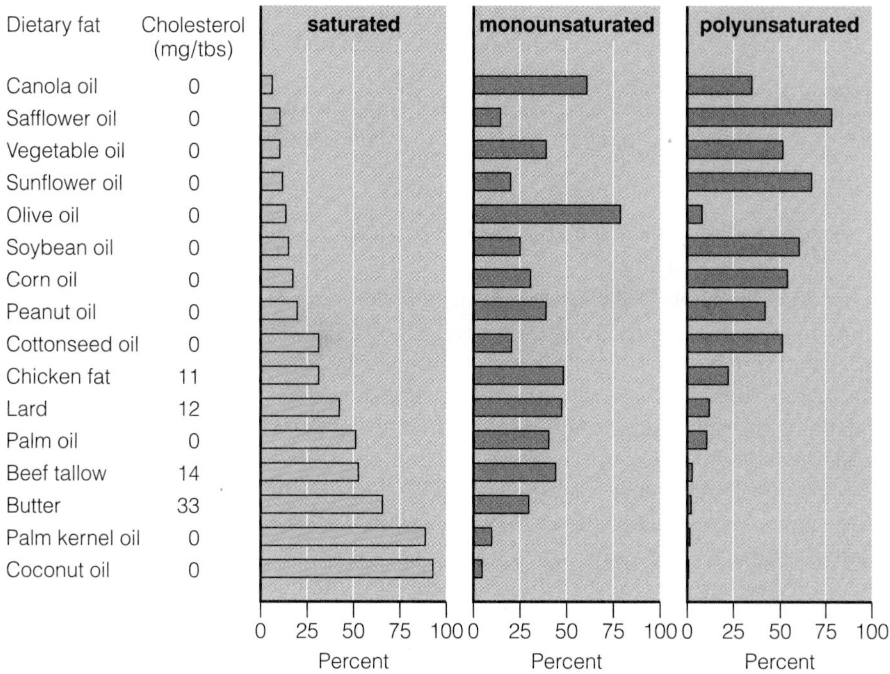

Dietary fat
Cholesterol (mg/tbs)
saturated
monounsaturated
polyunsaturated
Canola oil 0
Safflower oil 0
Vegetable oil 0
Sunflower oil 0
Olive oil 0
Soybean oil 0
Corn oil 0
Peanut oil 0
Cottonseed oil 0
Chicken fat 11
Lard 12
Palm oil 0
Beef tallow 14
Butter 33
Palm kernel oil 0
Coconut oil 0
0 25 50 75 100
Percent

Other Members of the Lipid Family

Thus far we have dealt with just one of the three classes of lipids—the triglycerides and their component fatty acids. These lipids represent 95 percent of all the lipids in the diet and in the body. The word *fat,* used properly, refers to the triglycerides.

The other two classes—phospholipids and sterols—merit a moment's attention, though, because they play important roles in the body. A phospholipid, like a triglyceride, consists of a molecule of glycerol with fatty acids attached, but it contains two, rather than three, fatty acids. In place of the third is a molecule containing phosphorus, which makes the phospholipid soluble in water, while its fatty acids make it soluble in fat. This versatility permits any phospholipid to play a role in keeping fats dispersed in water; it can serve as an **emulsifier.**

Food processors often blend fat with watery ingredients by way of **emulsification.** Some salad dressings separate to form two layers—vinegar on the bottom, oil on the top. Other dressings, such as mayonnaise, are also made from vinegar and oil but never separate. The difference lies in a special ingredient of mayonnaise, the emulsifier lecithin in egg yolks. Lecithin, a phospholipid, blends the vinegar and oil in a permanent emulsion.

Lecithins and other phospholipids also play key roles in the structure of cell membranes. Because phospholipids are emulsifiers, they have both water-loving and fat-loving characteristics, which enable them to help fats travel back and forth across the lipid-containing membranes of cells into the watery fluids on both sides. Almost magical health-promoting properties, such as the ability to lower blood cholesterol, are sometimes attributed to lecithin, but the people making the claims stand to gain from selling supplements. Although it is an important lipid to the body, lecithin has no special ability to promote health.

Sterols such as cholesterol are large, complicated molecules consisting of interconnected *rings* of carbon atoms with side chains of carbon, hydrogen, and oxygen attached. Cholesterol serves as the raw material for making another emulsifier, **bile,** which is important to digestion (see the next section for details). Other sterols are vitamin D, which is made from cholesterol, and several important hormones, the so-called steroid hormones, including the sex hormones.

emulsifier a substance that mixes with both fat and water and permanently disperses the fat in the water, forming an emulsion.

emulsification the process of mixing lipid with water by adding an emulsifier.

bile an emulsifier made by the liver from cholesterol and stored in the gallbladder. Bile does not digest fat as enzymes do but emulsifies it so that enzymes in the watery fluids may contact it and split the fatty acids from their glycerol for absorption.

Figure 5-6

Saturated Fats in the U.S. Diet

Note that fruits, grains, and vegetables are insignificant sources, unless saturated fats are intentionally added to them during preparation.

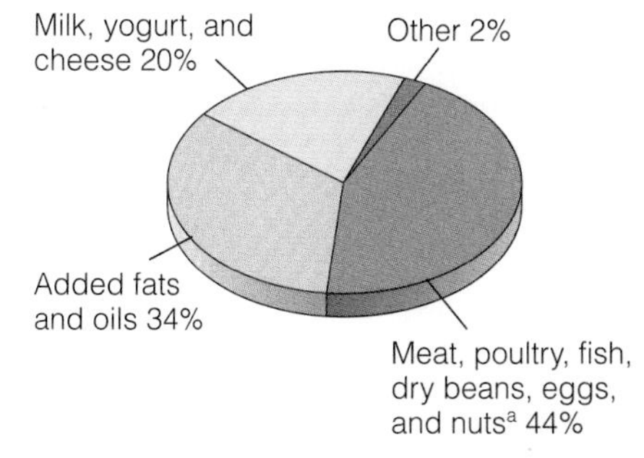

Milk, yogurt, and cheese 20%
Other 2%
Added fats and oils 34%
Meat, poultry, fish, dry beans, eggs, and nuts[a] 44%

[a]Eggs provide 2% of the 44%, and nuts and legumes provide another 2%.

monoglycerides (mon-oh-GLISS-er-ides) products of the digestion of lipids; consist of glycerol molecules with one fatty acid attached (*mono* means "one"; *glyceride* means "a compound of glycerol").

Cholesterol is an important sterol in the structure of brain and nerve cells. In fact, cholesterol is a part of every cell. Like lecithin, cholesterol can be made by the body, so it is not an essential nutrient. Though widespread in the body and necessary to its function, cholesterol is also the major part of the plaques that narrow the arteries in atherosclerosis, the underlying cause of heart attacks and strokes.

KEY POINT *Phospholipids, including lecithin, play key roles in cell membranes; sterols play roles as part of bile, vitamin D, the sex hormones, and other important compounds.*

Lipids in the Body

In handling lipids, the body faces a problem: how to thoroughly mix fats, which tend to separate from water, with its own watery fluids. The digestive system solves this problem through the use of bile, which emulsifies the fat in food in the watery digestive fluids. Thus, to digest fats, the digestive system first mixes them with its bile-containing digestive juices; once the fats are emulsified, the fat-digesting enzymes can break them down. After fats have been digested, they face another watery barrier, the watery layer of mucus that coats the absorptive lining of the digestive tract. Fats must traverse this layer to enter the cells of the digestive tract lining. Then the cells face another challenge: to package lipids so that they can travel in the watery fluids of the circulatory system. The next two sections describe the body's superb adaptations to meet all these needs for lipid digestion and transport.

Digestion of Fats

When you partake of animal products such as meat, fish, poultry, or eggs, you are eating fat and protein. When you eat oil-containing plant foods, such as nuts, coconut, or olives, you are eating fat and carbohydrate along with some protein. Of the fats and oils in foods, 95 percent are triglycerides that have been made in living animal or plant tissues, mostly from carbohydrate, the same way the human body makes them.

Food fat can end up in fat stores in the body, but first it has to be digested, absorbed, and transported to its cell destinations. A bite of food in the mouth first encounters the enzymes of saliva. One enzyme, produced by the tongue, acts on long-chain fatty acids, especially those of milk. The enzyme plays a major role in milk fat digestion in infants, but is thought to be of little importance to fat digestion in adults. Once the food has been chewed and swallowed, it travels to the stomach, where the fat separates from other components and floats as a layer on the top. Since fat does not mix with the stomach fluids, little fat digestion takes place.

By the time fat enters the small intestine, the gallbladder, which stores the liver's output of bile, has contracted and squirted its bile into the intestine. Bile mixes fat particles with watery fluid by emulsifying them (see Figure 5-7), suspending them in the fluid until the fat-digesting enzymes contributed by the pancreas can split them into smaller particles for absorption. A bile molecule, made from cholesterol, works because one of its ends attracts and holds fat, while the other end is attracted to and held by water.

Chapter 3 first described the action of bile and gave details of the digestive system.

People sometimes wonder how a person without a gallbladder can digest food. The gallbladder is just a storage organ. Without it, the liver still produces bile, but delivers it continuously into the small intestine. People who have had their gallbladders removed must reduce their fat intakes because they can no longer store bile and release it at mealtimes. As a result, their systems can handle only a little fat at a time until they can adjust to the change .

Once the intestine's contents are emulsified, fat-splitting enzymes act on triglycerides to split fatty acids from their glycerol backbones. Free fatty acids, glycerol, and **monoglycerides** cling together in balls surrounded by bile and are shuttled across the watery layer of mucus to the waiting absorptive cells of the intestinal villi. The bile may be absorbed and reused by the body, or it may exit with the feces as shown in Figure 4-4 in Chapter 4.

Figure 5-7

The Action of Bile in Fat Digestion

Detergents are emulsifiers and work the same way, which is why they are effective in removing grease spots from clothes. Molecule by molecule, the grease is dissolved out of the spot and suspended in the water, where it can be rinsed away.

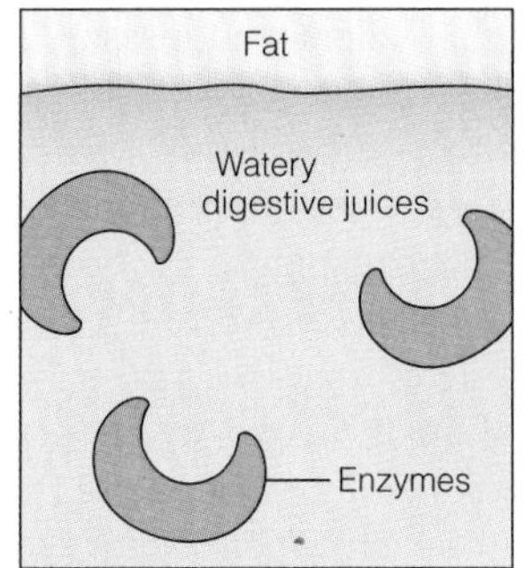
Fat
Watery digestive juices
Enzymes

In the stomach, the fat and watery digestive juices tend to separate. enzymes are in the water and can't get at the fat.

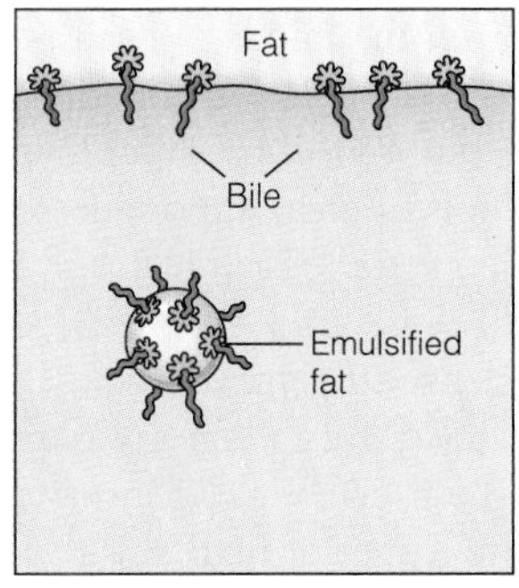
Fat
Bile
Emulsified fat

When fat enters the small intestine, the gallbladder secretes bile. Bile has an affinity for both fat and water, so it can bring the fat into the water.

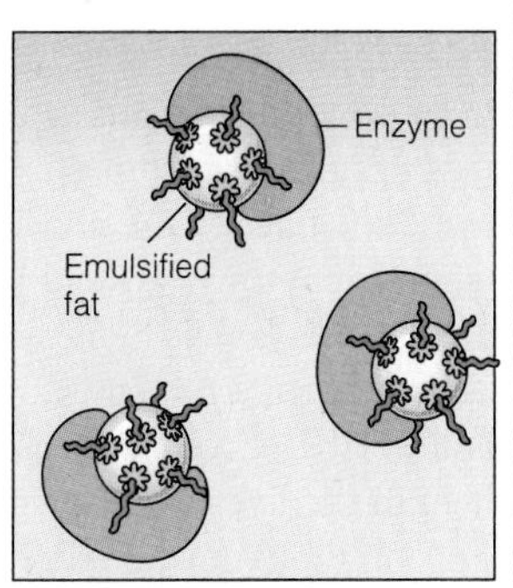
Enzyme
Emulsified fat

After emulsification, the enzymes have easy access to the fat droplets.

The small products of lipid digestion, glycerol and shorter-chain fatty acids, can pass directly through the cells of the digestive tract lining into the bloodstream. From there they can travel without help to the liver and to the tissues that need them.

The larger products of lipid digestion, monoglycerides and long-chain fatty acids, need some help to get to their destinations via the bloodstream. Once inside the intestinal cells, they are re-formed into triglycerides and incorporated into **chylomicrons,** clusters of proteins and the digested lipids. Chylomicrons are a class of **lipoproteins,** described later.

The digestive tract absorbs triglycerides from a meal with up to 98 percent efficiency. In other words, little fat is excreted by a healthy system. The process of fat digestion takes time, though, so the more fat taken in at a meal, the slower the digestive system action becomes. The efficient series of events just described is depicted in Figure 5-8 on the next page.

KEY POINT *In the stomach, fats separate from other food components. In the small intestine, bile emulsifies the fats, enzymes digest them, and the intestinal cells absorb them. Small lipids can travel alone in the blood after absorption, but large lipids must be incorporated into chylomicrons for transport.*

If Fat Floats, How Does It Travel around the Body in the Watery Blood?

Within the body, many fats travel from place to place in blood as passengers in lipoproteins. For example, the monoglycerides and long-chain fatty acids liberated from digested food fat are too large to be released directly into the bloodstream. Without some mechanism to keep them dispersed, these lipids would separate out and float in globules, disrupting the blood's normal functions. Therefore, before releasing the lipids, the intestinal cells allow them to cluster together, rejoin them as triglycerides, and combine them with protein to form the chylomicrons mentioned earlier. The protein and phospholipid in the clusters act as emulsifiers, attracting both water and fat. Their association with both substances enables chylomicrons to transport lipids in the watery body fluids. The tissues of the body can extract whatever fat they need from these clusters. The remnants are then picked up by the liver, which dismantles them and reuses their parts.

In addition to the chylomicrons, the body uses three other types of lipoproteins to carry fats: the **very-low-density lipoproteins (VLDL),** which carry triglycerides and other lipids made in the liver to the body cells for their use; the **low-density lipoproteins**

chylomicrons (KYE-low-MY-krons) clusters formed when lipids from a meal are combined with carrier proteins in the intestinal lining. Chylomicrons transport food fats through the watery body fluids to the liver and other tissues.

lipoproteins (LYE-poh-PRO-teens, LIH-poh-PRO-teens) clusters of lipids associated with protein, which serve as transport vehicles for lipids in blood and lymph. Major lipoprotein classes are the chylomicrons, the VLDL, the LDL, and the HDL.

very-low-density lipoproteins (VLDL) lipoproteins that transport triglycerides and other lipids from the liver to various tissues in the body.

low-density lipoproteins (LDL) lipoproteins that transport lipids from the liver to other tissues such as muscle and fat; contain a large proportion of cholesterol.

high-density lipoproteins (HDL) lipoproteins that return cholesterol from storage places to the liver for dismantling and disposal; contain a large proportion of protein.

(LDL), which are made from VLDL after they have donated much of their fat to body cells and picked up cholesterol; and the **high-density lipoproteins (HDL),** which carry cholesterol from body cells to the liver. The carrier proteins for both LDL and HDL are made in the liver, and both carry large quantities of cholesterol; the HDL also carry many phospholipids. Figure 5-9 depicts a typical lipoprotein and demonstrates how a lipoprotein's density changes with its lipid and protein contents.

Lipoproteins are very much on the minds of health-care providers who measure a person's blood lipid profile. They are interested not only in the types of fats in the blood (triglycerides and cholesterol) but also in the lipoproteins that carry them. The distinction between LDL and HDL is of great importance because it has implications for the health of the heart and blood vessels. Both LDL and HDL carry lipids in the blood, but LDL are larger, lighter, and more lipid filled; HDL are smaller, denser, and packaged with more protein. LDL deliver triglycerides and cholesterol from the liver to the tissues; HDL scavenge excess cholesterol and phospholipids from the tissues and return them to the liver for disposal. Both LDL and HDL carry cholesterol, but elevated LDL concentrations in the blood are a sign of high risk of heart attack, whereas elevated HDL concentrations are associated with a low risk. This is why some people refer to LDL as "bad" cholesterol, and HDL as "good" cholesterol. Keep in mind, though, that they carry the same kind of cholesterol—the difference to the heart between LDL and HDL lies in the tasks they perform, not in the type of cholesterol they carry. A later section examines the significance of lipoproteins to the health of the heart.

KEY POINT *Blood and other body fluids are watery, so fats need special transport vehicles to carry them around the body in these fluids. The chief lipoproteins are chylomicrons, VLDL, LDL, and HDL.*

Figure 5-8

The Process of Lipid Digestion and Absorption

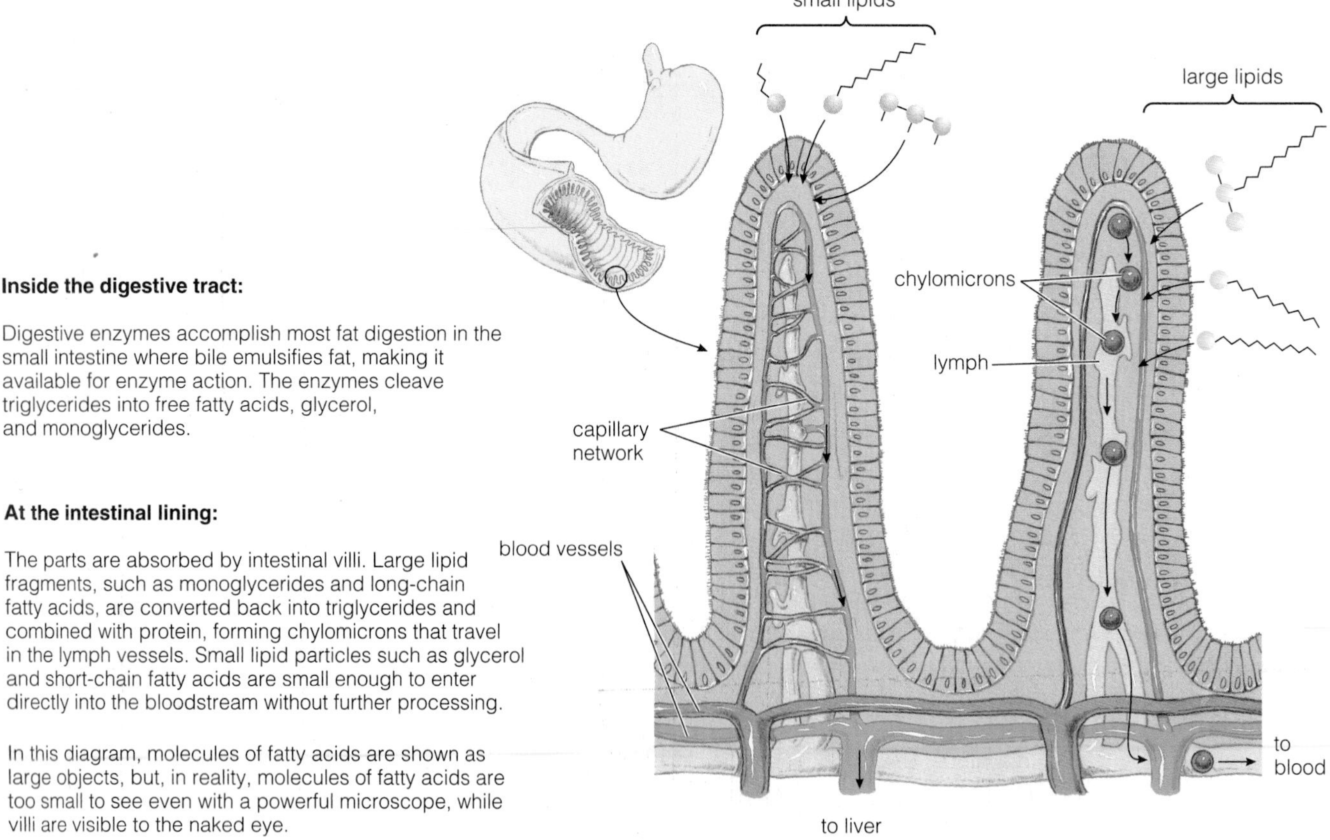
small lipids
large lipids
chylomicrons
lymph
capillary network
blood vessels
to liver
to blood
Inside the digestive tract:
Digestive enzymes accomplish most fat digestion in the small intestine where bile emulsifies fat, making it available for enzyme action. The enzymes cleave triglycerides into free fatty acids, glycerol, and monoglycerides.
At the intestinal lining:
The parts are absorbed by intestinal villi. Large lipid fragments, such as monoglycerides and long-chain fatty acids, are converted back into triglycerides and combined with protein, forming chylomicrons that travel in the lymph vessels. Small lipid particles such as glycerol and short-chain fatty acids are small enough to enter directly into the bloodstream without further processing.
In this diagram, molecules of fatty acids are shown as large objects, but, in reality, molecules of fatty acids are too small to see even with a powerful microscope, while villi are visible to the naked eye.

Figure 5-9

Lipoproteins

As the graph shows, the density of a lipoprotein is determined by its lipid-to-protein ratio. An LDL has a high ratio of lipid to protein (about 80 percent lipid to 20 percent protein) and is especially high in cholesterol. An HDL has more protein relative to its lipid content (about equal parts lipid and protein).

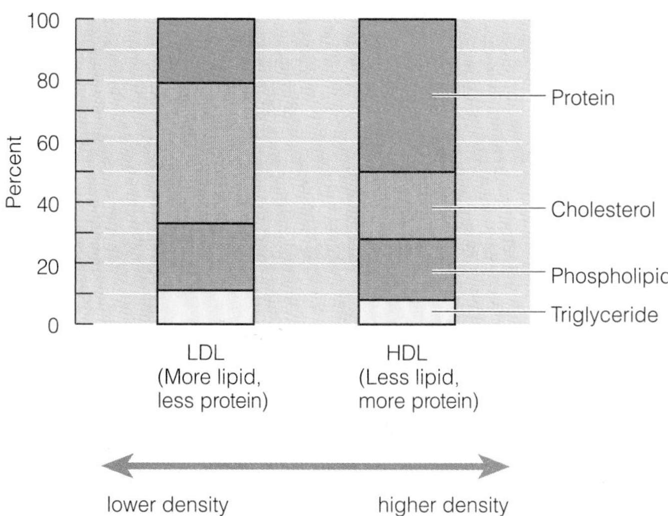

Percent
100
80
60
40
20
0
Protein
Cholesterol
Phospholipid
Triglyceride
LDL
(More lipid,
less protein)
HDL
(Less lipid,
more protein)
lower density
higher density

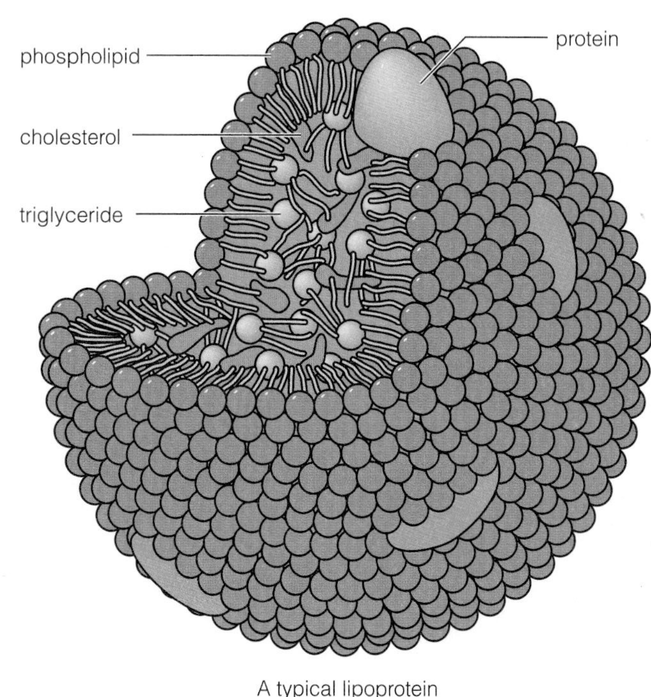

phospholipid
cholesterol
triglyceride
protein

A typical lipoprotein

How Can I Use My Stored Fat for Energy?

Excess fat carried in LDL is stored by the body's fat cells for later use. An earlier section described the body's remarkable ability to store excess energy as body fat. When a person's body starts to run out of fuel available from food, it begins to retrieve its stored fat to use for energy. It also draws on its stored glycogen, as the last chapter described. Fat cells respond to the call for energy by dismantling stored fat molecules and releasing fat components into the blood. Upon receiving these components, the energy-hungry cells break them down further into small fragments. Finally, each fat fragment is combined with a fragment derived from glucose, and the energy-releasing reaction continues, liberating energy, carbon dioxide, and water. The way to use stored fat for energy, then, is to create a demand for it in the tissues by decreasing intake of food energy, by increasing the body's expenditure of energy, or both.

Importantly, whenever body fat is broken down to provide energy, carbohydrate must be available as well. Without carbohydrate, ketosis will occur, as described in the last chapter, and products of incomplete fat breakdown (ketones) will appear in the blood and urine. Because this process and its consequences are so important in weight control, Chapter 9 describes them in greater detail.

The body can also store excess glucose as fat, but this conversion is not energy efficient. Figure 5-10 illustrates a simplified series of steps from carbohydrate to fat. As the figure shows, before excess glucose can be stored as fat, it must first be broken into tiny fragments and then reassembled into fatty acids, steps that require energy to perform. Fat goes through fewer chemical steps before storage. Thus, given the same number of calories from excess dietary fat or carbohydrate, the body stores more calories from the fat than from the carbohydrate. In short, you may get fatter on fat calories than on the same number of carbohydrate calories.

KEY POINT *When low on fuel, the body draws on its stored fat for energy. Glucose is necessary for the complete breakdown of fat; without carbohydrate, ketosis occurs.*

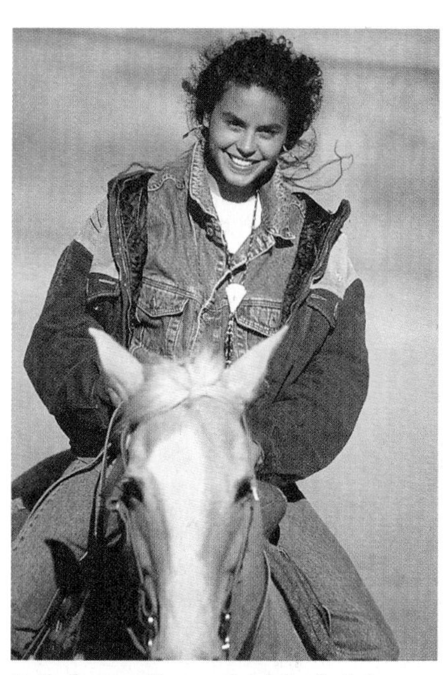

Body fat supplies much of the fuel these muscles need to do their work.

Another source of energy, alcohol, is the topic of this chapter's Controversy.

The American Heart Association Dietary Guidelines for Healthy American Adults are found in Table 11-10 of Chapter 11.

Figure 5-10

Glucose to Fat

Glucose can be used for energy, or it can be changed into fat and stored.

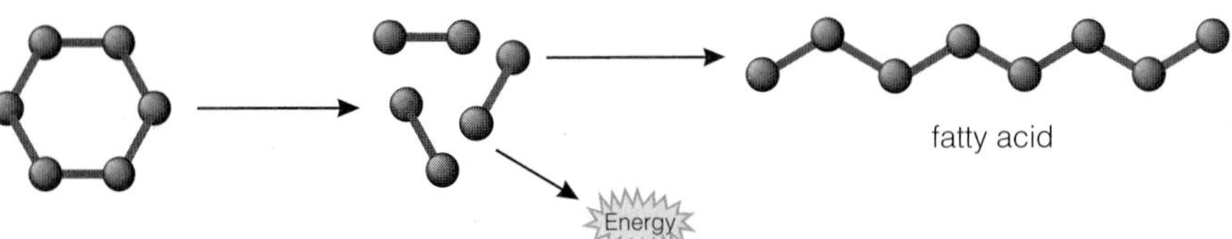
fatty acid
Energy

Glucose is broken down into fragments.

The fragments can provide immediate energy for the tissues.

Or, if the tissues need no more energy, the fragments can be reassembled, not back to glucose but into fatty acid chains.

How, Exactly, Do Fat and Cholesterol Affect People's Health?

High dietary fat intakes are associated with serious diseases. Obesity carries serious risks to health, and fat's energy density makes it likely that people who eat high-fat diets will exceed their energy needs and so gain weight. Furthermore, fat from food is stored with amazing efficiency in the body—its conversion to body fat costs little metabolic energy, unlike carbohydrate and protein from food, which must go through many metabolic steps before they are finally converted into fat that the body can store. People who eat high-fat diets often have more body fat than their energy intakes alone would predict. Chapter 9 comes back to the issue of the fattening power of fat and concludes that low-fat, high-carbohydrate diets are most appropriate for controlling body fatness.

Most importantly, the person who chooses a diet too high in certain fats may be inviting the risk of heart and artery disease, or CVD.[9] Heart disease is the number-one killer of adults in the United States. The person who eats a high-fat diet also incurs a greater-than-average risk of developing some forms of cancer, another leading killer disease. Much research has focused on the links between diet and disease, and later a whole chapter is devoted to these connections (Chapter 11). A few points about fats and heart health are presented here because they underlie dietary recommendations concerning fats (see Table 5-2).

Of great importance in regard to fat and disease is a medical test, the blood lipid profile, which reveals the amounts of various lipids, especially triglycerides and cholesterol, in the blood. It also identifies the protein carriers with which these lipids are traveling. The results of this test tell much about a person's risk of CVD.

Most important in regard to CVD is blood cholesterol.[†] A person's blood cholesterol concentration is considered to be a predictor of that person's likelihood of suffering a fatal heart attack or stroke, and the higher the cholesterol, the earlier the episode is expected to occur. Blood cholesterol is one of the three major risk factors for CVD (the other two are smoking and high blood pressure, or hypertension). The importance of blood cholesterol cannot be overemphasized.

Now, what does *food* cholesterol have to do with *blood* cholesterol? The answer may be, "Not as much as most people think." Most saturated food *fats* (triglycerides) raise blood cholesterol more than food *cholesterol* does. When told that cholesterol doesn't matter as much as fat, people may then jump to the wrong conclusion—that blood cholesterol doesn't matter. It does matter. High *blood* cholesterol is an indicator of risk for CVD. The main dietary factor associated with elevated blood cholesterol is a high *saturated fat* intake. In comparison, dietary cholesterol alone makes only a minor contribution.

[†] *Blood, plasma,* and *serum* cholesterol all refer to about the same thing; this book uses the term *blood* cholesterol. Plasma is blood with the cells removed; in serum the clotting factors are also removed. The concentration of cholesterol is not much altered by these treatments.

Table 5-2

Recommendations Concerning Intakes of Fats

1. Total Fat[a]

   Dietary Guidelines
   - Choose a diet low in fat.

   Daily Values[b]
   - 65 grams fat per day.

   Proposed Healthy People 2010[c]
   - Increase the proportion of people aged 2 and older who meet the average daily goal of no more than 30 percent of calories from fat.

   World Health Organization[d]
   - Lower limit for total fat intake: 15% of total calories from fat.[e]
   - Upper limit for total fat intake: 30% of total calories from fat.[f]

2. Saturated Fat

   Dietary Guidelines
   - Choose a diet low in saturated fat.

   Daily Values[b]
   - 20 grams of saturated fat per day.

   Proposed Healthy People 2010[c]
   - Increase the proportion of people aged 2 and older who meet the average daily goal of no more than 10 percent of calories from saturated fat.

   World Health Organization
   - Lower limit for saturated fat intake: 0% of total calories from saturated fat.
   - Upper limit for saturated fat intake: 10% of total calories from saturated fat.

3. Polyunsaturated Fatty Acids

   World Health Organization
   - Lower limit for polyunsaturated fat intake: 3% of total calories from polyunsaturated fatty acids.
   - Upper limit for polyunsaturated fat intake: 7% of total calories from polyunsaturated fatty acids.

4. Cholesterol

   Dietary Guidelines
   - Choose a diet low in cholesterol.

   Daily Values[b]
   - 300 milligrams cholesterol per day.

   World Health Organization
   - Lower limit for cholesterol intake: 0 milligrams cholesterol per day.
   - Upper limit for cholesterol intake: 300 milligrams cholesterol per day.

[a]Includes monounsaturated fatty acids.
[b]The Daily Values are for a 2,000-calorie diet.
[c]Other recommendations concerning lipid intakes may be included in the final version.
[d]WHO and FAO Joint Consultation, Fats and oils in human nutrition, *Nutrition Reviews* 53 (1995): 202–205.
[e]Except for women of reproductive age, who should consume at least 20% of energy from fat.
[f]Sedentary individuals. Active individuals may consume up to 35% of total energy from fat if saturated fat does not exceed 10% of calories and the diet is otherwise adequate.

Genetics modifies everyone's ability to handle food cholesterol somewhat.[10] About 10 percent of people exhibit little increase in their blood cholesterol even with a high dietary intake. About 10 percent more respond to the same diet with greatly increased blood cholesterol. A few individuals have inherited a total inability to clear from their blood the cholesterol they have eaten and absorbed. This condition is rare, but it is well known because studying it led to the discovery of how cholesterol is transported in the body. People with a genetic tendency toward high blood cholesterol must strictly limit fats and refrain from eating foods rich in cholesterol. Most people can eat limited amounts of eggs, liver, and other cholesterol-containing foods without fear of incurring high blood cholesterol. For most, moderation, not elimination, is key where cholesterol-containing foods are concerned.

Figure 5-11

Food Fat and Calories.

Fat hides calories in food. When you trim fat, you trim calories

Nutrition Facts

Amount Per Serving

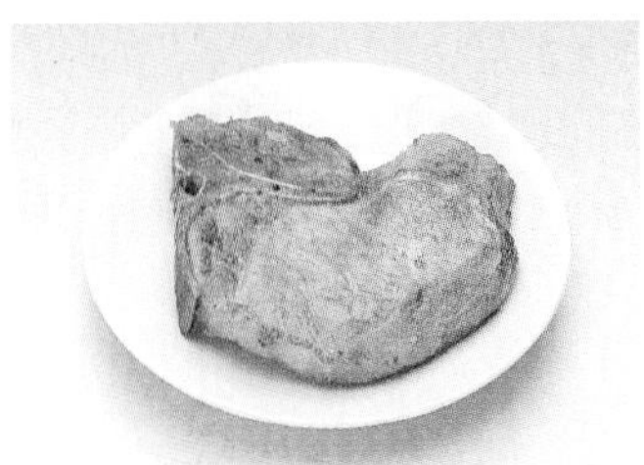

Pork chop (5 ounces) with ½ inch of fat	
Calories 450	Calories from Fat 315
	% Daily Value*
Total Fat 35g	**54%**
Saturated Fat 13g	**65%**

Potato (5 ounces) with 1 tablespoon butter and 1 tablespoon sour cream	
Calories 400	Calories from Fat 250
	% Daily Value*
Total Fat 28g	**43%**
Saturated Fat 18g	**90%**

Whole Milk
Whole Milk
Whole Milk
Whole Milk
Whole Milk
Whole Milk

Whole milk (1 cup)	
Calories 150	Calories from Fat 70
	% Daily Value*
Total Fat 8g	**12%**
Saturated Fat 5g	**25%**

Pork chop (4 ounces) with fat trimmed off	
Calories 230	Calories from Fat 100
	% Daily Value*
Total Fat 11g	**17%**
Saturated Fat 4g	**20%**

Plain potato (5 ounces)	
Calories 150	Calories from Fat 0
	% Daily Value*
Total Fat 0g	**0%**
Saturated Fat 0g	**0%**

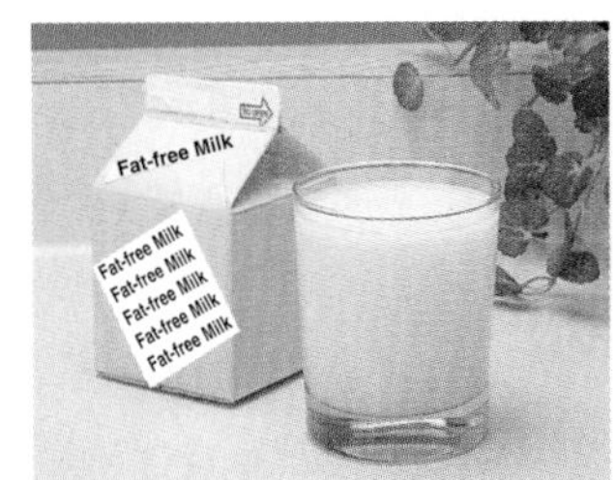
Fat-free Milk

Fat-free milk (1 cup)	
Calories 90	Calories from Fat 0
	% Daily Value*
Total Fat 0g	**0%**
Saturated Fat 0g	**0%**

To estimate the approximate number of fat grams allowed in a day to limit fat calories to 30% of the total, use this rule of thumb:

Estimate the total energy need (in calories)

Drop the last digit, and divide by 3.

For example, for a 2,300-calorie energy need:

$\frac{230}{3}$ = 77 g fat/day.

An effective dietary tactic against high blood cholesterol is to trim the fat, and especially the saturated fat, from foods. The photos of Figure 5-11 show that food trimmed of fat is also trimmed of much of its energy. A pork chop trimmed of its border of fat loses 220 calories. A plain baked potato has about 40 percent of the calories of one with butter and sour cream. Choosing fat-free milk over whole milk provides large savings of fat, saturated fat, and calories. The single most effective step you can take to reduce a food's potential for elevating blood cholesterol is to eat it without the fat.

KEY POINT *An important distinction: it is not the intake of cholesterol in foods but the saturated fat intake that is the major dietary factor that raises blood cholesterol. Elevated blood cholesterol is a risk factor for cardiovascular disease. Trimming fat from food trims calories and, often, saturated fat as well.*

What Is the Significance of LDL and HDL in Blood?

To repeat, cholesterol in foods contributes somewhat to cholesterol in the blood, and excesses of food cholesterol should be avoided. Dietary cholesterol is not as influential in raising blood cholesterol, however, as is dietary fat, especially saturated fat, which the body uses to make cholesterol. Now the link to the LDL can be explained. When a person's high blood cholesterol signifies a risk of heart disease, it is because the cholesterol, which is carried in LDL, is traveling to body tissues to be deposited there. If a person has high blood cholesterol in HDL, that is cause for celebration, not for concern. The vehicle matters.

Elevated LDL forecast heart and artery disease; elevated HDL signify a low disease risk. The rule of thumb is that a minimum of 35 milligrams HDL per deciliter of blood or plasma is associated with a low risk of heart attack. This amount of HDL seems to be especially predictive when compared with the total cholesterol count.[11] Another rule of thumb is that total blood cholesterol should be no more than about four times higher than HDL cholesterol for an acceptable level of risk.

Fortunately, the changes in diet that reduce blood cholesterol concentration mostly do so by reducing LDL. HDL concentration remains unaffected for the most part. The most influential dietary factor thought to raise blood cholesterol is a diet high in saturated fat. A person who attempts to lower blood cholesterol by reducing only the total fat in the diet may have little success.[12] Better to reduce total fat *and* replace cholesterol-raising saturated fat with monounsaturated or polyunsaturated fat. No beneficial change in blood lipids is seen when monounsaturated or polyunsaturated fat is *added* to a diet rich in saturated fat. That is, an eater of a bacon double cheeseburger cannot expect health protection from adding olive oil dressing to a side salad.

An important detail about LDL concerns its susceptibility to damage by **oxidation.**[13] Oxidation of the lipid part of LDL is thought to play a role in the injury of the arteries of the heart. **Antioxidant** nutrients, such as vitamin C and vitamin E, slow LDL oxidation. Other antioxidants found among the phytochemicals may also be helpful in this regard.

Some health authorities say all adults should take steps to reduce their blood cholesterol; others say that only those medically identified as at risk for heart disease should do so. In any case, it seems desirable for most people to limit saturated fat intake for health' sake. A diet that provides less than 30 percent of its calories from fat and less than 10 percent from saturated fat and that is rich in vegetables offers many advantages for health to anyone who eats it by supplying abundant nutrients and antioxidants along with beneficial fiber.

What about cholesterol intake? People respond differently. Dietary cholesterol has only a minimal effect on the blood cholesterol of about two-thirds of people. The other third must limit intakes to keep blood cholesterol from rising too high. Eggs, shellfish, liver, and other cholesterol-containing foods are nutritious, however. Cholesterol differs from salt and added fats and sugar in this respect: it cannot be omitted from the diet without omitting nutritious foods.

People can also take steps to raise HDL levels through exercise, a key weapon with wide-sweeping results in the fight against heart disease, as the Fitness for Life feature on the next page explains. Weight loss in the overweight and, for smokers, quitting smoking are other means of benefiting the health of the heart. Another factor probably unrelated to blood cholesterol but involving other effects on heart health is consumption of the essential fatty acids mentioned earlier. These relationships deserve some discussion.

KEY POINT *Dietary measures to lower LDL in the blood involve reducing saturated fat and substituting monounsaturated and polyunsaturated fats for saturated fat. A few people must also reduce cholesterol intake. Cholesterol-containing foods are nutritious and are best used in moderation.*

Essential Polyunsaturated Fatty Acids

The human body can use carbohydrate, fat, or protein to synthesize nearly all the fatty acids it needs. Two are well-known exceptions: **linoleic acid** and **linolenic acid.** These two polyunsaturated fatty acids, which the body needs for its basic functions, cannot be made from other substances in the body or from each other. They must be supplied by the diet and are therefore essential nutrients. These essential fatty acids are found in

oxidation interaction of a compound with oxygen; in this case, a damaging effect by a chemically reactive form of oxygen (Controversy 7 provides details).

antioxidant (anti-OX-ih-dant) a compound that protects other compounds from oxygen by itself reacting with oxygen (*anti* means "against"; *oxy* means "oxygen").

linoleic (lin-oh-LAY-ic) **acid** and **linolenic** (lin-oh-LEN-ic) **acid** polyunsaturated fatty acids that are essential nutrients for human beings.

Here's a trick:
Remember **H**DL is **H**ealthy.
LDL is **L**ess healthy.

Desirable Blood Lipid Values (mg/dL):

Total cholesterol <200.
LDL <130.
HDL >35.
Triglycerides <200.

Antioxidant nutrients are topics of Chapter 7 and its Controversy.

omega-6 fatty acid a polyunsaturated fatty acid with its endmost double bond six carbons from the end of the carbon chain. Also called n-6 fatty acid. Linoleic acid is an example.

FITNESS FOR LIFE

Why Exercise the Body for the Health of the Heart?

There's no doubt that, along with a carefully chosen diet, regular physical activity provides powerful disease protection for the heart and arteries. The heart, arteries, blood, and other body tissues respond when given the exercise they need by developing:

- Higher HDL concentration in the blood, an indicator of low risk of heart disease.
- Stronger muscles of the heart and arteries, needed to pump blood efficiently.
- Healthy body composition associated with low risk of heart disease.
- Reduced blood pressure; high pressure is a major risk factor for CVD.
- Improved insulin response, a factor associated with low risk of heart disease.
- Improved blood circulation to the heart's muscles; blocked vessels can result in heart attack.
- Greater heart volume; the heart pumps more blood with each beat, so its workload is reduced.

Is it any wonder that every leading authority lists physical activity among recommendations for maintenance of a healthy heart?

small amounts in the oils of plants and cold-water fish and are readily stored in the adult body. Both serve as raw materials from which the body makes hormonelike substances that regulate a wide range of body functions: blood pressure, blood clot formation, blood lipids, the immune response, the inflammation response to injury and infection, and many others.[‡14] Essential fatty acids also serve as structural parts of cell membranes, constitute a major part of the lipids of the brain and nerves, and are essential to normal growth in infants and children. To summarize then, essential polyunsaturated fatty acids:

- Provide raw materials for regulatory substances.
- Serve as structural parts of cell membranes.
- Constitute a major part of lipids in the brain and nerves.
- Are essential to normal growth.

To read about a possible relationship between polyunsaturated fatty acids and cancer, see Chapter 11.

To fail to obtain enough of the essential fatty acids is to invite ills of many kinds.

Deficiencies of Essential Fatty Acids A deficiency of an essential fatty acid in the diet leads to observable changes in cells, some more subtle than others. When the diet is deficient in *all* of the polyunsaturated fatty acids, symptoms of reproductive failure, skin abnormalities, and kidney and liver disorders appear. In infants and children, growth is retarded. Luckily, these extreme deficiency disorders are seldom seen except when intentionally induced in research. They sometimes do arise, however, on rare occasions when inadequate diets are provided by mistake. One such mistake is to feed hospital clients a formula lacking essential fatty acids through a vein for long periods. Another is to feed infants exclusively on a formula that lacks essential polyunsaturated fatty acids. Normal food, and especially a balanced diet that includes grains, seeds, nuts, leafy vegetables, and fish, supplies all the needed forms of fatty acids in abundance and prevents deficiencies.

The Omega-6 and Omega-3 Difference Linoleic acid is the primary member of a group of fatty acids named the **omega-6 fatty acid** family after their chemical structure. The body can convert linoleic acid to the other members of its omega-6 family, and these play active parts in body functioning. One plays a critical

‡The hormonelike derivatives referred to here are short-lived *eicosanoid* (eye-COSS-a-noid) compounds, such as prostaglandins and thromboxanes.

role in the cell membranes that define and protect each cell of the body. Any diet that contains vegetable oils, seeds, nuts, and whole-grain products supplies enough linoleic acid to meet the body's needs. Almost everyone eats enough.

Linolenic acid is the primary member of the **omega-3 fatty acid** family. This family has come to be appreciated for its role in health. Someone thought to ask why the native people of Greenland and Alaska, who eat a diet very high in fat, have such a low death rate from heart disease.[15] The trail led to the abundance of fish and other marine life that they eat, then to the oils in those fish, and finally to two omega-3 fatty acids, **EPA** and **DHA,** in the oils. Since then, research has revealed that these fatty acids make up a large part of the brain's thinking part, the cerebral cortex, and of the eye's main center of vision, the retina, and are necessary for the normal development of these organs. Additionally, omega-3 fatty acids are converted to hormonelike products that affect the heart and immune function.[16] Specifically, omega-3 fatty acids:

- Make up a large portion of the brain's cerebral cortex and are required for its development.
- Help to form the eye's retina and are required for development of normal vision.
- Are converted to hormonelike products that affect the heart and immune system.

These fatty acids can be made to some extent in the body from linolenic acid, or they can be derived from some foods, especially fatty fish.

Recommendations and Intakes No official intake recommendations for omega-6 and omega-3 fatty acids exist yet, but they soon will. Experts proclaim the U.S. diet deficient in omega-3 fatty acids, but excessive in omega-6.[17] They recommend more balanced intakes. Meanwhile their best dietary advice is to eat meals of fish two or three times a week, totaling about 10 ounces of fish, as well as small amounts of vegetable oils, to obtain the right balance between omega-3 and omega-6 intakes.[18] Even one fish meal per week has been associated with a reduced risk of heart attack, and people who make it a point to eat more than this amount suffer from strokes only half as often as those who eat no fish.[19] The ratio of omega-3 to omega-6 fatty acids should be about 1 to 4. This ratio may even turn out to be the key to human requirements; more omega-3 acids may not necessarily be better.[20]

Fish Oil Supplements Purchasing and taking fish oil supplements is not recommended, although many claims are made for their power to cure diseases. Supplement makers and others claim that fish oil capsules can cure arthritis, prevent cancer, reverse heart disease or boost the immune system, but although convincing evidence exists concerning heart disease prevention, other claims are based on only inconclusive preliminary research.[21] The Food and Drug Administration (FDA) does not allow labels to claim that fish oil supplements can prevent or cure diseases because these effects are not proved. In Canada, fish oil supplements require a physician's prescription.

Overdoses of fish oil may not be safe. The brain preferentially takes up omega-3 fatty acids with unknown results. They are also among the most vulnerable of the lipids to damage by oxidation, and people taking fish oil capsules may experience an increase in oxidative cell damage. Vitamin E in high doses can prevent the damage, but scientists are still investigating how much vitamin E is needed to do so.[22] One surprising finding is an increase in the number and size of tumors in rats fed fish oil when compared with rats fed corn oil or low-fat chow.[23] No one knows what such findings mean to people who are buying and taking untested supplements of fish oil, but the old axiom "let the buyer beware" seems to apply. A proven drawback is that fish oil supplements are made from fish skins and livers, which may have accumulated toxic concentrations of pesticides, heavy metals, and other ocean contaminants that may be further concentrated in the pills. Moreover, even without contamination, fish oil naturally contains high levels of the two most potentially toxic vitamins, A and D. Lastly, supplements of fish oil are expensive. So little is known about long-term effects of fish oil supplements that taking them is chancy. Better to go to the source for fish oils: eat fish.

omega-3 fatty acid a polyunsaturated fatty acid with its endmost double bond three carbons from the end of the carbon chain. Also called n-3 fatty acid. Linolenic acid is an example.

EPA, DHA eicosapentaenoic (EYE-cossa-PENTA-ee-NO-ick) acid, docosahexaenoic (DOE-cossa-HEXA-ee-NO-ick) acid; omega-3 fatty acids made from linolenic acid in the tissues of fish.

Here's where to find the omega fatty acids:

Omega-6

Linoleic acid:

- Leafy vegetables, seeds, nuts, grains, vegetable oils (corn, safflower, soybean, cottonseed, sesame, sunflower).

Omega-3

Linolenic acid:

- Fats and oils (canola, soybean, walnut, wheat germ, margarine and shortening made from canola and soybean oil).
- Nuts and seeds (butternuts, walnuts, soybean kernels).
- Vegetables (soybeans).

EPA and DHA:

- Human milk.
- Shellfish and fish[a] (mackerel, salmon, bluefish, mullet, sablefish, menhaden, anchovy, herring, lake trout, sardines, tuna).
- Or can be made from linolenic acid.

[a]All of these fish except tuna provide at least 1 gram of omega-3 fatty acids in 100 grams of fish (3.5 ounces); the fish oil content of each species varies with the season and site of harvest. Tuna is lower in omega-3 fatty acids, but because it is commonly consumed, its contribution can be significant.

More about vitamin E's antioxidant effects in Controversy 7; more about the safety of seafood in Chapter 14.

hydrogenation (high-dro-gen-AY-shun) the process of adding hydrogen to unsaturated fatty acids to make fat more solid and resistant to the chemical change of oxidation.

smoking point the temperature at which fat gives off an acrid blue gas.

KEY POINT *Two polyunsaturated fatty acids, linoleic acid (an omega-6 acid) and linolenic acid (an omega-3 acid), are essential nutrients used to make hormonelike substances that are prominent in the brain and in the retina of the eye and that perform many other functions. The omega-6 family includes linoleic acid; seed oils are rich sources. The omega-3 family includes linolenic acid, EPA, and DHA. Fish oils are rich sources of EPA and DHA.*

The Effects of Processing on Unsaturated Fats

Vegetable oils make up most of the added fat in the U.S. diet because fast-food chains use them for frying, food manufacturers add them to processed foods, and consumers tend to choose margarine over butter. Food manufacturers often process vegetable oils in ways that greatly change their effects on the body.

Consumers of vegetable oils may feel safe in choosing them because they are generally less saturated than animal fats. If consumers choose a liquid oil, they may be justified in feeling secure. If the choice is a processed food, however, their security may be questionable, especially if the word *hydrogenated* appears on the label's ingredient list.

What Is "Hydrogenated Vegetable Oil," and What's It Doing in My Peanut Butter?

When manufacturers process foods, they often alter the fatty acids in the fat (triglycerides) the foods contain through a process called **hydrogenation.** Hydrogenation of fats makes them stay fresher longer and also changes their physical properties.

Points of unsaturation in fatty acids are weak spots that are vulnerable to attack by oxygen. Oxidative damage is not confined to fats within body tissues but occurs anywhere oxygen mixes with fats. When the unsaturated points in the oils of food are oxidized, the oils become rancid and the food tastes "off."[24] This is why cooking oils should be stored in tightly covered containers that exclude air. If stored for long periods, they need refrigeration to retard oxidation.

One way to prevent spoilage of unsaturated fats and also to make them harder and more stable when heated to high temperatures is to change their fatty acids chemically by hydrogenation, as shown in Figure 5-12. When food producers want to use a polyunsaturated oil such as corn oil to make a spreadable margarine, for example, they hydrogenate it by forcing hydrogen into the oil. Some of the unsaturated fatty acids become more saturated as they accept the hydrogen, and the oil hardens. The product that results is more saturated and more spreadable than the original oil. It is also more resistant to damage from oxidation or breakdown from high cooking temperatures.

Once hydrogenated, oils lose their unsaturated character and the health benefits that go with it. An alternative to hydrogenation is to add a chemical preservative that will compete for oxygen and thus protect the oil. The additives are antioxidants, and they work just as vitamin E does, by reacting with oxygen before it can do damage. Examples are the well-known additives BHA and BHT[§] listed on snack food labels. Another alternative, already mentioned, is to keep the product refrigerated.

If you, the consumer, are looking for polyunsaturated oils to include in your diet, hydrogenated oils such as those in shortening, stick margarine, or peanut butter with added hydrogenated oils will not meet your need. Hydrogenated oils are easy to handle, easy to spread, and store well. For example,

Figure 5-12

How Hydrogenation Makes Fats More Saturated

Points of unsaturation are places on fatty acid chains where hydrogen is missing. The bonds that would normally be occupied by hydrogen in a saturated fatty acid are shared, reluctantly, as a double bond between two carbons that both carry a slightly negative charge.

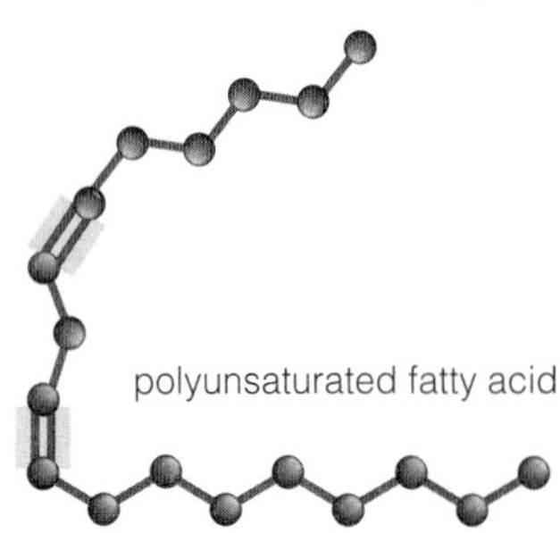
polyunsaturated fatty acid

When positively charged hydrogen is made available to one of those bonds, it readily accepts the hydrogen molecules and, in the process, becomes saturated. It no longer has a point of unsaturation.

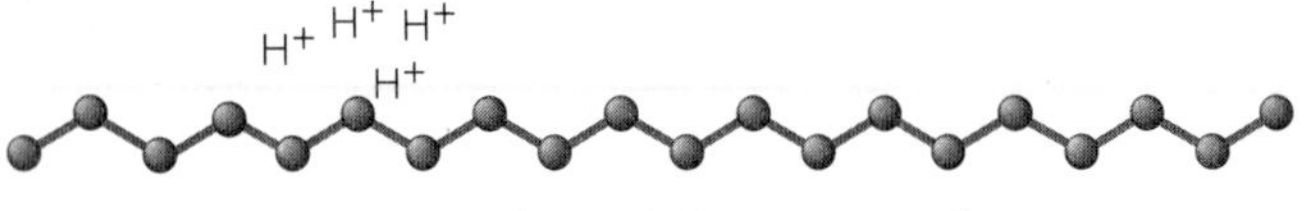
H+ H+ H+
H+
hydrogenated fatty acid (now saturated)

[§] BHA and BHT are butylated hydroxyanisole and butylated hydroxytoluene.

makers of peanut butter often replace some of the liquid oil from the ground peanuts with hydrogenated vegetable oils to create a creamy paste that does not separate. Peanut butters made without hydrogenated oils, the so-called "natural" types, separate into layers of liquid oil and peanut solids that must be stirred together before use. Hydrogenated oils also have a high **smoking point**, so they are suitable for purposes such as frying. Hydrogenated oils are more saturated than the oils from which they are made, however. Margarines that list liquid oil as the first ingredient are usually the most polyunsaturated, especially those made from highly unsaturated oils, such as safflower or canola. Margarines that are sold in tubs or squeeze bottles or labeled "soft" are sometimes less saturated than the stick varieties.

KEY POINT *Vegetable oils become more saturated when they are hydrogenated. Hydrogenated fats resist rancidity better, are firmer textured, and have a higher smoking point than unsaturated oils.*

trans-fatty acids fatty acids with unusual shapes that can arise when polyunsaturated oils are hydrogenated.

stanol esters compounds belonging to the sterol family of lipids, derived from plants, that have been shown experimentally to reduce blood cholesterol when consumed in place of other fats in a low-fat diet.

What Are *Trans*-Fatty Acids, and Are They Harmful?

Another concern about hydrogenation of fat centers around a change in chemical structure that occurs when polyunsaturated oils are hardened by hydrogenation. Some of the unsaturated fatty acids, instead of becoming saturated, end up changing their shapes (see Figure 5-13). This process creates unusual products that are not made by the body and that occur naturally in limited amounts, mainly in dairy foods and beef.[25] These changed fatty acids, or ***trans*-fatty acids,** may have implications for the body's health. Many researchers estimate that *trans*-fatty acids carry a risk to the health of the heart and arteries between that of saturated and unsaturated fats.[26] The risk arises because *trans*-fatty acids may elevate LDL cholesterol and lower beneficial HDL.[27] One study found a correlation between eating margarine, a source of *trans*-fatty acids, and increased incidence of cardiovascular disease.[28]

When processing changes omega-3 fatty acids into their *trans* counterparts, the eater derives none of their associated benefits. Finally, a high total fat consumption is also associated with cancer susceptibility, and *trans*-fatty acids contribute to the total fat intake. No strong evidence suggests that *trans*-fatty acids by *themselves* play a specific role in promoting or causing cancer, but research is continuing.[29]

When news of *trans*-fatty acids' possible effects on heart health first emerged, some people hastily switched from using margarine back to butter, believing oversimplified reports that margarine provided no heart health advantage over butter. It is true that most margarines and virtually all shortenings are made largely from hydrogenated fats and therefore are saturated and contain substantial *trans*-fatty acids—up to 40 percent. Some margarines, however, especially the soft or liquid varieties, are made from unhydrogenated oils. These have long proved to be less likely to elevate blood cholesterol than the saturated fats of butter. When oils (but not hydrogenated oils) are the first ingredient listed on a margarine label, that margarine is, in all probability, low in *trans*-fatty acids as well as in saturated fat.

In addition to soft and liquid margarine choices, consumers can now choose margarines containing few or no *trans*-fatty acids. Some of these also contain a newly approved ingredient, **stanol esters,** that reduces blood cholesterol when consumed in addition to a low-fat diet.* Stanol esters are not recognized by the intestine and therefore are not absorbed, and they also block the absorption of cholesterol.[30] Simply adding the margarine to a high-fat diet is unlikely to bring benefits, however. Stanol esters work only when people are also willing to cut their fat intakes as well. Drawbacks include the price (three or four times higher than regular margarine), a high fat content (the full fat kind equals the fat in regular margarine), and an unproved safety record (especially concerning daily use by healthy people of all ages).

Regular hard types of margarine contribute *trans*-fatty acids to the diet, but foods other than margarine contribute still more *trans*-fatty acids, and more total fat, too. Fast foods, chips, baked goods, and other commercially prepared foods are high in fats containing

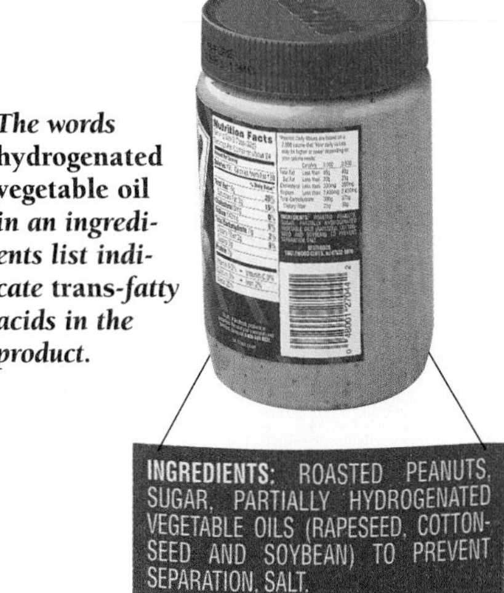
INGREDIENTS: ROASTED PEANUTS, SUGAR, PARTIALLY HYDROGENATED VEGETABLE OILS (RAPESEED, COTTONSEED AND SOYBEAN) TO PREVENT SEPARATION, SALT.

The words* hydrogenated vegetable oil *in an ingredients list indicate* trans-fatty *acids in the product.

Figure 5-13

A Trans Fatty Acid

Note that an unsaturated trans-*fatty acid is similar in shape to a saturated fatty acid (review Figure 5-12); it behaves similarly in the body, too.*

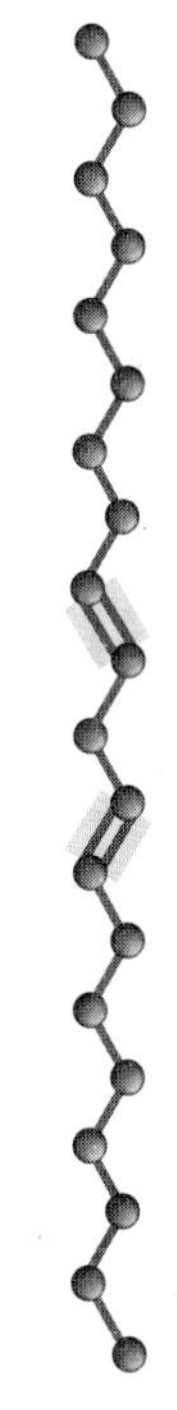

*The brand names of margarines with stanol esters that are currently on the market are *Benecol* and *Take Control*.

Some common food sources of *trans*-fatty acids:

- Most margarines, shortenings, and peanut butters.
- Fried fast foods.
- Salad dressing, mayonnaise.
- Biscuits, rolls, cakes, crackers.
- Corn snacks and chips.
- Other fried snacks and chips.
- Cookies, doughnuts.
- French fries, fried chicken or fish.

up to 50 percent *trans*-fatty acids. Overall, consumers are now eating more fats containing *trans*-fatty acids than ever before, amounting to about 3 percent of daily calories, and they are eating the *trans*-fatty acids in the form of processed foods (see the margin list).[31]

Food labels can be misleading in this regard. On Nutrition Facts panels, *trans*-fatty acids are counted with the polyunsaturated fats from which they arose, and not with the saturated fats whose health effects they mimic. Further, fast-food chains advertise foods fried in "vegetable oil" when that oil is a hydrogenated type containing abundant *trans*-fatty acids. Some experts are calling for a separate statement of *trans*-fatty acids on food labels to help consumers make informed choices.[32]

KEY POINT *The process of hydrogenation creates* trans-*fatty acids.* Trans-*fatty acids act somewhat like saturated fats in the body.*

Fat in the Diet

The remainder of this chapter shows you how to choose the right kinds of fat, and the right amounts, to provide optimal health and pleasure in eating. As you read, notice which foods offer unsaturated fat and which offer saturated fat. Your choices among them can make a difference in the unseen condition of your arteries.

Remember that some fat is necessary for health. People who take fat recommendations to an extreme and try to eliminate all traces of fat from food do so at their peril. Most adults need about 15 percent of their daily energy in the form of fat, an amount that is far, far smaller than the average intakes in this country.[33] Fat occupies, on average, 34 percent of total daily calories, down from 45 percent in 1965. At first glance, this appears to be a healthy trend—until the actual grams of fat and carbohydrate are inspected more closely. The total number of fat grams people take in has actually increased, and not decreased at all, but the carbohydrate grams increased even more during the same time period, causing energy intakes to rise. The result is a misleading drop in the percentage of fat calories from total energy intakes but not a drop in actual fat consumed.[34] Bottom line: most people must work to learn to recognize the fats in foods and reduce their intakes.

In the Food Guide Pyramid, two groups always contain fat (the fats and the meats and nuts), and two sometimes contain fat (the milk and milk products and the breads). Most unprocessed vegetables and fruits are fat-free, but two exceptions are rich in monounsaturated fat—avocados and olives. In their natural states, grains are like fruits and vegetables in that they contain little or no fat. Keep in mind that fats may be visible on foods, such as the fat trimmed from a steak, or they may be invisible, such as the fats in the marbling of meat, the fat ground into lunch meats, or the fats in avocados, biscuits, cheese, coconuts, other nuts, olives, or fried foods.**

Added Fats

A dollop of dessert topping, a spread of butter on bread, oil or shortening in a recipe, dressing on a salad—all of these are examples of *added* fats. All sorts of fats can be added to foods during commercial or home preparation or at the table. The following amounts of these fats contain about 5 grams of pure fat, providing 45 calories and negligible protein and carbohydrate:

- 1 teaspoon oil or shortening.
- 1½ teaspoon mayonnaise, butter, or margarine.
- 1 tablespoon regular salad dressing, cream cheese, or heavy cream.
- 1½ tablespoon sour cream.

These foods provide the majority of added fats to the diet. They are the hidden fats of fried foods or baked goods, sauces and mixed dishes, and dips and spreads.

KEY POINT *Fats added to foods during preparation or at the table are a major source of fat in the diet.*

Ten small olives or a sixth of an avocado each provide about 5 grams of mostly monounsaturated fat.

**Coconut is a nut, not a vegetable, although its oil is listed among vegetable oils.

CONSUMER CORNER

Fat Replacers

TODAY, SHOPPERS can choose from thousands of fat-reduced products. Many bakery goods, cheeses, frozen desserts, and other products made with **fat replacers** (see Table 5-3) now offer less than half a gram of fat in a serving. While some of these products contain **artificial fats,** others use conventional ingredients in unconventional ways to reduce fat and calories. Among the latter, manufacturers can:

- Add water or whip air into foods.
- Add fat-free milk to creamy foods.
- Use lean meats and soy protein to replace high-fat meats.
- Bake foods instead of frying them.

Common food ingredients such as fibers, sugars, or proteins can also take the place of fats in some foods. Products made this way still provide calories, but far fewer calories from fat.

Manufactured fat replacers consist of chemical derivatives of carbohydrate, protein, or fat, or modified versions of foods rich in those constituents (see Table 5-4). To gain the FDA's consent for the use of a new fat replacer in the food supply, U.S. manufacturers must prove that their fat replacer contributes little food energy, is nontoxic, is not stored in body tissues, and does not rob the body of needed nutrients. Olestra serves as an example of an artificial fat that has won FDA approval, but still awaits a final judgment that will determine its future: the approval of consumers.

An Artificial Fat: Olestra

Olestra, brand name Olean, is a member of the **sucrose polyester** chemical family. Chemically, olestra bears some resemblance to ordinary fat; it consists of a core molecule of the carbohydrate sucrose to which up to eight fatty acid molecules are bonded (see Figure 5-14). In comparison, ordinary triglycerides consist of a core of the carbohydrate glycerol to which three fatty acids are bonded. The human digestive enzymes that break down triglycerides in the digestive tract do not recognize the shape of the olestra molecule and so cannot split the fatty acids from their sucrose. All of the olestra eaten in a food remains undigested and passes through the digestive system intact.

Olestra's Advantages From some points of view, olestra is the most successful of the artificial fats, for its properties are identical to those of fats and oils when used in frying, cooking, and baking. It can be heated to frying temperatures without breaking down; it performs all of the functions of fat in cakes, pie crusts, and other baked goods; and most remarkably—aside from a slight aftertaste—it tastes like fat. Nevertheless, the wonders of olestra must be weighed against evidence concerning its safety.

Concerns about Olestra The company that invented olestra has been studying its safety for more than two decades. The results of the studies revealed that olestra is safe in most regards, but can cause digestive distress, nutrient losses, and losses of phytochemicals. The company has addressed each of these problems to the satisfaction of an FDA commission panel, which approved olestra for use in snack foods in 1996. Some nutrition experts disagree with the panel, however, and have come out strongly against olestra's use. Their concerns deserve a moment's attention.

The presence of indigestible olestra in the large intestine can potentially cause diarrhea, gas, cramping, and an urgent need for defecation in some people. Further, oily olestra can creep through the feces and leak uncontrollably from the anus, producing smelly dark yellow stains on underwear. The FDA commission that approved olestra decided that the digestive problems of olestra were unpleasant, but did not constitute a safety problem. The FDA

Table 5-3

Terms Related to Fat Replacers

- **artificial fats** zero-energy fat replacers that are chemically synthesized to mimic the sensory and cooking qualities of naturally occurring fats, but are totally or partially resistant to digestion. Also called *fat analogues.*
- **fat replacer** an ingredient that replaces some or all of the functions of fat and may or may not provide energy. Often used interchangeably with *fat substitute,* but the latter technically applies only to an ingredient that replaces all of the functions of fat and provides no energy.
- **olestra** a noncaloric artificial fat made from sucrose and fatty acids; formerly called *sucrose polyester.*
- **sucrose polyester** any of a family of compounds in which fatty acids are bonded with sugars or sugar alcohols. Olestra is an example.

Olestra and Triglyceride Compared

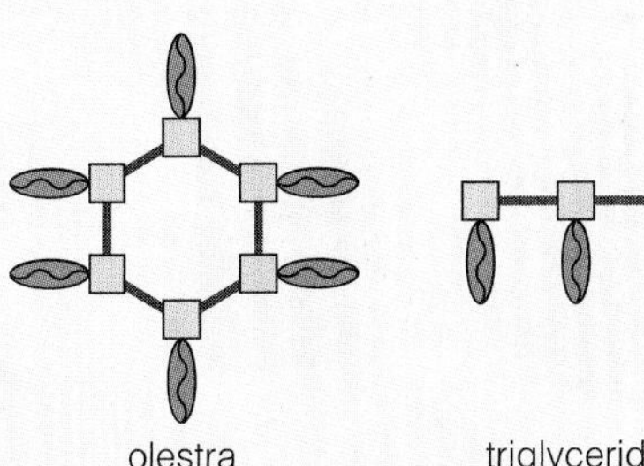
olestra
triglyceride

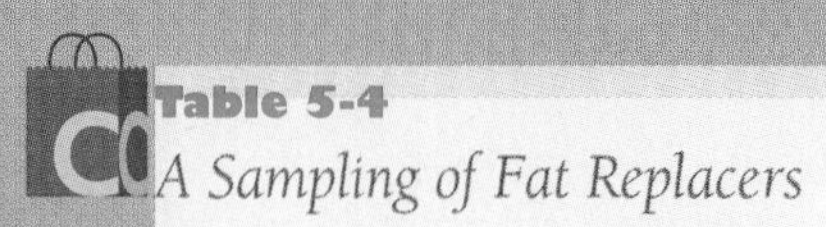

Table 5-4

A Sampling of Fat Replacers

Fat Replacers	Energy (cal/g)	Properties	Uses in Foods
Carbohydrate-Based Fat Replacers			
■ *Fruits* purees and pastes of apples, bananas, cherries, plums, or prunes; add bulk and tenderness to baked goods.	0	Replace bulk of fat; add moisture and tenderness.	Baked goods, candy, dairy products.
■ *Gels* derived from cellulose or starch to mimic the texture of fats in regular margarine and other products.	0–4[a]	Replaces bulk; lends thickness.	Fat-free margarines, salad dressing, frozen desserts.
■ *Gums* extracted from beans, sea vegetables or other sources.	0–4	Adds bulk; thickens salad dressings.	Salad dressing, processed meats, desserts.
■ *Maltodextrins* made from corn; is powdered and flavored to resemble the taste of butter.	1–4	Adds "buttery" flavor.	Butter-flavored "sprinkles" for melting on hot foods.
■ *Oatrim* derived from oat fiber; has the added advantage of providing satiety.	4	Creamy, replaces bulk of fat; can be used in baking but not frying.	Dips, dressings, baked goods.
■ *Z-trim* a modified form of insoluble fiber; is powdered and feels like fat in the mouth.	0	Creamy, replaces bulk of fat; can be used in baking but not frying.	Cheese, ground beef, chocolates, baked goods.
Fat-Based Fat Replacers			
■ *Olestra*[b] a noncaloric artificial fat made from sucrose and fatty acids; formerly called *sucrose polyester.*	0	Same properties as fats; heat stable in frying, cooking, and baking.	Potato chips, tortilla chips, crackers.
■ *Salatrim*[c] derived from fat and contains short- and long-chain fatty acids.	5	Same properties as fats; can be used in baking but not frying.	Chocolate coatings, dairy products, spreads.
Protein-Based Fat Replacers			
■ *Microparticulated protein*[d] processed from the proteins of milk or egg white into mistlike particles that roll over the tongue, making it feel and taste like fat.	4	Creamy; heat stable in some cooking and baking but not frying.	Ice cream, dairy products.

[a]Energy made available by action of colonic bacteria.
[b]Trade name: OLEAN.
[c]Trade name: Benefat.
[d]Trade names: Simplesse and K-Blazer.

a new experience in taste!
Nutrition Facts
let us know what you think!

requires olestra-containing foods to bear this warning:

> "**This Product Contains Olestra.** Olestra may cause abdominal cramping and loose stools. Olestra inhibits the absorption of some vitamins and other nutrients. Vitamins A, D, E, and K have been added."

When researchers recently provided olestra-containing snacks or ordinary snacks to some 3,000 volunteers, however, they found no significant increase in digestive distress with olestra.[1]

Olestra acts as a potent solvent for some of the fat-soluble substances in foods, such as the vitamins that dissolve in fat (vitamins A, D, E, and K). The absorption of these vitamins from a meal that includes olestra is greatly reduced because olestra dissolves them and carries them out of the digestive tract unabsorbed. To compensate for this effect, olestra is fortified with vitamins A, D, E, and K. The FDA ruled that fortification removes the threat of harm from malnutrition that olestra could otherwise cause.

An effect to which olestra's opponents object vigorously is the loss of health-promoting phytochemicals from foods. As an example, members of the carotene family, consumed over a lifetime, have been linked with prevention of a degenerative eye condition that affects many people as they age. One recent study showed that just 3 grams of olestra a day strongly reduced (by about 40 percent) the blood concentrations of lycopene, a phytochemical believed to defend the body against some forms of cancer (more on phytochemicals in foods in Controversy 11 in Chapter 11).[2] No studies exist to predict the effects, if any, of lifelong olestra exposure or the effects of olestra on growing children, but children often favor the foods in which olestra is allowed. Olestra's pros and cons are summed up in Figure 5-15.

FAT REPLACERS AND WEIGHT CONTROL

People hope, of course, that fat replacers will help fight both obesity and heart disease by lowering fat intakes. The question remains whether eating fat replacers will actually do these things because people may compensate for part or all of an energy reduction on one day by eating more energy in food on the next.[3] Also, a food made with fat replacers may be lower in fat than a full-fat counterpart, but not low enough to be called a "low-fat" food. Reduced-fat foods often deliver appreciable fat and as many calories as the regular product (compare the labels). And, if the U.S. experience with artificial sweeteners, described in Controversy 4, is any guide, consumers are likely to eat fat-replacer products *in addition* to other high-fat foods they prefer, negating the potential benefits of the fat replacers.

Used wisely, though, fat replacers may help some consumers achieve some of their dietary goals, especially when the user learns to cut fat from the diet in other ways as well.[4] The next section shows where the fat resides in foods, and the Food Feature gives practical advice about cutting fat in time-tested ways.

Figure 5-15
Olestra's Pros and Cons

Pros of Olestra

- Zero calories
- Zero fat and saturated fat
- Zero cholesterol
- Withstands frying
- Withstands baking
- Tastes like fat

Cons of Olestra

- Vitamin losses
- Phytochemical losses
- Possible digestive upset
- Possible anal leakage
- Slight aftertaste
- Expensive

Meat, Poultry, Fish, Dry Beans, Eggs, and Nuts

Meats conceal much of the fat—mostly saturated fat—that people unwittingly consume. To help people "see" the fat in meats, the exchange lists in Appendix D present the meats in four categories according to their fat contents: very lean, lean, medium-fat, and high-fat meats. Meats in all four categories contain about equal amounts of protein, but because their fat contents differ, their calorie amounts vary significantly. Figure 5-16 shows fat and calorie data for some ground meats.

According to the Food Guide Pyramid, a serving of meat amounts to just 2 or 3 ounces—very small by average consumption standards. A small, fast-food hamburger, for example, weighs about 3 ounces. A steak served in a restaurant averages 12 to 16 ounces, more than a whole day's meat allowance. Of course, your judgment of what is normal may differ from other people's, and you may have to weigh a serving or two of meat to see how much you are eating.

People think of meat as protein food, but calculation of its nutrient content reveals a surprising fact. A big (4-ounce), fast-food hamburger sandwich contains 23 grams of protein and 20 grams of fat. Because protein offers 4 calories per gram and fat offers 9, the sandwich provides 92 calories from protein and 180 calories from fat. The calo-

Figure 5-16

Fat in Ground Meats

Note that only the ground round, at 7 percent fat by weight, qualifies to bear the "lean" label. To be called "lean," products must contain fewer than 10 grams fat, 4 grams saturated fat, and 95 milligrams cholesterol per 100 grams of food. The numbers that qualify products to be called "extra lean" are, respectively, 5, 2, and 95. The red labels on these packages list rules for safe meat handling, explained in Chapter 14.

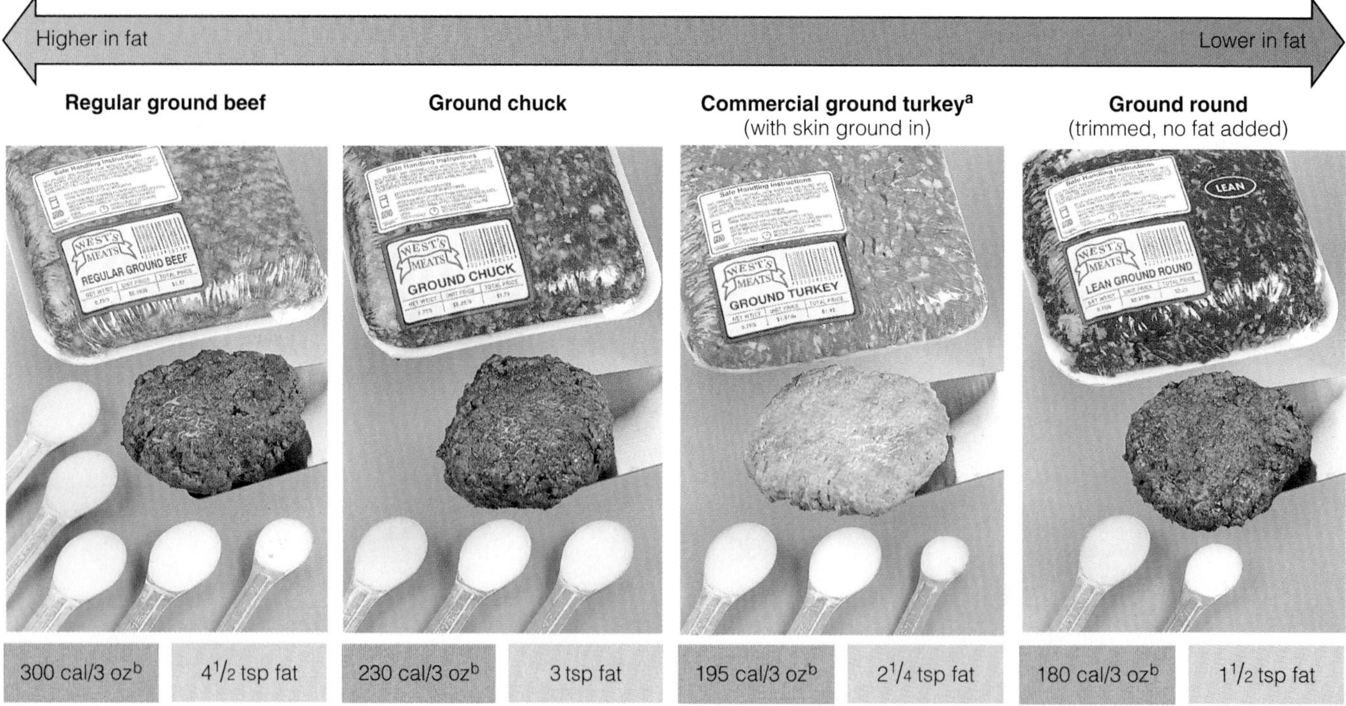

Higher in fat
Lower in fat
Regular ground beef
Ground chuck
Commercial ground turkey[a] (with skin ground in)
Ground round (trimmed, no fat added)
REGULAR GROUND BEEF
GROUND CHUCK
GROUND TURKEY
LEAN GROUND ROUND
LEAN
300 cal/3 oz[b]
4 1/2 tsp fat
230 cal/3 oz[b]
3 tsp fat
195 cal/3 oz[b]
2 1/4 tsp fat
180 cal/3 oz[b]
1 1/2 tsp fat

[a]Values for 3 ounces of cooked turkey breast ground without skin are 108 calories and ½ teaspoon fat (25% calories from fat). This type is not typically offered in many areas, but can be specially ordered from the butcher.

[b]The 3-ounce cooked serving used here may seem small to some, but it is the largest allowable meat serving according to the Daily Food Guide. Larger servings will, of course, provide more fat and calories than the values listed here.

rie total, counting carbohydrates from the bun and condiments, is over 400 calories, with more than 50 percent of them from fat. Hot dogs, fried chicken sandwiches, and fried fish sandwiches are also high-fat choices. Because so much of the energy in a meat eater's diet is hidden from view, people can easily overeat on high-fat food, making weight control difficult.

Animal breeders have been striving to produce beef and pork that are lower in fat.[35] This is a help to those shoppers who choose lean cuts: they get less fat in the same quantity of meat. When choosing beef or pork, look for lean cuts named *loin* or *round* from which the fat can be trimmed. Eat small portions, too.

Chicken and turkey meat is also naturally lean, but processing and frying add fats, especially in "patties," "nuggets," "fingers," or "wings." Chicken wings are mostly skin, and a chicken stores most of its fat just under its skin. The tastiest wing snacks have also been fried in cooking fat (often a saturated type), smothered with a buttery, spicy sauce, and then dipped in blue cheese dressing, making wings an extraordinarily high-fat snack. A person who snacks on wings should plan on eating low-fat foods at other meals to balance them out.

Watch out for ground turkey or chicken products. Many of these have the skin ground in to add moistness when cooked, and they can be much higher in fat than even lean beef, as Figure 5-16 has already shown. Table 2-10 on pp. 50 and 51 in Chapter 2 provided some definitions concerning the fat contents of meats.

KEY POINT *Meats account for a large proportion of the hidden fat in many people's diets. Most people consume meat in larger servings than those recommended.*

Milk, Yogurt, and Cheese

Some milk products contain fat. In homogenizing whole milk, milk processors blend in the cream, which otherwise would float and could be removed by skimming. A cup of whole milk, then, contains the protein and carbohydrate of fat-free milk, but in addition it contains about 60 extra calories from fat. A cup of reduced-fat (2 percent fat) milk falls between whole and fat-free, with 45 calories of fat. The fat of whole milk occupies only a teaspoon or two of the volume but nearly doubles the calories in the milk. Depending upon how much fat milk contains, it qualifies to bear one of the names listed in the margin (older names are listed in parentheses).

Milk's names:

- Milk, whole milk (whole milk).
- Reduced-fat, less-fat milk (2% milk).
- Low-fat milk (1% milk).
- Fat-free, zero-fat, or no-fat (skim or nonfat milk).

Milk and yogurt appear in the milk group, but cream and butter do not. Milk and yogurt are rich in calcium and protein, but cream and butter are not. Cream and butter are fats, as are whipped cream, sour cream, and cream cheese. That is why the food group that includes milk is carefully called the "milk, yogurt, and cheese group," and not the "dairy group." Figure 5-17 shows where the lipids are found in various kinds of milk, yogurt, and cheese.

KEY POINT *The choice between whole and fat-free milk products can make a large difference to the fat content of a diet.*

Bread, Cereal, Rice, and Pasta

Breads and cereals in their natural state are very low in fat, but they also may contain fat added during manufacturing, processing, or cooking. The fat in these foods can be particularly hard to detect, so diners must remember which foods stand out as being high in fat. Notable are granola and certain other ready-to-eat cereals, croissants, biscuits, cornbread, dinner rolls, fried rice, pasta with creamy or oily sauces, quick breads, snack and party crackers, muffins, pancakes, and homemade waffles. Packaged breakfast bars often resemble vitamin-fortified candy bars in their fat and sugar contents. Figure 5-18 shows the lipid contents of many grain products.

Figure 5-17

Lipids in Milk, Yogurt, and Cheese

Red shadows on labels indicate foods with higher lipid content that warrant moderation in their use. Green indicates lower-fat choices.

Nutrition Facts

Amount Per Serving

FAT FREE MILK
MILK
1% LOWFAT MILK
REDUCED FAT MILK
CHUNK

Fat-free, skim, zero-fat, no-fat, or nonfat milk, 8 oz (<0.5% fat by weight)

Calories 90	Calories from Fat 0
	% Daily Value*
Total Fat 0g	**0%**
Saturated Fat 0g	**0%**
Cholesterol 5mg	**2%**

Low-fat milk, 8 oz (1% fat by weight)

Calories 110	Calories from Fat 25
	% Daily Value*
Total Fat 2.5g	**4%**
Saturated Fat 1.5g	**8%**
Cholesterol 15mg	**5%**

Low-fat cheddar cheese, 1.5 oz

Calories 70	Calories from Fat 30
	% Daily Value*
Total Fat 3g	**5%**
Saturated Fat 2g	**10%**
Cholesterol 10mg	**3%**

Strawberry yogurt, 8 oz

Calories 250	Calories from Fat 40
	% Daily Value*
Total Fat 4g	**6%**
Saturated Fat 2.5g	**13%**
Cholesterol 15mg	**5%**

Whole milk, 8 oz (3.3% fat by weight)

Calories 150	Calories from Fat 70
	% Daily Value*
Total Fat 8g	**12%**
Saturated Fat 5g	**25%**
Cholesterol 35mg	**12%**

Reduced fat, less-fat milk, 8 oz (2% fat by weight)

Calories 120	Calories from Fat 45
	% Daily Value*
Total Fat 5g	**8%**
Saturated Fat 3g	**15%**
Cholesterol 20mg	**7%**

Cheddar cheese, 1.5 oz

Calories 170	Calories from Fat 130
	% Daily Value*
Total Fat 14g	**22%**
Saturated Fat 9g	**45%**
Cholesterol 45mg	**15%**

Low-fat strawberry yogurt, 8 oz

Calories 200	Calories from Fat 20
	% Daily Value*
Total Fat 2.5g	**4%**
Saturated Fat 1.5g	**8%**
Cholesterol 15mg	**5%**

KEY POINT *Fat in breads and cereals can be well hidden. Consumers must learn which foods of this group contain fats.*

Now that you know where the fats in foods are found, how can you reduce or eliminate them from your diet? The Food Feature that follows provides some pointers.

Figure 5-18

Lipids in Bread, Cereal, Rice, and Pasta

Red shadows on labels below indicate foods with higher lipid content that warrant moderation in their use. Green indicates lower-fat choices.

Nutrition Facts

Amount Per Serving

Low-fat granola, ½ c	
Calories 190	Calories from Fat 25
	% Daily Value*
Total Fat 3g	**5%**
Saturated Fat 0.5g	**3%**
Cholesterol 0mg	**0%**

A home-made waffle	
Calories 220	Calories from Fat 100
	% Daily Value*
Total Fat 11g	**17%**
Saturated Fat 2g	**10%**
Cholesterol 50mg	**17%**

A dinner roll	
Calories 80	Calories from Fat 20
	% Daily Value*
Total Fat 2g	**3%**
Saturated Fat 0g	**0%**
Cholesterol 0mg	**0%**

Crispy oat bran, ½ c	
Calories 130	Calories from Fat 35
	% Daily Value*
Total Fat 4g	**6%**
Saturated Fat 1.5g	**8%**
Cholesterol 0mg	**0%**

Fettuccine alfredo, ½ c	
Calories 250	Calories from Fat 130
	% Daily Value*
Total Fat 14g	**22%**
Saturated Fat 8g	**40%**
Cholesterol 60mg	**20%**

Buttery crackers, 4 crackers	
Calories 80	Calories from Fat 35
	% Daily Value*
Total Fat 4g	**6%**
Saturated Fat 1g	**5%**
Cholesterol 0mg	**0%**

A breakfast bar	
Calories 150	Calories from Fat 55
	% Daily Value*
Total Fat 6g	**9%**
Saturated Fat 2.5g	**13%**
Cholesterol 0mg	**0%**

Fried rice, ½ c[a]	
Calories 140	Calories from Fat 65
	% Daily Value*
Total Fat 7g	**11%**
Saturated Fat 1g	**5%**
Cholesterol 20mg	**7%**

A muffin	
Calories 170	Calories from Fat 65
	% Daily Value*
Total Fat 7g	**11%**
Saturated Fat 1.5g	**8%**
Cholesterol 25mg	**8%**

A biscuit	
Calories 190	Calories from Fat 80
	% Daily Value*
Total Fat 9g	**14%**
Saturated Fat 2.5g	**13%**
Cholesterol 0mg	**0%**

A croissant	
Calories 260	Calories from Fat 140
	% Daily Value*
Total Fat 16g	**25%**
Saturated Fat 10g	**50%**
Cholesterol 50mg	**17%**

[a]The fat content of fried rice varies by preparation method.

FOOD
FEATURE

Defensive Dining

To meet the most important recommendation of almost every nutrition authority—to reduce dietary fats—most people would have to make changes according to the following five principles. The changes would also lower intakes of saturated fat.

1. Eliminate much of the fat used as a seasoning and in cooking.
2. Cut down on intake of red meat.
3. Remove the fat from high-fat foods.
4. Replace high-fat foods with specially manufactured lower-fat versions of those foods.
5. Replace high-fat foods with naturally occurring low-fat alternatives.

With these principles in mind, you can begin to make choices about foods in your diet.

The first arena of choice that you as a consumer face is the grocery store. The right choices here can save many grams of fat at the dinner table. Food labels can reveal much about a processed food's fat content. Once you figure out whether a food is high in fat, the choice of whether to consume it depends on how you intend to use it in your diet: as a staple item, or as an occasional treat.

Once at home, one of the most effective steps for reducing fats is to eliminate fats used as seasonings. This means serving cooked vegetables without butter, bacon, or margarine; omitting high-fat gravies and sauces; and leaving off other last-minute fat additions. Butter and regular margarine contain the same number of calories (about 35 per teaspoon); diet margarine contains fewer calories because water, air, or fillers have been added. Imitation butter flavoring contains no fat and few calories.

For snacks, use an air popper for popcorn, and then add butter flavoring to the popcorn, if you like it. Keep that flavoring on hand together with other low-fat cooking substitutes such as diet margarine, low-fat salad dressings, fat-free sauce mixes or recipes, and nonstick spray for frying. To replace high-fat ingredients in recipes, check Table 5-5 for hints. These replacements will not change the taste or appearance of the finished product very much, but will dramatically lower its contents of fat and saturated fat.

If you must add fats, be sure that they are detectable in the food and that you enjoy them. For example, if you use strongly flavored fat, a little

Table 5-5

Substitutes for High-Fat Ingredients

Use	Instead of
Fat-free milk products	Whole-milk products
Evaporated fat-free ("skim") milk (canned)	Cream
Yogurt[a] or fat-free sour cream replacer	Sour cream
Reduced-calorie margarine; butter replacers	Butter
Wine, lemon juice, or broth	Butter
Fruit butters	Butter
Part-skim or fat-free ricotta; low-fat or fat-free cottage cheese[a]	Whole-milk ricotta
Part-skim, low-fat, or fat-free cheeses	Regular cheeses
1 tbs cornstarch (for thickening sauces)	1 egg yolk
Low-fat or fat-free mayonnaise	Regular mayonnaise
Low-fat or fat-free salad dressing (for salads and marinades)	Regular salad dressing
Water-packed canned fish and meats	Oil-packed fish and meats
Lean ground meat and grain mixture	Ground beef
Low-fat frozen yogurt or sherbet	Ice cream
Herbs, lemons, spices, fruits, liquid smoke flavoring, or ham-flavored boullion cubes	Butter, bacon, bacon fat

[a]If the recipe calls for the food to be boiled, the yogurt or cottage cheese must be stabilized with a small amount of cornstarch or flour.

goes a long way. Sesame oil, peanut butter, and the fats of strong cheeses are equal in calories to others, but they are so strongly flavored that you can use much less. Try small amounts of grated sapsago, romano, or other hard cheeses to replace larger amounts of less flavorful cheeses.

If you use oils, trade off among types to obtain the benefits different oils offer. Peanut and safflower oils are especially rich in vitamin E. Olive and canola oil present the heart health benefits associated with monounsaturates, mentioned earlier. Canola oil also contains omega-3 fatty acids. High temperatures, such as those used in frying, destroy omega-3 acids.

Here are some other tips to update old, high-fat recipes:

- Grill, roast, broil, boil, bake, stir-fry, microwave, or poach foods. Don't fry in fat.
- Add a little water or fat-free yogurt to thick, bottled salad dressings, and then apply them sparingly. They'll go farther this way, and you'll use less oil.
- Cut recipe amounts of meat in half; use only lean meats. Fill in the lost bulk with shredded vegetables, legumes, pasta, grains, or other low-fat items.
- Trim all visible fat and skin from meat and poultry.
- Refrigerate meat pan drippings and broth, and lift off the fat when it solidifies. Then add the defatted broth to a recipe.
- Make prepared mixes, such as rice or potato mixtures, without the fats called for on the label. The taste is practically unchanged.

All of these suggestions work well when a person carefully plans, selects, purchases, and prepares each meal with the loving attention it deserves. But in the real world, people sometimes fall behind schedule and don't have time to cook, so they eat fast food. Figure 5-19 shows that although some fast-food choices can be remarkably high in fat and calories, other choices can be reasonable.

Keep these facts about fast food in mind:

- Salads are a good choice. Avoid mixed salad bar items, such as macaroni salad. Use only about a quarter of the dressing provided or use low-fat dressing.
- If you are really hungry, order a small hamburger on the side. Hold the mayonnaise: use mustard or ketchup instead. A small bowl of chili or a plain baked potato can also satisfy a bigger appetite.
- Fried fish or chicken sandwiches are at least as high in fat as hamburgers. Broiled sandwiches are far less fatty—if you order them made without spreads, dressings, cheese, bacon, or mayonnaise.

Because fast foods are short on variety, let them be part of a lifestyle in which they complement the other parts. Eat differently, often, elsewhere.

By this time you may be wondering if you can realistically make all the changes recommended for your diet and keep high-fat foods completely under control. Be assured that most of the needed changes can easily become habits after a few repetitions. You need not give up all high-fat foods; you need only learn to exercise moderation. The famous French chef Julia Child makes this point about moderation:

> An imaginary shelf labeled INDULGENCES is a good idea. It contains the best butter, jumbo-size eggs, heavy cream, marbled steaks, sausages and pâtés, hollandaise and butter sauces, French butter-cream fillings, gooey chocolate cakes, and all those lovely items that demand disciplined rationing. Thus, with these items high up and almost out of reach, we are ever conscious that they are not everyday foods. They are for special occasions, and when that occasion comes we can enjoy every mouthful.
>
> *Julia Child, The Way to Cook,* 1989.

You decide what the treats should be and then choose them judiciously, just for pure pleasure. Meanwhile, make sure that your everyday, ordinary choices are those whole, nutrient-dense foods suggested throughout this book. That way you'll meet all your body's needs for nutrients and never feel deprived.

USDA facts:

- 70% of teenage males eat meals away from home each day.
- 57% of all Americans, and 40% of those over 60 years old, do so.
- The foods chosen away from home are higher in fat, saturated fat, and cholesterol and lower in vitamins and minerals, than meals eaten at home.

Figure 5-19
Fast-Food Choices

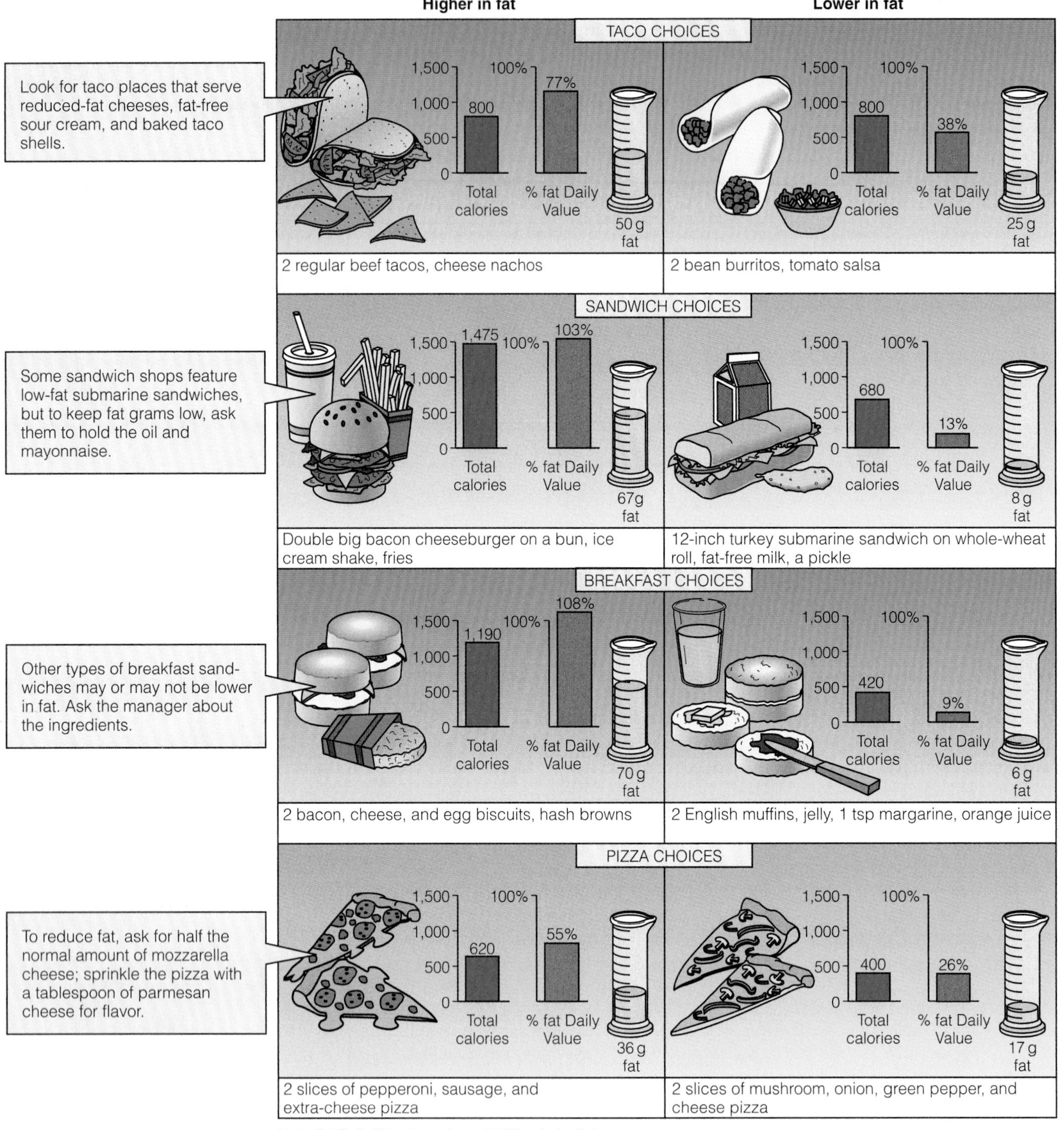
Higher in fat
Lower in fat
TACO CHOICES
Look for taco places that serve reduced-fat cheeses, fat-free sour cream, and baked taco shells.
1,500
1,000
500
0
800
Total calories
100%
77%
% fat Daily Value
50 g fat
2 regular beef tacos, cheese nachos
800
38%
25 g fat
2 bean burritos, tomato salsa
SANDWICH CHOICES
Some sandwich shops feature low-fat submarine sandwiches, but to keep fat grams low, ask them to hold the oil and mayonnaise.
1,475
103%
67g fat
Double big bacon cheeseburger on a bun, ice cream shake, fries
680
13%
8 g fat
12-inch turkey submarine sandwich on whole-wheat roll, fat-free milk, a pickle
BREAKFAST CHOICES
Other types of breakfast sandwiches may or may not be lower in fat. Ask the manager about the ingredients.
1,190
108%
70 g fat
2 bacon, cheese, and egg biscuits, hash browns
420
9%
6 g fat
2 English muffins, jelly, 1 tsp margarine, orange juice
PIZZA CHOICES
To reduce fat, ask for half the normal amount of mozzarella cheese; sprinkle the pizza with a tablespoon of parmesan cheese for flavor.
620
55%
36 g fat
2 slices of pepperoni, sausage, and extra-cheese pizza
400
26%
17 g fat
2 slices of mushroom, onion, green pepper, and cheese pizza
Note: Fat Daily Value based on a 2,000-calorie diet.

READ ABOUT FATS ON FOOD LABELS

Figure 5-20 presents three labels from packages of lasagna and asks you to compare the fat in one serving of each with your day's allowances for fat and saturated fat. Food labels make this comparison easy for people whose energy needs are about 2,000 calories a day. For those people, the Daily Values for fat and saturated fat are listed right on the food labels. Other people must perform a few calculations to arrive at meaningful numbers for themselves. This activity asks that you do three things:

- Calculate your own personal Daily Values for *total fat* and *saturated fat.*
- Calculate the percentage of your personal Daily Value for *total fat* contributed by each of the lasagnas.
- Calculate the percentage of your personal Daily Value for *saturated fat* contributed by each of the lasagnas.

Then it asks you some questions.

In calculating your Daily Values for fats, you will use three numbers. First is your recommended intake for energy (from the inside back cover, page Y). Second is the recommendation to consume no more than 30 percent of calories from fat, or 10 percent of calories from saturated fat. The third number arises from the calorie value of fats: 9 calories for every 1 gram of fat.

Find Your Personal Daily Values for Total Fat and Saturated Fat

Step 1. Calculate your personal Daily Value for total fat. On Form 5-1, section 1, part A, copy your energy intake recommendation from the inside back cover, page Y. Transfer your energy recommendation to part B and calculate 30 percent of it as shown. In part C, copy the answer of Part B and divide by 9 calories per gram to determine the number of grams of fat as shown. The answer is your personal Daily Value for total fat. Copy it into part D for later use.

Form 5-1

Calculate Your Daily Value for Fat and Saturated Fat

Section 1—Total Fat

A. Copy your energy recommendation in calories from inside back cover, page Y:
______________________ calories.
(energy recommendation)

B. Calculate the number of calories you can consume as fat in a day:
______________________ cal × .3 = ______________
(energy recommendation) (fat calories)

C. How many grams is this?
______________ cal ÷ 9 cal per gram = ______________ grams fat
(fat calories, from B)

D. Your personal Daily Value for total fat = ______________ g
(from C)

Section 2—Saturated Fat

A. Copy your energy recommendation from A in section 1: ______________________ calories.
(energy recommendation)

B. Calculate the number of calories you can consume as saturated fat each day:
______________________ cal × .1 = ______________________
(energy recommendation) (saturated fat calories)

C. How many grams is this?
______________ cal ÷ 9 cal per gram = ______________ grams saturated fat
(saturated fat calories from B)

D. Your personal Daily Value for saturated fat = ______________ g
(from C)

do it!

Figure 5-20

Comparison of Three Different Lasagnas

A

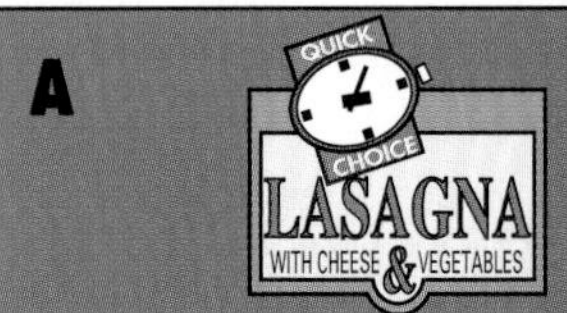
QUICK
CHOICE
LASAGNA
WITH CHEESE & VEGETABLES

Nutrition Facts

Serving size 10 1/2 oz (298g)
Servings per Package 1

Amount per serving

Calories 472 Calories from Fat 252

	% Daily Value*
Total Fat 28g	43%
Saturated Fat 16g	80%
Cholesterol 125mg	42%
Sodium 820mg	34%
Total Carbohydrate 35g	12%
Dietary fiber 4g	16%
Sugars 9g	
Protein 20g	

Vitamin A 25% • Vitamin C <2%
Calcium 50% • Iron 10%

*Percent Daily Values are based on a 2,000 calorie diet. Your daily values may be higher or lower depending on your calorie needs.

	Calories	2,000	2,500
Total Fat	Less than	65g	80g
Sat Fat	Less than	20g	25g
Cholesterol	Less than	300mg	300mg
Sodium	Less than	2,400mg	2,400mg
Total Carbohydrate		300g	375g
Dietary Fiber		25g	30g

Calories per gram
Fat 9 • Carbohydrate 4 • Protein 4

INGREDIENTS, Skim Milk, Ricotta Cheese (Whole milk,Cream, Skim Milk, Vinegar and Salt), Cooked Macaroni, Spinach, Parmesan Cheese, Carrots, Onions, Butter, Soybean Oil, Modified Cornstarch, Bread Crumbs (Enriched Bleached Wheat Flour, Sugar, Corn Syrup, Partially Hydrogenated Soybean Oil, Salt, Yeast, Calcium Propionate, Spice Extractives and BHT),Corn Syrup, Long Grain Rice Meal, Potato Flakes, Malt, Yeast, Vegetable Shortening (Partially Hydrogenated Soybean Oil), Salt, Calcium Propionate, Fat-Free Dry Milk Solids, Salt, Romano Cheese (made from Cow's Milk), Mushrooms, Sugar, Salt, Mono- and Diglycerides, Xanthan gum, Spices, Garlic Salt.

B

Home
Taste
Lasagna
WITH MEAT SAUCE

Nutrition Facts

Serving size 10 1/2 oz (298g)
Servings per Package 1

Amount per serving

Calories 361 Calories from Fat 117

	% Daily Value*
Total Fat 13g	20%
Saturated Fat 8g	40%
Cholesterol 87mg	29%
Sodium 860mg	36%
Total Carbohydrate 37g	12%
Dietary fiber 0g	
Sugars 8g	
Protein 26g	

Vitamin A 15% • Vitamin C 10%
Calcium 25 % • Iron 10%

*Percent Daily Values are based on a 2,000 calorie diet. Your daily values may be higher or lower depending on your calorie needs.

	Calories	2,000	2,500
Total Fat	Less than	65g	80g
Sat Fat	Less than	20g	25g
Cholesterol	Less than	300mg	300mg
Sodium	Less than	2,400mg	2,400mg
Total Carbohydrate		300g	375g
Dietary Fiber		25g	30g

Calories per gram
Fat 9 • Carbohydrate 4 • Protein 4

INGREDIENTS, Tomatoes, Cooked Macaroni Product, Dry Curd Cottage Cheese, Beef, Low-Moisture Part-Skim Mozzarella Cheese, Dehydrated Onions, Modified Cornstarch, Salt, Parmesan Cheese, Enriched Wheat Flour, Sugar, Spices, Tomato Flavor (Salt, Tomato Paste and Flavorings), Dehydrated Garlic.

C

SMART
Life
FLORENTINE
Lasagna

Nutrition Facts

Serving size 11 oz (312g)
Servings per Package 1

Amount per serving

Calories 217 Calories from Fat 9

	% Daily Value*
Total Fat 1g	2%
Saturated Fat 0g	0%
Cholesterol 10mg	3%
Sodium 500mg	21%
Total Carbohydrate 34g	11%
Dietary fiber 5g	20%
Sugars 10g	
Protein 18g	

Vitamin A 25% • Vitamin C 25%
Calcium 40 % • Iron 10%

*Percent daily values are based on a 2,000 calorie diet. Your daily values may be higher or lower depending on your calorie needs.

	Calories	2,000	2,500
Total Fat	Less than	65g	80g
Sat Fat	Less than	20g	25g
Cholesterol	Less than	300mg	300mg
Sodium	Less than	2,400mg	2,400mg
Total Carbohydrate		300g	375g
Dietary Fiber		25g	30g

Calories per gram
Fat 9 • Carbohydrate 4 • Protein 4

INGREDIENTS, Tomato Puree, Cooked Enriched Macaroni Product (Durum Semolina, [Niacin, Ferrous Sulfate, Thiamin Mononitrate, Riboflavin], Water, Egg White Solids, Disodium Phosphate, Powdered Cellulose, Soy Protein Isolate, Soy Protein, Vital Wheat Gluten, Guar Gum), Ricotta Cheese (Pasteurized Whey, Pasteurized Milk, Vinegar, Xanthum Gum) Tomatoes, Zucchini, Cheese (Pasteurized Skim-Milk, Water, Natural Flavors, Enzyme, Calcium Chloride, Salt and Vitamin A & D), Carrots, Spinach, Onions, Water, Mushrooms, Concentrated Dealcoholized Burgundy Wine, Sugar, Modified Food Starch, Salt, Spices, Microcrystalline Cellulose, Methylcellulose, Maltodextrin, Hydrolyzed Corn Protein, Xanthum Gum, Guar Gum, Autolyzed Yeast, Calcium Chloride, Citric Acid, Garlic Extractives, Dextrin.

Form 5-2

Compare Your Personal Daily Values with Fats in Three Lasagnas

A. Grams total fat per serving:

__________ __________ __________

(lasagna A) (lasagna B) (lasagna C)

Grams saturated fat per serving:

__________ __________ __________

(lasagna A) (lasagna B) (lasagna C)

B. Your personal Daily Value for total fat (from Form 5-1, section 1, part D) ________ g

Your Daily Value for saturated fat (from Form 5-1, section 2, part D) ________ g

C. What percentage of your Daily Value for total fat does a serving of each lasagna present?

Lasagna A ________ g ÷ ________ g × 100 = ________ % of personal Daily Value for total fat
(total fat) (total fat Daily Value)

Lasagna B ________ g ÷ ________ g × 100 = ________ % of personal Daily Value for total fat
(total fat) (total fat Daily Value)

Lasagna C ________ g ÷ ________ g × 100 = ________ % of personal Daily Value for total fat
(total fat) (total fat Daily Value)

D. What percentage of your Daily Value for saturated fat does a serving of each lasagna present?

Lasagna A ________ g ÷ ________ g × 100 = ________ % of personal Daily Value for saturated fat
(saturated fat) (saturated fat Daily Value)

Lasagna B ________ g ÷ ________ g × 100 = ________ % of personal Daily Value for saturated fat
(saturated fat) (saturated fat Daily Value)

Lasagna C ________ g ÷ ________ g × 100 = ________ % of personal Daily Value for saturated fat
(saturated fat) (saturated fat Daily Value)

Step 2. Calculate your personal Daily Value for saturated fat. On Form 5-1, section 2, part A, write in your personal energy recommendation (from part A of section 1.) Transfer your energy recommendation to part B and calculate 10 percent of it as shown. Transfer this number from part B to part C and divide by 9 calories per gram to find grams of saturated fat. The answer is your personal Daily Value for saturated fat. Copy it into part D for later use.

It's wise to memorize your Daily Value for fat and saturated fat and make food choices each day that do not exceed them. These are two of the most valuable numbers you can learn.

Compare the Fat in a Serving of each of Three Lasagnas with Your Daily Values

Step 3. Using Form 5-2, part A, copy the grams of total fat and saturated fat per serving of lasagna from the three package labels of Figure 5-20. Then copy to part B your personal Daily Values for both total fat and saturated fat from Form 5-1. Enter these values on part C of Form 5-2 and calculate the percentage of your Daily Value for total fat presented by each lasagna. Repeat the process for saturated fat in part D.

Analysis

Now use the information you have generated to respond to the following questions:

1. How many grams of fat can you consume in a day and not exceed 30 percent of calories from fat?
2. How many grams of saturated fat can you consume in a day and not exceed 10 percent of calories from saturated fat?
3. Which lasagna is highest in fat per serving? What percentage of your personal Daily Value for fat would this lasagna contribute to your day's intake?
4. What percentages of your personal Daily Value for fat do the other two lasagnas contribute?
5. If you ate a serving of the highest-fat lasagna, how could you avoid exceeding the recommended fat intake for the day?
6. If you substituted a serving of the lowest-fat lasagna for the highest-fat choice, what effects would this have on your other food choices and on your calorie and nutrient intakes that day?
7. How does the saturated fat in each of the three lasagnas compare with your personal Daily Value for saturated fat?

Consider the Ingredients

Read the ingredients list for each lasagna. Keep in mind that manufacturers list ingredients in descending order of their predominance in the food. Ingredients listed first are present in the largest quantities; those listed last are present in the smallest quantities. Now respond to the remaining questions:

8. Which ingredients contributed most to the total fat, saturated fat, and cholesterol in the highest-fat lasagna?
9. In the lowest-fat lasagna, which ingredients replaced or substituted for high-fat ingredients present in the other lasagnas? How did this help to lower the total fat content?
10. Do you agree or disagree with the following statement: "No food is good or bad based on its fat content alone." Justify your stance on this issue.

Take time to read labels, especially with regard to the fat contents of the foods you choose often. What you find there will often surprise you and may benefit your health.

Self Check

Answers to these Self Check questions are in Appendix G.

1. Which of the following is *not* one of the ways fats are useful in foods?
   a. Fats contribute to the taste and smell of foods.
   b. Fats carry fat-soluble vitamins.
   c. Fats provide a low-calorie source of energy compared to carbohydrates.
   d. Fats provide essential fatty acids.
2. Generally speaking, vegetable and fish oils are rich in:
   a. polyunsaturated fat
   b. saturated fat
   c. cholesterol
   d. *trans*-fatty acids
3. A benefit to health is seen when __________ is used in place of __________ in the diet.
   a. saturated fat/monounsaturated fat
   b. saturated fat/polyunsaturated fat
   c. monounsaturated fat/saturated fat
   d. polyunsaturated fat/cholesterol
4. Chylomicrons, a class of lipoproteins, are produced in the:
   a. gallbladder
   b. small intestinal cells
   c. large intestinal cells
   d. liver
5. Which foods from the breads, cereals, rice, and pasta group generally contain fat?
   a. biscuits
   b. muffins
   c. pasta
   d. (a) and (b)
6. Which of the following constitutes a serving of alcohol delivering ½ ounce of pure ethanol? (Read about alcohol in the Controversy on the next page.)
   a. 12 ounces of beer
   b. 10 ounces of wine cooler
   c. 3 to 4 ounces of wine
   d. all of the above
7. LDL deliver triglycerides and cholesterol from the liver to the body's tissues. T F
8. Given the same number of calories from excess dietary fat or carbohydrate, the body stores more calories from the fat than from the carbohydrate. T F
9. Consuming large amounts of *trans*-fatty acids lowers LDL cholesterol and thus lowers the risk of heart disease and heart attack. T F
10. When olestra is present in the digestive tract, it enhances the absorption of vitamin E. T F

NUTRITION ON THE NET

For further study of the topics of this chapter, access these websites and search for the phrases or words in quotation marks:

1. For the latest changes in chapter material, to find supplemental learning tools, or to access links to related websites, go to the "Nutrition: Concepts and Controversies" site at: **www.wadsworth.com/nutrition/prod/allprod.html**
2. For information about "cholesterol". **www/healthfinder.gov/searchoptions/topicsaz.htm**
3. The American Dietetic Association provides information on "fats, oils, and cholesterol": **www.eatright.org/cgi/search.cgi**
4. To learn more about "fat replacers": **ificinfo.health.org**
5. Search the Food and Drug Administration's site for "olestra": **www.fda.gov**
6. Search among thousands of current scientific and medical abstracts for any topic related to lipids at: **http://www.ncbi.nlm.nih.gov/PubMed/**

ALCOHOL AND NUTRITION

CONTROVERSY
5

On average, people in the United States consume from 6 to 10 percent of their total daily energy intake as alcohol. In the past month, an estimated 51 percent of people over 12 years of age drank alcohol.[1] Drinking habits span a wide spectrum: many adults drink no alcohol whatsoever, some take a glass of wine only with meals, others drink on social occasions, and still others take in huge quantities of alcohol daily because of a life-shattering addiction. A full third of U.S. college students are reported to be **binge drinkers,** and many pay a high price in terms of health and safety as a result of their episodes of heavy drinking.[2] People who are **moderate drinkers** usually consume the calories of alcohol in addition to their normal food intake.[3] Consequently, the energy of alcohol contributes to body fatness.[4] Alcohol is not just an energy source, however; it is also a psychoactive drug and a toxin to the body. At any level of intake, alcohol affects the body in ways that go far beyond its energy-yielding nature.

SOCIAL ASPECTS OF ALCOHOL

People naturally congregate to enjoy conversation and companionship, and it is only natural to offer beverages to companions. All beverages ease conversation, whether they contain alcohol or not. Still, many people are **social drinkers** who choose alcohol over cola, juice, milk, or coffee as a pleasant accompaniment to a meal, a drink of celebration, or a way to relax with friends. Taken in moderation, alcohol reduces inhibitions, and encourages social interactions, and produces feelings of **euphoria,** a pleasant sensation that people seek. The term *moderation* is important in this regard, for at intakes higher than this, alcohol worsens social interactions. An interesting side note: the nonalcoholic beers and wines now on the market also elevate mood and encourage social interaction, a testimony to the placebo effect at work.

In contrast to moderate social drinking, the effect of alcohol on **problem drinkers** or people with **alcoholism** is overwhelmingly negative. For these people, drinking alcohol brings irrational and often dangerous behavior such as driving a car while intoxicated, and regrettable human interactions such as arguments and violence. With continued drinking, such people face psychological depression, physical illness, severe malnutrition, and demoralizing erosion of self-esteem.

Moderation is not easily defined, for no single amount of alcohol per day is appropriate for everyone. Tolerances to alcohol differ. In general, women cannot handle as much alcohol as can men, and should not try to keep up drink for drink with male companions. Authorities have attempted to set limits at not more than two drinks a day for the average-sized, healthy man and not more than one drink a day for the average-sized, healthy woman. This amount is supposed to be enough to elevate mood without incurring long-term harm to health.

Doubtless some people could safely consume slightly more than the alcohol dose called moderate; others, especially those prone to alcohol addiction, could definitely not handle nearly so much without significant risk. If you think your own drinking might not be moderate or normal, if it has caused problems in your life, or if you feel guilty about your drinking, you may want to seek a professional evaluation.* Table C5-2 contrasts some behaviors of moderate drinkers with those of problem drinkers.

What Is Alcohol? In chemistry, the term *alcohol* refers to a class of chemical compounds whose names end in "-ol." The glycerol molecule of a triglyceride is an example. Alcohols affect living things profoundly, partly because they act as lipid solvents. Alcohols can easily penetrate a cell's outer lipid membrane, and once inside, they denature the cell's protein structures and kill the cell. Because some alcohols kill microbial cells, they make useful disinfectants and antiseptics.

The alcohol of alcoholic beverages, **ethanol,** is somewhat less toxic than others. Sufficiently diluted and taken in small enough doses, its action in the brain produces "euphoria." Used in this way, alcohol is a drug, and like many drugs, alcohol presents both benefits and hazards to the taker. Its effects depend largely on the quantity of alcohol consumed.

The Fattening Power of Alcohol As for the calories of alcohol, this energy source should probably be counted as fat in the diet because metabolic interactions occur between fat and alcohol in the body.[5] Presented with both fat and alcohol, the body stores the comparatively harmless fat and rids itself of the toxic alcohol by burning it off as fuel. Alcohol may promote fat storage particularly in the central abdominal area—the **"beer belly"** effect seen in moderate drinkers whose risks to the heart are described in Chapter 9.[6] Also, alcohol yields 7 calories of energy per gram to the body, so many alcoholic drinks are much more fattening than their nonalcoholic counterparts. A rule of

*The U.S. center for facts on alcohol is the National Clearinghouse for Alcohol and Drug Information: 1-800-729-6686.

Table C5-1

Alcohol and Drinking Terms

- **acetaldehyde** (ass-et-AL-deh-hide) a substance to which ethanol is metabolized on its way to becoming harmless waste products that can be excreted.
- **alcohol dehydrogenase** (ADH) an enzyme system that breaks down alcohol. The antidiuretic hormone listed below is also abbreviated ADH.
- **alcoholism** a dependency on alcohol marked by compulsive uncontrollable drinking with negative effects on physical health, family relationships, and social health.
- **antidiuretic hormone** (ADH) a hormone produced by the pituitary gland in response to dehydration (or a high sodium concentration in the blood). It stimulates the kidneys to reabsorb more water and so to excrete less. (This hormone should not be confused with the enzyme alcohol dehydrogenase, which is also abbreviated ADH.)
- **beer belly** central-body fatness associated with alcohol consumption.
- **binge drinker** a person who drinks 4 or more drinks in a short period.
- **cirrhosis** (seer-OH-sis) advanced liver disease, often associated with alcoholism, in which liver cells have died, hardened, turned an orange color, and permanently lost their function.
- **congeners** (CON-jen-ers) chemical substances other than alcohol that account for some of the physiological effects of alcoholic beverages, such as taste and after-effects.
- **drink** a dose of any alcoholic beverage that delivers ½ ounce of pure ethanol.
- **ethanol** the alcohol of alcoholic beverages, produced by the action of microorganisms on the carbohydrates of grape juice or other carbohydrate-containing fluids.
- **euphoria** an inflated sense of well-being and pleasure brought on by a moderate dose of alcohol and some other drugs.
- **fatty liver** an early stage of liver deterioration seen in several diseases, including kwashiorkor and alcoholic liver disease in which fat accumulates in the liver cells.
- **fibrosis** (fye-BROH-sis) an intermediate stage of alcoholic liver deterioration in which liver cells lose their function and assume the characteristics of connective tissue cells (fibers).
- **formaldehyde** a substance to which methanol is metabolized on the way to being converted to harmless waste products that can be excreted.
- **gout** (GOWT) accumulation of crystals of uric acid in the joints.
- **MEOS** (microsomal ethanol oxidizing system) a system of enzymes in the liver that oxidize not only alcohol but also several classes of drugs.
- **methanol** an alcohol produced in the body continually by all cells.
- **moderate drinker** a person who does not drink excessively and does not behave inappropriately because of alcohol. The person's health is not harmed by alcohol over the long term.
- **problem drinker** or **alcohol abuser** a person who suffers social, emotional, family, job-related or other problems because of alcohol. This person is on the way to alcoholism.
- **proof** a statement of the percentage of alcohol in an alcoholic beverage. Liquor that is 100 proof is 50% alcohol, 90 proof is 45%, and so forth.
- **social drinker** a person who drinks only on social occasions. Depending on how alcohol affects the person's life, the person may be a moderate drinker or a problem drinker.
- **urethane** a carcinogenic compound that commonly forms in alcoholic beverages.

thumb states that the calories in each ounce of ethanol in a drink represents about half an ounce of fat.[7] An observant reader, knowing that a gram of fat yields 9 calories, may wonder why a gram of alcohol, at 7 calories, is assigned only half the calorie value of fat. The answer is that the body rids itself of a small but measurable amount of alcohol by way of the breath and urine.

What Is a "Drink"? Alcoholic beverages contain a great deal of water and some other substances, as well as the alcohol ethanol. In wine, beer, and wine coolers, alcohol contributes a relatively low percentage of the beverage's volume. In contrast, whiskey, vodka, rum, and brandy may contain as much as 50 percent of their volume as alcohol. The percentage of alcohol is stated as **proof.** Proof equals twice the percentage of alcohol; for example, 100 proof liquor is 50 percent alcohol, 90 proof is 45 percent, and so forth. (Alcohol terms are defined in Table C5-1.)

A serving of alcoholic beverage, commonly called a **drink,** delivers ½ ounce of pure ethanol. Each of the following is considered a drink:

- 3 to 4 ounces wine.
- 10 ounces wine cooler.
- 12 ounces beer.
- 1½ ounce hard liquor (80 proof whiskey, gin, brandy, rum, vodka).

These standard measures may have little in common with the drinks served by enthusiastic bartenders, however. Many wine glasses easily hold 6 to 8 ounces of wine; a large beer stein can hold 16, 20, or even more ounces; a strong liquor drink may contain 2 or 3 ounces of various liquors.

DOES MODERATE ALCOHOL USE BENEFIT HEALTH?

One or two alcoholic beverages a day are credited with reducing the risk of death from heart disease in people over 60 years old who have an increased risk of heart disease.[8] Many studies support this effect of alcohol. However, the matter is not yet settled. Recently, researchers followed the alcohol intakes and health histories of almost 6,000 men for over 20 years.[9] The results showed no beneficial relationship between mortality from cardiovascular disease and any level of alcohol consumption, and an increased risk of death from all causes with more than 22 drinks per week. They also showed a strong correlation with mortality from stroke—men drinking more than 35 drinks a week had double the mortality from stroke compared with non-

Table C5-2

Behaviors Typical of Moderate Drinkers and Problem Drinkers

Moderate Drinkers Typically:	Problem Drinkers Typically:
■ Drink slowly, casually.	■ Gulp or "chug" drinks.
■ Eat food while drinking or beforehand.	■ Drink on an empty stomach.
■ Don't binge drink; know when to stop.	■ Binge drink; drink to get drunk.
■ Respect nondrinkers.	■ Pressure others to drink.
■ Avoid drinking when solving problems or making decisions.	■ Turn to alcohol when facing problems or decisions.
■ Do not admire or encourage drunkenness.	■ Consider drunks to be funny or admirable.
■ Remain peaceful, calm, and unchanged by drinking.	■ Become loud, angry, violent, or silent when drinking.
■ Cause no problems to others or themselves by drinking.	■ Physically or emotionally harm themselves, family members, or others when drinking.

drinkers, possibly because of elevated blood pressure.

Young people, especially, do not benefit their health by drinking, and in fact, they increase their risk of dying from any cause.[10] Young nondrinkers are found to have a lower risk of dying than even light drinkers (fewer than 15 drinks per month) of the same age.[11] Young women in particular should not be advised to drink alcohol for the sake of the heart. The risk of heart disease in women before menopause is low, but the risk of breast cancer is substantial, and alcohol in the amounts that have been said to benefit the hearts of older people raises the risk of breast cancer in young women.[12] Likewise, young adults would be ill advised to take up drinking for health—alcohol which is related to car crashes, homicides, and other violence that account for the great majority of deaths of people in this age group each year.

Red wine has also been credited with special health-supporting properties. In fact, the following two statements concerning wine and health were recently approved to appear on U.S. wine labels:

- "The proud people who made this wine encourage you to consult your family doctor about the health effects of wine consumption."
- "To learn the health effects of wine consumption, send for the Federal Government's Dietary Guidelines for Americans, Center for Nutrition Policy and Promotion, USDA, 1120 20th Street, NW, Washington, DC 20036 or visit its website."

These statements seem to promise that good news about wine and health may await the information seeker, but the science even on wine and health is mixed.[†] For example, the high potassium content of grape juice may lower high blood pressure, and this effect persists when the grape juice is made into wine. However, since alcohol in large amounts raises blood pressure, the grape juice may be more suitable than the wine for people with hypertension. Dealcoholized wine also facilitates the absorption of potassium, calcium, phosphorus, magnesium, and zinc. So does wine, but the alcohol in it promotes the *excretion* of these minerals, so again the dealcoholized version is preferred.

In addition to alcohol, wine contains phenols and other phytochemicals that may act as antioxidants that protect the cardiovascular system against damage from oxidation that leads to heart disease. Details about oxidation and phytochemicals are offered in Controversies 7 and 11. These protective chemicals of wine may explain the so-called French paradox: even though the French have many of the same risk factors as people in the United States, the wine-drinking population of France enjoys a lower incidence of heart disease.[13] People worried about oxidative damage to body systems may want to reconsider using alcoholic beverages as a source of phytochemicals, however. Alcohol itself may create oxidative stress damaging to the tissues of the liver and pancreas.[14]

[†]For more information on wine labels, visit the Internet website of the Bureau of Alcohol, Tobacco, and Firearms: www.atf.treas.gov/press/label_ab.htm.

Dealcoholized wine and purple grape juice contain similar phytochemicals to those of wine, but do not increase oxidative stress.[15]

Alcoholic beverages affect the appetite. Usually, they reduce it, making people unaware that they are hungry. But in people who are tense and unable to eat or in the elderly who have lost interest in food, small doses of wine taken 20 minutes before meals improve appetite. Certain compounds in the wine known as **congeners** are credited with this effect. For undernourished people and for people with severely depressed appetites, wine may facilitate eating even when psychotherapy fails to do so. Congeners are also involved in producing a hangover, as a later section notes.

Another example of the beneficial use of alcohol comes from research showing that moderate use of wine in later life improves morale, stimulates social interaction, and promotes restful sleep. In nursing homes, improved patient and staff relations have been attributed to greater self-esteem among elderly patients who drink moderate amounts of wine. Researchers hypothesize that chronic fatigue may be responsible for some behaviors associated with old age. The positive effects of wine on sleep may alleviate the fatigue, easing social interactions.

For people who choose to drink, a valid goal is learning to drink moderately. The next sections address that goal in the context of alcohol's physical effects.

ALCOHOL ENTERS THE BODY

From the moment an alcoholic beverage is swallowed, the body pays special attention to it. Unlike foods, which require digestion before they can be absorbed, the tiny alcohol molecules can diffuse right through the stomach walls and reach the brain within a minute. Ethanol is a toxin, and a too-high dose of alcohol triggers one of the body's primary defenses against poison—vomiting. Many times, though, alcohol arrives gradually and in a beverage dilute enough so that the vomiting reflex is delayed and the alcohol is absorbed.

A person can become intoxicated almost immediately when drinking, especially if the stomach is empty. When the stomach is full of food, molecules of alcohol have less chance of touching the stomach walls and diffusing through, so alcohol reaches the brain more gradually (see Figure C5-1). By the time the stomach contents are emptied into the small intestine, however, alcohol is absorbed rapidly whether food is present or not.

Figure C5-1

Food Slows Alcohol's Absorption

The alcohol in a stomach filled with food has a low probability of touching the walls and diffusing through. Food also holds alcohol in the stomach longer, slowing its entry into the highly absorptive small intestine.

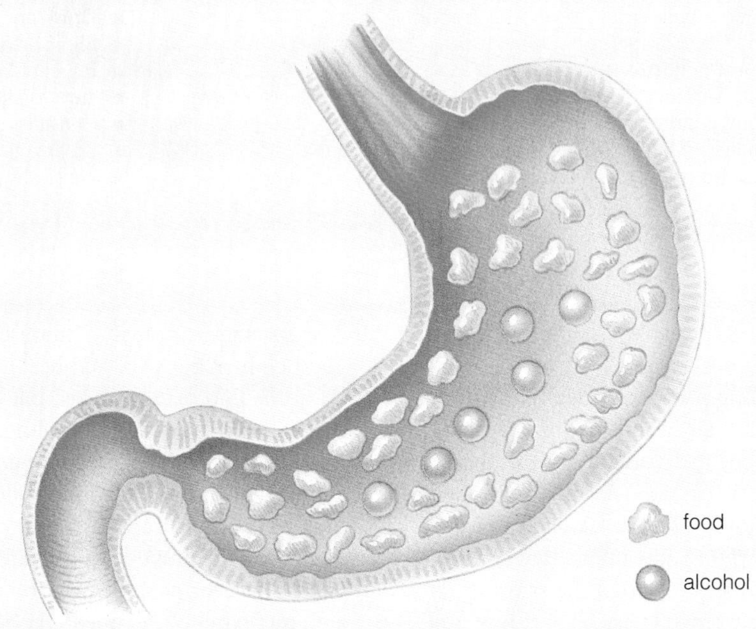
food
alcohol

A person who wants to drink socially and not become intoxicated should eat the snacks provided by the host (avoid the salty ones; they make you thirstier). Carbohydrate snacks slow alcohol absorption, and high-fat snacks help too because they slow peristalsis, keeping the alcohol in the stomach longer. Other tips include adding ice or water to drinks to dilute them and choosing nonalcoholic beverages first and then every other round to quench thirst.

If one drinks slowly enough, the alcohol, after absorption, will be collected by the liver and processed without much effect on other parts of the body. If one drinks more rapidly, however, some of the alcohol bypasses the liver and flows for a while through the rest of the body and the brain.

ALCOHOL ARRIVES IN THE BRAIN

Some people use alcohol as a kind of social anesthetic to help them relax or to relieve anxiety. One drink relieves inhibitions, and this gives people the impression that alcohol is a stimulant. Actually, it gives this impression by sedating *inhibitory* nerves, allowing excitatory nerves to take over. This effect is temporary. Ultimately, alcohol acts as a depressant and sedates all the nerve cells. Figure C5-2 shows alcohol's effects on the brain.

It is lucky that the brain centers respond to rising blood alcohol in the order shown because a person usually passes out before managing to drink a lethal dose. It is possible for a person to drink fast enough so the alcohol continues to be absorbed and its effects continue to accelerate after the person has gone to sleep. Every year, deaths that are attributed to this effect take place during drinking contests. Before passing out, the drinker drinks fast enough to receive a lethal dose. Table C5-3 on page 172 shows blood alcohol levels that correspond with progressively greater intoxication, and Table C5-4 shows brain responses that occur at these blood levels.

Binge Drinking Binge drinking (consuming 4 or more drinks in a short time) poses a serious health threat to college students, to others who engage in it, and to nearby nondrinkers, as well.[16] A binge is most likely to occur at a party, a sporting event, or another social occasion. Compared with nondrinkers or moderate drinkers, binge drinkers are more likely to damage property, to assault other people, to cause fatal automobile accidents, and to engage in risky unprotected and unplanned sexual intercourse.[17] Bingers rarely identify themselves as problem drinkers (refer back to Table C5-2) until their drinking behavior causes a crisis, such as a car crash, or until they've binged long enough to have caused substantial damage to their health. One target of such damage is the brain.

Alcohol's Effects on the Brain Brain cells are particularly sensitive to excessive exposure to alcohol. The brain shrinks, even in people who drink only moderately. The extent of the shrinkage is proportional to the amount drunk. Abstinence, together with good nutrition, reverses some of the brain damage, and possibly all of it, if heavy drinking has not continued for more than a few years. However, prolonged drinking beyond an individual's capacity to recover can do severe and irreversible harm to vision, memory, learning ability, and other functions.

Anyone who has had an alcoholic drink has experienced one of alcohol's physical effects: alcohol increases urine output. This is because alcohol depresses the brain's production of **antidiuretic hormone.** Loss of body water leads to thirst. The only fluid that will relieve dehydration is water, but if the only drinks available contain alcohol, each drink may worsen the thirst. The smart drinker, then, alternates alcoholic beverages with nonalcoholic choices and uses only the latter to quench thirst.

The water lost due to hormone depression takes with it important minerals, such as magnesium, potassium, calcium, and zinc, depleting the body's reserves. These minerals are vital to fluid balance and to nerve and muscle coordination. When drinking incurs mineral losses, the losses must be made up the next day if deficiencies are not to advance.

ALCOHOL ARRIVES IN THE LIVER

The capillaries that surround the digestive tract merge into veins that carry the alcohol-laden blood to the liver. Here the veins branch and rebranch into capillaries that touch every liver cell. The liver cells make nearly all of the body's alcohol-processing machinery, and the routing of blood through the liver allows the cells to go right to work on the alcohol. The liver's location at this point along the circulatory system enables it to remove toxic substances before they reach other body organs such as the heart and brain.

The Liver Metabolizes Alcohol The liver makes and maintains two sets of equipment for metabolizing alcohol. One is an enzyme that removes hydrogens from alcohol to break it down; the name, **alcohol dehydrogenase (ADH)**, almost says what it does.[††] This enzyme handles about 80 percent or more of the alcohol in the body. The other set of alcohol-metabolizing equipment is a chain of enzymes known as the **MEOS,** which is thought to handle about 10 percent of alcohol. The remaining 10 percent is excreted through the breath and in the urine. Because the alcohol in the breath is directly proportional to the alcohol in the blood, the breathalyzer test that law enforcement officers administer when someone may be driving under the influence of alcohol accurately reveals how intoxicated the person is.

The amount of alcohol a person's body can process in a given time is limited by the amount of ADH enzymes that reside in the liver. If more molecules of alcohol arrive at the liver cells than the enzymes can handle, the extra alcohol must wait. It circulates again and again through the brain, liver, and other organs until enzymes are available to degrade it.

Some ADH enzymes reside in the stomach and break down some alcohol before it enters the blood. Research shows that people with alcoholism make less stomach ADH than others and that women make less than men. Earlier, this Controversy warned that women should not to try to keep up with male

Figure C5-2

Alcohol's Effects on the Brain

When alcohol flows to the brain, it first sedates the frontal lobe, the reasoning part. As the alcohol molecules diffuse into the cells of this lobe, they interfere with reasoning and judgment.

With continued drinking, the speech and vision centers of the brain become sedated, and the area that governs reasoning becomes more incapacitated.

Still more drinking affects the cells of the brain responsible for large-muscle control; at this point people under the influence stagger or weave when they try to walk.

Finally the conscious brain becomes completely subdued, and the person passes out. Now the person can drink no more. This is fortunate because a higher dose would anesthetize the deepest brain centers that control breathing and heartbeat, causing death.

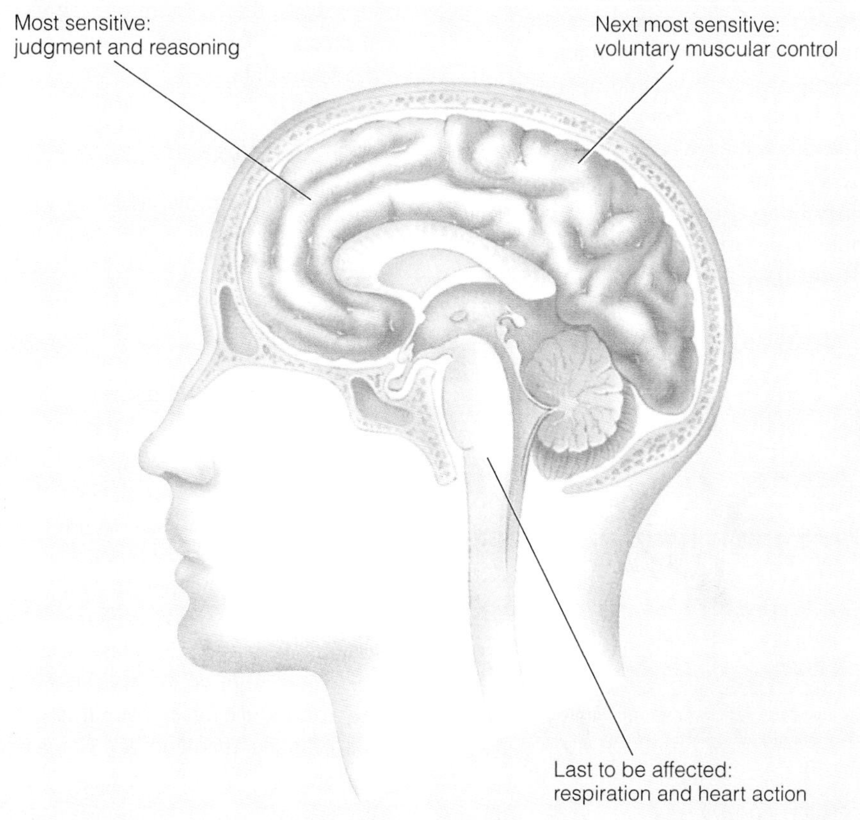
Most sensitive:
judgment and reasoning
Next most sensitive:
voluntary muscular control
Last to be affected:
respiration and heart action

[††]There are actually two ADH enzymes, each performing a specific task in alcohol breakdown.

Table C5-3

Alcohol Doses and Blood Levels

Number of Drinks[a]	Percent Blood Alcohol By Body Weight 100 lb	120 lb	150 lb	180 lb	200 lb
2	0.08	0.06	0.05	0.04	0.04
4	0.15	0.13	0.10	0.08	0.08
6	0.23	0.19	0.15	0.13	0.11
8	0.30	0.25	0.20	0.17	0.15
12	0.45	0.36	0.30	0.25	0.23
14	0.52	0.42	0.35	0.34	0.27

[a]Taken within an hour or so; each drink equal to 1/2 ounce pure ethanol.

drinkers, and here is the reason why: women absorb about one-third more alcohol than men do, even when the women are the same size as the men and drink the same amount of alcohol.

The amount of ADH enzymes present is also affected by whether a person eats. Fasting for as little as a day causes degradation of body proteins, including the ADH enzymes, and this can reduce the rate of alcohol metabolism by half. Prudent drinkers drink slowly, with food in their stomachs, to allow the alcohol molecules to move to the liver cells gradually enough for the enzymes to handle the load.

It takes about an hour and a half to metabolize one drink, depending on a person's body size, previous drinking experience, how recently the person has eaten, and the person's current state of health. The liver is the only organ that can dispose of significant quantities of alcohol, and its maximum rate of alcohol clearance cannot be speeded up. This explains why only time will restore sobriety. Walking will not because muscles cannot metabolize alcohol. Nor will drinking a cup of coffee help. Caffeine is a stimulant, but it won't speed up the metabolism of alcohol. The police say, ruefully, that a cup of coffee only makes a sleepy drunk into a wide-awake drunk. Table C5-5 on page 173 presents other alcohol myths.

Alcohol Affects Body Functions Upon exposure to alcohol, the liver speeds up its synthesis of fatty acids. Fat is known to accumulate in the livers of young men after a single night of heavy drinking and to remain there for more than a day. The first stage of liver deterioration seen in heavy drinkers is therefore known as **fatty liver;** it interferes with the distribution of nutrients and oxygen to the liver cells. If the condition lasts long enough, fibrous scar tissue invades the liver. This is the second stage of liver deterioration, called **fibrosis.** Fibrosis is reversible with good nutrition and abstinence from alcohol, but the next (last) stage, **cirrhosis,** is not. In cirrhosis, the liver cells harden, turn orange, and die, losing function forever. All of this points to the importance of moderation in the use of alcohol.

The presence of alcohol alters amino acid metabolism in the liver cells. Synthesis of some proteins important in the immune system slows down, weakening the body's defenses against infection. Synthesis of lipoproteins speeds up, increasing blood triglyceride and HDL levels. In addition, excess alcohol adds to the body's acid burden and interferes with normal uric acid metabolism, causing symptoms like those of **gout.**

The reproductive system is also vulnerable to alcohol's effects. Women who drink alcohol regularly may suffer increased menstrual discomfort, and heavy drinking may lead to infertility and spontaneous abortion.[18] Some evidence also leads researchers to suspect that alcohol may suppress the male reproductive hormone testosterone, leading to decreases in muscle and bone tissue, altered immunity, abnormal prostate gland, and a decreased reproductive ability.[19]

Table C5-4

Alcohol Blood Levels and Brain Responses

Blood Level (%)	Brain Response
0.05	Judgment impaired
0.10[a]	Emotional control impaired
0.15	Muscle coordination and reflexes impaired
0.20	Vision impaired
0.30	Drunk, lacking control
0.35	In a stupor
0.50–0.60	Loss of consciousness; death.

[a]The 0.10 percent level is the legal limit for intoxication according to most states' highway safety ordinances; however, driving ability is already impaired at blood alcohol levels lower than 0.10 percent.

THE HANGOVER

The hangover—the awful feeling of headache pain, unpleasant sensations in the mouth, and nausea that one has the morning after drinking too much—is a mild form of drug withdrawal. (The worst form is a delirium with severe tremors that presents a danger of death and demands medical management.) Hangovers are caused by several factors. One is the toxic effects of congeners, already mentioned, that accompany the alcohol in alcoholic beverages. The congeners in gin are different from those in vodka, which in turn are different from those in bourbon or rye whiskey. One particular kind of liquor may be more likely to produce a hangover at lower levels of intake than another. Congeners

Table C5-5

Myths and Truths Concerning Alcohol

Myth:	A shot of alcohol warms you up.
Truth:	Alcohol diverts blood flow to the skin making you *feel* warmer, but it actually cools the body.
Myth:	Wine and beer are mild; they do not lead to addiction.
Truth:	Wine and beer drinkers worldwide have high rates of death from alcohol-related illnesses. It's not what you drink, but how much, that makes the difference.
Myth:	Mixing drinks is what gives you a hangover.
Truth:	Too much alcohol in any form produces a hangover.
Myth:	Alcohol is a stimulant.
Truth:	Alcohol depresses the brain's activity.
Myth:	Alcohol is legal; therefore, it is not a drug.
Truth:	Alcohol is legal, but it alters body functions and is medically defined as a depressant drug.

are only one of several factors that produce hangovers, however, and mixing or switching drinks will not prevent them if too much is drunk.

Dehydration of the brain is a second factor: alcohol not only causes the body to lose water, but actually reduces the water content of the brain cells. When they rehydrate the morning after, nerve pain accompanies their swelling back to their normal size. Another contributor to the hangover is **formaldehyde,** the same chemical that medical laboratories use to preserve dead animals. Formaldehyde comes from **methanol,** an alcohol produced constantly by normal chemical processes in all the cells. Normally, a set of liver enzymes converts this methanol to formaldehyde, and then a second set immediately converts the formaldehyde to carbon dioxide and water, harmless waste products that can be excreted. But the same two sets of liver enzymes that do this are also used to process ethanol to its own intermediate waste product, **acetaldehyde,** and then to carbon dioxide and water. The enzymes prefer ethanol 20 times over methanol. Both alcohols are metabolized without delay until the excess acetaldehyde monopolizes the enzymes, leaving formaldehyde to wait for later detoxification. At that point, formaldehyde starts accumulating and the hangover begins.

Time alone is the cure for a hangover. Simple-minded remedies clearly will not work: vitamins, tranquilizers, aspirin, drinking more alcohol, breathing pure oxygen, exercising, eating, or drinking something awful are all useless. Fluid replacement can help to normalize the body's chemistry. The headache pain, unpleasantness in the mouth, and nausea of a hangover come simply from drinking too much.

ALCOHOL'S LONG-TERM EFFECTS

By far the longest-term effects of alcohol are those felt by the child of a woman who drinks during pregnancy. When a pregnant woman takes a drink, her fetus takes the same drink within minutes, and its body is defenseless against the effects. This topic is so important that it is given a space of its own in Chapter 12, where the recommendation is made that pregnant women should not drink at all. For nonpregnant adults, however, what are the effects of alcohol over the long term?

Left, normal liver; center, fatty liver; right, cirrhosis.

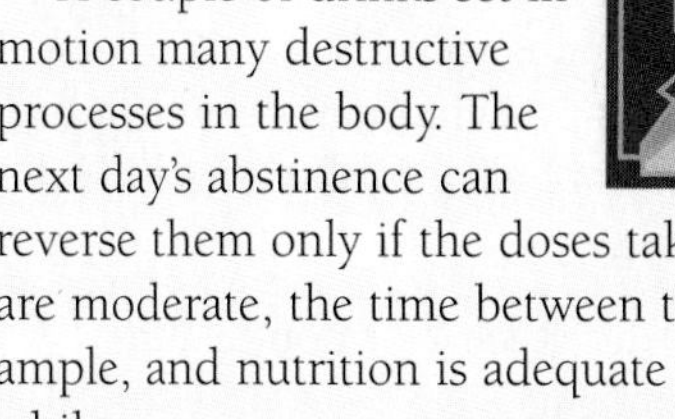

A couple of drinks set in motion many destructive processes in the body. The next day's abstinence can reverse them only if the doses taken are moderate, the time between them is ample, and nutrition is adequate meanwhile.

If the doses of alcohol are heavy, however, and the time between them is short, complete recovery cannot take place, and repeated onslaughts of alcohol gradually take a toll on the body. For example, alcohol is directly toxic to skeletal and cardiac muscle, causing weakness and deterioration that is greater, the larger the dose. Alcoholism makes heart disease likely, probably because chronic alcohol use raises blood pressure. At autopsy, the heart of a person with alcoholism appears bloated and weighs twice as much as a normal heart.

Alcohol attacks brain cells directly, and medical costs of treating the resulting dementia account for a substantial outlay of health dollars. Cirrhosis also develops after 10 to 20 years from the cumulative effects of frequent episodes of heavy drinking.

Alcohol abuse also leads to cancers of the breast, mouth, throat, esophagus, rectum, and lungs. Daily human exposure to ethanol ranks high among possible carcinogenic hazards. Alcohol seems to promote the development of cancer once the disease has started.

One recent study found a linear relationship between alcohol intake by women and cancer of the breast—women who drank less than 1 drink a day had a slightly elevated risk, and those who drank more increased their risk accordingly.[20] Cancer of the rectum occurs more often in those who drink more than 15 ounces of beer each day than in others. In the case of beer, alcohol may be acting together with other compounds formed during brewing to promote the cancer. One compound, **urethane,** is often found in some alcoholic beverages, especially flavored imported brandy. Urethane is known to cause cancer in animals, but the risk to

human beings is unknown.[21]

Other long-term effects of alcohol abuse include the following:

- Diabetes (type 2 or noninsulin-dependent).
- Ulcers and inflammation of the stomach and intestines.
- Nonviral hepatitis.
- Disease of the muscles of the heart.
- Severe psychological depression.
- Kidney, bladder, prostate, and pancreas damage.
- Skin rashes and sores.
- Impaired immune response.
- Deterioration of the testicles and adrenal glands.
- Feminization and sexual impotence in men.
- Brain disease and central nervous system damage.
- Impaired memory and balance.
- Malnutrition.
- Bone deterioration and osteoporosis.
- Increased risks of death from all causes.[22]

This list is by no means all-inclusive. Alcohol abuse exerts direct toxic effects on all body organs. All together, alcoholism costs our society an estimated $166 *billion* every year in medical services, lost wages, criminal offenses, auto crashes, and other losses.[23]

ALCOHOL'S EFFECT ON NUTRITION

Alcohol abuse also does damage indirectly via malnutrition. The more alcohol a person drinks, the less likely that he or she will eat enough food to obtain adequate nutrients. Like pure sugar and pure fat, alcohol is empty calories; it displaces nutrients. In a sense, every 150 calories spent on alcohol are going for a luxury item: the drinker receives no nutritional value in return. The more calories spent this way, the fewer are left to spend on nutritious foods. Table C5-6 shows the calorie amounts of typical alcoholic beverages.

Alcohol abuse also disrupts every tissue's metabolism of nutrients. Stomach cells oversecrete both acid and histamine, an agent of the immune system that produces inflammation. Beer in particular can irritate the stomach by stimulating it to release extra acid, opening the way for ulcers of the stomach and esophagus linings. Intestinal cells fail to absorb thiamin, folate, vitamin B_6, and other vitamins. Liver cells lose efficiency in activating vitamin D and alter their production and excretion of bile. Rod cells in the retina, which normally process vitamin A alcohol (retinol) to the form needed in vision, find themselves processing drinking alcohol instead. Liver cells, too, suffer a reduced capacity to process and use vitamin A.[24] The kidneys excrete magnesium, calcium, potassium, and zinc.

Alcohol's intermediate products interfere with vitamin B_6 metabolism, too. They dislodge the vitamin from its protective protein so that it is destroyed, leading to a deficiency that reduces production of red blood cells.

Most dramatic is alcohol's effect on folate. When an excess of alcohol is present, the body actively expels folate from all of its sites of action and storage. The liver, which normally contains enough folate to meet all needs, leaks its folate into the blood. As blood folate rises, the kidneys are deceived into excreting it, as if it were in excess. The intestine normally releases and retrieves folate continuously, but becomes so damaged by folate deficiency and alcohol toxicity that it fails to retrieve its own folate and misses out on any that may trickle in from food as well. Alcohol also interferes with the action of what little folate is left. This inhibits the production of new cells, especially the rapidly dividing cells of the intestine and the blood.

Nutrient deficiencies are thus a virtually inevitable consequence of alcohol abuse, not only because alcohol displaces food but also because alcohol directly interferes with the body's use of nutrients, making them ineffective even if they are present. Over a lifetime, excessive drinking brings about deficits of all the nutrients. People treated for alcohol addiction also need nutrition therapy to reverse deficiencies and even deficiency diseases rarely seen in others: night blindness, beriberi, pellagra, scurvy, and protein-energy malnutrition.

This discussion has touched on some of the ways alcohol affects health and nutrition. In contrast to some possible benefits of moderate alcohol consumption, excessive alcohol consumption presents a great potential for harm. Alcohol is guilty of contributing not only to deaths from health problems, but also to most of the other deaths of young people, including car crashes, falls, suicides, homicides, drownings, and other accidents.[25] The surest way to escape the harmful effects of alcohol is, of course, to refuse alcohol altogether. If you do drink, do so with care and strictly in moderation.

Table C5-6

Calories in Alcoholic Beverages and Mixers

Beverage	Amount (oz)	Energy (cal)
Beer	12	150
Light beer	12	100
Gin, rum, vodka, whiskey (86 proof)	1½	105
Dessert wine	3½	140
Table wine	3½	85
Tonic, ginger ale	8	80
Cola, root beer	8	100
Fruit-flavored soda, Tom Collins mix	8	115
Club soda, plain seltzer, diet drinks	8	1

6 The Proteins and Amino Acids

Sénèque Obin, 1893–1977, *Marché Poissons* (Fish Market) before 1957.

Contents

Frequently Asked Questions

proteins compounds composed of carbon, hydrogen, oxygen, and nitrogen and arranged as strands of amino acids. Some amino acids also contain the element sulfur.

amino (a-MEEN-o) **acids** the building blocks of protein. Each has an amine group at one end, an acid group at the other, and a distinctive side chain.

amine (a-MEEN) **group** the nitrogen-containing portion of an amino acid.

side chain the unique chemical structure attached to the backbone of each amino acid that differentiates one amino acid from another.

essential amino acids amino acids that either cannot be synthesized at all by the body or cannot be synthesized in amounts sufficient to meet physiological need. Also called *indispensable amino acids.*

THE PROTEINS are amazing, versatile, and vital cellular working molecules. Without them, life would not exist. First named 150 years ago after the Greek word *proteios* ("of prime importance"), **proteins** have revealed countless secrets of the life processes, and they account for many nutrition concerns. Why are certain chemical substances essential nutrients and not others? How do we grow? How do our bodies replace the materials they lose? How does blood clot? What gives us immunity? Understanding the nature of the proteins gives us many of the answers to these questions.

Some proteins are working proteins; others form structures. Working proteins include the body's enzymes, antibodies, transport vehicles, hormones, cellular "pumps," and oxygen carriers. Structural proteins include tendons and ligaments, scars, the cores of bones and teeth, the filaments of hair, the materials of nails, and more. All protein molecules have much in common.

The Structure of Proteins

The structure of proteins enables them to perform many vital functions. One key difference from carbohydrates and fats, which contain carbon, hydrogen, and oxygen atoms, is that proteins also contain nitrogen atoms. These nitrogen atoms give the name *amino* (nitrogen containing) to the **amino acids,** the building blocks of protein. Another key difference is that in contrast to the carbohydrates, whose repeating units, glucose molecules, are identical, the amino acids in a strand of protein are different from one another. A strand of amino acids that makes up a protein may contain 20 *different* kinds of amino acids.

Amino Acids

All amino acids have a simple chemical backbone consisting of a single carbon atom with both an **amine group** (the nitrogen-containing part) and an acid group attached to it. This backbone is the same for all amino acids. The differences among amino acids depend on a distinctive structure, the chemical **side chain,** that is also attached to the center carbon of the backbone (see Figure 6-1). It is the side chain that gives identity and chemical nature to each amino acid. About 20 amino acids with 20 different side chains make up most of the proteins of living tissue. Other rare amino acids appear in a few proteins.

The side chains make the amino acids differ in size, shape, and electrical charge. Some are negative, some are positive, and some have no charge (they are neutral). The first part of Figure 6-2 (page 178) is a diagram of three amino acids, each with a different side chain attached to its backbone. The rest of the figure shows how amino acids link to form protein strands. Long strands of amino acids form large protein molecules, and the side chains of the amino acids ultimately help to determine the molecules' shapes and behaviors.[1]

Hair, skin, eyesight, and the health of the whole body depend on protein from food.

Essential Amino Acids The body can make about half of the 20 amino acids for itself, given the needed parts: fragments derived from carbohydrate or fat to form the backbones and nitrogen from other sources to form the amine groups. The healthy adult body makes some other amino acids too slowly to meet its needs, however, or cannot make them at all. These are the **essential amino acids** (listed on page 178). Without these essential nutrients, the body cannot make the proteins it needs to do its work. Because the essential amino acids can only be obtained from foods, a person must eat often the foods that provide them.

Sometimes a nonessential amino acid becomes essential under special circumstances. For example, the body normally makes tyrosine (a nonessential amino acid) from the essential amino acid phenylalanine. But if the diet fails to supply enough phenylalanine, or if the body cannot make the conversion for some reason (as happens in the inherited disease phenylketonuria), then tyrosine becomes a **conditionally essential amino acid.**

conditionally essential amino acid an amino acid that is normally nonessential, but must be supplied by the diet in special circumstances when the need for it exceeds the the body's ability to produce it.

peptide bond a bond that connects one amino acid with another, forming a link in a protein chain.

Recycling Amino Acids The body not only makes some amino acids, but also breaks protein molecules apart and reuses their amino acids. Both food proteins, after digestion, and body proteins, when they have finished their cellular work, are dismantled to liberate their component amino acids. Pools of such amino acids provide the cells with raw materials from which they can build the protein molecules they need. Cells can also use the amino acids for energy and discard the nitrogen atoms as wastes. By reusing amino acids to build proteins, however, the body recycles and conserves a valuable commodity while easing its nitrogen disposal burden.

This recycling system also provides a sort of emergency fund of amino acids that tissues can draw on in times of fuel or protein deprivation. At such times, tissues break down their own proteins, sacrificing working molecules before the ends of their normal lifetimes, to supply energy and protein to the body's cells. The body employs a priority system in selecting the tissue proteins to dismantle—it uses the most dispensable ones first, such as the small proteins of the blood. It guards the protein structures of the heart and other organs until forced, by dire need, to relinquish them.

More about phenylketonuria, or PKU, in Controversy 4

Figure 6-1

An Amino Acid

The "backbone" is the same for all amino acids. The side chain differs from one amino acid to the next. The nitrogen is in the amine group.

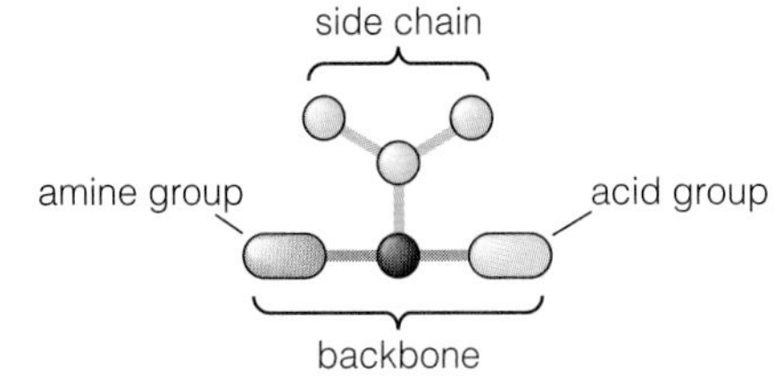
side chain
amine group
acid group
backbone

KEY POINT *Proteins are unique among the energy nutrients in that they possess nitrogen-containing amine groups and are composed of 20 different amino acid units. Some amino acids are essential, and some are essential only in special circumstances.*

How Do Amino Acids Build Proteins?

In the first step of making a protein, each amino acid is hooked to the next (as shown in Figure 6-2). A chemical bond, called a **peptide bond,** is formed between the amine group end of one amino acid and the acid group end of the next. The side chains bristle out from the backbone of the structure, and these give the protein molecule its unique character.

Taking Shape The strand of protein does not remain a straight chain. Figure 6-2 shows only the first step in making proteins, the linking of from several dozen to as many as 300 amino acid units with peptide bonds. The amino acids at different places along the strand are attracted to each other, and this attraction causes some segments of the strand to coil, somewhat like a metal spring. Also, each spot along the coiled strand is attracted to, or repelled from, other spots along its length. This causes the entire coil to fold this way and that, forming a globular structure, as shown in Figure 6-3 on the next page, or a fibrous structure (not shown).

The amino acids whose side chains are electrically charged are attracted to water. Therefore, in the body's watery fluids, they orient themselves on the outside of the protein structure. The amino acids whose side chains are neutral are repelled by water and are attracted to one another; these tuck themselves into the center, away from the body fluids. All these interactions among the amino acids and the surrounding fluids give each protein a unique architecture.

One final detail may be needed for the protein to become functional. Several strands may cluster together into a functioning unit; or a metal ion (mineral) or a vitamin may join to the unit and activate it.

Figure 6-2

Different Amino Acids Join Together

This is the basic process by which proteins are assembled.

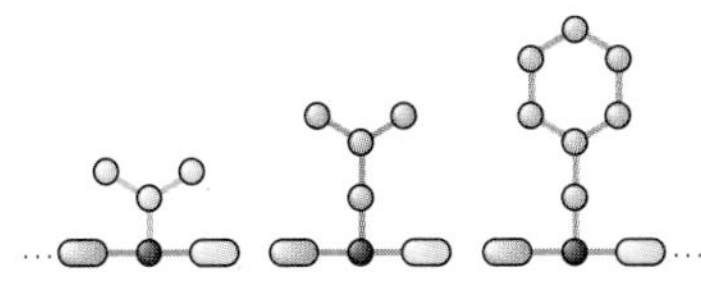
valine
leucine
tyrosine

Single amino acids with different side chains...

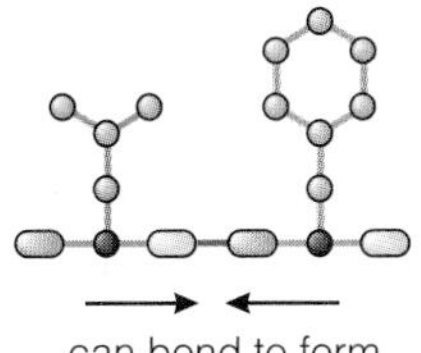

can bond to form...

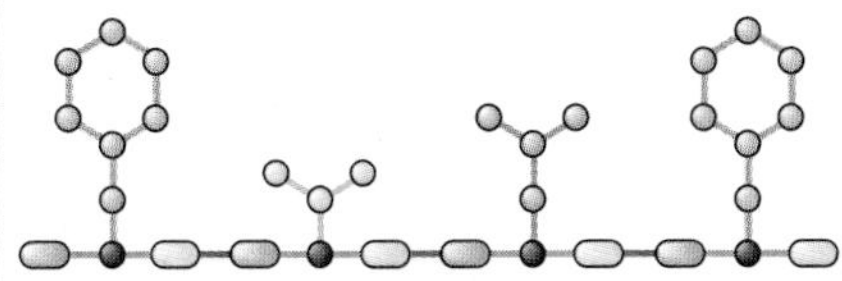

a strand of amino acids, part of a protein.

enzymes (EN-zimes) protein catalysts. A catalyst is a compound that facilitates a chemical reaction without itself being altered in the process.

hemoglobin the globular protein of red blood cells, whose iron atoms carry oxygen around the body. (More about hemoglobin in Chapter 8.)

The essential amino acids:

- Histidine.
- Isoleucine.
- Leucine.
- Lysine.
- Methionine.
- Phenylalanine.
- Threonine.
- Tryptophan.
- Valine.

Other amino acids important in nutrition:

- Alanine.
- Arginine.
- Asparagine.
- Aspartic acid.
- Cysteine.
- Glutamic acid.
- Glutamine.
- Glycine.
- Proline.
- Serine.
- Tyrosine.

NOTE: In special cases, some nonessential amino acids may become conditionally essential (see the text).

The Variety of Proteins

The dramatically different shapes of proteins enable them to perform different tasks in the body. Those of globular shape, such as some proteins of blood, are water soluble. Some are hollow balls, which can carry and store materials in their interiors. In some proteins, several coils of amino acids wind around each other and form ropelike fibers that can give strength and elasticity to body parts. Some, such as those that form tendons, are more than ten times as long as they are wide, forming stiff, rodlike structures that are somewhat insoluble in water and very strong. Still others act like glue. Among the most fascinating proteins are the **enzymes,** which act on other substances to change them chemically. The variety of proteins is endless. A model of a single large globular protein molecule, the **hemoglobin** that carries oxygen in the red blood cells, is shown in Figure 6-4.

The great variety of proteins in the world is possible because an infinite number of sequences of amino acids can be found. To understand how so many different proteins can be designed from only 20 or so amino acids, think of how many words are in an unabridged dictionary—all of them constructed from just 26 letters. The letters in a word must alternate between consonant and vowel sounds, but the amino acids in a protein need follow no such rules. Nor is there any restriction on the length of the chain of amino acids. Thus the number of possible proteins is much greater than the number of possible English words. A single human cell may contain as many as 10,000 different proteins, each one present in thousands of copies.

Inherited Amino Acid Sequences The sequences of amino acids that make up a protein molecule are specified by heredity. For each protein there is only

Figure 6-3

The Coiling and Folding of a Protein Molecule

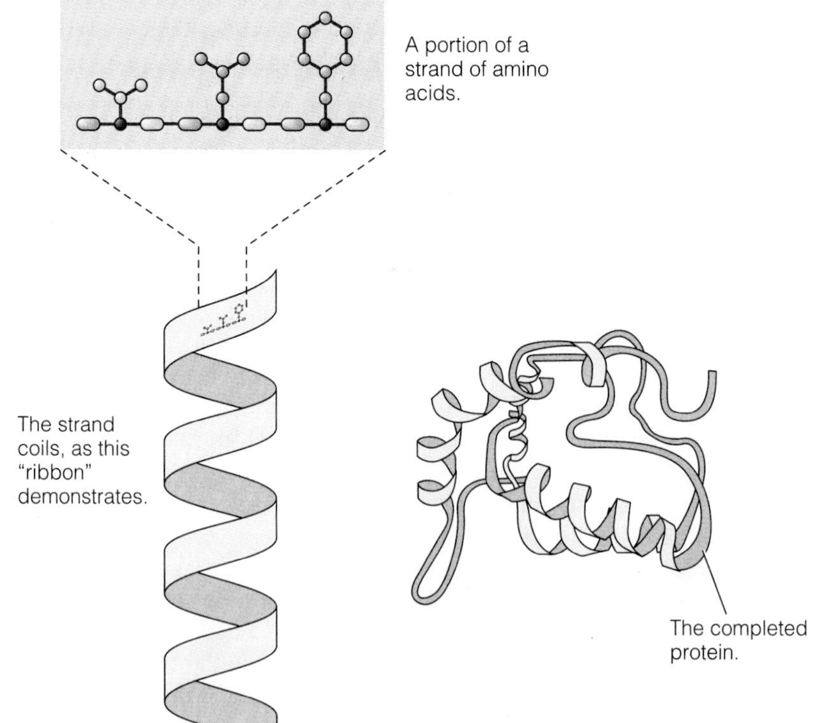
A portion of a strand of amino acids.
The strand coils, as this "ribbon" demonstrates.
The completed protein.
Coiling the strand. The strand of amino acids takes on a spring-like shape as their side chains variously attract and repel each other.
Folding the coil. Once coiled and folded, the protein may be functional as is, or it may need to join with other proteins or add a vitamin or mineral to become active.

one proper amino acid sequence. If a wrong amino acid is inserted, the result may be disastrous to health.

Sickle-cell disease, in which hemoglobin, the oxygen-carrying protein of the red blood cells, is abnormal, is an example of an inherited mistake in the amino acid sequence. Normal hemoglobin contains two kinds of chains. In sickle-cell hemoglobin, one of the chains is an exact copy of that in normal hemoglobin, but in the other chain, the sixth amino acid is valine, rather than glutamine as it should be. This replacement of one amino acid so alters the protein that it is unable to carry and release oxygen. The red blood cells collapse into crescent shapes instead of remaining disk shaped, as they normally do (see Figure 6-5). If too many abnormal, crescent-shaped cells appear in the blood, the result is illness and death. One can detect the disease by observing the altered red blood cells under a microscope.

Each person is different from any other human being. What makes you unique are minute differences in your body proteins. These differences are determined by the amino acid sequences of your proteins, which are written into the genetic code you inherited from your parents and they from theirs. At conception each person receives a unique combination of genes. The genes, passed down to a cell from its parent cell, direct the making of all the body's proteins, as shown in Figure 6-6. Notice that what the genes determine is the sequences of the amino acids in the finished protein.

Nutrients and Gene Expression When a cell makes a protein as shown in Figure 6-6, scientists say that the gene for that protein has been "expressed." The DNA for making every protein of the human body is contained within every cell nucleus, but not all cells make every type of protein, of course. Instead, cells specialize in making only the proteins typical of their cell types.* For example, only cells of the pancreas express the gene for the protein hormone insulin; in other cells, that gene is idle. Cells regulate the expression of genes, and so the synthesis of various proteins, in response to changing conditions within the body. Often, nutrients signal the need for more or less of a protein. For example, when the presence of glucose in the blood indicates that more of the hormone insulin is needed, pancreas cells speed up their production of this protein. Likewise, when the body's iron stores run low, bone marrow cells slow down their synthesis of hemoglobin. Nutrients often act in these ways to modulate gene expression, and later chapters provide more examples. The Fitness for Life feature on p. 181 addresses the question of whether increasing protein consumption can signal the genes in muscle cells to begin building larger muscles.

KEY POINT *Amino acids link into long strands that coil and fold to make a wide variety of different proteins. Each type of protein has a distinctive sequence of amino acids and so has great specificity.*

Denaturation of Proteins

Proteins can be denatured (distorted in shape) by heat, alcohol, acids, bases, or the salts of heavy metals. The **denaturation** of a protein is the first step in its destruction; thus these agents are dangerous because they damage the body's proteins. In digestion, however, denaturation is useful to the body. During the digestion of a food protein, the stomach acid opens up the protein's structure, permitting digestive enzymes to make contact with the peptide bonds and cleave them. Denaturation also occurs during the cooking of foods. Cooking an egg denatures the proteins of the egg and makes it firm. Perhaps more importantly, cooking denatures two proteins in raw eggs: one binds the B vitamin biotin and the mineral iron, and the other slows protein digestion. Thus cooking eggs liberates biotin and iron and aids digestion.

Many well-known poisons are salts of heavy metals like mercury and silver; these denature proteins wherever they touch them. The common first-aid remedy for swallowing a heavy-metal poison is to drink milk. The poison then acts on the protein of the milk rather than on the protein tissues of the mouth, esophagus, and

*Red blood cells lack nuclei and so also lack the DNA needed for making proteins.

denaturation the change in a protein's shape brought about by heat, acids, bases, alcohol, salts of heavy metals, or other agents.

Figure 6-4

The Protein Hemoglobin

The coiled and looped red structures are the globular proteins: the flat, jagged-edged objects are heme structures; the red center balls are iron atoms. This model represents a molecule of hemoglobin magnified 27 million times.

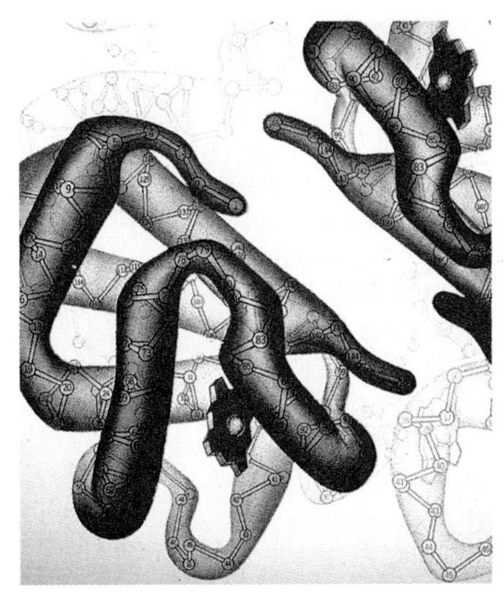

Figure 6-5

Normal Red Blood Cells and Sickle Cells

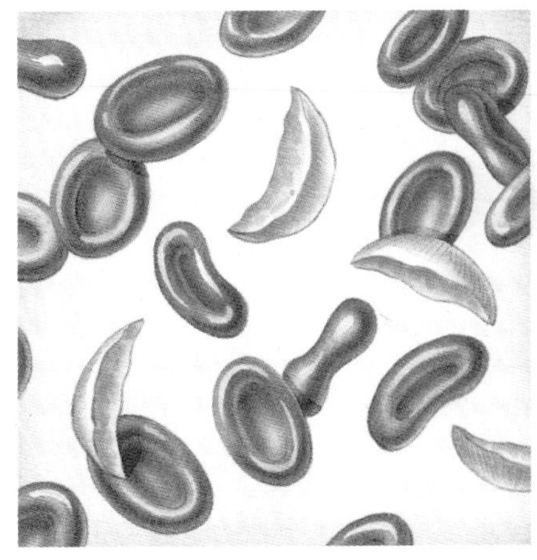

Figure 6-6

Protein Synthesis

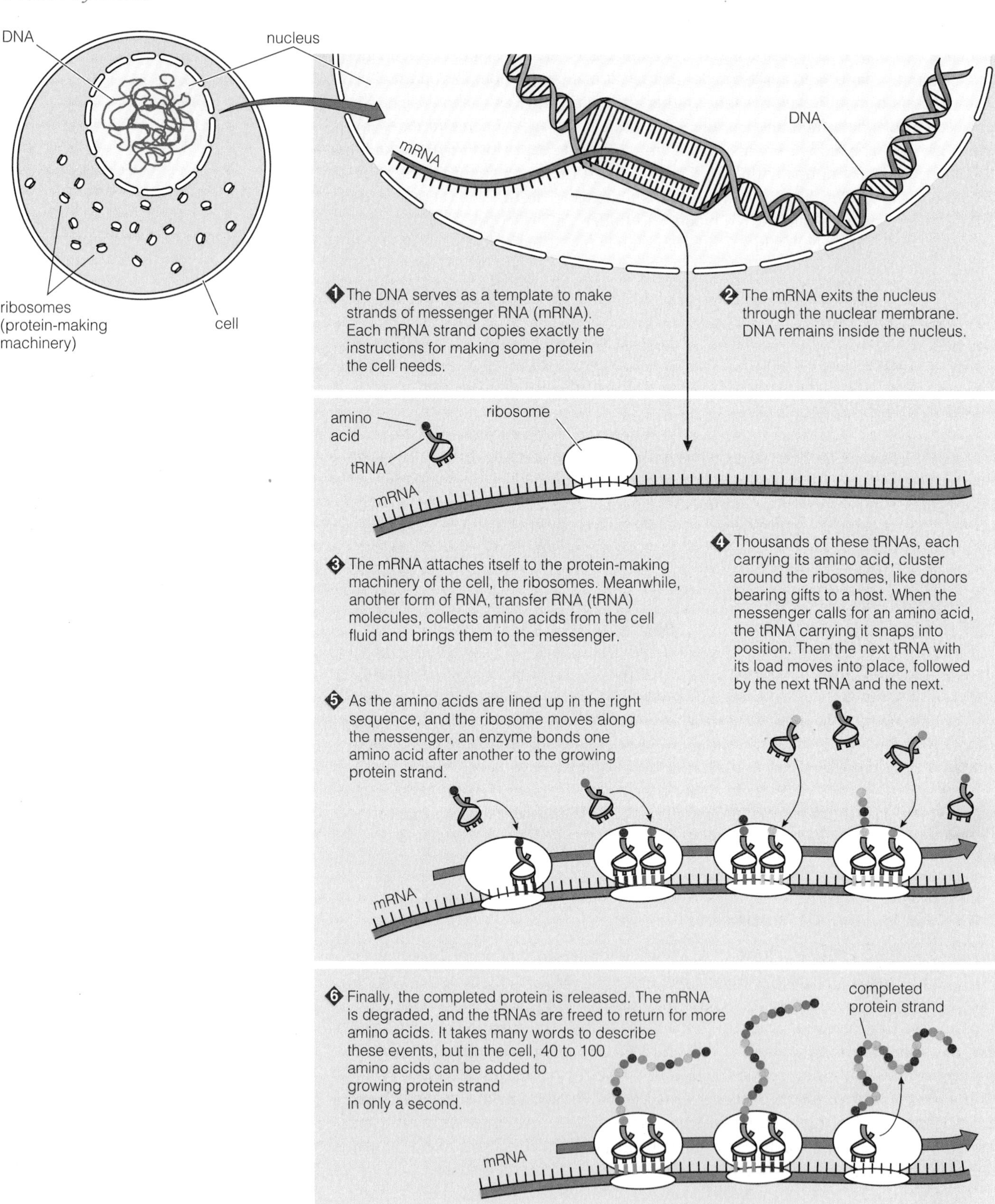
DNA
nucleus
ribosomes (protein-making machinery)
cell
mRNA
DNA
1 The DNA serves as a template to make strands of messenger RNA (mRNA). Each mRNA strand copies exactly the instructions for making some protein the cell needs.
2 The mRNA exits the nucleus through the nuclear membrane. DNA remains inside the nucleus.
amino acid
ribosome
tRNA
mRNA
3 The mRNA attaches itself to the protein-making machinery of the cell, the ribosomes. Meanwhile, another form of RNA, transfer RNA (tRNA) molecules, collects amino acids from the cell fluid and brings them to the messenger.
4 Thousands of these tRNAs, each carrying its amino acid, cluster around the ribosomes, like donors bearing gifts to a host. When the messenger calls for an amino acid, the tRNA carrying it snaps into position. Then the next tRNA with its load moves into place, followed by the next tRNA and the next.
5 As the amino acids are lined up in the right sequence, and the ribosome moves along the messenger, an enzyme bonds one amino acid after another to the growing protein strand.
mRNA
6 Finally, the completed protein is released. The mRNA is degraded, and the tRNAs are freed to return for more amino acids. It takes many words to describe these events, but in the cell, 40 to 100 amino acids can be added to growing protein strand in only a second.
completed protein strand
mRNA

Can Eating Extra Protein Make Muscles Grow Larger?

FITNESS FOR LIFE

Can athletes and fitness seekers stimulate their muscles to grow larger by consuming more protein? Or, to put it another way, will extra protein from food or supplements trigger the genes in muscle cells to build more muscle proteins?

The answers are "no" and a qualified "maybe," depending upon the athletic endeavor. Right away, it should be said that exercise, not excess dietary protein, triggers the building of muscle tissue. Exercise generates cellular messages that stimulate DNA to begin the process of building up muscle fibers (muscle fibers are made of protein). An excess of amino acids or other nutrients does not generate these messages.

Theoretically, it is possible for an athlete to choose a diet so monotonously high in carbohydrate-rich foods as to threaten protein malnutrition. This situation can easily be avoided just by including foods such as milk, eggs, legumes, and fish in balanced meals and by consuming enough total food to meet the increased energy needs of exercise. The great majority of people need not make a special effort to eat any particular foods or take supplements in order to obtain enough protein. Even most vegetarians can obtain the protein they need for sport from balanced diets made up of ordinary foods.[1] Over weeks or months, most nutritious diets end up supplying about 12 to 15 percent of their energy as protein, regardless of which foods are chosen.

In one exceptional case, a benefit is sometimes observed from taking in extra protein. That case is of a young male bodybuilder striving at competition level to build his muscle tissue to its physiological maximum limit.[2] In his case only, a protein intake of about double the recommendation (see the 1989 RDA on the inside front cover) for a male of the same age may be useful. Yet even this athlete probably needs to make no special effort to obtain protein. By meeting his large energy need with foods in a well-chosen diet, his increased protein need is amply provided—without supplements. To see how quickly protein grams add up with additional servings of ordinary foods, read the protein information on some food labels and pay attention to the protein values listed in the Food Feature and the Do It! sections of this chapter.

The path to bigger muscles, then, is rigorous physical training, with adequate protein from nutritious foods to support muscle growth. Food also supplies energy, fiber, water, other essential nutrients, and other constituents that active people need for health and performance—something supplements can never do. For the overwhelming majority of exercisers, adding excess protein or amino acids to an adequate diet will put on only pounds of body fat, not muscle.

stomach. Later, vomiting may be induced to expel the poison that has combined with the milk.

Chapter 10 discusses the nutrient needs of athletes in detail.

KEY POINT *Proteins can be denatured by heat, acids, bases, alcohol, or the salts of heavy metals. Denaturation begins the process of digesting food protein and can also destroy body proteins.*

Digestion and Absorption of Protein

Each protein is designed for a special purpose in a particular tissue of a specific kind of animal or plant. When a person eats food proteins, whether from cereals, vegetables, beef, fish, or cheese, the body must first alter them by breaking them down into amino acids; only then can it rearrange them into proteins with its own unique amino acid sequences.

Other than being crushed and moistened with saliva in the mouth, nothing happens to protein until it reaches the very strong acid of the stomach. There the acid helps to uncoil the protein's tangled strands so that molecules of the stomach's protein-digesting enzyme can attack the peptide bonds. You might expect that the stomach enzyme itself, being a protein, would be denatured by the stomach's acid. Unlike most enzymes,

dipeptides (dye-PEP-tides) protein fragments that are two amino acids long. A peptide is a strand of amino acids (*di* means "two").

tripeptides (try-PEP-tides) protein fragments that are three amino acids long (*tri* means "three").

polypeptides protein fragments of many (more than ten) amino acids bonded together (*poly* means "many"). A chain of between four and ten amino acids is called an oligopeptide.

Chapter 3 discussed the use of medicines to control the stomach's acidity and also defined pH as a measure of acidity. See p. 79.

though, the stomach enzyme functions best in an acid environment. Its job is to break other protein strands into smaller pieces. The stomach lining, which is also made partly of protein, is protected against attack by acid and enzymes by a coat of mucus, secreted by its cells.

Protein Digestion

The whole process of digestion is an ingenious solution to a complex problem. Proteins (enzymes), activated by acid, digest proteins from food, denatured by acid. The coating of mucus secreted by the stomach wall protects its proteins from attack by either acid or enzymes. The acid in the stomach is so strong (pH 1.5) that no food is acid enough to make it stronger; for comparison, the pH of pure vinegar is about 3. Obviously, the stomach is supposed to be acid to do its job.

By the time most proteins slip from the stomach into the small intestine, they are already broken into smaller pieces. Some are single amino acids; others are strands of two or three amino acids (**dipeptides** and **tripeptides**). The majority are longer chains (**polypeptides**), and a few are whole proteins. In the small intestine, alkaline juice from the pancreas neutralizes the acid delivered by the stomach. The pH rises to about 7 (neutral), enabling the next enzyme team to accomplish the final breakdown of the strands. Protein-digesting enzymes from the pancreas and intestine continue working until almost all pieces of protein are broken into small fragments or single amino acids. Figure 6-7 shows a dipeptide and a tripeptide, and the whole process is summarized in Figure 6-8.

KEY POINT *Digestion of protein involves denaturation by stomach acid, then enzymatic digestion in the stomach and small intestine to amino acids, dipeptides, and tripeptides.*

After Protein Is Digested, What Happens to the Amino Acids?

The cells all along the small intestine absorb single amino acids. As for dipeptides and tripeptides, the cells that line the small intestine have enzymes on their surfaces that split most of them into single amino acids, and the cells absorb them, too. Then the cells release all the single amino acids into the bloodstream. A few dipeptides, tripeptides, and even larger molecules can escape the digestive process altogether and cross the digestive tract wall to enter the bloodstream. It is thought that these larger particles may act as hormones to regulate body functions and provide the body with information about the environment. The larger molecules may also play a role in food allergy via the immune response.

The cells of the small intestine possess different sites for absorbing different types of amino acids. Amino acids of the same type compete for the same absorption sites. Consequently, when a person ingests a large dose of any single amino acid, that amino acid may limit absorption of others of its general type. The Consumer Corner, presented later, cautions against taking single amino acids as supplements, partly for this reason.

Once they are circulating in the bloodstream, amino acids are available to be taken up by any cell of the body. The body cells then use the amino acids to make proteins, either for their own use or for secretion into lymph or blood, or for other uses. Alternatively, the body cells can use amino acids for energy.

KEY POINT *The cells of the small intestine complete digestion, absorb amino acids and some larger peptides, and release them into the bloodstream.*

Figure 6-7

A Dipeptide and Tripeptide

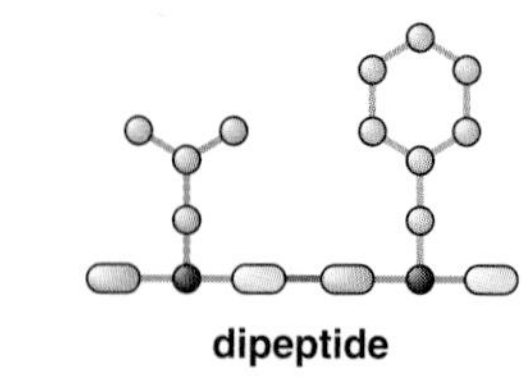
dipeptide

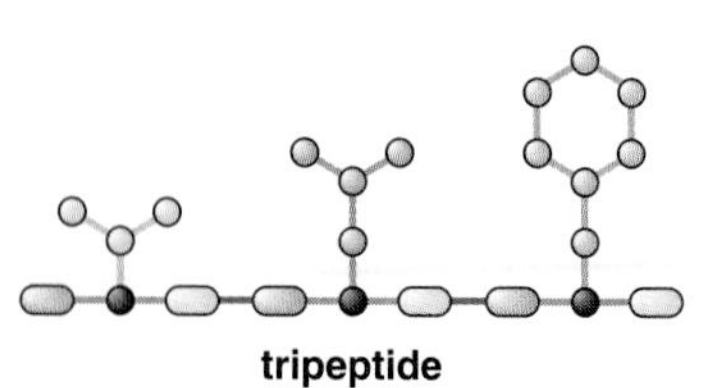
tripeptide

The Roles of Proteins in the Body

Only a sampling of the many roles proteins play can be described here, but these should serve to illustrate their versatility, uniqueness, and importance in the body. No wonder their discoverers called proteins the primary material of life.

Figure 6-8

How Protein in Food Becomes Amino Acids in the Body

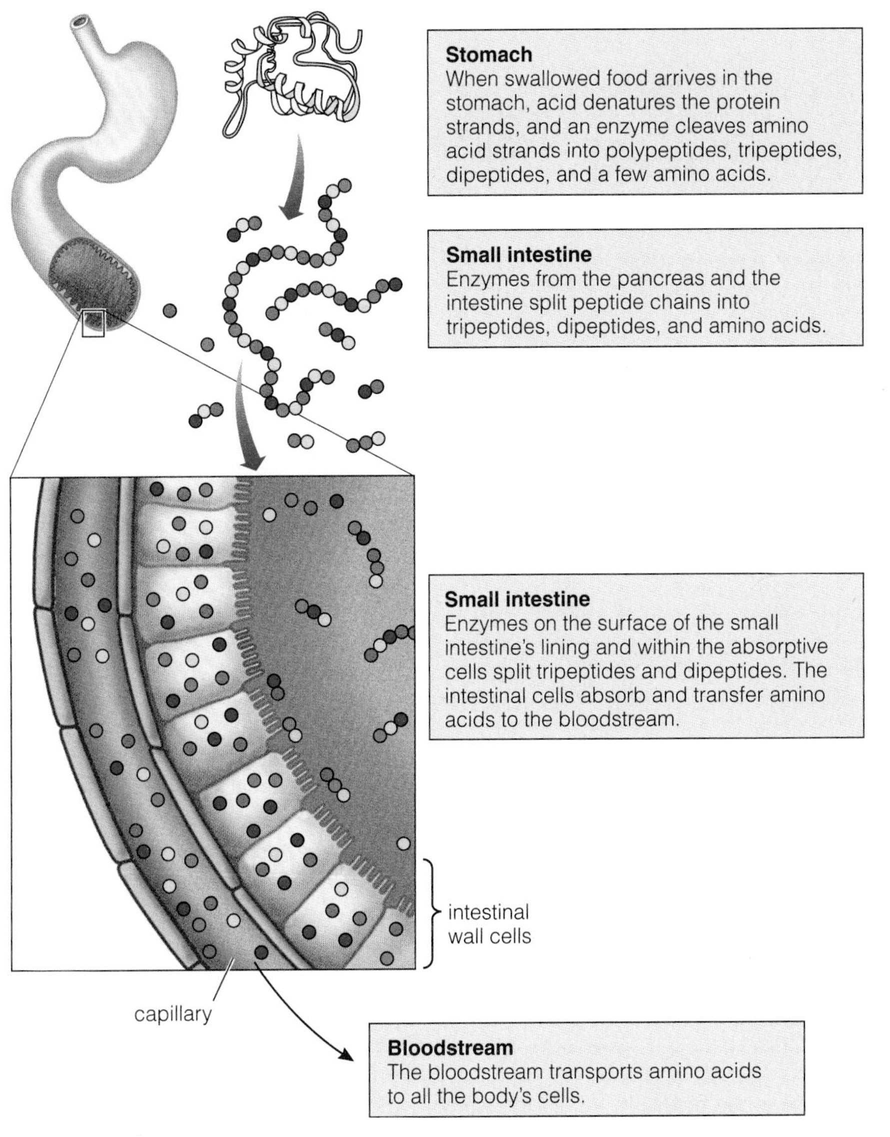
Stomach
When swallowed food arrives in the stomach, acid denatures the protein strands, and an enzyme cleaves amino acid strands into polypeptides, tripeptides, dipeptides, and a few amino acids.
Small intestine
Enzymes from the pancreas and the intestine split peptide chains into tripeptides, dipeptides, and amino acids.
Small intestine
Enzymes on the surface of the small intestine's lining and within the absorptive cells split tripeptides and dipeptides. The intestinal cells absorb and transfer amino acids to the bloodstream.
intestinal wall cells
capillary
Bloodstream
The bloodstream transports amino acids to all the body's cells.

Supporting Growth and Maintenance

Amino acids must be continuously available to build the proteins of new tissue. The new tissue may be in an embryo; in a growing child; in new blood needed to replace blood lost in burns, hemorrhage, or surgery; in the scar tissue that heals wounds; or in new hair and nails.

Less obvious is the protein that helps to replace worn-out cells in everyone's body all the time. Each of the millions of red blood cells lives for only three or four months. Then each must be replaced by a new cell produced by the bone marrow. The millions of cells that line the intestinal tract live for only three days; they are constantly being shed and replaced. The cells of the skin die and rub off, and new ones grow from underneath. Nearly all cells arise, live, and die this way, and while they are living, they constantly make and break down their proteins. Amino acids from food support all the new growth and maintenance of cells and the making of the working parts within them.

KEY POINT *The body needs amino acids to grow new cells and to replace worn-out ones.*

hormones as defined in Chapter 3, chemical messengers secreted by a number of body organs in response to conditions that require regulation. Each hormone affects a specific organ or tissue and elicits a specific response.

antibodies (AN-te-bod-ees) large proteins of the blood, produced by the immune system in response to an invasion of the body by foreign substances (antigens). Antibodies combine with and inactivate the antigens. Also defined in Chapter 3.

immunity specific disease resistance, derived from the immune system's memory of prior exposure to specific disease agents and its ability to mount a swift defense against them.

Controversy 13 presents some details about tryptophan's relationship to serotonin; the vitamin niacin is discussed in Chapter 7.

Figure 6-9

Enzyme Action

Enzymes are catalysts: they speed up reactions that would happen anyway, but much more slowly. This enzyme works by positioning two compounds, A and B, so that the reaction between them will be especially likely to take place.

Compounds A and B are attracted to the enzyme's active site and park there for a moment in the exact position that makes the reaction between them most likely to occur. They react by bonding together and leave the enzyme as the new compound, AB.

A single enzyme can facilitate several hundred such synthetic reactions in a second. Other enzymes break compounds apart into two or more products or rearrange the atoms in one compound to make another one.

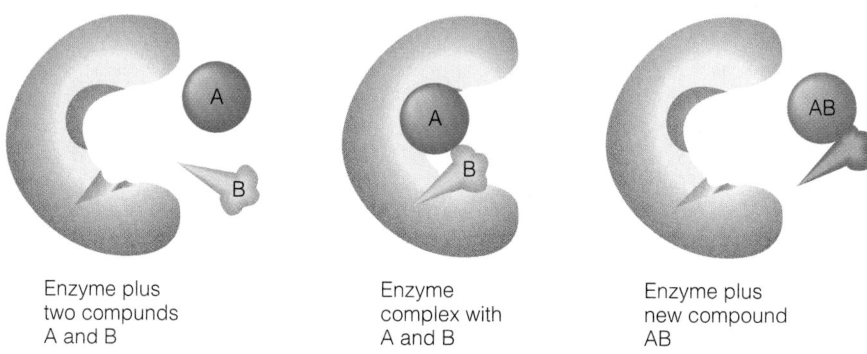
A
B
A
B
AB
Enzyme plus two compunds A and B
Enzyme complex with A and B
Enzyme plus new compound AB

Building Enzymes, Hormones, and Other Compounds

Enzymes are among the most important of the proteins formed in living cells. Thousands of enzymes reside inside a single cell, each one a catalyst that facilitates a specific chemical reaction. Figure 6-9 above shows how a hypothetical enzyme works.

The body's many **hormones** are messenger molecules, and some are made from amino acids. (Recall from Chapter 5 that some hormones are made from lipids.) Various body glands release hormones in response to changes in the internal environment. The hormones then elicit the responses necessary to restore normal conditions. Among the hormones made of amino acids is the thyroid hormone, which regulates the body's metabolism. An opposing pair of hormones, insulin and glucagon, maintain blood glucose levels, as described in Chapter 4. Many other hormones are at work in the body regulating equally critical body functions. For interest, Figure 6-10 shows how many amino acids are linked in sequence to form human insulin. It also shows how certain side groups attract one another to complete the insulin molecule and make it functional.

In addition to serving as building blocks for proteins, amino acids also perform tasks in the body as amino acids. For example, the amino acid tyrosine forms parts of the chemical messengers epinephrine and norepinephrine, which relay nervous system messages throughout the body. The body also uses tyrosine to make the brown pigment melanin, which is responsible for skin, hair, and eye color. It also becomes the hormone thyroxine, which helps to regulate the body's metabolic rate. The amino acid tryptophan serves as starting material for the neurotransmitter serotonin and the vitamin niacin.

KEY POINT *The body makes enzymes, hormones, and chemical messengers of the nervous system from its amino acids.*

Building Antibodies

Of all the proteins in living organisms, the **antibodies** best demonstrate that proteins are specific to one organism. Antibodies recognize every protein that belongs in "their" body and leave it alone, but they attack foreign particles (usually proteins) that invade the body. The foreign protein may be part of a bacterium, a virus, or a toxin, or it may be present in a food that causes allergy. The body, upon recognizing that it has been invaded, manufactures antibodies specially designed to inactivate the foreign protein.

Each antibody is designed to destroy one specific invader. An antibody active against one strain of influenza will be of no help to a person ill with another strain. Once the body has learned to make a particular antibody, it remembers. The next time the body encounters that same invader, it destroys the invader even more rapidly. In other words, the body develops **immunity** to the invader. This molecular memory underlies the principle of immunizations, injections of drugs made from destroyed and inactivated microbes or their products that activate the body's immune defenses. Some immunities are lifelong; others, such as that to tetanus, must be "boosted" at intervals.

KEY POINT *Antibodies are formed from amino acids to defend against foreign proteins and other foreign substances within the body.*

Maintaining Fluid and Electrolyte Balance

Proteins help to maintain the **fluid and electrolyte balance** by regulating the quantity of fluids in the compartments of the body. To remain alive, cells must contain a constant amount of fluid. Too much can cause them to rupture; too little would make them unable to function. Although water can diffuse freely into and out of cells, proteins cannot; and proteins attract water. By maintaining stores of internal proteins and also of some minerals, cells retain the fluid they need. The cells also keep the fluid volume constant in the spaces between them by secreting proteins (and minerals) into those spaces. Thus proper balance is maintained. Should this system begin to fail, too much fluid would collect outside the cells, causing **edema.**

Not only is the quantity of the body fluids vital to life, but so also is their composition. Transport proteins in the membranes of cells maintain this composition by continuously transferring substances into and out of cells. For example, sodium is concentrated outside the cells, and potassium is concentrated inside (see Figure 6-11). A disturbance of this balance can impair the action of the heart, lungs, and brain, triggering a major medical emergency. Cell proteins work constantly to avert such a disaster by holding fluids and electrolytes in their proper chambers.

KEY POINT *Proteins help to regulate the body's electrolytes and fluids.*

Maintaining Acid-Base Balance

Normal processes of the body continually produce **acids** and their opposite, **bases,** that must be carried by the blood to the organs of excretion. The blood must do this without allowing its own **acid-base balance** to be affected. This feat is another trick of the blood proteins, which act as **buffers** to maintain the blood's normal pH. The buffers pick up hydrogens (acid) when there are too many and release them again when there are too few. The secret is that negatively charged side chains of amino acids can accommodate additional hydrogens, which are positively charged, when necessary.

Blood pH is one of the most rigidly controlled conditions in the body. If it changes too much, the dangerous condition **acidosis** or the opposite, basic condition **alkalosis** can cause coma or death. These conditions are hazardous because of their effect on proteins. When the proteins' buffering capacity is filled—that is, when they have taken on all the acid hydrogens they can accommodate—additional acid pulls them out of shape, denaturing them and disrupting many body processes. Table 6-1 sums up the functions of protein we have discussed and adds several others.

KEY POINT *Proteins buffer the blood against excess acidity or alkalinity.*

fluid and electrolyte balance the distribution of fluid and dissolved particles among body compartments (see also Chapter 8).

edema (eh-DEEM-uh) swelling of body tissue caused by leakage of fluid from the blood vessels, seen in protein deficiency (among other conditions).

acids compounds that release hydrogens in a watery solution.

bases compounds that accept hydrogens from solutions.

acid-base balance equilibrium between acid and base concentrations in the body fluids.

buffers compounds that help keep a solution's acidity or alkalinity constant.

acidosis (acid-DOH-sis) blood acidity above normal, indicating excess acid (*osis* means "too much in the blood").

alkalosis (al-kah-LOH-sis) blood alkalinity above normal (*alka* means "base"; *osis* means "too much in the blood").

The control of water's location by electrolytes is discussed further in Chapter 8.

Figure 6-10

Amino Acid Sequence of Human Insulin

This picture shows a refinement of protein structure not mentioned in the text. The amino acid cysteine (cys) has a sulfur-containing side group. The sulfur groups on two cysteine molecules can bond together, creating a bridge between two protein strands or two parts of the same strand. Insulin contains three such bridges.

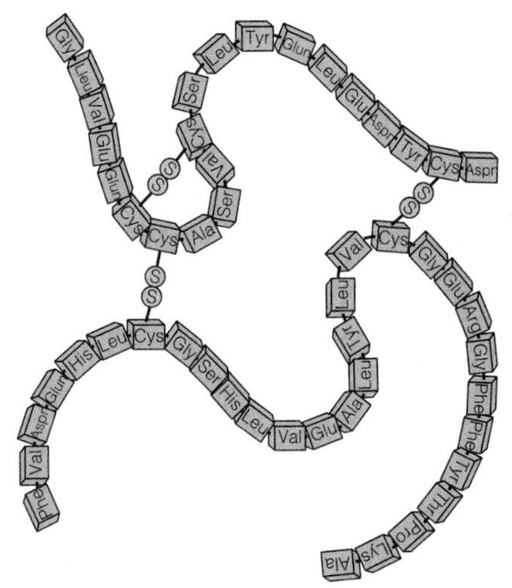
Gly
Ileu
Val
Glu
Glun
Cys
Cys
Ala
Ser
Ileu
Cys
Ser
Leu
Tyr
Glun
Leu
Glu
Aspn
Tyr
Cys
Aspn
S
S
S
S
S
S
Phe
Val
Aspn
Glun
His
Leu
Cys
Gly
Ser
His
Leu
Val
Glu
Ala
Leu
Tyr
Leu
Val
Cys
Gly
Glu
Arg
Gly
Phe
Phe
Tyr
Thr
Pro
Lys
Ala

Figure 6-11

Proteins Transport Substances Into and Out of Cells

A transport protein within a cell membrane acts as a sort of revolving door—it picks up substances on one side and flips them to the other side without leaving the membrane itself. This form of transporting substances into and out of cells is often called active transport. *The substances being transported here are sodium and potassium. The significance of sodium in fluid and electrolyte balance is discussed in Chapter 8.*

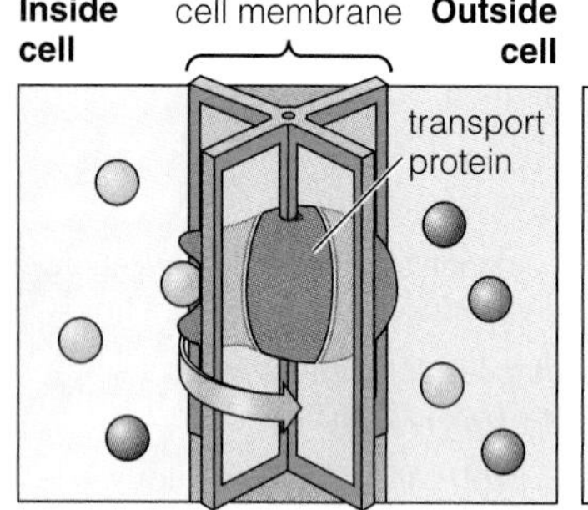
Inside cell
cell membrane
Outside cell
transport protein

Protein flips

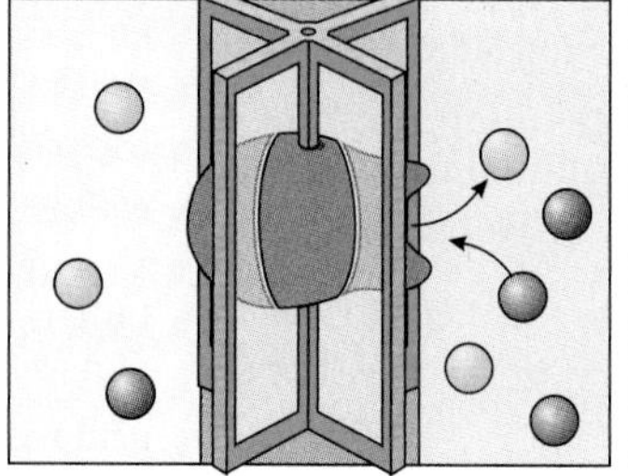

Molecules trade places

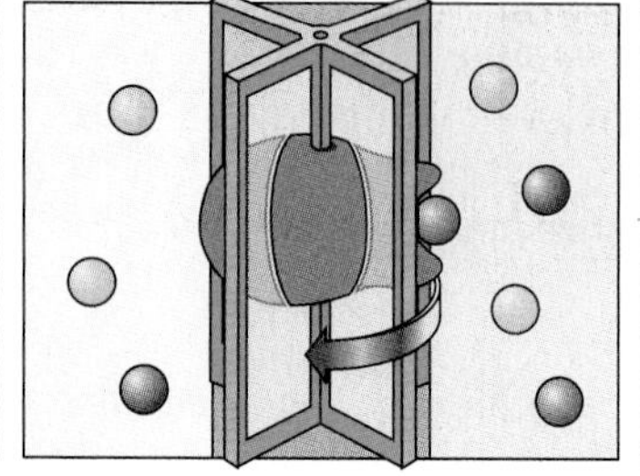

Protein flips

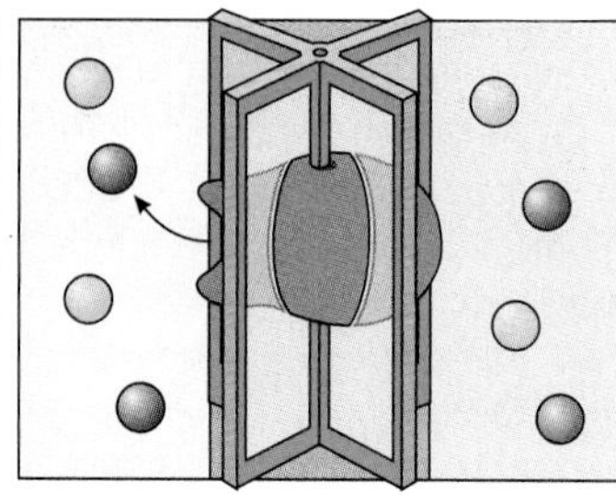

Balance restored

urea (yoo-REE-uh) the principal nitrogen-excretion product of metabolism; generated mostly by removal of amine groups from unneeded amino acids or from amino acids being sacrificed to a need for energy.

Providing Energy

Only protein can perform all the functions just described, but protein will be surrendered to provide energy if need be. The body must have energy to live from moment to moment, so obtaining that energy is a top priority.

When amino acids are degraded for energy, their amine groups are stripped off and used elsewhere or are incorporated by the liver into **urea** and sent to the kidneys for excretion in the urine. The fragments that remain are composed of carbon, hydrogen, and oxygen, as are carbohydrate and fat, and can be used to build those substances or can be metabolized like them.

Not only can amino acids supply energy, but many of them can be converted to glucose, as fatty acids can never be. Thus, if need be, protein can help to maintain a steady blood glucose level and so serve the glucose need of the brain.

The similarities and differences of the three energy-yielding nutrients should now be clear. Carbohydrate offers energy; fat offers concentrated energy; and protein, if needed, can offer energy plus nitrogen (see Figure 6-12).

Only if the protein-sparing energy from carbohydrate and fat is sufficient to power the cells will the amino acids be used for the work only they can perform—making proteins. The body does not make a specialized energy-storage compound from pro-

Table 6-1

Summary of Functions of Proteins

- **growth and maintenance** Proteins serve as building materials for growth and repair of body tissues.
- **enzymes** Proteins facilitate needed chemical reactions.
- **hormones** Proteins regulate body processes. Some hormones are proteins or are made from amino acids.
- **antibodies** Proteins form the immune system molecules that fight diseases.
- **fluid and electrolyte balance** Proteins help to maintain the fluid and mineral composition of various body fluids.
- **acid-base balance** Proteins help maintain the acid-base balance of various body fluids by acting as buffers.
- **energy** Proteins provide some fuel for the body's energy needs.
- **transportation** Proteins help transport needed substances, such as lipids, minerals, and oxygen, around the body.
- **blood clotting** Proteins provide the netting on which blood clots are built.
- **structural components** Proteins form integral parts of most body structures such as skin, tendons, ligaments, membranes, muscles, organs, and bones.

tein as it does from carbohydrate and fat. Glucose is stored as glycogen and fat as triglycerides, but body protein is available only as the active working molecular and structural components of the tissues. When the need becomes urgent, the body must dismantle its tissue proteins to obtain amino acids for energy. Each protein is taken in its own time: first, small proteins from the blood and liver; then, proteins from the muscles and other organs. Thus energy deficiency (starvation) always incurs wasting of lean body tissue as well as loss of fat.

If amino acids are oversupplied, the body cannot store them. It has no choice but to remove and excrete their amine groups and then convert the residues to glycogen or fat for energy storage.

KEY POINT *When insufficient carbohydrate and fat are consumed to meet the body's energy need, food protein and body protein are sacrificed to supply energy. The nitrogen part is removed from each amino acid, and the resulting fragment is oxidized for energy.*

The Fate of an Amino Acid

To review the body's handling of amino acids, let us follow the fate of an amino acid that was originally part of a protein-containing food. When the amino acid arrives in a cell, it may be used in several different ways, depending on the cell's needs at the time.

The amino acid may be used as is and become part of a growing protein. It may be altered somewhat to make another needed compound. Alternatively, the cell may dismantle the amino acid and use its amine group to build a different amino acid. The remainder may be used for fuel or, if not needed, converted to glucose or fat.

The same fate of being dismantled for fuel awaits the amino acid in a cell that is starved for energy but has no glucose or fatty acids. This amino acid may be needed to build a vital protein, but without energy, the cell will die. Therefore the cell strips the amino acid of its amine group (the nitrogen part) and uses the remainder of its structure for energy. The amine group is excreted from the cell and, finally, from the body in the urine.

Amino acids are also used for energy when the body has a surplus of amino acids and energy-yielding nutrients. In this case the cell takes the amino acid apart, excretes the amine group, converts the rest to fat, and then stores the fat in the fat cells.

Figure 6-12

Three Different Energy Sources

Carbohydrate offers energy; fat offers concentrated energy; and protein, if necessary, can offer energy plus nitrogen. The compounds at the left yield the 2-carbon fragments shown at the right. These fragments oxidize quickly in the presence of oxygen to yield carbon dioxide, water, and energy.

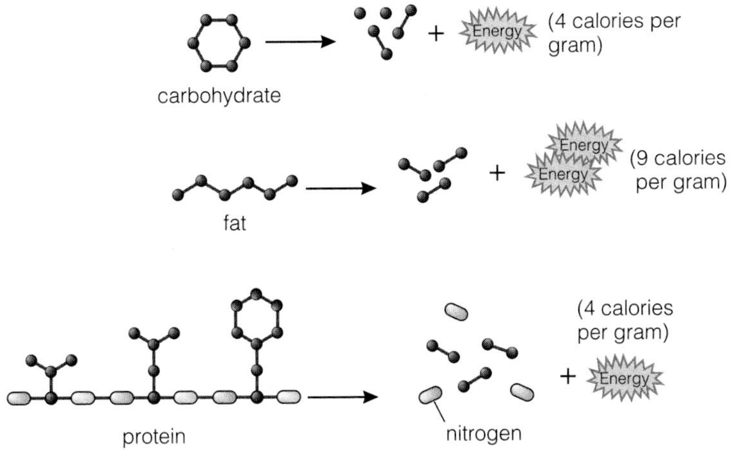

carbohydrate
Energy
(4 calories per gram)
fat
Energy
Energy
(9 calories per gram)
protein
nitrogen
Energy
(4 calories per gram)

legumes (leg-GOOMS, LEG-yooms) plants of the bean and pea family that have roots with nodules containing special bacteria. These bacteria can trap nitrogen from the air in the soil and make it into compounds that become part of the plant's seeds. The seeds are rich in protein compared with those of most other plant foods.

In summary, then, amino acids in a cell can be

- used to build proteins,
- converted to other small nitrogen-containing compounds such as the vitamin niacin, or
- converted to some other amino acids.

Stripped of their nitrogen, amino acids can be

- converted to glucose,
- burned as fuel, or
- stored as fat.

When not used to build protein or make other nitrogen-containing compounds, amino acids are "wasted" in a sense. This wasting occurs under any of four conditions:

1. When the body does not have enough energy from other sources.
2. When the body has more protein than it needs.
3. When the body has too much of any single amino acid, such as from a supplement.
4. When the diet supplies protein of low quality, with too few essential amino acids, as described in the next section.

Amino acids are wasted when:

- Energy is lacking.
- Protein is overabundant.
- An amino acid is oversupplied in supplement form.
- The quality of the diet's protein is too low (too few essential amino acids).

To prevent the wasting of dietary protein and permit the synthesis of needed body protein, three conditions must be met. First, the dietary protein must be adequate in quantity. Second, it must supply all essential amino acids in the proper amounts. Third, enough energy-yielding carbohydrate and fat must be present to permit the dietary protein to be used as such.

KEY POINT *Amino acids can be metabolized to protein, nitrogen plus energy, glucose, or fat. They will be metabolized to protein only if sufficient energy is present from other sources. The diet should supply all essential amino acids and a full measure of protein according to guidelines.*

Food Proteins: Quality, Use, and Need

The body responds to different proteins in different ways, depending on many factors: the body's state of health, the food source of the protein, its digestibility, the other nutrients taken with it, and its amino acid assortment. To know whether, say, 30 grams of a particular protein is enough to meet a person's daily needs, one must consider the effects of these other factors on the body's use of the protein.

Regarding a person's state of health, malnutrition or infection may greatly increase the need for protein while making it hard to eat even normal amounts of food. In malnutrition, secretion of digestive enzymes slows as the tract's lining degenerates, impairing protein digestion and absorption. When infection is present, extra protein is needed for enhanced immune functions.

Some concern exists about the formation of carcinogens in foods when meats are exposed to open flame—see Chapter 11.

Cooking with moist heat improves protein digestibility, whereas frying makes protein harder to digest.

Which Kinds of Protein-Rich Foods are Easiest to Digest and Use?

The digestibility of protein varies from food to food and affects protein quality profoundly. The protein of oats, for example, is less digestible than that of eggs. Generally, amino acids from animal proteins are most easily digested and absorbed (over 90 percent). Those from **legumes** are next (about 80 percent). Those from grains and other plant foods vary (from 60 to 90 percent). Cooking with moist heat generally improves protein digestibility, whereas dry heat methods can impair it.[2]

As for taking the other nutrients with protein, the need for carbohydrate and fat has already been emphasized. To be used efficiently, protein must also be accompanied by the full array of vitamins and minerals. In addition, as the next section

CONSUMER CORNER

Protein and Amino Acid Supplements

WHY DO PEOPLE take protein or amino acid supplements? Athletes take them to build muscle. Dieters take them to spare their bodies' protein while losing weight. People also take individual amino acids, mixtures of two or more amino acids, or products that combine amino acids with other nutrients. Some consumers believe the products will cure herpes, induce restful sleep, or relieve pain or depression. Do protein and amino acid supplements really do any of these things? Probably not. Are they safe? No.

Enthusiastic popular reports have led to widespread use of two amino acids. One is lysine, popularly recommended to prevent or relieve the infections that cause herpes sores on the mouth or genital organs. The other is tryptophan, popularly recommended to relieve pain, depression, and insomnia. Lysine does not cure herpes infections. Whether it reduces outbreaks, as popularly reported, is unknown because scientific studies are lacking.[1] If long-term use helps prevent outbreaks, it does so only in some individuals and with unknown associated risks. Tryptophan has some interesting effects with respect to pain and sleep in responsive individuals, as Controversy 13 explains later.

Some years ago, people who elected to take tryptophan developed a blood disorder (EMS, short for *eosinophilia-myalgia syndrome*), and at least 15 of the supplement takers died. Contaminants in the supplement were determined to be the cause of the disease, and the Food and Drug Administration (FDA) recalled tryptophan supplements and formulas to which it was added. Impurities of the same sort are still detected in some tryptophan-containing products currently on the market.[2] The lesson is that supplements of all sorts go to market in a largely untested state and consumers become unsuspecting experimental subjects. It is safer to derive amino acids from protein-rich foods taken with a little carbohydrate to facilitate their use. A glass of milk or a turkey sandwich is a good choice.

When used as a replacement for foods, protein supplements are often downright dangerous. The "liquid protein" diet, advocated some years ago for weight loss, caused the deaths of many users. Even the physician-supervised "protein-sparing" fast, also based on liquid protein, has caused abnormal heart rhythms (Chapter 9 has more about the effects of fasting). Processing can render some protein supplements less digestible than protein-rich food, and they cost more than food, too.

Amino acid supplements are also unnecessary. The body is designed to handle whole proteins best. It breaks them into manageable pieces (dipeptides and tripeptides), then splits these a few at a time, simultaneously absorbing them into the blood. This slow bit-by-bit absorption is ideal because groups of chemically similar amino acids compete for the carriers that absorb them into the blood. An excess of one amino acid can tie up a carrier and temporarily prevent the absorption of another similar amino acid. While carriers deal with an overdose of one or a few amino acids, some other needed amino acids may pass through the body unabsorbed. The result is a deficiency. The human body evolved without encountering the unbalanced arrays of highly concentrated amino acids found in supplements and therefore lacks the equipment to handle them.[3]

More about the effects of fasting in Chapter 9.

A decade ago, the FDA asked a panel of scientists from a well-known scientific research group to review the safety of amino acid supplements.[4] When the scientists searched the literature for well-controlled studies on the supplements, they found next to none, a situation that still exists today. The panel did find evidence of adverse health effects from amino acids, however, and therefore concluded that, without appropriate scientific research, no level of intake of these supplements could be considered safe. The panel also warned that any use of amino acids as dietary supplements is inappropriate for two reasons. First, some (serine and proline) present a high risk of toxicity. Second, no amino acid supplement performs any nutrient function in the human body.

The panel also singled out some groups of people whose growth or altered metabolism makes them especially likely to suffer harm from amino acid supplements:

- All women of childbearing age.
- Pregnant or lactating women.
- Infants, children, and adolescents.
- Elderly people.
- People with inborn errors of metabolism that affect their bodies' handling of amino acids.
- Smokers.
- People on low-protein diets.
- People with chronic or acute mental or physical illnesses who take amino acids without medical supervision.

Supplements may also be harmful to weight lifters and bodybuilders who may take frequent, massive amino acid doses, in the false belief that they offer benefits. As the Fitness for Life feature pointed out earlier, only muscle work builds muscle; protein supplements do not, and athletes do not need them. (Chapter 10 describes how muscles are built and the diet that best supports them.) Anyone considering taking amino acid supplements should check with a physician first.

Many of the chapters of this book present evidence on purified nutrients added to foods or taken singly. The Consumer Corner in Chapter 4 showed that a nutritionally inferior food (refined bread) enriched with a few added nutrients is still deficient in many others. The Consumer Corner in Chapter 5 showed that artificial fats can have side effects. The same is true of amino acids. Even with all that we know about science, it is hard to improve on nature.

makes clear, for the most efficient use of protein, a plentiful supply of all the needed amino acids is required.

KEY POINT *The body's use of a protein depends in part on the user's health and on the protein's digestibility. To be used efficiently, protein should be accompanied by all the other nutrients.*

Protein Quality

The quality of a food protein depends largely on its amino acid content. The cells, in making their own proteins, need a full array of amino acids from food, from their own **amino acid pools,** or from both. If a nonessential amino acid (that is, one the cell can make) is unavailable from food, the cell will synthesize it and continue attaching amino acids to the protein strands being manufactured. If an essential amino acid (one the cell cannot make) is missing from food, the cells begin to adjust their activities almost immediately. Within a single day of restricted essential amino acid intake, the cells begin to conserve it by limiting the breakdown of their working proteins and by reducing their use of amino acids for fuel.

amino acid pools amino acids dissolved in cellular fluid that provide cells with ready raw materials from which to build new proteins or other molecules.

Limiting Amino Acids Can Limit Protein Synthesis The measures just described help the cells to channel the available **limiting amino acid** to its wisest use: making new proteins. Even so, the normally fast rate of protein synthesis slows to a crawl, as the cells make do with the proteins on hand. When the limiting amino acid once again becomes available in abundance, the cells resume their normal protein-related activities. If the shortage becomes chronic, however, the cells begin to break down their protein-making machinery. Consequently, even when protein intakes become adequate, protein synthesis lags behind until the needed machinery can be rebuilt. Meanwhile, the cells function less and less effectively as their proteins wear out and are only partially replaced.

Thus a diet that is short in any of the essential amino acids limits protein synthesis. An earlier analogy likened amino acids to letters of the alphabet. To be meaningful, words must contain all the right letters. For example, a print shop that has no letter "N" cannot make personalized stationery for Jana Johnson. No matter how many J's, A's, O's, H's, and S's in the printer's possession, they cannot replace the missing N's. Likewise, in building a protein molecule, no amino acid can fill another's spot. If a cell that is building a protein cannot find a needed amino acid, synthesis stops, and the partial protein is released.

Partially completed proteins are not held for completion at a later time when the diet may improve. Rather, they are dismantled, and the component amino acids are returned to the circulation to be made available to other cells. If they are not soon inserted into protein, their amine groups are removed and excreted, and the residues are used for other purposes. The need that prompted the call for that particular protein will not be met. Since the other amino acids are wasted, the amine groups are excreted, and the body cannot resynthesize the amino acids later.

Mutual Supplementation and Complementary Proteins It follows that if all the essential amino acids are not consumed in proportion to the body's needs, the body's pools of essential amino acids will dwindle until body organs are compromised. Consuming the essential amino acids presents no problem to people who regularly eat **complete proteins,** such as those of meat, fish, poultry, cheese, eggs, milk, and many soybean products. The proteins of these foods contain ample amounts of all the essential amino acids. An equally sound choice is to eat two **incomplete protein** foods from plants so that each supplies the amino acids missing in the other. In this strategy, called **mutual supplementation,** the two protein-rich foods are combined to yield **complementary proteins** (see Table 6-2), or proteins containing all the essential amino acids in amounts sufficient to support health. This concept is illustrated in Figure 6-13. The two proteins need not even be eaten together, so long as the day's meals supply them both, and the diet provides enough energy and total protein from a variety of sources.[3]

Concern about the quality of individual food proteins is of only theoretical interest in settings where food is abundant. Most people in the United States and Canada eat a variety of nutritious foods to meet their energy needs—not just, say, cookies, potato chips, or alcoholic beverages. They would find it next to impossible *not* to meet their protein requirements, even if they were to eat no meat, fish, poultry, eggs, cheese, or soy products. Though *protein* is usually sufficient in North American diets, people must attend to *other* nutrients, as the next two chapters point out.

Protein quality can make the difference between health and disease when food energy intake is limited (where malnutrition is widespread) or when the selection of foods available is severely limited (where a single food such as potatoes or rice provides 90 percent of the calories). Even then, protein intake may be adequate, but it may not be. To be sure, in these cases, the primary food source of protein must be checked since its quality is crucial.

KEY POINT *A protein's amino acid assortment greatly influences its usefulness to the body. Proteins lacking needed amino acids can be used only if those amino acids are present from other sources.*

limiting amino acid an essential amino acid that is present in dietary protein in an insufficient amount, thereby limiting the body's ability to build protein.

complete proteins proteins containing all the essential amino acids in the right balance to meet human needs.

incomplete proteins proteins lacking, or low in, one or more of the essential amino acids.

mutual supplementation the strategy of combining two incomplete protein sources so that the amino acids in one food make up for those lacking in the other food. Such protein combinations are sometimes called *complementary proteins.*

complementary proteins two or more proteins whose amino acid assortments complement each other in such a way that the essential amino acids missing from one are supplied by the other.

Just as each letter of the alphabet is important in forming whole words, each amino acid must be available to build finished proteins.

BIG TIME PRINTING CO.
JA A JOH SO
ATTOR EY AT LAW

protein digestibility–corrected amino acid score (PDCAAS) a measuring tool used to determine protein quality. The PDCAAS reflects a protein's digestibility as well as the proportions of amino acids that it provides.

protein efficiency ratio (PER) a measure of protein quality assessed by determining how well a given protein supports weight gain in growing rats. The PER is used to judge the quality of protein in infant formulas and baby foods.

Table 6-2

Complementary Protein Combinations

Combine foods from two or more of these categories to obtain complete protein.

Grains
Barley Bulgur Cornmeal Oats Pasta Rice Whole-grain breads
Legumes
Dried beans Dried lentils Dried peas Peanuts Soy products
Seeds and Nuts
Cashews Nut butters Other nuts Sesame seeds Sunflower seeds Walnuts
Vegetables
Broccoli Leafy greens Other vegetables

Measuring Protein Quality

Researchers have developed many different methods of evaluating the quality of food protein. The most important one for consumers is the **protein digestibility–corrected amino acid score,** or **PDCAAS.** The protein values that U.S. consumers read on food labels are based on the PDCAAS. Another measure of protein quality, the **protein efficiency ratio (PER),** is also used for measuring the protein quality of infant formulas and baby foods.

The PDCAAS correction for protein digestibility is important. Simple measures of the total protein in foods are not useful by themselves—even animal hair or hooves would receive a top score by those measures alone. On the PDCAAS scale of 100 to 0, with 100 representing protein sources that are most readily digested and most perfectly balanced for meeting human needs, egg white, ground beef, chicken products, fat-free milk, and tuna fish all score 100. Soybean protein isn't far behind at 94. Most legumes rank in the 60s and 50s. The wheat protein, gluten, formed during bread making, ranks 25. Something interesting happens when pea flour (67) and whole-wheat flour (40) are combined: the score for the resulting flour is 82. Why? Mutual supplementation is at work. A person trying to choose between peanut butter and chili in the grocery store may have no use for the PDCAAS scoring method, but scientists who must establish adequacy of protein sources for human health worldwide rely on it heavily.[4] Of more relevance to the average well-fed North American, however, is the protein RDA, discussed next.

Figure 6-13

An Example of Mutual Supplementation

In general, legumes provide plenty of the amino acids isoleucine (Ile) and lysine (Lys), but fall short in methionine (Met) and tryptophan (Trp). Grains have the opposite strengths and weaknesses, making them a perfect match for legumes.

	Ile	Lys	Met	Trp
Legumes	■	■		
Grains			■	■
Together	■	■	■	■

KEY POINT *The quality of a protein is measured by its amino acids, by its digestibility, or by how well it supports growth.*

nitrogen balance the amount of nitrogen consumed compared with the amount excreted in a given time period.

How Much Protein Do People Really Need?

The 1989 RDA for protein stands as the current recommendation for protein intake. It is designed to cover the need to replace protein-containing tissue that people lose and wear out every day. Therefore it depends on body size: larger people have a higher protein need. The recommendation is also adjusted to cover additional needs for building new tissue and so is higher for growing children and pregnant and lactating women. The Canadian recommendation for protein is similar and is based on similar assumptions. Table 6-3 reviews the recommendations concerning dietary protein, first presented in Chapter 2. These ensure that the body is well supplied with the protein it needs.

Underlying the protein recommendation are **nitrogen balance** studies, which compare nitrogen lost by excretion with nitrogen eaten in food. In healthy adults, nitrogen-in (consumed) must equal nitrogen-out (excreted). Scientists measure the body's daily nitrogen losses in urine, feces, sweat, and skin under controlled conditions and then estimate the amount of protein needed to replace these losses.[†]

Under normal circumstances, healthy adults are in nitrogen equilibrium, or zero balance; that is, they have the same amount of total protein in their bodies at all times. When nitrogen-in exceeds nitrogen-out, people are said to be in positive nitrogen balance; somewhere in their bodies more proteins are being built than are being broken down and lost. When nitrogen-in is less than nitrogen-out, people are said to be in negative nitrogen balance; they are losing protein. Figure 6-14 illustrates these different states.

Growing children add new blood, bone, and muscle cells to their bodies every day. These cells contain protein, so children must have more protein, and therefore more nitrogen, in their bodies at the end of each day than they had at the beginning. A growing child is therefore in positive nitrogen balance. Similarly, when a woman is pregnant, she is, in essence, growing a new person; she, too, must be in positive nitrogen balance until after the birth when she once again reaches equilibrium.

[†]The average protein is 16 percent nitrogen by weight; that is, each 100 grams of protein contain 16 grams of nitrogen. As a rule of thumb, multiply the nitrogen's weight by 6.25 to estimate the protein's weight.

Table 6-3

Recommendations Concerning Intakes of Protein for Adults[a]

1989 Recommended Dietary Allowance (RDA)
- 0.8 gram protein per kilogram body weight per day.

Dietary Guidelines
- Every day eat 2 to 3 servings to total 4 to 9 ounces of cooked dry beans and peas, lean beef or other lean meats, poultry without the skin, fish and shellfish, and occasionally eggs and organ meats.
- Every day choose 2 to 3 servings of lowfat or fat-free milk, yogurt, or cheese.
- Eat a variety of foods to provide small amounts of protein from other sources.

Daily Values[b]
- 50 grams protein per day.

World Health Organization
- Lower limit: 10% of total calories from protein.
- Upper limit: 15% of total calories from protein.

[a]Protein recommendations for infants, children, and pregnant and lactating women are higher.
[b]The Daily Value is for a 2,000-calorie diet.

Figure 6-14

Nitrogen Balance

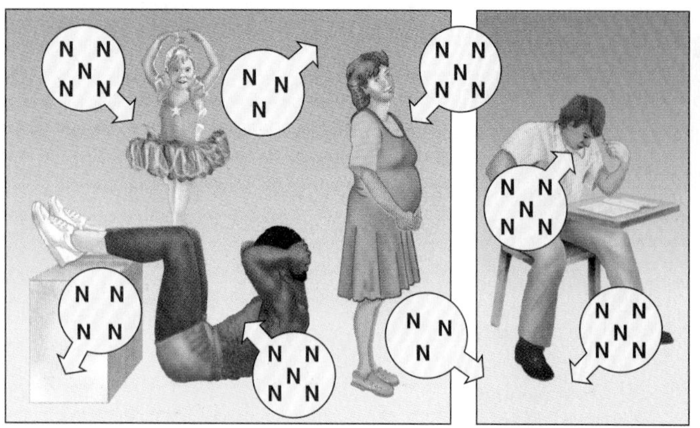
N N N N N
N N N
N N N N N
N N N N
N N N N N
N N N
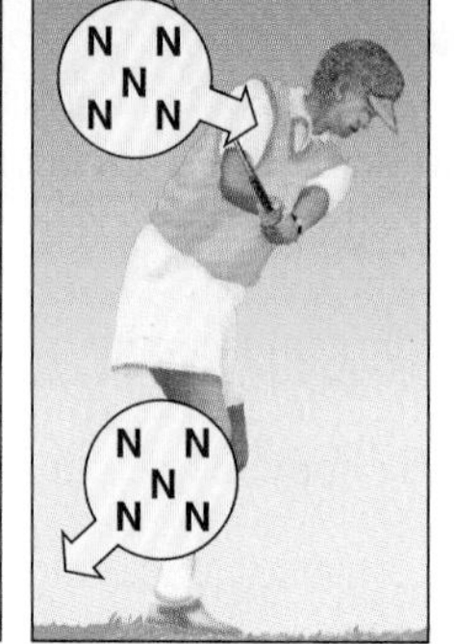
N N N N N
N N N N N

N N
N N N N N
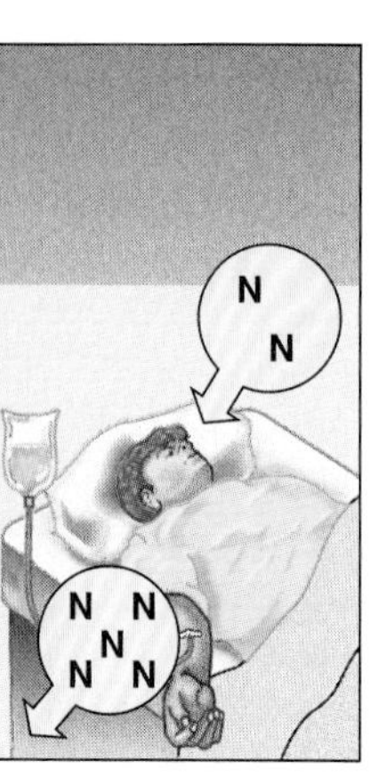
N N
N N N N N

Positive Nitrogen Balance
These people, a growing child, a person building muscle, and a pregnant woman, are all retaining more nitrogen than they are excreting.

Nitrogen Equilibrium
These people, a healthy college student and a young retiree, are in nitrogen equilibrium.

Negative Nitrogen Balance
These people, an astronaut and a surgery patient, are losing more nitrogen than they are taking in.

Negative nitrogen balance occurs when muscle or other protein tissue is broken down and lost.[5] Consider the situation of an ill person, for example. Illness or injury triggers the release of powerful messengers that signal the body to break down some of the less vital proteins, such as those of the skin.[‡] This action floods the blood with amino acids needed for building antibodies to fight the illness and for energy to fuel the body's defenses. The result is negative nitrogen balance. Astronauts, too, experience negative nitrogen balance. Nutritionists responsible for the welfare of astronauts must plan for the negative nitrogen balance that occurs after many days in space without gravity.[6] In the stress of space flight and with no need to support the body's weight against gravity, the astronauts' muscles waste and weaken.[7] To minimize the inevitable loss of muscle tissue, the astronauts must do special exercises in space.

For healthy adults, the 1989 RDA for protein has been set at 0.8 gram for each kilogram (or 2.2 pounds) of body weight. Athletes need slightly more, but the increased need is well covered by a regular diet (more on this in Chapter 10). For infants and growing children, the protein recommendation, like all nutrient recommendations, is higher per unit of body weight.

1989 RDA for protein (adult) = 0.8 g/kg.

To figure your protein need:

1. Find your body weight in pounds.
2. Convert pounds to kilograms (pounds divided by 2.2 equal kilograms).
3. Multiply kilograms by 0.8 to find total grams of protein recommended.

For example:

1. Weight = 110 lb.
2. 110 lb ÷ 2.2 = 50 kg.
3. 50 kg × 0.8 = 40 g.

Recommendations for protein intake assume that a normal diet will be mixed; that is, it will include a combination of animal and plant protein. Not all proteins are used with 100 percent efficiency, and individuals use protein with different efficiencies. Accordingly, the recommendation is quite generous, and many healthy people can consume less than this amount and still meet their bodies' protein needs. What this means in terms of food selections is presented in this chapter's Food Feature.

KEY POINT *Nitrogen balance compares nitrogen excreted from the body with nitrogen ingested in food. The amount of protein needed daily depends on size and stage of growth. The 1989 RDA for adults is 0.8 gram of protein per kilogram of body weight.*

Protein Deficiency and Excess

With all the attention that has been paid in recent years to the health effects of starch, sugars, fibers, fats, oils, and cholesterol, protein has been slighted. Protein deficiencies

‡The messengers are cytokines.

are well known because, together with energy deficiencies, they are the world's leading form of malnutrition. But the health effects of too much protein are far less well known. Both deficiency and excess are of concern.

protein-energy malnutrition (PEM) the world's most widespread malnutrition problem, including both marasmus and kwashiorkor and states in which they overlap; also called *protein-calorie malnutrition (PCM)*.

hunger the physiological craving for food; the progressive discomfort, illness, and pain resulting from the lack of food. (See also Chapters 9 and 15.)

marasmus (ma-RAZ-mus) the calorie-deficiency disease; starvation.

kwashiorkor (kwash-ee-OR-core, kwashee-or-CORE) a disease related to protein malnutrition, with a set of recognizable symptoms, such as edema.

dysentery (DISS-en-terry) an infection of the digestive tract that causes diarrhea.

What Happens When People Consume Too Little Protein?

Protein deficiency and energy deficiency go hand in hand. This combination—**protein-energy malnutrition (PEM)**—is the most widespread form of malnutrition in the world today. Over 500 million children face imminent starvation and suffer the effects of severe malnutrition and **hunger.** Most of the 33,000 children who die each day are malnourished.[8] PEM is prevalent in Africa, Central America, South America, the Middle East, and East and Southeast Asia, but developed countries including the United States are not immune to it.

PEM strikes early in childhood, but it endangers many adults as well. Inadequate food intake leads to poor growth in children and to weight loss and wasting in adults. Stunted growth due to PEM is easy to overlook because a small child can look perfectly normal. The small stature of children in impoverished nations was once thought to be a normal adaptation to the limited availability of food; now it is known to be an avoidable failure of growth due to a lack of food during the growing years.[9]

PEM seems to take two different forms, with some cases exhibiting a combination of the two. In one form, the person is shriveled and lean all over; in the other, a swollen belly and skin rash are present. In the combination, some features of each type are present. The two main forms of PEM have two different disease names: **marasmus** and **kwashiorkor,** respectively.[§10] Marasmus was once thought to be caused by energy deficiency and kwashiorkor by protein deficiency. In reality, though, marasmus reflects a chronic inadequate food intake and therefore inadequate energy, vitamins, and minerals as well as too little protein. Kwashiorkor may result from severe acute malnutrition, with too little protein to support body functions.[11]

Scant supplies of donated food save some from starvation, but many others go hungry.

Marasmus

Marasmus occurs most commonly in children from 6 to 18 months of age in overpopulated city slums. Children in impoverished nations subsist on a weak cereal drink with scant energy and protein of low quality; such food can barely sustain life, much less support growth. A starving child often looks like a wizened little old person—just skin and bones.

Without adequate nutrition, muscles, including the heart muscles, waste and weaken. Brain development is stunted and learning is impaired. Metabolism is so slow that body temperature is subnormal. There is little or no fat under the skin to insulate against cold. Hospital workers find that children with marasmus need to be wrapped up and kept warm. They also need love because they have often been deprived of parental attention as well as food.

Protein malnutrition impairs learning.

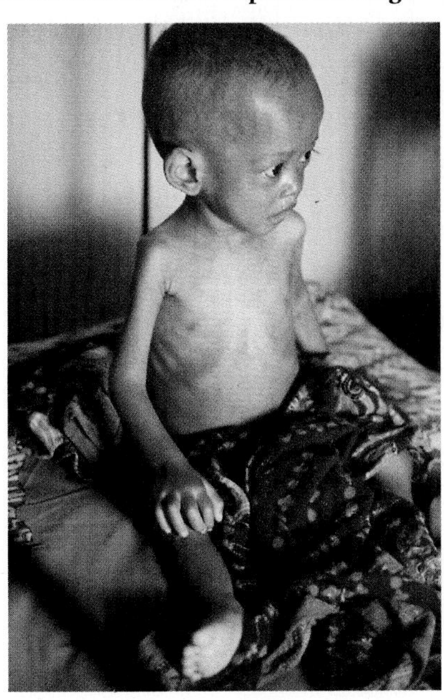

The starving child faces this threat to life by engaging in as little activity as possible—not even crying for food. The body collects all its forces to meet the crisis and so cuts down on any expenditure of protein not needed for the heart, lungs, and brain to function. Growth ceases; the child is no larger at age four than at age two. The skin loses its elasticity and moisture, so it tends to crack; when sores develop, they fail to heal. Digestive enzymes are in short supply, the digestive tract lining deteriorates, and absorption fails. The child can't assimilate what little food is eaten.

Blood proteins, including hemoglobin, are no longer produced, so the child becomes anemic and weak. The protein and energy needed for immune functions are lacking, which explains the high prevalence of infections in malnourished children. Antibodies to fight off invading bacteria are degraded to provide amino acids for other uses, leaving the child an easy target for infection.[12] Then **dysentery,** an infection of the digestive tract, causes diarrhea, further depleting the body of nutrients, especially minerals. Measles, which might make a healthy child sick for a week or two, kills a child with PEM

§A term gaining acceptance for use in place of kwashiorkor is *hypoalbuminemic-type PEM.*

within two or three days. In fact, infections that occur with malnutrition are responsible for two-thirds of the deaths of young children in developing countries.

Ultimately, marasmus progresses to the point of no return, when the body's machinery for protein synthesis, itself made of protein, has been degraded. At this point, attempts to correct the situation by giving food or protein fail to prevent death. If caught before this time, however, the starvation of a child may be reversed by careful nutrition therapy. The fluid balances are most critical. Diarrhea will have depleted the body's potassium and upset other electrolyte balances. The combination of electrolyte imbalances, anemia, fever, and infections often leads to heart failure and sudden death. Careful correction of fluid and electrolyte balances usually raises the blood pressure and strengthens the heartbeat within a few days. Later, fat-free milk, providing protein and carbohydrate, can safely be given; fat is introduced still later, when body protein is sufficient to provide carriers.

The term *electrolyte balance* refers to the proper concentrations of salts within the body fluids (see Chapter 8 for details).

Kwashiorkor Kwashiorkor is the Ghanaian name for "the evil spirit that infects the first child when the second child is born." In countries where kwashiorkor is prevalent, each baby is weaned from breast milk as soon as the next one comes along. The older baby no longer receives breast milk, which contains high-quality protein designed perfectly to support growth, but is given a watery cereal with scant protein of low quality. Small wonder the just-weaned child sickens when the new baby arrives.

Some kwashiorkor symptoms very much resemble those of marasmus (see Table 6-4), but often without severe wasting of body fat. Proteins and hormones that previously maintained fluid balance are now diminished, so fluid leaks out of the blood and accumulates in the belly and legs, causing edema, a distinguishing feature of kwashiorkor. The kwashiorkor victim's belly often bulges with a fatty liver, caused by lack of the protein carriers that transport fat out of the liver. The fatty liver loses some of its ability to clear poisons from the body, prolonging their toxic effects. Without sufficient tyrosine to make melanin, the child's hair loses its color; inadequate protein synthesis leaves the skin patchy and scaly; sores fail to heal.

Melanin, a brown pigment of hair, skin, and eyes, was mentioned earlier as a product made from tyrosine.

PEM at Home PEM is common among some groups in the United States and Canada: the poor living on U.S. Indian reservations, in inner cities, and in rural areas;

Table 6-4

Features of Marasmus and Kwashiorkor in Children

Separating PEM into two classifications oversimplifies the condition, but at the extremes, marasmus and kwashiorkor exhibit marked differences. Marasmus-kwashiorkor mix presents symptoms common to both marasmus and kwashiorkor. In all cases, children are likely to develop diarrhea, infections, and multiple nutrient deficiencies.

Marasmus	Kwashiorkor
Infancy (less than 2 yr)	Older infants and young children (1 to 3 yr)
Severe deprivation or impaired absorption of protein, energy, vitamins, and minerals	Inadequate protein intake or, more commonly, infections
Develops slowly; chronic PEM	Rapid onset; acute PEM
Severe weight loss	Some weight loss
Severe muscle wasting with fat loss	Some muscle wasting, with retention of some body fat
Growth: < 60% weight-for-age	Growth: 60 to 80% weight-for-age
No detectable edema	Edema
No fatty liver	Enlarged, fatty liver
Anxiety, apathy	Apathy, misery, irritability, sadness
Appetite may be normal or impaired	Loss of appetite
Hair is sparse, thin, and dry; easily pulled out	Hair is dry and brittle; easily pulled out; changes color; becomes straight
Skin is dry, thin, and wrinkled	Skin develops lesions

many elderly people; hungry and homeless children; and those suffering from the eating disorder anorexia nervosa.[13] People who are hospitalized for long periods are also at risk for PEM, as are those with wasting diseases such as cancer or AIDS and those addicted to drugs and alcohol. In a downward spiral, PEM and serious illness worsen each other, so treating the PEM often reduces medical complications and the suffering they may bring even when the underlying disease is untreatable.[14]

Today, millions of people who work to support their children earn so little that they cannot afford nutritious food—one child in eight under the age of 12 goes hungry. This situation presents a challenge to nutritionists. Hunger, especially in children, threatens everyone's future. Hungry children do not learn as well as fed children, nor are they competitive. They are ill more often, they have higher absentee rates from school, and when they attend, they cannot concentrate for long. The forces driving poverty and hunger will require many great minds working together to find solutions.

Turn to Chapter 15 for details concerning the causes of hunger at home and abroad.

KEY POINT *Protein-deficiency symptoms are always observed when either protein or energy is deficient. Extreme food-energy deficiency is marasmus; extreme protein deficiency is kwashiorkor. The two diseases overlap most of the time and together are called PEM.*

Is it Possible to Consume Too Much Protein?

While many of the world's people struggle to obtain enough food and enough protein to keep themselves alive, people in developed countries must consciously limit their protein intakes to avoid excesses. Overconsumption of protein offers no benefits and may pose health risks. For one thing, as mentioned before, protein-rich foods are often high-fat foods that contribute to obesity with its accompanying health risks. Furthermore, animal protein sources can be high in saturated fat, a known contributor to atherosclerosis and heart disease. Independently of saturated fat, however, animal protein itself may raise blood cholesterol, thus contributing to atherosclerosis and heart disease. In contrast, the protein of soybeans seems to lower blood cholesterol.[15]

Animals fed experimentally on high-protein diets often develop enlarged kidneys or livers. In human beings, a high-protein diet consumed over a lifetime worsens existing kidney problems and may also significantly alter the functioning of healthy kidneys.[16] One of the most effective treatments for people with established kidney problems may be to reduce protein intakes to slow down the progression of their disease.[17]

Evidence is mixed about whether high protein intakes from animal sources, especially when accompanied by low calcium intakes, can accelerate adult bone loss.[18] No doubt exists about the effect of feeding purified protein to human subjects—purified protein causes calcium to be spilled from the urine.[19] Also, eating diets high in animal protein, but not plant protein, correlates with a greater incidence of hip fractures in some populations.[20] In one population, however, the reverse is true—in malnourished elderly individuals, protein deficiency and hip fractures often occur together, and restoring protein can often improve bone status.[21]

1989 RDA for protein for women aged 19 to 24 = 46 g.

Average protein intake = 65 g.

1989 RDA for protein for men aged 19 to 24 = 58 g.

Average protein intake = 105 g.

Whether excess protein depletes bone minerals may depend in part upon the ratio of dietary calcium to protein.[22] An ideal ratio has not been established, but a woman whose intake just meets, but does not exceed, the recommendations for both nutrients has a calcium-to-protein ratio of 20 milligrams calcium to 1 gram protein.[23] For most women in the United States, however, average calcium intakes are lower and protein intakes are higher, yielding a 9-to-1 relationship that some theorize may lead to calcium losses that compromise bone health.[24] Other factors including genetic inheritance and regular physical activity also play roles in protecting the bones.[25]

U.S. protein recommendations set an upper limit for protein intake of no more than twice the 1989 RDA amount, and the World Health Organization (WHO) suggests an even more stringent upper limit of 15 percent of total calories. In a world where protein deficiency is such a threat to so many, it is ironic that some people in developed countries should be overconsuming protein.

The role of exercise in preventing bone loss is discussed in Controversy 8 and Chapter 10.

KEY POINT *Health risks may follow the overconsumption of protein-rich foods.*

FOOD FEATURE

Getting Enough, But Not Too Much Protein

It is clear by now that people in developed nations usually eat more than ample protein. The protein RDA is generous: it more than adequately covers the estimated needs of most people, even those with unusually high requirements.

Protein-Rich Foods

Foods in the meat, poultry, fish, dry beans, eggs, and nuts group and in the milk, yogurt, and cheese groups contribute an abundance of high-quality protein. Two others, the vegetables and bread, cereals, rice, and pasta groups, contribute smaller amounts of protein, but they can add up to significant quantities. What about the fruit group? Don't rely on fruit for protein—fruit contains only small amounts. Figure 6-15 shows the wide variety of foods that contribute most of the protein to the diet.

Protein is critical in nutrition, but too many protein-rich foods can displace other important foods from the diet. Foods richest in protein carry with them a characteristic array of vitamins and minerals, including vitamin B_{12} and iron, but they are notoriously lacking in others—vitamin C and folate, for example. In addition, many protein-rich foods such as meat are high in calories, and to overconsume them is to invite obesity.

With so many foods contributing to the protein of a day's meals, one can have confidence that one's protein intake is ample. With this knowledge, one can plan meatless or reduced-meat meals with pleasure. Of the many interesting, protein-rich meat alternates available, one has already been mentioned: the legumes.

Figure 6-15

Finding the Protein in Foods[a]

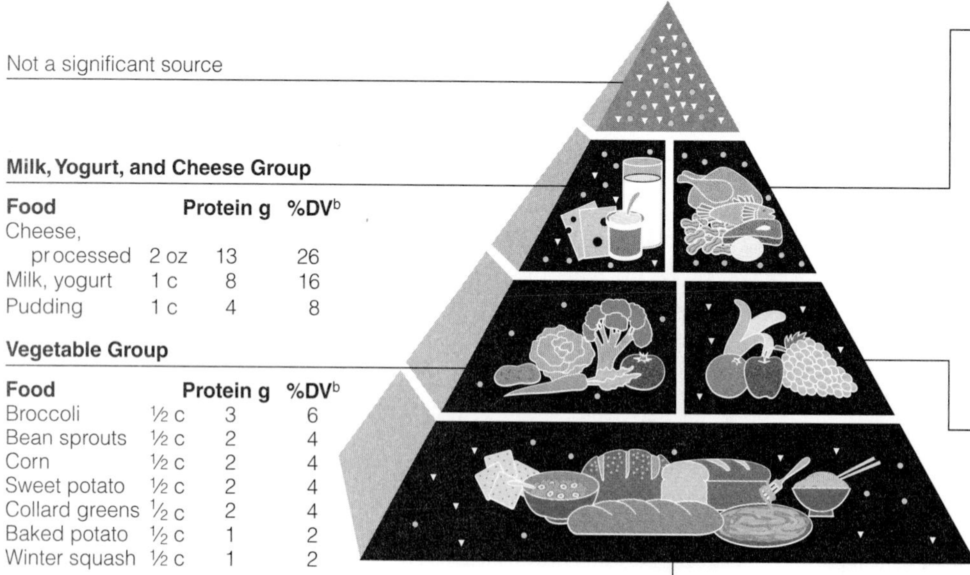

Not a significant source

Milk, Yogurt, and Cheese Group

Food		Protein g	%DV[b]
Cheese, processed	2 oz	13	26
Milk, yogurt	1 c	8	16
Pudding	1 c	4	8

Vegetable Group

Food		Protein g	%DV[b]
Broccoli	½ c	3	6
Bean sprouts	½ c	2	4
Corn	½ c	2	4
Sweet potato	½ c	2	4
Collard greens	½ c	2	4
Baked potato	½ c	1	2
Winter squash	½ c	1	2

Meat, Poultry, Fish, Dry Beans, Eggs, and Nuts Group

Food		Protein g	%DV[b]
Chicken breast	2 oz	18	36
Roast beef	2 oz	16	32
Pork meat	2 oz	16	32
Turkey leg	2 oz	16	32
Tuna	2 oz	14	28
Lentils, beans, peas	½ c	9	18
Peanut butter	2 tbs	8	16
Almonds	¼ c	7	14
Lunch meat	2 oz	7	14
Egg	1 lg	6	12
Hot dog	1 reg	6	12
Cashew nuts	¼ c	5	10

Fruit Group

Food		Protein g	%DV[b]
Avocado	½ c	2	4
Cantaloupe	½ c	1	2
Orange sections	½ c	1	2
Strawberries	½ c	0	0

Breads, Cereals, Rice, and Pasta Group

Food		Protein g	%DV[b]
Pancakes	2 sm	4	8
Bagel	½	4	8
Fried rice	½ c	3	6
Grain bread	1 sl	3	6
Noodles, pasta	½ c	3	6
Oatmeal	½ c	3	6
Barley	½ c	2	4
Cereal flakes	1 oz	2	4

[a] All foods are prepared and ready to eat.

[b] The Daily Value (DV) for protein is 50 g, and is based on an energy intake of 2,000 calories per day.

The Advantages of Legumes

The protein of some legumes is of a quality almost comparable to that of meat. In fact, for practical purposes, the quality of soy protein can be considered equivalent to that of meat. Figure 6-16 shows a legume plant's special root system that enables it to make abundant protein. Legumes are also excellent sources of fiber, many B vitamins, iron, calcium, and other minerals. On average, a cup of cooked legumes contains about 30 percent of the Daily Values for both protein and iron.[†] Like meats, though, legumes do not offer every nutrient, and they do not make a complete meal by themselves. They contain no vitamin A, vitamin C, or vitamin B_{12}, and their balance of amino acids can be much improved by using grains and other vegetables with them.

Soybeans are versatile legumes, and people make many products from them. One problem, however, is that the heavy use of soy products in place of meat inhibits iron absorption. The effect can be alleviated by using small amounts of meat and/or foods rich in vitamin C in the same meal with soy products. Vegetarians sometimes use convenience foods made from **textured vegetable protein** (soy protein) formulated to look and taste like hamburgers or breakfast sausages. Many of these are intended to match the known nutrient contents of animal-protein foods, but often they fall short.[*] A wise vegetarian would use such foods sparingly and learn to use combinations of whole foods to supply the needed nutrients.

Another form in which the nutrients of soybeans are available is as bean curd, or **tofu,** a staple used in many Asian dishes. Thanks to the use of calcium salts when some tofu is made, it can be high in calcium. Check the Nutrition Facts panel on the label.

The Food Features presented so far show that the recommendations for the three energy-yielding nutrients occur in balance with each other. If you reduce fat and increase carbohydrate, protein totals automatically come into line with the requirements. To help you accept that protein is abundant in most foods, the Do It! section that follows asks you to complete a day's meals and watch the protein grams add up.

[†]Data from the *Food Processor Plus,* ESHA research, version 7.11.

[*]In Canada, regulations govern the nutrient contents of such products.

textured vegetable protein processed soybean protein used in products formulated to look and taste like meat, fish, or poultry.

tofu (TOE-foo) a curd made from soybeans that is rich in protein, often rich in calcium, and variable in fat content; used in many Asian and vegetarian dishes in place of meat.

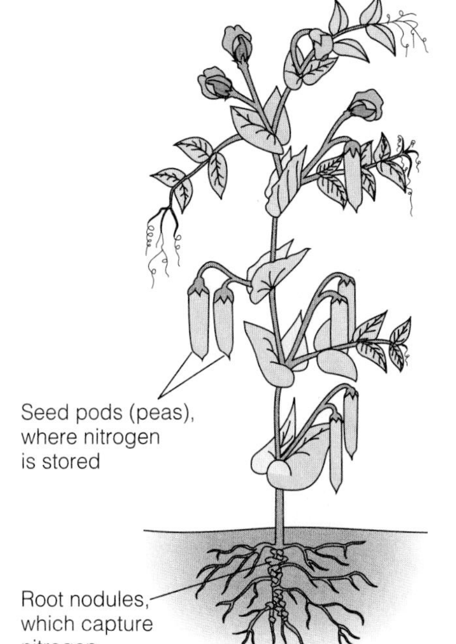
Seed pods (peas), where nitrogen is stored
Root nodules, which capture nitrogen

Figure 6-16

A Legume

The legumes include such plants as the kidney bean, soybean, garden pea, lentil, black-eyed pea, and lima bean. Bacteria in the root nodules can "fix" nitrogen from the air, contributing it to the beans. Ultimately, thanks to these bacteria, the plant accumulates more nitrogen than it can get from the soil and also leaves more nitrogen in the soil than it takes out. The legumes are so efficient at trapping nitrogen that farmers often grow them in rotation with other crops to fertilize fields. Legumes are shown among the meat alternates in Figure 2-5 of Chapter 2.

do it!

ADD UP THE PROTEIN IN A DAY'S MEALS

Consider the sources of protein in a day's meals. Look at the meals in Figure 6-17. Breakfast and lunch are given, but supper is yet to be planned. A simple breakfast of cereal, milk, and juice provides 14 grams of protein. Lunch is a bit heartier with a ham and cheese sandwich contributing most of its 18 protein grams. Now comes a puzzle—which supper to choose? After picking a supper from among the choices, check Table 6-5 to find out how much protein each supper contains (the totals are printed upside down to prevent you from peeking before you guess). Then compare the protein in the meals with your protein recommendation (use the 1989 RDA value from the inside front cover).

Two of the supper options are meatless (the spaghetti supper and the vegetable-rice supper), and one contains meat. To quickly assess the protein in such meals, remember that fruits provide only a little protein, but meats, milk, and cheeses are the richest sources, followed by legumes, grains, breads, and vegetables.

Settle on a supper choice and consider these questions:

1. Did the supper you chose add up to a protein total that meets, but does not exceed, your protein RDA? Which other suppers also qualify?
2. If your answer to the first part of question 1 was no, which foods might you substitute to achieve your goal without shorting yourself on nutrients? Hint: To keep from vastly overconsuming protein, you may need to restrict something, and an obvious "something" to restrict is some of the meat.

Figure 6-17

A Protein Puzzle

The protein values listed in this exercise are from the Food Processor Plus, 7.11, *a computerized diet analysis software program developed by ESHA Research.*

Breakfast
1 c orange juice
2 g protein
Cheerios cereal, 1 oz
4 g protein
1 c low-fat milk
8 g protein
Breakfast total = 14 g protein

Lunch
iced tea
0 g protein
ham and cheese sandwich (1 slice lunch meat; 1 slice cheese; ¾ c lettuce and tomato; 2 slices whole-wheat bread)
17 g protein
peaches ½ c
1 g protein
Lunch total = 18 g protein

So far, this day's meals have contributed 32 grams of protein. On this basis, what would you choose for supper? After choosing, turn Table 6-5 upside down to see how much protein each supper adds to the day's intake.

Supper A = ? protein

Supper B = ? protein

Supper C = ? protein

do it!

3. Make some educated guesses concerning the other two energy-yielding nutrients, fat and carbohydrate. Which foods contribute abundant fat and saturated fat to this day's meals? Which contribute carbohydrates and fiber?
4. Note that breakfast, though it contains no meat, provides almost as much protein as the ham and cheese sandwich at lunch. Which foods in this breakfast provide protein? Is the protein of each of these foods complete or incomplete? Is the total breakfast protein complete or incomplete? Why?
5. Which plant food shown in Figure 6-17 is the richest in protein? Which is next richest? Hint: If you have eliminated or are considering eliminating meat from your diet, read the Controversy that follows—it points out the pros and cons of both vegetarian and meat-containing diets.
6. If you were to design a day's meals around the lamb supper, yet did not want to consume too much protein, how would you change breakfast and lunch? What foods would you substitute for some of the protein-rich foods listed in Figure 6-17? Another hint: If you are considering doing away with the milk or cheese, remember that you must then provide other sources of calcium (you may reconsider this decision when you discover in Chapter 8 that few foods other than milk supply an abundance of calcium).

Table 6-5

Answers to Protein Puzzle

Supper Choice A
iced water/lemon
  0 g protein
garlic bread, 2 pieces
  6 g protein
large salad with ¼ c each garbanzo beans, artichoke and cucumber
  6 g protein
spaghetti, 1 c; parmesan cheese, 1 tbs
  13 g protein
sherbet, 1 c
  2 g protein
Totals:

Supper A total = 27 g protein
Entire day's protein = 59 g

Supper Choice B
iced tea/lemon
  0 g protein
tomato slices, ½ c
  1 g protein
grated cheese, 1 tbs
  3 g protein
mixed vegetables, 1 c
  4 g protein
brown rice, 1 c
  6 g protein
carrot cake, 1 pce
  4 g protein
Totals:

Supper B total = 18 g protein
Entire day's protein = 50 g

Supper Choice C
coffee, black, 1 c
  0 g protein
asparagus, ½ c
  2 g protein
potatoes au gratin, ½ c
  6 g protein
lamb chops, 2 oz
  35 g protein
sliced beets, ½ c
  1 g protein
bread pudding, ½ c
  7 g protein
Totals:

Supper C total = 51 g
Entire day's protein = 83 g

Self Check

Answers to these Self Check questions are in Appendix G.

1. The basic building blocks of protein are:
   a. glucose units
   b. amino acids
   c. side chains
   d. saturated bonds
2. Protein digestion begins in the:
   a. mouth
   b. stomach
   c. small intestine
   d. large intestine
3. Which of the following can form enzymes?
   a. carbohydrates
   b. lipids
   c. proteins
   d. (b) and (c)
4. For healthy adults, the 1989 RDA for protein has been set at:
   a. 0.8 gram per kilogram of body weight
   b. 2.2 pounds per kilogram of body weight
   c. 12 to 15 percent of total calories
   d. 100 grams per day
5. Which of the following conditions occur(s) less frequently in people who consume vegetarian diets? (Read about this in the upcoming Controversy.)
   a. obesity
   b. high blood pressure
   c. diverticular disease
   d. all of the above
6. Under certain circumstances, protein can be converted to glucose and so serve the energy needs of the brain. T F
7. Too little protein in the diet can have severe consequences, but excess protein has no adverse effects. T F
8. Although protein-energy malnutrition (PEM) is prevalent in developing nations, it is not seen in the United States. T F
9. Partially completed proteins are not held for completion at a later time when the diet may improve. T F
10. An example of a person in negative nitrogen balance is an astronaut. T F

NUTRITION ON THE NET

For further study of the topics of this chapter, access these websites and search for the phrases or words in quotation marks:

1. For the latest changes in chapter material, to find supplemental learning tools, or to access links to related websites, go to the "Nutrition: Concepts and Controversies" site at:
   www.wadsworth.com/nutrition/prod/allprod.html
2. To learn more about "protein" in foods, visit the American Dietetic Association (ADA) site at:
   www.eatright.org
3. Search the World Health Organization (WHO) site for information on "protein-energy malnutrition":
   www.who.org
4. For more on "vegetarian" diets, search the Food and Drug Administration (FDA) site for foods:
   www.fda.gov
5. Search among thousands of current scientific and medical abstracts for any topic related to protein at:
   http://www.ncbi.nlm.nih.gov/PubMed/

VEGETARIANS VERSUS MEAT EATERS: WHOSE DIET IS BEST?

CONTROVERSY
6

ONE YOUNG person rejects all animal products, shuns grains, and seeks out vegetables, fruits, and herbs. Another young person relishes meat at every meal and usually orders "a steak and potato: hold the rabbit food." These two have a lot more in common than either would probably believe. Both are extremists in their choices of foods. Both may be jeopardizing their health by their rigid, unbalanced eating styles.[1] But both **vegetarian** diets and meat-containing diets have elements in their favor, provided that they are not taken to extremes.

Some people mistakenly associate vegetarianism with a particular culture, but individuals choose it for many different reasons. Some believe that we should not kill animals to eat their meat. Some believe that we should not even partake of animal products such as milk, cheese, eggs, or honey or use items made from leather, wool, or silk. Many people are upset when livestock animals are treated inhumanely in feedlots and slaughterhouses. Some believe we should eat less meat for environmental reasons, or they may fear contracting food poisoning from ground meats. (The effects of food choices on the earth's resources are topics of Chapter 15 and its Controversy, and Chapter 14 provides the whole story on the threat from food poisoning.)

People who eat meat also do so for a variety of reasons. Some find that a hamburger makes a convenient lunch while providing a concentrated source of energy and nutrients. Others enjoy the taste of roasted chicken or beef stew. Others wouldn't know what to eat without meat; they are accustomed to seeing it on the plate.

Regardless of an individual's reasons for choosing a particular diet, these daily choices have implications for health.

This Controversy looks first at the positive health aspects of vegetarian diets, then at the positive aspects of meat eaters' diets. It concludes by showing how both types of eaters can maximize the benefits and minimize the risks of their diets.

POSITIVE HEALTH ASPECTS OF VEGETARIAN DIETS

A strong statement of support for some vegetarian eating styles comes from the *Dietary Guidelines for Americans,* which states that a vegetarian diet, well chosen, can confer excellent health on the eater.[2] This statement reflects strong evidence linking vegetarian diets with reduced incidences of chronic diseases.[3]

Such evidence, though abundant, is not easily obtained. It would be easy if vegerarians differed from others only in not consuming meat, but they often have *increased* intakes of fruits and vegetables as well, and these foods are the primary contributors of phytochemicals believed to reduce disease risks. Vegetarian diets may also contain more fiber, another factor associated with reduced disease risks.[4] Also, though there are exceptions, vegetarians typically use no tobacco, use alcohol in moderation if at all, and may be more physically active than other adults. Researchers must account for the effects of these lifestyle differences on disease development before they can see how health correlates with diet. Even then, *correlation* is not cause. Without more evidence, conclusions must be tentative.

Still, with all these qualifications, research findings are intriguing. They seem to indicate that a vegetarian diet may offer some protection against six conditions: obesity, diabetes, high blood pressure, heart disease, digestive disorders, and some forms of cancer. It matters, however, what form the vegetarian diet takes. Vegetarians differ, as Table C6-1 on the next page demonstrates. What is known about the relationships between vegetarianism and disease follows.

Obesity and Diabetes Vegetarians tend to be leaner than nonvegetarians.[5] Perhaps they consciously control their calorie intakes and make an effort to exercise regularly. Perhaps their diet, which tends to be high in fiber-rich bulky foods, is automatically lower in calories than the average diet based on meat.

Some vegetarians worry, and rightly so, about the health of their bones. Children who were fed a **macrobiotic diet,** an extreme form of vegetarian diet, were observed to have lower bone mineral density in adolescence, a time of great importance to the bones (see Controversy 8 and Chapter 13).[6] Nevertheless, parents who make sure that their vegetarian children consume milk, cheese, and other milk products regularly can be assured that their children's bones will receive the calcium they need.

The fattening power of fat may also be a factor in vegetarians' lower body weight: vegetarians tend to have low intakes of fat, which is efficiently stored in the body, and high intakes of carbohydrate, which costs more energy for its conversion to fat and so

yields less energy for storage.[7] Another possibility is that leanness may result from a higher metabolic rate (the rate at which the body burns fuel, see Chapter 9) in vegetarians compared with nonvegetarians.[8] This may occur because, when supplied with a mixed diet, the body preferentially burns off extra carbohydrate calories for fuel (increases carbohydrate metabolism) and stores the excess fat calories.[9] In any case, a healthy body weight combined with high intakes of complex carbohydrates and fiber reduces the risks of diabetes and several other obesity-related diseases. A limited body of research suggests a connection between meat-containing diets and increased incidences of diabetes, even without obesity.[10]

Table C6-1

Terms Used to Describe Vegetarians

- **fruitarian** includes only raw or dried fruits, seeds, and nuts in the diet.
- **lacto-ovo vegetarian** includes dairy products, eggs, vegetables, grains, legumes, fruits, and nuts; excludes flesh and seafood.
- **lacto-vegetarian** includes dairy products, vegetables, grains, legumes, fruits, and nuts; excludes flesh, seafood, and eggs.
- **macrobiotic diet** a vegan diet that progressively eliminates more and more foods. Ultimately, only brown rice and small amounts of water or herbal tea are consumed; taken to extremes, macrobiotic diets have resulted in malnutrition and even death.
- **ovo-vegetarian** includes eggs, vegetables, grains, legumes, fruits, and nuts; excludes flesh, seafood, and milk products.
- **partial vegetarian** includes seafood, poultry, eggs, dairy products, vegetables, grains, legumes, fruits, and nuts; excludes or strictly limits certain meats, such as red meats. Also called *semivegetarian.*
- **pesco-vegetarian** same as partial vegetarian, but eliminates poultry.
- **vegan** includes only food from plant sources: vegetables, grains, legumes, fruits, seeds, and nuts; also called *strict vegetarian.*
- **vegetarian** includes plant-based foods and eliminates some or all animal-derived foods.

Blood Pressure Vegetarians often are found to have lower blood pressure than nonvegetarians. Various combinations of lifestyle factors and diet seem to influence blood pressure. Among lifestyle factors, smoking and alcohol intake raise blood pressure, and exercise lowers it. Diet alone may be significant, however. A recent review of the literature in this area concluded that "there is now convincing evidence for a blood-pressure-lowering effect of . . . the type [of diet] eaten by some vegetarians."[11] The authors qualify this by saying that the effect is associated with diets low in fat and saturated fat and high in fiber, fruits, and vegetables; they also say that meat itself need not be totally excluded. It would be oversimplifying to say that including or excluding any one food or nutrient lowers blood pressure. Apparently, many factors act together.

Heart Disease Fewer vegetarians than meat eaters suffer from diseases of the heart and arteries or exhibit indicators of those diseases, even when the people being compared are all nonsmokers.[12] The dietary factor most directly related to coronary artery disease is saturated fat intake, but other factors may also play a role.[13] Research is currently concentrating on the sources of protein in the diet and on the antioxidant nutrients and phytochemicals found in plants (details on this line of research can be found in Controversies 7 and 11).[14]

When vegetarians are fed meat, which contains saturated fat, their lipid profiles change for the worse; when meat eaters are fed a low-fat vegetarian diet, their lipid profiles improve. One study compared two low-fat diets, one vegetarian and another containing lean meats. Both diets lowered blood cholesterol, but the vegetarian diet's effects were greater.

Researchers recently released some surprising findings from a 17-year study of dietary habits and causes of death among 11,000 health-conscious people.[15] About half of the study's subjects were vegetarians and half were meat eaters. The researchers reported a slight, but not statistically significant, correlation between lowered rates of heart disease and omission of meat from the diet. They reported a much stronger and more significant association between reduced risk of death from heart disease and the inclusion of fresh fruit and, to a lesser extent, green salads, regardless of meat consumption. Other researchers agree with this line of speculation: reduced risk of heart disease in vegetarians may result not entirely from what vegetarians *omit* from the diet but from what they *obtain*. Many vegetarians replace the meat in their diets with foods made with fruits, vegetables, whole grains, and soybeans. The case for benefits from consuming soy-based products, along with the other foods of vegetarians, is growing more solid each day.[16]

Protein itself may affect blood cholesterol. Experimental diets containing purified proteins from animal sources (milk, fish, and egg) raise blood cholesterol higher than do similar diets containing purified soybean protein. When rabbits were fed diets containing 50 percent of calories from purified animal protein along with high cholesterol, the animals suffered rapidly advancing atherosclerosis. Rabbits are naturally vegetarians, though, so this may not be a fair test of what happens in human beings. Also, the effect may be partly attributable to the heart-protecting effects of soy's accompanying phytochemicals. Clearly, though, soybean protein seems to lower cholesterol in people who consume soy products.[17]

People eat foods, not purified proteins, so the question of whether animal or vegetable proteins by themselves raise or lower blood cholesterol is academic. The purified proteins that experimentally raise blood cholesterol are milk protein (casein) and the protein of fish. If the whole foods are used instead of the isolated proteins, the contrast between animal and vegetable proteins is not seen. Milk *lowers* blood

A balanced meal need not include meat to be nutritious.

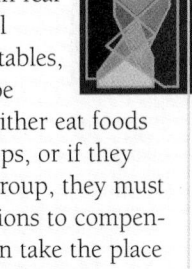

cholesterol just as soy does, and meals of fish provide benefits to heart health, as Chapter 5 made clear. Consumers of **vegan** diets who worry about a lack of omega-3 fatty acids (fish oil) in their foods might try the sources that fish themselves use: sea vegetables, which can be rich in omega-3 fatty acids.[18]

Against heart disease, then, a vegetarian diet that includes fruit, vegetables, and other nutritious foods may offer these advantages:

- Promotes leaner body composition and lower blood pressure.
- Provides less saturated fat.
- Provides more vegetable protein, fiber, antioxidant nutrients, and phytochemicals.

Digestive Disorders Constipation and diverticular disease are less common in people who consume high-fiber vegetarian or **partial vegetarian** diets than in people who consume typical meat-based diets. Chapter 4 presented possible ways fiber might influence the health of the digestive tract.

Cancer Seventh-Day Adventists, an often-studied health-conscious vegetarian group, enjoy a significantly lower cancer rate than the rest of the population, even when cancers linked to smoking and alcohol are taken out of the picture.[19] Their low cancer mortality may possibly be due to their low meat intakes, to their high intakes of fruits, vegetables, and cereal grains, to both, or to other lifestyle factors.

Some scientific findings support the idea that vegetarian diets may reduce the risks of colon cancer.[20] People with colon cancer seem to eat more meat, less fiber, and more saturated fat than others without colon cancer. Something about high-fat, high-protein, low-fiber diets creates an environment in the human colon that may promote the development of cancer. Such a diet has also been associated with a form of cancer of the lymphatic system.[21] In general, vegetarians tend to consume less fat and protein, and they seem to produce fewer carcinogens in the body than do meat eaters.[22] Further, they take in more carbohydrate, fiber, and water, and these dietary constituents add bulk to the stools, diluting any carcinogens that may be present. Meat also forms carcinogenic substances when it is browned during cooking (more about these in Chapter 11), and vegetarians are spared any risks these substances may present.[23]

Overall, then, many vegetarians have lower risks of developing obesity, diabetes, high blood pressure, heart disease, digestive disorders, and cancer than do meat eaters. Two million people in the United States follow vegetarian diets. If they plan their diets wisely, they obtain all the nutrients they need to support good health.

POSITIVE HEALTH ASPECTS OF THE MEAT EATER'S DIET

The meat lover introduced at the start of this Controversy was exaggerated to make a point. Those who really eat like that place themselves in immediate peril of malnutrition. In reality, few people shun all green and yellow vegetables, fruits, and grains. To be healthy, people must either eat foods from all five food groups, or if they omit foods from one group, they must make careful substitutions to compensate. No substitutes can take the place of fruits and vegetables (not even antioxidant supplements, as the next chapter points out). This section considers a balanced diet of which meat is a part.

Growth Meats, eggs, milk, and other foods from animal sources support growth well. Without them, children's growth often lags behind the growth of peers.[24] Populations existing on monotonous grain diets, either for reasons of strict meat taboos or of economic necessity, are often malnourished, as revealed by their short stature, low resistance to diseases, short life span, and high infant mortality. In developed nations, children whose parents are strict vegetarians often face three threats: lack of variety, reliance on nutritionally inadequate convenience foods, and a lack of exercise.[25]

Even in populations with somewhat more varied diets, the children who eat the most animal-derived products have been observed to grow the best.[26] Even when the protein amounts are equal, children whose protein intakes are from plant sources may not grow as well as those eating animal products.[27] Protein may not be the only nutrient affecting growth, however; families who can afford to buy animal products are also likely to consume a larger variety of fruits and vegetables. These foods provide the vitamins and minerals also needed for growth.[28]

Foods of plant origin generally offer much less energy for their bulk than do foods of animal origin. A child's small stomach can hold only so much food, and a vegetarian child may feel full before eating enough food to supply nutrients and energy sufficient to support growth. For obesity-prone adults, a bulky diet can be advantageous, but a child fed without meat, milk, or eggs may experience stunted growth that lasts a lifetime.

Are animal and dairy products superior to plants as protein-rich foods? It is true that meats, eggs, and dairy foods contain complete, more digestible proteins. Children of milk- and meat-eating populations are generally larger, fatter, and more resistant to infections than are those of grain-eating populations. They are also protected from the vitamin D–deficiency disease rickets, which is especially likely to strike vegans in cold climates who are rarely exposed to the sun.

It is also true, however, that meat itself is not necessary for children to achieve healthy growth. Many children grow normally when milk and eggs accompany a vegetarian diet and when knowledgeable adults plan and deliver the diet with care. An example is found among the children of a **lacto-ovo vegetarian** community in England: their growth is practically identical to the growth of children who eat meat.[29]

Support during Critical Times Meat eaters can generally rely on their diets during critical times of life. In contrast, a vegan woman who doesn't meet her nutrient needs may enter pregnancy too thin; have inadequate stores of iron, zinc, and vitamin B_{12}; and fail to gain enough weight during pregnancy to support the normal growth and development of her fetus.

The importance of careful planning for vegetarians is underscored by a severe, often irreversible, sometimes fatal disorder reported among breastfed infants of strict vegan mothers who fail to obtain sufficient vitamin B_{12}.[30] The infants exhibit a striking syndrome of body tremors and facial twitches involving the tongue and throat that combine to make nursing difficult. After some months, deprivation of vitamin B_{12} leads to severe psychomotor retardation of the infant, accompanied by shrinkage of the brain.[31] Sadly, in some cases, the retardation caused by the illness lingers on even after treatment with the missing vitamin.[32]

Unlike vegans, women who eat meat, eggs, and dairy products can be sure of receiving enough vitamin B_{12}, vitamin D, calcium, iron, and zinc, as well as protein, to support pregnancy and breastfeeding. Women following lacto-ovo vegetarian diets can also receive all of these nutrients in abundance, and they have the added advantage of habitually consuming more folate and other nutrients associated with vegetables and fruits than typical meat eaters do. The importance of adequate folate, especially for women in the childbearing years, will become evident in the next chapter.

Two meat servings of the size depicted here present the maximum daily meat intake suggested by the Daily Food Guide as health promoting.

OBTAINING NUTRIENTS

To obtain calcium, U.S. consumers who use no milk can purchase calcium-fortified soy milk or calcium-fortified orange juice. Alternatively, large servings of calcium-rich green vegetables such as broccoli or kale can make a sizable calcium donation to the diet. In addition, the nutrients iron, zinc, vitamin D, and vitamin B_{12} require special attention from strict vegetarians. Meat provides much of the iron and zinc in the meat eater's diet, and vitamin D and vitamin B_{12} are found reliably only in animal-derived foods. Compared with meat, vegetarian sources of iron and zinc, such as legumes, dark green, leafy vegetables, fortified cereals, and whole-grain breads and cereals, provide less of these minerals and in a less absorbable form.[33] To obtain enough iron and zinc, an emphasis on whole grains and legumes in the diet is important. A vegetarian relying heavily on white rice, refined white flour, and sweets for carbohydrates can easily obtain too few minerals for health. A strict vegetarian diet cannot meet vitamin D needs without a supplemental source or adequate exposure to sunlight (the next chapter provides details).[34] These substitutions take planning, but yield results.

Eggs, for those who eat them, can meet vitamin B_{12} needs; vegans, especially women of childbearing age, must take particular care to choose vitamin B_{12}–fortified cereals and other sources or must rely on supplements. Fermented plant products such as tempeh, made from soybeans, may contain some vitamin B_{12} contributed by the bacteria that did the fermenting, but unfortunately, much of the vitamin B_{12} in these foods may be inactive.

Have you noticed the lack of concern about protein for adult vegetarians? Protein is not the problem it was once thought to be for adults eating a varied diet. Even in vegans, protein deficiency is rare in those who consume adequate calories in various nutritious foods.[35]

Just as well-planned diets of both meat eaters and vegetarians can contribute to good health, so also can both diets damage it if haphazardly chosen. Both diets can be high in saturated fat and so pose a threat to the health of the heart. A vegetarian who dines on cheddar cheese, butter sauces, sour cream, and deep-fried vegetables invites the same health hazards as the overeater of high-fat meats. And both

Figure C6-1

Vegetarian Food Guide Pyramid[a]

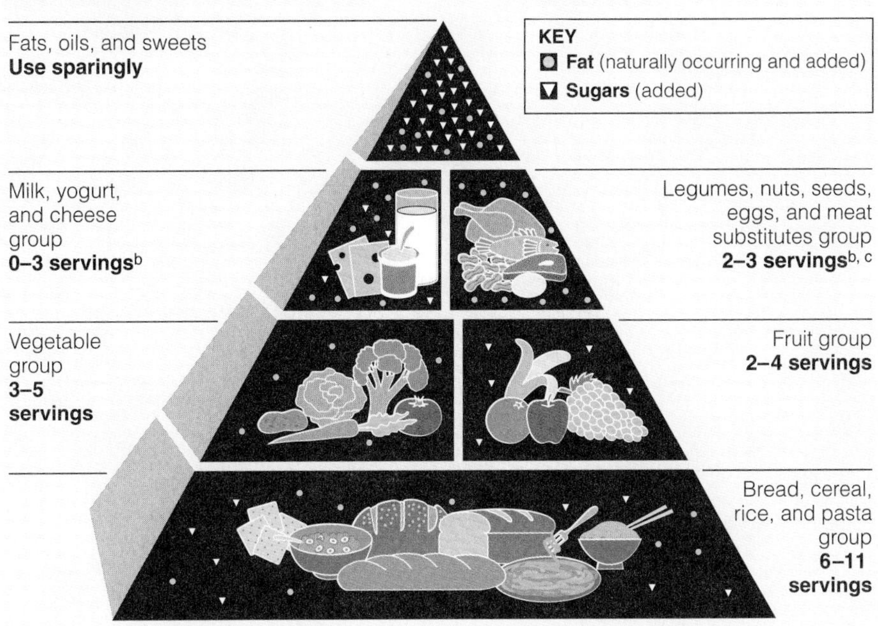

Fats, oils, and sweets
Use sparingly
KEY
Fat (naturally occurring and added)
Sugars (added)
Milk, yogurt, and cheese group
0–3 servings[b]
Legumes, nuts, seeds, eggs, and meat substitutes group
2–3 servings[b, c]
Vegetable group
3–5 servings
Fruit group
2–4 servings
Bread, cereal, rice, and pasta group
6–11 servings

[a] Serving sizes are listed in Figure 2-5 of Chapter 2.

[b] Vegans and other vegetarians who include no eggs, milk, or milk products in the diet must obtain calcium, vitamin D, and vitamin B_{12} from other sources. See text for details.

[c] Meat substitutes include soybean products, such as tofu, tempeh, and soy milk, and commercial meat replacers.

SOURCE: Adapted from the USDA Food Guide Pyramid and Food Guide Pyramid for Vegetarian Meal Planning in Position of the American Dietetic Association: Vegetarian diets, *Journal of the American Dietetic Association* 97 (1997): 1317–1321.

diets, if not properly balanced, can lack nutrients.

For both eaters, then, planning is the key to obtaining adequate nutrients. Those who eliminate meats can adapt the Food Guide Pyramid to their needs (see Figure C6-1).

CONCLUSION

This comparison has shown that both a meat eater's diet and a vegetarian's diet are best approached scientifically. Vegetarianism is not a religion like Buddhism or Hinduism; it is merely an eating plan that selects plant foods to deliver needed nutrients. Some people make much of the distinctions between types of vegetarians; although these distinctions are useful academically, they do not represent uncrossable lines. Some people use meat as a condiment or seasoning for vegetable or grain dishes. Some people eat meat only once a week and use plant protein foods the rest of the time. Many people rely mostly on milk products to meet their protein needs, but eat fish occasionally, and so forth. To force people into the categories of "vegetarians" and "meat eaters" leaves out all these in-between styles of eating that have much to recommend them.

To the person just beginning to study nutrition, consider adopting the attitude that the choice to make is not whether to be a meat eater or a vegetarian, but where along the spectrum to locate yourself. Your preferences, whatever they are, should be honored, and the only caveats are that you make your diet adequate, balanced, and varied, and that you use moderation when choosing foods high in saturated fat or calories.

7 The Vitamins

Mattie Lou O'Kelley 1908–1997, *Spring Vegetable Scene*, 1968.

SPRING 1968
M.L. O'KELLEY

Contents

Frequently Asked Questions

FAQ: Does eating carrots really protect the health of the eyes? p. 216

FAQ: How can people make a vitamin from sunlight? p. 217

FAQ: How are B vitamins related to heart disease? p. 232

FAQ: Is it possible to get too much vitamin C? p. 235

FAQ: Do I need to take a vitamin supplement? p. 237

FAQ: Which supplements are helpful, and which should I avoid? p. 238

FAQ: Supplements must be safe, or the government would not allow their sales, right? p. 240

vitamins organic compounds that are vital to life and indispensable to body functions, but are needed only in minute amounts; noncaloric essential nutrients.

precursors, provitamins compounds that can be converted into active vitamins.

AT THE BEGINNING OF the twentieth century, the romance and thrill of the discovery of the first **vitamins** captured the world's heart. People loved the vitamins. Catapulted from the shrouded mystery of folk cures into the technological era that brought us vitamin pills, they seemed a perfect answer to people who were looking for an easy route to good health. Only today are scientists beginning to uncover the complexity of the interactions of vitamins in the body and in foods.

It is easy to see why people were so impressed: with the discovery of each new vitamin, miraculous cures seemed to take place. In the usual scenario, a whole group of people were unable to walk (or were going blind or bleeding profusely) until an alert scientist stumbled onto the substance missing from their diets. The scientist usually confirmed the discovery by feeding vitamin-deficient feed to laboratory animals, which responded by becoming unable to walk (or going blind or bleeding profusely). When the missing ingredient was restored to their diet, they miraculously recovered. Miraculous cures of people followed as they, too, received the vitamins they lacked.

It took a sophisticated knowledge of chemistry and biology to isolate the vitamins and to learn their chemical structures. More scientific advances brought an understanding of the biological roles that vitamins play in maintaining health and preventing deficiency diseases. Now, our knowledge of the vitamins has entered a new era of hope and discovery.[1]

Today, people's excitement is growing as research hints that two of the major scourges of humankind, cardiovascular disease (CVD) and cancer, may be somehow linked with low intakes of vitamins. Research laboratories around the world are proposing new theories postulating relationships between vitamins and human health and disease. Can it be that vitamins will protect us from life-threatening diseases? We still have much to learn before this question is settled.

On hearing of the dramatic power of vitamins to cure deficiency diseases and learning that vitamins might possibly prevent chronic diseases as well, people can easily come to believe that vitamin pills will cure almost any ailment. Many people take supplements of vitamin C in the belief that they will cure a cold or the flu, a topic addressed in this chapter's Consumer Corner. This chapter's Controversy focuses on emerging knowledge about possible links between antioxidant nutrients and chronic diseases. Meanwhile, the media bombard us with a never-ending stream of overly simple claims for "miracle vitamins," and the supplement business is a *multibillion* dollar industry.

The only disease a vitamin can cure is the one caused by a deficiency of that vitamin.

For now, we can still say this with certainty: the only disease a vitamin will *cure* is the one caused by a deficiency of that vitamin. As for chronic disease prevention, the evidence is still emerging.

Definition and Classification of Vitamins

A child once defined a vitamin as "what, if you don't eat, you get sick." Although the grammar left something to be desired, the definition was accurate. Less imaginatively, a vitamin is defined as an essential, noncaloric, organic nutrient needed in tiny amounts in the diet. The role of many vitamins is to help make possible the processes by which other nutrients are digested, absorbed, and metabolized or built into body structures. Although small in size and quantity, the vitamins accomplish mighty tasks, some of which are still being discovered.

As they were discovered, the vitamins were named, and many were also given letters and numbers. This led to the confusing variety of vitamin names that still exists today. This chapter uses the names in Table 7-1; alternative names are given in Tables 7-5 and 7-6 at the end of the chapter.

Table 7-1

Vitamin Names[a]

Fat-Soluble Vitamins
- Vitamin A
- Vitamin D
- Vitamin E
- Vitamin K

Water-Soluble Vitamins
- B vitamins
  - Thiamin (B_1)
  - Riboflavin (B_2)
  - Niacin (B_3)
  - Folate[b]
  - Vitamin B_{12}
  - Vitamin B_6
  - Biotin
  - Pantothenic acid
- Vitamin C

[a] Vitamin names established by the International Union of Nutritional Sciences Committee on Nomenclature.
[b] Folate may be listed on food labels as *folic acid.*

Some of the vitamins occur in foods in a form known as **precursors,** or **provitamins.** Once inside the body, these are transformed chemically to one or more active vitamin forms. Thus, in measuring the amount of a vitamin found in food, it is often most accurate to count not only the amount of the true vitamin but also the vitamin activity potentially available from its precursors. Tables 7-5 and 7-6 specify which vitamins have precursors.

The vitamins fall naturally into two classes: fat soluble and water soluble. Solubility imparts to vitamins many of their characteristic behaviors and determines how they are absorbed into and transported around by the bloodstream, whether they can be stored in the body, and how easily they are lost from the body. In general, like other fats, fat-soluble vitamins are absorbed into the lymph and they travel in the blood in association with protein carriers. Fat-soluble vitamins can be stored with other lipids in fatty tissues, and because they are stored, some of them can build up to toxic concentrations. The water-soluble vitamins are generally absorbed directly into the bloodstream, where they travel freely. Most are not stored in tissues to any great extent; rather, excesses are excreted in the urine. Thus the risks of immediate toxicities are not as great as for fat-soluble vitamins, except in cases of extremely high doses. This chapter addresses first the fat-soluble vitamins and then the water-soluble ones. Some of the most important facts will be discussed separately for each vitamin, and the tables at the end of the chapter sum up the basic facts about all of them.

KEY POINT *Vitamins are essential, noncaloric nutrients, needed in tiny amounts in the diet, that help to drive cell processes. The fat-soluble vitamins are vitamins A, D, E, and K; the water-soluble vitamins are the B vitamins and vitamin C.*

beta-carotene an orange pigment with antioxidant activity; a vitamin A precursor made by plants and stored in human fat tissue.

retinol one of the active forms of vitamin A made from beta-carotene in animal and human bodies; an antioxidant nutrient. Other active forms are *retinal* and *retinoic acid.*

Vitamin intake recommendations (DRI or 1989 RDA) are found on the inside front cover.

Characteristics fat-soluble vitamins share:

- Dissolve in lipid.
- Require bile for absorption.
- Are stored in tissues.
- May be toxic in excess.

The Fat-Soluble Vitamins

The fat-soluble vitamins—A, D, E, and K—generally occur together in the fats and oils of foods. Like the lipids, these vitamins require bile for absorption. Once absorbed, they are stored in the liver and fatty tissues until the body needs them. For this reason the body can easily survive weeks of consuming foods that lack these vitamins, as long as the diet as a whole provides *average* amounts that approximate the recommended intakes. The capacity to be stored also sets the stage for toxic buildup, should an excess be taken in, especially in the form of supplements. Excesses of vitamins A, D, and K from supplements can reach toxic levels especially easily.

Deficiencies of the fat-soluble vitamins are likely when the diet is consistently low in them or when they are inadvertently lost from the digestive tract dissolved in undigested fat. We know that any disease that produces fat malabsorption (such as liver disease that prevents bile production) can bring about deficiencies of the fat-soluble vitamins. Deficiencies are also likely when people eat diets that are extraordinarily low in fat; such diets interfere with the absorption of these vitamins. Also, a person who uses mineral oil (which the body can't absorb) as a laxative or eats abundant foods containing the fat replacer olestra risks losing fat-soluble vitamins by excretion.

The roles that the fat-soluble vitamins play in the body are diverse. Vitamin A is, among many other things, a visual pigment. Vitamins A and D may act somewhat like hormones, directing cells to convert one substance to another, to store this, or to release that.[2] Vitamin E flows all over the body, preventing oxidative destruction of tissues. Vitamin K is necessary for blood to clot. Each is worth a book in itself.

Vitamins fall into two classes—fat soluble and water soluble.

A
E
D
K
Bs
C

Vitamin A

Vitamin A has the distinction of being the first fat-soluble vitamin to be recognized. Today, after almost a century of research, vitamin A and its plant-derived precursor, **beta-carotene,** are still very much a focus of research.

Vitamin A is certainly one of the most versatile vitamins, with roles in such diverse functions as vision, immune defenses, maintenance of body linings and skin, bone and body growth, normal cell development, and reproduction.[3] In short, vitamin A is needed everywhere. Three forms of vitamin A are active in the body; one of the active forms, **retinol,** is stored in the liver. The liver makes retinol available to the bloodstream

retina (RET-in-uh) the layer of light-sensitive nerve cells lining the back of the inside of the eye.

cornea (KOR-nee-uh) the hard, transparent membrane covering the outside of the eye.

rhodopsin (roh-DOP-sin) the light-sensitive pigment of the cells in the retina; it contains vitamin A (*rod* refers to the rod-shaped cells; *opsin* means "visual protein").

night blindness slow recovery of vision after exposure to flashes of bright light at night; an early symptom of vitamin A deficiency.

keratin (KERR-uh-tin) the normal protein of hair and nails.

keratinization accumulation of keratin in a tissue; a sign of vitamin A deficiency.

xerosis (zeer-OH-sis) drying of the cornea; a symptom of vitamin A deficiency.

xerophthalmia (ZEER-ahf-THALL-me-uh) hardening of the cornea of the eye in advanced vitamin A deficiency that can lead to blindness (*xero* means "dry"; *ophthalm* means "eye").

epithelial (ep-ith-THEE-lee-ull) **tissue** the layers of the body that serve as selective barriers to environmental factors. Examples are the cornea, the skin, the respiratory tract lining, and the lining of the digestive tract.

cell differentiation (dih-fer-en-she-AY-shun) the process by which immature cells are stimulated to mature and gain the ability to perform functions characteristic of their cell type.

and thereby to the body cells. The cells convert retinol to its other two active forms, retinal and retinoic acid, as needed.

A Jack of All Trades—Vitamin A Perhaps the most familiar function of vitamin A is in eyesight. Vitamin A plays two indispensable roles in the eye: in the events of light perception at the **retina** and in the maintenance of a healthy, crystal-clear outer window, the **cornea** (see the margin drawing on the next page).

When light falls on the eye, it passes through the clear cornea and strikes the cells of the retina, bleaching many molecules of the pigment **rhodopsin** that lie within them. Vitamin A is a part of the rhodopsin molecule. When bleaching occurs, the vitamin is broken off, initiating the signal that conveys the sensation of sight to the optic center in the brain. The vitamin then reunites with the pigment, but a little vitamin A is destroyed each time this reaction takes place, and fresh vitamin A arriving in the blood regenerates the supply. If the supply is low, a lag occurs before the eye can see again after a flash of bright light at night (see Figure 7-1). This lag in the recovery of night vision, termed **night blindness**, may indicate a vitamin A deficiency.[4] A bright flash of light can temporarily blind even normal, well-nourished eyes, but if you experience a long recovery period before vision returns, your health-care provider may want to check your vitamin A intake.

A deficiency of vitamin A that has progressed well beyond the night blindness stage may be reflected in an accumulation of a protein, **keratin,** that clouds the eye's outer vitamin A–dependent part, the cornea. The condition is known as **keratinization,** and, if the deficiency of vitamin A is not corrected, it can progress to **xerosis** (drying) and then to thickening and permanent blindness, **xerophthalmia.** Tragically, a half million vitamin A–deprived children become blind each year from this often preventable condition.[5] If the deficiency is discovered early, capsules containing 60,000 RE of vitamin A taken twice each year can reverse it. (RE, the unit of measure for vitamin A, is explained later.) Better still, a child fed fruits and vegetables regularly is virtually assured protection from ever incurring the problem.

Vitamin A is needed by all **epithelial tissue** (external skin and internal linings), not just by the cornea.[6] The skin and all of the protective linings of the lungs, intestines, vagina, urinary tract, and bladder serve as barriers to infection by bacteria and to damage from other sources. Vitamin A works at the genetic level to promote the process of **cell differentiation**, which allows each type of cell to mature so that it is capable of performing a particular function. For example, when goblet cells, a type of epithelial tissue cells, mature, they specialize in synthesizing and releasing mucus to protect those tissues from toxic particles, bacteria and other microbial invaders, and other potentially harmful substances.

If vitamin A is deficient, the differentiation and maturing process is impaired. Goblet cells, among others, fail to mature, then fail to make protective mucus, and eventually die off. Some of the cells in these areas are displaced by cells that secrete keratin, already mentioned. Keratin is the same protein that provides toughness in hair and fingernails. Keratin makes the surfaces dry, hard, cracked, and vulnerable to infection (see Figure 7-2). The dead cells accumulate on the tissue surface and become hosts to bacterial infection. In the cornea, as described, keratinization leads to xerophthalmia; in the lungs, the displacement of mucus-producing cells makes respiratory infections likely; in the vagina, the same process leads to vaginal infections.

Colorful foods often are rich in vitamins.

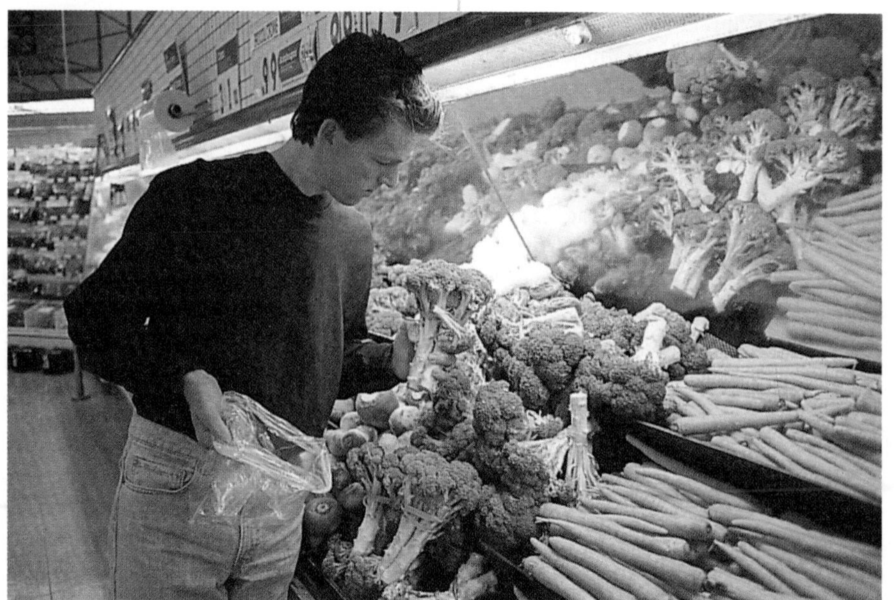

Figure 7-1

Night Blindness

This is one of the earliest signs of vitamin A deficiency.

In dim light, you can make out the details in this room.

A flash of bright light momentarily blinds you as the pigment in the retina is bleached.

You quickly recover and can see the details again in a few seconds.

With inadequate vitamin A, you do not recover but remain blind for many seconds; this is night blindness.

Vitamin A has gained a reputation as an "anti-infective" vitamin because so many of the body's defenses against infection depend on an adequate supply.[7] An emerging area of research concerns the roles of vitamin A in the regulation of the genes that produce proteins involved in immunity.[8] Without sufficient vitamin A, these complex genetic interactions produce an altered response to infection that weakens the body's defenses against disease.

When the defenses are weak, especially in vitamin A–deficient children, an illness such as measles can become severe. A downward spiral of malnutrition and infection can set in. The child's body must devote its scanty store of vitamin A to the immune system's fight against measles virus, but the infection causes vitamin A to be lost from the body.[9] As vitamin A dwindles, the infection worsens. More vitamin A is needed for the fight, but it is lacking, so the infection gains ground. Even if the child survives the measles infection, blindness is likely. The corneas, already damaged by the chronic vitamin A shortage, degenerate rapidly as their meager supply of vitamin A is diverted to the immune system. Blindness induced by vitamin A deficiency often follows bouts of infection.

An eye (sectioned).

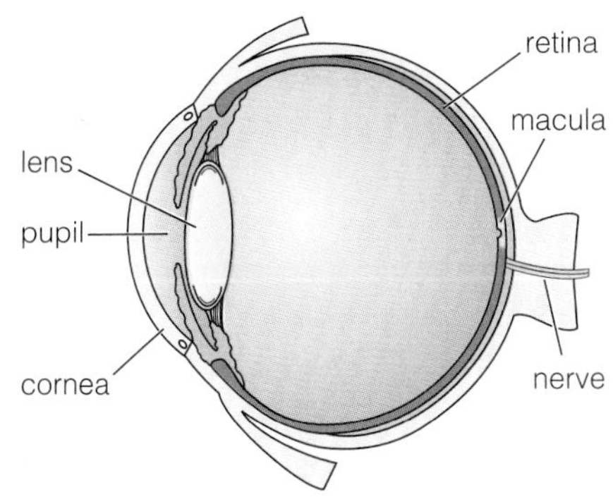
retina
macula
lens
pupil
cornea
nerve

Vitamin A also assists in bone growth. Normal children's bones grow longer, and the children grow taller, by remodeling each old bone into a new, bigger version. To do so, the body dismantles the old bone structures and replaces them with new, larger bone parts. Growth cannot take place just by adding on to the original small bone; vitamin A is needed in the critical dismantling steps.[10] In children, failure to grow is one of the first signs of poor vitamin A status. Restoring vitamin A to such children is imperative, but correcting dietary deficiencies may be more effective than giving vitamin A supplements alone; other nutrients from nutritious food are also needed for children to gain weight and grow taller.[11]

Vitamin A Deficiency around the World Although less frequently reported in developed countries, vitamin A deficiency has been a vast problem worldwide, placing a heavy burden on society. More than three million of the world's children suffer from signs of severe vitamin A deficiency—not only blindness but stunted growth with poor appetite. A staggering 275 million more children suffer from milder deficiency, sufficient to impair immunity and promote infections.[12] In just one country, Indonesia, vitamin A deficiency is responsible for the deaths of 150,000 preschool children each year. In other countries, the toll is many times greater. In some countries, supplements of vitamin A have reduced childhood death rates by as much as half.[13] The World Health Organization (WHO) and UNICEF (United Nations International Children's Emergency Fund) are working to eliminate

Figure 7-2

The Skin in Vitamin A Deficiency

The hard lumps reflect accumulations of keratin in the epithelial cells.

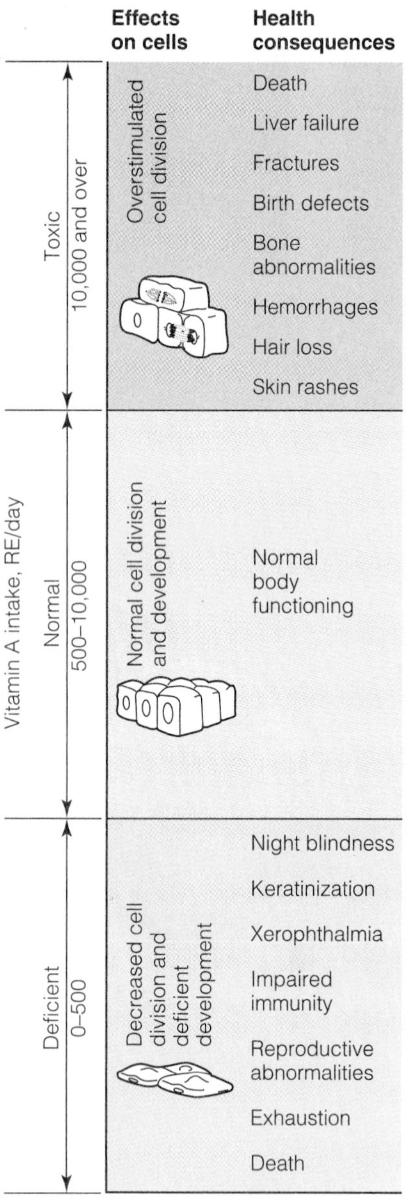

Figure 7-3

Vitamin A Deficiency and Toxicity

Danger lies both above and below a normal range of intakes of vitamin A.

Effects on cells
Health consequences
Vitamin A intake, RE/day
Toxic
10,000 and over
Overstimulated cell division
Death
Liver failure
Fractures
Birth defects
Bone abnormalities
Hemorrhages
Hair loss
Skin rashes
Normal
500–10,000
Normal cell division and development
Normal body functioning
Deficient
0–500
Decreased cell division and deficient development
Night blindness
Keratinization
Xerophthalmia
Impaired immunity
Reproductive abnormalities
Exhaustion
Death

vitamin A deficiency, a goal that will improve child survival throughout the developing world.

Vitamin A Toxicity As Figure 7-3 shows, for people who take excess vitamin A in supplements, toxicity presents a danger equal to that of deficiency. Toxicity's many symptoms include abdominal pain, hair loss, joint pain, stunted growth, bone and muscle soreness, cessation of menstruation, nausea, diarrhea, rashes, damage to the liver, and enlargement of the spleen. The earliest symptoms of overdoses are loss of appetite, blurred vision, growth failure in children, headache, itching of the skin, and irritability. Pregnant women should take note—chronic use of vitamin A supplements providing three to four times the amount recommended for pregnancy has been observed to cause malformations of the fetus.[14] Even a single massive vitamin A dose (100 times the need) will do so. Children, who often mistake chewable vitamin pills for candy, are also likely to be hurt from vitamin A excesses because they need less and are more sensitive to overdoses than adults are. Adolescents may take massive vitamin A doses in the mistaken belief that vitamin A can correct acne. True, an effective acne medicine, Accutane, is *derived* from vitamin A, but it is chemically altered and given in carefully controlled dosages. Vitamin A itself has no effect on acne.

Healthy people can eat vitamin A–rich foods in large amounts without risking toxicity, with the possible exception of liver. One report describes children falling ill after eating liver daily for years, but this is a medical rarity. Inuit people and Arctic explorers know that polar bear livers contain large enough amounts of the vitamin to be a dangerous food source because the bears eat fish whole including the livers. Likewise, when researchers fed pigs a chow made from the unusable parts of salmon, including livers, the animals stopped growing and fell ill from vitamin A toxicity.[15]

Sources of Vitamin A The foods that provide preformed active vitamin A are foods of animal origin. The richest sources, liver and fish oil, were already mentioned, but milk and milk products can also be good sources. Even butter and eggs provide some vitamin A to the diet. Plants contain no preformed vitamin A, but many vegetables and fruits contain the vitamin A precursor, beta-carotene. Snapshot 7-1 is the first of a series of figures that show a sampling of foods that provide more than 10 percent of the Daily Value for a vitamin in a standard-size serving and therefore qualify to be called "good" or "rich" sources.

The definitive fast-food meal—a hamburger, fries, and cola—lacks vitamin A. Many fast-food restaurants, however, now offer salads with cheese and carrots and other vitamin A–rich foods as alternatives or accompaniments to plain burgers. These selections greatly improve the nutritional quality of a fast-food meal.

As for vitamin A supplements, the National Research Council (NRC) and other nutrition agencies recommend that people avoid taking supplements that exceed intake recommendations. No official Tolerable Upper Intake Level for vitamin A has yet been set, but Table 7-3 later in the chapter lists a safe dose that will not be toxic even over a long period. But the best way to ensure a safe intake of vitamin A is to steer clear of supplements and obtain it instead from foods.

The current U.S. standard for vitamin A intake is the 1989 RDA, listed on the inside front cover, page B.

Vitamin A Recommendations The amount of vitamin A needed is proportional to a person's body weight. Although the vitamin A intake recommendation is given as a daily amount, the vitamin need not be consumed every day. An average intake that meets the daily need over several months is sufficient. According to the 1989 RDA, a man needs a daily average of about 1,000 **RE (retinol equivalents)**; a woman needs about 800 RE and more during lactation; children need less.

The effects of excessive vitamin A intakes during pregnancy are discussed in Chapter 12.

Vitamin A recommendations are expressed in RE, but some food tables and supplement labels still express vitamin A contents using a different unit, the **IU (international unit)**. Note that this book's Table of Food Composition (Appendix A) uses RE for your convenience, but be careful to notice whether other food tables or sup-

SNAPSHOT 7.1 Vitamin A and Beta-Carotene

These foods provide 10 percent or more of the vitamin A Daily Value (DV = 900 RE/day).[a, b]

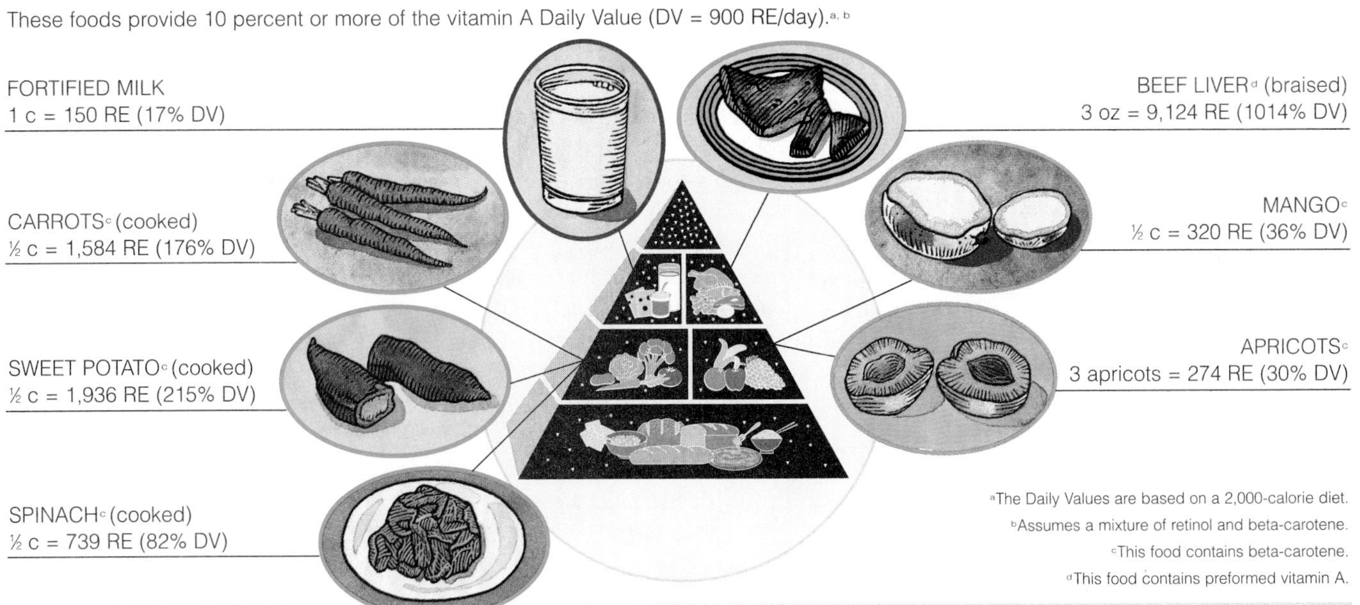

FORTIFIED MILK
1 c = 150 RE (17% DV)
CARROTS[c] (cooked)
½ c = 1,584 RE (176% DV)
SWEET POTATO[c] (cooked)
½ c = 1,936 RE (215% DV)
SPINACH[c] (cooked)
½ c = 739 RE (82% DV)
BEEF LIVER[d] (braised)
3 oz = 9,124 RE (1014% DV)
MANGO[c]
½ c = 320 RE (36% DV)
APRICOTS[c]
3 apricots = 274 RE (30% DV)

[a]The Daily Values are based on a 2,000-calorie diet.
[b]Assumes a mixture of retinol and beta-carotene.
[c]This food contains beta-carotene.
[d]This food contains preformed vitamin A.

plement labels use RE or IU. See the Aids to Calculations (Appendix C) for help in converting the units. When comparing vitamin A in foods, make sure that the amounts are all expressed in the same units.

KEY POINT *Vitamin A is essential to vision, integrity of epithelial tissue, bone growth, reproduction, and more. Vitamin A deficiency causes blindness, sickness, and death and is a major problem worldwide. Beef liver is a rich source of active vitamin A. Overdoses are possible and cause many serious symptoms. Foods are preferable to supplements for supplying vitamin A. Recommended intakes are expressed in RE, or retinol equivalents.*

Beta-Carotene Retinol in excess is toxic, but beta-carotene is not; it is not converted to retinol efficiently enough to cause toxicity symptoms. Beta-carotene has, however, been known to turn people bright yellow if they eat too much. Beta-carotene builds up in the fat just beneath the skin and imparts a yellow cast.

When beta-carotene is converted to retinol in the body, losses occur. This is why nutrition scientists do not express the amounts of beta-carotene in foods, but instead use the RE, which indicates how much retinol the body actually derives from a plant food after converting the beta-carotene. The body can make one unit of retinol from about two of beta-carotene.[16]

Many foods from plants contain beta-carotene. Some are such a bright orange color that they decorate the plate. Carrots, sweet potatoes, pumpkins, mango, cantaloupe, and apricots are all rich sources. Another colorful group, *dark* green vegetables, such as spinach, other greens, and broccoli, owe their color to chlorophyll and beta-carotene. The orange and green pigments together give a deep dark green color to the vegetables.

Other colorful vegetables, such as red cabbage, beets, and sweet corn, can fool you into thinking they contain beta-carotene, but these foods derive their colors from other pigments and are poor sources of beta-carotene. As for "white" plant foods such as grains and potatoes, they have none. Some confusion exists concerning the term *yam*. The white-fleshed Mexican root vegetable called "yam" is devoid of beta-carotene, but the orange-fleshed sweet potato termed *yam* in the United States is one of the richest

RE (retinol equivalent) a measure of vitamin A activity; the amount of retinol that the body will derive from a food containing vitamin A (preformed retinol) or its precursor beta-carotene.

IU (international unit) a measure of fat-soluble vitamin activity. For methods to convert IU to RE, see Aids to Calculations, Appendix C.

An acne medication and a wrinkle cream contain retinoic acid—see Chapter 13.

macular degeneration a common, progressive loss of function of the part of the retina that is most crucial to focused vision (the macula is shown on p. 213). This degeneration often leads to blindness.

rickets the vitamin D–deficiency disease in children; characterized by abnormal growth of bone and manifested in bowed legs or knock-knees, outward-bowed chest, and knobs on the ribs.

Child to parent: How do you know that carrots are good for my eyes? Exasperated parent: Did you ever see a rabbit wearing glasses?

Chapter 5 introduced antioxidants as substances that protect body compounds from damage by oxidation. Controversy 11 revisits the carotene family of phytochemicals and explores their disease-fighting potential.

Chapter 8 and Controversy 8 present more about bone minerals and their regulation, and about osteoporosis, the bone-weakening disease.

beta-carotene sources known. Recommendations state that a person should eat *deep* orange or *dark* green vegetables and fruits regularly.

Does Eating Carrots Really Protect the Health of the Eyes?

In plants, vitamin A exists only in its precursor forms. Beta-carotene, the most abundant of these precursors, has the highest vitamin A activity. For many years, scientists believed beta-carotene to be of interest solely as a vitamin A precursor, but now they also recognize beta-carotene and its other carotene relatives for their antioxidant actions in the body.

Studies of populations suggest that people whose diets are low in foods that contain beta-carotene have a high incidence of certain types of cancer. Likewise, a common form of blindness in the aged, **macular degeneration,** is linked with a lifelong diet that excludes foods rich in beta-carotene. Interestingly, research has not supported a protective effect of supplements of beta-carotene itself against these diseases.[17] Evidence is growing, however, to support the link between eating beta-carotene–rich *foods* regularly and low rates of eye diseases and other diseases. Such evidence has led the Dietary Reference Intakes (DRI) committee to consider establishing a separate DRI for beta-carotene and its relatives, should the committee find sufficient research to warrant doing so.[18] The Controversy following this chapter offers more explanation of beta-carotene's effects.

KEY POINT *The vitamin A precursor in plants, beta-carotene, is an effective antioxidant in the body. Brightly colored plant foods are richest in beta-carotene.*

Vitamin D

Vitamin D is different from all the other nutrients in the body in that the body can synthesize it with the help of sunlight. Therefore, in a sense, vitamin D is not an essential nutrient. Given enough sun each day, most people need consume no vitamin D at all from foods. Folk wisdom has long held that sunshine promotes health; only in the recent past did scientists work out the details of vitamin D synthesis. Now vitamin D is appreciated for many critical functions including assisting the immune system in fighting off infections.

Roles of Vitamin D The best-known role of vitamin D is as a member of a large and interacting team of nutrients and hormones that continuously maintain blood calcium levels and thereby bone integrity, which is especially important during growth. Many of these interactions take place at the genetic level of cellular function in ways that are just now beginning to be understood.[19] Vitamin D, along with its cast of other players, ensures that sufficient calcium and phosphorus are available in the blood to support the growing bone structure.

Calcium is also indispensable to the proper functioning of all tissues of the body; cells of muscles, nerves, glands, and others all draw calcium from the blood as they need it. The skeleton serves as a vast warehouse of stored calcium that can be tapped when the blood supply begins to fall even slightly. To raise the level of blood calcium, the body can draw from only two other places: the digestive tract, where food brings calcium in, and the kidneys, which can recycle calcium into the body from blood filtrate destined to become urine. When calcium is needed, vitamin D acts at all three locations to raise the blood calcium level.

Vitamin D functions as a hormone, a compound manufactured by one organ of the body that acts on other organs or tissues. In addition to its actions in the bones, intestines, and kidneys, vitamin D is known to play roles in the brain, pancreas, skin, reproductive organs, and some cancer cells, but these are not yet fully understood. Like vitamin A, vitamin D stimulates maturation of cells, including cells of the immune system.

Too Little Vitamin D—A Danger to Bones The most obvious sign of vitamin D deficiency is abnormality of the bones. The disease **rickets,** caused by vitamin D deficiency in children, has been recognized for several centuries. A child with rickets characteristically develops bowed legs from bones too weak to support the body weight and a protruding belly that results from lax abdominal muscles.

As early as the 1700s, rickets was known to be curable with cod-liver oil, which is rich in vitamin D. More than a hundred years later, a Polish physician linked sunlight exposure to prevention and cure of rickets. By the beginning of the twentieth century, enough was finally known about rickets to reproduce it in laboratory animals. Today, the bowed legs, knock-knees, and protruding (pigeon) chests of children with rickets are no longer common sights in the United States. Tragically, many children worldwide still suffer the ravages of rickets largely because of inadequate food due to poverty, combined with a lack of sunlight.

Adult rickets, or **osteomalacia,** occurs most often in women with low calcium intakes and little exposure to the sun (therefore little opportunity to make vitamin D) who have repeated pregnancies and then breastfeed their babies. Under these conditions, calcium is withdrawn from the bones, but is not picked up efficiently from the intestine or recycled by the kidneys. The bones of the legs and spine may soften to such an extent that a young woman who is tall and straight at the age of 20 years may, after several pregnancies, become bowlegged and bent by age 30.

osteomalacia (OS-tee-o-mal-AY-shuh) the vitamin D–deficiency disease in adults (*osteo* means "bone"; *mal* means "bad"). Symptoms include bending of the spine and bowing of the legs.

Too Much Vitamin D—A Danger to Soft Tissues Vitamin D is the most potentially toxic of all vitamins. Chronic ingestion of excesses can cause appetite loss, nausea, and vomiting.[20] A severe form of psychological depression may also result from the effects on the central nervous system. If overdoses continue, vitamin D raises the blood mineral level to dangerous extremes, forcing calcium to be deposited in soft tissues such as the heart and kidneys. Calcium deposited in critical organs may cause them to malfunction, with potentially serious consequences to health and life.[21]

The likeliest victims of vitamin D poisoning are infants whose well-intentioned but misguided parents think that if some is good, more is better. People who take supplements containing vitamin D may also easily overdose, not realizing that their tissues are building up stockpiles of the vitamin. Intakes of only five times the recommended amount have been associated with signs of vitamin D toxicity in young children and adults. Recently, some people fell ill with vitamin D toxicity, and two died, after drinking milk from a dairy that had mistakenly overfortified the milk with up to 500 times the standard requirement of vitamin D.[22] One survivor who was an infant at the time of the incident later developed dental problems. The vitamin D toxicity disturbed the normal development of her permanent teeth during their early development in the gums.[23] Such instances are rare, but the incident renewed awareness of the potential for harm from vitamin D and the need for close monitoring of those who fortify the nation's foods with vitamins. The DRI committee has set a Tolerable Upper Intake Level for vitamin D at 50 micrograms per day (2,000 IU on supplement labels).

This child has the vitamin D–deficiency disease rickets.

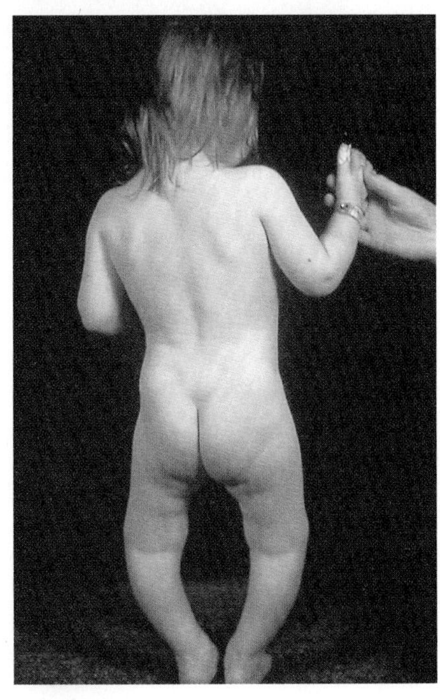

How can People Make a Vitamin from Sunlight? As mentioned, vitamin D is unique among vitamins because the body can make its own supply. When ultraviolet light from the sun shines on a cholesterol compound in human skin, the compound is transformed into a vitamin D precursor and is absorbed directly into the blood. Slowly, over the next day and a half or so, the liver and kidneys finish converting the precursor to the active form of vitamin D. Diseases that affect either the liver or the kidneys may impair the conversion of the inactive precursor to the active vitamin and therefore produce symptoms of vitamin D deficiency. Most of the world's population relies on natural exposure to sunlight to maintain adequate vitamin D nutrition.

Unlike concentrated supplements, sunlight presents no risk of vitamin D toxicity; the sun itself begins breaking down excess vitamin D made in the skin. Sunbathers run *other* risks, of course, such as premature wrinkling of the skin and the increased risk of skin cancer. Sunscreens with sun protection factors (SPF) of 8 and above can reduce these risks, but unfortunately they also prevent vitamin D synthesis.[24] In reality, though, production of vitamin D doesn't demand idle hours of sunbathing. Just being outdoors,

tocopherol (tuh-KOFF-er-all): a kind of alcohol. The active form of vitamin E is alpha-tocopherol.

Too much sun is dangerous—it may trigger the start of skin cancer.

even in lightweight clothing, is sufficient. The pigments of dark skin provide protection from ultraviolet radiation, but also reduce vitamin D synthesis.[25] Dark-skinned people require long exposure to direct sun (up to three hours, depending on the climate) for several days' worth of vitamin D, but light-skinned people need much less time (10 or 15 minutes). Thus, a person need only wait until enough time has elapsed to make some vitamin D and then apply sunscreen. Tanning booths may or may not promote vitamin D synthesis, but the Food and Drug Administration (FDA) has declared them risky because their unfiltered rays may promote skin cancer and damage the blood vessels and eyes. Daily doses of vitamin D are not necessary because the body stores enough vitamin D in its fat tissue to last through the dark winter months.

The ultraviolet rays of the sun that promote vitamin D synthesis cannot penetrate clouds, smoke, smog, heavy clothing, window glass, or even window screens. In the United States and Canada, almost all cases of rickets show up in dark-skinned people who live in smoggy northern cities or who lack exposure to sunlight. People who are housebound or institutionalized or who work at night may incur (over years) a vitamin D deficiency, as may elderly adults, who typically drink little milk, have limited exposure to sunlight, and lose efficiency in activating vitamin D as they age. Because of these risks, the DRI committe set recommended intakes for vitamin D that increase with age: 5 micrograms per day for adults 19 to 50 years, 10 micrograms for those 51 to 70 years, and 15 micrograms for those over 70.

Snapshot 7-2 shows the few significant food sources of vitamin D. Butter, cream, and fortified margarine contribute small amounts. In the United States and Canada, milk, whether fluid, dried, or evaporated, is fortified with vitamin D so that a daily quart (or liter) will supply the amount of the vitamin recommended for a young adult. That way young adults who drink the recommended 2 cups a day receive half their daily requirement; the other half comes from exposure to sunlight and other food sources. Children who drink 2 cups or more of milk a day will have a head start toward meeting their vitamin D needs for growth. Strict vegetarians, and especially their children, may have low vitamin D intakes because only two fortified plant sources exist: margarine and, in the United States, certain fortified cereals.

KEY POINT *Vitamin D raises blood minerals, notably calcium and phosphorus, permitting bone formation and maintenance. A deficiency can cause rickets in childhood or osteomalacia in later life. Vitamin D is the most toxic of all the vitamins, and excesses are dangerous or deadly. People exposed to the sun make vitamin D from a cholesterol-like compound in their skin; fortified milk is an important food source.*

The sunshine vitamin: vitamin D.

Vitamin E

More than 70 years ago, researchers discovered a compound in vegetable oils necessary for reproduction in rats. This compound was named **tocopherol** from *tokos*, a Greek word meaning "offspring." A few years later, the compound was named vitamin E. Eventually, four different tocopherol compounds were discovered, and each was designated by one of the first four letters of the Greek alphabet: alpha, beta, gamma, and delta.

The Extraordinary Bodyguard Vitamin E, because it functions as an antioxidant, serves as one of the body's main defenders against oxidative damage. By being oxidized itself, vitamin E protects the polyunsaturated fats and other vulnerable components of the cells and their membranes from destruction. Vitamin E protects all the cells' lipids and related compounds, such as vitamin A, from oxidation.

Vitamin E exerts an especially important antioxidant effect in the lungs, where the cells are exposed to high oxygen concentrations that can destroy molecules in their membranes. As the red blood cells carry oxygen from the lungs to other tissues, vitamin E protects their cell membranes, too. Vitamin E may also help defend against heart disease; this chapter's Controversy provides details.

SNAPSHOT 7.2 Vitamin D

These foods provide 10 percent or more of the vitamin D Daily Value (DV = 10 μg/day).[a]

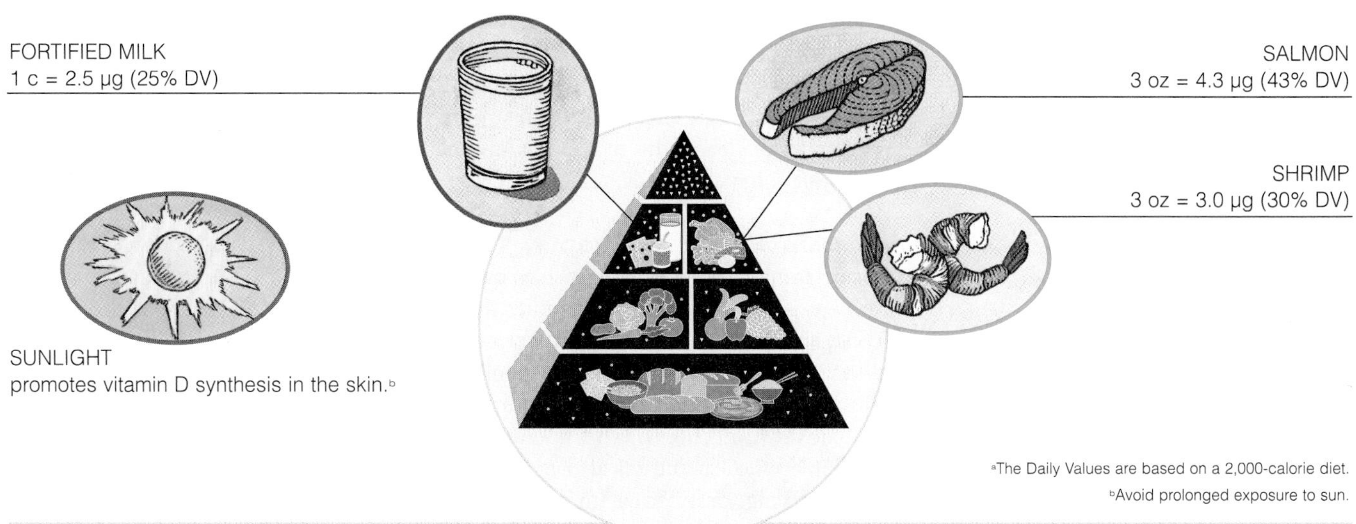
FORTIFIED MILK
1 c = 2.5 μg (25% DV)
SALMON
3 oz = 4.3 μg (43% DV)
SHRIMP
3 oz = 3.0 μg (30% DV)
SUNLIGHT
promotes vitamin D synthesis in the skin.[b]

[a]The Daily Values are based on a 2,000-calorie diet.

[b]Avoid prolonged exposure to sun.

Normal nerve development also depends on vitamin E. Vitamin E also protects the white blood cells that defend the body against disease and may play other roles in normal immunity. Supplements of the vitamin were found to improve the immune response in healthy elderly people, a group whose immunity is often impaired.[26]

Vitamin E Deficiency A deficiency of vitamin E produces a wide variety of symptoms in laboratory animals. Most of these symptoms have not been reproduced in human beings, however, despite many attempts. Three reasons have been given for this. First, the vitamin is so widespread in food that it is almost impossible to create a vitamin E–deficient diet. Second, the body stores so much vitamin E in its fatty tissues that a person could not eat a vitamin E–free diet for long enough to deplete these stores and produce a deficiency. Third, the cells recycle their working supply of vitamin E, using the same molecules over and over to ward off deficiency.

The classic vitamin E–deficiency symptom in human beings occurs in premature babies. Some of these babies are born before the transfer of the vitamin from the mother to the infant that takes place in the last weeks of pregnancy. Without sufficient vitamin E, the infant's red blood cells rupture **(erythrocyte hemolysis)**, and the infant becomes anemic. The few symptoms of vitamin E deficiency that have been observed in adults include loss of muscle coordination and reflexes with impaired movement, vision, and speech. All of these symptoms may be caused by oxidative damage; vitamin E treatment corrects them.[27]

An interesting line of vitamin E research concerns a possible effect on viruses residing in a vitamin E–deficient person. The research suggests that when body stores of vitamin E and the mineral selenium are low, viruses respond by becoming more virulent; that is, they cause more severe infections.[28] Even normally harmless viruses appear to undergo changes that make them more likely to cause diseases. No one yet knows the details of these effects, but they may be related to "oxidative stress" caused when oxidative activities outstrip the capacity of the tissues' antioxidant defenses.

In adults, vitamin E deficiency is usually associated with diseases, notably those that cause malabsorption of fat. These include disease or injury of the liver (which

erythrocyte (eh-REETH-ro-sight) **hemolysis** (HE-moh-LIE-sis, he-MOLL-ih-sis) rupture of the red blood cells, caused by vitamin E deficiency (*erythro* means "red"; *cyte* means "cell"; *hemo* means "blood"; *lysis* means "breaking").

makes bile, necessary for digestion of fat), the gallbladder (which delivers bile into the intestine), and the pancreas (which makes fat-digesting enzymes), as well as a number of hereditary diseases involving digestion and use of nutrients.

Though rare, vitamin E deficiencies may be seen in people without diseases. Deficiencies are most likely in those who for years eat diets extremely low in fat; use fat substitutes, such as diet margarines and salad dressings, as their only sources of fat; or consume diets composed largely of highly processed or "convenience" foods since vitamin E is destroyed by extensive heating in the processing of these foods.

Extravagant claims are being made that vitamin E cures all sorts of conditions because its deficiency affects animals' muscles and reproductive systems. Vitamin E deficiency does not affect the organs of human beings as it affects animals, however. Research has clearly discredited claims that vitamin E improves athletic endurance and skill, enhances sexual performance, or cures sexual dysfunction in males.

Some supplement labels list vitamin E in IU, or international units. To find IU, divide alpha TE by 1.5; to find alpha TE, multiply IU by 1.5.

For perspectives on possible risks and benefits of vitamin E supplements, see the Controversy.

Vitamin E Requirements, Toxicity, and Sources The 1989 RDA (inside front cover) for vitamin E is based on body size: it is 8 milligrams a day for women, 10 for men. Note that values for vitamin E in Appendix A and in the vitamin E Snapshot are given in units known as alpha TE (alpha-tocopherol equivalents). One of these units, 1 alpha TE, equals 1 milligram of active vitamin E. The need for vitamin E rises as people consume more polyunsaturated oil because the oil requires antioxidant protection by the vitamin. Luckily, most raw oils also contain vitamin E, so people who eat the oil also receive the vitamin. As mentioned, heat processing, such as frying, destroys vitamin E, as does oxidation, so most processed, fast, deep-fried, and convenience foods retain little intact vitamin E.

Ordinary supplemental doses of vitamin E taken over a period of months seem to have no adverse effects on health.[29] The medical literature contains isolated reports of adverse effects from very high doses in laboratory animals and occasional reports of nausea, intestinal distress, and other vague complaints in human beings. Large doses may augment the effects of anticoagulant medication used to oppose unwanted blood clotting; people taking such drugs risk uncontrollable bleeding when they also take large doses of vitamin E. For most individuals, however, daily doses below 800 milligrams alpha TE (530 IU) may be harmless.[30] We say "may be" because some preliminary reports link long-term use of even low doses of vitamin E supplements with brain hemorrhages, a form of stroke.

Vitamin E is widespread in foods. About 20 percent of the vitamin E people consume comes from vegetable oils and products made from them, salad dressings, and shortening (see Snapshot 7-3). Another 20 percent comes from fruits and vegetables although none of these is a good source by itself. Fortified cereals* and other grain products contribute about 15 percent of vitamin E in the diet, and meats, poultry, fish, eggs, nuts, and seeds contribute smaller percentages. Wheat germ and soybeans are good sources of vitamin E; animal fats, such as milk fat or the fat of meats, have almost none.

KEY POINT *Vitamin E acts as an antioxidant in cell membranes and is especially important for the integrity of cells that are constantly exposed to high oxygen concentrations, namely, the lungs and blood cells, both red and white. Vitamin E deficiency is rare in human beings, but it does occur in newborn premature infants. The vitamin is widely distributed in plant foods; it is destroyed by high heat; toxicity is rare.*

Vitamin K

Have you ever thought about how remarkable it is that blood can clot? The liquid turns solid in a life-saving series of reactions. If blood cannot clot, wounds will bleed for a dangerously long time; this is why hospitals measure the clotting time of a

*Cereals fortified with vitamin E may not be available in Canada.

SNAPSHOT 7.3 **Vitamin E**

These foods provide 10 percent or more of the vitamin E Daily Value (DV = 30 IU or 20 α-TE/day).[a]

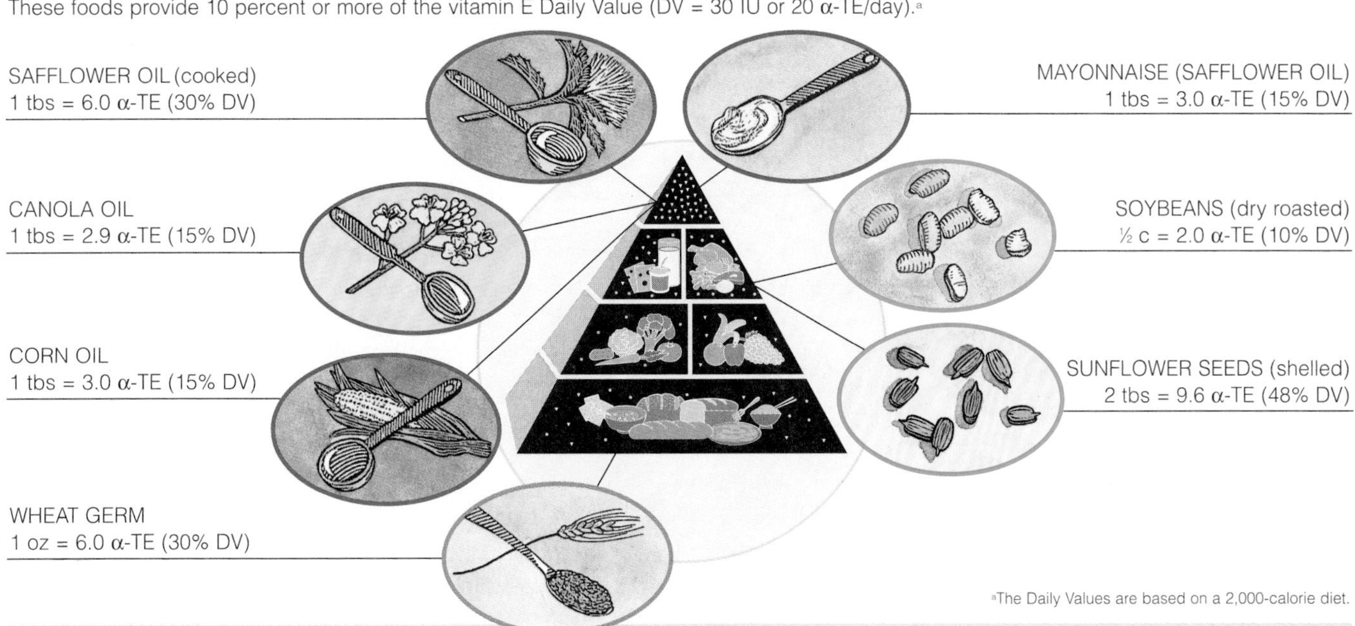
SAFFLOWER OIL (cooked)
1 tbs = 6.0 α-TE (30% DV)
CANOLA OIL
1 tbs = 2.9 α-TE (15% DV)
CORN OIL
1 tbs = 3.0 α-TE (15% DV)
WHEAT GERM
1 oz = 6.0 α-TE (30% DV)
MAYONNAISE (SAFFLOWER OIL)
1 tbs = 3.0 α-TE (15% DV)
SOYBEANS (dry roasted)
½ c = 2.0 α-TE (10% DV)
SUNFLOWER SEEDS (shelled)
2 tbs = 9.6 α-TE (48% DV)

[a]The Daily Values are based on a 2,000-calorie diet.

person's blood before surgery. Vitamin K, which is needed for the synthesis of proteins that help clot the blood, is sometimes administered before operations to reduce bleeding in surgery. Vitamin K may be of value at this time, but only if a vitamin K deficiency exists. Vitamin K does not improve clotting in those with other bleeding disorders, such as the genetic disease hemophilia.

K stands for the Danish word *koagulation* (clotting).

In some people with heart problems, there is a need to *prevent* the formation of clots within the circulatory system. This is popularly referred to as "thinning" the blood. One of the best-known medicines for this purpose is dicumarol, which interferes with the action of vitamin K in promoting clotting. Vitamin K therapy is necessary for people taking dicumarol if uncontrolled bleeding occurs.

Vitamin K is also necessary for the synthesis of a key protein needed in bone formation.[31] Together with the more famous bone vitamin, vitamin D, vitamin K ensures that the bones produce this protein normally so that bones can properly bind the minerals they need. Vitamin K intake may also play a part in reducing the risk of hip fracture: in a large study of women, those who ate abundant green vegetables, known sources of vitamin K, suffered hip fractures less often than those with lower intakes.[32]

Like vitamin D, vitamin K can be obtained from a nonfood source—in this case, the intestinal bacteria. Billions of bacteria normally reside in the intestines, and some of them synthesize vitamin K. The extent to which the body uses the vitamin K synthesized by these bacteria is not known, but it is thought that people may obtain about half of their daily needs from this source.

As Snapshot 7-4 shows, vitamin K's richest plant food sources are dark green, leafy vegetables, which provide from 50 to 800 micrograms per 3-ounce serving, and members of the cabbage family. There is also one rich animal food source, liver. Milk, other meats, eggs, cereals, and fruits provide smaller but still significant amounts. Tables of food composition do not include the vitamin K contents of foods because they are not well enough known.

Few U.S. adults are likely to experience vitamin K deficiency, even if they seldom eat vitamin K–rich foods. Exceptions are newborn infants whose intestinal tracts are

SNAPSHOT 7.4 Vitamin K[a]

The Daily Value (DV) for vitamin K is 80 µg/day.[b]

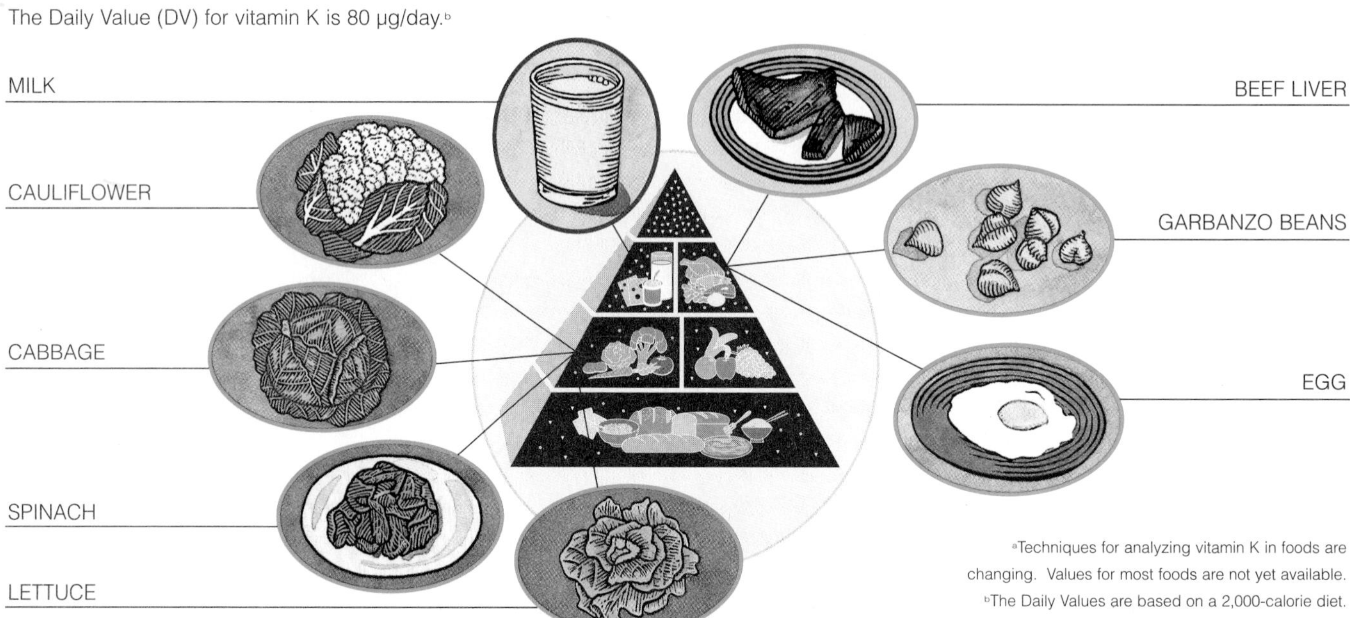
MILK
CAULIFLOWER
CABBAGE
SPINACH
LETTUCE
BEEF LIVER
GARBANZO BEANS
EGG

[a]Techniques for analyzing vitamin K in foods are changing. Values for most foods are not yet available.
[b]The Daily Values are based on a 2,000-calorie diet.

not yet inhabited by bacteria and people who have taken antibiotics that have killed their intestinal bacteria. Supplements of the vitamin are needed in these cases.

For others, vitamin K toxicity can result when supplements of a synthetic version of vitamin K are given, especially to infants or pregnant women. Toxicity induces breakage of the red blood cells and release of their pigment, which colors the skin yellow. A toxic dose of synthetic vitamin K causes the liver to release the blood cell pigment (bilirubin) into the blood (instead of excreting it into the bile) and leads to jaundice. When bilirubin invades the brain of an infant, the condition may lead to brain damage or death of the infant. Because vitamin K from supplements can easily reach toxic levels, it is available as a single vitamin only by prescription. The accompanying Fitness for Life feature addresses the question of whether physically active people need supplements of any vitamins.

KEY POINT *Vitamin K is necessary for blood to clot; deficiency causes uncontrolled bleeding. The bacterial inhabitants of the digestive tract produce vitamin K, and most people derive about half their requirement from them and half from food. Excesses are toxic.*

The Water-Soluble Vitamins

The water-soluble vitamins require special consideration in food preparation to avoid losing or destroying them. See the Food Feature of Chapter 14.

So far, this chapter has addressed the fat-soluble vitamins. All of the other vitamins, the B vitamins and vitamin C, are water soluble. Cooking and washing water can leach them out of foods. The body absorbs these vitamins easily and just as easily excretes them in the urine. Under ordinary circumstances, you need not be concerned about consuming modest excesses. Some of the water-soluble vitamins can remain in the lean tissues for a month or more, but these tissues are actively exchanging materials with the body fluids at all times. At any time, the vitamins may be picked up by the extracellular fluids, carried away by the blood, and excreted in the urine. Generally,

The Power of Vitamins

FITNESS FOR LIFE

Exercisers often harbor vague concerns about vitamins: Do their foods provide the vitamins that their muscles need to perform their best? There's no doubt that physically active people need vitamins. These metabolic servants are necessary to all physical activity—even getting up out of a chair. A look at the deficiency symptoms (listed in Tables 7-5 and 7-6)—anemia, fatigue, depression, and so forth—is enough to convince any active person that deficiencies are an enemy of performance. But as the Snapshots throughout this chapter demonstrate, rich vitamin sources occur in all of the food groups, so vitamins are easily obtained from a balanced diet of ordinary foods. In fact, the only foods with almost no vitamins are the refined sugars. Even oils extracted from vegetable sources often provide substantial vitamin E, and exercisers may need vitamin E to deal with the increased oxidation that results when exercise demands extra fuel.

The good news for exercisers is that even competitive athletes who choose their diets with reasonable care almost never suffer from vitamin deficiencies. The reason is elegantly simple. Exercise requires that people eat extra calories of food, and if that extra food is of the kind shown in this chapter's Snapshots—fruits, vegetables, milk, eggs, whole or enriched grains, lean meats, and even some oils—then obtaining the vitamins to support activity is a snap. One of the nicest things about understanding how activity and nutrition interact is that one can let go of unfounded fears about vitamins and other nutrients and concentrate on getting enough of the foods that meet all of the excercising body's needs.

you can make sure your three-day intake average meets the recommendation by choosing foods that are rich in water-soluble vitamins.

Foods never deliver toxic doses of the water-soluble vitamins, but the large doses concentrated in some vitamin supplements can reach toxic levels. Normally, though, the most likely hazard to the supplement taker is to the wallet. As one person aptly noted, "If you take supplements of the water-soluble vitamins, you may have the most expensive urine in town."

coenzyme (co-EN-zime) a small molecule that works with an enzyme to promote the enzyme's activity. Many coenzymes have B vitamins as part of their structure (*co* means "with").

The B Vitamins and Their Relatives

The B vitamins act as part of coenzymes. A **coenzyme** is a small molecule that combines with an enzyme to make it active. (Recall from Chapter 6 that enzymes are large proteins that do the body's building, dismantling, and other work.) Sometimes the vitamin part of the enzyme is the active site, where the chemical reaction takes place. The substance to be worked on is attracted to the active site and snaps into place; the reaction proceeds instantaneously. The shape of each enzyme predestines it to accomplish just one kind of job. Without its coenzyme, however, the enzyme is as useless as a car without wheels. Figure 7-4 shows how a coenzyme enables an enzyme to do its job.

Each B vitamin has its own special character, and the amount of detail known about each one is overwhelming. To simplify things, this introduction describes some of the ways the B vitamins work together as a group and emphasizes the consequences of deficiencies. The sections that follow present more details about the vitamins as individuals.

KEY POINT *As part of coenzymes, the B vitamins help enzymes do their jobs.*

Characteristics water-soluble vitamins share:

- Dissolve in water.
- Are easily absorbed and excreted.
- Are not stored extensively in tissues.
- Seldom reach toxic levels.

B Vitamin Roles in Metabolism

Figure 7-5 on page 225 shows some body organs and tissues in which the B vitamins help the body metabolize carbohydrates, lipids, and amino acids. The purpose of the

thiamin (THIGH-uh-min) a B vitamin involved in the body's use of fuels.

beriberi the thiamin-deficiency disease; characterized by loss of sensation in the hands and feet, muscular weakness, advancing paralysis, and abnormal heart action.

figure is not to present a detailed account of metabolism, which is best left to courses in biochemistry, but rather to give you an impression of where the B vitamins work together with enzymes in the metabolism of energy nutrients and in the making of new cells.

Many people mistakenly believe that B vitamins supply the body with energy. They do not, at least not directly. The energy-yielding nutrients, carbohydrate, fat, and protein, give the body fuel for energy; the B vitamins help the body use that fuel. More specifically, active forms of the B vitamins thiamin, riboflavin, niacin, pantothenic acid, and biotin participate in the release of energy from carbohydrate, fat, and protein. Vitamin B_6 helps the body use amino acids to make protein; the body then puts the protein to work in many ways—to build new tissues, to make hormones, to fight infections, or to serve as fuel for energy, to name only a few.

Folate and vitamin B_{12} help cells to multiply; this is especially important to cells with short life spans that must replace themselves rapidly. Such cells include both the red blood cells (which live an average of six weeks) and the cells that line the digestive tract (which replace themselves every three days). These cells deliver energy to all the others. In short, each and every B vitamin is involved, directly or indirectly, in energy metabolism.

Figure 7-4

Coenzyme Action

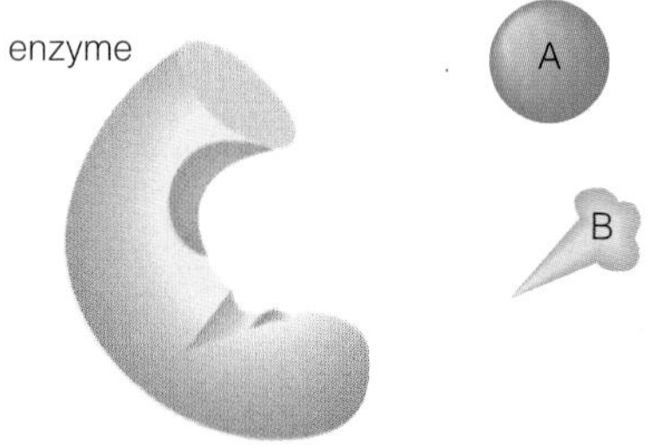
enzyme
A
B

Without the coenzyme, compounds A and B don't respond to the enzyme.

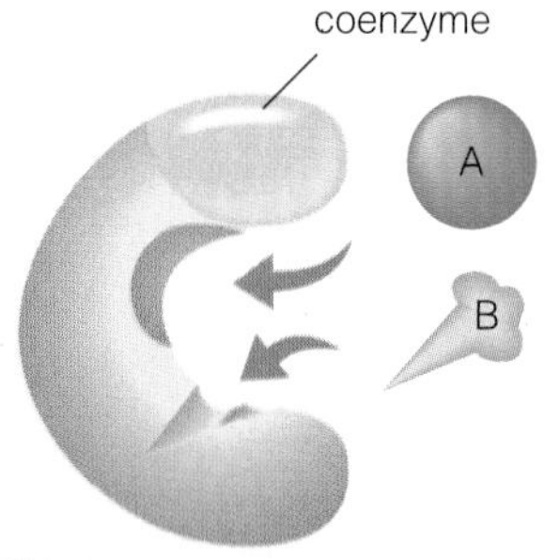
coenzyme
A
B

With the coenzyme in place, compounds A and B are attracted to the active site on the enzyme, and they react.

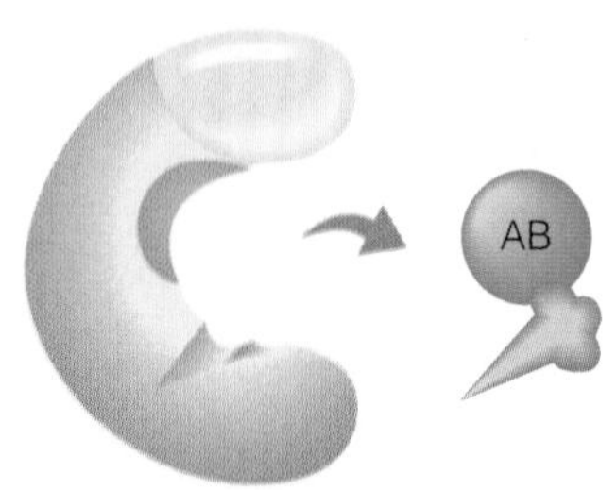
AB

The reaction is completed with the formation of a new product. In this case the product is AB.

KEY POINT *The B vitamins facilitate the work of every cell. Some help generate energy; others help make protein and new cells. B vitamins work everywhere in the body tissue to metabolize carbohydrate, fats, and protein.*

B Vitamin Deficiencies and Toxicities

As long as B vitamins are present, their presence is not felt. Only when they are missing does their absence manifest itself in a lack of energy and a multitude of other symptoms, as you can imagine after looking at Figure 7-5. The reactions by which B vitamins facilitate energy release take place in every cell, and no cell can do its work without energy. Thus, in a B vitamin deficiency, every cell is affected. Among the symptoms of B vitamin deficiencies are nausea, severe exhaustion, irritability, depression, forgetfulness, loss of appetite and weight, pain in muscles, impairment of the immune response, loss of control of the limbs, abnormal heart action, severe skin problems, teary or bloodshot eyes, and many more. Because cell renewal depends on energy and protein, and because these depend on the B vitamins, the digestive tract and the blood are invariably damaged. In children, full recovery may be impossible. In the case of a thiamin deficiency during growth, permanent brain damage can result.

In academic discussions of the vitamins, different sets of deficiency symptoms are given for each one. Actually, such clear-cut sets of symptoms are found only in laboratory animals that have been fed contrived diets that lack just one ingredient. In real life, a deficiency of any one B vitamin seldom shows up by itself because people don't eat nutrients singly; they eat foods that contain mixtures of nutrients. A deficiency of one B vitamin may appear to be responsible for a cluster of symptoms, but subtler, undetected deficiencies may accompany it. If treatment involves giving wholesome food rather than a single supplement, the subtler deficiencies will be corrected along with the major one. The symptoms of B vitamin deficiencies and toxicities are listed in Table 7-6 at the end of the chapter. The next few sections provide details about each of the B vitamins.

Thiamin and Riboflavin All cells use **thiamin**, which plays a critical role in their energy metabolism. Thiamin also occupies a special site on nerve cell membranes. Consequently, nerve processes and their responding tissues, the muscles, depend heavily on thiamin.

The classic thiamin-deficiency disease **beriberi** was first observed in East Asia, where rice provided 80 to 90 percent of the total calories most people consumed and was therefore their principal source of thiamin. When the custom of polishing rice (removing its brown coat, which contained the thiamin) became widespread, beriberi swept through the population like an epidemic. Scientists wasted years of time and

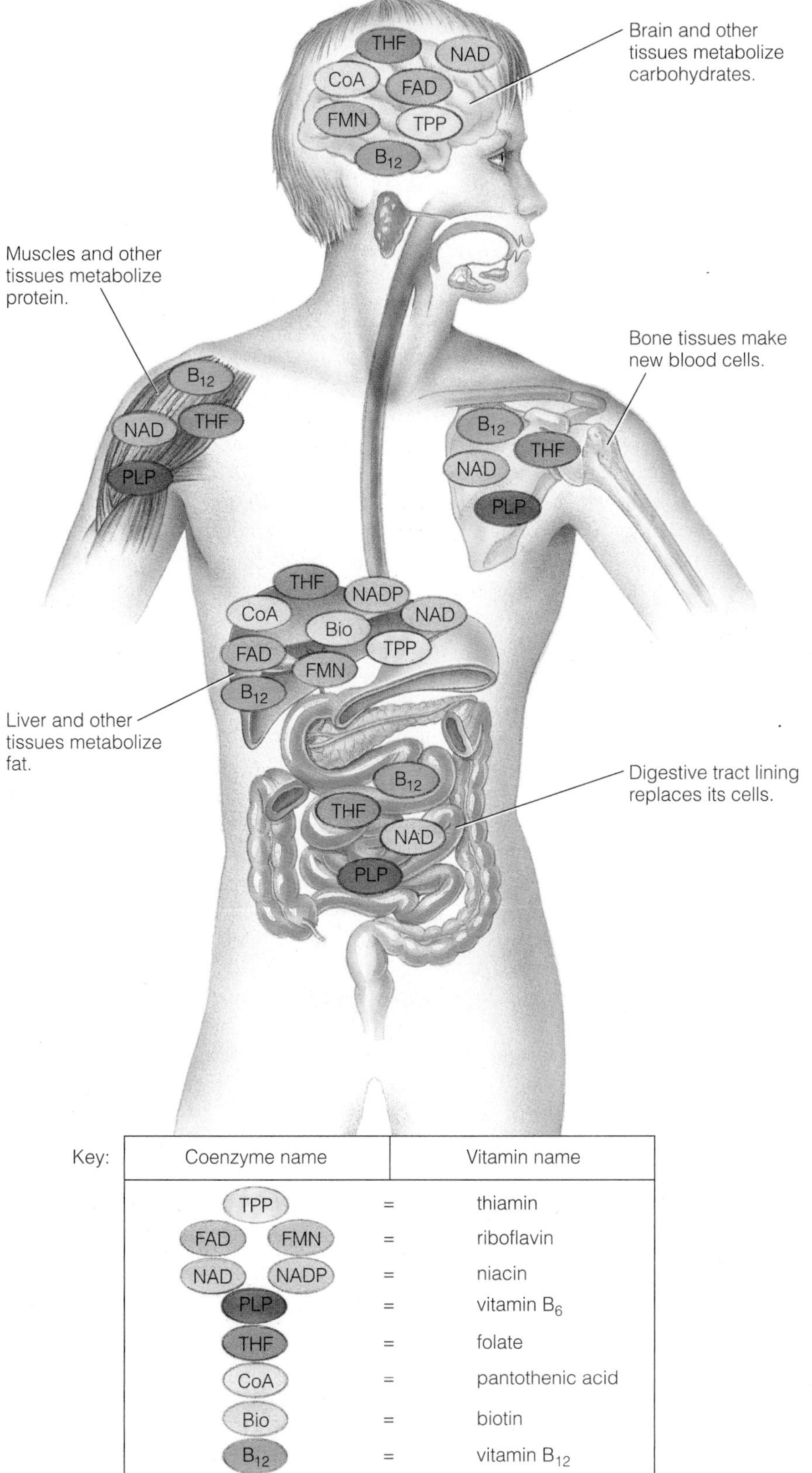
THF
NAD
CoA
FAD
FMN
TPP
B_{12}
Brain and other tissues metabolize carbohydrates.
Muscles and other tissues metabolize protein.
B_{12}
NAD
THF
PLP
Bone tissues make new blood cells.
B_{12}
THF
NAD
PLP
THF
NADP
CoA
Bio
NAD
FAD
TPP
FMN
B_{12}
Liver and other tissues metabolize fat.
B_{12}
Digestive tract lining replaces its cells.
THF
NAD
PLP
Key:

Coenzyme name		Vitamin name
TPP	=	thiamin
FAD FMN	=	riboflavin
NAD NADP	=	niacin
PLP	=	vitamin B_6
THF	=	folate
CoA	=	pantothenic acid
Bio	=	biotin
B_{12}	=	vitamin B_{12}

Figure 7-5

Some Roles of the B Vitamins in Metabolism: Examples

This figure does not attempt to teach intricate biochemical pathways or names of B vitamin–containing enzymes. Its sole purpose is to show a few of the many tissue functions that depend on B vitamin–containing enzymes. The B vitamins work in every cell, and this figure displays less than a thousandth of what they actually do.

Every B vitamin is part of one or more coenzymes that make possible the body's chemical work. For example, the niacin, thiamin, and riboflavin coenzymes are important in the energy pathways. The folate and vitamin B_{12} coenzymes are necessary for making RNA and DNA and thus new cells. The vitamin B_6 coenzyme is necessary for processing amino acids and, therefore, protein. Many other relationships are also critical to metabolism.

Wernicke-Korsakoff (VER-nih-kee KORE-sah-kof) **syndrome** a form of thiamin deficiency affecting the brain tissues. The syndrome is associated with alcohol abuse and is characterized by mental confusion and disorientation, loss of memory, jerky eye movements, and staggering gait.

riboflavin (RIBE-o-flay-vin) a B vitamin active in the body's energy-releasing mechanisms.

niacin a B vitamin needed in energy metabolism. Niacin can be eaten preformed or can be made in the body from tryptophan, one of the amino acids. Other forms of niacin are *nicotinic acid, niacinamide,* and *nicotinamide.*

pellagra (pell-AY-gra) the niacin-deficiency disease (*pellis* means "skin"; *agra* means "rough"). Symptoms include the "4 Ds": diarrhea, dermatitis, dementia, and, ultimately, death.

Table 7-6 on p. 245 lists the symptoms of riboflavin deficiency.

Figure 7-6

Beriberi

Beriberi takes two forms: wet beriberi, characterized by edema (fluid accumulation), and dry beriberi, without edema but with muscle wasting. This woman's leg retains the imprint of her physician's thumb, showing the edema of wet beriberi.

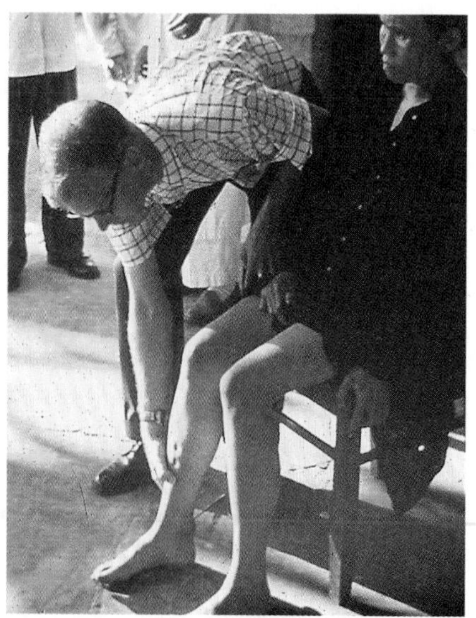

effort hunting for a microbial cause of beriberi before they realized that the cause was not something present in the environment but something absent from it. Figure 7-6 depicts beriberi and describes its two forms.

Just before 1900, an observant physician working in a prison in East Asia discovered that beriberi could be cured with proper diet. The physician noticed that the chickens at the prison had developed a stiffness and weakness similar to that of the prisoners who had beriberi. The chickens were being fed the rice left on prisoners' plates. When the rice bran, which had been discarded in the kitchen, was given to the chickens, their paralysis was cured. As might be expected, the physician met resistance when he tried to feed the rice bran, the "garbage," to the prisoners, but it worked—dramatically. Later, extracts of rice bran were used to prevent infantile beriberi; still later, thiamin was synthesized.

In developed countries today, abuse of alcohol often leads to a severe form of thiamin deficiency, **Wernicke-Korsakoff syndrome.**[33] Alcohol contributes energy, but carries almost no nutrients with it and often displaces food. In addition, alcohol impairs absorption of thiamin from the digestive tract and hastens its excretion in the urine, tripling the risk of deficiency. The syndrome is characterized by symptoms that are almost indistinguishable from alcohol abuse itself: mental confusion, disorientation, loss of memory, jerky eye movements, and a staggering gait. Unlike alcohol toxicity, the syndrome responds quickly to an injection of thiamin, and some experts recommend a precautionary dose for any patients suspected of having the syndrome.[34]

Thiamin occurs in small amounts in many nutritious foods. Ham and other pork products, leafy green vegetables, whole-grain cereals, and legumes are especially rich in thiamin (see Snapshot 7-5). People who keep empty-calorie foods to a minimum and include ten or more servings of nutritious foods each day will easily meet their thiamin needs. The DRI committee set the thiamin intake recommendation at 1.2 milligrams per day for men and at 1.1 milligrams per day for women. Pregnancy and lactation require somewhat more thiamin (see the DRI, inside front cover, page A).

Like thiamin, **riboflavin** plays a role in the energy metabolism of all cells. When thiamin is deficient, riboflavin may be lacking, too, but its deficiency symptoms may go undetected because those of thiamin deficiency are more severe.[35] Foods that remedy the thiamin deficiency invariably also contain some riboflavin, so they clear up both deficiencies. People obtain as much as half of their riboflavin from milk and milk products. Leafy green vegetables, whole-grain breads and cereals, and some meats contribute the rest of the riboflavin in people's diets (see Snapshot 7-6 on page 228).

Niacin The vitamin **niacin,** like thiamin and riboflavin, participates in the energy metabolism of every body cell. The niacin-deficiency disease **pellagra** appeared in Europe in the 1700s when corn from the New World came to be widely accepted as a staple food. In the early twentieth century in the United States, pellagra was devastating people's lives throughout the South and Midwest. Hundreds of thousands of pellagra victims were thought to be suffering from a contagious disease until this dietary deficiency was identified. The disease still occurs among poorly nourished people living in urban slums and particularly in those with alcohol addiction. Pellagra is also still common in parts of Africa and Asia.

Early workers seeking the cause of pellagra observed that well-fed people never got it. From there the researchers defined a diet that reliably produced the disease—one of cornmeal, salted pork fat, and molasses. Corn not only is low in protein, but also lacks tryptophan, the amino acid from which niacin is made. Salt pork is almost pure fat and contains too little protein to compensate; and molasses is virtually protein-free.

Figure 7-7 shows the skin disorder associated with pellagra. For comparison, a later figure (Figure 7-9 on page 231) shows a skin disorder associated with vitamin B_6 deficiency, a reminder that any nutrient deficiency affects the skin and all other cells. The skin just happens to be the organ you can see.

The key nutrient that prevents pellagra is niacin, but any protein containing sufficient amounts of the amino acid tryptophan will serve in its place. Tryptophan, which is abundant in almost all proteins (but is unavailable from the protein of corn), is converted to niacin in the body. In fact, it is possible to cure pellagra by administering

SNAPSHOT 7.5 Thiamin

These foods provide 10 percent or more of the thiamin Daily Value (DV = 1.5 mg/day).[a]

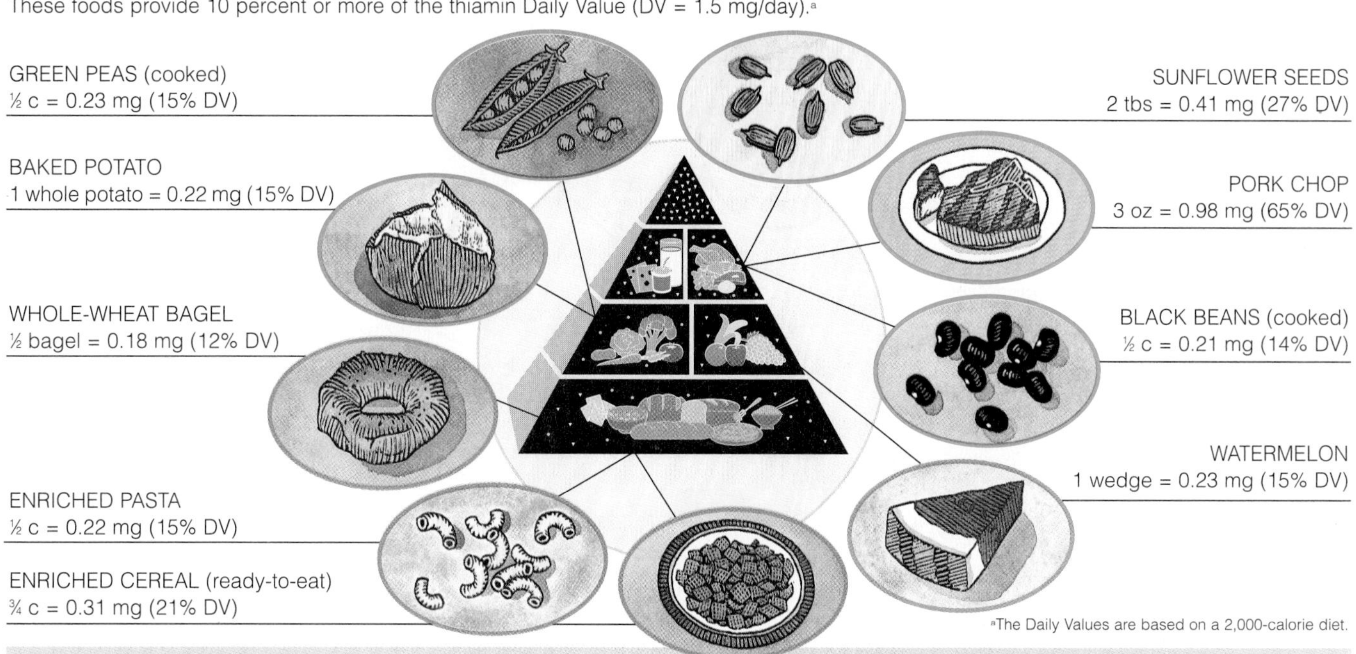

GREEN PEAS (cooked)
½ c = 0.23 mg (15% DV)
BAKED POTATO
1 whole potato = 0.22 mg (15% DV)
WHOLE-WHEAT BAGEL
½ bagel = 0.18 mg (12% DV)
ENRICHED PASTA
½ c = 0.22 mg (15% DV)
ENRICHED CEREAL (ready-to-eat)
¾ c = 0.31 mg (21% DV)
SUNFLOWER SEEDS
2 tbs = 0.41 mg (27% DV)
PORK CHOP
3 oz = 0.98 mg (65% DV)
BLACK BEANS (cooked)
½ c = 0.21 mg (14% DV)
WATERMELON
1 wedge = 0.23 mg (15% DV)

[a]The Daily Values are based on a 2,000-calorie diet.

tryptophan alone. Thus a person eating more than adequate protein (as most people in developed nations do) will not be deficient in niacin. The amount of niacin in a diet is stated in terms of **niacin equivalents,** a measure that takes available tryptophan into account. Snapshot 7-7 on page 229 shows some food sources of niacin.

Certain forms of niacin supplements in amounts ten times or more the recommendation cause "niacin flush," a dilation of the capillaries of the skin with perceptible tingling that, if intense, can be painful.[36] Physicians often administer large niacin doses as part of their arsenal of drugs against atherosclerosis.[37] Large doses of a form of niacin may also prove useful in preventing diabetes.[38] When used this way, niacin leaves the realm of nutrition to become a pharmacological agent, a drug. As with any drug, self-dosing with niacin is ill-advised; large doses may injure the liver, cause peptic ulcers or vision loss, or bring on any of a number of adverse effects.[39] For safety's sake, anyone taking large doses of niacin should do so only under the care of a physician.[40]

niacin equivalents the amount of niacin present in food, including the niacin that can theoretically be made from its precursor tryptophan that is present in the food.

folate (FOH-late) a B vitamin that acts as part of a coenzyme important in the manufacture of new cells. Other names for folate are *folacin* and *folic acid*.

Folate The vitamin **folate** is required to make all new cells. Folate helps synthesize the DNA needed for the new cells. Deficiencies may result from an inadequate intake or from illnesses that impair folate's absorption, increase its excretion, or otherwise increase the need for folate.

Because the blood cells and digestive tract cells divide most rapidly, they are most vulnerable to deficiency. As a result, deficiencies of folate cause anemia and abnormal digestive function. In the United States, a significant number of cases of folate-deficiency anemia occur yearly. This anemia is also related to the anemia of vitamin B_{12} malabsorption; see Figure 7-8, page 230. Folate deficiency may also elevate a woman's risk for cervical cancer, a major cause of cancer deaths among women worldwide.[41]

The DRI committee advises all women of childbearing age to consume 400 micrograms of synthetic folate each day, in addition to the folate that occurs naturally in their foods.[42] The reason for this recommendation is that folate deficiency is associated with a group of devastating birth defects known as neural tube defects. These defects affect 1 in every 1,000 births, making them the second most common form of birth defect after Down syndrome. Neural tube defects range from slight

Figure 7-7
Pellagra

The typical dermatitis of pellagra develops on skin that is exposed to light.

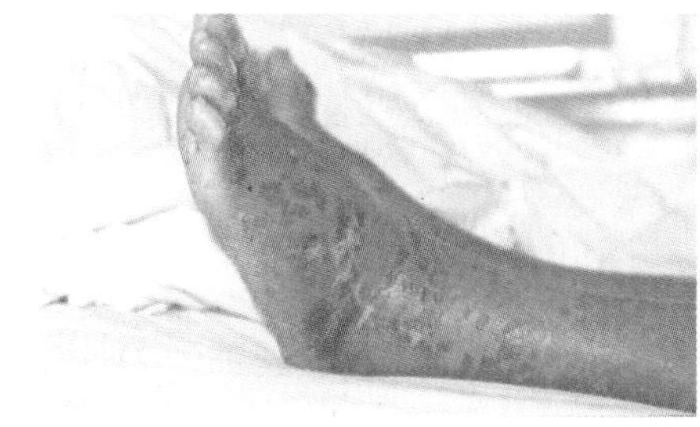

SNAPSHOT 7.6 **Riboflavin**

These foods provide 10 percent or more of the riboflavin Daily Value (DV = 1.7 mg/day).[a]

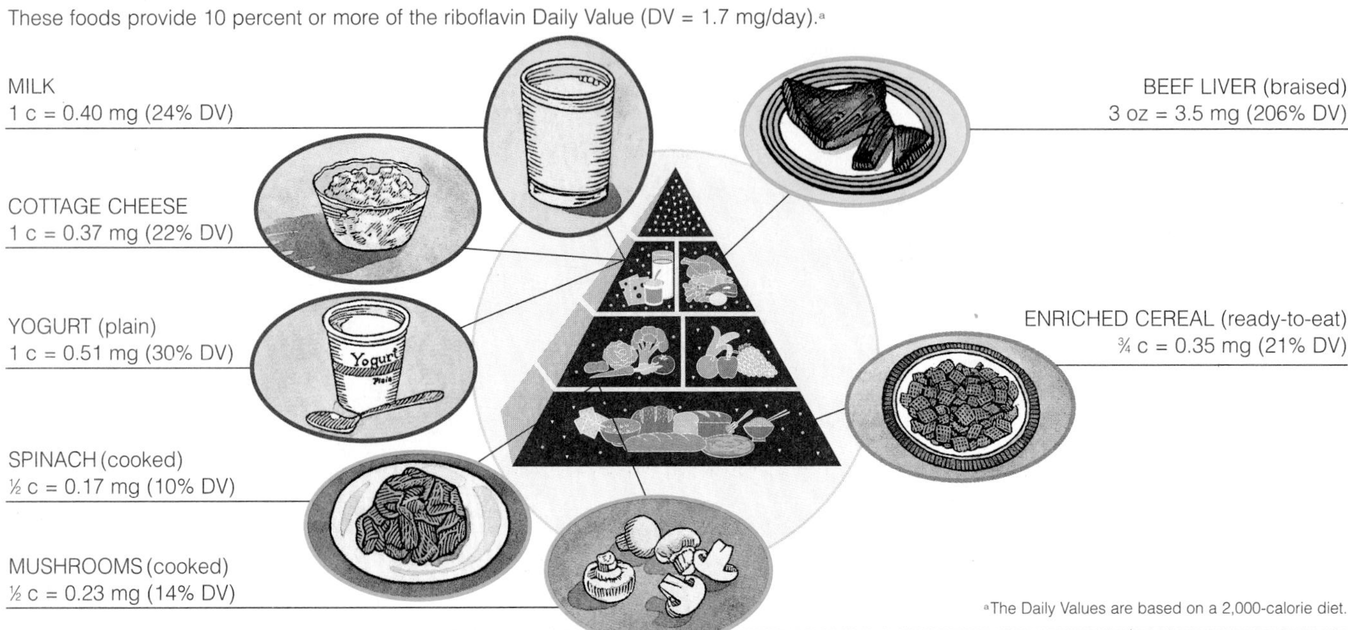

MILK
1 c = 0.40 mg (24% DV)
COTTAGE CHEESE
1 c = 0.37 mg (22% DV)
YOGURT (plain)
1 c = 0.51 mg (30% DV)
SPINACH (cooked)
½ c = 0.17 mg (10% DV)
MUSHROOMS (cooked)
½ c = 0.23 mg (14% DV)
BEEF LIVER (braised)
3 oz = 3.5 mg (206% DV)
ENRICHED CEREAL (ready-to-eat)
¾ c = 0.35 mg (21% DV)

[a]The Daily Values are based on a 2,000-calorie diet.

Folate (micrograms) in enriched foods:

- 1 serving highly fortified cereal: 400 μg (680 μg DFE).
- 1 serving most other cereals: 100 μg (170 μg DFE).
- ½ c cooked pasta: 50 μg (85 μg DFE).
- ½ c cooked rice: 40 μg (68 μg DFE).
- ½ large bagel: 25 μg (43 μg DFE).
- 1 slice bread (any kind): 15–25 μg (26–43 μg DFE).

The B vitamins thiamin, niacin, riboflavin, and folate (as folic acid) are among the enrichment nutrients added to grain foods such as breads and cereals sold in the United States. Chapter 4 presented more details on enrichment of grain foods.

Dietary Folate Equivalent (DFE) a unit of measure expressing the amount of folate available to the body from naturally occurring sources. The measure mathematically equalizes the difference in absorption between less absorbable food folate and highly absorbable synthetic folate added to enriched foods and found in supplements.

problems in the spine to mental retardation, severely diminished brain size, and death shortly after birth.

Neural tube defects arise in the first days or weeks of pregnancy, long before most women even suspect that they are pregnant. Unfortunately, most women eat too few fruits and vegetables to supply even half the needed folate. The FDA therefore ordered fortification of all enriched grain products with an especially absorbable synthetic form of folate.[43] Because common foods such as bread are now fortified with folate, women's intakes *before* pregnancy, and during the first critical days of pregnancy, are increasing. If women take in sufficient folate from these sources and predictions hold true, the nation should see a significant drop in the incidence of neural tube defects (about half are attributable to causes other than folate deficiency) and possibly in some other birth defects and some miscarriages as well.[44]

Still unanswered are concerns about folate's ability to mask deficiencies of vitamin B_{12} (more about this effect later). The possibility also exists that, once in the blood, excess folate may negate actions of some anticancer drugs that work by blocking the activities of folate in rapidly dividing cancer cells. Other unwanted effects on populations other than women of childbearing age are also possible; and time will tell whether the benefits of folate enrichment outweigh the risks.[45]

The difference in absorption between naturally occurring food folate and the synthetic folate added to enriched foods and found in supplements necessitated a new unit of measurement for folate: the **Dietary Folate Equivalent**, or **DFE.**[46] The DFE converts all forms of dietary folate into units that are equivalent to the folate in foods. Currently, most labels and tables of food composition still express folate values in micrograms; Appendix C offers a conversion factor for calculating DFE.

Folate's name is derived from the word *foliage*, and as that implies, folate is naturally abundant in leafy green vegetables such as spinach and turnip greens (see Snapshot 7-8, page 230). Fresh, uncooked vegetables and fruits are the best natural sources because the heat of cooking and the oxidation that occurs during storage destroy as

SNAPSHOT 7.7 Niacin[a]

These foods provide 10 percent or more of the niacin Daily Value (DV = 20 mg/day).[b]

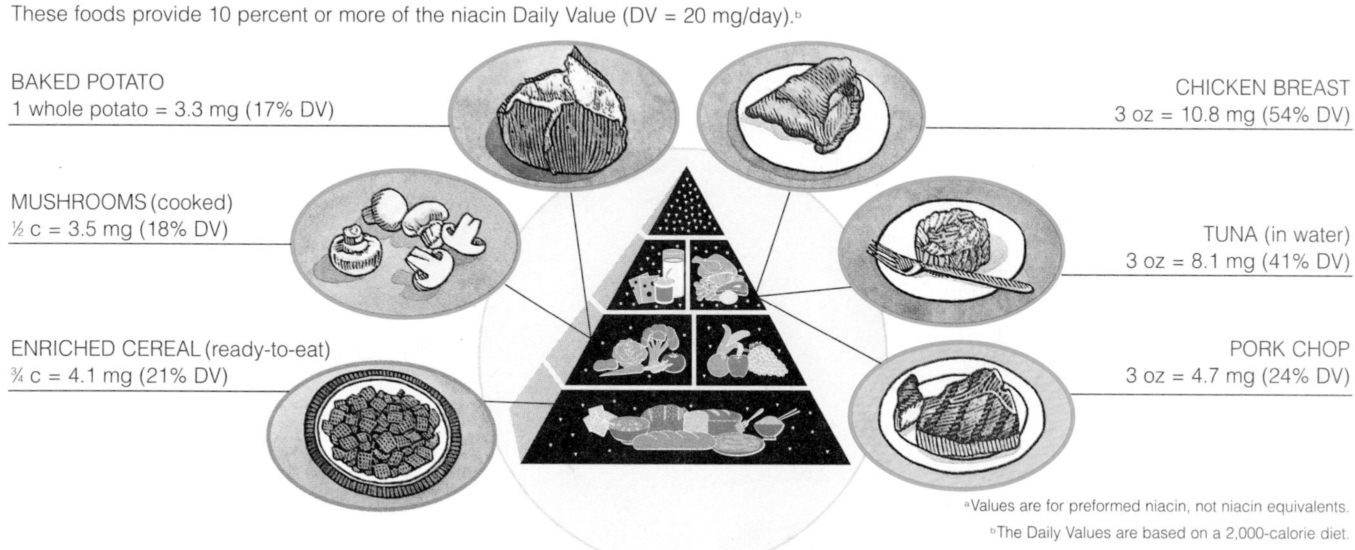
BAKED POTATO
1 whole potato = 3.3 mg (17% DV)
MUSHROOMS (cooked)
½ c = 3.5 mg (18% DV)
ENRICHED CEREAL (ready-to-eat)
¾ c = 4.1 mg (21% DV)
CHICKEN BREAST
3 oz = 10.8 mg (54% DV)
TUNA (in water)
3 oz = 8.1 mg (41% DV)
PORK CHOP
3 oz = 4.7 mg (24% DV)

[a]Values are for preformed niacin, not niacin equivalents.

[b]The Daily Values are based on a 2,000-calorie diet.

much as half the folate in foods. Eggs also contain some folate. Orange juice and legumes also contain folate, but they contain factors that may interfere with folate absorption, so their usefulness as folate contributors may be limited. Milk may enhance the absorption of folate, although it is unclear which constituent in milk may do so.

Of all the vitamins, folate seems to be most likely to interact with medications. Ten major groups of drugs, including antacids and aspirin and its relatives, have been shown to interfere with the body's use of folate. Occasional use of these drugs to relieve headache or upset stomach presents no concern, but frequent users may need to attend to their folate intakes. These include people with chronic pain or ulcers who rely heavily on aspirin or antacids as well as those who smoke or take oral contraceptives or anticonvulsant medications.

Vitamin B_{12} **Vitamin B_{12}** helps folate make red blood cells. By itself vitamin B_{12} also serves the body by helping to maintain the sheaths that surround and protect nerve fibers. It may also influence the cells that build bone tissue.

The absorption of vitamin B_{12} requires an **intrinsic factor,** a compound made by the stomach. The design for this factor is carried in the genes. With the help of the stomach's acid, intrinsic factor attaches to the vitamin; the complex then passes to the small intestine and is absorbed into the bloodstream. A few people have an inherited defect in the gene for intrinsic factor, which makes vitamin B_{12} absorption abnormal, beginning in mid-adulthood; absorption may decline with age for other reasons (see Chapter 13). Without normal absorption of vitamin B_{12} from food, they develop deficiency symptoms. In this case or in the case of stomach injury that limits production of intrinsic factor, vitamin B_{12} must be supplied by injection to bypass the defective absorptive system. The vitamin B_{12} deficiency caused by lack of intrinsic factor is known as **pernicious anemia.**

Without sufficient vitamin B_{12}, nerves become damaged and folate fails to do its blood-building work, so vitamin B_{12} deficiency causes an anemia identical to that caused by folate deficiency. The blood symptoms of a deficiency of either folate or vitamin B_{12} include the presence of large, immature red blood cells (see Figure 7-8). Administering extra folate will often clear up this blood condition, but will allow the

vitamin B_{12} a B vitamin that enables folate to get into cells and also helps maintain the sheath around nerve cells. Vitamin B_{12}'s scientific name, not often used, is *cyanocobalamin.*

intrinsic factor a factor found inside a system. The intrinsic factor necessary to prevent pernicious anemia is now known to be a compound that helps in the absorption of vitamin B_{12}.

pernicious (per-NISH-us) **anemia** a vitamin B_{12}-deficiency disease, caused by lack of intrinsic factor and characterized by large, immature red blood cells and damage to the nervous system (*pernicious* means "highly injurious or destructive").

SNAPSHOT 7.8 Folate[a]

These foods provide 10 percent or more of the folate Daily Value (DV = 400 μg/day).[b]

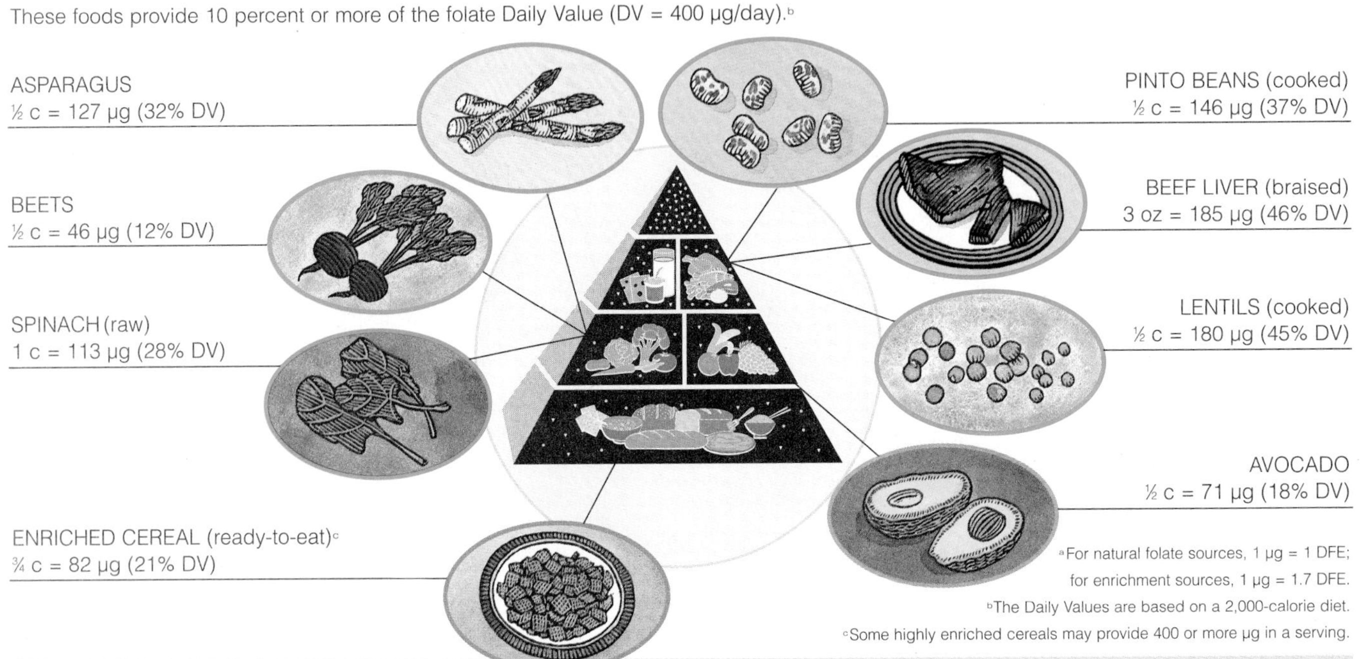
ASPARAGUS
½ c = 127 μg (32% DV)
BEETS
½ c = 46 μg (12% DV)
SPINACH (raw)
1 c = 113 μg (28% DV)
ENRICHED CEREAL (ready-to-eat)[c]
¾ c = 82 μg (21% DV)
PINTO BEANS (cooked)
½ c = 146 μg (37% DV)
BEEF LIVER (braised)
3 oz = 185 μg (46% DV)
LENTILS (cooked)
½ c = 180 μg (45% DV)
AVOCADO
½ c = 71 μg (18% DV)

[a]For natural folate sources, 1 μg = 1 DFE; for enrichment sources, 1 μg = 1.7 DFE.
[b]The Daily Values are based on a 2,000-calorie diet.
[c]Some highly enriched cereals may provide 400 or more μg in a serving.

Figure 7-8

Blood Cells of Pernicious Anemia and Normal Cells

The top photo shows the large, immature red blood cells of pernicious anemia; the bottom photo shows normal cells. The abnormal red blood cells produced in folate deficiency are indistinguishable from those of vitamin B_{12} deficiency because the two vitamins work closely together in red blood cell production.

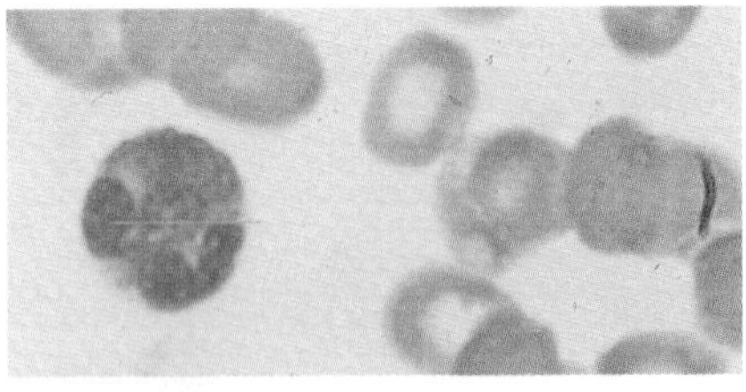

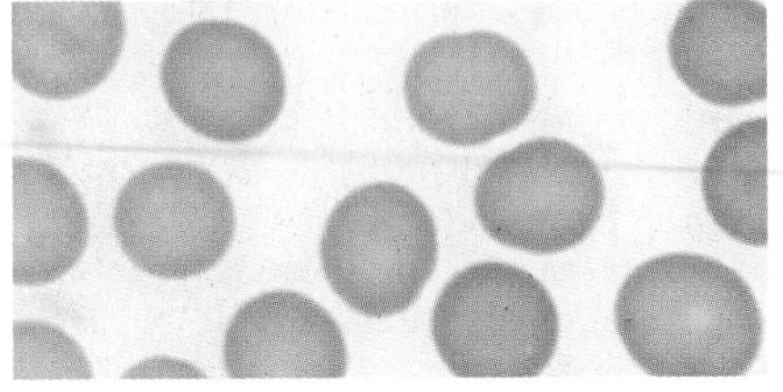

deficiency of vitamin B_{12} to continue undetected.[47] Vitamin B_{12}'s other functions then become compromised, and the results can be devastating: damaged nerve sheaths, creeping paralysis, and general malfunctioning of nerves and muscles.

A physician may notice signs of the B_{12} problem, but it is hard to diagnose correctly. More likely, the damage will proceed unchecked. In an effort to prevent excessive folate intakes that could mask symptoms of a vitamin B_{12} deficiency, the FDA specifies exact amounts of folate that may be added to enriched foods.

As Snapshot 7-9 shows, vitamin B_{12} is present only in foods of animal origin, not in foods from plants. Consequently, the uninformed, strict vegetarian is at special risk. People who give up all foods of animal origin may not show signs of deficiency right away because up to five years' worth of vitamin B_{12} can be stored in the body; but eventually signs will develop. A pregnant or lactating woman who is eating such a diet should be aware that her infant can develop a vitamin B_{12} deficiency, even if the mother appears healthy. A deficiency of this vitamin can cause irreversible nervous system damage in the fetus. The birth of an infant with nerve problems can be the mother's first clue to the deficiency. All strict vegetarians, and especially pregnant women, must be sure to use vitamin B_{12}–fortified products, such as vitamin B_{12}–fortified soy "milk,"or to take the appropriate supplements.

The way folate masks the anemia of vitamin B_{12} deficiency underlines a point already made several times. It takes a skilled professional to correctly diagnose a nutrient deficiency or imbalance, and you clearly take a serious risk when you diagnose yourself or listen to self-proclaimed experts. A second point should also be underlined here. Since vitamin B_{12} deficiency in the body may be caused by either a lack of the vitamin in the diet or a lack of the intrinsic factor necessary to absorb the vitamin, a change in diet alone may not correct the deficiency, another reason for seeking professional diagnosis of physical symptoms.

SNAPSHOT 7.9 Vitamin B_{12}

These foods provide 10 percent or more of the vitamin B_{12} Daily Value (DV = 6 μg/day).[a]

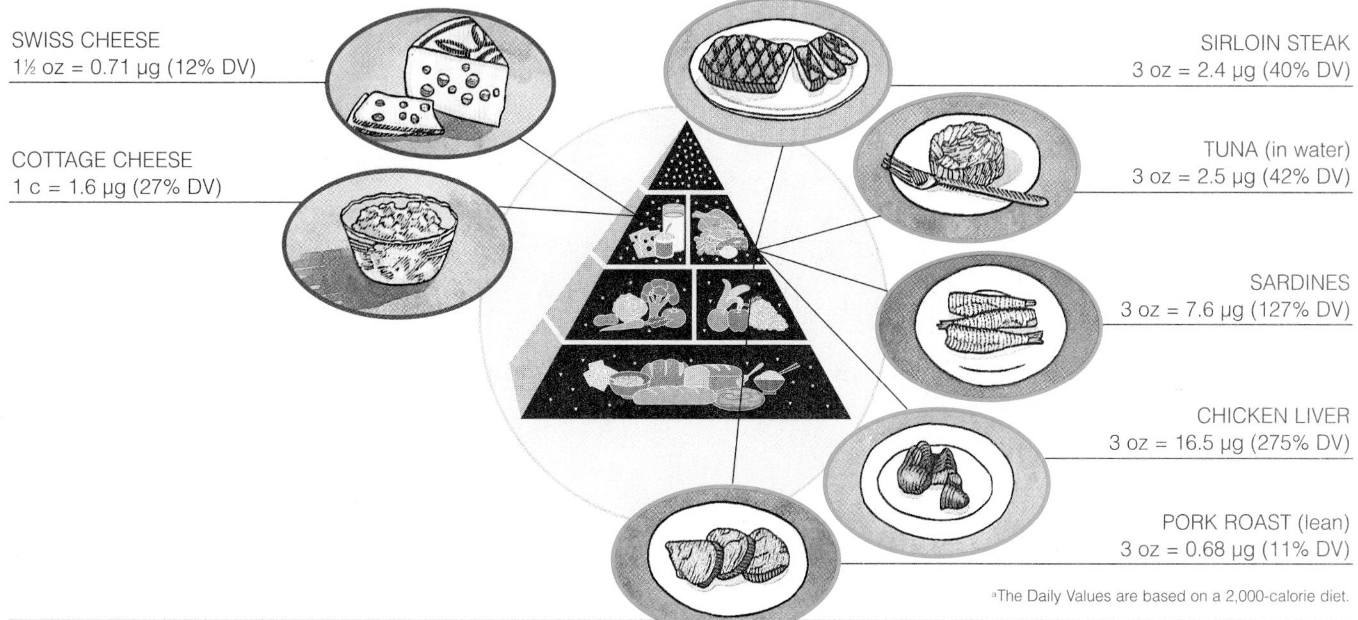
SWISS CHEESE
1½ oz = 0.71 μg (12% DV)
COTTAGE CHEESE
1 c = 1.6 μg (27% DV)
SIRLOIN STEAK
3 oz = 2.4 μg (40% DV)
TUNA (in water)
3 oz = 2.5 μg (42% DV)
SARDINES
3 oz = 7.6 μg (127% DV)
CHICKEN LIVER
3 oz = 16.5 μg (275% DV)
PORK ROAST (lean)
3 oz = 0.68 μg (11% DV)

[a]The Daily Values are based on a 2,000-calorie diet.

Vitamin B_6 In the cells, **vitamin B_6** helps to convert one kind of amino acid, which cells have in abundance, to others that the cells lack. It also aids in the conversion of tryptophan to niacin and plays important roles in the synthesis of hemoglobin. Vitamin B_6 also assists in releasing stored glucose from glycogen and thus contributes to the regulation of blood glucose. During the last decade or so, vitamin B_6 research has revealed roles for the vitamin in immune function and steroid hormone activity.[48] In addition, vitamin B_6 is critical to the developing brain and nervous system of a fetus. Without enough of the vitamin during this stage, the child often suffers behaviorally later on.[49]

vitamin B_6 a B vitamin needed in protein metabolism. Its three active forms are *pyridoxine, pyridoxal,* and *pyridoxamine.*

Because of these diverse functions, vitamin B_6 deficiency is expressed in general symptoms, such as weakness, irritability, and insomnia. Other symptoms include the greasy dermatitis depicted in Figure 7-9, anemia, and, in advanced cases of deficiency, convulsions. A shortage of vitamin B_6 also weakens the immune response.[50] Low vitamin B_6 intakes may also be related to increased incidence of heart disease.[51]

Figure 7-9

Vitamin B_6 Deficiency

In this dermatitis, the skin is greasy and flaky, unlike the skin affected by the dermatitis of pellagra.

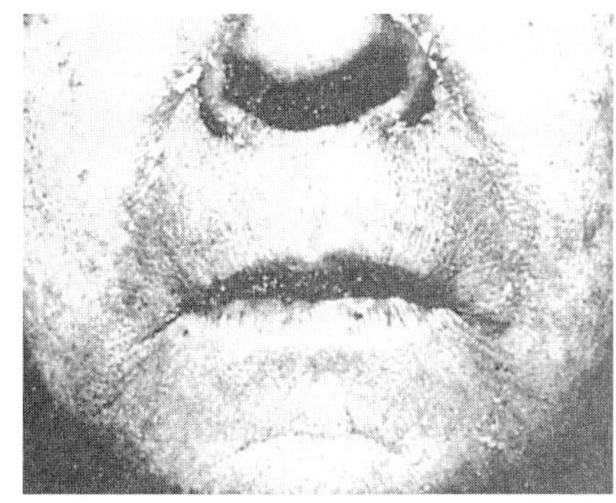

Large doses of vitamin B_6 can be dangerous. Years ago it was generally believed that, like most of the other water-soluble vitamins, vitamin B_6 could not reach toxic concentrations in the body. This belief changed when a report told of women who took more than 2 grams of vitamin B_6 daily (the recommendation for women is less than 2 *milligrams*) for two months or more in an attempt to cure the symptoms of premenstrual syndrome (PMS). The women developed numb feet, then lost sensation in their hands, and eventually became unable to work. Later, in some cases, their mouths became numb. Since the first report of vitamin B_6 toxicity, researchers have seen toxicity symptoms in more than 100 women who took vitamin B_6 supplements for more than five years. The women recovered after they stopped taking the supplements. The potential toxicity of vitamin B_6 is yet another reason why people should not self-diagnose and self-prescribe high doses of vitamins for their illnesses. Table 7-3 on p. 239 lists some potentially safe intakes for vitamins, and Table 7-6 on p. 246

SNAPSHOT 7.10 Vitamin B_6

These foods provide 10 percent or more of the vitamin B_6 Daily Value (DV = 2 mg/day).[a]

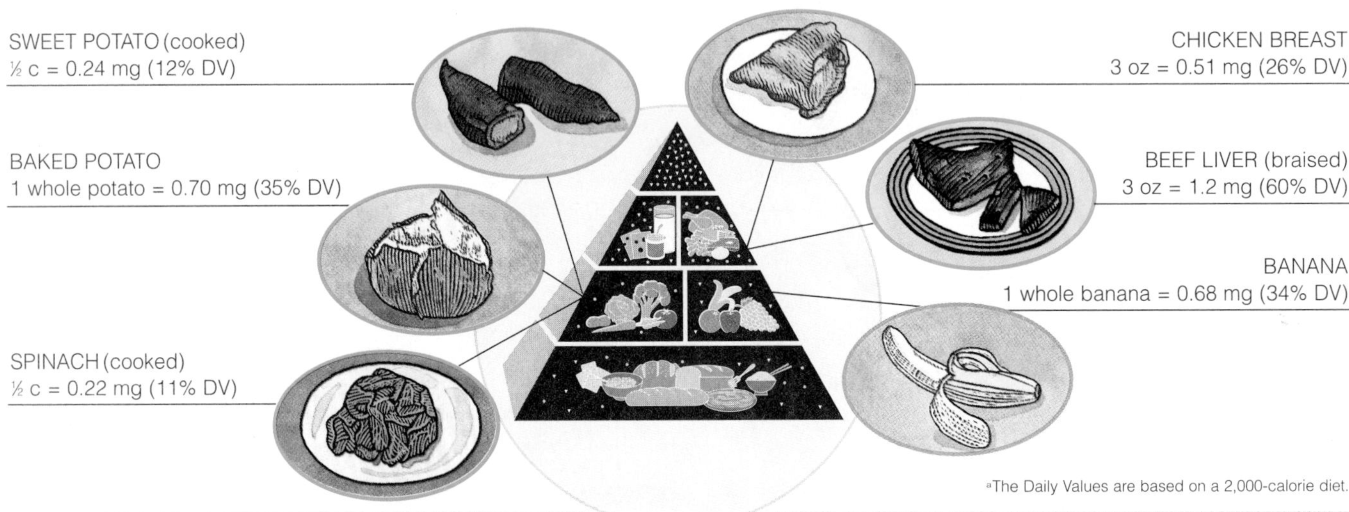

SWEET POTATO (cooked)
½ c = 0.24 mg (12% DV)
BAKED POTATO
1 whole potato = 0.70 mg (35% DV)
SPINACH (cooked)
½ c = 0.22 mg (11% DV)
CHICKEN BREAST
3 oz = 0.51 mg (26% DV)
BEEF LIVER (braised)
3 oz = 1.2 mg (60% DV)
BANANA
1 whole banana = 0.68 mg (34% DV)

[a]The Daily Values are based on a 2,000-calorie diet.

The links between PMS and vitamin B_6 are explored in Chapter 13.

lists common deficiency and toxicity symptoms of vitamin B_6 and other water-soluble vitamins.

Vitamin B_6 plays so many roles in protein metabolism that the body's requirement for vitamin B_6 is roughly proportional to protein intakes.[52] The DRI committee set the vitamin B_6 intake recommendation high enough to cover most people's needs, regardless of differences in protein intakes (see the inside front cover, page A).[53] Meats, fish, and poultry (protein-rich foods), potatoes, leafy green vegetables, and some fruits are good sources of vitamin B_6 (see Snapshot 7-10).

How Are B Vitamins Related to Heart Disease? A theory proposing that a deficiency of the B vitamin folate is a risk factor for cardiovascular disease is gaining strength.[54] Along with deficiencies of vitamin B_6 and possibly other B vitamins, a lack of folate is often associated with the buildup of the amino acid homocysteine in the blood. Elevated homocysteine, in turn, correlates with a high incidence of heart and artery disease.[55] People with high intakes of folate are reported to have significantly less cardiovascular disease than those with low intakes.[56] When researchers gave supplements of B vitamins (folate, vitamin B_6, and vitamin B_{12}) to a group of healthy men, their homocysteine levels dropped significantly.[57] Researchers hope that fortified food may provide enough folate to achieve the same effect. Indeed, a recent analysis of people's folate and homocysteine status before and after folate fortification shows measurable improvements since folate fortification was implemented.[58] Whether reduced rates of cardiovascular disease will follow, however, remains to be seen. Some researchers feel ready to advise everyone to take a standard multivitamin tablet or eat a serving of highly fortified cereal each day to provide an additional 400 micrograms of folate.[†59] This amount of folate does not have an unblemished safety record, however. A handful of studies report interference with zinc metabolism.[60]

†Read cereal labels to identify those supplying 400 micrograms of folate.

Furthermore, not every study has consistently found a relationship between high folate intakes and lower rates of heart disease, and none has produced evidence that giving B vitamins to a population reduces their incidence of heart disease.[61] Thus, a cautious approach is in order. Also, some groups of people seem to clear homocysteine from the blood more readily than others regardless of their vitamin status, leading researchers to postulate that genetics may also play a role.[62] Supplementation in people with such inborn protection would probably not reduce their risks from blood homocysteine, but would still expose them to whatever risks are posed by supplements.

Biotin and Pantothenic Acid Two other B vitamins, **biotin** and **pantothenic acid,** are, like thiamin, riboflavin, and niacin, important in energy metabolism. Biotin is a cofactor for several enzymes active in the metabolism of carbohydrate, fat, and protein. Pantothenic acid was first recognized as a substance that stimulates growth. Pantothenic acid is a component of a key coenzyme that makes possible the release of energy from the energy nutrients. It also participates in more than 100 different steps in the synthesis of lipids, neurotransmitters, steroid hormones, and hemoglobin.

Although rare diseases may precipitate deficiencies of biotin and pantothenic acid, both vitamins are widespread in foods. A steady diet of raw egg whites, which contain a protein that binds biotin, can produce biotin deficiency, but a person would have to consume more than two dozen egg whites daily to produce the effect. Cooking eggs denatures the protein. Healthy people eating ordinary diets are not at risk for deficiencies.

KEY POINT *Historically, famous B vitamin–deficiency diseases are beriberi (thiamin), pellagra (niacin), and pernicious anemia (vitamin B_{12}). Pellagra can be prevented by adequate protein because the amino acid tryptophan can be converted to niacin in the body. A high intake of folate can mask the blood symptom of vitamin B_{12} deficiency but will not prevent the associated nerve damage. Vitamin B_6 is important in amino acid metabolism and can be toxic in excess. Biotin and pantothenic acid are important to the body and are abundant in food.*

Non-B Vitamins In addition to the B vitamins just discussed, a few compounds that are topics of debate among researchers deserve mention. **Choline** could be considered an essential nutrient because when the diet is devoid of choline, the body cannot make enough of the compound to meet its need. Luckily, choline is common in foods, and deficiencies are practically unheard of outside the laboratory.

The compounds **carnitine, inositol,** and **lipoic acid** might appropriately be called *nonvitamins* because they are not essential nutrients for human beings. Carnitine, sometimes called "vitamin B_T," is an important piece of cell machinery, but it is not a vitamin. Although deficiencies can be induced in laboratory animals for experimental purposes, these substances are abundant in ordinary foods. Even if these compounds were essential in human nutrition, supplements would be unnecessary for healthy people eating a balanced diet. Vitamin companies often include these substances to make their formulas appear more "complete," but there is no physiological reason to do so.

In addition to carnitine, inositol, and lipoic acid, other substances have been mistakenly thought essential in human nutrition because they are needed for growth by bacteria or other life-forms. These substances include PABA (para-aminobenzoic acid), bioflavonoids ("vitamin P" or hesperidin), and ubiquinone (coenzyme Q). Other names you may hear are "vitamin B_{15}," or pangamic acid (a hoax); "vitamin B_{17}" (laetrile or amygdalin, not a cancer cure and not a vitamin by any stretch of the imagination); and more.

KEY POINT *Choline is needed in the diet, but it is not a vitamin and deficiencies are unheard of outside the laboratory. Many other substances that people claim are B vitamins are not. Among these substances are carnitine, inositol, and lipoic acid.*

biotin (BY-o-tin) a B vitamin; a coenzyme necessary for fat synthesis and other metabolic reactions.

pantothenic (PAN-to-THEN-ic) **acid** a B vitamin.

choline (KOH-leen) a nonessential nutrient used to make the phospholipid lecithin and other molecules.

carnitine a nonessential nutrient that functions in cellular activities.

inositol (in-OSS-ih-tall) a nonessential nutrient found in cell membranes.

lipoic (lip-OH-ic) **acid** a nonessential nutrient.

The DRI recommended intakes for biotin and pantothenic acid are listed on the inside front cover, page A.

Links between choline and brain function are discussed in Chapter 13.

scurvy the vitamin C-deficiency disease.

ascorbic acid one of the active forms of vitamin C (the other is *dehydroascorbic* acid); an antioxidant nutrient.

prooxidant a compound that triggers reactions involving oxygen.

collagen (COLL-a-jen) the chief protein of most connective tissues, including scars, ligaments, and tendons, and the underlying matrix on which bones and teeth are built.

Vitamin C

Two hundred odd years ago, any person who joined the crew of a seagoing ship knew they had only half a chance of returning alive—not because they might be slain by pirates or die in a storm but because they might contract **scurvy**, a dreaded disease that might kill as many as two-thirds of a ship's crew on a long voyage. Only ships that sailed on short voyages, especially around the Mediterranean Sea, were safe from this disease. It was not known at the time that the special hazard of long ocean voyages was that the ship's cook used up the fresh fruits and vegetables early and relied for the duration of the voyage on cereals and live animals.

The first nutrition experiment to be conducted on human beings was devised nearly 250 years ago to find a cure for scurvy. A physician of the time divided some British sailors with scurvy into groups. Each group received a different test substance: vinegar, sulfuric acid, seawater, oranges, or lemons. Those receiving the citrus fruits were cured within a short time. Sadly, it was 50 years before the British navy made use of the information and required all its vessels to provide lime juice to every sailor daily. The term *limey* was applied to the British sailors in mockery because of this requirement. The name later given to the vitamin, **ascorbic acid,** literally means "no-scurvy acid."

The Work of Vitamin C Since vitamin C is also a water-soluble vitamin, you might expect its mode of action to resemble that of the B vitamins. To some extent, in some situations, vitamin C does help a specific enzyme perform its job, just as the B vitamins do. In other cases, vitamin C acts in a more general way, as an antioxidant. Many substances found in foods and important in the body can be destroyed by oxidation. Vitamin C protects them from oxidation by being oxidized itself. In the intestines, vitamin C protects iron from oxidation and so promotes its absorption. In the blood, vitamin C protects sensitive blood constituents from oxidation, and helps to protect vitamin E. Red blood cells are thought to recycle the "used" vitamin C back to the active form to conserve the supply.[63] The antioxidant roles of vitamin C are the focus of extensive study, especially in relation to disease prevention. In test tubes, however, high concentrations of vitamin C have the opposite effect; that is, they act as a **prooxidant**. The questions of whether this effect occurs in the human body and, if so, what it may mean to health remain unanswered.

Vitamin C is required for the production and maintenance of **collagen,** a protein substance that forms the base for all connective tissues in the body: bones, teeth, skin, and tendons. Collagen forms the scar tissue that heals wounds, the reinforcing structure that mends fractures, and the supporting material of capillaries that prevents bruises.

Vitamin C also enhances the immune response and so protects against infection. The vitamin is also important to the production of thyroxine, the hormone that regulates basal metabolic rate and body temperature. The relationship between vitamin C and the common cold is the topic of this chapter's Consumer Corner.

Long voyages without fresh fruits and vegetables spelled death by scurvy for the crew.

The Need for Vitamin C The adult 1989 RDA (inside front cover, page B) for vitamin C of 60 milligrams is midway between two extremes. At one extreme is the requirement, 10 milligrams per day, which is enough to prevent the symptoms of scurvy from appearing. At the other extreme is the amount at which the body's pool of vitamin C is full to overflowing: about 100 milligrams per day. The recommendation for smokers is set at the high end, 100 milligrams, because this amount is needed to maintain blood levels comparable to those of nonsmokers. Other authorities have set different standards. For example, Canada recommends 30 milligrams per day and Japan 100.

Cigarette smoking, among its many harmful effects, interferes with the use of vitamin C. Smokers, and even "passive smokers" who live and work with smokers, therefore need more vitamin C than others.[64] Consumption of extra vitamin C can normalize blood levels, but it cannot protect against the damage caused by exposure to tobacco smoke.

Most of the symptoms of scurvy can be attributed to the breakdown of collagen in the absence of vitamin C: loss of appetite, growth cessation, tenderness to touch, weakness, bleeding gums (shown in Figure 7-10), loose teeth, swollen ankles and wrists, and tiny red spots in the skin where blood has leaked out of capillaries. One symptom, anemia, reflects an important role already mentioned—that vitamin C helps the body to absorb and use iron.

In the United States, scurvy is seldom seen today except in infants who are fed only cow's milk, the elderly, and people addicted to alcohol or other drugs. Breast milk and infant formula supply enough vitamin C, but infants who are fed cow's milk and receive no vitamin C in formula, fruit juice, or other outside sources are at risk. Low intakes of fruits and vegetables and a poor appetite for food in general lead to low vitamin C intakes and are not uncommon among people 65 years of age and older.[65]

Is It Possible to Get Too Much Vitamin C? The easy availability of vitamin C in pill form and the publication of books recommending vitamin C as a "nutraceutical" treatment to prevent and cure colds and cancer have led thousands of people to take large doses of vitamin C. These "volunteer" subjects enabled researchers to study potential toxic effects of large vitamin C doses. Effects that are theoretically possible (but have not been seen with intakes as high as 3 grams a day) include formation of kidney stones, alteration of the acid-base balance, and interference with the action of vitamin E. One effect that has been observed with a 2-gram dose is alteration of the insulin response to carbohydrate in people with otherwise normal glucose tolerances.[66]

Other adverse effects may include nausea, abdominal cramps, excessive gas, and diarrhea. Several instances of interference with medical regimens are known. Large amounts of vitamin C excreted in the urine can obscure the results of tests used to detect diabetes, giving a false positive result in some instances and a false negative result in others. Vitamin C in amounts over 250 milligrams can also produce false negative results on tests for blood in the digestive system, masking the presence of potentially dangerous medical conditions.[67] People taking medications to prevent blood clotting may unwittingly undo the effect if they also take massive doses of vitamin C.[68] Vitamin C supplements in any dosage may be dangerous for people with an overload of iron in the blood because vitamin C increases iron absorption from the intestine and releases iron from storage.[69]

The published research on large doses of vitamin C reveals few instances in which consuming more than 100 to 300 milligrams a day is beneficial. Adults may not be taking major risks if they dose themselves with a gram a day, but doses approaching 10 grams can be expected to be unsafe. In short, the range of safe vitamin C intakes seems to be broad. Between the absolute minimum of 10 milligrams a day and the conservative maximum of 1,000 milligrams (1 gram), nearly everyone should be able to find a suitable intake.[70] People who venture outside these limits do so at their own risk. Vitamin C from food sources, such as those shown in Snapshot 7-11 on page 237, is always safe.

When people dose themselves with supplements, they leave the realm of nutrition and enter that of pharmacology. Like drugs, large doses of nutrients can have medicinal effects on the body and can present serious side effects as well. The next section discusses supplements as sources of nutrients. The Controversy takes up topics of interest to those who would seek medicinal effects from nutrients.

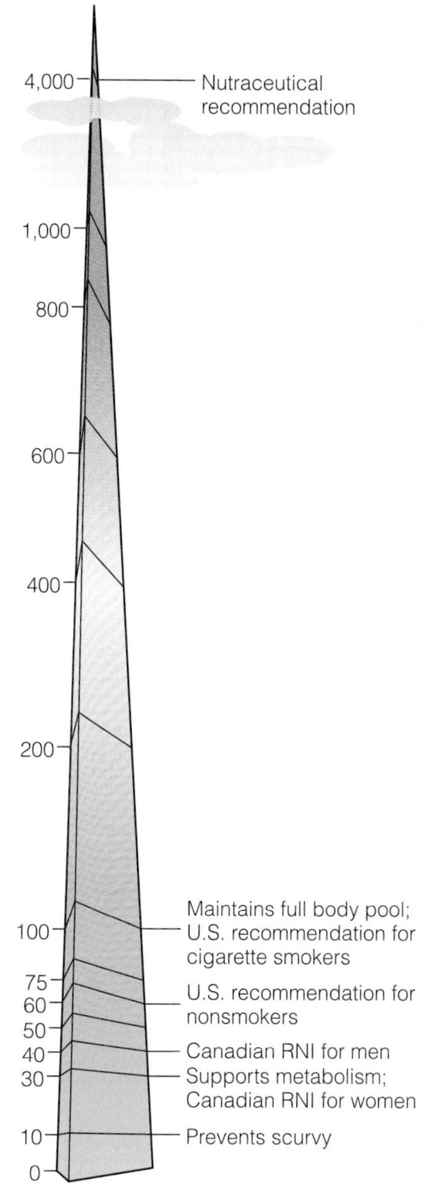

4,000
Nutraceutical recommendation
1,000
800
600
400
200
100
Maintains full body pool; U.S. recommendation for cigarette smokers
75
60
U.S. recommendation for nonsmokers
50
40
Canadian RNI for men
30
Supports metabolism; Canadian RNI for women
10
Prevents scurvy
0

Figure 7-10

Scurvy

Vitamin C deficiency causes the breakdown of collagen, which supports the teeth.

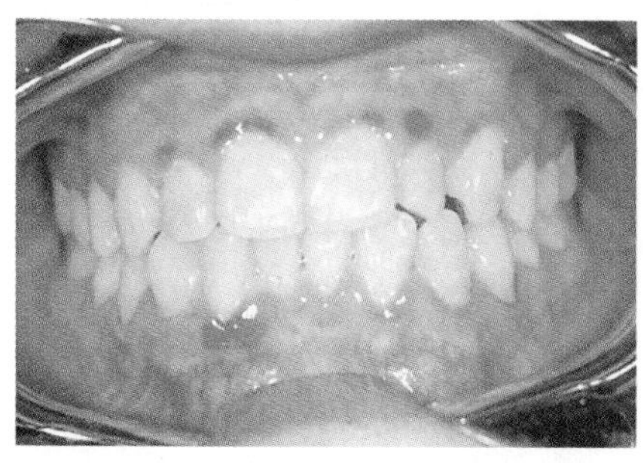

KEY POINT *Vitamin C, an antioxidant, helps to maintain the connective tissue protein collagen, protects against infection, and helps in iron absorption. The theory that vitamin C prevents or cures colds or cancer is not well supported by research. High vitamin C doses may be hazardous. Ample vitamin C can be obtained from foods.*

Vitamin Supplements

Who should take vitamin supplements? Almost 40 percent of the population do so regularly, collectively spending billions of dollars a year on them. But who really needs

CONSUMER CORNER

Vitamin C and the Common Cold

NO STUDY TO date has conclusively proved that vitamin C can prevent colds or reduce their severity. For years, people have claimed that taking supplements of vitamin C helps to cure their colds, but researchers have found little support for such claims in the literature.[1] Why, then, do so many people continue buying and taking the supplements to relieve colds? Does any research support such use?

Classic literature on this topic is mixed. More than 20 years ago, claims by Linus Pauling, a famous Nobel Prize winner and vocal supporter of vitamin C supplements, were all but discounted by the scientific community because of negative conclusions reported in a major review of the literature.[2] In particular, research suggested that regular ingestion of vitamin C supplements is not effective in reducing the number of colds people suffer each year.[3]

Lately, though, researchers have been taking a second look. An exception to the general conclusion that vitamin C has no effect came recently from a pair of researchers who revisited the cornerstone review of the literature on vitamin C and colds. By making some changes in the underlying assumptions and methodology of the original review, these researchers concluded that vitamin C in amounts up to 1 gram per day may indeed shorten the duration of a cold by about one day and reduce the severity of its symptoms by about 23%.[4] This effect may be greater in children than in adults; in adults, even more vitamin C (2 or more grams a day) may be required to produce an effect.[5]

It could turn out that vitamin C's antioxidant or other activities boost the body's immunity or somehow improve its defenses. In a test tube, antioxidant nutrients, including vitamin C, stimulate cells of the immune system to move and work more efficiently.[6] In people, vitamin C (2 grams taken daily for two weeks) has also been observed to reduce blood histamine.[7] Anyone who has ever had a cold knows the effects of histamine: sneezing, a runny or stuffy nose, and swollen sinuses. Antihistamines, a group of drugs used in cold medications, provide relief from just those symptoms. In druglike doses, vitamin C may work like a weak antihistamine by deactivating histamine.

Can vitamin C ease the suffering of a person with a cold?

One other effect is hard at work with supplements of all kinds. The placebo effect, first defined in Chapter 1, was demonstrated vividly with regard to vitamin C and colds. Half of the experimental subjects in one study received a placebo but thought they were receiving vitamin C. These subjects had fewer colds than the group who had received vitamin C but thought they were receiving the placebo. At work was the powerful healing effect of faith—the placebo effect.

Much more research is required concerning vitamin C and the common cold before any recommendations are possible. One thing is certain, though—no drug is risk-free, and vitamin C in large doses qualifies as a drug that may have side effects, as described in this chapter.[8]

SNAPSHOT 7.11 Vitamin C

These foods provide 10 percent or more of the vitamin C Daily Value (DV = 60 mg/day).[a]

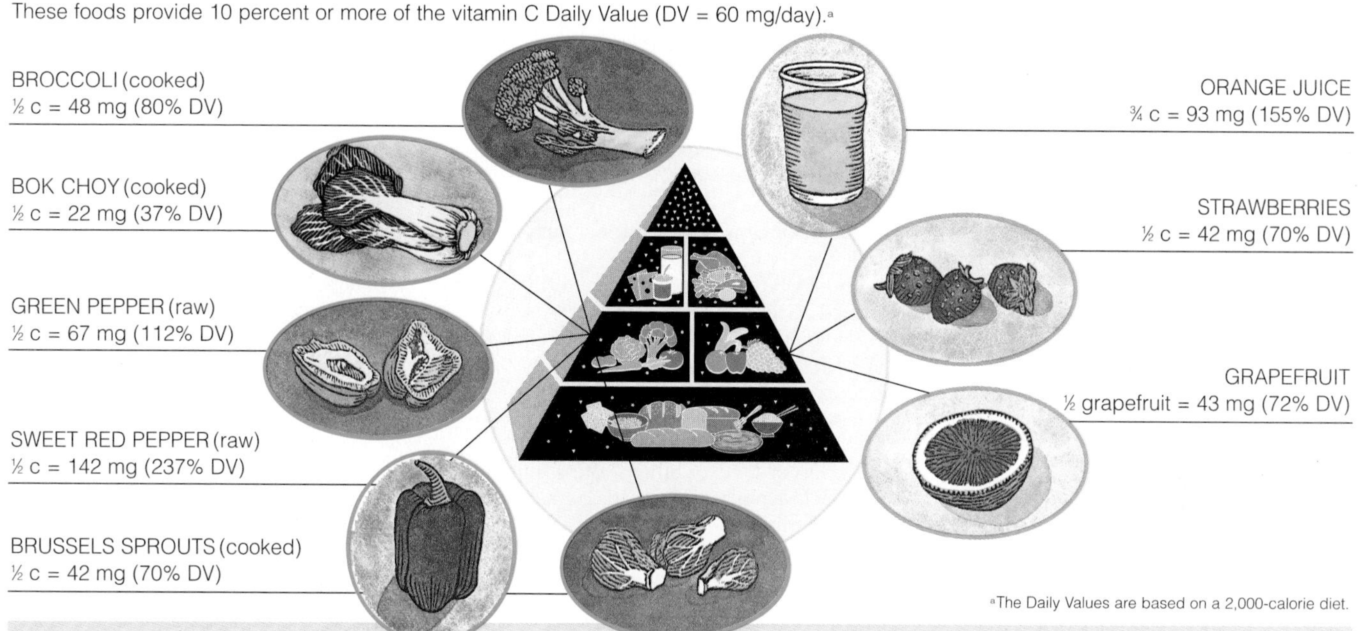
BROCCOLI (cooked)
½ c = 48 mg (80% DV)
BOK CHOY (cooked)
½ c = 22 mg (37% DV)
GREEN PEPPER (raw)
½ c = 67 mg (112% DV)
SWEET RED PEPPER (raw)
½ c = 142 mg (237% DV)
BRUSSELS SPROUTS (cooked)
½ c = 42 mg (70% DV)
ORANGE JUICE
¾ c = 93 mg (155% DV)
STRAWBERRIES
½ c = 42 mg (70% DV)
GRAPEFRUIT
½ grapefruit = 43 mg (72% DV)

[a]The Daily Values are based on a 2,000-calorie diet.

supplements? No one? Everyone? Some people? Which people? And which supplements should they take? People need a few facts about supplements before they open their wallets and buy trouble.

 Do I Need to Take a Vitamin Supplement?

When people think of supplements, they often think of vitamins, but minerals are important, too, of course. People whose diets lack several vitamins, for whatever reason, probably lack several minerals as well. In some cases, supplements may be appropriate.

People who may need supplements fall into several groups. Some people's diets put them at risk of developing **subclinical,** or **marginal, deficiencies:** states of unwellness shy of the classical, full-blown nutrient deficiencies. In contrast to the classical deficiencies, which are easy to recognize, subclinical deficiencies are subtle and easy to overlook.[71] The first remedy should be to attempt to improve the diet so that it supplies the needed nutrients. Nutrient supplements may also be appropriate in special cases.

Consider a woman who loses a lot of blood and therefore a lot of iron and other blood-building nutrients in menstruation each month. This woman may be able to eat in such a way as to make up all nutrient losses except that of iron, and for iron, she may need a supplement prescribed by a health-care provider. Table 7-2 points out valid and invalid reasons for taking supplements.

KEY POINT *People who routinely fail to obtain the recommended amounts of vitamins and minerals from the diet and people with special needs, such as those who are pregnant or elderly, may be at risk for deficiencies and may benefit from a multivitamin-mineral supplement.*

subclinical, or **marginal, deficiency** a nutrient deficiency that has no detectable (clinical) symptoms. The term is often used to scare consumers into buying unneeded nutrient supplements.

Table 7-2

Some Valid and Invalid Reasons for Taking Supplements

These People May Need Supplements:	These People Don't:
■ People with nutrient deficiencies. ■ Women in their childbearing years (they may need more folate than foods alone can supply to reduce risk of neural tube defects in infants). ■ Pregnant or lactating women (they may need iron and folate). ■ Newborns (they are routinely given a vitamin K dose). ■ Infants (they may need various supplements, see Chapter 12). ■ Those who are lactose intolerant (they need calcium to forestall osteoporosis). ■ Habitual dieters (they may eat insufficient food). ■ Elderly people (they may choose poorly, have trouble chewing, or absorb or metabolize less efficiently; see Chapter 13). ■ Victims of AIDS or other wasting illnesses (they lose nutrients faster than foods can supply them). ■ Those addicted to drugs or alcohol (they absorb fewer and excrete more nutrients; nutrients cannot undo damage from drugs or alcohol). ■ Those recovering from surgery, burns, injury, or illness (they need extra nutrients to help regenerate tissues). ■ Strict vegetarians (they may need vitamin B_{12}, vitamin D, iron, and zinc). ■ People taking medications that interfere with the body's use of nutrients.	■ Those who feel insecure about the amounts of nutrients in the food supply. ■ Those who feel tired and falsely believe that supplements can provide energy. ■ Those who believe that supplements will help them cope with stress. ■ Those who wish to build lean body tissue without physical work or wrongly believe that supplements will build muscles faster than work and diet alone. ■ Those who want to prevent or cure self-diagnosed conditions, from the common cold to cancer. ■ Those with kidney or liver diseases because these disorders can increase vitamin toxicity. ■ Those who hope that excess nutrients will produce mysterious, though beneficial, reactions in the body. ■ Those who are taking certain medications because supplements may interfere with the action of the medications. ■ Those who smoke and take beta-carotene supplements because such supplements may possibly increase the risk of lung cancer in smokers.

Which Supplements Are Helpful, and Which Should I Avoid?

When people self-prescribe supplements, they have to choose doses. "I'll take one of these a day," a person may say, or "I'll take two of these and three of those a day." Whatever the choice, the higher the dose, the greater the risk of toxicity. People's tolerances for high doses of nutrients vary. Amounts that some can tolerate may not be safe for others, and no one knows who falls into which category. Toxic overdoses of vitamins and minerals may be more common than we realize.

Even more worrisome than the possibility of short-term, acute overdoses is the potential for chronic, low-level nutrient toxicity in which the subtle effects develop slowly and go unrecognized. For example, a woman took just 1,000 RE (5,000 IU) of vitamin A a day, an amount typically found in vitamin-mineral supplements—but she took this dose daily for ten years. Then she was diagnosed with liver disease. Only when she discontinued the supplement did the condition clear up.[72] Vitamin A reliably produces liver injury at doses greater than 10,000 RE, and even at 5,000 RE, abnormal levels of liver enzymes are detectable in the blood.[73] In view of the potential hazards that supplements present, some authorities believe they should be required to bear warning labels, but such labels have not been seriously considered.

It is impossible to say precisely how much of a nutrient is too much for every person because people's tolerances vary. Assuming, however, that it is best to err on the conservative side, Table 7-3 presents suggested limits for vitamin and mineral doses in supplements for daily use.

Another potential problem is that supplements may lull their takers into a false sense of security. A person may eat irresponsibly, thinking, "My supplement will cover my needs." More often, supplements supply precisely the nutrients people need least—those they consume in food—while failing to provide those missing from the diet.

Another problem is **bioavailability.** In general, nutrients are absorbed best from foods in which they are dispersed among other ingredients that facilitate their absorption. In contrast, nutrients taken in pure, concentrated form are likely to interfere with the absorption of other nutrients. Minerals provide examples: zinc hinders copper and calcium absorption, iron hinders zinc absorption, calcium hinders magnesium and iron absorption, and so on. Interactions between vitamins also exist.

bioavailability absorbability; the individual differences in the proportion of a nutrient that is available for absorption from various sources.

Table 7-3

Vitamin and Mineral Supplement Doses (Adults)

Nutrient	Probable Safe Intake/Day	Average Multivitamin-Mineral Supplement	Single-Nutrient Supplement
Vitamins			
Vitamin A	3,000 µg RE (10,000 IU)	5,000 IU	8,000 to 10,000 IU
Vitamin D	50 µg (2,000 IU)	400 IU	400 IU
Vitamin E	200 to 800 mg α-TE (130 to 530 IU)	30 IU	100 to 1,000 IU
Thiamin	—[a]	1.5 mg	50 mg
Riboflavin	—[a]	1.7 mg	25 mg
Niacin (as niacinamide)	35 mg	20 mg	100 to 500 mg
Vitamin B_6	100 mg	2 mg	100 to 200 mg
Folate	1,000 µg	400 µg	400 µg
Vitamin B_{12}	—[a]	6 µg	100 to 1,000 µg
Pantothenic acid	—[a]	10 mg	100 to 500 mg
Biotin	—[a]	30 µg	300 to 600 µg
Vitamin C	1,000 mg	50 mg	500 to 2,000 mg
Choline	3,500 mg	10 mg	250 mg
Minerals			
Calcium	2,500 mg	160 mg	250 to 600 mg
Phosphorus	4,000 mg	110 mg	—[c]
Magnesium	350 mg	100 mg	250 mg
Iron	10 to 40 mg	18 mg	18 to 30 mg
Zinc	10 to 25 mg	15 mg	10 to 100 mg
Iodine	—[b]	150 µg	—[c]
Selenium	200 µg	10 µg	50 to 200 µg
Fluoride	10 mg	—	—[c]

[a] Thiamin, riboflavin, vitamin B_{12}, pantothenic acid, and biotin have been evaluated by the DRI committee for Upper Tolerable Intake Levels, but none were established for these nutrients because of insufficient data. No adverse effects have been reported with intakes of these nutrients at levels typical of supplements, but caution is still advised, because the potential for harm accompanies excessive intakes.

[b] Not recommended in supplemental form.

[c] Available as a single supplement by prescription.

SOURCES: Values for vitamin C are from M. Levine and coauthors, Criteria and recommendations for vitamin C intake, *Journal of the American Medical Association* 281 (1999): 1415–1428; Values for niacin, vitamin B_6, folate, and choline reflect Tolerable Upper Intake Levels as established in Committee on Dietary Reference Intakes, *Dietary Reference Intakes for Thiamin, Riboflavin, Niacin, Vitamin B_6, Folate, Vitamin B_{12}, Pantothenic Acid, Biotin, and Choline* (Washington, D.C.: National Academy Press, 1998); values for vitamin D, calcium, phosphorus, magnesium, and fluoride reflect Tolerable Upper Intake Levels as established in Committee on Dietary Reference Intakes, *Dietary Reference Intakes for Calcium, Phosphorus, Magnesium, Vitamin D, and Fluoride* (Washington, D.C.: National Academy Press, 1997); values for vitamin A, vitamin C, and selenium have been adapted from J. N. Hathcock, Vitamins and minerals: Efficacy and safety, *American Journal of Clinical Nutrition* 66 (1997): 427–437; values for vitamin E were adapted from R. J. Sokol, Vitamin E, in *Present Knowledge in Nutrition*, eds. E. E. Ziegler and L. J. Filer (Washington, D.C.: International Life Sciences Institute Press, 1996), pp. 130–136.

dietary supplement as defined by DSHEA a product, other than tobacco, that is added to the diet and contains one of the following ingredients: a vitamin, mineral, herb, botanical (plant extract), amino acid, metabolite, constituent, or extract, or a combination of any of these ingredients.

Another concern is not about known nutrients, but about the landslide of other substances legally sold in the United States as "dietary supplements" (Table 7-4 provides a sampling). A quirk of U.S. labeling laws allows almost any substance, including herbs, amino acids, dried organ tissues of animals, microorganisms, or even concentrated hormones, to be sold freely over-the-counter, even though little is known about the product's effects or safety.

KEY POINT *Before you decide to take supplements, make sure to recognize the potential for toxicity as well as the associated limitations.*

Supplements Must Be Safe, or the Government Would Not Allow Their Sales, Right?

The Dietary Supplement Health and Education Act (DSHEA) of 1994 was intended to enable consumers to make informed choices about nutrient supplements without much regulation from the government. The act subjects supplements to some of the same general labeling requirements that apply to foods: they must provide nutrient information (see Figure 7-11), they may not make unapproved health claims, and they may not claim to diagnose or treat specific illnesses. Labels may, however, describe the roles of a nutrient or other substance in the body, state how the com-

Table 7-4

A Sample of Substances Sold as Dietary Supplements

According to legal definitions, all of these substances qualify as dietary supplements, even though some appear to have the effects of drugs, not nutrients. Table 11-9 of Chapter 11 defines many more medicinal herbs sold as dietary supplements.

- **DHEA**[a] a hormone secretion of the adrenal gland whose level falls with advancing age. DHEA may protect antioxidant nutrients. Theories that DHEA might stimulate hormone-responsive cancers such as breast or prostate are unproved. Real DHEA is available only by prescription; herbal DHEA imitator for sale in health food stores is not active in the body. No safety information exists.
- **desiccated liver** a powder sold in health-food stores and supposed to contain in concentrated form all the nutrients found in liver (*desiccated* means "totally dried").
- **ephedrine** One of a group of compounds with dangerous amphetamine-like stimulant effects; commonly added to herbal preparations such as Ma huang, to weight-loss products, and to products claimed to imitate the effects of illegal drugs of abuse. The most severe reported side effects of ephedrine include sudden death, heart attack, and stroke; other reported effects include chest pain, dizziness, fatigue, headache, insomnia, nausea, psychosis, seizure, tremor, and vomiting. The World Health Organization has called for worldwide controls on ephedrine-containing products.
- **garlic oil** an extract of garlic; may or may not contain the chemicals associated with garlic; claims for health benefits unproved.
- **green pills, fruit pills** pills containing dehydrated, crushed vegetable or fruit matter. An advertisement may claim that each pill equals a *pound* of fresh produce, but in reality a pill may equal one small forkful—minus nutrient losses incurred in processing.
- **kelp tablets** tablets made from dehydrated kelp, a kind of seaweed used by the Japanese as a foodstuff.
- **ma huang** an evergreen plant derivative that supposedly boosts energy and helps with weight control; Ma huang contains ephedrine (see above), especially dangerous in combination with kola nut or other caffeine-containing substances.
- **melatonin** a hormone of the pineal gland believed to help regulate the body's daily rhythms, to reverse the effects of jet lag, and to promote sleep. Claims for life extension or enhancement of sexual prowess are without merit.
- **nutritional yeast** a preparation of yeast cells, often praised for its high nutrient content. Yeast is a source of B vitamins, as are many other foods. Also called *brewer's yeast;* not the yeast used in baking.
- 1,000 others

[a]Dehydroepiandrosterone

pound performs its function, and indicate that consuming the compound is associated with general well-being. The DSHEA also defined the term **dietary supplement**, but the definition is so broad as to be almost meaningless.[75]

In effect, the DSHEA resulted in the deregulation of the supplement industry.[76] Unlike food and drugs, supplements now need no government approval before entering the market. Most people still believe that the FDA is watching out for their safety concerning supplements, but this is no longer the case. Manufacturers alone decide whether their products are safe and effective. The FDA has the burden of proving that a supplement ingredient is unsafe and should be removed from the market, but there are so many ingredients that the FDA lacks the resources to test them all and remove any that are unsafe from the market before they cause harm. Chapter 9 tells the story of **ephedrine** (see Table 7-4), a dangerous ingredient still available in weight-loss products; Chapter 11 describes other herbs and dietary supplements.[77]

In view of all the negatives and uncertainties associated with supplement taking, several nutrition societies have indicated that most healthy people should not use supplements. These experts say that if a person's diet is inadequate, the remedy is to improve food choices and eating patterns. Failing this, supplements rank a distant second choice.

KEY POINT *Ingredients of dietary supplements are not tested or approved by the FDA before marketing, but their labels must meet certain criteria.*

Selection of a Multinutrient Supplement

Now the question is, do *you* need a supplement? As the preceding sections have indicated, if you choose to take one, you do so at some risk. If you fall into one of the categories already noted, however, and if you cannot meet your nutrient needs from foods, a supplement containing nutrients only may be in order. Avoid "dietary supplements" containing herbs and other unnecessary ingredients.

The next question is, which supplement to choose? Do you need one—"For vitality!" "Infants only!" "Time release!" "Stress formula!"—as the ads claim? It doesn't take much time in front of a supplement counter to see that competition for consumer dollars is fierce and not always highly ethical.

The first step in escaping the clutches of the health hustlers is to use your imagination and simply white out the label picture of sexy people on the beach and the meaningless, glittering generalities like "new and improved." Now all you have left is the list of ingredients, the form they are in, and the price—the plain facts.

You have two basic questions to answer. The first question: What form do you want—chewable, liquid, or pills? If you'd rather drink your vitamins and minerals than chew them, fine. If you choose a fortified liquid or bar-type "energy" meal replacer, you must then proportionately reduce the calories you consume as food, or you may gain unwanted weight. If you choose chewable pills, be aware that vitamin C can erode tooth enamel. Swallow promptly and flush the teeth with a drink of water.

The second question: Who are you? What vitamins and minerals do you need? The nutrient intake recommendations listed in the tables on the inside front cover and the Canadian recommendations listed in Appendix B are the standards appropriate for virtually all reasonably healthy people.

Generally, an appropriate supplement provides all the vitamins and minerals in amounts smaller than, equal to, or very close to the intake recommendations. Avoid any preparation that, in a daily

Figure 7-11

A Supplement Label

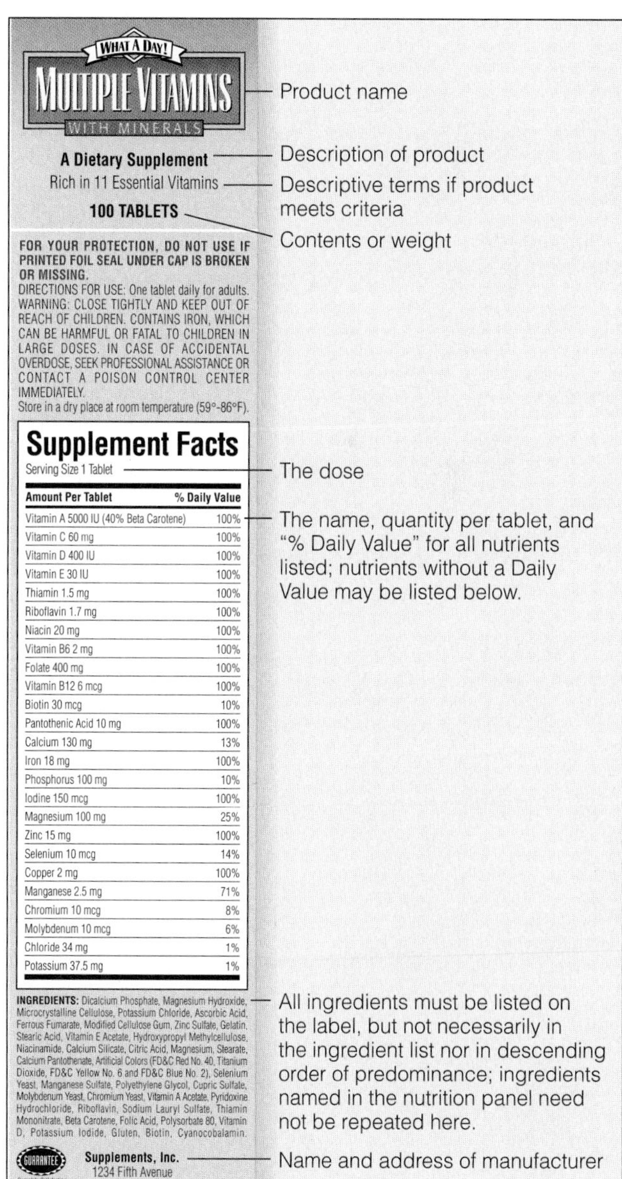
WHAT A DAY!
MULTIPLE VITAMINS
WITH MINERALS
Product name
A Dietary Supplement
Description of product
Rich in 11 Essential Vitamins
Descriptive terms if product meets criteria
100 TABLETS
Contents or weight
FOR YOUR PROTECTION, DO NOT USE IF PRINTED FOIL SEAL UNDER CAP IS BROKEN OR MISSING.
DIRECTIONS FOR USE: One tablet daily for adults.
WARNING: CLOSE TIGHTLY AND KEEP OUT OF REACH OF CHILDREN. CONTAINS IRON, WHICH CAN BE HARMFUL OR FATAL TO CHILDREN IN LARGE DOSES. IN CASE OF ACCIDENTAL OVERDOSE, SEEK PROFESSIONAL ASSISTANCE OR CONTACT A POISON CONTROL CENTER IMMEDIATELY.
Store in a dry place at room temperature (59°-86°F).
Supplement Facts
Serving Size 1 Tablet
The dose
Amount Per Tablet % Daily Value
Vitamin A 5000 IU (40% Beta Carotene) 100%
Vitamin C 60 mg 100%
Vitamin D 400 IU 100%
Vitamin E 30 IU 100%
Thiamin 1.5 mg 100%
Riboflavin 1.7 mg 100%
Niacin 20 mg 100%
Vitamin B6 2 mg 100%
Folate 400 mg 100%
Vitamin B12 6 mcg 100%
Biotin 30 mcg 10%
Pantothenic Acid 10 mg 100%
Calcium 130 mg 13%
Iron 18 mg 100%
Phosphorus 100 mg 10%
Iodine 150 mcg 100%
Magnesium 100 mg 25%
Zinc 15 mg 100%
Selenium 10 mcg 14%
Copper 2 mg 100%
Manganese 2.5 mg 71%
Chromium 10 mcg 8%
Molybdenum 10 mcg 6%
Chloride 34 mg 1%
Potassium 37.5 mg 1%
The name, quantity per tablet, and "% Daily Value" for all nutrients listed; nutrients without a Daily Value may be listed below.
INGREDIENTS: Dicalcium Phosphate, Magnesium Hydroxide, Microcrystalline Cellulose, Potassium Chloride, Ascorbic Acid, Ferrous Fumarate, Modified Cellulose Gum, Zinc Sulfate, Gelatin, Stearic Acid, Vitamin E Acetate, Hydroxypropyl Methylcellulose, Niacinamide, Calcium Silicate, Citric Acid, Magnesium, Stearate, Calcium Pantothenate, Artificial Colors (FD&C Red No. 40, Titanium Dioxide, FD&C Yellow No. 6 and FD&C Blue No. 2), Selenium Yeast, Manganese Sulfate, Polyethylene Glycol, Cupric Sulfate, Molybdenum Yeast, Chromium Yeast, Vitamin A Acetate, Pyridoxine Hydrochloride, Riboflavin, Sodium Lauryl Sulfate, Thiamin Mononitrate, Beta Carotene, Folic Acid, Polysorbate 80, Vitamin D, Potassium Iodide, Gluten, Biotin, Cyanocobalamin.
All ingredients must be listed on the label, but not necessarily in the ingredient list nor in descending order of predominance; ingredients named in the nutrition panel need not be repeated here.
GUARANTEE
Complete Satisfaction or Your Money Back
Supplements, Inc.
1234 Fifth Avenue
Anywhere, USA
Name and address of manufacturer

To get the most from a supplement of vitamins and minerals:

- Take the supplement with food. A full stomach retains the pill and dissolves it with its churning action.
- If you take an iron supplement, choose foods that will assist in its absorption, such as meats, fish, or poultry, or foods that contain vitamin C.

dose, provides more than the recommended amount of vitamin A, vitamin D, or any mineral or more than ten times the recommendation for any nutrient. A warning: Expect to reject about 80 percent of available preparations when you choose according to these criteria; be choosy where your health is concerned. Avoid these:

- High doses of iron (more than 10 milligrams per day) except for menstruating women. People who menstruate need more iron, but people who don't, don't.
- "Organic"or "natural" types of preparations with added substances. They are no better than standard types, but they cost much more.[78]
- "High-potency" or "therapeutic dose" supplements. More is not better.
- Items not needed in human nutrition, such as carnitine and inositol. These particular items won't harm you, but they reveal a marketing strategy that makes the whole mix suspect. The manufacturer wants you to believe that its pills contain the latest "new" nutrient that other brands omit, but, in fact, for every valid discovery of this kind, there are 999,999 frauds.
- "Time release." Medications such as some antibiotics or pain relievers often must be sustained at a steady concentration in the blood to be effective, but nutrients are incorporated into the tissues where they are needed whenever they arrive.
- "Stress formulas." Although the stress response depends on certain B vitamins and vitamin C, the recommended amount provides all that is needed of these nutrients. If you are under stress (and who isn't?), generous servings of fruits and vegetables will more than cover your need.
- Pills containing extracts of parsley, alfalfa, and other vegetable components.
- Geriatric "tonics." They are generally poor in vitamins and minerals and yet may be so high in alcohol as to threaten inebriation.
- Any supplement sold with claims that today's foods lack the nutrients they once contained. Truth: Plants make vitamins for their own needs, not ours. A plant lacking a mineral or failing to make a needed vitamin dies before it can bear food for our consumption.

Ironically, people in developed nations are far more likely to suffer from overnutrition and poor lifestyle choices than from nutrient deficiencies. People wish that simply swallowing vitamin pills might boost their health. The truth—that they need to improve their eating and exercise habits—is harder to swallow.

KEY POINT *If you feel a supplement is needed, then remember to examine the ingredients and choose one that satisfies your needs.*

This chapter has addressed all 13 of the vitamins. It sums up the basic facts about each one in Tables 7-5 and 7-6.

Table 7-5

The Fat-Soluble Vitamins—Functions, Deficiencies, and Toxicities

Vitamin A

OTHER NAMES	DEFICIENCY SYMPTOMS	TOXICITY SYMPTOMS
Retinol, retinal, retinoic acid; main precursor is beta-carotene	**Blood/Circulatory System**	
CHIEF FUNCTIONS IN THE BODY	Anemia (small-cell type)[a]	Red blood cell breakage, nosebleeds
Vision; health of cornea, epithelial cells, mucous membranes; skin health; bone and tooth growth; reproduction; hormone synthesis and regulation; immunity	**Bones/Teeth**	
Beta-carotene: antioxidant	Cessation of bone growth, painful joints; impaired enamel formation, cracks in teeth, tendency toward tooth decay	Bone pain; growth retardation; increased pressure inside skull mimicking brain tumor; headaches
DEFICIENCY DISEASE NAME	**Digestive System**	
Hypovitaminosis A	Diarrhea, changes in intestinal and other body linings	Abdominal cramps and pain, nausea, vomiting, diarrhea, weight loss
SIGNIFICANT SOURCES	**Immune System**	
Retinol: fortified milk, cheese, cream, butter, fortified margarine, eggs, liver	Depression; frequent respiratory, digestive, bladder, vaginal, and other infections	Overreactivity
Beta-carotene: spinach and other dark, leafy greens; broccoli; deep orange fruits (apricots, cantaloupe) and vegetables (squash, carrots, sweet potatoes, pumpkin)	**Nervous/Muscular Systems**	
	Night blindness (retinal)	Blurred vision, pain in calves, fatigue, irritability, loss of appetite
	Skin and Cornea	
	Keratinization, corneal degeneration leading to blindness,[b] rashes	Dry skin, rashes, loss of hair
	Other	
	Kidney stones, impaired growth	Cessation of menstruation, liver and spleen enlargement

Vitamin D

OTHER NAMES	DEFICIENCY SYMPTOMS	TOXICITY SYMPTOMS
Calciferol, cholecalciferol, dihydroxy vitamin D; precursor is cholesterol	**Blood/Circulatory System**	
		Raised blood calcium
CHIEF FUNCTIONS IN THE BODY	**Bones/Teeth**	
Mineralization of bones (raises blood calcium and phosphorus via absorption from digestive tract and by withdrawing calcium from bones and stimulating retention by kidneys)	Abnormal growth, misshapen bones (bowing of legs), soft bones, joint pain, malformed teeth	
DEFICIENCY DISEASE NAME	**Nervous System**	
Rickets, osteomalacia	Muscle spasms	Excessive thirst, headaches, irritability, loss of appetite, weakness, nausea
SIGNIFICANT SOURCES	**Other**	
Self-synthesis with sunlight; fortified milk or margarine, eggs, liver, sardines		Kidney stones, stones in arteries, mental and physical retardation

(Continued on next page)

[a]Small-cell anemia is termed *microcytic anemia;* large-cell type is *macrocytic* or *megaloblastic anemia.*

[b]Corneal degeneration progresses from *keratinization* (hardening) to *xerosis* (drying) to *xerophthalomia* (thickening, opacity, and irreversible blindness).

Table 7-5

The Fat-Soluble Vitamins—Functions, Deficiencies, and Toxicities (Continued)

Vitamin E

		DEFICIENCY SYMPTOMS	TOXICITY SYMPTOMS
OTHER NAMES: Alpha-tocopherol, tocopherol	Blood/Circulatory System	Red blood cell breakage, anemia	Augments the effects of anticlotting medication
CHIEF FUNCTIONS IN THE BODY: Antioxidant (detoxification of strong oxidants), stabilization of cell membranes, regulation of oxidation reactions, protection of PUFA and vitamin A	Digestive System		General discomfort
DEFICIENCY DISEASE NAME: (No name)	Nervous/Muscular Systems	Degeneration, weakness, difficulty walking, leg cramps	
SIGNIFICANT SOURCES: Polyunsaturated plant oils (margarine, salad dressings, shortenings), green and leafy vegetables, wheat germ, whole-grain products, nuts, seeds	Other	Fibrocystic breast disease	

Vitamin K

		DEFICIENCY SYMPTOMS	TOXICITY SYMPTOMS
OTHER NAMES: Phylloquinone, naphthoquinone	Blood/Circulatory System	Hemorrhage	Interference with anticlotting medication; vitamin K analogues may cause jaundice
CHIEF FUNCTIONS IN THE BODY: Synthesis of blood-clotting proteins and a blood protein that regulates blood calcium			
DEFICIENCY DISEASE NAME: (No name)			
SIGNIFICANT SOURCES: Bacterial synthesis in the digestive tract; liver, green leafy vegetables, cabbage-type vegetables, milk			

Table 7-6

The Water-Soluble Vitamins—Functions, Deficiencies, and Toxicities

Thiamin

OTHER NAMES
Vitamin B_1

CHIEF FUNCTIONS IN THE BODY
Part of a coenzyme used in energy metabolism, supports normal appetite and nervous system function

DEFICIENCY DISEASE NAME
Beriberi

SIGNIFICANT SOURCES
Occurs in all nutritious foods in moderate amounts; pork, ham, bacon, liver, whole grains, legumes, nuts

	DEFICIENCY SYMPTOMS	TOXICITY SYMPTOMS
Blood/Circulatory System	Edema, enlarged heart, abnormal heart rhythms, heart failure	(No symptoms reported)
Nervous/Muscular Systems	Degeneration, wasting, weakness, pain, low morale, difficulty walking, loss of reflexes, mental confusion, paralysis	(No symptoms reported)

Riboflavin

OTHER NAMES
Vitamin B_2

CHIEF FUNCTIONS IN THE BODY
Part of a coenzyme used in energy metabolism, supports normal vision and skin health

DEFICIENCY DISEASE NAME
Ariboflavinosis

SIGNIFICANT SOURCES
Milk, yogurt, cottage cheese, meat, leafy green vegetables, whole-grain or enriched breads and cereals

	DEFICIENCY SYMPTOMS	TOXICITY SYMPTOMS
Mouth, Gums, Tongue	Cracks at corners of mouth,[a] magenta tongue	(No symptoms reported)
Nervous System and Eyes	Hypersensitivity to light,[b] reddening of cornea	(No symptoms reported)
Other	Skin rash	(No symptoms reported)

Niacin

OTHER NAMES
Nicotinic acid, nicotinamide, niacinamide, vitamin B_3; precursor is dietary tryptophan

CHIEF FUNCTIONS IN THE BODY
Part of a coenzyme used in energy metabolism; supports health of skin, nervous system, and digestive system

DEFICIENCY DISEASE NAME
Pellagra

SIGNIFICANT SOURCES
Milk, eggs, meat, poultry, fish, whole-grain and enriched breads and cereals, nuts, and all protein-containing foods

	DEFICIENCY SYMPTOMS	TOXICITY SYMPTOMS
Digestive System	Diarrhea	Diarrhea, heartburn, nausea, ulcer, irritation, vomiting
Mouth, Gums, Tongue	Black, smooth tongue[c]	
Nervous System	Irritability, loss of appetite, weakness, dizziness, mental confusion progressing to psychosis or delirium	Fainting, dizziness
Skin	Flaky skin rash on areas exposed to sun	Painful flush and rash, sweating
Other		Abnormal liver function, low blood pressure

(Continued on next page)

[a]Cracks at the corners of the mouth are termed *cheilosis* (kee-LOH-sis).
[b]Hypersensitivity to light is *photophobia*.
[c]Smoothness of the tongue is caused by loss of its surface structures and is termed *glossitis* (gloss-EYE-tis).

Table 7-6

The Water-Soluble Vitamins—Functions, Deficiencies, and Toxicities (Continued)

Vitamin B_6

	DEFICIENCY SYMPTOMS		TOXICITY SYMPTOMS
OTHER NAMES Pyridoxine, pyridoxal, pyridoxamine	Anemia (small-cell type)[d]	**Blood/Circulatory System**	Bloating
CHIEF FUNCTIONS IN THE BODY Part of a coenzyme used in amino acid and fatty acid metabolism, helps convert tryptophan to niacin, helps make red blood cells	Smooth tongue[c]	**Mouth, Gums, Tongue**	
	Abnormal brain wave pattern, irritability, muscle twitching, convulsions	**Nervous/Muscular Systems**	Depression, fatigue, impaired memory, irritability, headaches, numbness, damage to nerves, difficulty walking, loss of reflexes, weakness, restlessness
DEFICIENCY DISEASE NAME (No Name)			
SIGNIFICANT SOURCES Green and leafy vegetables, meats, fish, poultry, shellfish, legumes, fruits, whole grains	Irritation of sweat glands, rashes, greasy dermatitis	**Skin**	
	Kidney stones	**Other**	

Folate

	DEFICIENCY SYMPTOMS		TOXICITY SYMPTOMS
OTHER NAMES Folic acid, folacin, pteroyglutamic acid	Anemia (large-cell type)[d]	**Blood/Circulatory System**	Masks vitamin B_{12} deficiency
CHIEF FUNCTIONS IN THE BODY Part of a coenzyme needed for new cell synthesis	Heartburn, diarrhea, constipation	**Digestive System**	
DEFICIENCY DISEASE NAME (No Name)	Suppression, frequent infections	**Immune System**	
SIGNIFICANT SOURCES Leafy green vegetables, legumes, seeds, liver, enriched breads, cereal, pasta, and grains	Smooth red tongue[c]	**Mouth, Gums, Tongue**	
	Depression, mental confusion, fainting	**Nervous System**	

Vitamin B_{12}

	DEFICIENCY SYMPTOMS		TOXICITY SYMPTOMS
OTHER NAMES Cyanocobalamin	Anemia (large-cell type)[d]	**Blood/Circulatory System**	(No toxicity symptoms known)
CHIEF FUNCTIONS IN THE BODY Part of a coenzyme used in new cell synthesis, helps maintain nerve cells	Smooth tongue[c]	**Mouth, Gums, Tongue**	
DEFICIENCY DISEASE NAME (No Name)[e]	Fatigue, degeneration progressing to paralysis	**Nervous System**	
SIGNIFICANT SOURCES Animal products (meat, fish, poultry, milk, cheese, eggs)	Hypersensitivity	**Skin**	

Table 7-6

The Water-Soluble Vitamins—Functions, Deficiencies, and Toxicities (Continued)

Pantothenic Acid

OTHER NAMES / CHIEF FUNCTIONS / DEFICIENCY DISEASE NAME / SIGNIFICANT SOURCES		DEFICIENCY SYMPTOMS	TOXICITY SYMPTOMS
OTHER NAMES: (None)	**Digestive System**	Vomiting, intestinal distress	
CHIEF FUNCTIONS IN THE BODY: Part of a coenzyme used in energy metabolism	**Nervous System**	Insomnia, fatigue	
DEFICIENCY DISEASE NAME: (No name)	**Other**		Water retention (infrequent)
SIGNIFICANT SOURCES: Widespread in foods			

Biotin

OTHER NAMES / CHIEF FUNCTIONS / DEFICIENCY DISEASE NAME / SIGNIFICANT SOURCES		DEFICIENCY SYMPTOMS	TOXICITY SYMPTOMS
OTHER NAMES: (None)	**Blood/Circulatory System**	Abnormal heart action	(No toxicity symptoms reported)
CHIEF FUNCTIONS IN THE BODY: A cofactor for several enzymes used in energy metabolism, fat synthesis, amino acid metabolism, and glycogen synthesis	**Digestive System**	Loss of appetite, nausea	
DEFICIENCY DISEASE NAME: (No name)	**Nervous/Muscular Systems**	Depression, muscle pain, weakness, fatigue	
SIGNIFICANT SOURCES: Widespread in foods	**Skin**	Drying, rash, loss of hair	

Vitamin C

OTHER NAMES / CHIEF FUNCTIONS / DEFICIENCY DISEASE NAME / SIGNIFICANT SOURCES		DEFICIENCY SYMPTOMS	TOXICITY SYMPTOMS
OTHER NAMES: Ascorbic acid	**Blood/Circulatory System**	Anemia (small-cell type),[d] pinpoint hemorrhages	
CHIEF FUNCTIONS IN THE BODY: Collagen synthesis (strengthens blood vessel walls, forms scar tissue, matrix for bone growth), antioxidant, thyroxine synthesis, amino acid metabolism, strengthens resistance to infection, helps in absorption of iron	**Digestive System**		Nausea, abdominal cramps, diarrhea, excessive urination
DEFICIENCY DISEASE NAME: Scurvy	**Immune System**	Suppression, frequent infections	
SIGNIFICANT SOURCES: Citrus fruits, cabbage-type vegetables, dark green vegetables, cantaloupe, strawberries, peppers, lettuce, tomatoes, potatoes, papayas, mangoes	**Mouth, Gums, Tongue**	Bleeding gums, loosened teeth	
	Nervous/Muscular Systems	Muscle degeneration and pain, hysteria, depression	Headache, fatigue, insomnia
	Skeletal System	Bone fragility, joint pain	
	Skin	Rough skin, blotchy bruises	Rashes
	Other	Failure of wounds to heal	Interference with medical tests; aggravation of gout symptoms; deficiency symptoms may appear at first on withdrawal of high doses

[a]Cracks at the corners of the mouth are termed *cheilosis* (kee-LOH-sis).

[b]Hypersensitivity to light is *photophobia*.

[c]Smoothness of the tongue is caused by loss of its surface structures and is termed *glossitis* (gloss-EYE-tis).

[d]Small-cell anemia is termed *microcytic anemia;* large-cell type is *macrocytic* or *megaloblastic anemia.*

[e]The name *pernicious anemia* refers to the vitamin B_{12} deficiency caused by lack of intrinsic factor, but not to that caused by inadequate dietary intake.

FOOD
FEATURE

Choosing Foods Rich in Vitamins

Most people, upon learning how important the vitamins are to their health, want to choose foods that are vitamin-rich. A way to identify such foods is to look at the vitamins and calories in single servings. This method is shown in Table 7-7. Figure 7-12 depicts information similar to that of Table 7-7, but shows at a glance some foods that are rich sources of a particular vitamin and others that are poor sources. The serving sizes in Figure 7-12 are those recommended by the Food Guide Pyramid. For example, 3 ounces is used for most meats. The vegetable serving size is ½ cup. The colors of the bars represent the various food groups.

Which Foods Should I Choose?

Some people, after viewing a figure such as this, believe that to meet their vitamin needs, they must memorize the richest sources of each vitamin and include those foods daily. This notion is false and can lead people to limit the variety of foods they choose while overemphasizing the components of a few foods. Though it is reassuring to know that your carrot-raisin salad at lunch provided more than the entire Daily Value amount for vitamin A, it is a mistake to think that you must then select equally rich sources of all the other vitamins. Such rich sources do not exist for many vitamins. Rather, foods work in harmony to provide most nutrients. For example, a baked potato, though not a star performer among vitamin C providers, contributes substantially to a day's need for this nutrient and contributes some thiamin, too. By the end of the day, assuming that your food choices were made with reasonable care, the bits of thiamin, vitamin B_6, and vitamin C from each serving of food have accumulated to make a more-than-adequate total diet.

A Variety of Foods Works Best

With a few exceptions, nutritious foods generally provide small quantities of thiamin, as shown in Figure 7-12 on page 250. A few meats are exceptionally good thiamin sources; these are members of the pork family. As you can see in the graph, one small pork chop (275 calories) provides over half of the Daily Value for thiamin, but again, this does not suggest that you eat pork every day. Legumes and grains are also good, low-fat sources, and they provide beneficial fiber and nutrients lacking from meats. Beans lack the vitamin B_{12} provided by meats, however. Peanut butter is a good source of thiamin, as it is of most B vitamins, but its high fat and calorie contents call for moderation in its use.

The vitamin B_6 data provide another insight to support the argument for variety. From just the few foods listed here, you can see that no one source can provide the whole day's requirement, but that a variety of meats, fish, and poultry along with potatoes and a few other vegetables and fruits can work together to supply it.

The last two graphs of Figure 7-12 show sources of folate and vitamin C. These nutrients are both richly supplied by fruits and vegetables. The richest source of either may be only a moderate source of the other, but the recommended servings of fruits and vegetables in food group plans cover both needs amply. As for vitamin E, vegetable oils are the richest sources. Some vegetables, nuts, and fruits contribute some vitamin E, too.

Are you ready to try out what you've learned about the food sources of vitamins? The Do It! section that follows gives you a chance to test your skills.

Table 7-7

Vitamin Contents of Restaurant Meals

	Energy (cal)	Vitamin A (RE)	Thiamin (mg)	Niacin (mg)	Vitamin C (mg)	Folate (µg)	Vitamin B_{12} (µg)
Breakfast Foods							
Hotcakes with syrup and butter, scrambled egg and sausage patty	1,140	150	0.41	4	0	32	1.7
Egg, ham, and cheese muffin	290	100	0.49	3	0	30	0.7
Oatmeal, brown sugar	107	2	0.13	0	0	5	0
Cornflakes	110	375	0.87	5	15	100	0
Hash browned potatoes	130	0	0.08	1	3	8	0
2 small cinnamon sweet rolls	300	50	0.25	2	2	20	0.1
Large blueberry muffin	400	10	0.2	2	2	20	0.1
English muffin	130	0	0.25	2	0	20	0.1
Orange juice	80	40	0.17	1	90	60	0
Milk (low-fat)	120	140	0.1	0	2	10	0.8
Lunch Foods							
Homemade chili/crackers	350	150	0.26	5	25	40	0.5
Cold cut hoagie sandwich/chips	460	80	1.0	6	12	55	1.1
Peanut butter and jelly sandwich on whole wheat; fruit cocktail	450	30	0.3	7	3	62	0
Tuna sandwich on white; banana	470	40	0.32	6	11	49	0.6
Chef's salad with cheese, ham, turkey, and dressing/crackers	580	140	0.43	7	16	100	0.8
Fat-free milk	85	150	0.1	0	0	10	0.9
Apple juice	120	0	0.05	0	2	0	0
Supper Foods							
New England boiled dinner (corned beef, potatoes, brussels sprouts)	440	150	0.27	5	80	70	1.4
Vegetable plate	400	680	0.26	2	53	120	0.1
Spaghetti and meatballs; small salad	600	220	0.38	6	25	50	1.3
Fried fish, tartar sauce, corn, and macaroni salad	510	70	0.26	3	6	83	1
Rolls	100	0	0.15	1	0	0	0
Garlic bread	190	4	0.4	3	0	0	0
Corn muffins	300	0	0.12	2	0	0	0
Lemon pie	350	70	0.16	1	4	11	0.2
Chocolate cake	250	20	0.02	0	0	5	0.1

Figure 7-12

Food Sources of Vitamins Selected to Show a Range of Values

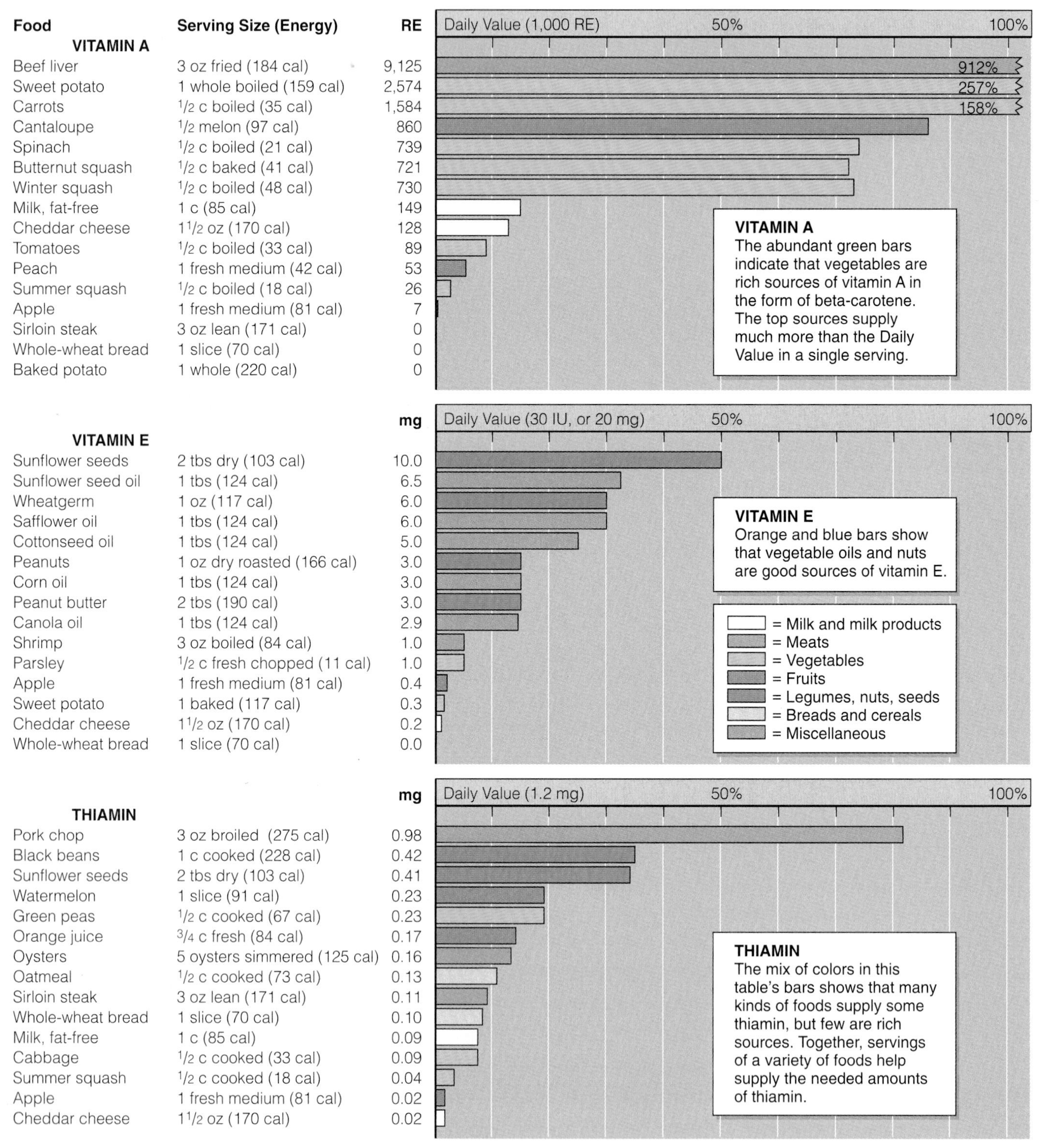

Food	Serving Size (Energy)	RE
VITAMIN A		
Beef liver	3 oz fried (184 cal)	9,125
Sweet potato	1 whole boiled (159 cal)	2,574
Carrots	1/2 c boiled (35 cal)	1,584
Cantaloupe	1/2 melon (97 cal)	860
Spinach	1/2 c boiled (21 cal)	739
Butternut squash	1/2 c baked (41 cal)	721
Winter squash	1/2 c boiled (48 cal)	730
Milk, fat-free	1 c (85 cal)	149
Cheddar cheese	1 1/2 oz (170 cal)	128
Tomatoes	1/2 c boiled (33 cal)	89
Peach	1 fresh medium (42 cal)	53
Summer squash	1/2 c boiled (18 cal)	26
Apple	1 fresh medium (81 cal)	7
Sirloin steak	3 oz lean (171 cal)	0
Whole-wheat bread	1 slice (70 cal)	0
Baked potato	1 whole (220 cal)	0

Food	Serving Size (Energy)	mg
VITAMIN E		
Sunflower seeds	2 tbs dry (103 cal)	10.0
Sunflower seed oil	1 tbs (124 cal)	6.5
Wheatgerm	1 oz (117 cal)	6.0
Safflower oil	1 tbs (124 cal)	6.0
Cottonseed oil	1 tbs (124 cal)	5.0
Peanuts	1 oz dry roasted (166 cal)	3.0
Corn oil	1 tbs (124 cal)	3.0
Peanut butter	2 tbs (190 cal)	3.0
Canola oil	1 tbs (124 cal)	2.9
Shrimp	3 oz boiled (84 cal)	1.0
Parsley	1/2 c fresh chopped (11 cal)	1.0
Apple	1 fresh medium (81 cal)	0.4
Sweet potato	1 baked (117 cal)	0.3
Cheddar cheese	1 1/2 oz (170 cal)	0.2
Whole-wheat bread	1 slice (70 cal)	0.0

Food	Serving Size (Energy)	mg
THIAMIN		
Pork chop	3 oz broiled (275 cal)	0.98
Black beans	1 c cooked (228 cal)	0.42
Sunflower seeds	2 tbs dry (103 cal)	0.41
Watermelon	1 slice (91 cal)	0.23
Green peas	1/2 c cooked (67 cal)	0.23
Orange juice	3/4 c fresh (84 cal)	0.17
Oysters	5 oysters simmered (125 cal)	0.16
Oatmeal	1/2 c cooked (73 cal)	0.13
Sirloin steak	3 oz lean (171 cal)	0.11
Whole-wheat bread	1 slice (70 cal)	0.10
Milk, fat-free	1 c (85 cal)	0.09
Cabbage	1/2 c cooked (33 cal)	0.09
Summer squash	1/2 c cooked (18 cal)	0.04
Apple	1 fresh medium (81 cal)	0.02
Cheddar cheese	1 1/2 oz (170 cal)	0.02

Daily Value (1,000 RE)
50%
100%
912%
257%
158%
VITAMIN A
The abundant green bars indicate that vegetables are rich sources of vitamin A in the form of beta-carotene. The top sources supply much more than the Daily Value in a single serving.
Daily Value (30 IU, or 20 mg)
50%
100%
VITAMIN E
Orange and blue bars show that vegetable oils and nuts are good sources of vitamin E.
= Milk and milk products
= Meats
= Vegetables
= Fruits
= Legumes, nuts, seeds
= Breads and cereals
= Miscellaneous
Daily Value (1.2 mg)
50%
100%
THIAMIN
The mix of colors in this table's bars shows that many kinds of foods supply some thiamin, but few are rich sources. Together, servings of a variety of foods help supply the needed amounts of thiamin.

Figure 7-12

Food Sources of Vitamins Selected to Show a Range of Values (Continued)

Food	Serving Size (Energy)	mg
VITAMIN B_6		
Baked potato	1 whole (220 cal)	0.70
Banana	1 peeled (109 cal)	0.68
Turkey breast	3 oz (133 cal)	0.46
Watermelon	1 slice (91 cal)	0.41
Sirloin steak	3 oz lean (171 cal)	0.38
Pork roast	3 oz lean (175 cal)	0.29
Spinach	1/2 c cooked (21 cal)	0.22
Salmon	3 oz broiled/baked (183 cal)	0.19
Navy beans	1/2 c cooked (129 cal)	0.15
Broccoli	1/2 c cooked (22 cal)	0.11
Milk, fat-free	1 c (85 cal)	0.10
Orange juice	3/4 c fresh (84 cal)	0.08
Apple	1 fresh medium (81 cal)	0.07
Summer squash	1/2 c boiled (18 cal)	0.06
Whole-wheat bread	1 slice (69 cal)	0.05
Cheddar cheese	1 1/2 oz (170 cal)	0.03

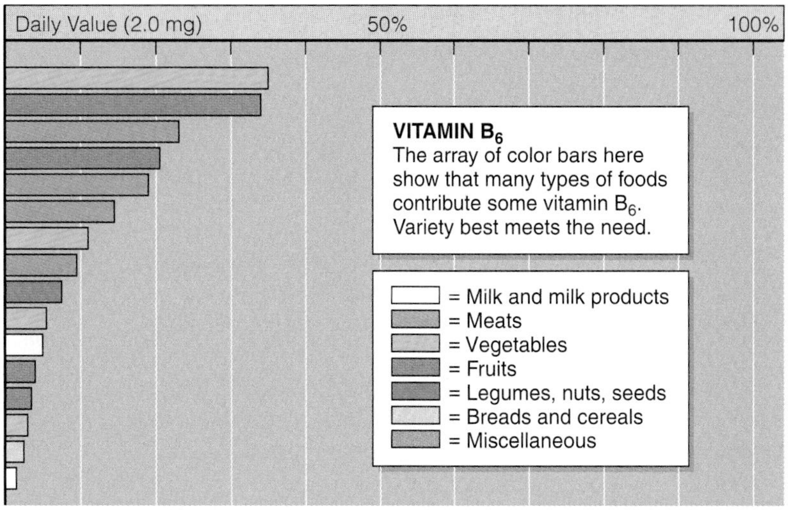

Daily Value (2.0 mg) 50% 100%

VITAMIN B_6
The array of color bars here show that many types of foods contribute some vitamin B_6. Variety best meets the need.

= Milk and milk products
= Meats
= Vegetables
= Fruits
= Legumes, nuts, seeds
= Breads and cereals
= Miscellaneous

Food	Serving Size (Energy)	µg
FOLATE		
Beef liver	3 oz fried (184 cal)	187
Spinach	1/2 c cooked (21 cal)	131
Asparagus	4 spears cooked (14 cal)	88
Turnip greens	1/2 c cooked (15 cal)	85
Winter squash	1/2 c cooked (48 cal)	69
Beets	1/2 c cooked (37 cal)	68
Orange juice	3/4 c fresh (84 cal)	57
Cantaloupe	1/2 melon (97 cal)	47
Broccoli	1/2 c cooked (22 cal)	39
Lima beans	1/2 c cooked (85 cal)	18
Summer squash	1/2 c cooked (18 cal)	18
Whole-wheat bread[a]	1 slice (70 cal)	14
Milk, fat-free	1 c (85 cal)	13
Sirloin steak	3 oz lean (171 cal)	8
Cheddar cheese	1 1/2 oz (170 cal)	5
Apple	1 fresh medium (81 cal)	4

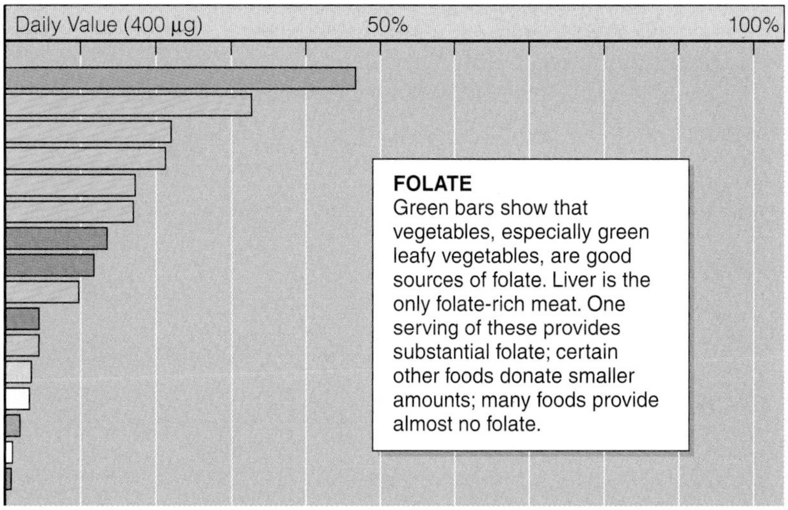

Daily Value (400 µg) 50% 100%

FOLATE
Green bars show that vegetables, especially green leafy vegetables, are good sources of folate. Liver is the only folate-rich meat. One serving of these provides substantial folate; certain other foods donate smaller amounts; many foods provide almost no folate.

Food	Serving Size (Energy)	mg
VITAMIN C		
Cantaloupe	1/2 melon (97 cal)	116
Orange juice	3/4 c fresh (84 cal)	93
Green peppers	1/2 c (20 cal)	67
Broccoli	1/2 c cooked (26 cal)	48
Brussels sprouts	1/2 c cooked (30 cal)	48
Tomato juice	3/4 c canned (31 cal)	33
Baked potato	1 whole (220 cal)	26
Cabbage	1/2 c cooked (17 cal)	15
Apple	1 fresh medium (81 cal)	8
Oysters	3 oz (69 cal)	7
Milk, fat-free	1 c (85 cal)	2
Whole-wheat bread	1 slice (69 cal)	0
Sirloin steak	3 oz lean (171 cal)	0
Cheddar cheese	1 oz (170 cal)	0

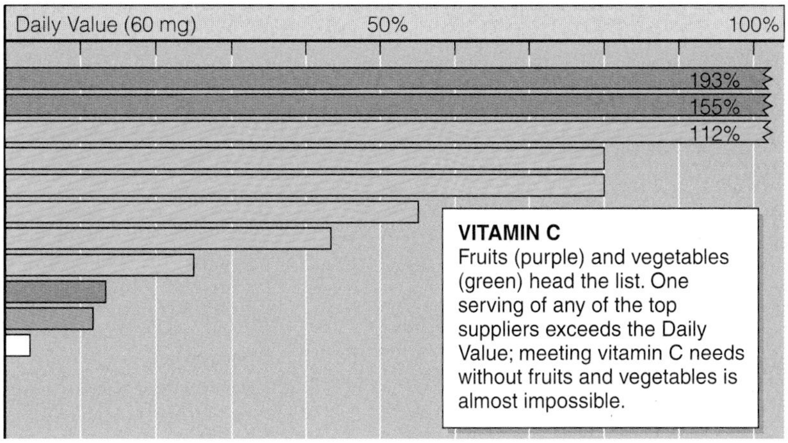

Daily Value (60 mg) 50% 100%

193% 155% 112%

VITAMIN C
Fruits (purple) and vegetables (green) head the list. One serving of any of the top suppliers exceeds the Daily Value; meeting vitamin C needs without fruits and vegetables is almost impossible.

[a]Unenriched.

do it!

FIND THE VITAMINS ON A MENU

Can you meet your vitamin needs when you eat in restaurants? One way is to learn to identify the foods on restaurant menus that are rich sources of vitamins. Assume you are spending a day on the road and have to eat all three of the day's meals in restaurants. Read over the three menus in Figure 7-13 and create a meal from each one, using the foods offered. Note that, just like real menus, these menus lack serving size information. For this exercise, take for granted that the serving sizes of foods agree with those in the Food Guide Pyramid. (Never assume this about real menus, however—commercial serving sizes vary enormously.) Most vegetable and grain servings in the meals listed in Figure 7-13 are ½-cup servings. Those identified as "large" are 1-cup servings; meats are 2- to 3-ounce portions; milk is an 8-ounce serving, and so forth.

Step 1. On a copy of Form 7-1, record your food choices down the left-hand column.

Step 2. Consult Table 7-7 on p. 249, earlier, to determine the values for calories and nutrients in the day's meals. Fill in the values for the foods you listed on Form 7-1. Coffee, tea, and water contribute negligible energy and vitamins; use zeros for their values on Form 7-1.

Step 3. Enter the nutrient intake recommendations that apply to your age and gender (see the inside front cover for nutrients, inside back cover for energy, or the Canadian appendix, Appendix B) in the spaces provided on Form 7-1.

Step 4. Divide the day's total intakes by the recommendation amounts and multiply by 100 to determine the percentages of your nutrient and energy needs contributed by the foods chosen this day.

- Example: If total vitamin C intake equals 40 milligrams and your vitamin C recommendation equals 60 milligrams, then $(40 \div 60) \times 100 = 67\%$. The meals met two-thirds of the daily recommended amount of vitamin C.

Repeat this process for all five vitamins and for energy.

Figure 7-13

Restaurant Menus

Welcome to
BURGER DOODLE
★ ★ ★

★ BREAKFAST
- Hotcakes, eggs, and sausage ....
- Egg, ham, and cheese muffin ....
- Oatmeal with brown sugar ....
- Cornflakes ....

★ SIDES
- Hash browned potatoes ....
- A pair of cinnamon sweet rolls ....
- Large blueberry muffin ....
- Plain English muffin ....

★ BEVERAGES
- Orange juice ....
- Low-fat milk ....
- Coffee or tea ....

The Box Lunch
EXPRESS

Soups
- Homemade chili with crackers ....

Sandwiches
- Cold cut hoagie sandwich with chips ....
- Peanut butter and jelly sandwich on whole-wheat bread, served with fruit cocktail ....
- Tuna sandwich on white bread, served with a fresh banana ....

Salads
- Large chef's salad with cheese, ham, and turkey with crackers ....

Beverages
- Fat-free milk ....
- Apple juice..........
- Sparkling water..........

Elf's Cafeteria
Blue Plate Specials

Specials

New England boiled dinner: corned beef, large serving of potatoes, and brussels sprouts with margarine

Southern style vegetable plate: steamed yellow squash, collard greens, fried okra, candied yams, and fried green tomatoes

Large spaghetti & meatballs with small salad

Fried fish, tartar sauce, macaroni, salad, and corn on the cob with margarine

Choice of 2 rolls, 1 slice garlic bread, or 2 corn muffins

Beverages
Tea or coffee

Desserts
Lemon pie or chocolate cake

Form 7-1

Vitamin Tally

	Energy (cal)	Vitamin A (RE)	Thiamin (mg)	Niacin (mg)	Vitamin C (mg)	Folate (μg)	Vitamin B_{12} (μg)
Breakfast:							
Lunch:							
Supper:							
Day's totals							
Energy and vitamin recommendations							
% of recommendations							

Analysis

Answer the following questions:

1. How did the day's totals for calories and vitamins compare with your recommended amounts? Did the day's meals meet or exceed your need for energy or any of these vitamins? Which ones?

2. Did the meals present too little of energy or any of the nutrients listed here? Which ones? What changes in your choices among those foods would have improved the vitamin totals for the day?

3. Which foods listed on the menus contributed significant amounts of vitamin A? Remembering that brightly colored fruits and vegetables offer the vitamin A precursor beta-carotene, name any foods in your meal choices that provide beta-carotene (turn back to Snapshot 7-1 on p. 215 for hints). Which foods contributed little or no vitamin A in any form?

4. Did your choices supply enough folate to meet your requirement? Remembering that young women are urged to obtain 400 micrograms of folate each day, what percentage of your folate need did the day's meals provide? Which individual foods were richest in folate? Turn to the Table of Food Composition, Appendix A at the back of the book, and look down the folate column of the fruits and vegetables sections. Identify some foods that, in a single serving, provide 10 percent or more of the 400 micrograms of folate recommended for women.

5. What are the sources of niacin in the day's meals? Rich sources are shown in Snapshot 7-7 on page 229.

6. What about vitamin C? What percentage of your daily need of vitamin C did these meals provide? Which individual foods were the main contributors? To what food groups do they belong?

7. How did your total energy intake compare with your energy need? Express your answer as a percentage, and use it to answer the next two questions.

8. Which meals are vitamin "bargains"? Compare the vitamin contents (as a percentage of your need) with the

energy contents of the chosen meals (also as a percentage of your need). Which are the most vitamin-dense, providing the most vitamins for the fewest calories?

9. Which of the breakfast choices is highest in vitamin A? Which is highest in folate and niacin? What characteristics of this food make its vitamin content so high?

10. Which foods on the menus, though low in vitamins, possess other valuable constituents that make them desirable as part of a health-promoting diet?

Should you wonder about the vitamins in other foods you choose, you can find their vitamin and other nutrient contents in several references in this book. First, the vitamin Snapshots appearing throughout this chapter depict the richest vitamin sources; second, the Daily Food Guide, on pp. 40 and 41, notes which vitamins characterize foods of each group; and third, the Table of Food Composition, Appendix A at the back of the book, provides actual values for vitamins in about 2,000 foods. With these references, you can judge the vitamin values of foods on virtually any menu.

Self Check

Answers to these Self Check questions are in Appendix G.

1. Which of the following vitamins are fat soluble?
   a. vitamins B, C, D, and E
   b. vitamins B, C, and E
   c. vitamins A, C, E, and K
   d. vitamins A, D, E, and K
2. Night blindness and xerophthalmia are the result of a deficiency of which vitamin?
   a. niacin
   b. vitamin C
   c. vitamin A
   d. vitamin K
3. Which of the following foods is (are) rich in beta-carotene?
   a. sweet potatoes
   b. pumpkin
   c. cantaloupe
   d. all of the above
4. A deficiency of vitamin D may result in which disease?
   a. rickets
   b. beriberi
   c. scurvy
   d. pellagra
5. Which vitamin(s) is (are) present only in foods of animal origin?
   a. the active form of vitamin A
   b. vitamin B_{12}
   c. riboflavin
   d. (a) and (b)
6. Which of the following describes the water-soluble vitamins?
   a. vitamins D and E
   b. frequently toxic
   c. stored extensively in tissues
   d. easily absorbed and excreted
7. Almost any substance, including herbs, amino acids, dried animal organ tissues, microorganisms, and hormones can be sold freely over-the-counter in the United States. T F
8. In general, nutrients are absorbed equally well from foods and supplements. T F
9. People in developed nations are more likely to suffer from overnutrition than from nutrient deficiencies. T F
10. The best way to consume a diet rich in vitamins is to eat a variety of nutrient-dense foods every day. T F
11. The body's defenses against free-radical damage include vitamin E, vitamin C, beta-carotene, and phytochemicals. (Read about these in the upcoming Controversy.) T F

NUTRITION ON THE NET

For further study of the topics of this chapter, access these websites and search for the phrases or words in quotation marks:

1. For the latest changes in chapter material, to find supplemental learning tools, or to access links to related websites, go to the "Nutrition: Concepts and Controversies" site at:
   www.wadsworth.com/nutrition/prod/allprod.html
2. Search the American Dietetic Association site for more information on "vitamins":
   www.eatright.org
3. For worldwide information on "vitamin deficiency":
   www.who.int/home/search
4. Visit the U.S. Government site for information on "folate":
   www.healthfinder.gov/searchoptions/topicsaz.htm
5. To learn more about "neural tube defects":
   www.sbaa.org/
6. For more details on "vitamin and mineral supplements":
   vm.cfsan.fda.gov/
7. The National 5 A Day Program provides information on fruits and vegetables to provide vitamins:
   www.dcpc.nci.nih.gov/5aday
8. Search among thousands of current scientific and medical abstracts for any topic related to vitamins at:
   http://www.ncbi.nlm.nih.gov/PubMed/

ANTIOXIDANT NUTRIENTS: MAGIC BULLETS?

CONTROVERSY
7

LATELY, "ANTIOXIDANT vitamins" have become household words. On hearing about these metabolic busybodies, people want to know whether they are receiving enough of them. Knowledgeable people claim that *foods* rich in these nutrients and other **dietary antioxidants** help to prevent disease. Others claim that *supplements* are more reliable allies. Consumers can tell who is right by weighing the evidence on both sides. This Controversy offers a way to score foods versus supplements as sources of these beneficial compounds. To anticipate the punch line, the antioxidant vitamins and the mineral selenium are only a few among thousands of health-promoting constituents of foods, about which Controversy 11 has much more to say. Still, these nutrients are of great interest in themselves.

Most of the results presented here are preliminary. More research is needed to confirm or deny their possible implications. Supplement makers rightly claim that "scientific evidence backs up the claims we make for our products," but almost none of this evidence makes a complete case for taking supplements. For example, epidemiological studies suggest interesting general trends, but have no specific implications for individuals. They must be followed with laboratory work and human intervention studies for a complete picture. So proceed with caution and demand rigorous, repeated testing before you consider a finding confirmed—but do proceed. The evidence that lies before you holds secrets that are evolving toward being tomorrow's established nutrition concepts.

FREE RADICALS AND DISEASE

The body's cells use oxygen to produce energy. In the process, oxygen sometimes reacts with body compounds to produce highly unstable molecules known as **free radicals** (see Table C7-1 for the definition of this and related terms). In addition to normal body processes, environmental factors such as radiation, pollution, tobacco smoke, and others can act as **oxidants** and cause free-radical formation (see Table C7-2).[1] The trouble begins when free radicals in the body exceed its defenses against them, a condition known as **oxidative stress.** The body's level of oxidative stress is hard to quantify, but researchers are proposing ways of doing so.[2]

A free radical is a molecule with one or more unpaired **electrons.*** An electron without a partner is unstable and highly reactive. To regain its stability, the free radical quickly finds a stable but vulnerable compound from which to steal an electron. With the loss of an electron, the formerly stable molecule becomes a free radical itself and steals an electron from some other nearby molecule. Thus an electron-snatching chain reaction is under way.

Unopposed, free radicals are like sparks, starting wildfires that lead to widespread damage by oxidative stress. Free-radical damage commonly disrupts unsaturated fatty acids in cell membranes, damaging the membranes' ability to transport substances into and out of cells. Free radicals also cause damage to cell proteins, altering their functions, and to DNA, disrupting all cells that inherit the damaged DNA.

It's an exaggeration to say that the connections between oxidative stress and diseases are endless, but today's research has identified links with the development of over 200 diseases. Among them are age-related blindness, arthritis, some cancers, cardiovascular disease, cataracts, diabetes, and kidney disease.[3] In diabetes, oxidative stress reduces cell membrane responsiveness to insulin and damages arteries and

*Oxygen-derived free radicals are common in the human body. Examples are superoxide radical (O_2•−), hydroxyl radical (OH •), and nitric oxide (NO •). The dots in the symbols represent the unpaired electrons. Scientists sometimes use the term *reactive oxygen species (ROS)* to describe all of these compounds.

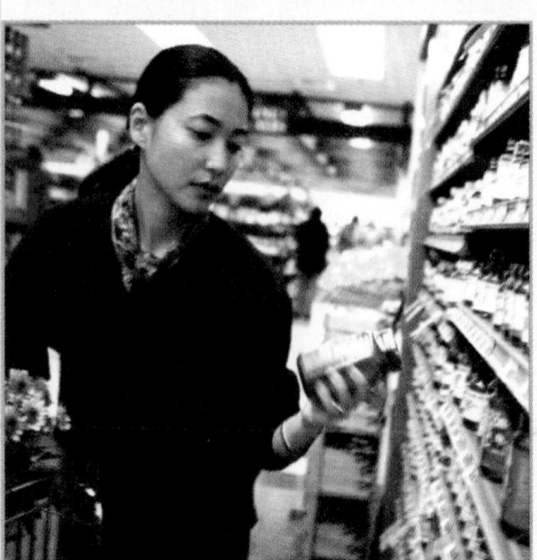

Table C7-1

Antioxidant Terms

- **dietary antioxidant** a compound typically found in foods that significantly decreases the adverse effects of oxidants on human physical functions. The antioxidant vitamins are vitamin E, vitamin C, and beta-carotene, the plant precursor of vitamin A.
- **electron** part of an atom; a negatively charged particle. Stable atoms (and molecules, which are made of atoms) have even numbers of electrons in pairs. An atom or molecule with an unpaired electron is an unstable *free radical.*
- **free radical** an atom or molecule with one or more unpaired electrons that make it unstable and highly reactive.
- **oxidant** a compound (such as oxygen itself) that oxidizes other compounds. Compounds that prevent oxidation are called *anti*oxidants, whereas those that promote it are called *pro*oxidants (*anti* means "against"; *pro* means "for").
- **oxidative stress** damage inflicted on living systems by free radicals.

Table C7-2

Factors That Increase Free-Radical Formation

Body Factors	Environmental Factors
Energy metabolism	Air pollution
Diabetes	Asbestos
Exercise	High levels of vitamin C
Acute illness	High levels of oxygen
Immune response	Radioactive emissions (for example, from radon gas)
Injury	Some herbicides
Obesity	Tobacco smoke
Other diseases	Trace minerals (iron, copper)
Other metabolic reactions	Ultraviolet light rays

SOURCE: Data from B. N. Ames and coauthors, Oxidants, antioxidants, and the degenerative disease of aging, *Proceedings of the National Academy of Sciences* 90 (1993): 7915–7920.

kidney tissues, producing severe vascular and kidney disease.[4] Uncontrolled diabetes, in turn, causes more oxidative stress, creating a cycle of worsening disease. Even the effects of physical aging are thought by some to be the result of years of free-radical damage. No wonder antioxidants seem to be part of the answer, whatever the question about nutrition.

Free radicals are not all bad. In fact, their destructive properties are put to good use by some cells of the immune system. These cells make free radicals, to use as ammunition in an "oxidative burst" that lays waste to viruses and bacteria that might otherwise cause diseases. Partly for this reason, infections cause a detectable increase in free-radical activity all over the body.

THE BODY'S DEFENSES AGAINST FREE RADICALS

The body's two main systems of defense against damage from free radicals are its reserves of antioxidants and its enzyme systems that oppose oxidation.* These defense systems try to handle all free radicals, but they are not 100 percent effective. If insufficient radical-fighting agents are present in the body, if free radicals become excessive, or if the body's repair systems cannot undo all of the damage, health problems may develop. Unrepaired damage accumulates as people age.

*Internal enzyme systems include an enzyme-selenium complex and the *superoxide dismutase* (*SOD*) system. Chapter 8 offers more on minerals and oxidation.

Antioxidant Vitamins The body maintains pools of the antioxidant vitamins: vitamin E, vitamin C, and the vitamin A precursor beta-carotene. These vitamins actively scavenge and quench free radicals, becoming oxidized (and inactive) themselves in the process. Once oxidized, they can to some extent be regenerated to become active antioxidants again, but some are dismantled and discarded. Free radicals attack the body continuously, so to maintain defenses, a person's supplies of antioxidants must be replenished as rapidly as they are used up.

Among the antioxidant vitamins, vitamin E and beta-carotene defend the body's lipids. Vitamin E efficiently breaks the free-radical chain reaction at a rate 200 times faster than BHT,† a commercial antioxidant added to baked goods to prevent rancidity from fat oxidation. Vitamin C protects the body's watery components, such as the fluid of the blood, against free-radical attacks. Vitamin C seems especially adept at neutralizing free radicals from polluted air and cigarette smoke; it also has the knack of restoring oxidized vitamin E to its active state. An example of how these defenders prevent tissue damage from free radicals appears in Figure C7-1 on page 258.

If extra antioxidants are present in the tissues, then free radicals are less likely to damage the cells (see Figure C7-2). This line of thinking is used to sell millions of dollars worth of antioxidant supplements to consumers each year, although, in fact, protection may be conferred, not by these particular nutrients, but by *other* factors, such as phytochemicals, in foods that contain them.[5] With this uncertainty, we begin our tally of points for foods versus supplements at zero points for both.

†BHT is butylated hydroxytoluene.

Phytochemicals Some phytochemicals have antioxidant activity, although they work in many other ways, too. One phytochemical filters incoming light, another slows blood clotting, some act as hormones, others inhibit harmful chemical reactions, still others stimulate immunity, and many act by mechanisms as yet unknown. The evidence on phytochemicals concludes the core chapters of this book, in Controversy 11, but even those who never read it can arrive at a conclusion based on what is be presented here. Foods or supplements: Which is the better choice? Phytochemicals score a major point in favor of foods: foods,1; supplements, 0.

The Internal Defense System In addition to using antioxidant compounds from foods, the body defends itself against the free-radical threat by using a powerful system of cellular enzymes that neutralize the free radicals they encounter. These enzymes are proteins whose concentrations are controlled both by inherited genes and by influences affecting those genes. A group of enzymes contain the mineral selenium; another enzyme is *superoxide dismutase (SOD),* which has been intensively studied and even purified for use as an anti-aging supplement. Because these enzymes are proteins, however, they are useless as dietary supplements. Enzymes taken by mouth are digested in the stomach and small intestine long before they reach the bloodstream. In our scoring of foods versus supplements, then, supplements still score 0 points.

DO ANTIOXIDANT NUTRIENTS PROTECT AGAINST CANCER?

Cancers arise when cellular DNA is damaged—sometimes by free-radical attacks.

If antioxidant nutrients protect DNA from this damage, then they probably reduce cancer risks. The strongest evidence that they do is found in studies of populations with high intakes of vegetables and fruits rich in antioxidant nutrients.[6] These populations most often have low rates of cancer.[7] Laboratory studies with animals and with cell cultures seem to support such findings.

Beta-Carotene and Its Relatives People with high cancer rates have been found to consume the fewest vegetables and fruits, especially those containing beta-carotene and its relatives. This evidence was strengthened by findings of researchers who collected human blood samples and found that low concentrations of beta-carotene consistently correlated with the development of cancers of the breast, cervix, and lung.[8] For a while, people were convinced that this evidence provided conclusive proof of beta-carotene's protective anticancer effect. In fact, many people bought and took beta-carotene supplements long before they were proved safe and effective.

Support for beta-carotene supplements soon crumbled, however.[9] One study declared them useless in preventing a condition that leads to colon cancer.[10] A surprise result came from another study designed to determine the benefits of beta-carotene supplements on the incidence of lung cancer among smokers.[11] The researchers expected to see a beneficial effect, but it never appeared.[12] All major clinical trials of beta-carotene supplements had to be stopped when 28 percent more of the participants taking beta-carotene developed lung cancer than did the members of the control group who took placebos. The researchers are continuing to observe the experimental group to detect any possible long-term adverse effects of taking beta-carotene. Another long-term study reported no differences in disease rates between physicians who took beta-carotene for 14 years and a matched group who took placebos.[13] What exactly these results mean to the individual supplement taker is unknown, but overall the studies seem to indicate no benefit, and a possible risk to smokers, from taking supplements of beta-carotene.

These results presented an apparent contradiction: beta-carotene in the diet and elevated beta-carotene in the blood are associated with lower cancer incidence, but the taking of beta-carotene supplements is not. Researchers quickly discovered why, however. Beta-carotene itself is just one of the antioxidant nutrients present in the foods chosen for study. Along with beta-carotene, the foods contained other essential nutrients, not only the vitamins named here, and not only antioxidants, but nutrients that fight cancer in other ways. These include vitamin A itself, vitamin B_6, pantothenic acid, vitamin B_{12}, zinc, iron, copper, selenium, and more.

Besides all of these nutrients, hundreds of phytochemicals are present in foods, too. Health effects attributed to beta-carotene may, in reality, be the work of one or a number of *nonnutrient* compounds in fruits and vegetables.[14] Beta-carotene may simply tag along, serving as a marker for one of its relatives or for one or more as-yet-unknown phytochemicals that are actually responsible for the effect. Foods now have 2 points in their favor; supplements still score 0.

Vitamins C and E When people's diets include foods rich in vitamin C, they seem to develop fewer cancers of the mouth, larynx, and esophagus. A

Figure C7-1

The Theory of Antioxidants and Disease

Free-radical formation occurs during metabolic processes, and it accelerates when diseases or other stresses strike.

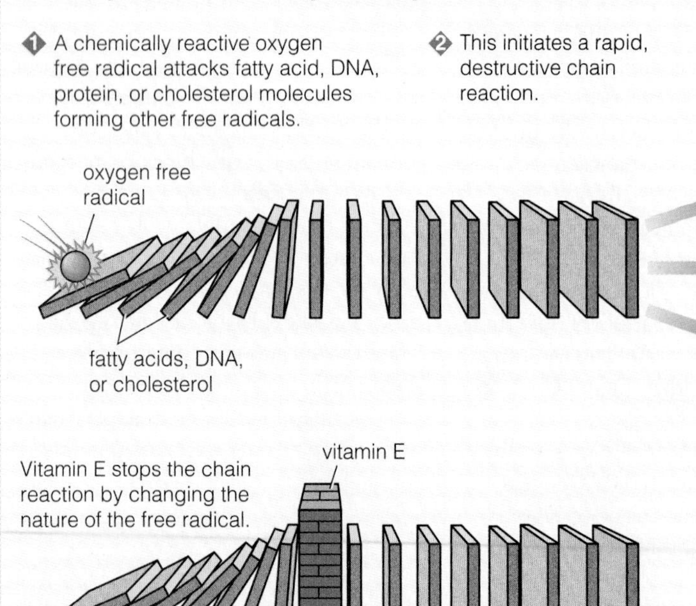

1 A chemically reactive oxygen free radical attacks fatty acid, DNA, protein, or cholesterol molecules forming other free radicals.
oxygen free radical
fatty acids, DNA, or cholesterol
2 This initiates a rapid, destructive chain reaction.
3 The result is injury to tissues and the formation of more free radicals:
damage to cell membrane lipids and proteins, disabling them
precancerous changes in DNA
oxidation of blood cholesterol initiating steps leading to heart disease
4 And ultimately, diseases and tissue aging:
cancer
heart disease
macular degeneration
other diseases
aging
Vitamin E stops the chain reaction by changing the nature of the free radical.
vitamin E

Figure C7-2

The Antioxidant Theory of Disease Prevention

A. *Normally, the body's antioxidant enzymes, vitamins, and other molecules are sufficient to neutralize free-radical molecules before they do much damage.*
B. *An increased free-radical load can overwhelm the body's antioxidant systems. Free radicals then damage the tissues.*
C. *The body stocked with extra antioxidants is best equipped to handle an increase in free radicals. The two ways to obtain more antioxidants are to build them into cells by exercising, which stimulates production of more antioxidant enzymes, and to eat them in foods or supplements, which supply antioxidant vitamins and phytochemicals.*

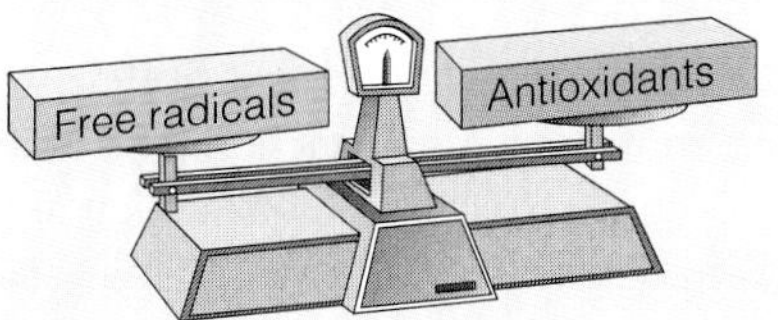
Free radicals
Antioxidants

(a) Balanced system

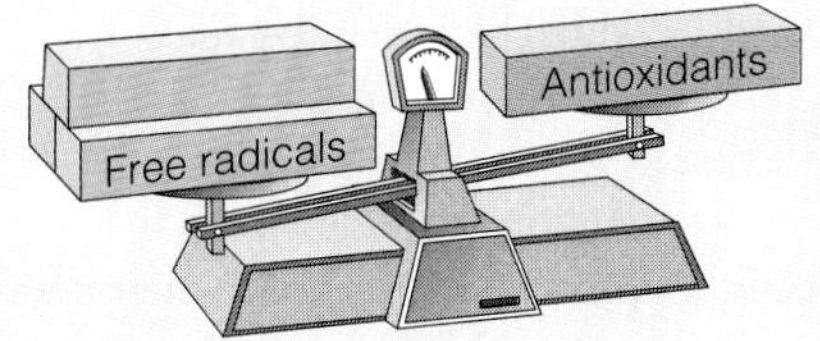
Free radicals
Antioxidants

(b) Additional free-radical load—damage

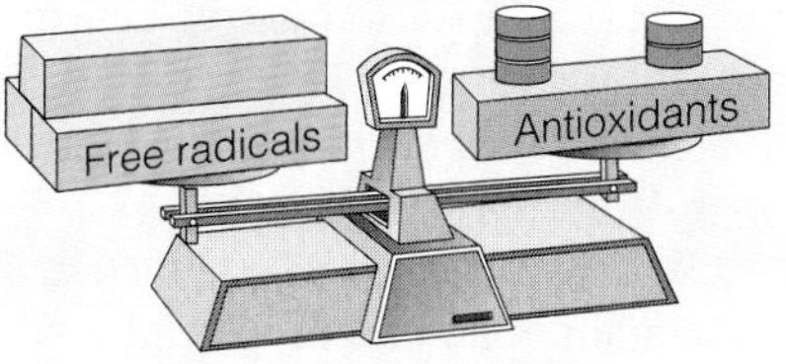
Free radicals
Antioxidants

(c) Additional antioxidants added and balance restored

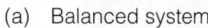

dozen or so studies have confirmed this relationship. Like beta-carotene, however, vitamin C occurs in foods together with other cancer-fighting constituents. For example, broccoli and its sprouted seeds, leafy greens, and citrus fruits, which are all vitamin C–rich foods, also contain powerful phytochemicals thought to be active against cancer. They are also fiber-rich and low in fat—two other dietary characteristics believed to reduce cancer risk. Like beta-carotene, then, vitamin C may simply be a marker for a diet rich in fruits and vegetables. If so, the taking of vitamin C pills alone will do nothing to prevent cancer. Foods, 3; supplements, still 0.

The final vitamin of note on the list of vitamins against cancer is vitamin E. A study reported a reduced risk of melanoma, a deadly skin cancer, in people with high intakes of *foods* rich in vitamin E.[15] Studies have generally found vitamin E *supplements* useless against both colon and lung cancers.[16] Foods, 4; supplements, still 0.

Selenium The selenium content of food depends partly on the selenium content of the soil, which varies from region to region. In low-selenium areas of the United States, people suffer from higher rates of some cancers.[17] Studies of animals confirm these results and suggest that selenium may play active roles in cancer prevention. A recent experiment with human beings yielded interesting results. In a study of over 1,300 people with a history of skin cancer, half of the subjects were given 200 micrograms of selenium over several years while the others received a placebo during the same period.[18] The researchers found that the incidence of recurrence of skin cancer was about the same between the two groups. However, the selenium-treated group had fewer cancers of the prostate, colon, and lung. In light of these seeming benefits, the researchers stopped their study to allow all study participants access to the selenium treatment.

Selenium supporters hail this study as proof that selenium supplements prevent cancer, but a problem remains unresolved. Though the selenium takers did indeed suffer fewer cancers of some types, their rate of death from all causes was virtually identical to that of the placebo group. Dr. Victor Herbert, a researcher famous for his anti-supplement stance, concluded that the selenium given in the study must have *increased* the rate of deadly diseases other than cancer, thus equalizing the overall death rates between the two groups.[19] What those diseases might have been and how they may have been affected by selenium supplements remains a mystery.

In any case, an adequate intake of selenium is not to be neglected, especially by men. In another study, men who tested low on selenium were reported as being more likely to develop prostate cancer than men whose selenium stores were full.[20] No one knows for certain whether selenium's link to cancer results from its role in enzymatic antioxidant activity, is a function of improved immunity, or whether selenium may have a direct inhibiting effect on cancer cell development. Research is advancing knowledge along these lines.[21]

So far, the evidence is a long way from supporting the taking of selenium supplements, and they are not without risks. In doses above about 750 micrograms per day, selenium is toxic.[22] In animals grazing on selenium-rich pastures, toxicity has caused hoof loss and nerve and muscle damage known as "blind staggers." In people, toxicities have caused nausea, loss of hair and fingernails, and nerve damage. Meanwhile, no harm can come from including in the diet nutritious foods such as fish and whole grains, which are especially reliable sources of selenium.

The Food and Drug Administration (FDA) agrees that foods, not supplements, provide anticancer benefits. After reviewing the evidence, the agency concluded that *diets* high in fruits and vegetables, which are particularly good sources of beta-carotene and vitamin C, are strongly associated with reduced risks of several types of cancer. The reductions in risk could not be attributed solely to the named vitamins. Accordingly, the FDA rejected a request by the supplement industry to allow health claims on the labels of antioxidant supplements and ruled that food labels may make health claims only in terms of fruits and vegetables

and cancer. From the evidence so far, then, give supplements another zero, but score another point for foods. Foods, 5; supplements, 0.

EVIDENCE CONCERNING AGE-RELATED BLINDNESS

The leading cause of age-related blindness in the United States is macular degeneration, already mentioned in Chapter 7. This form of blindness occurs when the macula, located at the focal center of the retina, loses integrity. This causes the loss of the most important field of vision, the area in focus (peripheral vision remains unimpaired). This blindness has, until now, been untreatable and unpreventable, but the possibility of prevention has been suggested by a study of people's diets. Researchers have reported a 43 percent reduced incidence of macular degeneration in those consuming diets high in carotenoids.[23] Beta-carotene is not, however, the carotenoid credited with the effect. Other nonnutritive carotenoids are more likely at work in this regard. These carotenoids, supplied by fruits and vegetables, have the ability to filter out damaging light rays before they can harm the macula. The same foods often also provide vitamin C, another protector of eye health.[24] No food or supplement so far protects the eyes from destructive effects inflicted by smoking, however.[25] Foods now have 6 points; supplements, still 0.

DO ANTIOXIDANT NUTRIENTS PROTECT AGAINST HEART DISEASE?

Antioxidant nutrients, especially vitamin E, may help protect against cardiovascular disease.[26] A theory about how vitamin E might accomplish this involves oxidation of blood cholesterol. Cholesterol carried in low-density lipoproteins (LDL) in the blood correlates directly with cardiovascular disease. Further, most of the cholesterol scientists have collected from damaged arteries has turned out to be oxidized cholesterol. The theory suggests that LDL undergo oxidation by free radicals inside the artery wall, thereby promoting the formation of artery-clogging plaques.[27]

Vitamin E and Heart Disease

Vitamin E may offer some protection against heart disease by protecting LDL from oxidation.[28] Some research supports this possibility. Scientists selected groups of men in 16 European regions where rates of death from heart disease varied sixfold. The researchers compared the plasma vitamin E, cholesterol, and blood pressure among the men from each region. The men with the lowest vitamin E values died more often from heart disease. The correlation of heart disease mortality with low vitamin E was even stronger than with high cholesterol or high blood pressure, supporting the "antioxidant hypothesis" of heart disease. The authors cautioned, though, that although this evidence was suggestive, it was also indirect. The question remained whether the vitamin E in the blood was "the" agent that should get the credit, or whether it came as part of a package (such as foods) containing an unknown active agent.

Evidence bearing on this question comes from two large epidemiological studies that used large-dose vitamin E supplements. The supplements appeared to be associated with a significantly reduced risk of heart disease.[29] This correlation remained strong after the researchers controlled for heart disease risks and for other dietary antioxidants. In another study, researchers tracked changes in the arteries of men with heart disease. In men who were taking cholesterol-lowering drugs as well as at least 100 milligrams of vitamin E a day and whose arteries were only mildly damaged, the disease progressed more slowly than in others. Men with advanced disease and those not taking cholesterol-lowering drugs gained no benefit from vitamin E.[30] Another study of women suggests that a diet of foods rich in vitamin E is associated with fewer heart disease deaths.[31]

An experimental study recently added another piece of evidence to the vitamin E–heart disease puzzle. Researchers at Cambridge University, England, who are conducting a long-term clinical experiment, have released data supporting the idea that vitamin E may prevent heart attacks in people with diagnosed advanced heart disease.[32] Ten months into the study, the researchers could already see that people who received doses of 400 and 800 milligrams of vitamin E per day for two years suffered fewer *heart attacks* than others like them who were not given vitamin E. Total *deaths* from heart disease, however, were unaffected by vitamin E.

Vitamin E may also protect the arteries from changes in functioning that are thought to be caused by oxidation and that precede heart disease.[33] Fat in the diet causes the linings of the arteries to constrict, an effect observed after even a single high-fat meal. When a high-fat meal is accompanied by supplementation with vitamins E and C, the excess oxidative activity that typically follows a high-fat meal is eliminated.

By whatever mechanism, high doses of vitamin E, a nutrient, produce health effects like those of a drug. Because both supplements and foods have been observed to coincide with reduced heart disease, the rationale for taking supplements of vitamin E gains credibility. At last, we can give a point to supplements, specifically those of vitamin E, but vitamin E–rich foods must receive another point, too. Foods, 7; supplements, 1.

Vitamin C and Heart Disease

Vitamin C may or may not affect susceptibility to heart disease; research results are mixed. Logic suggests teamwork between vitamin C and vitamin E in defending LDL against oxidation: vitamin C defends against free radicals in the watery compartments of cells, vitamin E in lipid environments.[34] Also, as mentioned, vitamin C regenerates vitamin E from its oxidized form, making it available to act again as an antioxidant.[35] Some studies also suggest that vitamin C works with vitamin E to reduce the damage from an atherogenic diet, that is, a diet likely to lead to atherosclerosis.[36] It may also raise HDL, lower total cholesterol, and improve blood pressure.[37] These findings might support a point for vitamin C supplements, but consider these results from a recent study of over 1,600 men.[38] After measuring the men's vitamin C status and tracking them for an average of five years, researchers observed a striking correlation between low blood vitamin C values and heart attacks. Mild vitamin C deficiency increased a man's risk of heart attack by two and a half times. However, two other factors measured in this study also correlated with both low blood vitamin C and heart attack—a low intake of fruits and vegetables and low blood carotene levels. From these results, then, it is impossible to isolate vitamin C as the only factor at play. Something else lacking from a diet low in fruits and vegetables could have elevated the men's risks. It's easy to get enough vitamin C from fruits and vegetables, and these foods supply an almost unimaginable host of other nutrients and phytochemicals (see Controversy 11), many with the potential to protect the heart. A few servings a day of vitamin C–rich foods can make high-dose vitamin C supplements, along with their associated risks, unnecessary.[39] Perhaps both could receive a point. Foods, now 8; supplements, 2.

PERSONAL STRATEGY ON VITAMIN SUPPLEMENTS

To this point, foods have beaten vitamin supplements by a score of 8 to 2. Much more evidence in favor of foods over supplements arises from study of the phytochemicals, still to be discussed in Chapter 11. Still, a wrap-up can be based on what has been presented here.

Should We Take Supplements, Just to Be Safe? Supplement manufacturers have proclaimed single antioxidant *pills* as the new magic bullets against aging, disease, and even death itself. Dr. Victor Herbert, the scientist introduced earlier, energetically opposes these claims. He points to a host of side effects that might endanger supplement takers' health. Among these effects are the following:

- Vitamin E supplements, taken over a period of time, may increase the risk of brain hemorrhage (a form of stroke).
- Vitamin E supplements delay blood clotting.
- Vitamin E supplements may worsen autoimmune diseases, such as asthma or rheumatoid arthritis.
- Vitamin C supplements enhance iron absorption, making iron overload likely in some people.
- Daily supplements of vitamin E, beta-carotene, or both do not reduce the incidence of lung cancer among smokers, and beta-carotene may increase it.[40]
- Selenium supplements are toxic in the ways already mentioned.

For the latest cancer fighters, visit your local produce center.

Dr. Herbert makes the point that while orange juice and pills may both contain vitamin C, the orange juice presents a balanced array of chemicals that modulate vitamin C's effects. The pill provides only vitamin C, a lone chemical. And in general, although fruits and vegetables rich in antioxidant nutrients have been associated with a diminished risk of many cancers, supplements of beta-carotene and vitamins C and E have not always proved beneficial.

The struggle for truth continues. Members of the Food and Nutrition Board of the National Research Council have broadened the definition of their intake recommendations to not only prevent classic deficiency diseases, but to help protect against chronic diseases as well. Meanwhile, most scientists agree that it is too early to recommend that people start taking antioxidant supplements now, even those of vitamin E. The risks are real, and clinical studies to quantify them and clarify the benefits will take several years to complete.

Should We Try to Eat More Antioxidant-Rich Foods? Foods deliver thousands of chemicals other than the handful we call nutrients. Anyone judging a food solely by its nutrient content might scorn a turnip root or a radish because these foods appear almost devoid of nutrients in a nutrient-composition table such as the one in this book's Appendix A. But turnip roots and radishes are full of phytochemicals that enhance their value to the body. Researchers must be careful in crediting a particular health benefit to a nutrient when the source of that nutrient is whole food.

It seems reasonable, for now, to conclude by recommending this personal strategy. Don't try to single out only a few magic nutrients to take as supplements. Instead, invest energy in eating a wide variety of fruits and vegetables in generous quantities every day. It is one of the most important favors you can do yourself, and the benefits are well backed by research.

8 *Water and Minerals*

Teresa Fasolino, *Rice Paddy,* date unknown

Contents

Frequently Asked Questions

minerals naturally occurring, inorganic, homogeneous substances; chemical elements.

major minerals essential mineral nutrients found in the human body in amounts larger than 5 grams.

trace minerals essential mineral nutrients found in the human body in amounts less than 5 grams.

"ASHES TO ASHES and dust to dust": This familiar biblical quotation reminds us of our mortality. Perhaps we need this reminder to put our own importance into perspective. It is true that when the life force leaves the body, what is left behind ultimately becomes nothing but a small pile of ashes. Carbohydrates, proteins, fats, vitamins, and water are present at first, but they soon disappear.

The carbon atoms in all the carbohydrates, fats, proteins, and vitamins combine with oxygen to produce carbon dioxide, which vanishes into the air; the hydrogens and oxygens of those compounds unite to form water; and this water, along with the water that was a large part of the body weight, evaporates. The ashes that are left behind are the **minerals,** a small pile that weighs only about 5 pounds. The pile is not impressive in size, but when you consider the tasks these minerals perform, you realize their great importance in living tissue.

Consider calcium and phosphorus. If you could separate these two minerals from the rest of the pile, you would take away about three-fourths of the total. Crystals made of these two minerals, plus a few others, form the structure of the bones and so provide the architecture of the skeleton.

By running a magnet through the pile that remains, you would pick up the iron. It would not fill a teaspoon, but it consists of billions and billions of iron atoms. As part of hemoglobin, these iron atoms have the special property of being able to attach to oxygen and make it available at the sites inside the cells where metabolic work is taking place.

If you now extracted all the other minerals, leaving only copper and iodine in the pile of ashes, you would want to close the windows first. A slight breeze would blow these remaining bits of dust away. Yet the amount of copper remaining in the dust is necessary for iron to hold and to release oxygen, and iodine is the critical mineral in the thyroid hormones. Figure 8-1 shows the amounts of the **major minerals** and a few of the **trace minerals** in the human body. Other minerals such as gold and aluminum, though present in the body, are not known to be nutrients.

That we distinguish between the major and the trace minerals doesn't mean that one group is more important in the body than the other. A daily deficiency of a few micrograms of iodine is just as serious as a deficiency of several hundred milligrams of calcium. Major minerals and trace minerals all play specific roles. Because the major minerals are present in larger total quantities, however, they influence the body fluids, thereby affecting the whole body in a general way.

A person can drink pure water, but in the body, that water mingles with minerals to become fluids in which all life processes take place. This chapter begins with a discussion of water—the most indispensable nutrient of all—and the major minerals that characterize the body's fluids and regulate their distribution within the body. Then the chapter discusses the specialized roles of the minerals.

Water is the most indispensable nutrient.

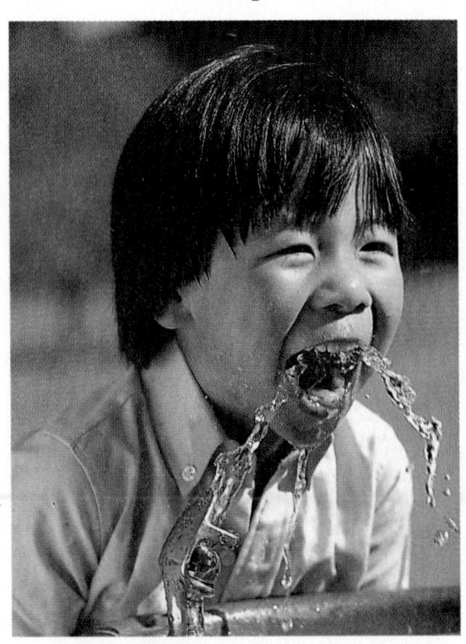

Water

You began as a single cell bathed in a nourishing fluid. As you became a beautifully organized, air-breathing body of trillions of cells, each of your cells had to remain next to water to stay alive. Water brings to each cell the exact ingredients the cell requires and carries away the end products of its life-sustaining reactions.

Water in the body is not simply a river coursing through the arteries, capillaries, and veins. Some of the water is part of the chemical structure of compounds that form the cells, tissues, and organs of the body. For example, proteins hold water molecules within them. This water is locked in and is not readily available for any other use. Water also participates actively in many chemical reactions.

What Does Water Do for My Body?

As the medium for the body's traffic of nutrients and waste products, water is nearly a universal solvent. Luckily for our physical integrity, this is not quite the case, but

Figure 8-1

Minerals in a 60-Kilogram (132-Pound) Person

The major minerals are those present in amounts larger than 5 grams (a teaspoon). The essential trace minerals number a dozen or more: only four are shown. A pound is about 454 grams; thus only calcium and phosphorus appear in amounts larger than a pound.

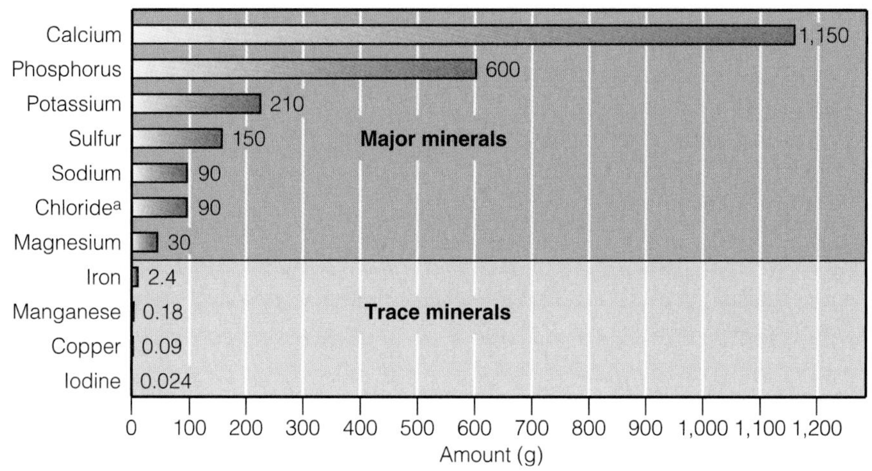
Calcium
Phosphorus
Potassium
Sulfur
Sodium
Chloride[a]
Magnesium
Iron
Manganese
Copper
Iodine
1,150
600
210
150
90
90
30
2.4
0.18
0.09
0.024
Major minerals
Trace minerals
0
100
200
300
400
500
600
700
800
900
1,000
1,100
1,200
Amount (g)

[a]Chlorine appears in the body as the chloride ion.

water does dissolve amino acids, glucose, minerals, and many other substances needed by the cells. Fatty substances are specially packaged with water-soluble proteins so that they, too, can travel freely in the blood and lymph. The water of the body fluids is thus the transport vehicle for all the nutrients.

Boasting scientist: "I'm working on discovering the universal solvent." Skeptic: "Is that so? Well, when you've got it, what are you going to keep it in?"

Water is also the body's cleansing agent. Small molecules, such as the nitrogen wastes generated during protein metabolism, dissolve in the watery blood and must be removed before they build up to toxic concentrations. The kidneys filter these wastes from the blood and excrete them, mixed with water, as urine. When the kidneys become diseased, as can happen in diabetes, toxins can build to life-threatening levels. A machine must then take over the task of cleansing the blood by filtering wastes into water contained in the machine.*

Another important characteristic of water is its incompressibility. Its molecules resist being crowded together. Thanks to this characteristic, water can act as a lubricant and a cushion for the joints. For the same reason, it can protect a sensitive tissue such as the spinal cord from shock. The fluid that fills the eye serves in a similar way to keep optimal pressure on the retina and lens. From the start of human life, a fetus is cushioned against shock by the bag of amniotic fluid in which it develops into an infant. Water also lubricates the digestive tract and all tissues that are moistened with mucus.

Still another of water's special features is its heat-regulating capacity. This characteristic of water is familiar to coastal dwellers, who know that land near large bodies of water is protected from wide variations in temperature from day to night. Water itself changes temperature slowly; at night, when the land cools, the water gradually gives up its heat to the air, moderating the coolness of the night. In contrast, desert temperatures vary widely from day to night because there is no water to moderate them. In the body, water plays a similar role—it helps to maintain body temperature.

Human life begins in water.

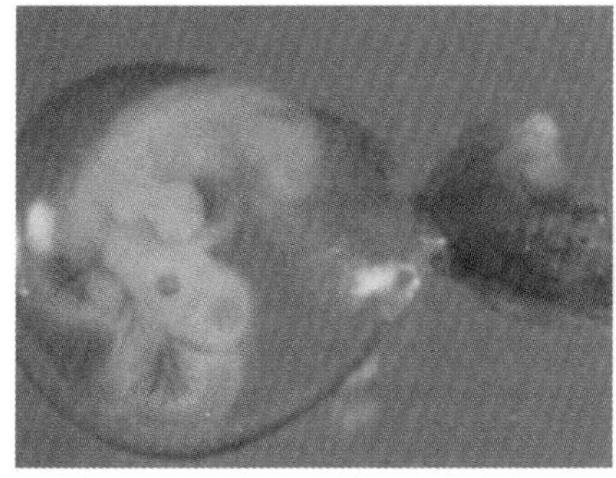

The water of sweat is the body's coolant. Heat, produced as a by-product of energy metabolism, can build up dangerously in the body. To rid itself of excess heat, the body routes its blood supply through the capillaries just under the skin. At the same time, the skin secretes sweat and its water evaporates. Converting water to vapor takes energy; therefore, as sweat evaporates, heat energy dissipates, cooling the skin and the underlying blood. The cooled blood then flows back to cool the body's core. Sweat is

*The machine that cleanses the blood is a kidney dialysis machine.

water balance the balance between water intake and water excretion, which keeps the body's water content constant.

dehydration loss of water. The symptoms progress rapidly, from thirst to weakness to exhaustion and delirium, and end in death.

water intoxication the rare condition in which body water content is too high. Symptoms are headache, muscular weakness, lack of concentration, poor memory, and loss of appetite.

constantly evaporating from the skin, usually in slight amounts that go unnoticed; thus the skin is a major organ through which water is lost from the body. To sum up, water:

- Carries nutrients throughout the body.
- Cleanses the blood of wastes.
- Serves as the solvent for minerals, vitamins, amino acids, glucose, and other small molecules.
- Actively participates in many chemical reactions.
- Acts as a lubricant around joints.
- Serves as a shock absorber inside the eyes, spinal cord, joints, and amniotic sac surrounding a fetus in the womb.
- Aids in maintaining the body's temperature.

KEY POINT *Water acts as a solvent, provides the medium for transportation, participates in chemical reactions, provides lubrication and shock protection, and aids in temperature regulation in the human body.*

The Body's Water Balance

Factors that increase water needs:

- Alcohol or caffeine consumption.
- Diseases that disturb water balance, such as diabetes.
- Exercise (see Chapter 10).
- Forced air environments, such as airplanes or sealed buildings.
- Heated environments.
- Hot weather.
- Increased dietary fiber, protein, salt, or sugar.
- Medications (diuretics).
- Pregnancy and breastfeeding (Chapter 12).
- Prolonged diarrhea, vomiting, or fever.
- Surgery, blood loss, or burns.
- Very young or old age.

Water makes up about 60 percent of the body's weight. It is such an integral part of us that people seldom are conscious of water's importance, unless they are deprived of it. You can survive a deficiency of any of the other nutrients for a long time, in some cases for months or years, but you can survive only a few days without water. Since the body must excrete at least a pint of water a day to cleanse its fluids, a person must consume at least a pint each day to avoid life-threatening losses, that is, to maintain **water balance.**

The total amount of fluid in the body is kept constant by delicate balancing mechanisms. Imbalances can occur, such as **dehydration** and **water intoxication,** but the balance is restored to normal as promptly as the body can manage it. Both intake and excretion are controlled to maintain water balance.

The amount of the body's water varies by pounds at a time, especially in women who retain water at the menses. Anyone who eats a meal high in salt can temporarily increase the body's water content, which the body sheds over the next day or so as the sodium is excreted. These temporary fluctuations in body water also cause changes in body weight on the scales. People who gain or lose water weight may believe the change reflects a change in body fat, but fat weight takes days or weeks to change noticeably, whereas water weight can change overnight.

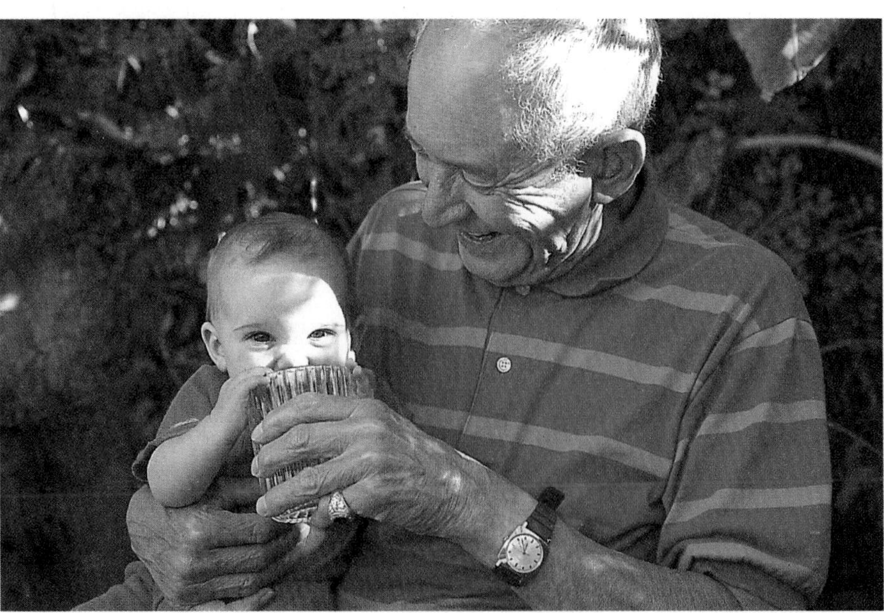

An extra drink of water benefits both young and old.

KEY POINT *Water makes up about 60 percent of the body's weight. A change in the body's water content can bring a change in body weight.*

Quenching Thirst and Balancing Losses

Thirst and satiety govern water intake. When the blood is too concentrated (having lost water but not salt and other dissolved substances), the molecules and particles in the blood attract water out of the salivary glands. The mouth becomes dry as a result, and you drink to wet your mouth. The brain center known as the hypothalamus (described in Chapter 3) also monitors the concentration of the blood. When the blood is too concentrated, or when the blood volume or pressure is too low, the hypothalamus initiates impulses that stimulate drinking behavior. The hypothalamus also calls forth a hormone from the pituitary gland that directs the kidneys to shift water back into the bloodstream from the pool destined for excretion. The kidneys themselves also respond to the sodium concentration in the blood passing through them and secrete regulatory substances of their own. The net result is that the more water the body needs, the less it excretes. Figure 8-2 shows how intake and excretion naturally balance out.

The hypothalamus and the pituitary gland were shown in Figure 3-5 of Chapter 3.

Thirst lags behind a lack of water. A water deficiency that develops slowly can switch on drinking behavior in time to prevent serious dehydration, but one that develops quickly may not. When too much water is lost from the body and is not replaced, dehydration can threaten survival. A first sign of dehydration is thirst, the signal that the body has lost up to 2 cups of its total fluid. Rather than waiting until thirst sets in, people should drink regularly throughout the day. But suppose a person is unable to obtain fluid or, as in many elderly people, fails to perceive the thirst message. With a loss of just 5 percent of body fluid, perceptible symptoms appear: headache, fatigue, confusion or forgetfulness, and an elevated heart rate. Instead of "wasting" any of its precious water in sweat, the dehydrated body diverts most of its water into the blood vessels to maintain the life-supporting blood pressure. Meanwhile, body heat builds up because sweating has ceased, creating the possibility of serious consequences (see Table 8-1).

KEY POINT *Water losses from the body necessitate intake equal to output to maintain balance. The brain regulates water intake; the brain and kidneys regulate water excretion. Dehydration can have serious consequences.*

Figure 8-2

Water Balance

Water enters the body in liquids and foods, and some water is created in the body as a by-product of metabolic processes. Water leaves the body through the evaporation of sweat, in the moisture of exhaled breath, in the urine, and in the feces.

Water input (Total = 1,450–2,800 ml)

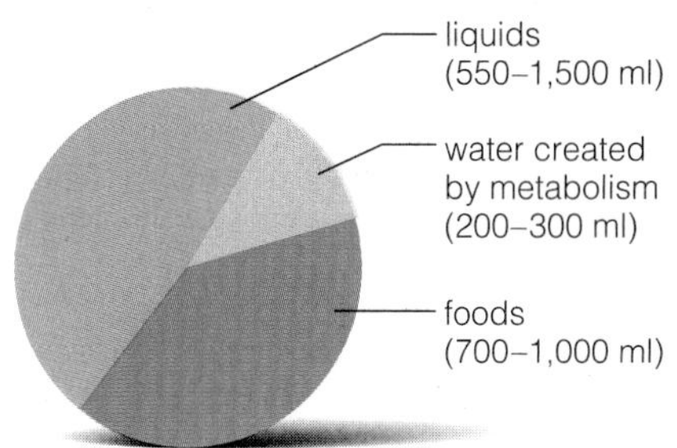
liquids (550–1,500 ml)
water created by metabolism (200–300 ml)
foods (700–1,000 ml)

Water output (Total = 1,450–2,800 ml)

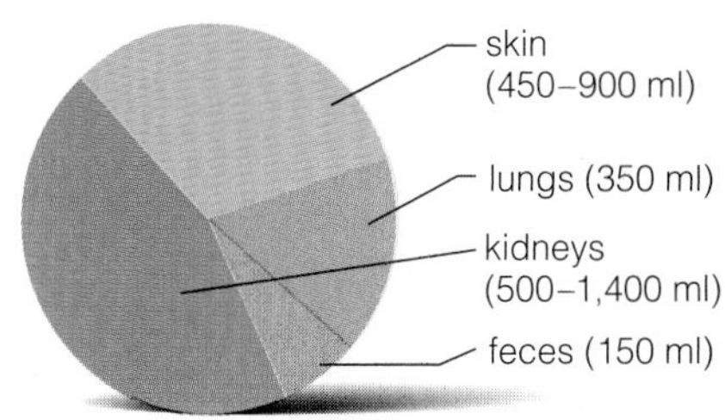
skin (450–900 ml)
lungs (350 ml)
kidneys (500–1,400 ml)
feces (150 ml)

Water Recommendations and Sources

Water needs vary greatly depending on the foods a person eats, the environmental temperature and humidity, the person's activity level, and other factors. Under normal dietary and environmental conditions, adults need between 1 and 1½ milliliters of

Table 8-1

Signs of Mild and Severe Dehydration

Mild (< 5% Loss of Body Weight)	Severe (> 5% Loss of Body Weight)
Thirst	Pale skin
Sudden weight loss	Bluish lips and fingertips
Rough dry skin	Confusion; disorientation
Dry mouth, throat, body linings	Rapid, shallow breathing
Rapid pulse	Weak, rapid, irregular pulse
Low blood pressure	Thickening of blood
Lack of energy; weakness	Shock; seizures
Impaired kidney function	Coma; death
Reduced quantity of urine; concentrated urine	

diuretics (dye-you-RET-ics) compounds, usually medications, causing increased urinary water excretion; "water pills."

hard water water with high calcium and magnesium concentrations.

soft water water with a high sodium concentration.

bottled water drinking water sold in bottles.

salts compounds composed of charged particles (ions). An example is potassium chloride (K^+Cl^-).

ions (EYE-ons) electrically charged particles, such as sodium (positively charged) or chloride (negatively charged).

electrolytes compounds that partly dissociate in water to form ions, such as the potassium ion (K^+) and the chloride ion (Cl^-).

Water content of various foods and beverages:

- 100% = water, diet soft drinks, seltzer (unflavored), plain tea.
- 95–99% = sugar-free gelatin dessert, clear broth, Chinese cabbage, celery, cucumber, lettuce, summer squash, decaffeinated black coffee.
- 90–94% = Gatorade, grapefruit, fresh strawberries, broccoli, tomato.
- 80–89% = sugar-sweetened soft drinks, milk, yogurt, egg white, fruit juices, low-fat cottage cheese, fresh apple, carrot.
- 60–79% = low-calorie mayonnaise, instant pudding, banana, shrimp, lean steak, pork chop, baked potato.
- 40–59% = diet margarine, sausage, chicken, macaroni and cheese.
- 20–39% = bread, cake, cheddar cheese, bagel, cooked oatmeal.
- 10–19% = butter, margarine, regular mayonnaise, cooked rice.
- 5–9% = peanut butter, popcorn.
- 1–4% = ready-to-eat cereals, pretzels.
- 0% = cooking oils, meat fats, shortening, white sugar.

Lead poisoning is especially harmful to children (see Chapter 13).

water from all sources for each calorie spent in the day. For the person who expends about 2,000 calories a day, this works out to a fluid intake of about 2 to 3 liters (about 8 to 12 cups). Sweating increases water needs.

In addition to water itself and other beverages made of water, nearly all foods contain water. Most fruits and vegetables contain large quantities of water, up to 95 percent of their volume; many meats and cheeses contain at least 50 percent. Also, the energy-yielding nutrients in foods release additional water as the body breaks them down. Beverages containing alcohol or caffeine have a negative effect on the body's water balance—they are **diuretics,** compounds that cause water excretion. Many people who drink beer or coffee, then, may end up with a net fluid loss rather than a gain, to the detriment of the body's fluid balance.[1] Better choices are foods and beverages that contain abundant water without unwanted alcohol or caffeine (see the list in the margin). The Table of Food Composition, Appendix A, lists the water contents of most other foods and beverages.

Water naturally occurs as **hard water** or **soft water,** a distinction that affects health with regard to three minerals. Hard water has high concentrations of calcium and magnesium. Soft water's principal mineral is sodium. In practical terms, soft water makes more bubbles with less soap; hard water leaves a ring on the tub, a jumble of rocklike crystals in the teakettle, and a gray residue in the wash. Soft water may seem more desirable, and homeowners even purchase water softeners that remove magnesium and calcium and replace them with sodium. However, soft water appears to aggravate hypertension and heart disease in areas where it is used. Hard water may oppose these conditions.

Soft water also more easily dissolves certain metals, such as cadmium and lead, from pipes. Cadmium is not an essential nutrient. In fact, it can harm the body, affecting enzymes by displacing zinc from its normal sites of action and disturbing iron and copper transfer during pregnancy.[2] Cadmium is also suspected of promoting hypertension. Lead is another toxic metal, and the body seems to absorb it more readily from soft water than from hard water, possibly because the calcium in hard water protects against its absorption. Old plumbing may contain cadmium or lead. People who live in old buildings should run the cold water tap a minute to flush out harmful minerals before drawing water for the first use in the morning and whenever no water has been drawn during the previous six hours.

Many people turn to **bottled water** as an alternative to tap water. As the Consumer Corner points out, though, bottled water may or may not contain more health-promoting mineral arrays than ordinary tap water does.

KEY POINT *Hard water is high in calcium and magnesium. Soft water is high in sodium, and it dissolves cadmium and lead from pipes.*

Body Fluids and Minerals

Much of the body's water weight is contained inside the cells, and some water also bathes the outsides of the cells. The remainder fills the blood vessels. Special provisions are needed to ensure that the cells do not collapse when water leaves them or swell up when too much water enters them. The cells cannot regulate the amount of water directly by pumping it in and out because water slips across membranes freely. They can, however, pump minerals across their membranes. The major minerals form **salts** that dissolve in the body fluids; the cells direct where the salts go; and this determines where the fluids flow because water follows salt.

When mineral (or other) salts dissolve in water, they separate into single, electrically charged particles known as **ions.** Unlike pure water, which conducts electricity poorly, ions dissolved in water carry electrical current. For this reason, the electrically charged ions are called **electrolytes.** As Figure 8-3 on page 272 shows, when dissolved particles, such as electrolytes, are present in unequal concentrations on either side of a water-permeable membrane, water flows toward the more concentrated side to equalize the

CONSUMER CORNER

Which Type of Water Is Safest?

MANY PEOPLE, knowing that contaminated water can be injurious to health, are concerned about the safety of their water supplies. Households, traffic, industry, and agriculture all add pollutants to environmental water and thereby degrade its quality. Hundreds of contaminants, including disease-causing bacteria and viruses from human wastes, toxic pollutants from highway fuel runoff, spills and heavy metals from industry, and organic chemicals such as pesticides from agriculture, have been detected in public drinking water.

Treatment by public water systems can at least partly remove some of these hazards. The treatment includes the addition of a disinfectant (usually chlorine) to kill most microorganisms. Private well water is usually not chlorinated or cleansed, so the 40 million Americans who drink water from private wells are especially likely to encounter microorganisms in their water. All public drinking water must be tested regularly for contamination, and the Environmental Protection Agency (EPA) is responsible for ensuring that public water systems meet minimum standards for protection of public health.

The law mandates that customers of public water utilities receive a yearly statement, written in plain language, that names the chemicals and bacteria found in local water. This document makes fascinating reading for those interested in how pure their tap water is. The law also requires a utility to notify the public within 24 hours of discovering any dangerous contaminants in drinking water. The intent is to reduce the threat from the most harmful contaminants such as *Cryptosporidium,* a parasite common in lakes and rivers that invaded the public water supply and caused 400,000 people living in Milwaukee, Wisconsin to fall ill several years ago. Effects of the infection ranged from digestive distress to death. Water authorities monitor the water for this microbe, which often survives the killing effects of chlorine. Nevertheless, the incident stands as testimony that even our sophisticated water system cannot always guarantee 100 percent safety, especially during floods, drought, excessive runoff, or chemical spills, because of increased demand for water and budget constraints.* Some people fear that chlorine itself presents a danger to health, and several preliminary studies of tap water use have issued disturbing results. Researchers have noted a statistical correlation between consuming chlorinated surface water and a substantial increase in the likelihood of developing colon and other cancers.[2] Earlier experiments revealed that by-products in chlorinated water cause cancer in laboratory animals.

Although most investigators acknowledge a connection between consumption of chlorinated drinking water and cancer incidence, they also passionately defend chlorination as a benefit to public health. In areas of the world without chlorination, an estimated 25,000 people die each day from

*Concerned consumers can call the Safe Drinking Water Hotline toll-free at 1-800-426-4791, or ask experts water safety questions by e-mail at: *hotline-sdwa@epamail.epa.gov,* or visit the EPA's Drinking Water Homepage at *http://www.epa.gov.*

diseases caused by organisms carried by water and easily killed by chlorine.

Substitutes for chlorine exist, but they may create their own by-products; no practical substitute has proved safer than chlorine. Under development is a treatment to destroy organic compounds and living organisms in water by bombarding them with beams of high-energy electrons, but this method is years away from approval for use.[3] In the meanwhile, what is a consumer to drink?

One option may be to purify tap water with home purifying equipment, which ranges in price from about $20 to $800. Some home systems may do an adequate job of removing lead, but others only improve the water's taste. Many are not designed to remove dangerous microorganisms that are left unaffected by chlorine. Each system has advantages and drawbacks, and all require periodic maintenance or filter replacements that vary in price. Unfortunately, not all companies or representatives are legitimate. Some perform water tests that yield dramatic-appearing but meaningless results to sell unneeded systems. Verify all claims of contamination with local water agencies before buying any purifying system.

Another alternative to tap water is to use bottled water. About 1 in 15 households uses bottled water as the main drinking water source, believing it to be safer than tap water and therefore worth its substantial price.

Unfortunately, bottled water is just as vulnerable to contamination as tap water is. After all, whether it comes from the tap or is poured from a bottle, all water comes from the same sources, **surface water** and **ground water** (see Table 8-2). Each of these sources supplies water for about half of the population.

Surface water comes from lakes, rivers, and reservoirs and provides drinking water for most of the nation's major cities. Surface water is easily contaminated by acid rain, runoff from highways, pesticides, fertilizer, animal wastes, and industrial wastes that run directly from pavements, septic tanks, farmlands, and industrial areas into streams that feed surface water bodies.

Surface water generally moves faster than ground water and stays aboveground where aeration and exposure to sunlight can cleanse it. The plants and microorganisms that live in surface water also filter it. These processes can remove some contaminants, but others stay in the water.

Ground water comes from **aquifers,** underground rock formations saturated with water. People in rural areas rely mostly on ground water pumped from private wells. Ground water is susceptible to contamination from hazardous waste sites, dumps, oil and gas pipelines, and landfills, as well as downward seepage from surface water bodies. Ground water moves slowly and is not aerated or exposed to sunlight, so contaminants break down more slowly than in surface water. To reach the aquifer, though, ground water must "percolate," or seep, through soil, sand, or rock, which filters out some contaminants.

Bottled water is classed as a food, so the Food and Drug Administration (FDA) regulates it.[4] The FDA requires yearly tests of bottled water to ensure that it meets the same standards as those set for the purity and sanitation of U.S. tap water.[5] As a rule, bottled water is low in fluoride, a mineral important to the health of teeth and bones (see later sections).[6]

Overwhelmingly, the people who buy bottled water say that it tastes better than the water from their taps. Most water-bottling plants disinfect their products with ozone, a form of oxygen that, unlike chlorine, leaves no flavor or odor in the water. Another

Surface water is easily contaminated by acid rain, pesticides, and other pollutants that fall or wash into streams, rivers, and lakes.

Table 8-2

Water Sources

- **aquifers** underground rock formations containing water that can be drawn to the surface for use.
- **ground water** water that comes from underground aquifers.
- **surface water** water that comes from lakes, rivers, and reservoirs.

reason people choose bottled water is to avoid public water that consistently tests positive for one or more chemicals or other contaminants.

As a consumer, what should you look for when buying bottled water? Look for the trademark of the International Bottled Water Association (IBWA). The IBWA supports the FDA's regulations and enforcement efforts. Determine the water's source. If the bottled water comes from a public source, what treatment processes were used to remove contaminants? If the bottled water comes from a spring or stream, where is it located? Is the area agricultural, residential, industrial, or undeveloped? Bottled water from municipal sources must be labeled as such unless it has been distilled or otherwise purified. Table 8-3 defines some terms you may see on labels. What you are unlikely to find on the label, however, is the water's mineral content. The best choice of bottled water is one rich in magnesium and calcium, but low in sodium.[7] Some water manufacturers will provide mineral information, if a consumer requests it.

If your water is dispensed from a water cooler, cleanse the cooler once a month by running half a gallon of white vinegar through it. Remove the vinegar residue by rinsing the cooler with 4 or 5 gallons of tap water. The microbial content of water coolers has been found to be considerably higher than that recommended by the government. Regular cleaning reduces bacterial and mold growths that can cause serious infection and disease in those who ingest water contaminated with them.

Sales of bottled water show little sign of slowing down as more and more people question the safety of their water. Before you spend your money, though, be sure what you are getting in return.

Table 8-3

Water Terms That May Appear on Labels

- **artesian water** water drawn from a well that taps a confined aquifer in which the water is under pressure.
- **carbonated water** water that contains carbon dioxide gas, either naturally occurring or added, that causes bubbles to form in it; also called *bubbling* or *sparkling* water. Seltzer, soda, and tonic waters are legally soft drinks and are not regulated as water.
- **distilled water** water that has been vaporized and recondensed, leaving it free of dissolved minerals.
- **filtered water** water treated by filtration, usually through *activated carbon filters* that reduce the lead in tap water, or by *reverse osmosis* units that force pressurized water across a membrane removing lead, arsenic, and some microorganisms from tap water.
- **mineral water** water from a spring or well that typically contains 250 to 500 parts per million (ppm) of minerals. Minerals give water a distinctive flavor. Many mineral waters are high in sodium.
- **natural water** water obtained from a spring or well that is certified to be safe and sanitary. The mineral content may not be changed, but the water may be treated in other ways such as with ozone or by filtration.
- **public water** water from a municipal or county water system that has been treated and disinfected.
- **purified water** water that has been treated by distillation or other physical or chemical processes that remove dissolved solids. Because purified water contains no minerals or contaminants, it is useful for medical and research purposes.
- **spring water** water originating from an underground spring or well. It may be bubbly (carbonated), or "flat" or "still," meaning not carbonated. Brand names such as "Spring Pure" do not necessarily mean that the water comes from a spring.
- **well water** water drawn from ground water by tapping into an aquifer.

fluid and electrolyte balance maintenance of the proper amounts and kinds of fluids and minerals in each compartment of the body.

fluid and electrolyte imbalance failure to maintain the proper amount and kind of fluid in every body compartment; a medical emergency.

acid-base balance maintenance of the proper degree of acidity in each of the body's fluids.

buffers molecules that can help to keep the pH of a solution from changing by gathering or releasing H ions.

Figure 8-3

How Electrolytes Govern Water Flow

Water flows in the direction of the more highly concentrated solution.

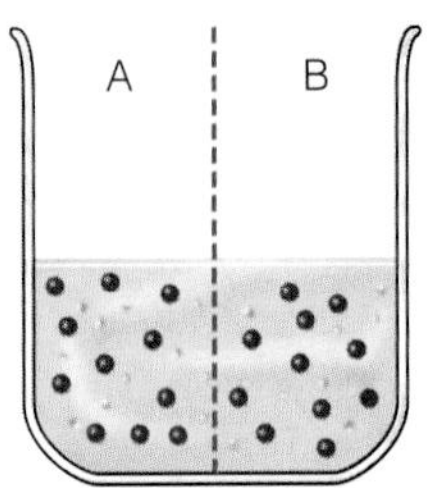
A
B

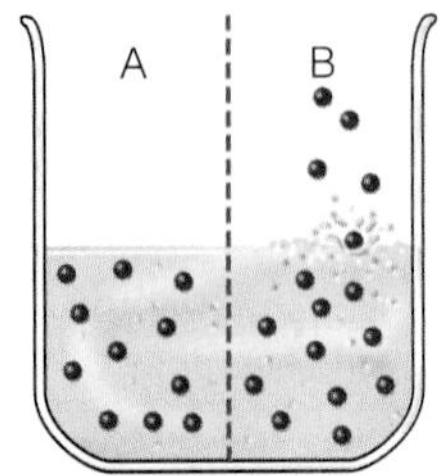
A
B

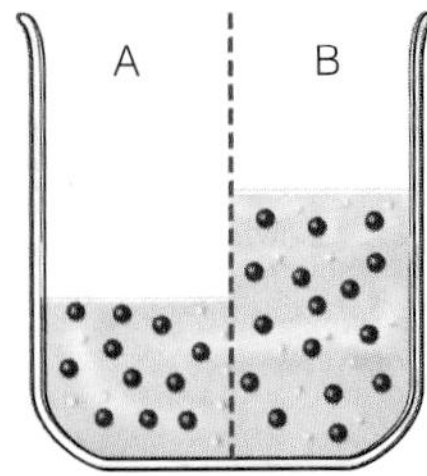
A
B

1. With equal numbers of dissolved particles on both sides of a water-permeable divider, water levels remain equal.
2. Now additional particles are added to increase the concentration on side B. Particles cannot flow across the divider (in the case of fluid inside and outside a cell, the divider is a cell membrane).
3. Water can flow both ways across the divider, but tends to move from side A to side B, where there is a greater concentration of dissolved particles. The *volume* of water increases on side B, and the *concentrations* on sides A and B become equal.

concentrations. Cells and their surrounding fluids go through the same process. Think of a cell as a sack made of a water-permeable membrane. The sack is filled with watery fluid, sealed, and suspended in a dilute solution of salts and other dissolved particles. Water flows freely between the fluids inside and outside the cell, but generally moves from the more dilute solution toward the more concentrated one. To control the flow of this water, the body must spend energy moving its electrolytes from one compartment to another. Figure 6-11 of Chapter 6 showed that proteins form the pumps that move mineral ions across cell membranes. The result is **fluid and electrolyte balance,** the proper amount and kind of fluid in every body compartment.

If something happens to overwhelm the fluid balance, severe illness can develop quickly because fluid can shift rapidly from one compartment to another. For example, in vomiting or diarrhea, the loss of water from the intestinal tract pulls fluid from between the cells in every part of the body. Fluid then leaves the cell interiors to restore balance. Meanwhile the kidneys detect the water loss and attempt to retrieve water from the pool destined for excretion. To do this, they raise the sodium concentration outside the cells, and this pulls still more water out of them. The result is the very serious condition of **fluid and electrolyte imbalance.** Water and minerals lost in vomiting or diarrhea ultimately come from every body cell. This loss disrupts the heartbeat and threatens life. It is a cause of death among those with eating disorders.

Controversy 10 describes the problems of eating disorders.

The minerals help manage still another balancing act, the **acid-base balance,** or pH, already mentioned in Chapters 3 and 6. In pure water, a small percentage of water molecules (H_2O) exist as positive (H) and negative (OH) ions, but they exist in equilibrium—the positive charges exactly equal the negatives. When dissolved in watery body fluids, some of the major minerals give rise to acids (H, or hydrogen, ions), and others to bases (OH). Excess H ions in a solution make it an acid; they lower the pH. Excess OH ions in a solution make it a base; they raise the pH.

Figure 3-10 of Chapter 3 showed where common substances fall along the pH scale; Figure 3-3 depicts fluid movement in and around cells.

It is critical to life that the body carefully maintains its fluids at a nearly constant pH. Even slight changes in pH drastically change the structure and chemical functions of most biologically important molecules. The body's proteins and some of its mineral salts help prevent changes in the acid-base balance of its fluids by serving as **buffers**—molecules that gather up or release H ions as needed to maintain the correct pH. The kidneys help to control the pH balance by excreting more or less acid (H ions). The lungs also help by excreting more or less carbon dioxide. (Dissolved in the blood, carbon dioxide forms an acid, carbonic acid.) The tight control of the acid-base balance permits all other life processes to take place.

KEY POINT *Electrolytes help keep fluids in their proper compartments and buffer these fluids, permitting all life processes to take place.*

hydroxyapatite (hi-DROX-ee-APP-uh-tight) the chief crystal of bone, formed from calcium and phosphorus.

The Major Minerals

Major minerals:

- Calcium.
- Chloride.
- Magnesium.
- Phosphorus.
- Potassium.
- Sodium.
- Sulfur.

Though all the major minerals help to maintain the balances just described, each also plays some special roles of its own. These roles are described in the following sections and are summarized in Table 8-11 on pp. 298–299.

Calcium

As Figure 8-1 showed, calcium is by far the most abundant mineral in the body. Nearly all (99 percent) of the body's calcium is stored in the bones, where it plays two important roles. First, it is an integral part of bone structure. Second, bone calcium serves as a bank that can release calcium to the body fluids if even the slightest drop in blood calcium concentration occurs. Many people have the idea that, once deposited in bone, calcium (together with the other minerals of bone) stays there forever—that once a bone is built, it is inert, like a rock. Not so. The minerals of bones are in constant flux, with formation and dissolution taking place every minute of the day and night.

Calcium and phosphorus are essential to the formation of bone (see Figure 8-4). As bones begin to form, calcium phosphate salts crystallize on a foundation material composed of the protein collagen. The resulting **hydroxyapatite** crystals invade the collagen and gradually lend more and more rigidity to the maturing bones until they are able to support the weight they will have to carry. Thus the long leg bones of children can support their weight by the time they have learned to walk. During and after

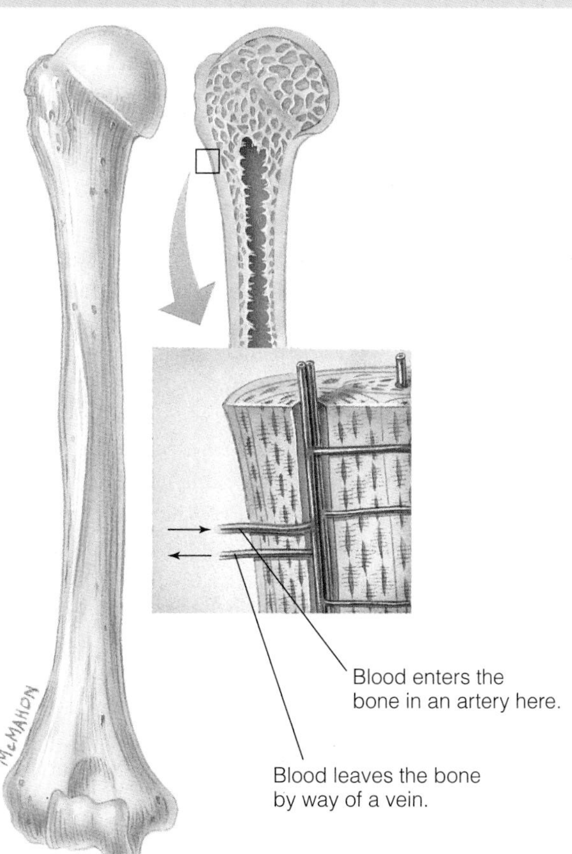
Blood travels in capillaries throughout the bone. It brings nutrients to the cells that maintain the bone's structure, and carries away waste materials from those cells. It picks up and deposits minerals as instructed by hormones.
This bone derives its structural strength from the lacy network of crystals that lie along the bone's lines of stress. If minerals are withdrawn to cover deficits elsewhere in the body, the bone will grow weak and ultimately will bend or crumble.
Blood enters the bone in an artery here.
Blood leaves the bone by way of a vein.
McMAHON

Figure 8-4
A Bone

fluorapatite (floor-APP-uh-tight) a crystal of bones and teeth, formed when fluoride displaces the hydroxy portion of hydroxyapatite. Fluorapatite resists being dissolved back into body fluid.

osteoporosis (OSS-tee-oh-pore-OH-sis) a reduction of the bone mass of older persons in which the bones become porous and fragile (*osteo* means "bones"; *poros* means "porous"); also known as **adult bone loss.**

peak bone mass the highest attainable bone density for an individual; developed during the first three decades of life.

the bone-strengthening processes, fluoride may displace the "hydroxy" parts of these crystals, making **fluorapatite**. Fluorapatite resists bone-dismantling forces to help maintain bone integrity.

The formation of teeth follows a pattern similar to that of bones. Hydroxyapatite crystals form on a collagen matrix to create the dentin that gives strength to the teeth (see Figure 8-5). Calcification of the "baby" teeth occurs in the gums during the latter half of the infant's time in the womb. The calcification of the permanent teeth takes place during early childhood, up to about the age of three; that of the "wisdom" teeth begins at about the age of ten. The turnover of minerals in teeth is not as rapid as in bone, but some withdrawal and redepositing do take place throughout life. As in bone, fluoride hardens and stabilizes the crystals of teeth and makes the enamel resistant to decay.

Calcium in Body Fluids Only about 1 percent of the body's calcium is in the fluid that bathes and fills the cells, but this tiny amount plays major roles:

- It regulates the transport of ions across cell membranes and is particularly important in nerve transmission.
- It helps maintain normal blood pressure (see Chapter 11).
- It is essential for muscle contraction and therefore for the heartbeat.
- It allows secretion of hormones, digestive enzymes, and neurotransmitters.
- It plays an essential role in the clotting of blood.

Because of its importance, blood calcium is tightly controlled.

Calcium Balance Cells need continuous access to calcium, so the body maintains a constant calcium concentration in the blood. The skeleton serves as a bank from which the blood can borrow and return calcium as needed. Blood calcium is regulated, not by a person's daily calcium intake, but by hormones sensitive to blood calcium.† This means that you can go without adequate dietary calcium for years and not suffer noticeable symptoms. Only later in life do you suddenly discover that your calcium savings account has dwindled to the point at which the integrity of your skeleton can no longer be maintained. Throughout your adult years, you have been developing the fragile bones of **osteoporosis,** or **adult bone loss.** Osteoporosis constitutes a major health problem for many older people whose bones suddenly begin to shatter. The problem and its possible causes and prevention are the topics of this chapter's Controversy.

Bone loss in adulthood is widespread. To protect against these losses, high calcium intakes early in life are recommended. A calcium-poor diet during the growing years may prevent a person's achievement of maximum **peak bone mass.**[3] Too little calcium packed into the skeleton during childhood and young adulthood strongly predicts susceptibility to osteoporosis later in adulthood.

The body is sensitive to an increased need for calcium, although it sends no signals to the conscious brain indicating calcium need. Instead, the body quietly increases the absorption of calcium from the intestine and prevents its loss from the kidneys, thus conserving it. For example, more calcium is needed for growth, so infants and children absorb up to 75 percent of ingested calcium and pregnant women absorb about 50 percent. The body of an adolescent girl hungers for calcium, absorbing and retaining more calcium from each meal than does the body of an adult.[4] Adults absorb about 30 percent. The body also absorbs a higher percentage of calcium when less total calcium is provided in the diet. Deprived of calcium for months or years, an adult may double the calcium absorbed; when supplied for years with abundant calcium, the same person may absorb only about one-third the normal amount. These adjustments take time, though. A person accustomed to high calcium intakes who suddenly cuts back is likely to lose calcium from bone stores until the body adapts to the lower intake.

Figure 8-5

A Tooth

The inner layer of dentin is bonelike material that forms on a protein (collagen) matrix. The outer layer of enamel is harder than bone. Both dentin and enamel contain hydroxyapatite crystals (made of calcium and phosphorus). The crystals of enamel may become even harder when exposed to the trace mineral fluoride.

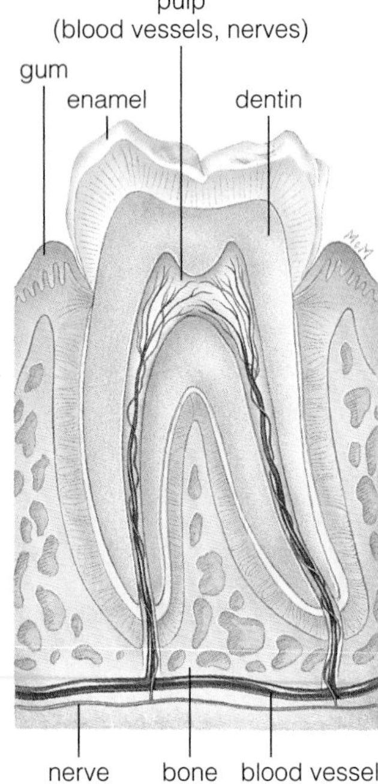
pulp
(blood vessels, nerves)
gum
enamel
dentin
nerve
bone
blood vessel

†Calcitonin, made in the thyroid gland, is secreted whenever the calcium concentration in the blood rises too high. It acts to stop withdrawal from bone and to slow absorption somewhat from the intestine. Parathormone, from the parathyroid glands, has the opposite effect.

How Much Calcium Do I Need? Setting recommended intakes for calcium is difficult because absorption varies not only with age, but also with a person's vitamin D status, the calcium content of the diet, and calcium binders in foods. The DRI committee took such variations into account when setting recommendations for calcium and established an amount that is observed to produce maximum calcium retention. At lower intakes, the body does not store calcium to capacity; at greater intakes, the excess calcium is excreted and so wasted.

Recommended intakes are high for young people and also for those in old age because calcium absorption declines with age. The high intake recommendations are perhaps appropriate because people need calcium throughout life. They develop their peak bone mass during their young years. After 26 years of age or so, the skeleton no longer adds significantly to bone density. After about 40 years of age, regardless of calcium intake, bones begin to lose density, but the loss can be slowed somewhat by a diet high in calcium. Thus obtaining enough calcium during the young years of life ensures that the skeleton will start out with peak bone density, and obtaining enough later on helps to minimize bone losses throughout life. Table 8-4 offers the DRI recommendations and other calcium goals. Everyone needs to meet their calcium needs, and the Food Feature later in the chapter suggests ways to do so. Snapshot 8-1 provides a look at some foods that are good or excellent sources of calcium.

In vitamin D deficiency:

- Rickets causes the bones of children to be soft and malformed.
- Osteomalacia causes the bones of adults to soften and bend.

In osteoporosis:

- Bones of older adults become brittle and fragile.

The importance of vitamin D in calcium absorption was described in Chapter 7.

KEY POINT *Calcium makes up bone and tooth structure and plays roles in nerve transmission, muscle contraction, and blood clotting. Calcium absorption rises when there is a dietary deficiency or an increased need such as during growth.*

Phosphorus

Phosphorus is the second most abundant mineral in the body. About 85 percent of it is found combined with calcium in the crystals of the bones and teeth. The rest is everywhere else you might look in the body. The concentration of phosphorus in the blood is less than half that of calcium, but its functions are wide reaching and critical to life.

Phosphorus salts are critical buffers, helping to maintain the acid-base balance of cellular fluids. Each cell also depends on phosphorus as part of its DNA and RNA, the cells' genetic material. Thus phosphorus is essential for growth and renewal of tissues. In metabolism of energy nutrients, phosphorus compounds carry, store, and release energy, and they assist many enzymes and vitamins in extracting the energy from nutrients. Recall from Chapter 5 that phosphorus also forms part of the molecules of the phospholipids, lipids that are principal components of cell membranes. Phosphorus is present in some proteins, too. With all of these critical roles to its credit, people might

The mineral is *phosphorus*. The adjective form is spelled with an -ous (as in *phosphorous salts*).

Table 8-4

Calcium Intake Recommendations

Proposed Healthy People 2010

- Increase to at least 90 percent the proportion of people aged 2 and older who meet the DRI dietary recommendations for calcium.

DRI Recommended Intakes[a]

- Adolescents: 1,300 milligrams per day.
- Women and men (19–50 years): 1,000 milligrams per day.
- Women and men (51 years and older): 1,200 milligrams per day.

World Health Organization

- 400 to 500 milligrams per day.

[a] For values for other groups, see the inside front cover; Canadian values are listed in Appendix B.

SNAPSHOT 8.1 Calcium

All except broccoli provide 10 percent or more of the calcium Daily Value (DV = 1,000 mg/day).[a]

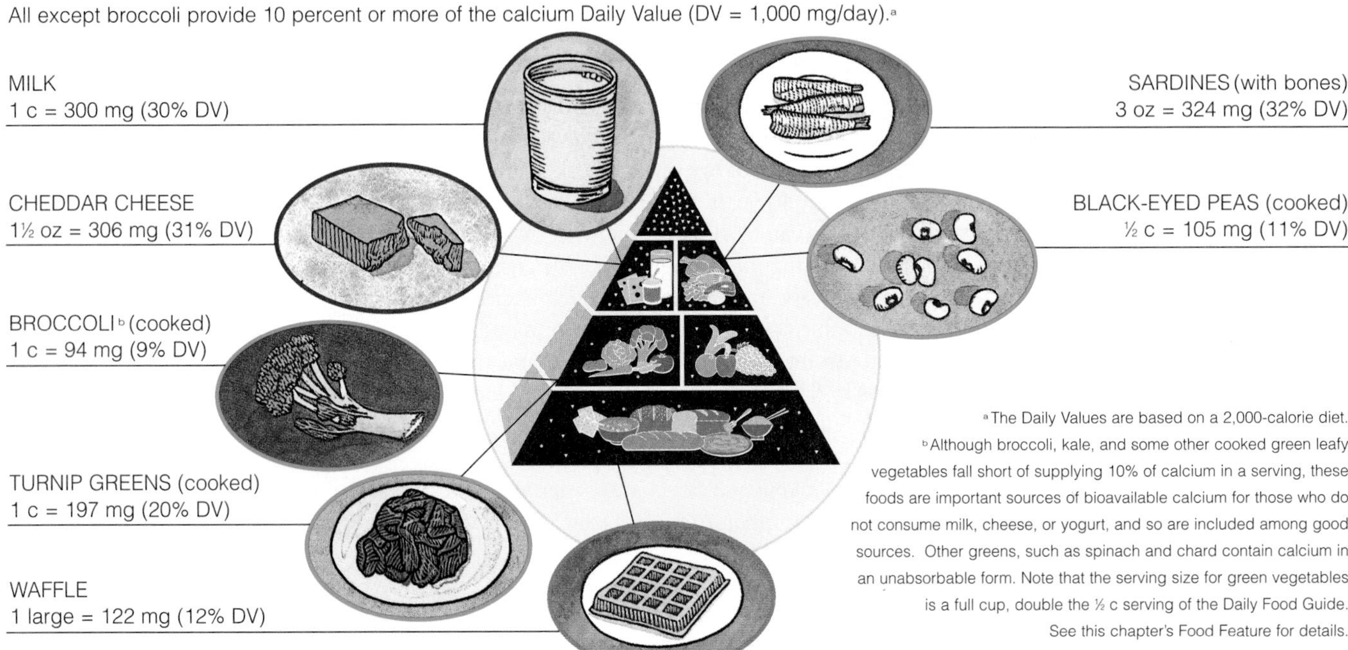

MILK
1 c = 300 mg (30% DV)
CHEDDAR CHEESE
1½ oz = 306 mg (31% DV)
BROCCOLI[b] (cooked)
1 c = 94 mg (9% DV)
TURNIP GREENS (cooked)
1 c = 197 mg (20% DV)
WAFFLE
1 large = 122 mg (12% DV)
SARDINES (with bones)
3 oz = 324 mg (32% DV)
BLACK-EYED PEAS (cooked)
½ c = 105 mg (11% DV)

[a] The Daily Values are based on a 2,000-calorie diet.

[b] Although broccoli, kale, and some other cooked green leafy vegetables fall short of supplying 10% of calcium in a serving, these foods are important sources of bioavailable calcium for those who do not consume milk, cheese, or yogurt, and so are included among good sources. Other greens, such as spinach and chard contain calcium in an unabsorbable form. Note that the serving size for green vegetables is a full cup, double the ½ c serving of the Daily Food Guide. See this chapter's Food Feature for details.

worry about getting enough phosphorus in the diet. Luckily, needs for phosphorus are easily met by almost any diet, and deficiencies are unknown.

As Snapshot 8-2 shows, animal protein is the best source of phosphorus. The reason is that phosphorus is abundant in the cells of animals. In the past, researchers emphasized the importance of an ideal calcium-to-phosphorus ratio to support the body's absorption and use of calcium. A wide range of ratios has since proved to have no effect on the body's calcium balance, and total intakes of phosphorus and calcium, not their ratio, seem to be of importance.[5]

KEY POINT *Most of the phosphorus in the body is in the bones and teeth. Phosphorus in the blood helps maintain acid-base balance, is part of the genetic material in cells, assists in energy metabolism, and is part of cell membranes. Under normal circumstances, deficiencies of phosphorus are unknown.*

Magnesium

Magnesium barely qualifies as a major mineral: only about 1 ounce is present in the body of a 130-pound person, over half of it in the bones.[6] Most of the rest is in the muscles, heart, liver, and other soft tissues, with only 1 percent in the body fluids. The supply of magnesium in the bones can be tapped to maintain a constant blood level whenever dietary intake falls too low. The kidneys can also act to conserve magnesium.

Like phosphorus, magnesium is critical to many cell functions. It assists in the operation of more than 300 enzymes, is needed for the release and use of energy from the energy-yielding nutrients, and directly affects the metabolism of potassium, calcium, and vitamin D. Magnesium acts in the cells of all the soft tissues, where it is part of the protein-making machinery and is necessary for the release of energy. Magnesium works with calcium in contracting and relaxing muscles: calcium promotes contraction, and magnesium helps the muscles relax afterward, so both are

SNAPSHOT 8.2 Phosphorus

These foods provide 10 percent or more of the phosphorus Daily Value (DV = 1,000 mg/day).[a]

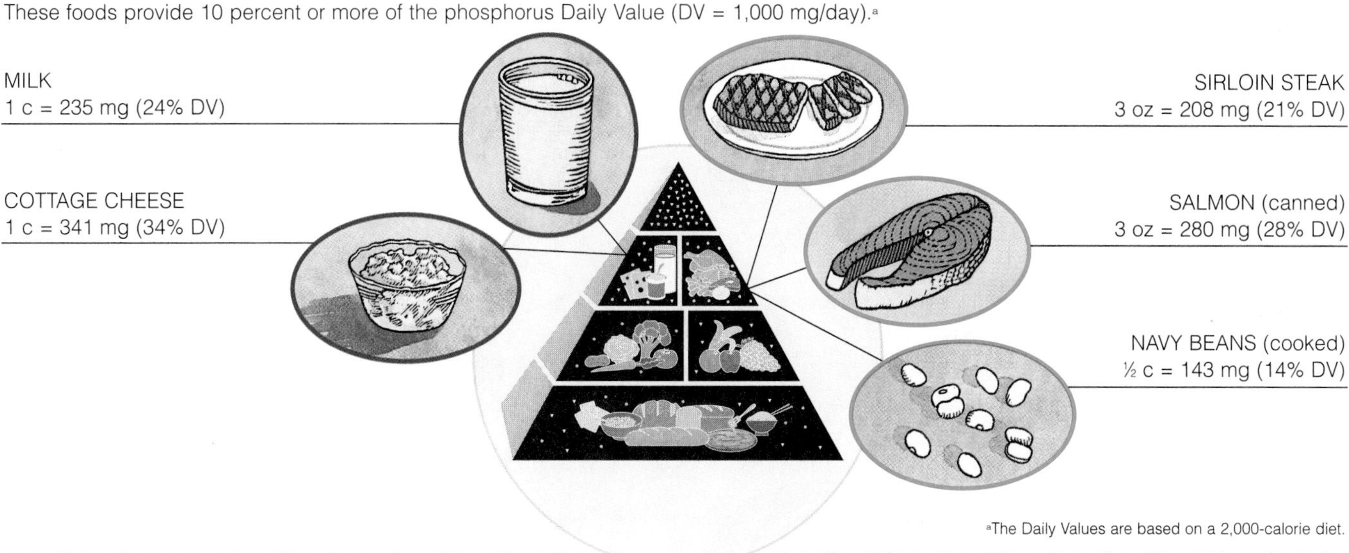

MILK
1 c = 235 mg (24% DV)
COTTAGE CHEESE
1 c = 341 mg (34% DV)
SIRLOIN STEAK
3 oz = 208 mg (21% DV)
SALMON (canned)
3 oz = 280 mg (28% DV)
NAVY BEANS (cooked)
½ c = 143 mg (14% DV)

[a]The Daily Values are based on a 2,000-calorie diet.

needed for proper functioning. In the teeth, magnesium promotes resistance to tooth decay by holding calcium in tooth enamel.

A magnesium deficiency may occur as a result of inadequate intake, vomiting, diarrhea, alcoholism, or protein malnutrition. It may also occur in hospital clients who have been fed magnesium-poor fluids through a vein for too long or in persons using diuretics. People whose drinking water has a high magnesium content experience a lower incidence of sudden death from heart failure than other people. It seems likely that magnesium deficiency makes the heart unable to stop itself from spasms once it starts. Magnesium deficiency may also be related to cardiovascular disease, heart attack, and high blood pressure.[7] A deficiency also causes hallucinations that can be mistaken for mental illness or drunkenness. Although intakes are often below those recommended, overt deficiency symptoms are rare in normal, healthy people.

Most people in the United States receive only about three-quarters of the magnesium they need from their diets. Snapshot 8-3 shows magnesium-rich foods. Magnesium is easily washed and peeled away from foods during processing, so slightly processed or unprocessed foods are the best choices. In various parts of the country, water can contribute significantly to magnesium intakes, so people living in those regions need less from food.

Magnesium toxicities are most often reported in older people who abuse magnesium-containing laxatives, antacids, and other medications. The consequences can be severe: diarrhea, acid-base imbalance, kidney impairment, lack of coordination, confusion, coma, and, in extreme cases, death from heart failure.[8] For safety, use magnesium-containing laxatives with discretion.

KEY POINT *Most of the body's magnesium is in the bones and can be drawn out for all the cells to use in building protein and using energy. Most people in the United States fail to obtain enough magnesium from their food.*

Sodium

Salt has been known and valued throughout recorded history. The biblical saying "You are the salt of the earth" means that a person is valuable. If "you are not worth your

SNAPSHOT 8.3 Magnesium

These foods provide 10 percent or more of the magnesium Daily Value (DV = 400 mg/day).[a]

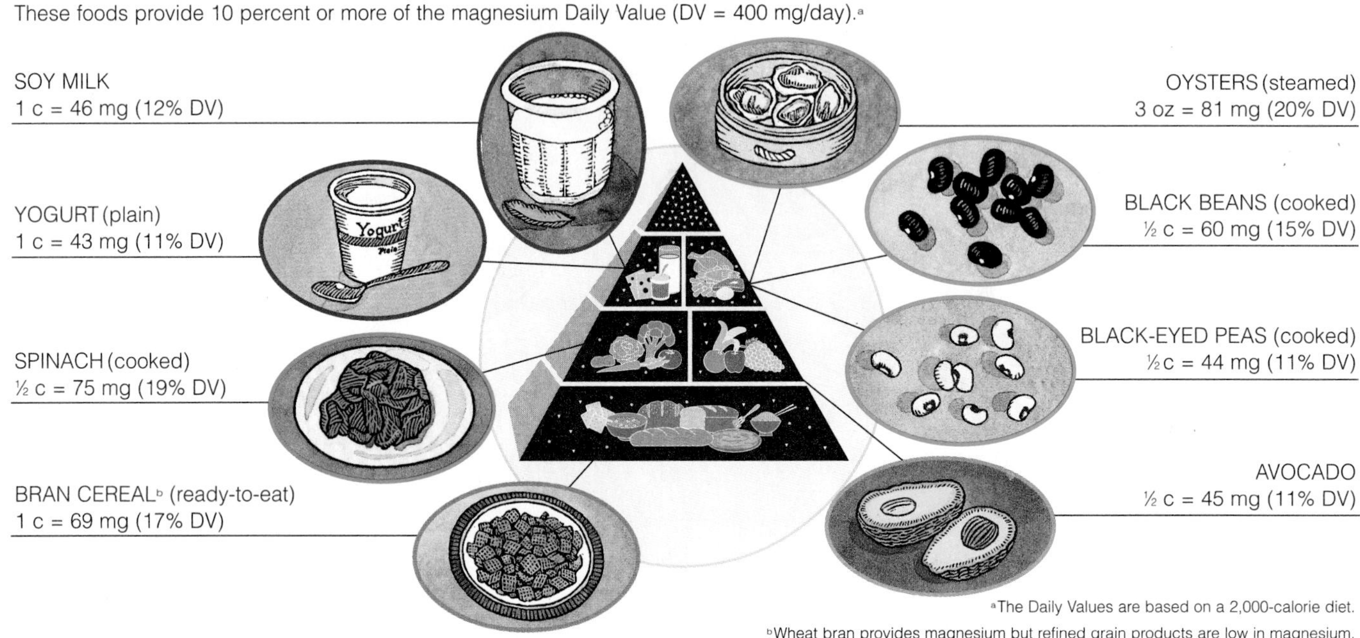
SOY MILK
1 c = 46 mg (12% DV)
YOGURT (plain)
1 c = 43 mg (11% DV)
SPINACH (cooked)
½ c = 75 mg (19% DV)
BRAN CEREAL[b] (ready-to-eat)
1 c = 69 mg (17% DV)
OYSTERS (steamed)
3 oz = 81 mg (20% DV)
BLACK BEANS (cooked)
½ c = 60 mg (15% DV)
BLACK-EYED PEAS (cooked)
½ c = 44 mg (11% DV)
AVOCADO
½ c = 45 mg (11% DV)
Yogurt

[a]The Daily Values are based on a 2,000-calorie diet.

[b]Wheat bran provides magnesium but refined grain products are low in magnesium.

To the chemist, a salt results from the neutralization of an acid and a base. Sodium chloride, table salt, results from the reaction between hydrochloric acid and the base sodium hydroxide. The positive sodium ion unites with the negative chloride ion to form the salt. The positive hydrogen ion unites with the negative hydroxide ion to form water.

Base + acid = salt + water.

Sodium hydroxide + hydrochloric acid = sodium chloride + water.

For a brief summary of the kidneys' actions, see Chapter 3.

salt," you are worthless. Even our word *salary* comes from the Latin word for *salt*. Sodium is the positive ion in the compound sodium chloride (table salt) and contributes 40 percent of its weight. Thus a person who consumes a gram of salt consumes 400 milligrams of sodium. As already mentioned, sodium is the chief ion used to maintain the volume of fluid outside cells. Sodium also helps maintain acid-base balance and is essential to muscle contraction and nerve transmission. About 30 to 40 percent of the body's sodium is thought to be stored on the surface of the bone crystals, where the body can easily draw on it to replenish the blood concentration, if necessary.

A deficiency of sodium would be harmful, but few diets lack sodium. Foods usually include more salt than is needed, and the body absorbs it freely. The kidneys filter the surplus out of the blood into the urine. They can also sensitively conserve sodium. In the rare event of a deficiency, they can return to the bloodstream the exact amount needed. Normally, the amount of sodium you excrete in a day equals the amount you have ingested that day.

What Is "Water Weight," and What Does It Have to Do with Eating Salty Food? If blood sodium rises, as it will after a person eats salted foods, thirst ensures that the person will drink water until the sodium-to-water ratio is restored. Then the kidneys excrete the extra water along with the extra sodium.

Dieters sometimes think that eating too much salt or drinking too much water will make them gain weight, but they do not gain fat, of course. They gain water, but they excrete this excess water immediately. Excess salt is excreted as soon as enough water is drunk to carry the salt out of the body. From this perspective, then, the way to keep body salt (and "water weight") under control is to drink more, not less, water.

If blood sodium drops, body water is lost, and both water and sodium must be replenished to avert an emergency. Overly strict use of low-sodium diets in the treatment of hypertension, kidney disease, or heart disease can deplete the body of needed sodium; so can vomiting, diarrhea, or extremely heavy sweating. As Chapter 10 makes clear, the sodium lost by way of normal sweating due to exercise is easily replaced later in the day with ordinary foods.

Sodium Intakes No known human diet lacks sodium. For this reason, no intake recommendation has been set. Instead, the minimum sodium requirement for U.S. adults is estimated to be 500 milligrams, 115 milligrams in Canada. Both these amounts are provided by a diet of plain foods with no salt added. The World Health Organization emphasizes moderation as its key concern about sodium (see Table 8-5).

Cultures vary in their use of salt. Men in the United States consume an average of 3,300 milligrams of sodium, or more than 8 grams of salt, a day (see Figure 8-6, in the margin). Asian people, whose staple sauces and flavorings are based on soy sauce and monosodium glutamate (MSG or Accent), may consume the equivalent of about 30 to 40 grams of salt per day.

Does Everyone Need to Cut Back on Sodium Intake?

Often, communities with high intakes of salt experience high rates of **hypertension** and cerebral hemorrhage, a hypertension-related stroke. Over 30 years of observational evidence has suggested a relationship between elevated blood pressure and sodium intakes of over 2,400 milligrams. Recently, however, some have begun to question the advice for everyone to reduce dietary sodium to reduce hypertension and cardiovascular disease risks.[9]

hypertension high blood pressure.

A concern is that, while some people with hypertension respond to reduced sodium intakes with lowered blood pressure (they are salt-sensitive), other individuals do not (they are not salt-sensitive). The critics rightly point out that *non* salt-sensitive people with hypertension will likely not benefit from restricting dietary sodium and salt as much as they might from taking other steps to bring their pressure down. For example, weight loss in overweight people reliably reduces blood pressure. Also, according to the DASH study, increasing fruit and vegetable intakes, increasing fish and low-fat dairy intakes, and reducing fat intake, while only slightly reducing sodium, may bring about a healthy reduction in blood pressure.[10]

DASH stands for Dietary Approaches to Stop Hypertension. The study included more than 450 adults with hypertension at research centers around the nation. More about the DASH diet in Chapter 11.

One study seemed to indicate that high sodium intakes do not correlate with increased risk of death from cardiovascular diseases.[11] This study has been criticized, however, because it may have included subjects who had previously been diagnosed with heart disease and so had reduced their sodium intakes; including them may have skewed the study's results.[12]

There is no question that the risk of death from heart disease climbs steadily with increased blood pressure.[13] The question is whether reducing sodium and salt will reduce blood pressure in a population. It is true that salt sensitivity varies widely among people, but many authorities believe that a population-wide reduction in blood pressure is attainable by holding salt intakes to the recommended level. This action, they say, can reasonably be expected to reduce the rate of death from cardiovascular

Figure 8-6
Sodium Intakes of U.S. Adults

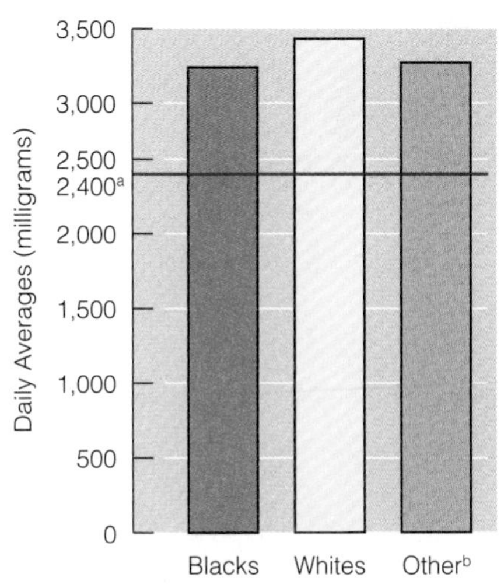

Daily Averages (milligrams)
3,500
3,000
2,500
2,400[a]
2,000
1,500
1,000
500
0
Blacks
Whites
Other[b]

[a]Recommended maximum intake.
[b]Except Alaskan natives who average over 4,500 milligrams per day.
SOURCE: Adapted from USDA, *Nutrition Insights*, Issue 3, May 1997.

Table 8-5
Salt and Sodium Intake Guidelines

Estimated Safe and Adequate Daily Intakes
- Adolescents and adults: 500 milligrams per day.

Proposed Healthy People 2010
- Increase to at least 65 percent the proportion of people aged 2 and older who meet the Daily Value of 2,400 mg or less of sodium consistent with the *Dietary Guidelines*.[a]

World Health Organization
- Upper limit: 6 grams salt from mixed food sources per day. Lower limit: not defined.

Dietary Guidelines
- Choose a diet moderate in salt and sodium.

[a] Fruit and vegetable intake also affects blood pressure.

disease. Research is now under way to help redefine some of these relationships, but both the USDA and the committee developing the Healthy People 2010 Objectives for the Nation have restated their support of the *Dietary Guideline* that urges people to choose a diet moderate in salt and sodium.[14] Many Americans may have much to gain in terms of cardiovascular health and nothing to lose from cutting back on salt as part of an overall dietary strategy to reduce blood pressure.[15]

More on factors relating to hypertension in Chapter 11.

Cut down on salt by minimizing intakes of:

- Foods prepared in brine (pickles, olives, sauerkraut).
- Salted or smoked meats (bologna, corned or chipped beef, franks, ham, lunch meats, salt pork, sausages, bacon).
- Salted or smoked fish (anchovies, caviar, salted dried cod, pickled herring, canned sardines, smoked salmon).
- Salty snacks (potato chips, pretzels, salted popcorn, salted nuts, most crackers).
- Fast foods that do not bear a label stating *healthy* or *low-sodium* (pizza, chicken nuggets or wings, fish and chips, tacos, sausage biscuits, fried chicken, convenience dinners, frozen TV dinners, canned pastas).
- Bouillon cubes, horseradish, mustard, seasoned salts, sauces (barbeque, ketchup, soy, Worcestershire).
- Cheeses, especially processed types.
- Canned and instant soups.

Here's a short cut: look for the words *healthy* or *low-sodium* on the labels of packaged foods.

Controlling Salt Intake The connection between salt and blood pressure in salt-sensitive people is direct: the more salt they eat, the higher their blood pressure goes. People tending toward salt sensitivity usually include those with kidney disease, those of African descent, those whose parents had high blood pressure, and anyone over 50 because salt sensitivity becomes more pronounced in older age. As mentioned, salt responders benefit from mild salt restriction, but other people may not.[16] It seems, then, that only salt responders should be advised to cut down on salt. The trouble is, it is impossible to tell to which group people belong before they become seriously ill with hypertension.

Other valid reasons exist for most people to hold their salt intakes at or below the recommended maximum. For example, older people without clinical hypertension often die of stroke, and reducing dietary sodium may lower their blood pressure enough to reduce their stroke risk.[17] Excess sodium in the diet causes increased calcium excretion—just the wrong effect for preserving the integrity of the bones. Excessive salt may also directly stress a weakened heart or aggravate kidney problems.[18] Asians' high salt intakes have been suggested as a possible cause for their greatly elevated rate of stomach cancer.[19]

For all these reasons, cutting down on salt and sodium in the diet may be wise, and the margin offers tips for doing so. Foods eaten without salt may seem less tasty at first, but with repetition, tastes adjust and the natural flavor becomes the preferred taste. Also, the guideline is to reduce, not eliminate, salt intake.

Table 8-6

Processing Reduces Potassium, Increases Sodium in Foods

Food	Potassium (mg)	Sodium (mg)	Ratio
Milk Products			
Milk (whole), 1 c	371	120	3:1
Chocolate pudding (home cooked), 1 c	506	274	2:1
Chocolate pudding (instant), 1 c	488	834	1:2
Meats			
Beef roast (cooked), 3 oz	250	53	5:1
Corned beef (canned), 3 oz	115	855	1:7
Frankfurter, 1 large	95	638	1:7
Chipped beef, 3 oz	377	2,953	1:8
Vegetables			
Corn (cooked), 1 c	242	8	30:1
Creamed corn (canned), 1 c	390	572	1:2
Cornflakes, 1 c	25	300	1:12
Fruit			
Peaches (fresh), 1	193	0	171:1
Peaches (canned), 1 c	241	16	15:1
Peach pie, 1 piece	131	253	1:2
Grains			
Whole-wheat flour, 1 c	486	6	81:1
Shredded wheat cereal, 1 c	155	4	39:1
Whole-wheat bread, 1 slice	71	148	1:2
Wheat crackers, 4	16	70	1:4

An obvious step in controlling salt intake is to control the saltshaker, but this source may contribute as little as 15 percent of the total salt consumed. A more productive step may be to cut down on processed and fast foods, the source of almost 75 percent of salt in the U.S. diet. Notice, too, in Table 8-6, that the least processed foods in each food group are not only lowest in sodium but also highest in potassium. Low potassium intakes are thought to play an important role in the development of hypertension.[20] The advice to increase potassium often goes hand in hand with reducing sodium. Figure 8-7 and this Chapter's Do It! section identify some sodium sources in the U.S. diet.

KEY POINT *Sodium is the main positively charged ion outside the body's cells. Sodium attracts water. Thus too much sodium (or salt) may aggravate hypertension. Diets rarely lack sodium.*

Potassium

Potassium is the principal positively charged ion inside body cells. It plays a major role in maintaining fluid and electrolyte balance and cell integrity. It is also critical to maintaining the heartbeat. The sudden deaths that occur during fasting or severe diarrhea and in children with kwashiorkor or people with eating disorders are thought to be due to heart failure caused by potassium loss.

Kwashiorkor is described in Chapter 6.

Unlike sodium, potassium may exert a positive effect against hypertension and related ills. See Chapter 11 for details.

Dehydration leads to potassium loss from inside cells. This condition is especially dangerous because when brain cells lose potassium, the victim loses the ability to notice the need for water. For this reason, adults are warned not to take diuretics (water pills) that cause potassium loss except under a physician's supervision. When taking such diuretics, a person should alert all other health-care providers to their use. Any physician prescribing diuretics will tell the client to eat potassium-rich foods to compensate for the losses. Depending on the diuretic, the physician may also advise a lower sodium intake.

A dietary deficiency of potassium is unlikely in healthy people, although a low potassium intake is possible with a steady diet of highly processed foods. Because potassium is found inside all living cells and because cells remain intact unless foods are processed, the richest sources of potassium are *fresh* foods of all kinds (see Snapshot 8-4 on the next page). Most whole vegetables and fruits are outstanding.

Figure 8-7

Sources of Sodium in the U.S. Diet

3 mg (1 cob)
50 mg (½ c)
120 mg (1 c)
60 mg (1 egg)
50 mg (3 oz)
20 mg (1 small potato)

Unprocessed foods that are low in sodium contribute less than 10 percent of the total sodium in the U.S. diet.

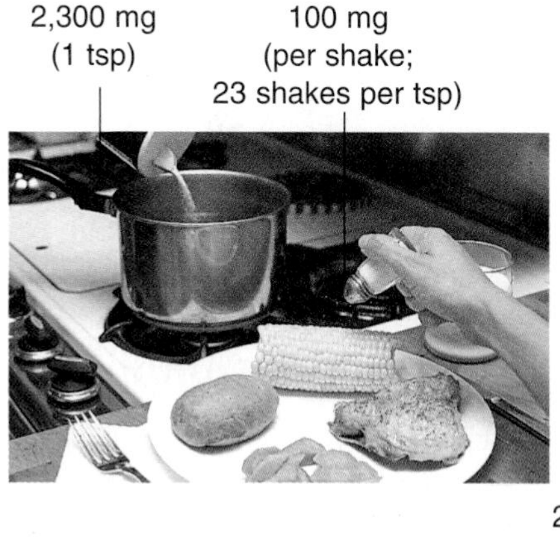

2,300 mg (1 tsp)
100 mg (per shake; 23 shakes per tsp)

Salt added at home, in cooking or at the table, contributes 15 percent of the total sodium in the U.S. diet.

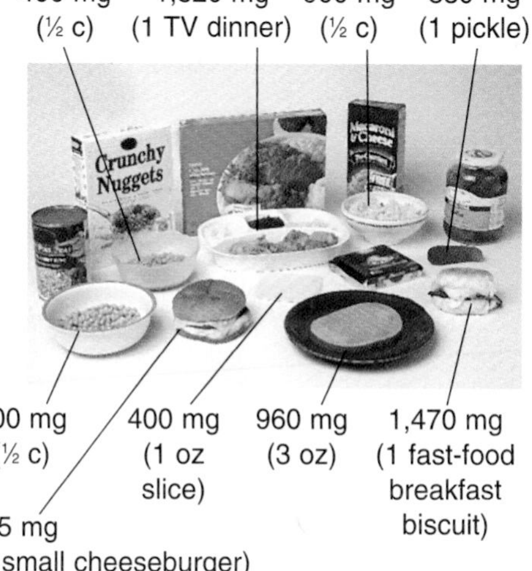

400 mg (½ c)
1,820 mg (1 TV dinner)
900 mg (½ c)
830 mg (1 pickle)
Crunchy Nuggets
200 mg (½ c)
400 mg (1 oz slice)
960 mg (3 oz)
1,470 mg (1 fast-food breakfast biscuit)
725 mg (1 small cheeseburger)

Processed foods such as these contribute 75 percent of the sodium in the U.S. diet.

SNAPSHOT 8.4 Potassium

These foods provide 10 percent or more of the potassium Daily Value (DV = 3,500 mg/day).[a]

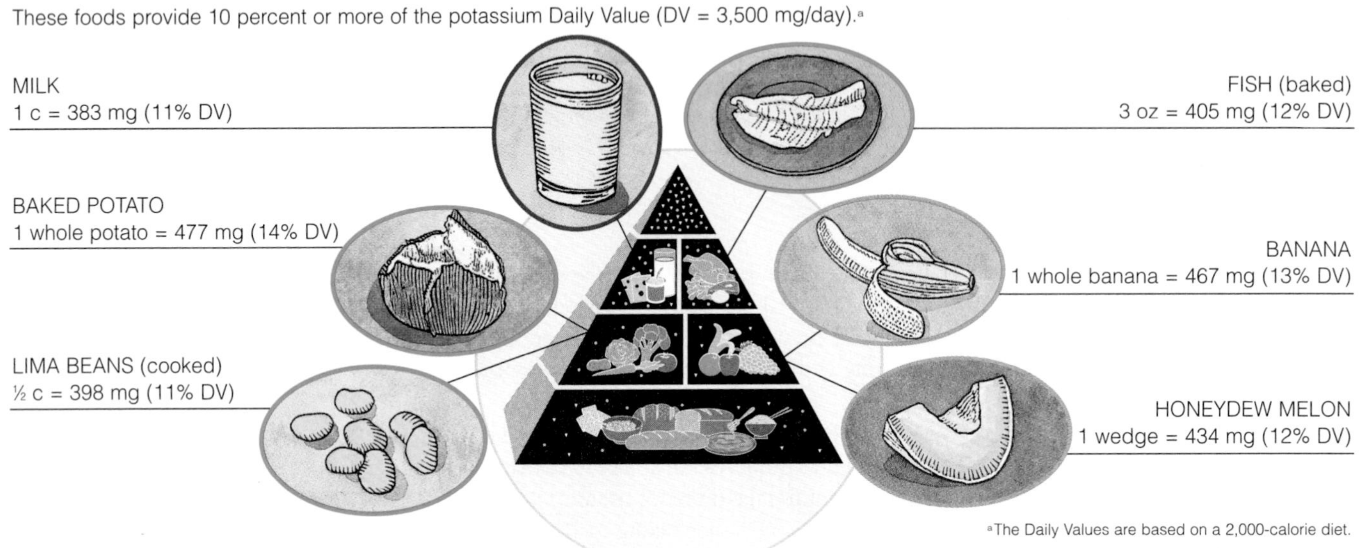
MILK
1 c = 383 mg (11% DV)
BAKED POTATO
1 whole potato = 477 mg (14% DV)
LIMA BEANS (cooked)
½ c = 398 mg (11% DV)
FISH (baked)
3 oz = 405 mg (12% DV)
BANANA
1 whole banana = 467 mg (13% DV)
HONEYDEW MELON
1 wedge = 434 mg (12% DV)

[a]The Daily Values are based on a 2,000-calorie diet.

Bananas, despite their fame as the richest potassium source, are just one among many rich sources. Bananas, however, are available everywhere, are easy to chew, and have a sweet taste that almost everyone likes, so health-care providers often recommend them to enhance potassium intake.

Potassium chloride pills are available over the counter and are sold in health food stores without a warning label, but they should not be used except on a physician's advice. Potassium overdoses normally are not life-threatening as long as they are taken by mouth because the presence of excess potassium in the stomach triggers a vomiting reflex that expels the unwanted substance. A person with a weak heart, however, should not be put through this trauma, and a baby may not be able to withstand it. Several infants have died when well-meaning parents overdosed them with potassium supplements. Potassium from foods is safe, but potassium injected into a vein can stop the heart.

KEY POINT *Potassium, the major positive ion inside cells, is important in many metabolic functions. Fresh foods are the best sources of potassium. Diuretics can deplete the body's potassium and so can be dangerous; potassium excess can also be dangerous.*

Chloride and Sulfur

The chloride ion is a major negative ion in the body. In the fluids outside the cells, it accompanies sodium; inside the cells, it occurs primarily in association with potassium. Thus it helps to maintain the crucial fluid balances (acid-base and electrolyte balances) mentioned earlier in the discussion of water. The chloride ion also plays a special role as part of the hydrochloric acid that maintains the strong acidity of the stomach. The principal food source of chloride is salt, both added and naturally occurring in foods. In its elemental form, chlorine forms a deadly green gas; dissolved in fluid, chlorine can be useful as a disinfectant, but it must be handled carefully.

As for sulfur, the body does not use it by itself as a nutrient, but it is present in essential nutrients that the body does use, such as thiamin and all proteins. Sulfur plays its most important role in helping strands of protein to assume a functional shape. Skin, hair, and nails contain some of the body's more rigid proteins, which have high sulfur contents.

There is no recommended intake for sulfur, and deficiencies are unknown. The summary table at the end of this chapter presents the main facts about the major minerals.

KEY POINT *Chloride is the body's major negative ion inside and outside cells. It is essential to the acid-base balance and is part of the stomach's hydrochloric acid, which is necessary to digest protein. Sulfur is also considered a major mineral, although it occurs only as part of other compounds such as protein.*

goiter (GOY-ter) enlargement of the thyroid gland due to iodine deficiency is *simple goiter;* enlargement due to an excess is *toxic goiter.*

cretinism (CREE-tin-ism) severe mental and physical retardation of an infant caused by the mother's iodine deficiency during pregnancy.

The Trace Minerals

An obstacle to determining the precise roles of the trace elements is the difficulty of providing an experimental diet lacking in the one element under study. Thus research in this area is limited mostly to the study of laboratory animals, which can be fed highly refined, purified diets in environments free of all contamination. New laboratory techniques have enabled scientists to detect minerals in smaller and smaller quantities in living cells, and research is now rapidly expanding our knowledge about them. Whole books have been published on the trace minerals alone. As Table 8-7 shows, intake recommendations for human beings have been established for some trace elements. Others are recognized as essential nutrients for some animals, but have not been proved to be required for human beings.

Iodine

Iodine is needed by the body in an infinitesimally small quantity, but its principal role in human nutrition makes obtaining this amount critical. Iodine is a part of thyroxine, the hormone responsible for regulating the basal metabolic rate. Iodine must be available for thyroxine to be synthesized.

When the iodine concentration of the blood is low, the cells of the thyroid gland enlarge in an attempt to trap as many particles of iodine as possible. Sometimes the gland enlarges until it makes a visible lump in the neck, a **goiter.** People with iodine deficiency this severe become sluggish and gain weight. In a pregnant woman, severe iodine deficiency causes extreme and irreversible mental and physical retardation of the infant known as **cretinism.** Much of the mental retardation can be averted if the woman's deficiency is detected and treated within the first six months of pregnancy, but if treatment comes too late or not at all, the child may spend a lifetime with an IQ as low as 20 (100 is average).[21] Iodine deficiency is one of the world's most common and most preventable causes of mental retardation.[22] In developing nations, both cretinism and goiter pose problems of enormous proportions that researchers hope may be remediable through adding iodine to community water supplies.[23]

The iodine in food varies. Generally, it reflects the soil in which plants are grown or on which animals graze. Iodine is plentiful in the ocean, so seafood is a completely dependable source. In the central parts of the United States that were never under the ocean, the soil is poor in iodine. In those areas, the use of iodized salt and the consumption of foods shipped in from iodine-rich areas have wiped out the iodine deficiency that once was widespread. Surprisingly, sea salt delivers little iodine to the eater because iodine becomes a gas and flies off into the air during the salt-drying process. In the United States, salt labels state whether the salt is iodized; in Canada, all table salt is iodized.

Excessive intakes of iodine can cause an enlargement of the thyroid gland resembling goiter, which in infants can block the airways and cause suffocation. U.S. intakes are several times the recommended intake of 150 micrograms. The toxic level at which detectable harm results is thought to be over 2,000 micrograms per day for an adult. Like chlorine and fluorine, iodine is a deadly poison in large amounts.

Much of the excess iodine in U.S. diets today comes from fast-food establishments, which use iodized salt with a liberal hand, and from bakery products and milk. The

Table 8-7
Trace Minerals

Human Intake Recommendations Established
Iodine
Iron
Zinc
Selenium
Fluoride
Chromium
Copper
Manganese
Molybdenum
Known Essential for Animals; Human Requirements under Study
Arsenic
Boron
Nickel
Silicon
Known Essential for Some Animals; No Evidence That Intake by Humans Is Ever Limiting
Cobalt

The evidence for requirements and essentiality is weak for the trace minerals cadmium, lead, lithium, tin, and vanadium.

hemoglobin (HEEM-oh-globe-in) the oxygen-carrying protein of the blood; found in the red blood cells (*hemo* means "blood"; *globin* means "spherical protein").

myoglobin (MYE-oh-globe-in) the oxygen-holding protein of the muscles (*myo* means "muscle").

iron deficiency the condition of having depleted iron stores, which, at the extreme, causes iron-deficiency anemia.

iron-deficiency anemia a form of anemia caused by iron deficiency and characterized by red blood cell shrinkage and color loss. Accompanying symptoms are weakness, apathy, headaches, pallor, intolerance to cold, and inability to pay attention. (For other anemias, see the index.)

anemia the condition of inadequate or impaired red blood cells; a reduced number or volume of red blood cells along with too little hemoglobin in the blood. The red blood cells may be immature and therefore too large or too small to function properly. Anemia can result from blood loss, excessive red blood cell destruction, defective red blood cell formation, and many nutrient deficiencies. Anemia is not a disease, but a symptom of another problem; its name literally means "too little blood."

In iodine deficiency, the thyroid gland enlarges—a condition known as simple goiter.

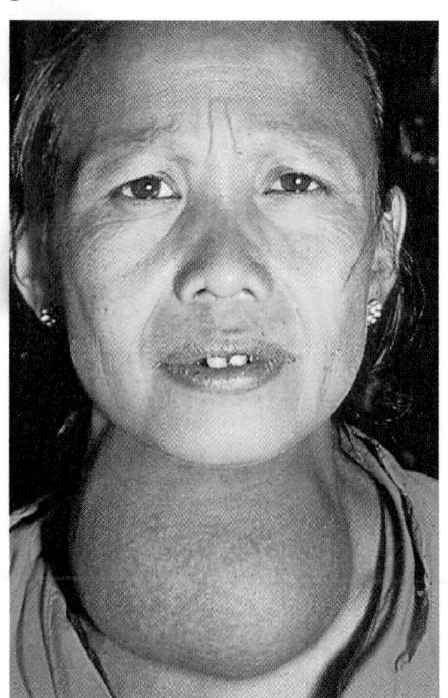

baking industry uses iodine-containing dough conditioners, and most dairies use iodine to disinfect milking equipment. One cup of milk supplies nearly half of one day's recommended intake of iodine, and less than a half-teaspoon of iodized salt meets the entire recommendation. The dairy and baking industries have been reducing their use of iodine compounds, but the emergence of this problem points to a need for continued surveillance of the food supply.

KEY POINT *Iodine is part of the hormone thyroxine, which influences energy metabolism. The deficiency diseases are goiter and cretinism. Iodine occurs naturally in seafood and in foods grown on land that was once covered by oceans; it is an additive in milk and bakery products. Large amounts are poisonous.*

Iron

Every living cell, whether plant or animal, contains iron. Most of the iron in the body is a component of two proteins: **hemoglobin** in red blood cells and **myoglobin** in muscle cells. Hemoglobin in the blood carries oxygen from the lungs to tissues throughout the body. Myoglobin carries and stores oxygen for the muscles. Both hemoglobin and myoglobin contain iron, which helps them to hold and carry oxygen and then release it.

All the body's cells need oxygen to help them handle the carbon and hydrogen atoms they release as they break down energy nutrients. The oxygen combines with these atoms to form the waste products carbon dioxide and water; thus the body constantly needs fresh oxygen and nutrients to keep the cells going. As cells use up and excrete their oxygen (as carbon dioxide and water), red blood cells shuttle between the metabolizing tissues and the lungs to bring in fresh oxygen supplies. Besides helping hemoglobin to carry oxygen and myoglobin to hold it in muscles, iron helps many enzymes in energy pathways to use oxygen. Iron is also needed to make new cells, amino acids, hormones, and neurotransmitters.

Iron is clearly the body's gold, a precious mineral to be hoarded. The liver packs iron sent from the bone marrow into new red blood cells, also from the bone marrow, and ships them out to the blood. Red blood cells live for about three to four months. When they die, the spleen and liver break them down, salvage their iron, and send it back to the bone marrow to be kept for reuse. Only tiny losses of iron occur in nail clippings, hair cuttings, and shed skin cells. If bleeding occurs, the loss of blood can cause significant iron loss from the body.

The body has special provisions for obtaining iron. Normally, only about 10 to 15 percent of dietary iron is absorbed; but if the body's supply of iron is diminished or if the need increases for any reason (such as pregnancy), absorption increases.

What Happens When Iron Is Lacking? If absorption cannot compensate for losses or low dietary intakes, then iron stores are used up and iron deficiency sets in. Iron deficiency and anemia are not one and the same, though they often occur together. The distinction between **iron deficiency** and **iron-deficiency anemia** is a matter of degree. People may be iron deficient, meaning that they have depleted iron stores without being anemic, or they may be iron deficient *and* anemic. With regard to iron, the term **anemia** refers to severe depletion of iron stores resulting in low blood hemoglobin.

The body that has been severely deprived of iron becomes unable to make enough hemoglobin to fill its new blood cells, and anemia results. A sample of iron-deficient blood examined under the microscope shows cells that are smaller and lighter red than normal (see Figure 8-8). The undersized cells contain too little hemoglobin and thus deliver too little oxygen to the tissues. The diminished supply of oxygen limits the cells' energy metabolism, so a person with iron-deficiency anemia lacks "get up and go." Tiredness, apathy, and a tendency to feel cold all reflect the energy deficiency of iron-deficiency anemia.

Long before the red blood cells are affected, though, people exhibit the impact of iron deficiency in their behavior. Even at slightly lowered iron levels, physical work capacity and productivity are impaired. With reduced energy available to work, play, think, or learn, people simply do these things less. Because they work and play less, they become less physically fit. Many of the symptoms associated with iron deficiency are easily mistaken for behavioral or motivational problems. Children deprived of iron become restless, irritable, and unable to pay attention, and they may fall behind their peers academically. Such symptoms disappear when iron intake improves.

A curious symptom seen in some people with iron deficiency is an appetite for ice, clay, paste, and other nonnutritious substances. Such people have been known to eat as many as eight trays of ice in a day. This consumption of nonfood substances, most often observed in poverty-stricken women and children, has been given the name **pica.** In many cases, pica clears up dramatically within days after iron is given, even before the red blood cells have had a chance to respond. Other times, pica is unresponsive to iron.

What Causes Iron Deficiency? Iron deficiency is usually caused by malnutrition, that is, inadequate iron intake, either from sheer lack of food or from high consumption of the wrong foods. In the Western world, overconsuming foods rich in sugar and fat and poor in nutrients is often responsible for low iron intakes.

Among nonnutritional causes of anemia, blood loss is the primary one. About 80 percent of the iron in the body is in the blood, so anyone who loses blood loses iron. Because of menstrual losses, women need one and a half times as much iron as men do. Women are especially vulnerable to iron deficiency because they not only need more iron than men but they also, on average, eat less food. The information about iron in foods, later in this section, is especially important for most women.

In developing countries, parasitic infections of the digestive tract cause people to lose blood daily. For their entire lives, they may feel tired and listless but never know why. Ulcers and other sores of the digestive tract can also cause blood loss severe enough to cause anemia.[24]

Worldwide and in the United States, iron deficiency is the most common nutrient deficiency.[25] Iron-deficiency anemia affects an estimated 40 percent of the world's population, with the highest prevalence in developing countries.[26] Older infants, young children, and pregnant women are especially vulnerable. Happily, the iron status of U.S. infants and young children has improved over the last decade, thanks to more widespread breastfeeding, which promotes iron absorption, and greater use of iron-fortified infant formula and cereals. For low-income families, the Special Supplemental Food Program for Women, Infants, and Children (WIC) provides coupons redeemable for foods high in iron, giving another boost to the iron status of many U.S. children.

Can a Person Take in Too Much Iron? Iron is toxic in large amounts, and once inside the body, it is difficult to excrete. The body's defense against iron poisoning is a control system: the intestinal cells trap some of the iron and hold it within their boundaries. When they are shed, these cells carry out of the intestinal tract the excess iron that they collected during their brief lives.

Some individuals, most often men; are poorly defended against iron toxicity. Once considered rare, **iron overload** has emerged as an important iron disorder that seems to have increased in frequency over the last few decades. Iron overload is caused by a hereditary defect that causes the intestine to helplessly absorb excess iron. Tissue damage occurs, especially in iron-storing organs such as the liver. Infections are also likely because bacteria thrive on iron-rich blood. The effects are most severe in alcohol abusers because alcohol damages the intestine, impairing its defense against absorbing too much iron.

The body guards against iron's renegade nature. Left free, iron is a powerful oxidant that can start free-radical reactions that damage cellular structures.[27] Protein carriers guard the body's iron molecules and keep them away from vulnerable body compounds, thereby preventing damage. Iron's actions are thus tightly controlled.[28]

An interesting study of Finnish men conducted in the early 1990s suggested a link between increased risk of heart disease and elevated iron stores.[29] In that study, elevated

pica (PIE-ka) a craving for nonfood substances. Also known as *geophagia* (gee-oh-FAY-gee-uh) when referring to clay eating, and *pagophagia* (pag-oh-FAY-gee-uh) when referring to ice craving (*geo* means "earth"; *pago* means "frost"; *phagia* means "to eat").

iron overload the state of having more iron in the body than it needs or can handle. Too much iron is toxic and can damage the liver.

Feeling fatigued, weak, and apathetic is a sign that something is wrong. It is not a sign that you necessarily need iron or other supplements. Three actions are called for: first, get your diet in order; second, get some exercise (see the Fitness for Life section); then, if symptoms persist for more than a week or two, consult a physician for a diagnosis.

Figure 8-8

Normal and Anemic Blood Cells

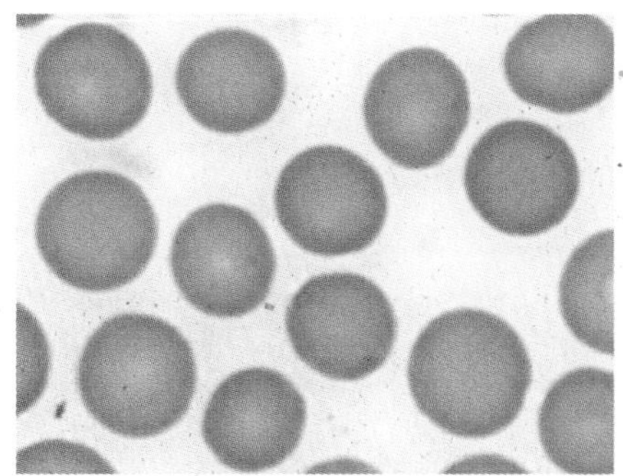

Normal red blood cells. Both size and color are normal.

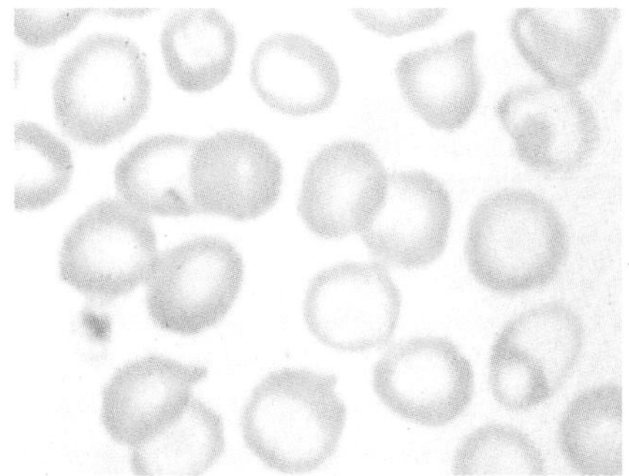

Blood cells in iron-deficiency anemia. These cells are small and pale because they contain less hemoglobin.

heme (HEEM): the iron-containing portion of the hemoglobin and myoglobin molecules.

Table 8-11, pp. 298–299, summarizes the effects of iron toxicity.

serum ferritin (the protein carrying iron in the blood) doubled the risk of heart attack. Research results have been mixed since then. One group of researchers fed rats chow laced with increasing doses of dietary iron. The chows high in iron caused increases in the oxidation of low-density lipoproteins (LDL) in the rats' blood, a process believed important in heart disease development.[30] When researchers searched for a link between high iron intakes and elevated heart disease in human populations, however, they found none.[31] More will be known as research progresses.

Other links may exist between iron and disease states. For example, a link between iron in foods and cancer of the colon may be a possibility.[32] Iron supplements may also stunt normal growth in healthy, well-fed children.[33] Much more work is needed to clarify these associations, however.

An argument against widespread fortification of foods with iron is that it might put more people at risk of iron overload. The U.S. population's love of vitamin C supplements may be worsening problems associated with excess iron because vitamin C greatly enhances iron absorption.[34] In any case, widespread iron fortification of foods would make it difficult for susceptible people to follow a low-iron diet.

Be forewarned against unnecessarily taking supplements that contain iron, and keep such pills safely out of children's reach. Iron supplements are the number one cause of fatal accidental poisonings among U.S. children under three years old. The FDA has called for a label warning parents that "iron may be lethal to children."[35] If you are feeling tired, do not automatically reach for an iron supplement; as the Fitness for Life feature on page 289 explains, what you may need is exercise.

Iron Recommendations and Sources Men need 10 milligrams of iron each day and so do women past age 51. For women of childbearing age, the recommendation is higher—15 milligrams—to replace menstrual losses. During pregnancy, a woman needs double this amount, or 30 milligrams. Adult men rarely experience iron-deficiency anemia. Should a man have a low hemoglobin concentration, his health-care provider is alerted to examine him for a blood-loss site. Table 8-8 sums up iron recommendations.

To meet iron needs, it is best to rely on foods because the iron from supplements is much less well absorbed than that from food. The usual Western mixed diet, however, provides only about 5 to 6 milligrams of iron in each 1,000 calories, not enough for some people. An adult male who eats 2,500 calories or more a day has no trouble meeting his need of 10 milligrams, but a woman who eats fewer calories and needs more iron understandably has trouble meeting her need. To meet it, she must select high-iron, low-calorie foods from each food group.

Iron occurs in two forms in foods. Some is bound into **heme**, the iron-containing part of hemoglobin and myoglobin in meat, poultry, and fish. Some is nonheme iron, the kind in foods from plants and the nonheme iron in meats. The form affects absorption.

Table 8-8

Iron Intake Recommendations (U.S.)[a]

1989 RDA

- Men (19–50 years) 10 milligrams per day.
  (51 years and older) 10 milligrams per day.
- Women (19–50 years) 15 milligrams per day.
  (51 years and older) 10 milligrams per day.

Proposed Healthy People 2010

- Reduce iron deficiency among children and females of childbearing age.
- Reduce anemia among low-income pregnant women to 23 percent.

[a]Canadian values are listed in Canadiana, Appendix B.

Absorbing Iron Heme iron is much more reliably absorbed than is nonheme iron. Healthy people with adequate iron stores absorb heme iron at a rate of about 23 percent over a wide range of meat intakes. People absorb nonheme iron at rates of 2 to 20 percent, depending on dietary factors and iron stores.

Meat, fish, and poultry contain a factor **(MFP factor)** other than heme that promotes the absorption of nonheme iron from other foods eaten with it. Vitamin C is also a potent promoter of iron absorption and can triple nonheme iron absorption from foods eaten in the same meal. A system of calculating the amount of iron absorbed from a meal, based on these factors, is presented in Table 8-9.

Some factors impair iron absorption. These include tea, coffee, the calcium and phosphorus in milk, and the **phytates** and **tannins** that accompany fiber in whole-grain cereals. Ordinary black tea is so efficient at reducing absorption of iron that clinical dietitians advise people with iron overload to drink it with their meals. For those who need more iron, the opposite advice applies—don't drink tea with food. Snapshot 8-5 shows iron amounts in foods that are good or excellent sources of iron.

The amount of iron ultimately absorbed from a meal depends on the interaction between promoters of iron absorption and inhibitors. When you eat meat with legumes (for example, ham and beans or chili with beans and meat), MFP factor

MFP factor a factor (identity unknown) present in meat, fish, and poultry that enhances the absorption of nonheme iron present in the same foods or in other foods eaten at the same time.

phytates compounds present in plant foods (particularly whole grains) that bind iron and prevent its absorption.

tannins compounds in tea (especially black tea) and coffee that bind iron. Tannins also denature proteins.

Dietary factors that increase iron absorption:

- Vitamin C.
- MFP factor.

Factors that hinder iron absorption:

- Tea.
- Coffee.
- Calcium and phosphorus.
- Phytates, tannins, and fiber.

Table 8-9

Calculation of Iron Absorbed from Meals

Three factors go into the calculation of the amount of iron absorbed from a meal: (1) how much of the iron in the meal was heme iron and how much was nonheme iron; (2) how much vitamin C was in the meal; and (3) how much total meat, fish, and poultry (MFP factor) was consumed. (It is assumed your iron stores are moderate; otherwise, you'd have to take this into consideration, too.) Write down the foods you eat at a typical meal, look up their iron content in the Table of Food Composition, Appendix A, and then answer these questions:

1. How much iron was from animal tissues (MFP)? ________mg.
2. 40% of (1), on the average, is heme iron: (1) _________ mg × 0.40 = _________ mg heme iron.
3. How much iron was from other sources? ___________mg.
4. This (3), plus 60% of (1), is nonheme iron: (3) _______ mg + 0.60 × (1) ___________ mg = _________ mg nonheme iron.
5. How much vitamin C was in the meal? Less than 25 mg is low; 25 to 75 mg is medium; more than 75 mg is high. _________ mg.
6. How much MFP factor was in the meal? Less than 1 oz lean MFP is low; 1 to 3 oz is medium; more than 3 oz is high. _______ oz.
7. Now calculate the heme iron absorbed. You absorbed 23% of the heme iron, or (2) _________mg × 0.23 = _________ mg heme iron absorbed.
8. Now, take your best score from (5) and (6). If either vitamin C or MFP factor was high or if both were medium, the availability of your nonheme iron was high. If neither was high, but one was medium, the availability of your nonheme iron was medium. If both were low, your nonheme iron had poor availability. You absorbed:
   - High availability: 8% of the nonheme iron.
   - Medium availabilty: 5% of the nonheme iron.
   - Poor availability: 3% of the nonheme iron.
9. Now calculate the nonheme iron absorbed. You absorbed _______ % of the nonheme iron, or (4) _________ mg × _________ = _________ mg nonheme iron absorbed.
10. Add the heme and nonheme iron from (7) and (9) together:
    - ________ mg heme iron absorbed.
    - ________ mg nonheme iron absorbed.

    Total = ________ mg iron absorbed.

Recommendations assume you will absorb 10% of the iron you ingest. Thus, if you are a man over age 18 or a woman over age 50 (recommended intake 10 mg), you need to absorb 1 mg per day. If you are a woman 11 to 50 years old (recommended intake 15 mg), you need to absorb 1.5 mg per day. If you have higher menstrual losses than the average woman, you may need still more.

This chili dinner provides iron and MFP factor from meat, iron from legumes, and vitamin C from tomatoes. The combination of heme iron, nonheme iron, MFP factor, and vitamin C helps to achieve maximum iron absorption.

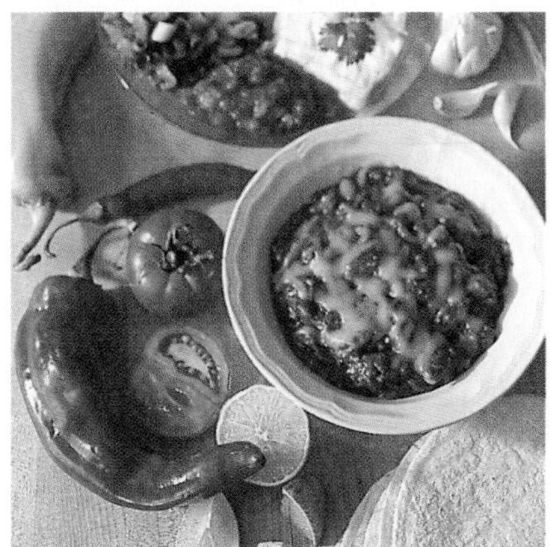

SNAPSHOT 8.5 Iron

These foods provide 10 percent or more of the iron Daily Value (DV = 18 mg/day).[a, b]

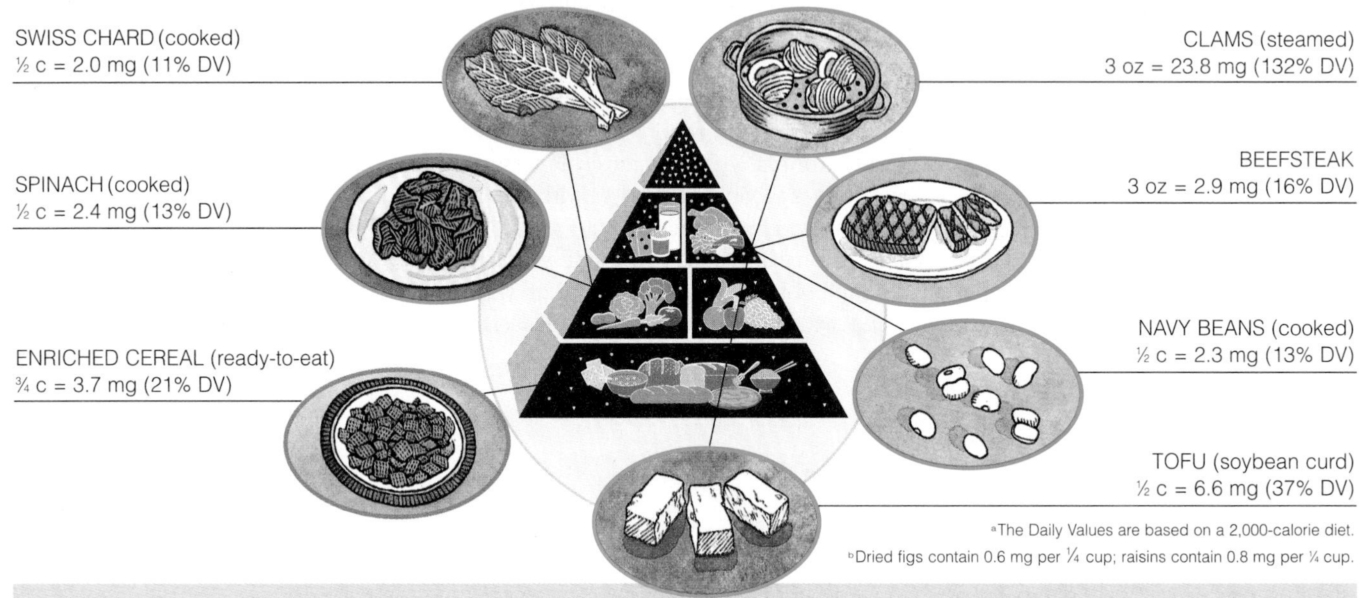

SWISS CHARD (cooked)
½ c = 2.0 mg (11% DV)
SPINACH (cooked)
½ c = 2.4 mg (13% DV)
ENRICHED CEREAL (ready-to-eat)
¾ c = 3.7 mg (21% DV)
CLAMS (steamed)
3 oz = 23.8 mg (132% DV)
BEEFSTEAK
3 oz = 2.9 mg (16% DV)
NAVY BEANS (cooked)
½ c = 2.3 mg (13% DV)
TOFU (soybean curd)
½ c = 6.6 mg (37% DV)

[a]The Daily Values are based on a 2,000-calorie diet.

[b]Dried figs contain 0.6 mg per ¼ cup; raisins contain 0.8 mg per ¼ cup.

The old-fashioned iron skillet adds supplemental iron to foods.

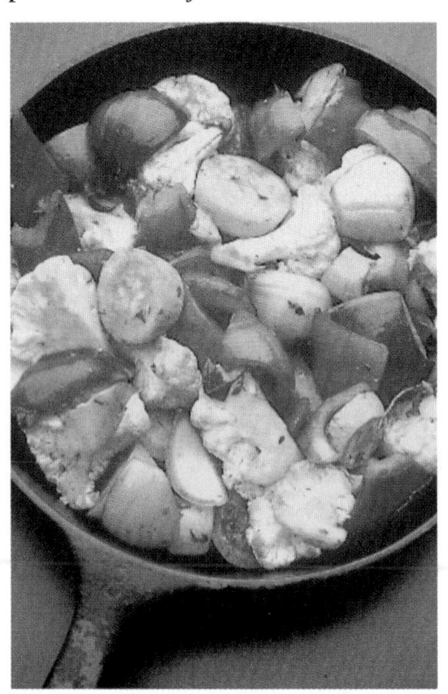

enhances iron absorption from both. The vitamin C from a slice of tomato and a leaf of lettuce in a sandwich will enhance iron absorption from the bread. The meat and tomato in spaghetti sauce help the eater absorb the iron from the spaghetti. A sauce cooked in an iron pan draws iron from this source, too.

Foods cooked in iron pans contain iron salts somewhat like those in supplements. The iron content of 100 grams of spaghetti sauce simmered in a glass dish is 3 milligrams, but it is 87 milligrams when the sauce is cooked in a black iron skillet. Even in the short time it takes to scramble eggs, a cook can triple the eggs' iron content by scrambling them in an iron pan. Similarly, dried peaches and raisins contain more iron than the fresh fruit because they are dried in iron pans. This iron salt is not as well absorbed as iron from meat, but some does get into the body, especially if the meal also contains MFP factor or vitamin C.

KEY POINT *Most iron in the body is contained in hemoglobin and myoglobin or occurs as part of enzymes in the energy-yielding pathways. Iron-deficiency anemia is a problem worldwide; too much iron is toxic. Iron is lost through menstruation and other bleeding; the shedding of intestinal cells protects against overload. For maximum iron absorption, use meat, other iron sources, and vitamin C together.*

Zinc

Zinc occurs in a very small quantity in the body, but works with proteins in every organ. It helps more than 50 enzymes to:[‡]

[‡]Estimates of as many as 200 zinc-containing metalloenzymes exist, such estimates distinguish between animal, microbial, and plant forms of the same enzyme. Estimates in the 50 to 70 range count the same enzyme from different lifeforms only once.

Iron or Exercise for More Energy?

FITNESS FOR LIFE

Feeling fatigued and worn-out can take the joy out of life. People suffering from chronic fatigue naturally seek out ways of increasing their vigor. On hearing about the symptoms of iron deficiency, tired people may assume that they need to take iron supplements to restore their pep. True, iron deficiency causes fatigue, but so do many other nutrient deficiencies and toxicities, including iron overload. A person who feels fatigued may also need to correct sleep or exercise habits. Getting to bed on time and taking steps to ensure sound sleep (see Controversy 13) are obvious remedies for too little rest. As for an exercise deficiency, few realize that too little exercise is as exhausting over weeks and months as too much—the less you do, the less you're able to do, and the more you feel fatigued throughout the day. The condition has even been given a name: "sedentary inertia."

Many people fear that exercise will worsen their fatigue. In truth, a new workout regimen can increase the need for sleep for a while, and often encourages sound sleep. Soon, as the body adapts to the new demands and increases its capacity for activity, the person feels more energetic than before. No longer taxed by everyday exertions, the physically active person becomes more able to do what needs to be done, all through the day.

Fatigue is a symptom of so many physical and psychological conditions that persistent or severe fatigue demands a diagnosis by a physician. But if you generally feel tired and unmotivated, and you rarely meet the guidelines for activity (see Chapter 2), then try gradually increasing your daily physical activity. It could be the best antidote for exercise deficiency fatigue.

Chapter 2 presented a "Physical Activity Pyramid" to serve as a guide in increasing activity.

- Make parts of the cells' genetic material.
- Make heme in hemoglobin.
- Help the pancreas with its digestive functions.
- Help metabolize carbohydrate, protein, and fat.
- Liberate vitamin A from storage in the liver.
- Dispose of damaging free radicals.

Zinc also affects behavior and learning, assists in immune function, and is essential to wound healing, sperm production, taste perception, fetal development, and growth and development in children. Zinc is needed to produce the active form of vitamin A in visual pigments. When zinc deficiency occurs, it impairs all these and other functions.[36] Even a mild zinc deficiency can result in impaired immunity, abnormal taste, and abnormal vision in the dark.

Problem: Too Little Zinc Zinc deficiency in human beings was first reported in the 1960s from studies with growing children and adolescent boys in the Middle East. The native diets were typically low in animal protein and high in whole grains and beans; consequently, they were high in fiber and phytates, which bind zinc as well as iron. Furthermore, the bread was not the **leavened** type; in leavened bread, yeast breaks down phytates as the bread rises.

leavened (LEV-end) literally, "lightened" by yeast cells, which digest some carbohydrate components of the dough and leave behind bubbles of gas that make the bread rise.

Since the first reports, zinc deficiency has been recognized elsewhere, and it is known to affect much more than growth. It alters digestive function profoundly and causes diarrhea, which worsens the malnutrition already present, with respect not only to zinc but to all nutrients. It drastically impairs the immune response, making infections likely.[37] Infections of the intestinal tract worsen malnutrition, including zinc malnutrition. Even mild zinc deficiency, brought on after one month of consuming a low-zinc diet, causes imbalances in the body's immune system that can increase susceptibility to infections.[38]

Normal vitamin metabolism depends on zinc, so zinc-deficiency symptoms often include vitamin-deficiency symptoms. Zinc deficiency also disturbs thyroid function and slows the body's energy metabolism. It causes loss of appetite and slows wound

How old does the boy in the picture appear to be? He is 17 years old but is only 4 feet tall, the height of a seven-year-old in the United States. His genitalia are like those of a six-year-old. The retardation is rightly ascribed to zinc deficiency because it is partially reversible when zinc is restored to the diet. The photo was taken in Egypt.

healing. In laboratory animals, a mild deficiency may reduce physical activity and attention span.[39] In fact, the symptoms are so pervasive that when faced with zinc deficiency, physicians are more likely to diagnose it as general malnutrition and sickness than as zinc deficiency.

Although severe zinc deficiencies are not widespread in developed countries, they occur among some groups, including pregnant women, young children, the elderly, and the poor. Among these people, poor growth, poor appetite, and impaired taste sensitivity may indicate zinc deficiency. When pediatricians or other health workers evaluating children's health note poor growth accompanied by poor appetite, they should think zinc.

Problem: Too Much Zinc Zinc is toxic in large quantities, and zinc supplements can cause serious illness or even death in high enough doses. Doses of zinc only a few milligrams above the recommended intake, especially when taken regularly over time, block copper absorption and lower the body's copper content, an effect that, in animals, leads to degeneration of the heart muscle.[40] In high doses, zinc may reduce the concentration of beneficial HDL in the blood.

High doses of zinc can also inhibit iron absorption from the digestive tract. A protein in the blood that carries iron from the digestive tract to tissues that need it also carries some zinc. If this protein is burdened with excess zinc, little or no room is left for iron to be picked up from the intestine. The opposite is also true; too much iron leaves little room for zinc to be picked up, thus impairing zinc absorption. Zinc and iron are often found together in foods, but food sources are safe and never cause imbalances in the body. Supplements, in contrast, can easily do so.

Unlike excess iron, excess zinc has a normal escape route from the body. The pancreas secretes zinc-rich juices into the digestive tract, and some of these are excreted. Still, large overdoses from zinc supplements can overwhelm the escape route and cause toxicity. Some experts consider small doses of zinc to be relatively nontoxic compared to other more hazardous trace minerals, but zinc supplements should be approached with caution.[41] In particular, zinc lozenges sold for relief of sore throats and colds may be safe if used sparingly. Mixed results are achieved when people use zinc lozenges against cold symptoms, however, so sufferers may be tempted to take higher and higher doses when the lozenges fail to deliver relief.[42] Most people find the taste of zinc lozenges objectionable, and this alone may protect some consumers from overdosing on zinc from this source.

Meats, shellfish, and poultry are among top providers of zinc (see Snapshot 8-6). Among plant sources, some legumes and whole grains are rich in zinc, but the zinc is not as well absorbed as from meat. Most people probably do not meet the recommended 15 milligrams per day for men and 12 milligrams per day for women; the average daily intake for men and women falls short of the recommendations by about 2 milligrams each.[43] Vegetarians are advised to eat varied diets that include whole-grain breads well leavened with yeast, which helps make zinc available for absorption.

KEY POINT *Zinc assists enzymes in all cells. Deficiencies in children cause growth retardation with sexual immaturity. Zinc supplements can reach toxic doses, but zinc in foods is nontoxic. Animal foods are the best sources.*

Selenium

Selenium has been attracting much attention for its role in antioxidant enzyme activity in protecting vulnerable body chemicals from oxidation. Selenium assists a group of enzymes that, in concert with vitamin E, works to prevent the formation of free radicals and prevent oxidative harm to cells and tissues.[44] The question of whether selenium protects against the development of some forms of cancers is currently under investigation, and some, but not all, preliminary results seem encouraging.[45] Among the recent discoveries about selenium is that it plays a role in activating thyroid hormone, the hormone that regulates the body's rate of metabolism.[46]

SNAPSHOT 8.6 Zinc

These foods provide 10 percent or more of the zinc Daily Value (DV = 15 mg/day).[a]

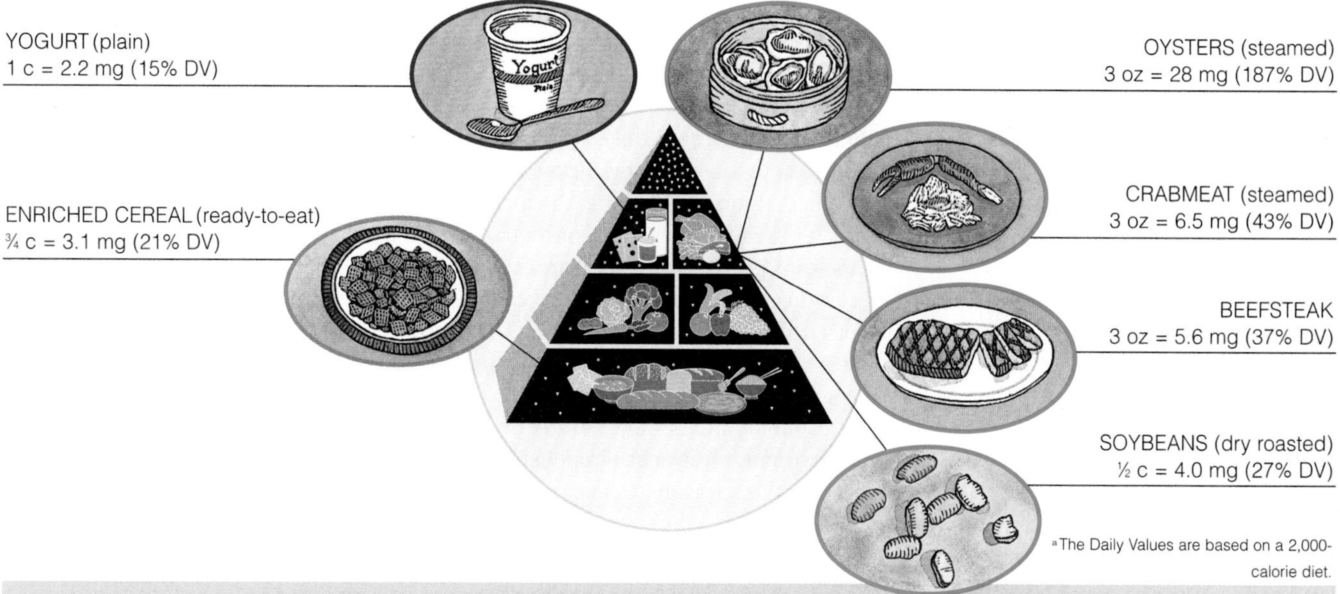
YOGURT (plain)
1 c = 2.2 mg (15% DV)
OYSTERS (steamed)
3 oz = 28 mg (187% DV)
CRABMEAT (steamed)
3 oz = 6.5 mg (43% DV)
ENRICHED CEREAL (ready-to-eat)
¾ c = 3.1 mg (21% DV)
BEEFSTEAK
3 oz = 5.6 mg (37% DV)
SOYBEANS (dry roasted)
½ c = 4.0 mg (27% DV)
Yogurt

[a]The Daily Values are based on a 2,000-calorie diet.

A deficiency of selenium can open the way for a specific type of heart disease (unrelated to the heart disease discussed in Chapters 5 and 11). The condition, first identified in China among people from areas with selenium-deficient soils, prompted researchers to give this mineral its rightful place among the essential nutrients (see the inside front cover, page B, for selenium's intake recommendation). Foods grown on U.S. and Canadian soils supply plenty of selenium.

Anyone who eats a normal diet composed of mostly unprocessed foods need not worry about selenium. It is widely distributed in foods such as meats and shellfish and in vegetables and grains grown on selenium-rich soil. From current indications, it is likely that most people in the United States receive plenty of selenium, partly because they eat supermarket foods transported from many regions and partly because they eat meat, a food rich in selenium.

Toxicity is possible, especially when people take selenium supplements over a long period. Selenium toxicity brings on symptoms such as hair loss, diarrhea, and nerve abnormalities. Selenium supplements taken inappropriately as an anticancer agent make selenium toxicity likely.

KEY POINT *Selenium works with vitamin E to protect body compounds from oxidation. A deficiency induces a disease of the heart. Deficiencies in developed countries are rare, but toxicities occur from overuse of supplements.*

Fluoride

Fluoride is not essential to life, but it is beneficial in the diet because of its ability to inhibit the development of dental caries in both children and adults.[47] Only a trace of fluoride occurs in the human body, but with this amount the crystalline deposits in bones and teeth are larger and more perfectly formed. As mentioned earlier, fluoride replaces the hydroxy portion of hydroxyapatite, forming the more decay-resistant fluorapatite.

fluorosis (floor-OH-sis) discoloration of the teeth due to ingestion of too much fluoride during tooth development.

Drinking water is the usual source of fluoride. In communities where the water contains too much—2 to 8 parts per million—discoloration of the teeth, or **fluorosis,** may occur. In recent years, widespread availability of fluoridated toothpaste, mouthwash, foods made with fluoridated water, and fluoride-containing supplements has led to an increase in the mildest form of fluorosis.[48] In this condition, characteristic white spots form in the tooth enamel. A more severe form of fluorosis is shown in Figure 8-9. Fluorosis is irreversible and occurs only during tooth development, never after teeth have formed. To prevent fluorosis, people in areas with fluoridated water may wish to limit other sources for infants and young children, and especially to avoid giving them fluoride supplements unless prescribed by a physician. Fluoride supplements are no longer recommended for children from birth, and the dosages for older children have been reduced.[49] Young children should be told not to swallow their toothpaste when brushing the teeth to avoid excessive fluoride intake.

Where fluoride is lacking, the incidence of dental decay is very high. Fluoridation of water to raise its fluoride concentration to 1 part per million is recommended as an important public health measure. Those fortunate enough to have had sufficient fluoride during the tooth-forming years of infancy and childhood are protected from tooth decay throughout life. Figure 8-10 shows the extent of fluoridation nationwide; states that have adopted fluoridation in more than half of their counties are shown in color.

Figure 8-9

Fluorosis

The brown mottled stains on these teeth indicate exposure to high concentrations of fluoride during development.

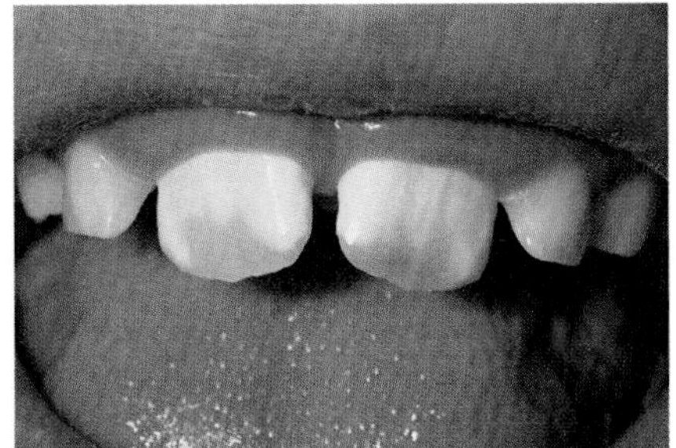

Despite fluoride's value, strong disagreement often surrounds the decision to fluoridate community water. Proponents argue that fluoridation is an obvious, safe, and cost-effective measure to help prevent dental caries in the young.[50] Opponents argue that altering the community water supply is "unnatural" and deprives its consumers of the freedom to refuse to take fluoride. They may claim that communities using fluoridated water have an increased cancer rate, but studies show no connection.

Opponents of fluoridation also fear accidental overdoses. In fact, serious cases of fluoride poisoning have occurred when public water systems failed and allowed fluoride to reach toxic concentrations.[51] Such incidents make a strong case for vigorous monitoring of public water supplies.

On the basis of the accumulated evidence of its beneficial effects, fluoridation has been endorsed by the National Institute of Dental Health, the American Dietetic Association, the American Medical Association, the National Cancer Institute, and the National Nutrition Consortium. The allegation that it causes cancer has no basis in fact and has been refuted by the National Cancer Institute, the American Cancer Society, and the National Institute of Dental Research.

KEY POINT *Fluoride stabilizes bones and makes teeth resistant to decay. Excess fluoride discolors teeth; large doses are toxic.*

Chromium

Chromium works closely with the hormone insulin to regulate and release energy from glucose. When chromium is lacking, insulin action is impaired, resulting in a diabetes-like condition of high blood glucose that resolves with chromium supplementation. Supplements of chromium cannot cure the common forms of diabetes, of course, but people with diabetes who eat low-chromium diets may make their condition worse.[52] Diets high in simple sugars deplete the body's supply of chromium.

People who hope that chromium-containing supplements will build extra muscle tissue or melt off body fat or ward off its regain are in for a disappointment.[53] Chromium supplements have been reported to slightly increase lean body mass in laboratory animals and sometimes in human beings tested under laboratory conditions. These results have led to exaggerated claims for chromium's ability to bring about weight loss and muscle gain. Follow-up studies so far have shown no effect of chromium on body fat and lean tissue. If such an effect turns up, it is likely to be minor in comparison with those of diet and exercise. Likewise, chromium supplementation does not lower blood cholesterol in people, as is often implied by "information" from supplement marketers.

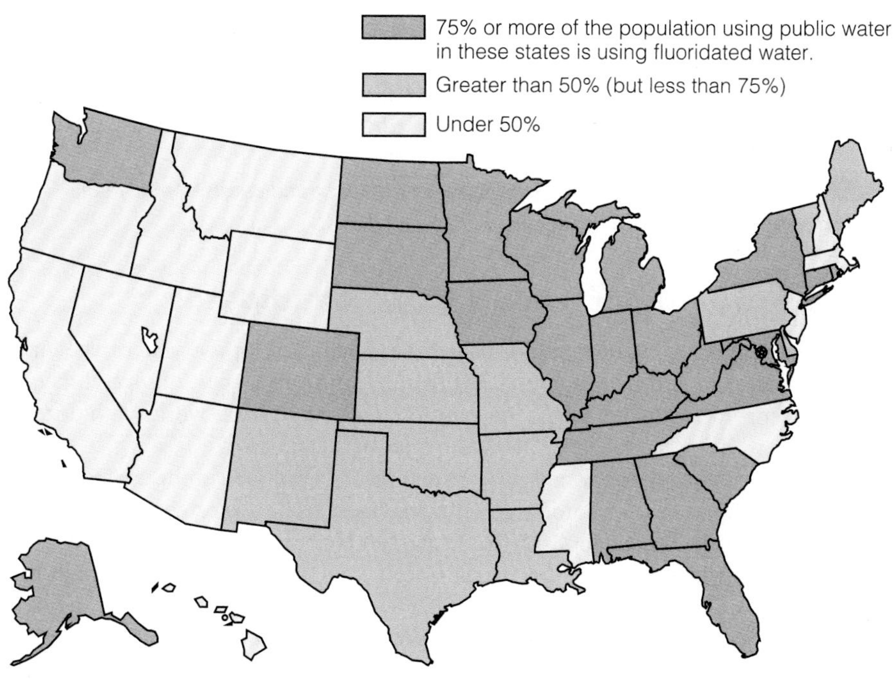
75% or more of the population using public water in these states is using fluoridated water.
Greater than 50% (but less than 75%)
Under 50%

SOURCE: Fluoridation Census 1989 Summary, U.S. Department of Health and Human Services, Public Health Service, Centers for Disease Control and Prevention, National Center for Prevention Services, Division of Oral Health, Atlanta, GA, April 1993.

Figure 8-10
Fluoridation in the United States

Chromium compounds used in industry are known carcinogens and are responsible for many cases of cancer in exposed workers.[54] The form of chromium in foods and supplements seems to be relatively nontoxic by comparison, and amounts of about 200 micrograms per day seem to be safe.[55] Taking in much more than this amount is ill-advised, however.

Chromium from food is found in complexes with other compounds that make it easily controlled and used by the body. Researchers use the terms *biologically active chromium* or *glucose tolerance factor* to describe these chromium-containing compounds.[56]

Although chromium is present in a variety of foods, it is estimated that 90 percent of U.S. adults consume less than the recommended minimum intake of 50 micrograms a day. Chromium is easily lost during food processing, as are other trace minerals. As people find themselves more pressed for time and depending more heavily on refined foods, chromium deficiencies become more likely. The best chromium food sources are liver, whole grains, nuts, and cheeses.

KEY POINT *Chromium works with the hormone insulin to control blood glucose concentrations. Many adults in the United States have low dietary intakes of chromium.*

Copper

One of copper's most vital roles is to help form hemoglobin and collagen. Many enzymes depend on copper for its oxygen-handling ability. Copper, like iron, assists in reactions leading to the release of energy. It also works with proteins to regulate the activity of certain genes.[57] One copper-dependent enzyme helps to control damage from free-radical activity in the tissues.[§] Some researchers are investigating the possibility that a low-copper diet may contribute to heart disease by suppressing the activity of this enzyme.

Copper deficiency is rare but not unknown. It has been seen in severely malnourished infants fed a copper-poor formula.[58] It can severely disturb growth and metabolism. In adults, it may impair immunity and blood flow through the arteries.[59]

[§]The enzyme is superoxide dismutase, first mentioned in Controversy 7.

Excess zinc interferes with copper absorption and can cause deficiency.[60] Copper toxicity from foods is unlikely, but supplements can cause it. The best food sources of copper include organ meats, seafood, nuts, and seeds. Water may also supply copper, especially where copper plumbing pipes are used.[61]

KEY POINT *Copper is needed to form hemoglobin and collagen and assists in many other body processes. Copper deficiency is rare.*

Other Trace Minerals and Some Candidates

Daily dietary intake recommendations have been established for two other trace minerals, molybdenum and manganese. Molybdenum functions as part of several metal-containing enzymes, some of which are giant proteins. Manganese works with dozens of different enzymes that facilitate many different body processes.

Several other trace minerals are now recognized as important to health. Research suggests that a low intake of boron may enhance susceptibility to osteoporosis by way of its effects on calcium metabolism. The richest food sources of boron are noncitrus fruits, leafy vegetables, nuts, and legumes. Cobalt is recognized as the mineral in the large vitamin B_{12} molecule; the alternative name for vitamin B_{12}, cobalamin, reflects cobalt's presence. Nickel is important for the health of many body tissues; deficiencies harm the liver and other organs. Silicon is known to be involved in bone calcification, at least in animals. The future may reveal key roles played by many other trace minerals including barium, cadmium, lead, lithium, mercury, silver, tin, and vanadium. Even arsenic, a known poison and carcinogen, may turn out to be essential in tiny quantities.

All trace minerals are toxic in excess. The hazards of overdoses are among the chief risks faced by people who take multiple nutrient supplements. The way to obtain the trace minerals is from food, which is not hard to do. You need only eat a variety of whole foods in the amounts recommended in the Food Guide Pyramid. Some claim that organically grown foods contain more trace minerals than those grown with chemical fertilizers. Organic fertilizers do contain more trace minerals than do refined chemical fertilizers, and plants do take up some of the minerals they are given, so this claim may be valid.

As research on the trace minerals continues, many interactions among them are also coming to light. An excess of one may cause a deficiency of another. A slight manganese overload, for example, may aggravate an iron deficiency. A deficiency of one mineral may open the way for another to cause a toxic reaction. Iron deficiency, for example, makes the body much more susceptible to lead poisoning than it normally is. Good food sources of one are poor food sources of another, and factors that cooperate with some trace elements oppose others. Vitamin C, for example, enhances the absorption of iron and depresses that of copper. The continuous outpouring of new information about the trace minerals is a sign that we have much more to learn. Table 8-11 on pp. 298–299 sums up what this chapter has said about the minerals and fills in some additional information.

KEY POINT *Many different trace elements play important roles in the body. All of the trace minerals are toxic in excess.*

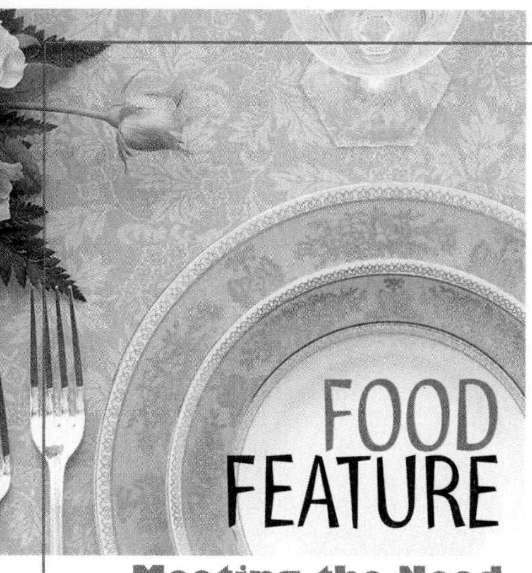

FOOD FEATURE

Meeting the Need for Calcium

kefir a yogurt-based beverage.

nori a type of seaweed popular in Asian, particularly Japanese, cooking.

stone-ground flour flour made by grinding kernels of grain between heavy wheels made of limestone, a kind of rock derived from the shells and bones of marine animals. As the stones scrape together, bits of the limestone mix with the flour, enriching it with calcium.

Table 8-10

Recommended Fluid Milk Intakes

Age	Recommended U.S. Daily Intake
Children	2 cups
Teenagers	3 cups
Adults	2 cups
Pregnant or lactating women	3 cups
Pregnant or lactating teens	4 cups

MOST PEOPLE consume far less than the recommended amount of calcium. The average woman meets just a third of her requirement; men do somewhat better, with intakes close to three-fourths of the recommendation. Low calcium intakes are associated with all sorts of major illnesses, including adult bone loss (see the following Controversy for details), high blood pressure and colon cancer (see Chapter 11), kidney stones, and even lead poisoning. This Food Feature focuses attention on the sources of calcium in the diet. Despite these benefits, consumption of one of the best sources of calcium—milk—has been decreasing in recent years while consumption of other beverages, such as soft drinks has been dramatically increasing. One national dairy group calls the situation a "national calcium crisis," and 250 representatives of many national health and nutrition organizations agree.[62] Representatives of these groups joined together at a "Calcium Summit" where they identified these goals:

- Educate professionals about the nation's low calcium intakes.
- Educate consumers with consistent messages on the importance of calcium.
- Develop strategies for making milk consumption more appealing and convenient to the U.S. population.

This Food Feature focuses attention on milk and other sources of calcium in the diet.

Milk, Yogurt, and Cheese Group

Milk and milk products are traditional sources of calcium for people who can tolerate them. Table 8-10 shows the current milk recommendations that help to meet the calcium needs of various age groups. People who do not use milk because of lactose intolerance, dislike, or allergy must obtain calcium from other sources. Care is needed, though; *wise* substitutions must be made. Most of milk's many relatives are recommended choices: yogurt, **kefir**, buttermilk, cheese (especially the low-fat or fat-free varieties), and, for people who can afford the calories, ice milk. Cottage cheese and frozen yogurt desserts contain about half the calcium of milk, with 2 cups being equivalent in calcium to 1 cup of milk. Butter, cream, and cream cheese are almost pure fat and contain negligible calcium.

If no milk product is acceptable as is, consider tinkering with it to make it work. Add cocoa to milk, fruit to yogurt, or fat-free milk powder to any dish. By the way, the cocoa powder added to make chocolate milk contains a small amount of oxalic acid, a compound that binds with some of milk's calcium and inhibits its absorption. The effect is probably insignificant, however—only slightly less calcium is abosrbed from a glass of chocolate milk than from white milk. Sugar lends both sweetness and calories to chocolate milk, so mix your chocolate milk at home where you control the amount of sugary chocolate added to the milk.

Vegetables

Among vegetables, rutabaga, broccoli, beet greens, turnip greens, mustard greens, bok choy (a Chinese cabbage), and kale are good sources of available calcium. So are collard greens, green cabbage, kohlrabi, watercress, parsley, and probably some seaweeds, such as the **nori** popular in Japanese cookery. Certain other foods, including spinach, Swiss chard, and rhubarb, appear equal to milk in calcium content but actually provide no calcium, or very little, to the body because they contain binders that prevent calcium's absorption (see Figure 8-11 on the next page). Of course, the presence of calcium binders does not make spinach an inferior food. Dark greens of all kinds are superb sources of riboflavin and virtually indispensable for the vegan or anyone else who does not drink milk. Spinach is also rich in iron, beta-carotene, and dozens of other essential nutrients and potentially helpful phytochemicals. Just don't rely on it for calcium.

Calcium in Other Foods

For the many people who cannot use milk and milk products, oysters are a rich source of calcium. So are small fish such as canned sardines or other canned fishes prepared with their bones. Stocks

Label reader's tip: To convert % Daily Value for calcium to milligrams, drop the percent sign and add a zero.* Example: 40% Daily Value = 400 mg.

*NOTE: This tip works for calcium, but not for other minerals.

or extracts made from bones are another rich source. The Vietnamese people's tradition of making such a stock helps account for their adequate calcium intake without the use of milk. To make a high-calcium extract, soak cracked bones from chicken, turkey, pork, or fish in vinegar; then slowly boil the bones until they become soft, indicating that they have released their calcium into the acid medium. By this time, most of the vinegar taste will have boiled off. Use the stock in place of water to cook soup, vegetables, rice, or stew. One *tablespoon* of such stock may contain more than 100 milligrams of calcium.

Calcium-Fortified Foods

Next in order of preference among nonmilk sources of calcium are foods that contain large amounts of calcium salts by an accident of processing or by intentional fortification. In the processed category are bean curd (tofu: calcium salt is often used to coagulate it); canned tomatoes (firming agents donate 63 milligrams per cup of tomatoes); **stone-ground flour** or self-rising flour; stone-ground whole or self-rising cornmeal; and blackstrap molasses.

Some food products available to U.S. consumers are specially fortified to add calcium to people's diets. The richest in calcium is high-calcium milk itself, that is, milk with extra calcium added; it provides more calcium per cup than any natural milk, 500 milligrams per 8 ounces. Then comes calcium-fortified orange juice, with 300 milligrams per 8 ounces, a good choice because the bioavailability of its calcium compares favorably with that of milk. Calcium-fortified soy milk can also be prepared so that it contains more calcium than whole cow's milk. Soy-based infant formula is fortified with calcium, and no law prevents adults from using it in cooking for themselves.

Finally, there are supplements intended to meet calcium needs without regard to needs for energy or other nutrients. Most people who take calcium supplements do so in hopes of warding off osteoporosis. As Controversy 8 points out, however, supplements are not magic bullets against bone loss.

Figure 8-11

Calcium Absorption from Food Sources

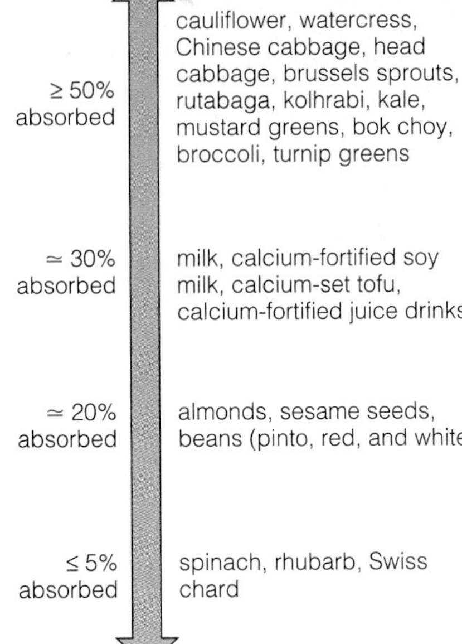
≥ 50% absorbed
cauliflower, watercress, Chinese cabbage, head cabbage, brussels sprouts, rutabaga, kolhrabi, kale, mustard greens, bok choy, broccoli, turnip greens
≈ 30% absorbed
milk, calcium-fortified soy milk, calcium-set tofu, calcium-fortified juice drinks
≈ 20% absorbed
almonds, sesame seeds, beans (pinto, red, and white)
≤ 5% absorbed
spinach, rhubarb, Swiss chard

SOURCE: Data from C. M. Weaver and K. L. Plawecki, Dietary calcium: Adequacy of a vegetarian diet, *American Journal of Clinical Nutrition* 59 (1994): S1238–S1241; R. P. Heaney and coauthors, Absorbability of calcium from *Brassica* vegetables: Broccoli, bok choy, and kale, *Journal of the American Dietetic Association* 93 (1993): 1378–1380.

Making Meals Rich in Calcium

The following are some tips for including calcium-rich foods in your day's meals. Many cooks slip extra calcium into meals by sprinkling a tablespoon or two of fat-free dry milk into almost everything. The added calorie value is small, changes to taste and texture of the dish are practically nil, but each 2 tablespoons adds about 100 extra milligrams of calcium and moves people closer to meeting the recommendation to obtain about 2 servings of milk each day (see Figure 8-12). Here are some additional tips:

At Breakfast

- Serve tea or coffee with milk, hot or iced.
- Choose cereals, hot or cold, with milk.

Calcium in a delicious form.

- Cook hot cereals with milk instead of water; then mix in 2 tablespoons of fat-free dry milk.
- Make muffins or quick breads with milk and extra powdered milk.
- Add milk to scrambled eggs.
- Moisten cereals with flavored yogurt.
- Choose calcium-fortified orange juice.

At Lunch

- Add low-fat cheeses to sandwiches, burgers, or salads.
- Use a variety of green vegetables, such as Swiss chard or kale, in salads and on sandwiches.
- Stuff potatoes with broccoli and low-fat cheese.
- Try pasta such as ravioli stuffed with low-fat ricotta cheese instead of meat.
- Sprinkle parmesan cheese on pasta salads.
- Mix the mashed bones of canned salmon into salmon salad or patties.
- Eat sardines with their bones.
- Use fat-free milk or calcium-fortified soy milk as a beverage.
- Marinate cabbage shreds or broccoli spears in low-fat Italian dressing for an interesting salad.
- Choose coleslaw over potato and macaroni salads.

At Supper

- Add fat-free powdered milk to almost anything—meat loaf, sauces, gravies, soups, stuffings, casseroles, blended beverages, puddings, quick breads, cookies, brownies. Be creative.
- Choose frozen yogurt, ice milk, or custards for dessert.
- Toss a handful of thinly sliced green vegetables, such as kale or young turnip greens, with hot pasta dishes; the greens wilt pleasingly in the steam of the freshly cooked pasta.
- Serve a green vegetable every night and try new ones—how about kohlrabi? It tastes delicious when boiled like broccoli.
- Learn to stir-fry Chinese cabbage and other Asian foods.
- Try some tofu (the calcium-set kind); this versatile food has inspired whole cookbooks devoted to creative uses.

People who learn to identify and regularly choose rich calcium sources among foods are rewarded with many benefits to their health. The Do It! section at the chapter's end provides practice at finding calcium and other minerals in foods when you are in a hurry—from a convenience store.

Figure 8-12

Milk, Yogurt, and Cheese Group: Food Supply Servings, 1970–1996[a]

On average, people in the United States fall far short of meeting the recommendation to obtain 2 or 3 servings of milk, yogurt, or cheese each day. The picture is worse for the dark green vegetables that supply calcium—only 3 percent of the vegetables consumed each day meet this description.

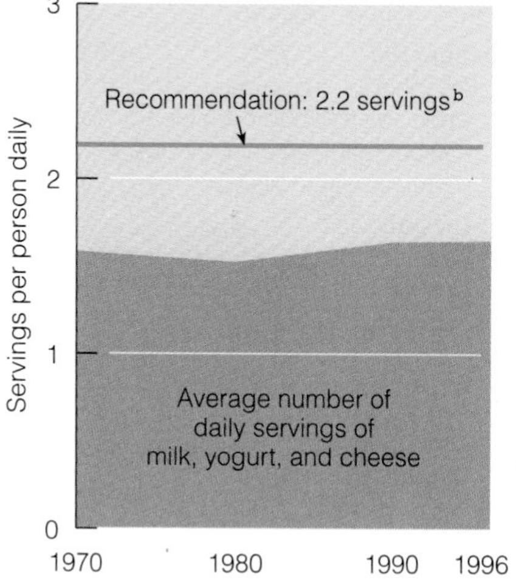
Servings per person daily
3
2
1
0
Recommendation: 2.2 servings[b]
Average number of daily servings of milk, yogurt, and cheese
1970
1980
1990
1996

[a]Includes all forms of milk, yogurt, cheese, and frozen dairy desserts.

[b]Recommended servings based on weighted average of recommended servings for different age groups of the U.S. population, excluding the higher needs of pregnant and lactating women.

SOURCE: U.S. Department of Agriculture, Economic Research Service, 1998.

Table 8-11

The Minerals—A Summary

Mineral and Chief Functions in the Body	Deficiency Symptoms	Toxicity Symptoms	Significant Sources
MAJOR MINERALS			
Calcium			
The principal mineral of bones and teeth. Also acts in normal muscle contraction and relaxation, nerve functioning, blood clotting, blood pressure, and immune defenses.	Stunted growth in children; adult bone loss (osteoporosis).	Excess calcium is excreted except in hormonal imbalance states (not caused by nutritional deficiency).	Milk and milk products, oysters, small fish (with bones), tofu (bean curd), greens, legumes.
Phosphorus			
Phosphorus is important in cells' genetic material, in cell membranes as phospholipids, in energy transfer, and in buffering systems.	Appetite loss, bone pain, muscle weakness, impaired growth, and rickets in infants.[a]	Excess phosphorus may cause calcium excretion.	All animal tissues.
Magnesium			
A factor involved in bone mineralization, the building of protein, enzyme action, normal muscular contraction, transmission of nerve impulses, and maintenance of teeth.	Weakness; muscle twitches; appetite loss; confusion; depressed pancreatic hormone secretion; if extreme, convulsions, bizarre movements (especially of eyes and face), hallucinations, and difficulty in swallowing. In children, growth failure.[b]	Excess magnesium from abuse of laxatives and other medications by the elderly or those with kidney disease has caused confusion, lack of muscle coordination, coma, and death.	Nuts, legumes, whole grains, dark green vegetables, seafoods, chocolate, cocoa.
Sodium			
Sodium, chloride, and potassium (electrolytes) maintain cells' normal fluid balance and acid-base balance in the body. Sodium is critical to nerve impulse transmission.	Muscle cramps, mental apathy, loss of appetite.	Hypertension.	Salt, soy sauce, processed foods.
Potassium			
Potassium facilitates reactions, including the making of protein; the maintenance of fluid and electrolyte balance; the support of cell integrity; the transmission of nerve impulses; and the contraction of muscles, including the heart.	Deficiency accompanies dehydration; causes muscular weakness, paralysis, and confusion; can cause death.	Causes muscular weakness; triggers vomiting; if given into a vein, can stop the heart.	All whole foods: meats, milk, fruits, vegetables, grains, legumes.
Chloride			
Chloride is also part of the hydrochloric acid found in the stomach, necessary for proper digestion.	Growth failure in children; muscle cramps, mental apathy, loss of appetite; can cause death (uncommon).	Normally harmless (the gas chlorine is a poison but evaporates from water); can cause vomiting.	Salt, soy sauce; moderate quantities in whole, unprocessed foods, large amounts in processed foods.

[a]Seen only rarely in infants fed phosphorus-free formula or in adults taking medications that interact with phosphorus.

[b]A still more severe deficiency causes tetany, an extreme, prolonged contraction of the muscles similar to that caused by low blood calcium.

Table 8-11

The Minerals—A Summary continued

Mineral and Chief Functions in the Body	Deficiency Symptoms	Toxicity Symptoms	Significant Sources
MAJOR MINERALS			
Sulfur			
A component of certain amino acids; part of the vitamins biotin and thiamin and the hormone insulin; combines with toxic substances to form harmless compounds; stabilizes protein shape by forming sulfur-sulfur bridges (see Figure 6-10 in Chapter 6).	None known; protein deficiency would occur first.	Would occur only if sulfur amino acids were eaten in excess; this (in animals) depresses growth.	All protein-containing foods.
TRACE MINERALS			
Iodine			
A component of the thyroid hormone thyroxine, which helps to regulate growth, development, and metabolic rate.	Goiter, cretinism.	Depressed thyroid activity; goiter-like thyroid enlargement.	Iodized salt; seafood; bread; plants grown in most parts of the country and animals fed those plants.
Iron			
Part of the protein hemoglobin, which carries oxygen in the blood; part of the protein myoglobin in muscles, which makes oxygen available for muscle contraction; necessary for the use of energy.	Anemia: weakness, pallor, headaches, reduced resistance to infection, inability to concentrate, lowered cold tolerance.	Iron overload: fatigue, infections, liver injury, possible increased risk of colon cancer, growth retardation in children, acidosis, bloody stools, shock.	Red meats, fish, poultry, shellfish, eggs, legumes, dried fruits.
Zinc			
Part of insulin and many enzymes; involved in making genetic material and proteins, immune reactions, transport of vitamin A, taste perception, wound healing, the making of sperm, and normal fetal development.	Growth failure in children, dermatitis, sexual retardation, loss of taste, poor wound healing.	Fever, nausea, vomiting, diarrhea, muscle incoordination, dizziness, anemia, accelerated atherosclerosis, kidney failure.	Protein-containing foods: meats, fish, shellfish, poultry, grains, vegetables.
Selenium			
Part of an enzyme that breaks down reactive chemicals that harm cells; works with vitamin E.	Muscle degeneration and pain, cataracts, depressed sperm production, fragile red blood cells, pancreas damage, heart damage, growth failure in children.	Nausea; abdominal pain; nail and hair changes; nerve, liver, and muscle damage.	Seafoods, organ meats; other meats, grains, and vegetables depending on soil conditions.
Fluoride			
Helps form bones and teeth; confers decay resistance on teeth.	Susceptibility to tooth decay.	Fluorosis (discoloration) of teeth, nausea, vomiting, diarrhea, chest pain, itching.	Drinking water if fluoride containing or fluoridated; tea; seafood.
Chromium			
Associated with insulin; needed for energy release from glucose.	Abnormal glucose metabolism.	Possible link with muscle degeneration.	Meat, unrefined grains; vegetable oils.
Copper			
Helps form hemoglobin; part of several enzymes.	Anemia; poor wound healing.	Vomiting, diarrhea.	Meat, drinking water.

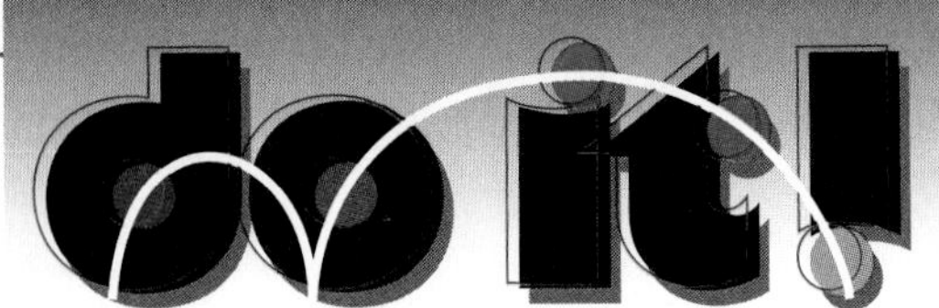

FIND THE MINERALS IN SNACK FOODS

In an ideal world, each person would set aside time every day for planning, shopping, and cooking nutritious meals. In the real world, you've had nothing to eat, your research papers are due, your room is topsy turvy, and your car needs fuel as you're rushing to class. A convenience store seems like a good idea; you can fill up your car and grab a bite to eat in one stop. What you grab makes a difference to your day's calorie and mineral intakes, however—to the benefit or detriment of your nutritional health.

The shelves of convenience stores are lined with all sorts of edible items, and luckily, their labels list their contents of energy, calcium, iron, and sodium. For the sake of learning, we've added two other minerals to this exercise: magnesium and zinc.

List Your Nutrient Intake Goals and Choose Some Foods

Step 1. List your intake goals for energy and iron, zinc, and magnesium (see the inside front cover, or Appendix B for Canadian recommendations) as indicated on Form 8-1. For sodium, use the recommended upper limit of 2,400 milligrams as your target for the day.

Step 2. Scan the variety of items on the convenience store shelves represented in Figure 8-13. Choose a snack and enter your choices on Form 8-1. The portion sizes are those

Figure 8-13

Snacks at the Fuel 'n' Feed

Snacks at the Fuel 'n' Feed	
Sandwiches/ Canned foods	**Snacks**
Cheese pizza slice	Apple pie, packaged
Ham biscuit	Banana
Pork and beans	Cheese slice
Roast beef sandwich	Chocolate candy bar
Sardines	Dill pickle
Vienna sausage	Dried fruit mix
	Fig bar cookies
	Ice cream and sherbet bar
	Potato chips, small bag
	Pretzels, regular
Beverages	Pretzels, unsalted
Milk, 1% fat	Roasted almonds
Chocolate milk, 2% fat	Sunflower seeds, shelled
Cola	Wheat crackers
Pineapple orange juice	Yogurt, low fat, with fruit

Form 8-1

Energy and Mineral Tally

Your Snack Foods	Energy (cal)	Calcium (mg)	Iron (mg)	Magnesium (mg)	Potassium (mg)	Sodium (mg)	Zinc (mg)
Snack totals							
Personal intake goals[a]						2,400	
% of recommendation							

[a]See Inside front cover or Appendix B.

commonly used for such foods, and not those recommended in the Food Guide Pyramid. For example, a can of vienna sausages contains 5 ounces of sausages, not the recommended 3 ounces.

Record Energy and Mineral Values

Step 3. Obtain energy and mineral values for your snack choices from Table 8-12 and enter these values on Form 8-1. Total the seven columns for energy and minerals.

Step 4. Calculate the percentages of energy and mineral requirements contributed by this snack. Divide each nutrient total by the recommendation for that nutrient, and multiply by 100. Example: a snack providing 40 milligrams of calcium would meet 4 percent of an intake recommendation of 1,000 milligrams (40 ÷ 1,000) × 100 = 4%.

Analysis

Now answer these questions:

1. What percentage of your energy need did the snack contribute? Did the snack also contribute a proportional amount of calcium, iron, magnesium, or zinc?
2. What about sodium? Did the sodium in this snack exceed 10 percent of the maximum of 2,400 milligrams? Did it exceed 30 percent? If so, look for low-sodium supper choices to round out the day. Consult Figure 8-7 to review high- and low-sodium foods.
3. Which of the foods listed in Table 8-12 are the best "bargains" in terms of sodium and other minerals? For example, a food that supplies much of the daily sodium allowance and few other nutrients may support nutrition goals less well than a similar food with less

Table 8-12

Energy and Mineral Contents of Selected Convenience Store Foods[a]

	Energy (cal)	Calcium (mg)	Iron (mg)	Magnesium (mg)	Potassium (mg)	Sodium (mg)	Zinc (mg)
Cheese pizza, slice	243	184	0.7	16	209	467	0.8
Ham biscuit	386	160	2.7	23	210	1432	1.7
Vienna sausages, 5 oz	395	14	1.3	10	86	1347	2.3
Sardines, 3¾ oz	221	405	3.1	41	338	536	1.4
Pork and beans, 5 oz	139	80	4.7	50	425	624	8.4
Roast beef sandwich	318	47	3.1	22	367	1252	3.0
Ice cream and sherbet bar	92	62	0.1	7	102	43	0.4
Potato chips, salt and vinegar, 1 oz	150	7	0.5	19	360	380	0.3
Pretzels, unsalted, 1 oz	110	10	1.2	10	41	60	0.2
Pretzels, salted, 1 oz	108	10	1.2	10	41	486	0.2
Fig bar cookie, 4	195	36	1.3	15	116	214	0.2
Chocolate candy bar, 1½ oz	226	84	0.6	26	169	36	0.6
Wheat crackers, 1 oz	134	14	1.3	18	84	225	0.5
Apple pie, fast-food type	225	6	1.0	6	67	179	0.2
Low-fat chocolate milk, 2% fat, 1 c	179	285	0.6	33	423	151	1.0
Pineapple orange drink, 12 oz	170	17	0.9	20	156	10	0.2
Cola, 12 oz	186	14	0.1	5	3	18	0.0
Milk, 1% fat, 1 c	102	300	0.1	34	381	123	1.0
Dill pickle	12	6	0.3	7	75	833	0.1
Low-fat yogurt, with fruit, 8 oz	232	345	0.2	33	478	133	1.7
Sunflower seeds, shelled, 3¾ oz	654	123	7.2	375	513	3	5.4
Dried fruit mix, 3¾ oz	258	40	2.9	41	846	19	0.5
Banana	105	7	0.4	33	451	1	0.2
Cheese slice, 1 oz	69	121	0.2	6	59	250	0.6
Roasted almonds, salted, 3¾ oz	660	326	4.0	273	629	828	3.3

[a]Data from the Food Processor 7.11 diet analysis program, ESHA, 1996.

sodium. List some high-sodium and low-sodium choices from among the foods.

4. Did you find any "bargains" in terms of energy and minerals among the foods? For example, calcium and potassium are delivered by many foods of varying calorie values. List snack foods rich in needed minerals, yet relatively low in calories.
5. Where is the calcium in your snack foods? List the snack foods on your list that supply substantial calcium (10 percent of the recommendation). If your choices lack calcium, look back to the menu on the previous page and list some calcium sources you find there.
6. Can a person looking for iron do well in a convenience store? Which foods or ingredients contribute iron? See Snapshot 8-5 for good iron sources.
7. Is the iron in some of these snack foods more absorbable than in others? Explain. What constituents of these foods can enhance iron's absorption?
8. Did any of your choices provide 10 percent or more of your need for magnesium? Magnesium is easily lost in processing foods and so is difficult to obtain from highly processed snacks.
9. Consider the zinc in your snack. Did any foods provide 10 percent or more of the daily recommendation? If so, which ones? If not, consult Snapshot 8-6 for ideas about the kinds of foods that are good sources of zinc, and list some snack foods likely to contain significant amounts.
10. What might a whole day's mineral intake look like if a person ate convenience foods at every meal?

Americans are snacking more as the pace of life quickens. The next chapter makes clear that your choices among foods to snack on can make a difference of several hundred calories a day to energy intakes.

Self Check

Answers to these Self Check questions are in Appendix G.

1. Water excretion is governed by the:
   a. liver
   b. kidneys
   c. brain
   d. (b) and (c)
2. Compared with hard water, soft water:
   a. better supports the health of the heart
   b. contains fewer harmful dissolved minerals
   c. is higher in sodium
   d. leaves a gray residue in the wash
3. Which two minerals are the major constituents of bone?
   a. calcium and zinc
   b. phosphorus and calcium
   c. sodium and magnesium
   d. selenium and calcium
4. A deficiency of ________________ is one of the world's most common preventable causes of mental retardation.
   a. zinc
   b. magnesium
   c. selenium
   d. iodine
5. Which mineral in excess is the number one cause of fatal accidental poisonings of U.S. children under three years old?
   a. iron
   b. sodium
   c. chloride
   d. potassium
6. Bottled water must meet higher standards for purity and sanitation than U.S. tap water. T F
7. You can survive being deprived of water for about a week. T F
8. The best way to control salt intake is to cut down on processed and fast foods. T F
9. Electrolytes help keep fluids in their proper compartments in the body. T F
10. The dairy foods butter, cream, and cream cheese are good sources of calcium whereas vegetables such as broccoli are poor sources. T F
11. Actions to prevent osteoporosis are best begun in middle age, when the bones are ceasing their growth. True or false? (This is addressed in the upcoming Controversy) T F

NUTRITION ON THE NET

For further study of the topics of this chapter, access these websites and search for the phrases or words in quotation marks:

1. For the latest changes in chapter material, to find supplemental learning tools, or access links to related websites, go to the "Nutrition: Concepts and Controversies" site at:
   www.wadsworth.com/nutrition/prod/allprod.html
2. To find out more about the importance of water:
   ificinfo.health.org/insight/waterref.htm
3. Visit the American Dietetic Association site for information on "minerals":
   www.eatright.org
4. To find out more about sodium in foods:
   www.fda.gov/fdac/foodlabel/sodium.html
5. For more on iron:
   www.healthfinder.gov/searchoptions/topicsaz.htm
6. For details concerning iron overload:
   www.ironoverload.org
7. To learn more about osteoporosis and related bone diseases:
   www.osteo.org
8. Search among thousands of current scientific and medical abstracts for any topic related to minerals at:
   http://www.ncbi.nlm.nih.gov/PubMed/

CONTROVERSY 8

OSTEOPOROSIS AND CALCIUM

OSTEOPOROSIS is one of the most prevalent of the degenerative diseases.[1] It affects more than 25 million people in the United States alone. Most are women, but men are far from immune. Osteoporosis causes much illness and many deaths among men each year.[2]

In 1998, a million and a half people suffered bone breaks of the hips, pelvis, legs, arms, hands, and ankles attributable to osteoporosis.[3] Of these, hip fractures prove most serious. The break is rarely clean; the bone explodes into fragments so numerous and scattered that they cannot be reassembled. Just removing them is a struggle, and replacing them with an artificial joint requires major surgery. About 300,000 people, 90 percent of them over age 50, suffer hip fractures each year.[4] Many never walk or live independently again. About a fifth die from related complications within a year.

Osteoporosis sets in during the later years, but it develops much earlier.[5] Silently, relentlessly, it saps bone strength with no warning symptoms until late in life. Exceedingly few people are aware in advance that osteoporosis is sapping the strength of their bones, then, suddenly, dramatically, someone's hip gives way.[6] People say, "She fell and broke her hip," but in fact the hip may have been so fragile that it broke *before* she fell. Just stepping off the curb may jar a bone enough to shatter it, if the bone is sufficiently porous from loss of minerals.

The causes of osteoporosis are tangled, and although evidence is mounting that insufficient dietary calcium influences the development of osteoporosis, other factors are also major potential players. This Controversy addresses several questions about osteoporosis: What is it? Who gets it? What factors increase the risk? What can people do to reduce their risks? And where does calcium fit into the picture?

THE PROBLEM OF OSTEOPOROSIS

To understand how the skeleton loses minerals in later years, you must first know a few things about bones. Table C8-1 offers definitions of relevant terms. The photograph on this page shows a human leg bone sliced lengthwise, exposing the lattice of calcium-containing crystals (the **trabecular bone**) inside. These lacy crystals, part of the body's calcium bank, are tapped to raise blood calcium when the supply from the day's diet runs short; they are redeposited in bone when dietary calcium is plentiful. Invested as savings during the milk-drinking years of childhood and young adulthood, these deposits provide a nearly inexhaustible fund of calcium.

In contrast to trabecular bone, **cortical bone** is the dense, ivorylike bone that forms the exterior shell of a bone and the shaft of a long bone (look closely at the photograph). Both types of bone are crucial to overall bone strength. Cortical bone forms a sturdy outer wall, and trabecular bone provides strength along the lines of stress.

The differences between the two types of bone are meaningful with regard to osteoporosis. Trabecular bone is generously supplied with blood vessels and is more metabolically active than is cortical bone. Trabecular bone is also more sensitive to hormones that govern calcium deposits and withdrawals from day to day. Cortical bone's calcium can be withdrawn, but slowly. Trabecular bone readily gives up its minerals at the necessary rate whenever blood calcium needs replenishing. Losses of trabecular bone begin to be significant for men and women in their mid-20's, although losses can occur anytime calcium withdrawals exceed calcium deposits. Cortical bone loss begins at about age 40; bone tissue dwindles steadily thereafter.

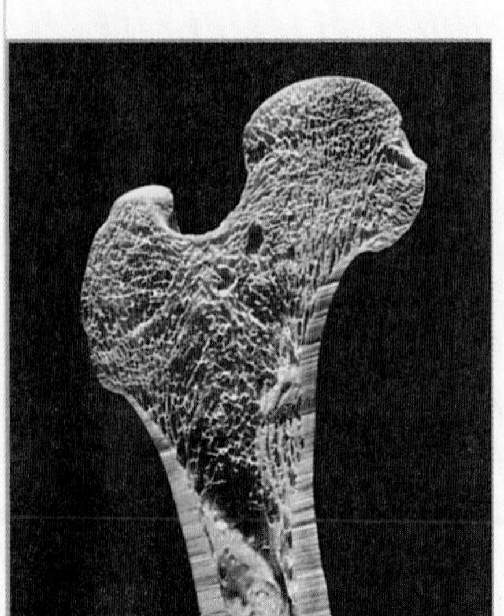

Table C8-1

Osteoporosis Terms

- **bone density** a measure of bone strength; the degree of mineralization of the bone matrix.
- **cortical bone** the ivorylike outer bone layer that forms a shell surrounding trabecular bone and that comprises the shaft of a long bone.
- **trabecular** (tra-BECK-you-lar) **bone** the weblike structure composed of calcium-containing crystals inside a bone's solid outer shell. It provides strength and acts as a calcium storage bank.
- **type I osteoporosis** osteoporosis characterized by rapid bone losses, primarily of trabecular bone.
- **type II osteoporosis** osteoporosis characterized by gradual losses of both trabecular and cortical bone.

Menopause, when a woman's estrogen secretion sharply declines and menstruation ceases, produces a surge of bone loss. This surge tapers off after a few years so that, once again, women's losses equal those sustained by men of the same age. Losses of bone minerals continue throughout the rest of a woman's life, but not at the free-fall pace of the menopause years.

Researchers have associated the losses of trabecular and cortical bone with two types of osteoporosis, types I and II, identified by the bone breaks they produce. People with **type I osteoporosis** lose mostly trabecular bone (see Figure C8-1), sometimes at three times the expected rate or even faster, and bone breaks become frequent when the victim passes the age of 65 years. In this condition, trabecular bones become so fragile that the body's weight can overburden the spine; vertebrae may suddenly disintegrate and crush down, painfully pinching major nerves. Wrists may break as trabecula-rich bone ends weaken, and teeth may loosen or fall out as the trabecular bone of the jaw recedes.[7] Women are most often the victims of type I osteoporosis, six to one over men.

In **type II osteoporosis,** the calcium of both cortical and trabecular bone is drawn out of storage, but slowly over the years. As old age approaches, the vertebrae may compress into wedge shapes, usually painlessly, forming what is insensitively called "dowager's hump," the posture seen in many older people as they "grow shorter." Figure C8-2 on page 306 shows the effect of the compression of spinal bone on a woman's height and posture. Because the cortical shell as well as the trabecular interior weakens, breaks most often occur in the hip, as described earlier. A woman is twice as likely as a man to suffer type II osteoporosis, probably due to the process set in motion years before at menopause.

Scientists searching for ways to prevent osteoporosis must first discover its causes. One group has amassed an osteoporosis database that includes hundreds of thousands of women who have suffered from osteoporosis.[8] The researchers hope one day to identify the most influential risk factors in the development of osteoporosis. Right now, many findings seem to conflict with each other, and others simply raise more questions. Nevertheless, some areas of agreement have been reached. Whether a person develops osteoporosis seems to depend partly on heredity and partly on the environment, including nutrition. The sections that follow discuss the factors thought to be the main determinants of **bone density.**

AGE, CALCIUM, AND VITAMIN D

Two major life stages have been established as critical in the development of osteoporosis.[9] The first is the bone-acquiring stage during adolescence and young childhood. The second stage is the bone-losing decades following menopause.The bones gain strength and density all through the growing years and into young adulthood. As people age, the cells that build bone gradually become less active, but those that dismantle bone continue working. Because the body depends on bone tissue to supply calcium throughout life, bone mass diminishes and bones lose strength and density.

One factor that likely affects bone withdrawal and deposition is calcium nutrition during childhood and early adult life. Preteen children who consume extra calcium together with adequate vitamin D lay more calcium into the structure of their bones than children with less adequate intakes. The same two nutrients, when supplemented in the later years, produce small but beneficial effects on the bone mass and fracture rates of older people.[10]

When people reach the bone-losing years of middle age, those who formed dense bones during youth have the advantage. They have more bone tissue starting out and can lose more before beginning to suffer ill effects. Figure C8-3 on page 307 demonstrates this affect. Therefore, whatever factors contribute to the building of strong bones in youth, these same factors, including calcium nutrition, become protective against osteoporosis much later.

Another factor, which becomes important in later life, is calcium absorption. Absorption declines after about the age of 65 years, probably because the kidneys do not activate vitamin D as well as they did earlier. Also, sunlight is

Figure C8-1

Losses of Trabecular Bone

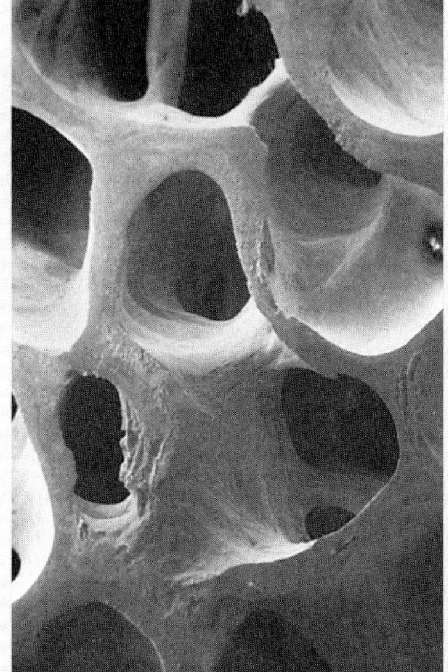

Electron micrograph of healthy trabeculae.

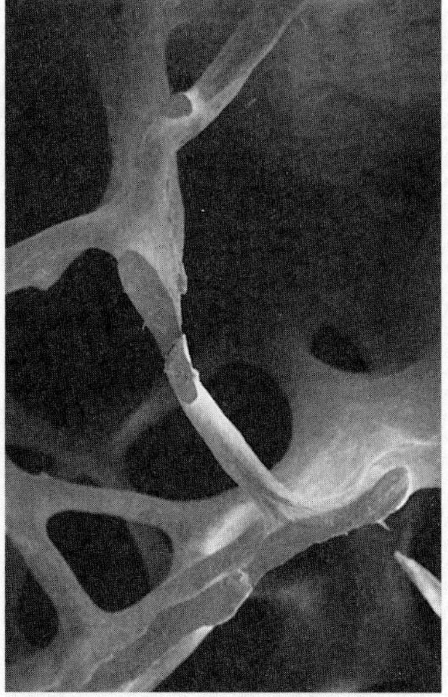

Electron micrograph of trabecular bone affected by osteoporosis.

needed to form vitamin D, and many older people fail to go outdoors into the sunshine. Some of the hormones that regulate bone maintenance and calcium metabolism also change with age and accelerate bone mineral withdrawal.* Older people typically take in less calcium and vitamin D than others, and they absorb less calcium, too. According to one leading expert, we could cut our osteoporosis problem by half simply by ensuring that everyone receives adequate calcium and vitamin D.[11]

GENDER AND HORMONES

After age itself, experiencing menopause is, for women, the next-strongest predictor of loss of bone density with aging. As mentioned, estrogen secretion declines at menopause in women at about age 50, but cases exist in which *young* women's ovaries fail to produce enough estrogen to maintain menstruation. These young women also undergo rapid bone losses. In some, diseased ovaries have been removed; in others, the ovaries produce inadequate estrogen because the women overexercise and restrict their body weights unreasonably, (details in Controversy 10).[12] Estrogen prescription drugs, and even estrogen-containing oral contraceptives, can help nonmenstruating women to prevent further bone loss and reduce fractures with little risk to health.[13]

If estrogen deficiency is a major cause of osteoporosis in women, what is the explanation for the pattern seen in men? Men produce only a little estrogen, yet they are normally more resistant to osteoporosis than are women. Does the male sex hormone testosterone play roles in osteoporosis? Perhaps so, because men reliably suffer more fractures after undergoing removal of diseased testes or when their testes lose function with aging. Testosterone replacement therapy may benefit the bones of men with reduced sex hormone status.[14]

Hormones slow calcium losses from bones, but a new drug reverses bone loss to some degree.†[15] It works in both men and women by inhibiting the activities of the bone-dismantling cells, thus allowing the bone-building cells to slowly shore up bone tissue with new calcium deposits. In a clinical study, the drug increased bone mass by 8 percent over three years accompanied by a 30 to 40 percent reduction in fractures. The hope now is for a drug that can stimulate the bone-building cells to work faster to restore bone strength.

*Among the hormones suggested as influential in governing calcium balance are parathyroid hormone and calcitonin.

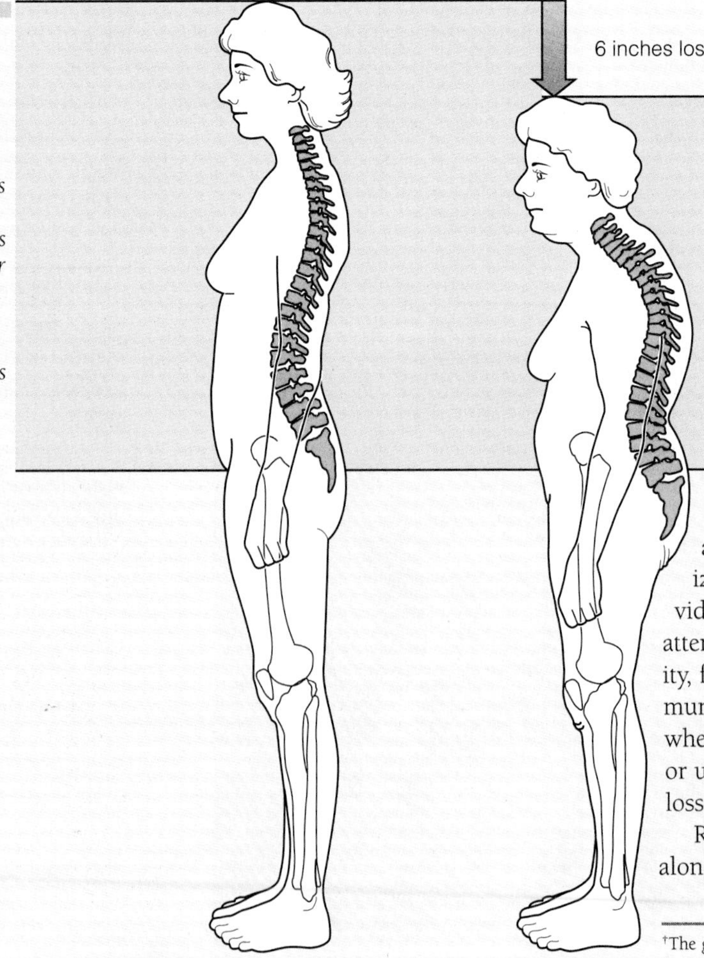

6 inches lost
50 years old
80 years old

Figure C8-2

Loss of Height in a Woman Caused by Osteoporosis

The woman on the left is about 50 years old. On the right, she is 80 years old. Her legs have not grown shorter; only her back has lost length, due to collapse of her spinal bones (vertebrae). When collapsed vertebrae cannot protect the spinal nerves, the pressure of bones pinching the nerves causes excruciating pain.

INHERITED BONE DIFFERENCES

Studies of mothers and daughters and of twins confirm that heredity plays a role in bone density.[16] Most likely, inheritance influences both the maximum bone mass possible during growth and the extent of bone loss during menopause. The extent to which a given genetic potential is realized, however, depends on individual life experiences. Those who attend to nutrition and physical activity, for example, attain their maximum bone density during growth, whereas those who overuse alcohol or use tobacco accelerate their bone losses.

Risks of osteoporosis also run along racial lines.[17] People of African

†The generic name of the drug is aldendronate, marketed as Fosamax.

Figure C8-3

Two Women's Bone Mass History Compared

Woman A entered adulthood with enough calcium in her bones to last a lifetime. Woman B had less bone mass starting out and so suffered ill effects from bone loss later on.

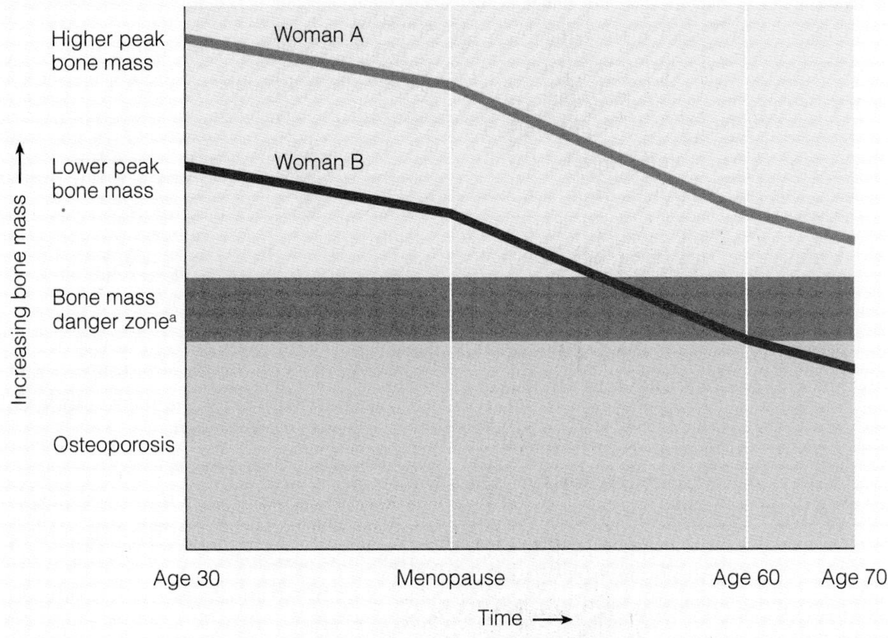
Higher peak bone mass
Woman A
Lower peak bone mass
Woman B
Increasing bone mass ⟶
Bone mass danger zone[a]
Osteoporosis
Age 30
Menopause
Age 60
Age 70
Time ⟶

[a]People with a moderate degree of bone mass reduction are said to have osteopenia and are at increased risk of fractures.
SOURCE: Data from Standing Committee on the Scientific Evaluation of Dietary Reference Intakes, Food and Nutrition Board, Institute of Medicine, *Dietary Reference Intakes for Calcium, Phosphorus, Magnesium, Vitamin D, and Fluoride* (Washington, D.C.: National Academy Press, 1997), pp. 4–10.

extraction have denser bones than do those of northern European descent, and these differences are dramatically evident even before birth in X-ray images of fetuses. This difference holds true for both sexes of all ages and expresses itself in a much lower total rate of osteoporosis among Africans. Hip fractures, for example, are reported to be about three times more likely in 80-year-old northern European women than in African women of the same age.

Other ethnic groups have lower bone densities than do northern Europeans. Asians from China and Japan, Mexican Americans, Hispanic people from Central and South America, and Inuit people from St. Lawrence Island all have lower bone density than do people of northern European descent. Although those lower densities might suggest that these groups would suffer more bone fractures, the picture is not that tidy. Chinese people living in Singapore have low bone density, but their hip fracture rates are among the lowest in the world.

An interesting study of Chinese women found them to have exceptionally potent intestinal absorption of calcium—two to three times that of people of northern European or even African descent.[18] Further, the Chinese women's rate of calcium absorption remained unchanged throughout menopause and into old age. Thus, it may be that the underlying causes of osteoporosis vary by race, and that recommendations will be most effective when they fit this reality.

These studies of populations demonstrate that although a person's genes may lay the groundwork for a likely outcome, environmental factors influence the genes' ultimate expression. It appears that calcium nutrition may be one of those environmental factors, but there are others, such as physical activity, body weight, smoking, alcohol use, and protein intake. Importantly, all of these factors are under a person's own control.

PHYSICAL ACTIVITY

Though you can't do much about your genetic inheritance, you can control your physical activity. It has long been known that when people lie idle—for example, when they are confined to bed—the bones lose strength just as the muscles do.[19] Astronauts who live without gravity for days or weeks at a time also experience remarkably rapid and extensive bone losses.

Muscle strength and bone strength usually go together, and muscle use seems to promote bone strength.[20] When cross sections of bones of sedentary and active people are compared, the active bones are denser by far. The hormones that promote synthesis of new muscle tissue also favor the building of bone. Also, increasing flexibility and muscle strength improves balance and helps to prevent falls from occurring.[21]

To keep the bones healthy, then, and to prevent falls, include weight-bearing exercises such as calisthenics, dancing,

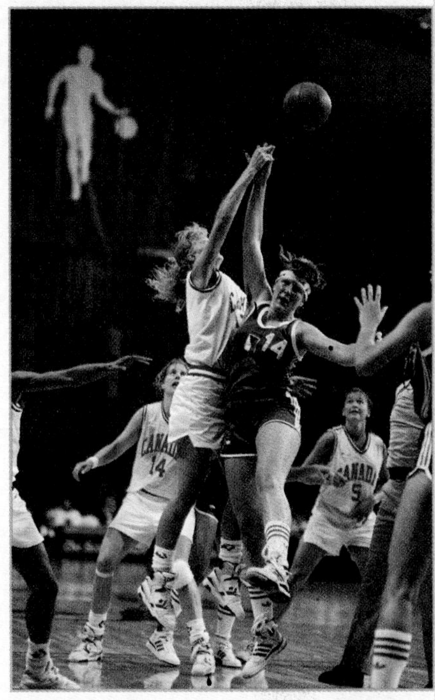

These college women are putting bone in the bank.

jogging, vigorous walking, or weight training every day.[22] Even sports or gardening, perfomed with vigor, may build bone strength, and moderate daily exercise for women of all ages is strongly advised.[23]

BODY WEIGHT

Heavier body weights, and to some degree higher body fatness, stress bones and promote their maintenance; osteoporosis is most often associated with underweight.[24] Women who are thin through life, and especially those who lose 10 percent or more of their body weight after the age of menopause, face a hip fracture rate twice as high as that of most other women. Also, the type of diabetes associated with slender body structure alters the body's handling of calcium and magnesium with possible harmful effects on the health of the bones.

SMOKING, ALCOHOL, AND CAFFEINE

Smoking seems to be hard on the bones. Women who smoke lose 5 to 10 percent more of their bone density than nonsmokers by the time they reach menopause.[25] Surgeons report that, compared with nonsmokers, users of tobacco suffer more bone degeneration and spinal injuries, and surgeries to repair their backs are less often successful.[26] Researchers speculate that the weakened bones may be related to the lower body weights of smokers or to early menopause in female smokers.[27]

People who are addicted to alcohol also experience relatively more frequent fractures. It may be that because alcohol (a diuretic) causes fluid excretion, it induces excessive calcium losses through the urine. Heavy drinking may also upset the hormonal balance required for healthy bones or it may restrict reproduction of the bone-building cells, diminishing their numbers.[28] An earlier suggestion that moderate drinking may be protective of the bones has been dismissed by research—a lifetime of moderate drinking is clearly linked with reduced bone mass.[29]

Heavy users of caffeinated beverages, such as coffee, tea, or colas, should be aware that although some evidence suggests a link between caffeine use and osteoporosis, other findings do not point to caffeine use as a risk factor.[30] It may be that caffeine exerts an effect only when calcium intakes are low.[31]

Table C8-2 summarizes the risk factors covered so far and includes some others, among them, high protein and low calcium intakes, discussed next. The more risk factors that apply to you, the greater your chances of developing osteoporosis in the future and the more seriously you should take the advice offered in a later section of this Controversy.

Table C8-2

Risk and Protective Factors That Correlate with Osteoporosis

Risk Factors	Protective Factors
HIGH CORRELATION	
Advanced age	Black race
Alcoholism, heavy drinking	Estrogens, long-term use
Chronic steroid use	
Female gender	
Rheumatoid arthritis	
Surgical removal of ovaries or testes	
Thinness or weight loss	
White race	
MODERATE CORRELATION	
Chronic thyroid hormone use	Having given birth
Cigarette smoking	High body weight
Diabetes (insulin-dependent, type 1)	High-calcium diet
Early menopause	Regular physical activity
Excessive antacid use	
Family history of osteoporosis	
Low-calcium diet	
Sedentary lifestyle	
Vitamin D deficiency	
PROBABLY IMPORTANT BUT NOT YET PROVED	
Alcohol taken in moderation	Adequate vitamin K intake
Caffeine intake	Low-sodium diet (later years)
High-fiber diet	
High-protein diet	
Lactose intolerance	

PROTEIN, SODIUM, AND VITAMIN K

Researchers have discovered that extra dietary protein causes the body to excrete calcium in the urine. This often repeated finding has led to the suspicion that a lifetime of consuming excess dietary protein may accelerate bone loss.

Does a high-protein diet cause bone loss? One reason that it may is the tendency for a high-protein diet to create excess acid during its metabolism.[32] When presented with excess acidity, the body must quickly neutralize it and so withdraws calcium carbonate from the bones and uses the carbonate part to buffer the acid. The remaining calcium is excreted in the urine. Consistently, when people are fed increasing amounts of protein, the calcium balance tips toward the negative (they lose more calcium than they take in). In one study, women consuming more than 95 grams of protein from food each day suffered more forearm fractures than women consuming less than 68 grams of protein per day.[33] The highest recommended protein intake is 65 grams per day, and the upper safe limit for protein intake by women established by the World Health

Organization is around 83 grams per day.

Researchers who hold protein harmless in regard to the bones cite other studies suggesting that increased intakes of protein-rich foods do not always produce negative calcium balance. This may be because extra calcium often follows protein into the diet, and the extra calcium helps to oppose withdrawal of calcium from the skeleton.[34] Other studies find that protein from animal, but not vegetable, sources increases calcium in the urine.[35] The reason may be that the metabolism of vegetables and fruits generates alkali instead of acid and may therefore help conserve calcium.

Of course, consuming too little dietary protein is never a good idea either. (Chapter 6 made clear many reasons why this is so). In fact, protein deprivation also stimulates calcium losses. These facts seem to conflict at first glance, but they really just demonstrate a sound nutrition principle—that a happy medium is best.

Cola beverages and other soft drinks have been studied for their effects on calcium. In rats, feeding colas, but not other soft drinks, results in significant urinary losses of calcium.[36] In children, an unexplained tendency to develop bone problems such as stress fractures have been observed to accompany high intakes of cola beverages or fruit juices, but not other soft drinks.[37] Today's adolescents are taking in less milk and less calcium, while consuming more and more soft drinks than ever before—a trend that may bode ill for future bone health.[38]

Another line of research suggests a link between bone loss and sodium intakes. Most studies confirm that with increasing dietary sodium, urinary calcium loss also increases.[39] One group of researchers found that women who kept sodium intakes low did not lose bone tissue from their hips, but those who consumed more sodium did lose bone.[40] Another group found no direct effect of sodium on the bones, but noted that urinary calcium losses increased by 73 percent in young women with sodium sensitivity.[41] How-ever, these researchers found no indication of increased bone losses among subjects who lost calcium. Still, it may be that a diet plan for the bones should both increase calcium intakes in the early years and reduce sodium intakes in the later years, especially in older people with osteoporosis.[42]

Recently, vitamin K has been suggested as important in sustaining the health of the bones through life.[43] Vitamin K plays important roles in the production of at least one bone protein that participates in bone maintenance.[‡44] In a study of female athletes in whom strenuous exercise had lowered estrogen production, an increased vitamin K intake reduced markers of bone loss and increased markers of bone formation.[45] It was as though their reduced estrogen status somehow made the women *functionally* deficient in vitamin K, even though their diets met adequacy recommendations. Also, people with hip fractures have often been found to have lower intakes of dietary vitamin K.[46] More investigation is needed, but it may turn out that obtaining enough vitamin K in foods such as lettuce and other green vegetables is as important to bone health as consuming enough vitamin D or calcium.

CALCIUM RECOMMENDATIONS

The DRI committee recommends 1,300 milligrams of calcium, the amount in about 4 cups of milk, each day for everyone 9 through 18 years of age. Unfortunately, few girls meet the recommendation for calcium during their bone-forming years. (Boys generally obtain intakes close to those recommended because they eat more food.) As adults, women rarely meet the recommended 1,000 to 1,2000 milligrams from food within their energy allowances.

How should all this calcium be obtained? Consider a bit of hard-won advice: use foods if at all possible and supplements only when advised to do so by a physician. People can best support their bones' health by following the recommendations of Table C8-3. Calcium supplements cannot equal any of the actions listed in the table. No one should be led to think that calcium supplements are risk-free, or that popping pills can take the place of sound food choices and other healthy

Table C8-3

A Lifetime Plan for Healthy Bones

Age	Action
0–18	Use milk as the primary beverage to meet the need for calcium within a balanced diet that provides all nutrients; play actively in sports or other activities; limit television; do not start smoking or drinking alcohol; drink fluoridated water.
19–25	Choose milk as the primary beverage, or if milk causes distress, include other calcium sources; commit to a lifelong program of physical activity; do not smoke or drink alcohol—if you have started, quit; drink fluoridated water.
26–50	Continue as for 19- to 25-year-olds. At menopause, women should be evaluated for possible estrogen replacement therapy. Obtain the recommended amount of calcium from food. Take calcium and fluoride supplements only if prescribed by a physician.
51 and above	Continue as for 19- to 25-year-olds; continue following a physician's advice concerning estrogen and supplements. Continue striving to meet the calcium need from diet, and continue bone-strengthening exercises.

‡The vitamin K-dependent bone protein is osteocalcin.

habits. For those who desire details on supplements, however, a discussion follows.

A PERSPECTIVE ON CALCIUM SUPPLEMENTS

As important as calcium intake may be, bone loss is not a calcium-deficiency disease comparable to iron-deficiency anemia. In iron-deficiency anemia, high iron intakes reliably reverse the condition. With respect to calcium balance, though, calcium intakes alone do little or nothing to reverse bone loss. During the menopausal years, calcium supplements of 1 gram may slow, but cannot fully prevent, the inevitable bone loss.[47]

Calcium supplements are part of standard therapy for already developed osteoporosis. Taking self-prescribed calcium supplements entails possible risks, though (see Table C8-4).[48] If these risks are deemed acceptable, the consumer must decide among the many products available for purchase. Regular vitamin-mineral pills contain little or no calcium. The label may list a few milligrams of calcium, but the amounts are insignificant when compared with the calcium requirement.

Table C8-4

Calcium Supplement Pitfalls to Avoid

People who take calcium supplements risk:

- *Impaired iron status.* Calcium inhibits iron absorption.
- *Accelerated calcium loss.* Calcium-containing antacids that also contain aluminum and magnesium hydroxide cause a net calcium loss.
- *Urinary tract stones or kidney damage in susceptible individuals.* People who have a history of kidney stones should be monitored by a physician and choose calcium citrate if they must take supplements.
- *Exposure to contaminants.* Some preparations of bone meal and dolomites are contaminated with hazardous amounts of arsenic, cadmium, mercury, and lead.
- *Vitamin D toxicity.* Vitamin D, which is present in many calcium supplements, can be toxic. Users must eliminate other concentrated vitamin D sources.
- *Excess blood calcium.* This complication is seen only with doses of calcium fourfold or more greater than customarily prescribed.
- *Milk alkali syndrome.* This condition is rare, but not absent. It is characterized by high blood calcium, metabolic alkalosis, and renal failure. Early symptoms include irritability, headaches, and apathy.
- *Other nutrient interactions.* Calcium inhibits absorption of magnesium, phosphorus, and zinc.
- *Drug interactions.* Calcium and tetracycline form an insoluble complex that impairs both mineral and drug absorption.
- *GI distress.* Constipation, intestinal bloating, and excess gas are common.

Supplements are available in three forms. Simplest are the purified **calcium compounds,** such as calcium carbonate, citrate, gluconate, lactate, malate, or phosphate, and compounds of calcium with amino acids (called **amino acid chelates**). Then there are mixtures of calcium with other compounds, such as calcium carbonate with magnesium carbonate, with aluminum salts (as in some **antacids**), or with vitamin D. Then there are powdered, calcium-rich materials such as **bone meal, powdered bone, oyster shell,** or **dolomite** (limestone). See Table C8-5 for supplement terms.

The first question to ask is how well the body absorbs and uses the calcium from various supplements. Based on research to date, many people seem to absorb calcium reasonably well—and about as well as from milk—from amino acid chelates and from calcium compounds such as calcium phosphate dibasic, calcium acetate, calcium carbonate, calcium citrate, calcium gluconate, and calcium lactate. People absorb calcium less well from a mixture of calcium and magnesium carbonates, from oyster shell calcium fortified with inorganic magnesium, from a chelated calcium-magnesium combination, and from calcium carbonate fortified with vitamins and iron. Some people absorb calcium better from milk and milk products than from even the most absorbable supplements.

The next question to ask is how much calcium the supplement provides. Healthy people have consumed up to 2 grams daily of calcium without problems.[49] To be safe, though, supplements should provide less than this, since foods also provide calcium. Read the label to find out how much a dose supplies. Calcium carbonate is 40 percent elemental calcium, whereas calcium gluconate is only 9 percent. The user should select a low-dose supplement and take it several times a day rather than taking a large-dose supplement all at once.[50] Divided doses can improve a day's total absorption up to 20 percent.

Then consider that when manufacturers compress large quantities of calcium into small pills, the stomach acid has difficulty penetrating the pill. To test a supplement's absorbability, drop it into a 6-ounce cup of vinegar, and stir occasionally. A high-quality formulation will dissolve within half an hour. The chewable kind, because they are broken into bits before swallowing, may be less prone to this problem.

Think one more time, then, before you commit yourself to taking supplements for calcium. The Consensus Conference on Osteoporosis recommends milk. The American Society for Bone and Mineral Research recommends foods as a source of calcium in preference to supplements. The *Diet and Health* report from the National Academy of Sciences concludes that foods are best.[51] The authors of this book are so impressed with the importance of using abundant, calcium-rich foods that they have worked out ways to do so at every meal. Seldom do nutritionists agree so unanimously.

Table C8-5

Calcium Supplement Terms

- **amino acid chelates** (KEY-lates) compounds of minerals (such as calcium) combined with amino acids in a form that favors their absorption. A *chelating agent* is a molecule that surrounds another molecule and can then either promote or prevent its movement from place to place (*chele* means "claw").
- **antacids** acid-buffering agents used to counter excess acidity in the stomach. Calcium-containing preparations (such as Tums) contain available calcium. Antacids with aluminum or magnesium hydroxides (such as Rolaids) can accelerate calcium losses.
- **bone meal** or **powdered bones** crushed or ground bone preparations intended to supply calcium to the diet. Calcium from bone is not well absorbed and is often contaminated with toxic materials such as arsenic, mercury, lead, and cadmium.
- **calcium compounds** the simplest forms of purified calcium. They include calcium carbonate, citrate, gluconate, lactate, malate, and phosphate. These supplements vary in the amount of calcium they contain, so read the labels carefully. A 500-milligram tablet of calcium gluconate may provide only 45 milligrams of calcium, for example.
- **dolomite** a compound of minerals (calcium magnesium carbonate) found in limestone and marble. Dolomite is powdered and is sold as a calcium-magnesium supplement, but may be contaminated with toxic minerals, is not well absorbed, and interacts adversely with absorption of other essential minerals.
- **oyster shell** a product made from the powdered shells of oysters that is sold as a calcium supplement, but is not well absorbed by the digestive system.

9 Energy Balance and Healthy Body Weight

Helen Berggruen
Lake Balaton Sanitorium, 1994

Contents

Frequently Asked Questions

body composition: the proportions of muscle, bone, fat, and other tissue that make up a person's total body weight.

adipose tissue the body's fat tissue, consisting of massses of fat-storing cells and blood vessels to nourish them.

Are you pleased with your body weight? If you answered yes, you are a rare individual. Nearly all people in our society think they should weigh more or less (mostly less) than they do. Usually, their primary concern is appearance, but they often perceive, correctly, that physical health is also somehow related to weight. At the extremes, both overweight and underweight present definite health risks.

People also think of their weight as something they should control, once and for all. Three misconceptions frustrate their efforts, however. The first is the focus on weight; the second is the focus on *controlling* weight, and the third is the focus on short-term endeavors. Simply put, it isn't your weight you need to control; it's the fat in your body in proportion to the lean—your **body composition.** And it isn't possible to control body composition directly; it is only possible to control your *behavior.* Furthermore, sporadic bursts of activity, such as "dieting," are not effective; the behaviors that achieve and maintain a healthy body weight take a lifetime of commitment.

This chapter has several missions. First, it presents the problems associated with deficient and excessive body fatness. Next, it suggests how to judge body weight on the sound basis of health. Finally, it reveals the recommended strategies for coping with body weight. It starts by considering how the body manages its energy budget.

Energy Balance

What happens inside the body when you eat too much or too little food? The body ends up with an unbalanced energy budget—you have taken in more or less food energy than you spent. The mechanisms by which the body handles its energy underlie changes that occur in body composition.

When more food energy is consumed than is needed, excess fat enters the fat cells in the body's **adipose tissue** for storage. When energy supplies run low, stored fat is withdrawn. The daily energy balance can therefore be stated like this:

> Change in energy stores equals food energy taken in minus energy spent on metabolism and muscle activities.

More simply:

> Change in energy stores = energy in − energy out.

Too much or too little fat on the body today does not necessarily reflect today's energy budget, of course. Small imbalances in the energy budget compound over time.[1]

1 lb body fat = 3,500 cal.

Energy In

The energy in foods and beverages is the only contributor to the "energy in" side of the energy balance equation. Before you can decide how much food energy you need in a day, you must first become familiar with the amounts of energy in foods and beverages. One way to do this is to look up calorie amounts associated with foods and beverages in the Table of Food Composition (Appendix A). Many people now have access to computer programs that provide this information in the blink of an eye. However they are derived, the numbers are always fascinating to those concerned with managing body weight.

As examples of the calories associated with food portions, an apple gives you 125 calories from carbohydrate; a regular-size candy bar gives you about 250 calories mostly from fat and carbohydrate. You may already know that for each 3,500 calories you eat in excess of expenditures, you store approximately 1 pound of body fat.*

KEY POINT *The "Energy In" side of the body's energy budget is measured in calories taken in each day in the form of foods and beverages. Calories in foods and beverages can be obtained from published tables or computer diet analysis programs.*

*Pure fat is worth 9 calories per gram. A pound of it (450 grams), then, would store 4,050 calories. A pound of *body* fat is not pure fat, though; it contains water, protein, and other materials—hence the lower calorie value.

Energy Out

Though the energy present in a serving of food or in a day's meals can easily be estimated by consulting a table of food composition or a computer, no easy method exists for determining the energy an individual spends, and therefore needs. Many authorities have published recommended energy intakes for various age-sex groups in their populations, and these are found on the inside back cover, p. Y or in Appendix B. These recommendations are useful for population studies, but energy needs vary so widely among individuals in a group that it is impossible to guess any individual person's need without knowing something about the person's lifestyle and metabolism.

The recommendations for energy intake are based on average people. For example, an intake of 2,200 calories per day is recommended for a woman who is assumed to be 20 years old, standing 5 feet 5 inches tall, weighing about 128 pounds, of average body fatness, and engaging in light activity. The man, needing 2,900 calories per day, is a healthy 20-year-old of average body fatness who stands 5 feet 10 inches tall, weighs 160 pounds, and is lightly active. Taller people need proportionately more energy than shorter people to balance their energy budgets because their greater surface area allows more energy to escape as heat. Older people generally need less than younger people due to slowed metabolism and reduced muscle mass that occur partly because of reduced physical activity.[2] On average, the energy need diminishes by 5 percent per decade beyond the age of 30 years.

In reality, though, no one is average and people's energy needs vary widely. In any group of 20 similar people with similar activity levels, one may expend twice as much energy per day as another. A 60-year-old person who bikes, swims, or jogs each day may need as many calories as another person of 30. Clearly, with such a wide range of variation, a necessary step in determining any person's energy need is to study that person.

One way to estimate your energy needs is to monitor your food intake and body weight over a period of time in which your activities are typical of your lifestyle. If you keep an accurate record of all the foods and beverages you consume for a week or two and if your weight has not changed during the past few months, you can conclude that your energy budget is balanced. At least three days of honest record keeping are necessary because intakes fluctuate from day to day. A week or two is better still. (On about half the days, you eat less food energy than the average; on the other half, more.)

An alternative method of determining energy need is based on energy output. To estimate output, a person must compute the amount of the two major components of energy expenditure and then add them together. This method leaves out a third energy component, the body's metabolic response to food. About 5 to 10 percent of a meal's energy value is used up in stepped-up metabolism in the five or so hours after the meal; this category of energy expenditure is called the **thermic effect of food.** Although this amount of energy could affect expenditures over the long run, most experts believe its effects are negligible.[3] For our purposes, it can be ignored.

The two major ways in which the body spends energy are (1) to fuel its **basal metabolism** and (2) to fuel its **voluntary activities** (see Figure 9-1). Basal metabolism generates energy to support the body's work that goes on all the time without our conscious awareness.

Basal metabolism consumes a surprisingly large amount of fuel, and the **basal metabolic rate (BMR)** varies from person to person.[4] A person whose total energy needs are 2,000 calories a day spends as many as 1,200 to 1,400 of them to support basal metabolism. The hormone thyroxine directly controls basal metabolism—the less secreted, the lower the energy requirements for basal functions. Many other factors also affect the BMR (see Table 9-1 on the next page).

People often want to know how they can speed up their metabolism to promote fat loss. You cannot speed up your BMR much today. You can, however, amplify the second component of your energy expenditure, your voluntary activities. If you do, you will spend more calories today, and if you keep doing so day after day, your BMR will also increase. Lean tissue is more metabolically active than fat tissue, so a way to speed up your BMR to the maximum possible rate is to make endurance and strength-building

thermic effect of food (TEF) the body's speeded-up metabolism in response to having eaten a meal. Also called *diet-induced thermogenesis*.

basal metabolism the sum total of all the involuntary activities that are necessary to sustain life, including circulation, respiration, temperature maintenance, hormone secretion, nerve activity, and new tissue synthesis, but excluding digestion and voluntary activities. Basal metabolism is the largest component of the average person's daily energy expenditure.

voluntary activities intentional activities (such as walking, sitting, or running) conducted by voluntary muscles.

basal metabolic rate (BMR) the rate at which the body uses energy to support its basal metabolism.

The recommendations for energy are presented inside the back cover; Canadian energy allowances are presented in Appendix B.

For both women and men, *light* activity means sleeping or lying down for eight hours a day, sitting for seven hours, standing for five, walking for two, and spending two hours in light physical activity.

Balancing food energy intake with physical activity can add to life's enjoyment.

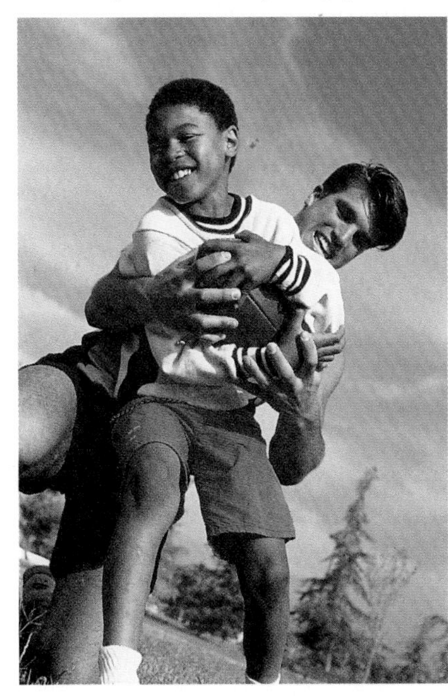

Figure 9-1

Components of Energy Expenditure

Generally, basal metabolism represents a person's largest expenditure of energy, followed by exercise and the thermic effect of foods. Of these categories, energy spent in physical activity is most responsive to voluntary control.

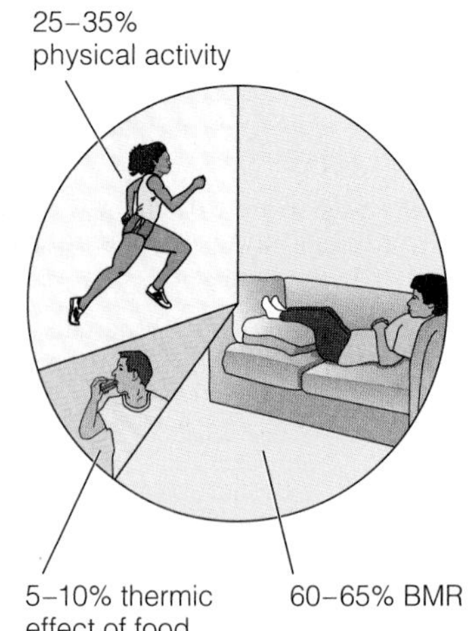
25–35%
physical activity
5–10% thermic
effect of food
60–65% BMR

exercise a daily habit so that your body composition becomes as lean as possible.[5] A warning: some ads for weight-loss diets claim that eating certain foods can elevate the BMR and thus promote weight loss. This claim is false. Any meal promotes a temporary stepped-up energy expenditure in the form of the thermic effect of food; in the context of a mixed diet, the differences among foods are not large enough to be worth notice.

As for fuel for voluntary activities, the amount of energy you spend in exercise depends somewhat on your personal style. In general, the heavier the weight of the body parts you move in your activity and the longer the time you invest, the more calories you spend. So important to energy balance is physical activity that much of Chapter 10 is devoted to presenting details of how the body spends its energy during activity. The next section shows how to calculate an approximation of your daily energy output.

KEY POINT *Two major components of the "energy out" side of the body's energy budget are basal metabolism and voluntary activities. A third component of energy expenditures is the thermic effect of food.*

How Many Calories Do I Need Each Day?

Simply put, you need enough to cover your energy expenditure. Here is one way to estimate your total energy expenditure. First, estimate the two major components of energy expenditure; then, add them together. The first component is the energy spent in basal metabolism. Follow these steps. Use the BMR factor 1.0 calorie per kilogram of body weight per hour for men or 0.9 for women; the factors differ because men usually have more muscles (metabolically active tissue) than women do. Example (for a 150-pound man):

1. Change pounds to kilograms:
   150 pounds ÷ 2.2 pounds per kilogram = 68 kilograms.
2. Multiply weight in kilograms by the BMR factor:
   68 kilograms × 1 calorie per kilogram per hour = 68 calories per hour.
3. Multiply the calories used in one hour by the hours in a day:
   68 calories per hour × 24 hours per day = 1,632 calories per day.

The second major component of energy expenditure, physical activity, is calculated by multiplying the BMR calories by a percentage that varies by activity level. These percentages are estimates or approximations of energy expenditure based on the amount of muscular work a person typically performs in a day:

Table 9-1

Factors That Affect the BMR

Factor	Effect on BMR
Age	The BMR is higher in youth; as lean body mass declines with age, the BMR slows. Continued physical activity may prevent some of this decline.
Height	Tall people have a larger surface area, so their BMRs are higher.
Growth	Children and pregnant women have higher BMRs.
Body composition	The more lean tissue, the higher the BMR.
Fever	Fever raises the BMR.
Stress	Stress hormones raise the BMR.
Environmental temperature	Adjusting to either heat or cold raises the BMR.
Fasting/starvation	Fasting/starvation hormones lower the BMR.
Malnutrition	Malnutrition lowers the BMR.
Thyroxine	The thyroid hormone thyroxine is a key BMR regulator; the more thyroxine produced, the higher the BMR.

- Sedentary lifestyle: Men, 25 to 40 percent; women, 25 to 35 percent.
- Light activity: Men, 50 to 70 percent; women, 40 to 60 percent.
- Moderate activity: Men, 65 to 80 percent; women, 50 to 70 percent.
- Heavy activity: Men, 90 to 120 percent; women, 80 to 100 percent.
- Exceptional activity: Men, 130 to 145 percent; women, 110 to 130 percent.†

To select the activity level appropriate for you, consult the list in the margin. Remember to think in terms of the amount of *muscular* work performed; don't confuse being *busy* with being *active*. Table 9-5 later in this chapter provides more exact energy costs per minute of activity based on body weight and Chapter 10 comes back to activity for control of body fatness.

Calculate your energy expenditure using both the upper and lower ends of the range of percentages given for your gender and activity level. Suppose the 150-pound man used as an example earlier is a student who bikes about ten minutes a day and walks to classes but otherwise sits and studies. He falls into the light activity category, so we can estimate the range of energy he needs by multiplying his BMR calories per day by both 50 and 70 percent:

1,632 calories per day × 0.50 = 816 calories per day.
1,632 calories per day × 0.70 = 1,142 calories per day.

The man needs from 816 to 1,142 calories per day for his activities. Now total the metabolic and activity components, first using the lower number for activity energy, then using the higher number. In a day, the man in our example spends either:

1,632 calories per day + 816 calories per day = 2,448 calories per day

or

1,632 calories per day + 1,142 calories per day = 2,774 calories per day.

Express the man's needs as a range of rounded values: 2,400 to 2,800 calories per day.

KEY POINT *To estimate the energy spent on basal metabolism, use the factor 1.0 calorie (for men) or 0.9 calorie (for women) per kilogram of body weight per hour for a 24-hour period. Then add a percentage of this amount depending on how much daily muscular activity the person engages in.*

- *Sedentary:* You sit down most of the day and drive or ride whenever possible.
- *Light activity:* You move around some of the time, as a teacher might during working hours.
- *Moderate activity:* You engage in some intentional exercise, such as an hour of jogging four or five times a week, or your occupation calls for some physical work.
- *Heavy activity:* Your job requires much physical labor, such as a roofer or a carpenter.
- *Exceptional activity:* The exceptional category is reserved for those few who spend many hours a day in intense physical training, such as professional or college athletes during their seasons.

Table 9-2

Recommendations concerning Body Weight

Dietary Guidelines
- Balance the food you eat with physical activity—maintain or improve your weight.

Healthy People 2010
- Increase to 60 percent the prevalence of healthy weight (BMI from 19.0 to 25.0) among all people age 20 or older.
- Reduce to less than 15 percent the prevalence of BMI of 30.0 or above in people age 20 and older.

NOTE: BMI values are on the inside back cover.

The Problems of Too Little or Too Much Body Fat

Both deficient and excessive body fat present health risks.[6] In the United States, too little body fat is not a widespread problem. Obesity, in contrast, is considered by experts to be an escalating epidemic.[7] Figure 9-2 shows that too many people have enough body fat to slip into the weight range that incurs health risks. An estimated 97 million U.S. adults (almost 55 percent) are overweight, while 22 percent of adults are dangerously obese.[8] Additionally one of every seven children and one of every nine teenagers in the United States are overweight.[9] The *Healthy People 2010* objectives include mandates to reduce obesity and overweight (see Table 9-2). To accomplish these goals, however, will require first that the population reverse its current trend toward gaining body weight.

Figure 9-2

Obesity Among Adults in 1988–1994 and the 2010 Objective

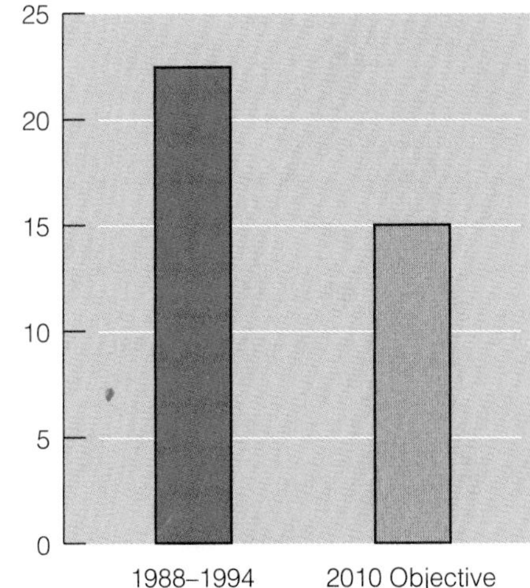
25
20
15
10
5
0
1988–1994
2010 Objective

NOTE: Obesity is defined as having a BMI at or above 30.0 for people aged 20 years and older. A later section explains BMI.

What Are the Risks from Underweight?

It has long been known that thin people will die first during a siege or in a famine. A fact not always recognized, even by health-care providers, is that overly thin people are also at a disadvantage in the hospital, where they may have to refrain from eating food for days at a time so that they can undergo tests or surgery and their nutrient status

†Percentages are derived from the RDA (1989) formula for energy expenditure allowing a 15 to 30 percent range.

wasting the progressive, relentless loss of the body's tissues that accompanies certain diseases and shortens survival time.

obesity overfatness with adverse health effects, as determined by reliable measures and interpreted with good medical judgment. Obesity is sometimes defined as a body mass index of 30 or higher (see p. 320).

visceral fat fat stored within the abdominal cavity in association with the internal abdominal organs. Also called *intra-abdominal fat*.

central obesity excess fat in the abdomen and around the trunk.

"Obesity itself has become a lifelong disease, not a cosmetic issue, not a moral judgment—and it is becoming a dangerous epidemic."—Robert H. Eckel, M.D., American Heart Association, 1998.

Figure 9-3

Visceral Fat and Subcutaneous Fat

The fat lying deep within the body's abdominal cavity may pose an especially high risk to health.

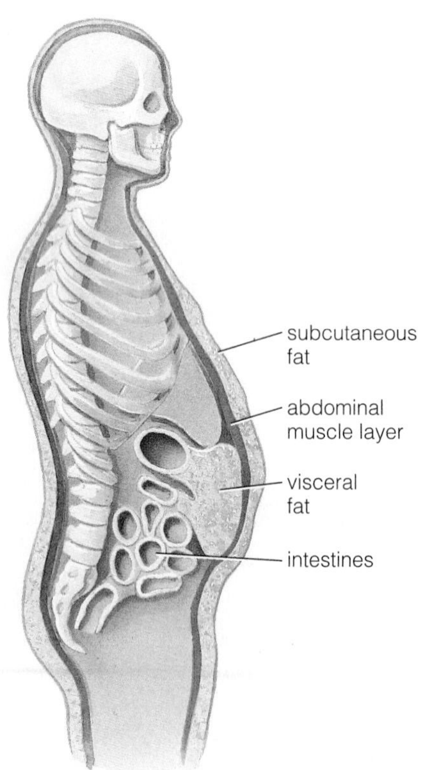
subcutaneous fat
abdominal muscle layer
visceral fat
intestines

can easily deteriorate.[10] Underweight also increases the risk for any person fighting a **wasting** disease. In fact, people with cancer often die, not from the cancer itself, but from starvation. Thus excessively underweight people are urged to gain body fat as an energy reserve and as protection for the bones and to acquire protective amounts of all the nutrients that can be stored.

KEY POINT *Deficient body fatness threatens survival during a famine or in wasting diseases.*

How Fat Is Too Fat?

People want to know exactly how fat is too fat for health, but no one cutoff point is appropriate for everyone. Obesity has recently been recognized as a major risk factor for cardiovascular disease, and evidence also suggests that being even mildly or moderately overweight also increases the risk.[11] This assumes, of course, that excess weight is composed of fat, not muscle. A few obese people, however, remain healthy and live long despite their excess body fatness. Perhaps a person's physical fitness, despite body fatness, lends some protection against early death from disease.[12] In fact, in some populations of the world, the longest-lived people seem to be those on the heavy end of the normal weight range; whether their high weights are due to extra fat or muscle is not known. Genetics and other risk factors such as smoking may also help to determine who among the overweight are most likely to stay well, and these factors must be taken into account when deciding whether an individual is overfat.

For most people, however, many studies confirm that excessive body fat substantially increases the risk of diseases and premature death.[13] In the United States, more than 70 percent of obese people suffer from at least one other major health problem.[14] As one example, excess weight may cause up to half of all cases of hypertension, thereby contributing to an increased risk of stroke.[15] Often a loss of just a few pounds can normalize the blood pressure of an overfat person; some people with hypertension can tell you at exactly what weight their blood pressure begins to rise.

Obesity as a Chronic Disease The health risks of overfatness are so many that **obesity** has been declared a chronic disease.[16] In addition to cardiovascular disease and hypertension just mentioned, being overfat also triples a person's risk of developing diabetes and all of its associated ills. Even a modest weight gain during adulthood may increase the risk of diabetes among women.[17] If hypertension, cardiovascular disease, or diabetes runs in your family, you urgently need to attend to controlling body fatness.

Obese adults are also threatened by other risks. Among them are high blood lipids, sleep apnea (abnormal pauses in breathing during sleep), osteoarthritis, abdominal hernias, some cancers, varicose veins, gout, gallbladder disease, arthritis, respiratory problems (including Pickwickian syndrome, a breathing blockage linked to sudden death), liver malfunction, complications in pregnancy and surgery, flat feet, and even a high accident rate. Moreover, after the effects of diagnosed diseases are taken into account, the risk of death from other causes remains almost twice as high for people with lifelong obesity as for others.[18]

Risks from Central Obesity Even more than total fatness, fat that collects deep within the central abdominal area of the body, called **visceral fat**, may be especially dangerous with regard to risks of diabetes, stroke, hypertension, and coronary artery disease (see Figure 9-3). In fact, the risk of death from all causes may be higher in those with **central obesity** than in those whose fat accumulates elsewhere in the body. The health risks of obesity seem to run on a continuum: normal weight brings no extra risk, central obesity carries severe risks, and other forms of obesity fall somewhere in between.[19]

Why should fat in the abdomen bring extra risk to the heart? Some researchers suspect that differences in fat mobility are part of the explanation. Visceral fat, which is readily released into the bloodstream, may make a significant contribution to the blood's daily burden of cholesterol-carrying lipoproteins, LDL, thereby increasing heart disease risk.[20] Fat layers lying just beneath the skin **(subcutaneous fat)** of the abdomen, thighs, hips, and legs also release fat, but sluggishly.[21] Subcutaneous fat tends to stay longer in storage and so, theoretically, may contribute less to blood lipids.

Men of all ages and women who are past menopause are more prone to develop the "apple" profile of central obesity, whereas women in their reproductive years develop more of a "pear" profile (fat around the hips and thighs). Some women change profile at menopause, and lifelong "pears" may suddenly face increased risks of diseases that accompany excess visceral fat. Smokers, too, may carry more of their body fat centrally. Although a smoker may weigh less than the average nonsmoker, the smoker's waist measurement may be greater, leading to the theory that smoking may directly affect body fat distribution. Two other factors may also affect body fat distribution. Moderate-to-high intakes of alcohol have a positive association with central obesity, and higher exercise levels have a negative association. A later section explains how to judge whether a person carries too much fat around the middle.

A further word about smoking: weight gain is often a concern for people who are thinking of quitting smoking, and some adolescents, especially girls, may even take up smoking to control weight. Smokers do tend to weigh less than nonsmokers, and many gain weight when they stop smoking.[22] Nicotine blunts feelings of hunger, so when hunger strikes, a smoker can reach for a cigarette instead of food. The best advice to smokers wanting to quit seems to be to adjust diet and exercise habits to maintain weight during and after cessation. The best advice to a person flirting with the idea of taking up smoking for weight control is don't do it—many thousands of people who became addicted as teenagers die from tobacco-related illnesses each year.

subcutaneous fat fat stored directly under the skin (*sub* means "beneath"; *cutaneous* refers to the skin).

body mass index (BMI) an indicator of obesity, calculated by dividing the weight of a person by the square of the person's height.

Smoking may keep some people's weight down, but at what cost?

- Heart disease.
- Cancer.
- Osteoporosis.
- Chronic lung diseases.
- Shortened life span.
- Low-birthweight babies.
- Miscarriage.
- Sudden infant death.
- Many others.

Social and Economic Costs of Obesity Though some overfat people seem to escape health problems, no one who is fat in our society quite escapes the social and economic handicaps. Our society places enormous value on thinness, especially for women, and fat people are less sought after for romance, less often hired, and less often admitted to college.[23] They pay higher insurance premiums, and they pay more for clothing.[24] Psychologically, too, fat people are made to feel rejected and embarrassed, and this diminishes self-esteem.

Medical advice urges all overweight people to reduce their fatness to reduce associated risks to health and life.[25] However, such advice may be easier to give than to follow. Almost everyone who "goes on a diet" and loses weight regains the weight in short order. This and other risks of dieting have led some experts to question traditional approaches. This chapter's Controversy explores this debate from both points of view.

KEY POINT *Overfatness presents social and economic handicaps as well as a host of physical ills. Obesity has been named a chronic disease. Central obesity may be more hazardous to health than other forms of obesity.*

Body Weight versus Body Fatness

An easy way to determine whether a given person has a healthy body weight is to measure the person's height and weight and then compare the results to a list of suggested weights for heights deemed by nutrition authorities to reflect health. Table 9-3 offers one such list. Weight-for-height tables, though easy to use, inexpensive, and accessible, are not the most accurate method of evaluation, however. Instead, experts use guidelines involving evaluations in three areas.[26] The first is a person's **BMI,** or **body mass index.** The BMI, which defines average relative weight for height in people older than 20 years, usually correlates with body fatness and degree of disease risks (see the inside back cover). The second indicator is the waist circumference, indicating the degree of visceral fatness in proportion to body fatness. These two measurements are discussed in the next section. The third area of concern is the person's disease risk profile, which takes into account, for example, whether the person has hypertension, type 2 diabetes, elevated lipoproteins, whether the person is a

Table 9-3

Suggested Weights for Adults of All Ages: 1995 Guidelines

Height[*]	Weight (lb)[a] MIDPOINT	RANGE[b]
4′10″	105	91–119
4′11″	109	94–124
5′0″	112	97–128
5′1″	116	101–132
5′2″	120	104–137
5′3″	124	107–141
5′4″	128	111–146
5′5″	132	114–150
5′6″	136	118–155
5′7″	140	121–160
5′8″	144	125–164
5′9″	149	129–169
5′10″	153	132–174
5′11″	157	136–179
6′0″	162	140–184
6′1″	166	144–189
6′2″	171	148–195
6′3″	176	152–200
6′4″	180	156–205
6′5″	185	160–211
6′6″	190	164–216

[a]Without shoes or clothes.
[b]Higher weights within the ranges generally apply to men, and lower weights to women, because men tend to have more muscle and bone.
SOURCE: Report of the Dietary Guidelines Advisory Committee on the Dietary Guidelines for Americans, 1995.

Finding your BMI:

$$BMI = \frac{\text{weight (kg)}}{\text{height (m)}^2}$$

or

$$BMI = \frac{\text{weight (lb)}}{\text{height (in)}^2} \times 705$$

Example: A 5′10″ person weighing 150 pounds has a BMI of 21.6.

$$BMI = \frac{150}{70^2} \times 705$$

$$BMI = \frac{150}{4{,}900} \times 705$$

$$BMI = .0306 \times 705$$

$$BMI = 21.6 \text{ (rounded)}$$

smoker, and so forth. The more risk factors present, the greater the urgency to control body fatness.

Body Mass Index

BMI values correlate significantly with body fatness, and experts use them to help evaluate a person's health risks associated with underweight or overweight. According to recent guidelines from the National Heart, Lung, and Blood Institute of the National Institutes of Health, overweight for adults is defined as BMI of 25 through 29.9 and obesity as BMI equal to or greater than 30. As Table 9-4 shows, overweight conditions present health risks with increasing severity: overweight presents a somewhat elevated risk of disease, and in class III obesity, the risks to health are extreme. The inside back cover of this book provides an easy way to find and evaluate BMI in adults and adolescents. Currently, researchers are working to develop BMI values for young children.[27]

The BMI values are probably most valuable for evaluating degrees of obesity and are less useful for evaluating nonobese people's body fatness. The BMI values have two major drawbacks: they fail to indicate how much of the weight is fat and where that fat is located. These drawbacks make the BMI unsuitable for use with:

- Athletes (because their highly developed musculature falsely increases their BMI values).
- Pregnant and lactating women (because their increased weight is normal during childbearing).
- Adults over 65 (because BMI values are based on data collected from younger people and because people "grow shorter" with age).

The photo of the bodybuilder in the margin of the next page underscores this point: with a BMI over 30, he would be classified as obese by BMI standards alone. On further investigation, however, we find that his percentage of body fat is well below average and his waist circumference is within a healthy range. A diagnosis of obesity or overweight requires more than a BMI value. One must also learn something about the person's body composition and fat distribution. There is no easy way to look inside a person to measure bones and muscles, but indirect measures can reveal some clues.

KEY POINT *The body mass index mathematically correlates heights and weights with risks to health. It is especially useful for evaluating health risks of obesity but fails to measure body composition or fat distribution.*

Table 9-4

BMI Values and Health Risks[a]

BMI	Obesity Class	Risks to Health
<18.5	Underweight	The lower the BMI, the greater the risk
18.5 through 24.9	Normal	Very low risk
25.0 through 29.9	Overweight	Increased risk/high risk[b]
30.0 through 34.9	Class I obesity	High risk/very high risk[b]
35.0 through 39.9	Class II obesity	Very high risk
40.0 or above	Class III obesity	Extreme risk

[a]Risk for type 2 diabetes, hypertension, and cardiovascular disease.
[b]The lower risk applies to men with a waist circumference of 102 centimeters (40 inches) or less and women with a waist circumference of 88 centimeters (35 inches) or less. The higher risk applies to those with a waist circumference above these values.
SOURCE: Data from National Heart, Lung, and Blood Institute Expert Panel, Executive summary of the clinical guidelines on the identification, evaluation, and treatment of overwieght and obesity in adults, *Journal of the American Dietetic Association* 98 (1998): 1178–1191.

Estimating Body Fatness

A person who stands about 5 feet 10 inches tall and weighs 150 pounds carries about 90 of those pounds as water and 30 as fat. The other 30 pounds are the so-called lean tissues: muscles; organs such as the heart, brain, and liver; and the bones of the skeleton. Stripped of water and fat, then, the person weighs only 30 pounds![‡] This lean tissue is vital to health. The person who seeks to lose weight wants, of course, to lose fat, not this precious lean tissue. And for someone who wants to gain weight, it is desirable to gain lean and fat in proportion, not just fat.

Several laboratory techniques for estimating body fatness have been used for years. These include, among others:

- *Anthropometry*. Measurements such as the **fatfold test** or waist circumference can be taken (see Figures 9-4 and 9-5 on page 322).[28]
- *Density* (the measurement of body weight compared with volume). Lean tissue is denser than fat tissue, so the denser a person's body is, the more lean tissue it must contain. Density can be determined by **underwater weighing.**
- *Conductivity*. Only lean tissue and water conduct electrical current; **bioelectrical impedance** measures how well a tiny harmless electrical charge is conducted through the lean tissue of the body and so reflects the body's contents of lean tissue, including water. A drawback is that temporary changes in body water content can produce changes in the measurement.

Other sophisticated techniques are available. Some can not only estimate lean versus fat tissue but also determine where the fat is located. Even these methods do not provide perfect analyses of some people, however. For example, the body fat of African American women may be underestimated in laboratory studies because standard methods do not account for variability of body composition among ethnic groups.

Fatfold measurements provide an accurate estimate of total body fat and a fair assessment of the fat's location.[29] About half of the fat in the body lies directly beneath the skin, so the thickness of this subcutaneous fat is assumed to reflect total body fat. Measures taken from central-body sites (around the abdomen) better reflect changes in fatness than those taken from upper sites (arm and back). Fatfold measurements taken without skill and training are often inaccurate, however.

Waist circumference serves as an indicator of visceral fatness. Previously, a ratio of waist to hip measurements served this purpose, but waist circumference alone has been deemed a valid indicator for both men and women. Above a certain waist circumference, disease risks rise—even for some whose BMI values fall within the normal range.[30] For a fair indication of whether you develop fat centrally, measure your waist as shown in Figure 9-5; then compare your measurement with these cutoff points:

- Men: 102 centimeters (40 inches).
- Women: 88 entimeters (35 inches).

Anyone with a waist measurement larger than these standards may carry an increased risk of disease along with the extra girth.

KEY POINT *An assessor can determine the percentage of fat in a person's body by measuring fatfolds, body density, or other parameters. Distribution of fat can be estimated by determining the waist circumference.*

Defining Priorities

Even after you have a body fatness estimate, questions arise. What is the "ideal" amount of fat for a body to have? To answer this question, one must first decide, ideal for what? If the answer is "society's approval," be aware that fashion is fickle. Body

‡For a healthy person 5 feet tall and weighing 100 pounds, the comparable figures would be 60 pounds of water, 20 pounds of fat, 20 pounds of lean.

fatfold test measurement of the thickness of a fold of skin on the back of the arm (over the triceps muscle), below the shoulder blade (subscapular), or in other places, using a caliper (depicted in Figure 9-3). Also called *skinfold test*.

underwater weighing a measure of density and volume used to determine body fat content.

bioelectrical impedance a technique to measure body fatness by measuring the body's electrical conductivity.

The BMI standards are not accurate for athletes. At 6 feet 3 inches tall and 245 pounds, Mike O'Hearn would be judged to be obese by BMI standards alone. Further measures reveal that his body contains only 8 percent of its weight as fat, less than the average fat percentage for men, and that his waist circumference is within a healthy range.

Figure 9-4

Three Methods of Assessing Body Fatness

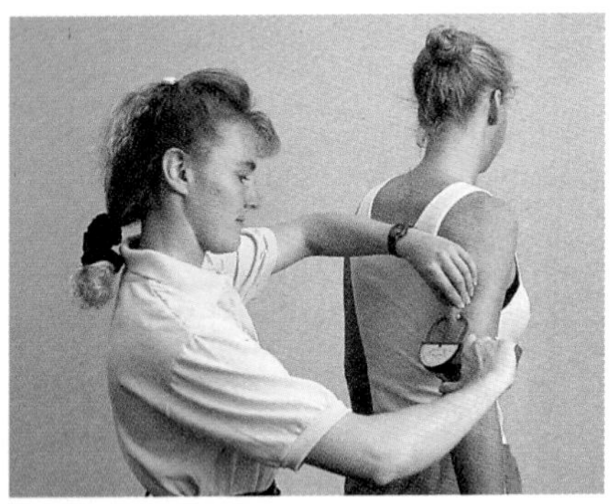

A fatfold measures can yield accurate results when taken by a trained technician using a calibrated metal caliper. (This procedure is painless.)

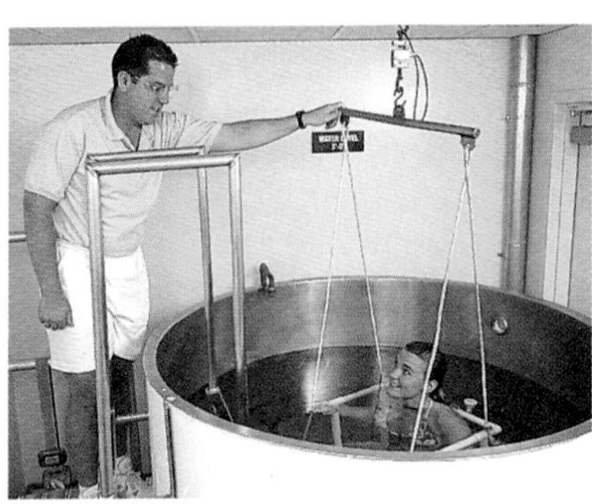

Underwater weighing determines body volume and density; measurements reflect the percentage of body fat in experimental subjects.

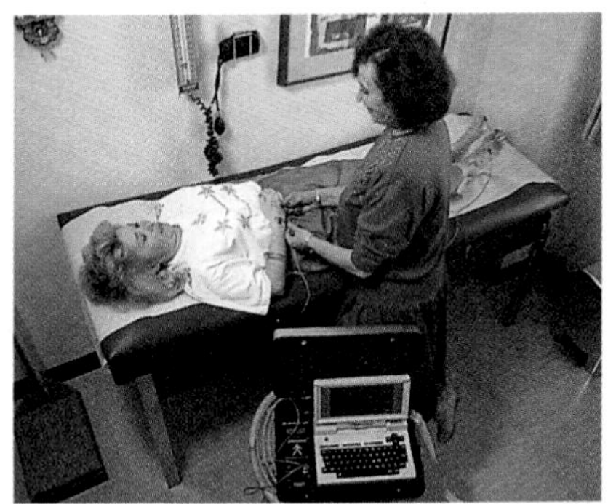

Bioelectrical impedance is simple, painless, and accurate when properly administered; the method determines body fatness by measuring conductivity. Lean tissue conducts a mild electric current; fat tissue does not.

shapes valued for looks may have little to do with health, and many of the most popular body shapes are not achievable goals for most people.

If the answer is "health," then the ideal depends partly on who you are. A man of normal body composition may have, on average, 15 percent of body weight as fat, and a woman may have 20 percent. Researchers draw the line when body fat exceeds 22 percent in young men, 25 percent in older men, 32 percent in younger women, and 35 percent in older women; age 40 is the dividing line between younger and older.[31]

Because people have different lifestyles and are at various stages of their lives, different standards must be used for different people, which requires applying judgment. For example, competitive endurance athletes need just enough body fat to provide fuel, insulate the body, and permit normal fat-soluble hormone activity, but not so much fat as to weigh them down. An Alaskan fisherman, in contrast, needs a blanket of extra fat to insulate against the cold. For a woman starting pregnancy, the ideal percentage of body fat may be different again; the outcome of pregnancy is compromised if the woman begins it with too little body fat. Below a threshold for body fat content set by heredity, some individuals become infertile, develop depression or abnormal hunger regulation, or become unable to keep warm. These thresholds are not the same for each function or in all individuals, and much remains to be learned about them.

The person seeking a single, authoritative answer to the question "How much should I weigh?" is bound to be disappointed. No one can tell you exactly how much you should weigh; but with health as a value, at least you have a starting framework. Your weight should fall within the range that best supports your health.

KEY POINT *No single body weight suits everyone; different people have different needs.*

Figure 9-5

Measuring Waist Circumference

Using a nonstretching tape measure, measure the body around the point near the belly button. A healthy waist circumference for men is no larger than 102 centimeters (40 inches); for women, no larger than 88 centimeters (35 inches).

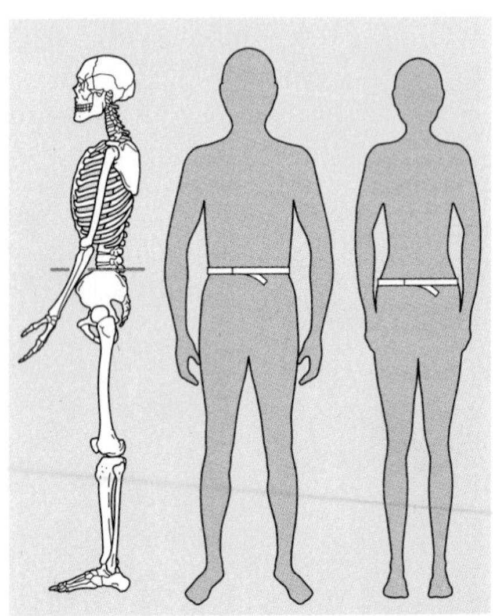

The Mystery of Obesity

Why do some people get fat? Why do some get thin? And most amazingly, why do some people stay at the same weight year after year? Is weight controlled by hereditary, metabolic factors? Or is it controlled by environmental influences? Is it a matter of eating behaviors—and if so, what controls these behaviors, internal signals or a person's free will? The section sorts through the pieces of the obesity puzzle, but no law says that only one cause must prevail. In all likelihood, internal and external factors operate together and in different combinations in different people. We begin by examining the appetite and its controls.

Why Did I Eat That?

Obesity researchers are interested in why people eat and what they eat, and especially why some people overeat. Overeating may or may not explain some people's obesity, but scientists hope that by discovering how food intake is regulated in the body, they may devise new effective treatments for obesity.[32] Eating behavior seems to be regulated by a series of signals that fall into two broad functional categories: "go" mechanisms that stimulate eating and "stop" mechanisms that signal the body to cease or refrain from eating. One view of the process of food intake regulation is summarized in Figure 9-6 on the next page.

"Go" Signals—Hunger and Appetite Most people recognize **hunger** as an irritating feeling that prompts them to search out food and to start eating. Hunger is the physiological response to a need for food triggered by chemical messengers originating and acting in the brain, especially in the brain's hypothalamus.[33] One such messenger is the neurotransmitter **neuropeptide Y (NPY),** which is suspected of being a strong stimulant of the appetite, especially for carbohydrate-rich foods (see the top part of Figure 9-6). Hunger can also be affected by a number of other brain chemicals and other factors, including the nutrients present in the bloodstream, the size and composition of the preceding meal, customary eating patterns, the weather (heat reduces food intakes; cold increases it), exercise, sex hormones, and physical and mental disease states. Typically, hunger makes itself known roughly four to six hours after eating, when the food has left the stomach and much of it has been absorbed by the intestine.

The body's hunger response adapts quickly to changes in food intake. A person who restricts the amount of food consumed at each meal may feel extra hungry for a few days, but then hunger may diminish for a time. This dimming of hunger represents an adaptation to smaller quantities of food. During this period a large meal may make the person feel uncomfortably full, partly because the stomach's capacity has adapted to a smaller quantity of food.[34] It is now that a dieter may report that "My stomach has shrunk," but the stomach organ itself doesn't shrink except in cases of chronic starvation.

This may seem to be good news for dieters, but beware: at some point in food deprivation, hunger almost invariably returns with a vengeance, and can lead to bouts of overeating that more than make up for the calories lost during the deprivation period. Be aware, too, that the stomach's capacity can also gradually adapt to larger and larger quantities of food, until a meal of normal size no longer feels satisfying. This observation may partly explain why obesity is on the rise: Popular demand has led to larger meals of giant burgers, super-sized fries, and huge candy bars and soft drinks, and many people's stomachs have adapted to accommodate them.

Like hunger, **appetite** also initiates eating, but unlike hunger, appetite is learned. When the two coincide, appetite intensifies hunger, but a person may experience appetite without hunger. An example of the latter is the effect of seeing and smelling a freshly baked apple pie after finishing a big meal—despite an already overfull stomach, the appetite for the pie can be strong. In contrast, a person may feel hungry but have no appetite for food such as when under stress or ill. Other factors affecting appetite include:

- Learned preferences, aversions, and timings (cravings for favorite foods, fear of new foods, eating according to the clock).
- Environmental conditions (people prefer hot foods in cold weather and vice versa).
- **Endorphins** (the brain's pleasure chemicals that enhance the desire for the taste of delicious foods and may be triggered by the smell, sight, or taste of foods; as mentioned in the top part of Figure 9-6).
- Inborn appetites (inborn preferences for fatty, salty, and sweet tastes).

hunger the physiological need to eat, experienced as a drive for obtaining food; an unpleasant sensation that demands relief.

neuropeptide Y (NPY) a neurotransmitter whose functions in the brain's hypothalamus include the stimulation of appetite, especially appetite for carbohydrate-rich foods.

appetite the psychological desire to eat; a learned motivation and a positive sensation that accompanies the sight, smell, or thought of appealing foods.

endorphins, endogenous opiates compounds of the brain whose actions mimic those of opiate drugs (morphine, heroin) in reducing pain and producing pleasure. In appetite control, endorphins are released on seeing, smelling, or tasting delicious food and are believed to enhance the drive to eat or continue eating.

Hunger is physical and asks, "Is there anything to eat?" Appetite is psychological and asks, "What do I feel like eating?"

Chapter 3 described the brain's hypothalamus.

Figure 2-7 in Chapter 2 demonstrated how portion sizes have increased over recent decades.

satiation (SAY-she-AY-shun) the perception of fullness that builds throughout a meal, eventually reaching the degree of fullness and satisfaction that halts eating. Satiation generally determines how much food is consumed at one sitting.

satiety (sah-TIE-eh-tee) the perception of fullness that lingers in the hours after a meal and inhibits eating until the next mealtime. Satiety generally determines the length of time between meals. .

leptin an appetite-suppressing hormone produced in the fat cells that conveys information about body fatness to the brain; believed to be involved in the maintenance of body composition (*leptos* means "slender").

- Social interactions (cultural or religious acceptability of foods, companionship).
- Some disease states (obesity may be associated with increased taste sensitivity, whereas colds, flu, and zinc deficiency reduce taste sensitivity).
- Some drugs (appetite stimulants or depressants affect food intake).

"Stop" Signals—Satiation and Satiety At some point during a meal, the brain receives messages from several sources that enough food has been eaten. This condition is called **satiation,** and it originates from the presence of food in the GI tract (consult the lower part of Figure 9-6). When the stomach stretches, nerve receptors in the stomach fire, sending the signal to the brain that the stomach is full and causing the person to stop eating.[35] As nutrients from the meal enter the small intestine, they stimulate receptor nerves and trigger the release of hormones that provide the brain with information about the meal just eaten.[§] The brain also detects absorbed nutrients passing by in the blood and releases neurotransmitters that suppress food intake in response.[36] Together, stomach distension, nutrients in the small intestine, and hormonal and neural

[§]One such hormone is cholecystokinin.

Figure 9-6

A Cascade of Regulation: Hunger, Appetite, Satiation, and Satiety

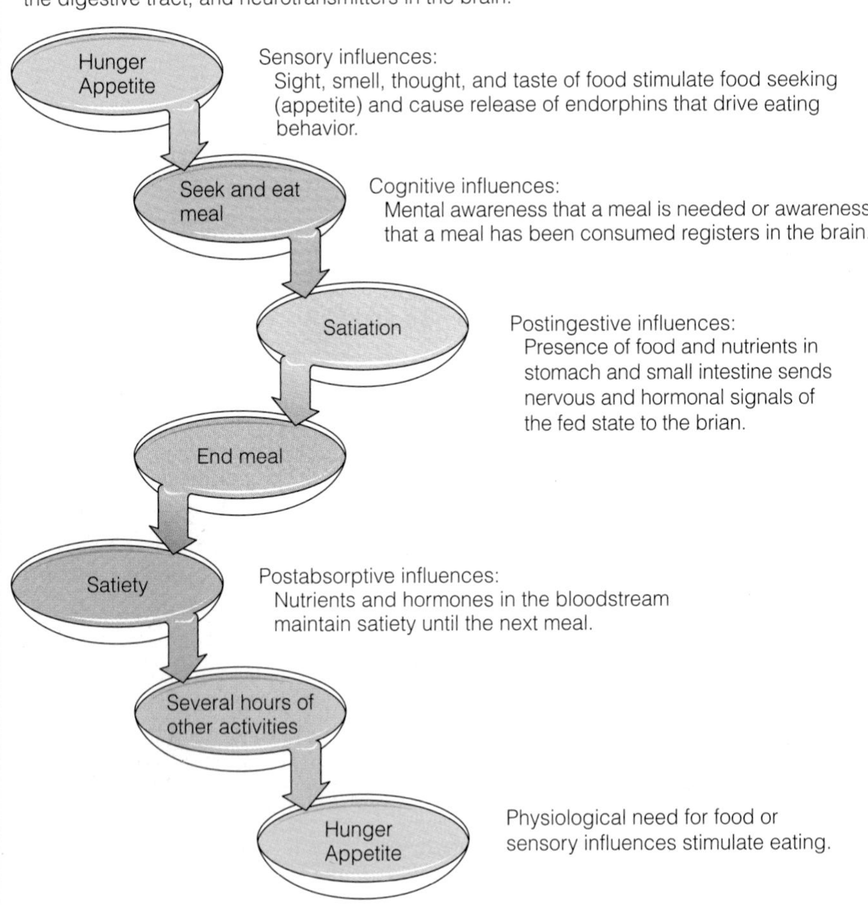
Physiological need for food:
Hunger is produced by hormones, absence of food from the digestive tract, and neurotransmitters in the brain.
Hunger
Appetite
Sensory influences:
Sight, smell, thought, and taste of food stimulate food seeking (appetite) and cause release of endorphins that drive eating behavior.
Seek and eat meal
Cognitive influences:
Mental awareness that a meal is needed or awareness that a meal has been consumed registers in the brain.
Satiation
Postingestive influences:
Presence of food and nutrients in stomach and small intestine sends nervous and hormonal signals of the fed state to the brian.
End meal
Satiety
Postabsorptive influences:
Nutrients and hormones in the bloodstream maintain satiety until the next meal.
Several hours of other activities
Hunger
Appetite
Physiological need for food or sensory influences stimulate eating.

SOURCE: Adapted from ideas in J. E. Blundell and coauthors, Control of human appetite: Implications for the intake of dietary fat, *Annual Review of Nutrition* 16 (1996): 285–319.

signals inform the brain's hypothalamus about the size and nature of the meal. The response: satiation occurs, the eater feels full, and the meal ends.

After a meal, the feeling of **satiety** continues to suppress hunger and allows for a period of some hours in which the person is free to dance, study, converse, wonder, fall in love, and concentrate on endeavors other than eating. Whereas satiation informs the body when to stop eating, satiety allows the body to stay stopped for a while. One powerful source of satiety is **leptin,** a hormone produced by the adipose tissue and directly linked to both appetite and body fatness. The next section discusses leptin and its relationship to obesity.

A lack of satiety between meals can cause problems when people attempt to reduce their food intakes. A person who suffers through hours of annoying hunger pains may overeat regrettably when mealtime arrives, setting up a destructive cycle of starving and binge eating, with no weight loss to show for the effort. The choice of foods may affect satiety—some foods seem to sustain feelings of satiety for longer periods than others.[37] In trying to rank individual foods by their ability to produce satiety, researchers fed subjects food portions that were equal in calories, and then questioned subjects about feelings of hunger later on. The satiety value of white bread was set at 100 and other foods were ranked in relation to it. The results, shown in Figure 9-7, seem to indicate that, in general, foods high in fiber or protein sustain satiety for longer than those high in fat, white flour, or sugar. Plain boiled potatoes ranked highest, but the reasons why are unclear. Keep in mind that the researchers tested portions that were equal in calories, and that fat rich foods pack many calories in a small bulk of food. It takes more than a slice of bread to deliver the calories in just a third of a croissant, for example. It may be that bulky foods kept people feeling full longer simply because they presented more food to the digestive tract. More research is needed to clarify these intriguing, but preliminary, findings.

Although interesting and important, findings about appetite regulation do not fully explain why some people gain too much body fatness while others stay lean. Many other forces are also at work, including a person's genetic inheritance.

Figure 9-7
Satiety Scores of Foods

Higher satiety

Food	Score
Boiled white potatoes	323
Baked fish	225
Oatmeal with milk[a]	209
Orange, apple	200
Whole-grain pasta	188
Beefsteak, baked beans	170
Popcorn, eggs, bran cereal with milk[a]	150
Brown rice or white rice	135
White bread	100
Snack chips, ice cream	94
Candy bar	70
Cake, doughnuts	67
Croissant	47

Lower satiety

[a]Cereals were served with 1.5% fat milk.
SOURCE: Data from S. H. A. Holt and coauthors, A satiety index of common foods, *European Journal of Clinical Nutrition* 49 (1995): 675–690.

KEY POINT *Food intake is regulated by hunger, appetite, satiation, and satiety. Discoveries of neural and hormonal regulators of eating behaviors may lead to new effective pharmacological treatments of obesity.*

Genetics and Obesity

A person's genetic makeup almost certainly influences the body's tendency to consume or store too much energy or burn too little.[38] For a person who has one obese parent, the chance of becoming obese is 60 percent; if both parents are obese, the probability may rise to as high as 90 percent. Researchers have also found that adopted children tend to be similar in weight to their biological parents, not to their adoptive parents. Studies of twins yield similar findings: identical twins are twice as likely to weigh the same as fraternal twins—even when reared apart. These findings support a role for genetics in a person's susceptibility to obesity.[39]

Clearly, something affects a person's tendency to gain or lose weight when eating more or less food energy than needed.[40] When given an extra 1,000 calories of food a day for 100 days, some people gain 30 pounds, but others gain less than 10 pounds. Similarly, some people lose more weight faster than others on comparable exercise regimens. The efforts of researchers around the world are focused on the genetic links to obesity in hopes of finding new effective treatments. The following sections explore the main areas of

The mouse on the right is genetically obese—it lacks the gene for producing leptin. The mouse on the left is also genetically obese but remains lean because it receives leptin.

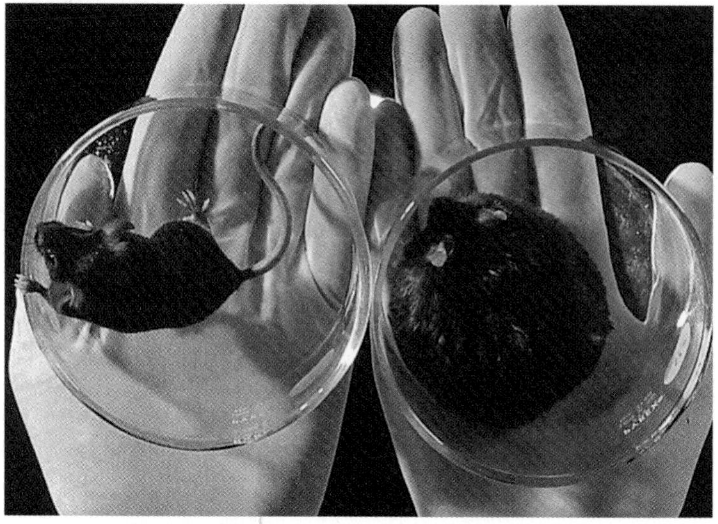

set-point theory the theory that the body tends to maintain a certain weight by means of its own internal controls.

interest. The first concerns one of the products of the genes—leptin, the protein hormone already mentioned as a possible influence in obesity development.

Leptin Imagine an appetite suppressant designed to work in perfect harmony with the brain's appetite-regulating chemistry, while having no adverse effects on other tissues. Further, the tissue ultimately controlled by this compound, the adipose tissue, also produces it, affording precise feedback to the mechanism. When the body gains fatness, this triggers greater production of the appetite suppressant; less food is consumed, and weight loss ensues. Fat losses bring the opposite effect. This fanciful description is not far off the mark from a true accounting of the workings of the hormone leptin. And it performs beautifully—at least in mice.

The genetic code used for making leptin is held in an obesity-related gene, called the *ob* gene.[**41] Mice with a defective *ob* gene, like the round one already shown in the margin of the previous page, fail to produce leptin even when grossly overweight and often end up weighing three times as much as normal mice.[42] The thin mouse is of the same strain—it is also genetically obese—but has received injections of leptin, and so remains lean. Figure 9-8 demonstrates the influences of leptin: appetite falls off, energy expenditures increase, the fat cells store less fat, and some fat cells also self-destruct, trimming both the body's stored fat and the tissue that contains it.[43]

Leptin's discovery, and then the further discovery of its responding site in the human brain, opened a whole new chapter in the investigation into a genetic basis for obesity.[44] Leptin suppresses the action of the neurotransmitter NPY, the strongest appetite stimulant in the brain. Why not just give leptin injections to overweight people to suppress their appetites and increase their energy expenditures? As it turns out, most obese people already have elevated leptin concentrations in their blood, and giving them more has no effect on appetite.[45] Just a tiny percentage of obese people are exceptions: they fail to produce leptin at all.[46] In these few, leptin injections would be expected to restore normal appetite control and body fatness.

Logically, it would seem, then, that most obese people, with their large amounts of fat tissue and thus greater concentrations of leptin, should feel *less* hungry than people of normal fatness. This bit of logic fails to hold up under research scrutiny, however. Fat people feel hungry and stay fat despite their elevated leptin levels. It seems likely that as fatness increases and leptin rises, the brain's leptin receptors may become less responsive to leptin in much the same way as the body cells of people with type 2 diabetes become unresponsive to insulin. In fact, one line of research suggests a possible link between increased leptin and development of type 2 diabetes.[47]

Recently, other regulatory roles of leptin around the body have been recognized. For example, leptin may selectively reduce centrally located fat.[48] Leptin may also inform the female reproductive system about body fat reserves; stimulate growth of new blood vessels, especially in the cornea of the eye; act on bone marrow cells to enhance their maturation into specialized cells; promote formation of red blood cells; and help support a normal immune response.[49] The actions of leptin on the body tissues reach far beyond controlling body weight and so may prohibit using leptin as an anti-obesity drug.

Obviously, leptin is not the final answer to human obesity as once was hoped. The complexity of obesity demands more than one approach. Potentially, though, by harnessing the activities of the *ob* genes and their protein products, researchers may discover new, effective methods for treating obesity—an outcome much desired, but so far not feasible.

Set-Point Theory The discovery of leptin and the appetite-regulating system may support the long-popular **set-point theory** of obesity. As mentioned, earlier, most people who lose weight on reducing diets later quickly regain all the lost weight. This phenomenon seems to suggest that somehow the body chooses a weight

**Another genetic link to obesity is the gene that codes for a protein known as *mahogany,* also thought to suppress obesity.

Figure 9-8

A Model of Leptin's Effects on Energy Balance and Body Fatness

A. In positive energy balance, body fat accumulates and production and release of leptin increases. The higher blood leptin concentration then suppresses hunger, decreases food intake, and steps up energy expenditure, reducing the body's energy excess.

B. In negative energy balance, body fat stores dwindle and production and release of leptin slows. Less circulating leptin means increased hunger, increased food intake, and decreased energy expenditures, and correction of the energy deficit.

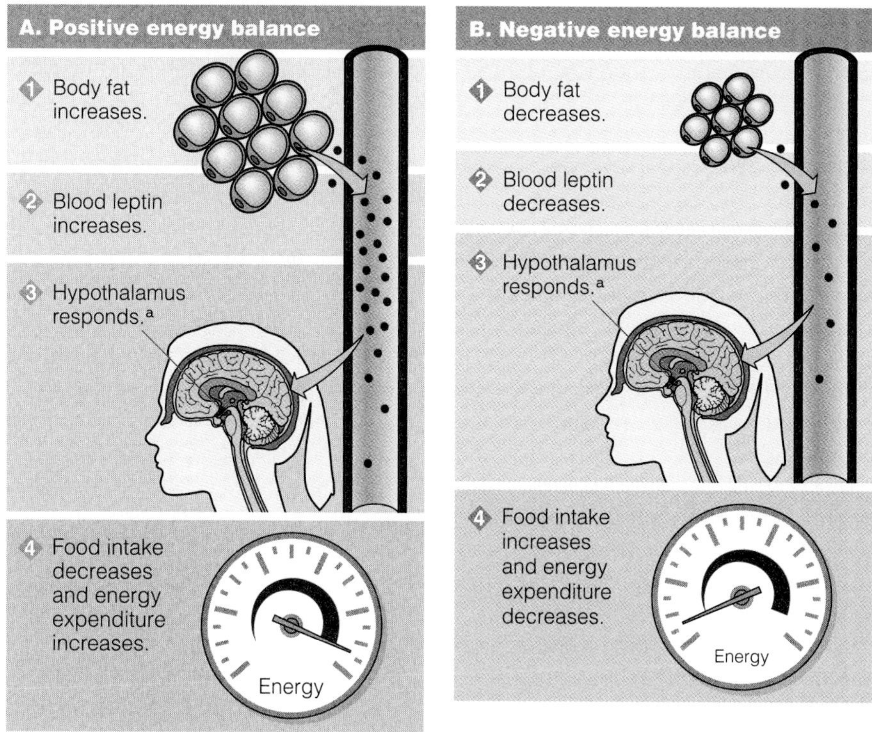

A. Positive energy balance
1 Body fat increases.
2 Blood leptin increases.
3 Hypothalamus responds.a
4 Food intake decreases and energy expenditure increases.
Energy
B. Negative energy balance
1 Body fat decreases.
2 Blood leptin decreases.
3 Hypothalamus responds.a
4 Food intake increases and energy expenditure decreases.
Energy

[a]The brain's hypothalamus was introduced in Chapter 3.

that it wants to be and defends that weight by regulating eating behaviors and hormonal actions.[50]

Unfortunately for those trying to lose weight or maintain a loss, evidence is piling up to suggest that biological systems, and not human will, may determine a body weight set point. Just as a thermostat setting triggers a heater to run when air temperature falls and to turn off when warmth is restored, whenever weight is lost or gained, the set-point mechanism seems to trigger a change in metabolic energy expenditure in the direction that restores the initial body weight.[51] These changes in energy expenditure are greater than those predicted based on body composition and may help to explain why it is so difficult for obese people to maintain weight losses.

Enzyme Theory Strong evidence also links fat storage with elevated concentrations of the enzyme that enables fat cells to store triglycerides. Concentrations of this enzyme, **LPL** or **lipoprotein lipase**, increase as cells become enlarged with fat. The more LPL, the more easily fat cells store lipid, and the more likely the body will remain obese. An interesting question relating to the set-point theory just described is whether some people's fat cells contain elevated concentrations of LPL *before* the onset of obesity. If so, this situation might partly explain why obesity tends to run in families, for the making of all enzymes including LPL is governed by the genes.

LPL (lipoprotein lipase) an enzyme mounted on the surfaces of fat cells that splits triglycerides in the blood into fatty acids and glycerol to be absorbed into the cells for reassembly and storage.

thermogenesis the generation and release of body heat associated with the breakdown of body fuels.

brown fat adipose tissue abundant in hibernating animals and human infants. Brown fat cells are packed with pigmented, energy-burning enzymes that release heat rather than accomplishing other tasks. These enzymes give the cells a darkened appearance under a microscope.

adaptive thermogenesis adjustments in energy expenditure related to changes in environment such as cold and to physiological events such as underfeeding or trauma.

Fat Cell Number Theory Another cause of obesity may be the development of excess fat cells during childhood. The amount of fat on a person's body reflects both the *number* and the *size* of fat cells. The number of fat cells increases during the growing years and then levels off during adulthood. Fat cell number increases more rapidly in obese children than in lean children, and obese children entering their teen years may already have as many fat cells as do adults of normal weight.

Fat cells of obese people also contain more LPL, so they are likely to reach a large size quickly. Therefore, obesity reflects not only more and larger fat cells, but more efficient ones, too. For this reason, people with obesity may encounter extra difficulty in trying to lose weight. They may also tend to regain lost weight rapidly. Obesity-fighting measures may therefore be most effective during the growing years when fat cell number is increasing.

Thermogenesis Theories Genes code for proteins involved in energy metabolism. Some of these proteins control the body's manufacturing of heat, or **thermogenesis.** For years, researchers have focused efforts on a tissue that specializes in converting energy to heat—**brown fat.** Regular white fat cells store energy in fat's chemical bonds and have a slow metabolic rate; brown fat cells break those bonds and actively release their stored energy as heat. Brown fat is more abundant and more active in lean animals than in fat ones. It is theorized that a person whose fat burns off too little energy or a person who has less than the normal amount of brown fat would tend to store more white fat than other people store. Not much brown fat occurs in adult human beings, however.

Lately, researchers have identified a system of thermogenesis enzymes that occurs not only in brown fat but in many tissues distributed throughout the body, such as the muscles, spleen, and bone marrow.[52] These enzymes, like those of brown fat, produce heat during exposure to cold temperature. They also adapt to other changing conditions, such as physical conditioning, overfeeding, starvation, trauma, or other types of stress, producing more or less heat as needed. Known as **adaptive thermogenesis,** this heat production partly reflects the work of the body in building the tissues, enzymes, and hormones necessary to meet the new demand. Another way of producing heat wastes energy; the tissues break down fuel for heat without building or doing anything. They can speed up or reduce heat production to spend or conserve energy whenever the body's energy supply is too high or too low. Weight-loss dieters are all too familiar with this counterproductive effect: their efforts at reducing food energy intake are soon met by reduced heat production and slowed metabolism.[53] Researchers speculate that as body fat diminishes, this powerful energy conservation system acts to prevent further losses and to restore the body fatness to its previous level.

Yet another form of thermogenesis, the thermic effect of food mentioned earlier, varies between obese and nonobese people. In lean people who have just eaten a meal, energy use speeds up for a while and then drops back to normal. In many obese people, no change in energy use occurs after eating.

So far no one has shown conclusively that overweight people expend less energy overall than normal-weight people, and in fact, the opposite seems to hold true. Overweight people seem to spend more energy each day, not less, than do people of normal weight. This is probably because heavier bodies require more energy to move and to maintain themselves.

KEY POINT *Theories about genetic causes of obesity include obesity genes and leptin theory, set-point theory, enzyme theory, fat cell theory, and theories centered around thermogenesis.*

Do People's Behaviors Cause Obesity?

A different line of research looks at whether obesity is determined by behavioral responses to environmental stimuli. These researchers ask what external forces might cause people to eat more food than they need.

External Cue Theory Being creatures of free will, people can override signals of satiety and hunger and eat whenever they wish, especially when presented with circumstances that stimulate them to do so. Almost everyone has had the experience of walking into a store for some small item, not feeling particularly hungry, and walking out snacking on a favorite treat. No doubt about it, most people will eat when presented with delectable foods, even if they are not hungry. A classic experiment showed that even animals respond in this way. Normal-weight rats rapidly became obese when fed "cafeteria style" on a variety of rich, palatable foods. Rats are known to maintain a precise, healthy weight when fed a standard diet of rat chow. People, too, may be prone to gain weight when their diets provide many kinds of rich, palatable foods. One study found a positive correlation between high body fatness and a diet presenting a wide variety of sweets, snacks, condiments, and main dishes.[54] The opposite was also true: subjects consuming a wide variety of vegetables, but not many treats, had lower body fatness.

One food constituent stands out in being perceived as palatable—fat. Not only does fat deliver more than twice the calories, gram for gram, as protein and carbohydrate, it is also stored preferentially by the body and with great efficiency. Of the three energy nutrients, fat also stimulates the least energy expenditure in diet-induced thermogenesis and may be the least satiating and thus lead to overconsumption of calories at each meal. A person whose diet contains much fat is often one who battles against overweight.[55]

Eating behavior also occurs in response to complex human sensations such as yearning, craving, addiction, or compulsion. For an emotionally insecure person, eating when lonely may be less threatening than calling a friend and risking rejection. Some people experience food cravings when feeling down or depressed. Food picks them up for a while. Some respond to other external stimuli such as the time of day ("I'm not hungry, but it's time for lunch.").

Any kind of stress can cause overeating, perhaps because **arousal** feelings are mistaken for hunger. ("What do I do when I'm grieving? Eat. What do I do when I'm celebrating? Eat!") The opposite can also be true, however. Some people undereat or cannot eat at all when under stress. Though all of these behaviors can easily lead to the overconsumption of food energy, they cannot fully explain obesity development because even thin people are susceptible to them.

arousal heightened activity of certain brain centers associated with attention, excitement, and anxiety.

Physical Inactivity Although, as mentioned, overweight people may spend more energy than normal-weight people in their regular daily activities, they less often engage in exercise.[56] Physical activity clearly reduces body fatness, builds healthy lean tissue, and reduces disease risks, yet the United States seems locked in what has been called an epidemic of inactivity.[57] For many people, television watching has all but replaced outdoor work and play as the major leisure time activity. In addition, sponsors run advertisements for delicious, high-fat (and low-nutrient) foods designed to spur the appetites of viewers. One study showed that in children, obesity increases by 2 percent per hour of television watching per day. Another revealed that watching television costs *less* energy than simply doing nothing.[58] The Fitness for Life feature near here underscores the importance of physical activity in weight management.

Physical activity can help to regulate the appetite, help overweight people lose fat, and help underweight people gain muscle.

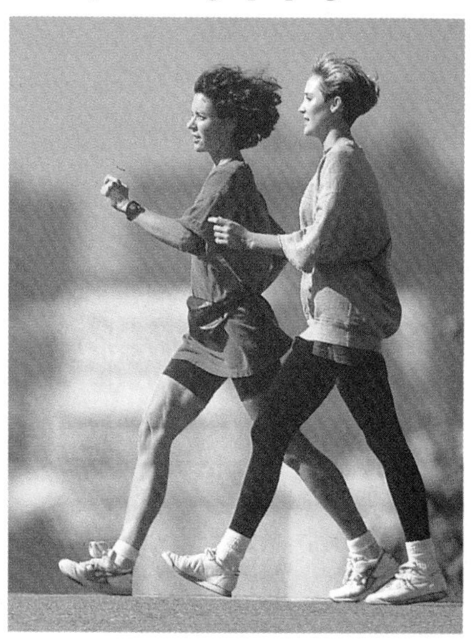

End of Story? Anyone involved in a good mystery wants to know how it ends. In the case of the causes of obesity, no one yet knows which of the suspects is the real culprit, and until evidence proves otherwise, any or all may be guilty as charged. A number of factors no doubt contribute to any one person's obesity, and the best treatment for most overweight people boils down to control in three areas: diet therapy, physical activity, behavior modification.[59]

Later sections focus on these areas. For people who have a BMI of 30 or over or who have many disease risk factors, weight-loss drugs may be appropriate as a means

Reminder: The three lifestyle components leading to healthy body weight are diet, physical activity, and behavior modification.

of achieving initial weight-loss goals; for those with a BMI of 40 or over, surgical approaches may be indicated to speed reversal of obesity's risks to health. In all cases, however, regain of lost weight is almost a certainty unless changes in food intake, physical activity, and behaviors become ingrained as a new way of living. The section following the Fitness for Life feature explains the mechanics of how the body gains or loses weight.

KEY POINT *Studies of human behavior identify stimuli that lead to eating and exercise habits. Physical inactivity is clearly linked with overfatness. Obesity treatment involves diet, exercise, and behavior modification.*

FITNESS FOR LIFE

Physical Activity for Building Up and Trimming Down

People trying to build up, slim down, or maintain weight (and doesn't that include everybody?) can rely on physical activity to help them achieve their goals. For those who are reducing, it can ensure that weight loss is mostly fat loss; for those who wish to gain, it can ensure that the gain is a healthy blend of muscle and fat. For the weight reducer, physical activity:

- Increases energy expenditure.
- Increases resting metabolic rate (slightly) over the long term.
- Accelerates loss of body fat.
- Helps regulate appetite.
- Helps control stress and stress-induced overeating or undereating with subsequent indulgences.
- Enhances self-esteem.

To help with fat loss, the activity must involve the voluntary moving of muscles, not passive motion such as being jiggled by a machine at a health spa or being massaged. The more muscles you move, and the more time you spend moving them, the more calories you spend and the more muscle tissue you build. Exercise using approximately 150 calories of energy per day, or about 1,000 calories per week, is considered to be moderate activity.[1] Table 9-5 shows the approximate calorie costs of various activities.

Some people believe that physical activity must be long and arduous to achieve fat loss or build muscle. Not so. A brisk, hour-long walk each day assists those who want to lose body fat. The greater the intensity, the more calories burned, of course, so an hour-long jog helps even more. In a given time frame, total energy spent, as well as total energy used from fat, is greater with higher-intensity activity.[2] To lose fat, then, expend as much energy as your time and stamina allow at first, and increase from there.

To build muscle and increase body weight, strengthening activities, such as calisthenics or weight training, can add bulk to muscles and lean tissue to the body. Be sure to work all the body's muscle groups to balance their strength, and start slowly with light weights. Starting out too fast and too hard invites failure. Build gradually to support the ultimate goal of sustaining a lifetime of physical activity.

Another strategy that trims fat and builds lean tissue is to incorporate more physical activity into your daily schedule in many simple, small-scale ways. Park the car at the far end of the parking lot; use the stairs instead of the elevator; work in the garden; work your abdominal muscles while you stand in line; fidget while sitting down; tighten your buttocks each time you get up from your chair. Individually, these activities make only tiny differences, but over a year's time the total becomes significant.

Physical activity for weight loss or maintenance:

1. Choose moderate activities.
2. Move large muscle groups.
3. Invest longer times in physical activity.
4. Adopt informal strategies to be more active.

Physical activity for building body mass:

1. Choose strength-building exercises.
2. Use a balanced exercise routine.
3. Perform exercises with increasing intensity.
4. Adopt informal strategies to be more active.

The next chapter provides details about exercising and weight maintenance.

Table 9-5

Energy Demands of Activities

Activity	Energy per Pound of Body Weight per Minute	Body Weight (lb) 110	125	150	175	200
	cal/lb/min[a]			cal/min[b]		
Aerobic dance (vigorous)	.062	6.8	7.8	9.3	10.9	12.4
Basketball (vigorous, full court)	.097	10.7	12.1	14.6	17.0	19.4
Bicycling						
13 miles per hour	.045	5.0	5.6	6.8	7.9	9.0
15 miles per hour	.049	5.4	6.1	7.4	8.6	9.8
17 miles per hour	.057	6.3	7.1	8.6	10.0	11.4
19 miles per hour	.076	8.4	9.5	11.4	13.3	15.2
21 miles per hour	.090	9.9	11.3	13.5	15.8	18.0
23 miles per hour	.109	12.0	13.6	16.4	19.0	21.8
25 miles per hour	.139	15.3	17.4	20.9	24.3	27.8
Canoeing (flat water, moderate pace)	.045	5.0	5.6	6.8	7.9	9.0
Cross-country skiing (8 miles per hour)	.104	11.4	13.0	15.6	18.2	20.8
Golf (carrying clubs)	.045	5.0	5.6	6.8	7.9	9.0
Handball	.078	8.6	9.8	11.7	13.7	15.6
Horseback riding (trot)	.052	5.7	6.5	7.8	9.1	10.4
Rowing (vigorous)	.097	10.7	12.1	14.6	17.0	19.4
Running						
5 miles per hour	.061	6.7	7.6	9.2	10.7	12.2
6 miles per hour	.074	8.1	9.2	11.1	13.0	14.8
7.5 miles per hour	.094	10.3	11.8	14.1	16.4	18.8
9 miles per hour	.103	11.3	12.9	15.5	18.0	20.6
10 miles per hour	.114	12.5	14.3	17.1	20.0	22.9
11 miles per hour	.131	14.4	16.4	19.7	22.9	26.2
Soccer (vigorous)	.097	10.7	12.1	14.6	17.0	19.4
Studying	.011	1.2	1.4	1.7	1.9	2.2
Swimming						
20 yards per minute	.032	3.5	4.0	4.8	5.6	6.4
45 yards per minute	.058	6.4	7.3	8.7	10.2	11.6
50 yards per minute	.070	7.7	8.8	10.5	12.3	14.0
Table tennis (skilled)	.045	5.0	5.6	6.8	7.9	9.0
Tennis (beginner)	.032	3.5	4.0	4.8	5.6	6.4
Walking (brisk pace)						
3.5 miles per hour	.035	3.9	4.4	5.2	6.1	7.0
4.5 miles per hour	.048	5.3	6.0	7.2	8.4	9.6
Wheeling self in wheelchair[c]	.030	3.3	3.75	4.5	5.25	6
Wheelchair basketball[c]	.084	9.2	10.5	12.6	14.7	16.8

[a]Use this column if you want to calculate calories spent for your own exact body weight. Multiply cal/lb/min by your exact weight and then multiply that number by the number of minutes spent in the activity. For example, if you weigh 142 pounds, and you want to know how many calories you spent doing 30 minutes of vigorous aerobic dance: .062 × 142 = 8.8 calories per minute. 8.8 × 30 (minutes) = 264 total calories spent.

[b]Use this column if you weigh 110, 125, 150, 175, or 200 pounds. This eliminates the need to calculate from column 1.

[c]Wheelchair values estimated from data in Table IV-4 of the National Heart, Lung, and Blood Institute Expert Panel, National Institutes of Health, *Clinical Guidelines on the Identification, Evaluation, and Treatment of Overweight and Obesity in Adults* (Washington, D.C: Government Printing Office, 1998).

How the Body Loses and Gains Weight

The balance between the energy you take in and the energy you spend determines whether you will gain, lose, or maintain body *fat*. When you step on the scale and note a change in body *weight* of a pound or two, however, this may not indicate a change in body fat. A change in weight can reflect shifts in body fluid content, in bone minerals, in lean tissues such as muscles, or in the contents of the bladder or digestive tract. It often correlates with the time of day: people generally weigh the least before breakfast. It is important for people concerned with weight control to realize that quick, large changes in weight are usually not changes in fat alone, or even at all.

The type of tissue lost or gained depends on how the person goes about losing or gaining it. To lose fluid, for example, one can take a "water pill" (diuretic), causing the kidneys to siphon extra water from the blood into the urine. Or one can engage in intense exercise while wearing heavy clothing in hot weather and lose abundant fluid in sweat. (Both practices are dangerous, incidentally, and are not being recommended here.) To gain water weight, a person can overconsume salt and water; for a few hours, the body will retain water until it manages to excrete the salt. (This, too, is not recommended.) Most quick weight-change schemes promote large changes in body fluids that register dramatic, but temporary, changes on the scale and accomplish little weight change in the long run.

Chapter 8 gave details about the body's water balance.

Moderate Weight Loss versus Rapid Weight Loss

When you eat less food energy than you need, your body draws on its stored fuel to keep going. It is a great advantage to be able to eat periodically, store fuel, and then use up that fuel between meals. The between-meal interval is normally about 4 to 6 waking hours—about the length of time the body takes to use up most of the available liver glycogen—or 12 to 14 hours at night, when body systems slow down and the need is less.

If a person exercises appropriately, moderately restricts calories, and consumes an otherwise balanced diet that meets protein and carbohydrate needs, the body will be forced to use up its stored fat for energy. Gradual weight loss will occur. This is preferred to rapid weight loss because lean body mass is spared and fat is lost.

The Body's Response to Fasting If a person doesn't eat for, say, three whole days or a week, then the body makes one adjustment after another. Soon, the liver's glycogen is essentially exhausted. Where, then, can the body obtain *glucose* to keep its nervous system going? Not from the muscles' glycogen because that is reserved for the muscles' own use. The underfed body must turn to the protein in its own lean tissues.

An alternative source of *energy* might be the abundant fat stores most people carry, but these are of no use to the nervous system. The muscles, heart, and other organs use fat as fuel, but at this stage the nervous system needs glucose. Most importantly, the body's major fuel, fat, cannot be converted to glucose—the body lacks enzymes for this conversion.[††] The body does, however, possess enzymes that can convert protein to glucose. Therefore, body proteins are sacrificed to supply raw materials from which to make glucose.

If the body were to continue to consume its lean tissue unchecked, death would ensue within about ten days. After all, in addition to skeletal muscle, the blood proteins, liver, heart muscle, and lung tissue—all vital tissues—are being burned as fuel. (In fact, fasting or starving people remain alive only until their stores of fat are gone or until half their lean tissue is gone, whichever comes first.) To prevent this, the body plays its last ace: it begins converting fat into compounds that the nervous system can

In early food deprivation:

- The nervous system cannot use fat as fuel; it can only use glucose.
- Body fat cannot be converted to glucose.
- Body protein can be converted to glucose.

In later food deprivation:

- Ketone bodies help feed the nervous system and so help spare tissue protein.

[††] Glycerol, which makes up 5 percent of fat, can yield glucose but is a negligible source.

adapt for use and so forestall the end. This process is ketosis, first mentioned in Chapter 4 as an adaptation to prolonged fasting or carbohydrate deprivation.

In ketosis, instead of breaking down fat molecules to carbon dioxide and water as it normally does, the body takes partially broken-down fat fragments and combines them to make **ketone bodies,** compounds that are normally rare in the blood. It converts some amino acids—those that cannot be used to make glucose—to ketone bodies, too. These ketone bodies circulate in the bloodstream and help to feed the brain, since about half of the brain's cells can make the enzymes needed to use ketone bodies for energy. After about ten days of fasting, the brain and nervous system can meet most of their energy needs using ketone bodies.

ketone bodies acidic compounds derived from fat and certain amino acids. Normally rare in the blood, they help to feed the brain during times when too little carbohydrate is available. Also defined in Chapter 4.

Thus, indirectly, the nervous system begins to feed on the body's fat stores. Ketosis reduces the nervous system's need for glucose, spares the muscle and other lean tissue from being devoured quickly, and prolongs the starving person's life. Thanks to ketosis, a healthy person starting with average body fat content can live totally deprived of food for as long as six to eight weeks. Figure 9-9 on page 334 reviews how energy is used during both feasting and fasting.

Respected, wise people in many cultures have practiced fasting as a periodic discipline. Clearly, the body tolerates short-term fasting, although there is no evidence that the body becomes internally "cleansed," as some believe. Ketosis may harm the body by upsetting the acid-base balance of the blood and by promoting mineral losses in the urine. Strong evidence also indicates that food deprivation leads to a tendency to overeat or even binge on food when food becomes available.[60] The effect seems to last beyond the point when weight is restored to normal. In addition, people with eating disorders (see Chapter 10's Controversy) often report that a fast or a severely restricted diet heralded the beginning of their loss of control over eating.

Side effects of low carbohydrate diets may include:

- Nausea.
- Fatigue.
- Constipation or diarrhea.
- Low blood pressure.
- Elevated uric acid (contributor to gout, a painful condition of the joints).
- Foul taste in the mouth.
- Bad breath.

For the person who wants to lose weight, fasting is not the best way. The body's lean tissues continue to be degraded. The body is deprived of nutrients it needs to assemble new enzymes, red and white blood cells, and other vital components. The body also slows its metabolism to conserve energy. A diet only moderately restricted in calories has actually been observed to promote a greater rate of *weight* loss, a faster rate of *fat* loss, and the retention of more lean tissue than a severely restricted fast.[61]

The Body's Response to Carbohydrate Restriction Any diet too low in carbohydrate will bring about responses that are similar to fasting. Many low-carbohydrate diets have been promoted to the public in many different guises. Each diet has enjoyed a surge of popularity thanks largely to a sizable initial weight loss. These diets are designed to throw a person into ketosis. The sales pitch is that "you'll never feel hungry" and that "you'll lose weight fast—faster than you would on any ordinary diet." Both claims are true, but also misleading. Loss of appetite accompanies any low-calorie diet. Severe calorie restriction means loss of water and lean tissue, and the water is rapidly regained when people begin eating normally again. People listed in a national registry of those who successfully lost an average of 30 pounds and maintained those losses for five years did so by eating more, not less, carbohydrate than those who were unsuccessful at weight loss or maintenance.[62] They also consumed substantially less fat.

Names of some low-carbohydrate diets: Atkins New Diet Revolution, Calories Don't Count Diet, Drinking Man's Diet, Mayo Diet, Protein-Sparing Fast, Scarsdale Diet, Ski Team Diet, Stillman Diet, and the Zone Diet. New ones keep coming out under new names, but they are essentially the same diet.

Even if a person can stick to a low-carbohydrate diet long enough to lose some body fat, the loss results from reduced intakes of food and calories, and not from any metabolic hocus-pocus as claimed in popular diet books. It's simple arithmetic: stripped of its carbohydrate-rich beans, tortilla wrapper, and chopped vegetables, a burrito is reduced to a tiny pile of ground beef. Likewise, a steak dinner emerges as a lone piece of meat and a green salad without the calories (and nutrients) of a whole-grain roll, fresh fruit, milk, or even a baked potato.

People who lose weight and keep it off report eating more, not less, carbohydrate-rich foods.

Aside from the obvious threats to nutrition, many hazards accompany low-carbohydrate diets: extraordinarily high intakes of cholesterol, fat, and saturated fat set the stage for high blood cholesterol; too little carbohydrate can bring on hypoglycemia; no milk or fruits and minimal vegetables rob the body of needed nutrients and phytochemicals; a lack of fiber may invite digestive tract ailments ranging from constipation to colon cancer. Some low-carbohydrate diets, particularly those called protein-sparing

Figure 9-9

Feasting and Fasting

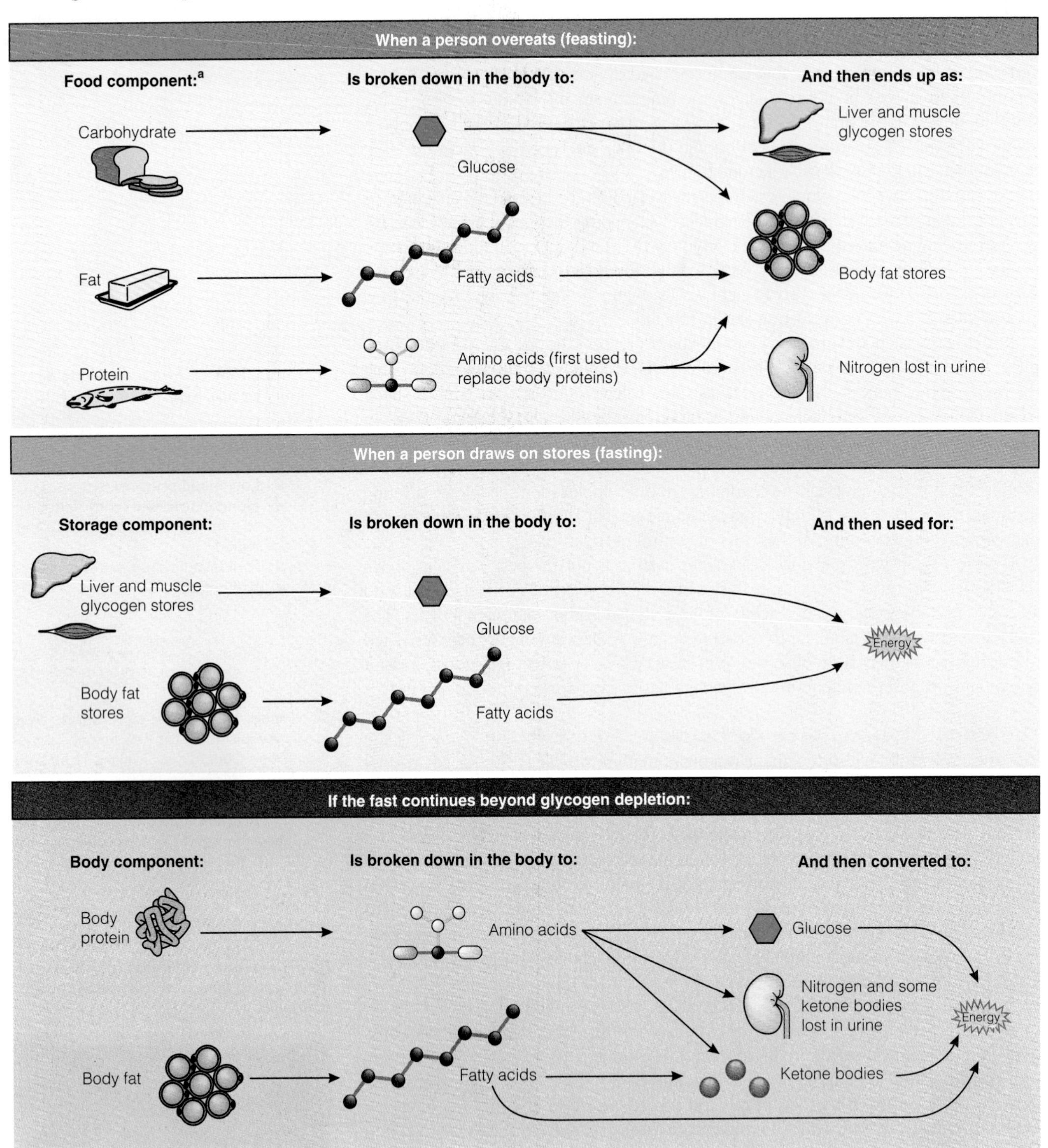

When a person overeats (feasting):
Food component:[a]
Is broken down in the body to:
And then ends up as:
Carbohydrate
Glucose
Liver and muscle glycogen stores
Fat
Fatty acids
Body fat stores
Protein
Amino acids (first used to replace body proteins)
Nitrogen lost in urine
When a person draws on stores (fasting):
Storage component:
Is broken down in the body to:
And then used for:
Liver and muscle glycogen stores
Glucose
Energy
Body fat stores
Fatty acids
If the fast continues beyond glycogen depletion:
Body component:
Is broken down in the body to:
And then converted to:
Body protein
Amino acids
Glucose
Nitrogen and some ketone bodies lost in urine
Energy
Body fat
Fatty acids
Ketone bodies

[a]Alcohol is not included because it is a toxin and not a nutrient, but it does contribute energy to the body. After detoxifying the alcohol, the body uses the remaining two-carbon fragments to build fatty acids and stores them as fat.

fasts, have caused heart failure. These diets are never recommended by knowledgeable practitioners.

KEY POINT *When energy balance is negative, glycogen returns glucose to the body. When glycogen runs out, body protein is called upon for glucose. Fat also supplies fuel as fatty acids. If glucose runs out, fat supplies fuel as ketone bodies, but ketosis can be dangerous. Both fasts and low-carbohydrate diets are ill-advised. People who successfully lose weight and keep it off eat abundant carbohydrates.*

Weight Gain

What happens inside the body when a person does not use up all of the food energy taken in? What does the body do with the excess? Previous chapters have already provided the answer; the energy-yielding nutrients contribute to body stores as follows:

- Carbohydrate (other than fiber) is broken down to sugars for absorption. In the body tissues, excesses of these may be built up to *glycogen* and stored, burned off as heat, or converted to *fat* and stored.
- Fat is broken down to glycerol and fatty acids for absorption. Inside the body, these are especially easily stored as body *fat*.
- Protein is broken down to amino acids for absorption. Inside the body, these may be used to replace lost body *protein* and, in a person who is exercising, to build new muscle and other lean tissue. This protein must be functioning protein; excess protein is not passively stored. Excess amino acids can have their nitrogen removed and be used for energy or be converted to *glucose* or *fat*, mostly fat.
- Alcohol is easily absorbed intact and is converted into body fat for storage.

Note that although three kinds of energy-yielding nutrients and alcohol may enter the body, they become only two kinds of energy stores: glycogen and fat. Glycogen stores amount to about three-fourths of a pound; fat stores can, of course, amount to many pounds. Note, too, that when excess protein is converted to fat, it cannot be recovered later as protein because the nitrogen is stripped from the amino acids and excreted in the urine. Thus, if you eat enough of any food, whether it's steak, brownies, or baked beans, any excess will be turned to fat within hours. Weight gain comes from spending less food energy than is taken in. Weight may be gained as body fat or as lean tissue, depending largely upon whether the eater is also exercising.

Chapter 10 further discusses muscle gains in response to exercise.

As for ethanol, the alcohol of alcoholic beverages, it has been shown to slow down the body's use of fat for fuel by as much as a third, causing more fat to be stored. The storage is primarily in the visceral fat tissue of the "beer drinker's belly" and also on the thighs, legs, or anywhere the person tends to store surplus fat. Alcohol therefore is fattening, both through the calories it provides and through its effects on fat metabolism.[‡‡] The obvious conclusion is that weight control and daily alcohol intake cannot coexist.[63]

It is worth emphasizing these points by repeating them:

- Any food can make you fat if you eat enough of it. A net excess of energy is almost all stored in the body as fat in fat tissue.
- Fat, as opposed to carbohydrate or protein, from food is especially easy for the body to store as fat tissue.
- Protein is not stored in the body except in response to exercise; it is present only as working tissue. Excess protein is converted to fat and stored as such.
- Alcohol both delivers calories and encourages storage of body fat.
- Too little physical activity encourages body fat accumulation.

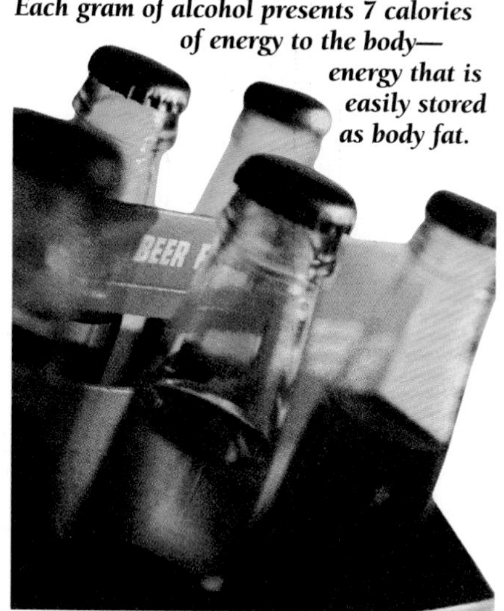
BEER

Each gram of alcohol presents 7 calories of energy to the body—energy that is easily stored as body fat.

[‡‡]People addicted to alcohol are often overly thin because of diseased organs, depressed appetite, and subsequent malnutrition.

KEY POINT *When energy balance is positive, carbohydrate is converted to glycogen or fat, protein is converted to fat, and food fat is stored as fat. Alcohol delivers calories and encourages fat storage.*

Achieving and Maintaining a Healthy Body Weight

Before setting out to change your body weight, think about your motivation for doing so. Many people in our society are dissatisfied with their body weights, not because of potential health risks but because their weight fails to meet society's idea of attractiveness. Yet people arrive in this world with varying weight tendencies; just as some tend to be tall and others short, some tend to be lean and others stout. No one expects tall people to grow shorter or short people to grow taller to become "normal."

The adoption of health as the ideal, rather than some ill-conceived image of beauty, could avert much misery. Fashion ideals can be out of reach for most people. The truth is that the human body is not infinitely malleable—few overweight people will ever become rail-thin, even with the right diet, exercise habits, and behaviors. Likewise, most underweight people will remain on the slim side even after putting on some heft. Sadly, many people adhere to the false belief that happiness, career advancements, wealth, and companionship would be within their reach, if only their bodies were of the "right" size.

Beyond issues of style, people who are overweight may have major health concerns. For them, the benefits of weight loss are clear: fewer heart attacks, fewer strokes, less cancer, less diabetes, and a greater chance at a longer life than otherwise would be expected.

What Diet Strategies Are Best for a Healthy Body Weight?

This section reveals diet-related changes that most often lead to successful weight change and maintenance. It is written in terms of advice to "you." The idea is not to put you under pressure to take the advice personally but to give you the illusion of listening in on a conversation in which an overweight person is benefiting from competent dietary counseling.

Keeping Records The best diet is one tailored to your individual needs. To that end, keeping records is of great importance. Recording your food intake and exercise habits can help you to spot trends and identify areas needing improvement. To monitor changes in body composition, body weight can serve as a quick indicator of changes in body fatness. In addition to weight, waist circumference can be measured to track changes in central adiposity.

Setting Goals No matter what your purpose for controlling your weight, setting goals helps to achieve the desired result. For example, three broad goals have been identified as important for obese people who must reduce their weight to reduce their disease risks:

1. Reduce body weight by about 10 percent over half a year's time.
2. Maintain a lower body weight over the long term.
3. At a minimum, prevent further weight gain.[64]

Whether your goals involve weight gain, loss, or maintenance, set broad goals in terms of end results that you wish to achieve, and then set smaller step goals in terms of the dietary, physical activity, and behavioral changes necessary to achieve the desired result. These changes do not produce a dramatic weight loss overnight, but the person who faithfully employs them can lose a pound or two of body fat each week, safely and effectively. Losses greater or faster than this are not recommended because they are almost invariably followed by just as rapid regain. Also, the person who reduces rapidly and restricts fats excessively may suffer from stones forming in the gallbladder or from dangerous electrolyte imbalances.[65] It's better to take your time and achieve a lasting change.

In addition to quick-fix diets, weight-loss gimmicks abound in the marketplace (see the margin). Most are ineffective, and some are truly dangerous. As for drugs and surgery, these are reserved for efforts at saving the lives of obese people at critical risk, as the Consumer Corner points out. How, then, can a person go about choosing a diet to help in weight loss? Table 9-6 provides a way to choose a safe, effective weight-loss plan according to standard nutrition principles, and the next section provides details.

Planning a Diet for Weight Loss and Maintenance No particular food plan is magical, and no particular food must be either included or excluded. You are the one who will have to live with the plan, so you had better be the one to design it. Remember, you are not "going on a diet" because you will not be "going off" your eating plan. Instead, you are adopting a healthy eating plan for life, so it must consist of foods that you like, that are readily available, and that you can afford.

Ineffective or dangerous weight-loss gimmicks:

- Diet pills.
- Expanding pills.
- Glucomannan, bee pollen, spirulina.
- Herbs, such as ma huang, ephedrine, and dieter's teas.
- Hormones.
- Laxatives.
- Liposuction, lipectomy.
- Massages, muscle stimulators.
- Spa belts, rollers, saunas, whirlpools.
- Thigh-reducing cream.

Table 9-6

Rating Sound and Unsound Weight-Loss Schemes

Start by giving each diet or program 160 points. Subtract points as instructed, whenever a plan falls short of ideals. A plan that loses more than 20 points might still be of value, but deserves careful scrutiny.

Does the diet or program:

1. Provide a reasonable number of calories (not fewer than 1,200 calories for an average-size person)? If not, give it a minus 10.
2. Provide enough, but not too much, protein (at least the recommended intake, but not more than twice that much)? If no, minus 10.
3. Provide enough fat for balance but not so much fat as to go against current recommendations (about 30% of calories from fat)? If no, minus 10.
4. Provide enough carbohydrate to spare protein and prevent ketosis (100 grams of carbohydrate for the average-size person)? Is it mostly complex carbohydrate (not more than 10% of the calories as concentrated sugar)? If no to either or both, minus 10.
5. Offer a balanced assortment of vitamins and minerals by including foods from all food groups? If it omits a food group (for example, meats), does it provide a suitable food (not supplement) substitute? Count five food groups in all: milk/milk products; meat/fish/poultry/eggs/legumes; fruits; vegetables; and breads/cereals/grains. For *each* food group omitted and not adequately substituted for, subtract 10 points.
6. Offer variety, in the sense that different foods can be selected each day? If you'd class it as boring or monotonous, give it a minus 10.
7. Consist of ordinary foods that are available locally (for example, in the main grocery stores) at the prices people normally pay? Or does the dieter have to buy special, expensive, or unusual foods to adhere to the diet? If you would class it as "bizarre" or "requiring special foods," minus 10.
8. Promise dramatic, rapid weight loss (substantially more than 1% of total body weight per week)? If yes, minus 10.
9. Encourage permanent, realistic lifestyle changes, including regular exercise and the behavioral changes needed for weight maintenance? If not, minus 10.
10. Misrepresent salespeople as "counselors" supposedly qualified to give guidance in nutrition and/or general health without a profit motive, or collect large sums of money at the start, or require that clients sign contracts for expensive, long-term programs? If so, minus 10.
11. Fail to inform clients about the risks associated with weight loss in general or the specific program being promoted? If so, minus 10.
12. Promote unproven or spurious weight-loss aids such as human chorionic gonadotrophin hormone (HCG), starch blockers, diuretics, sauna belts, body wraps, passive exercise, ear stapling, acupuncture, electric muscle stimulating (EMS) devices, spirulina, amino acid supplements (e.g., arginine, ornithine), glucomannan, appetite suppressants, "unique" ingredients, and so forth? If so, minus 10.

CONSUMER CORNER

Surgery and Drugs for Weight Loss

PEOPLE WHO suffer from obesity may seek help from surgery, prescription or over-the-counter pills, and other products. Weight-loss help from surgery is available for those who suffer from serious diseases brought on by their severe obesity. Such people are likely to die early unless drastic measures are taken to improve their weight.[1] In people who face elevated risks of chronic diseases, such as heart disease or diabetes, and who have not succeeded in weight loss unaided, prescription medications may be effective, although the war chest of effective pharmaceutical agents has diminished in recent years.[2]

Two surgical procedures performed today—**gastric bypass** and **gastroplasty**—seem to offer some hope. (Table 9-7 defines these and other terms related to weight-loss surgery and drugs.) In bypass surgery, the digestive tract is shortened by joining a section of the small intestine directly to the stomach, this procedure can reduce risks associated with coronary artery disease. In gastroplasty (sometimes called stomach stapling), the stomach's volume is constricted; this can also reverse a trend toward worsening health as a result of obesity. In one study, about a third of those who received surgical procedures achieved long-term success, defined as five years of reduced weight.[3] The other two-thirds of surgical patients, though, failed to lose weight (although they reported improved health and attitude), failed to maintain their losses, or did not survive the surgery. Other studies report that up to half of patients may benefit from surgery.

Early reports of a new procedure, **gastric banding,** seem promising.[4] The new technique may be safer than the traditional surgeries because the surgeon makes only two tiny incisions in the abdomen and through them applies a restricting band or pouch around the stomach organ. So far, patients receiving this type of surgery seem to have less pain, faster recovery, and fewer complications than those having more extensive surgeries, and many lose weight because the band causes them to feel full after eating just a small amount of food.

Surgeons carefully screen obesity candidates for surgery and accept only the very obese (BMI greater than 40) whose need for weight loss outweighs the threat of possible risks from surgery

Table 9-7

Terms Related to Weight-Loss Surgery and Drugs

- **gastric banding** a surgical means of producing weight loss by restricting stomach size with a constricting band or pouch; used in people whose severe obesity brings extreme health risks.
- **gastric bypass** surgery that reroutes food from the stomach to the lower part of the small intestine; creates a chronic, lifelong state of malabsorption by preventing normal digestion and absorption of nutrients.
- **gastroplasty** surgery that partitions the stomach by stapling off a "pouch" or otherwise modifying the stomach, thereby reducing total food intake.
- **phenylpropanolamine (PPA)** a stimulant of the sympathetic nervous system used as a weight-loss agent and available in over-the-counter medications.
- **serotonin** a compound related in structure to (and made from) the amino acid tryptophan. It serves as one of the brain's principal neurotransmitters.

and whose motivation to make the necessary lifestyle changes is sufficient for success. During surgery, obese people face more infections, more respiratory problems, more blood clots, more difficulties in anesthesia, and slower wound healing than do normal-weight patients. In addition to immediate risks of surgery, the person may face lifelong problems of severe diarrhea, frequent vomiting, intolerance to sweets and milk, dehydration, limited dietary intake, and multiple nutrient deficiencies even with medical follow-up care.[5] Staples pull out, stomach pouches enlarge, and gastric banding must sometimes be repeated with repeated risks. Lifelong medical supervision is critical for those who choose the surgical route, but for up to half of those who choose it, the benefits of weight loss prove worth the risks.

Possibly less risky than surgery for help in weight loss are prescription drugs. Most prescription drugs cause either a drop in food intake by suppressing appetite or an increase in the use of fat for fuel through speeded-up metabolism. Though some drugs have succeeded in promoting modest weight loss in the first six months of treatment, their long-term effectiveness is uncertain, and their long-term use involves unknown risks.[6] This uncertainty places practitioners in a bind because treatment of obesity, once begun, is a long-term endeavor, not something to be accomplished in just a few months and then abandoned.[7]

The now infamous appetite-suppressing drugs, phentermine and fenfluramine (phen-fen for short), provide a dramatic example of the potential problems. These drugs suppress appetite by increasing an appetite-suppressing neurotransmitter, **serotonin,** in the brain. Soon after the approval of these drugs, it was discovered that people taking them, whether for relief of obesity or to gain fashionable figures, were likely to develop a dangerous condition of elevated blood pressure in the lungs (pulmonary hypertension) or to sustain damage to the valves of the heart.[8] Both conditions are life-threatening. Within a year the Food and Drug Administration (FDA) had determined that phen-fen posed an unacceptable risk and recalled the drugs from the market. Anyone who has used phen-fen is urged to seek evaluation by a physician, for the damage to the valves of the heart often remains symptomless.

Another appetite-suppressing drug, sibutramine, remains in use.* Sibutramine works on the brain's neurotransmitters, too, but in other ways. It raises blood pressure, so the FDA cautions those with hypertension against using it. Anyone taking sibutramine should monitor blood pressure carefully. As more becomes known about the molecular chemistry of appetite control, safer appetite-suppressing drugs may be developed.

The drug orlistat, recently approved by the FDA, takes a different approach to weight loss.† Not an appetite suppressant, orlistat inhibits the production of fat-digesting enzymes in the pancreas and so reduces fat absorption by about 30 percent.[9] As a result of absorbing less fat, people often lose weight.[10] The problem with undigested fat is that it must exit the digestive tract intact, carrying with it dissolved fat-soluble vitamins and phytochemicals that would otherwise have been absorbed by the body. Possible effects of orlistat resemble those attributed to the artificial fat olestra—diarrhea, digestive distress, and leakage of smelly, undigested fat from the anus.[11]

For most people, surgery or prescription drugs are not the answer. What about over-the-counter weight-loss pills? Two ingredients are approved by the FDA for inclusion in over-the-counter weight-loss drugs. One, **phenylpropanolamine (PPA),** also found in cold medications, carries a side effect of elevated blood pressure.[12] PPA makes some people feel jumpy because one of its actions is to trigger the body's stress response. Currently under investigation are the deaths of three apparently healthy women from brain hemorrhage

*Sibutramine's trade name is Meridia.

†Orlistat's trade name is XENICAL.

within several hours after taking a prescribed dose of PPA.[13] People taking PPA should know their blood pressure and monitor it carefully. The other approved ingredient, the anesthetic benzocaine (usually in gum or candy form), numbs the taste buds and supposedly reduces the desire to taste food. These two drugs are still under study by the FDA and may be removed from shelves if they prove ineffective or unsafe.

As for water pills (diuretics), they do nothing to solve a fat problem, as the chapter made clear. Likewise, a "thigh cream" on the market contains an asthma medication that two obesity researchers, citing two small unpublished studies, claim shrinks fat on women's thighs. This cream is considered a cosmetic, not a drug, so no proof of effectiveness is required.

Wildly popular but unproved for effectiveness or safety are herbal weight-loss products sold as "dietary supplements" instead of drugs and therefore not subject to rigorous FDA scrutiny. People are drawn to these products because they believe, falsely, that herbs grown in nature are never harmful to the body. Of course, this is not the case—many poisonous herbs and toxins, such as belladonna and hemlock, are found in nature. Furthermore, because weight-loss herbs are marketed as "dietary supplements," manufacturers need not present a shred of evidence of their safety or effectiveness to the FDA. Evidence about safety is gathered only through reports of actual consumers who sicken or even die after purchasing and using herbal remedies. A good example is ephedrine, which showed promise as a weight-loss drug in a limited number of studies. Without further ado, manufacturers rushed to market with ma huang or "herbal fen-phen," ephedrine-containing herbs combined with caffeine from coffee bean extract or aspirin as willow extract. Since that time, more than 800 victims have reported ill effects, ranging from headaches and vomiting to heart attacks and brain hemorrhage.[14] More than 15 people have died. As of this writing, Canada has banned ma huang, and the FDA has issued a warning against ephedrine-containing weight-loss herbs and products. The World Health Organization has called for regulation and control of these products. Nevertheless, they remain freely available for purchase by unsuspecting consumers.[15] Don't take them. The risks of doing so are too high.

In addition to ephedrine, herbal preparations containing senna, aloe, rhubarb root, cascara, castor oil, or buckthorn are sold as "dieter's tea" because their laxative effect may cause a temporary water loss of a pound or two. Users commonly report nausea, vomiting, diarrhea, cramping, and fainting. Such "teas" may have contributed to the deaths of four women who used them and also drastically reduced their food intakes.[16] Some herbs may turn out to be useful for some purposes, but those sold for weight loss clearly fail the safety test.

In the end, the only means of reducing body fat is to shift the energy budget from positive to negative, to take less energy in and put more energy out. Nonprescription diet pills, diuretics, illegal hormones, fat-melting creams, and other gimmicks are useless. They enjoy brisk sales only because the lifelong effort required for weight control is difficult and people can be lured by promises of quick and easy weight loss.

Guidelines for a weight-loss diet are outlined in Table 9-8. For all but the severely obese, a deficit of 500 to 1,000 calories per day will produce the desired loss. This amounts to an intake of about 1,200 calories per day for most people.[66] Figure 9-10 suggests daily food servings to build a balanced 1,200-calorie diet. Diets providing energy intakes lower than about 800 calories, the so-called very-low-calorie diets (VLCD), are notoriously unsuccessful at achieving lasting weight loss and can be dangerous, and so are not recommended.[67]

If you plan resolutely to include the number of servings of food from each food group that you need each day, you will find that you will have little appetite left for

Table 9-8

Recommendations for a Weight-Loss Diet

Nutrient	Recommended Intake
Calories[a]	Approximately 500 to 1,000 calories per day reduction from usual intake
Total fat	30% or less of total calories
Saturated fatty acids[b]	8 to 10% of total calories
Monounsaturated fatty acids	Up to 15% of total calories
Polyunsaturated fatty acids	Up to 10% of total calories
Cholesterol[b]	300 mg or less per day
Protein[c]	Approximately 15% of total calories
Carbohydrate[d]	55% or more of total calories
Sodium chloride	No more than 2,400 mg of sodium or approximately 6 g of sodium chloride (salt) per day
Calcium	1,000 to 1,500 mg per day
Fiber[d]	20 to 30 g per day

[a]For people with BMI of 35 or above. For those with BMI in the range of 27 to 35, a decrease of 300 to 500 calories per day will result in ½ to 1 pound lost per week.
[b]People with high blood cholesterol should aim for less than 7 percent calories from saturated fat and 200 milligrams cholesterol per day.
[c]Protein should be derived from plant sources and lean sources of animal protein.
[d]Carbohydrates and fiber should be derived from vegetables, fruits, and whole grains.
SOURCE: National Heart, Lung, and Blood Institute Expert Panel, National Institutes of Health, *Clinincal Guidelines on the Identification, Evaluation, and Treatment of Overweight and Obesity in Adults* (Washington, D.C.: Government Printing Office, 1998), p. 74.

high-fat or empty-calorie foods. Foods such as fruits, vegetables, and whole grains are high in carbohydrates and fiber, low in fat, and take a lot of chewing, too. Crunchy, wholesome, unprocessed or lightly processed foods offer bulk and satiety for far fewer calories than smooth, refined foods. Limit, but don't eliminate, lean meats or other low-fat protein sources: an ounce of lean ham contains more calories than an ounce of bread, but the ham produces greater satiety.

Remember to pay careful attention to portion sizes. The monstrous helpings served by restaurants and others are the enemy of the person controlling weight. Especially don't lose track of the fat in foods. Fat calories add up quickly and probably contribute more to body fat stores than do carbohydrate calories, and fat contributes little to the eater's feelings of fullness during a meal.[68] The Food Feature of Chapter 5 is an excellent resource for the person trying to control fat. Beware, though, of overdoing foods that are manufactured to be low in fat but make up the calories with sugar. Many people have found out the hard way that reducing fat without restricting calories does not produce weight loss. They may eat less fat, but calorie intakes remain high or even increase. (Compare the calories of yogurts in Figure 9-11 on the next page.) Read labels, compare the calories per serving among similar foods, and choose the lowest-calorie item of the bunch.

For those who drink alcoholic beverages, cutting down or eliminating alcohol is an obvious way to eliminate many unneeded calories and to enhance the use of fat for energy. This step can also make room in the diet for more nutritious foods.

Three meals a day is standard in our society, but no law says you can't have four or five—only be sure they are smaller, of course. People who eat small, frequent meals are reported as successful at weight loss and maintenance.[69] Make sure that mild hunger, not appetite, is prompting you to eat. Eat regularly, and eat before you become extremely hungry. When you do decide to eat, eat the entire meal you have planned for yourself. Then don't eat again until the next meal or snack. Save calorie-free or favorite foods or beverages for a planned snack at the end of the day if you need insurance against late-evening hunger.

One meal you should strive to include is breakfast. Much evidence supports the health effects of breakfast, and people who eat breakfast seem to need fewer snacks and consume less fat all day long.[70]

Figure 9-10

Suggested Daily Servings for an Adequate 1,200-Calorie Diet[a]

Choose the lowest-calorie, fat-free, or lowest-fat options from each group. Strictly limit serving sizes to those specified in Figure 2-5 of Chapter 2. To further reduce calories, reduce servings of fats and added sugars.

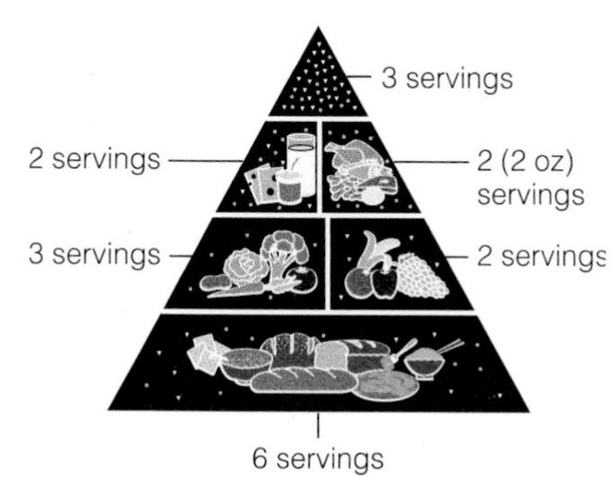
3 servings
2 servings
2 (2 oz) servings
3 servings
2 servings
6 servings

[a]Assumes no alcohol intake.

Figure 9-11

Comparing Yogurts: Fat, Carbohydrates, Calories

The calorie values of the whole-milk and fat-free yogurts are almost the same. When sugar is controlled as well as the fat, calories are significantly reduced, so the wise calorie watcher pays attention to more than just fat.

Whole Milk Fruit Yogurt

Nutrition Facts	
Serving Size 1 container (170g)	
Amount Per Serving	
Calories 175	Calories from Fat 15
	% Daily Value*
Total Fat 1.5g	3%
Cholesterol 10mg	3%
Sodium 105mg	4%
Total Carbohydrate 33g	11%
Sugars 27g	
Protein 7g	16%

Fat-Free Fruit Yogurt

Nutrition Facts	
Serving Size 1 container (170g)	
Amount Per Serving	
Calories 160	Calories from Fat 0
	% Daily Value*
Total Fat 0g	0%
Cholesterol <5mg	1%
Sodium 105mg	4%
Total Carbohydrate 33g	11%
Sugars 27g	
Protein 7g	16%

Fat-Free, Reduced Sugar Fruit Yogurt

Nutrition Facts	
Serving Size 1 container (170g)	
Amount Per Serving	
Calories 90	Calories from Fat 0
	% Daily Value*
Total Fat 0g	0%
Cholesterol <5mg	1%
Sodium 105mg	4%
Total Carbohydrate 16g	5%
Sugars 8g	
Protein 7g	16%

KEY POINT *To achieve and maintain a healthy body weight, set realistic goals, keep records, progress slowly, and avoid using pills, herbs, or gimmicks that may endanger health. Make the diet adequate, limit fat and calories, reduce alcohol, and eat regularly.*

What Diet Strategies Can Help Me to Gain Weight?

Should an underweight person try to gain weight? Not necessarily. If you are healthy at your present weight, stay there. If your physician has advised you to gain, if you are excessively tired, if you are unable to keep warm, if you fall into the "underweight" category of the BMI table (see inside back cover), or if, for women, you have missed at least three consecutive menstrual periods, you may be in danger from a too-low body weight.

For those who wish to gain, a healthful weight gain can be achieved through physical activity, particularly strength training (see Chapter 10 for details), combined with a high-calorie diet. Diet alone can bring about weight gain, but the gain will be mostly fat. For someone facing a wasting disease, the gain of fat tissue may be a welcome sign of improvement. For an athlete, however, such a gain can impair performance. Therefore, for most people, physical activity is an essential component of a sound weight-gain plan. Many an underweight person has simply been too busy (for months) to eat or to exercise enough to gain or to maintain weight.

As important to weight gain as exercise are the calories to support that activity—otherwise you will lose weight (body fat). If you eat just enough to fuel the activity, you will build muscle, but at the expense of body fat; that is, fat will be burned to support the muscle building. If you eat more, you will gain both muscle and fat. To gain a pound of muscle and fat requires taking in about 3,000 extra calories.[§§] Conventional advice on diet to the person building muscle is to eat about 700 to 1,000 calories a day above normal energy needs; this is enough to support both the added activity and the formation of new muscle.

The weight gainer needs nutritious calorie-dense foods. No matter how many sticks of celery you consume, you won't gain weight because celery simply doesn't offer enough calories. Calorie-dense foods (the very ones the weight-loss dieter is trying to avoid) are high in fat, but if they are contributing energy that will be spent building new tissue and if the fat is mostly unsaturated, they will not contribute to heart disease. Choose peanut butter instead of lean meat, avocado instead of cucumber, olives instead of pickles, whole-wheat muffins instead of whole-wheat bread, and milkshakes instead of milk. When you do eat celery, stuff it with tuna salad (use oil-packed tuna); choose flavored coffee drinks over plain coffee; use olive oil or canola oil dressings on salads, whipped toppings on fruit, soft or liquid margarine on potatoes, and the like. Because fat contains more than twice as many calories per teaspoon as sugar, it adds calories without adding much bulk, and its energy is in a form that is easy for the body to store.

Expect to feel full, sometimes even uncomfortably so. Most underweight individuals are accustomed to small quantities of food. When they begin eating significantly more food, they complain of uncomfortable fullness. This feeling is normal, and it passes as the stomach gradually adapts to the extra food.

Eat frequently. Make three sandwiches in the morning and eat them between classes in addition to the day's three regular meals. Spend time making foods appealing—the more varied and palatable, the better. If you fill up fast during a meal, start with the main course or a meat- or cheese-filled appetizer, not carrot sticks. Drink between meals, not with them, to save space for higher-calorie foods. Make milkshakes of milk, frozen bananas, and flavorings to drink between meals. Always finish with dessert. Be aware that most "weight-gain" supplements designed to add body weight are useless without physical activity and confer no special benefits on the taker. Of body weight gained in a day, only a half ounce to an ounce is protein tissue, so no special protein supplements can help speed weight gain. Ordinary food in abundance along with exercise to work the nutrients into place supports efforts to gain weight.

[§§]Theoretically, it takes an excess of 2,000 to 2,500 calories to gain a pound of pure lean tissue and about 3,500 calories to gain a pound of fat.

Smoking tobacco depresses the appetite and makes taste buds and olfactory (smelling) organs less sensitive. A person who smokes should quit before trying to gain weight. Quitters find that appetite picks up, food tastes and smells better, and the body reaps additional benefits too numerous to mention.

For the person trying to gain weight, physical activity can ensure a healthy gain. The Fitness for Life feature already explained how important physical activity can be to improving body composition. Changes in daily habits such as playing sports or working out must become as routine as brushing your teeth; they are not temporary measures to "solve" a weight problem. The Food Feature explores how, exactly, a person can modify daily behaviors into healthy lifelong habits.

KEY POINT *Weight gain requires a diet of calorie-dense foods, eaten frequently through the day.*

self-efficacy a person's belief in his or her ability to succeed in an undertaking.

weight cycling repeated rounds of weight loss and subsequent regain, with reduced ability to lose weight with each attempt. Also called *yo-yo dieting*.

People's hormone status also affects their ability to gain weight, but taking steroids and other drugs to enhance weight gain is a bad idea—see Chapter 10.

Once I've Changed My Weight, How Can I Stay Changed?

One reason gimmicks fail in weight control is that they fail to produce lasting change. Millions have experienced the frustration of achieving a desired change in weight only to see their hard work visibly slipping away: "I have lost 200 pounds, but I was never more than 20 pounds overweight." Disappointment, frustration, and self-condemnation are common in dieters who find they have slipped back to their original weight or even higher. What makes the difference between a successful, long-term weight-control program and one that doesn't stick? How can one go about maintaining a healthy body weight?

A key to weight maintenance is acceptance of the chronic nature of weight management. Acceptance of weight control as a lifelong endeavor, and not a goal to be achieved and then forgotten, helps prepare the mind for making permanent changes. For example, people who maintain a loss continue to employ the behaviors that reduce calorie intakes and increase expenditures through exercise.[71] They cultivate the habits and attributes of people who maintain a healthy weight, such as eating diets higher in carbohydrate, lower in fat, and higher in fruits and vegetables than average.[72] They also often eat several small meals throughout the day before their hunger becomes severe.[73] Those who maintain healthy weight also:

- Are more physically active than the average person.[74]
- Monitor fat grams, caloric intake, and body weight.
- Believe they have the ability to control their weight, an attribute known as **self-efficacy**, even in the face of previous failures.
- Develop social support systems.
- Eat controlled portions at planned times, and eat them at a leisurely pace.
- Eat high-fiber foods, and consume sufficient water each day.

Also important are realistic expectations regarding body size and shape. The importance of exercise cannot be overstated. When people try to lose body fat or maintain a loss without physical activity, they fight an uphill battle.[75] Those who endeavor to lose weight without exercise often become trapped in **weight cycling**, endless repeating rounds of weight loss and regain sometimes called "yo-yo" dieting. In fact, previous weight-cycling history can predict a person's success (or lack thereof) in maintaining weight.[76]

Self-acceptance also predicts success, while self-hate predicts failure. A paradox of behavior change is that it takes self-acceptance (loving the overweight self) to lay the foundation for changing that self. Once body weight is changed, further improvements are seen in many areas of life.[77] Self-acceptance is the basis of a beneficial cycle.

KEY POINT *People who succeed at maintaining lost weight keep to their eating routines, keep exercising, and keep track of calorie and fat intakes and body weight. The more traits related to positive self-image and self-efficacy a person possesses or cultivates, the more likely that person will succeed.*

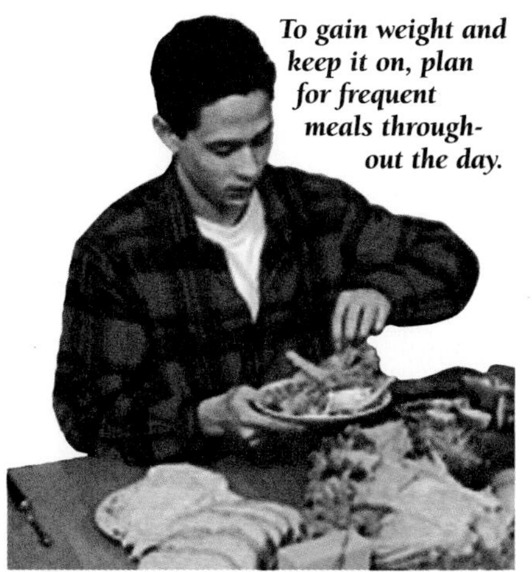
To gain weight and keep it on, plan for frequent meals throughout the day.

FOOD
FEATURE

Behavior Modification for Weight Control

SUPPORTING BOTH diet and exercise is the technique of **behavior modification**, which cements into place all the behaviors that lead to and perpetuate the desired body composition. Behavior modification is based on the knowledge that habits drive behaviors.

How Does Behavior Modification Work?

The following example illustrates behavior modification at work. Suppose a friend tells you about a shortcut to class. To take it, you must make a left-hand turn at a corner where you now turn right. You decide to try the shortcut the next day, but when you arrive at the familiar corner, you turn right as always. Not until you arrive at class do you realize that you failed to turn left, as you had planned. You can learn to turn left, of course, but at first you will have to make an effort to remember to do so. After a while, the new behavior will become as automatic as the old one was.

For those striving to lose weight, learning to say "No, thank you" might be among the first habits to establish. Learning not to "clean your plate" might be another. Once you identify the behaviors you need to change, do not attempt to modify all of them at once. No one who attempts too many changes at one time is successful. Set your priorities and begin with a behavior you can handle; then practice it until it becomes habitual and automatic. Then select another.

Applying Behavior Modification

Behavior researchers have identified six elements of behavior modification to use to replace old eating habits with new ones:

1. Eliminate inappropriate eating cues.
2. Suppress the cues you cannot eliminate.
3. Strengthen cues to appropriate eating and exercise.
4. Repeat the desired eating and exercise behaviors.
5. Arrange or emphasize negative consequences of inappropriate eating.
6. Arrange or emphasize positive consequences of appropriate eating and exercise behaviors.

Table 9-9 provides specific examples of how these six elements can be put into action. Before doing so, however, you must establish a baseline, a record of your present eating behaviors against which to measure future progress. Keep a diary so that you can learn what particular eating stimuli, or cues, affect you.

To begin, set about eliminating or suppressing the cues that prompt you to eat inappropriately. An overeater's life may include many such cues: watching television, talking on the telephone, entering a convenience store, studying late at night, and many more. Resolve that you will no longer respond to such cues by eating. Respond only to one set of cues designed by you, in one particular place in one particular room. If some cues to inappropriate eating behavior cannot be eliminated, suppress them, as described in Table 9-9; then strengthen the appropriate cues, and reward yourself for doing so. The list in the margin suggests some activities and rewards to substitute for eating.

As you progress in your new behaviors of physical activity and sensible eating, enjoy your new, emerging self. Get in touch with and reach out to your fit and healthy self, and help that self feel welcome in the light of day.

behavior modification alteration of behavior using methods based on the theory that actions can be controlled by manipulating the environmental factors that cue, or trigger, the actions.

Table 9-9

Applying Behavior Modification to Control Body Fatness

1. Eliminate inappropriate eating cues:
   - Don't buy problem foods.
   - Eat only in one room at the designated time.
   - Shop when not hungry.
   - Turn off television food commercials.
   - Avoid vending machines, fast-food restaurants, and convenience stores.
2. Suppress the cues you cannot eliminate:
   - Serve individual plates; don't serve "family style."
   - Make small portions look large by spreading them over the plate.
   - Create obstacles to consuming problem foods—wrap them and freeze them, making them less quickly accessible.
   - Control deprivation; plan and eat regular meals.
3. Strengthen cues to appropriate behaviors:
   - Share appropriate foods with others.
   - Store appropriate foods in convenient spots in the refrigerator.
   - Learn appropriate portion sizes.
   - Plan appropriate snacks.
   - Keep sports and play equipment by the door.
4. Repeat desired behaviors:
   - Slow down eating—put down utensils between bites.
   - Always use utensils.
   - Leave some food on your plate.
   - Move more—shake a leg, pace, stretch often.
   - Join groups of active people and participate.
5. Arrange negative consequences for negative behavior:
   - Ask that others respond neutrally to your deviations (make no comments—even negative attention is a reward).
   - If you slip, don't punish yourself.
6. Reward yourself personally and immediately for positive behaviors:
   - Buy tickets to sports events, movies, concerts, or other nonfood amusement.
   - Indulge in a new small purchase.
   - Get a massage; buy some flowers.
   - Take a bubble bath; read a good book.
   - Join a card game; listen to music.
   - Praise yourself; visit friends.
   - Nap; relax.

Activities and rewards to substitute for eating:

- Attending sporting events.
- Enjoying leisure activities.
- Exercising or playing sports.
- Gardening.
- Getting praise from others.
- Going to a movie or play.
- Listening to music.
- Napping.
- Praising yourself.
- Reading.
- Receiving token rewards (stickers, stars).
- Redecorating.
- Relaxing.
- Saving money for future treats.
- Shopping.
- Taking a bubble bath.
- Telephoning.
- Tidying your room or house.
- Vacationing.
- Working on hobbies or crafts.

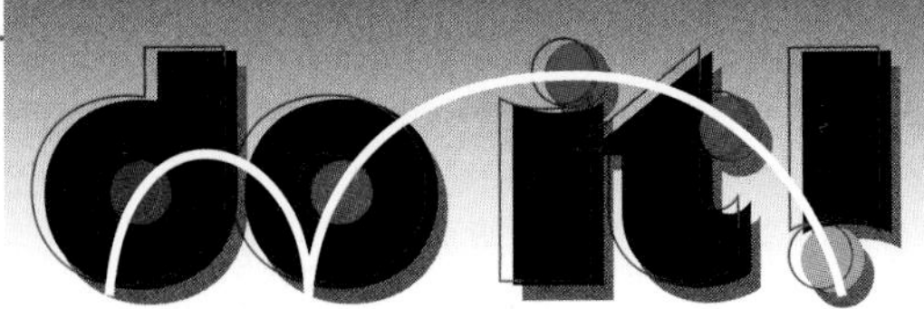
do it!

CONTROL THE CALORIES IN A DAY'S MEALS

This exercise speaks to the person who wants to control calorie intakes to control body fatness. To choose foods to meet nutrient needs while staying within a calorie limit, a diet planner can choose any or all of these four lines of action:

- *Cut down food portions,* if they are significantly larger than those recommended in the Food Guide Pyramid.
- *Eliminate* high-calorie, low-nutrient foods from the everyday diet (save for special treats).
- *Remove the high-calorie constituents* from most foods (trim fat from meat, choose fat-free milk).
- *Replace* high-calorie foods with lower-calorie versions, either naturally occurring or manufactured.

Fat is a main target for cutting calories because of its high calorie density. When using manufactured low-fat and fat-free items, the consumer must read labels carefully, however. These products may be lower in *fat,* but some contain as many *calories* as the originals.

Read Figure 9-12

The meals shown in Figure 9-12 represent one person's attempt to control calories. Shown on the left are the meals this person started with; the number of calories was too large to permit weight maintenance. On the right side of Figure 9-12, the meals have been modified to reduce their calories while still presenting the minimum number of servings from each food group recommended in the Food Guide Pyramid. Note that through these changes, the person has already reduced calories by almost 1,000. To reduce calories further, however,

Figure 9-12

Calories in Two Sets of Meals

About 3,200 cal

2% Milk, 1 c, 121 cal
Orange juice, 1 c 112 cal
Waffles, 2 each, 185 cal
Margarine, 2 tsp, 68 cal
Syrup, 4 tbs, 210 cal
Banana slices, ½ c, 69 cal

About 2,300 cal

2% Milk, 1 c, 121 cal
Orange juice, ¾ c, 84 cal
Waffle, 1 each, 93 cal
Margarine, 1 tsp, 34 cal
Syrup, 2 tbs, 105 cal
Banana slices, ½ c, 69 cal

2% Milk, 1 c, 121 cal
Hamburger, quarter pound, 415 cal
French fries, large (about 50), 448 cal
Ketchup, 2 tbs, 32 cal
Apple pie, 1 each, 225 cal

BIGGIE

2% Milk, 1 c, 121 cal
Hamburger, small, 266 cal
Green salad, 1 c, with light dressing, 1 tbs, 67 cal; croutons, ½ c, 50 cal
French fries, regular serving, 207 cal
Ketchup, 1 tbs, 16 cal
Gelatin dessert with fruit, 73 cal

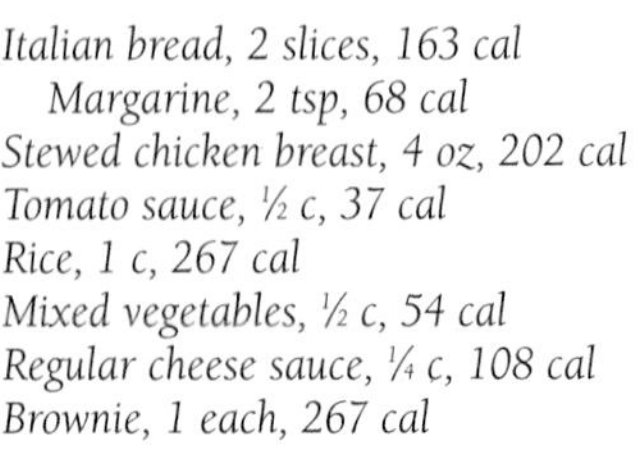

Italian bread, 2 slices, 163 cal
Margarine, 2 tsp, 68 cal
Stewed chicken breast, 4 oz, 202 cal
Tomato sauce, ½ c, 37 cal
Rice, 1 c, 267 cal
Mixed vegetables, ½ c, 54 cal
Regular cheese sauce, ¼ c, 108 cal
Brownie, 1 each, 267 cal

Italian bread, 1 slice, 82 cal
Margarine, 1 tsp, 34 cal
Stewed chicken breast, 4 oz, 202 cal
Tomato sauce, ½ c, 37 cal
Rice, 1 c, 267 cal
Mixed vegetables, ½ c, 54 cal
Low-fat cheese sauce, ¼ c, 85 cal
Brownie, 1 each, 267 cal

takes more careful thought. The person must cut calories while still meeting the minimum number of servings from the food groups to maintain nutrient adequacy.

The three top contributors of both calories and fat in the high-calorie meals are:

french fries > large hamburger > brownie

These foods are worth considering when cutting calories. Also, remember from Chapter 5's Food Feature that breads, cereals, baked goods, rice, and pasta also vary in their fat and sugar contents and, therefore, in calories. For example, the waffles depicted in the breakfast of Figure 9-12 contribute more calories than plain bread does because waffle mix contains fat; and the syrup served on top provides more calories than the waffles do.

Some 400 calories were trimmed from the 3,200-calorie meals by reducing sweets—reducing syrup served at breakfast and eliminating apple pie from lunch. (The planner kept the brownie, however—pleasure matters, too!) These two actions alone, repeated each day for one month, produce a calorie reduction more than sufficient to make a 3½-pound difference in the person's body weight.

Reduce Calories Further

Try your hand at cutting calories further to 1,800 or even 1,600 calories for the day. The only "must" in cutting calories is to make the diet adequate. Replace some of the higher calorie choices with better calorie bargains. Make substitutions with an eye for adequacy: for example, substituting diet cola for one of the two milk servings would compromise calcium adequacy and so is not allowable.

Step 1. Use Form 9-1, the column marked *Lower-Calorie Choices,* to record your changes in the 2,300-calorie day's meals (already listed for convenience at the left of the form).

Form 9-1

Try Your Hand: Reduce Calories Further

2,300-Calorie Day	Lower-Calorie Choices	Calories Saved	Food Group Servings: Name of Group	Food Group Servings: Number of Servings
Breakfast				
2% Milk, 1 c, 121 cal				
Orange juice, ¾ c, 84 cal				
Waffle, 1 each, 93 cal				
Margarine, 1 tsp, 34 cal				
Syrup, 2 tbs, 105 cal				
Banana slices, ½ c, 69 cal				
Lunch				
2% Milk, 1 c, 121 cal				
Hamburger, small, 2 oz, 266 cal				
Green salad, 1 c,				
with 1 tbs light dressing, 67 cal				
croutons, ½ c, 50 cal				
French fries, regular, 207 cal (about 30)				
Ketchup, 1 tbs, 16 cal				
Gelatin dessert with fruit, 73 cal				
Dinner				
Italian bread, 1 slice, 82 cal				
Margarine, 1 tsp, 34 cal				
Stewed chicken breast, 4 oz, 202 cal				
Tomato sauce, ½ c, 37 cal				
Rice, 1 c, 267 cal				
Mixed vegetables, ½ c, 54 cal				
Low-fat cheese sauce, ¼ c, 85 cal				
Brownie, 1 each, 267 cal				
	Total calories Saved:		2,300 – ________ (calories saved)	= ________ (new day's total calories)

Food group totals:
Milk, yogurt, cheese ________ Fruit ________ Vegetables ________ Meats and alternates ________
Bread, cereal, rice, pasta ________

Step 2. Record the calorie savings. As you make changes, turn to the Table of Food Composition, Appendix A, to find calorie values for foods you propose as substitutions for the originals. There is no need to look up every food on the menu—just the substitutes for those you choose to change. Subtract the calorie value of the new food from that of the original and write the calorie differences in the *Calories Saved* column.

Step 3. Add the *Calories Saved* column to obtain your total calorie savings for the day.

Step 4. Subtract the total savings from the original total of 2,300 calories to find the calorie value of the new day's meals.

Step 5. Assign each food from the new menu to its appropriate food group, and estimate the number of servings it represents. (Tips on how to estimate servings were given in Chapter 2's Do It! section.) Write the food group names on the left-hand side of the *Food Group Servings* column of Form 9-1. On the right-hand side of the same column write your estimate of the number of servings each food item represents. Total the day's servings from each group, and write the totals in the spaces provided at the bottom of the form.

Analysis

Answer the following questions:

1. How many total calories did your changes save?
2. Assuming that a pound of body weight is worth 3,500 calories, how much weight would a dieter theoretically lose in a month by cutting every day's calories to this extent?
3. By what methods (see the four-item list in the first paragraph of this Do It!) did you reduce calories?
4. Did you remove high-calorie constituents from any foods? Which ones?
5. Which high-calorie foods did you replace with lower-calorie ones? Try to judge how these changes may have affected the saturated fat, vitamin, mineral, or fiber content of the meals; write your responses.
6. Which of the changes most significantly reduced calories in the meals? Which changed calories least?
7. Did you find it necessary to replace the fast-food lunch in the diet? Why or why not?
8. Were you able to cut calories significantly while still meeting the minimum number of servings from each food group? If not, which groups fell short? List ways of adjusting your choices to include the missing foods.
9. Are the reduced-calorie meals appealing? If not, how can you include more appealing foods without increasing the calorie values?
10. Did your meals include any sweets or other treats? If so, which ones? If not, why not?

When you develop skill in making these sorts of changes, they tend to come to mind whenever the opportunity arises. Choosing foods with an eye for their contributions of calories and nutrients becomes a natural part of living. In case you are curious about how the authors might reduce calories, Table 9-10 shows our ideas for changes.

Table 9-10

Our Answers to Figure 9-11: New Calorie Level = About 1,900 Calories

Food	Changes	Calories Saved
Orange juice	Same	0
Milk	Replace with fat-free	35
Waffle	Same	0
Margarine/syrup	Replace with light margarine	12
Banana	Same	0
Milk	Replace with fat-free	35
Hamburger	Same	0
French fries	Omit	207
Ketchup	Same	0
Salad	Same	0
Croutons	Same	0
Gelatin	Omit	73
Italian bread with margarine	Same	0
Chicken	Same	0
Rice	Same	0
Tomato sauce	Same	0
Mixed vegetables	Same	0
Low-fat cheese sauce	Omit	85
Brownie	Same	0
Total calories saved	447	

Food group totals
Milk, yogurt, cheese 2 Fruit 2
Vegetables 4 Meats and alternates 2
Bread, cereal, rice, pasta 7

Self Check

Answers to these Self Check questions are in Appendix G.

1. Which of the following statements about basal metabolic rate (BMR) is correct?
   a. The more fat tissue, the higher the BMR.
   b. The more thyroxine produced, the higher the BMR.
   c. Fever lowers the BMR.
   d. Pregnant women have lower BMRs.
2. Which of the following is a health risk associated with excessive body fat?
   a. high blood lipids
   b. diabetes
   c. gallbladder disease
   d. all of the above
3. Body density (the measurement of body weight compared with volume) is determined by which technique?
   a. fatfold test
   b. bioelectrical impedance
   c. underwater weighing
   d. all of the above
4. The obesity theory that suggests that the body chooses to be at a specific weight is the:
   a. set-point theory
   b. enzyme theory
   c. fat cell theory
   d. external cue theory
5. Which of the following is a recommended weight-loss strategy?
   a. muscle stimulators
   b. stomach stapling
   c. herbs containing ephedrine
   d. none of the above
6. Which of the following is a possible physical consequence of fasting?
   a. loss of lean body tissues
   b. lasting weight loss
   c. body cleansing
   d. all of the above
7. The thermic effect of food plays a major role in energy expenditure. True or false? T F
8. If you bike about ten minutes a day and walk to classes but otherwise sit and study, you are considered lightly active. True or false? T F
9. The BMI standard is an excellent tool for evaluating obesity in athletes and the elderly. True or false? T F
10. A person wishing to gain weight will gain faster by consuming protein supplements in addition to ordinary foods. T F

NUTRITION ON THE NET

For further study of the topics of this chapter, access these websites and search for the phrases or words in quotation marks:

1. For the latest changes in chapter material, to find supplemental learning tools, or access links to related websites, go to the "Nutrition: Concepts and Controversies" site at:
   www.wadsworth.com/nutrition/prod/allprod.html
2. For more information on the number of calories in foods and beverages; type in the name of the food or beverage at:
   www.nal.usda.gov/fnic/foodcomp
3. Visit the Shape Up America site for information on body weight and body fat:
   www.shapeup.org
4. Visit the U.S. Government website for additional information on "obesity":
   www.healthfinder.gov/searchoptions/topicsaz.htm
5. To learn more about nutrition and weight maintenance:
   www.niddk.nig.gov/health/nutrit/nutrit.htm
6. The Center for Drug Evaluation and Research provides information about "drugs for weight loss":
   www.fda.gov/cder
7. Search the Mayo Clinic's Nutrition Center site for information on the "treatment of obesity":
   www.mayohealth.org
8. Search among thousands of current scientific and medical abstracts for any topic related to obesity at:
   http://www.ncbi.nlm.nih.gov/PubMed/

CONTROVERSY

9

THE DIET DEBATE: WHO CHOOSES, WHO PROFITS, AND WHAT ARE THE RISKS?

SOME LEADERS in obesity research are questioning the very foundations of obesity treatment. The issues dividing the experts center on how, when, and even whether to advise weight-loss dieting for overfat clients.

Do all obese clients benefit from routine advice to lose weight? Or is this "one size fits all" approach now obsolete? The question arises from research that suggests that diets almost always fail in the long term. Right away, it should be said that weight loss is possible and that the techniques presented in Chapter 9 are valid. The chapter also made clear, however, that controlling body composition is a lifelong effort that involves more than diet alone. Lost fat can be regained when overweight people "go on a diet," lose weight, but then return to old eating patterns and sedentary lifestyles. Unrealistic expectations are another problem. Practically no dieters will end up looking like fashion models despite sales pitches from diet programs. A more attainable and sustainable goal is to lose enough body fat to prevent diseases or regain health. This Controversy centers on the question of who, exactly, should lose body fat for health's sake; it also introduces an alternative for healthy overfat people: self-acceptance.

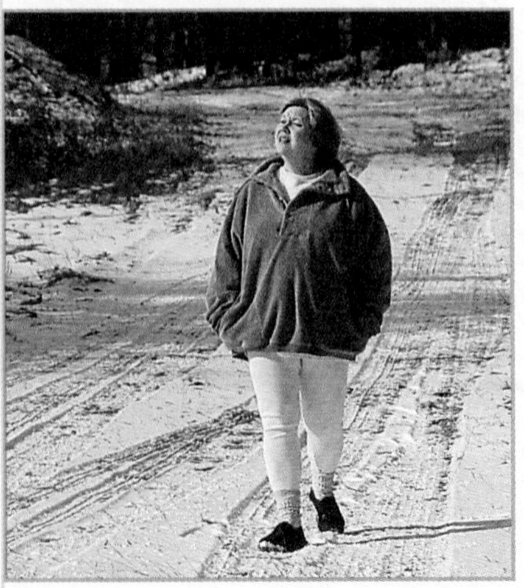

THE DECISION TO TREAT OBESITY

Those in favor of aggressive treatment of obesity are still in the majority. All major governmental dietary guidelines mention weight control as a health-supporting ideal. The experts cite many studies showing that with obesity come increased disease risks and that weight loss reduces those risks.[1] What good would it do to treat a person's Type 2 diabetes, say researchers, without treating the obesity underlying the disorder? The treatment may reduce the ravages of the present symptoms, but more symptoms will surely develop unless truly effective and lasting treatment for the predisposing condition is offered. According to one estimate, prevention of obesity could have saved over $45 billion in medical costs in one year alone.[2] The definition of obesity is important in this regard. Who should be told to lose weight? At what point does extra fat present a health risk?

When researchers followed the weights, health status, and mortality rates of women over a period of 16 years, they found that women who had gained 22 pounds or more since age 18 were nearly twice as likely to die from heart disease as women who gained little or no weight.[3] This risk tripled for women gaining more than 44 pounds during the study period. Women who had a body mass index (BMI) below 19 and weighed 15 percent *below* the average U.S. weights had the lowest risk of death from all causes. A related study found similar trends in men.

The women's study findings are presented in graph A of Figure C9-1. Observe that the correlation is direct: the higher the BMI, the higher the mortality rate. An interesting note is that the data exclude smokers. Typical results of data not excluding smokers are shown in graph B of the figure. Because smoking contributes to both low body weight and early death, including smokers elevates the risk of mortality at the lowest BMI values. One might have concluded, then, that thinness itself is not a risk factor for early death, but that smoking is. However, even when smokers are excluded from the data, the mortality risks of being underweight still appear as severe as the risks of extreme obesity.[4]

Does obesity cause diseases and early death, then? Remember that in science correlation is not cause. Yes, obesity is often found in those who suffer from the conditions mentioned in the chapter, but this does not prove that obesity *causes* those conditions. It is as likely that a high-fat diet, excess caloric intake, or lack of exercise causes both the diseases and the obesity.[5] Or it may be that a hereditary factor predisposes people to developing both the diseases and the obesity (some evidence exists to support this idea).[6] To say that obesity is a direct cause of disease, and must therefore be treated to prevent disease, is to disobey the laws of scientific reasoning. In fact, the results of recent surveys appear to oppose this view: cardiovascular disease incidence *declined* in recent years, a period when obesity was on the *rise*.[7]

Other medical authorities counter these arguments by pointing to proved benefits from even 10 to 20 pounds of weight loss in obese patients.[8] Indisputably, many of the complications of obesity such as glucose tolerance and insulin resistance, hypertension, hyperlipidemia, and sleep disturbances show rapid improvement with even small losses. Thus weight loss in the obese who suffer these conditions offers practical benefits, and whether these benefits are brought about by reduction in body fatness or adoption of healthy lifestyle habits seems immaterial.

Other researchers point out, however, that a large subset of the obese population, termed the "healthy

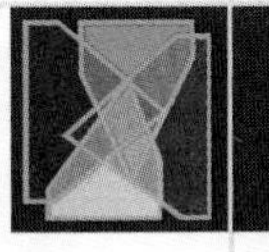

obese," is not at increased risk of disease. These people live long and remain healthy. For them, the risks associated with weight-loss dieting may far outweigh any health benefits they can expect from weight loss.

A larger question may be whether the weight loss and benefit correlation holds as true for the mildly obese as for the very obese. Obese people may carry 80, 100, or even more pounds of excess fat and suffer health hazards, but is a seemingly healthy person who carries 20 or 30 extra pounds also inviting those hazards? Should such a person be advised by medical and nutrition experts to lose weight? A recent study suggests that although risks from obesity are severe up to age 30, from that age to the mid-70s, they may become less severe.[9] Findings like these call into question the validity of making blanket weight-loss recommendations to all overweight people.

DIET SAFETY—PROVED OR JUST IMPLIED?

When consumers are offered a medical treatment, they expect to be fully informed of any risks involved in undertaking it. This expectation is reasonable because the FDA requires proof that treatments such as medicines and surgery not only work, but are safe. Unfortunately, this is not true of many diets, which pose risks but do not inform users about them, a practice that many believe to be unethical. Overall, the multibillion dollar weight-loss industry remains only loosely regulated, if at all, and often places consumers at significant health risk.

The risks associated with weight loss are more serious than most dieters would suspect. Linked with fluctuations in body weight, but not specific to any one mode of achieving weight loss, may be:

- A doubled risk of developing gallstones, especially when food intake is severely restricted (but successful weight loss and maintenance reduce this risk).
- An increased risk of death from all causes (but successful weight loss and maintenance reduce the risk).[10]
- An increased risk of bone loss and osteoporosis.[11]
- Slowed reaction time (a driving hazard).[12]
- Diminished concentration, poor attention span (possibly due to impairment of blood iron status).[13]
- Increased moodiness, grumpy mood.[14]
- An increased likelihood of overeating and binge eating.[15]
- An increased chance of developing an eating disorder.[16]

In the short run, weight loss may reduce blood pressure and normalize blood lipids and glucose metabolism. In the long run, the benefits of weight loss may not translate into longer life.[17]

Some diet programs are straightforward about the potential risks as well as the benefits that participants can expect to encounter. Other weight-loss centers intentionally hide both risks and costs associated with the programs while exaggerating their efficacy. A movement to subject the weight-loss industry to strong regulation is gaining support in the United States, and such controls have already been adopted in some other locations.[18] Table C9-1 presents a list of guidelines for weight-management programs recommended by the American Heart Association.

Figure C9-1

Mortality Data: The Effects of Body Weight and Smoking

A. Women who never smoked (1,499 deaths)

Risk of death from all causes: 0.0, 0.5, 1.0, 1.5, 2.0, 2.5

Body mass index: < 19.0, 19.0–21.9, 22.0–24.9, 25.0–26.9, 27.0–28.9, 29.0–31.9, ≥ 32

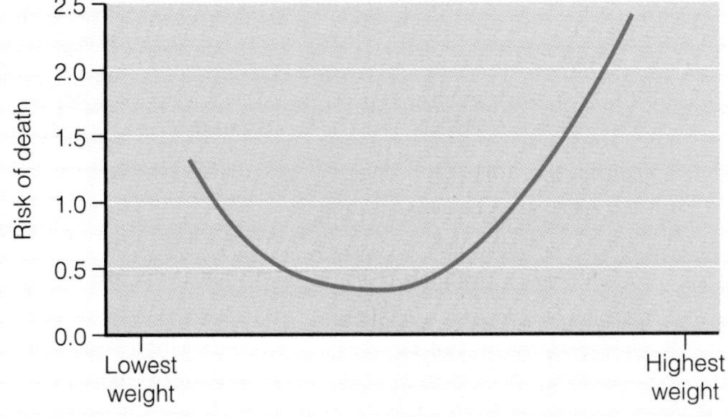
B. Data that include smokers
Risk of death
2.5
2.0
1.5
1.0
0.5
0.0
Lowest weight
Highest weight
Body mass index

A. These are actual data for mortality and body mass index (BMI) that exclude smokers. There is a direct linear relationship between higher BMI and higher mortality.

B. Data for mortality and body weight that include smokers often generate a J-shaped graph like this one. Smokers weigh less, but they die sooner than others, so their deaths may elevate the left-hand side of the curve.

SOURCE: Adapted from J. E. Manson and coauthors, *New England Journal of Medicine*, 1995, vol. 333, pp. 677–685. Copyright 1995. Massachusetts Medical Society. All rights reserved.

MEDICAL ETHICS, WEIGHT LOSS, AND PROFITS

Obesity is considered to be a chronic disease, like diabetes or heart and artery disease. Most physicians feel compelled by medical ethics to offer treatment for all diseases. Many fear that to delay offering treatment could constitute medical malpractice.[19] Not only do obese people who lose weight experience reduction of indicators of disease risks, but they may acquire a new sense of self-esteem and improved quality of life. Certainly, to give people the opportunity to achieve these benefits would be desirable.

Besides, weight-loss advocates argue, if doctors withdraw their medical support for weight control, quacks will rush to fill the void. Physicians may not be able to dictate to society or even to individuals what an ideal weight should be, but they should stay involved with weight control and strive to guide their clients toward healthy habits.

While no one can argue against treating a disease that causes misery and illness, questioners point out that weight-loss efforts involving diets, even medically supervised diets, are not effective—they almost always fail to produce long-term results. Thus it is unethical and misleading to promote these diets as effective treatments. If the FDA were asked to approve the sale of a medical drug with the failure record of weight-loss diets and programs, the agency would ban the drug as ineffective. If obesity is treated as a medical problem, shouldn't prescribed treatments be required to meet the criteria that other medical treatments must meet?

Unquestionably, the huge potential for profit drives many programs. An estimated 30 to 40 percent of all adult U.S. women (and 20 to 25 percent of U.S. men) are trying to lose weight at any given time and spending $30 to $40 billion each year to do so. Such income potential is bound to attract those looking for "get-rich-quick" schemes. In fact, the FDA names weight-loss scams as one of the leading forms of fraud in the United States. People have come to attach so many unproved benefits to weight loss that they are willing to risk huge sums for the slightest chance of success. These factors—a hard-to-change condition, a willingness to believe in unproved benefits, and the ability to pay—create a fertile field for profiteers who continue, year after year, to rake in huge sums while hiding the improbable odds of success.

Table C9-1

American Heart Association Weight-Management Program Guidelines

Properly constructed weight-loss programs provide:

- Participant information about the program format, its costs, and potential risks; the qualifications of its professionals, the expected results, and the time frame both for reaching goals and for follow-up.
- Qualified experts with credentials in nutrition, exercise, and behavior change.
- Medical screening for conditions that may make weight loss risky, and physician evaluation of those identified as having such conditions.
- Reasonable weight-loss goals to support specific health targets, such as reduced risk from cardiovascular disease.
- Nutrition, exercise, and behavior components, specifically designed by experts to meet individual needs.
- A maintenance program of at least 2 years.
- Follow-up at 2 and 5 years.

NOTE: The full text of these guidelines is available on the Internet at www.americanheart.org/Scientific/statements

SOURCE: Adapted from AHA Medical/Scientific Statement, *American Heart Association Guidelines for Weight Management Programs for Healthy Adults,* available from Office of Scientific Affairs, AHA, 7272 Greenville Ave., Dallas, TX 75231–4596.

ATTITUDES TOWARD BODY FATNESS

People in the United States today value slenderness highly. The image of a successful, healthy, well-adjusted person almost invariably includes a slender body form. Irrationally, many people equate slenderness with happiness, intelligence, psychological stability, harmony in relationships, success in the workplace, and many other valued attributes that have little if anything to do with body size or weight. Some claim we emphasize slenderness, especially in women, not for the sake of their health, but for other people's viewing pleasure.

Another widespread misconception holds that obese people are to blame for their overfatness; they face unspoken accusations of laziness, slothfulness, and self-indulgence. Prejudice and discrimination against overweight people, especially overweight women, cause them great psychological stress. Overweight people in our society readily assume the burden of responsibility for their fatness and feel personally guilty when weight-loss diets fail to produce promised results.

Especially damaging are the attitudes of some physicians and other health professionals toward the overweight. A story is told of a woman who had this problem:

> The woman had complained to her family physician for years about her indigestion and diarrhea. He always reminded her that she was overweight and if she wouldn't eat so much, her digestion would be fine. Finally, she was diagnosed by a specialist as having a gluten [wheat] intolerance. Although her health is better today, she is still overweight and is embarrassed to see her family doctor for her regular checkup.

Not only physicians, but other health-care providers including dietitians who run weight-loss clinics can be moralistic in their approach to the obese. They present weight loss as an easily achievable goal, a grossly misleading message. They may regard those who do not become slender as "deviant" and as failures. In reality, such people are among the overwhelming majority who have repeatedly

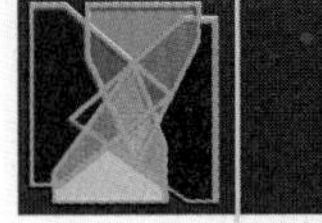

Figure C9-2

Do I Need to Lose Weight?

The "branches" of this tree display signs indicating that loss of weight may benefit health. The more factors that are present, the higher the risk of health problems in the future.

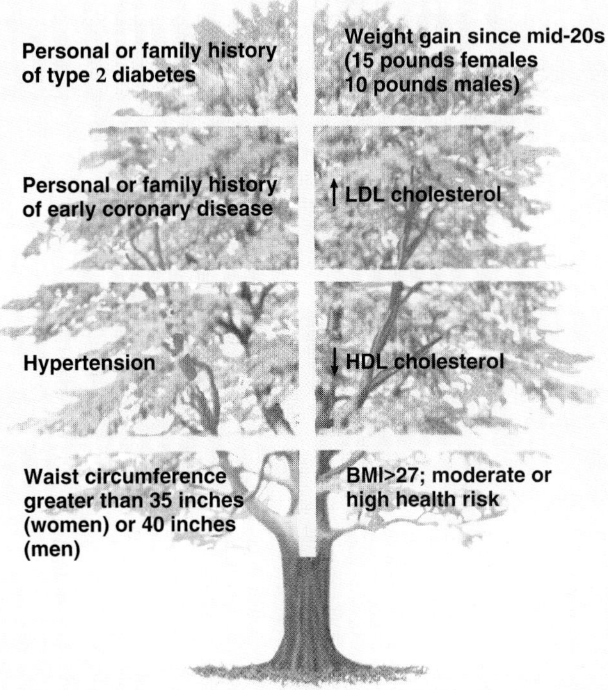

Personal or family history of type 2 diabetes
Weight gain since mid-20s (15 pounds females 10 pounds males)
Personal or family history of early coronary disease
↑ LDL cholesterol
Hypertension
↓ HDL cholesterol
Waist circumference greater than 35 inches (women) or 40 inches (men)
BMI>27; moderate or high health risk

proved that it is really the weight-loss diet schemes that are the failures.

TOWARD DEVELOPING ANSWERS

In searching for appropriate solutions to the problem of obesity, researchers have looked back to the population a century ago. In those days, significantly fewer people were overweight, and dieting was practically unknown. Physical activity was necessarily a part of life, for elevators, automobiles, and other labor-saving devices were yet to be invented.

With the advent of fast food, obesity became more common in the U.S. population. It is known that adopting a calorie-controlled low-fat, high-carbohydrate eating plan for life can significantly push body composition toward the lean. Equally important, or more important, is an active lifestyle. Improvements in fitness improve a person's odds for remaining healthy, and an active lifestyle may also prevent substantial weight gain and obesity as a person ages.[20]

Should obese people make efforts to lose weight? The answer is "yes," if risks of illness are present (see Figure C9-2 to evaluate the risks). What weight to aim for is a trickier question. An immediate goal might be to stop gaining. Then, focusing on healthy behaviors rather than weight loss itself may be a key.

People with none of the risk factors of Figure C9-2 should seriously question the desirability of losing weight. People who hold to thin ideals face a danger to their own and others' self-esteem and well-being. Fashion models whose careers depend on body shape often suffer from eating disorders. Teenage girls across the country feel compelled to diet, ignorant of their peril. To break free from these dangerous ideals requires that people revamp old ways of thinking and accept themselves and others regardless of body weight. Table C9-2 offers tips to people who are working to accept their body weights rather than to change them.

A safe option exists for those who would like to improve their body composition without attempting weight loss. The low-fat, adequate diet advocated by the *Dietary Guidelines* and the Daily Food Guide can make a difference for those who are accustomed to eating high-fat diets. Taken with the routine of behavior modification and physical activity described in Chapter 9, the change can help produce fitness while removing the issue of weight from the realm of success or failure. This course of action is not only life-changing for the people who adopt it, but it also sends an important message of self-acceptance to a younger generation.

Table C9-2

Tips for Accepting a Healthy Body Weight

- Adopt a new value system. Value yourself and others for human attributes other than body weight. Realize that prejudging people by weight is as harmful as prejudging them by race, religion, or gender.
- Use supportive, nonjudgmental descriptions of your body; never use degrading negative descriptions.
- Take compliments seriously. Positive comments from others probably reflect an objective viewpoint.
- Avoid frequent checking of your weight or appearance; focus on your whole self including your intelligence, social grace, and professional and scholastic accomplishments.
- Accept that no magic diet exists, and that it is *diets* that fail, not the people who try to use them.
- Stop dieting to lose weight. Adopt healthy eating and exercise habits.
- Memorize and employ the Food Guide Pyramid. Never restrict food intake beyond the minimum levels that meet nutrient needs.
- Become physically active, not because it will help you get thin but because it will enhance your health.
- Seek support from loved ones. Tell them of your plan for a healthy life in the body you have been given.
- Seek professional counseling, *not* from a weight-loss counselor, but from someone who can help you make gains in self-esteem without weight as a factor.
- Join with others to fight weight discrimination and fashion stereotypes. (Search your local paper, or see the Nutrition Resources Appendix for names of groups.)

10 Nutrients, Physical Activity, and the Body's Responses

Fernand Leger, 1881–1955, *The Cyclists*, 1944.

Contents

Frequently Asked Questions

training regular practice of an activity, which leads to physical adaptations of the body with improvement in flexibility, strength, or endurance.

flexibility the capacity of the joints to move through a full range of motion; the ability to bend and recover without injury.

muscle strength the ability of muscles to work against resistance.

muscle endurance the ability of a muscle to contract repeatedly within a given time without becoming exhausted.

cardiorespiratory endurance the ability to perform large-muscle dynamic exercise of moderate-to-high intensity for prolonged periods.

IN THE BODY, nutrition and physical activity go hand in hand. The working body demands all three energy-yielding nutrients to fuel activity: carbohydrate, lipids, and protein all contribute to body fuel. The body also needs protein and a host of supporting nutrients to build lean tissue. Physical activity, in turn, benefits the body's nutrition by helping to regulate the use of fuels, by pushing the body composition toward the lean, and by increasing the daily calorie allowance, and with more calories comes more nutrients and other beneficial constituents of foods.

Luckily for most people, it isn't necessary to run marathons to reap the health rewards of physical activity.[1] In fact, people who regularly engage in just moderate physical activity live longer on average than those who are physically inactive.[2] A sedentary lifestyle ranks with such powerful risk factors as smoking and obesity for developing the major killer diseases of our time—cardiovascular disease, some forms of cancer, stroke, diabetes, and hypertension.[3] Chapter 1 listed many other benefits as well.

For many people, then, the health benefits of regular, moderate physical activity are reward enough. Others, however, seek the kinds and amount of physical activity that will not only benefit health, but improve physical fitness or their performance in sports. In 1998, the American College of Sports Medicine (ACSM) put forth recommendations for the quantity and quality of physical activity needed to develop and maintain fitness in healthy adults (see Table 10-1).[4]

Maybe you're already physically fit. If so, the following description applies to you. You are graceful and move with ease. You are strong and meet physical challenges without strain. You have endurance, and your energy lasts for hours. You meet daily physical challenges and have plenty of energy in reserve. What's more, you are prepared to meet mental and emotional challenges, too, for physical fitness also supports mental and emotional energy and resilience.

The body benefits from physical activity, but so does the mind—see the Fitness for Life feature of Chapter 1.

For those just beginning to increase fitness, be assured that improvement is not only possible, but is an inevitable result of becoming more active. As you improve your physical fitness, you not only *feel* better and stronger, but you *look* better, too. Physically fit people walk with confidence and purpose because posture and self-image improve along with physical fitness. Socially, groups that work out together bring companionship and the fun of belonging. The kinds and amounts of physical activity that improve physical fitness also provide still greater health benefits (further reduction of cardiovascular disease risk and improved body composition, for example).[5]

This chapter is written for athletes and for exercisers who train like athletes. Casual athletes (those who compete only with their own goals) and competitive athletes (those who compete with others) are cut from the same cloth as far as their food and fluid needs are concerned. To understand the interactions between physical activity and nutrition, you must first know a few things about physical fitness and **training.**

KEY POINT *Physical activity and fitness benefits people's physical, psychological, and social well-being and improves their resistance to disease. Physical activity to improve physical fitness eases the tasks of daily living and offers additional personal benefits.*

People's bodies are shaped by the activities they perform.

The Essentials of Fitness and Training

To be physically fit, you need to develop your own potential along several lines. You need to achieve enough of the four components of fitness—**flexibility, muscle strength, muscle endurance,** and **cardiorespiratory endurance**—to allow you to meet the everyday demands of life, with some to spare and to achieve a reasonable body composition.

Table 10-1

Physical Activity Guidelines

Guidelines for developing and maintaining *physical fitness:*

- Frequency of activity: 3 to 5 days per week.
- Intensity of activity: 55 to 90% of maximum heart rate.
- Duration of activity: 20 to 60 minutes of continuous activity.
- Mode of activity: Any activity that uses large-muscle groups.
- Resistance activity: Strength training of moderate intensity at least two times per week.
- Flexibility activity: Stretching of the major muscle groups two to three times per week.

 How Do My Muscles Become Physically Fit?

People shape their bodies by what they choose to do and not do. Muscle cells and tissues respond to an **overload** of physical activity by gaining strength and size, a response called **hypertrophy**. The opposite is also true: if not called on to perform, muscles dwindle and weaken, a response called **atrophy.** Thus cyclists often have well-developed legs but less arm or chest strength; a tennis player may have one arm that is superbly strong, while the other is just average. A variety of physical activities will produce the best overall fitness. To this end, people are told to work different muscle groups from day to day. For balanced fitness, stretching enhances flexibility, weight training and calisthenics develop muscle strength and endurance, and **aerobic** activity improves cardiorespiratory endurance. It makes sense to give muscles a rest, too, because it takes a day or two to replenish muscle fuel supplies and to repair wear and tear incurred through physical activity.

Periodic rest also gives muscles time to adapt to an activity. During rest, muscles build more of the equipment required to perform the activity that preceded the rest. The muscle cells of a superbly trained weight lifter, for example, store extra granules of glycogen, build up strong connective tissues, and add bulk to the special proteins that contract the muscles, thereby increasing the muscles' ability to perform.* In the same way, the muscle cells of a distance swimmer develop huge stocks of **myoglobin,** the muscles' oxygen-handling protein, and other equipment needed to burn fat and to sustain prolonged exertion. Therefore, if you wish to become a better jogger, swimmer, or biker, you should train mostly by jogging, swimming, or biking. Your performance will improve as your muscles develop the specific equipment they need to do the activity. Keep in mind that although everyone's muscles adapt to physical activity to some degree, true champions in sports are born with the genetic potential to excel while others are not. This fact cannot be changed even through hard work.

KEY POINT *The components of fitness are flexibility, muscle strength, muscle endurance, and cardiorespiratory endurance. To build fitness, a person must engage in physical activity. Muscles adapt to activities they are called upon to perform.*

Can I Expect any Health Benefits from Weight Training?

Weight training builds lean body mass, develops strength and endurance of muscles, and benefits health and overall fitness. Strong muscles in the back and abdomen improve posture and reduce the risk of back injury. Weight training can also help prevent the decline in physical mobility that often accompanies aging.[6] Older adults who participate in weight training programs not only gain muscle strength, but also improve

*All muscles contain a variety of muscle fibers, but there are two main types—slow-twitch (also called *red fibers*) and fast-twitch (also called *white fibers*). Slow-twitch fibers contain extra metabolic equipment to perform fat-burning aerobic work; the fast-twitch type store extra glycogen for anaerobic work. Muscle fibers of one type take on some of the characteristics of the other as an adaptation to exercise.

overload an extra physical demand placed on the body; an increase in the frequency, duration, or intensity of an activity. A principle of training is that for a body system to improve, it must be worked at frequencies, durations, or intensities that increase by increments.

hypertrophy (high-PURR-tro-fee) an increase in size (for example, of a muscle) in response to use.

atrophy (AT-tro-fee) a decrease in size (for example, of a muscle) because of disuse.

aerobic (air-ROE-bic) requiring oxygen. Aerobic activity strengthens the heart and lungs by requiring them to work harder than normal to deliver oxygen to the tissues.

myoglobin the muscles' iron-containing protein that stores and releases oxygen in response to the muscles' energy needs.

weight training (also called **resistance training**) the use of free weights or weight machines to provide resistance for developing muscle strength and endurance. A person's own body weight may also be used to provide resistance as when a person does push-ups, pull-ups, or sit-ups.

cardiac output the volume of blood discharged by the heart each minute.

stroke volume the amount of oxygenated blood ejected from the heart toward body tissues at each beat.

their muscle endurance, which enables them to walk significantly longer before exhaustion. Leg strength and walking endurance are powerful indicators of an older adult's physical abilities.[7]

Weight training to improve muscle strength and endurance also helps maximize and maintain bone mass.[8] For example, young women participating in a weight training program were able to increase the bone density of their spines.[9]

Depending on the technique, weight training can emphasize either muscle strength or muscle endurance. To emphasize muscle strength, combine high resistance (heavy weight) with a low number of repetitions. To emphasize muscle endurance, combine less resistance (lighter weight) with more repetitions.

Weight training enhances performance in other sports, too. Swimmers can develop a more efficient stroke and tennis players a more powerful serve, when they train with weights.

KEY POINT *Weight training offers health and fitness benefits to adults. Weight training improves older adults' physical mobility and helps maximize and maintain bone mass.*

How, Exactly, Does Cardiorespiratory Training Benefit the Heart?

Everyone has felt the heart beat pick up its pace during physical activity. Cardiorespiratory endurance determines how long a person can remain active with an elevated heart rate—it is the ability of the heart and lungs to sustain a given physical demand. Working muscles need abundant oxygen to produce energy, and the heart and lungs work together to provide that oxygen. Cardiorespiratory endurance training, therefore, is aerobic. As the body adapts to the demands of aerobic activity, cardiorespiratory endurance improves—the body delivers oxygen more efficiently. With cardiorespiratory endurance, the total blood volume and number of red blood cells increase, so the blood can carry more oxygen. The heart muscle becomes stronger and larger, and each beat empties the heart's chambers more completely, so the heart pumps more blood per beat. This makes fewer beats necessary, so the pulse rate falls. The muscles that inflate and deflate the lungs gain strength and endurance, thereby allowing breathing to become more efficient. Blood moves easily through the blood vessels because the muscles of the heart contract powerfully, and contraction of the skeletal muscles pushes the blood through the veins. Such improvements keep resting blood pressure normal. The improvements that come with cardiorespiratory endurance also raise blood HDL, the lipoprotein associated with lower heart disease risk.[10]

The importance of HDL to heart health is a topic of the next chapter.

Cardiorespiratory endurance is characterized by:

- increased **cardiac output** and oxygen delivery.
- increased heart strength and **stroke volume.**
- slowed resting pulse.
- increased breathing efficiency.
- improved circulation.
- and reduced blood pressure.

Figure 10-1 shows the major relationships among the heart, lungs, and muscles.

Which activities produce these beneficial changes? Effective activities elevate the heart rate, are sustained for longer than 20 minutes, and use most of the large-muscle groups of the body (legs, buttocks, and abdomen). Examples are swimming, cross-country skiing, rowing, fast walking, jogging, fast bicycling, soccer, hockey, basketball, water polo, lacrosse, and rugby.

An informal pulse check can give you some indication of how conditioned your heart is. The average resting pulse rate for adults is around 70 beats per minute, but the rate can be higher or lower. Active people can have resting pulse rates of 50 or even lower. To take your pulse, follow the directions in the margin. The rest of this

To take your resting pulse:

Using a watch or clock with a second hand, place your hand over your heart or your finger firmly over an artery at the underside of the wrist or side of the throat under the jawbone. Start counting your pulse at a convenient second, and continue counting for ten seconds. If a heartbeat occurs exactly on the tenth second, count it as one-half beat. Multiply by 6 to obtain the beats per minute. To ensure a true count:

- Use only fingers, not your thumb, on the pulse point (the thumb has a pulse of its own).
- Press just firmly enough to feel the pulse. Too much pressure can interfere with the pulse rhythm.

Figure 10-1

Delivery of Oxygen by the Heart and Lungs to the Muscles

The more fit a muscle is, the more oxygen it draws from the blood. This oxygen comes from the lungs, so the person with more fit muscles extracts oxygen from inhaled air more efficiently than a person with less fit muscles. The cardiovascular system responds to increased demand for oxygen by building up its capacity to deliver oxygen. Researchers can measure cardiovascular fitness by measuring the amount of oxygen a person consumes per minute while working out. This measure of fitness is called **VO_2 max.**

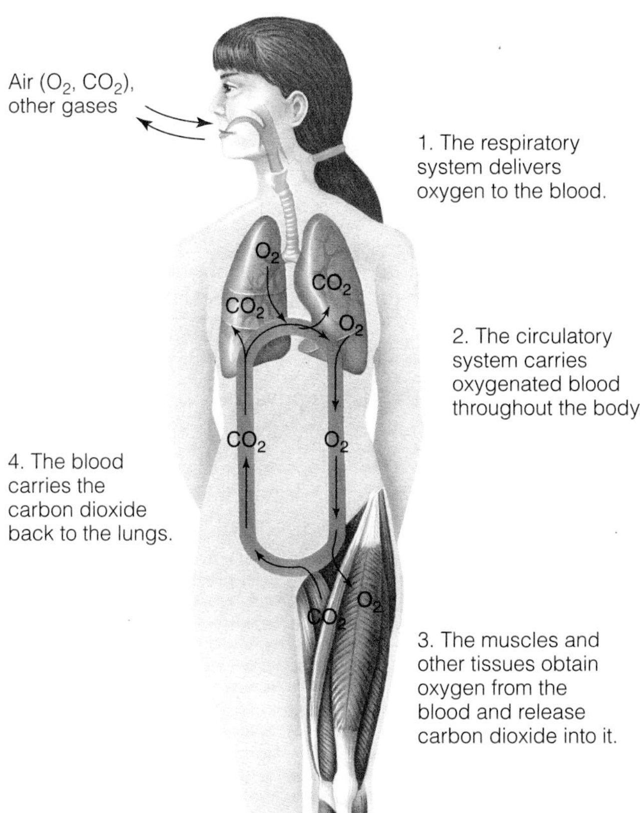
Air (O_2, CO_2), other gases
1. The respiratory system delivers oxygen to the blood.
2. The circulatory system carries oxygenated blood throughout the body.
3. The muscles and other tissues obtain oxygen from the blood and release carbon dioxide into it.
4. The blood carries the carbon dioxide back to the lungs.
O_2
CO_2

VO_2 max the maximum rate of oxygen consumption by an individual (measured at sea level).

chapter describes the interactions between nutrients and physical activity. Nutrition alone cannot endow you with fitness or athletic ability, but along with the right mental attitude, it can complement the effort you put forth to obtain them. Conversely, unwise food selections can stand in your way.

KEY POINT *Cardiorespiratory endurance training enhances the ability of the heart and lungs to deliver oxygen to the muscles. With cardiorespiratory endurance training, the heart becomes stronger, breathing becomes more efficient, and the health of the entire body improves.*

The Active Body's Use of Fuels

The fuels that support physical activity are glucose (from carbohydrate), fatty acids (from fat), and, to a small extent, amino acids (from protein). The body uses different mixtures of fuels at different times depending on the intensity and duration of its activities and also depending on its own prior training.

During rest, the body derives a little more than half of its energy from fatty acids, most of the rest from glucose, and a little from amino acids. During physical activity, the body adjusts its fuel mix. The stored glucose of muscle glycogen is a major fuel for physical activity. In the early minutes of an activity, muscle glycogen provides the

Epinephrine was discussed and defined in Chapter 3. You may recall that it is the major hormone that elicits the body's stress response, mobilizing fuels and readying the body for action.

majority of energy the muscles use to go into action. As activity continues, messenger molecules, including the hormone epinephrine, flow into the bloodstream to signal the liver and fat cells to liberate their stored energy nutrients, primarily glucose and fatty acids. Thus hormones set the table for the muscles' energy feast, and the muscles help themselves to the fuels passing by in the blood.

Glucose Use and Storage

Both the liver and muscles store glucose as glycogen; the liver can also make glucose from fragments of other nutrients. It has been said that muscles hoard their glycogen stores—they do not release their glucose into the bloodstream to share with other body tissues, as the liver does. This is fortunate. A muscle that shared its glycogen reserves with other tissues might lack glucose at a critical time, say, when running from danger. A muscle that conserves its glycogen is prepared to act in emergencies because muscle glucose is the fuel for quick action. Later on, as activity continues, glucose from the liver's stored glycogen and dietary glucose absorbed from the digestive tract also become important sources of fuel for muscle activity.

The body constantly uses and replenishes its glycogen. The more carbohydrate a person eats, the more glycogen muscles store, and the longer the stores will last to support physical activity.

A classic report compared fuel use during physical activity by three groups of runners, each on a different diet. For several days before testing, one of the groups ate a normal mixed diet (55 percent of calories from carbohydrate); a second group ate a high-carbohydrate diet (83 percent of calories from carbohydrate); and the third group ate a high-fat diet (94 percent of calories from fat). As Figure 10-2 shows, the high-carbohydrate diet enabled the athletes to work longer before exhaustion. This study and many others that followed established that a high-carbohydrate diet enhances an athlete's endurance by ensuring ample glycogen stores.

Figure 10-2

The Effect of Diet on Physical Endurance

A high-carbohydrate diet can triple an athlete's endurance.

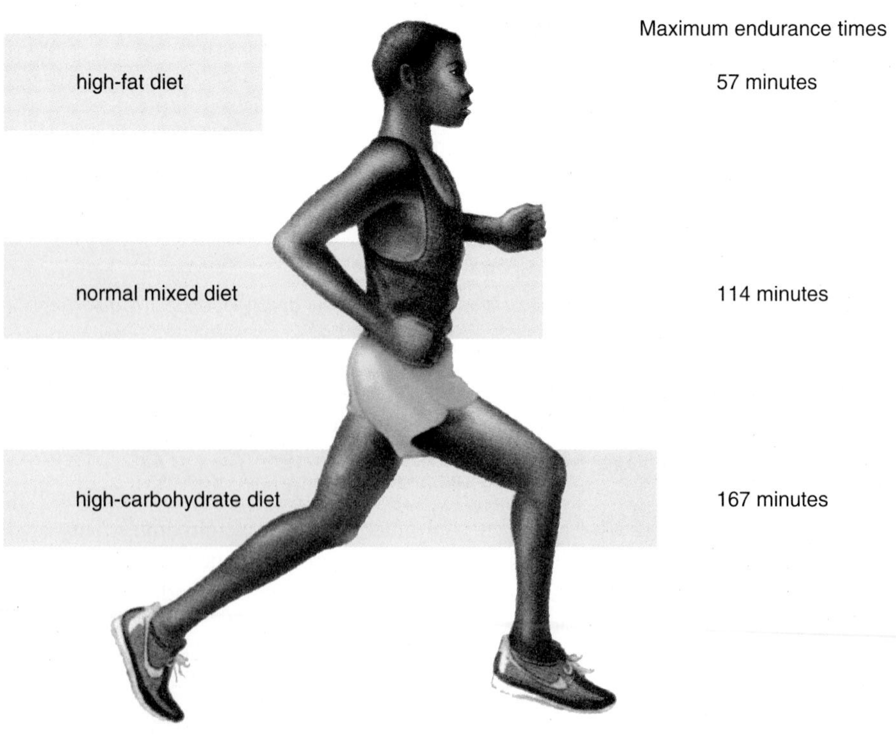
Maximum endurance times
high-fat diet
57 minutes
normal mixed diet
114 minutes
high-carbohydrate diet
167 minutes

KEY POINT *Glucose is supplied by dietary carbohydrate or made by the liver. It is stored in both liver and muscle tissue as glycogen. Total glycogen stores affect an athlete's endurance.*

Activity Intensity, Glucose Use, and Glycogen Stores

The body's glycogen stores are much more limited than its fat. A person with 30 pounds of body fat to spare may have only a pound or so of muscle and liver glycogen to draw on. How long an exercising person's glycogen will last depends not only on diet but also on the intensity of the activity. The most intense activities—the kind that make it difficult "to catch your breath," such as a quarter-mile run—use glycogen quickly. Less intense activities, such as jogging, during which breathing is steady and easy, use glycogen more slowly. Thus competitive athletes demand much more from their glycogen stores than do casual joggers. Joggers still use glycogen, however, and eventually they can run out of it. Glycogen depletion usually occurs after about two hours of vigorous activity.[†] [11]

Aerobic Use of Glucose During *moderate* physical activity, the lungs and circulatory system have no trouble keeping up with the muscles' need for oxygen. The individual breathes easily, and the heart beats at a faster pace than at rest but steadily—the activity is aerobic. As Figure 10-3 shows, during aerobic activity muscles extract their energy from both glucose and fatty acids when both are present together with oxygen. In this way, a little glucose helps to metabolize a lot of fat. Fat yields a lot of energy, so moderate aerobic activity conserves glycogen stores.

Anaerobic Use of Glucose Intense activity presents a different picture. The heart and lungs can provide only so much oxygen only so fast. When muscle exertion is so great that the demand for energy outstrips the oxygen supply, aerobic metabolism cannot sufficiently meet energy needs. This means that fat cannot be used because oxygen is required for its breakdown. Instead, muscles must begin to rely more heavily on glucose, which can be partially broken down by **anaerobic** metabolism. Thus the muscles begin drawing more heavily on their limited glycogen supply.

As the upper portion of Figure 10-3 on the next page shows, glucose can yield some energy in anaerobic metabolism, but not as much as in aerobic metabolism. Anaerobic breakdown of glycogen yields energy to muscle tissue when energy demands outstrip the body's ability to provide energy aerobically as, for example, during intense activity. Thus anaerobic metabolism supplies energy, but it does so by lavishly spending the muscles' glycogen reserves.

Lactic Acid Anaerobic breakdown of glucose produces **lactic acid**, fragments of glucose molecules that accumulate in the tissues and blood. When the nervous and hormonal systems detect these fragments in the blood, they respond by speeding up the heart and lungs to draw in more oxygen and break down the fragments. At some point, however, the heart and lungs are no longer able to keep up, and lactic acid accumulates. If you exercise intensely, you may have to slow down or even stop to "catch your breath" (replenish your oxygen supply). Then your body begins relying on aerobic metabolism once more. At this point, lactic acid is burned for fuel or used by the liver to regenerate glucose.

Lactic acid causes burning muscle pain, followed within seconds by a type of muscle fatigue. A strategy for dealing with lactic acid is to relax the muscles at every opportunity during activity so that the circulating blood can carry away the lactic acid and bring in more oxygen to sustain aerobic metabolism.

KEY POINT *The more intense an activity, the more glucose it demands. During anaerobic metabolism, the body spends glucose rapidly and accumulates lactic acid.*

[†] Here "vigorous exercise" means exercise at 75 percent of VO_2 max.

anaerobic (AN-air-ROE-bic) not requiring oxygen. Anaerobic activity may require strength but does not work the heart and lungs very hard for a sustained period.

lactic acid a product of the incomplete breakdown of glucose during anaerobic metabolism. When oxygen becomes available, lactic acid can be completely broken down for energy or converted back to glucose.

Figure 10-3

Glucose and Fatty Acids in Their Energy-Releasing Pathways

Glucose is partially broken down under anaerobic conditions. It yields some quick energy and leaves behind some fragments that must await the renewed availability of oxygen (aerobic conditions) to be broken down completely to carbon dioxide, water, and energy. Fat can enter the energy cycle only in the presence of oxygen.

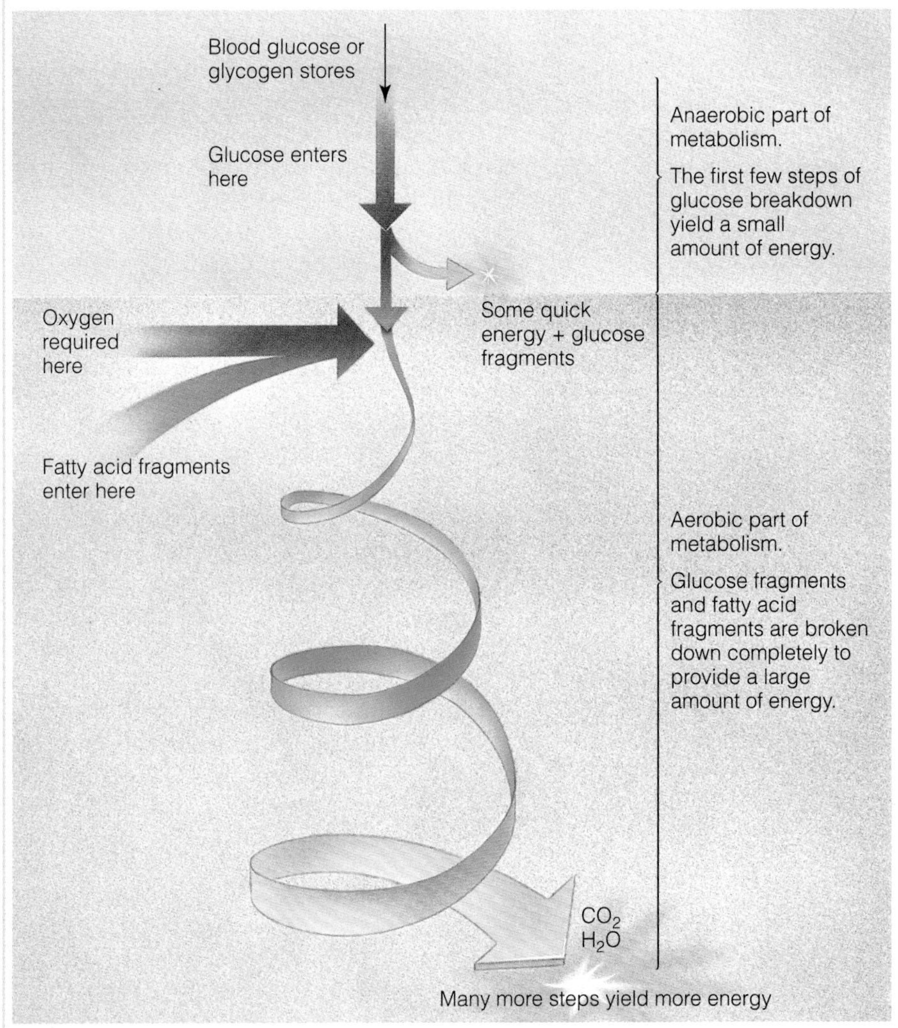

Blood glucose or glycogen stores
Glucose enters here
Anaerobic part of metabolism.
The first few steps of glucose breakdown yield a small amount of energy.
Oxygen required here
Some quick energy + glucose fragments
Fatty acid fragments enter here
Aerobic part of metabolism.
Glucose fragments and fatty acid fragments are broken down completely to provide a large amount of energy.
CO_2
H_2O
Many more steps yield more energy

Activity Duration and Glucose Use

Glucose use during physical activity depends on the *duration* of the activity as well as its *intensity*. In the first 10 minutes or so of an activity, the active muscles rely almost completely on their own stores of glycogen. Within the first 20 minutes or so of moderate activity, a person uses up about one-fifth of the available glycogen. As the muscles devour their own glycogen, they become ravenous for more glucose and increase their uptake of blood glucose dramatically.[12] If you test a person's blood glucose during moderate activity, you will find that it declines slightly, reflecting the muscles' use of blood glucose.

A person who continues exercising moderately for longer than 20 minutes begins to use less glucose and more fat for fuel. Still, glucose use continues, and if the activity goes on for long enough and at a high enough intensity, muscle and liver glycogen stores will run out almost completely (see Figure 10-4). Physical activity can continue for a short time thereafter only because the liver scrambles to produce some glucose from available lactic acid and certain amino acids. This minimum amount of glucose

Figure 10-4

Glycogen Depletion in Cyclists

After three and a half hours of constant cycling, muscle glycogen is used up, but the demand for glucose fuel declines only slightly (dotted line).

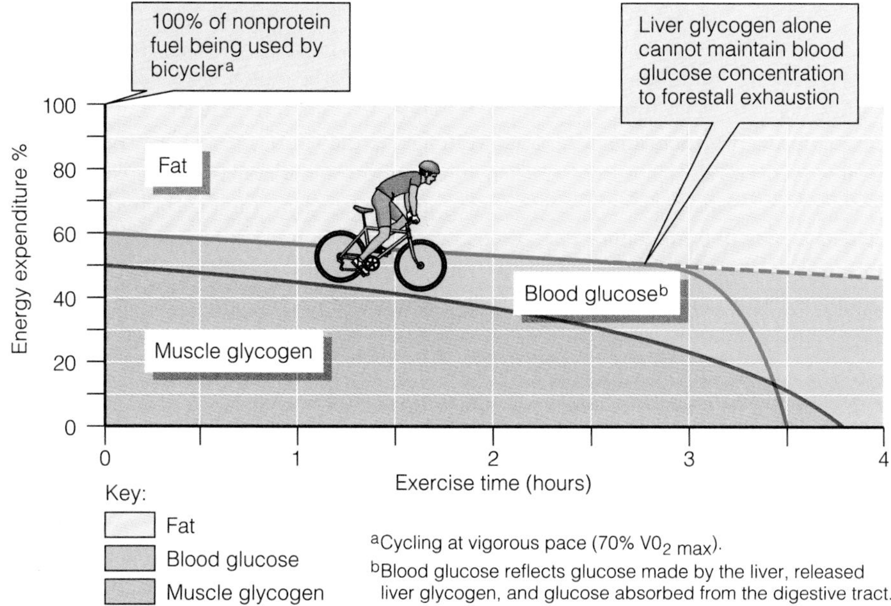

100% of nonprotein fuel being used by bicycler[a]
Liver glycogen alone cannot maintain blood glucose concentration to forestall exhaustion
Energy expenditure %
100
80
60
40
20
0
Fat
Blood glucose[b]
Muscle glycogen
0
1
2
3
4
Exercise time (hours)
Key:
Fat
Blood glucose
Muscle glycogen
[a]Cycling at vigorous pace (70% VO_2 max).
[b]Blood glucose reflects glucose made by the liver, released liver glycogen, and glucose absorbed from the digestive tract.

may briefly forestall exhaustion, but when hypoglycemia accompanies glycogen depletion, it brings nervous system function almost to a halt, making activity impossible. This is what marathon runners call "hitting the wall."

Maintaining Blood Glucose for Activity To postpone exhaustion, endurance athletes must try to maintain their blood glucose concentrations for as long as they can. Three dietary strategies and one training strategy may help maintain glucose concentrations. One diet strategy is to eat a high-carbohydrate diet on a daily basis (see this chapter's Food Feature). Another is to take in some glucose during the activity, usually in fluid (see the next section). The third is to eat carbohydrate-rich foods after activity to boost the storage of glycogen. As for training, the strategy involves training the muscles to store as much glycogen as they can, while supplying enough dietary glucose to enable them to do so (this strategy is called *carbohydrate loading,* described in a later section).

Glucose during Activity Glucose ingested before or during a long-duration competition, makes its way from the digestive tract to the working muscles. This external source of glucose augments dwindling internal glucose supplies from the muscle and liver glycogen stores.[13] Especially during games such as soccer or hockey, which last for hours and demand repeated bursts of intense activity, athletes may benefit from carbohydrate-containing drinks taken during the activity.

Before concluding that sugar might be good for your own performance, consider first whether you engage in *endurance* activity. Do you run, swim, bike, or ski nonstop at a rapid pace for more than an hour at a time, or do you compete in games lasting for hours? If not, the sugar picture changes. For an everyday jog or swim lasting less than 60 minutes, sugar probably won't help performance, and it may do no harm, either. Even in athletes, extra carbohydrate does not benefit those who engage in sports in which fatigue is unrelated to blood glucose, such as 100-meter sprinting, baseball, casual basketball, and weight lifting.[14]

Four strategies can help to maintain blood glucose to support sports performance (for endurance athletes only):

1. Eat a high-carbohydrate diet regularly.
2. Take glucose (usually in diluted fruit juice or other sweet beverages) during endurance activity.
3. Eat carbohydrate-rich foods after performance.
4. Train the muscles to maximize glycogen stores.

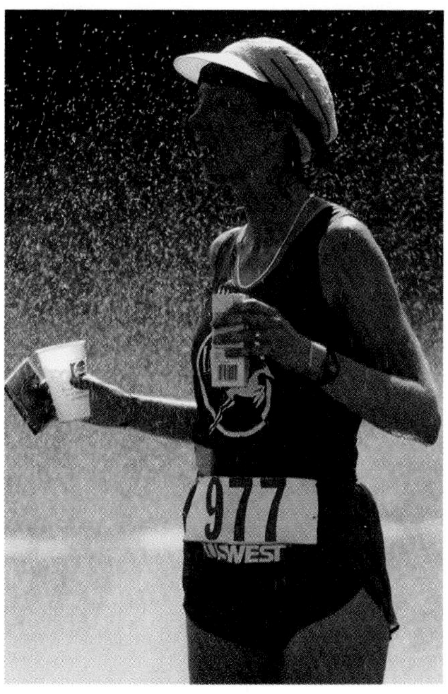

977
USWEST

Those who compete in endurance activities require fluid and carbohydrate fuel.

carbohydrate loading a regimen of exhausting exercise, followed by eating a high-carbohydrate diet, that enables muscles to temporarily store glycogen beyond their normal capacity; also called *glycogen loading* or *glycogen supercompensation*.

To make glycogen, muscles need carbohydrate, but they also need rest. Vary daily activity routines to work different muscles on different days.

Carbohydrate Loading Athletes whose sports routinely exhaust their glycogen stores sometimes use a technique called **carbohydrate loading** to trick their muscles into storing extra glycogen before competition. During the first 4 days of the week before competition, the athlete trains moderately hard (1 to 2 hours per day) and eats a diet that is moderate in carbohydrate. During the 3 days before competition, the athlete gradually cuts back on activity and eats a very-high-carbohydrate diet (about 8 grams of carbohydrate per kilogram of body weight or 70 percent of total energy intake).[‡15] By manipulating activity and carbohydrate, the athlete packs in extra glycogen to fuel activity lasting 90 minutes or longer at a stretch. In a hot climate, extra stored glycogen confers an additional advantage on the endurance athlete. As glycogen breaks down, it releases water, which helps to meet the athlete's fluid needs.

A simpler measure can also be used The athlete can eat a high-carbohydrate meal, such as a glass of orange juice and some crackers, toast, or cereal, within 2 hours after physical activity. This accelerates the rate of glycogen storage by 300 percent.[16] Timing is important—eating the meal after two hours have passed reduces the glycogen synthesis rate by almost half.

KEY POINT *Physical activity of long duration places demands on the body's glycogen stores. Carbohydrate ingested before and during long-duration activity may help to forestall hypoglycemia and fatigue. Carbohydrate loading is a regimen of physical activity and diet that enables an athlete's muscles to store larger-than-normal amounts of glycogen to extend endurance.*

Degree of Training Affects Glycogen Use

Training also affects glycogen use during activity. Muscles that deplete their glycogen stores through work adapt to store greater amounts of glycogen to support that work. As you know, the more glycogen muscles store, the longer the stores last during physical activity.

Muscles make still another adaptation to training that affects glycogen use during activity. Trained muscles burn more fat, and at higher intensities, than untrained muscles, so they require less glucose to perform the same amount of work.[17] A person attempting an activity for the first time uses up much more glucose per minute than an athlete who is trained to perform it. A trained person can work at high intensities for longer periods than an untrained person while using the same amount of glycogen.

Factors that affect glucose use during physical activity:

- Carbohydrate intake.
- Intensity and duration of activity.
- Degree of training.

People with diabetes should know that the moderating effect of physical training on glucose metabolism may have implications for them. Those who must take insulin or insulin-eliciting drugs sometimes find that as their muscles adapt to physical activity, they can reduce their daily drug doses. Physical activity may also improve type 2 diabetes by helping the body lose excess fat. For those with type 1 diabetes, physical activity has been shown to lower the risk of cardiovascular disease, increase insulin sensitivity, lower blood pressure, and improve blood lipids.[18]

Chapter 4 described the action of insulin on blood sugar.

KEY POINT *Highly trained muscles use less glucose and more fat than do untrained muscles to perform the same work, so their glycogen lasts longer.*

To Burn More Fat during Activity, Should Athletes Eat More Fat?

An athlete who eats a fat-rich diet with little carbohydrate will burn more fat during activity but will sacrifice endurance, as Figure 10-2 showed. The importance of a

‡Percentage of energy intake is meaningful only when total energy intake is known. Consider that at high energy intakes (say 5,000 calories per day), even a moderate carbohydrate diet (40 percent of energy intake) supplies 500 grams of carbohydrate—enough for a 137-pound athlete in heavy training. By comparison, at a moderate energy intake (2,000 calories per day), a high carbohydrate intake (70 percent of energy intake) supplies 350 grams—plenty of carbohydrate for most people, but not enough for athletes in heavy training.

high-carbohydrate diet for endurance has long been recognized. Of course, extremes in diet such as severely limiting fats or eliminating them from the diet are always detrimental to nutrition and do the athlete no service.[19] Indeed, limited research has found that a high-fat diet eaten for less than one month may actually benefit performance in some athletes.[20]

Some words of caution are in order to those who may begin to wonder if a high-fat diet might improve their own athletic performance. In endurance activity, the athlete's oxygen requirement is high but is made higher still by a high-fat diet. Fat, remember, can yield energy only in the presence of oxygen, that is, aerobically (refer back to Figure 10-3). Extra fat in the diet, therefore, places extra demands on the pumping heart to supply the needed oxygen to the working muscles.[21] Carbohydrate fuel, in contrast, yields some energy without oxygen. In addition, high-fat diets carry risks of heart disease. Physical activity offers some protection against cardiovascular disease, but even athletes can suffer heart attacks and strokes. Most nutrition experts agree that the potential for adverse health effects from prolonged high-fat diets makes them an unwise choice for athletes.

Body fat stores are more important as fuel for activity than is fat in the diet. Unlike the body's glycogen stores, which are limited, fat stores can fuel hours of activity without running out; body fat is (theoretically) an unlimited source of energy. Even the lean bodies of elite runners carry enough fat to fuel several marathon runs.

Early in activity, muscles begin to draw on fatty acids from two sources—fats stored within the working muscles and fats from fat deposits such as the fat under the skin. Areas that have the most fat to spare donate the greatest amounts of fatty acids to the blood (although they may not be the areas that you would choose to lose fat from). This is why "spot reducing" doesn't work: muscles do not own the fat that surrounds them. Fat cells release fatty acids into the blood for all the muscles to share. Proof of this is found in a tennis player's arms: the fatfolds measure the same in both arms, even though one arm has better-developed muscles than the other.

Intensity and Duration Affect Fat Use The *intensity* of physical activity also affects the percentage of energy contributed by fat. As mentioned, fat can be broken down for energy in only one way—by aerobic metabolism. When the intensity of activity becomes so great that energy demands surpass the ability to provide energy aerobically, the body cannot burn more fat. Instead, it burns more glucose.

The *duration* of activity also matters to fat use. At the start of activity, the blood fatty acid concentration falls, but a few minutes into an activity, the neurotransmitter norepinephrine signals the fat cells to break apart their stored triglycerides and to liberate fatty acids into the blood. After about 20 minutes of activity, the blood fatty acid concentration rises above the normal resting concentration. It is only during this phase of sustained, submaximal activity, after the first 20 minutes, that the fat cells begin to shrink in size as they empty out their fat stores. Intense, prolonged activity can have other effects as well, as the accompanying Fitness for Life feature explains.

Factors that affect fat use during physical activity:

- Fat intake.
- Intensity and duration of the activity.
- Degree of training.

Degree of Training Affects Fat Use Training—repeated aerobic activity—stimulates the muscles to develop more fat-burning enzymes. Aerobically trained muscles burn fat more readily than untrained muscles. With aerobic training, the heart and lungs also become stronger and better able to deliver oxygen to the muscles during high-intensity activities. This improved oxygen supply also enables the muscles to burn more fat.

KEY POINT *Athletes who eat high-fat diets may burn more fat during endurance activity, but the risks to health outweigh any possible performance benefits. The intensity and duration of activity, as well as the degree of training, affect fat use.*

Can Physical Training Speed Up an Athlete's Metabolism?

FITNESS FOR LIFE

Athletes in training, whether endurance athletes or power athletes, expend huge amounts of energy each day while practicing their chosen activity. Common sense tells us that the harder an athlete works, the more energy the athlete spends. But what about *after* the work is done? Does the athlete continue to spend more energy at rest than a sedentary person or a casual exerciser? Research suggests that the answer may be yes, for a limited time after intense, prolonged activity. For example, resistance training can elevate an athlete's resting metabolic rate by increasing the body's lean tissues, which are more metabolically active than fat. Intense endurance activity (at greater than 70 percent of maximal oxygen consumption), however, seems to increase resting metabolic rate independently of any change in muscle mass. The rise in resting metabolic rate may last anywhere from minutes to hours depending on the intensity and duration of the activity and the training status of the athlete. The bottom line is that athletes need more energy to cover their increased expenditures both on and off the playing field. They need more nutrients, too, to help use that energy. The Food Feature at the end of this chapter tells how to include more servings of nutritious foods each day to cover both needs amply.

Using Protein and Amino Acids for Building Muscles and to Fuel Activity

Athletes use protein to build and maintain muscle and other lean tissue structures and, to a small extent, to fuel activity. The body handles protein differently during activity than during rest, however.

Protein for Building Muscle Tissue In the hours of rest that follow physical activity, muscles speed up their rate of protein synthesis—they build more of the proteins they need to perform the activity. Additionally, whenever the body rebuilds a part of itself, it must tear down the old structures to make way for the new ones. Physical activity, with just a slight overload, calls into action both the protein-dismantling and the protein-synthesizing equipment of each muscle cell that work together to remodel muscles.

Physical activity itself triggers the building of muscle proteins.

Dietary protein provides the needed amino acids for synthesis of new muscle proteins. As Chapter 6 pointed out, however, the true director of synthesis of muscle protein is physical activity itself. Repeated activity signals the muscle cells' genetic material to begin producing more of the proteins needed to perform the work at hand.

The genetic protein-making equipment inside the nuclei of muscle cells seems to "know" when proteins are needed. Furthermore, it knows *which* proteins are needed to support each type of physical activity. Apparently, the intensity and pattern of muscle contractions initiate signals that direct the muscles' genetic material to make particular proteins. For example, a weight lifter's workout sends the information that muscle fibers need added bulk for strength and more enzymes for making and using glycogen. A jogger's workout stimulates production of proteins needed for aerobic oxidation of fat and glucose. Muscle cells are exquisitely responsive to the need for proteins, and they build them conservatively only as needed.

Finally, after muscle cells have made all the decisions about which proteins to build and when, protein nutrition comes into play. During active muscle-building phases of training, a weight lifter might add to existing muscle mass between ¼ ounce and 1 ounce (between 7 and 28 grams) of protein each day. This extra protein comes from ordinary food.

Protein Used for Fuel Not only do athletes retain more protein, but they also use a little more protein as fuel.[22] Studies of nitrogen balance show that the body speeds up its use of amino acids for energy during physical activity, just as it speeds up its use of glucose and fatty acids. Protein contributes about 10 percent of the total fuel used, both during activity and during rest.

Diet Affects Protein Use during Activity The factors that regulate how much protein is used during activity seem to be the same three that regulate the use of glucose and fat. One factor is diet—a carbohydrate-rich diet spares protein from being used as fuel. Some amino acids can be converted into glucose when needed. Others, the **branched-chain amino acids,** can stand in for glucose in energy pathways. If the diet is low in carbohydrate, much more protein will be used in place of glucose.

branched-chain amino acids amino acids that, unlike the others, can provide energy directly to muscle tissue: leucine, isoleucine, and valine.

Factors that affect protein use during physical activity:

- Carbohydrate intake.
- Intensity and duration of the activity.
- Degree of training.

Intensity and Duration Affect Protein Use Second, the intensity and duration of the activity also affect protein use.[23] Endurance athletes who train for over an hour a day, engaging in aerobic activity of moderate intensity and long duration, may deplete their glycogen stores by the end of their training and become more dependent on body protein for energy. The protein needs of bodybuilders and weight lifters are higher than those of sedentary people, but not as high as the protein intakes many bodybuilders consume.

Degree of Training Affects Protein Use Finally, the degree of training also affects the use of protein. The better trained the athlete, the less protein used during activity at a given intensity.

How Much Protein Should an Athlete Consume?

Most athletes probably need somewhat more protein than do sedentary people. In the United States, however, average protein intakes are high enough to cover even the needs of most athletes. Therefore, athletes in training should attend to protein needs, but should back up the protein with ample carbohydrate. Otherwise, they will burn off as fuel the very protein they wish to retain in muscle.

A joint position paper from the American Dietetic Association (ADA) and the Dietitions of Canada (DC) recommends 1.0 to 1.5 grams of protein per kilogram of body weight each day, an amount somewhat higher than the 0.8 gram of protein per kilogram of body weight recommended for sedentary people.[24] Another authority suggests different protein intakes for different athletes.[25] Table 10-2 lists some recommendations and translates them into daily intakes for athletes.

Table 10-2

Recommended Protein Intakes for Athletes

	Recommendations (g/kg/day)	Protein Intakes (g/day)	
		MALES	FEMALES
1989 RDA for adults	0.8	56	44
ADA/CDA recommended intake	1.0–1.5	70–105	55–83
Recommended intake for power (strength or speed) athletes	1.7–1.8	119–126	94–99
Recommended intake for endurance athletes	1.2–1.4	84–98	66–77
U.S. average intake		95	65

NOTE: Daily protein intakes are based on a 70-kilogram (154-pound) man and 55-kilogram (121-pound) woman.
SOURCES: Committee on Dietary Allowances, *Recommended Dietary Allowances,* 10th ed. (Washington, D.C.: *National Academy Press,* 1989); Position of the American Dietetic Association and the Canadian Dietetic Association; Nutrition for physical fitness and athletic performance for adults, *Journal of the American Dietetic Association* 93 (1993): 691–695; P. W. R. Lemon, Effect of exercise on protein requirements, in C. Williams and J. T. Devlin, eds., *Foods, Nutrition and Sports Performance: An International Scientific Consensus* (London: E & FN Spon, 1992), pp. 65–86.

After considering these recommendations, athletes may still wonder whether their diet provides the protein they need. This chapter's Food Feature answers questions about choosing a performance diet. Meanwhile, relax. Athletes who eat a balanced, high-carbohydrate diet that provides enough total energy also consume enough protein.

KEY POINT *Physical activity stimulates muscle cells to break down and synthesize protein, resulting in muscle adaptation to activity. Athletes use protein both for building muscle tissue and for energy. Diet, intensity and duration of activity, and training affect protein use during activity.*

Vitamins and Minerals—Keys to Performance

Many vitamins and minerals assist in releasing energy from fuels and transporting oxygen. In addition, vitamin C is needed for the formation of the protein collagen, the foundation material of bones and the cartilage that forms the linings of the joints and other connective tissues. Folate and vitamin B_{12} help build the red blood cells that carry oxygen to working muscles. Calcium and magnesium help make muscles contract, and so on. Do active people need extra nutrients to support their work? Do they need supplements?

Do Nutrient Supplements Benefit Athletic Performance?

An estimated 84 percent of world-class athletes take nutrient supplements. Many other athletes also take supplements in the hope of improving their performance.

The summary tables listing functions of vitamins and minerals begin on pp. 243 and 298.

Thiamin, Riboflavin, and Niacin This vitamin trio plays key roles in energy release. Scientists have concluded, however, that extra amounts of these vitamins from supplements do not benefit performance.[26] An adequate diet supplies all the thiamin, riboflavin, and niacin needed by an active person, even an athlete. Athletes, with their greater energy needs, eat more food, so most athletes' diets are adequate in these vitamins. Extra amounts provide no competitive advantage.

Indeed, niacin in excess of the recommended amount may affect performance adversely: excess niacin suppresses the release of fatty acids and thus forces muscles to use extra glycogen during physical activity.[27] Thus glycogen may be depleted more rapidly, making the work seem more difficult.

These comments should not be interpreted to mean that thiamin, riboflavin, niacin, or any other vitamins are not important. The words *adequate diet* are weighty in this regard. They mean that athletes should obtain most of the extra energy they need from nutrient-dense foods, not fats, sweets, or highly refined foods. Anyone who consumes a diet of relatively empty-calorie foods risks becoming vitamin deficient. But athletes who use mostly nutrient-dense foods to provide energy exceed recommendations by far, not only for thiamin and riboflavin, but for other vitamins and minerals as well.[28]

Vitamin B_6 and Vitamin B_{12} Vitamin B_6 plays key roles in the release of energy from nutrients, in the liberation of glucose from glycogen, and in the formation of hemoglobin. Thus sellers of supplements claim that vitamin B_6 pills promote athletic performance, but scientific research proves otherwise.[29] To ensure that the diet is adequate in vitamin B_6, a person need only include some leafy green vegetables, meats, fish, legumes, fruits, and whole grains. Megadoses of vitamin B_6 provide no additional benefit, and large doses can be toxic.

The belief that vitamin B_{12} supplementation will enhance performance stems from its role in the production of red blood cells. Anemias of all kinds rob the blood of its oxygen-carrying capacity by reducing the number and impairing the function of circulating red blood cells. Vitamin B_{12} deficiency causes anemia, but so do iron and folate deficiencies (and others, see the margin). Chances are that a diet low enough in vitamin B_{12} to bring on anemia will be low in other nutrients as well. A person with so poor a diet does not need to take pills; the person needs to eat right. For a well-nourished athlete, any perceived benefits from vitamin B_{12} supplements or shots taken before competition are based on psychology, not physiology.

stress fracture a bone injury or break caused by the stress of exercise on the bone surface.

amenorrhea the absence or cessation of menstruation.

Nutrients necessary to ward off anemias include vitamins A, B_6, B_{12}, and folate and the minerals iron, zinc, copper, and magnesium along with protein—in short, the perfect mix of nutrients that occurs naturally in whole, nutrient-dense foods.

Vitamins C and E, the Antioxidants During prolonged, high-intensity physical activity, the muscles' consumption of oxygen increases tenfold or more, enhancing the production of damaging free radicals in the body.[30] Vitamin E is a potent fat-soluble antioxidant that vigorously defends cell membranes against oxidative damage.[31] Vitamin C, a water-soluble antioxidant, helps regenerate vitamin E. Vitamin E seems to be the most important antioxidant related to physical activity. Many athletes are taking antioxidant supplements, particularly vitamin E, in hopes of preventing oxidative damage to muscles. The results of some research lend support to this practice.[32] Research suggests that supplementation either with vitamins C and E together, or with vitamin E alone, offers protection against exercise-induced oxidative stress.[33] Supplement doses in these studies varied considerably, however, and no one yet knows the precise dose that will offer the greatest benefits with the least risk of toxicity. Furthermore, although research suggests that antioxidant supplements may offer protection against oxidative stress, there is little evidence that these supplements can improve performance.[34] Clearly, more research is needed before drawing conclusions about antioxidant supplements for athletes.

Chapter 7 specified the risks of toxicity from vitamin pills and supplements.

Stringent weight requirements pose a risk of developing eating disorders. See this chapter's Controversy.

So far, research indicates that, with the possible exception of vitamin E, the working body gains no benefit from vitamin supplements.[35] To meet recommendations for vitamins, athletes need only consume sufficient nutrient-dense food—and most athletes certainly can do this. Athletes who must lose weight to meet low body-weight requirements, however, may consume so little food that they fail to obtain all the nutrients they need. The practice of "making weight" is opposed by many health-minded groups, but for athletes who choose this course, a single daily multivitamin-mineral tablet that provides no more than the recommended amounts of nutrients may be beneficial.

KEY POINT *Vitamins are essential for releasing the energy trapped in energy-yielding nutrients and for other functions that support physical activity. Active people can meet their vitamin needs if they eat enough nutrient-dense foods to meet their energy needs.*

Female athlete triad:

- Disordered eating.
- Amenorrhea.
- Osteoporosis.

Physical Activity and Bone Loss

Osteoporosis, the condition of reduced bone mass, increases susceptibility to bone damage, including **stress fractures.** Controversy 8 pointed out that moderate physical activity and adequate calcium intakes protect against bone loss. Extremes in physical activity, however, may be detrimental to bone health, at least in some young women and especially in adolescent girls.[36] Many young women athletes and dancers restrict energy intakes to meet the weight guidelines of their sport and thus have calcium intakes below the recommendations.[37] Such young women risk developing a potentially fatal triad of medical problems: abnormal eating behaviors, **amenorrhea,** and osteoporosis.[38] These three associated disorders, called the "female athlete triad," are discussed in Controversy 10.

KEY POINT *Moderate physical activity strengthens the bones, but young women athletes who train strenuously, become amenorrheic, and practice abnormal eating behaviors are susceptible to stress fractures and osteoporosis.*

Foods like these are packed with the nutrients that active people need.

Iron and Performance

Endurance athletes, and especially women athletes, are prone to iron deficiency. Physical activity may impair iron status in any of several ways. For one, iron may be excreted in sweat.[39] For another, iron may be lost through red blood cell destruction; blood cells are squashed when body tissues (such as the soles of the feet) make high-impact contact with an unyielding surface (such as the ground). In addition, physical activity may cause small blood losses through the digestive tract, at least in some athletes. Perhaps more significant than losses are the high iron demands by muscles to make the iron-containing molecules of aerobic metabolism. Habitually low intakes of iron-rich foods as well as increased losses and extra demands may contribute to iron deficiency in young women athletes.[40] Vegetarian women athletes may be especially vulnerable to iron insufficiency.[41]

Iron deficiency impairs performance because iron helps deliver the muscles' oxygen. Insufficient oxygen delivery reduces aerobic work capacity, so the person tires easily. Whether marginal deficiency without clinical signs of anemia hinders physical performance is a point of debate among researchers.[42]

Early in training, athletes may develop low blood hemoglobin for a while. This condition, sometimes called "sports anemia," probably reflects a normal adaptation to physical activity. Aerobic training promotes increases in the fluid of the blood; with more fluid, the red blood cell count in a unit of blood drops. True iron-deficiency anemia requires treatment with prescribed iron supplements, but sports anemia goes away by itself, even with continued training.

The best strategy concerning iron may be to determine individual needs. Many menstruating women probably border on iron deficiency even without the additional iron demand and losses incurred by physical activity. Teens of both sexes, because they are growing, have high iron needs, too. Especially for women and teens, then, prescribed supplements may be needed to correct a deficiency of iron that is confirmed by tests. (Medical testing is needed to eliminate nondietary causes of anemia, such as internal bleeding or cancer.)

KEY POINT *Iron-deficiency anemia impairs physical performance because iron is the blood's oxygen handler. Sports anemia is probably a harmless temporary adaptation to physical activity.*

Women athletes may be at special risk of iron deficiency.

Other Minerals

Three trace minerals—chromium, zinc, and copper—have specific roles in physical activity. The excretion of all three accelerates during physical training, but research has not yet pinpointed the nutrition implications of this finding.[43] Chromium has received a lot of media attention as a purported muscle builder or fat burner for athletes. The Consumer Corner discusses chromium picolinate and the claims made for it.

During physical activity, the body loses electrolytes—the minerals sodium, potassium, chloride, and magnesium—in sweat. Beginners lose sodium, potassium, and chloride (but not magnesium) to a much greater extent than do trained athletes. As the body adapts to physical activity, it becomes better at conserving these electrolytes.

Magnesium losses in sweat are no greater for untrained individuals than for trained athletes. Both lose more magnesium in urine than in sweat, and some studies suggest that physical activity accelerates magnesium excretion.[44] Perhaps working muscles take the mineral from storage for their use and then release it; then the kidneys excrete it.

Wise athletes plan to obtain enough magnesium. They know that a magnesium deficiency has been shown to hinder the muscle gains associated with a given amount of training. Magnesium is abundant in leafy vegetables, legumes, and whole-wheat products. Other foods offer small but significant amounts. Convenience and highly processed snack foods lack magnesium.

Normally, potassium remains safely inside the cells where it does its work. In prolonged dehydration from profuse sweating, it may migrate outside the cells and be lost by excretion in the urine. Even so, potassium is easily replaced with just a few servings of fresh fruits and vegetables. Avoid potassium supplements unless prescribed by a physician; although they improve some conditions, they worsen others. Most times, a regular diet supplies all the electrolytes athletes need.

Athletes comprise a huge and favorable market for the supplement industry, and they are one of the groups most often victimized by frauds. The following Consumer Corner touches on some of the most common schemes aimed at athletes and warns of the dangers of using steroids and other drugs.

KEY POINT *The body adapts to compensate for sweat losses of electrolytes, but urinary magnesium and potassium losses may persist. Athletes are advised to use foods, not supplements, to make up for these losses.*

heat stroke an acute and life-threatening reaction to heat buildup in the body.

hypothermia a below-normal body temperature.

Fluids and Temperature Regulation in Physical Activity

The body's need for water far surpasses its need for any other nutrient. If the body loses too much water, as in dehydration, its life-supporting chemistry is compromised.

The exercising body loses water primarily via sweat; second to that, breathing costs water, exhaled as vapor. During physical activity, both routes can be significant, and dehydration is a real threat. The first symptom of dehydration is fatigue. A water loss of even 1 to 2 percent of body weight can reduce a person's capacity to do muscular work.[45] A person with a water loss of about 7 percent is likely to collapse.[46] The athlete who arrives at an event even slightly dehydrated starts out at a competitive disadvantage.

Temperature Regulation

Chapter 8 pointed out that sweat cools the body. The conversion of water to vapor uses up a great deal of heat, so as sweat evaporates, it cools the skin's surface and the blood flowing beneath it.

In hot, humid weather, sweat may fail to evaporate because the surrounding air is already laden with water. Little cooling takes place and body heat builds up. In such conditions, athletes must take precautions to avoid **heat stroke.** Heat stroke is an especially dangerous accumulation of body heat with accompanying loss of body fluid. Three measures to prevent heat stroke are to drink enough fluid before and during the activity, to rest in the shade when tired, and to wear lightweight clothing that encourages evaporation.[47] Hence the rubber or heavy suits sold with promises of weight loss during physical activity are dangerous because they promote profuse sweating, prevent sweat evaporation, and invite heat stroke. If you experience any of the symptoms of heat stroke listed in the margin, stop your activity, sip cold fluids, seek shade, and ask for help. The condition demands medical attention; it can kill.

Symptoms of heat stroke: headache, nausea, dizziness, clumsiness, stumbling, sudden cessation of sweating (hot, dry skin), internal (rectal) temperature above 104° Fahrenheit, and confusion or loss of consciousness.

In cold weather, **hypothermia,** or loss of body heat, can pose as serious a threat as heat stroke does in hot weather. Inexperienced runners participating in long races on cold or wet, chilly days are especially vulnerable to hypothermia. Slow runners can produce too little heat to keep warm, especially if their clothing is inadequate. Early symptoms of hypothermia include shivering and euphoria. As body temperature continues to fall, shivering may stop, and weakness, disorientation, and apathy may set in. People with these symptoms soon become helpless to protect themselves from further body heat losses. Even in cold weather, the body still sweats and needs fluids, but the fluids should be warm or at room temperature, not cold.

KEY POINT *Evaporation of sweat cools the body. Heat stroke can be a threat to physically active people in hot, humid weather. Hypothermia threatens those who excercise in the cold.*

CONSUMER CORNER

Steroids and "Ergogenic" Aids

ATHLETES CAN BE sitting ducks for quacks. An endless array of scams are aimed at them: protein powders and amino acid supplements, vitamin and mineral supplements, steroid replacers, "muscle-building" powders, and electrolyte pills. Some athletes take dangerous illegal drugs to try to gain a competitive edge. Others use sodium bicarbonate, caffeine, or other products. Table 10-3 describes these and many other so-called **ergogenic aids** used by athletes. The term *ergogenic* implies that such products have special work-enhancing powers, but no food or supplement is really ergogenic.

An athlete who takes a nutrient supplement or other substance to improve performance cannot be sure that it will deliver on the promises made for it. A variety of supplements make claims based on misunderstood or misinterpreted nutrition principles. The claims may sound good, but they have no factual basis. In some cases, little or no information is available about the ingredients of the supplements. In other cases, dosage levels for listed ingredients are not given, or the suggested doses are extremely high. Rarely, if ever, do the products mention possible side effects or offer warnings to pregnant women or people with hypertension or other conditions that might contraindicate their use. The findings of a survey of advertisements in a dozen popular health and bodybuilding magazines underscore these points.[1] Researchers identified over 300 products containing 235 different ingredients advertised as beneficial, mostly for muscle growth. None had been scientifically proved effective.

Even worse, someone considering using illegal drugs such as steroids should be aware that such drugs can contain anything, even poisons, because no one tests illegal drugs for safety. Among the most dangerous products sold to athletes are steroids, other hormones, amphetamines, cocaine, muscle relaxants, tranquilizers, barbiturates, diuretics, and even veterinary drugs.

Of the hormones, **anabolic steroid hormones** are made naturally by the testes and adrenal cortex in men and by the adrenal cortex in women. The steroid drugs some athletes take are synthetic varieties that combine the masculinizing effects of male hormones and the growth stimulation of the adrenal steroids. In the body, the steroids produce accelerated muscle bulking in response to physical activity in both men and women. Injections of these hormones produce muscle size and strength far beyond that attainable by training alone, but at the price of great risks to health, as Figure 10-5 on page 374 demonstrates. The American Academy of Pediatrics and the American College of Sports Medicine condemn athletes' use of anabolic steroids, and the International Olympic Committee bans their use.[2]

Some athletes use DHEA as an alternative to anabolic steroids. DHEA is a hormone (dehydroepiandrosterone), made in the adrenal glands, that serves as a precursor to the male hormone testosterone. Advertisements claim it "burns fat," "builds muscle," and "slows aging," but evidence to support such claims is lacking.

DHEA's short-term side effects include oily skin, acne, body hair growth, liver enlargement, and aggressive behavior.[3] Long-term effects of DHEA use remain to be seen and may take years to become evident. The potential for harm from DHEA supplements is great, and athletes, as well as others, should avoid it. DHEA is banned by the International Olympic Committee and the National Collegiate Athletic Association.

Though not a steroid, **growth hormone** can induce huge body size and is less readily detected in drug tests than steroids. Its abuse can result in a condition known as acromegaly, characterized by a widened jawline, widened nose, protruding brow, buck teeth, weakened heart walls, an enlarged heart, and an increased likelihood of death before age 50.[4] Athletes who have paid the price of hormone abuse, even some for whom the drugs made careers in sports possible, have come forward to warn young athletes away from growth hormones. They say that even the rewards of success in sports are not worth the side effects of the drugs.

Growth hormone "stimulators," such as the amino acids ornithine and arginine, are useless in the form sold to athletes. In laboratory studies, huge doses of these amino acids do stimulate growth hormone release, but the effective dose would be too dangerous to take. There are safe ways to maximize growth hormone production, however. One is rest. Growth hormone is released during sleep, especially after physical activity, so getting enough rest and adequate training are effective.

Extracted herb and insect sterols are hawked as legal substitutes for steroid drugs. Sellers falsely claim that these substances contain hormones or that they enhance the body's natural ability to make anabolic hormones. In some cases, the substances actually are plant or insect sterols, but the body cannot convert them to human steroids. None of these products has any proven anabolic activity, nor can any of them strengthen muscles; in fact, they may

Table 10-3

Products Athletes Use

- **anabolic steroid hormones** chemical messengers related to the male sex hormone testosterone that stimulate building up of body tissues (*anabolic* means "promoting growth;" *sterol* refers to compounds chemically related to cholesterol).
- **androstendione** a precursor of testosterone that elevates both testosterone and estrogen in the blood of both males and females.
- **arginine** a nonessential amino acid falsely promoted as enhancing the secretion of human growth hormone, the breakdown of fat, and the development of muscle.
- **bee pollen** a product consisting of bee saliva, plant nectar, and pollen that confers no benefit on athletes and may cause an allergic reaction in individuals sensitive to it.
- **boron** a nonessential mineral that is promoted as a "natural" steroid replacement.
- **caffeine** a stimulant that in small amounts may produce alertness and reduced reaction time in some people, but that also creates fluid losses. Overdoses cause headaches, trembling, an abnormally fast heart rate, and other undesirable effects. More about caffeine appears later in this chapter and in Controversy 12.
- **carnitine** a nitrogen-containing compound formed in the body from lysine and methionine that helps transport fatty acids across the mitochondrial membrane. Carnitine is claimed to "burn" fat and spare glycogen during endurance events, but it does neither.
- **cell salts** a mineral preparation supposedly prepared from living cells.
- **chaparral** an herb, promoted as an antioxidant (see also Table 11-9, p. 416).
- **chromium picolinate** a trace element supplement; falsely promoted to increase lean body mass, enhance energy, and burn fat.
- **coenzyme Q10** a lipid found in cells (mitochondria) that has been shown to improve exercise performance in heart disease patients, but is not effective in improving performance of healthy athletes.
- **creatine** a nitrogen-containing compound that combines with phosphate to burn a high-energy compound stored in muscle. Claims that creatine safely enhances energy and stimulates muscle growth are unconfirmed.
- **DHEA (dehydroepiandrosterone)** a hormone made in the adrenal glands that serves as a precursor to the male hormone testosterone; falsely promoted as burning fat, building muscle, and slowing aging.
- **DNA** and **RNA (deoxyribonucleic acid** and **ribonucleic acid)** the genetic materials of cells necessary in protein synthesis; falsely promoted as ergogenic aids.
- **ergogenic aids** products that supposedly enhance performance, although none actually do so; the term *ergogenic* implies "energy giving" (*ergo* means "*work*"; *genic* means "give rise to").
- **ginseng** a plant whose extract supposedly boosts energy (see Table 11-9, p. 416).
- **glycine** a nonessential amino acid, promoted as an ergogenic aid because it is a precursor of the high-energy compound phosphocreatine. Other amino acids that are commonly packaged for athletes but are equally useless include ornithine, arginine, lysine, and the branched-chain amino acids.
- **growth hormone** a hormone produced by the brain's pituitary gland that regulates normal growth and development (see text discussion); also called *somatotropin*.
- **growth hormone releasers** herbs or pills that supposedly regulate hormones; falsely promoted as enhancing athletic performance.
- **guarana** a reddish berry found in Brazil's Amazon valley that contains seven times as much caffeine as its relative the coffee bean. It is used as an ingredient in carbonated sodas and taken in powder or tablet form to enhance speed and endurance and serve as an aphrodisiac, a "cardiac tonic," an "intestinal disinfectant," and a smart drug that supposedly improves memory and concentration and wards off senility. High doses may stress the heart and can cause panic attacks.
- **inosine** an organic chemical that is falsely said to "activate cells, produce energy, and facilitate exercise." Studies have shown that it actually reduces the endurance of runners.
- **ma huang** an herbal preparation sold with promises of weight loss and increased energy, but that contains ephedrine, a cardiac stimulant with serious adverse effects (see Table 7-4, p. 240).
- **octacosanol** an alcohol extracted from wheat germ, often falsely promoted as enhancing athletic performance.
- **ornithine** a nonessential amino acid falsely promoted as enhancing the secretion of human growth hormone, the breakdown of fat, and the development of muscle.
- **phosphate salt** a salt that has been demonstrated to raise the concentration of a metabolically important compound (diphosphoglycerate) in red blood cells and enhance the cells' potential to deliver oxygen to muscle cells. The salts may cause calcium losses from the bones if taken in excess.
- **plant sterols** lipid extracts of plants, called ferulic acid, oryzanol, phytosterols, or "adaptogens," marketed with false claims that they contain hormones or enhance hormonal activity.
- **pyruvate** a 3-carbon compound derived during the metabolism of glucose, certain amino acids, and glycerol; falsely promoted as burning fat and enhancing endurance. Common side effects include intestinal gas and diarrhea.
- **royal jelly** a substance produced by worker bees and fed to the queen bee; often falsely promoted as enhancing athletic performance.
- **sodium bicarbonate** baking soda; an alkaline salt believed to neutralize blood lactic acid and thereby reduce pain and enhance possible workload. "Soda loading" may cause intestinal bloating and diarrhea.
- **superoxide dismutase (SOD)** an enzyme that protects cells from oxidation. When it is taken orally, the body digests and inactivates this protein; it is useless to athletes.

contain natural toxins. Controversy 7 first made this point, but it is worth repeating: don't make the mistake of equating "natural" with "harmless."

It also bears repeating that amino acid supplements can be dangerous (see the Consumer Corner in Chapter 6). Healthy athletes never need them. Advertisers point to research that identifies the branched-chain amino acids as a source of fuel for the exercising body, but fail to mention that when amino acids are needed, the muscles have plenty on hand. Any diet low in carbohydrate or calories seems to activate or assist an enzyme that breaks down branched-chain amino acids for energy. Otherwise, branched-chain amino acids are conserved. Studies to determine whether supplements of branched-chain amino acids enhance

Figure 10-5

Physical Risks of Taking Steroid Hormone Drugs

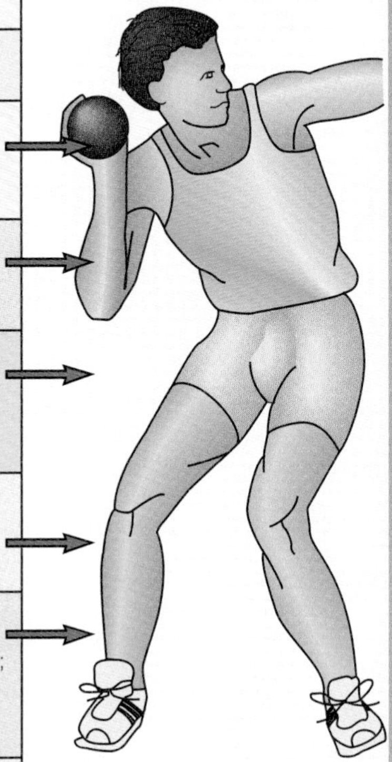

Mind	Extreme aggression with hostility ("steroid rage"); mood swings; anxiety; dizziness; drowsiness; unpredictability; psychotic depression; personality changes; suicidal thoughts
Face and Hair	Swollen appearance; greasy skin; severe, scarring acne; mouth and tongue soreness; yellowing of whites of eyes; In females, male-pattern hair loss and increased growth of face and body hair
Voice	In females, irreversible deepening of voice
Chest	In males, breathing difficulty; breathing stoppage; breast development In females, breast atrophy
Heart	Heart disease; elevated or reduced heart rate; heart attack; stroke; hypertension; increased LDL; drastic reduction in HDL
Abdominal Organs	Nausea; vomiting; bloody diarrhea; pain; liver tumors (possibly cancerous); liver damage, disease, or rupture leading to fatal liver failure (peliosis hepatitis)[a]; kidney stones and damage; frequent urination; possible rupture of aneurysm or hemorrhage
Blood	Blood clots; high risk of blood poisoning; those who share needles risk contracting HIV (the AIDS virus) or other disease-causing organisms
Reproductive System	In males, permanent shrinkage of testes; prostate enlargement with increased risk of cancer; sexual dysfunction; loss of fertility; excessive and painful erection; In females, loss of menstruation and fertility; permanent enlargement of external genitalia
Muscles, Bones, and Connective Tissues	Increased susceptability to injury with delayed recovery times; cramps, tremors; seizure-like movements; injury at injection site; in adolescents, failure to grow to normal height
Other	Fatigue; increased risk of cancer

[a]In peliosis hepatitis, excess buildup of bile causes destruction of liver cells. Blood pools form and liver failure causes death.

endurance have produced conflicting results. Some studies suggest improved endurance, some report no effect on endurance, and still others show a detrimental effect.[5] Large doses of branched-chain amino acids can, however, raise plasma ammonia concentrations, which can cause fatigue and can be toxic to the brain.[6] The wise athlete, then, takes no amino acid supplements, but eats a diet adequate in carbohydrate and energy.

Chromium is an essential trace mineral involved in carbohydrate and lipid metabolism. Advertisements in bodybuilding magazines claim that chromium picolinate, which is supposed to be more easily absorbed than chromium alone, builds muscle, enhances energy, and burns fat. Such claims derive from one or two initial studies reporting that men who weight trained while taking chromium picolinate supplements increased lean body mass and reduced body fat.[7] Most subsequent studies of chromium picolinate and strength training, however, show no effects of chromium picolinate supplementation on strength, lean body mass, or body fat.[8] A dangerous condition of muscle degeneration was reported in an athlete ingesting 1,200 micrograms of chromium picolinate over two days' time.[9] No one knows for sure whether the chromium was the cause of the disease, but the doctors strongly suspect that it was.

Interest in—and use of—creatine monohydrate supplements to enhance performance during intense activity has grown dramatically in the last few years.[10] Power athletes such as weight

lifters use creatine supplements in the belief that they enhance stores of the high-energy compound creatine phosphate (or phosphocreatine) in muscles. Theoretically, the more creatine phosphate in muscles, the higher the intensity at which an athlete can train.

The results of some studies suggest that creatine supplementation may somewhat enhance performance of high-intensity strength activity such as weight lifting.[11] Other findings, however, find no such effect of creatine supplements on strength performance but research is continuing.[12]

New studies on creatine are published almost daily, but definitive answers to such questions as "Does creatine supplementation enhance sports performance?" and "Is creatine supplementation safe?" are not close at hand. For example, although some research does suggest creatine can improve certain kinds of performance, most such studies are conducted in laboratories using small numbers of subjects and, as research protocol dictates, controlled dosages.[13] Still lacking are studies of athletes using commercially available creatine products in the doses athletes actually consume. One group of researchers studied the performance of U.S. Navy SEALS.[14] Navy SEALS need both strength and endurance to perform the extremely demanding physical tasks their jobs often require. The researchers timed the performance of 24 SEALS on a four-station obstacle course. The SEALS were randomly assigned to receive either a creatine supplement or a placebo. The group receiving creatine did not exhibit improved performance over SEALS receiving the placebo. In another study outside the laboratory setting, creatine did improve the strength of trained weight lifters.[15] Thus the question of whether short-term creatine supplementation improves athletic performance remains to be answered.

The question of whether creatine supplements are safe also cannot yet be answered with certainty.[16] Even short-term (5 to 7 days) creatine supplementation may pose risks to athletes with kidney disease or other conditions, and studies on the safety of long-term creatine supplementation are not available.

Some medical and fitness experts voice concern that like many performance enhancement supplements, creatine is being taken in huge doses (5 to 30 grams per day) before evidence of its value has been ascertained.[17] Creatine from foods, even in diets high in red meat, a creatine-rich food, does not approach the amount that athletes take in supplement form. Athletes who take megadoses of creatine risk possible long-term side effects such as organ and muscle damage. Despite the uncertainties, creatine supplements are not illegal in international competition.

The overwhelming majority of schemes touted for athletes are frauds.* The placebo effect is strongly at work, however. When you hear reports of a performance boost from a new concoction, give it time. Chances are that the effect came from the power of the mind over the body. Incidentally, don't discount that power—it is formidable. You can use it by visualizing yourself as a winner in your sport. You don't have to rely on useless supplements for an extra edge because you already have a real one—your mind.

*If you have questions about a fitness product, book, or program, write to the American College of Sports Medicine at the address in the Nutrition Resources, Appendix E.

Fluid Needs during Physical Activity

Endurance athletes can lose 2 or more quarts of fluid in every hour of activity, but the digestive system can absorb only about a quart or so an hour. Hence the athlete must hydrate before and rehydrate during and after activity to replace it all. Even then, in hot weather the digestive tract may not be able to absorb enough water fast enough to keep up with an athlete's sweat losses, and some degree of dehydration becomes inevitable. Wise athletes preparing for competition drink extra fluids in the last few days of training before the event. The extra fluid is not stored in the body, but drinking extra ensures maximum tissue hydration at the start of the event. Any coach or athlete who withholds fluids during practice for any reason takes a great risk and is subject to sanctions by the ACSM.

Athletes who rely on thirst to govern fluid intake can easily become dehydrated. During activity thirst becomes detectable only *after* fluid stores are depleted. Don't wait to feel thirsty before drinking. Table 10-4 presents one schedule of hydration for physical activity. To find out how much water you need to replenish losses, weigh yourself before and after the activity. The difference is all water. Two cups (16 ounces) of fluid weigh about a pound.

What is the best fluid to support physical activity? Surprisingly, the best drink for most active bodies is just plain cool water, for two reasons: (1) water rapidly leaves the digestive tract to enter the tissues, and (2) it cools the body from the inside out.

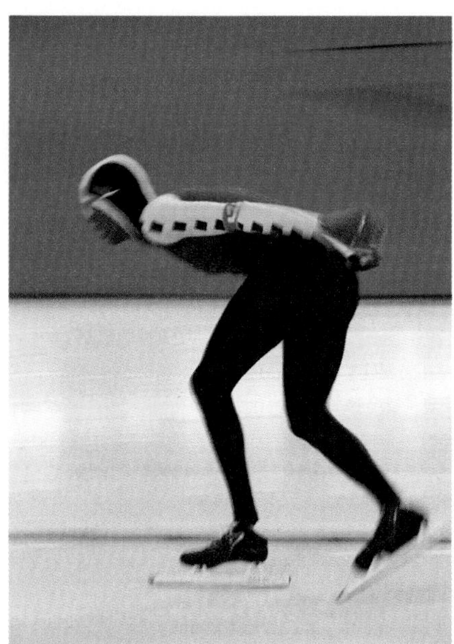

Active people need extra fluid, even in cold weather.

Fluid replacement tips:[a]

- To insure adequate fluid intake without being distracted during an event: before the event, fill a 32 ounce (4 cup) water bottle and place 2 colored rubber bands about equal distance from the top. Finish off the first segment of the bottle in the first 30 minutes of activity; finish the next segment in the next 30 minutes, and the remainder in the next. Have someone refill the bottle if activity lasts longer than 90 minutes.
- The urine of a person who is adequately hydrated is the color of pale lemonade. Urine the color of apple juice indicates slight dehydration.

[a]Ideas from J. Berning, nutrition professor and sports nutrition consultant, personal communication, 1999.

Table 10-4

Schedule of Hydration before, during, and after Physical Activity

When to Drink	Total Amount of Fluid (Consume in 1 cup Servings)
Day of event	Extra fluid throughout the day
2 hours before exercise	About 3 c
10 to 15 minutes before exercise	About 2 c
Every 15 to 30 minutes during exercise	4–8 oz (about 1 qt in 60 to 90 minutes)
After exercise	Replace each pound of body weight lost with 2 c fluid

As mentioned earlier, however, endurance athletes are an exception: they need more from their fluids than water alone. The first priority for endurance athletes should always be to replace fluids to prevent life-threatening heat stroke.[48] But endurance athletes also need carbohydrate to supplement their limited glycogen stores, so glucose is important, too.

Many good-tasting drinks are marketed for athletes and active people. Manufacturers reason that if a drink tastes good, people will drink more, thereby ensuring adequate hydration. Research backs up such reasoning, too. Fluids that are flavored, sweetened, and cool stimulate fluid intake.[49] The drinks also can provide a psychological edge to people who associate them with success in sports. Although you may hear much promotion of these drinks from their manufacturers, none of these companies is likely to run ads about the possible advantages of water because they aren't selling water. The next section compares these two drink options objectively.

KEY POINT *Physically active people lose fluids and must replace them to avoid dehydration. Thirst indicates that water loss has already occurred.*

What Do Sports Drinks Have to Offer?

First, and most important, of course, sports drinks offer fluids to help offset the loss of fluids during physical activity. As discussed earlier, however, plain water can do this, too.

Second, sports drinks supply glucose. A beverage that supplies glucose in some form can be useful during endurance activity lasting 60 minutes or more or during prolonged competitive games that demand repeated intermittent activity.[50] Not just any sweet beverage can meet this need, however, because a carbohydrate concentration greater than 10 percent can delay fluid emptying from the stomach and thereby slow down the delivery of water to the tissues. Most sports drinks contain about 7 percent glucose—a safe level for fluid transport.

Third, sports drinks offer sodium and other electrolytes to help replace those lost during physical activity. Sodium in sports drinks also helps to accelerate the rate of fluid absorption from the digestive tract.

Most athletes do not need to replace the minerals lost in sweat immediately; a meal eaten within hours of competition replaces these minerals soon enough. Most sports drinks are relatively low in sodium, however, so healthy people who choose to use these beverages run little risk of excessive intake.

In strenuous world-class competitions lasting for many hot, humid days, heavy sweating coupled with drinking large amounts of plain water has been reported to dangerously dilute blood sodium. If an athlete works up a drenching sweat, exceeding 5 to 10 pounds a day (or 3 percent of body weight), for several consecutive days, electrolyte replacement is advised. Athletes who *compete* for longer than six to eight hours each day may especially need to replace sodium.[51]

Equally as effective as commercial sports drinks but much less expensive is a homemade mixture of one-third teaspoon of salt (to provide sodium chloride) and 1 cup of sugar-sweetened fruit juice (to provide potassium and glucose) per quart of water. Avoid electrolyte or salt tablets; they increase potassium losses, can irritate the stomach, cause vomiting, and always pull water out of the tissues into the digestive tract at first.

KEY POINT *Sports drinks offer fluid and glucose for athletes who compete in endurance or intense intermittent activities. For most athletes, electrolyte and mineral repletion is best accomplished by eating a balanced diet, not by taking mineral-containing drinks. In the heat, however, electrolyte-containing fluids may benefit athletes who are training or competing.*

Other Beverage Choices

Some drinks, such as iced tea, deliver caffeine along with fluid. Moderate doses of caffeine (2 milligrams per pound of body weight or about the amount in 2 cups of coffee) one hour prior to activity seem to assist some people's athletic performance. Theoretically, caffeine may stimulate the release of fatty acids into the blood early in activity, thus conserving glycogen. Better than caffeine for this purpose, though, is a warm-up activity. Light activity initiates fat release, but unlike caffeine, it also warms the muscles and connective tissues, making them flexible and resistant to injury.

Caffeine also has adverse effects, including stomach upset, nervousness, sleeplessness, irritability, headaches, and diarrhea. It has been shown to constrict the arteries and raise blood pressure above normal. Constricted arteries make the heart work harder to pump blood to working muscles, an effect detrimental to sports performance.

In a hot environment, caffeine's diuretic effect is potentially hazardous, too. Physical activity slows the excretion of caffeine, prolonging its effects, so beverages containing caffeine should be used in moderation and in addition to other fluids, not as substitutes for them. In college, national, and international athletic competitions, the use of caffeine is forbidden in amounts greater than about 800 milligrams, the equivalent of drinking 5 or 6 cups of strong, brewed coffee in a two hour period before the event.

Carbonated beverages are not the best choice for meeting an athlete's fluid needs. Although they are composed largely of water, the air bubbles from the carbonation take up room in the stomach that might otherwise be filled with fluid that the athlete can absorb.

Athletes, like others, sometimes drink beverages that contain alcohol, but these beverages are inappropriate as fluid replacements. Like caffeine, alcohol is a diuretic. Both substances promote the excretion of water; of vitamins such as thiamin, riboflavin, and folate; and of minerals such as calcium, magnesium, and potassium—exactly the wrong effects for fluid balance and nutrition. It is hard to overstate alcohol's detrimental effects on physical activity. It impairs temperature regulation, making hypothermia or heat stroke much more likely. It alters perceptions and slows reaction time. It depletes strength and endurance and deprives people of their judgment, thereby compromising their safety in sports. Many sports-related fatalities and injuries each year involve alcohol or other drugs.

KEY POINT *Caffeine-containing drinks within limits may not impair performance, but water and fruit juice are preferred. Alcohol use can impair performance in many ways and is not recommended.*

Controversy 12 provides a discussion of caffeine's effects and sources.

Beer facts:

- Beer is not carbohydrate-rich. Beer is calorie-rich, but only one-third of its calories are from carbohydrates. The other two-thirds are from alcohol.
- Beer is mineral-poor. Beer contains a few minerals, but to replace those lost in sweat, athletes need good sources such as fruit juices.
- Beer is vitamin-poor. Beer contains tiny traces of some B vitamins, but it cannot compete with rich food sources.
- Beer causes fluid losses. Beer is a fluid, but alcohol is a diuretic and causes the body to lose more fluid in urine than is provided by the beer.

Read about alcohol's effects on the brain in Controversy 5.

FOOD FEATURE

Choosing a Performance Diet

Compare and decide which best meets your needs:

- 1 sandwich of 2 slices bologna, 2 slices white bread, 2 tbs mayonnaise (525 calories, 9% protein, 23% carbohydrate, 68% fat).

or

- 2 sandwiches of 2 slices lean ham, 4 slices whole-wheat bread, 2 tsp mayonnaise (503 calories, 20% protein, 51% carbohydrate, 29% fat).

Small daily choices, when made consistently, enhance an athlete's nutritional health.

THERE IS NO one diet that must be eaten to support an athlete's performance. Indeed, many different diets can be excellent for athletes. However, food choices must obey the rules for diet planning.

Nutrient Density

First, athletes need a diet composed mostly of nutrient-dense foods, the kind that supply a maximum of vitamins and minerals for the energy they provide. When athletes eat mostly refined, processed foods that have suffered nutrient losses and contain added sugar and fat, nutrition status suffers. Even if foods are fortified or enriched, manufacturers cannot replace the whole range of nutrients and nonnutrients lost in refining. Consider, for example, that manufacturers mill out much of a food's original magnesium and chromium but do not replace them. This doesn't mean that athletes can never choose a white bread, bologna, and mayonnaise sandwich but only that later they should eat a large salad or big portions of vegetables and whole grains and drink a glass of milk to compensate. The nutrient-dense foods will provide the magnesium and chromium; the bologna sandwich provided extra energy, mostly from fat.

Balance

Athletes must eat for energy, and their energy needs may be immense. Athletes need full glycogen stores, and they need to strive to prevent heart disease and cancer by limiting fat and saturated fat. Simply stated, a diet that is high in carbohydrate (60 to 70 percent of total calories), low in fat (20 to 25 percent), and adequate in protein (12 to 15 percent) is best for all these purposes. Even if the athlete does not compete in glycogen-depleting events, such a diet will provide adequate fiber while supplying abundant nutrients and energy.

With these principles in mind, compare the two 500-calorie sandwich meals in the margin. The trick to getting enough carbohydrate energy is easy, at least in theory: just reduce the amount of fat and meat in a meal, and let carbohydrate-rich foods fill in for them.

Adding carbohydrate-rich foods is a sound and reasonable option for increasing energy intake, up to a point. It becomes unreasonable when the person cannot eat enough food to meet energy needs. At that point, the person can cram more food energy into the diet only by using refined sugars and fats or liquid meals. Still, these energy-rich additions must be superimposed on nutrient-rich choices; energy alone is not enough.

Protein

In addition to carbohydrate, athletes need protein. What quantities of what kinds of foods supply enough protein to meet the needs of athletes? Meats and milk products head the list of protein-rich foods, but suggesting that athletes eat more than the recommended servings of meat would be shortsighted advice for many reasons. Athletes must protect themselves from heart disease, and even lean meats contain fat, much of it saturated fat. Besides, the extra servings of carbohydrate-rich foods such as legumes, grains, and vegetables that an athlete needs to meet energy requirements also boost protein intakes.

Earlier in this chapter, Table 10-2 showed some possible protein intakes for a 55-kilogram female athlete or a 70-kilogram male athlete based on recommendations of various authorities. It is likely that an athlete weighing 70 kilograms who engages in vigorous physical activity on a daily basis could require 3,000 to 5,000 calories per day. As a rule of thumb, endurance athletes should aim for an average intake of 50 calories per kilogram (2.2 pounds) of body weight. Others may need more. To meet such an energy requirement, an athlete should select from a variety of nutrient-dense foods. Figure 10-6 provides an example of how foods that provide the extra nutrients athletes need can be added to regular meal selections to attain a 3,300-calorie diet. These meals supply over 130 grams of protein, more than the highest recommended intake for an athlete. For those with reasonable diets, protein is rarely a problem.

The meals in Figure 10-6 provide 63 percent of their calories from carbohydrate. Athletes who train exhaustively for endurance events may want

Figure 10-6
An Athlete's Meals

Regular Meals		Modifications	Athlete's Meals
Breakfast: 1 c shredded wheat. 1 c 1% low-fat milk. 1 small banana. 1 c orange juice		The regular breakfast *plus*: 2 pieces whole-wheat toast. 1/2 c orange juice. 4 tsp jelly.	
Lunch: 1 turkey sandwich. 1 c 1% low-fat milk.		The regular lunch *plus*: 1 turkey sandwich. 1/2 c 1% low-fat milk. Large bunch of grapes.	
Snack: 2 c plain popcorn. A smoothie made from: 1 1/2 c apple juice. 1 1/2 frozen banana.		The regular snack *plus*: 1 c popcorn.	
Dinner: Salad: 1 c spinach, carrots, and mushrooms. 1/2 c garbanzo beans. 1 tbs sunflower seeds. 1 tbs ranch dressing. 1 c spaghetti with meat sauce. 1 c green beans. 1 slice Italian bread. 2 tsp butter. 1 1/4 c strawberries. 1 c 1% low-fat milk.		The regular dinner *plus*: 1 corn on the cob. 1 slice Italian bread. 2 tsp butter. 1 piece angel food cake. 1 tbs whipping cream.	
Total cal: 2,600 62% cal from carbohydrate 23% cal from fat 15% cal from protein		Total cal: 3,300 63% cal from carbohydrate 22% cal from fat 15% cal from protein All vitamin and mineral intakes exceed the recommendations for both men and women.	

to aim for somewhat higher carbohydrate levels—from 65 to 75 percent. Notice that breakfast, though light in fat, is filling and hearty. Current thinking supports the idea that athletes benefit from such a morning start. If you train early in the morning, try splitting breakfast into two parts. An hour or so before training, eat some toast, juice, and fruit. Later, after your workout, come back for the cereal and milk.

Planning an Athlete's Meals

Table 10-5 shows some sample food patterns for athletes at various high energy and high carbohydrate intakes.

Table 10-5

High-Carbohydrate Food Patterns for Athletes

	Number of Servings for a Daily Energy Intake of:					
Food Group	1,500 cal	2,000 cal	2,500 cal	3,000 cal	3,500 cal	4,000[a] cal
Milk	3	3	4	4	4	4
Fruit	5	6	7	9	10	12
Vegetable	3	3	3	5	6	7
Grain	7	11	16	18	20	24
Fat[b]	2	3	5	6	8	10
Meat (ounces)	5	5	5	5	6	6
Percent carbohydrate:	58%	58%	63%	64%	60%	62%

[a]A way to add more energy to the diet without adding much bulk is to snack on milkshakes or "complete meal" liquid supplements (see the text).

[b]A fat serving is 1 teaspoon of butter, margarine, oil, or the equivalent.

pregame meal a meal eaten three to four hours before athletic competition.

Looking for an amino acid supplement that rates a perfect score of 100 for protein quality? Try 1 ounce of chicken breast—it provides almost 10,000 mg of amino acids in perfect complement for use by the human body.

Good choices for pregame meals:

- Apricot nectar, pineapple juice, grape juice, banana, toast with jam or jelly, pancakes with syrup, baked white or sweet potatoes, pasta with steamed vegetables, lentils or other peas or beans, raisins, figs, dates, frozen yogurt, graham crackers, sponge cake, angel food cake.

Not recommended:

- Stuffing, muffins, biscuits, croissants, french fries, onion rings, potato chips, meats, cheese, pies, ice cream, eggnog, creams, nuts, butter, gravy, mayonnaise, salad dressing, frosted cakes.

These plans are effective only if the user chooses foods to provide nutrients as well as energy: extra milk for calcium and riboflavin; many servings of fruit for folate and vitamin C; energy-rich vegetables such as sweet potatoes, peas, and legumes; modest portions of lean meat for iron and other vitamins and minerals; and whole grains for B vitamins, magnesium, zinc, and chromium. In addition, these foods provide plenty of electrolytes.

A trick used by professional sports nutritionists to maximize athletes' intakes of energy and carbohydrates is to make sure that vegetable and fruit choices are as dense as possible in both nutrients and energy. A whole cupful of iceberg lettuce supplies few calories or nutrients, but a half-cup portion of cooked sweet potatoes is a powerhouse of vitamins, minerals, and carbohydrate energy. Similarly, it takes a whole cup of cubed melon to equal the calories and carbohydrate in a half-cup of canned fruit. Small choices like these, made consistently, can contribute significantly to nutrient, energy, and carbohydrate intakes.

Before competition, athletes may eat particular foods or practice rituals that convey psychological advantages. One eats steak the night before; another spoons up honey at the start of the event. As long as these foods or rituals remain harmless, they should be respected. Still, science has recommendations for the **pregame meal.** The foods should be carbohydrate-rich and the meal light (300 to 800 calories). It should be easy to digest and should contain fluids. Breads, potatoes, pasta, and fruit juices—carbohydrate-rich foods low in fat, protein, and fiber—form the basis of the pregame meal. Bulky, fiber-rich foods such as raw vegetables or high-fiber cereals, although usually desirable, are best avoided just before competition. Such foods can cause stomach discomfort during performance. The competitor should finish eating three to four hours before competition to allow time for the stomach to empty before exertion.

What about drinks or candylike sport bars claiming to provide "complete" nutrition? These mixtures of carbohydrate, protein (usually amino acids), fat, some fiber, and certain vitamins and minerals usually taste good and provide additional food energy before a game or for those needing to gain weight. They fall short of providing "complete" nutrition, however, since they lack many of real food's nutrients and the nonnutrients that benefit health. These products provide no special advantage for active people except one—they are easy to eat in the hours before competition. However, they are expensive. Concerning "complete" drinks, Table 10-6 demonstrates that there is no point in paying high prices for fancy brand-name drinks.

Table 10-6

Commercial and Homemade Meal Replacers Compared

	Cost (U.S.)	Energy (cal)	Protein (g)	Carbohydrate (g)	Fat (g)
12-ounce commercial liquid meal replacer[a]	about $2 per serving	360	15 (17% of calories)	55 (61%)	9 (22%)
12-ounce homemade milkshake[b]	about 50¢ per serving	330	15 (18% of calories)	53 (63%)	7 (19%)

[a]Average values for three commercial formulas.

[b]Home recipe: 8 oz fat-free milk, 4 oz ice milk, 3 heaping tsp malted milk powder. For even higher carbohydrate and calories values, blend in ½ mashed banana or ½ c other fruit. For athletes with lactose intolerance, use lactose-reduced milk or soy milk and chocolate or other flavored syrup, with mashed banana or other fruit blended in.

Homemade shakes are inexpensive and easy to prepare, and they perform every bit as well as do commercial products. Don't drop a raw egg in the blender, though, because raw eggs often carry bacteria that cause food poisoning.[52]

The person who wants to excel physically will apply the most accurate nutrition knowledge along with dedication to rigorous training. A diet that provides ample fluid and consists of a variety of nutrient-dense foods in quantities to meet energy needs will enhance not only athletic performance but overall health as well. Training and genetics being equal, who would win a competition—the person who habitually consumes less than the amounts of nutrients needed or one who arrives at the event with a long history of full nutrient stores and well-met metabolic needs?

DETECT FITNESS DECEPTION

Chapter 1 and Controversy 1 offered ways to distinguish between valid nutrition information and nutrition fraud, and the Consumer Corner in this chapter warned you about ergogenic products and the deceptive tactics some advertisers use to sell these products. Here is another chance to practice your deception detection skills: Browse the aisles of your local health food store or flip through the pages of any popular bodybuilding or fitness magazine. Look for products or ads for products that claim to provide amazing health or fitness benefits (see Figure 10-7). Then answer these questions:

1. What kinds of product descriptors are used on the labels or in the advertisements? Turn back to Figure C1-1 in Controversy 1 to get some ideas. What information do you gain from phrases such as "most scientifically advanced fat-burning formula" or "new cutting-edge formula" or "puts more meat in your muscle"?
2. Are you familiar with the ingredients in fitness products? Go to a store that sells supplements and read the ingredient lists or descriptions on some of these products. Do you know the health effects of each ingredient? For example, many product advertisements or labels boast that they contain steroidlike ingredients to enhance muscle growth and strength. The "steroidlike" ingredients are either plant or insect steroids whose effects on human beings can be nonexistent or even dangerous.
3. Do some products contain amino acids? Are these needed by healthy athletes?
4. Are the dosage levels for the ingredients given? If you are thinking about using a product, how much of each ingredient is appropriate? Compare vitamin and mineral amounts with recommendations; amounts between 50 and 150 percent of the recommendation for each nutrient reflect ranges commonly found in foods. Such amounts are compatible with the body's normal handling of nutrients. For other ingredients, call the National Institute of Health Information Center, Office of Alternative Medicine: (301) 402-2466.

Figure 10-7
Deceptive Fitness Claims

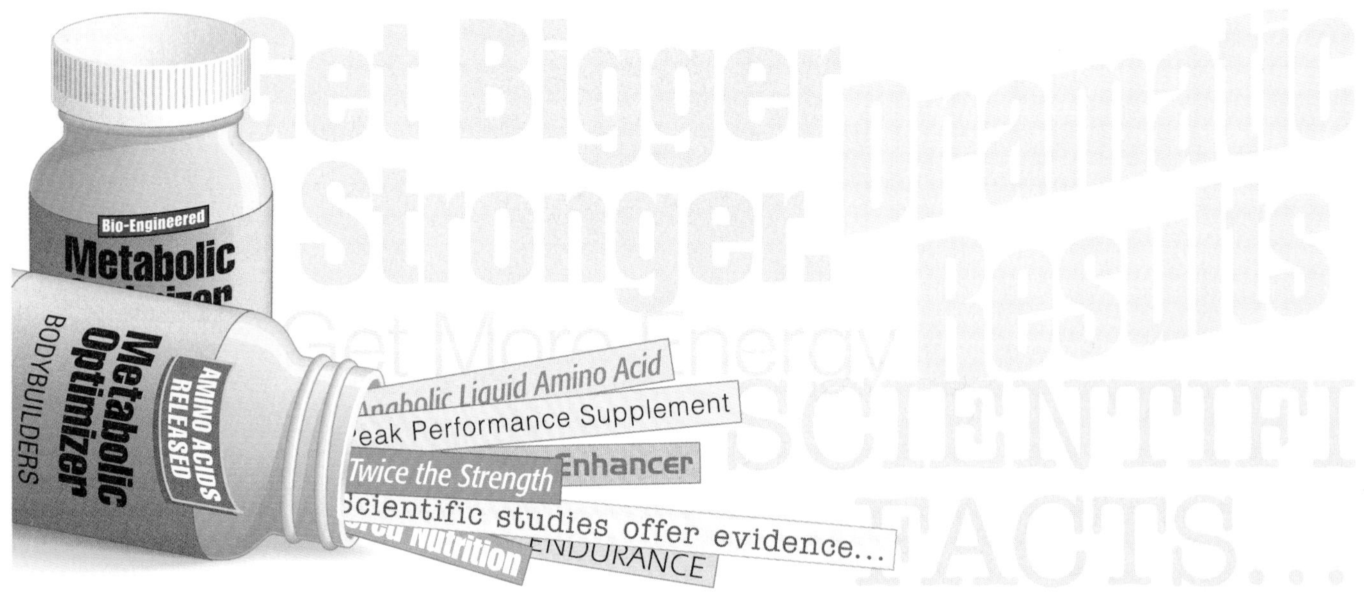
Get Bigger
Stronger.
Dramatic
Results
Get More Energy
SCIENTIFI
FACTS...
Bio-Engineered
Metabolic
AMINO ACIDS RELEASED
Metabolic Optimizer
BODYBUILDERS
Peak Performance Supplement
Enhancer
Twice the Strength
Scientific studies offer evidence...
ENDURANCE

Self Check

Answers to these Self Check questions are in Appendix G.

1. Which of the following provides most of the energy the muscles use in the early minutes of activity?
   a. fat
   b. protein
   c. glycogen
   d. (b) and (c)
2. Which diet has been shown to increase an athlete's endurance?
   a. high-fat diet
   b. high-carbohydrate diet
   c. normal mixed diet
   d. Diet has not been shown to have any effect.
3. Which of the following stimulates synthesis of muscle cell protein?
   a. physical activity
   b. a high-carbohydrate diet
   c. a high-protein diet
   d. amino acid supplementation
4. Which of the following has been proved to impart work-enhancing powers to healthy athletes?
   a. chromium picolinate
   b. DNA and RNA supplements
   c. vitamin E
   d. none of the above
5. What effect does alcohol have on the exercising body?
   a. impairs temperature regulation
   b. acts as a diuretic
   c. enhances performance
   d. (a) and (b)
6. The guidelines for developing and maintaining physical fitness are the same as the guidelines for obtaining health benefits. T F
7. The average resting pulse rate for adults is around 70 beats per minute, but the rate is higher in people who are physically fit. T F
8. Research does not support the idea that active people need supplements of vitamins to perform their best. T F
9. It is best for an athlete to drink extra fluids in the last few days of training before an event to ensure proper hydration. T F
10. Anorexia nervosa occurs only in women. (Read about this in the upcoming Controversy.) T F

NUTRITION ON THE NET

For further study of the topics of this chapter, access these websites and search for the phrases or words in quotation marks:

1. For the latest changes in chapter material, to find supplemental learning tools, or to access links to related websites, go to the "Nutrition Concepts and Controversies" site at:
   www.wadsworth.com/nutrition/prod/allprod.html
2. Search the American College of Sports Medicine site for information on "physical fitness":
   www.acsm.org
3. Explore the many resources offered on the Nutrition and Physical Activity site fromthe Centers for Disease Control and Prevention:
   www.cdc.gov/nccdphp/dnpa/
4. Visit the U.S. Government site for the Surgeon General's Report on Physical Activity:
   www.cdc.gov/nccdphp/sgr/sgr.htm
5. To learn more about the President's Council on Physical Fitness and Sports:
   www.whitehouse.gov/WH/PCPFS/html/fitnet.html
6. Visit the Shape Up America site at:
   www.shapeup.org
7. For information on sports drinks visit the Gatorade Sports Science Institute site at:
   www.gssiweb.com
8. Search among thousands of current scientific and medical abstracts for any topic related to exercise physiology at:
   http://www.ncbi.nlm.nih.gov/PubMed/

CONTROVERSY 10

WHAT CAUSES EATING DISORDERS?

An estimated 2 million people in the United States, primarily girls and women, suffer from the **eating disorders, anorexia nervosa** and **bulimia nervosa.** Many more suffer from **binge eating disorder** or other related conditions that do not meet the strict criteria for anorexia nervosa or bulimia nervosa but still imperil the sufferer's well-being. Certain characteristics of disordered eating such as restrained eating, binge eating, purging, fear of fatness, and distorted body image may be extraordinarily common among middle-class adolescent girls. In one survey of high-schoolers across the nation, 59 percent of girls and 23 percent of boys reported having dieted to lose weight.[1] Another survey found that only 8 percent of the students at a large university were overweight according to BMI standards, yet a majority of the students reported themselves as overweight and a full half of those who were *underweight* according to their BMI made the same claim.[2] In most other societies, these behaviors and attitudes are much less prevalent.

Why do so many people in our society suffer from eating disorders? Excessive pressure to be thin is at least partly to blame. When low body weight becomes an important goal, people begin to view normal healthy body weight as too fat, and they take unhealthy actions to lose weight.[3] Overly severe restriction of food intake may create intense hunger that leads to binges. Research confirms this theory, showing that unhealthy or dangerous diets predict binge eating in adolescent girls.[4] Energy restriction followed by bingeing can set in motion a pattern of repeated weight losses and gains—weight cycling.[5]

Young girls who attempt extreme weight loss may be expressing negative emotions other than dissatisfaction with body weight. They may have learned to identify discomforts such as anger, jealousy, or disappointment as "feeling fat."[6] They may also be depressed or suffer social anxiety. As weight loss and maintenance become more and more a focus, psychological problems worsen and the likelihood of developing full-blown eating disorders intensifies.

Athletes are at special risk for eating disorders. Many women athletes and dancers appear healthy but in fact may easily develop the three associated medical problems of the **female athlete triad:** disordered eating, amenorrhea (cessation of menstruation), and osteoporosis.[7] Table C10-1 defines this and other eating disorder terms.

THE FEMALE ATHLETE TRIAD

At age 14, Suzanne was a top contender for a spot on the state gymnastics team. Each day her coach reminded team members that they must weigh no more than a few ounces above their assigned weights in order to qualify for competition. The coach chastised gymnasts who gained weight. Convinced that the less she weighed the better she would perform, Suzanne weighed herself several times a day to make sure that she had not exceeded her 80-pound limit. Suzanne dieted and exercised to an extreme, and unlike many of her friends, she never began to menstruate. A few months before her fifteenth birthday, Suzanne's coach dropped her back to the second-level team. Suzanne blamed her poor performance on a slow-healing stress fracture. Mentally stressed and physically exhausted, she quit gymnastics and began overeating between periods of self-starvation. Suzanne had developed the dangerous combination of problems that characterize the female

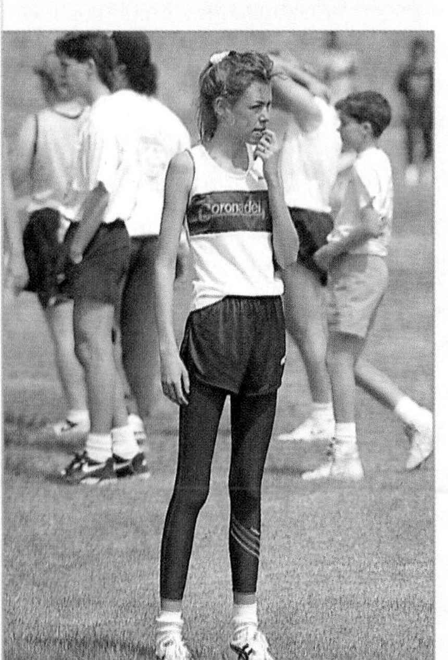

Table C10-1

Eating Disorder Terms

- **anorexia nervosa** an eating disorder characterized by a refusal to maintain a minimally normal body weight, self-starvation to the extreme, and a disturbed perception of body weight and shape; seen (usually) in teenage girls and young women (*anorexia* means "without appetite"; *nervos* means "of nervous origin").
- **binge eating disorder** an eating disorder whose criteria are similar to those of bulimia nervosa, excluding purging or other compensatory behaviors.
- **bulimia** (byoo-LEEM-ee-uh) **nervosa** recurring episodes of binge eating combined with a morbid fear of becoming fat; usually followed by self-induced vomiting or purging.
- **cathartic** a strong laxative.
- **cognitive therapy** psychological therapy aimed at changing undesirable behaviors by changing underlying thought processes contributing to these behaviors; in anorexia, a goal is to replace false beliefs about body weight, eating, and self-worth with health-promoting beliefs.
- **eating disorder** a disturbance in eating behavior that jeopardizes a person's physical or psychological health.
- **emetic** (em-ETT-ic) an agent that causes vomiting.
- **female athlete triad** a potentially fatal triad of medical problems seen in women athletes: disordered eating, amenorrhea, and osteoporosis.

athlete triad—disordered eating, amenorrhea, and weakening of the bones.

Disordered Eating Part of the reason many athletic women engage in disordered eating behaviors may be that they and their coaches have embraced unsuitable weight standards. An athlete's body must be heavier for a given height than a nonathlete's body because the athlete's body is denser; it contains more healthy muscle and bone tissue and less fat. Upon consulting a standard weight-for-height table and falling on the heavy side for height, an athlete can easily be misled into believing that he or she is too fat. Chapter 9 underscored this point with a photo of a champion bodybuilder who would be classified as obese by BMI standards. Weight standards such as BMI work well for others, but they are inappropriate for athletes. Body composition measures such as fatfold measures yield more useful information.

Another reason so many young athletes severely restrict their eating is that an ultra slim appearance has traditionally been considered aesthetically appealing in certain activities.[8] Male athletes, especially wrestlers and gymnasts, may be caught up in unhealthy weight loss as well, but research shows that female athletes and dancers, especially young ones, are most at risk.[9] Possible risk factors for the female athlete triad of disorders include the following:

- Young age (adolescence).
- Pressure to excel at a chosen sport.
- Focus on achieving or maintaining an "ideal " body weight or body fat percentage.
- Participation in endurance sports or competitions where performance is judged on aesthetic appeal such as ballet dancing, gymnastics, or figure skating.
- Dieting, even moderate dieting, undertaken at an early age.
- Unsupervised dieting.

Amenorrhea The prevalence of amenorrhea among premenopausal women in the United States is about 2 to 5 percent overall, but among female athletes, it may be as high as 66 percent.[10] Contrary to previous notions, amenorrhea is *not* a normal adaptation to strenuous physical training: it is a symptom of something going wrong. Amenorrhea is characterized by low blood estrogen, infertility, and bone mineral losses. Some research indicates that depleted body fat contributes to amenorrhea, but studies conflict about whether having a certain percentage of body fat is critical for normal menstruation in athletes. However it develops, amenorrhea in athletes threatens the integrity of the bones, immediately and in the future.

Osteoporosis For most people, weight-bearing exercise helps to protect bones against the calcium losses of osteoporosis. For young women with anorexia nervosa, however, strenuous activity can imperil their bones.[11] Vigorous training, along with low food energy intakes and other life stresses, may reduce estrogen levels and greatly increase the risks of stress fractures today and of osteoporosis in later life.[12] Stress fracture, a serious form of bone injury, occurs commonly among dancers and athletes with low body weights and disordered eating. Many underweight young athletes have bones like those of postmenopausal women, and they may never recover their lost bone even after diagnosis and treatment.[13] This makes preventing such losses critical for protecting lifelong bone health. Athletes, especially young females, should be encouraged to optimize their nutrient intakes by consuming at least 1,300 milligrams of calcium each day, and to modify activity so that they expend no more energy than they consume. Women with bulimia rarely cease menstruating and so may be spared this loss of bone integrity.[14]

Making Weight Only female athletes face the threats of the female athlete triad, but many male athletes and dancers also face pressure to achieve a certain body weight and many develop eating disorders. On average, male teenagers carry about 15 percent of body weight as fat, but some high school athletes strive for only 5 percent body fatness. Wrestlers, for example, are required to "make weight" to compete in the lowest possible weight class to face the smallest possible opponents. To that end, wrestlers starve themselves, don rubber suits, sweat in steam rooms, and take diuretics to shed water weight before weighing in for competition. These practices were responsible for the deaths of three college athletes in 1997 and have caused untold misery and harm to many others.[15] A sad irony is that these practices actually compromise athletic performance. The diminished anaerobic strength, reduced endurance, decreased oxygen capacity, and general weakness caused by food deprivation and dehydration can hobble performance, an effect lasting days after food and water are replenished.[16]

Preventing Eating Disorders in Athletes To prevent eating disorders in athletes and dancers, both the performers and their coaches must be educated about links between inappropriate body-weight ideals, improper weight-loss techniques, eating disorder development, effective sports nutrition, and safe weight-control methods. Young people naturally search for identity during the teen years and will often follow the advice of a person in authority without question. Therefore, coaches and dance instructors should never encourage unhealthy weight loss to qualify for competition or to conform to idealistic artistic standards. To do so is to risk eroding the teenager's self-esteem and interfering with the normal development of identity, two important defenses against eating disorders. Unquestionably, old standards based on slim appearance or low body weight should be replaced with performance-based standards. Table C10-2 provides some suggestions to help athletes and dancers protect themselves against developing eating disorders. The next sections describe eating disorders that anyone, athlete or nonathlete, may experience.

ANOREXIA NERVOSA

Julie is 18 years old and is a superachiever in school. She watches her diet with great care, and she exercises daily, maintaining a heroic schedule of self-discipline. She is thin, but she is determined to lose more weight. She is 5 feet 6 inches tall and weighs 85 pounds. She has anorexia nervosa.

Characteristics of Anorexia Nervosa

Julie is unaware that she is undernourished, and she sees no need to obtain treatment. She stopped menstruating several months ago and is moody and chronically depressed. She insists that she is too fat, although her eyes are sunk in deep hollows in her face. Although she is close to physical exhaustion, she no longer sleeps easily. Her family is concerned, and although reluctant to push her, they have finally insisted that she see a psychiatrist. Julie's psychiatrist has prescribed group therapy as a start, but warns that if Julie does not begin to gain weight soon, she will need to be hospitalized.

Most anorexia nervosa victims come from middle- or upper-class families. Men account for only 1 or 2 in 20 cases in the general population, although among male athletes and dancers the incidence may approach that of women.[17] Men with eating disorders face many of the same physical ills as women, and their diagnosis and treatment are the same.

Central to the diagnosis of anorexia nervosa is a distorted body image that overestimates body fatness.[18] When Julie looks at herself in the mirror, she sees her 85-pound body as fat. The more Julie overestimates her body size, the more resistant she is to treatment, and the more unwilling to examine her faulty values and misconceptions. Malnutrition itself is known to affect brain functioning and judgment in this way. People with anorexia nervosa cannot recognize it in themselves; only professionals can diagnose it. Table C10-3 shows the criteria that experts use.

Table C10-2

Tips for Combating Eating Disorders

General Guidelines

- Never restrict food servings to below the numbers suggested for adequacy by the Daily Food Guide.
- Eat frequently. People often do not eat frequent meals because of time constraints, but eating can be incorporated into other activities, such as snacking while studying or commuting. The person who eats frequently never gets so hungry as to allow hunger to dictate food choices.
- If not at a healthy weight, establish a reasonable weight goal based on a healthy body composition. (Chapter 9 provides help in doing so.)
- Allow a reasonable time to achieve the goal. A reasonable rate for losing excess fat is about 1% of body weight per week.
- Establish a weight-maintenance support group with people who share interests.

Specific Guidelines for Athletes and Dancers

- Remember that eating disorders impair physical performance. Seek confidential help in obtaining treatment if needed.
- Restrict weight-loss activities to the off-season.
- Focus on proper nutrition as an important facet of your training, as important as proper technique.

The Role of the Family

Certain family attitudes, and especially parental attitudes, stand accused of contributing to eating disorders. Families of persons with anorexia nervosa are likely to be critical and to overvalue outward appearances while undervaluing inner self-esteem. Parents may oppose one another's authority and vacillate between defending the anorexic child's behavior and condemning it, confusing the child and disrupting normal parental control. In the extreme, parents may even be sexually abusive or abusive in other ways.[19]

Julie is a perfectionist, just as her parents are: she identifies so strongly with her parents' ideals and goals that she cannot get in touch with her own identity. She is respectful of authority, but sometimes feels like a robot, and she may act that way, too: polite but controlled, rigid, and unspontaneous.[20] For Julie, rejecting food is a way of gaining control.

Although families of children with anorexia nervosa often have problems, blame is a useless concept, and parents may suffer deeply from being blamed for their child's illness. In truth, no one knows what causes anorexia nervosa, and it may turn out that the cause is a physical one, and not related to parenting. Rather than judging parents, a more useful tactic is to identify the family's strong points and resources and to prepare them for the job of helping their ill child benefit from treatment.

Self-Starvation

How can a person as thin as Julie continue to starve herself? Julie uses tremendous discipline to strictly limit her portions of low-calorie foods. She will deny her hunger, and having become accustomed to so little food, she feels full after eating only a half-dozen carrot sticks. She can recite the calorie contents of dozens of foods and the calorie costs of as many exercises. If she feels that she has gained an ounce of weight, she runs or jumps rope until she is sure she has exercised it off. She drinks water incessantly to fill her stomach, risking dangerous mineral imbalances.[21] If she fears that the food energy she has eaten exceeds the exercise she has done, she takes laxatives to hasten the passage of food from her system, not knowing that laxatives reduce water absorption, but not food energy absorption. Her other methods of staying thin are so effective that she is unaware that laxatives have no effect on body fat. She is desperately hungry. In fact, she is starving, but she doesn't eat because her need for self-control dominates other needs.

Many people, on learning of this disorder, say they wish they had "a touch" of it to get thin. They mistakenly think that people with anorexia nervosa feel no hunger. They also fail to recognize the pain of the associated psychological and physical trauma.

Physical Perils

Anorexia nervosa damages the body much as other forms

Table C10-3

Criteria for Diagnosis of Anorexia Nervosa

A person with anorexia nervosa demonstrates the following:

A. Refusal to maintain body weight at or above a minimal normal weight for age and height, e.g., weight loss leading to maintenance of body weight less than 85% of that expected; or failure to make expected weight gain during period of growth, leading to body weight less than 85% of that expected.
B. Intense fear of gaining weight or becoming fat, even though underweight.
C. Disturbance in the way in which one's body weight or shape is experienced; undue influence of body weight or shape on self-evaluation, or denial of the seriousness of the current low body weight.
D. In females past puberty, amenorrhea, i.e., the absence of at least three consecutive menstrual cycles. (A woman is considered to have amenorhea if her periods occur only following hormone, e.g., estrogen, administration.)

Two types of anerexia nervosa include:

- Restricting type: during the episode of anorexia nervosa, the person does not regularly engage in binge eating or purging behavior (i.e., self-induced vomiting or the misuse of laxatives, diuretics, or enemas).
- Binge eating/purging type: during the episode of anorexia nervosa, the person regularly engages in binge eating or purging behavior (i.e., self-induced vomiting or the misuse of laxatives, diuretics, or enemas).

SOURCE: Reprinted with permission from American Psychiatric Association, *Diagnostic and Statistical Manual of Mental Disorders,* 4th ed. (Washington, D.C.: American Psychiatric Association, 1994).

of starvation do. In young people, growth ceases and normal development falters. They lose so much lean tissue that basal metabolic rate slows. In athletes, the loss of lean tissue handicaps physical performance.[22] The heart pumps inefficiently and irregularly, the heart muscle becomes weak and thin, the heart chambers diminish in size, and the blood pressure falls. Electrolytes that help to regulate the heartbeat go out of balance. Many deaths in people with anorexia are due to heart failure.

Starvation brings other physical consequences as well: significant losses of brain tissues, impaired immune response, anemia, and a loss of digestive function that worsens malnutrition.[23] Digestive functioning becomes sluggish, the stomach empties slowly, and the lining of the intestinal tract shrinks. The ailing digestive tract fails to digest food adequately, even if the victim does eat. The pancreas slows its production of digestive enzymes. Diarrhea sets in, further worsening malnutrition.

Starvation also brings altered blood lipids, high concentrations of vitamin A and vitamin E in the blood, low blood proteins, an imbalance among the hormones of growth and development, dry skin, abnormal nerve functioning, low body temperature, and the development of fine body hair (the body's attempt to keep warm). The electrical activity of the brain becomes abnormal, and insomnia is common. In the growing years, growth slows or ceases and development is impaired. In adulthood, both women and men lose their sex drives. Mothers with anorexia nervosa may severely underfeed their children who then fail to grow and suffer the other harms typical of starvation.[24]

Treatment of Anorexia Nervosa Treatment of anorexia nervosa requires a multidisciplinary approach that addresses two sets of issues and behaviors: those relating to food and weight and those involving relationships with oneself and others.[25] Teams of physicians, nurses, psychiatrists, family therapists, and dietitians work together to treat people with anorexia nervosa. Appropriate diet is crucial for normalizing body weight and must be tailored individually to each client's needs.[26] Clients are seldom willing to eat for themselves, but if they are, chances are they can recover without other interventions.

Professionals classify clients based on the risks posed by the degree of malnutrition present.* Clients with low risks may benefit from family counseling, **cognitive therapy,** behavior modification, and nutrition guidance; those with greater risks may also need other forms of psychotherapy and supplemental formulas to provide extra nutrients and energy. High-risk clients may require involuntary hospitalization and may need to be force-fed by tube at first to forestall death.[27] This step causes psychological trauma.[28] Drugs are commonly prescribed, but to date, they play a limited role in treatment.

Stopping weight loss is a first goal of treatment; establishing regular eating patterns is another. At first, progress is slow partly owing to a speeded-up metabolic rate and an increased thermic response to food that occur upon refeeding.[29] Further, as small gains of body fat occur, blood values of the appetite-suppressing hormone leptin begin creeping up, too, causing researchers to speculate that leptin may contribute to difficulties in weight restoration.[30]

Few people with anorexia nervosa seek treatment on their own, and denial makes treatment difficult. Almost half of the women who are treated can maintain their body weight within 15 percent of a healthy weight; at that weight, many of them begin menstruating again. The other half have poor or fair outcomes of treatment, and two-thirds of those treated continue a mental battle with recurring morbid thoughts about food and body weight.[31] Many relapse into abnormal eating behaviors.[32] About 5 percent die during treatment, 1 percent by suicide.

Before drawing conclusions about someone who is extremely thin, remember that diagnosis of anorexia nervosa requires professional assessment. People seeking help with anorexia nervosa, either for themselves or for others, can call the National Anorexic Aid Society hotline.†

*Indicators of malnutrition include a low percentage of body fat, low blood proteins, and impaired immune response.

†Phone numbers, addresses, and Internet addresses are in Appendix E.

BULIMIA NERVOSA

Sophia is a 20-year-old airline stewardess, and although her body weight is healthy, she thinks constantly about food. She alternately starves herself and then secretly binges; when she has eaten too much, she vomits. Few people would fail to recognize that these symptoms signify bulimia nervosa.

Characteristics of Bulimia Nervosa Bulimia nervosa is distinct from anorexia nervosa and is much more prevalent, although the true incidence is difficult to establish. People with bulimia nervosa often suffer in secret and, when asked, may deny the existence of a problem. More men suffer from bulimia nervosa than from anorexia nervosa, but bulimia nervosa is still most common in women. Based on a questionnaire, one study estimates that 19 percent of female college students experience bulimic symptoms.[33] A true diagnosis of bulimia nervosa is based on the criteria listed in Table C10-4.

Like the typical person with bulimia nervosa, Sophia is single, female, and white. She is well educated and close to her ideal body weight, although her weight fluctuates over a range of 10 pounds or so every few weeks. As a flight attendant, she is required to weigh no more than a certain cutoff weight, slightly below the weight that her body maintains naturally.

Sophia seldom lets her bulimia nervosa interfere with her work or other activities, although a third of all bulimics do. From early childhood she has been a high achiever but emotionally dependent on her parents. As a young teen, Sophia cycled on and off crash diets. She feels anxious at social events and cannot easily establish close relationships. She is usually depressed, is often impulsive, and has low self-esteem. When crisis hits, Sophia responds by replaying events, worrying excessively, and blaming herself but never asking for help—behaviors that are barriers to effective coping.[34]

Women with anorexia nervosa see themselves as fat, even when they are dangerously underweight.

The Role of the Family Families of bulimic people are observed to be externally controlling, but emotionally uninvolved with their children, resulting in a stifling negative self-image.[35] Dieting, arguments, criticism of body shape or weight, minimal affection, and high expectations with little support are common in the families of people with bulimia. Such attitudes make perfectionism likely. The unrealistic expectation of perfection is followed by devastating self-blame and criticism when the unrealistic standard is not met. Typically, the family has "secrets " that are hidden from outsiders. Bulimic women who report having been abused sexually or physically by family members or friends may continually suffer a sense of being unable to gain control.[36]

Should the member with bulimia nervosa begin making the needed changes toward recovery, others in the family may feel threatened. Family cooperation is important, however, because making changes within a family requires effort from everyone. Such effort is well spent, for changing destructive family interactions can greatly benefit the person who has begun to fight against bulimia nervosa.

Binge Eating and Purging A bulimic binge is unlike normal eating, and the food is not consumed for its nutritional value. During a binge, Sophia's eating is accelerated by her hunger from previous calorie restriction. She may take in anywhere from 1,000 to many thousands of calories of easy-to-eat, low-fiber, smooth-textured, high-fat, and, especially, high-carbohydrate foods. Typically, she chooses cookies, cakes, and ice cream; and she eats the entire bag of cookies, the whole cake, and every spoonful in a carton of ice cream. By the end of the binge, she has vastly overcorrected for her attempts at calorie restriction at other times.[37]

The binge is a compulsion and usually occurs in several stages: "anticipation and planning, anxiety, urgency to begin, rapid and uncontrollable consumption of food, relief and relaxation, disappointment, and finally shame or disgust." Then, to purge the food from her body, she may use a **cathartic**—a strong laxative that can injure the lower

Table C10-4

Criteria for Diagnosis of Bulimia Nervosa

A person with bulimia nervosa demonstrates the following:

A. Recurrent episodes of binge eating. An episode of binge eating is characterized by both of the following:
   1. Eating, in a discrete period of time (e.g., within any two-hour period), an amount of food that is definiely larger than most people would eat during a similar period of time and under similar circumstances, and,
   2. A sense of lack of control over eating during the episode (e.g., a feeling that one cannot stop eating or control what or how much one is eating).

B. Recurrent inappropriate compensatory behavior in order to prevent weight gain, such as self-induced vomiting; misuse of laxatives, diuretics, enemas, or other medications; fasting; or excessive exercise.

C. Binge eating and inappropriate compensatory behaviors that both occur, on average, at least twice a week for three months.

D. Self-evaluation unduly influenced by body shape and weight.

E. The disturbance does not occur exclusively during episodes of anorexia nervosa.

Two types:

- Purging type: the person regularly engages in self-induced vomiting or the misuse of laxatives, diuretics, or enemas.
- Nonpurging type: the person uses other inappropriate compensatory behaviors, such as fasting or excessive exercise, but does not regularly engage in self-induced vomiting or the misuse of laxatives, diuretics, or enemas.

SOURCE: Reprinted with permission from American Psychiatric Association, *Diagnostic and Statistical Manual of Mental Disorders*, 4th ed. (Washington, D.C.: American Psychiatric Association, 1994).

intestinal tract. Or she may induce vomiting, using an **emetic**—a drug intended as first aid for poisoning. After the binge she pays the price with hands scraped raw against the teeth during induced vomiting, swollen neck glands and reddened eyes from straining to vomit, and the bloating, fatigue, headache, nausea, and pain that follow.

Physical and Psychological Perils On first glance, purging seems to offer a quick and easy solution to the problems of unwanted calories and body weight. Bingeing and purging have serious physical consequences, however. Fluid and electrolyte imbalances caused by vomiting or diarrhea can lead to abnormal heart rhythms and injury to the kidneys. Urinary tract infections can lead to kidney failure. Vomiting causes irritation and infection of the pharynx, esophagus, and salivary glands; erosion of the teeth; and dental caries. The esophagus or stomach may rupture or tear. Overuse of emetics can lead to death by heart failure.

A person may consume up to 10,000 calories during an eating binge.

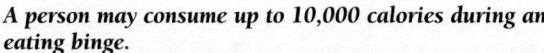

cola

Unlike Julie, Sophia is aware that her behavior is abnormal, and she is deeply ashamed of it. She wants to recover, and this makes recovery more likely for her than for Julie, who clings to denial. Sophia tends to be passive and to look to others, primarily men, for confirmation of her sense of worth. When she experiences rejection, either in reality or in her imagination, her bulimia nervosa becomes worse. If Sophia's depression deepens, she may seek solace in drug or alcohol abuse or other addictive behaviors. Many studies show a link between bulimia nervosa and drug and alcohol dependency.[38]

Treatment of Bulimia Nervosa To gain control over food and establish regular eating patterns requires adherence to a structured eating plan. Restrictive dieting is forbidden, for it almost always precedes and may even trigger binges. Steady maintenance of weight and prevention of relapse into cyclic gains and losses, are the goals. Many a former bulimia nervosa sufferer has taken a major step toward recovery by learning to consistently eat enough food to satisfy hunger needs (at least 1,600 calories a day). Table C10-5 offers some ways to begin correcting the eating problems of bulimia nervosa. About half of women receiving a diagnosis of bulimia may recover completely after five to ten years, with or without treatment, but treatment probably speeds the recovery process.[39]

BINGE EATING DISORDER

Anorexia nervosa and bulimia nervosa are distinct eating disorders, yet they sometimes overlap in important ways. People with both conditions share an overconcern with body weight and the tendency to drastically undereat. Both may purge. The two disorders can also appear in the same person, or one can lead to the other. Other people have eating disorders that fall short of anorexia nervosa or bulimia nervosa, but share some of their features, such as fear of body fatness. One such condition is binge eating disorder (defined earlier in Table C10-1).

Up to half of all people who restrict eating to lose weight end up periodically binge eating without purging. This describes about one-third of obese people who regularly engage in binge eating. Obesity itself, however, does not constitute an eating disorder. Table C10-6 lists the official diagnostic criteria for binge eating disorder.

Clinicians note differences between people with bulimia nervosa and those with binge eating disorder. Binge eaters rarely purge, they consume less during a binge, and they are less restrained during nonbinge eating. Similarities also exist, including feeling out of control, or feeling disgusted, depressed, embarrassed, or guilty after bingeing.

Binge eating behavior may respond more readily to treatment than other eating disorders, and resolving such behaviors can be a first step to authentic weight control. Successful treatment also improves physical health, mental health, and the chances of breaking the cycle of rapid weight losses and gains.

Table C10-5

Diet Strategies for Combating Bulimia Nervosa

- Avoid finger foods; eat foods that require the use of utensils.
- Enhance satiety by eating warm foods.
- Include vegetables, salad, and/or fruit at meals to prolong eating time.
- Choose whole-grain and high-fiber breads and cereals to maximize bulk.
- Eat a well-balanced diet and meals consisting of a variety of foods.
- Use foods that are naturally divided into portions, such as potatoes (rather than rice or pasta); 4- and 8-ounce containers of yogurt, ice cream, or cottage cheese; precut steak or chicken parts; and frozen entrées.
- Include foods containing ample complex carbohydrates (for satiety) and some fat (to slow gastric emptying).
- Eat meals and snacks sitting down.
- Plan meals and snacks, and record plans in a food diary prior to eating.

EATING DISORDERS IN SOCIETY

Eating disorders seem to have complex causes. Some may have hereditary components; some may have psychological components that our society brings out. Proof that society plays a role in eating disorders is found in their demographic distribution: they are known only in developed nations, and they become more prevalent as wealth increases and food becomes plentiful.

A food-centered society that favors thinness puts people in a bind. Families may encourage hearty eating and socializing around the dinner table, and family members and guests are obliged to indulge. Restaurants and others provide enormous helpings of foods as their standard-sized meals. A child raised in such a setting may see little alternative but to indulge in vast quantities of food and then vomit, crash diet, or fast to "undo" possible weight gain.

No doubt our society sets unrealistic ideals for body weight, especially for women, and devalues those who do not conform to them. Magazines and other media convey the message that to be thin is to be happy; eating disorders are not a form of rebellion against these unrealistic ideas, but rather an exaggerated acceptance of them. Even professionals, including physicians and dietitians, can suffer from these disorders or tend to praise people for losing weight and to suggest weight loss to people who do not need it. As a result, at so tender an age as 11 or 12, beautifully growing, normal-weight girls fear that they are too fat. Most are "on diets," and many are poorly nourished.[40] Some eat too little food to support normal growth; thus they miss out on their adolescent growth spurts and may never catch up.

Societal pressure to be thin is no doubt a factor in the development of eating disorders. Most experts, agree, however, that the disorders are multifactorial: sociocultural, psychological, and probably also neurochemical. New research has uncovered potential roles for the brain's neurotransmitters in the development of eating disorders and is raising the hope that effective pharmacological treatments may soon be available.[41]

Perhaps a young person's best defense against these disorders is to learn about their own normal, expected growth patterns, especially the characteristic weight gain of adolescence (see Chapter 13), and to learn respect for the inherent wisdom of the body. When people discover and honor the body's real needs, they become unwilling to sacrifice health for conformity. The author Eda LeShan, once a slave to bulimic behavior, achieved this inner ideal and described her recovery from overeating: "Deep inside there had always been a small child begging for my attention. . . . All I gave her was food. Now I give her love."[42]

Table C10-6

Criteria for Diagnosis of Binge Eating Disorder

A person with a binge eating disorder demonstrates the following:

A. Recurrent episodes of binge eating. An episode of binge eating is characterized by both of the following:
   1. Eating, in a discrete period of time (e.g., within any two-hour period) an amount of food that is definitely larger than most people would eat in a similar period of time under similar circumstances.
   2. A sense of lack of control over eating during the episode (e.g., a feeling that one cannot stop eating or control what or how much one is eating).

B. Binge eating episodes are associated with at least three of the following:
   1. Eating much more rapidly than normal.
   2. Eating until feeling uncomfortably full.
   3. Eating large amounts of food when not feeling physically hungry.
   4. Eating alone because of being embarrassed by how much one is eating.
   5. Feeling disgusted with oneself, depressed, or very guilty after overeating.

C. The binge eating causes marked distress.

D. The binge eating occurs, on average, at least twice a week for six months.

E. The binge eating is not associated with the regular use of inappropriate compensatory behaviors (e.g., purging, fasting, excessive exercise) and does not occur exclusively during the course of anorexia nervosa or bulimia nervosa.

SOURCE: Reprinted with permission from American Psychiatric Association, *Diagnostic and Statistical Manual of Mental Disorders*, 4th ed. (Washington, D.C.: American Psychiatric Association, 1994).

11 Diet and Health

Pierre Auguste Renoir 1841–1919,
The Luncheon of the Boating Party, 1881.

Contents

Frequently Asked Questions

infectious diseases diseases caused by bacteria, viruses, parasites, and other microbes, which can be transmitted from one person to another through air, water, or food; by contact; or through vector organisms such as mosquitoes or fleas.

degenerative diseases chronic, irreversible diseases characterized by degeneration of body organs due in part to such personal lifestyle elements as poor food choices, smoking, alcohol use, and lack of physical activity. Also called *lifestyle diseases, chronic diseases,* or the *diseases of old age.*

AIDS acquired immune deficiency syndrome; caused by infection with HIV, a virus that is transmitted primarily by sexual contact, contact with infected blood, needles shared among drug users, or fluids transferred from an infected mother to her fetus or infant.

If there is any deficiency in food or exercise the body will fall sick.—Hippocrates, a Greek physician, c. 400 B.C.

Deficiencies (↓) and toxicities (↑) known to impair immunity:

- Protein (↓).
- Energy (↓).
- Vitamin A (↓↑).
- Vitamin E (↓).
- Vitamin D (↓).
- B vitamins (↓).
- Folate (↓).
- Vitamin C (↓).
- Iron (↓↑).
- Zinc (↓↑).
- Copper (↓).
- Magnesium (↓).
- Selenium (↓).

THE DISEASES that afflict people around the world are of two main kinds: **infectious diseases** and **degenerative diseases.*** Infectious diseases such as tuberculosis, smallpox, and polio have been widespread among humankind since before the dawn of history, and they strike people of all ages. In any civilization not well defended against them, infectious diseases can cut life so short that the average person dies at age 20, 30, or 40, instead of living out the full potential life span of 70 or 80 years or more.

With the advent of vaccines and antibiotics, many people have become complacent about infectious diseases. This is a serious error. New microorganisms such as the virus known as HIV, which causes **AIDS,** are emerging and others are developing resistance to antibiotic therapy and threaten health and life across the globe. Tuberculosis, once declared a controlled disease, has overpowered formerly curative antibiotics and has produced a worldwide epidemic that some have called a modern-day plague.[1]

Degenerative diseases remain the leading causes of death and illness among people in developed nations; examples are diabetes, cancer, heart disease, and osteoporosis. Ironically, the longer a person dodges life's other perils, the more likely that these diseases will take their toll. They become prevalent in a population only when infectious diseases are kept firmly enough in check that people survive to older ages.

Degenerative diseases arise, not from simple infection, but from a mixture of three sets of factors. Two of these, hereditary susceptibility and prior disease, people cannot control. The other set of factors, daily life choices, people themselves can control directly. Young people can choose whether to nourish their bodies well, to smoke, to exercise, or to abuse alcohol. As people age, their bodies accumulate impacts from these choices, and in the later years these impacts can make the difference between a life of health or one of chronic disability. Degenerative diseases are often called the *chronic* diseases.

Today, we rely on public health measures, such as disinfected water supplies, and medical care to reduce the likelihood of infectious diseases. Still, people are exposed to millions of microbes each day, and although nutrition cannot directly prevent or cure infectious diseases, it can strengthen or weaken the body's defenses against them. With respect to prevention of chronic diseases, choices people make about their nutrition can help at least to postpone them and sometimes to avoid them altogether. Figure 11-1 shows how deaths from today's top ten killers of adults have changed over time. Much of this chapter is devoted to diseases in which nutrition plays important roles.

Nutrition and Immunity

Without your awareness, your immune system guards continuously against thousands of enemy attacks mounted against you by microorganisms and cancer cells. If your immune system falters, you become vulnerable to disease-causing agents, and disease invariably follows.

These facts can help to underscore nutrition's importance to immunity:

- Both excessive intakes and deficient intakes of almost all vitamins and minerals are associated with impaired immune response.
- Immune tissues are among the first to be impaired in the course of a nutrient deficiency or toxicity.[2]
- Some deficiencies are more immediately harmful to immunity than others; the speed of the impact is affected by whether another nutrient can perform some of the metabolic tasks of the missing nutrient, how severe the deficiency is, whether an infection has already taken hold, and how old the person is.
- The risk of sickness and death increases dramatically when medical tests of a malnourished person indicate weakened immunity.

Malnutrition often worsens diseases, which, in turn, worsen malnutrition.[3] The cycle often begins when impaired immunity opens the way for diseases; then diseases

*The term *disease* is also used to refer to conditions such as birth defects, alcoholism, obesity, mental disorders, and others.

Figure 11-1

Ten Leading Causes of Death—United States, 1997, Age-Adjusted Rates[a, b]

The causes identified with the red bars are related to nutrition; those with green bars are alcohol-related. See Controversy 5.

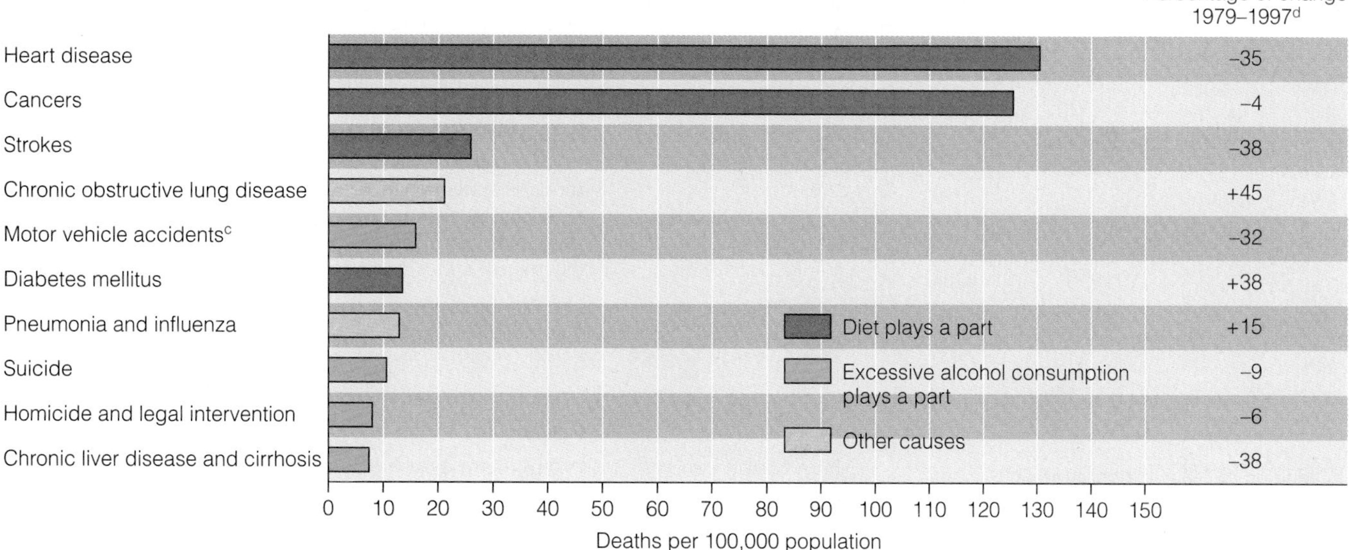

Percentage of change 1979–1997d
Heart disease
Cancers
Strokes
Chronic obstructive lung disease
Motor vehicle accidentsc
Diabetes mellitus
Pneumonia and influenza
Suicide
Homicide and legal intervention
Chronic liver disease and cirrhosis
–35
–4
–38
+45
–32
+38
+15
–9
–6
–38
Diet plays a part
Excessive alcohol consumption plays a part
Other causes
0 10 20 30 40 50 60 70 80 90 100 110 120 130 140 150
Deaths per 100,000 population

[a]Rates are age-adjusted to the 1940 U.S. population; these data appear lower than gross mortality data used previously because rates of death have declined and because they have been standardized to allow meaningful comparison between the rates of demographic groups and over time.

[b]Five other leading causes of death are kidney disease, Alzheimer's disease, septicemia, human immuno deficiency virus infections, and atherosclerosis.

[c]Accidents are the leading cause of death among people aged 15–24, followed by homicide, suicide, cancer, heart disease, birth defects, lung disease, pneumonia and influenza, and stroke. Alcohol contributes to about half of all accident fatalities. Age-adjusted death rates for 1997, percentage of change in age-adjusted death rates for the 15 leading causes of death, 1996–1997 and 1979–1997, and ratio of age-adjusted death rates, by sex and race of decedent, 1997—United States, *Morbidity and Mortality Weekly Report* 48 (1999): 664.

[d]Rounded values.

SOURCE: Data from National Center for Health Statistics.

impair food assimilation; and nutrition status suffers further. Drugs become necessary and most of them impair nutrition status (see Controversy 12). Other treatments, such as surgery, take a further toll. Thus disease and poor nutrition together form a downward spiral that must be broken for recovery to occur (see Figure 11-2).

Certain groups of people are especially likely to be caught in the downward spiral of malnutrition and weakened immunity. Among them are people who restrict their food intakes, whether because of lack of appetite, eating disorders, desire for weight loss, or any other reason. Also susceptible are those who fit one or more, or even all four, of these descriptions: they are very young or old, poor, hospitalized, or malnourished.

Protein-energy malnutrition (PEM) is especially destructive to various immune system organs and tissues. Table 11-1 shows PEM's effects on body defenses. Listed first are the body's initial barriers to infection—the skin and the mucous membranes. The digestive system has an especially active defense force. The mucous membranes of the digestive system are heavily laced with active immune tissues. These tissues both work at the absorptive site and form cells that travel to other organs, such as the liver, pancreas, mammary glands, and uterus.

In PEM, indispensable tissues and cells of the immune system dwindle in size and number, opening the whole body to infection. The skin and body linings, the first line of defense against infections, become thinner as their connective tissue is broken down, and so become less of a barrier to agents of disease. The number of antibodies normally present in secretions of the lungs and digestive tract diminishes, opening the way for repeated lung and digestive tract infections. Infectious agents that are normally barred from the body are allowed to enter, and the defensive responses mounted against these invaders are weak.

Figure 11-2

Nutrition

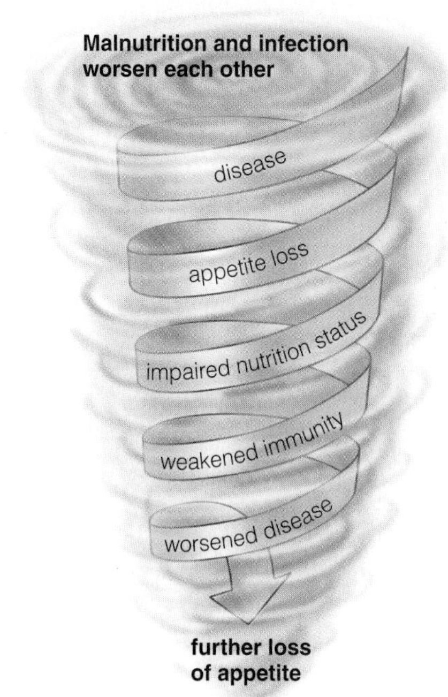

Malnutrition and infection worsen each other
disease
appetite loss
impaired nutrition status
weakened immunity
worsened disease
further loss of appetite

risk factors factors known to be related to (or correlated with) diseases but not proved to be causal.

Table 11-1

Effects of Protein-Energy Malnutrition (PEM) on the Body's Defense Systems

System Component	Effects of PEM
Skin	Thinned, with less connective tissue to serve as a barrier for protection of underlying tissues; delayed skin sensitivity reaction to antigens
Digestive tract and other body linings	Antibody secretions and immune cell number reduced
Lymph tissues	Immune system organs[a] reduced in size; cells of immune defense depleted
General response	Invader kill time prolonged; circulating immune cells reduced; antibody response impaired

[a]Thymus gland, lymph nodes, and spleen.

In addition to malnutrition, another important concern for people with AIDS is food safety. Common food bacteria, such as *Salmonella*, can easily overwhelm an overly compromised immune system. Cleanliness and thorough cooking are protective. General information on food safety can be found in Chapter 14.

The great majority of HIV infections are preventable. The authors urge you to seek out information about AIDS prevention and to heed it.

Some immune cells and their functions were described in Chapter 3, and Chapter 6 provided details about PEM.

Malnutrition can result not only from a lack of availability of food, but from diseases, such as AIDS and cancer, and their treatments. These alter the appetite and metabolism, causing a wasting away of the body's tissues similar to that seen in the last stages of starvation. For people with AIDS, the severity of their wasting or deficiencies of vitamins or minerals can determine the duration of their survival, making them critically in need of medical nutrition therapy.[4] Nutrients cannot cure or reverse the progression of AIDS, of course, but an adequate diet may help to improve responses to drug therapy, reduce duration of hospital stays, and promote greater independence with an improved quality of life overall.[5] And along with diet, exercise to strengthen the muscles may hold wasting to a minimum.[6]

A deficiency or a toxicity of even a single nutrient can seriously weaken even a healthy person's immune defenses. For example, in vitamin A deficiency, the body's skin and membranous linings become unhealthy and unable to ward off infectious organisms. A vitamin C deficiency robs white blood cells of their killing power. Too little vitamin E impairs many aspects of immunity.[7] Deficient and excessive zinc both impair immunity.[8] The obvious conclusion is that a well-balanced diet holds the key to maintaining the best possible immune system support. The rest of this chapter is devoted to the diet-related factors that affect the degenerative diseases that develop over a lifetime.

KEY POINT *Adequate nutrition is a key player in maintaining a healthy immune system to defend against infectious diseases. Medical nutrition therapy can improve the course of wasting diseases. Both excessive and deficient nutrients can harm the immune system.*

Immunity

The ideal situation in which nutrition is the cornerstone of immunity against disease

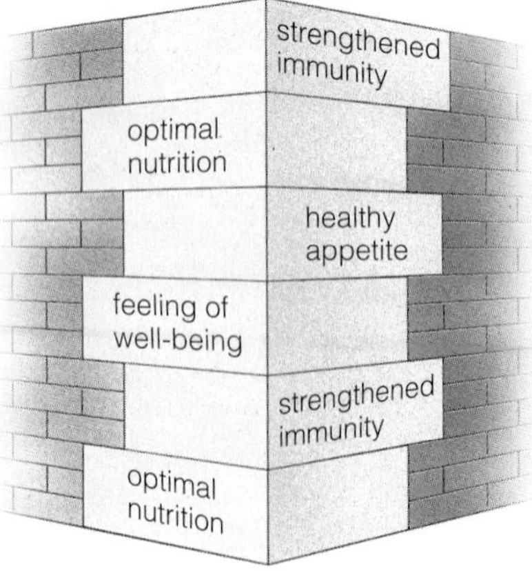
strengthened immunity
optimal nutrition
healthy appetite
feeling of well-being
strengthened immunity
optimal nutrition

Lifestyle Choices and Risks of Degenerative Disease

In contrast to the infectious diseases, each of which has a distinct microbial cause such as a bacterium or virus, the degenerative diseases of adulthood tend to have clusters of suspected contributors known as **risk factors.** Among them are environmental, behavioral, social, and genetic factors that tend to occur in clusters and interact with each other. In many cases, one disease or condition intensifies the risk of another. Risk factors show a correlation with a disease, and although they are candidates for causes, they have not yet been voted in or out. We can say with confidence that a virus causes influenza, but we cannot name the cause of heart disease with such confidence.

An analogy may help clarify the concept of risk factors. A risk factor is like a person who is often seen lurking around the scene of a particular type of crime, say, arson. The police may suspect that person of setting fires, but it may very well be that another, sneakier individual who goes unnoticed is actually pouring the fuel and lighting the match. The evidence against the known suspect is only circumstantial. The police can be sure of guilt only when they observe the criminal in the act. Risk factors have not yet been caught in the act of causing diseases (the mechanisms are still largely unknown). The presence of risk factors often predicts the occurrence of diseases, however, and researchers are working on theories of how risk factors may be related to disease causation.

People's behaviors, including food behaviors, underlie many risk factors.[9] The choice to eat a diet high in fat and calories, for example, is a choice to risk becoming obese and contracting cancer, hypertension, diabetes, atherosclerosis, diverticulosis, or other diseases. Figure 11-3 shows connections among some of the risk factors associated with today's major degenerative diseases and highlights the diet-related behaviors that contribute to them.

The exact contribution diet makes to each disease is hard to estimate. Many experts believe that diet accounts for about a third of all cases of coronary heart disease. Diet's link to cancer incidence is harder to pin down because cancer's different forms associate with different dietary factors. Other important risk factors for cancer include tobacco use, alcohol abuse, exposure to radiation and to environmental and other contamination, and advanced age. General trends, however, support many links between diet and cancer, and the evidence in some cases is overwhelming.[10] People can control their own food choices. If a dietary change can't hurt and might help, why not make it?

Making some choices, such as not smoking, is of importance to most people's health. Making other choices, such as those relating to diet, is doubtless more important for some people than for others because some people are genetically predisposed to certain diseases.[11] To begin deciding whether certain diet recommendations are especially important to you, you should search your family's medical history for diseases common to your

Figure 11-3

Diet/Lifestyle Risk Factors and Chronic Diseases

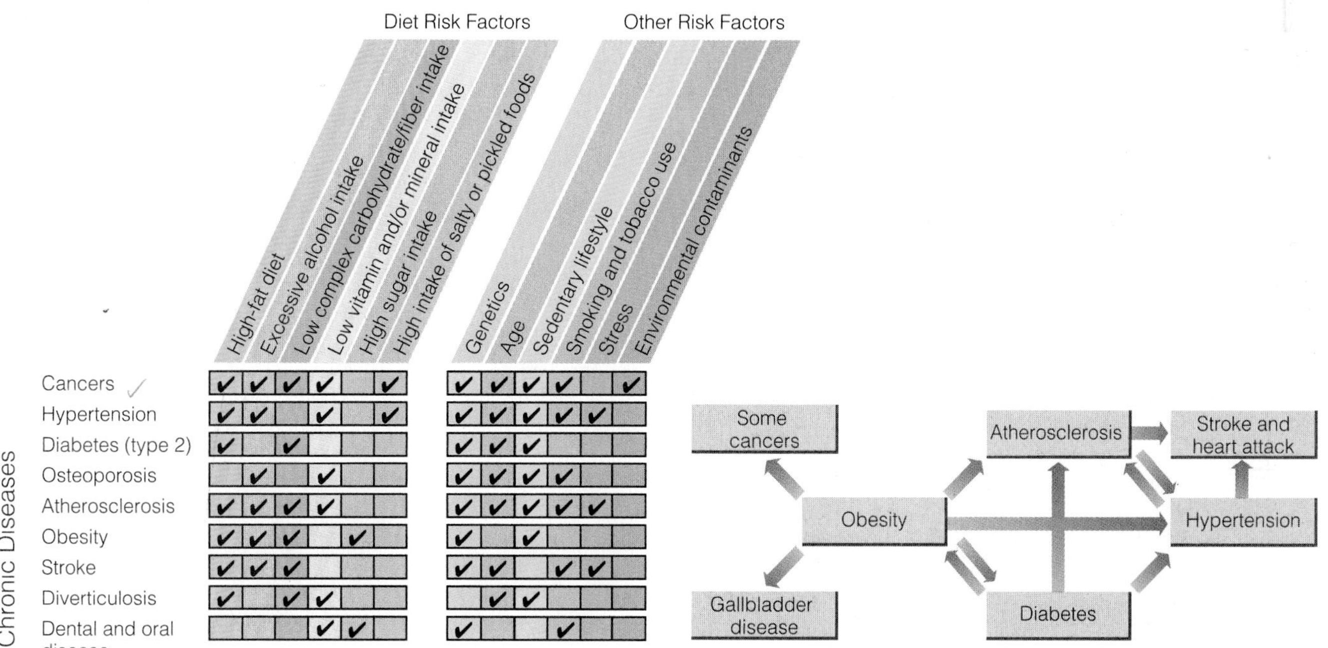

Diet Risk Factors
Other Risk Factors
High-fat diet
Excessive alcohol intake
Low complex carbohydrate/fiber intake
Low vitamin and/or mineral intake
High sugar intake
High intake of salty or pickled foods
Genetics
Age
Sedentary lifestyle
Smoking and tobacco use
Stress
Environmental contaminants
Chronic Diseases
Cancers
Hypertension
Diabetes (type 2)
Osteoporosis
Atherosclerosis
Obesity
Stroke
Diverticulosis
Dental and oral disease
Some cancers
Obesity
Gallbladder disease
Atherosclerosis
Stroke and heart attack
Hypertension
Diabetes

This chart shows that the same risk factor can affect many chronic diseases. Notice, for example, how many diseases have been linked to a high-fat diet. The chart also shows that a particular disease, such as atherosclerosis, may have several risk factors.

This flow chart shows that many of these conditions are themselves risk factors for other chronic diseases. For example, a person with diabetes is likely to develop atherosclerosis and hypertension. These two conditions, in turn, worsen each other. Notice how all of these chronic diseases are linked to obesity.

forebears. Any condition that shows up in several close blood relatives may be a special concern for you. Also find out, after your next physical examination, which test results are out of line. Family history and lab test results together are powerful predictors of disease. Table 11-2 presents a summary of the signs to watch for in both categories.

Accepting that you have certain unchangeable " givens," you can look to the things you can change and choose the most influential among them. For example, a person whose parents, grandparents, or other close blood relatives suffered from diabetes and heart disease is urgently advised to avoid becoming obese and not to smoke. A person who has hypertension is urged to control weight, to exercise regularly, to eat a nutritious diet, to control salt intake, and not to smoke. The guidelines in the Food Feature toward the end of this chapter can benefit most people, while presenting the smallest possible risk to health.

KEY POINT *The same diet and lifestyle risk factors may contribute to several degenerative diseases. A person's family history can reveal strategies for disease prevention.*

Nutrition and Atherosclerosis

Currently, our major cause of death in men and women over 50 is disease of the heart and blood vessels (cardiovascular disease, henceforth abbreviated, CVD). CVD accounts for more of the world's deaths each year than any other single cause, mostly by way of heart attacks and strokes. Efforts to fight CVD have led to valuable discoveries and public education. In 1998, the American Heart Association defined these five major CVD risk factors: obesity, smoking, high blood pressure, high blood cholesterol, and lack of exercise. Many people have changed their lifestyles accordingly.[12] Many have quit smoking or have refrained from starting. Many have been willing to change their diets, consuming less fat, less saturated fat, less cholesterol, less salt, more fruits and vegetables, and more fiber. People are still reluctant to exercise, however. Figure 11-4 provides a method to rate your own risks of developing CVD.

The rate of CVD has fallen steadily and substantially since 1960, but it still remains our leading cause of death and disability in adults.[13] How can people minimize their risks of CVD? Or, more positively, what actions can they take to help them hold on to their cardiovascular health and vigor through life?

Table 11-2
Family and Medical Indicators for Increased Disease Risk

Nutrition Changes Recommended for All People	This is Especially Important if Your Family History Indicates:	And/or if Your Medical History Indicates:
Reduce consumption of fat (especially saturated fat) and cholesterol. Achieve and maintain a desirable body weight. Increase consumption of complex carbohydrates and fiber.	Diabetes, obesity, cancer, or any form of cardiovascular disease (atherosclerosis, hypertension, heart attacks, strokes)	Glucose intolerance, high blood cholesterol or triglycerides, hypertension
Reduce intake of salt/sodium.	Hypertension, diabetes, or any form of cardiovascular disease (atherosclerosis, hypertension, heart attacks, strokes)	Hypertension
Drink alcohol in moderation, if at all.	Liver disease (cirrhosis), cancer, any form of cardiovascular disease (atherosclerosis, hypertension, heart attacks, strokes),[a] osteoporosis	Glucose intolerance, high blood cholesterol or triglycerides, hypertension, any sign of adult bone loss

[a]Moderate alcohol intakes may reduce cardiovascular disease risks in some people. Alcohol excesses injure the heart.

Figure 11-4

Assess Your Heart Disease Risk

Do you know your heart disease risk score? Respond to the statements below, and score yourself as directed. Be aware that a high risk score does not mean you will develop heart disease, but it should warn you of the possibility. Consult your physician if you have questions about your score results.

In each category, circle the number next to the statement that's most true for you.

Cigarette Smoking

I never smoked or stopped smoking three or more years ago.	1
I don't smoke but live and/or work with smokers.	2
I stopped smoking within the past three years.	3
I smoke regularly.	4
I smoke regularly and live and/or work with other smokers.	5

Total Blood Cholesterol

Use the number from your most recent blood cholesterol measurement:

Less than 160	1
160–199	2
Don't know	3
200–239	4
240 or higher	5

HDL Cholesterol

Use the number from your most recent HDL cholesterol measurement:

Over 60	1
56–60	2
Don't know	3
35–55	4
Less than 35	5

Systolic Blood Pressure

Use the first (highest) number from your most recent blood pressure measurement:

Less than 120	1
120–139	2
Don't know	3
140–159	4
160 or higher	5

Excess Body Weight

I am within 10 pounds of my desirable weight.	1
I am 10–20 pounds above my desirable weight.	2
I am 21–30 pounds above my desirable weight.	3
I am 31–50 pounds above my desirable weight.	4
I am more than 50 pounds above my desirable weight.	5

Physical Activity

Determine which statements best describe your usual level of physical activity:

A: Highly Active

My job requires very hard physical labor (such as digging or loading heavy objects) at least four hours a day

or

I do vigorous activities (jogging, cycling, swimming, etc.) at least three times a week for 30 minutes or more

or

I do at least one hour of moderate activity such as brisk walking at least four days a week.

B: Moderately Active

My job requires that I walk, lift, carry, or do other moderately hard work for several hours a day (day-care worker, stock clerk, or busboy/waitress)

or

I spend much of my leisure time doing moderate activities (dancing, gardening, walking, or housework).

C: Inactive

My job requires that I sit at a desk most of the day

and

Much of my leisure time is spent in sedentary activities (watching TV, reading, etc.)

and

I seldom work up a sweat, and I cannot walk fast without having to stop to catch my breath.

Now circle the number that best describes your level of physical activity:

A: Highly Active	1
Between A and B	2
B: Moderately Active	3
Between B and C	4
C: Inactive	5

Scoring Your Heart Attack Risk

To learn your estimated risk, add the six numbers you've circled.

If Your Total Score Is:	*Your Heart Attack Risk Is:*
6–13	*Low*
14–22	*Moderate*
23–30	*High*

SOURCE: American Heart Association.

The twin demons that lead to most CVD are **atherosclerosis** and **hypertension.** Atherosclerosis is the common form of hardening of the arteries; hypertension is high blood pressure; and each makes the other worse. The remainder of this section and the next on hypertension describe these relationships.

atherosclerosis (ath-er-oh-scler-OH-sis) the most common form of cardiovascular disease; characterized by plaques along the inner walls of the arteries (*scleros* means "hard"; *osis* means "too much"). The term *arteriosclerosis* refers to all forms of hardening of the arteries and includes some rare diseases.

hypertension high blood pressure.

plaques (PLACKS) mounds of lipid material mixed with smooth muscle cells and calcium that develop in the artery walls in atherosclerosis (*placken* means "patch"). The same word is also used to describe the accumulation of a different kind of deposits on teeth, which promote dental caries.

How Atherosclerosis Develops

No one is free of atherosclerosis. The question is not whether you have it but how far advanced it is and what you can do to retard or reverse it. Atherosclerosis usually begins with the accumulation of soft, fatty streaks along the inner walls of the arteries, especially at branch points.[14] These gradually enlarge and become hardened **plaques** that damage artery walls, making them inelastic and narrowing the passage through them (see Figure 11-5). Most people have well-developed plaques by the time they reach age 30.

aneurysm (AN-you-rism) the ballooning out of an artery wall at a point that is weakened by deterioration.

aorta (ay-OR-tuh) the large, primary artery that conducts blood from the heart to the body's smaller arteries.

platelets tiny cell-like fragments in the blood, important in blood clot formation (*platelet* means "little plate").

What, exactly, causes the plaques to form is a subject of intense scientific investigation. A suspected first step may have to do with an oxidative change that occurs in LDL particles when the body runs low on antioxidant nutrients, especially vitamin E. A destructive sequence may begin when the lipids of LDL lack adequate antioxidants to protect them and become oxidized to form dangerous free-radical compounds—a condition known as *oxidative stress*.[15] Inside the artery wall, immune cells (phagocytes, see Chapter 3) are attracted to and engulf the oxidized LDL. Once engorged with oxidized LDL, the immune cells become known as foam cells, which themselves become sources of oxidation that attract more and more fresh immune scavengers to the scene. Smooth muscle cells of the arterial wall proliferate, mixing with the foam cells to form hardened areas of plaque. The process is repeated until many inner artery walls become virtually covered with disfiguring plaque.[16]

Normally, the arteries expand with each heartbeat to accommodate the pulses of blood that flow through them. Arteries hardened and narrowed by plaques cannot expand, however, so the blood pressure rises. The increased pressure damages the artery walls further and strains the heart. Plaques are especially likely to form at damage sites, so the development of atherosclerosis becomes a self-accelerating process.

As pressure builds up in an artery, the arterial wall may become weakened and balloon out, forming an **aneurysm.** An aneurysm can burst, and in a major artery such as the **aorta,** this leads to massive bleeding and death.

Figure 11-5

The Formation of Plaques in Atherosclerosis

When plaques have covered 60 percent of the coronary artery walls, the critical phase of heart disease begins.

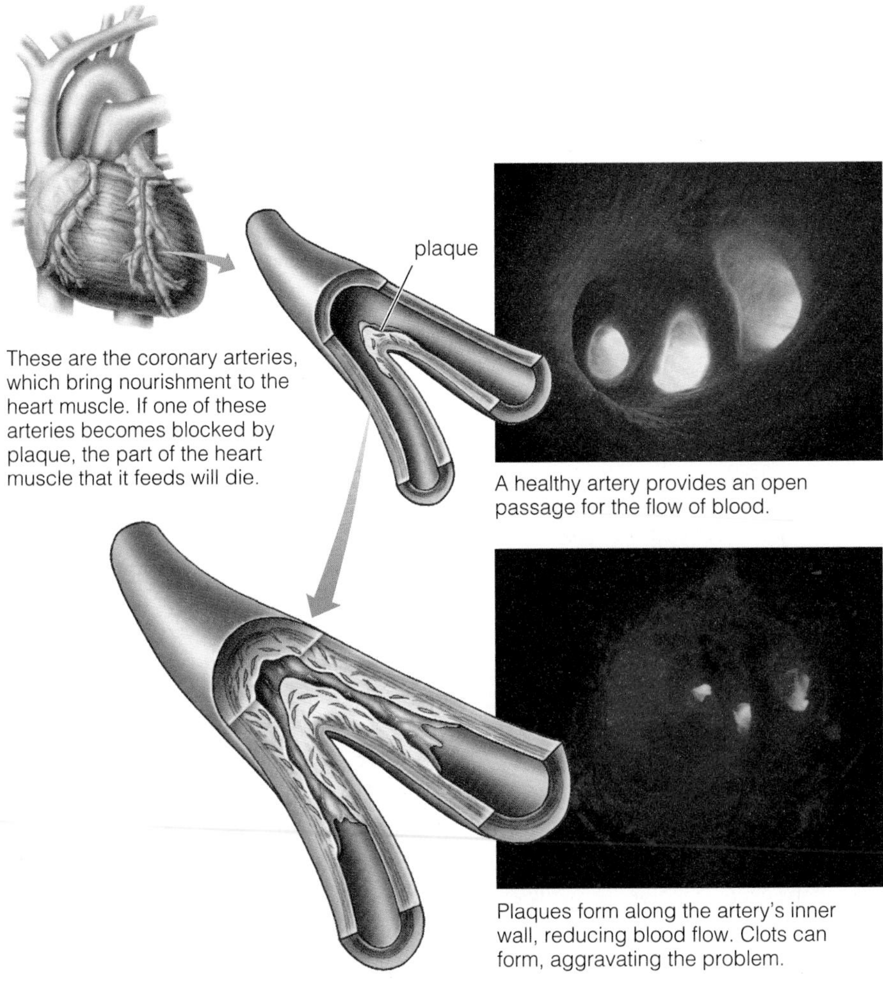
plaque
These are the coronary arteries, which bring nourishment to the heart muscle. If one of these arteries becomes blocked by plaque, the part of the heart muscle that it feeds will die.
A healthy artery provides an open passage for the flow of blood.
Plaques form along the artery's inner wall, reducing blood flow. Clots can form, aggravating the problem.

Abnormal blood clotting can also threaten life. Clots form and dissolve in the blood all the time, and the balance between these processes ensures that clots do no harm. That balance is disturbed in atherosclerosis. Small, cell-like bodies in the blood, known as **platelets,** normally cause clots to form whenever they encounter injuries in blood vessels. In atherosclerosis, the platelets respond to plaques as they do to injuries and form unneeded clots. Platelets also release substances that enlarge plaques. Opposing platelet action are the active products of omega-3 fatty acids. A diet lacking the seafoods that contain these essential fatty acids may contribute to clot formation.[17]

A clot, once formed, may remain attached to a plaque in an artery and gradually grow until it shuts off the blood supply to the surrounding tissue. That tissue may die slowly and be replaced by nonfunctional scar tissue. The stationary clot is called a **thrombus.** When it has grown large enough to close off a blood vessel, it is a **thrombosis.** A clot can also break loose, becoming an **embolus,** and travel along the system until it reaches an artery too small to allow its passage. There the clot becomes stuck and is referred to as an **embolism.** The tissues fed by this artery will be robbed of oxygen and nutrients and will die suddenly. Such a clot can lodge in an artery of the heart, causing sudden death of part of the heart muscle, a **heart attack.** A clot may also lodge in an artery of the brain, killing a portion of brain tissue, a **stroke.**

On many occasions heart attacks and strokes occur with no apparent blockage. An artery may go into spasms, restricting or cutting off the blood supply to a portion of the heart muscle or brain. Much research today is devoted to finding out what causes plaques to form, what causes arteries to go into spasms, what governs the activities of platelets, and why the body allows clots to form unopposed by clot-dissolving cleanup activity.

Hypertension worsens atherosclerosis. A stiffened artery, already strained by each pulse of blood surging through it, is stressed still more by high internal pressure. Injuries multiply, more plaques grow, and more weakened vessels become likely to burst and bleed.

Atherosclerosis also worsens hypertension. Since hardened arteries cannot expand, the heart's beats raise the blood pressure. Hardened arteries also fail to let blood flow freely through the kidneys, which control blood pressure. The kidneys sense the reduced flow of blood and respond as if the blood pressure were too low; they take steps to raise it further (see the discussion of hypertension later in the chapter).

KEY POINT *Plaques of atherosclerosis induce hypertension and trigger abnormal blood clotting, leading to heart attacks or strokes. Abnormal vessel spasms can also cause heart attacks and strokes.*

Risk Factors for CVD

Table 11-3 on the next page lists the factors known to increase the risk of developing CVD. Most people reaching middle age exhibit at least one of these factors besides gender, and many have several factors, silently increasing their risks of CVD.[18] Figure 11-6 shows how rates of heart attacks for both men and women rise with the number of risk factors. It befits a nutrition book to focus on dietary strategies to reduce these risks, but it should be noted that diet is not the only, and perhaps not even the most important, factor in development of CVD.

The big *diet-related* risk factors for CVD are type 2 diabetes (discussed in Chapter 4), obesity (discussed in Chapter 9), high blood cholesterol (discussed next), and hypertension (discussed after cholesterol). People found to have all four of this "deadly quartet" of risk factors are said to have **insulin resistance syndrome** and face a high likelihood of developing cardiovascular disease.[†19] Preventing these four conditions is important because each elevates CVD risk independently, and when they occur together, they synergistically elevate the risk.[20]

In diabetes, blood vessels often become blocked and circulation diminishes. More than 80 percent of people with diabetes die of CVD, usually from heart attacks. Even without diabetes, people whose insulin values are high may face an elevated risk of

[†]Insulin resistance syndrome is also called "syndrome X."

thrombus a stationary clot.

thrombosis a thrombus that has grown enough to close off a blood vessel. A *coronary thrombosis* is the closing off of a vessel that feeds the heart muscle. A *cerebral thrombosis* is the closing off of a vessel that feeds the brain (*coronary* means "crowning" [the heart]; *thrombo* means "clot"; the cerebrum is part of the brain).

embolus (EM-boh-luss) a thrombus that breaks loose (*embol* means "to insert").

embolism an embolus that causes sudden closure of a blood vessel.

heart attack the event in which the vessels that feed the heart muscle become closed off by an embolism, thrombus, or other cause with resulting sudden tissue death. A heart attack is also called a *myocardial infarction* (*myo* means "muscle"; *cardial* means "of the heart"; *infarct* means "tissue death").

stroke the sudden shutting off of the blood flow to the brain by a thrombus, embolism, or the bursting of a vessel (hemorrhage).

insulin resistance syndrome a combination of four risk factors—diabetes, obesity, hypertension, and high blood cholesterol—that greatly increase a person's risk of developing CVD. Also called *syndrome X.*

Proposed Healthy People 2010 Objective: Reduce coronary heart disease deaths to no more than 51 per 100,000 people.

Chapter 5 described the effects of omega-6 and omega-3 fatty acids on heart health and identified some food sources of each.

A blood clot in an artery, like the fatal heart embolism shown, blocks the blood flow to tissues fed by that artery.

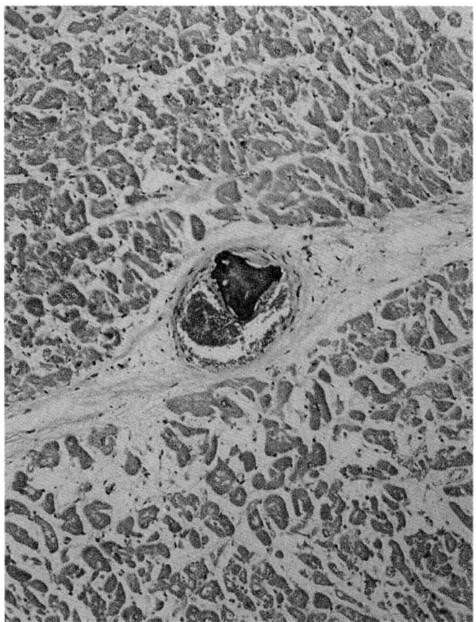

Table 11-3

Factors Contributing to CVD Risk

- Smoking[a]
- Hypertension[a]
- High LDL cholesterol[a]
- Low HDL cholesterol
- Obesity, especially central obesity, as described in Chapter 9[a]
- Glucose intolerance (diabetes)
- Lack of exercise[a]
- Heredity (history of CVD in family members younger than age 55 for males, 65 for females)
- Male gender (after age 45)
- Menopause in women

[a]The American Heart Association includes these factors as major risk factors.

heart disease.[21] About a third of middle-aged men may have a symptomless type of insulin resistance that greatly increases the likelihood of a heart attack—*without* elevated cholesterol values that might otherwise warn them.[22]

Table 11-4 shows the standards by which blood lipids, blood pressure, and obesity are evaluated. Almost half of all deaths from CVD occur among men with blood cholesterol in the borderline-high range, so clearly only the lowest values, if any, are "safe." The moral of the story: everyone, even those with normal cholesterol values, should take seriously diet and exercise advice for reducing CVD risk.

As for triglycerides, this measurement is often elevated in the blood of people with CVD. In association with other risk factors, such as diabetes, central obesity, artery disease, hypertension, or kidney disease, elevated triglycerides are thought to worsen atherosclerosis and accelerate clotting activity while slowing clot destruction in the blood.[23] A recent study hints that elevated triglycerides alone may also raise CVD risk, but more evidence is needed to bear this out.[24] Other factors emerging as important in CVD research are the antioxidant nutrients, such as vitamin E, mentioned earlier.[25] Detailed information about antioxidants was presented in Controversy 7.

What Is the Significance of High Blood Cholesterol? High *blood* cholesterol, particularly when the ratio of LDL to HDL is high, predicts CVD. Generally, cholesterol carried in LDL correlates *directly* with risk of heart disease, whereas that carried in HDL correlates *inversely* with risk (see Figure 11-7). High total cholesterol generally reflects elevated LDL. A population whose average blood cholesterol is 10 percent lower than another population's will suffer one-third less CVD; a 30 percent difference in blood cholesterol predicts a CVD rate that is four times lower.

Diet and Blood Cholesterol Now, how does *diet* relate to high blood cholesterol? Diet relates in two ways: first, a diet high in saturated fat and *trans*-fatty acids contributes to high blood cholesterol, and second, reducing those fats in the diet lowers blood cholesterol and may reduce the rate of CVD.[26]

Generally, wherever in the world diets are high in saturated fat and low in fish, fruits, and vegetables, blood cholesterol is high and heart disease takes a great toll on

Figure 11-7

LDL to HDL Ratio and Risk of Heart Disease

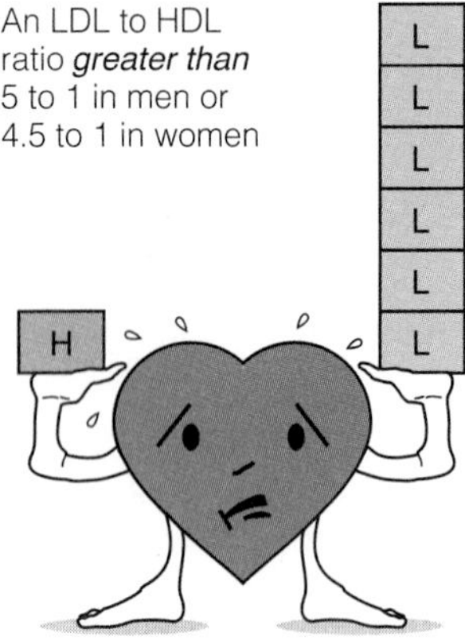
An LDL to HDL ratio *greater than* 5 to 1 in men or 4.5 to 1 in women
H
L
L
L
L
L
L
Increased risk of heart disease

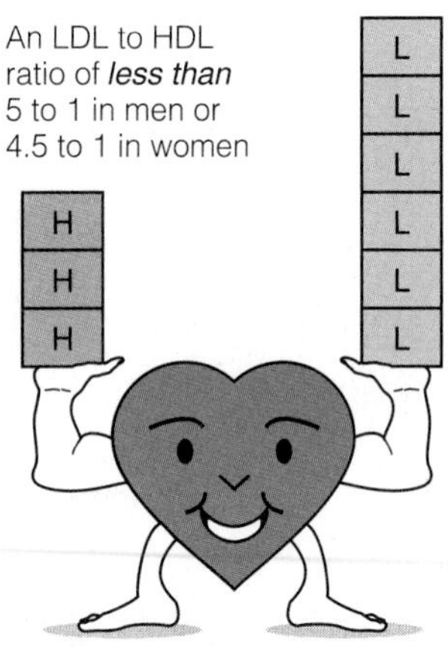
An LDL to HDL ratio of *less than* 5 to 1 in men or 4.5 to 1 in women
H
H
H
L
L
L
L
L
L
Reduced risk of heart disease

Figure 11-6

How Risk Factors Compound a Person's Risk of Heart Attack

This graph shows how risk of heart attack rises in people with more than one risk factor. In the graph, "high cholesterol" is 260 or above, and "high blood pressure" is 150 or above (systolic pressure—the first figure in a blood pressure reading).

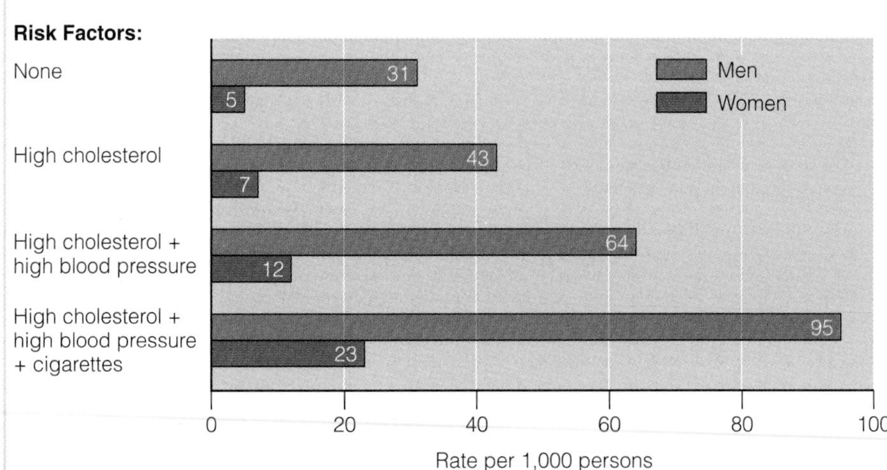
Risk Factors:
None
31
5
High cholesterol
43
7
High cholesterol + high blood pressure
64
12
High cholesterol + high blood pressure + cigarettes
95
23
Men
Women
0
20
40
60
80
100
Rate per 1,000 persons

SOURCE: Framingham Heart Study. Personal communication, Thomas Thom, National Heart, Lung, and Blood Institute.

Table 11-4

U.S. Standards for Blood Pressure, Obesity, Blood Cholesterol, and Triglycerides

Blood Pressure
Systolic and diastolic pressure:[a]
<120 and <80 = optimal.
<130 and <85 = normal.
130–139 or 85–89 = high-normal.
140–159 or 90–99 = mild.
160–179 or 100–109 = moderate hypertension.
≤180 or ≤110 = severe hypertension.

Obesity
Body mass index: > 30.

Total Cholesterol
< 200 mg/dL = desirable.
200–239 mg/dL = borderline high.[b]
≥240 mg/dL = high.[b]

HDL Cholesterol, LDL to HDL Ratio
HDL: ≤35 mg/dL indicates risk.
LDL to HDL ratio: >5 for men or >4.5 for women indicates risk.

LDL Cholesterol
<130 mg/dL = desirable.
130–159 mg/dL = borderline high.
≥160 mg/dL = high.

Triglycerides (Fasting)[c]
>200 mg/dL may indicate risk in those with other risk factors (see text).

[a]The diastolic pressure is the lower of the two numbers in the blood pressure reading—for example, the 60 in 105/60; recently, systolic pressure has also been identified as predictive of heart attack and stroke.
[b]About half of U.S. citizens have cholesterol readings in these ranges, according to the Third Report on Nutrition Monitoring in the United States.
[c]High triglycerides do not normally indicate direct risk, but may reflect lipoprotein abnormalities associated with CVD. High triglycerides also occur in conditions such as kidney disease and diabetes, which suggest a high CVD risk.

health and life. Conversely, wherever dietary fat consists mostly of monounsaturated fats with abundant fish, fruits, and vegetables, blood cholesterol and the rate of death from heart disease are low.

More about the Mediterranean diet in Controversy 2; food sources of saturated fat and *trans*-fatty acids were listed in Chapter 5.

The bulk of research supports the idea that lowering saturated fat intakes will lead to lower blood cholesterol and reduced heart disease risks.[27] In addition, restricting intakes of *trans*-fatty acids along with saturated fats seems important for reducing heart disease risk.[28] Most authorities agree that for people living in the United States and Canada, saturated fat in the diet should account for no more than 10 percent of calories. The American Heart Association recommends limiting daily intakes of fats and oils to about 5 to 8 teaspoons to hold intakes of both saturated fat and *trans*-fatty acids to a level likely to be safe for most healthy people.[29]

Recommendations for U.S. and Canadian citizens also urge that total fat be held to no more than 30 percent of calories and that the cholesterol intake from food be limited to 300 milligrams a day. These measures may be more important for some people than others. The links between intakes of total dietary fat, dietary cholesterol, and high blood cholesterol are not as firm as the links between saturated fat, *trans*-fatty acids, blood cholesterol, and CVD. Data on the people of Mediterranean countries illustrate that diets high in total fat can coexist with low rates of heart disease so long as the diet is rich in fish, fruits, and vegetables and the fat is of the monounsaturated type (Controversy 2 provided details).[30]

When diets are rich in vegetables and fruits, life expectancies are long.

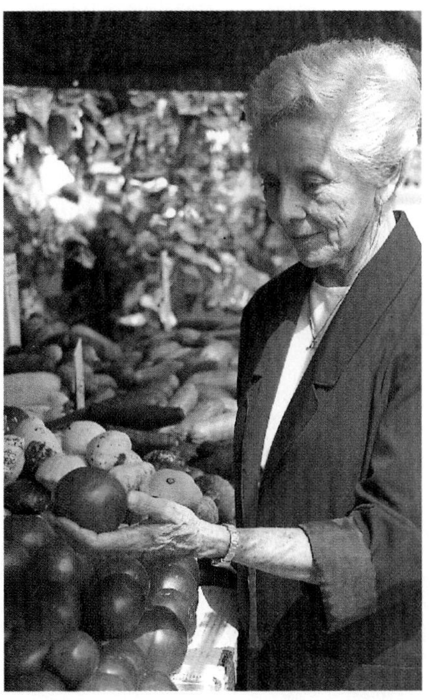

Most people in the United States, however, eat diets rich in meats and hydrogenated fats, so if they reduce the *total* fat in their diets, no doubt their *saturated* fat and their *trans*-fatty acids, which act much like saturated fats with respect to blood cholesterol, will follow suit. Table 11-5 on the next page presents diet adjustments to lower blood cholesterol in two steps. Step 1 is to reduce risk in everyone over 2 years of age; Step 2 is for people with high risks or already-diagnosed CVD.

Chapter 5 first mentioned a new form of margarine with stanol esters added. Stanol esters, compounds derived from the fiber cellulose, block absorption of cholesterol from the intestine and so may lower blood cholesterol by about 7 to 10 percent. These margarines may turn out to be of some use to those with elevated blood cholesterol but only in the context of a low-fat diet and other cholesterol-controlling measures. Also, anyone who does not have elevated cholesterol would be unwise to use such margarines for prevention—their other effects on the body have yet to be proved as safe. Particularly, modified functional foods that contain novel ingredients like stanol esters should not be fed to children.

Previous chapters have already mentioned other dietary factors that seem to lower blood cholesterol or improve heart disease risks in other ways. These include:

- Soluble fibers of cereals, fruits, legumes, and other vegetables.[31]
- Omega-3 fatty acids of fish, which may also help prevent blood clots.
- Weight loss in the overweight, which brings improved blood pressure and blood lipid values.[32]
- Vitamin E and other antioxidant nutrients, which may slow plaque formation, lower the risk of heart attack in people with CVD, and reduce oxidative stress.
- Possibly, the vitamins folate, vitamin B_6, and B_{12}, which may also turn out to fight heart disease.[33] These vitamins play roles in clearing from the blood an amino acid[‡] that may promote both the plaques of atherosclerosis and blood clots.[34]
- A diet based on foods rich in antioxidants and other phytochemicals—that is, one based upon legumes, vegetables, and fruits—which repeatedly turns up in research as related to low risk for CVD and many other diseases.[35]

Controversies 7 and 11 explore the link between reduced disease risks and antioxidant nutrients and phytochemicals in foods.

Many aspects of life probably affect heart health, but the focus of this section has been on blood cholesterol. To return to the main points: (1) high blood cholesterol indicates a risk of heart disease, and (2) it is possible to lower blood cholesterol, in part, by controlling dietary saturated fat. If people lower their blood cholesterol, they will reduce their risk of heart disease.

KEY POINT *Diet-related risk factors for cardiovascular disease include type 2 diabetes, obesity, high blood cholesterol, and hypertension, or the combination of all four called insulin resistance syndrome. Dietary measures to reduce fat, saturated fat, and cholesterol intakes are part of the first line of treatment for high blood cholesterol. Diets rich in fruits and vegetables are also important.*

What Else, Besides Controlling Diet, Can I Do to Reduce My Risk of CVD?

Diet alone may not be enough to reduce CVD risk. Physical activity can amplify a low-fat diet's benefits, and moderate use of alcohol may also be protective.

Physical Activity Physical activity merits special attention. Endurance activities, such as brisk walking or jogging, are effective in lowering LDL and raising HDL concentrations. *Aerobic* exercise, particularly when combined with a low-fat diet, may even help to reverse atherosclerosis. In fact, men with moderate-to-high levels of cardiovascular fitness (determined by a treadmill test) were found to have lower risk of dying from heart attack than men who were less fit, even when the fit men had other risk factors such as high blood cholesterol.[36] Some forms of weight training, if undertaken regularly, may also be protective by elevating blood HDL concentrations somewhat. If pursued daily, even 30 minutes of light exercise, performed at intervals throughout the day, improves the odds against heart disease considerably. The Fitness for Life feature near here offers suggestions for incorporating physical activity into your daily routine.

The benefits of regular exercise are many. In addition to helping normalize blood lipids, regular exercise can

Table 11-5
LDL-Lowering Diet

	Step 1	Step 2
Energy	Energy should be adequate to achieve or maintain desirable weight in both Step 1 and Step 2.	
Total fat[a]	<30%	<30%
Saturated fat[a]	8–10%	<7%
Polyunsaturated fat[a]	<10%	<10%
Monounsaturated fat[a]	10–15%	10–15%
Cholesterol	<300 mg/day	<200 mg/day

[a]Fat amounts are expressed as percentages of total food energy, assuming energy intake is adequate to achieve and maintain desirable weight.
SOURCE: Adapted from The Expert Panel, Summary of the second report of the National Cholesterol Education Program (NCEP) Expert Panel on Detention, Evaluation, and Treatment of High Blood Cholesterol in Adults (Adult Treatment Panel II), *Journal of the American Medical Association* 269 (1993): 3015–3023.

[‡]Homocysteine (mentioned in Chapter 7) derived from the amino acid methionine.

strengthen the heart and blood vessels, alter body composition in favor of lean over fat tissue, lower blood pressure, improve insulin response, and expand the volume of blood the heart can pump to the tissues at each beat and so reduce the heart's workload. Physical activity also stimulates development of new arteries to nourish the heart muscle, and this may be a factor in the excellent recovery seen in some heart attack victims who exercise. These changes are so beneficial that some experts believe that physical activity should be the primary focus of cardiovascular disease prevention efforts.[37]

Both physical activity and the weight loss it induces raise HDL concentrations, and the effects of these two factors are additive. If exercise also reduces of central obesity, the result is exceptionally beneficial because central obesity is an important determinant of CVD risk.[38]

Diet helps a little, physical activity helps more, and the combination is better still. People in a clinical setting have been able to reduce plaque buildup in their arteries by following a strict plan combining an extremely low-fat diet (less than 10 percent of calories from fat), no smoking, stress management, and exercise.[39] Without this program, atherosclerosis would likely have progressed; instead, it regressed and allowed the participants to avoid or postpone heart surgery. Some people do not respond favorably to such lifestyle changes, however, and for them, medication and surgery can be life-saving.

Alcohol Consumption People ask whether moderate consumption of alcohol will reduce their CVD risk. Some research indicates that middle-aged or older people who drink one or two drinks a day (with no binge drinking) suffer fewer fatalities from heart disease compared with abstainers of the same ages.[40] Researchers theorize that moderate alcohol intakes may elevate a form of HDL in the blood, reduce the blood's tendency to clot, or suppress the growth of a constituent of plaques and so reduce heart disease risks.[41] However, a recent study described in Controversy 5 opposes the idea of a

Learn to recognize the signs of a heart attack. Should they occur, call emergency 911.

- Pressure, pain, squeezing, or fullness in the chest.
- Chest pain spreading to the shoulders, neck, or arms.
- Chest pain with light-headedness, fainting, sweating, nausea, or shortness of breath.

Call 911 also for these sudden warning signs of stroke:

- Numbness or weakness of face or extremities, especially on one side.
- Confusion, trouble speaking or understanding.
- Trouble seeing or walking, loss of balance or coordination.
- Severe headache with no known cause.

Prompt treatment is most effective.

Obesity worsens many disease risks (Chapter 9). Chapter 10 specified exercise guidelines for health.

Ways to Include Physical Activity in a Day

FITNESS FOR LIFE

By now, you know the good news about physical activity—it helps to support health, ward off diseases, and keep body fat within bounds. Instead of spending more time thinking about the benefits of physical activity, why not tie up your shoes, head out the door, and get going? Here are some ideas:

- Coach a sport.
- Garden.
- Hike, bike, or walk to nearby stores.
- Mow, trim, and rake by hand.
- Park a block from your destination and walk.
- Play a sport.
- Play with children.
- Take classes for credit in dancing, sports, conditioning, or swimming.
- Take the stairs, not the elevator.
- Walk a dog.
- Wash your car with extra vigor, or bend and stretch to wash your toes in the bath.
- Work out at a fitness club.
- Work out with friends who help one another stay fit.
- Many others—be imaginative.

Also, try these:

- Give two labor-saving devices to charity.
- Lift small hand weights while talking on the phone or watching TV.
- Stretch often during the day.

Now that you know what to do—go ahead and do it!

heart benefit from alcohol—in this study, moderate drinkers suffered no fewer heart disease deaths than did abstainers.[42] The potential relationship between moderate alcohol intake and the health of the heart has yet to be clearly defined.

Heavy alcohol use and abuse are also known to elevate blood pressure, to damage the heart muscle, and to have many other deleterious effects on the body's organs. In fact, heart attacks among apparently healthy young people have been associated with alcohol intoxication from heavy weekend drinking.[43] Heavy drinking (three or more drinks a day) on a regular basis increases the risk of death from breast cancer and other causes, and some evidence suggests that even moderate drinking may do so.[44] A later section examines alcohol's link with cancer in more detail. The bottom line for young people is that, for them, the risks of consuming alcohol greatly outweigh any benefit to the heart, and they do their health no favor by drinking alcohol.

More details about alcohol's effects on the body are in Chapter 5's Controversy.

Proposed Healthy People 2010 objectives:

- Reduce the mean serum cholesterol level among adults to no more than 193 mg/dL.
- Reduce to 16 percent the percentage of adults with high blood pressure.
- Increase to at least 95 percent the proportion of adults who can state whether their blood pressure is normal or high.

More Strategies Drug therapy can lower blood cholesterol, but it also presents risks and side effects that accumulate during years of therapy.[45] Cholesterol-lowering drugs also seem to work best in association with other efforts, such as proper diet and exercise.

Periodically, the media repopularize the idea that the vitamin niacin can lower blood cholesterol. Experimentally, pharmaceutical doses of a form of niacin act like a drug in lowering blood cholesterol and prolonging life, but other drugs effective for this purpose have fewer side effects. Ordinary niacin supplements are useless in lowering blood cholesterol.

Although diet and exercise are not the easy route to heart health that everyone hopes for, they form a powerful combination for improving health. Weight control may not lower blood cholesterol, but it may reduce blood pressure (see the next section). So will eating a diet low in fat, restricted in cholesterol, and high in complex carbohydrates with lots of whole grains, fruits, and vegetables. And even if the diet high in complex carbohydrates does not lower cholesterol or blood pressure, it will help by normalizing blood glucose (diabetes). Remember, diabetes is a major risk factor for CVD. A meal of fish each week may help by favoring the right fatty acid balance so that clot formation is unlikely. The pattern of protection from the recommended diet and exercise regimen becomes clear—the effects of each small choice add to the beneficial whole. While you are at it, don't smoke. Relax. Meditate or pray. Control stress.[46] Play. Happy people have lower blood cholesterol levels.

KEY POINT *Physical activity can reduce CVD risk. Moderate alcohol intake is also associated with reduced risk, but its use can be problematic.*

Nutrition and Hypertension

The most effective single step you can take against hypertension is to learn your own blood pressure.

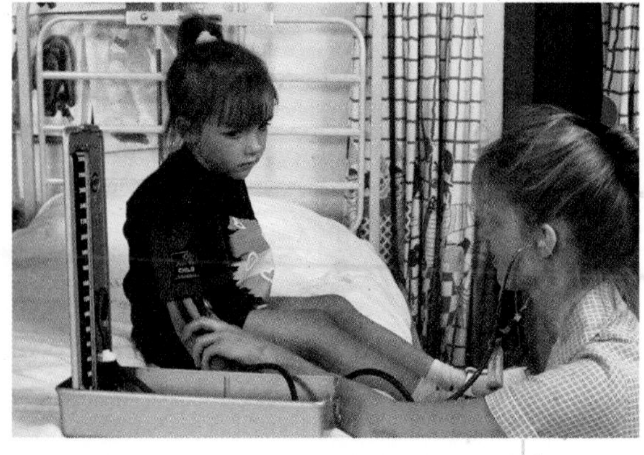

People with low blood pressure, unless it is extremely low, generally enjoy a long life expectancy and low heart disease risk. When blood pressure is high, it threatens to impair the quality of life and strikes many people down before their time. Chronic high blood pressure, or hypertension, remains one of the most prevalent forms of cardiovascular disease, affecting about a quarter of the entire U.S. adult population.[47] It contributes to half a million strokes and to over a million heart attacks each year. The higher above normal the blood pressure, the greater the risk of heart disease. Paired with atherosclerosis, as it often is, hypertension is especially threatening.

You cannot tell if you have high blood pressure; it presents no symptoms you can feel. The most effective single step you can take to protect yourself from hypertension is to find out whether you have it. At checkup time, a health-care professional can take an accurate resting blood pressure reading. Self-test machines in drugstores and other places are often inaccurate. If your resting blood pressure is above normal, the reading should be repeated before confirming the diagnosis of hypertension. Thereafter blood pressure should be checked at regular intervals.

When blood pressure is measured, two numbers are important: the pressure during contraction of the heart's ventricles (large pumping chambers) and the pressure during their relaxation. The numbers are given as a fraction, with the top number representing the **systolic pressure** (ventricular contraction) and the bottom number the **diastolic pressure** (relaxation). Return to Table 11-4 to see how to interpret your resting blood pressure.

Resting blood pressure should ideally be 120 over 80 or lower, but less than 140 over 90 is also considered normal. Above this level the risks of heart attacks and strokes increase in direct proportion to increasing blood pressure.

systolic (sis-TOL-ik) **pressure** the first figure in a blood pressure reading (the "dub" of the heartbeat), which reflects arterial pressure caused by the contraction of the heart's left ventricle.

diastolic (dye-as-TOL-ik) **pressure** the second figure in a blood pressure reading (the "lub" of the heartbeat), which reflects the arterial pressure when the heart is between beats.

KEY POINT *Hypertension is silent, progressively worsens atherosclerosis, and makes heart attacks and strokes likely. All adults should know their blood pressure.*

How Does Blood Pressure Work in the Body, and What Makes It Too High?

Blood pressure is vital to life. It pushes the blood through the major arteries into smaller arteries and finally into tiny capillaries whose thin walls permit exchange of fluids between the blood and the tissues (see Figure 11-8). When the pressure is right, the cells receive a constant supply of nutrients and oxygen and can release their wastes.

Figure 11-8

The Blood Pressure

Three major factors contribute to the pressure inside an artery. For one, the heart pushes blood into the artery. For another, the small-diameter arteries and capillaries at the other end resist the blood's flow (peripheral resistance). Third, the volume of fluid in the circulatory system, which depends on the number of dissolved particles in that fluid, adds pressure.

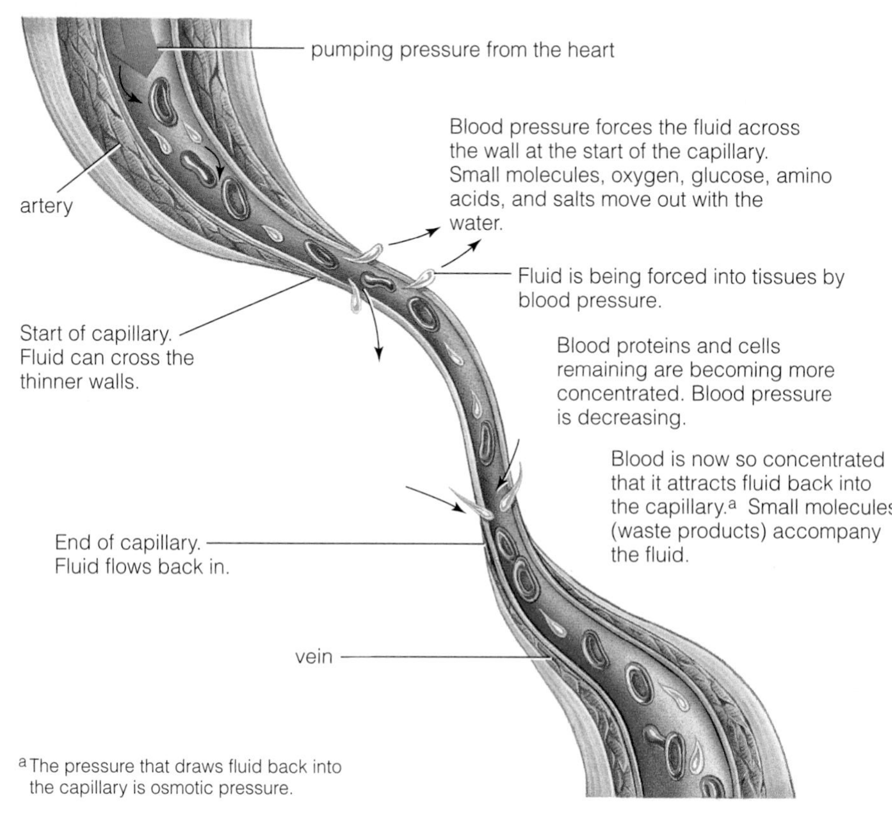

pumping pressure from the heart
Blood pressure forces the fluid across the wall at the start of the capillary. Small molecules, oxygen, glucose, amino acids, and salts move out with the water.
artery
Fluid is being forced into tissues by blood pressure.
Start of capillary. Fluid can cross the thinner walls.
Blood proteins and cells remaining are becoming more concentrated. Blood pressure is decreasing.
Blood is now so concentrated that it attracts fluid back into the capillary.[a] Small molecules (waste products) accompany the fluid.
End of capillary. Fluid flows back in.
vein

[a] The pressure that draws fluid back into the capillary is osmotic pressure.

The Role of the Kidneys The kidneys depend on the blood pressure to help them filter waste materials out of the blood into the urine. (The pressure has to be high enough to force the blood's fluid out of the capillaries into the kidneys' filtering networks.) If the blood pressure is too low, the kidneys set in motion actions to increase it; they send hormones to constrict the peripheral blood vessels and bring about the retention of water and salt in the body. Dehydration sets these actions in motion, and in this case they are beneficial because when the blood volume is low, higher blood pressure is needed to deliver substances to the tissues. By constricting the blood vessels and conserving water and sodium, the kidneys ensure that normal blood pressure is maintained until the dehydrated person can drink water.

As mentioned in an earlier section, atherosclerosis also sets this process in motion, however, and this is not beneficial. By obstructing blood vessels, atherosclerosis fools the kidneys: they react as if there were a water deficiency. The kidneys raise the blood pressure high enough so that they will get the blood they need, but in the process they may make the pressure too high for the arteries and heart to withstand. Hypertension also aggravates atherosclerosis by mechanically injuring the artery linings, making plaques likely to form; plaques restrict blood flow to the kidneys; this may raise the blood pressure still further; and the problem snowballs.

The Roles of Risk Factors Primary among the risk factors that precipitate or aggravate hypertension are atherosclerosis, obesity (particularly central obesity), and insulin resistance (which leads to type 2 diabetes).[48] Excess adipose tissue means miles of extra capillaries through which the blood must be pumped.[49] Strain on the heart's pump, the left ventricle, can enlarge and weaken it, until it finally fails (heart failure). Pressure in the aorta may cause it to balloon out and burst (aneurysm). Pressure in the small arteries of the brain may make them burst and bleed (hemorrhage, a form of stroke). The kidneys can also be damaged when the heart cannot pump enough blood through them (kidney failure). Fluid may then accumulate in the body, straining the heart further and making it hard to breathe (congestive heart failure).

Epidemiological studies have identified several other risk factors that predict hypertension. One is age: most people who develop hypertension do so in their 50s and 60s. Another is heredity: a family history of hypertension and heart disease raises the risk of developing hypertension two to five times, and people of African American descent are likely to develop more severe hypertension earlier in life than people of European or Asian descent. As yet unidentified environmental factors in the United States may also favor the development of hypertension because African Americans living in the United States have higher rates than Africans living in Africa. Hypertension has also been observed to bear some relation to insulin resistance, and measures to prevent diabetes no doubt also protect against hypertension.

The rate of hypertension has been rising steadily over the past four decades.[50] While researchers continue looking for the cause or causes, clearly it is urgent to do what we can to prevent hypertension. Failing in that, we must make every effort to detect and treat it. Even mild hypertension can be dangerous, but individuals who adhere to treatment are less likely to suffer illness or early death.

KEY POINT *Atherosclerosis, obesity, insulin resistance, age, family background, and race contribute to hypertension risks. Prevention and treatment both deserve high-priority efforts.*

How Does Nutrition Affect Hypertension?

Some people need medications to bring their blood pressure down, but diet and exercise alone can bring improvements for many and prevent hypertension for many others.[51] For some, both drugs and diet and exercise are suggested at first until some progress has been made; then the drugs can be stopped. This section focuses on diet.

Salt (Sodium) and Prevention A controversy surrounds the relationship between excessive salt and sodium in the diet and hypertension. The benefit from reducing salt intake in *treatment* of hypertension is not questioned. For about half of people with hypertension, a lower salt (or sodium) intake leads to a reduction in blood pressure. Such people are said to be salt-sensitive, and they often fit one or more of these descriptions: have a family history of hypertension, are African American, have kidney problems or diabetes, are older, or have experienced sustained psychological stress.[52] As for the role of salt in *prevention* of hypertension, the evidence is mixed. One major worldwide epidemiological study revealed that, without question, blood pressure rises whenever sodium or salt intakes increase.[53] Other studies, though, seem to suggest that the degree of sodium restriction needed to produce a meaningful reduction in blood pressure is neither achievable nor sustainable in our population.[54] Nevertheless, the consensus of many professionals and agencies is that enough evidence is available to recommend that everyone moderately restrict salt intake to the level suggested by the *Dietary Guidelines for Americans*—that is, no more than 2,400 mg of sodium (6 grams salt) per day. (Chapter 8 showed how).[55] They reason that, at worst, such a diet cannot be harmful, and it may help some people to avoid hypertension.[56]

Details concerning sodium, salt-sensitivity, and hypertension were presented in Chapter 8.

Assuredly, salt avoidance prevents hypertension in salt-sensitive people. For others, other nutrition-related factors are important, most notably, losing weight for those who are overweight, using moderation with regard to alcohol consumption, and increasing intakes of fruit, vegetables, fish, and low-fat dairy products, and reducing intakes of fat. Exercise and nutrients such as calcium, magnesium, and others also play roles. A blanket recommendation for prevention of hypertension, then, would center on controlling weight, obtaining a balanced diet, exercising, and reducing intakes of alcohol and, possibly, of salt.

Weight Control and Exercise For people who are overweight and hypertensive, a weight loss of as little as 10 pounds may significantly lower blood pressure. Those who are using drugs to control their blood pressure can often cut down their doses if they lose weight.[57]

Moderate physical activity helps in weight loss and also helps to reduce hypertension directly.[58] The right kind of physical activity, regularly undertaken, can lower blood pressure in almost everyone, even in those without hypertension. The "right kind" of activity is the same kind observed to increase blood HDL and lower LDL, that is, the aerobic kind recommended for cardiovascular fitness (Chapter 10 provided details). Physical activity also changes the hormonal climate in which the body does its work. By reducing stress, physical activity reduces the secretion of stress hormones, and this lowers blood pressure. Physical activity also redistributes body water and eases transit of the blood through the peripheral arteries.

Alcohol In moderate doses, alcohol initially relaxes the peripheral arteries and so reduces blood pressure, but high doses clearly raise blood pressure. Hypertension is common among people with alcoholism. The hypertension is apparently caused directly by the alcohol, and it leads to cardiovascular disease, the same as hypertension caused by any other factor. Furthermore, alcohol may cause strokes—even *without* hypertension.[59] The *Dietary Guidelines* urge moderation for those who drink alcohol. *Moderation* means no more than a drink a day for women or two drinks a day for men, an amount that seems safe relative to blood pressure.[60]

Calcium, Magnesium, Potassium, and Vitamin C Other dietary factors may help to regulate blood pressure.[61] A diet providing enough calcium is certainly one such factor. Little doubt remains of calcium's ability to reduce blood pressure in both healthy people and those with hypertension.[62] In fact, adding calcium-rich foods may lower blood pressure even in salt-sensitive people whose sodium intakes remain high.[63] Anyone concerned about blood pressure would do well to think of ways to include calcium-rich foods in the diet.

Adequate potassium[§] and magnesium also appear to help prevent and treat hypertension in certain populations. Diets low in potassium are often associated with

[§]People using diuretics to control hypertension should know that some cause potassium excretion and can induce a deficiency. Those using these drugs must be particularly careful to include rich sources of potassium in their daily diets.

cancer a disease in which cells multiply out of control and disrupt normal functioning of one or more organs.

carcinogen (car-SIN-oh-jen) a cancer-causing substance (*carcin* means "cancer"; *gen* means "gives rise to").

initiation an event, probably occurring in a cell's genetic material, caused by radiation or by a chemical carcinogen that can give rise to cancer.

promoters factors that do not initiate cancer but speed up its development once initiation has taken place.

metastasis (meh-TASS-ta-sis) movement of cancer cells from one body part to another, usually by way of the body fluids.

hypertension, whereas high-potassium diets appear to both prevent and correct hypertension.[64] Magnesium deficiency causes the walls of the arteries and capillaries to constrict and so may raise the blood pressure. Similarly, vitamin C adequacy seems to help normalize blood pressure, while vitamin C deficiency may tend to raise it.[65]

How can people be sure of getting all of the nutrients needed to keep blood pressure low? The best answer may be to consume a low-fat diet with abundant fruits, vegetables, and low-fat dairy products that provide the needed magnesium, potassium, vitamin C, and calcium.[66] One such diet, known as DASH (Dietary Approaches to Stop Hypertension, mentioned in Chapter 8) recommends *twice* the servings of fruits and vegetables of the Food Guide Pyramid, provides 30 percent of its calories from fat, and meets other Dietary Guidelines besides (see Table 11-6 in the margin).

Other dietary factors may affect blood pressure in one way or another. Roles for cadmium, selenium, lead, caffeine, protein, and fat are currently under study. The Food Feature of this chapter provides more detail on dietary measures that help support normal blood pressure.

KEY POINT *For most people,weight reduction, exercise, restricted alcohol use, and a diet that provides adequate nutrients work to keep blood pressure normal. For some, salt restriction is also required. A diet high in fruits, vegetables, fish, and low-fat dairy products and low in fat may help to lower blood pressure.*

Table 11-6

DASH Diet and the Food Guide Pyramid Compared

Food Group	DASH	Pyramid
Grains	7–8	6–11
Vegetables	4–5	3–5
Fruits	4–5	2–4
Milk (fat-free/low-fat)	2–3	2–3
Meat (lean)	2 or less	2–3

SOURCE: E. N. Whitney and S. R. Rolfes, *Understanding Nutrition* (Belmont, Calif.: Wadsworth, 1999), p. 575.

Nutrition and Cancer

One out of every four people in the United States will eventually contract **cancer**, and an estimated 20 to 50 percent of these cancers are attributable to diet.[67] Dietary fat, alcohol, excess calories, and low intakes of fruits and vegetables are thought to be especially important in the occurrence of cancer, but diet relates to cancer in several ways. It is important to get them all in perspective. Constituents in foods may be cancer causing, cancer promoting, or protective against cancer. Also, for the person who has cancer, diet can make a crucial difference in recovery.

Of course, nondiet factors are also important in the development of cancer. A very few cancers are known to be caused by genetics alone and will appear regardless of lifestyle choices. Other cancers are related to environmental factors other than diet, including smoking, exposure to sun, and exposure to water and air pollution or other toxic chemicals. Lack of physical activity may also play a role in the development of some forms of cancer.[68] To give some idea of the extent of these relationships, Table 11-7 lists some of the factors linked to particular kinds of cancers. The emphasis here is on diet, of course.

How Does Cancer Develop?

Cancer is thought to develop through the following steps (illustrated in Figure 11-9):

1. Exposure to a **carcinogen.**
2. Entry of the carcinogen into a cell.
3. **Initiation** of cancer as the carcinogen probably alters the cell's genetic material in some way.
4. Acceleration by other carcinogens, called **promoters,** so that the cell begins to multiply out of control.
5. Spreading of cancer cells via blood and lymph (**metastasis**); disruption of normal body functions.

Researchers think that the first three steps, which culminate with initiation, are key to cancer prevention. On hearing this, many people mistakenly believe that they should avoid eating all foods that contain carcinogens. Doing so would be impossible, however, because most carcinogens occur naturally among thousands of other chemicals and nutrients the body needs. Luckily, the body is well equipped to deal with the minute amounts of carcinogens naturally occurring in foods, such as those listed in the margin.

Selected chemicals and carcinogens occurring naturally in breakfast foods:

- Coffee: acetaldehyde, acetic acid, acetone, atractylosides, butanol, cafestol palmitate, chlorogenic acid, dimethyl sulfide, ethanol, furan, furfural, guaiacol, hydrogen sulfide, isoprene, methanol, methyl butanol, methyl formate, methyl glyoxal, propionaldehyde, pyridine, 1,3,7,-trimethylxanthine.
- Toast and coffee cake: acetic acid, acetone, butyric acid, caprionic acid, ethyl acetate, ethyl ketone, ethyl lactate, methyl ethyl ketone, propionic acid, valeric acid.

NOTE: Consuming coffee, toast, and coffee cake does not elevate a person's risk of developing cancer.

Table 11-7

Some Factors Associated with Cancer

Cancer Sites	Incidence Associated with:	Protective Effect Associated with:
Bladder cancer	Weak associations with coffee, artificial sweeteners, and alcohol; stronger associations with cigarette smoking, chlorinated drinking water	Fruits and vegetables, especially green and yellow ones; adequate fluid intake
Breast cancer	High intakes of food energy and possibly alcohol; sedentary lifestyle; probably not associated with dietary fat	Fruits and vegetables, especially green and yellow ones; soybeans and soy products; physical activity
Cervical cancer	Folate deficiency	Adequate folate intake
Colorectal cancer	High intakes of fat (particularly saturated fat), meat, and alcohol (especially beer); low intakes of fiber, folate, and vegetables; inactivity	Vegetables; calcium, vitamin D, and dairy intake; whole wheat, wheat bran, and other fiber-rich foods; physical activity
Esophageal and mouth cancers	High alcohol, tobacco, and especially combined use; use of preserved foods (such as pickles); low intakes of vitamins and minerals; high intakes of vitamin A supplements	Fruits and vegetables
Liver cancer	Infection with hepatitis virus; high intakes of alcohol; iron overload or other toxicity	
Lung cancer	Supplements of beta carotene (in smokers)	Fruits and vegetables
Ovarian cancer	No dietary risk factors established; inversely correlated with oral contraceptive use	Fruits and vegetables, especially green ones
Pancreatic and lung cancer	No dietary risk factors established; correlated with cigatette smoking and air pollution	Fruits and vegetables, especially green and yellow ones
Prostate cancer	High intakes of fats, especially saturated fats from meats	Fruits and vegetables, especially green and yellow ones; soybeans and soy products; flax seed; adequate selenium intake
Stomach cancer	High intakes of smoke- or salt-preserved foods (such as dried, salted fish); low intakes of fresh fruits and vegetables; infection with ulcer-causing bacteria	Fresh fruits and vegetables, especially tomatoes

NOTE: Findings based on epidemiological studies.

SOURCES: M. C. Jansen and coauthors, Dietary fiber and plant foods in relation to colorectal cancer mortality: The Seven Countries Study, *International Journal of Cancer* 81 (1999): 174–179; B. S. Reddy, Role of dietary fiber in colon cancer; An overview, *American Journal of Medicine* 106 (1999): S50–S51; J. A. Baron and coauthors, Calcium supplements for the prevention of colorectal adenomas, *New England Journal of Medicine* 340 (1990): 101–107: G. J. Handelman, High-dose, vitamin supplements for cigarette smokers: Caution is indicated, *Nutrition Reviews* 55 (1997): 369–370; D. J. Hunter and coauthors, Cohort studies of fat intake and the risk of breast cancer:—A pooled analysis, *New England Journal of Medicine* 334 (1996): 356–361; E. Giovannucci and coauthors, Intake of carotinoids and retinol in relation to risk of prostate cancer, *Journal of the National Cancer Institute* 87 (1995): 1767–1776; Potential mechanisms for food-related carcinogens and anticarcinogens: A scientific status summary by the Institute of Food Technologists' Expert Panel on Food Safety and Nutrition, *Food Technology* 47 (1993); 105–118.

Figure 11-9

Cancer Development

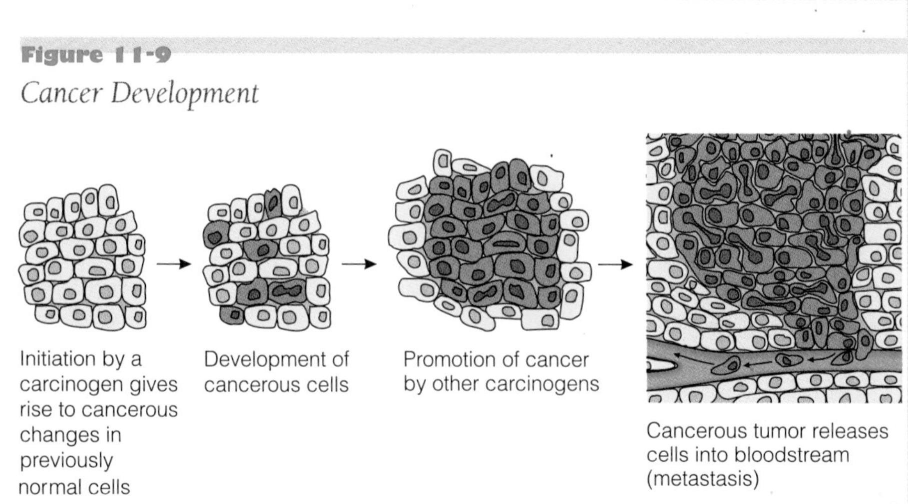
Initiation by a carcinogen gives rise to cancerous changes in previously normal cells
Development of cancerous cells
Promotion of cancer by other carcinogens
Cancerous tumor releases cells into bloodstream (metastasis)

Many people suspect food additives of being carcinogenic, but additives are held to strict standards, and none of those used in the United States cause cancer. (Details concerning saccharin are found in Controversy 4.) Contaminants that enter foods by accident or toxins that arise naturally, however, may be powerful carcinogens, or they may be converted to carcinogens by the body's attempts to metabolize them. Luckily, these constituents are present in foods in doses well below those that may pose significant cancer risks to consumers.[69]

The incidence of certain cancers varies both by region and by ethnic group. For example, Japanese citizens develop more stomach cancers and fewer colon cancers than do U.S. citizens. When Japanese people move to the United States, however, their children develop both stomach and colon cancers at the same rates as do native-born U.S. children. Japan and the United States are both industrial countries, and their environmental pollution rates are similar. Even so, something in the environment must account for the changed cancer pattern in immigrants, and an obvious candidate is diet. The traditional Japanese diet is rich in vegetables and low in fat, two characteristics of diets associated with low rates of colon cancer. The Japanese diet also contains many salted and pickled foods associated with stomach cancer. Traditional Japanese foods are not widely available in the United States, so immigrants largely adopt U.S.-style food choices.

Chapter 14 comes back to the topic of the safety of the U.S. food supply.

KEY POINT *Cancer develops in steps including initiation and promotion, which are thought to be influenced by diet. The body is equipped to handle tiny doses of carcinogens that occur naturally in foods. Populations with high vegetable and grain intakes generally have low rates of cancer.*

How Powerful Is Diet in Reducing a Person's Risk of Developing Cancer?

From the evidence so far, it is almost certain that diet affects cancer rates in the world's people both for the worse and for the better.[70] According to one estimate, deaths from cancers of the colon, prostate, pancreas, and breast might be reduced a full 50 percent, if everyone would adopt a diet that supports good health.[71]

Significantly, studies of populations suggest that low rates of many kinds of cancer correlate with intakes of fiber-rich vegetables and whole grains. Case-control studies, in which researchers can control some of the variables, support the population studies and implicate fat in cancer causation and fruits, vegetables, and whole grains, and especially whole wheat, in its prevention.[72] A related finding is that vegetarians have lower mortality rates from cancer than the rest of the population, even when cancers linked to smoking and alcohol are taken out of the picture.

The following paragraphs explore what is known about the effects of nutrients and fiber on cancer development. This chapter's Controversy delves into an important and rapidly changing frontier of research: the effects of the phytochemicals of fruits, whole grains (and especially whole wheat), herbs, and vegetables on cancer prevention.

Simple advice can be powerful. Think Food Guide Pyramid.

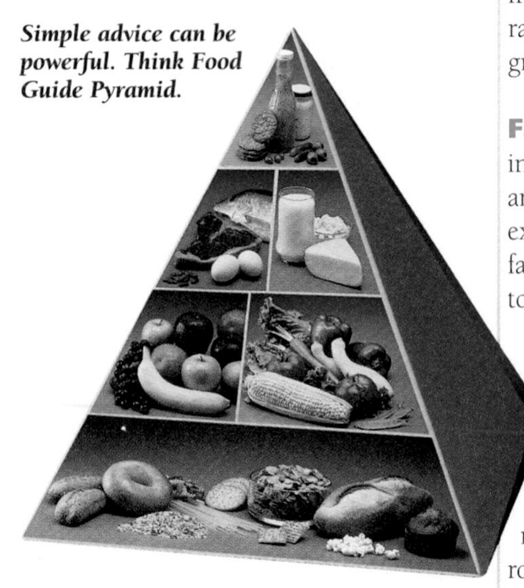

Fat and Fatty Acids Laboratory studies using animals suggest that high fat intakes may correlate with development of cancer. Simply feeding fat to experimental animals is not enough to get tumors started, however; an experimenter must also expose the animals to a known carcinogen. After that exposure, animals fed the high-fat diet develop more cancers faster than animals fed low-fat diets.[73] Thus fat appears to be a stronger cancer promoter than initiator.

In human beings, diets high in fat and cholesterol are positively associated with many forms of cancer.[74] Dietary fat is extremely calorie dense, and energy itself promotes cancer, so researchers must untangle the effects of fat alone from those of the energy content of the diet.[75] A diet high in saturated fat may increase the risk of cancer of the lymph organs and digestive tract organs.[76] Overall, research indicates that breast cancer is probably unrelated to dietary fat.[77] Emerging explanations for rising breast cancer rates (see Figure 11-10) include exposure to environmental pollutants, such as pesticides that mimic estrogen in the body.

One theory of how dietary fat might be related to certain cancers blames the tendency of food fat to oxidize when exposed to high cooking temperatures. In the body, oxidized fat compounds may set up a condition of oxidative stress that may trigger cancerous changes in the tissues of the colon and rectum.[78] Alternatively, a high-fat diet may promote cancer by eliciting the secretion of certain hormones that favor development of certain cancers. Fat also stimulates bile secretion, and organisms in the colon may then convert the bile into cancer-causing compounds.

It may not be fat in general but certain forms of fat that have these effects. Research findings in this area sometimes seem to conflict, however. For example, some findings appear to implicate linoleic acid, the essential omega-6 fatty acid of vegetable oils, in the promotion of cancer, yet a modified form of linoleic acid found only in food from animal sources seems protective against cancers.[**79] Saturated fat seems to end up in the colon in greater concentrations than an equal amount of unsaturated oils or omega-3 fatty acids, and fat in the feces is related to colon carcinogenesis.[80] Further, some evidence suggests that omega-3 fatty acids from fish may protect against cancer, not promote it.[81] In any case, moderation in fat intakes remains a sound principle.

Alcohol, Smoked Foods, and Meats Cancers of the head and neck seem to correlate strongly with the combination of alcohol and tobacco use and with low intakes of green and yellow fruits and vegetables. Alcohol intake alone is associated with cancers of the mouth, throat, and breast, and alcoholism often damages the liver and increases the risk of liver cancer.

Smoke generated from burning wood or charcoal, like smoke from burning tobacco, is made up of a multitude of chemical substances, some of which initiate cancer.[††] Some carcinogens from smoke settle on food during cooking; others form when meat fats or added oils land on the coals and then vaporize, creating carcinogens that rise and stick to the food. Just the process of browning foods triggers chemical changes among the sugars and amino acids of foods such as steak that are typically seared in cooking. These changes often improve the flavor, aroma, and appearance of foods, but they may also create carcinogens.[82] Eating smoked, grilled, charbroiled, or browned foods introduces the carcinogens into the digestive system, where they may affect the tissues of the intestinal lining. Once the compounds are absorbed and enter the body's tissues, however, they are quickly captured and detoxified by the body's competent detoxifying system.

Evidence from population studies spanning 13 countries over a period of almost 20 years supports the theory that diets high in meat, and particularly red meat, are related to a greater risk of developing colon cancer.[83] Remember, however, that even strong correlation is not causation; although certain foods may appear at the scene of the colon cancer crime, no one knows whether eating such foods actually causes cancer or whether some other feature of a meat-containing diet is at fault. Still, a health-savvy diner chooses meats and grilled, fried, highly browned, and smoked foods in moderation.

Food Energy When calorie intakes are reduced, cancer rates fall. In animal experiments, this **caloric effect** holds true regardless of the energy source: excess calories from carbohydrates, fat, or protein all raise cancer rates.[84] When researchers establish a cancer-causing condition and then restrict the energy in laboratory animals' feed, the onset of cancer in the restricted animals is delayed beyond the time when animals on normal feed have died. At the moment, no experimental evidence exists showing this effect in human subjects, but some population observations seem to imply that the effect seen in animals may hold true for human beings as well.[85]

The processes by which excess calories may promote cancer development remain obscure, but some researchers have a hunch that the hormones produced by the kidneys' adrenal gland may be involved. High calorie intakes stimulate the release of these hormones, which cause inflammation, and inflammation stimulates the growth of

**The linoleic compound referred to here is known by the acronym CLA for conjugated (dienoic) linoleic acid.

††The carcinogens of greatest concern are some of those called *polycyclic aromatic hydrocarbons*.

Figure 11-10

New Cases of Cancer, United States, 1999

In 1999, 175,000 women were diagnosed with breast cancer. Incidence of breast cancer has been rising steadily since 1940, when about 55,000 women were diagnosed with breast cancer.

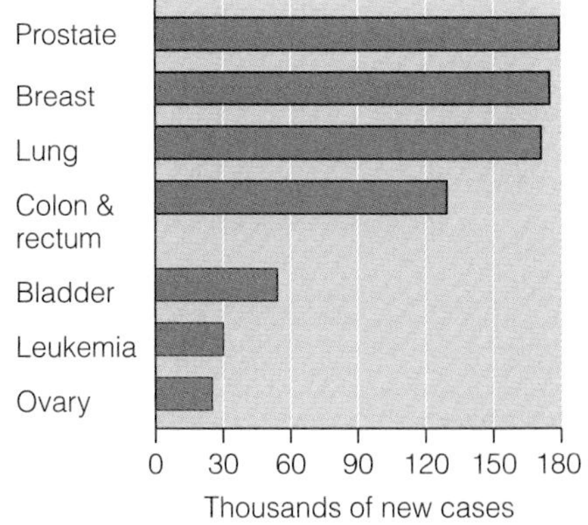
Prostate
Breast
Lung
Colon & rectum
Bladder
Leukemia
Ovary
0 30 60 90 120 150 180
Thousands of new cases

*Projected figures.

SOURCE: Data from American Cancer Society, *Cancer Facts and Figures—1999*, available on the Internet at www.cancer.org/statistics

Chapter 5's Food Feature offers suggestions for cutting fat from the diet, and Figure 5-5 shows the fatty acid breakdown of common fats.

Controversy 5 addressed the topic of alcoholic beverages and cancer risks.

To minimize risks from carcinogens formed during cooking:

- When grilling, line the grill with foil, or wrap the food in foil to minimize the formation of carcinogenic compounds.
- Take care not to burn foods while cooking by any method.
- Marinating meats beforehand may help to reduce the carcinogens formed during grilling.
- Limit intakes of fried, browned, and broiled foods.
- Limit intakes of smoked foods.

caloric effect the drop in cancer incidence seen whenever intake of food energy (calories) is restricted.

carcinogenesis the origination or beginning of cancer.

tumors. Restricting energy intakes inhibits adrenal hormone release. Another idea is that obesity, and the insulin resistance it fosters, may promote the development of cancer.[86] Other theories propose other mechanisms. Importantly, a high-calorie diet can potentiate (augment) the damaging actions of other carcinogens that may be present in the tissues, making the advice to consume diets moderate in energy particularly important.

Fiber and Fluid Many studies have shown that a diet with ample high-fiber foods helps to protect against some forms of cancer.[87] One recent report made headlines by contradicting long-standing evidence for this protective effect—the researchers detected no effect of fiber on rates of colon and rectal cancer—but other research has reaffirmed the original finding that a fiber-rich diet may indeed afford protection.[88] It may do so by promoting the excretion of bile from the body, by absorbing toxins and carrying them out of the body, by generating beneficial hormonelike fragments within the colon, by scavenging free-radical compounds in the feces, or by stimulating the body's immune system to oppose **carcinogenesis.**[89] Fiber may be especially important for preventing cancers of the colon, rectum, and possibly breast, but some features of a high-fiber diet other than fiber itself, such as abundant fruits and vegetables, may also help fight other forms of cancer. High-fiber diets that are also high in both fat and calories seem not to protect against cancer risks.

If a fat-rich, calorie-dense diet is implicated in causation of certain cancers and if a vegetable-rich diet is associated with prevention, then one would expect vegetarians to have a lower incidence of those cancers. They do, as the many studies cited in Controversy 6 have shown.

One type of cancer, bladder cancer, may be related to intake of fluids. Men who drink about 10 cups of fluid a day have been reported to develop substantially less bladder cancer than those drinking only about half this amount.[90] The most probable explanation for this effect involves carcinogens that form naturally in urine. A greater fluid intake dilutes the carcinogens and causes more frequent urination, thus reducing the likelihood that carcinogens will interact with the tissues of the bladder. Plain water seems most beneficial in this regard, but almost any kind of fluid, save one kind, will do. The exception is alcoholic beverages, which do not lower bladder cancer risks.

Smoked, grilled, charbroiled, or browned foods introduce carcinogens into the digestive system.

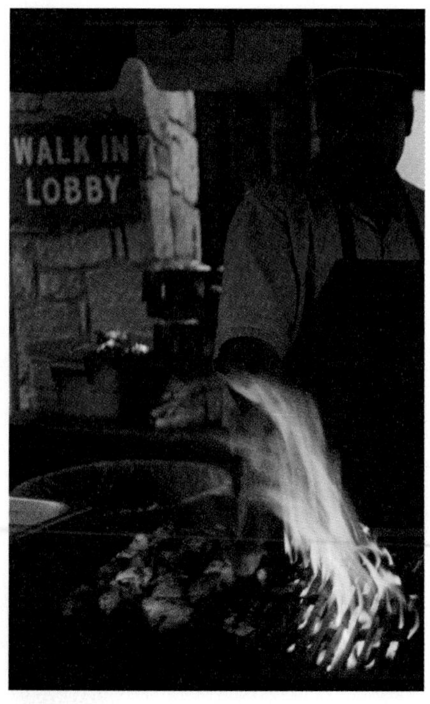
WALK IN
LOBBY

Folate and Other Vitamins Vitamin E, vitamin C, and beta-carotene received attention in Controversy 7, which included a discussion of their antioxidant roles and cancer-fighting effects. Other vitamins may fight against cancer in other ways, for example, as antipromoters. These include vitamins B_6 and B_{12} and pantothenic acid. Vitamin A regulates aspects of cell division and communication that go awry in cancer. It also helps to maintain the immune system. Immune cells can often identify cancerous cells and destroy them before cancer can develop.

Folate is known to play a special role with respect to cervical cancer and may fight other cancers as well. Cervical cancer presents a major health threat for women worldwide. Each year 50,000 new cases of cervical cancer are diagnosed in the United States, and many more cases of early precancerous changes known as cell dysplasia are treated. The underlying cause of this ailment is an often symptomless, sexually transmitted virus, human papilloma virus (HPV), but inadequate dietary folate may be related to the activation of the virus. For this reason alone, all sexually active women should attend to their folate needs. Many other reasons have appeared in previous chapters.

Calcium and Other Minerals Some evidence suggests that a high-calcium diet may help to prevent colon cancer. In a large, long-term study, people who developed colon cancer were found to have consumed slightly less calcium and vitamin D than people who did not develop the cancer. Other studies attempting to confirm this finding have obtained mixed results, but when calcium intakes of populations are compared, the trend appears consistent. Populations consuming more calcium are seen to develop less colon cancer even after researchers subtract the effects of dietary fat from their analysis. In animal studies, calcium seems to protect the colon lining from some of the effects of a high-fat diet. These studies have not yet proved that dietary calcium prevents colon cancer, but with all the other points in calcium's favor, prudence dictates that everyone should arrange to meet calcium needs every day.

Other minerals are thought to play roles in cancer prevention, perhaps by helping antioxidant enzymes. These include zinc, iron, copper, selenium, and probably more.

Foods Themselves In the end, whole foods, not single nutrients, may be most influential on cancer development. For example, many vegetables are known to contain combinations of phytochemicals known as **antipromoters**, substances that oppose cancer. One way antipromoters may protect against cancer is by acting as mild toxins that force the body to build up its arsenal of carcinogen-destroying enzymes. Then, when a potent carcinogen arrives, the prepared body deals with it swiftly. Almost without exception, studies find that infrequent use of green and yellow fruits and vegetables and citrus fruits correlates with cancers of many types. Specifically, infrequent use of **cruciferous vegetables**—broccoli, brussels sprouts, cabbage, cauliflower, turnips, and the like—is common in colon cancer victims (see the margin). One recent review of the literature found an almost unheard-of perfect association between reduced incidence of lung cancer and diets high in fruits and vegetables—every study included in the review reported a protective effect.[91] Incidence of stomach cancer, too, correlates with too few vegetables in the diet: in one study, with vegetables in general; in another, with fresh vegetables; and in others, with lettuce and other fresh greens or vegetables containing vitamin C. Herbs and spices often contain such compounds, too, and their medicinal use is growing rapidly among those who search for natural remedies to ailments through alternative medicine (see the Consumer Corner near here).

Foods that contain phytochemicals are believed to promote health and fight diseases. Such foods, often called *functional foods,* have been recognized as potentially beneficial by the National Academy of Sciences. According to the academy, a functional food is any of a number of "potentially healthful products, [that include] any modified food or food ingredient, that may provide a health benefit beyond the traditional nutrients it contains."[92] This chapter's Controversy explores the state of the science concerning functional foods and their phytochemical constituents.

KEY POINT *Diets high in certain fats and red meats are associated with cancer development. Foods containing fiber, vitamin C, beta-carotene, many other vitamins and minerals, phytochemicals, and an ample intake of fluid, are thought to be protective.*

antipromoters compounds in foods that act in several ways to oppose the formation of cancer.

cruciferous vegetables vegetables with cross-shaped blossoms. Their intake is associated with low cancer rates in human populations. Examples include broccoli, brussels sprouts, cabbage, cauliflower, rutabagas, and turnips.

Cruciferous vegetables belong to the cabbage family: bok choy, broccoli, broccoli sprouts, brussels sprouts, cabbages (all sorts), cauliflower, greens (collard, mustard, turnip), kale, kohlrabi, rutabaga, and turnip root.

Conclusion

Nutrition is often associated with promoting health and medicine with fighting disease, but if we ever believed that a clear line separated nutrition and medicine, we surely can do so no longer.[93] Every major agency involved with health recommends a healthful diet as part of a lifestyle that provides the best possible chance for a long and healthy life.

This chapter has summarized the major forms of disease and their links with nutrition. You may have noticed a philosophical shift from previous chapters. There, we could say "a deficiency of nutrient X causes disease Y." Here we could only cite theories and discuss research that illuminates current thinking. We can say with certainty, for example, that "a diet lacking vitamin C causes scurvy," but to say that a low-fiber diet that lacks vegetables causes cancer would be inaccurate. We can say only that, as a general trend, people who eat few vegetables suffer more often from cancer. We can, however, recommend behaviors that are prudent and reduce the likelihood of illness. The Food Feature presents these recommendations.

This chapter concludes this book's treatment of normal adult nutrition. The next two chapters are about nutrition's contribution to health in each stage of the life cycle from pregnancy and infancy to old age.

Often, it is foods like these, not individual chemicals, that lower people's cancer rates.

CO

CONSUMER CORNER

Complementary, Alternative, and Herbal Medicine

WHERE DO YOU turn for help when illness strikes? Do you see a physician who offers treatment methods sanctioned by the established medical community? Or do you seek out an herbalist, an acupuncturist, or another practitioner of **alternative therapy** (see Table 11-8)? Recently, scientists have been taking a close look at alternative therapies, which attract millions of people and billions of health-care dollars each year.

Unlike conventional therapies, alternative therapies:

- Generally have not been well established by scientific experimentation to be safe and effective—they are used but unproven.
- Are not taught by most medical schools in the United States.
- Are not reimbursable by most health insurance providers in the United States.

In 1992, the prestigious National Institutes of Health established its National Center for Complementary and Alternative Medicine to investigate the efficacy of alternative therapies. Indeed, some of today's proven medical practices started out as alternative medicine. For example, cancer radiation therapy was once an unconventional therapy, but it quickly proved its clinical value and became part of the mainstream cancer treatment.

Critics charge that tax money might be better spent elsewhere, but the office has released a few interesting findings.[1] Among them is a statement of possible support for use of **acupuncture** for quelling nausea associated with surgery, cancer chemotherapy, and pregnancy and for relieving pain associated with dental procedures. Acupuncture, when performed by skilled practitioners under sterile conditions, presents virtually no risk to the user, a claim that cannot be made for many standard drugs administered for relief of nausea and pain.

Many people seek out alternative therapies because they distrust standard medical practices and desire more "natural" treatments that they hope will be safer. This hope is often ill-founded, however. Often, so little scientific evidence exists about the use of alternative therapies that no reasonable conclusions about their safety or efficacy can be drawn. Almost anyone can claim to be an expert in a "new" or "natural" therapy, even though no clinical evidence supports the efficacy of that therapy. Many practitioners display false credentials and act knowledgeable, when in reality, they are untrained (see Controversy 1). For example, purveyors of alternative medicine almost always suggest taking vitamins, but few have the authentic training in nutrition that would qualify them to make such recommendations.

Stories abound that credit alternative therapies with miraculous cures of diseases. The listener may think that unless the speaker is lying, the therapies really do cure the diseases. But a third option also exists: remember from Chapter 1 that giving a placebo medication very often brings about a physical healing when the patient believes in the treatment.[2] The placebo effect powerfully augments a person's ability to heal when a therapy is convincingly administered. No doubt many an alternative therapy has helped people to recover by engaging the ability of the body to heal itself.

This is not to say that all alternative medicine is placebo based or that none has authentic medicinal value. Of special interest in this regard is **herbal medicine.** Since the dawn of humankind herbs have been harvested for use as medicines. Dozens of herbs, when investigated scientifically, are found to contain effective natural drugs. For example, ancient people greatly valued the resin they called myrrh, which has recently been discovered to contain an analgesic (pain-killing) compound.[3] Willow bark contains aspirin; the herb valerian contains a tranquilizing oil; senna leaves produce a powerful laxative. A constituent of green tea leaves may even induce cancer cells to self-destruct.[4] Penicillin is a dramatic example of a life-saving drug harnessed from a wild mold. Medicinal herbs are showing up on grocery shelves in "functional foods," making consumer education about their effects a pressing need.

Beneficial compounds from wild species form the basis of more than half of our modern medicines. By analyzing these compounds, pharmaceutical labs can synthesize pure forms of the drugs. Unlike herbs and wild species, which vary from batch to batch, these synthesized medicines are supplied in predictable dosages. By creating synthetic drugs, we are also able to conserve endangered species. Consider that producing one 300-milligram dose of the anticancer drug Taxol took all of the bark from a 40-foot-tall, 100-year-old yew tree, until scientists learned how to synthesize it.[5] A larger concern is the rampant extinction of species on earth. Many yet undiscovered cures may be forever lost as wild species die out as their habitats are destroyed, long before their secrets are revealed to medicine.

Herbal medicine has some serious drawbacks. For one thing, few herbalists prescribing herbs have the under-

Table 11-8

Alternative Therapy Terms[a]

- **acupuncture** (AK-you-PUNK-cher) a technique that involves piercing the skin with long thin needles at specific anatomical points to relieve pain or illness. Acupuncture sometimes uses heat, pressure, friction, suction, or electromagnetic energy to stimulate the points.
- **alternative therapy** any approach to medical diagnosis and treatment not fully accepted by the established medical community and as such, not widely taught at U.S. medical schools or practiced in U.S. hospitals; also called *adjunctive, unconventional,* or *unorthodox* therapy.
- **aroma therapy** a technique that uses oil extracts from plants and flowers (usually applied by massage or baths) to try to enhance physical, psychological, and spiritual health.
- **ayurveda** (EYE-your-VAY-dah) a traditional Hindu system of using herbs, diet, meditation, massage, and yoga to stimulate health.
- **bioelectromagnetic medical applications** the use of electrical energy, magnetic energy, or both in an attempt to stimulate bone repair, wound healing, and tissue regeneration.
- **biofeedback** the use of sensors to convey information about heart rate, blood pressure, skin temperature, muscle relaxation, and the like to enable a person to learn how to consciously control these medically important functions.
- **biofield therapeutics** a manual healing method that supposedly directs a healing force from an outside source (commonly God or another supernatural being) through the practitioner and into the client's body; also called *Reike* (ray-kee), or "laying on of hands."
- **chelation therapy** the use of ethylene diamine tetraacetic acid (EDTA), supposedly to bind with metallic ions and heal the body by removing toxic metals.
- **chiropractic** (KYE-roe-PRAK-tik) a manual healing method of manipulating vertebrae to relieve musculoskeletal pain.
- **complementary medicine** therapy involving techniques of mainstream medicine or alternative therapies, whichever produces the desired outcome, or employs both at the same time.
- **faith healing** the practice of invoking divine intervention without the use of medical, surgical, or other traditional therapy.
- **herbal medicine** the use of herbs and other natural substances with the intention of preventing or curing diseases.
- **homeopathic** (HOME-ee-oh-PATH-ick) **medicine** a practice based on the theory that "like cures like," that is, that substances that cause symptoms in healthy people can cure those symptoms when given in very dilute amounts (*homeo* means "like"; *pathos* means "suffering").
- **hypnotherapy** a technique that uses hypnosis and the power of suggestion in the attempt to improve health behaviors, relieve pain, and heal.
- **imagery** a technique that guides clients to achieve a desired physical, emotional, or spiritual state by visualizing themselves in the state.
- **iridology** the study of changes in the iris of the eye and their purported relationships to disease.
- **naturopathic medicine** a system that integrates traditional medicine with botanical medicine, clinical nutrition, homeopathy, acupuncture, East Asian medicine, hydrotherapy, and manipulative therapy.
- **orthomolecular medicine** the use of large doses of vitamins to attempt treatment of chronic disease.

[a]This is a partial list of popular alternative therapies.

standing of botany, pharmacology, or human physiology necessary to use these drugs effectively and safely. Instead they rely on hearsay and folklore. Dangerous mistakes with herbs are extraordinarily likely. For example, most mint is safe when brewed as tea, but some may contain highly toxic pennyroyal oil. Folk medicine urges parents to soothe a colicky baby with mint tea, but one concoction laden with pennyroyal was recently blamed for liver and neurological injuries to at least two infants, one of whom died.[6]

In addition to cases of mistaken identity, purity of herbal products can be a problem. Quantities of mercury and arsenic detected in traditional Chinese herb balls for treating fever, rheumatism, and cataracts have exceeded the Environmental Protection Agency's maximum allowable levels by 20,000 and 1,000 times, respectively.[7]

Another problem is lack of information about which herbs to use and *not* to use and when. For example, no one really knows if St. John's wort, said to have a calming effect, is safe to take during pregnancy. Also, interactions with medications are common, but labels of herb bottles typically bear no warnings to that effect.[8] Foxglove leaves contain digoxin, a compound that modifies the heart's action; a person taking cardiac medication who also decides to take foxglove may be headed for disaster from the combined effect on the heart.

Some food manufacturers add medicinal herbs to common foods and can legally make "structure and function" claims on the labels. A claim such as "helps maintain healthy cholesterol levels" can appear on labels without FDA approval, so long as a disclaimer stating that the FDA has not evaluated the statement also appears. A problem with kava kava-containing corn chips, tomato soup with St. John's wort, or gingko biloba chocolate chewies, is that parents can easily believe that these foods are superior for health and so feed them to their children. Such foods are most probably of no benefit to children, but may incur substantial risks to their health.

Additionally, herbal medicines are sold as "dietary supplements" instead of drugs, a step that allows their labels to make an almost unbelievable array of unsubstantiated claims. Not surprisingly, when a label claims that an herbal product "may" strengthen immunity, support eyesight, or maintain heart

health, consumers believe that taking the products will provide those benefits. Beware. Herbal medicine labels, like the food labels just described, may legally include unproved claims as long as the label also includes the words "Has not been evaluated by the FDA." Table 11-9 lists some additional herbs and their potential actions.*

Another huge source of misinformation about herbs is the Internet. People tend to believe what they find there, but many of the testimonials and advice from "doctors" are intended to market products, and not to educate consumers. All of the Internet cautions of Controversy 1 apply; just because you see something on the "Net" doesn't necessarily make it so.

At present, scientific knowledge concerning alternative medicine is growing but with the exception of acupuncture for the purposes already noted, most such therapies are unexplored. However, a growing number of health-care professionals are trying out alternative therapies and incorporating the helpful ones into their practices. This open-minded approach, termed **complementary medicine,** takes advantage of the best of both kinds of medicine. As more becomes known, no doubt more beneficial therapies will be ushered into mainstream medicine. Until then, consumers are left in the dark about the potential risks and benefits of choosing alternative therapies.

*A reliable source of information about herbs is V. Tyler, *The Honest Herbal* (New York: Pharmaceutical Products Press). Look for the latest edition.

Table 11-9

Selected Herbs and Their Effects

- **aloe** a tropical plant with widely claimed value as a topical treatment for minor skin injury. Some scientific evidence supports this claim; evidence against its use in severe wounds also exists.
- **belladonna** any part of the deadly nightshade plant; a fatal poison.
- **cat's claw** an herb from the rain forests of Brazil and Peru; claimed, but not proved, to be an "all-purpose" remedy.
- **chamomile** flowers that may provide some limited medical value in soothing menstrual, intestinal, and stomach discomforts.
- **chaparral** an herbal product made from ground leaves of the creosote bush and sold in tea or capsule form; supposedly, this herb has antioxidant effects, delays aging, "cleanses" the bloodstream, and treats skin conditions—all unproven claims. Chaparral has been found to cause acute toxic hepatitis, a severe liver illness.
- **comfrey** leaves and roots of the comfrey plant; believed, but not proved, to have drug effects. Comfrey contains cancer-causing chemicals.
- **echinacea** an herb popular before the advent of antibiotics for its "anti-infectious" properties and as an all-purpose remedy, especially for colds and allergy and for healing of wounds. A small body of research seems to lend preliminary support for some of the claims, but also points to an insecticidal property, leading to questions about safety. Also called *cone-flower.*
- **feverfew** an herb sold as a migraine headache preventive. Some evidence exists to support this claim.
- **foxglove** a plant that contains a substance used in the heart medicine digoxin.
- **ginkgo biloba** an extract of a tree of the same name, claimed to enhance mental alertness, but not proved to be effective or safe.
- **ginseng** (JIN-seng) a plant root containing chemicals that have stimulant drug effects. *Ginseng abuse syndrome* is a group of symptoms associated with the overuse of ginseng, including high blood pressure, insomnia, nervousness, confusion, and depression.
- **hemlock** any part of the hemlock plant, which causes severe pain, convulsions, and death within 15 minutes.
- **kava-kava** the root of a tropical pepper plant, often brewed as a a tea consumed for its calming effects. Limited scientific research supports the effectiveness of kava-kava for treating anxiety. Adverse effects include skin rash, metabolic abnormalities, elevated blood cholesterol, and, possibly, lethargy and mental disorientation.
- **kombucha** a product of fermentation of sugar-sweetened tea by various yeasts and bacteria. Proclaimed as a treatment for everything from AIDS to cancer but lacking scientific evidence. Microorganisms in home-brewed teas have caused serious illnesses in people with weakened immunity. Also known as *Manchurian tea, mushroom tea,* or *Kargasok tea.*
- **kudzu** a weedy vine, whose roots are harvested and used by Chinese herbalists as a treatment for alcoholism. Kudzu reportedly reduces alcohol absorption by up to 50 percent in rats.
- **medicinal herbs** nonwoody plants, plant parts, or extracts valued by some people for their medicinal qualities, both proved and unproved.
- **sassafras** root bark from the sassafras tree; once used in beverages but now banned as an ingredient in foods or beverages because it contains cancer-causing chemicals.
- **St. John's wort** an herb containing psychoactive substances that has been used for centuries to treat depression, insomnia, bedwetting, and "nervous conditions." Most scientific reports find St. John's wort equal in effectiveness to standard antidepressant medication for relief of depression. Long-term safety, however, has not been established.
- **valerian** a preparation of the root of an herb used as a sedative and sleep agent. Safety and effectiveness of valerian have not been scientifically established.
- **witch hazel** leaves or bark of a witch hazel tree; not proved to have healing powers.

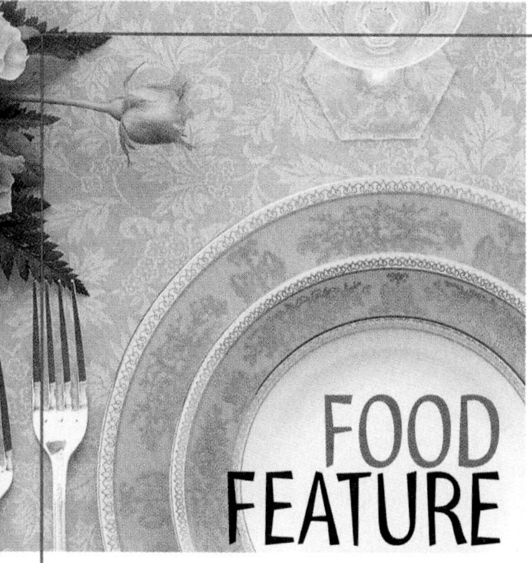
FOOD
FEATURE

Diet as Preventive Medicine

If you find yourself saying, "I know I should eat well, but I'm too busy" (or too fond of fast food, or have too little money, or a dozen other excuses) take note:

- *No time.* Everyone is busy. In truth, eating well takes little time. Convenience packages of frozen vegetables, jars of pasta sauce, and prepared salads are abundant in markets today and take no longer to pick up than snack chips and colas.
- *Love fast food.* Occasional fast-food meals can support health, if you choose wisely (see Chapter 5).
- *Too little money.* Eating right costs no more than eating poorly. Chips, colas, fast food, doughnuts, and premium ice cream cost *more* than foods such as fruits, vegetables, legumes, cereals, and milk. And serious illness costs more than a well person can imagine.
- *Overeat.* Everyone blows it once in a while. An occasional splurge, say, once a month, is part of moderation.
- *Take vitamins instead.* Vitamin pills cannot make up for consistently poor food choices. Food constituents such as fiber and phytochemicals are also important to good health.
- *Love sweets.* If your sweet tooth takes control, know that occasional sweets are an acceptable, and even desirable, part of a balanced diet.

SOURCE: Ideas adapted from Seven excuses for not eating better, *Tufts University Healthletter,* December 1998, p. 8.

A REMARK BY A former surgeon general is worth repeating: for the two out of every three Americans who do not smoke or drink excessively, " your choice of diet can influence your long-term health prospects more than any other action you might take."[94] Indeed, healthy young adults today are privileged to be the first generation in history who can know enough now to lay the foundation for healthy later years through a lifetime of proper nutrition. Figure 11-11 illustrates this point.

Dietary Guidelines for Disease Prevention

An early chapter of this book presented dietary guidelines for the prevention of diseases. Chapters that followed focused on the "whys" and "hows" of those guidelines. This Food Feature comes full circle to revisit the guidelines with a broader and deeper understanding of their significance. As is already clear, not all of the diet recommendations that follow apply equally to all of the diseases, but fortunately for the consumer, most of the recommendations that help prevent one disease help with others as well.

The American Heart Association and American Cancer Society offer suggestions specifically for disease prevention. Table 11-10 shows how similar is the advice from the two sources and clinches the argument that it's time to get busy putting the recommendations into practice. (For those who would argue, " yes, but . . ." see the margin.) The following paragraphs review the specifics.

Reduce Fat Intake

Primary among the recommendations is to reduce fat intake. The Food Feature of Chapter 5 showed how to keep total fat down by selecting low-fat foods. If fat is to account for less than 30 percent of calories, then it is especially important to limit pure fat foods such as sour cream, butter, and margarine; high-fat foods such as mayonnaise, cheese, and cream cheese; and foods high in hidden fat

Figure 11-11

Proper Nutrition Shields against Diseases

A well-chosen diet can protect your health.

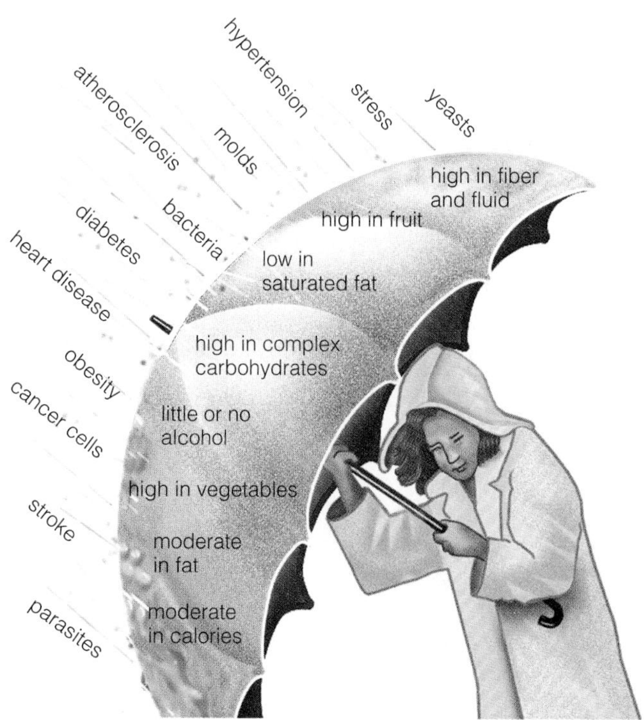
hypertension
stress
yeasts
atherosclerosis
molds
high in fiber
and fluid
high in fruit
diabetes
bacteria
low in
saturated fat
heart disease
high in complex
carbohydrates
obesity
cancer cells
little or no
alcohol
high in vegetables
stroke
moderate
in fat
parasites
moderate
in calories

SOURCE: Adapted from an idea in R. K. Chandra, Nutrition and the immune system: An introduction, *American Journal of Clinical Nutrition* 66 (1997): S460–S463.

Table 11-10

Dietary Guidelines for Disease Prevention

American Heart Association Dietary Guidelines for Healthy American Adults, 1996[a]	American Cancer Society Guidelines on Diet, Nutrition and Cancer Prevention, 1996
■ Eat a variety of foods.	■ Choose most of the foods you eat from plant sources. Eat five or more servings of fruits and vegetables each day. Eat other foods from plant sources, such as breads, cereals, grain products, rice, pasta, or beans several times each day.
■ Balance food intake with physical activity and maintain or reduce weight.	
■ Choose a diet low in fat, saturated fatty acids, and cholesterol.	■ Limit your intake of high-fat foods, particularly from animal sources. Choose foods low in fat. Limit consumption of meats, especially high-fat meats.
■ Choose a diet with plenty of vegetables, fruits, and whole-grain products.	
■ Choose a diet moderate in sugar.	■ Be physically active: achieve and maintain a healthy weight. Be at least moderately active for 30 minutes or more on most days of the week. Stay within your healthy weight range.
■ Use salt and sodium in moderation.	
■ If you drink alcohol, do so in moderation.	■ Limit consumption of alcoholic beverages, if you drink at all.

[a]The American Heart Association also recommends stopping smoking and reducing weight in those who are overweight.

SOURCES: Adapted from American Heart Association Nutrition Committee, Dietary Guidelines for healthy American adults, *Circulation* 94 (1996): 1795–1800, and from American Cancer Society, Nutrition and diet, *Prevention and Risk Factors,* 1998, available from www.cancer.org or upon request from the American Cancer Society at (800) ACS-2345.

Exercise regularly, all your life.

such as convenience foods, foods with sauces, fried foods, fat-marbled meat cuts, sausages, ground beef, whole milk, and others. For each 1,000 calories of food, allow a maximum of about 30 grams of fat. Shop for foods whose labels indicate no more than 3 grams of fat per 100 calories.

When you must add fat, use olive oil or canola oil, since these are high in monounsaturated fatty acids, but use them, like all fats, sparingly. Eat meals of fish regularly, especially fatty fish such as salmon, to balance your intakes of omega-6 and omega-3 fatty acids. Consult the Table of Food Composition, Appendix A, for further details about the fatty acid contents of your favorite foods.

Include Fruits and Vegetables

Every legitimate source of dietary advice urges people to include a variety of fruits and vegetables in the diet, not just for nutrients but also for the phytochemicals that promote health. Vow to try a new fruit or vegetable each week. Who knows? Some of the foods still waiting for you on the produce shelves may become your favorites. An adventurous spirit is a plus in this regard. For example, most people are not familiar with soybeans and soy products, but soy milk is a natural addition to cereals, casseroles, or hot beverages; textured soy protein products can replace part of the hamburger in any recipe; tofu makes delicious puddings; and soybeans themselves are good in any recipe calling for beans. Read some cookbooks for ideas on incorporating other new foods into the diet.

Go for Variety

Eat foods high in potassium (whole foods), high in calcium and magnesium (milk products and appropriate substitutes), low in fat, high in fiber (whole grains, legumes, vegetables, and fruits), and ample in fluids. If you are prone to hypertension, experts advise that you eat less salt.

Advice not to let your diet become monotonous is based on an important concept in the prevention of cancer initiation—dilution. Whenever you switch from food to food, you are diluting whatever is in one food with what is in the others. It is safe to eat *some* salt-cured foods or smoked or grilled meats, but don't eat them all the time. One recent study found that omission of several food groups from the diet brought extra risks of both cancer and cardiovascular disease.[95]

Be Physically Active

In addition to making wise food choices, maintain a proven program of weight control. Expend energy, so as to earn the right to eat more nutrient-dense foods; that is, be physically active. If the threat of CVD doesn't motivate you, then exercise to improve your self-image, to improve your morale, or to make friends—but do exercise.

In the end, people's choices are based on their own likes and dislikes within the limits that their own lives impose on them. Whoever you are, we encourage you to take the time to work out ways of making your diet meet the guidelines known to support health, at least on most days. If you include fruits, vegetables, and high-fiber grains and control your fat intake, you can feel confident that you are supporting your health. Take time to enjoy your meals, too: the sights, smells, and tastes of good foods are among life's greatest pleasures. Joy, even the simple joy of eating, contributes to a healthy life.

PRACTICE EYEBALLING A MEAL

Selecting nutritious foods from among pictures in a textbook is easy when the nutrient values of the foods appear on the page. It is more difficult to select foods in the real world with nothing to go on but appearance. This Do It! section offers the chance to hone your "eyeballing" skills for judging meals according to nutrient ideals such as the *Dietary Guidelines*.

Step 1. Study the suppers shown in Figure 11-12.

Step 2. Consider each dietary goal of Form 11-1 individually, and judge each of the suppers according to that goal.

Step 3. Write into the blanks on Form 11-1 the letters that identify the three meals that you think are highest or lowest in the characteristics named. (For example, for "Lowest in calories," you might write in "A B C".) Instructions and hints follow.

- *Calories.* List the three meals *lowest* in calories (or if you need to gain weight, list those highest in calories). Without nutrient data, this question might stump even the most skilled eyeballer. The person who can identify fats in foods (see Chapter 5's Food Feature) can "see" excess calories in foods right away. Added sugar adds calories, too (Chapter 4's Food Feature can help). Don't forget about portion sizes—too much of almost anything pushes the calories of a meal into the high ranges.
- *Fat, saturated fat, and cholesterol.* Select the three meals likely to be *lowest* in fat, those lowest in saturated fat, and those lowest in cholesterol. These ingredients can be as obvious on the plate as pats of margarine and added salad dressings, or they can be hidden in dishes with fat-laden sauces, cheeses, high-fat meats, creamy sauces, baked goods, toppings, and crusts. Chapter 5's Food Feature identifies the lipids in common fatty foods.
- *Sodium.* Select the meals you think are lowest in sodium. As mentioned in Chapter 8 (p. 281), foods that contain the most salt, and therefore are highest in sodium, are often the most processed foods such as lunch meats, canned soups, chips, and condiments. It isn't always possible to guess which foods contain salt, but it's a safe bet that most farm-fresh, unprocessed foods are low in salt, unless salt was added during preparation.
- *Vitamin A and calcium.* Pick out the three meals highest in vitamin A and the three highest in calcium. Foods rich in these nutrients are shown in the Snapshots on pp. 215 and 276.
- *Fiber.* Select the three meals highest in fiber. Fiber follows fruits, vegetables, and whole grains.
- *Variety.* Choose the meals that provide the greatest variety of foods. Of all the nutrition recommendations, the one concerning variety of foods may be hardest to measure. The U.S. government's *Healthy Eating Index* suggests a minimum of eight different types of food in a day. Other experts believe that more variety best supports health.

Step 4. Score yourself. Compare your answers with the actual ranking of the meals as shown in Table 11-11. Score as instructed on Form 11-1, and then add the column to obtain a total score. Interpret your score as suggested at the bottom of Form 11-1.

Figure 11-12

Six Suppers

Meal A
roasted chicken breast with skin, 1 average
creamed corn, ½ c
mashed potatoes, 1 c, with ¼ c gravy
dinner roll with 1 tsp margarine
fruit punch, 10 oz

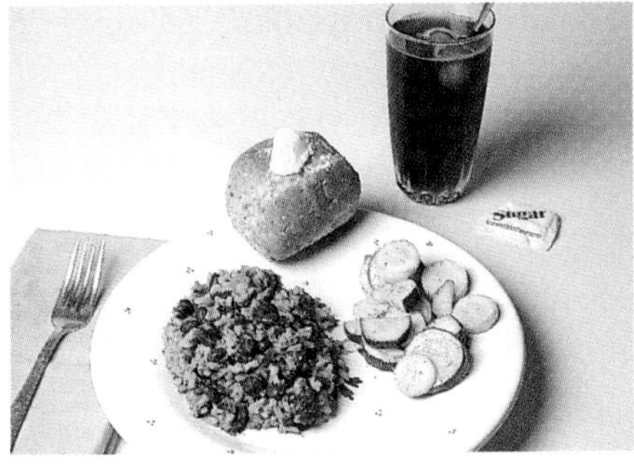

Sugar

Meal B
red beans and rice, 1 c
zucchini and yellow squash mix, ½ c
whole-wheat roll with 1 tsp margarine
sweetened iced tea, 10 oz

Meal C
extra cheese, sausage, and pepperoni pizza, 3 slices
lettuce and tomato salad, 1½ c with 2 tbs dressing
cola, 10 oz

Meal D
macaroni and cheese, 1½ c
green peas, canned, ½ c
iced water

Meal E
pot roast, 2½ oz, with ½ c gravy
potatoes, 1 c
onions and celery mix, 1 c
whole-wheat roll with 1 tsp margarine
tomato, ½ c
fat-free milk, 1 c

Meal F
taco salad, fast food
lemon-lime soda pop, 10 oz

Form 11-1

Dietary Recommendations Scoreboard

Write into the blanks the letters that identify the three meals that you judge to be highest or lowest in each characteristic.

The Three Meals	Your Score
Lowest in calories ______ ______ ______	______________
Lowest in fat ______ ______ ______	______________
Lowest in saturated fat ______ ______ ______	______________
Lowest in cholesterol ______ ______ ______	______________
Lowest in salt (sodium) ______ ______ ______	______________
Highest in vitamin A ______ ______ ______	______________
Highest in calcium ______ ______ ______	______________
Highest in fiber ______ ______ ______	______________
Highest in variety ______ ______ ______	______________
Total:	______________

Compare each of your responses with the ranking of meals in the charts of Table 11-11. Scoring: If you correctly identified the top three meals (in any order) for the category, give yourself three points. If you correctly identified two of the top-ranking meals, give yourself two points. For one meal, you get one point. If your total score adds up to be:

- 22 or more points—you have excellent eyeballing skills.
- 12–21 points—you are well on your way to developing your skills.
- 11 or fewer—you should identify your weakest areas and reread the corresponding material, especially the Food Features, in previous chapters. Then try this exercise again.

Analysis

Answer the following questions:

1. Did you guess which meals were highest in calories, fats, salt, and the rest? For example, logic might predict that a taco salad, because of the word *salad,* would be low in fat and calories, but high-fat ingredients such as sour cream, cheddar cheese, and commercial ground beef change the nutrient picture dramatically.
2. Which areas did you find to be the hardest to judge? How can you hone your eyeballing skills in these problem areas? What information are you lacking?
3. What do you notice about meals that are lowest in fat? Do they often meet more than one of the nutrition goals of this exercise?
4. In Table 11-11, which meal fell into the top three most often? What were the flaws in that meal?
5. Look at the meals that fell at the very bottom of the various lists. Give some reasons why the meals fell so far from the nutrition goals in terms of the foods they included or lacked.
6. Does any meal rank at the top of the list for one goal and at the bottom of the list for another? Which one? What does this mean to the eater? Should you avoid such a meal, or might it bring benefits when used in moderation?

Keep in mind that not all the *Dietary Guidelines* focus on food; some also mention that being physically active and limiting alcohol are crucial to overall health. As you learn to identify meals that best meet your needs, choosing them will become second nature.

Table 11-11

Ranking Meals According to Dietary Recommendations

Important: The meal that comes closest to meeting each dietary ideal ranks highest on that chart; in other words, the meal *highest* in fiber is at the top of its list, as is the meal *lowest* in sodium, because both meals come closest to their ideals.

Ideal: Moderate in Calories[a]	Ideal: 30% or Fewer Calories from Fat	Ideal: 10% or Fewer Calories from Saturated Fat
MEAL	MEAL	MEAL
B — Lowest in calories	B — Lowest in fat	B — Lowest in saturated fat
E	A	A
D	E	E
A	D	F
F	C	C
C — Highest in calories	F — Highest in fat	D — Highest in saturated fat
Ideal: Moderate in Cholesterol[b]	**Ideal: Moderate in Sodium[c]**	**Ideal: 25 Grams or More of Fiber**
MEAL	MEAL	MEAL
B — Lowest in cholesterol	B — Lowest in sodium	B — Highest in fiber
E	F	F
A	A	E
D	D	C
F	E	A
C — Highest in cholesterol	C — Highest in sodium	D — Lowest in fiber
Ideal: Provides Significant Vitamin A[d]	**Ideal: Provides Significant Calcium[e]**	**Ideal: More Variety**
MEAL	MEAL	MEAL
F — Highest in vitamin A	C — Highest in calcium	E — Highest in variety
C	D	C
D	E	A
E	F	B
A	A	F
B — Lowest in vitamin A	B — Lowest in calcium	D — Lowest in variety

[a] This ranking reflects a need to reduce calorie intakes. A person needing to increase calorie intakes might want to change the order of this list to give higher ranks to higher-calorie meals that also meet other guidelines.
[b] Ideal for cholesterol: a day's meals should contain 300 milligrams or fewer of cholesterol.
[c] Ideal for sodium: a day's meals should contain 2,400 milligrams or fewer of sodium.
[d] Ideal for vitamin A: a day's meals should contain 800 to 1,000 RE vitamin A.
[e] Ideal for calcium: a day's meals should contain 800 to 1,200 milligrams calcium.

Self Check

Answers to these Self Check questions are in Appendix G.

1. Most people have well-developed plaques in their arteries by the time they reach the age of ______.
   a. 20 years
   b. 30 years
   c. 40 years
   d. 50 years
2. Which of the following dietary factors appears to influence heart disease risk the most?
   a. sodium
   b. cholesterol
   c. saturated fat
   d. total fat
3. Which of the following dietary factors may help to regulate blood pressure?
   a. calcium
   b. magnesium
   c. potassium
   d. all of the above
4. Which type of cancer is associated with overuse of alcohol?
   a. mouth
   b. breast
   c. throat
   d. all of the above
5. In general, the following can be said of the phytochemicals; (Read about this in the upcoming Controversy.)
   a. They alter body functions only weakly, if at all
   b. They are safe because they occur naturally
   c. They are most effective when consumed in foods, not supplements
   d. None of the above
6. The best way to plan a diet to support the immune system is to exceed the recommended intake for each nutrient. T F
7. The main diet-related risk factors for cardiovascular disease are type 2 diabetes, obesity, high blood cholesterol, and hypertension. T F
8. Resting blood pressure should ideally be 120 over 80 or lower. T F
9. Hypertension is more severe and occurs earlier in life among people of European or Asian descent than among African Americans. T F
10. Vegetarians and meat eaters have the same mortality rates from cancer. T F

NUTRITION ON THE NET

For further study of the topics of this chapter, access these websites and search for the phrases or words in quotation marks:

1. For the latest changes in chapter material, to find supplemental learning tools, or access links to related websites, go to the "Nutrition: Concepts and Controversies" site at: **www.wadsworth.com/nutrition/prod/allprod.html**
2. Information about "AIDS" can be located at: **www.nih.gov/od/oar**
3. To learn more about "chronic diseases" and their relation to diet: **www.cdc.gov**
4. For information about atherosclerosis and cardiovascular disease, search the American Health Association site at: **www.amhrt.org**
5. For information about stroke and related topics: **www.stroke.org**
6. For information about blood pressure, obesity, and blood cholesterol, visit the National Heart, Lung, and Blood Institute site at: **www.nhlbi.nih.gov/nhlbi/nhlbi.htm**
7. For more information about diet and hypertension: **www.healthfinder.gov/searchoptions/topicsaz.htm**
8. For more information on cancer development and prevention, contact the American Cancer Society site at: **www.cancer.org**
9. To learn more about alternative medicine: **http://nccam.nih.gov**
10. Search among thousands of current scientific and medical abstracts for any topic related to diseases and nutrition including individual phytochemicals, at: **www.ncbi.nlm.nih.gov/PubMed/**

CONTROVERSY 11

PHYTOCHEMICALS: CAN THEY PREVENT CHRONIC DISEASES?

MANY HEALTH-conscious people anxiously await news about the **phytochemicals** in foods. Each new scientific revelation about these fascinating substances sparks new enthusiasm in the media and among consumers. Headlines promise remarkable benefits: "Miracle Tomato Chemical **Lycopene** Prevents Prostate Cancer," "Eat Soybeans and Prevent Breast Cancer," "Drink Up!! Red Wine **Flavonoids** Cure Heart Disease," "50 **Functional Food** Cures You Cannot Afford to Miss."

Are phytochemicals simple, safe, and effective weapons against diseases? Or are the headlines exaggerated to sell newspapers and magazines? The aim of this Controversy is to help the reader answer these questions. It begins by presenting some basic facts about the phytochemicals (Table C11-1 defines some terms). Then, using just a few of the phytochemicals of current scientific interest as examples, it presents some general concepts to keep in mind as you encounter news about phytochemicals in the popular media. The Controversy ends with some reasonable and safe personal strategies regarding phytochemicals based on current scientific understandings.

Some warnings are in order at the beginning. Research on most phytochemicals is scanty at present, but each year many studies emerge from the nation's laboratories to add to our knowledge. Only a few of the tens of thousands of phytochemicals known to exist have been researched at all as of this writing. Therefore, some of the details presented in this Controversy may change, and indeed, some may have already changed by the time you read these pages. Still, our basic understandings remain solid, and the reader can build upon this foundation of knowledge. Another caution: Although phytochemicals do appear to play important roles in human health, virtually no safety studies exist to support the taking of purified phytochemicals. That is, a phytochemical in *food* found to benefit the body in some way may well harm it in some other way when taken in purified form. The conclusion of this Controversy and of every other valid assessment of the phytochemicals is that foods, not supplements, are the best and safest source of these substances.

CLASSIFICATION OF THE PHYTOCHEMICALS

Phytochemicals are not classified as nutrients because they do not meet the classic definition for a nutrient as a substance that is indispensable to the body for energy or building materials. Nevertheless, as mentioned, evidence is mounting that phytochemicals may perform other important functions related to prevention of diseases.

Chemists classify the phytochemicals according to their chemical features. People interested primarily in the health effects of the phytochemicals, however, have given some of them names that reflect their effects on the body. The dual names of these phytochemicals can lead to some confusion. For example, **genistein**, a phytochemical of soybeans and the subject of intensive study for its possible anticancer and other health effects, belongs to the broad chemical group *flavonoids* and the subgroup known as *isoflavones*.[1] Functionally, however, genistein is often identified as a member of a group known as **phytosterols** or *phytoestrogens* to reflect its hormone-related effects on the body. Table C11-2 presents the names, possible effects and food sources of some of the better-known phytochemicals that often appear in scientific and popular reporting.

Table C11-1

Phytochemical Terms

- **carotenoids** (CARE-oh-ten-oyds) a group of pigments in foods, ranging from light yellow to reddish orange in color.
- **flavonoids** (FLAY-vone-oyds) members of a larger chemical group *polyphenols,* flavonoids are a group of compounds that include some colorful pigments in foods (*flavus* means "yellow").
- **functional foods** a general term for foods with beneficial physiological or psychological effects beyond their traditional nutrient functions. Also called *medical foods, foods for medical purposes,* or other names with no legal or scientific definition. Also defined in Chapter 1.
- **genistein** (GEN-ih-steen) a phytosterol found primarily in soybeans that both mimicks and blocks the action of estrogen in the body.
- **green tea** tea brewed from unfermented tea leaves; green tea contains substantial amounts of phytochemicals that have antioxidant characteristics.
- **indoles** nitrogen-containing compounds found in cabbages and other cruciferous vegetables.
- **isothiocynates** a group of organosulfur compounds that lend aroma to cabbages and other cruciferous vegetables. Sulforaphane of broccoli is one of these.
- **lycopene** (LYE-koh-peen) a carotenoid pigment responsible for the red color of tomatoes and a few other red-hued fruits.
- **monoterpenes** substances in the oils of orange peels and other citrus oils that confer their distinctive aromas. Some monoterpenes are also found in other fruit, such as cherries.
- **organosulfur compounds** a large group of sulfur-containing phytochemicals, including some common to the onion, leek, chive, shallot, and garlic family that are responsible for their pungent flavors and fragrances.
- **phenolic** (fen-OL-ick) **acids** members of a larger chemical group *polyphenols,* phenolic acids are organic acids occuring in a variety of foods; many are related to tannins, substances that produce an astringent sensation in the mouth, familiar to drinkers of tea.
- **phytochemicals** (FIGH-toe-CHEM-icals) biologically active compounds of plants believed to confer resistance to diseases on the eater; also defined in Chapter 1 (*phyto* means "plant").
- **phytosterols** (FIGH-toe-STEER-ols) compounds of plants structurally similar to mammalian steroid hormones, such as the female sex hormone estrogen. Also called *phytopharmaceuticals.*

THE STATE OF KNOWLEDGE CONCERNING PHYTOCHEMICALS

Though still in an early stage, research on phytochemicals is progressing.[2] One day soon, according to the committee developing the DRI recommendations, daily intake recommendations for some phytochemicals will very likely appear alongside those for the vitamins and minerals.[3] By making this statement, the DRI committee recognizes that phytochemicals may have powerful physiological effects, capable of changing the basic functions of living cells at a metabolic level thus affecting

Table C11-2

A Sampling of Phytochemicals: Classification, Possible Health Effects, and Food Sources

Name or Class	Possible Effects	Food Sources
Capsaicin	Modulates blood clotting, possibly reducing the risk of fatal clots in heart and artery disease.	Hot peppers
Carotenoids (include beta-carotene, lycopene, and hundreds of related compounds)[a]	May act as antioxidants, possibly reducing risks of cancer and other diseases.	Deeply pigmented fruits and vegetables (apricots, broccoli, cantaloupe, carrots, pumpkin, spinach, sweet potatoes, tomatoes)
Curcumin	May inhibit enzymes that activate carcinogens.	Tumeric, a yellow-colored spice.
Flavonoids (include flavones, flavonols, isoflavones, catechin, and others)[b,c]	Many flavonoids may act as antioxidants; scavenge carcinogens; bind to nitrates in the stomach, preventing conversion to nitrosamines; inhibit cell proliferation.	Berries, black tea, celery, citrus fruits, green tea, olives, onions, oregano, purple grapes, purple grape juice, soybeans and soy products, vegetables, whole wheat, wine
Indoles[d]	May trigger production of enzymes that block DNA damage from carcinogens; may inhibit estrogen action.	Broccoli and other cruciferous vegetables (brussels sprouts, cabbage, cauliflower), horseradish, mustard greens
Isothiocyanates (including sulforaphane)	Inhibit enzymes that activate carcinogens; trigger production of enzymes that detoxify carcinogens.	Broccoli and other cruciferous vegetables (brussels sprouts, cabbage, cauliflower), horeseradish, mustard greens
Lignans[e]	Block estrogen activity in cells, possibly reducing the risk of cancer of the breast, colon, ovaries, and prostate.	Flaxseed and its oil, whole grains
Monoterpenes (include limonene)	May trigger enzyme production to detoxify carcinogens; inhibit cancer promotion and cell proliferation.	Citrus fruit peels and oils
Organosulfur compounds	May speed production of carcinogen-destroying enzymes; slow production of carcinogen-activating enzymes.	Chives, garlic, leeks, onions
Phenolic acids[c]	May trigger enzyme production to make carcinogens water soluble, facilitating excretion.	Coffee beans, fruits (apples, blueberries, cherries, grapes, oranges, pears, prunes), oats, potatoes, soybeans
Phytic acid	Binds to minerals, preventing free-radical formation, possibly reducing cancer risk.	Whole grains
Phytosterols (genistein and diadzein)	Estrogen inhibition may produce these actions: inhibit cell replication in GI tract; reduce risk of breast, colon, ovarian, prostate, and other estrogen-sensitive cancers; reduce cancer cell survival. Estrogen mimicking may reduce risk of osteoporosis.	Soybeans, soy flour, soy milk, tofu, textured vegetable protein, other legume products
Protease inhibitors	May suppress enzyme production in cancer cells, slowing tumor growth; inhibit hormone binding; inhibit malignant changes in cells.	Broccoli sprouts, potatoes, soybeans and other legumes, soy products
Saponins	May interfere with DNA replication, preventing cancer cells from multiplying; stimulate immune response.	Alfalfa sprouts, other sprouts, green vegetables, potatoes, tomatoes
Tannins[c]	May inhibit carcinogen activation and cancer promotion; act as antioxidants.	Black-eyed peas, grapes, lentils, red and white wine, tea

[a]Other carotenoids include alph-carotene, beta-cryptoxanthin, lutein, and zeaxanthin.
[b]Other flavonoids of interest include ellegic acid and ferulic acid; see also *phytosterols.*
[c]A subset of the larger group *phenolic phytochemicals.*
[d]Indoles include dithiothiones, isothiocyantes, and others.
[e]Lignans act as phytosterols, but their food sources are limited.

the likelihood of diseases. Phytochemicals may act in these ways to reduce disease risks:

- Reduce formation of atherosclerosis by acting as antioxidants in much the same ways as the antioxidant nutrients already discussed in Controversy 7.
- Increase the activity of enzymes that destroy carcinogens or suppress enzymes that form or activate carcinogens. Phytochemicals may modify the genetic expression of such enzymes.
- Stimulate the immune system's response to abnormal cells or invaders.
- Stimulate spontaneous death of cancer cells.[*4]
- Interfere with hormonal stimulation of some cancers.
- Reduce the likelihood of infection by acting as antimicrobial agents.

Many of these functions have implications for the development of heart disease, cancer, and other diseases. Should phytochemicals prove reliable in reducing rates of these killers, the potential for benefits to the nation would be enormous. Cutting the number of cases of cancer and heart disease in half would save more than 100 *billion* dollars in the United States alone each year and would prevent immeasurable personal suffering.[5] There is no question that reducing chronic diseases is a good idea—it's a stated goal of the governments of virtually all developed countries. But can phytochemicals play significant roles in achieving this goal? And to what extent should health seekers place their prevention eggs in the phytochemical basket? The next few sections provide some clues.

The Scientists' Approach to Phytochemicals Scientists often become interested in a food constituent because someone notices a connection between the presence of the substance in a population's diet and a reduced incidence of disease. To try to discover the biochemical effects of the constituent on living tissues, scientists expose animals, tissue samples, or human cell cultures to disease-causing conditions. Then they add the substance under study to some of the samples, but not to others and observe any differences between the two. If differences are apparent, then the scientists devise further studies to try to isolate the precise effects of the substance on disease development.

Once the mechanism is known, researchers say that "biological plausibility" exists for a theory that a particular phytochemical may reduce disease risks. Today, research on many phytochemicals is reaching this stage—laboratory studies are being conducted to find biological plausibility for the roles of phytochemicals in disease prevention, but no solid evidence yet exists for an effect on human health.

Much more research is needed to determine whether a phytochemical may actually protect free-living people from diseases. Supplement promoters, however, typically jump ahead of research to make claims that the phytochemicals that they sell prevent or cure diseases. For example, many claims are made concerning an ordinary food, garlic, and its extracts that are for sale as dietary supplements.

What the Scientists Say about Garlic Compounds in garlic, **organosulfur compounds,** appear to suppress a set of enzymes that activate carcinogen, thereby reducing cancers in laboratory animals. These enzymes typically convert nitrogen-containing dietary compounds into *nitrosamines*—compounds known to be carcinogenic.[6] Nitrosamines damage DNA in animal cells and are suspected of triggering cancerous changes in human beings. In the presence of garlic, the enzymes slow their nitrosamine production. No one yet knows whether eating garlic or taking garlic supplements affects cancer in human beings. The studies to settle this issue are lacking. Some evidence suggests a possibility of a role for garlic or its derivatives as cancer-quelling agents, but this research is in its infancy.[7]

Although studies on heart disease and garlic as food seem promising, research on garlic *supplements* has been disappointing. Intakes of naturally occurring garlic often correlate with improved blood lipids in subjects given garlic, but this effect is not reliably produced when preparations, such as powder and oil, are administered to people with high blood cholesterol.[8] From the scientists' view, then, much more information is needed before garlic or products made from it can be recommended as protection against diseases.[9] No one can say for certain whether large doses of concentrated chemicals from garlic or of any other phytochemical substances, may improve a person's health or injure it.

What Supplement Supporters Say about Garlic Users and sellers of supplements argue that the evidence cited above is good enough to recommend that people take supplements containing garlic. They want to reap immediately the benefits that they believe these products offer and see no need to wait for studies to prove their efficacy and safety. Besides, they say, consuming garlic must be completely safe because people have been doing so for tens of thousands of years; clearly, the body is accustomed to handling the substances in garlic. Accordingly, they reason, supplements containing purified chemicals from garlic must be safe as well.

Such lines of thinking raise concerns among scientists, however. They point out that although the body may have received the chemicals of garlic diluted in natural foods, it has never encountered them in concentrated form, and their effects may not be the same.

The case of garlic demonstrates three important points about phytochemicals that constitute the main points of the remainder of this Controversy:

1. Phytochemicals can alter body functions, sometimes powerfully so.
2. Caution is needed because evidence for safety of isolated phytochemicals in human beings is lacking.[10]
3. The best-known, most effective, and safest sources for phytochemicals are foods, not supplements.

Right now, the best advice might be to enjoy garlic in the traditional cuisines of

*The spontaneous death of cells is *apoptosis* (pronounced A-poh-TOE-sis or A-pop-TOE-sis); some phytochemicals restrict proliferation of cancer cells and initiate apoptosis in those cells.

the many places where rates of heart disease and some cancers are low, such as China and Greece. Consumed in moderation, foods prepared with garlic may contribute to health in yet-undiscovered ways and are likely to do no harm. In fact, if they are rich in vegetables, such foods may bring into play a virtual army of potentially health-defending phytochemicals, as the next few sections show.

PHYTOSTEROLS: ALTERED BODY FUNCTIONING

Compared with women in the West, Asian women living in Asia suffer more rarely from osteoporosis, breast cancer, and symptoms related to menopause. Among men, Asians have lower rates of cancer of the prostate. When Asians move to the United States and adopt Western diets and habits, however, their risks soon approach those of native Westerners. Researchers who study this phenomenon have examined Asian diets and have concluded that Asians with high intakes of soy products, such as soybeans, tofu, and soy milk, have the lowest rates of these diseases.

Soybeans are rich sources of *phytosterols,* a fascinating group of phytochemicals. Phytosterols are chemically related to steroid hormones of the human body, but come from plants. They weakly mimic or modulate the effects of the hormones estrogen and progesterone in the human body.

Because of the actions of the phytosterols, soybeans and products made from them are emerging as leaders among potential functional foods that correlate with low rates of cancer, and especially cancer of the breast, prostate, and other hormone-sensitive organs.[11] Phytosterols in soy foods may also improve the condition of the arteries, and soy protein is known to reduce blood LDL concentrations. Therefore, in theory, soy foods may help to reduce the risk of heart attack.[12] Indeed, the rate of heart attack is much lower in soy-eating Asian people than in westerners.[13]

Clearly, the phytosterols of soy foods are potentially beneficial, but no one can yet say which of soy's many phytosterols might have these effects. An especially interesting and well-researched possibility is the soy phytosterol genistein. Genistein, an isoflavone of soybeans, is renowned for its roles in both mimicking and opposing the more potent sex hormones of the body and is thought to slow the growth of certain cancers.

The body readily absorbs genistein from food. Combined with a protein, it travels in the bloodstream until cells that recognize it, such as those of the breast, prostate, brain, or uterus, selectively harvest it from the blood.[14] Such cells collect genistein because they are "estrogen-sensitive"; that is, they are equipped to recognize the hormone estrogen, and genistein weakly mimics estrogen in the body.

At high concentrations, genistein prevents cancer cell proliferation and reduces mammary gland cancers in laboratory animals. In this case, genistein is believed to oppose estrogen's actions, altering a woman's monthly hormonal cycle in ways that may reduce her risk of breast cancer.[15]

From observations in Asian women, researchers are inclined to guess that phytosterols like genistein work in two ways—mimicking and opposing estrogen—in the body. When natural estrogen is lacking, such as after menopause, phytosterols may step in to stimulate estrogen-dependent tissues. By way of this action, phytosterols may help to prevent the rapid bone losses of the menopause years that lead to osteoporosis in later life, along with the most common symptoms of menopause, such as mood swings and the sensation of elevated body temperature often called "hot flashes."

At lower concentrations, genistein mimicks some unwelcome effects of estrogen, triggering rapid division in breast cancer cells in a laboratory dish. This finding demonstrates the wisdom of waiting until science clarifies the effects of phytochemicals before seeking out supplements containing untested doses of them.

The paired, opposing actions of genistein should send a red flag of warning against taking genistein supplements. They may *promote* cancer instead of protecting against it.

How can people receive the benefits of genistein without incurring risks? Clearly, eating a diet that includes daily soybean products is harmless and may even be helpful, if research on soybeans holds true. Unfortunately, studies of isolated soy supplements have been disappointing. So far, supplements do not seem to produce the beneficial effects observed in people who consume soy foods.

The dual possibility that phytosterols may stimulate or suppress cancers raises a safety issue that applies to all of the phytochemicals. Until research determines them safe, people who take supplements of the phytochemicals are taking chances with their health. Next, details about two carotenoids offer support for the benefits of whole foods, and cautions about supplements.

CAROTENOIDS: EVIDENCE FOR CAUTION

In areas where many people suffer from cancers of the esophagus, prostate, or stomach, those who eat tomatoes often, say, about five tomato-containing meals per week, are less likely than others to suffer from those cancers.[16] Among the likely candidates for this effect is one of the **carotenoids,** beta-carotene's 600 or so relatives.[17] This particular carotenoid, *lycopene*, is a pigment with antioxidant activity that is responsible for the red color of guava, papaya, pink grapefruit, tomatoes, and watermelon. Lycopene is especially abundant in tomatoes and cooked tomato products.[18] In fact, the name *lycopene* comes from the Latin name for tomatoes, *Lycopersicon.*

In test tubes, lycopene inhibits the reproduction of cancer cells. Studies of human populations indicate that lycopene may have the same effect in living human beings.[19] In a case-controlled study, African American women with low lycopene intakes ran a three to four times greater risk of developing cancerous changes of the cervix than similar women with high lycopene intakes.[20]

An interesting aspect of lycopene is its sensitivity to the harmful, cancer-causing ultraviolet (UV) rays of the sun. When exposed to UV light in the laboratory, the skin loses more than a third of its lycopene concentration.[21] Researchers suspect that lycopene

molecules are destroyed as they absorb and render harmless UV radiation, thereby protecting other vulnerable molecules of the skin from the sun's carcinogenic effects.

Lycopene-rich foods are especially easy to love: pizza sauce, spaghetti sauce, sloppy joes, tomato salads, and other tomato-based foods present the body with lycopene in abundance.

Though evidence in favor of foods containing lycopene as anticancer agents, seems strong, not a shred of evidence supports the idea that consuming lyco-pene in the form of a purified supplement would be of benefit. In fact, if a lesson can be learned from experience with lycopene's chemical cousin, beta-carotene, then the only safe option is to avoid concentrated doses of any of the carotenoids until more is known about their effects. Recall from Controversy 7 that supplements of beta-carotene were once believed to have an antioxidant effect that people hoped could prevent cancer. When tested empirically, however, beta-carotene supplements not only did not prevent cancers, they greatly increased the risk of developing lung cancer among smokers. Luckily, carotenoid-rich foods eaten in the context of balanced diet, and in close to their natural state, never cause such problems.

Scientific data also do not support the originally-suspected role for beta-carotene from supplements as defender of the body against oxidation imbalances.[22] In fact, there is no evidence that any one of the 600 or so carotenoids is essential in fighting diseases: people around the world eat many different fruits and vegetables with widely varying mixtures of the carotenoids. The relationship that remains true is that fruit and vegetables rich in carotenoids, regardless of their carotenoid mix, are associated with similar degrees of reduction of disease risks.

FLAVONOIDS: WHOLE-FOOD BENEFITS

It has been observed that people who eat mostly foods in their natural state are healthier than people who rely mostly on highly refined foods. These health effects have often been credited to the fiber in whole foods, for all of the reasons made clear in Chapter 4. Often researchers find that the benefits of whole foods exceed those of fiber alone. Some credit the *flavonoids*, a large group of phytochemicals found concentrated in fruits and vegetables, many herbs, spices, and whole grains and in tea and red wine. Many are candidates for research concerning potential health-promoting qualities.[23]

In one large study that spanned seven countries, researchers reported that deaths from heart disease were inversely related to intakes of flavonoid-containing foods.[24] The more flavonoid-containing foods and beverages present in the diets of people of each country, the lower their rates of heart disease. Other studies have yielded similar results.

How flavonoids may contribute to reduced risk of death from heart attack is not at all clear, but some likely mechanisms are being investigated.[25] Many flavonoids are strong antioxidants that may protect LDL in the blood against oxidation.[26] In addition, flavonoids may reduce blood platelet "stickiness," a term used to descibe the platelets' tendency to cause blood clotting, a major cause of heart attack and strokes.[27] Additionally, flavonoids may reduce inflammation of diseased arteries and promote relaxation of the smooth muscles of the arteries, thereby lowering blood pressure.[28] All of these potential actions of flavonoids are under investigation, but none is yet proved.

Flavonoids impart a bitter taste to foods that consumers may dislike. This prompts manufacturers to refine the foods to reduce flavonoids. Flavonoids in the bran of wheat kernels, for example, confer a strong taste on whole-wheat baked goods and pastas. Another example is white wine—winemakers remove the skins of red grapes to lighten the flavor and the color of the wine, a process that also removes the wine's potentially beneficial flavonoids. Red wine (but not white wine), dealcoholized red wine, black and green tea, purple grape juice (but not white grape juice), grapefruit, other fruits and juices, and many kinds of vegetables all derive a bitter tang from flavonoids that many people dislike.[29] If preliminary research results hold true, everyone might be wise to cultivate a liking for the interesting flavors of foods and beverages rich in flavonoids.

The body absorbs only limited amounts of flavonoids from most foods, though onions are a notable exception—their flavonoids are relatively easily absorbed.[30] Some others do get in, but frequent repeated exposure is needed to accumulate most flavonoids in the blood. Because the body excludes all but a fraction of the flavonoids in a meal, it seems prudent to include flavonoid-rich vegetables, fruits, juices, and whole grains throughout the day. Prudence also dictates avoiding supplements of flavonoids until science uncovers possible reasons why the body may limit their absorption; they may pose unsuspected hazards.

PHYTOCHEMICAL RECOMMENDATIONS

As mentioned, the DRI committee has not yet made recommendations for the intake of phytochemicals. Nevertheless, the committee recognizes the hundreds of studies that overwhelmingly suggest that the phytochemicals in plant-derived foods, such as cabbages, celery, cucumbers, endive, parsley, radishes, and soybeans and other legumes, have the potential to help people to stay well.[31] Scientists credit an astounding number of fruits and vegetables with potential anticancer activity (see Table C11-2, earlier). Based on the accumulated data, researchers speculate that people might cut their risks of some cancers by half simply by meeting the current recommendation to consume five or more servings of fruits and vegetables each day.[32]

The Key: A Variety of Fruits, Vegetables, and Other Whole Foods An important conclusion for most people to have gleaned from the existing research seems to be that an intake of fruits and vegetables in general, and not of any single fruit or vegetable, is what correlates best with cancer prevention. It isn't enough to add some daily green beans to an otherwise vegetable-poor diet (although this is certainly a first step in the right direction). What is needed is a variety of fruits, vegetables, whole grains, and legumes each day.[33] As Table C11-2 pointed out, many phytochemicals in fruits and vegetables may work in many different ways. Variety matters.

An exception to the rule may be soybeans and foods made from soybeans. No other food is known to supply the phytosterols of soybeans, and they are rich in low-fat protein and fiber and supply some needed vitamins and minerals. Evidence associating low disease risks with a diet rich in soy food has gained enough strength to suggest that many people may benefit from consuming more of these foods in their diets.

Not all of the phytochemicals' effects have been pinned down yet, nor have the effects of the specific chemicals been completely teased out of the evidence on foods themselves.[34] For example, **green tea** contains hundreds of phytochemicals and seems to protect against cancer in some who drink it regularly, but no one knows exactly which chemical in tea deserves the credit.[35] Many bits of evidence exist on this phytochemical or that one, but a unified and complete picture based on research is lacking.

Choosing and Preparing Food for Maximum Phytochemicals

Chapter 14 discusses how to choose foods with the most nutrients and how to conserve those nutrients once they are in your kitchen. Many of the principles you will read about there apply to phytochemicals as well. Here are some general rules about obtaining and conserving phytochemicals in foods:

- Choose fruits and vegetables harvested at the fully ripe stage. Avoid those harvested green and ripened chemically or in greenhouses; some phytochemicals require time and direct sunshine to develop fully.[36]
- Try marmalades and chutneys that include the peels of citrus fruits to obtain the **monoterpenes** concentrated in the oils of citrus peels. Toss well-washed rinds of lemons (preferably, the kind grown without pesticides) into lemonade.
- Quickly steam, broil, sauté, stir-fry, or microwave vegetables to conserve phytochemicals. Avoid long boiling times that can destroy heat-sensitive phytochemicals and dissolve others into the cooking water that is discarded.
- Eat some raw and some cooked vegetables and fruits each day.
- Store fruit and vegetables in the refrigerator to reduce enzymatic or oxidative destruction of phytochemicals.
- Try to find interesting new foods to enjoy. Particularly, try soy milk on cereal, tofu on a sandwhich, or frozen vegetarian "burgers" and main dishes made with the soy product called texturized vegetable protein to obtain the phytochemicals in soybeans.

An interesting tip about brewing tea for phytochemical content has recently emerged: to obtain the maximum phytochemicals from black or green tea, use boiling water and allow the tea to steep for five minutes. Researchers determined that 84 percent of the total maximum potential antioxidant concentration is transferred to the beverage within the first five minutes of brewing.[37] Additional tips for consuming phytochemicals are listed in Table C11-3.

Table C11-3

Tips for Consuming Phytochemicals

- Eat more fruit. The average U.S. diet provides little more than one serving a day. Remember to choose juices and raw, dried, or cooked fruits and vegetables at mealtimes as well as for snacks. Choose dried fruit in place of candy.
- Double the normal serving size of vegetables to 1 cup.
- Use herbs and spices in cooking. Cookbooks offer ways to include parsley, basil, garlic, hot peppers, oregano, and other beneficial seasonings.
- Replace some of the meat in the diet with grains, legumes, and vegetables. Oatmeal, soy meat replacer, or grated carrots mixed with ground meat and seasonings make a luscious, nutritious meat loaf, for example.
- Add grated vegetables to other familiar foods: carrots in chili or meatballs, celery and squash in spaghetti sauce, etc.
- Try a new fruit, vegetable, or whole grain each week. Walk through vegetable aisles and visit farmers' markets. Read recipes. Try tofu, soy milk, or soybeans in cooking.

WHAT ABOUT PHYTOCHEMICAL SUPPLEMENTS AND FUNCTIONAL FOODS?

Each of the phytochemicals listed in Table C11-2 works in characteristic ways in the laboratory, for example, by inducing or inhibiting various enzymes, acting as an antioxidant, or prompting cancer cells to die by their own internal mechanisms.[38] Further, each kind of food possesses its own characteristic array of phytochemicals—citrus fruits contain limonene, deeply pigmented vegetables are rich in carotenoids, and flaxseed is the richest source of lignans. Broccoli alone may contain as many as 10,000 phytochemicals and metabolites with potential activity in the body.[39]

Manufacturers of food products hope to capture this knowledge and put it to use by adding one or more of these chemicals to their recipes, thereby creating functional foods. Opponents to the idea of isolating phytochemicals and adding them to foods question the motivations for doing so. According to these critics, the research has not matured nearly enough to pinpoint appropriate additions of isolated phytochemicals to the diet, and further based on the beta-carotene experience described earlier, doing so may be risky in unimaginable ways. Also, they question whether manufacturers should be allowed to douse fried snack chips or cup cakes with phytochemicals and then label them as "functional" foods, implying that they are supportive of health. They say that the pressure to approve such foods has more to do with healthy profits for the manufacturers than with the health of the population.[40]

Problems exist concerning phytochemicals, and they are serious. The small quantities of the combinations of phytochemicals occurring in whole foods provide a variety of health benefits and can be consumed safely every day.

Large doses of purified phytochemicals, however, may produce only single effects and may be potentially toxic. Even overusing some foods might be hazardous to some people. People with kidney disease, for example, should avoid large doses of parsley. Parsley contains carotenoids and other beneficial phytochemicals, but it also contains a diuretic that can irritate the tissues of the kidneys when consumed in high doses.[41]

To date, the bulk of research seems *not* to support the addition of phytochemicals to the diet through supplementation or fortification because the benefits of doing so are not proved, and the adverse effects are not known.[42]

Of course, ignoring these observations, many manufacturers have already marketed broccoli extracts, spinach chemical pills, parsley pills, and vegetable or fruit pills and powders as supplements. Such products imply that they will deliver the same benefits as many servings of vegetables, but they fall far short—they contain the equivalent of just a few spoonfuls.[43]

CONCLUSIONS

Many more phytochemicals than those mentioned in this Controversy are under study for their disease-fighting potential, and much more can be said about the fascinating research concerning them. Other sections of this book, such as Controversy 7, provide some details, but it is up to you, the student, to continue reading about the results of ongoing phytochemical research with a scientific eye.

For now, take another look at Table C11-2. Focus on the right-hand column, the food sources, and note what a tremendous argument they make for emphasizing plant foods in the diet. They represent a great variety of plant-derived foods. Consumers could never combine dietary supplements into a feast like the one that plant foods provide.

Given that the U.S. population consumes just a fraction of the recommended five servings of fruits and vegetables each day, a reasonable goal is to make sure that the diet you choose provides at least this minimum amount.[44] Study the photo of Figure C11-1, identify foods that you like, and vow to eat them more often. Then identify foods new to you, and vow to try new ones each week. This way, you'll be assured of receiving the many health benefits that fruits, vegetables, and other foods can deliver.

Clearly, no magic bullets yet exist to undo the damage from an otherwise haphazardly chosen diet. One thing is certain: the debate can be expected to continue as scientists and commercial ventures wrangle over how phytochemicals should be used and regulated in the future.

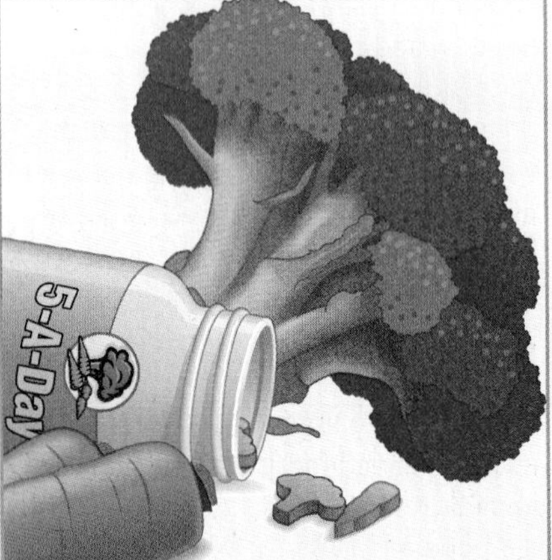
5-A-Day

Costs per equivalent of a half-cup vegetable serving: Pills, about $2.00. Vegetables, about 25¢. Who needs pills? Go for the vegetables.

Figure C11-1

Get to Know Some Players in the Phytochemical Game

The foods depicted in the photo appear in bold print in the list to the left.

apricot, asparagus, barley, basil, berries, ***bok choy, broccoli, broccoli rabe,*** *brown rice,* ***brussels sprouts, cabbage*** *(all types), cantaloupe, carrots, cauliflower, celery, chives,* ***citrus fruit,*** *cucumber,* ***fennel,*** *flaxseed, garlic, ginger, green onions, guava,* ***kale,*** *kohlrabi,* ***leafy greens, lemon,*** *lettuce (dark green),* ***mango,*** *oats, onion, orange, oregano,* ***papaya,*** *parsley,* ***parsnip,*** *potato (white),* ***rutabaga,*** *soybeans, soy products, spinach,* ***squash (summer), squash (winter), sweet potato,*** *tangerine, tarragon, tea (green and black), thyme,* ***tomato,*** *turmeric,* ***turnip roots,*** *whole wheat.*

12 Life Cycle Nutrition: Mother and Infant

Michael Escoffery, *Mother and Son* 1996.

Contents

Frequently Asked Questions

FAQ: How much weight should a woman expect to gain during pregnancy? p. 440

FAQ: Why do some women crave pickles and ice cream while others can't keep anything down? p. 442

FAQ: What behaviors or substances should pregnant women avoid? p. 443

FAQ: Why is breast milk so good for babies? p. 448

FAQ: Are there any situations in which a woman should not breastfeed? p. 453

low birthweight a birthweight of less than 5½ pounds (2,500 grams); used as a predictor of probable health problems in the newborn and as a probable indicator of poor nutrition status of the mother before and/or during pregnancy. Low-birthweight infants are of two different types. Some are premature; they are born early and are the right size for their gestational age. Others have suffered growth failure in the uterus; they may or may not be born early, but they are small for gestational age (small for date).

THOUGH ALL PEOPLE need the same nutrients, the amounts they need change as they move through life. This chapter is the first of a two-chapter segment on life's changing nutrient needs. It focuses on the two life stages that are arguably the most important to an individual's lifelong health—pregnancy and infancy.

Pregnancy: The Impact of Nutrition on the Future

We normally think of our nutrition as personal, affecting only our own lives. The woman who is pregnant, or who will be, must understand that her nutrition today will be critical to the health of her child for years to come. The nutrition demands of pregnancy are extraordinary because the growth of a new person requires all the nutrients, and extra amounts of many of them.

Preparing for Pregnancy

Before she becomes pregnant, a woman must establish eating habits that will optimally nourish both the growing fetus and herself. She must be well nourished at the outset because early in pregnancy the embryo undergoes rapid and significant developmental changes that depend on her prior nutrition status.

Fathers-to-be also are wise to consider their eating and other habits. Limited evidence suggests that men who consume too few fruits and vegetables containing vitamin C or who drink too much alcohol in the weeks before conception may sustain damage to their sperm's genetic material. This damage can cause birth defects in future children.

Underweight is defined as BMI <19.8. Obese is defined as BMI >29 (see Table 12-5 on page 440).

Prepregnancy Weight Prior to pregnancy, all women should strive for appropriate body weights. This is especially important for underweight women. An underweight woman who fails to gain adequately during pregnancy is most likely to bear a baby with a dangerously low birthweight. Infant birthweight is the most potent single indicator of an infant's future health status. A **low-birthweight** baby, defined as one who weighs less than 5½ pounds (2,500 grams), is nearly 40 times more likely to die in the first year of life than is a normal-weight baby. Underweight women are therefore advised to try to gain weight before becoming pregnant or to strive to gain adequately during pregnancy.

Nutritional deficiency, coupled with low birthweight, is the underlying cause of more than half of all the deaths worldwide of children under five years of age. In 1997, the U.S. infant mortality rate was the lowest the nation has ever recorded: 7.1 deaths per 1,000 live births.[1] This rate is still higher than that of some other developed countries, but as part of a significant steady decline for two decades, it stands as a tribute to public health efforts aimed at reducing infant deaths.

Neural tube defects and gestational diabetes are discussed in later sections.

Both parents can prepare in advance for a healthy pregnancy.

Obese women, too, are urged to attain healthy weights before pregnancy. The infant of an obese mother may be larger than normal and born late, or it may be large in size even if born prematurely. In the latter case, the baby may not be recognized as premature and may not receive the special care it requires from medical staff. Maternal obesity also may double the risk for neural tube defects in the infant.[2] Also, obese pregnant women more often suffer gestational diabetes, hypertension, and infections after the birth than do women of healthy weight.[3] The birth itself may be more likely to require drugs to induce labor or require surgical intervention. An appropriate goal for the obese woman who wishes to become pregnant is to attain a prepregnancy body weight low enough to minimize her medical risks.

A Healthy Placenta and Other Organs A major reason the mother's nutrition before pregnancy is so crucial is that it determines whether her **uterus** will be able to support the growth of a healthy **placenta** during the first month of **gestation.** If the placenta works perfectly, the fetus wants for nothing; if it doesn't, no alternative source of sustenance is available and the fetus will fail to thrive.[4] The placenta is shown in Figure 12-1; it is a sort of cushion of tissue in which the mother's and baby's blood vessels intertwine and exchange materials. The two bloods never mix, but nutrients and oxygen cross from the mother's blood into the baby's blood while wastes move out of the baby's blood, ultimately to be excreted by the mother. The **amniotic sac** forms to cradle the baby, cushioning it with fluids.

Far from being passive in its transport of molecules, the placenta is a highly metabolic organ with some 60 sets of enzymes of its own. It actively gathers up hormones, nutrients, and protein molecules such as antibodies and transfers them into the fetal bloodstream. It also produces hormones that maintain pregnancy and prepare the mother's breasts for **lactation.**

If the mother's nutrient stores are inadequate during the period when the body is preparing to develop the placenta, then the placenta will never develop properly. As a consequence, no matter how well she eats later, the fetus will not receive optimal nourishment. The infant is likely to be a low-birthweight baby with all of the associated risks. After getting such a poor start on life, children may be ill equipped, even as adults, to store sufficient nutrients, and a girl may later be unable to grow an adequate placenta. In turn, she may bear an infant who is unable to reach full potential.

Not all cases of low birthweight reflect poor nutrition. Other factors associated with low birthweight are heredity, disease conditions, smoking, and drug (including alcohol) use during pregnancy. Even with optimal nutrition and health during pregnancy, some women give birth to small infants for reasons unknown. Still, poor nutrition is the major factor in low birthweight, and generally, it is an avoidable one as later sections make clear.

uterus (YOO-ter-us) the womb, the muscular organ within which the infant develops before birth.

placenta (pla-SEN-tuh) the organ that develops inside the uterus in early pregnancy in which maternal and fetal blood circulate in close proximity and exchange materials. The fetus receives nutrients and oxygen across the placenta; the mother's blood picks up carbon dioxide and other waste materials to be excreted via her lungs and kidneys.

gestation the period of about 40 weeks (three trimesters) from conception to birth; the term of a pregnancy.

amniotic (am-nee-OTT-ic) **sac** the "bag of waters" in the uterus in which the fetus floats.

lactation production and secretion of breast milk for the purpose of nourishing an infant.

Figure 12-1

The Placenta

The placenta is composed of spongy tissue in which fetal blood and maternal blood flow side by side, each in its own vessels. The maternal blood transfers oxygen and nutrients to the fetus's blood and picks up fetal wastes to be excreted by the mother. Thus the placenta performs the nutritive, respiratory, and excretory functions that the fetus's digestive system, lungs, and kidneys will provide after birth.

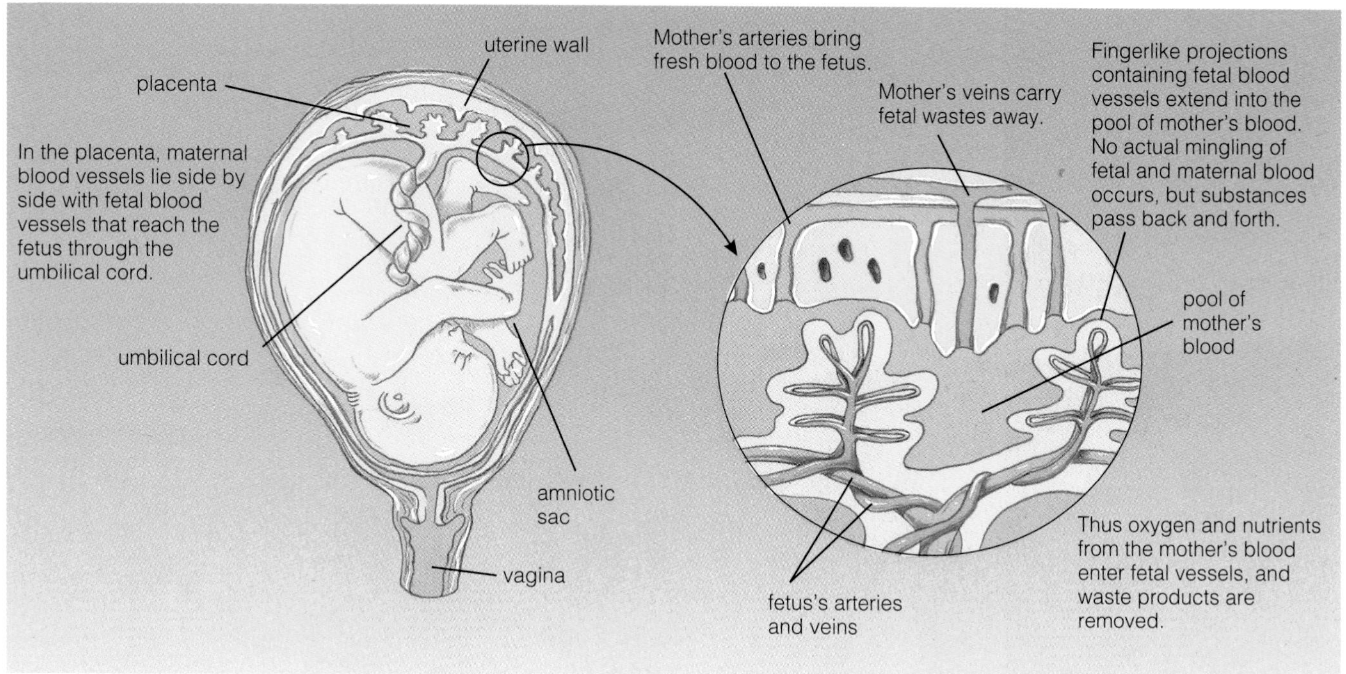
uterine wall
placenta
Mother's arteries bring fresh blood to the fetus.
Mother's veins carry fetal wastes away.
Fingerlike projections containing fetal blood vessels extend into the pool of mother's blood. No actual mingling of fetal and maternal blood occurs, but substances pass back and forth.
In the placenta, maternal blood vessels lie side by side with fetal blood vessels that reach the fetus through the umbilical cord.
pool of mother's blood
umbilical cord
amniotic sac
vagina
fetus's arteries and veins
Thus oxygen and nutrients from the mother's blood enter fetal vessels, and waste products are removed.

implantation the stage of development, during the first two weeks after conception, in which the fertilized egg (fertilized ovum or zygote) embeds itself in the wall of the uterus and begins to develop.

ovum the egg, produced by the mother, that unites with a sperm from the father to produce a new individual.

zygote (ZYE-goat) the term that describes the product of the union of ovum and sperm during the first two weeks after fertilization.

critical period a finite period during development in which certain events may occur that will have irreversible effects on later developmental stages. A critical period is usually a period of cell division in a body organ.

embryo (EM-bree-oh) the stage of human gestation from the third to the eighth week after conception.

fetus (FEET-us) the stage of human gestation from eight weeks after conception until the birth of an infant.

KEY POINT *Adequate nutrition before pregnancy establishes physical readiness and nutrient stores to support fetal growth. Both underweight and overweight women should strive for appropriate body weights before pregnancy. Babies who weigh less than 5½ pounds at birth face greater health risks than normal-weight babies. The healthy development of the placenta depends on adequate nutrition before pregnancy.*

The Events of Pregnancy

On **implantation** of the newly fertilized **ovum** in the uterine wall, a placenta begins to grow inside the uterus. During the two weeks following fertilization, the **zygote** divides into many cells, and these cells sort themselves into three layers. Minimal growth in size takes place at this time, but it is a **critical period** developmentally. Adverse influences such as smoking, drug abuse, and malnutrition at this time lead to failure to implant or to abnormalities such as neural tube defects that can cause loss of the zygote, possibly even before the woman knows she is pregnant. Both mother and child will benefit most from an optimal supply of nutrients uncontaminated by other materials.

The Embryo During the next six weeks of development, the **embryo** registers astonishing physical changes (see Figure 12-2). At eight weeks, the **fetus** has a complete central nervous system, a beating heart, a fully formed digestive system, well-defined fingers and toes, and the beginnings of facial features.

In the last seven months of pregnancy, the fetal period, the size of the fetus increases tremendously. Critical periods of cell division and development occur in organ after organ. The amniotic sac fills with fluid and the mother's body changes. The uterus and its supporting muscles increase in size, the breasts may become tender and full, the nipples may darken in preparation for lactation, and the mother's blood volume increases by half to accommodate the added load of materials it must carry. Gestation lasts approximately 40 weeks and ends with the birth of the infant.

Figure 12-2

Stages of Embryonic and Fetal Development

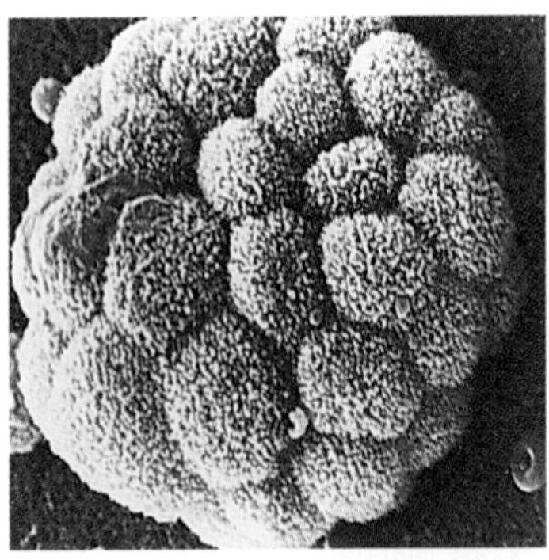

(1) *A newly fertilized ovum is about the size of the period at the end of this sentence. This zygote at less than one week after fertilization is not much bigger and is ready for implantation.*

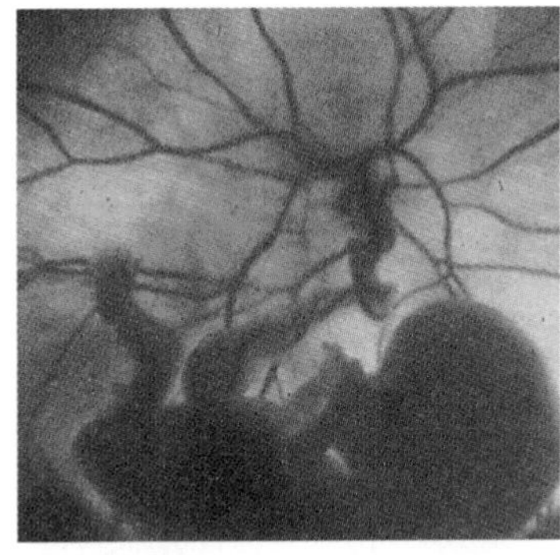

(3) *A fetus after 11 weeks of development is just over an inch long. Notice the umbilical cord and blood vessels connecting the fetus with the placenta.*

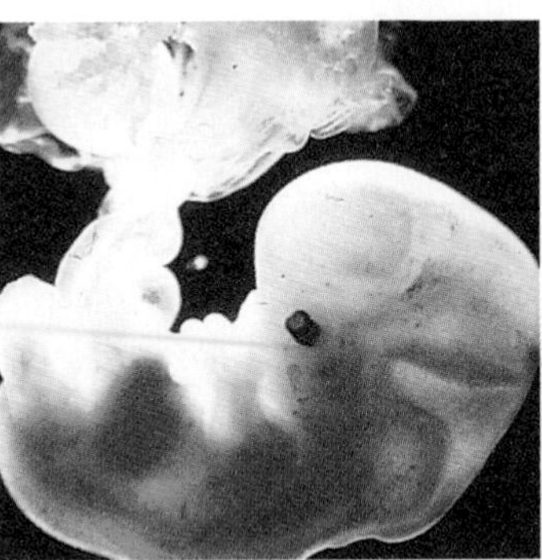

(2) *After implantation, the placenta develops and begins to provide nourishment to the developing embryo. An embryo five weeks after fertilization is about ½ inch long.*

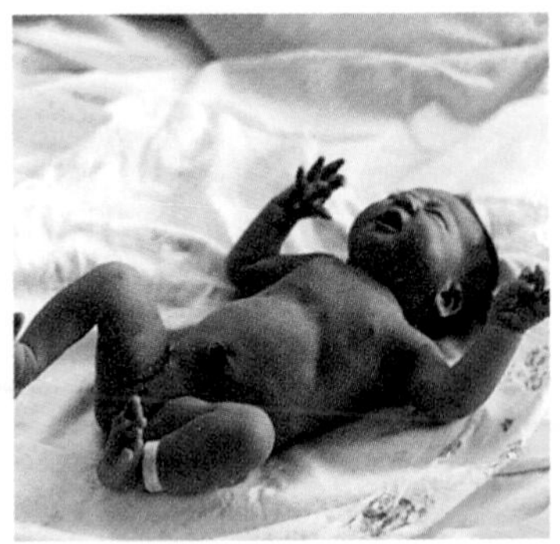

(4) *A newborn infant after nine months of development measures close to 20 inches in length. From eight weeks to term, this infant grew 20 times longer and 50 times heavier.*

Critical Periods Each organ and tissue type grows with its own characteristic pattern and timing. Each organ depends most on its supply of nutrients during its own intensive growth period. For example, the fetus's heart and brain are well developed at 14 weeks; the lungs, 10 weeks later. Therefore early malnutrition impairs the heart and brain; later malnutrition impairs the lungs.

Developments that occur during a critical period can take place only at that time and at no other. Whatever nutrients and other environmental conditions are necessary during this period must be supplied on time if the organ is to reach its full potential. If the development of an organ is limited during a critical period, recovery is impossible.[5] Thus early malnutrition often does irreversible damage, although this may not become fully apparent until the person has matured and may never be attributed to events of pregnancy. When people having such a poor start on life reach adulthood, they may be vulnerable to infections, and some may have high risks of diabetes, hypertension, stroke, or heart disease.[6] Table 12-1 provides a list of factors that make nutrient deficiencies likely during pregnancy. Notice that young age heads the list; a later section explains why pregnant adolescents are especially prone to malnutrition.

The effects of malnutrition during critical periods are seen in neural tube defects of the nervous system (explained later), in the short height of people who were undernourished in their early years, and in the poor dental health of children whose mothers were malnourished during pregnancy. The effects of malnutrition during critical periods are irreversible. There is no second chance to provide nutrients. No matter how abundant and nourishing the food, if it is fed after the critical time, it fails to remedy harm already done.

KEY POINT *Maternal nutrition before and during pregnancy affects both present and future development of the infant. Placental development, implantation, and early critical periods depend on nutrient supply and influence future growth, health, and developmental events.*

Increased Need for Nutrition

Nutrient needs during periods of intensive growth are greater than at any other time and are greater for certain nutrients than for others. The nutrient needs of pregnancy are shown in Figure 12-3.

Energy, Protein, and Fat A pregnant woman needs extra food energy, but only a little extra—300 calories above the allowance for nonpregnant women—and only during the second and third **trimesters.** A woman can easily get 300 calories with just one extra serving from each of the five food groups—a slice of bread, a serving of vegetables, an ounce of lean meat, a piece of fruit, and a cup of fat-free milk (see Table 12-2 on page 437). Pregnant teenagers, underweight women, and physically active women may require more. This increment of extra energy is needed to spare protein for its all-important tissue-building work.

The increase recommended for protein is greater than for energy: from about 45 to 50 grams of protein per day for a nonpregnant woman to about 60 grams for a pregnant woman. Most women in the United States, however, need not add protein-rich foods to their diets because they already exceed the recommended protein intake for pregnancy. Besides, extra servings of almost any calorie-yielding food add protein to the diet. Excess protein may also have adverse effects, as Chapter 6 explained.

Some vegetarian women limit or omit protein-rich meats, eggs, and dairy products from their diets. For them, meeting the recommendation for food energy each day and including several generous servings of plant-protein foods such as legumes, tofu, whole grains, nuts, and seeds are imperative steps. Use of high-protein supplements during pregnancy can be harmful and is discouraged. All pregnant women need generous amounts of carbohydrate-rich foods to spare their protein and to provide energy.

The high nutrient requirements of pregnancy leave little room in the diet for excess energy from added purified fats such as oil, margarine, and butter. The essential fatty acids, however, are important to the growth of the fetus and are regarded by some as

trimester one-third of gestation, about 13 to 14 weeks.

Neural tube defects were first discussed in Chapter 7.

Nutrient intake recommendations for pregnant women are listed on the inside front cover.

Recommended protein intake: 60 grams/day.

Recommended carbohydrate intake: about 50% of energy intake. In a 2,000-calorie/day intake, this represents 1,000 calories of carbohydrate, or about 250 grams.

Table 12-1

Factors Placing Pregnant Women at Nutritional Risk

Women likely to develop nutrient deficiencies include those who:

- Are young (adolescents).
- Have had many recent previous pregnancies. (This depletes maternal nutrient stores.)
- Lack nutrition knowledge, have too little money to purchase adequate food, or have too little family support.
- Ordinarily consume an inadequate diet due to food faddism, preferences, weight-loss "dieting," uninformed vegetarianism, other limited food choices, or other reasons.
- Smoke cigarettes or abuse alcohol or drugs.
- Are lactose intolerant or suffer chronic health conditions requiring special diets.
- Are underweight or overweight at conception.
- Are carrying twins or triplets.
- Gain insufficient or excessive weight during pregnancy.
- Have a low level of education.

essential nutrients in early human development.[7] The brain is largely made of lipid material, and it depends heavily on products of both omega-3 and omega-6 fatty acids for its growth, function, and structure. If a mother-to-be regularly eats a diet that includes seafood, she receives a balance of the essential fatty acids and their derivatives. This benefits her pregnancy and afterward her infant by way of her milk. Supplements of fish oil are not recommended, however, both because they may carry concentrated toxins and because high fish oil intakes seem to alter the course of pregnancy and labor with unknown effects.

Figure 12-3

Comparison of Nutrient Recommendations for Nonpregnant, Pregnant, and Lactating Women

For actual values, turn to the tables on the inside front and back covers.

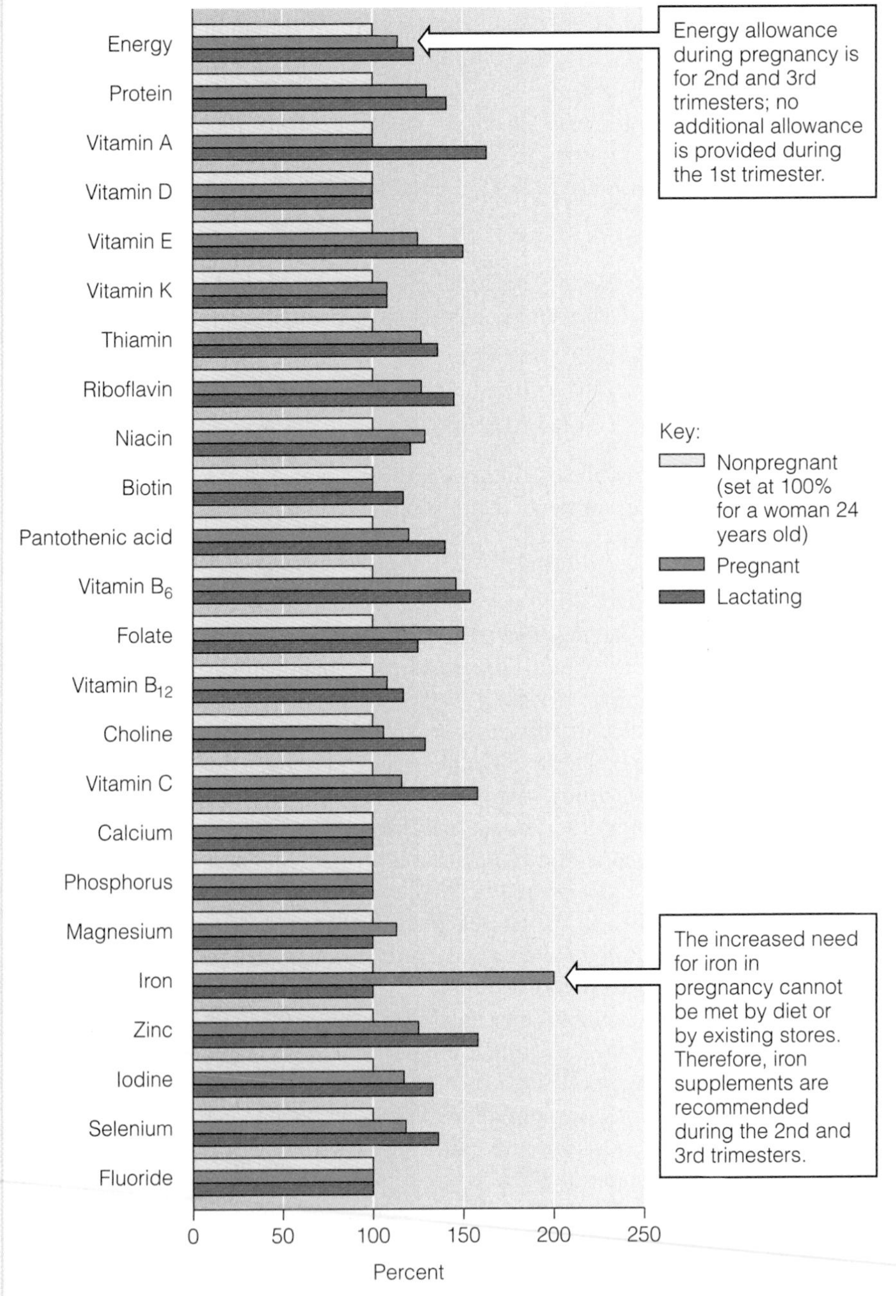

Energy
Protein
Vitamin A
Vitamin D
Vitamin E
Vitamin K
Thiamin
Riboflavin
Niacin
Biotin
Pantothenic acid
Vitamin B_6
Folate
Vitamin B_{12}
Choline
Vitamin C
Calcium
Phosphorus
Magnesium
Iron
Zinc
Iodine
Selenium
Fluoride
0
50
100
150
200
250
Percent
Energy allowance during pregnancy is for 2nd and 3rd trimesters; no additional allowance is provided during the 1st trimester.
Key:
Nonpregnant (set at 100% for a woman 24 years old)
Pregnant
Lactating
The increased need for iron in pregnancy cannot be met by diet or by existing stores. Therefore, iron supplements are recommended during the 2nd and 3rd trimesters.

Of Special Interest: Folate and Vitamin B_{12} The vitamins required for rapid cell proliferation—folate and vitamin B_{12}—are needed in large amounts during pregnancy. New cells are laid down at a tremendous pace as the fetus grows and develops. At the same time, the number of the mother's red blood cells rises, so the recommendation for folate during pregnancy increases from 400 to 600 micrograms a day.

As described in Chapter 7, folate plays an important role in preventing neural tube defects. To review, the early weeks of pregnancy are a critical period for the formation and closure of the **neural tube** that will later develop to form the brain and spinal cord. By the time a woman suspects she is pregnant, usually around the sixth week, the embryo's neural tube is supposed to have closed. A **neural tube defect** occurs when the tube fails to close properly. In the United States, about 4,000 pregnancies each year are affected by neural tube defects.[8] When the upper end of the neural tube fails to close, a rare but lethal neural tube defect known as **anencephaly** occurs. All infants with anencephaly die shortly after birth. When the lower end of the neural tube fails to close, **spina bifida**, a common neural tube defect, occurs (see Figure 12-4). In an infant with spina bifida, the spinal cord and backbone do not develop normally. The membranes covering the spinal cord often protrude as a sac, and sometimes a portion of the spinal cord is contained in the sac. Spina bifida is accompanied by varying degrees of paralysis, depending on the extent of spinal cord damage. Mild cases may not even be noticed, but severe cases lead to death. Common problems include clubfoot, dislocated hip, kidney disorders, curvature of the spine, muscle weakness, mental handicaps, and motor and sensory losses.

Fortunately, women today have several options for obtaining the folate that can reduce the risk of these defects, and it's important that they do so before they become pregnant.[9] Women who are capable of becoming pregnant can:

- Choose a diet of foods naturally rich in folate, but few women obtain the folate they need this way. Natural sources contribute about 200 micrograms of folate to most women's intakes, but the need is much higher.
- Take a daily supplement containing 400 micrograms of folic acid. This ensures an adequate intake.
- Eat a daily serving of a fortified breakfast cereal that provides 400 micrograms of folic acid (see Table 12-3 on the next page).[10] This also ensures an adequate intake.
- Choose a variety of foods naturally rich in folate together with foods enriched with folic acid (see Table 12-3 on the next page).

Supplements and fortified foods offer women a convenient way to ensure sufficient folate regularly and continuously enough to benefit pregnancy.[11] Furthermore, the synthetic form of folate, *folic acid*, in supplements and fortified food is better absorbed than the naturally occurring folate in foods. Thus folate status improves more with intakes of supplements or fortified foods than with intakes of only natural sources of folate. The foods that naturally contain folate are still important, however, because

neural tube the embryonic tissue that later forms the brain and spinal cord.

neural tube defects a group of nervous system abnormalities caused by interruption of the normal early development of the neural tube.

anencephaly (an-en-SEFF-ah-lee) a severe neural tube defect in which the brain fails to form; anencephaly leads to death soon after birth.

spina bifida (SPY-na BIFF-ih-duh) one of the most common types of neural tube defects; the infant is born with gaps in the bones of the spine, leaving the spinal cord protected only by a sheath of skin in those spots, or with no protection at all. The spinal cord may bulge and protrude through the gaps in the vertebral column.

Table 12-2

Daily Food Guide for Pregnant and Lactating Women

Food Group	Number of Servings[a]	
	ADULTS	PREGNANT OR LACTATING WOMEN
Breads/cereals/rice/pasta	6 to 11	7 to 11
Vegetables	3 to 5	4 to 5
Fruits	2 to 4	3 to 4
Meat/meat alternates	2 to 3	3
Milk/milk products	2	3 to 4

[a]Figure 2–5 in Chapter 2 provides examples of foods in each group and serving sizes.

Figure 12-4

Spina Bifida—A Neural Tube Defect

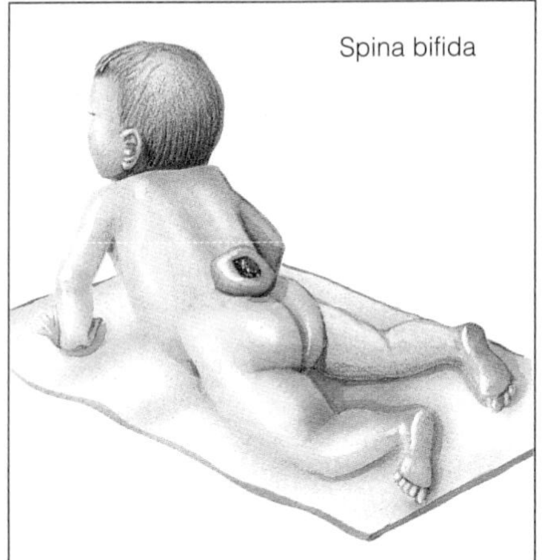
Spina bifida

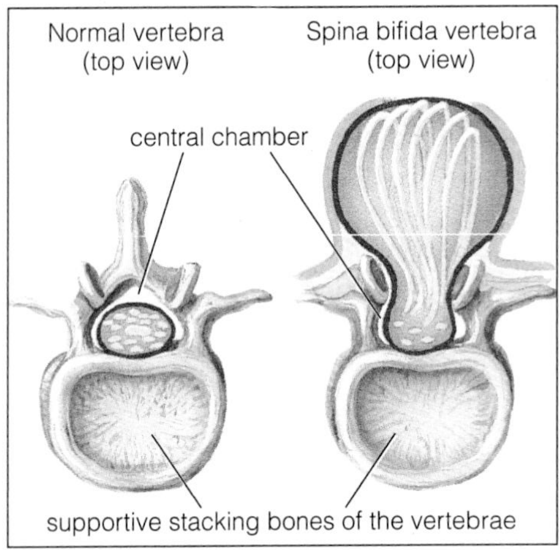
Normal vertebra
(top view)
Spina bifida vertebra
(top view)
central chamber
supportive stacking bones of the vertebrae

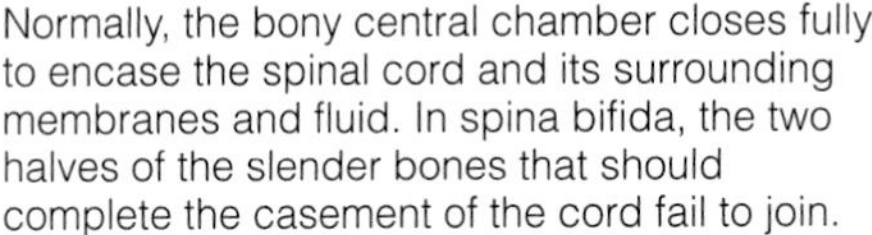

Normally, the bony central chamber closes fully to encase the spinal cord and its surrounding membranes and fluid. In spina bifida, the two halves of the slender bones that should complete the casement of the cord fail to join.

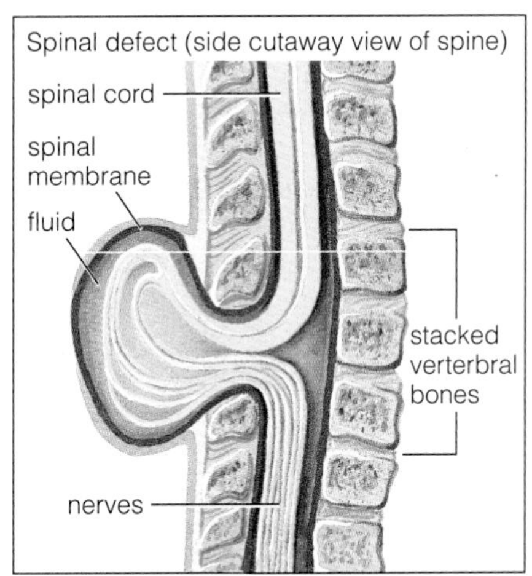
Spinal defect (side cutaway view of spine)
spinal cord
spinal
membrane
fluid
stacked
verterbral
bones
nerves

In the serious form shown here, membranes and fluid have bulged through the gap and nerves are exposed, invariably leading to some degree of paralysis and often to mental retardation.

they are rich sources of other vitamins, minerals, fiber, and the phytochemicals thought to protect against heart disease, cancer, and other diseases.

As of 1999, all refined grain products (cereal, pasta, flour, rolls, buns, farina, grits, cornmeal, and rice) in the United States are fortified with folic acid.* Folate fortification is expected to prevent half of all neural tube defects that occur each year. Folate fortification, however, does raise some safety concerns. The pregnant woman needs a greater amount of vitamin B_{12} to assist folate in the manufacture of new cells. High intakes of folate complicate the diagnosis of a vitamin B_{12} deficiency. For this reason, folate intakes should not exceed 1 milligram per day.[12]

People who eat meat, eggs, or dairy products receive all the vitamin B_{12} they need, even for pregnancy. Those who exclude all animal products from the diet, however, need vitamin B_{12}-fortified soy milk or supplements.

*These products must be fortified with 1.4 milligrams of folate per 100 grams of food.

Table 12-3

Rich Folate Sources[a]

Natural Folate Sources	Fortified Folate Sources
Liver (3 oz) 185 µg	Multi-Grain Cheerios Plus cereal (1 c) 400 µg[b]
Lentils (½ c) 180 µg	Product 19 cereal (1 c) 400 µg[b]
Chickpeas or pinto beans (½ c) 145 µg	Total cereal (1 c) 400 µg[b]
Asparagus (½ c) 125 µg	Pasta, cooked (1 c) 110 µg
Spinach (1 c raw) 115 µg	Rice, cooked (1 c) 80 µg
Avocado (½ c) 70 µg	Bagel (1 small whole) 50 µg
Orange juice (1 c) 60 µg	Waffles, frozen (2) 40 µg
Beets (½c) 46 µg	Bread, white (1 slice) 20 µg

[a]Folate amounts for these and 2,000 other foods are listed in the Table of Food Composition in Appendix A.
[b]Folate in cereals varies; read the Nutrition Facts panel of the label.

Calcium and Other Minerals Among the minerals, calcium, phosphorus, and magnesium are in great demand during pregnancy because they are involved in building the skeleton. Insufficient intakes may result in abnormal development of fetal bones and teeth. Intestinal absorption of calcium doubles early in pregnancy, and the mineral is stored in the mother's bones. Later, when the fetal bones begin to calcify, the mother's bone stores are drawn upon, and there is a dramatic shift of calcium across the placenta. Whether calcium added to the mother's bones early in pregnancy is withdrawn to build the fetus's bones later is unclear.[13] In the final weeks of pregnancy, more than 300 milligrams of calcium a day are transferred to the fetus. Recommendations to ensure an adequate calcium intake during pregnancy are aimed at conserving the mother's bone mass while supplying fetal needs.

For women whose prepregnancy calcium intakes are below recommendations, as most are, increased calcium intakes may be especially important. Because bones are still actively depositing minerals until about age 25, adequate calcium is especially important for young women. Pregnant women under age 25 who consume less than 600 milligrams of calcium a day need to increase their intakes of milk, cheese, yogurt, and other calcium-rich foods. Alternatively, and less preferably, they may need a daily supplement of 600 milligrams of calcium. The DRI recommendation for calcium intake is the same for nonpregnant and pregnant women in the same age group. Magnesium for bone and tissue growth is needed during pregnancy in amounts slightly higher than recommendations for nonpregnant women.

The body conserves iron even more than usual during pregnancy. Menstruation ceases, and absorption of iron increases up to threefold. Despite these conservation measures, iron stores dwindle because the developing fetus draws on its mother's iron stores to create stores of its own to carry it through the first three to six months of life. Also, maternal blood losses are inevitable at birth, especially during a delivery by **cesarean section**, and these losses can further drain the mother's iron supply. Few women enter pregnancy with adequate stores to meet pregnancy demands, so a daily iron supplement containing 30 milligrams is recommended during the second and third trimesters for all pregnant women.[14] Ideally, supplements plus food intakes supply pregnant women with the iron they need.

Nutrient Supplements Women who make wise food choices during pregnancy can meet most of their nutrient needs except for iron. As already discussed, iron supplements are recommended during the second and third trimesters for all pregnant women. Daily multivitamin-mineral supplements are also recommended for women who do not eat adequately and for those in high-risk groups: women carrying multiple fetuses, cigarette smokers, and alcohol and drug abusers. The use of prenatal supplements may help to reduce the risks of preterm delivery and low infant birthweights.[15] Table 12-4 lists recommended amounts of supplements for pregnant women at nutritional risk.

Food Assistance Programs Pregnancy is clearly a time of increased nutrient needs. A woman of limited financial means may need help in obtaining the food and the nutrition counseling that she needs. At the federal level, the **Special Supplemental Food Program for Women, Infants, and Children (WIC)** provides nutrition education and vouchers redeemable for nutritious foods to low-income pregnant and breastfeeding women and their children. WIC provides food vouchers for milk and cheese, iron-fortified cereals, fruit or vegetable juices, eggs, dried beans, and peanut butter. These foods provide nutrients often lacking in diets of low-income women and children (calcium, iron, vitamins A and C, and protein). For infants given formula, WIC also provides iron-fortified infant formula.

Participation in the WIC program benefits both the iron status and the growth and development of infants and children. WIC participation during pregnancy has been shown to reduce the risks of delivering preterm or low-birthweight infants.[16]

Federal food stamps can also help to stretch the pregnant woman's grocery dollars. Her own community may provide educational services and materials, including

cesarean (see-ZAIR-ee-un) **section** surgical childbirth, in which the infant is taken through an incision in the woman's abdomen.

Special Supplemental Food Program for Women, Infants, and Children (WIC) a USDA program to provide nutrition support to low-income women who are pregnant or who have infants or preschool children. WIC offers coupons redeemable for specific foods to supply the nutrients deemed most needed for growth and development.

In midpregnancy, hemoglobin values below 12 grams are not unusual, and 10.5 grams is where the line defining "too low" is often drawn.

Table 12-4

Nutrient Supplements for Pregnancy[a]

Nutrient	Amount
Folate	300 µg
Vitamin B_6	2 mg
Vitamin C	50 mg
Vitamin D	5 µg
Calcium	250 mg
Copper	2 mg
Iron	30 mg
Zinc	15 mg

[a]For pregnant women at nutritional risk (see Table 12-1).
SOURCE: Reprinted with permission from *Nutrition during Pregnancy* © 1990 by the National Academy of Sciences. Published by National Academy Press, Washington, D.C.

Table 12-5

Recommended Weight Gains for Pregnancy

- Underweight women: 28 to 40 lb
- Normal-weight women: 25 to 35 lb
- Overweight women: 15 to 25 lb
- Obese women: 13 lb minimum

NOTE: Underweight is defined as BMI <19.8; normal weight, as BMI 19.8 to 26.0; overweight as BMI 26.0 to 29.0; and obese as BMI >29.0 (BMI standards are on the inside back cover).

SOURCE: Committee on Nutritional Status during Pregnancy and Lactation, Food and Nutrition Board, *Nutrition during Pregnancy* (Washington, D.C.: National Academy Press, 1990), pp.10, 12.

nutrition, food budgeting, and shopping information, through the local agricultural extension service. Organizations such as the American Dietetic Association, the American Diabetes Association, and local hospitals may also provide nutrition information.

KEY POINT *Pregnancy brings physiological adjustments that demand increased intakes of energy and nutrients. A daily iron supplement is recommended for all pregnant women during the second and third trimesters. Food assistance programs such as WIC can benefit pregnant women of limited financial means.*

How Much Weight Should a Woman Expect to Gain during Pregnancy?

The pregnant woman must gain a certain amount of weight during pregnancy as a defense against bearing a low-birthweight baby. Ideally, she will have begun her pregnancy at the appropriate weight for her height, but even more importantly, she will gain enough weight based on her prepregnancy body mass index (BMI). Table 12-5 presents recommended weight gains for pregnancy. For the normal-weight woman, the ideal pattern is about 2 to 4 pounds during the first trimester and a pound per week thereafter.

Dieting during pregnancy is not recommended. Even an obese woman should gain about 15 pounds for the best chances of delivering a healthy infant. Weight gain must be especially generous to meet the needs of a teenager who is still growing herself. Women who are carrying twins should strive for a weight gain of 35 to 45 pounds. Women have been known to exceed or undershoot the recommended limits in pregnancy without ill effects, but the best chances of health are predicted by recommended weight gains. A sudden, large weight gain is always a danger signal: it may indicate the onset of preeclampsia. See the section entitled "Troubleshooting" later on.

The weight the pregnant woman puts on is nearly all lean tissue: placenta, uterus, blood, milk-producing glands, and, of course, the baby itself (see Figure 12-5). The fat she gains is needed later for lactation. Physical activity can help a pregnant woman cope with the extra weight, as the accompanying Fitness for Life feature explains. Some weight is lost at delivery, but many women retain a few pounds with each pregnancy.

KEY POINT *Weight gain is essential for a healthy pregnancy. A woman's prepregnancy BMI, her own nutrient needs, and the number of fetuses she is carrying help to determine an appropriate weight gain.*

Pregnant women can enjoy the benefits of physical activity.

FITNESS FOR LIFE

Should Pregnant Women Be Physically Active?

Physical activity is important to the pregnant woman, not only to help her carry the extra weight of pregnancy without strain, but also to help ease her upcoming childbirth. Staying active can improve the fitness of the mother-to-be, prevent complications in pregnancy, facilitate labor, and reduce psychological stress.[1] Women who remain active during pregnancy report fewer discomforts throughout their pregnancies.[2] Pregnant women should take care in choosing their physical activities, however. Pregnant women should participate in "low-impact" activities and avoid sports in which they might fall or be hit by other people or objects. As is true for everyone, the frequency, duration, and intensity of the activity affect the likelihood of the benefits or risks.[3] A pregnant woman should consult her health-care provider before taking up additional activity. A few guidelines are offered in Table 12-6.

Figure 12-5

Components of Weight Gain during Pregnancy

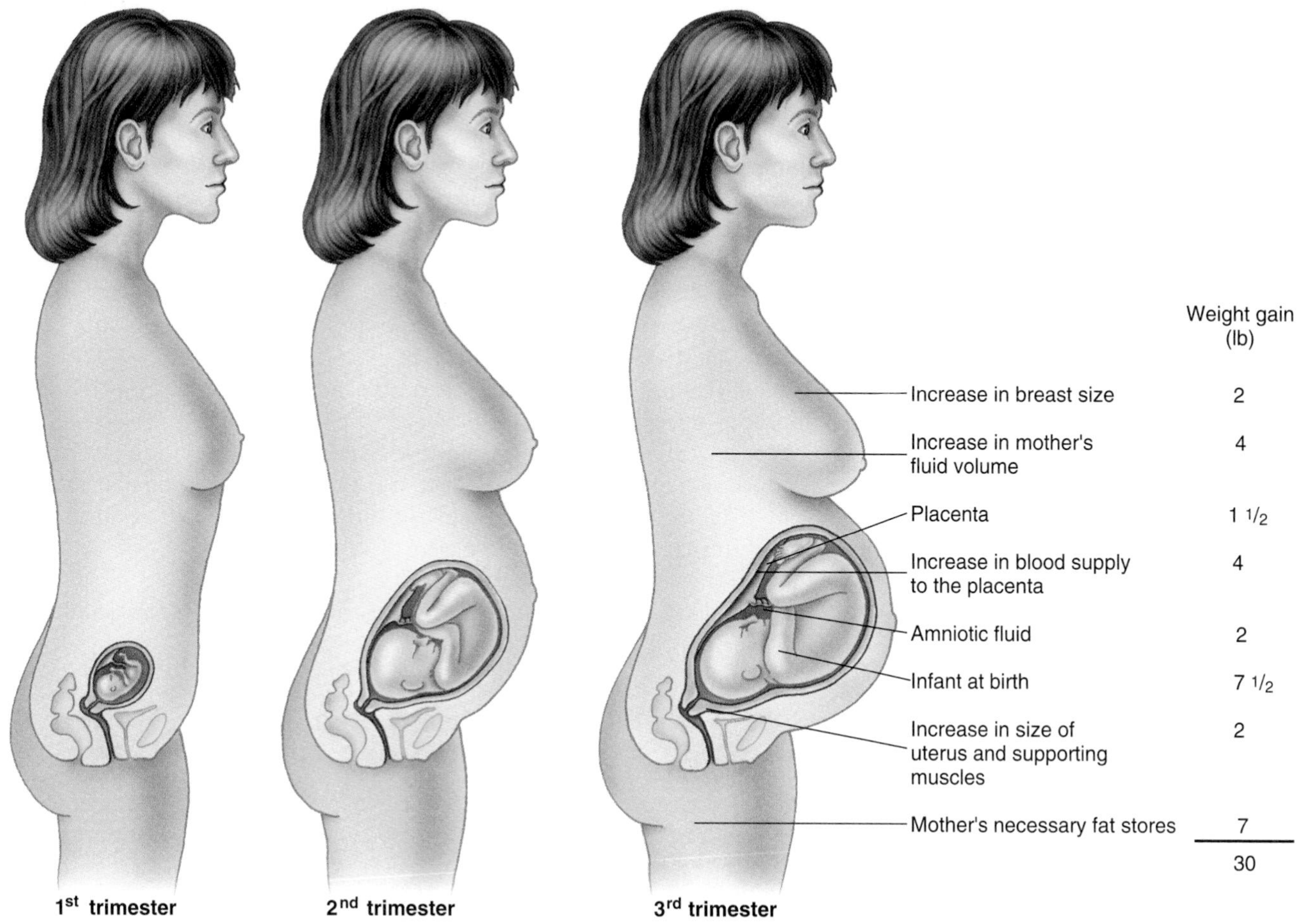
Weight gain (lb)
Increase in breast size 2
Increase in mother's fluid volume 4
Placenta 1 1/2
Increase in blood supply to the placenta 4
Amniotic fluid 2
Infant at birth 7 1/2
Increase in size of uterus and supporting muscles 2
Mother's necessary fat stores 7
30
1st trimester
2nd trimester
3rd trimester

Table 12-6

Guidelines for Physical Activity during Pregnancy

- Be physically active on a regular basis (at least three times a week), not intermittently.
- Stop exercising if you feel overheated.
- Drink plenty of fluids before, during, and after physical activity.
- Avoid exerting yourself in hot, humid weather; avoid overheating.
- Avoid jarring or jerky motions.
- Avoid any activity that has the potential to cause even mild abdominal trauma.
- Avoid prolonged periods of standing still.
- Discontinue any activity that causes discomfort.
- Do not exercise while lying on your back after the fourth month.
- Do not allow your heart rate to exceed 150 beats per minute.
- Eat enough to support the energy needs of pregnancy and physical activity.

Teen Pregnancy

Each year in the United States, about one million adolescent girls become pregnant.[17] Of these, about half choose to continue their pregnancies. A pregnant adolescent presents a special case of intense nutrient needs. Even when not pregnant, a teenager is hard-pressed to meet her nutrient needs. Many teenage girls, especially the youngest ones, have not had time to store the nutrients needed to support their own rapid growth and development, much less nutrients needed to support pregnancy and the developing fetus. Many teens enter pregnancy deficient in several vitamins and minerals, including vitamins A and D, folate, iron, calcium, and zinc. Nutrient shortages place both mother and infant at risk. Pregnant teenagers have more miscarriages, premature births, stillbirths, and low-birthweight infants than do pregnant adult women. Their greatest risk, though, is death of the infant; mothers under age 16 bear more babies who die within the first year than do women in any other age group. Clearly, teenage pregnancy is a major public health problem.

To support the needs of both mother and fetus, a pregnant teenager with a BMI in the normal range is encouraged to gain about 35 pounds or so. Teenagers who gain less have smaller infants with associated risks. Adequate nutrition can substantially improve the health of the mother and infant; it is an indispensable component of prenatal care.[18] Table 12-7 provides a guide to the number of food servings recommended for pregnant and lactating teenagers.

KEY POINT *Of all the population groups, pregnant teenage girls have the highest nutrient needs and an increased likelihood of having problem pregnancies.*

Why Do Some Women Crave Pickles and Ice Cream While Others Can't Keep Anything Down?

Does pregnancy give a woman the right to demand pickles and ice cream at 2 A.M.? Perhaps not, at least not for nutrition's sake. Food cravings and aversions during pregnancy, though common, do not seem to reflect real physiological needs. In other words, a woman who craves pickles is not likely to be in need of salt. Food cravings and aversions that arise during pregnancy are usually due to changes in taste and smell sensitivities, and they quickly disappear after the baby's birth.

Sometimes cravings may occur in women with nutrient-poor diets. A pregnant woman who is deficient in iron, zinc, or other nutrients may crave and eat clay, ice, cornstarch, and other nonnutritious substances, but this does not prove that the deficiency caused the craving. The practice is pica (first mentioned in Chapter 8). Such cravings are not adaptive; the substances the woman craves do not deliver the nutrients she needs. In fact, clay and other substances can cling to the intestinal wall and form a barrier that interferes with normal nutrient absorption.

Table 12-7

Daily Food Guide for Pregnant and Lactating Teenagers

Food Group	Number of servings[a]	
	TEENAGERS	PREGNANT OR LACTATING TEENAGERS
Breads/cereals/rice/pasta	6 to 11	7 to 11
Vegetables	3 to 5	4 to 5
Fruits	2 to 4	3 to 4
Meat/meat alternates	2 to 3	3
Milk/milk products	3	4

[a] Figure 2-5 provided details concerning serving sizes and foods within the groups listed here.

The nausea of "morning" (actually, anytime) sickness seems unavoidable because it arises from the hormonal changes of early pregnancy. Many women complain that smells, especially cooking smells, make them sick. Thus minimizing odors is a key to alleviating morning sickness. Sipping on carbonated drinks and nibbling soda crackers or other salty snack foods before getting out of bed can sometimes alleviate nausea.[19] Some women may do well to simply eat what they desire whenever they feel hungry. Table 12-8 offers some other suggestions, but morning sickness can be persistent. If morning sickness interferes with normal eating for more than a week or two, the woman should seek medical advice to prevent nutrient deficiencies.

Later, as the hormones of pregnancy alter her muscle tone and the thriving fetus crowds her intestinal organs, an expectant mother may complain of heartburn or constipation. Many women find that raising the head of the bed with two or three pillows helps to relieve nighttime heartburn. A high-fiber diet, physical activity, and a plentiful water intake will help relieve constipation. The pregnant woman should use laxatives or heartburn medication only if her physician prescribes them.

KEY POINT *Food cravings usually do not reflect physiological needs, and some may interfere with nutrition. Nausea arises from normal hormonal changes of pregnancy.*

What Behaviors or Substances Should Pregnant Women Avoid?

Some substances in a woman's diet and environment can be harmful, and their potential impact is too great to ignore. Of these, alcohol predominates and is the topic of the next section. A few others also deserve some discussion.

Cigarette Smoking Cigarette (and cigar) smoking is clearly a harmful practice. Smoking adversely affects the pregnant woman's nutrition status, which, in turn, impairs fetal nutrition. Smokers tend to have lower intakes of dietary fiber, vitamin A, beta-carotene, folate, and vitamin C. Oxidants in cigarette smoke accelerate vitamin C metabolism and deplete smokers' body stores of this antioxidant, further compromising smokers' vitamin C status. Smoking also restricts the blood supply to the growing fetus

Table 12-8

Tips for Relieving Common Discomforts of Pregnancy

To alleviate the nausea of pregnancy:

- On waking, arise slowly.
- Eat dry toast or crackers.
- Chew gum or suck hard candies.
- Eat small, frequent meals whenever hunger strikes.
- Avoid foods with offensive odors.
- When nauseated, drink no citrus juice, water, milk, coffee, or tea.

To prevent or alleviate constipation:

- Eat foods high in fiber.
- Exercise daily.
- Drink at least 8 glasses of liquids a day.
- Respond promply to the urge to defecate.
- Use laxatives only as prescribed by a physician; avoid mineral oil—it carries needed fat-soluble vitamins out of the body.

To prevent or relieve heartburn:

- Eat small, frequent meals.
- Drink liquids between meals.
- Avoid spicy or greasy foods.
- Sit up while eating.
- Wait an hour after eating before lying down.
- Wait 2 hours after eating before exercising.

environmental tobacco smoke (ETS) the combination of exhaled smoke (mainstream smoke) and smoke from lighted cigarettes, pipes, or cigars (sidestream smoke) that enters the air and may be inhaled by other people.

and so limits the delivery of oxygen and nutrients and the removal of wastes. It slows growth, thus retarding physical development of the fetus, and it may cause behavioral or intellectual problems later on.[20] A mother who smokes is more likely to have a complicated birth, and her infant is more likely to be of low birthweight.[21] Of all preventable causes of low birthweight in the United States, smoking has the greatest impact.

Research suggests that even in women who do not smoke, exposure to **environmental tobacco smoke** (ETS, or secondhand smoke) during pregnancy increases the risk of low birthweight.[22] For an infant of a smoker, ETS exposure increases the likelihood of sudden infant death syndrome (SIDS), the unexplained deaths that sometimes occur in otherwise healthy infants.[23] Constituents of cigarette smoke such as nicotine, cyanide, and others are directly toxic to a fetus and to the infant later on. With great urgency, the surgeon general has warned that parental smoking can kill an otherwise healthy fetus or newborn.

Medications and Illicit Drugs Other drugs taken during pregnancy can cause serious birth defects. The use of medications not prescribed by a physician, even over-the-counter drugs, herbal preparations, or high-dose vitamin supplements, is inadvisable.[24] Women are advised to take medications only if their physicians deem it necessary to protect their life and health. Drug labels warn: "As with any drug, if you are pregnant or nursing a baby, seek the advice of a health professional before using this product." For aspirin and ibuprofen, an additional warning immediately follows: "It is especially important not to use aspirin (or ibuprofen) during the last three months of pregnancy unless specifically directed to do so by a doctor because it may cause problems in the unborn child or excessive bleeding during delivery." Such warnings should be taken seriously.

Research shows that mothers who abuse drugs such as marijuana and cocaine during pregnancy inflict serious health consequences, including nervous system disorders, on their future infants.[25] Crack and other forms of cocaine pose hazards to infants who may face low-birthweight complications, heartbeat abnormalities, the pain of withdrawal, and even death as they first experience life outside the womb. Some effects of other drugs of abuse on the fetus are listed in the margin; the effects of the drug alcohol are so profound and widespread that they are addressed in a section of their own.

Fetal effects of abused drugs:

- Amphetamines: Suspected nervous system damage; behavioral abnormalities.
- Barbiturates: Drug withdrawal symptoms in the newborn, lasting up to six months.
- Cocaine: Uncontrolled jerking motions; paralysis; permanent mental and physical damage.
- Marijuana: Short-term irritability at birth.
- Opiates (including heroin): Drug withdrawal symptoms in the newborn; permanent learning disability (attention deficit hyperactivity disorder).

Vitamin-Mineral Megadoses Many vitamins are toxic when taken in excess and the minerals are even more so. Among vitamins, a single massive dose of preformed vitamin A (100 times the recommended intake) has caused birth defects. Chronic use of lower doses of vitamin A supplements (three to four times the recommended intake) may also cause birth defects.[26] Intakes before the seventh week of pregnancy appear to be the most damaging. For this reason, vitamin A is not given as a supplement in the first trimester of pregnancy unless there is evidence of deficiency, which is rare. Women taking supplements should take heed—experts urge pregnant women not to exceed three times the recommended daily intake of vitamin A.

Dieting Dieting, even for short periods, is also hazardous during pregnancy. Low-carbohydrate diets or fasts that cause ketosis deprive the growing brain of needed glucose and may impair its development. Such diets are also likely to be deficient in other nutrients vital to fetal growth. Energy restriction during pregnancy is dangerous, regardless of the woman's prepregnancy weight or the amount of weight gained in the previous month.

Caffeine Caffeine crosses the placenta, and the fetus has only a limited ability to metabolize it. No firm limit for caffeine intake is yet available. So far, research studies have not proved that caffeine causes birth defects in human beings (as it does in animals). Some evidence suggests that moderate-to-heavy use (more than 300 milligrams per day—the equivalent of 2 to 3 cups of coffee) may lower infant birthweight.[27] However, another well-controlled study of caffeine's effects on pregnancy found that 3 cups of coffee daily is not a risk factor for growth retardation.[28] In light of such conflicting evidence, it seems most sensible to limit caffeine consumption to the equivalent of one cup of coffee or two 12-ounce cola beverages a day. Caffeine amounts in food and beverages are listed in this chapter's Controversy, on page 466.

KEY POINT *Abstaining from smoking and other drugs, avoiding large doses of nutrients, refraining from dieting, and limiting caffeine use are recommended during pregnancy.*

Apgar score a system of scoring an infant's physical condition right after birth. Heart rate, respiration, muscle tone, response to stimuli, and color are ranked 0, 1, or 2. A low score indicates that medical attention is required to facilitate survival.

fetal alcohol syndrome (FAS) the cluster of symptoms seen in an infant or child whose mother consumed excess alcohol during her pregnancy. FAS includes, but is not limited to, brain damage, growth retardation, mental retardation, and facial abnormalities.

Drinking during Pregnancy

Alcohol is arguably the most hazardous drug to future generations because it is legally available, heavily promoted, and widely abused. Society often sends mixed messages concerning alcohol. Beverage companies promote an image of drinkers as wealthy, healthy, young, and active. Opposing this image, health authorities warn that alcohol may have adverse effects, especially during pregnancy (see Figure 12-6). Every container of beer, wine, or liquor for sale in the United States is required to warn pregnant women of the danger of drinking during pregnancy. In the past, many women who say they would have ceased drinking during pregnancy had they known the danger, unwittingly damaged their infants. Women of childbearing age need to know about alcohol's effects.

Alcohol's Effects

Oxygen is indispensable on a minute-to-minute basis to the development of the fetus's central nervous system. A sudden dose of alcohol can halt the delivery of oxygen through the umbilical cord. Alcohol also slows cell division, reducing the number of cells produced and inflicting abnormalities on those that are produced.[29] During the first month of pregnancy, even a few minutes of alcohol exposure can exert a major effect on the fetal brain, which at that time is growing at the rate of 100,000 new brain cells a minute. Alcohol also interferes with placental transport of nutrients to the fetus and can cause malnutrition in the mother; then, all of malnutrition's harmful effects compound the effects of the alcohol.

Controversy 5 defined "a drink" as:

- 3 to 4 ounces wine.
- 10 ounces wine cooler.
- 12 ounces beer.
- 1 ounce hard liquor.

KEY POINT *Alcohol limits oxygen delivery to the fetus, slows cell division, and reduces the number of cells organs produce. Alcoholic beverages must bear warnings to pregnant women.*

Figure 12-6

Mixed Messages in Alcohol Advertisements

Labels on alcoholic beverages often display "healthy" images, but their warnings tell the truth.

Tropical Breeze
Rum Cooler
GOVERNMENT WARNING:
ACCORDING TO THE SURGEON GENERAL WOMEN SHOULD NOT DRINK ALCOHOLIC BEVERAGES DURING PREGNANCY BECAUSE OF THE RISK OF BIRTH DEFECTS.

Fetal Alcohol Syndrome

Drinking alcohol during pregnancy threatens the fetus with irreversible brain damage, growth retardation, mental retardation, facial abnormalities, vision abnormalities, low **Apgar scores**, and more than 40 identifiable health problems—a cluster of symptoms known as **fetal alcohol syndrome** or **FAS**.[30] The fetal brain is extremely vulnerable to a glucose or oxygen deficit, and alcohol causes both by disrupting placental functioning. In addition, alcohol itself crosses the placenta freely and is directly toxic to the defenseless fetal brain and nervous system. The result is permanent brain damage and lifelong mental retardation. FAS is preventable by limiting alcohol intake during pregnancy, but once present, it is incurable.

Between 1979 and 1993, incidence of FAS increased sixfold.[31] About a fifth of women continue drinking alcohol after they learn that they are pregnant. For women who want to drink during their pregnancies, then, the important question is how much alcohol is too much.

Clearly, 3 ounces of alcohol (about 6 drinks) a day is too much early in pregnancy, even if the woman stops drinking immediately after she learns that she is pregnant. Birth defects have been observed in the children of some women who drank 2 ounces (4 drinks) of alcohol daily during pregnancy. Low birthweight has been observed in infants born to some women who drank 1 ounce (2 drinks) per day during pregnancy. At that level of alcohol intake, a sizable and significant increase in the rate of spontaneous abortions occurs; the reason is unclear, but perhaps the alcohol poisons the fetus or causes the placenta to detach. FAS is also known to occur with as few as 2 drinks a day.

fetal alcohol effect (FAE) partial abnormalities from prenatal alcohol exposure, not sufficient for diagnosis with FAS, but impairing to the child. Also called *alcohol-related birth defects (ARBD)* or *subclinical FAS.*

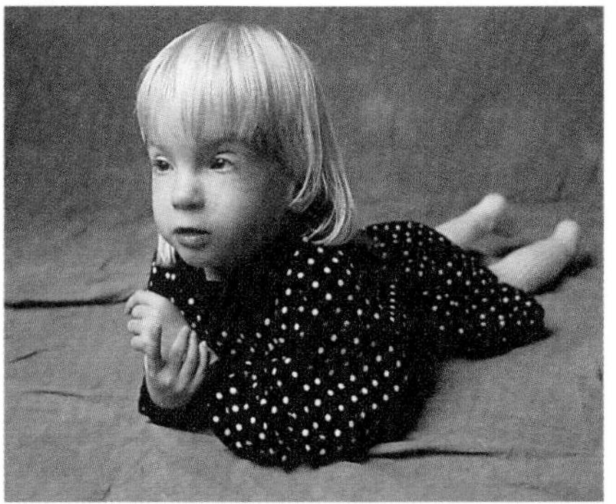

A child with FAS.

The pattern of a woman's drinking may be as important as the average alcohol intake. For example, a woman whose average intake was only 1 ounce of alcohol a day might not drink at all during the week, but then have 14 drinks each weekend. Thus the fetus might be intermittently exposed to high alcohol doses. No matter what the intake or pattern, the most severe impact is likely to occur in the first month, before the woman may be aware that she is pregnant.

Research using animals shows that one-fifth of the amount of alcohol needed to produce major, outwardly visible defects will surely produce learning impairment in the offspring, a condition known as **fetal alcohol effect (FAE)**. Some children show no outward sign of the impairment, but the damage is there on the inside.[32] Others may be short in stature or display subtle facial abnormalities. Most perform poorly in school and in social interactions and suffer a subtle form of brain damage. Anyone exposed to alcohol before birth may always respond differently to it, and also to certain drugs, than if no exposure had occurred. Even before fertilization, alcohol may damage the ovum or sperm in the mother- or father-to-be, and so lead to abnormalities in the child.

The children born with alcohol damage remain damaged. They may live, but they never fully recover. Figure 12-7 shows the facial abnormalities of FAS, which are easy to depict. A visual picture of the internal harm is impossible, but it is that damage that virtually seals the fate of the child for life. About 3 in every 1,000 children are victims

Figure 12-7

Typical Facial Characteristics of FAS

The severe facial abnormalties shown here are just outward signs of severe mental impairments and internal organ damage. These defects, though hidden, may create major health problems later.

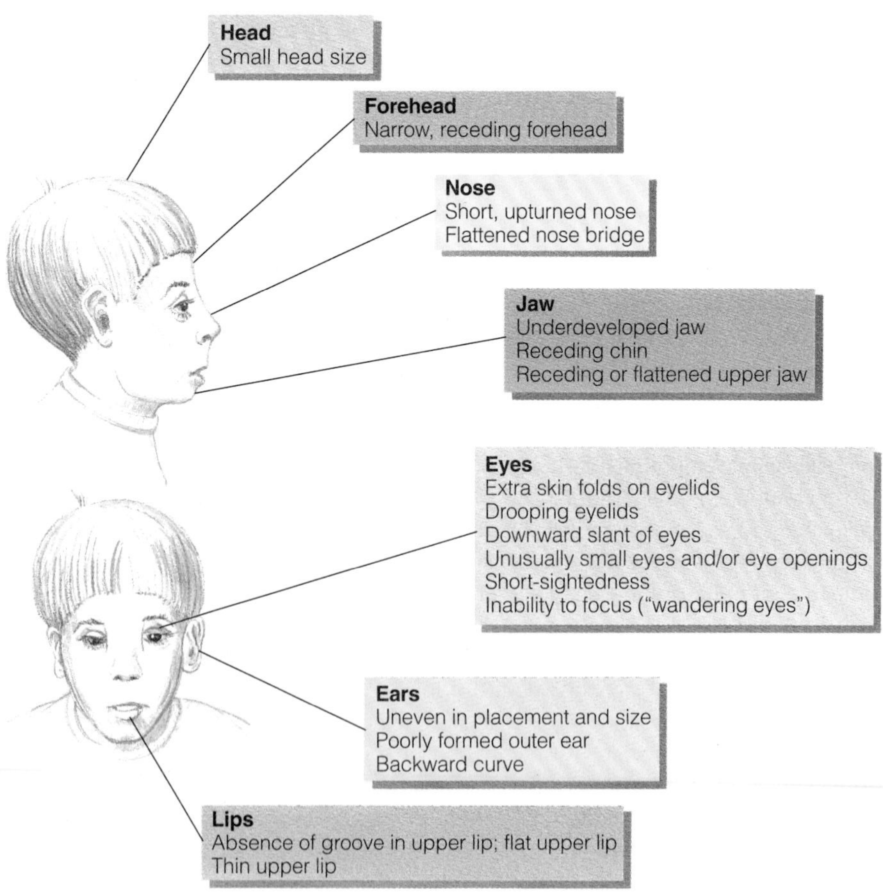
Head
Small head size
Forehead
Narrow, receding forehead
Nose
Short, upturned nose
Flattened nose bridge
Jaw
Underdeveloped jaw
Receding chin
Receding or flattened upper jaw
Eyes
Extra skin folds on eyelids
Drooping eyelids
Downward slant of eyes
Unusually small eyes and/or eye openings
Short-sightedness
Inability to focus ("wandering eyes")
Ears
Uneven in placement and size
Poorly formed outer ear
Backward curve
Lips
Absence of groove in upper lip; flat upper lip
Thin upper lip

of this preventable damage, making FAS the leading known cause of mental retardation in the world. Moreover, for every baby diagnosed with FAS, 3 or 4 with FAE may go undiagnosed until problems develop later in the preschool years. Upon reaching adulthood, such children are ill equipped for employment, relationships, and the other facets of life most adults take for granted.

KEY POINT *The birth defects of fetal alcohol syndrome arise from severe damage to the fetus caused by alcohol. A lesser condition, fetal alcohol effect, may be harder to diagnose, but also robs the child of a normal life.*

Experts' Advice

The American Academy of Pediatrics takes the position that women should stop drinking as soon as they *plan* to become pregnant.[33] As mentioned, this step is important for fathers-to-be as well. It is important to know, though, that a woman who has drunk heavily during the first two-thirds of her pregnancy can still prevent some organ damage by stopping heavy drinking during the third trimester.

Experts have not always agreed that women need to abstain totally from using alcohol during pregnancy. Nevertheless, researchers looking for a "safe" intake limit have come full circle to concede that abstinence from alcohol is the best policy for pregnant women. The authors of this book do, too. It is a personal choice, but if we had it to make, we would give up even the pleasure of wine with meals "for the duration." After the birth of our healthy baby, if we chose to drink at all, we would celebrate with one glass of the finest champagne.

KEY POINT *Abstinence from or strict restriction of alcohol is critical to prevent irreversible damage to the fetus.*

Troubleshooting

Just as adequate nutrition and normal weight gain support the health of the mother and growth of the fetus, maternal diseases detract from health and growth. If discovered early, many diseases can be controlled—another reason early prenatal care is recommended. Some additional measures can help women to avoid the most common problems encountered during pregnancy.

Gestational Diabetes Pregnancy precipitates the onset of diabetes in some women. This condition, known as **gestational diabetes**, is a common medical complication of pregnancy.[34] With gestational diabetes, blood glucose becomes abnormal during pregnancy, but it usually returns to normal after the infant is born. In about one-third of such cases, however, diabetes becomes permanent. Without proper management, gestational diabetes can lead to fetal or infant sickness and death. Properly managed, it will cause no harm at all except that surgical birth may be necessary.[35] The American Diabetes Association recommends that women be screened for diabetes between 24 and 28 weeks' gestation, with the exception of women who meet specific criteria such as no family history of diabetes.[36]

Preeclampsia A certain degree of **edema** is to be expected in late pregnancy, and some women also develop hypertension during that time. If a rise in blood pressure is mild, it may subside after childbirth and cause no harm. In some cases, however, hypertension may signal the onset of **preeclampsia**, a condition characterized not only by high blood pressure, but by protein in the urine and fluid retention (edema). The edema of preeclampsia is a severe, whole-body edema, distinct from the localized fluid retention women normally experience late in pregnancy. The normal edema of pregnancy is a response to gravity; fluid from blood pools in the ankles. The edema of preeclampsia causes swelling of the face and hands as well as of the feet and ankles.

gestational diabetes abnormal glucose tolerance appearing during pregnancy, with subsequent return to normal after the end of pregnancy.

edema accumulation of fluid in the tissues (also defined in Chapter 6).

preeclampsia a potentially dangerous condition during pregnancy characterized by edema, hypertension, and protein in the urine.

Warning signs of preeclampsia:

- Headaches.
- Swelling, especially facial swelling.
- Dizziness.
- Blurred vision.
- Sudden weight gain.

Preeclampsia affects almost all of the mother's organs—the circulatory system, liver, kidneys, and brain. If the condition progresses, she may experience convulsions; when this occurs, the condition is called eclampsia. Maternal mortality during pregnancy is rare in developed countries, but eclampsia is the most common cause. Preeclampsia demands prompt medical attention. Treatment focuses on regulating blood pressure and preventing convulsions.

Pregnancy is a time of adjustment to major changes, physical, social, emotional, and financial. The couple who are expecting a baby will have to change their lifestyles as they take on the responsibility of caring for a child. Ideally, the mother will start developing this sense of responsibility by caring for herself during pregnancy. The expectant parents need support in thinking of themselves as important people with a new and challenging task that they can and will perform well.

KEY POINT *Common medical problems associated with pregnancy are gestational diabetes and preclampsia. These should be managed to minimize associated risks.*

Breastfeeding

As the time of childbirth nears, a woman must decide whether she will feed her baby breast milk or formula. Before she makes this choice, she should be aware of some things about breastfeeding. Both the American Academy of Pediatrics (AAP) and the Canadian Pediatric Society stand behind this statement: "Breastfeeding is strongly recommended for full term infants, except in the few instances where specific contraindications exist." The American Dietetic Association advocates breastfeeding for the nutritional health it confers on the infant as well as for the physiological, social, economic, and other benefits it gives to the mother.[37] The AAP recommends that infants receive breast milk for at least the first 12 months of life.[38] All other legitimate nutrition authorities share this view, but some makers of baby formula try to convince women otherwise, as the Consumer Corner points out.

Why Is Breast Milk So Good for Babies?

Breastfeeding is a natural extension of pregnancy—the mother's body continues to nourish the infant.

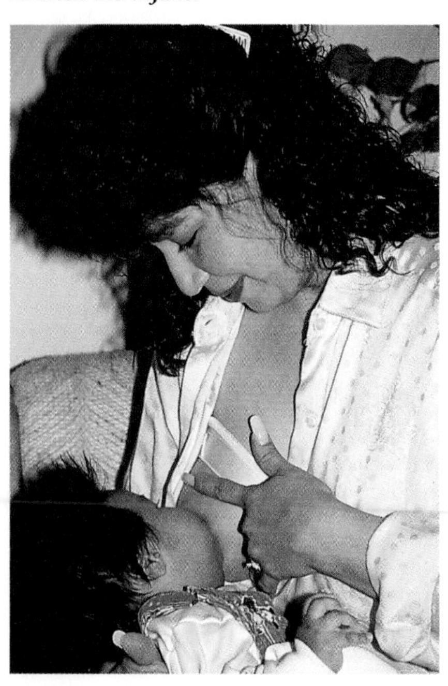

Breast milk is tailor-made to meet the nutrient needs of the human infant.[39] Its carbohydrate is lactose, and its fat provides a generous portion of the essential omega-6 fatty acid linoleic acid and its products. In addition, a mother who consumes food rich in omega-3 fatty acids will pass these beneficial nutrients on to her child through her milk. The protein of breast milk is especially digestible and usable to support tissue growth. Breast milk contains fat-digesting enzymes that help ensure efficient fat absorption by the infant. Breast milk also conveys information to the infant's body about its environment by way of antibodies, whole proteins, and other constituents.

Immune Factors in Breast Milk Breast milk offers the infant unsurpassed protection against infection.[40] This protection includes antiviral and antibacterial agents and infection inhibitors. Some of these immune molecules are proteins that the infant absorbs whole, but the greatest protection may occur in the milk itself. For example, immune factors in breast milk interfere with growth of bacteria that could otherwise attack the infant's vulnerable digestive tract linings. Breastfed babies are less prone to develop stomach and intestinal disorders during the first few months of life and so experience less vomiting and diarrhea than formula-fed babies do.[41] In fact, research shows that breast milk contains not only antibodies against the most common cause of diarrhea in infants and young children but also another factor that binds to, and inhibits replication of, the infective agent.[†42] Breastfeeding reduces the

† The most common cause of diarrhea in the United States is rotavirus. More children are hospitalized for rotavirus infection than for any other single cause.

severity and duration of symptoms associated with this infection. Breastfeeding also protects against other common illnesses of infancy such as middle ear infection and respiratory illness.[43]

Chapter 4 first mentioned research showing that children with diabetes (type 1) almost always have antibodies to cow's milk protein in their pancreatic tissues.[44] This finding leads some to suspect that early feeding of cow's milk formula may set the stage for abnormal immune functioning that causes diabetes later on. Others say a virus, not milk protein, initiates the immune changes associated with diabetes. In any case, breastfeeding prevents early exposure to cow's milk protein and also confers extra immune cells that destroy the kind of virus suspected of involvement in diabetes.

During the first two or three days of lactation, the breasts produce **colostrum**, a premilk substance containing antibodies and white cells from the mother's blood. Because it contains immunity factors, colostrum helps protect the newborn infant from those infections against which the mother has developed immunity, precisely those in the environment likely to infect the infant.[45] Maternal antibodies from colostrum inactivate harmful bacteria within the infant's digestive tract. Later, breast milk also delivers antibodies, although not as many as colostrum.

Certain factors in colostrum and breast milk favor the growth of "friendly" bacteria in the infant's digestive tract, preventing other, harmful bacteria from thriving there.[‡] Hormones and other factors present in colostrum and breast milk stimulate the development of the infant's digestive tract. Worn cells in the infant's digestive tract are promptly replaced, facilitating the tract's functioning. Clearly, breast milk is a very special substance.

colostrum (co-LAHS-trum) a milklike secretion from the breasts during the first day or so after delivery before milk appears; rich in protective factors.

alpha-lactalbumin (lact-AL-byoo-min) the chief protein in human breast milk. The chief protein in cow's milk is *casein* (CAY-seen).

lactoferrin (lack-toe-FERR-in) a factor in breast milk that binds iron and keeps it from supporting the growth of the infant's intestinal bacteria.

Nutrients in Breast Milk

Breast milk changes in composition throughout lactation. Milk from the mother of a premature infant meets the developmental needs of a preterm infant in ways that full-term mother's milk cannot match. For example, the milk for a premature infant provides more protein in less volume, just the right mix to support the rapid growth required to help a premature infant survive its first critical weeks. Some preliminary research even suggests that preterm infants who are breastfed may have an intellectual advantage over formula-fed preterm infants.[46] More research is needed to support or refute this idea, however. Breast milk composition keeps on changing from early to late infancy to meet the infant's changing energy and nutrient needs.

The protein in breast milk is largely **alpha-lactalbumin**, a protein the human infant can easily digest. Another breast milk protein, **lactoferrin**, indirectly benefits the baby's iron nutrition and also acts as an antibacterial agent. Lactoferrin is an iron-gathering compound that helps absorb iron into the infant's bloodstream, keeps intestinal bacteria from getting enough iron to grow out of control, and also works directly to kill some bacteria.

The vitamin content of the breast milk of a well-nourished mother is ample. Even vi-tamin C, for which cow's milk is a poor source, is supplied generously by the breast milk of such a mother. The concentration of vitamin D in breast milk is low, but this is not a threat to light-skinned infants who are taken out into the sunshine regularly. A dark-skinned infant, or one who has little exposure to sunlight, however, may not make enough vitamin D to prevent rickets. Because so many variables exist regarding vitamin D and sunlight exposure, the AAP recommends vitamin D supplementation beginning at birth for breastfed babies who do not receive sufficient exposure to sunlight.

As for minerals, the 2-to-1 calcium-to-phosphorus ratio of breast milk is ideal for calcium absorption, and both of these minerals, along with magnesium, support the rate of growth expected in a human infant. Breast milk is also low in sodium. The limited amount of iron in breast milk is highly absorbable, and its zinc, too, is absorbed better than from cow's milk, thanks to the presence of a zinc-binding protein.

‡The "friendly" bacteria are the *Lactobacillus bifidus* type.

certified lactation consultant a health-care provider, often a registered nurse, with specialized training in breast and infant anatomy and physiology who teaches the mechanics of breastfeeding to new mothers. Certification is granted after passing a standardized post-training examination.

An exclusively breastfed baby does not need supplements except possibly for vitamin D and, after four months, iron. Before four months, supplemental iron is unnecessary. Although breast milk contains little total iron, the absorption of that iron is excellent. As lactation progresses, the iron in breast milk dwindles further, making iron a concern for the four-to-six month old. Most babies are born with enough iron in their livers to last about half a year, and iron deficiency is rarely seen in very young infants. Certainly, by six months, it seems desirable to begin feeding the breastfed infant iron-fortified cereals. If the water supply is severely deficient in fluoride, both breastfed and formula-fed infants require fluoride supplementation after six months of age.

KEY POINT *Breast milk is normally the ideal food for infants with the needed nutrients in the right proportions and also protective factors. It is especially valuable for premature infants.*

Concerns for the Breastfeeding Mother

Toward the end of her pregnancy, a woman who plans to breastfeed her baby should begin to prepare. No elaborate or expensive preparations are needed, but the expectant mother might want to read at least one of the many handbooks available on breastfeeding or consult a **certified lactation consultant,** employed at many hospitals.[§] One way to prepare is to learn what dietary changes are needed. Adequate nutrition is essential to successful lactation; without it, lactation may falter.

In rare cases, women produce too little milk to nourish their infants adequately. Severe consequences, including infant dehydration, malnutrition, and brain damage, can occur should the condition go undetected for long. Early warning signs of insufficient milk are dry diapers (a well-fed infant wets about six diapers a day) and infrequent bowel movements. In such cases, formula feeding is essential.

A nursing mother produces about 25 ounces of milk a day, more or less, depending primarily on the infant's demand for milk. Producing this milk costs a woman almost 650 calories per day above her regular need during the first six months of lactation. To meet this energy need, the woman is advised to eat an extra 500 calories of food each day. The other 150 calories may be drawn from the fat stores she accumulated during pregnancy. Energy needs for women who are breastfeeding exclusively range from 2,500 to 3,300 calories a day, depending on physical activity.[47] Severe energy restriction hinders milk production and can compromise the mother's health.[48]

The food energy consumed by the nursing mother should carry with it abundant nutrients. Figure 12-3 (page 436) showed a lactating woman's nutrient needs and those of a nonpregnant woman, and Table 12-2 on page 437 suggested a food pattern that meets them.

Some infants may be sensitive to foods such as cow's milk, onions, or garlic in the mother's diet and become uncomfortable when she eats them. Nursing mothers should not automatically avoid such foods, however. A mother who is nursing her baby is advised to eat whatever nutritious foods she chooses. Then, if a particular food seems to cause the infant discomfort, she can try eliminating that food from her diet for a few days and see if the problem goes away.

The volume of breast milk produced depends on how much milk the baby demands, not on how much fluid the mother drinks. The nursing mother is nevertheless advised to drink at least 2 quarts of liquids each day to protect herself from dehydration. To help themselves remember to drink enough liquid, many women make a habit of drinking a glass of milk, juice, or water each time the baby nurses as well as at mealtimes.

A question often raised is whether a mother's milk may lack a nutrient if she fails to get enough in her diet. The answer differs from one nutrient to the next, but in general, the effect of nutritional deprivation of the mother is to reduce the quantity, not the quality, of her milk. For protein, carbohydrate, and most minerals, the milk of a healthy mother has a fairly constant composition. Any excess water-soluble vitamins

[§] The LaLeche League is an international organization that helps women with breastfeeding concerns. See Appendix E for the address.

CONSUMER CORNER

Formula's Advertising Advantage

THERE IS A STRONG scientific consensus that breastfeeding is preferable for most infants. Why, then, do so many women who could breastfeed their infants choose formula?

Certainly, most women are free to choose whatever feeding method best suits their needs. For only a few is breastfeeding either prohibited for medical reasons or medically indicated for special needs of the infant. Some women may find the time and logistics of breastfeeding burdensome, especially if they work outside the home. For many women, though, the decision to forgo breastfeeding is influenced not only by medical and personal concerns but also by aggressive advertising of formulas.[1] The ads can lead women to believe that formula is just as good for infants as human milk.

Advertisers of infant formulas often strive to create the illusion that formula is identical to human milk. In reality, no formula can match the nutrients, agents of immunity, and environmental information conveyed to infants through human milk. The ads are convincing, though: "Like mother's milk, our formula provides complete nutrition" or "Why trust anything but our brand? It's scientifically formulated to meet your baby's needs." These ads imply, falsely, that breast milk is "unscientific," unknown, and therefore untrustworthy.

To augment their market share, formula sellers give coupons and samples of free formula to pregnant women who are deciding whether to breastfeed. After childbirth, women in the hospital receive "goody bags" with more coupons to tempt them to go and receive their "formula gifts." Later, drugstores dispense still more coupons whenever computerized cash registers ring up items related to breastfeeding, such as pads that protect clothing from milk. And still more coupons arrive by mail a couple of months later, at a time when many women give up breastfeeding, even though authorities urge continued breastfeeding for several more months. Aggressive marketing tactics such as these can undermine a woman's confidence concerning her breastfeeding choice.

Dismissal of new mothers and newborns from hospitals within a day or two after delivery has had a negative impact on breastfeeding. Early releases limit the professional support and lactation education available to new mothers in hospitals within a day or two after delivery. Many hospitals employ certified lactation consultants who specialize in helping new mothers establish a healthy breastfeeding relationship with their newborns. Such counseling can have considerable influence on a woman's decision to initiate and sustain breastfeeding, but with early release, many mothers are lucky to see such a counselor by the time they leave the hospital.[2]

National efforts to promote breastfeeding seem to be working, at least to some extent. An encouraging trend of breastfeeding initiation rates is emerging, with almost 60 percent of women initiating breastfeeding in 1995, up from 50 percent in 1990. Few infants, however, are breastfed beyond about two months of age.[3]

Formula-fed infants in developed nations are generally healthy and usually grow normally, but they miss out on the breastfeeding advantages described in the text.[4] In developing nations, however, the consequence of choosing not to breastfeed can be tragic. Feeding formula is often fatal to the infant in nations where poverty

limits access to formula mixes, clean water is unavailable for safe formula preparation, and medical help is limited. The World Health Organization (WHO) strongly supports breastfeeding for the world's infants in its "baby friendly" initiative and actively opposes the marketing of infant formulas to new mothers.[5] Table 12-9 lists important provisions of WHO's code of ethics for formula makers worldwide.

Women should, of course, be free to choose between breast and bottle, but the decision is important and should be made by carefully weighing valid factual information. The choice should not be influenced by sophisticated advertising ploys created by an industry that profits from formula sales.

Table 12-9

Ten Provisions of the International Code for Marketing Breastmilk Substitutes

- NO advertising of any of these products to the public.
- NO free samples to mothers.
- NO promotion of products, including the distribution of free or low-cost supplies, in health-care facilities.
- NO company sales representatives to advise mothers.
- NO gifts or personal samples to health workers.
- NO words or pictures idealizing artificial feeding, or pictures of infants on labels of infant milk containers.
- Information to health workers should be scientific and factual.
- ALL information on artificial infant feeding, including that on labels, should explain the benefits of breastfeeding and the costs and hazards associated with artificial feeding.
- Unsuitable products, such as sweetened condensed milk, should not be promoted for babies.
- Manufacturers and distributors should comply with the *Code's* provisions even if countries have not adopted laws or other measures.

the mother takes in are excreted in the urine; the body does not release them into the milk. The amounts of fat-soluble vitamins in human milk, however, are affected by the mother's excessive or deficient intakes. For example, large doses of vitamin A correspondingly raise the concentration of this vitamin in breast milk. Vitamin supplementation of undernourished women appears to help normalize the vitamin concentrations in their milk and may be beneficial.

Another question often raised is whether breastfeeding promotes a more rapid loss of the extra body fat accumulated during pregnancy. In one study, researchers found that women who breastfed or combined breastfeeding and formula feeding experienced faster loss of body weight during the first month after delivery than those who fed formula alone.[49] Results of other studies about the relationship between feeding method and loss of body fat and body weight are inconsistent. In most studies where breastfeeding duration was three months or longer, researchers found that lactation accelerated a woman's weight loss.[50] This does not mean that a breastfeeding woman can eat unlimited food and still effortlessly return to prepregnancy weight. Breastfeeding costs energy, true, but carefully chosen programs of diet and physical activity are still the cornerstones of weight control. Physical activity in particular helps to reduce body fatness and improve fitness while having little effect on a woman's milk production or her infant's weight gain.[51] A gradual weight loss (1 pound per week) is safe and does not reduce milk output. Too large an energy deficit, however, especially soon after birth, will inhibit lactation.

The lactating woman needs extra fluid and enough energy and nutrients to make sufficient milk each day. Malnutrition most often diminishes the quantity of the milk produced without altering quality. Lactation facilitates loss of the extra fat gained during pregnancy.

Are There Any Situations in Which a Woman Should Not Breastfeed?

Some substances impair maternal milk production or enter breast milk and interfere with infant development. Some medical conditions prohibit breastfeeding.

Alcohol, Other Drugs, and Environmental Contaminants Alcohol easily enters breast milk and can adversely affect production, volume, composition, and ejection of breast milk as well as overwhelm an infant's immature alcohol-degrading system.[52] Alcohol concentration peaks within one hour after ingestion of even moderate amounts (equivalent to a can of beer). This amount may alter the taste of the milk to the disapproval of the nursing infant, who may, in protest, drink less milk than normal.

Similarly, excess caffeine can make a baby jittery and wakeful. As during pregnancy, caffeine consumption should be moderate. As for cigarette smoke, research shows that lactating women who smoke produce less milk, and milk with a lower fat content, than mothers who do not smoke. Consequently, their infants gain less weight than infants of nonsmokers. Smoking also exerts the numerous harmful effects mentioned earlier on both mother and child.

If a nursing mother must take medication that is secreted in breast milk and is known to affect the infant, then breastfeeding must be put off for the duration of treatment. Meanwhile, the flow of milk can be sustained by pumping the breasts and discarding the milk. Many prescription medications do not reach nursing babies in sufficient quantities to affect them adversely and so have no impact on breastfeeding. Other drugs are not at all compatible with breastfeeding either because they are secreted into the milk and can harm the infant or because they suppress lactation.[53] A nursing mother should consult with the prescribing physician before taking medicines. Drug addicts, including alcohol abusers, are capable of taking such high doses that their infants can become addicts by way of breast milk. In these cases, too, breastfeeding is contraindicated.

Many women wonder about using oral contraceptives during lactation. One type that combines the hormones estrogen and progestin seems to suppress milk output, lower the nitrogen content of the milk, and shorten the duration of breastfeeding. In contrast, progestin-only pills have no effect on breast milk or breastfeeding and are considered appropriate for lactating women.

For more about contaminants and nutrition, turn to Chapter 14.

A woman sometimes hesitates to breastfeed because she has heard that environmental contaminants may enter breast milk and harm her infant. Although some contaminants do enter breast milk, others may be filtered out of the milk. Formula-fed infants consume a great deal of tap water because formula is made with water, so they receive directly any contaminants that may be in the water supply. The decision whether to breastfeed on this basis might best be made after consultation with a physician or dietitian familiar with the local circumstances.

Maternal Illness If a woman has an ordinary cold, she can go on nursing without worry. The infant will probably catch it from her anyway, and thanks to immunological protection, a breastfed baby may be less susceptible than a formula-fed baby would be. If a woman has a serious communicable disease such as tuberculosis or hepatitis, then mother and baby have to be separated. Breastfeeding may be continued by pumping the mother's breasts several times a day and letting the baby drink the milk from a bottle (see the margin for tips for safe handling).

For safe breast milk storage:

- Wash hands thoroughly before pumping.
- Clean pumping equipment according to manufacturer's directions.
- Sterilize bottles, nipples, and rings before using.
- Refrigerate milk to be fed within 48 hours.
- Freeze milk to be stored longer than 48 hours.
- Thaw milk gently on defrost cycle of microwave or in refrigerator.
- Do not refreeze thawed milk.

The human immunodeficiency virus (HIV), responsible for causing AIDS, can be passed from an infected mother to her infant during pregnancy, at birth, or through breastfeeding, so women in developed countries who have tested positive for HIV should not breastfeed. They should choose a safe alternative feeding method, such as breast milk from a milk bank.[54] Milk banks in the United States pasteurize donated human milk and make it available to infants who lack access to milk from their own mothers. Pasteurization destroys harmful organisms, such as HIV, but leaves intact most of the beneficial constituents of the milk.[55]

In developing countries, the feeding of inappropriate or contaminated formulas causes 1.5 million infant deaths each year, so WHO and UNICEF urge mothers to breastfeed irrespective of HIV infection. For these infants, the protection of being breastfed outweighs the risk of HIV transmission.[56]

KEY POINT *Breastfeeding may be inadvisable if milk is contaminated with alcohol, drugs, or environmental pollutants. Most ordinary infections such as colds have no effect on breastfeeding.*

Feeding the Infant

For the first four to six months of life, the infant drinks only breast milk or formula, but later becomes able to handle other foods. Early nutrition affects later development, and early feedings establish eating habits that influence nutrition throughout life.

Trends change and experts argue the fine points, but nourishing a baby is relatively simple. Common sense in the selection of infant foods and a nurturing, relaxed environment go far to promote the infant's well-being.

Nutrient Needs

A baby grows faster during the first year of life than ever again, as Figure 12-8 shows. Pediatricians carefully monitor the growth of infants and children because growth directly reflects their nutrition status. The infant's birthweight doubles around four months of age and triples by the age of one year. (If a 150-pound adult were to grow like this, the person's weight would increase to 450 pounds in a single year.) By the end of the first year, the growth rate slows considerably.

The rapid growth and metabolism of the infant demand an ample supply of all the nutrients. Of special importance during infancy are the energy nutrients and the vitamins and minerals critical to the growth process, such as vitamin A, vitamin D, calcium, and iron.

Because they are small, babies need smaller total amounts of these nutrients than adults do, but as a percentage of body weight, babies need over twice as much of most nutrients. Infants require about 100 calories per kilogram of body weight per day; most adults require fewer than 40. Figure 12-9 compares a five-month-old baby's needs (per unit of body weight) with those of an adult man. As you can see, some of the differences are extraordinary. Sometime around six months of age, energy needs begin to increase less rapidly as the growth rate begins to slow down, but some of the energy saved by slower growth is spent in increased activity. When their growth slows, infants spontaneously reduce their energy intakes. Thus parents should expect their babies to adjust their food intakes downward when appropriate and should not force or coax them to eat more.

Figure 12-8

Weight Gain of Human Infants and Children In the First Five Years of Life

The colored vertical bars show how the yearly increase in weight gain slows its pace over the years.

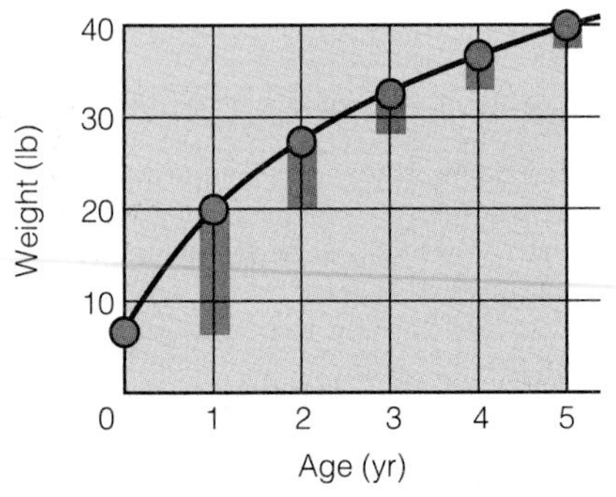

40
30
20
10
Weight (lb)
0
1
2
3
4
5
Age (yr)

Vitamin K nutrition for newborns presents a unique case. A newborn's digestive tract is sterile, and vitamin K–producing bacteria take weeks to establish themselves in the baby's intestines. To prevent uncontrolled bleeding in the newborn, the AAP recommends that a single dose of vitamin K be given at birth.[57]

The most important nutrient of all, for infants as for everyone, is the one easiest to forget: water. The younger a child is, the more of its body weight is water and the faster the water is lost and replaced. Proportionately more of an infant's body water than an adult's is between the cells and in the vascular space, and this water is easy to lose. Conditions that cause fluid loss, such as hot weather, vomiting, diarrhea, or sweating, can rapidly propel an infant into life-threatening dehydration.

Figure 12-9

Nutrient Recommendations for a Five-Month-Old Infant and an Adult Male Compared on the Basis of Body Weight

Infants may be relatively small and inactive, but they use large amounts of energy and nutrients in proportion to their body size to keep all their metabolic processes going.

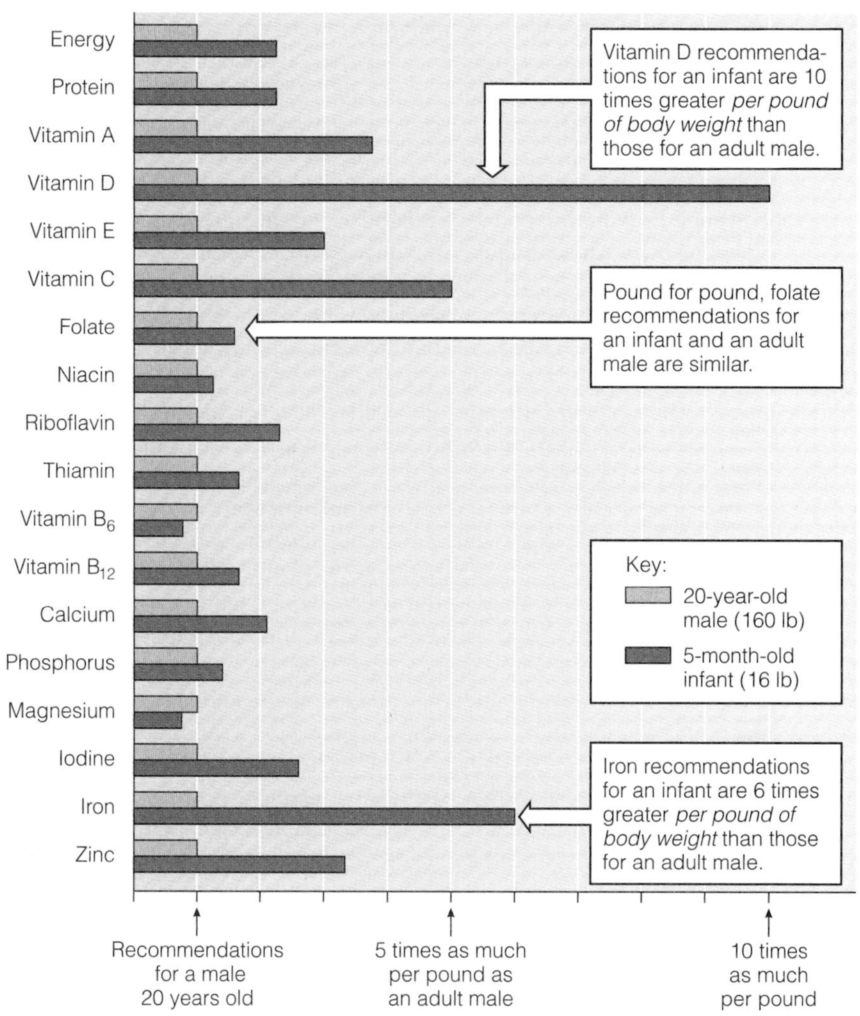

Energy
Protein
Vitamin A
Vitamin D
Vitamin E
Vitamin C
Folate
Niacin
Riboflavin
Thiamin
Vitamin B6
Vitamin B12
Calcium
Phosphorus
Magnesium
Iodine
Iron
Zinc
Vitamin D recommendations for an infant are 10 times greater per pound of body weight than those for an adult male.
Pound for pound, folate recommendations for an infant and an adult male are similar.
Key:
20-year-old male (160 lb)
5-month-old infant (16 lb)
Iron recommendations for an infant are 6 times greater per pound of body weight than those for an adult male.
Recommendations for a male 20 years old
5 times as much per pound as an adult male
10 times as much per pound

In early infancy, breast milk or infant formula normally provides enough water for a healthy infant to replace water losses from the skin, lungs, feces, and urine. When the older infant starts eating solid foods, additional water is required.

KEY POINT *Infants' rapid growth and development depend heavily on adequate nutrient supplies. Adequate water is also crucial.*

Formula Feeding and Weaning to Milk

The type of milk the infant receives and the age at which solid foods are introduced are major areas of concern in infant nutrition research. Under most circumstances, a woman can freely choose to feed breast milk or formula; either one will meet the infant's nutrient needs. If the family has a low income, however, or if other factors threaten the baby's health, then the advantages of breastfeeding tip the balance in its favor.

The substitution of formula feeding for breastfeeding involves striving to copy nature as closely as possible. Human milk and cow's milk differ; cow's milk is significantly

After six months of age, the energy saved by slower growth is spent on increased activity.

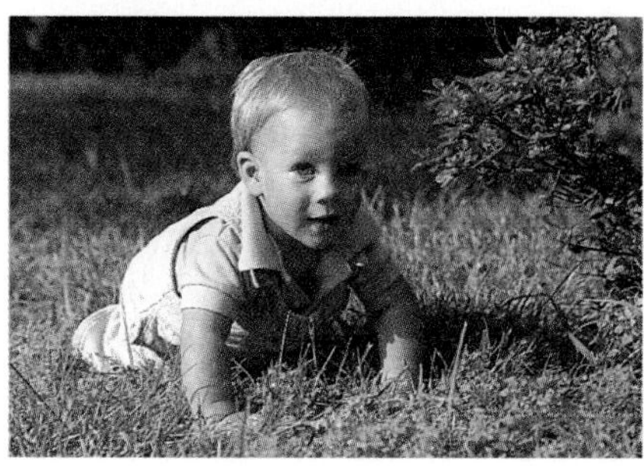

Formula options:

- Liquid concentrate (inexpensive, relatively easy)—mix with equal part water.
- Powdered formula (cheapest, lightest for travel)—follow label directions.
- Ready-to-feed (easiest, most expensive)—pour directly into clean bottles.
- Never an option—whole cow's milk before 12 months of age.

higher in protein, calcium, and phosphorus, for example, to support the calf's faster growth rate. A formula can be prepared from cow's milk that does not differ significantly from human milk in these respects; the formula makers first dilute the milk and then add carbohydrate and nutrients to make the proportions comparable to those of human milk (see Table 12-10 for a comparison of human milk and standard formulas).

Standard formulas are inappropriate for some infants. For example, premature babies require special formulas. Infants allergic to milk protein can drink special formulas based on soy protein. Soy formulas are lactose-free and so can be used for infants with lactose intolerance as well. They are also useful as an alternative to milk-based formulas for vegetarian families. For infants with other special needs, many other variations are available.

The AAP recommends iron-fortified formulas for all formula-fed infants. Low-iron formulas have no role in infant feeding. Use of iron-fortified formulas has risen in recent decades and is credited with the decline of iron-deficiency anemia in U.S. infants.

Formula feeding offers an acceptable alternative to breastfeeding. Nourishment for the infant from formula is adequate, and a mother can choose this course with confidence. One advantage is that parents can see how much milk the baby drinks during feedings. Also, other family members can participate in feeding sessions, giving them a chance to develop the special closeness that feeding fosters. Mothers who resume employment early after giving birth may choose formula for their infants, but they have another option. Breast milk can be pumped into bottles and given to the baby in day care. At home, mothers may breastfeed as usual. Many mothers use both methods—they breastfeed at first but wean to formula later on.

For as long as breast milk or formula is the baby's major food, unmodified cow's milk is an inappropriate replacement, primarily because milk provides little iron and vitamin C. Plain, unmodified cow's milk (including whole, reduced-fat, low-fat, fat-free, or evaporated milk) is not recommended before the baby's first birthday. The infant's digestive tract may be sensitive to the protein content and, if so, may bleed and worsen iron deficiency. Thus cow's milk both causes iron loss and fails to replace iron. Also, the infant's immature kidneys are stressed by plain cow's milk. Once the baby is obtaining at least two-thirds of total daily food energy from a balanced mixture of cereals, vegetables, fruits, and other foods (usually after 12 months of age), whole cow's milk, fortified with vitamins A and D, is an acceptable accompanying beverage. Reduced-fat milk is not recommended before the age of two years. Table 12-11 defines some terms applied to types of milk.

The infant thrives on formula offered with affection.

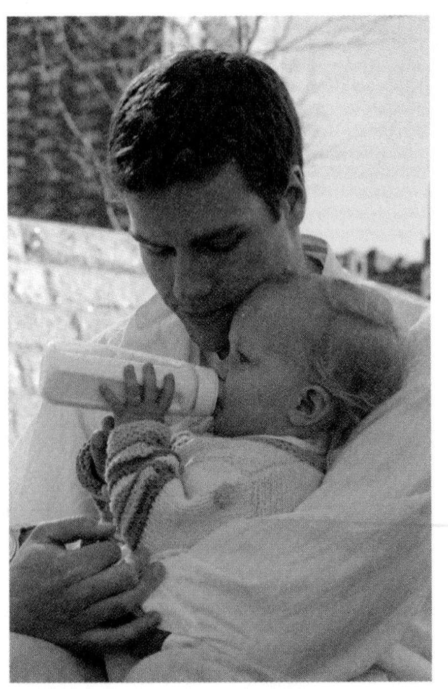

KEY POINT *Infant formulas are designed to resemble breast milk and must meet an AAP standard for nutrient composition. Special formulas are available for premature babies, allergic babies, and others. Formula should be replaced with milk only after the baby's first birthday.*

Table 12-10

Human Milk Compared with Infant Formula for Selected Nutrients

Content	Mature Human Milk	Fortified Infant Formula
Energy (cal/100 ml)	64	67
Protein (% of cal)	6	9
Fat (% of cal)	40–50	50
Carbohydrate (% of cal)	41	42
Iron (mg/L)	0.5	1.5–12
Vitamin A (µg/L)	675	660
Niacin (mg/L)	1.5	7.5
Vitamin D (µg/L)	2.2	41
Inositol (mg/L)	149	32

SOURCE: American Academy of Pediatrics, Committee on Nutrition, *Pediatric Nutrition Handbook*, 3rd ed., ed. L. A. Barness (Elk Grove, Ill.: American Academy of Pediatrics, 1993), Appendix E.

An Infant's First Foods

Foods can be introduced into a baby's diet as the baby becomes physically ready to handle them. This readiness develops in stages. A newborn baby can swallow only liquids that are well back in the throat. Later (at four months or so), the baby's tongue can move against the palate to swallow semisolid food such as cooked cereal. Still later, the first teeth erupt, but not until sometime during the second year can a baby begin to handle chewy food. The stomach and intestines are immature at first; they can digest milk sugar (lactose) but not starch. At about four months, most babies can begin to digest starchy foods.

The Need for Water The baby's kidneys are unable to concentrate waste efficiently, so a baby must excrete relatively more water than an adult to carry off a comparable amount of waste. This means that the risk of dehydration is higher for infants than for adults, and it becomes even greater once solid foods are introduced. Foods high in protein or electrolytes such as meat and eggs can promote dehydration if offered without water. Water should be offered to infants regularly once they are eating solid food. If the weather is hot and the mother is thirsty, her infant probably is, too. Infants cannot tell you what they are crying for; remember that they may need plain water, and let them drink it until they quench their thirst.

When to Introduce Solid Food The timing for adding solid foods to a baby's diet depends on several factors. Formula or breast milk alone is sufficient until age four to six months. Babies who are ready for solid foods thrive on receiving them and develop new skills through handling the foods. Indications of readiness for solid foods include:

- The infant can sit with support and can control its head movements.
- The infant is about six months old.

All babies develop according to their own schedules, and although Table 12-12 on the next page presents a suggested sequence, individuality is important. Three considerations are relevant: the baby's nutrient needs, the baby's physical readiness to handle different forms of foods, and the need to detect and control allergic reactions. With respect to nutrient needs, the nutrient needed most is iron, then vitamin C.

Foods to Provide Iron and Vitamin C Iron deficiency is prevalent in children between the ages of six months and three years due to their rapid growth rate and the significant place that milk has in their diets. Excessive milk consumption (more than 3½ cups a day) displaces iron-rich foods and can lead to iron-deficiency anemia, popularly called **milk anemia**.

Iron ranks highest on the list of nutrients most needing attention in infant nutrition. A baby's stored iron supply from before birth runs out after the birthweight doubles, long before the end of the first year. Breast milk or formula with iron for formula-fed babies, then iron-fortified cereals, and then meat or meat alternates such as legumes are recommended. Once babies are eating iron-fortified cereals, parents or caregivers should begin selecting vitamin C–rich foods to go with meals to enhance absorption. The best sources of vitamin C are fruits and vegetables.

Fruit juices should be diluted and served in cup, not a bottle, once the infant is six months of age or older. Juices should be used moderately, so as not to displace other foods.

Physical Readiness for Solid Foods Foods introduced at the right times contribute to an infant's physical development. For example, experience with solid food at four to six months, when swallowing ability is developing, helps to desensitize the gag reflex. When the baby can sit up, can handle finger foods, and is teething, hard crackers and other hard finger foods may be introduced under the watchful eye of an adult. These foods promote the development of manual dexterity and control of

milk anemia iron-deficiency anemia caused by drinking so much milk that iron-rich foods are displaced from the diet.

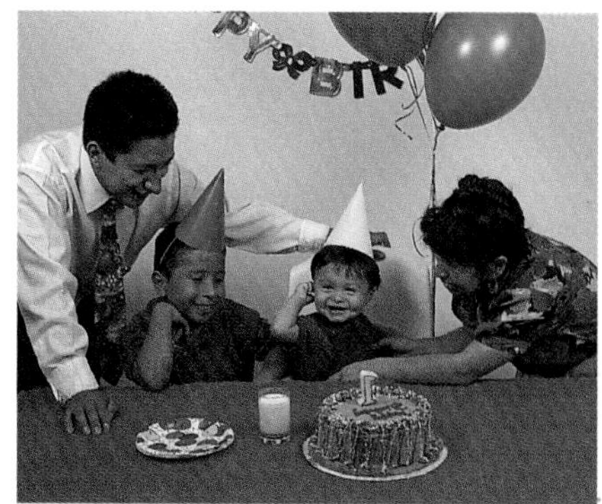

A first birthday party and the possibility of tasting whole, unmodified cow's milk for the first time.

Table 12-11

Milk Terms

- **casein** or **sodium caseinate** the principal protein of cow's milk. Another milk protein found in human milk's whey is **lactalbumin**.
- **evaporated milk** milk concentrated to half volume by evaporation. Adding water reconstitutes the milk; the taste is altered by the processing, however.
- **homogenized milk** milk treated to mix the fat evenly with the watery part (fat ordinarily floats to the top as cream). Heated milk is forced under high pressure through small openings to emulsify the fat.
- **pasteurized milk** milk that is heat treated to eliminate disease-causing microbes and to reduce its total bacterial count to an acceptable level.
- **powdered milk** dehydrated milk solids. Some powdered milks rehydrate easily (instant milk); others require extensive blending. Both whole and fat-free milk can be powdered.
- **whey** the liquid that remains after milk has coagulated (see also *casein*).
- **whole milk** full-fat cow's milk.

Table 12-12

Infant Feeding Skills and Recommended Foods

Age (mo)	Feeding Skill	Foods Introduced into the Diet
0–4	Turns head toward any object that brushes cheek. Initially swallows using back of tongue; gradually begins to swallow using front of tongue as well. Strong reflex (extrusion) to push food out during first 2 to 3 months.	Feed breast milk or infant formula.
4–6	Extrusion reflex diminishes, and the ability to swallow nonliquid foods develops. Indicates desire for food by opening mouth and leaning forward. Indicates satiety or disinterest by turning away and leaning back. Sits erect with support at 6 months. Begins chewing action. Brings hand to mouth. Grasps objects with palm of hand.	Begin iron-fortified cereal mixed with breast milk, formula, or water. Begin pureed vegetables and fruits.
6–8	Able to feed self with fingers. Develops pincher (finger to thumb) grasp. Begins to drink from cup.	Begin breads and other cereals and mashed vegetables and fruits. Begin plain, unsweetened fruit juices from cup.
8–10	Begins to hold own bottle. Reaches for and grabs food and spoon. Sits unsupported.	Begin yogurt. Begin pieces of soft, cooked vegetables and fruit from table. Gradually begin finely cut meats, fish, casseroles, cheese, eggs, and legumes.
10–12	Begins to master spoon, but still spills some.	Include at least 4 servings of breads and cereals from table, in addition to infant cereal; at least 2 servings of fruits and 3 servings of vegetables; and 2 servings of meat, fish, poultry, eggs, or legumes.[a]

[a]Serving sizes for infants and young children are smaller than those for an adult. For example, a serving might be ½ slice of bread instead of 1 slice, or ¼ cup rice instead of ½ cup.

SOURCE: Adapted in part from American Academy of Pediatrics, Committee on Nutrition, *Pediatric Nutrition Handbook*, 3rd ed., ed. L. A. Barness (Elk Grove Village, Ill.: American Academy of Pediatrics, 1993), pp. 23–33.

Foods such as iron-fortified cereals and formulas, mashed legumes, and strained meats provide iron.

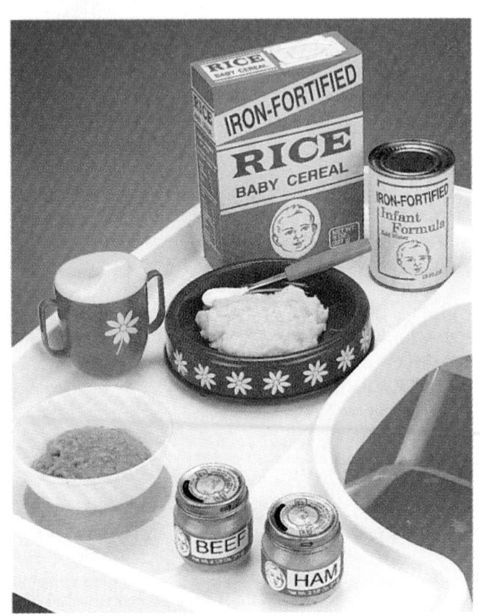
RICE
IRON-FORTIFIED
RICE
BABY CEREAL
IRON-FORTIFIED
Infant
Formula
BEEF
HAM

the jaw muscles, but the caregiver must make sure that the infant does not choke on them. Hard crackers that melt slowly to a mush that is easy to swallow are best. Babies and even young children can easily choke on popcorn, nuts, hot dogs, raw carrots, whole grapes, and hard candy; these foods are not worth the risk.

Some parents want to feed solids as early as possible on the theory that "stuffing the baby" at bedtime will promote sleeping through the night. There is no proof for this theory. Babies start to sleep through the night whenever they are ready, no matter when solid foods are introduced. By three months, most are sleeping adequately regardless.

Food Allergies New foods should be introduced one at a time so that allergies or other sensitivities can be detected. For example, when fortified baby cereals are introduced, try rice cereal first for several days; it causes allergy least often. Try wheat-containing cereal last; it is a common offender. Egg whites, soy products, peanut products, cow's milk, and citrus fruits are introduced still later for the same reason. If a food causes an allergic reaction (irritability due to skin rash, digestive upset, or respiratory discomfort), discontinue its use before going on to the next food. About nine times out of ten, the allergy won't be evident immediately but will manifest itself in vague symptoms occurring up to five days after the offending food is eaten. Wait a month or two to try the food again; many sensitivities disappear with maturity. If the family history indicates allergies, apply extra caution in introducing new foods. Parents or caregivers who detect allergies early in an infant's life can spare the whole family much grief.

Chapter 13 offers more information on allergies.

Choice of Infant Foods Baby foods commercially prepared in the United States and Canada are safe, and except for mixed dinners with added starch fillers and heavily sweetened desserts, they generally have high nutrient density. Brands vary in their use of starch and sugar—the ingredient lists distinguish one from another. Parents or caregivers should not feed directly from the jar but should remove portions to a dish for feeding so as not to contaminate the unused food that will be stored in the jar.

Appendix A includes the nutrient composition of many commercial baby foods.

An alternative to commercial baby food is to process a small portion of the family's table food in a blender, food processor, or baby food grinder. This necessitates cooking without salt or sugar, though, as the best baby food manufacturers do. The adults can season their own food after taking out the baby's portion. Pureed food can be frozen in an ice cube tray to yield a dozen or so servings that can be quickly thawed, heated, and served on a busy day.

Foods to Omit Sweets of any kind (including baby food "desserts") have no place in a baby's diet. The added food energy they contribute can promote obesity, and they convey few or no nutrients to support growth. Canned vegetables are inappropriate for babies; they often contain too much salt. Also, awareness of food poisoning and precautions against it are imperative. Honey should never be fed to infants because of the risk of botulism.

Chapter 14 provides details about botulism.

Foods at One Year For the baby weaned to whole milk after one year of age, whole milk can supply most of the nutrients the infant needs; 2 to 3½ cups a day meet those needs. Other foods—meat and meat alternates, iron-fortified cereal, enriched or whole-grain bread, fruits and vegetables—should be supplied in variety and in amounts sufficient to round out total energy needs. Ideally, the one-year-old sits at the table, eats many of the same foods everyone else eats, and drinks liquids from a cup, not a bottle. A meal plan that meets the requirements for a one-year-old is shown in Table 12-13.

KEY POINT *Solid food additions to a baby's diet should begin at about six months and should be governed by the baby's nutrient needs and readiness to eat. By one year, the baby should be receiving foods from all food groups.*

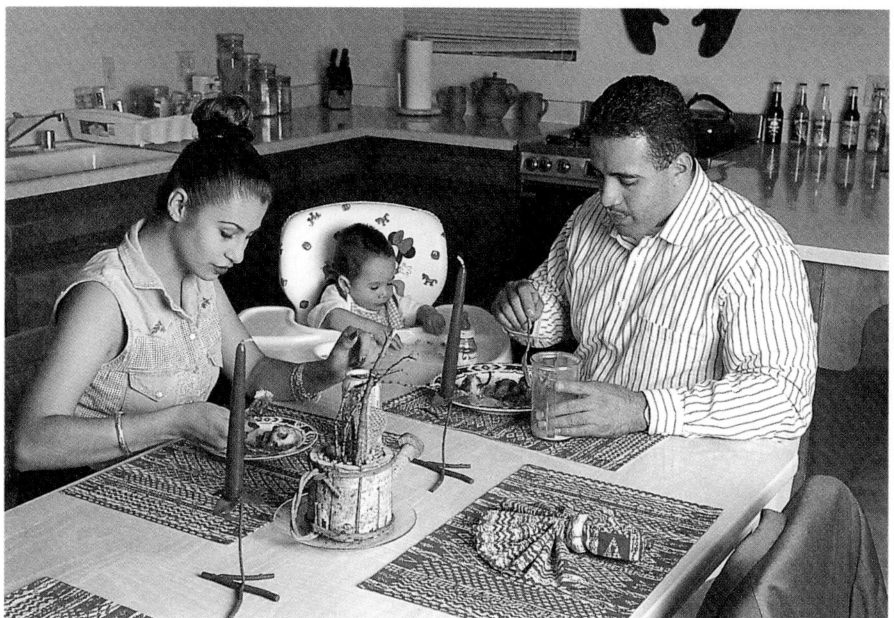

Children love to eat what their families eat.

Table 12-13

Meal Plan for a One-Year Old

Breakfast	Afternoon snack
½ c whole milk	½ c whole milk
½ c iron-fortified cereal	½ slice toast
½ c orange juice	1 tbs peanut butter
Morning snack	**Dinner**
½ c yogurt	1 c whole milk
1 to 2 tbs fruit[a]	2 oz chopped meat or well-cooked mashed legumes
Lunch	½ c potato, rice, or pasta
½ c whole milk	2 to 3 tbs vegetables[b]
2 to 3 tbs vegetables[b]	2 to 3 tbs fruit[a]
1 egg or ¼ c tofu	
½ c noodles	

[a]Include citrus fruits, melons, and berries.
[b]Include dark green, leafy, and deep yellow vegetables.

Looking Ahead

The first year of a baby's life is the time to lay the foundation for future health. From the nutrition standpoint, the problems most common in later years are obesity and dental disease. Prevention of obesity may also help prevent the obesity-related diseases: atherosclerosis, diabetes, and cancer.

Probably the most important single measure to undertake during the first year is to encourage eating habits that will support continued normal weight as the child grows. This means introducing a variety of nutritious foods in an inviting way, not forcing the baby to finish the bottle or baby food jar, avoiding concentrated sweets and empty-calorie foods, and encouraging physical activity. Parents should not teach babies to seek food as a reward, to expect food as comfort for unhappiness, or to associate food deprivation with punishment. If they cry for thirst, give them water, not milk or juice. If they cry for companionship, pick them up, don't feed them. If they are hungry, by all means, feed them appropriately. More pointers are offered in this chapter's Food Feature.

Nursing bottle syndrome in an early stage.

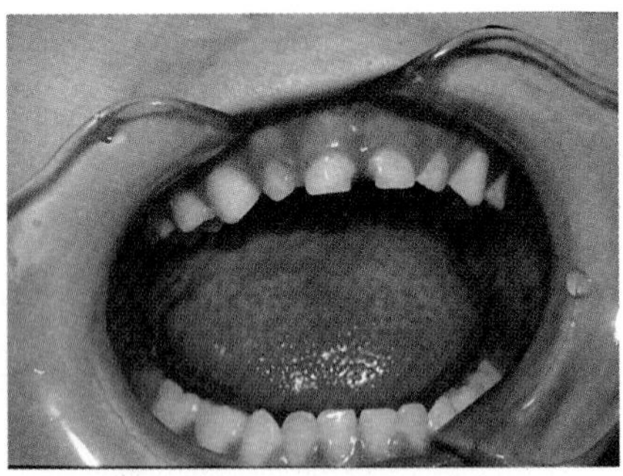

An irrational fear of obesity leads some parents to underfeed their infants, depriving them of the energy and nutrients they need to grow. Others wonder if they should feed their infants a low-fat diet to reduce heart disease risk, but the AAP recommends fat intake of 40 to 50 percent of total calories for infants. A diet too low in fat often hinders growth and development even when energy from carbohydrate and protein is ample. With rare exceptions, to be identified by physicians, babies from age one to two years need the food energy and fat of whole milk. They also need frequent servings of food containing the essential fatty acids.

Nursing bottle syndrome, an extreme example. The lower teeth have decayed all the way to the gum line.

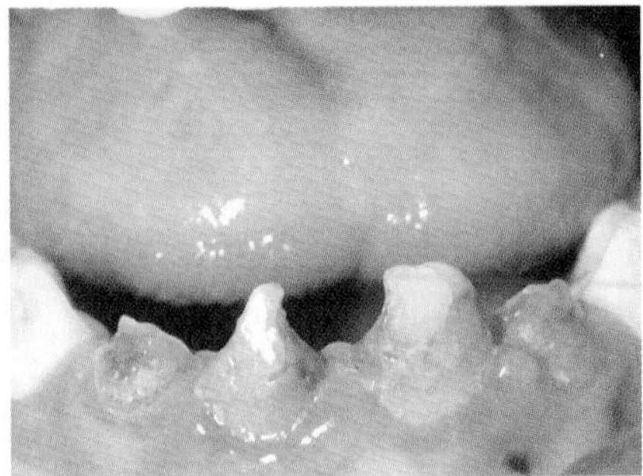

The same strategies promote normal dental development: supplying nutritious foods, avoiding sweets, and discouraging the association of food with reward or comfort. In addition, dentists strongly discourage the practice of giving a baby a bottle as a pacifier. Sucking for long periods of time pushes the normal jawline out of shape and causes a bucktoothed profile: protruding upper and receding lower teeth. Furthermore, prolonged sucking on a bottle of milk or juice bathes the upper teeth in a carbohydrate-rich fluid that favors the growth of decay-producing bacteria. The bacteria produce acid that dissolves tooth material. Babies regularly put to bed with a bottle sometimes have teeth that are decayed all the way to the gum line, a condition known as nursing bottle syndrome, shown in the margin photos.

KEY POINT *The early feeding of the infant lays the foundation for lifelong eating habits. It is desirable to foster preferences that will support normal development throughout life and that may help to ward off common lifestyle diseases.*

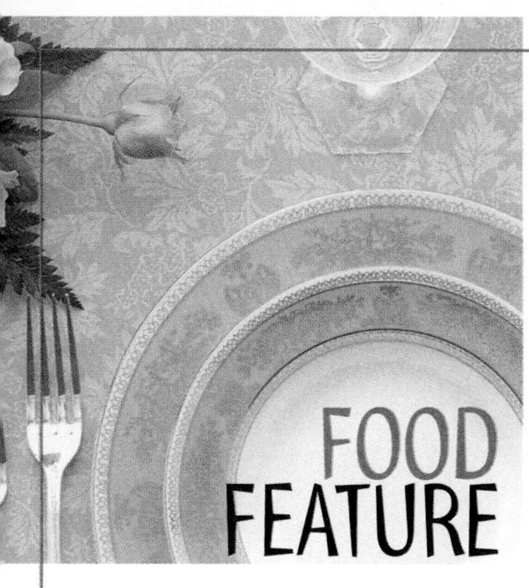
FOOD
FEATURE

Mealtimes with Infants

THE WISE PARENT or caregiver of a one-year-old offers nutrition and affection together. It is literally true that "feeding with love" produces better growth in both weight and height than feeding the same food in an emotionally negative climate.

Foster a Sense of Autonomy

The person feeding a one-year-old has to be aware that the child's exploring and experimenting are normal and desirable behaviors. The child is developing a sense of autonomy that, if allowed to develop, will provide the foundation for later assertiveness in choosing when and how much to eat and when to stop eating. The child's self-direction, if consistently overridden, can later turn into shame and self-doubt.

Some Feeding Guidelines

In light of the developmental and nutrient needs of one-year-olds and in the face of their often contrary and willful behavior, a few feeding guidelines may be helpful:

- *Discourage unacceptable behavior (such as standing at the table or throwing food) by removing the child from the table to wait until later to eat.* Be consistent and firm, not punitive. The child will soon learn to sit and eat.
- *Let the child explore and enjoy food.* This may mean the child eats with fingers for a while. Use of the spoon will come in time.
- *Don't force food on children.* Provide children with nutritious foods, and let them choose which ones and how much they will eat. Gradually, they will acquire a taste for different foods. If children refuse milk, provide cheese, cream soups, and yogurt.
- *Limit sweets strictly.* Infants have little room in their 1,000-calorie daily energy allowance for empty-calorie sweets, except occasionally.

These recommendations reflect a spirit of tolerance that serves the best interest of the infant emotionally as well as physically. This attitude, carried throughout childhood, helps the child to develop a healthy relationship with food. The next chapter finishes the story of growth and nutrition.

WITHDRAWN
FROM
UNIVERSITIES
AT
MEDWAY
LIBRARY

Self Check

Answers to these Self Check questions are in Appendix G.

1. Only a slightly increased intake is recommended during pregnancy for:
   a. folate
   b. iron
   c. energy
   d. protein
2. A deficiency of which nutrient appears to be related to an increased risk of neural tube defects in the newborn?
   a. vitamin B_6
   b. folate
   c. calcium
   d. niacin
3. Which of the following may be hazardous during pregnancy?
   a. vitamin A supplementation
   b. dieting for weight loss
   c. drinking alcohol
   d. all of the above
4. Breastfed infants may need supplements of:
   a. fluoride, iron, and vitamin D
   b. zinc, iron, and vitamin C
   c. vitamin E, calcium, and fluoride
   d. vitamin K, magnesium, and potassium
5. Breastfeeding is contraindicated if a woman has:
   a. AIDS
   b. hepatitis
   c. tuberculosis
   d. all of the above
6. A major reason why a woman's nutrition before pregnancy is crucial is that it determines whether her uterus will support the growth of a normal placenta. T F
7. Fetal alcohol syndrome (FAS) is the leading known cause of mental retardation in the world. T F
8. In general, the effect of nutritional deprivation on a breastfeeding mother is to reduce the quality of her milk. T F
9. A sure way to get a baby to sleep through the night is to feed solid foods as soon as the baby can swallow them. T F
10. Caffeine seems relatively harmless to normal adults when used in moderation (the equivalent of two average-sized cups of coffee a day). (Read about this in the upcoming Controversy.) T F

NUTRITION ON THE NET

For further study of the topics of this chapter, access these websites and search for the phrases or words in quotation marks:

1. For the latest changes in chapter material, to find supplemental learning tools, or access links to related websites, go to the "Nutrition: Concepts and Controversies" site at: **www.wadsworth.com/nutrition/prod/allprod.html/**
2. Search the Mayo Clinic Children's Health Center site for additional information on "pregnancy": **www.mayohealth.org**
3. Visit the U.S. Government website to learn more about "birth defects," "pregnancy," "adolescent pregnancy," "maternal and child health," "breastfeeding," "infants," "baby bottle tooth decay," or "premature birth": **www.healthfinder.gov/searchoptions/topicsaz.htm**
4. The March of Dimes provides information about birth defects: **www.modimes.org**
5. For information about food assistance programs: **www.fns.usda.gov/fns**
6. Search the American Dietetic Association's site to learn more about "pregnancy": **www.eatright.org**
7. Visit the American College of Obstetricians and Gynecologists site for additional information about nutrition during pregnancy: **www.acog.org**
8. Additional information about fetal alcohol syndrome can be located at: **www.nofas.org**
9. The American Diabetes Association's site offers information about "gestational diabetes": **www.diabetes.org**
10. Visit the American Academy of Allergy, Asthma, and Immunology site for more information on food allergies: **www.aaaai.org**
11. Search the LaLeche League International website for additional information about "breastfeeding": **www.lalecheleague.org**
12. Search among thousands of current scientific and medical abstracts for any topic related to pregnancy and lactation at: **www.ncbi.nlm.nih.gov/PubMed/**

MEDICINES, OTHER DRUGS, AND NUTRITION

CONTROVERSY 12

A 45-YEAR-OLD Chicago business executive attempts to give up smoking with the help of nicotine gum. She decides to replace smoking breaks with beverage breaks and begins drinking frequent servings of tomato juice, coffee, and colas. She is discouraged when her stomach becomes upset and her craving for tobacco continues unabated despite the nicotine gum. Problem: nutrient-drug interaction.

A 14-year-old girl develops frequent and prolonged respiratory infections. Over the past six months, she has suffered constant fatigue despite adequate sleep, has had trouble completing school assignments, and has given up playing volleyball because she runs out of energy on the court. During the same six months, she has been taking huge doses of antacid pills each day because she heard this was a sure way to lose weight. Her pediatrician has diagnosed iron-deficiency anemia. Problem: nutrient-drug interaction.

A 30-year-old schoolteacher who takes antidepressant medication attends a faculty wine and cheese party. After sampling the cheese with a glass or two of red wine, his face becomes flushed. His behavior prompts others to drive him home. In the early morning hours, he awakens with severe dizziness, a migraine headache, vomiting, and trembling. An ambulance delivers him to an emergency room where a physician takes swift action to save his life. Problem: nutrient-drug interaction.

MEDICINES AND NUTRITION

People sometimes think that medical drugs do only good, not harm. As the opening stories illustrate, both prescription and over-the-counter (OTC) medicines can and do cause harm when they interact with the body's normal use of nutrients. Alcohol is infamous for its interactions with nutrients, but other drugs also interact with nutrition in the following ways:

- Foods can delay or prevent drug absorption.
- Drugs can modify taste, appetite, or food intake.
- Drugs can delay or prevent nutrient absorption.
- Nutrients can interfere with drug action, metabolism, or excretion.
- Drugs can interfere with nutrient action or excretion.

Not all the interactions discussed occur every time a person takes a drug. Some people are more vulnerable than others to nutrient-drug interactions. The potential for undesirable nutrient-drug interactions is greatest for those who:

- Take drugs (or medicines) for long times.
- Take two or more drugs at the same time.
- Are poorly nourished to begin with or are not eating well.

Absorption of Drugs and Nutrients The business executive described earlier felt the effects of the type of interaction mentioned first in the list above. Acid from tomato juice, coffee, and colas she drank before chewing nicotine gum kept the nicotine from being absorbed through the lining of her mouth. With this route blocked, the nicotine traveled to her stomach, remained unabsorbed, and caused nausea. Other foods and beverages can have similar effects (see Table C12-1). Once identified, the problem is easy to prevent by waiting to eat or drink until after chewing the gum.

Drugs can also interfere with the small intestine's absorption of nutrients, particularly minerals. This explains the experience of the tired 14-year-old. Her overuse of antacids eliminated the stomach's normal acidity, on which iron absorption depends. The medicine bound tightly to the iron molecules, forming an insoluble, unabsorbable complex. Her iron stores already bordered on deficiency, as iron stores for young girls typically do, so her misuse of antacids pushed her over the edge into frank deficiency.

Chronic laxative use can also lead to malnutrition. Laxatives can carry nutrients through the intestines so rapidly that many vitamins have no time to be absorbed. The laxative mineral oil, which the body cannot absorb, can rob a person of fat-soluble vitamins. Vitamin D deficiencies can occur this way; calcium, too, may be excreted with the oil, accelerating adult bone loss.

Table C12-1

Foods and Beverages That Limit the Effectiveness of Nicotine Gum

- Apple juice
- Beer
- Coffee
- Colas
- Grape juice
- Ketchup
- Lemon-lime soda
- Mustard
- Orange juice
- Pineapple juice
- Soy sauce
- Tomato juice

Metabolic Interactions and Nutrient Excretion

As for the teacher who landed in the emergency room, he was taking an antidepressant medicine, one of the monoamine oxidase inhibitors (MAOI). At the party, he suffered a dangerous chemical interaction between the medicine and the compound tyramine in his cheese and wine. Tyramine is produced during the fermenting process in cheese and wine manufacturing.

The MAOI medication works by depressing the activity of enzymes that destroy the brain neurotransmitter dopamine. With less enzyme activity, more dopamine is left, and depression lifts. At the same time, the drug also depresses enzymes in the liver that destroy tyramine. Ordinarily, the man's liver would have quickly destroyed the tyramine from the cheese and wine. But due to the MAOI medication, tyramine built up too high in the man's body and caused the potentially fatal reaction.

Other culprits that affect the metabolism of medication include grapefruit juice and one of the most popular herbal supplements in the United States, Ginkgo biloba. Something in the grapefruit juice suppresses an enzyme responsible for breaking down more than 20 kinds of medical drugs.[1] With less drug breakdown, doses build up in the blood to levels that can have undesirable effects on the body. For example, in a drinker of grapefruit juice, a normal dosage of the blood-thinning drug coumarin can lead to dangerously prolonged bleeding and delayed clotting of blood. As for Ginkgo biloba, many people take it as a supplement in hopes of improving memory, but this effect is unproved. Takers may not know that it has been found to stimulate the activity of liver enzymes responsible for metabolizing many medications, and so may diminish their effects.[2]

Foods can slow down the absorption of drugs in the digestive tract.

DIGESTIVE
TRACT

Drugs often cause nutrient losses, too. Many people take large quantities of aspirin, easily 10 to 12 tablets each day, to relieve the pain of arthritis, backaches, and headaches. This much aspirin can speed up blood loss from the stomach by as much as ten times, enough to cause iron-deficiency anemia in some people. People who take aspirin regularly should make sure they eat iron-rich foods regularly as well. Table C12-2 lists some examples of other possible nutrient-drug interactions, including both prescription and OTC medications. Some details of the more common interactions follow.

Oral Contraceptives and Estrogen Millions of women use oral contraceptives, daily doses of hormones that prevent pregnancy. The case of the oral contraceptives illustrates that interactions between drugs and nutrients can be complex.

Each nutrient responds differently to oral contraceptive use (see Table C12-2). The vitamin B_{12} status of oral contraceptive users may be slightly lower than in others.[3] Beta-carotene values may also be reduced, leading researchers to wonder whether the lower levels of this antioxidant might influence some disease risks.[4] Vitamin D levels, on the other hand, may be higher in oral contraceptive users, with unknown effects.[5] At first glance these findings might seem to indicate that women using oral contraceptives are on their way to suffering deficiencies of some nutrients and have somehow enlarged their body stores of others. The research in the area has yielded conflicting results, however, so any such assumptions would be premature.

Significantly, oral contraceptives alter blood lipids, possibly making cardiovascular disease more likely to occur in menstruating women.[6] This effect poses very little risk for young healthy women who do not smoke.[7] Beyond about age 35, however, most oral contraceptives raise total cholesterol and triglyceride concentrations and lower HDL, amplifying the risk of stroke and heart disease. A few women using oral contraceptives also experience mild hypertension.

Some women lose weight when taking oral contraceptives, but others may gain as much as 20 pounds or more from fat deposited in the hips, thighs, and breasts or from retained fluid. Some lean tissue is also deposited in response to an androgenic (steroid) effect of the pills. Sometimes a switch to another form of pill can normalize body weight.

As with oral contraceptives, women's responses to estrogen replacement drugs must be assessed individually. Some women may suffer edema because estrogen promotes sodium conservation by the kidneys. Sodium restriction can correct this condition. Others may develop abnormally low blood folate or vitamin B_6, indicating a need to include more vitamin-rich, nutrient-dense foods in the diet. All women taking estrogen should be aware that vitamin C doses of a gram or more may elevate serum estrogen and falsely suggest that a lower dose is needed.

If a women taking any form of estrogen thinks she may have a nutrient deficiency, she should refrain from taking individual supplements and seek testing and a diagnosis from a healthcare professional to rule out other causes of her symptoms. For most women a nutritious diet is all that is needed. If a woman feels compelled to take a supplement, however, a standard multivitamin-mineral supplement is probably harmless, as long as it accompanies a well-balanced diet. Chapter 7 showed how to select a supplement.

CAFFEINE

The well-known "wake-up" effect of caffeine is the primary reason people in every society use it in some form. Compared with the drugs discussed so far, though, caffeine's interactions with foods and nutrients are subtle. And yet in one important way caffeine's relationship to nutrition is more notable. Caffeine is so widespread among foods, beverages, and medications that people may be unaware that they are consuming it. Table C12-3 lists the caffeine contents of many beverages and foods. Parents should take note: caffeine is present in chocolate bars, colas, and other foods children favor. Children are especially sensitive to caffeine's effects because they are small and, at first, not adapted to its use.

Caffeine is the most popular and widely consumed drug in the United States. One in three U.S. citizens consumes about 200 milligrams of caffeine per day (as in 2 small cups of coffee), but many others consume much more. Many people's intake patterns fulfill some of the accepted criteria for a diagnosis of drug

Table C12-2

Nutrition Effects of a Few Commonly Used Drugs

Medicines and Caffeine	Effects on Absorption	Effects on Excretion	Effects on Metabolism
Antacids (aluminum containing)	Reduce iron absorption	Increase calcium and phosphorus excretion	May accelerate destruction of thiamin
Antibiotics (long-term usage)	Reduce absorption of fats, amino acids, folate, fat-soluble vitamins, vitamin B_{12}, calcium, copper, iron, magnesium, potassium, phosphate, zinc	Increase excretion of folate, niacin, potassium, riboflavin, vitamin C	Destroy vitamin K–producing bacteria and reduce vitamin K production
Aspirin (large doses, long-term usage)	Lowers blood concentration of folate	Increases excretion of thiamin, vitamin C, vitamin K; causes iron and potassium losses through blood loss	
Caffeine		Increases secretion of small amounts of calcium and magnesium	Stimulates release of fatty acids into the blood
Diuretics		Raise blood calcium and zinc; lower blood folate, chloride, magnesium, phosphorus, potassium, vitamin B_{12}; increase excretion of calcium, sodium, thiamin, potassium, chloride, magnesium	Interfere with storage of zinc
Laxatives (effects vary with type)	Reduce absorption of glucose, fat, carotene, vitamin D, other fat-soluble vitamins, calcium, phosphate, potassium	Increase excretion of all unabsorbed nutrients	
Oral contraceptives	Reduce absorption of folate, may improve absorption of calcium	Cause sodium retention	Raise blood vitamin A, vitamin D, copper, iron; may lower blood beta-carotene, riboflavin, vitamin B_6, vitamin B_{12}, vitamin C; may elevate requirements for riboflavin and vitamin B_6; alter blood lipids elevating risk of heart disease in smokers and older women
Estrogen replacement therapy	May reduce absorption of folate	Causes sodium retention	May raise blood glucose, triglycerides, vitamin A, vitamin E, copper, and iron; may lower blood vitamin C, folate, vitamin B_6, riboflavin, calcium, magnesium, and zinc

SOURCES: Data from Z. M. Pronsky, *Food Medication Interactions*, 9th ed. (Pottstown, Pa.: Food-Medication Interactions, 1995); G. Berg, L. Kohlmeier, and H. Brenner, Use of oral contraceptives and serum beta-carotene, *European Journal of Clinical Nutrition* 51 (1997): 181–187; S. S. Harris and B. Dawson-Hughes, The association of oral contraceptive use with plasma 25-hydroxyvitamin D levels, *Journal of the American College of Nutrition* 17 (1998): 282–284; S. M. Vaziri and coauthors, The impact of female hormone usage on the lipid profile: The Framingham Offspring Study, *Archives of Internal Medicine* 153 (1993): 2200–2206.

dependence.[8] Many OTC cold and headache remedies contain caffeine because, in addition to being a mild pain reliever in its own right, caffeine remedies the headache caused by caffeine withdrawal that no other pain reliever can touch.

Caffeine is a true stimulant drug. Like all stimulants, it increases the respiratory rate, heart rate, blood pressure, and secretion of stress and other hormones. A moderate dose of caffeine may speed up metabolic energy expenditures for several hours, and it stimulates the digestive tract, promoting efficient elimination. Because caffeine is a diuretic, it promotes water loss from the body as well. People who are accustomed to caffeine report that the caffeine in a cup of coffee or tea in the morning decreases feelings of fatigue and improves mental alertness.[9] More caffeine than this amount probably confers no additional advantages and may present some small risk to health.[10]

Despite caffeine's tremendous popularity, many people today are using less because they fear possible harms to their health. Research in the last decade has yielded sporadic reports linking caffeine to health problems such as cancer, birth defects, and hypertension. Much other research, however, refutes any links between caffeine and cancer or birth defects and finds only weak links between caffeine and hypertension.[11]

C12-3

Caffeine Content of Beverages and Foods

Drinks and Foods	Average (mg)	Range (mg)
Coffee (5 oz cup)		
Brewed, drip method	130	110–150
Brewed, percolator	94	64–124
Instant	74	40–108
Instant "lite"	30	no data
Decaffeinated, brewed or instant	3	1–5
Tea (5 oz cup)		
Brewed, major U.S. brands	40	20–90
Brewed, imported brands	60	25–110
Instant	30	25–50
Iced (12 oz glass)	70	67–76
Herb teas (caffeine-free)	0	0
Soft drinks (12 oz can)		
Dr. Pepper		40
Colas and cherry colas:		
Regular		30–46
Diet		2–58
Clear and caffeine-free		0–trace
Extra caffeine (Jolt)		75–100
Mountain Dew, Mello Yello		52
Big Red		38
Fresca, 7-Up, Sprite, Squirt, Sunkist Orange, seltzers, root beers		0
Cocoa beverage (5 oz cup)	4	2–20
Chocolate milk beverage (8 oz)	5	2–7
Milk chocolate candy (1 oz)	6	1–15
Dark chocolate, semisweet (1 oz)	20	5–35
Baker's chocolate (1 oz)	26	26
Chocolate-flavored syrup (1 oz)	4	4
Carob	0	0

NOTE: Many over-the-counter medications such as pain relievers and cold medicines also contain caffeine. Their labels must list the milligram amounts of caffeine per dose of medicine. Read medicine labels carefully.

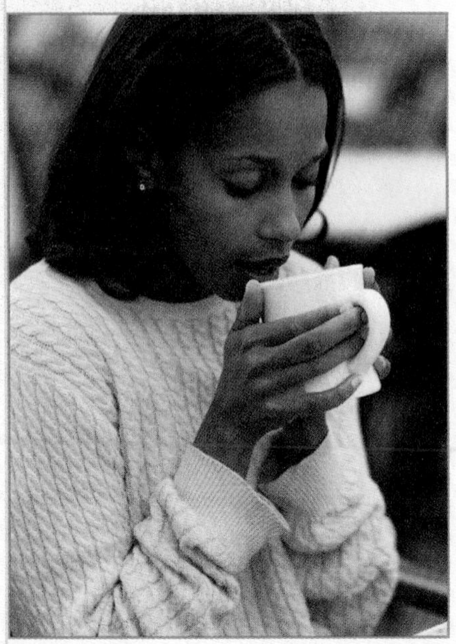

Caffeine seems relatively harmless when used in moderation (again, as in 2 cups of coffee a day). In fact, a recent study reports that men consuming 2 to 3 cups of coffee a day run a 40 percent lower risk of developing active gallstone disease than men who avoid coffee.[12] However, *cola* consumption may increase the risk.[13] In higher doses, caffeine can cause symptoms associated with anxiety: sweating, tenseness, and inability to concentrate. High doses may also accelerate bone loss in women past mid-life. Caffeine may also contribute to painful but benign fibrocystic breast disease.

If you like caffeine-containing foods or beverages, the most reasonable approach may be to limit your intake to the equivalent of about 2 small cups of coffee per day. For most people this is enough to reduce drowsiness and sharpen awareness without paying too high a price. Pregnant women, especially, should exercise moderation in using caffeine, and parents should monitor and control their children's intakes.

TOBACCO

Cigarette and other tobacco use causes thousands of people to suffer from cancer and other diseases of the cardiovascular, digestive, and respiratory systems. These effects are beyond nutrition's scope, but smoking does depress hunger and body fatness and change nutrient status, and the nutrition effects are also linked to lung cancer. Chapter 9 provided details on smoking and body fatness.

Nutrient intakes of smokers and nonsmokers differ. Smokers have lower intakes of dietary fiber, vitamins, and minerals, even when their energy intakes are quire similar to those of nonsmokers. The association between smoking and low vitamin intake may be noteworthy, considering the altered metabolism of vitamin C in smokers and their lower blood values for a number of nutrients. The research has just begun on many nutrients, but much is known about vitamin C.

Research shows that the vitamin C requirement of smokers exceeds that of nonsmokers. Smokers break down vitamin C faster and so must take in more vitamin C–containing foods to achieve steady body pools comparable to those of nonsmokers. It is estimated that the vitamin C requirement of smokers may be twice as high as that of nonsmokers. The evidence for this is so strong that the vitamin C recommendation is set higher for smokers than for nonsmokers.

ILLICIT DRUGS

People know that illicit drugs are harmful, but many choose to abuse them anyway in spite of the risks. Like OTC and prescription drugs, illegal drugs modify body functions. They are unlike medicines, however, in that no watchdog agency such as the Food and Drug Administration monitors them for safety, effectiveness, or even purity.

Smoking a marijuana cigarette affects several senses including the sense of taste. It produces an enhanced enjoyment of eating, especially of sweets, commonly known as "the munchies." Why or how this effect occurs is not known. Despite higher food intakes, marijuana abusers often consume fewer nutrients than do nonabusers because the extra foods they choose tend to be high-calorie, low-nutrient snack foods. Besides the nutrition effects, regular marijuana users face the same risk of lung cancer as people who smoke a pack of cigarettes a day.

Cocaine elicits effects such as intense euphoria, restlessness, heightened self-confidence, irritability, insomnia, and loss of appetite. Weight loss is a common side effect, and cocaine abusers often develop eating disorders. Repeated use can cause a rapid heart rate, irregular heartbeats, heart attacks, and death.

Unlike marijuana, cocaine causes serious malnutrition in those who use it. The craving of the drug replaces hunger; the stronger the craving for cocaine, the less a drug abuser wants nutritious food. Rats given unlimited access to cocaine will choose the drug over food until they die of starvation. The effects of the other addictive drugs vary in degree but are similar in kind to those of cocaine. A few are listed in Table C12-4. Drug abusers face multiple nutrition problems, and an important aspect of addiction recovery is the identification and correction of nutrition problems.

Table C12-4

Nutrition Effects of Four Nonmedical Drugs

Drug of Abuse	Possible Effects on Nutrition Status
Cocaine	Reduces intakes of nutritious foods; increases intakes of alcohol, coffee, and fat; may induce or aggravate eating disorders
Heroin	Heightens and delays insulin response to glucose; reduces intakes of nutritious foods
Marijuana	Increases intakes of foods, especially sweets; may cause weight gain
Nicotine	Reduces intake of sweet foods and water; increases intakes of fat; reduces fetal weight; lowers blood concentration of beta-carotene.

SOURCES: Data from M. E. Mohs, R. R. Watson, and T. Leonard-Green, Nutritional effects of marijuana, heroin, cocaine, and nicotine, *Journal of the American Dietetic Association* 90 (1990): 1261–1267; G. van Poppel, S. Spanhaak, and T. Ockhuizen, Effects of beta carotene on immunological indexes in healthy male smokers, *American Journal of Clinical Nutrition* 57 (1993): 402–407.

PERSONAL STRATEGY

In conclusion, when you need to take a medicine, do so wisely. Ask your physician, pharmacist, or other health-care provider for specific instructions about the doses, times, and how to take the medication—for example, with meals or on an empty stomach. If you notice new symptoms or if a drug seems not to be working well, consult your physician. The only instruction people need about illicit drugs is to avoid them altogether for countless reasons. As for smoking and chewing tobacco, the same advice applies: don't take these habits up, or if you already have, take steps to quit. For drugs with lesser consequences to health, such as caffeine, use moderation.

Try to live life in a way that requires less chemical assistance. If sleepy, try a 15-minute nap or meditation instead of a 15-minute coffee break. The coffee will stimulate your nerves for an hour, but the alternatives will refresh your attitude for the rest of the day. If you suffer constipation, try getting enough exercise, fiber, and water for a few days. Chances are that a laxative will be unnecessary. The strategy being suggested here is to take control of your body, allowing your reliable, self-healing nature to make fine adjustments that you need not force with chemicals. Bodies have few requests: adequate nutrition, rest, exercise, and hygiene. Give your body what it asks for, and let it function naturally, day-to-day, without interference from drugs.

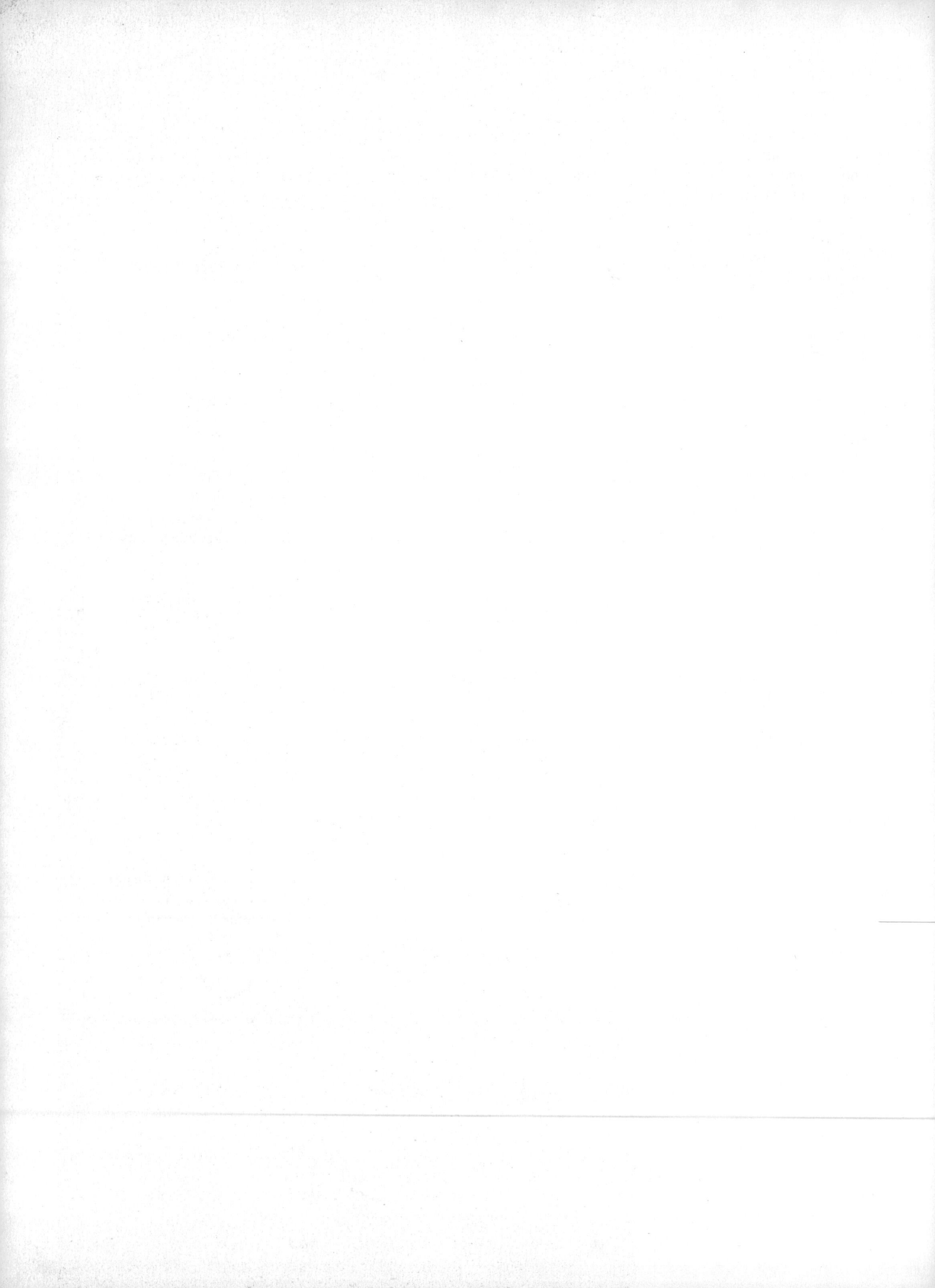

13 Child, Teen, and Older Adult

Diego Rivera, 1886–1957,
El Pan Nuestro de Cada Dia, 1923–28.

Contents

Frequently Asked Questions

FAQ: Can nutrient deficiencies impair a child's thinking or cause misbehavior? p. 473

FAQ: Does diet affect hyperactivity? p. 477

FAQ: Is breakfast really the most important meal of the day for children? p. 480

FAQ: How nourishing are the lunches served at school? p. 480

FAQ: Can nutrition help people to live longer? p. 493

FAQ: Can supplements affect the course of Alzheimer's disease ? p. 494

To grow and to function well in the adult world, children need a solid background of sound eating habits. These habits begin during babyhood with the introduction of solid foods, as described in the last chapter. But at that point the person's nutrition story has just begun; the plot thickens. Nutrient needs change steadily throughout life into old age, depending on the rate of growth, gender, activities, and many other factors. Nutrient needs also vary from individual to individual, but generalizations are possible and useful.

Early and Middle Childhood

Imagine growing 10 inches taller in just one year, as the average healthy baby does during the first dramatic year of life. At age one, infants have just learned to stand and toddle, and growth has slowed by half; by two years, they can take long strides with solid confidence and are learning to run, jump, and climb. These new accomplishments are possible thanks to the accumulation of a larger mass and greater density of bone and muscle tissue. The same trends, a lengthening of the long bones and an increase in musculature, continue until adolescence, though unevenly and more slowly.

Growth and Nutrient Needs of Young Children

An infant's appetite decreases markedly near the first birthday, in line with the great reduction in growth rate. Thereafter the appetite fluctuates. At times children seem to be insatiable, and at other times they seem to live on air and water. Parents and other caregivers need not worry about this: a child will need and demand more food during periods of rapid growth than during slow periods. The perfection of appetite regulation in children of normal weight guarantees that their overall energy intakes will be right for each stage of growth. One caution: some children may overeat in response to external cues, disregarding internal satiety signals and thereby inviting the onset of obesity.

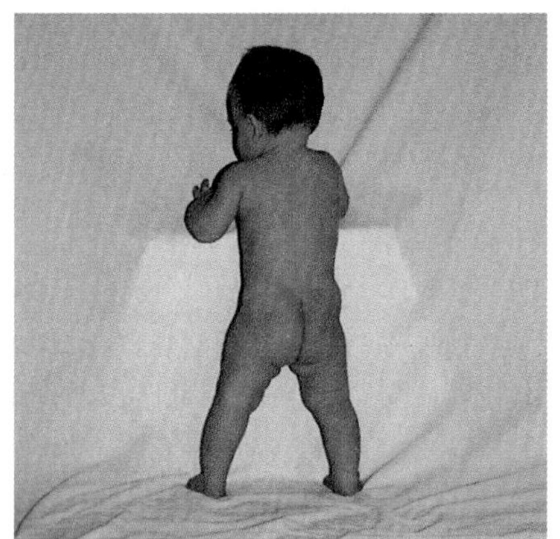

The body shape of a one-year-old (above) changes dramatically by age two (below). The two-year-old has lost much baby fat; the muscles (especially in the back, buttocks, and legs) have firmed and strengthened; and the leg bones have lengthened.

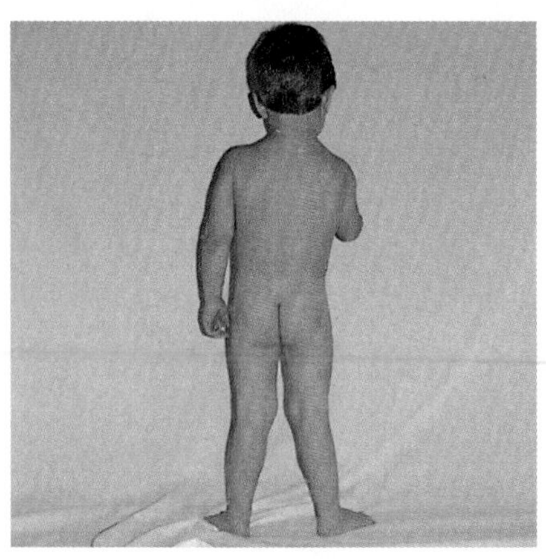

A one-year-old child needs perhaps 1,000 calories a day; a three-year-old needs 300 calories more. The next seven years add 700 more calories for a total of about 2,000 calories a day. Though total energy needs have doubled by age ten, the child's energy need per pound of body weight has steadily declined. More active children of any age need more energy because they spend more, and an inactive child can become obese even when eating less than average.

Growth enlarges the demand for all the nutrients per pound of body weight. On this basis, a five-year-old's need for, say, vitamin A is about double the need of an adult man (see the margin of the next page). Before the adolescent growth spurt, children accumulate stores of nutrients that they will need in the years ahead. Then, when they take off on that growth spurt and their nutrient intakes cannot meet the demands of rapid growth, they draw on the nutrients they stored earlier. This is especially true of calcium; the denser the bones are in childhood, the better prepared they will be to support teen growth and still withstand the inevitable bone losses of later life.

One means of providing these nutrients to children is to follow the Food Guide Pyramid for Young Children, shown in Figure 13-1. Notice that this pyramid differs from the one for adults (see Chapter 2) in that a set number of servings, rather than a range, is suggested from each food group. Children 2 to 6 years old need at least the specified number of servings from each food group to meet their nutrient needs, but the serving sizes should vary according to age. Generally, children in the 4- to 6-year age group can eat the serving sizes recommended for adults. Caretakers of children 2 to 3 years old should offer servings that are about two-thirds the size of an adult serving. To satisfy the larger appeties of older children and adolescents, caretakers should provide additional servings of the same nutritious foods for the additional energy and nutrients growing children need.

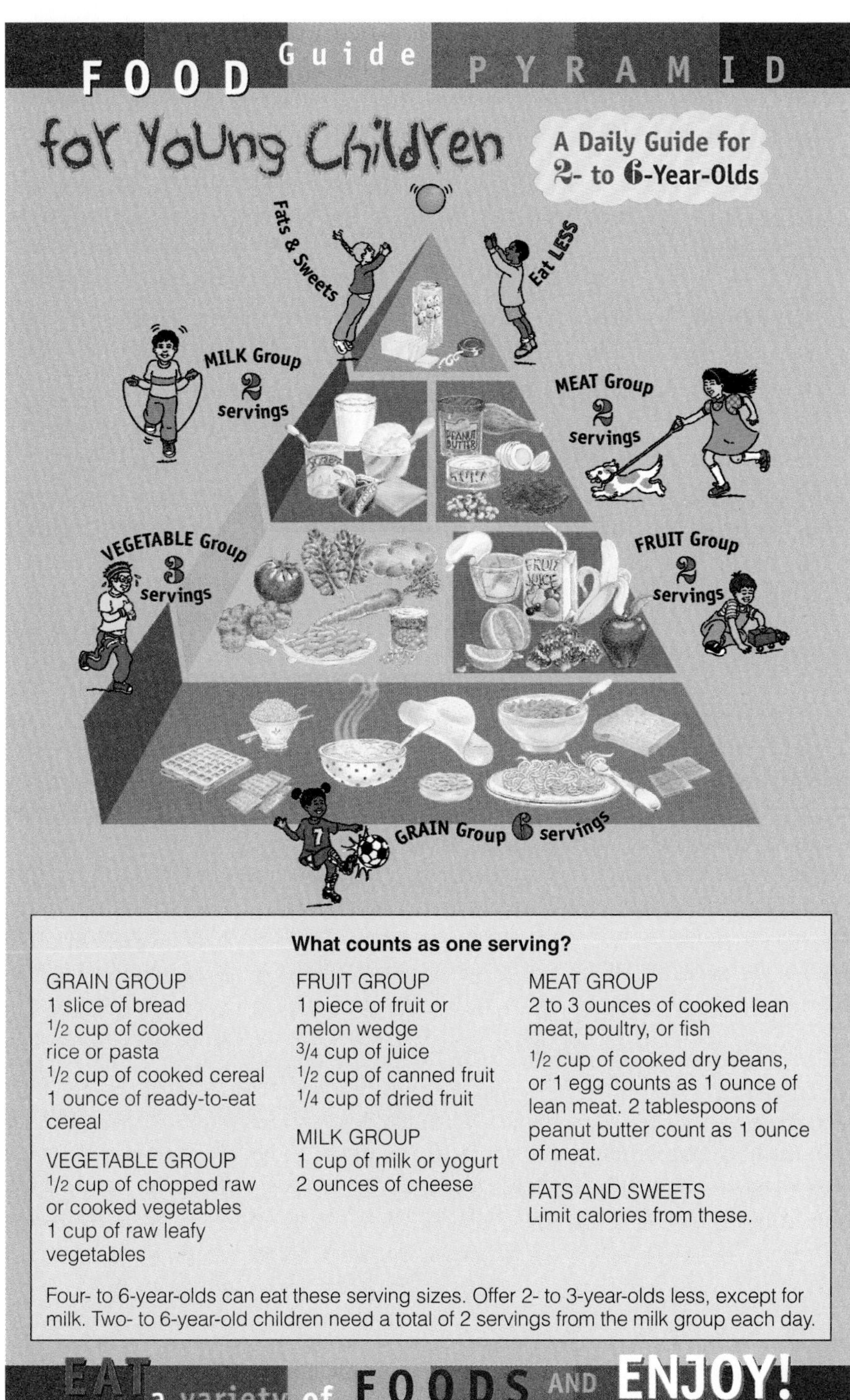

FOOD Guide PYRAMID
for Young Children
A Daily Guide for
2- to 6-Year-Olds
Fats & Sweets
Eat LESS
MILK Group
2
servings
MEAT Group
2
servings
VEGETABLE Group
3
servings
FRUIT Group
2
servings
GRAIN Group 6 servings
What counts as one serving?
GRAIN GROUP
1 slice of bread
1/2 cup of cooked
rice or pasta
1/2 cup of cooked cereal
1 ounce of ready-to-eat
cereal
VEGETABLE GROUP
1/2 cup of chopped raw
or cooked vegetables
1 cup of raw leafy
vegetables
FRUIT GROUP
1 piece of fruit or
melon wedge
3/4 cup of juice
1/2 cup of canned fruit
1/4 cup of dried fruit
MILK GROUP
1 cup of milk or yogurt
2 ounces of cheese
MEAT GROUP
2 to 3 ounces of cooked lean
meat, poultry, or fish
1/2 cup of cooked dry beans,
or 1 egg counts as 1 ounce of
lean meat. 2 tablespoons of
peanut butter count as 1 ounce
of meat.
FATS AND SWEETS
Limit calories from these.
Four- to 6-year-olds can eat these serving sizes. Offer 2- to 3-year-olds less, except for
milk. Two- to 6-year-old children need a total of 2 servings from the milk group each day.
EAT a variety of FOODS AND ENJOY!

Figure 13-1

Food Guide Pyramid for Young Children

SOURCE: USDA Center for Nutrition and Policy Promotion, March 1999, Program AID 1649.

Careful food selection is essential to ensure that a child receives the right amounts of nutrients. When a child consistently skips breakfast or is allowed to choose sugary foods (candy or marshmallows) in place of nourishing ones (whole-grain cereals), it is virtually certain that the child will fail to get enough of several nutrients. The nutrients missed from a skipped breakfast won't be "made up" at lunch and dinner but will be left out completely that day. A child can't be trusted to choose nutritious foods on the basis of taste alone; the preference for sweets is inborn, as Chapter 3 made clear.

A 174-pound adult male needs 1,000 RE of vitamin A. That is 5.75 RE per pound. A 44-pound five-year-old needs 500 RE of vitamin A. That is 11.4 RE per pound.

Active, normal-weight children may enjoy occasional treats of high-calorie but nutritious foods. From the milk group, ice cream or pudding is good now and then;

from the bread group, whole-grain or enriched cakes, cookies, or even doughnuts are an acceptable addition to a balanced diet. These foods are made from milk and grain, they carry valuable nutrients, and they encourage a child to learn, appropriately, that eating is fun. Should a child regularly eat large quantities of these treats, however, the only possible outcomes are nutrient deficiencies, obesity, or both. Parents of wandering elementary school children should be aware that they may be spending pocket money at nearby stores and filling up on sweets. The next section presents more tips on the role of parents in feeding children.

KEY POINT *Children's nutrient needs reflect their stage of growth. Positive parental guidance can help establish food patterns that provide adequate nourishment for growth without obesity.*

Mealtimes and Snacking

The childhood years are a parent's last chance to influence food choices. Appropriate eating habits and attitudes toward food ensure positive development during growth and may help future adults emerge with healthy eating habits to reduce risks of degenerative diseases in later life.

Children's Preferences Children naturally like nutritious foods in all the food groups, with one exception—vegetables, which some young children frequently refuse. Here presentation may be the key.

Many children prefer vegetables that are mild flavored, slightly undercooked and crunchy, bright in color, and easy to eat. Cooked foods should be served warm, not hot, because a child's mouth is much more sensitive than an adult's. The mild flavors of carrots, peas, and corn are often preferred over sharper tasing broccoli or turnips because a child has more taste buds. Smooth foods such as grits, oatmeal, mashed potatoes, and pea soup should have no lumps in them. Children prefer familiar foods; fear of new foods is practically universal among children. Suggesting, rather than commanding, that a child try small amounts of new foods at the beginning of a meal, when the child is hungry, seems to work best.

Choking A child may make no sound when choking, so an adult should keep an eye on children when they are eating. Encouraging the child to sit when eating is a good practice; choking is more likely when children are running or reclining. Round foods such as grapes, nuts, hard candies, and pieces of hot dog can easily become lodged in a child's small windpipe. Other potentially dangerous foods include tough meat, popcorn, chips, and peanut butter eaten by the spoonful.

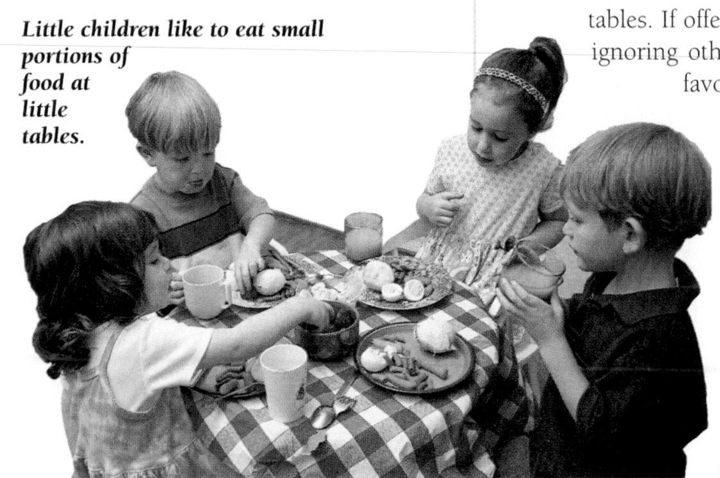
Little children like to eat small portions of food at little tables.

Portion Sizes Little children like to eat small portions of food at little tables. If offered large portions, children may well fill up on favorite foods, ignoring others. Toddlers often go on food jags, eating only one or two favored foods. The best way to handle food jags lasting a week or so is to make no response, since two-year-olds regard any form of attention as a reward. After two weeks of indulging the jag, try serving tiny portions of many foods, including the favored items. Distract the child with friends at meals, and make other foods as attractive as possible.

Remember, too, that just as parents are entitled to their likes and dislikes, a child who genuinely and consistently rejects a food should be allowed the same privilege. Also, children should be believed when they say they are full: the "clean-your-plate" dictum should be stamped out for all time. Children who are forced to override their own satiety signals are essentially in training for obesity. Encourage children to listen to their bodies, and do not make an issue of food acceptance. The parent is responsible for *what* the child is offered to eat, but the child is responsible for *how much* and even *whether* to eat.

Snacking and Other Healthy Habits Parents may find that their children often snack so much that they are not very hungry at mealtimes. This need not be a problem as long as children know how to snack. Nutritious snacks can meet the same needs as nutritious small meals do. Keep snack foods simple and readily available. Milk, cheese, crackers, fruit, vegetable sticks, yogurt, peanut butter sandwiches, and whole-grain cereal can all satisfy children and help meet nutrient and fiber needs.

A bright, unhurried atmosphere free of conflict is conducive to good appetite. Parents who serve meals in a relaxed and casual manner, without anxiety, provide a climate in which a child can learn to enjoy eating. Parents who beg, cajole, and demand that their children eat set up power struggles. A child may find mealtimes unbearable if they are accompanied by a barrage of accusations—"Susie, your hands are filthy . . . your report card . . . and clean your plate!" The child's stomach recoils as both body and mind react to stress of this kind.

Children love to be included in meal preparation, and they like to eat foods they helped to prepare. Children as young as age two can develop new skills by helping out (see Table 13-1 in the margin). A positive experience is most likely when tasks match children's developmental abilities and are undertaken in a spirit of enthusiasm and enjoyment, not criticism or drudgery. Praise for a job well done (or at least well attempted) expands a child's sense of pride and helps to develop skills and positive feelings toward healthy foods.

Many parents may overlook perhaps the single most important influence on their child's food habits—their own habits. Parents who don't prepare, serve, and eat carrots shouldn't be surprised when their child refuses to eat carrots.[1] A child learns much through imitation. Parents set an irresistible example by enjoying nutritious foods at meals and snacks.

While preparing, serving, and enjoying food, caretakers can promote not only physical but also emotional growth at every stage of a child's life. It is important for parents to help their youngsters to remember that they are good kids. What they do may sometimes be unacceptable, but they are still normal, healthy, growing, fine human beings.

KEY POINT *Healthy eating habits and positive relationships with food are learned in childhood. Parents teach children best by example.*

Table 13-1

Food Skills of Preschoolers[a]

Age 1–2 years, when large muscles develop, the child:

- uses short-shanked spoon.
- helps feed self.
- lifts and drinks from cup.
- helps scrub, tear, break, or dip foods.

Age 3 years, when medium hand muscles develop, the child:

- spears food with fork.
- feeds self independently.
- helps wrap, pour, mix, shake, or spread foods.
- helps crack nuts with supervision.

Age 4 years, when small finger muscles develop, the child:

- uses all utensils and napkin.
- helps roll, juice, mash, or peel foods.
- crack egg shells.

Age 5 years, when fine coordination of fingers and hands develops, the child:

- helps measure, grind, cut, and grate.
- uses hand-cranked egg beater with supervision.

[a]These ages are approximate. Healthy, normal children develop at their own pace.

SOURCES: Adapted from M. Sigman-Grant, Feeding preschoolers: Balancing nutrition and developmental needs, *Nutrition Today,* July/August 1992, pp. 13–17; A. A.. Hertzler, Preschoolers' food handling skills—Motor development, *Journal of Nutrition Education* 21 (1989): 100B–100C.

Can Nutrient Deficiencies Impair a Child's Thinking or Cause Misbehavior?

A child who suffers from nutrient deficiencies exhibits physical and behavioral symptoms: the child is sick and out of sorts. Diet-behavior connections are of keen interest to caretakers who both feed children and live with them.

Deficiencies of protein, energy, vitamin A, iron, and zinc plague children the world over. In developing nations, such deficiencies cause or contribute to nearly half the deaths of children under four and inflict blindness, stunted growth, and vulnerability to infections on millions more.

In developed countries such as the United States and Canada, most deficiencies have subtle, even unnoticeable, effects. A study of British children found that about 40 percent of them had intakes of less than half the recommended amounts of folate, vitamin D, calcium, iron, magnesium, selenium, zinc, and many other minerals. The researchers gave multinutrient supplements to some of the children and later administered intelligence tests to all of them. Those who had received the supplements

Controversy 13 provides details concerning the mental symptoms of anemia.

scored significantly higher on the tests than the others did. The researchers took the findings to mean that although children may be well nourished in terms of protein and some vitamins, as they are in the United States, brain function may be sensitive to borderline deficiencies of many other nutrients, a conclusion supported by many previous findings.[2]

Iron deficiency remains common in children and adolescents despite iron fortification of foods and other programs to combat this deficiency.[3] Reducing the incidence of iron deficiency remains a top priority of U.S. policymakers.[4] Besides carrying oxygen in the blood, iron works as part of large molecules to release energy within cells. A lack of iron not only causes an energy crisis but also directly affects behavior, mood, attention span, and learning ability.[5] Iron plays key roles in many molecules of the brain and nervous system. Some research links iron deficiency in infancy or early childhood with later development of mental retardation severe enough to require special education services, although no cause-and-effect evidence yet exists.[6] In laboratory animals, deficiencies of iron have caused abnormal metabolism in neurotransmitters, notably those that regulate the ability to pay attention, which is crucial to learning.

Iron deficiency is usually diagnosed by a deficit of iron in the *blood,* after anemia has developed. A child's *brain,* however, is sensitive to slightly lowered iron concentrations long before the blood effects appear. It is difficult to distinguish the effects of iron deficiency from those of other factors in children's lives, but studies have found connections between iron deficiency and behavior. Iron deficiency seems to manifest itself in a lowering of the motivation to persist in intellectually challenging tasks, a shortening of the attention span, and a reduction of overall intellectual performance. A child with such symptoms may be irritable, aggressive, and disagreeable or sad and withdrawn. One might label such a child "hyperactive," "depressed," or "unlikable," but these traits may not be purely psychological; they may arise from malnutrition. Inspection of a disruptive or apathetic child's diet by a qualified health-care professional can identify these reversible problems, and additions to the diet can correct them. Table 13-2 lists some iron-rich foods children often like to eat. Only a health-care provider should make the decision to give iron supplements, of course, and if used, supplements should be kept out of children's reach. Iron toxicity is a leading cause of poisoning each year in toddlers and other children who accidentally ingest iron pills.

Table 13-2

Iron-Rich Foods Kids Like[a]

Breads, Cereals, and Grains
Canned macaroni (½ c)
Canned spaghetti (½ c)
Cream of wheat (¼ c)
Fortified dry cereals (1 oz)[b]
Noodles, rice, or barley (½ c)
Tortillas (1 flour, 2 corn)
Whole-wheat, enriched, or fortified bread (1 slice)
Vegetables
Baked flavored potato skins (½ skin)
Cooked mushrooms (½ c)
Cooked mung bean sprouts or snow peas (½ c)
Green peas (½ c)
Mixed vegetable juice (1 c)
Fruits
Apple juice (1 c)
Canned plums (3 plums)
Cooked dried apricots (¼ c)
Dried peaches (4 halves)
Raisins (1 tbs)
Meats and Legumes
Bean dip (¼ c)
Canned pork and beans (⅓ c)
Mild chili or other bean/meat dishes (¼ c)
Liverwurst (½ oz)
Meat casseroles (½ c)
Peanut butter and jelly sandwich (½ sandwich)
Lean roast beef or cooked ground beef (1 oz)
Sloppy joes (½ sandwich)

[a]Each serving provides at least 1 milligram iron, or one-tenth of a child's iron RDA. Vitamin C–rich foods included with these snacks increase iron absorption.
[b]Some fortified breakfast cereals contain more than 10 milligrams iron per half-cup serving (read the labels).
SOURCE: Many of these ideas reflect data in A. A. Hertzler, Children's food patterns—A review: I. Food preferences and feeding problems, *Journal of the American Dietetic Association* 83 (1983): 551–554.

KEY POINT *The detrimental effects of nutrient deficiencies in children of developed nations can be subtle. Iron deficiency is the most widespread nutrition problem of children and causes abnormalities in both physical health and behavior. Iron toxicity is a major form of poisoning in children.*

The Problem of Lead

Another form of metal poisoning arises from ingestion of lead. Lead poisoning can cause iron-deficiency anemia. Conversely, a child with iron-deficiency anemia is three times as likely to have elevated blood lead as a child with normal iron status. Adequate calcium may also slow lead's absorption or interfere with its toxic effects in the body.[7] Often, lead poisoning occurs because babies like to explore, and they put everything into their mouths, including things that may harm them, such as chips of old paint, pieces of metal, and other unlikely substances. These are normal baby activities, but they may be silently increasing blood lead levels until toxic concentrations have built up. Not until much later, after lead toxicity has set in, do caretakers notice unusual symptoms.

Joey's Story Joey was such a child. This normal-appearing baby grew up in an industrial city where dust settled on his playthings, sprinkling lead from industrial emissions into his environment. He loved to taste everything—pets, toys, the spindles of old painted railings—within his reach. And his mother often mixed his formula with the first water from the tap, water that had spent the night absorbing lead from the old building's lead pipes.

Joey became a cautious, quiet preschooler who clung to stair railings as he slowly climbed up and down. He was late in walking and talking, small for his age, seldom played vigorously, and was prone to diarrhea, irritability, and lethargy. Finally, a pediatrician detected lead toxicity in Joey's blood and started treating him with lead-scavenging drugs. Except for persistent, minor learning disabilities, Joey is now growing normally and playing vigorously.

Once diagnosed, lead toxicity is easy to treat. The trick is to identify it before too much damage has set in. For kids like Joey, the truth can easily come too late, as even one year of lead exposure can permanently impair the brain, nervous system, and psychological functioning. Older children with high blood lead also suffer physical complaints and are often delinquent, aggressive, and distractible.[8]

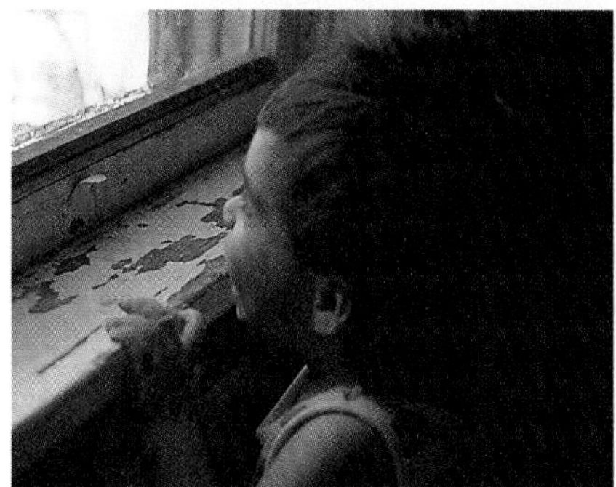

Old paint is the main source of lead in most children's lives.

The Nature of Lead Lead is an indestructible metal element; the body cannot alter it. Because it is chemically similar to nutrient minerals like iron, calcium, and zinc, lead displaces these minerals from their sites of action, but then is unable to perform their biological functions. Consequently, lead interferes with many of the body's systems, particularly the vulnerable tissues of the nervous system, kidneys, blood, and bone marrow. During pregnancy, lead crosses the placenta to inflict severe damage on the fetal nervous system. Infants and young children absorb five to ten times as much lead as do adults.

The Environmental Protection Agency (EPA) provides this toll-free hotline for lead information: 1-800-LEAD-FYI (1-800-532-3394).

Alternative agricultural and industrial processes that can help to reduce environmental lead and other contaminants are discussed in Chapter 15.

A Public Health Success A ban on leaded gasoline and reductions in other uses of lead have dramatically reduced the amount of lead in the environment.[9] The result has been a gratifying decline in children's blood lead concentrations since the 1970s (see Figure 13-2). The decline follows exactly the nation's reduction in use of leaded gasoline, leaded house paint, and lead-soldered food cans. A nationwide lead-monitoring system is now in place, and aggressive community programs are testing and treating children for lead poisoning.

Lead is still a problem, however, especially among children of low-income families. Over a million children have blood lead concentrations high enough to harm their health.[10] Lead is found in layers of old paint that find their way into children's mouths, and it is still discharged into their environment, contaminating the soil or water near their homes. Some tips for avoiding lead toxicity are offered in Table 13-3.

Figure 13-2

Reduction in Children's Blood Lead Concentrations over Two Decades

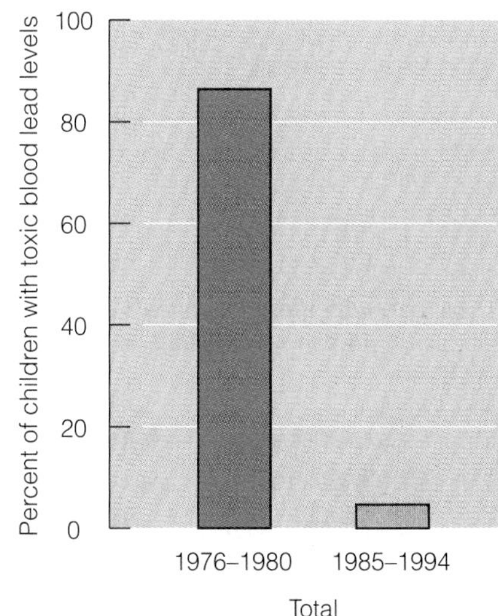
Percent of children with toxic blood lead levels
100
80
60
40
20
0
1976–1980
1985–1994
Total

SOURCE: Data from the Centers for Disease Control and Prevention, National Center for Heath Statistics, *National Health and Nutrition Examination Survey II* (conducted 1976–1980) and *National Health and Nutrition Examination Survey III* (conducted 1988–1994).

Table 13-3

Steps to Prevent Lead Poisoning

To protect children:

- If your home was built before 1978, wash floors, windowsills, and other surfaces weekly with warm water and detergent to remove dust released by old lead paint; clean up flaking paint chips immediately.
- Feed children balanced, timely meals with ample iron and calcium.
- Prevent children from chewing on old painted surfaces.
- Wash children's hands, bottles, and toys often.
- Wipe soil off shoes before entering the home.
- Ask a pediatrician whether your child should be tested for lead.

To safeguard yourself:

- Avoid daily use of handmade, imported, or old ceramic mugs or pitchers for hot or acidic beverages, such as juices, coffee, or tea. Commercially made U.S. ceramic, porcelain, and glass dishes or cups are safe. If ceramic dishes or cups become chalky, use them for decorative purposes only.
- Do not use lead crystal decanters for storing alcoholic or other beverages.
- If your home is old and may have lead pipes, run the water for a minute before using, especially before the first use in the morning.
- Remove lead foil from wine bottles, and wipe the mouth of the bottle before pouring.

allergy an immune reaction to a foreign substance, such as a component of food. Also called *hypersensitivity* by researchers.

antigen a substance foreign to the body that elicits the formation of antibodies or an inflammation reaction from immune system cells. Food antigens are usually glycoproteins (large proteins with glucose molecules attached). Inflammation consists of local swelling and irritation and attracts white blood cells to the site.

antibodies as defined in Chapter 6, large protein molecules that are produced in response to the presence of antigens and then inactivate the antigens.

histamine a substance that participates in causing inflammation; produced by cells of the immune system as part of a local immune reaction to an antigen.

food intolerance an adverse effect of a food or food additive not involving the immune response.

anaphylactic (an-AFF-ill-LAC-tic) **shock** a life-threatening whole-body allergic reaction to an offending substance.

A concern exists about allergic reactions to new genetically engineered foods. See Controversy 14.

Warning signs of allergic anaphylactic shock: itching tongue and tightness in the throat, abdominal pain, itchy and blotchy skin, nausea, vomiting, diarrhea, inflamed nasal membranes, chest pain, swelling, low blood pressure, shock, and respiratory arrest.

KEY POINT *Lead poisoning has declined dramatically over the past two decades, but when it strikes, it can inflict severe, irreparable damage on growing children. More awareness of the remaining sources of lead poisoning can help to reduce the present rate of its occurrence.*

Food Allergy, Intolerance, and Aversion

Food **allergy** is frequently blamed for physical and behavioral abnormalities in children. In truth, the prevalence of food allergy among children is far less (up to 8 percent) than many grown-ups believe.[11] Among adults, allergies are even less common, occurring in 2 percent of the population.

A true food allergy occurs when a whole food protein or other large molecule enters the body tissues. Recall that most large molecules of food are normally dismantled to smaller ones in the digestive tract before absorption. Some, however, are not digested but enter the bloodstream whole. Once they are inside, the body's immune system reacts to undigested food proteins or other large molecules as it does to any other **antigen**: it releases **antibodies, histamine,** or other defensive agents. A problem not involving the immune system that results from exposure to food substances is known as a **food intolerance.**

Almost 75 percent of food allergies are caused by just three foods: eggs, peanuts, and milk. The other 25 percent are caused by a variety of foods from almonds to yeast breads. The life-threatening reaction of **anaphylactic shock** is most often caused by peanuts, tree nuts, fish, or shellfish. Food manufacturers are becoming more aware of the problems these foods pose and are taking pains to prevent cross-contamination during production. For example, equipment used for making peanut butter is dissembled and thoroughly cleaned before using it to pulverize another kind of food, say, cashew nuts.[12] That way, a person with peanut allergy needn't worry about traces of peanut allergen being present in cashew butter. When cross-contamination is likely, food labels must state that the product may contain an allergy-producing food.

Detecting Food Allergy

Allergies may have one or two components. They always involve antibodies; they may or may not involve symptoms. A person may produce antibodies *without* exhibiting any symptoms or may produce antibodies *and* exhibit symptoms. Symptoms without antibody production, however, are not due to allergy. This means that allergies cannot be diagnosed from symptoms alone; they have to be diagnosed by testing for antibodies.

A food allergy can produce many different symptoms. Most frequently, food allergies show up on the skin as hives, swelling, and rashes.[13] In the digestive tract, a food allergy may cause cramping, constipation, bloating, nausea, diarrhea, or vomiting; in the lungs, it can cause asthma; it can also cause a runny nose or irritated, reddened eyes.[14] Most dangerous is the severe, generalized allergic reaction known as anaphylactic shock.

Allergic reactions to food can occur with different timings; symptoms may appear within minutes or up to 24 hours later. Identifying a food that causes an immediate allergic reaction is relatively easy because symptoms correlate closely with the time of eating the food. If the reaction is delayed, though, identifying the offending food is more difficult because many other foods will have been eaten by the time the symptoms have appeared. Many people are allergic to just one food, but some are allergic to many.

Allergy Testing

A skin prick test and a double-blind food challenge under medical supervision can confirm a true food allergy.[15] These tests are time-consuming and often expensive, however, so people—and even physicians—often try to guess the cause of an adverse reaction and use the term *food allergy* loosely. A parent whose child has any kind of discomfort after eating, such as stomachache, headache, pain, rapid

These normally wholesome foods are most likely to induce symptoms in people with allergies.

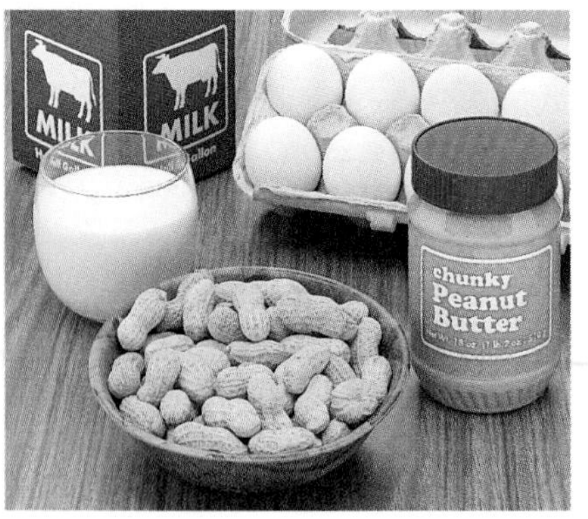
MILK
MILK
Chunky
Peanut
Butter

pulse rate, nausea, wheezing, hives, bronchial irritation, or cough, may decide that an allergy is responsible, when in fact the cause is something else entirely. Only careful, skilled testing by a physician can distinguish the many possibilities, but such testing is seldom done.

Because reliable tests for food allergy are inconvenient and expensive, people are tempted to believe quacks offering quick and easy but sophisticated-sounding laboratory work. For example, "cytotoxic testing" involves mixing blood with foods to see what blood cells "react" to. As you might guess, this test is invalid for detecting allergy because isolated blood cells are cut off from the body's immune system, which produces the allergic response. Other terms indicating allergy quackery are *brain allergy* and *metabolic rejectivity syndrome.*

Food Aversion A **food aversion,** an intense dislike of a food, may be a biological response to a food that once caused trouble. Children's food aversions may be the result of nature's efforts to protect them from allergic or other adverse reactions. Parents are advised to watch for signs of food dislikes and to take them seriously. Such a dislike may turn out to be a whim or fancy, but it may turn out to be an allergy or other valid reason to avoid a certain food. In any case, don't prejudge. Test. Then, if an important staple food must be excluded from the diet, find other foods to provide the omitted nutrients and ensure the child's continued good nutrition.

Allergies are often blamed when behavior problems arise, but children who are sick from any cause are likely to be cranky. Evidence does not support the hypothesis that allergy can cause misbehavior without other symptoms. The next section singles out a type of misbehavior that is not caused by foods.

KEY POINT *Food allergies cause illness, but diagnosis is difficult. Tests are imperative to determine whether allergy exists. Food aversions can be related to food allergies or to adverse reactions to food.*

food aversion an intense dislike of a food, possibly biological in nature, resulting from an illness or other negative experience associated with that food.

hyperactivity (in children) a syndrome characterized by inattention, impulsiveness, and excess motor activity; usually diagnosed before age seven, lasts six months or more, and usually does not entail mental illness or mental retardation. Properly called *attention-deficit/hyperactivity disorder (ADHD)* and may be associated with minimal brain damage.

learning disabilities a group of conditions resulting in an altered ability to learn basic cognitive skills such as reading, writing, and mathematics.

Does Diet Affect Hyperactivity?

Hyperactivity, or attention-deficit/hyperactivity disorder (ADHD), is a kind of **learning disability** that occurs in 5 to 10 percent of young, school-aged children, that is, in 2 or 3 in a classroom of 30 children. It can delay growth, lead to academic failure, and cause major behavioral problems.[16] ADHD is characterized by the chronic inability to pay attention, along with overly active behavior and poor impulse control (see the margin). Although some children improve with age, many reach the college years or adulthood before they receive a diagnosis. When that day finally arrives, they breathe a sigh of relief on learning that their problem has a name and that the possibility of treatment exists.

Food allergies have been blamed for ADHD. Research to date is not promising, but studies continue.[17] Research has also all but dismissed the idea that sugar makes children hyperactive (see Controversy 4 for details). One study did find an association between doses of the food colorant tartrazine and increased irritability, restlessness, and sleep disturbances in a small percentage of hyperactive children.[18] Parents who wish to avoid tartrazine can find it listed with the ingredients on labels of the processed foods that contain it.

Physicians often diagnose ADHD by conducting trials with stimulant drugs. Stimulants normally speed up people's activity, but they calm down children with ADHD. The reason is unclear, but the drugs may stimulate centers in the brain that control behavior. Clearly ADHD is complex, but many parents hope that a new diet or some other simple solution may improve children's behavior. Unfounded dietary "treatments" may seem to help for a while due to the placebo effect, but they fail to provide lasting cures.

Common sense says that all children at times get unruly and "hyper."

A child with ADHD often:

- Has a short attention span, even while playing.
- Has trouble with tasks that require sustained mental effort.
- Has trouble learning and earns failing grades in school.
- Has poor impulse control and acts physically or verbally before thinking.
- Angers friends by not taking turns or playing by the rules.
- Runs instead of walks, climbs instead of sitting still, talks excessively.

These symptoms are clues to ADHD but they are not enough to make a diagnosis. Tests by a knowledgeable clinician are needed for diagnosis and treatment.

The placebo effect was defined earlier. It is the healing effect produced by faith in a treatment, rather than by the treatment itself.

Controversy 12 presented a table of the caffeine in some foods and beverages.

There are many normal, everyday causes of such behavior:

- Too much caffeine from colas or chocolate.
- Desire for attention.
- Lack of sleep.
- Overstimulation.
- Too much television.
- Lack of exercise.
- Chronic hunger.[19]

A child who often fills up on colas and chocolate, misses lunch, becomes too cranky to nap, misses out on outdoor play, and spends hours in front of a television suffers stresses that trigger chronic patterns of crankiness. This cycle of tension and fatigue resolves itself when the caretakers begin insisting on regular hours of sleep, regular mealtimes, and regular outdoor exercise. As for chronic hunger of children, the issues are many, and are topics of Chapter 15.

KEY POINT *Hyperactivity, properly named attention-deficit/hyperactivity disorder (ADHD), is not caused by poor nutrition; temporary "hyper" behavior may reflect excess caffeine consumption or inconsistent care. A wise parent will limit children's caffeine intakes and meet their needs for structure to prevent tension and fatigue.*

Table 13-4

The Caries Potential of Foods

Low Caries Potential

These foods are less damaging to teeth:

- Eggs, legumes
- Fresh fruit, fruits packed in water
- Lean meats, fish, poultry
- Milk, cheese, plain yogurt
- Most cooked and raw vegetables
- Pizza
- Popcorn, pretzels
- Sugarless gum and candy,[a] diet soft drinks
- Toast, hard rolls, bagels

High Caries Potential

Brush teeth especially well and quickly after eating these foods:

- Cakes, muffins, doughnuts, pies
- Candied sweet potatoes
- Chocolate milk
- Cookies, granola bars, crackers
- Dried fruits (raisins, figs, dates)
- Frozen or flavored yogurt
- Fruit juices or drinks
- Fruits in syrup
- Glazed carrots
- Ice cream or ice milk
- Jams, jellies, preserves
- Lunch meats with added sugar
- Meats with sugary glazes
- Oatmeal, oat cereals, oatmeal baked goods[b]
- Peanut butter with added sugar
- Potato and other snack chips
- Ready-to-eat sugared cereals
- Sugared gum, soft drinks, candies, honey, sugar, molasses, syrups
- Toaster pastries

[a]Cariogenic bacteria cannot efficiently metabolize the sugar alcohols in these products, so they do not contribute to dental caries.

[b]The soluble fiber in oats makes this grain particularly sticky and therefore cariogenic.

Television and Children's Nutrition

Television has adverse effects on children's nutrition. On average, children in the United States spend more time watching television than they do attending school; over 25 percent of children watch four or more hours of television every day, and 67 percent watch two or more hours every day.[20] Television exerts four major kinds of impacts on children's nutrition. First, television viewing requires no energy. It seems to reduce the metabolic rate to a level below that of rest, requiring even less energy than daydreaming. The effect may be most pronounced in obese children.[21] Second, it consumes time that could be spent in energetic play. Third, watching television correlates with between-meal snacking and with buying and eating the calorically dense foods most heavily advertised on children's programs. Children who watch more than two hours of television per day may also have higher serum cholesterol than do more active children.[22] Fourth, it encourages food behaviors that damage dental health.

Children who watch hours of television a day are prone to frequent snacking on high-sugar foods, a major factor in dental caries development. Sticky, high-carbohydrate snack foods cling to the teeth and provide an ideal environment for the growth of mouth bacteria that cause caries. What child can resist the delicious-looking, sugar-filled fun foods that dance across the television screen? Television commercials have no stake in promoting dental health—their goal is to persuade children to buy and eat sugary foods. Parents must combat this influence by helping children to do the following:

- Limit between-meal snacking.
- Brush and floss daily, and brush or rinse after eating meals and snacks.
- Choose foods that don't stick to teeth and are swallowed quickly.
- Snack on crisp or fibrous foods to stimulate the release and rinsing action of saliva.

Table 13-4 in the margin lists foods that promote dental health and those that require speedy removal from the teeth.

Evidence from these many points of view suggests that television's effects on children's nutritional health are negative. It seems prudent, therefore, to advise parents to limit children's television viewing time to one or two hours a day or less. Replace television viewing with reading, developing hobbies, cooking with a parent, biking or playing ball, visiting a relative or friend, working around the house, or any of a thousand other activities.

KEY POINT *Television viewing can contribute to obesity through lack of exercise and overconsumption of snacks. Television advertising of sugary foods promotes sugar consumption and tooth decay.*

Childhood Obesity and Disease in Later Life

Children are heavier today than they were 20 or so years ago. Since the late 1970s, the prevalence of overweight has almost doubled for children.[23] This pattern is a secular trend; that is, it cannot be explained by genetics. Diet and physical activity are most likely responsible.

Are children today consuming significantly more calories than in the past? No, children's energy intake has remained relatively stable over the past 15 years. On average, their intake of fat, as a percentage of calories, has even declined slightly. Of course, some children do eat extraordinarily high-fat diets, and these children often have overweight parents who choose high-fat diets for themselves.[24] Such findings confirm the significant role of parents in teaching children about food choices and providing powerful role models that children readily imitate.

A more likely explanation for the fattening of the nation's children, however, is that they have grown more sedentary.[25] A child who spends hours in front of a television or computer monitor can become obese even while eating fewer calories of food than a more active peer. A study of 1,000 teenagers revealed that compared with inactive teens, active teens weighed less, smoked less, ate a diet lower in saturated fats, and had healthier blood lipid profiles. Further, those who were initially described as inactive were still inactive six years later.[26] The reverse was also true: those who started out physically active remained so. The message seems clear: children who are active are less likely to become overweight and more likely to choose healthy behaviors throughout life.

Chapter 2 offered some guidelines for activity.

Prevention of childhood obesity seems critical because overweight children are especially likely to suffer from heart and artery disease later in life. Often, an obese child will also have high blood cholesterol, high blood pressure, and diabetes, which together may predict an elevated risk of heart disease in adult life.[27] Though some experts recommend regular cholesterol screening for all children, others think such measures are unjustified and favor testing only those whose parents or grandparents developed cardiovascular disease. The American Heart Association makes these recommendations for all children over the age of about 2 years and adolescents:

The combination of these four conditions—obesity, hyperlipidemia, hypertension, and insulin resistance—is known as *insulin resistance syndrome* or *syndrome X*. The risks it poses to health were described in Chapter 11.

- Adequate nutrition should be achieved by eating a wide variety of foods.
- Energy (calories) should be adequate to support growth and development and to reach or maintain desirable body weight.
- The following pattern of nutrient intake is recommended:
  - Saturated fatty acids—less than 10 percent of total calories.
  - Total fat—an *average* of no more than 30 percent of total calories.
  - Dietary cholesterol—less than 300 milligrams per day.[28]

No harm can come to children over the age of 2 who are encouraged to eat a variety of foods, to reach or maintain a desirable weight, and, within reason, to obtain enough fiber and limit saturated fat and cholesterol intakes. One easy way to determine the fiber needs of children is the "age plus 5" method: add 5 to the child's age to obtain the approximate number of fiber grams needed per day.

Habits instilled in childhood, such as overeating on high-fat foods and following a sedentary lifestyle, are especially difficult to change in adulthood. The margin lists some suggestions for establishing positive behaviors early on that can help combat obesity. Weight gain in overweight children must be sensitively addressed, however. Children are impressionable and can easily come to believe that their worth or lovability is somehow tied to their weight. Some parents fail to realize that society's ideal of slimness can be perilously close to starvation. Adolescents, especially, can be led to make all sorts of unhealthy weight-loss attempts. Even healthy adolescents without diagnosable eating disorders have been observed to stunt their own growth through "dieting." Adolescents who "break the rules" with extreme weight-loss behaviors have

Here are some tips for preventing childhood obesity. Encourage children to:

- Eat slowly.
- Pause and enjoy their table companions.
- Play vigorously outdoors every day.
- Select appropriate food portions.
- Select lower-calorie snacks.
- Stop eating when they are full.

And then set the example for them to follow.

also been observed to engage in other health-compromising behaviors, such as using alcohol, tobacco, and other drugs.[29] It may be that a low sense of self-worth opens the door to taking risks of all sorts with little regard for personal well-being.

The child who is already obese needs careful management. Weight loss ordinarily is not recommended because as mentioned, diet restriction can easily interfere with normal growth. Instead, the aim is to hold the child's weight steady over time while the child grows taller. The goal is to support normal lean body development, while letting children "grow out" of their obesity. By feeding the child balanced meals, restricting treats, and boosting the child's physical activity, that goal is often accomplished.

KEY POINT *The nation's children are growing fatter and face advancing risks of diseases. Childhood obesity demands careful management.*

Breakfast ideas for rushed mornings:

- Make ahead and freeze sandwiches to thaw and serve with juice. Fillings may include peanut butter, low-fat cream cheese, other cheeses, jams, fruit slices, or meats. Or use flour tortillas with cheese; roll up, wrap, and freeze for later heating in a toaster oven or microwave oven.
- Teach school-aged children to help themselves to dry cereals, milk, and juice. Keep unbreakable bowls and cups in low cupboards, and keep milk and juice in small unbreakable pitchers on a low refrigerator shelf.
- Keep a bowl of fresh fruit and small containers of shelled nuts, trail mix (the kind without candy), or roasted peanuts for grabbing. Granola or other grain cereal poured into an 8-ounce yogurt tub is easy to eat on the run. So are plain toasted whole-grain frozen waffles—no syrup needed.
- Untraditional choices are often acceptable. Purchase or make ahead enough carrot sticks to divide among several containers; serve with yogurt or bean dip. Leftover casseroles, stews, or pasta dishes are nutritious choices that children can eat hot or cold.

Is Breakfast Really the Most Important Meal of the Day for Children?

Elders have long held that breakfast is the most important meal of the day, and for children, this bit of wisdom is now backed by science. A nutritious breakfast that includes cereal is a central feature of a diet that meets the needs of children and supports their healthy growth and development.[30]

Children who eat no breakfast perform poorly in tasks requiring concentration; their attention spans are shorter, they achieve lower test scores, and they are tardy or absent more often than their well-fed peers. Common sense tells us that it is unreasonable to expect anyone to study and learn when no fuel has been provided. Even children who have eaten breakfast suffer from distracting hunger by late morning. Chronically underfed children suffer all the more.[31]

The U.S. government funds several programs to provide nutritious, high-quality meals, including breakfast, to children at school. Schools that begin to participate in the federal school breakfast program observe higher achievement test scores and lower tardiness and absence rates. One Canadian study found that an astonishing 15 to 20 percent of children attended school once or more each week without eating breakfast and just 30 percent consumed adequate daily servings from all food groups.[32]

KEY POINT *Breakfast is critical to school performance. Not all children start the day with an adequate breakfast, but school breakfast programs help to fill the need for some.*

How Nourishing Are the Lunches Served at School?

For the past 50 years, lunches served at school have been meeting many of the nutrient needs of the nation's children. Today, 66 percent of children from age 6 to 11 years partake of school lunches providing servings of milk, protein-rich foods (meat, poultry, fish, cheese, eggs, legumes, or peanut butter), vegetables, fruits, and breads or other grain foods each day.[33] The lunches are designed to provide at least a third of the recommended intake for each of the nutrients. Table 13-5 shows school lunch patterns for different ages.

Parents often rely on school lunches to meet a significant part of their children's nutrient needs on school days. Indeed, students who regularly eat school lunches have higher intakes of energy and nutrients than students who do not. Children don't always like what they are served, though, and school lunch programs must strike a balance between what children want to eat and what will nourish them and guard their health.

Many schoolchildren in the United States have significant risk factors for developing cardiovascular disease.[34] In an effort to help reduce their risk, the U.S. Department of Agriculture (USDA) has ruled that all government-funded meals served at schools must follow the *Dietary Guidelines for Americans.*[35] This admirable attitude leaves many schools with a problem, however. Though a school's own cafeteria may serve such meals, private vendors offer other, unregulated meals, even fast foods, side-by-side with the school lunches. U.S. children develop a taste for fat and salt early in

Table 13-5
School Lunch Patterns for Different Ages

Food Group	Preschool (Age) 1 to 2	Preschool (Age) 3 to 4	Grade School Through High School (Grade)[a] K to 3	4 to 6	7 to 12
Milk					
1 serving of fluid milk[b]	¾ c	¾ c	1 c	1 c	1 c
Meat or Meat Alternate					
1 serving:					
Lean meat, poultry, or fish	1 oz	1½ oz	1½ oz	2 oz	3 oz
Cheese	1 oz	1½ oz	1½ oz	2 oz	3 oz
Large egg(s)	½	¾	¾	1	1½
Cooked dry beans or peas	¼ c	⅜ c	⅜ c	½ c	¾ c
Peanut butter	2 tbs	3 tbs	3 tbs	4 tbs	6 tbs
Peanuts, soynuts, tree nuts, or seeds[c]	½ oz	¾ oz	¾ oz	1 oz	1½ oz
Vegetable and/or Fruit					
2 or more servings, both to total	½ c	½ c	½ c	¾ c (plus ½ c extra over a week)	¾ c
Bread or Bread Alternate Servings[d]	5 per week (minimum ½ per day)	8 per week (minimum 1 per day)	8 per week (minimum 1 per day)	8 per week (minimum 1 per day)	10 per week (minimum 1 per day)

[a]These patterns may be used so long as the meals served meet the *Dietary Guidelines for Americans* and provide one-third of the child's recommendations for nutrients.
[b]Whole milk and unflavored low-fat milk must be offered; flavored milks or fat-free milk may also be offered.
[c]These foods may meet no more than one-half a serving of meat and must be accompanied by other meat or alternate in the meal.
[d]A serving is 1 slice of whole-grain or enriched bread; a whole-grain or enriched biscuit, roll, muffin, or the like; or ½ cup cooked rice, pasta, or other grain.
SOURCE: U.S. Department of Agriculture, 1998.

life and thus may reject nutritious school meals when offered a choice of meals higher in fat and salt. Children receive a mixed message when they are left on their own to choose between the health-supporting school lunch and high-fat, high-salt, low–nutrient density foods that their taste buds may prefer.[36]

Some children in high schools also face the additional option of choosing soft drinks, frozen confections, candies, and other low-nutrient treats from school snack bars, vending machines, or school stores.[37] No federal laws exist to restrict sales of these items to schoolchildren. The administrators of the school lunch program have tried to outlaw such sales on school grounds, but have been defeated by the powerful lobbying efforts of industries that reap huge profits from children's pocket money. Given a choice, though, many children still select nutritious snacks, such as yogurt, milk, or fruit juices, when these foods are also made available.

KEY POINT *School lunches are designed to meet at least a third of the daily nutrients needed by growing children and to stay within limits set by the* Dietary Guidelines for Americans. *Vending machines, fast food, school stores, and snack bars tempt schoolchildren with foods high in fats and sugars.*

The American Dietetic Association has set nutrition standards for child-care programs. Among them, meal plans should:

- Be nutritionally adequate.
- Involve parents in planning.
- Meet the Dietary Guidelines for Americans.
- Follow recommended meal patterns while respecting cultural and ethnic differences.
- Minimize added fat, sugar, and sodium.
- Emphasize fresh fruit, fresh and frozen vegetables, and whole grains.
- Respect children's small appetites.

SOURCE: Position of the American Dietetic Association: Nutrition standards for child-care programs, *Journal of the American Dietetic Association* 99 (1999): 981–988.

Nutrition Education

Coincident with the school lunch program is the Nutrition Education and Training (NET) program available to all public schools. The major goals of NET are these:

- To encourage lifelong healthy eating habits in children and parents by teaching them about the roles of nutrition in supporting health.
- To educate school foodservice personnel and others in the preparation of nutritious meals and snacks that meet the *Dietary Guidelines for Americans.*

epiphyseal (eh-PIFF-ih-seal) **plate** a thick, cartilage-like layer that forms new cells that are eventually calcified, lengthening the bone (*epiphysis* means "growing" in Greek).

- To support classroom nutrition education.
- To provide educators with resources and ideas to combine classroom and mealtime experiences into an effective nutrition curriculum.

Nutritionists and others regularly lobby their legislators to fund NET, so the program remains in effect but it operates on a shoestring.* As children grow, they will need nutrition knowledge to enable them to make healthy food choices, as the choices become theirs to make.

KEY POINT *Schools share with families the responsibility of offering nutrition education to children.*

Nutritious snacks play an important role in an active teen's diet.

The Teen Years

Teenagers are not fed; they eat. The teen years bring a search for identity, which is acquired largely through trial and error. At the same time, teens face tremendous pressures from peers and the media, especially regarding body image. Many teens readily adopt fads and scams offering promises of slenderness, good-looking muscles, freedom from acne, or control over symptoms that may accompany menstruation. Nutrient needs are high during adolescence, and choices made during the teen years profoundly affect health, both now and in the future.

Growth and Nutrient Needs of Teenagers

With the onset of adolescence, needs for all nutrients become greater than at any other time of life except during pregnancy and lactation. The need for iron is especially great to support menstruation in girls and to develop lean body mass in boys.

Adolescence and the Bones Adolescence is a crucial time for bone development. The bones are growing longer at a rapid rate (see Figure 13-3) thanks to a special bone organ, the **epiphyseal plate**, that disappears as a teenager reaches adult height. At the same time, the bones aqre gaining density, laying down the calcium that may make the difference between weak and strong bones later in life. Low calcium intakes are all too common, and especially when paired with physical inactivity, these low intakes may compromise the development of peak bone mass.[38] The requirement for calcium is high during these years, but teenagers often choose soft drinks over milk (see Table 13-6).[39] Soft drinks, when chosen as the primary beverage, may affect the density of the bones because they displace milk from the diet. Milk chosen instead of soft drinks as the primary beverage provides calcium and other bone-building nutrients as well.[40] The attainment of maximal bone mass during the young years is considered the best protection against age-related bone loss and fractures in later life.

Table 13-6

Soft Drink Consumption of U.S. Adolescents

In general, soft drink consumption is inversely associated with the consumption of the nutrients in milk and fruit juices.

This Percentage of Adolescents:	Consumes This Many Soft Drinks Each Day:
22%	More than 26 oz (more than 3¼ c or 2 cans)
28%	13 to 26 oz (1¾ c to 3¼ c or 2 cans)
32%	Up to 13 oz (about 1¾ c or 1 can)
18%	None

SOURCE: Data from L. Harnack, J. Stang, and M. Story, Soft drink consumption among U.S. children and adolescents: Nutritional consequences, *Journal of the American Dietetic Association* 99 (1999): 436–441.

The Body Changes of Adolescence Teenagers' rates and patterns of growth vary tremendously. This intensive growth period, or the adolescent growth spurt, brings hormonal changes that profoundly affect every organ of the body, including the brain. Girls' growth spurts begin at 10 or 11 years of age and peak at about 12 years. Boys' growth spurts begin at 12 or 13 years and peak at about 14 years, slowing down at about 19. Growth charts used to track children's normal height and

*To express your support for nutrition education in schools, write to the president of the United States and your state legislators and ask that NET be securely funded.

Figure 13-3

Growth of Long Bones

Bones grow longer as new cartilage cells accumulate at the top portion of the epiphyseal plate and older cartilage cells at the bottom of the plate are calcified.

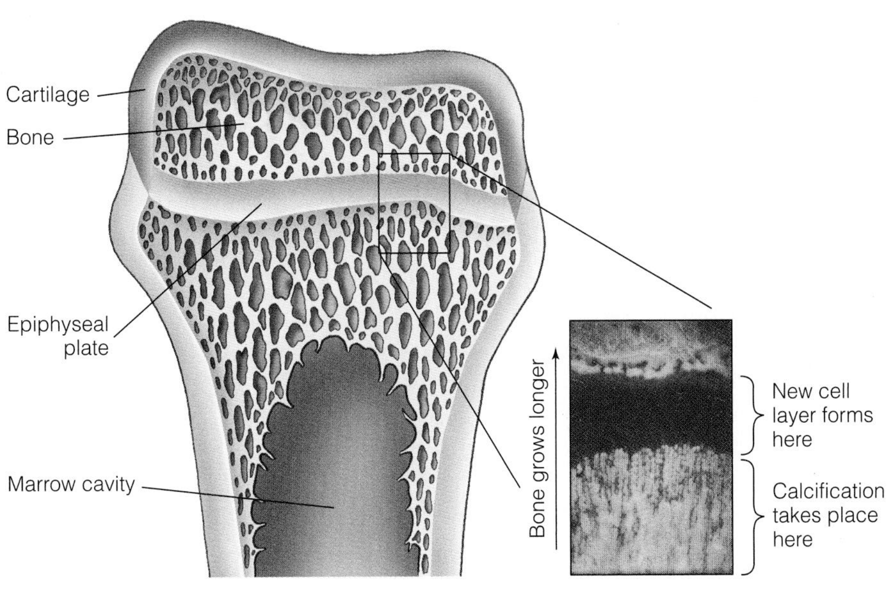

Cartilage
Bone
Epiphyseal plate
Marrow cavity
Bone grows longer
New cell layer forms here
Calcification takes place here

puberty the period in life when a person develops sexual maturity and the ability to reproduce.

premenstrual syndrome (PMS) a cluster of symptoms that some women experience prior to and during menstruation. They include, among others, abdominal cramps, back pain, swelling, headache, painful breasts, and mood changes.

weight gains by age don't fit teens very well, but height and weight charts meant for adults fit even less well. Two boys of the same age may vary in height by a foot, but if both have been growing steadily, each is fulfilling his genetic destiny according to an inborn schedule of events. Parents should watch only for reasonably smooth progress; to apply external standards that a child cannot "live up to" is to invite a lasting diminished self-image. The only way to be sure that a teenager is growing satisfactorily is to compare each new height and weight measure with her or his own measures taken earlier. Health-care providers also compare measures of the changes of **puberty** with standard rating scales.

The energy needs of adolescents vary tremendously. An active, rapidly growing boy of 15 may need 4,000 calories or more a day just to maintain his weight, but an inactive girl of the same age who is growing slowly may need less than 2,000 calories to keep from becoming obese. The insidious problem of obesity may first become apparent in adolescence, mostly in girls, and may last a lifetime. Girls normally develop a somewhat higher percentage of body fat than boys do, a fact that causes much needless worry about becoming overweight. Teen athletes especially need energy and nutrients, and the nutrition advice to athletes in Chapter 10 is especially important for them.

One of the many changes girls face as they become women is the onset of menstruation. The hormones that regulate the menstrual cycle powerfully affect not just the uterus and the ovaries but metabolic rate, glucose tolerance, appetite, food intake, mood, and behavior. Most women live easily with the cyclic rhythm of the menstrual cycle, but some are afflicted with physical and emotional pain prior to menstruation, a condition called **premenstrual syndrome**, or **PMS.** This chapter's Consumer Corner offers more on PMS.

Food sources of iron and calcium are listed in Chapter 8, and peak bone mass was discussed in Controversy 8.

KEY POINT *Nutrient and energy needs of teens vary with gender, body size, and activity level. Growth patterns vary widely.*

CONSUMER CORNER

Nutrition and PMS

A WOMAN SUFFERING from PMS may complain of any or all of the following symptoms: cramps and aches in the abdomen, back pain, headache, acne, swelling of the face and limbs associated with water retention, food cravings (especially for chocolate and other sweets), abnormal thirst, pain and lumps in the breasts, diarrhea, and mood changes, including both nervousness and depression. The symptoms can range from mild complaints to a severe, life-altering form—premenstrual dysphoric disorder—that poses major problems to the women who suffer from it and to the clinicians trying to treat it. A diagnosis of PMS is based on the timing and pattern of symptoms reported daily in records kept by women who seek help. Many physicians, however, may be unaware of diagnostic techniques and offer less than helpful treatments for PMS.[1] This finding is not surprising for scientific literature concerning PMS is in an early stage of development, and no consensus exists.

One of the candidates as a cause of PMS is altered response to the two major regulatory hormones of the menstrual cycle, estrogen and progesterone.[2] In particular, the hormone estrogen may affect the brain's neurotransmitters and, in turn, alter mood. One of the symptoms of PMS is depression, a mood disorder that many people, both male and female, experience under a wide variety of conditions. Adequate serotonin in the brain buoys a person's mood. Conversely, deficient serotonin creates a dysphoric state (depressed mood) and is related to major depressive disorders. In fact, some of the most effective antidepressants work by elevating the brain's serotonin levels (more about serotonin in this chapter's Controversy). In PMS, the natural rise and fall of estrogen levels in the blood may affect the activities of serotonin in the brain during the last half of each menstrual cycle. Mood often improves with administration of hormones in the form of oral contraceptives to eliminate estrogen's peaks and valleys.[3]

The major connections between PMS and nutrition concern energy metabolism and vitamin and mineral status. During the two weeks prior to menstruation, two things are believed to happen that may affect a woman's energy metabolism:

- The basal metabolic rate during sleep speeds up, although the daytime rate may not change.[4]
- Appetite and calorie intakes may increase.

Most studies seem to indicate that women take in an average of 300 calories a day more during the ten days prior to menstruation than during the ten days after it. As a consequence, women who wish to control their weight may find it relatively easy to restrict calories during the two weeks following menstruation. During the two weeks before the next menstruation, however, they may find it harder to limit calories because they are fighting a natural, hormone-governed increase in appetite.

A particularly common symptom of PMS is sodium retention, with the water retention that accompanies it. Some doctors prescribe diuretics to get rid of the excess sodium and water, with mixed results. The placebo effect is extraordinarily powerful in PMS. Often a drug or other treatment that first appears to relieve PMS in the short term proves useless in the long term

because the early relief attributed to the treatment was in fact brought about by the placebo effect. Diuretic therapy causes loss of minerals such as potassium, possibly making PMS symptoms worse. Also, if women do retain sodium and water just before menstruation, their effect may be normal and desirable.

The role of vitamin B_6 in PMS has been heavily researched. The logic of ascribing PMS to a vitamin B_6 deficiency is that women with PMS may have abnormal levels of hormones that require vitamin B_6 for their action. In trials, a few subjects with PMS have responded favorably to treatment with vitamin B_6, but the improvement is not statistically meaningful. No need exists for megadoses of vitamin B_6, and the hazards associated with such doses are well documented (see Table 7-6 of Chapter 7).

Vitamin E deficiency is another possible contributor to PMS. One double-blind, placebo-controlled study of 75 women suggested that supplemental vitamin E brought relief from sore breasts associated with PMS, when the placebo did not. Some women without PMS also have sore breasts, however, and they, too, can sometimes be relieved by vitamin E. Possibly, then, vitamin E deficiency does not cause PMS, but can worsen symptoms associated with the menstrual period.

In a recent randomized, double-blind, placebo-controlled study, women were given a daily supplement of 200 milligrams of magnesium or a placebo. The magnesium seemed effective in reducing weight gain, swelling and water retention, and breast tenderness.[5] In another well-controlled study, calcium supplements produced a 50 percent reduction in PMS-related moodiness, water retention, food cravings, and pain.[6] Few women in the United States consume enough magnesium or calcium from their diets, so an obvious first step for women with PMS is to make sure that the food they choose provides all of the nutrients they need.

Tea consumption has been strongly linked with PMS. Women who drink the most tea seem to have the worst symptoms. Which component of tea—the caffeine, pigments, or other substances—is responsible is not known, but evidence indicates a role for caffeine. Data from questionnaires administered to more than 800 women correlated caffeine intakes with PMS in a linear fashion: the more caffeine-containing beverages the women reported drinking, up to 10 cups per day, the more symptoms of PMS they reported suffering. Slight increases in PMS symptoms accompanied even one cup of a caffeinated beverage a day. Thus any woman who finds menstrual symptoms troublesome may want to try a caffeine-free lifestyle for a while and see if her symptoms improve.

One thing seems clear: the woman with PMS should look to her total lifestyle, of which diet is only a part. Adequate sleep and physical activity help, and controlling stress may be important.[7] She should also moderate her intakes of caffeine, salt, alcohol, and any other abusable substances. Finally, she should watch out for snake-oil salespeople selling PMS "cures," for they are everywhere.

Eating Patterns and Food Choices

With a multitude of after-school, social, and job activities, teenagers often fall into irregular eating habits and miss out on nutrients. Ideally, the adult becomes a **gatekeeper**, controlling availability but not intakes of food in the teenager's environment. In households where all the adults work outside the home, teens may be asked to perform some of the gatekeeper's roles, such as shopping for groceries or choosing fast foods or prepared foods. Though this arrangement may work well for some, teens typically turn a deaf ear to adults' attempts at persuasion to choose nutritious foods. Instead the teens listen to their taste buds and repeatedly select familiar favorite foods in a monotonous diet that lacks the nutrients they need.[41]

gatekeeper with respect to nutrition, a key person who controls other people's access to foods and thereby affects their nutrition profoundly. Examples are the spouse who buys and cooks the food, the parent who feeds the children, and the caretaker in a day-care center.

The nutritive values of selected fast foods are presented in the Table of Food Composition, Appendix A.

Wise gatekeepers set examples to follow and provide access to nutritious foods that are low in sugar and fat. They make sure that their teenage sons and daughters and their friends find plenty of nutritious, easy-to-grab food in the refrigerator (meats for sandwiches, raw vegetables, milk, fruit and fruit juices) and more in the cupboards (breads, peanut butter, nuts, popcorn, cereals).

On the average, about a fourth of a teenager's total daily energy intake comes from snacks. They are one way that teens with irregular schedules can gain all the protein, thiamin, riboflavin, vitamin B_6, magnesium, and zinc that they need. Regardless of how often teenagers eat, their diets are commonly too high in fat, sodium, and protein, and too low in fiber to support the future health of their arteries.[42] Their calcium intakes often fall short unless they snack on dairy products, and they often fail to obtain enough iron and vitamin A. For iron and other nutrients, a teen might snack on iron-containing meat sandwiches, low-fat bran muffins, or tortillas with spicy bean spread along with a

acne chronic inflammation of the skin's follicles and oil-producing glands, which leads to an accumulation of oils inside the ducts that surround hairs; usually associated with the maturation of young adults.

glass of orange juice to help maximize the iron's absorption. For vitamin A, why not carrot sticks, mixed vegetable juice, cantaloupe, or some dried apricots?

Inevitably, teenagers do a lot of eating away from home. They love fast food, and fortunately, some fast-food establishments are offering nutritious choices alongside their standard fat-laden fare. This is a positive trend, but teenagers must choose the foods they need from among those offered. The gatekeeper can help by presenting the needed nutrition information in a way that is meaningful to the individual teen. Teens who are prone to gain weight will often open their ears to news about the fat and calorie contents of fast foods. Others attend best to information about the negative effects of an ill-chosen diet on sport performance.

Teenagers are intensely involved in day-to-day life with their peers and in preparation for their future lives as adults. The gatekeeper can set an example, provide an environment with plenty of nutritious foods, keep lines of communication open, and stand by with reliable nutrition information and advice, but the rest is up to the teens themselves. Ultimately, they make the choices.

KEY POINT *With planning, the gatekeeper can encourage teens to meet nutrient requirements by providing nutritious snacks.*

Acne

No one knows why some people get **acne** while others do not, but heredity plays a role—acne runs in families. The hormones of adolescence also play a role by stimulating the glands in the skin. The skin's natural oil is made in deep glands and is supposed to flow out through tiny ducts to the skin's surface. In acne, the ducts become clogged, and oily secretions build up in the ducts.

One medical treatment for acne is to apply a vitamin A relative, retinoic acid or Retin A, directly to the skin. This loosens the plugs that form in the ducts, allowing the oil to flow normally, but the acid may burn the skin and cause pimples to form, making the acne look worse at first. Retin A is also available in topical wrinkle creams for older skin because retinoic acid can make fine lines and wrinkles less obvious in some people.

Prescribed antibiotic pills and ointments work for some, and antibiotic ointments do not burn the skin. Some antibiotic creams contain a form of the mineral zinc because it improves their staying power on the skin and may reduce inflammation.[43] However, zinc supplements taken orally are not of benefit against acne in well-nourished people. The oral prescription medicine Accutane is made from vitamin A, but it is much more powerful than the vitamin itself and is effective against the deep lesions of cystic acne. Accutane is highly toxic and causes serious birth defects in the infants of women who have taken it during their pregnancies. Women with acne who wish to use Accutane should use contraception diligently before beginning treatment and for a time after treatment has ceased.

Although medicines made from vitamin A are successful in treating acne, vitamin A itself has no effect, and supplements of the vitamin can be toxic. Quacks remain undaunted by these facts, though, and market vitamin A supplements to people hoping to cure acne. Of course, a certain amount of vitamin A is essential for healthy skin, but too much can damage the body.

Among foods charged with aggravating acne are chocolate, cola beverages, fatty or greasy foods, milk, nuts, sugar, and foods or salt containing iodine. None of these factors has been proved to worsen acne, and two, chocolate and sugar, have been shown not to worsen it. Psychological stress, though, clearly worsens acne. Vacations from school often bring acne relief. Sun and swimming also help, perhaps because they are relaxing and also because the sun's rays kill bacteria and water cleanses the skin. Too much sun exposure in a teen may make skin cancer likely in later life, however.

One remedy always works: time. While waiting, attend to basic needs. Petal-smooth, healthy skin reflects a tended, cared-for body whose owner provides it with nutrients and fluids to sustain it, exercise to stimulate it, and rest to restore its cells.

KEY POINT *Althought no foods have been proved to aggravate acne, stress can worsen it. Supplements are useless against acne, but relief from stress, sunlight, and proven medications can help.*

life expectancy the average number of years lived by people in a given society.

life span the maximum number of years of life attainable by a member of a species.

longevity long duration of life.

The Later Years

This looks like a section about older people, but it is relevant even if the reader is only 20 years old. How you live and think at 20 years of age can profoundly affect the quality of your life at 60 or 80 years. Without realizing it, most people hold a stereotype, largely negative, of what it is like to be old—and then, later, they become that way. An old saying has it that "as the twig is bent, so grows the tree." Unlike a tree, however, you can bend your own twig.

Before you will adopt nutrition behaviors that will enhance your health in old age, you must accept on a personal level that you yourself are aging. People who fear age may make the mistake of equating age with disease. Given time, aging affects everyone, but disease can strike anyone at any stage of the life cycle. To learn what negative and positive views you hold about aging, try answering the questions in the margin. Your answers reveal not only what you think of older people now but also what will probably become of you. You may wish to review some of the reasons for your answers and change your beliefs if they are not supported by science. When older adults were asked to give tips to younger people on how to live life fully in the later years, they offered the ten suggestions in the margin.[44]

The majority of the U.S. population is now middle-aged, and as that group ages, the ratio of old people to young people is growing larger, a trend called the "graying" of America (see Figure 13-4 on the next page). Since 1950, the number of people over age 65 has doubled, and people over 85 years old are the fastest-growing age group.

In the United States, the **life expectancy** at birth is 79 years for women and 73 years for men, up from about 50 years in 1900.[45] Once a person survives the perils of youth and middle age to reach age 80, the average woman can expect nine more years of life and the average man seven more years.[46] Thanks largely to advances in medical science, including antibiotics and other treatments, the life expectancy almost doubled in the twentieth century. Still, the biological schedule that we call aging cuts off life at a genetically fixed point in time. The **life span** (the maximum length of life possible for a species) of human beings, 130 years, is probably the upper limit of human **longevity.**[47] Even this limit may one day be challenged with advances of medical and genetic technologies.[48] One caution: to date, scientists who study the aging process have found no specific diet or nutrient supplement that will prolong life, despite hundreds of unproven claims to the contrary.

KEY POINT *Life expectancy for U.S. adults has increased in the twentieth century and may increase further with genetic and medical advances.*

How will you age?

- In what ways do you expect your appearance to change as you age?
- What physical activities do you see yourself enjoying at age 70?
- What will be your financial status? Will you be independent?
- What will your sex life be like? Will others see you as sexy?
- How many friends will you have? What will you do together?
- Will you be happy? Cheerful? Curious? Depressed? Uninterested in life or new things?

Tips for productive aging:

1. Simplify your life; identify priorities and set limits.
2. Pay attention to yourself—your body, your mind, and your spirit.
3. Continue to teach, continue to learn; take up leisure activities (painting, woodwork).
4. Let yourself laugh and cry; be flexible; learn to navigate change.
5. Be charitable; make it a practice to give (wisdom, experience, money, time, yourself).
6. Be financially astute; invest early for retirement.
7. Get a life; you'll live better in retirement if you do.
8. Practice good nutrition and exercise.
9. Think about your past and future; deal with your mortality.
10. Be involved; be positive; link with others.

SOURCE: Adapted with permission from H. Kerschner and J. M. Pegues, Productive aging: A quality of life agenda, *Journal of the American Dietetic Association* 98 (1998): 1445–1448.

Nutrition in the Later Years

Knowledge of nutrition in older adults has grown considerably in the last decade. Nutrient needs become more individual with age, depending on genetics and individual medical history. For example, one person's stomach acid secretion, which helps in iron absorption, may decline, so that person may need more iron. Another person may excrete more folate due to past liver disease and thus need a higher dose. Table 13-7 lists some changes that can affect nutrition. Nutrition, in turn, affects the aging body profoundly.

Figure 13-4

U.S. Population Growth, 1960 to 1990

The "oldest old"—those over 85 years—are the fastest-growing age group in the United States. Between 1960 and 1990, the U.S. population grew 39 percent, but the population of those over 85 more than doubled. An estimated 25,000 Americans now living are 100 years old or older.

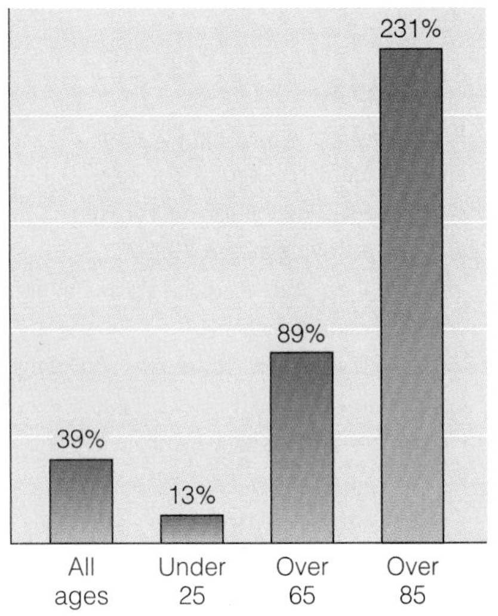

231%
89%
39%
13%
All ages
Under 25
Over 65
Over 85

The new DRI nutrient intake standards provide separate recommendations for those 51 to 70 years and for those 70 and older. See the inside front cover, page A.

Body mass index was discussed in Chapter 9.

KEY POINT *Nutrient needs change with age. Individual histories strongly influence older people's nutrient needs.*

Energy and Activity

Energy needs often decrease with advancing age. One reason is that the number of active cells in each organ decreases, reducing the body's overall metabolic rate, although much of this loss may not be inevitable. Another reason is that older people usually reduce their physical activity and so their lean tissue diminishes. For people older than 70 years, the best health and lowest risk of death have been observed in those who maintain a body mass index (BMI), ranging from about 25 to 32 on the BMI scale.[49] This is considerably higher than the BMI considered optimal for younger people. After about the age of 50 years, the intake recommendation for energy assumes about a 5 percent reduction in energy output per decade (see the inside back cover). For those who must limit energy intake, there is little leeway in the diet for foods of low nutrient density such as sugars, fats, and, of course, alcohol.

Current thinking seems to refute the idea that declining energy needs are entirely unavoidable. Physical activity, along with a diet adequate in nutrients and rich in phytochemicals, probably holds part of the key not only to maintaining energy needs but to upholding many other functions as well.[50] Adequate blood levels of nutrients are associated with a healthy immune response, a function often diminished in the aged.[51] Physical activity and diet also seem effective to some degree against a destructive spiral of sedentary behavior and mental and physical losses in the elderly that one expert has called "the dwindles."[52] The "dwindles" refers to a complex of interacting failures in the elderly including:

- Weight loss.
- Diminished mental function.
- Decreased physical ability to function
- Social withdrawal.
- Malnutrition.

The Fitness for Life Feature near here emphasizes the importance of physical activity to maintaining body tissue integrity throughout life.

KEY POINT *Energy needs decrease with age, but exercise burns off excess fuel, maintains lean tissue, and brings health benefits.*

Table 13-7

Examples of Physical Changes of Aging That Affect Nutrition

Digestive Tract	Intestines lose muscle strength resulting in sluggish motility that leads to constipation. Stomach inflammation, abnormal bacterial growth, and greatly reduced acid output impair digestion and absorption. Pain may cause food avoidance or reduced intake.
Hormones	For example, the pancreas secretes less insulin and cells become less responsive, causing abnormal glucose metabolism.
Mouth	Tooth loss, gum disease, and reduced salivary output impede chewing and swallowing. Choking may become likely; pain may cause avoidance of hard-to-chew foods.
Sensory Organs	Diminished senses of smell and taste can reduce appetite; diminished sight can make food shopping and preparation difficult.
Body Composition	Weight loss and decline in lean body mas lead to lowered energy requirements. May be preventable or reversible through physical activity.

Fitness in the Later Years

FITNESS FOR LIFE

It would be hard to overstate the importance of physical activity during the later years. Active older adults have greater flexibility and endurance, have a better sense of balance, suffer fewer falls and broken bones, have greater blood flow to the brain, have stronger immune systems, enjoy better overall health, and even live longer than their couch-loving peers.[1] A nutrition expert puts the importance of exercise to the elderly like this:

> We now know that physically active elders can build and rebuild muscle mass. Even the frail elderly can improve function by a remarkable 200 percent on a short, focused exercise regimen. No single feature of aging can more dramatically affect basal metabolism, insulin sensitivity, calorie intake, appetite, breathing, ambulation, mobility, and independence than muscle mass.[2]

Even institutionalized people in their *nineties* have been able to gain muscle bulk and strength to regain balance, and to add some pep to their walking steps after just eight weeks of weight training.

The photos below emphasize this point: they compare cross sections of the thigh of a young woman and of an old woman to demonstrate the muscle loss that sedentary aging can bring. Strength training helps to prevent at least some of this muscle loss, and recovery of lost muscle is possible when older poeple take up exercise.[3] Also, a person spending energy in physical activity can afford to eat more food, along with its associated nutrients. Some unscrupulous practitioners try to sell elderly people a "shortcut" to muscle and bone tissue retention in the form of growth hormone (GH) "therapy." Although secretion of GH does decline with age and the changes of aging are the opposite of the physiologic effects of GH, science does not yet support its usefulness or safety in reversing the tissue loss of aging.[4]

Any exercise, even a ten-minute walk a day, provides a benefit. Older people should feel free to exercise in their own way, at their own pace. They should not hold themselves to standards set in younger days. Although great achievements are possible and improvements are inevitable, an aging person unavoidably loses some capacity to perform exercise.

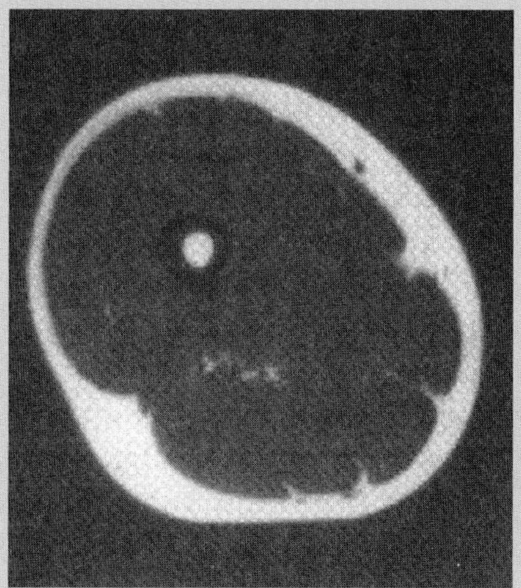

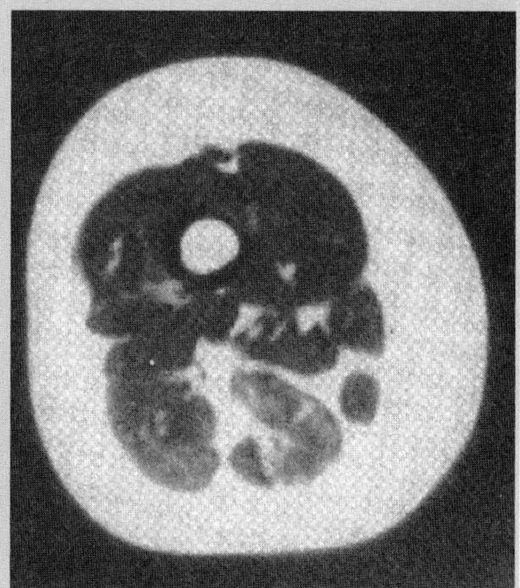

Cross sections of two thighs. These two women's thighs may appear to be about the same size from the outside, but the younger woman's thigh (left) is dense with muscle tissue. The older woman's thigh (right) has lost muscle and gained fat, changes that may be largely preventable with strength-building physical activities.

Carbohydrates and Fiber

The recommendation to obtain 6 to 11 servings of breads, grains, or pasta is appropriate for older people. It is especially wise to choose the majority of those servings from whole grains. With age, fiber takes on extra importance for its role against constipation, a common complaint among older adults and especially among nursing

arthritis a usually painful inflammation of joints caused by many conditions, including infections, metabolic disturbances, or injury; usually results in altered joint structure and loss of function.

gout a painful form of arthritis resulting from a metabolic abnormality in which excessive amounts of the waste product uric acid collect in the blood and uric acid salt is deposited as crystals in the joints.

Bogus or unproven arthritis treatments:

- Alfalfa tea.
- Aloe vera liquid.
- Any of the amino acids.
- Burdock root.
- Calcium.
- Celery juice.
- Copper or copper complexes.
- Dimethyl sulfoxide (DMSO).
- Fasting.
- Fresh fruit.
- Honey.
- Inositol.
- Kelp.
- Lecithin.
- Melatonin.
- Para-aminobenzoic acid (PABA).
- Raw liver.
- Selenium.
- Superoxide dismutase (SOD).
- Most vitamin or mineral supplements.
- Watercress.
- Yeast.
- Zinc.
- 100 other substances.

home residents. Older adults generally do not obtain the recommended daily 27 to 40 grams of fiber. When low fiber intakes are combined with low fluid intakes, inadequate exercise, and constipating medications, constipation becomes almost inevitable.

KEY POINT *Generous carbohydrate intakes are recommended for older adults. Including fiber in the diet is important to avoid constipation.*

Fats and Arthritis

Older adults must limit the overall fat in the diet for many reasons. Not only are the foods lowest in fat often richest in vitamins, minerals, and phytochemicals, but as Chapter 11 made clear, a diet high in certain fats is associated with many diseases. A high-fat diet also correlates closely with obesity, which, in turn, is a risk factor for developing **arthritis,** the painful deterioration and swelling of the joints that constitutes the leading cause of physical limitation in the United States.[53] Loss of body weight often brings relief, especially from the most common form of arthritis, osteoarthritis, and especially in the knees.[54]

Two kinds of arthritis afflict the bones: osteoarthritis and rheumatoid arthritis. Osteoarthritis affects millions of older adults, setting in from unknown causes as people age. During movement, the ends of normal bones are protected by small sacs of fluid that act as lubricants. With arthritis, the sacs erode, cartilage and bone ends disintegrate, and joints become malformed and painful to move. Beyond weight loss, nutrition does not seem to play a role in the causation of osteoarthritis, but high dietary intakes of vitamins E and C may help to slow its progression and ease pain somewhat once it has started.[55] Conversely, low intake of vitamin D may speed its progression.

Rheumatoid arthritis can strike at any age. It probably arises from a malfunction of the immune system—the immune system mistakenly attacks the bone coverings as if they were made of foreign tissue. In some cases, certain foods such as milk may possibly stimulate the immune attack, but for the most part, links with nutrition center around dietary antioxidants and the omega-3 fatty acid, EPA, found in fish oil.[56] EPA may interfere with activities of the fatty acid–derived hormonelike chemicals involved in inflammation. Alternatively, perhaps antioxidants in vegetables and fruits interfere with inflammation by reducing oxidative stress in the joints. In any case, supplements of vitamin E may help to reduce oxidative stress, but they do not improve active cases of rheumatoid arthritis. The same diet recommended for heart health—one low in fats and high in fruit, vegetables, and oils from fish—may help prevent or reduce the inflammation in the joints that makes arthritis so painful.

One form of arthritis known as **gout** worsens when sufferers consume foods that are high in purines, compounds that cause crystals of uric acid to form in the joints. Some of the best sources of omega-3 fatty acids, unfortunately, are also the highest in purines—among them, sardines, herring, anchovies, mackerel, and other fish and shellfish. This makes attempts at self-diagnosis and treatment of arthritis especially unwise. Though some with arthritis may benefit from increased fish intakes, others may make themselves worse.

No one universally effective diet for arthritis relief, and especially for osteoarthritis relief, is known. Many *ineffective* or unproven "cures" are sold, however, as the margin list shows. The safest bet for those with arthritis is to obtain a medical diagnosis and treatment.

KEY POINT *A diet high in fruits and vegetables and low in fats of meats and dairy products may improve some symptoms of arthritis. Omega-3 fatty acids may also have a positive effect. Foods high in purines can worsen the arthritis of gout.*

Protein

Protein needs of older people seem to remain about the same as for young adults. Too much protein, though, can be hard on the kidneys of older adults because of the extra burden of excreting its nitrogen. Which protein-rich foods elders choose to eat takes on extra importance, too. Some older people have lost their teeth, so chewing tough foods is next to impossible. They need soft cooked beans or meats or chopped foods. Individuals with chronic constipation, heart disease, or diabetes may receive benefits from fiber-rich low-fat legumes and grains as sources of protein. Such foods are easy to chew and can help stretch limited food budgets as well.

cataracts (CAT-uh-racts) thickening of the lens of the eye that can lead to blindness. Cataracts can be caused by injury, viral infection, toxic substances, genetic disorders, and, possibly, some nutrient deficiencies or imbalances.

KEY POINT *Protein needs remain about the same through adult life, but choosing low-fat fiber-rich protein foods may help control other health problems.*

Vitamins

Vitamin A stands alone among the vitamins, in that its absorption appears to increase with aging. For this reason researchers have proposed lowering the vitamin A requirement for aged populations. Some resist such a change, though, because foods containing vitamin A and its precursor beta-carotene are under study for preventing oxidative damage to body tissues, an effect described in Chapter 7.

Older adults face a greater risk of vitamin D deficiency than younger people do. Many older adults drink little or no vitamin D–fortified milk, and many go day after day with no exposure to sunlight, especially if they reside in nursing homes. As people age, vitamin D synthesis declines fourfold, setting the stage for deficiency. Thus, the recommendation for vitamin D intake for the elderly was recently doubled to 10 micrograms daily. Every elderly person should obtain this amount of vitamin D and get outside more often or even just sit by an open window some of the time.[57]

The Committee on DRI has recommended that adults aged 51 years and older should obtain 2.4 micrograms of vitamin B_{12} daily *and* that vitamin B_{12} fortified foods (such as fortified cereals) or supplements be used to meet much of the DRI recommended intake.[58] The committee's recommendation reflects the finding that between 10 to 30 percent of people older than 50 years lose the ability to produce enough stomach acid to make the protein-bound form of vitamin B_{12} available for absorption. The synthetic vitamin B_{12} is reliably absorbed, however, and much misery among elderly people could be averted by preventing deficiencies of vitamin B_{12}.

The macula of the eye was described in Chapter 7.

Of particular interest in aging are theories linking nutrients and phytochemicals in foods to age-related changes in the eyes. One such theory concerns the leading cause of permanent blindness in people over age 60—macular degeneration, already described in Controversy 7. People with lifelong high intakes of vegetables, particularly dark green leafy vegetables such as spinach and collard greens, rarely suffer from macular degeneration. Dark green leafy vegetables are rich in certain nonnutrient carotenoids that may protect the eyes from this destructive disease.[‡59]

The other theory concerns **cataracts.** A cataract is a thickening of the lens that impairs vision and ultimately leads to blindness. Cataracts can occur even in well-nourished individuals due to injury or other trauma, but most cataracts are vaguely called senile cataracts, meaning "caused by aging." Only 5 percent of people younger than 50 years have cataracts; by age 65, the percentage jumps to over 50 percent.

The lens of the eye is easily oxidized. People who eat few fruits and green vegetables obtain too few antioxidants, both nutrients and phytochemicals, and this causes oxidative stress in the eyes that puts them at risk of developing cataracts.[60] People taking supplements of vitamins C and E seem less likely to develop cataracts.[61] Sadly, many people needlessly endanger their vision by shunning the dark green leafy vegetables that may protect against both macular degeneration and cataracts.

Energy like this requires continued physical activity and all the nutrients to support it.

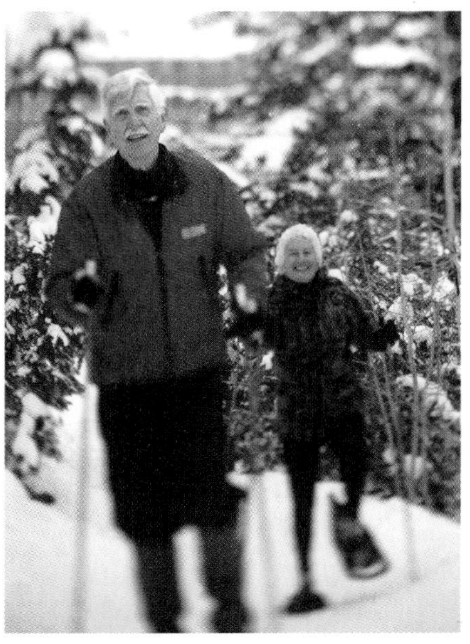

KEY POINT *Vitamin A absorption increases with aging. Older people suffer more from vitamin D deficiency than do young people. Cataracts and macular degeneration often occur among those with low fruit and vegetable intakes.*

‡The carotenoids are lutein and zeaxanthin, which help to form pigments of the macula of the eye.

I,LL TAKE
ONE SIX-PACK
PLEASE!
H_2O H_2O H_2O
Adults of all ages need 6 to 8 glasses of water each day.

Water and the Minerals

Dehydration is a major risk for older adults, who may not notice or pay attention to their thirst. With age, the thirst mechanism may become imprecise, and older people may go for long periods without drinking fluids. The kidneys also gradually lose the ability to efficiently recapture water before it is lost as urine. This causes some problems and worsens others, such as dehydration, constipation, and other intestinal problems. In a person with asthma, dehydration thickens mucus in the lungs, which may then block airways. Urinary tract infections, pneumonia, pressure ulcers (bed sores), and mental confusion can result.[62] Regardless of age, adults need to drink 6 to 8 glasses of water each day. A person we know uses this trick to ensure getting enough water: He keeps six inexpensive 8-ounce cups in the cupboard. Through the day he uses each one to drink water only once and then collects them in the dish drain. In the afternoon he checks the cupboard and makes sure to drink from any remaining cups. For him, drinking water has become a habit, and seldom are any cups left in the cupboard after supper.

Iron Among the minerals, iron deserves mention. Iron-deficiency anemia is less common in older adults than in younger people. In fact, iron status generally improves in later life, especially for women when menstruation ceases. Iron deficiency still occurs in some elderly people, however, especially in those with low food energy intakes. Aside from diet, other factors in many older people's lives make iron deficiency likely:

- Chronic blood loss from ulcers, hemorrhoids, or the like.
- Poor iron absorption due to reduced stomach acid secretion.
- Antacid use, which interferes with iron absorption.
- Use of medicines that cause blood loss, including anticoagulants, aspirin, and arthritis medicines.

Older people take more medicines than others, and drug and nutrient interactions are common.

These foods provide iron and zinc together: meat, poultry, liver, oysters, whole grains, fortified breakfast cereals,* and legumes.

*Cereals fortified with iron and zinc may not be available in Canada.

Zinc Zinc deficiencies are common in older people. Many older adults do not get the zinc they need, and many miss the mark by more than half.[63] Zinc deficiency, in turn, may depress the appetite and blunt the sense of taste, thereby leading to low food intakes and worsening of zinc status. Many medications interfere with the body's absorption or use of zinc, and elderly people often take many medicines, worsening the zinc nutrition picture.

The bright side of the zinc story is that some healthy older adults may need less than they did when they were younger. Research on zinc supplements demonstrates that nutrient supplements taken by the elderly can bring unexpected results. Researchers studying the immune response of elderly people sometimes observe a reduced immune response, sometimes an enhanced response, and sometimes no effect in those given supplements of zinc.[64] Vitamin A has been seen to depress the immunity of elders, while vitamin E may enhance it.[65] Overall, elderly people often benefit from a balanced low-dose vitamin and mineral supplement. Such supplements supply many of the needed minerals along with the vitamins often lacking in older people's diets without presenting too much of any one nutrient.[66] Many times, those taking such supplements suffer fewer sicknesses caused by infection.[67]

Calcium-rich foods are listed in Chapter 8, and Controversy 8 discusses osteoporosis.

Calcium Many previous sections of this book have emphasized the importance of abundant dietary calcium throughout life to protect against osteoporosis later on. The calcium intakes of many people, especially women, in the United States are well below the recommended amount. If fresh milk causes stomach discomfort, as the majority of older people report, then lactose-modified milk or other calcium-rich foods should take its place.

A summary of the effects of aging on nutrient needs appears in Table 13-8. As people live longer, attention to nutrition concerns can help to ensure the best possible quality of life.

Table 13-8

Summary of Nutrient Concerns in Aging

Nutrient	Effects of Aging	Comments
Energy	Need decreases.	Physical activity moderates the decline.
Fiber	Low intakes make constipation likely.	Inadequate water intakes and physical activity, along with some medications, compound the problem.
Protein	Needs stay the same.	Choices of low-fat, high-fiber legumes and grains meet both protein and other needs.
Vitamin A	Absorption increases.	Supplements normally not needed.
Vitamin D	Increased likelihood of inadequate intake; skin synthesis declines.	Daily moderate exposure to sunlight may be of benefit.
Vitamin B_{12}	Malabsorption of some forms.	Foods fortified with synthetic vitamin B_{12} or a low-dose supplement may be of benefit in addition to a balanced diet.
Water	Lack of thirst and increased urine output make dehydration likely.	Mild dehydration is a common cause of confusion.
Iron	In women, status improves after menopause; deficiencies linked to chronic blood losses and low stomach acid output.	Stomach acid required for absorption; antacid or other medicine use may aggravate iron deficiency; vitamin C and meat enhance absorption.
Zinc	Intakes are often inadequate and absorption may be poor, but needs may also increase.	Medications interfere with absorption; deficiency may depress appetite and sense of taste.
Calcium	Intakes may be low; osteoporosis becomes common.	Lactose intolerance commonly prevents milk intake; substitutes are needed.

Aging alters vitamin and mineral needs. Some needs rise while others decline.

Can Nutrition Help People to Live Longer?

The evidence concerning nutrition and longevity is intriguing. One approach studies the combined effects of nutrition and other lifestyle habits on aging. In a classic study, researchers in California observed nearly 7,000 adults and noticed that some were young for their ages, while others were old for their ages.[68] To find out what made the difference, the researchers focused on health habits and identified six factors that affect physiological age. Three of the six factors were related to nutrition:

- Abstinence from, or moderation in, alcohol use.
- Regular meals.
- Weight control.

The others were regular adequate sleep, abstinence from smoking, and regular physical activity. The physical health of those who reported all six positive health practices was comparable to that of people 30 years younger who reported few or none. Numerous studies have since confirmed the benefits of these six factors. The findings suggest that even though people cannot alter the year of their birth, they can alter the probable length and quality of their lives. The pair of tables in the margins of this page and the next, Tables 13-9 and 13-10, list some changes of aging that are unpreventable and also some that may yield to lifestyle influences.

The first evidence that diet might extend the length of life came more than half a century ago from experiments on rats. Researchers fed young rats diets adequate in all nutrients but short in energy. The rats stopped growing. Then the researchers increased the energy, and growth resumed. Meanwhile, control rats were allowed to eat and grow normally. Many of the rats in the energy-deficient group died young from the effects of malnutrition. A few survivors, however, lived an extraordinarily long time, and they developed the diseases of old age later in life, even though they had suffered malformations and stunting that did not improve with normal feed. These rats remained alive far beyond the normal life span for such animals.

In the study just described, food restriction was begun as soon as the animals were weaned (at three weeks). Later studies repeated these findings using more moderate

Table 13-9

Changes with Age You Probably Must Accept

These changes are probably beyond your control:

- ✔ Graying of hair
- ✔ Balding
- ✔ Some drying and wrinkling of skin
- ✔ Impairment of near vision
- ✔ Some loss of hearing
- ✔ Reduced taste and smell sensitivity
- ✔ Reduced touch sensitivity
- ✔ Slowed reactions (reflexes)
- ✔ Slowed mental function
- ✔ Diminished visual memory
- ✔ Menopause (women)
- ✔ Loss of fertility (men)
- ✔ Loss of joint elasticity

senile dementia the loss of brain function beyond the normal loss of physical adeptness and memory that occurs with aging.

Differences in maximum life span between animals eating normally and those that are energy restricted:

- *Rats*:
  Normal diet, 33 months.
  Restricted diet, 47 months.
- *Spiders*:
  Normal diet, 100 days.
  Restricted diet, 139 days.
- *Single-celled animals (protozoans)*:
  Normal diet, 13 days.
  Restricted diet, 25 days.

SOURCE: R. Weindruch, Caloric restriction and aging, *Scientific American*, January 1996, pp. 46–52.

energy restriction with adequate nutrient intakes that do not inflict physical malformations. In fact, restricted rats seem to retain youthfulness longer and develop fewer of the factors associated with chronic diseases.[69] Energy restriction in adult animals has been reported to lower blood pressure and blood glucose and to improve the insulin response to glucose. Interestingly, extending life span appears to depend on restricting energy and not on body fatness. Genetically obese rats live longer when energy is restricted, even though their body fat remains similar to that of nonobese rats allowed to eat freely.[70] Evidence from other species (see the margin) suggests that this effect spans many biological systems.

The obvious question is whether findings on aging taken from animal studies can be applied to human beings. Many of the physiological responses seen in rats during energy restriction do seem to occur in human beings who moderately restrict energy intakes. In experiments, when men of normal weight cut back on their usual energy intake by 20 percent, their body weight, body fat, and blood pressure dropped, and their HDL cholesterol rose—favorable changes for preventing obesity and chronic diseases.[71] Much more evidence must be collected, however, before such findings can be safely applied to the general population.

Investigators have proposed several mechanisms to explain how energy restriction prolongs life in rats, but none has been proved. Already mentioned is a delay in the onset of age-related diseases that seems to be related to restricted energy intake. Food restriction also slows the metabolic rate and helps to control blood glucose. A free-radical hypothesis of aging blames damage from oxidative stress for the physical deterioration associated with aging.[72] Research in support of this theory points out that the body's internal antioxidant enzymes diminish with age and that many "age-related" degenerative diseases are linked to free-radical damage.[73] This line of research is, unfortunately, the basis for a storm of anti-aging hoaxes in which elderly people are duped into spending their limited resources on worthless pills, supplements, and treatments with promises of life extension. Better to spend money on fresh fruit and green and yellow vegetables, known to provide the kinds of antioxidants the body can use, and on milk, lean meats, and whole grains for many more nutrients that support health throughout life.

KEY POINT *Lifestyle factors can make a difference in aging. In rats, food energy deprivation may lengthen the lives of individuals who survive the treatment. Claims for life extension through antioxidants or other supplements are common hoaxes.*

Table 13-10

Changes with Age You Probably Can Slow or Prevent

By exercising, eating an adequate diet, reducing stress, and planning ahead, you may be able to slow or prevent:

- ✔ Wrinkling of skin due to sun damage
- ✔ Some forms of mental confusion
- ✔ Raised blood pressure
- ✔ Speeded-up resting heart rate
- ✔ Reduced breathing capacity and oxygen uptake
- ✔ Increased body fatness
- ✔ Raised blood cholesterol
- ✔ Slowed energy metabolism
- ✔ Decreased maximum work rate
- ✔ Loss of sexual functioning
- ✔ Loss of joint flexibility
- ✔ Oral health: loss of teeth, gum disease
- ✔ Bone loss
- ✔ Digestive problems, constipation

Can Supplements Affect the Course of Alzheimer's Disease?

Alzheimer's disease is now the third costliest health problem in the United States, following heart disease and cancer.[74] In Alzheimer's disease, the most prevalent form of **senile dementia,** the brain deteriorates abnormally with brain cell death occurring in areas of the brain that coordinate memory and cognition. Dementia from conditions such as Alzheimer's may rob 6 to 10 percent of U.S. adults of productive life by age 65—and the rate doubles when milder cases are added to the count.[75] A cluster of symptoms justifies diagnosis: losses of memory and reasoning power, loss of the ability to communicate, and loss of physical capabilities, finally causing death. More research is needed, and quickly, to find solutions.

Nutrition bears only weak links to Alzheimer's disease, mostly centering on a buildup of metals, including aluminum in the brain tissue. Researchers often find elevated levels of copper, iron, and zinc in the brain tissues of those with Alzheimer's and theorize that these metals may accelerate the progression of the disease, possibly by increasing oxidative stress.[76] Some research casts doubt on the idea that supplements of zinc or other trace minerals can worsen Alzheimer's, but other research lends support to the theory. To err on the side of safety, food sources, not concentrated supplements, of trace minerals may be advisable for people with the disease.[77] As for the mineral aluminum, a causal connection seems unlikely. Brain aluminum in people with Alzheimer's exceeds normal brain aluminum by some 10 to 30 times. Still, blood and hair aluminum remains normal, indicating that the accumulation is caused by something in the brain itself, not by high aluminum in the diet. Other metals follow similar trends. Thus the high levels of trace elements in the brain must be more a result, than a cause, of the disease.

A new connection to a human gene that makes part of a lipoprotein* has sparked new hope for prevention of Alzheimer's.[78] Ultimately, researchers hope to use this genetic finding to develop an early test to identify people who are prone to develop Alzheimer's.

The brain of a person with Alzheimer's has an abnormally small amount of an enzyme that makes a compound from choline that is essential to memory.** To date, oral supplements of choline or lecithin (which contains choline, first mentioned in Chapter 5) have had no consistent effect on memory, mental functioning, or the progression of Alzheimer's. Some evidence suggests that vitamin E may help to slow its progression, but only in a dose (2,000 milligrams per day) high enough to adversely affect some people, especially those taking blood-thinning medications or those prone to strokes.[79]

Some degree of memory loss is often simply a function of aging, and is termed *benign* (meaning *harmless*) *senecent* (meaning *of aging*) *forgetfulness*. Occasional forgetful moments do not generally forecast the development of Alzheimer's disease in an older person.

Supplements of vitamin E may be safe up to 800 milligrams a day. See Chapter 7 for details.

"Smart" drugs, drinks, and supplements, sold with promises of brainpower enhancement, are discussed in this chapter's Controversy.

People wonder if alternative medicines, and particularly Gingko biloba, might reverse the mental impairment of their loved ones with Alzheimer's. To date, no proven benefits are available from herbs or other remedies, but claims from quacks are all too commonplace. The results of one small, yearlong study seemed to indicate a modest benefit in cognitive and social functioning in Alzheimer's patients given an extract of Ginkgo biloba.[80] This study has lead to all sorts of claims for the herb, including the cure and prevention of Alzheimer's disease. However, the findings did not indicate a cure or prevention by any stretch of the imagination, and the small beneficial effect reported there has yet to be supported or refuted in further research.

More promising are drugs used to treat Alzheimer's disease. Drugs that slow down the destruction of the choline compound just mentioned improve memory and other aspects of cognition.[81] Others seem to slow the advance of the disease in about 20 percent of users, but cannot reverse the damage already done. Meanwhile, other drugs, such as the nicotine of tobacco and estrogen hormone therapy, seem to favorably influence older people's ability to remember. Scientists also seem closer to developing a vaccine that may one day be able to prevent brain damage from Alzheimer's.[82]

Nutrient deficiencies, especially those that continue over many years, may contribute to losses of memory and thinking ability that some older adults experience. Subtle vitamin deficits also impair cognition and are now under study.[83] Such deficiencies are not believed to cause Alzheimer's disease, and they can be largely reversed with diet.

KEY POINT *Alzheimer's disease causes some degree of brain deterioration in as many as one-fifth of people past age 65. A genetic link to Alzheimer's may open the way to prevention. Current treatment helps only marginally; dietary aluminum is probably unrelated.*

Food Choices of Older Adults

Results of national surveys help indicate what older adults are eating. Many older people seem to have heard and heeded nutrition messages. They have cut down on saturated fats in dairy foods and meats and are eating slightly more vegetables and whole-grain breads. Smart marketers appeal to this growing group, many of whom are willing and able to spend more money on food than are people of other ages. Store shelves now prominently display good-tasting, low-fat, nutritious foods in easy-to-open, single-serving packages with labels that are easy to read. Many nutrient and antioxidant supplements are marketed for older adults. Whether to take a supplement is a personal choice, but as mentioned earlier, evidence supports the idea that a single low-dose multivitamin-mineral tablet a day can improve resistance to disease in the elderly. The best choice may be a supplement low in vitamin A, with ample amounts of all of the other vitamins for which daily intake recommendations are made (see the inside front cover).

---

*The gene associated with Alzheimer's is apo E4, one of three apolipoprotein E varieties. A report on the genetic and other aspects of Alzheimer's is available from Alzheimer's Disease Education and Referral Center, P.O. Box 8250, Silver Springs, MD 20907-8250.

**The memory compound made from choline is acetylcholine.

The DETERMINE predictors of malnutrition in the elderly:

- **D**isease.
- **E**ating poorly.
- **T**ooth loss or oral pain.
- **E**conomic hardship.
- **R**educed social contact.
- **M**ultiple medications.
- **I**nvoluntary weight loss or gain.
- **N**eed of assistance with self-care.
- **E**lderly person older than 80 years.

The Senior Nutrition Program is part of the Child and Adult Care Food Program (CACFP) designed to help public and private nonresidential child and adult day care programs provide nutritious meals to those younger than age 12 or older than 65, or people with disabilities.

Sources of support for the elderly:

- Social Security.
- Food Stamps.
- Senior Nutrition Program.
- Meals on Wheels.

Obstacles to Adequacy The food choices and eating habits of older adults are affected not just by preference but by the changes that accompany the experience of aging in our society. Whether people live alone, with others, or in institutions affects the way they eat. Men living alone, for example, are likely to consume poorer-quality diets than those living with spouses. Some older people suffer medical conditions that affect nutrition. They may have difficulty chewing because of tooth loss, or they may have lost a form of taste sensitivity so that they no longer seek a wide variety of foods.[84]

Two other factors also seem to make older people vulnerable to malnutrition: use of multiple medications and abuse of alcohol. People over age 65 take about a fourth of all the medications, both prescription and over-the-counter, sold in the United States. Typically, an older person receives prescriptions from several physicians, usually specialists, who are not aware that other drugs have been prescribed. Though medications may enable a person with health problems to live longer and more comfortably, they may also pose a threat to nutrition status because they interact with nutrients, depress the appetite, or alter the perception of taste (see Controversy 12).

The incidence of alcoholism, alcohol abuse, or problem drinking among the elderly in the United States is estimated at between 2 and 10 percent.[85] Evidence is also mounting that loneliness, isolation, and depression in the elderly accompany the overuse of alcohol. It isn't possible to say whether the depression or the alcohol abuse comes first, for each worsens the other, and both detract from nutrient intakes. Table 13-11 and the margin list provide means of identifying those who might be at risk for malnutrition.

Programs That Help For older people who lack funds to buy nutritious foods, federal programs can be of at least some help. Social Security provides an income to retired people over age 62 who paid in to the system during their working years. The Food Stamp program assists the very poor by supplementing their monthly food budgets with paper coupons, or electronic transfers of benefits to a card similar to a credit card, redeemable for food. The Senior Nutrition Program provides nutritious meals, social interaction, education and shopping assistance, counseling and referral to other needed services, and transportation to the elderly. An estimated 25 percent of the nation's elderly poor benefit from meals provided by the program.[86] Many people say that the shared midday meals this program provides are the high point of their day. They enjoy gathering with friends to share both conversation and nutritious meals. For the homebound, Meals on Wheels volunteers deliver meals to the door, a benefit even though the recipients miss out on the social atmosphere of the congregate program.

Table 13-11

Nutrition Screening Initiative Checklist for Older Americans

Circle the number to the right if the statement applies to you.

Statement	Yes
I have an illness or condition that makes me eat different kinds and/or amounts of food.	2
I eat fewer than 2 meals per day.	3
I eat few fruits or vegetables and use few milk products.	2
I have 3 or more drinks of beer, liquor, or wine almost every day.	2
I have tooth or mouth problems that make it hard for me to eat.	2
I don't always have enough money to buy the food I need.	4
I eat alone most of the time.	1
I take 3 or more different prescribed or over-the-counter drugs a day.	1
Without wanting to, I have lost or gained 10 pounds in the last 6 months.	2
I am not always physically able to shop, cook, and/or feed myself.	2
Total	

Score:
0–2: Good. Recheck your score in 6 months.
3–5: Moderate nutritional risk. Visit your local office on aging, senior nutrition program, senior citizens center, or health department for tips on improving eating habits.
6 or more: High nutritional risk. See your doctor, dietitian, or other health-care professional for help in improving your nutrition status.

NOTE: The Nutrition Screening Initiative is part of a national effort to identify and treat nutrition problems in older Americans.

Nutritionists are wise not to focus solely on nutrient and food intakes of the elderly because social interactions may be as important. A professor of psychiatry wrote perceptively of many elderly people, using the pronoun "he" to mean a typical elderly person:

> It is not what the older person eats but with whom that will be the deciding factor in proper care for him. The oft-repeated complaint of the older patient that he has little incentive to prepare food only for himself is not merely a statement of fact but also a rebuke to the questioner for failing to perceive his isolation and aloneness and to realize that food . . . for one's self lacks the condiment of another's presence which can transform the simplest fare to the ceremonial act with all its shared meaning.[87]

The need for companionship during meals is as great today as it was when these words were written.

Nutrition knowledge meets health in the real world of cooking, cleaning, and shopping, but many older people, even able-bodied ones with financial resources, find themselves unable to perform these tasks. For anyone living alone and for those of advanced age especially, it is important to work through the problems that food preparation presents.[88] This chapter's Food Feature presents some ideas.

KEY POINT *Food choices of the elderly are affected by aging, altered health status, and changed life circumstances. Assistance programs can help both by providing nutritious meals and by easing financial problems. Social stimulation also helps people eat well by relieving loneliness.*

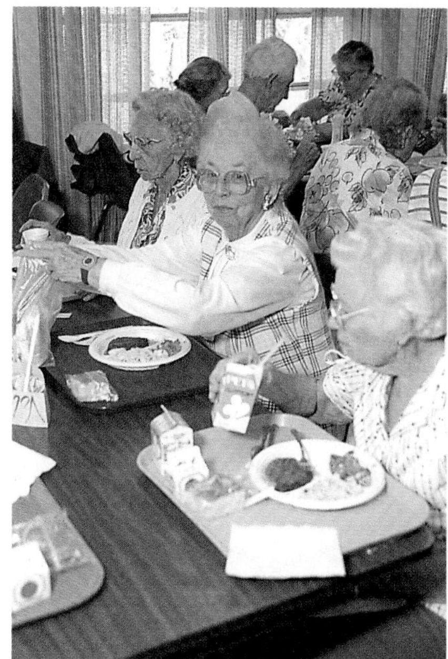

Shared meals can be the high point of the day.

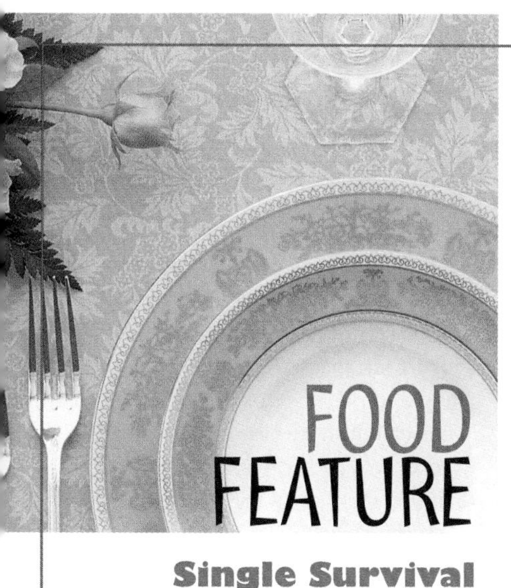

FOOD FEATURE

Single Survival and Nutrition on the Run

SINGLES OF ALL ages face problems when it comes to feeding themselves wisely. Their concerns range from selection of fast foods to the purchasing, storing, and preparing of food from the grocery store. Whether a single person is a busy student in a college dormitory, an elderly person in a retirement apartment, or a professional in an efficiency suite, the problems of preparing nourishing meals are often the same. Many such people live in places without kitchens and freezers, and for them, purchasing and storing foods can be especially problematic. Following is a collection of ideas gathered from single people who have devised answers to some of these problems.

Is Eating in Restaurants the Answer?

For the single person as for others, restaurants mean convenience. On average, almost 40 percent of a U.S. household's food budget is spent on foods prepared and eaten away from home. Even food consumed at home is not always prepared there. In any given month, up to 70 percent of households eat food that has been purchased from a carry-out restaurant.[89] Restaurant foods may be the quickest, easiest, and least taxing way to satisfy hunger at mealtime, but can they meet your body's nutrient needs or support health as well as homemade foods?

The answer may be "perhaps," with the condition that the diner makes the effort to meet nutritional needs when dining out.[90] Although a few chefs and restaurant owners are concerned with the nutritional health of their patrons, more often diners are on their own with regard to nutrition concerns. Restaurant foods are often overly endowed with calories, fat, saturated fat, and salt, yet not overly generous with needed constituents such as fiber, iron, or calcium.[91] In addition, as Chapter 2 pointed out, generous restaurant portions often equal three or more standard-size servings. Nevertheless, if you are willing to restrict your portions to sizes that do not exceed your energy needs, ask that excess portions be placed in take-out containers, and make judicious choices of foods that stay within intake guidelines for fat and salt, then restaurants can provide both convenience and nutrition. The Food Feature of Chapter 5 made many specific suggestions for ordering fast food and other foods with an eye to keeping fat intakes within bounds, and Chapter 8 showed where salt arises in the diet.

Grocery Store Take-Out Choices

The demand for take-out delicatessen-style foods from grocery stores has grown enormously. Such foods offer convenience—they can be purchased while shopping for other items—and a modicum of control over their nutrient contents. An additional attraction: they often cost substantially less than similar foods from restaurants. A bonus is that you specify the amount you need and then portion it onto your plate at home—you control how much you will buy and eat.

Depending on your choices, grocery store take-out can be an excellent bargain in terms of nutrient density, too. Choose from among roast chicken, smoked seafood, pasta with tomato sauce, steamed vegetables, precut salads and fruit without dressings, cooked beans, and plain baked potatoes for convenient nutrient bargains. These foods and others like them are prepared with a minimum of added fats and generally have no fatty gravy or sauce. Be aware, however, that some grocery store take-out foods rival fast foods in their calories, fat, and salt. Limit such choices as stuffing, macaroni and cheese, meat loaf and gravy, vegetables with creamy sauces, mayonnaise-dressed mixed salads, and, unless you plan to remove its crunchy skin before eating, fried chicken.

Buy only what you will use.

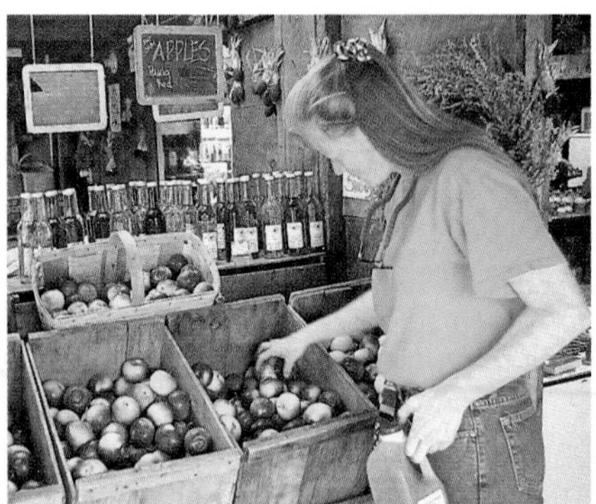
APPLES

More Grocery Store Know-How

In raw and fresh grocery store departments and in frozen food aisles, singles often face problems. Large packages of meat and vegetables are often suitable for a family of four or more, and even a head of lettuce can spoil before one person can use it all.

Buy only what you will use. Don't be timid about asking the grocer to break open a family-sized package of wrapped meat or fresh vegetables. Look for bags of prepared salad greens to take the place of lettuce in both salads and sandwiches. Prepared salads and other small-size containers of food may be expensive, but it is also expensive to let the unused portion of a large-size container spoil. Buy only three pieces of each kind of fresh fruit: a ripe one, a medium-ripe one, and a green one. Eat the first right away and the second soon, and let the last one ripen to eat days later.

Think up a variety of ways to use a vegetable that you must buy in large quantity. For example, you can divide a head of cauliflower into thirds. Cook one-third and eat it as a hot vegetable. Toss another third into a salad dressing marinade for use as an appetizer. Save the rest to use raw in salad.

Buy fresh milk in the sizes you can best use. If your grocer doesn't carry pints or quarts of milk, try a convenience store. If you eat lunch in a cafeteria, buy two pints of milk—one to drink and one to take home and store. Buy a loaf of bread and immediately store half, well wrapped, in the freezer (not the refrigerator, which will make it stale).

Food Preparation Hints

A wise person once said, "An hour spent organizing can save three hours

Time-saving tips to turn convenience foods into nutritious meals:

- Add extra nutrients and a fresh flavor to canned stews and soups by tossing in some frozen ready-to-use mixed vegetables. Choose vegetables frozen without salty, fatty sauces—prepared foods generally contain enough salt to season the whole dish including added vegetables.
- Buy frozen vegetables in a bag, toss in a variety of herbs, and use as needed. Vary your choices to prevent boredom.
- When grilling burgers, wrap a mixture of frozen broccoli, onion, and carrots in a foil packet with a tablespoon of Italian dressing and grill alongside the meat for seasoned grilled vegetables.
- Use canned fruits in their own juices as desserts. Toss in some frozen berries or peach slices and top with flavored yogurt for an instant fruit salad.
- Prepared rice or noodle dishes are convenient, but those claiming to contain broccoli, spinach, or other vegetables really contain just a trifle—not nearly enough to qualify as a serving of vegetable. Pump up the nutrient value by adding a half-cup of your frozen vegetables per serving of pasta or rice just before cooking.
- Purchase frozen onion, mushroom, and pepper mixtures to embellish jarred spaghetti sauce or small frozen pizzas. Top with parmesan cheese.
- Use frozen shredded potatoes, sold for hash browns, in soups or stews, or mix with a handful of shredded reduced-fat cheese or a can of fat-free "cream of anything" soup and bake for a quick and hearty casserole.

later on."** This holds true in food preparation. For shelf-stable items, prepare a space for rows of glass jars (jars from spaghetti sauce, applesauce, or other foods work well). Use the jars to store pasta, rice, lentils, other dry beans, flour, cornbread or biscuit mix, dry fat-free milk, and cereal, to name only a few possibilities. Light destroys riboflavin, so use opaque jars for enriched pasta and dry milk. Cut the directions-for-use label from the package of each item and store it in the jar. Place each jar, tightly sealed, in the freezer for a few days to kill any eggs or organisms before storing it on the shelf. Then the jars will keep bugs out of the foods indefinitely. The jars make an attractive display and will remind you of possibilities for variety in your menus.

Experiment with stir-fried foods. A large fry pan works well to stir-fry foods on a modern range. A variety of vegetables and meats can be enjoyed this way; inexpensive vegetables such as cabbage and celery are delicious when crisp cooked in a little oil with soy sauce or lemon juice added. Interesting frozen vegetable mixtures are available in larger grocery stores. Cooked, leftover vegetables can be dropped into a stir-fry at the last minute. A bonus of a stir-fried meal is that you'll have only one pan to wash.

Make mixtures using what you have on hand. A thick stew prepared from any leftover vegetables and bits of meat, with some added frozen onions, peppers, celery, and potatoes, makes a complete and balanced meal, except for milk. If you like creamed gravy, add some fat-free dry milk to your stew.

If you can afford a microwave oven, buy one. Cooking times are quick, and you'll use fewer pots and pans. Freeze or refrigerate meals in microwavable containers to reheat whenever you like. Be sure to use containers designed for the purpose, however. Margarine tubs, plastic bowls, and storage bags and containers can release potentially harmful chemicals into food when they are heated in the microwave oven. Use glass, or buy plastic containers that are labeled as safe for microwaving.

Depending on your freezer space, make a regular-size recipe of a dish that takes time to prepare: a casserole, vegetable pie, or meat loaf. Freeze individual portions in containers that can be heated later. Be sure to date these so you will use the oldest first.

Dealing with Loneliness

For nutrition's sake, it is important to attend to loneliness at mealtimes. The person who is living alone must learn to connect food with socializing. Cook for yourself with the idea that you will invite guests, and make enough food so that you can enjoy the leftovers later on. If you know an older person who eats alone, you can bet that person would love to join you for a meal now and then. Invite the person often.

**That wise person was Eva May Hamilton, one of the original authors of this textbook.

Invite guests to share a meal.

Self Check

Answers to the Self Check questions are in Appendix G.

1. Which of the following can contribute to choking in children?
   a. peanut butter eaten by itself
   b. reclining while eating
   c. popcorn and nuts
   d. all of the above
2. Which of the following is most commonly deficient in children and adolescents?
   a. folate
   b. zinc
   c. iron
   d. vitamin D
3. An allergic reaction to a food always involves:
   a. anaphylactic shock
   b. dislike of a particular food
   c. antibodies
   d. all of the above
4. Which of the following may alleviate symptoms of PMS?
   a. adequate vitamin E and vitamin B_6
   b. exercise
   c. omitting caffeine from the diet
   d. all of the above
5. Which of the following have been shown to improve acne?
   a. avoiding chocolate and fatty foods
   b. retinoic acid or Retin A
   c. vitamin A supplements
   d. all of the above
6. Research to date supports the idea that food allergies or intolerances are common causes of hyperactivity in children. T F
7. Children who watch more than two hours of television per day may be prone to develop elevated serum cholesterol and dental caries. T F
8. Dietary fat may play a role in causing arthritis. T F
9. Vitamin A absorption decreases with age. T F
10. Some nutrient deficiencies can cause mental symptoms. (Read about this in the upcoming Controversy.) T F

NUTRITION ON THE NET

For further study of the topics of this chapter, access these websites and search for the phrases or words in quotation marks:

1. For the latest changes in chapter material, to find supplemental learning tools, or access links to related websites, go to the "Nutrition: Concepts and Controversies" site at:
   www.wadsworth.com/nutrition/prod/allprod.html
2. For information about the "nutrient needs of young children and teenagers," visit the American Dietetic Association at:
   www.eatright.org
3. To learn more about food allergies, access the Food Allergy Network at:
   www.foodallergy.org
4. Search the U.S. Government site for more information about "hyperactivity" at:
   www.healthfinder.gov/searchoptions/topicsaz.htm
5. For more information about childhood obesity and development of disease:
   www.kidshealth.org
6. Information regarding the "National School Lunch Program" can be found at:
   www.fns.usda.gov/fns
7. The National Institute of Child Health and Development can provide information about bone development at:
   www.nih.gov/nichd
8. To learn more about the aging process, contact:
   www.nih.gov/nia
9. For additional information about "arthritis" and "Alzheimer's":
   www.healthfinder.gov/searchoptions/topicsaz.htm
10. To obtain information about nutrition problems and malnutrition in older adults, search for the "Nutrition Screening Initiative" at:
    www.aafp.org
11. Search among thousands of current scientific and medical abstracts for any topic related to "child, adolescent, or elderly nutrition" at:
    www.ncbi.nlm.nih.gov/PubMed/

FOOD, MIND, AND MEMORY

CONTROVERSY 13

WHY DO YOU feel like eating a steak at one meal and a doughnut at another? Why do you feel sleepy after lunch and not after dinner? Do some foods help you to think or to remember? Human behavior and the brain are still largely uncharted territory, and research is uncovering layer after layer of complexity. Researchers no longer doubt that intakes of food and nutrients affect the mind and memory, and they are beginning to understand how and why. [1] The neurosciences are sure to be the focus of many future nutrition investigations.

THE BRAIN AND ITS NEUROTRANSMITTERS

The brain has special needs. Encased in the hard, bony, inelastic helmet of its protective skull, the brain cannot expand and contract as can, say, the liver or adipose tissue. It cannot store its own reserve supply of glycogen or fat because those fuels take up space. It cannot store oxygen to oxidize those fuels or nutrients to help it do so. Therefore the brain must depend on the passing blood supply for both its fuels and its oxygen.

Furthermore, its needs for those substances are extraordinary. Although it accounts for only 2 percent of an adult's body weight, at any given time the brain contains 15 percent of the body's blood, and it devours 20 to 30 percent of the fuels that support the basal metabolism. Should the blood deliver too little oxygen or glucose, the brain's cells would cease communicating with each other (see Figure C13-1 on the next page) and coma would occur within minutes. Should the blood supply be interrupted altogether, coma would ensue within 10 seconds.

Nutrients of all kinds are crucial to brain function, and the blood supply must deliver these, too. For example, the brain requires amino acids to make its messenger molecules, some 30 to 40 **neurotransmitters** and related compounds. The brain also needs electrically charged minerals to help transmit its electrical impulses, vitamins and other minerals to facilitate these processes, lipids to repair its cell membranes, and water to maintain the fluids in which many of the chemical processes take place. Table C13-1 defines some terms relating to brain function.

The brain is extremely sensitive to fluctuations in its own internal chemical composition. To keep its internal environment constant, it has its own molecular sieve to filter from the blood the fluid and chemicals it needs: the **blood-brain barrier** (see Table C13-1). No matter how widely the chemical composition of the blood may fluctuate, the brain's internal milieu hardly changes at all. Because of its dependence on blood-borne fuels, oxygen, and nutrients, the brain monitors the blood closely and sends messages to other organs to signal its need for these substances. At one time the brain may need glucose; at another, amino acids. Animals regulate their intakes of protein and carbohydrate, each proportional to the other, in response to promptings from the brain. People probably do this, too, unconsciously.

A day's intake of lipids, vitamins, and minerals is unlikely to affect the brain's functioning immediately, although intakes of these nutrients probably do affect it over time. Amino acids, in contrast, are used to form neurotransmitters the same day they are eaten, so their effects are seen within minutes or hours of ingestion. Scientists exploring the effects of nutrients on the brain are especially interested in these associations:

- Some amino acids serve as the starting material from which some neurotransmitters are built.
- Vitamins and minerals assist enzymes in the syntheses of neurotransmitters.

Table C13-1
Brain Terms

- **blood-brain barrier** a barrier composed of the cells lining the blood vessels in the brain. These cells are so tightly glued to each other that blood-borne substances cannot get into the brain between the cells, but only by crossing the cell bodies themselves. Thus the cells, using all their sophisticated equipment, can screen substances for entry.
- **catecholamines** neurotransmitters made from the amino acid tyrosine: dopamine, epinephrine, and norepinephrine.
- **neurotransmitter** a chemical messenger released by a nerve cell when that cell is firing (conducting a nerve impulse). The neurotransmitter diffuses to the next nerve cell and alters the membrane of that cell, making it either less or more likely to fire. Exposed to enough neurotransmitter molecules, the next nerve cell will fire.
- **precursor control** control of a compound's synthesis by the availability of that compound's precursor. The more precursor there is, the more of the compound is made.
- **serotonin** a compound related in structure to (and made from) the amino acid tryptophan. It serves as one of the brain's principal neurotransmitters.

If the quantity of a nutrient eaten is to affect the brain directly, the nutrient must be free to come and go as it pleases; its brain concentration must fluctuate in response to diet. Most amino acids never exceed a given level in the brain no matter how much is consumed, but the brain's regulatory amino acids cross the blood-brain barrier freely. Furthermore, once in the brain, these precursor nutrients exert **precursor control**; that is, the brain responds to larger or smaller amounts of them by making larger or smaller amounts of neurotransmitters from them. Thus the food a person eats can influence brain chemistry by changing the rates at which the brain makes its neurotransmitters. These facts link eating directly to brain chemistry, and thereby, as you will see in a moment, the eater's mood and other sensations.

One neurotransmitter whose brain concentration is especially sensitive to changes in precursor supply has been studied in depth: **serotonin,** whose precursor is the amino acid tryptophan. Similarly, a set of neurotransmitters, the **catecholamines,** depend on the availability of their precursor amino acid, tyrosine. The discussion that follows centers on serotonin because more details about it are known.

Tryptophan to Serotonin

Ordinarily meals of the kind people eat every day raise or lower the concentration of serotonin in the brain, depending on the meal's protein and carbohydrate content.[2] Serotonin release, in turn, affects sensations and mood, so the ingredients of meals may have real effects on how people feel afterward.[3] Many people report feeling relaxed, peaceful, or

Figure C13-1

Communication within the Brain

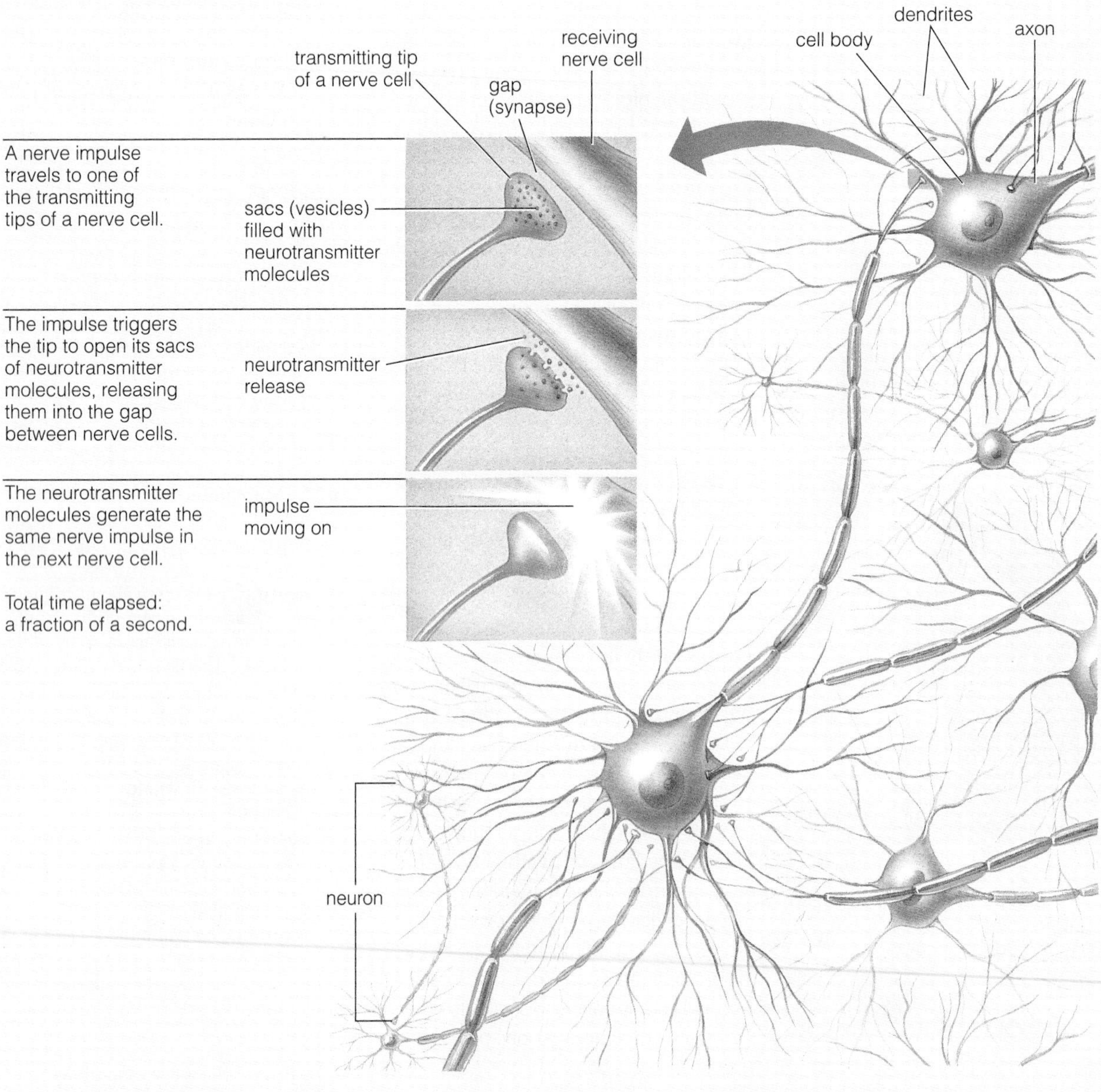

transmitting tip of a nerve cell
receiving nerve cell
gap (synapse)
cell body
dendrites
axon
A nerve impulse travels to one of the transmitting tips of a nerve cell.
sacs (vesicles) filled with neurotransmitter molecules
The impulse triggers the tip to open its sacs of neurotransmitter molecules, releasing them into the gap between nerve cells.
neurotransmitter release
The neurotransmitter molecules generate the same nerve impulse in the next nerve cell.
impulse moving on
Total time elapsed: a fraction of a second.
neuron

sleepy after eating certain foods; these sensations are sometimes attributed to serotonin's effects on the brain. Research shows that a lack of tryptophan flowing into the brain can manifest itself in wakefulness, depressed mood, a tendency to startle, and an enhanced sensitivity to pain. Animals that have been made tryptophan-deficient exhibit these symptoms, and when given tryptophan, they return to normal as their brain serotonin is restored. Tests show that tryptophan reduces sensitivity to pain in people, too.[4]

The amount of tryptophan that enters the brain depends not only on the amount of tryptophan the person eats but also on the total protein and carbohydrates eaten with it. If tryptophan is taken as a single amino acid, then brain serotonin increases proportionately. Normally, however, whole proteins, not just tryptophan, are eaten. In this case, some of the other large amino acids in the proteins compete with tryptophan for entry into the brain because they use the same transport mechanism to get across the blood-brain barrier. In this situation, tryptophan fails to enter the brain in increased quantities and so does not effectively enhance brain serotonin synthesis.

If carbohydrate is fed instead of protein, however, the carbohydrate can "help" to deliver tryptophan, already in the bloodstream, to the brain because it elicits the secretion of the hormone insulin. Insulin drives the *other* amino acids, but not tryptophan, into *body* cells, leaving the tryptophan free to enter the brain without competition. Thus, paradoxically, food high in carbohydrate—*not* food high in protein—eases tryptophan's transport into the brain and so promotes serotonin synthesis.

Choosing a meal to raise brain serotonin represents a problem, though. The amount of protein in most mixed meals, even those high in carbohydrate, is generally more than sufficient to block the tryptophan-delivering effect of carbohydrate.[5]

Researchers speculate that people's food choices may be partly governed by the brain's reactions to prior foods eaten. Thus, when a meal elicits top mental performance, the brain of the eater may be "trained" to prefer those foods and so to choose them often.[6] That could partly explain why, in the great majority of societies in the world, people seem naturally to prefer mixed meals over single foods. It may be that meals with mixed protein and carbohydrate sources provide just the right balance to support brain function.

Serotonin and Appetite An understanding of the absorption of tryptophan by the brain may help explain how animals and human beings, depending on what kinds of foods they have eaten last, seem to know what foods to choose next time to ensure balance. According to one theory, by raising brain serotonin, a high-carbohydrate meal satisfies a need and so reduces the urge to eat more carbohydrate. Therefore a person or animal who has eaten plenty of carbohydrate will seek out more protein at the next meal. A high-protein meal creates a serotonin deficit, awakens the carbohydrate craving, and once again leads to the consumption of carbohydrate-containing foods.

Though not everyone agrees, a chief appeal of this theory is that it seems to account for what is often observed to happen with low-carbohydrate diets. The more dieters try to restrict carbohydrate, the more they seem to crave it. The effect is accentuated if they are insulin resistant, as is likely in obese people. When an insulin-resistant person eats carbohydrate, insulin's normal actions do not follow, and the cells continue to hunger for glucose. Furthermore, brain serotonin does not rise, and so the carbohydrate craving is intensified.

Another theory links serotonin to obesity. The theory gains support from the finding that serotonin is often below average in the brains of obese people and of normal-weight people who report craving carbohydrate-rich food.[7] Though direct cause-and-effect conclusions are not yet possible, the authors of the study suggest that high-carbohydrate weight-loss diets may be the most successful because they work *with* many people's brain chemistry rather than against it.

Dieting in general seems to disrupt mental functioning somewhat.[8] During dieting, people are easily distracted from tasks requiring vigilance and have slower reaction times. They also score lower on memory tests than at times of normal eating. Whether these effects are related to neurotransmitter synthesis or are equally likely to result from any sort of change in regular eating habits remains a mystery.

Some evidence weakens the theory that links serotonin to the appetite for carbohydrate. When animals were injected with tryptophan, which should stimulate protein intake according to the theory, no preference for protein or carbohydrate was observed.[9] Ongoing research should ultimately help to untangle this problem, but it may well first become more knotty.

Serotonin, Mood, and Sleep Some people say they feel anxious, tense, and somewhat depressed before eating carbohydrate and feel the opposite afterward. The amino acid tryptophan given by itself often has similar effects, consistent with the notion that it is the indirect agent of carbohydrate's effect. When tryptophan is restricted in the diet, some people report depressed feelings.[10] When tryptophan is restored, depressed feelings lift, but in their place come drowsiness, clumsiness, and mental slowing.[11]

Reports exist of disturbed serotonin metabolism in people with major depression. Researchers interested in studying the effects of food on people's emotions fed tryptophan-deficient diets to two groups of young men: one with long family histories of depression and a matched (control) group with no reported depression in the family.[12] As blood tryptophan concentrations dropped, the men with family histories of depression scored significantly lower on a mood scale, indicating depression. None of the controls scored lower on the test—their moods hadn't changed. Major depression is a serious condition that is not reversible through diet

alone. Still, those who tend to be depressed may be more sensitive to the effects of serotonin than are others. Such people would do well to eat balanced meals at regular intervals to supply the brain with the materials needed to make serotonin.

Carbohydrate or tryptophan may also induce fatigue or sleepiness. The tryptophan effect is particularly well known from at least 50 studies in people and animals. Carbohydrate or tryptophan can also increase the error rate in performance tests. Elevated brain serotonin is known to reduce aggression in rats, and it may have that effect on people, too.

Studies demonstrate clearly that single amino acids, when administered alone, act like drugs in the body. These effects are interesting, but they are not well characterized, and people should not dose themselves with amino acids seeking mental effects. The Consumer Corner of Chapter 6 warned that to do so is to imperil health.

EFFECTS OF OTHER NUTRIENTS AND FOODS ON THE BRAIN

Nutrients other than amino acids are also involved in the synthesis of neurotransmitters. Iron is needed in one of the first steps of neurotransmitter synthesis. Vitamin B_6 and riboflavin are needed in later steps. These nutrients are but three among many; deficiencies of them are reflected in depressed or otherwise disturbed mood. Deficiencies of many nutrients also cause anemia, which produces mental symptoms of its own (see Table C13-2). Some mental effects of nutrient deficiencies become apparent only in severe deficiency states, but others such as fatigue, depressed mood, or impaired learning can be among the first symptoms of a developing deficiency.[13] Administration of the missing nutrient rapidly reverses these effects.

An interesting idea still under study concerns the possibility that lipids may affect human emotions, too. Several years ago, a research team examined mortality data from subjects participating in ongoing cholesterol-lowering studies. People receiving cholesterol-lowering drugs or dietary treatments were twice as likely as controls to die from suicide or violence. The possibility that lowering blood cholesterol might cause negative emotion gained strength with the report of a study of monkeys fed diets either high or low in fat and cholesterol. Monkeys fed the low-fat, low-cholesterol diet were rated as significantly more aggressive and less social and had lower brain serotonin than controls.[14] Low serum cholesterol in human beings has been linked with both suicide and aggression.[15] In contrast, fish oil was shown to reduce aggression in college students under the stress of final exams.[16]

A number of plausible mechanisms exist for the associations just described, but none has yet been proved. Heart disease brought on by high blood cholesterol is a proven threat to health and life; people receiving treatments for high blood cholesterol should make no changes until more studies make clear the implications of these early findings.

CAN CERTAIN FOODS OR "SMART" SUPPLEMENTS ENHANCE BRAIN FUNCTIONING?

If deficiencies of nutrients can cause mental disturbances, can extra amounts of some nutrients make brain function excel? The idea has appeal, especially to businesspeople, students, textbook authors, medical workers, and others who must remain alert despite long hours, little sleep, or international travel.

Table C13-2

The Mental Symptoms of Anemia

Apathy, listlessness
Behavior disturbances
Clumsiness
Hyperactivity
Irritability
Lack of appetite
Learning disorders (vocabulary, perception)
Low scores on latency and associative reactions
Lowered IQ
Reduced physical work capacity
Repetitive hand and foot movements
Shortened attention span

NOTE: These symptoms are not caused by anemia itself but by iron deficiency in the brain. Children with much more severe anemias from other causes, such as sickle-cell anemia and thalassemia, show no reduction in IQ when compared with children without anemia.

Purveyors of "smart" drugs, supplements, and drinks say these products can speed thinking and learning, jump-start a failing memory, and reverse aging processes.[17] Some drugs sold with these claims are being tested for reversal of the mental deterioration of Alzheimer's disease common in old age. Some are available by prescription, but people have to order others, often over the Internet, from foreign countries because they are not approved for use in the United States. The idea seems to be that if a drug can reverse or slow deterioration in a diseased brain, then perhaps it can boost the thinking power of a normal brain. An expert in the field of neurobiology of learning and memory commented concisely on smart drugs: "I think they are silly."[18]

Experts at the Food and Drug Administration (FDA) have warned that many of the smart products produce well-known side effects such as gastrointestinal distress, headaches, ulcers of the nasal cavity, and insomnia. They warn that long-term effects of many drugs are not known.[19]

Among herbs, an extract of the evergreen Gingko biloba is sold as a "brainpower" enhancer. Some studies support a theory that the herb's antioxidant effects may bring slight improvement to the mental functioning of the impaired aging brain.[20] Unknown is whether the herb can boost functioning in a healthy brain and whether long-term use is safe. Until more is known, it cannot be recommended.

The supplements and drinks, available from shopping mall juice bars and health food stores, not only claim to make people smarter, but also are promoted as a legal high (euphoria), or as legal stimulants or depressants, to people too young to buy alcohol. Made mostly of amino acids, vitamins, choline, and lecithin, none of these drinks or any other known nutrient supplement produces euphoria, but the placebo effect can do so.

Users tell convincing stories about the effects of these drinks, but none of the products has been shown to improve intelligence in clinical trials. Researchers who attempt to measure people's feelings of being smart, witty, energetic, or able to remember run into problems. Measurements of this sort are always clouded by the placebo effect and wishful thinking. Also, a person with slight nutrient deficiencies may well respond favorably to a potion that provides the missing nutrients.

Other mental effects of nutrients may exist as well. Some inquiries have suggested a role for ingestion of carbohydrate on the formation of memory. It may be that blood glucose somehow sparks the formation of memory or improves recall. The very act of eating also enhances memory, although the details of how it does so are not known. People given a snack during a learning task exhibit better recall later than when no snack is given. Hungry mice, fed immediately after learning a task, later remember how to perform the task better than do mice fed before the learning session. This effect may be a survival adaptation; in the wild, individuals who learn something new and obtain food as a result benefit from remembering the new behavior. Table C13-3 offers a playful memory quiz, just for fun.

A curious response to food is the almost universal love for the taste of chocolate, which some claim is a craving akin to an addiction. Most chocolate lovers say that eating chocolate lifts their spirits.[21] Chocolate contains phytochemicals, such as caffeine (a central nervous system stimulant), theobromine (another stimulant), and phenylethylamine (a biologically active amine), all of which could conceivably affect mood. One group of researchers suggested that one amine in chocolate may activate some of the same brain areas that marijuana targets.[22] Of course, chocolate has none of the mind-altering effects of marijuana, but if it stimulates the brain in ways that enhance the sensory qualities of food, as marijuana is known to do, this effect might make chocolate seem extra delicious. The vast majority of studies of chocolate's effects have not found chocolate to affect the brain's functioning or lift mood, however, doubtless a disappointment to chocoholics everywhere.[23]

APPLICATIONS

Can a person choose foods to maximize the brain's performance? With some qualifications, the answer seems to be that food choices probably can minimize diet-induced sluggishness, clumsiness, or poorer-than-normal memory capacity. The opposite food choices may also bring on those effects in some people who want to go to sleep.

A student who wishes to perform optimally on an examination, for example, may be wise to consider carefully the foods chosen for breakfast or lunch before the exam. It should be said that normal variations in meals presenting mixtures of carbohydrate, fat, and protein are unlikely to produce any noticeable effects on mental functioning. That is, mixed meals generally support mental functioning well. When the eater strays far from normal eating, however, brain functioning can change.[24]

If our exam-taking student tends to nod off during exams, then it might be best to avoid foods extremely high in carbohydrate and low in protein in the hours before the test.[25] Foods to avoid might include plain white rice or potatoes (eaten alone), rolls, bread, muffins, pasta with plain tomato sauce, pancakes or waffles with syrup, fruit and juices, and a jelly or jam sandwich. Otherwise, the carbohydrate might speed up the brain's production of serotonin, producing mental grogginess. As already mentioned, carbohydrate stimulates insulin release and triggers the synthesis of serotonin. Serotonin's calming, sleep-inducing effect is exactly the wrong effect for the test taker who needs to be alert.

However, if our test taker is the nervous type who may suffer from poor performance due to stress, such foods may be just what's needed to lend a calming influence for clear thinking.[26] Foods that provide protein along with carbohydrate do not induce serotonin synthesis, so regular, mixed meals may have no effect in this regard.

Research, as well as common sense, tells us that breakfast is of prime importance to a person taking an exam in the morning hours, while lunch is linked to mental performance in the afternoon. Though any effects from food on the neurotransmitters of the brain take at least an hour to become manifest, food in the digestive tract seems to have an immediate, unexplained effect on performance. No one knows why, but a common finding is the "post-lunch dip"—an early afternoon period of less-than optimal mental performance. Paying attention to mental tasks can seem harder after lunch, especially after a large meal, but the effect is by no means universally observed. A prudent action for those taking tests in the early afternoon would be to eat lightly at the lunch hour and save heavier eating for after the test.

If the student is accustomed to taking caffeine in beverages, then it may be especially important to do so on exam day. Persons missing their normal caffeine intakes experience headaches, drowsiness, and fatigue that interfere with mental tasks. For people not accustomed to caffeine, its addition will not improve performance. Alcohol, of course, worsens mental performance.

Table C13-3

Matching Memory Test

Here's a typical short-term memory test. Carefully read through this list of 15 foods just once. Concentrate on each word. Then turn the page and write down as many of the items as you can remember.

onions	shrimp	mangoes
plums	tonic water	pasta
eggs	mayonnaise	ham
blackberries	basil	brownies
hazelnuts	zucchini	oatmeal

How did you do? The average 18- to 39-year-old can remember 10 of the items. It's nine for the average 40- to 59-year old, eight for the average 60- to 69-year-old, and seven if you're 70 or older.

SOURCE: The Memory Assessment Clinic, Bethesda, Maryland.

An interesting finding is that mental performance seems most adversely affected by food that is different in composition from that ordinarily eaten.[27] Among people who eat breakfasts, those eating foods that differ from the norm perform worse than those eating foods of more typical composition. The advice to be gleaned from this finding is, simply, don't rock the boat on examination day. Stick to normal mealtime choices to sidestep adverse effects on mental functioning.

For those seeking the calming effect of serotonin to induce sleep, the opposite advice holds. Foods that provide less than 1 gram of protein per 100 calories are the best choices for this purpose.[28] Note that the foods already named for inducing serotonin carry a lot of calories. A person who eats an extra meal before bedtime must reduce intakes at other meals or risk gaining body fat. Ample daily exercise can both burn off calories and help induce healthy sleep.

Finally, it goes without saying that the diet should be adequate to supply all the precursor and supporting nutrients needed for mental functioning. No manipulation of energy nutrients will correct less-than-optimal brain functioning if needed nutrients are missing. Fluid is also important. Even slight dehydration can cause confusion.

With all of its unsolved mysteries, the human brain remains fascinating to researchers. The ideas presented here, and others like them, offer plenty of grist for the mills of researchers for many years to come.

14 Food Safety and Food Technology

Roosevelt, *Landscape with Royal Palms*, 1952.

Contents

Frequently Asked Questions

food-borne illness illness transmitted to human beings through food and water; caused by a poisonous substance (*food intoxication*) or an infectious agent (*food-borne infection*). Also called *food poisoning.*

hazard a state of danger; used to refer to any circumstance in which harm is possible under normal conditions of use.

safety the practical certainty that injury will not result from the use of a substance.

CONSUMERS HAVE questions about their food. Are today's food products nutritious? Are they safe and free from contamination? Are the additives in them safe? And who is looking out for these consumer concerns?

The Food and Drug Administration (FDA) is the major agency charged with monitoring the food supply, but other agencies are involved as well (see Table 14-1). The following list indicates the areas of concern the FDA has identified in our food supply. The first, microbial **food-borne illness,** commonly called *food poisoning,* constitutes a true **hazard** and so is of most concern. The last, food additives, is of least concern. The others fall somewhere in between.

1. *Microbial food-borne illness.* This affects the most people every year.
2. *Natural toxins in foods.* These constitute a hazard whenever people consume single foods either by choice (fad diets) or by necessity (poverty).
3. *Residues in food.*
   a. Environmental contaminants (other than pesticides) such as household and industrial chemicals. These are increasing yearly in number and concentration, and their impacts are hard to foresee and to forestall.
   b. Pesticides. These are a subclass of environmental contaminants, but are listed separately because they are applied intentionally to foods and so, in theory, can be controlled.
   c. Animal drugs. These include metabolically active proteins that increase growth or milk production in food animals and dairy cows.
4. *Nutrients in foods.* These require close attention as more and more artificially constituted foods appear on the market.
5. *Intentional food additives.* These are listed last because so much is known about them that they pose virtually no hazard to consumers and because their use is well regulated.

These concerns are remarkably similar to those of other nations around the world.

In our free market, where food companies compete for sales, the consumer enjoys the safest, most pleasing, and most abundant food supply in the world. With this benefit comes the consumer's responsibility of distinguishing between foods with a good **safety** record and foods that may pose a hazard. Often foods are safe until they are mishandled.

This chapter provides the information consumers need to purchase, handle, and use foods with confidence. It begins with the most pressing concern of the FDA, food producers, and food consumers alike: food-borne illness.

With the privilege of abundance comes the responsibility to choose wisely.

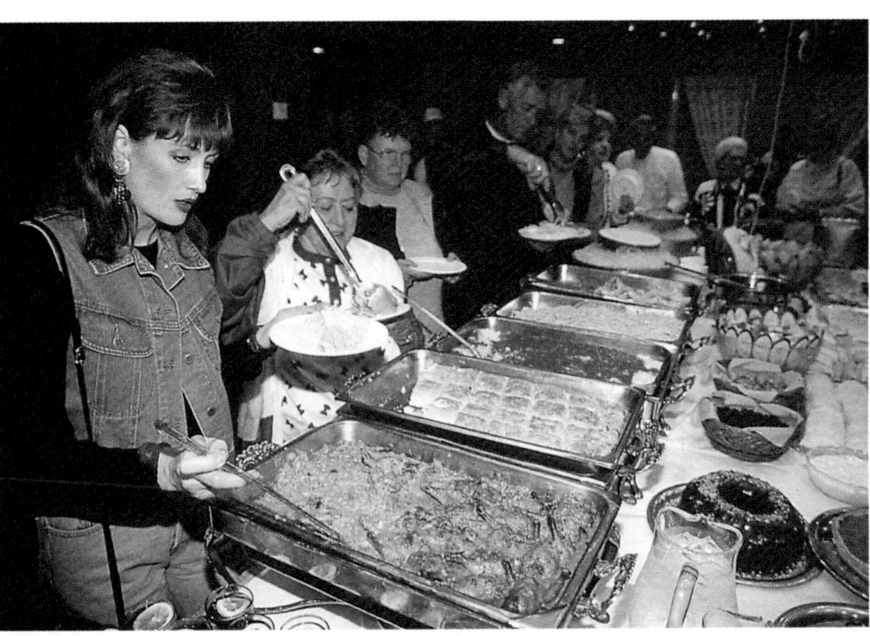

Table 14-1

Agencies That Monitor the U.S. Food Supply

- **CDC (Centers for Disease Control and Prevention)** a branch of the Department of Health and Human Services that is responsible, among other things, for monitoring food-borne diseases.
- **EPA (Environmental Protection Agency)** a federal agency that is responsible, among other things, for regulating pesticides and establishing water quality standards.
- **FDA (Food and Drug Administration)** a part of the Department of Health and Human Services' Public Health Service that is responsible for ensuring the safety and wholesomeness of all foods sold in interstate commerce except meat, poultry, and eggs (which are under the jurisdiction of the USDA); inspecting food plants and imported foods; and setting standards for food consumption.
- **USDA (U.S. Department of Agriculture)** the federal agency responsible for enforcing standards for the wholesomeness and quality of meat, poultry, and eggs produced in the United States; conducting nutrition research; and educating the public about nutrition.
- **WHO (World Health Organization)** an international agency that, among other responsibilities, develops standards to regulate pesticide use. A related organization is the FAO (Food and Agricultural Organization).

enterotoxins poisons that act upon mucous membranes, such as those of the digestive tract.

neurotoxins poisons that act upon the cells of the nervous system.

botulism an often-fatal food poisoning caused by botulinum toxin, a toxin produced by the *Clostridium botulinum* bacterium that grows without oxygen in nonacidic canned foods.

Microbes and Food Safety

The FDA lists food-borne illness as the leading food safety concern because episodes of food poisoning far outnumber any other kind of food contamination. Each year in the United States, an estimated 81 million people become ill from food-borne diseases, and about 9,000 of them die.[1] By taking a few preventive steps, however, people can minimize their chances of contracting food-borne illnesses.

How Do Microorganisms in Food Cause Illness in the Body?

Food-borne illness can be caused either by infection or by intoxication. Microorganisms such as *Salmonella* varieties that occur in foods commonly infect the human body themselves. Other microorganisms in foods produce **enterotoxins** or **neurotoxins** in foods or in the human digestive tract that are then absorbed into the body. Bacteria may multiply or create toxins in food during improper preparation or storage or within the digestive tract after a person eats contaminated food. If you experience the digestive tract disturbances listed in Table 14-2 (on the next page) as the major or only symptoms of your next bout of "flu," chances are excellent that what you really have is a food-borne illness. For people who are otherwise ill or malnourished, have a compromised immune system, or are very old or very young, even relatively mild disturbances can be fatal.

The symptoms of one neurotoxin stand out as severe and commonly fatal—those of **botulism.** Botulism is caused by the toxin of the *Clostridium botulinum* bacterium, which grows in improperly canned (and especially home-canned) foods, improperly prepared or stored vacuum-packed foods, or oils flavored with herbs, garlic, vegetables, or other flavoring agents and stored at room temperature. The microbe grows only in the absence of oxygen, in low-acid conditions, and at temperatures that support growth of most bacteria—40° to 120° Fahrenheit.[2]

Botulism danger signs constitute a true medical emergency (see the margin). Even with medical assistance, survivors can suffer symptoms for months, years, or a lifetime. So potent is the botulinum toxin that an amount as tiny as a single grain of salt can kill several people within an hour. The botulinum toxin is destroyed by heat, so canned foods that contain the toxin can be rendered harmless by boiling them for ten minutes. Food can be canned safely at home, but only if proper canning techniques are followed to the letter.*

For safety, when making flavored oils, wash and dry the herbs before adding them to the oil and keep the oil refrigerated.

Warning signs of botulism:

- Double vision.
- Weak muscles.
- Difficulty swallowing.
- Difficulty breathing.

*Complete, up-to-date, safe home-canning instructions are available in the USDA's 172-page *Complete Guide to Home Canning*, available for $11.00 from the Superintendent of Documents, Government Printing Office, Washington, DC 20402.

Table 14-2

Food-Borne Illnesses

Disease and Organism That Causes It	Most Frequent Food Source	Onset and General Symptoms	Prevention Methods
FOOD-BORNE INFECTIONS			
Campylobacteriosis *Campylobacter jejuni* bacterium	Raw poultry, beef, lamb, unpasteurized milk (foods of animal origin eaten raw or undercooked or recontaminated after cooking).	Onset: 2 to 5 days. Diarrhea, nausea, vomiting, abdominal cramps, fever; sometimes bloody stools; lasts 7 to 10 days.	Cook foods thoroughly; use pasteurized milk; use sanitary food-handling methods.
Giardiasis *Giardia lamblia* protozoa	Contaminated water; uncooked foods.	Onset: 5 to 25 days. Diarrhea (but occasionally constipation), abdominal pain, gas, abdominal distension, digestive disturbances, anorexia, nausea, and vomiting.	Use sanitary food-handling methods; avoid raw fruits and vegetables where protozoa are endemic; dispose of sewage properly.
Hepatitis Hepatitis A virus	Undercooked or raw shellfish.	Onset: 15 to 50 days (28 to 30 days average). Inflammation of the liver with tiredness; nausea, vomiting, or indigestion; jaundice (yellowed skin and eyes from buildup of wastes); muscle pain.	Cook foods thoroughly.
Listeriosis *Listeria monocytogenes* bacterium	Raw meat and seafood, raw milk, and soft cheeses.	Onset: 7 to 30 days. Mimics flu; blood poisoning, complications in pregnancy, and meningitis (stiff neck, severe headache, and fever).	Use sanitary food-handling methods; cook foods thoroughly; use pasteurized milk.
Perfringens food poisoning *Clostridium perfringens* bacterium	Meats and meat products stored at between 120° and 130°F.	Onset: 8 to 12 hr (usually 12). Abdominal pain, diarrhea, nausea, and vomiting; symptoms last a day or less and are usually mild; can be serious in old or weak people.	Use sanitary food-handling methods; cook foods thoroughly; refrigerate foods promptly and properly.
Salmonellosis *Salmonella* bacteria	Raw or undercooked eggs, meats, poultry, milk and other dairy products, shrimp, frog legs, yeast, coconut, pasta, and chocolate.	Onset: 6 to 48 hr. Nausea, fever, chills, vomiting, abdominal cramps, diarrhea, and headache; can be fatal.	Use sanitary food-handling methods; use pasteurized milk; cook foods thoroughly refrigerate foods promptly and properly.

KEY POINT *Each year in the United States, many millions of people suffer from mild to life-threatening symptoms caused by food-borne illness.*

Get medical help when these symptoms occur:

- Bloody stools.
- Headache accompanied by muscle stiffness and fever.
- Rapid heart rate, fainting, dizziness.
- Fever of longer than 24 hours' duration.
- Diarrhea of more than 3 days' duration.
- Numbness, muscle weakness, tingling sensations in the skin.

Food Safety in the Marketplace

The overwhelming majority of food-poisoning cases result from errors consumers make in handling foods *after* purchase. Though commercially prepared food is usually safe, rare accidents do occur, and they can affect many people at once. Such episodes attract the attention of reporters of the daily news. Milk producers, for example, rely on **pasteurization,** a process of heating milk to kill many disease-causing organisms and make milk safe for consumption. When, on occasion, a major dairy develops flaws in its pasteurization system, tens of thousands of cases of food-borne illness may result and are sure to be reported by the media.

Table 14-2

Food-Borne Illnesses Continued

Disease and Organism That Causes It	Most Frequent Food Source	Onset and General Symptoms	Prevention Methods
FOOD-BORNE INFECTIONS Continued			
E. coli *Escherichia coli* bacteria	The most serious disease is caused by undercooked ground beef, unpasterized milk and milk products, contaminated water, sprouts, and person-to person contact.	Onset: 12 to 72 hr. Severe bloody diarrhea, abdominal cramps, acute kidney failure; can be fatal.	Cook ground beef thoroughly; avoid raw milk and milk products; use sanitary food-handling methods; avoid dark or slimy-looking sprouts and keep fresh sprouts refrigerated; use treated, boiled, or bottled water.
Traveler's diarrhea A variety of microorganisms	Contaminated water, undercooked ground beef, raw foods, imported unpasteurized soft cheeses.	Onset: 12 to 18 hr. Loose and watery stools, nausea, bloating, and abdominal cramps.	Cook foods thoroughly; use safe, treated water and pasteurized milk; wash fruits and vegetables.
Trichinosis *Trichinella spiralis* parasite	Raw or undercooked pork or wild game (bear). Worms burrow through the body tissues to reach muscle tissue where they remain alive.	Onset: 24 hr. Abdominal pain, nausea, vomiting, diarrhea, and fever. One to two weeks later, muscle pain, low-grade fever, pain on breathing, edema (swelling), skin eruptions, loss of appetite, and weight loss. Drug therapy kills the worms and deaths are rare.	Cook foods thoroughly.
FOOD INTOXICATIONS			
Botulism Botulinum toxin (produced by the *Clostridium botulinum* bacterium)	Anaerobic environment of low acidity (canned corn, peppers, green beans, soups, beets, asparagus, mushrooms, ripe olives, spinach, tuna, chicken, chicken liver, liver paté, luncheon meats, ham, sausage, stuffed eggplant, herb-flavored oils, lobster, and smoked and salted fish).	Onset: 4 to 36 hr. Nervous system symptoms, including double vision, inability to swallow, speech difficulty, and progressive paralysis of the respiratory system; often fatal; leaves prolonged symptoms in survivors.	Use proper canning methods for low-acid foods; avoid commercially prepared foods with leaky seals or with bent, bulging, or broken cans.
Staphylococcal food poisoning Staphylococcal toxin (produced by the *Staphylococcus aureus* bacterium)	Toxin produced in meats, poultry, egg products, tuna, potato, and macaroni salads, and cream-filled pastries.	Onset: ½ to 8 hr. Diarrhea, nausea, vomiting, abdominal cramps, and fatigue; mimics flu; lasts 24 to 48 hr; rarely fatal.	Use sanitary food-handling methods; cook food thoroughly; refrigerate foods promptly and properly.

Attention on *E. coli* In the mid-1990s, a fast-food restaurant chain in the Northwest served undercooked hamburgers tainted with a particularly dangerous strain of *E. coli* bacteria, the 0157:H7 strain. As a result, four lives were lost, and hundreds of other patrons were stricken with serious illness. News coverage of this incident focused the national spotlight on two important food safety issues: live, disease-causing organisms of many types are routinely found in raw meats, and thorough cooking is necessary to make animal-derived foods safe. Revelations about *E. coli* 0157:H7 have led to a much needed overhaul of the country's mechanisms for ensuring meat safety.

Though usually associated with undercooked beef, outbreaks of *E. coli* 0157:H7 have also been traced to the consumption of commercially grown alfalfa sprouts. Some

pasteurization the treatment of milk with heat sufficient to kill certain pathogens (disease-causing microbes) commonly transmitted through milk; not a sterilization process. Pasteurized milk retains bacteria that cause milk spoilage. Raw milk, even if labeled "certified," transmits many food-borne diseases to people each year and should be avoided.

Hazard Analysis Critical Control Point (HACCP) a systematic plan to identify and correct potential microbial hazards in the manufacturing, distribution, and commercial use of food products.

biotechnology the science that manipulates biological systems or organisms to modify their products or components or create new products (see the chapter Controversy).

biosensor a genetically altered microbe that provides a rapid, low-cost, and accurate test for toxic products of microbial agents in foods.

people, such as those with liver disease or compromised immunity, should probably avoid raw sprouts altogether, but most people can eat sprouts safely if they select fresh-looking packages of crisp, green, living sprouts.[3]

Industry Controls One result of the media reports and public concern about food-borne illness is a law requiring that producers of meat, poultry, and seafood employ an effective prevention method, a **Hazard Analysis Critical Control Point (HACCP)** plan.[4] Each producer must review its processes to identify "critical control points" where the risk of food contamination is high and then develop and implement a HACCP plan to prevent loss of control at those critical points. For many years, meat and seafood inspectors relied on their senses of sight, smell, and touch to detect bad meat and seafood. Unfortunately, human senses cannot detect dangerous organisms until after the food has begun to decay. Thanks to advances in **biotechnology** better tests to ensure the purity of foods, and especially of poultry, meats, and seafoods, are now available. Geneticists are now able to tailor the DNA inside a living bacterial cell to yield a **biosensor** organism. The biosensor can detect chemicals that disease-causing microorganisms create in foods. These sensitive chemical tests for microbial contamination have now replaced less reliable methods and are used to verify that each food-producing company's HACCP plan is effective. The results of the HACCP system seem promising. Since its implementation, *Salmonella* contamination of poultry, ground beef, and pork has decreased by almost 50 percent, 40 percent, and 25 percent, respectively.[5] Encouraging data from the Centers for Disease Control and Prevention also show a decline in the overall incidence of *Salmonella* and *Campylobacter* infections, two of the most common causes of food-borne disease in the United States.[6]

Luckily, large-scale commercial incidents, though dramatic, make up only a fraction of the nation's total food-poisoning cases each year. Most cases arise from one person's error in a small setting and affect just a few victims. Some people have come to accept a yearly bout or two of intestinal illness as inevitable, but in truth, these illnesses can and should be prevented. To protect themselves, consumers need to learn how to select, prepare, and store food safely. See the margin for some food safety myths than can make consumers sick.

More on biotechnology in this chapter's Controversy.

Myths that could make consumers sick:

- "If it tastes okay, it's safe to eat."
- "We have always handled our food this way and nothing has ever happened."
- "I sampled it a couple of hours ago and didn't get sick, so it should be safe to eat."

Consumer Protection Canned and packaged foods sold in grocery stores are easily controlled, but rare accidents do happen. Batch numbering makes it possible to recall contaminated foods through public announcements via newspapers, television, and radio, and the FDA monitors large suppliers. You can help protect yourself, too. Carefully inspect the seals and wrappers of packages. Reject leaking or bulging cans. Many jars have safety "buttons," areas of the lid designed to pop up once the jar is opened; make sure that they are firmly sealed. If a package on the shelf looks ragged, soiled, or punctured, do not buy the product; turn it in to the store manager. A badly dented can or a mangled package is useless in protecting food from microorganisms, insects, spoilage, or even vandals. Frozen foods should be solidly frozen, and those in a chest-type freezer case should be stored below the frost line.

Raw foods from the grocery store, especially meats, poultry, eggs, and seafood, contain microbes, as all things do. Whether the microbes from these sources will multiply and cause illness can be largely a matter of what you do or fail to do in your own kitchen.

KEY POINT *Industry employs sound practices to safeguard the commercial food supply from microbial threats. Still, incidents of commercial food-borne illness have caused widespread harm to health.*

Food Safety in the Kitchen

In the home, concerns regarding food safety revolve around three main functions: food storage, food handling, and cooking. Just for fun, take the food safety quiz in Table 14-3 to see how well you follow food safety rules. Then read on to learn how you can make the meals from your kitchen as safe as they can be.

Table 14-3

Can You Pass the Kitchen Food Safety Quiz?

1. The temperature of the refrigerator in my home is:
   A. 50°F (10° Celsius).
   B. 40°F (5°C).
   C. I don't know; I don't own a refrigerator thermometer.
2. The last time we had leftover cooked stew or other meaty food, the food was:
   A. cooled to room temperature, then put in the refrigerator.
   B. put in the refrigerator immediately after the food was served.
   C. left at room temperature overnight or longer.
3. If a cutting board is used in my home to cut raw meat, poultry, or fish and it is going to be used to chop another food, the board is:
   A. reused as is.
   B. wiped with a damp cloth.
   C. washed with soap and water.
   D. washed with soap and hot water and then sanitized.
4. The last time I had a hamburger, I ate it:
   A. rare.
   B. medium.
   C. well-done.
5. The last time there was cookie dough where I live, the dough was:
   A. made with raw eggs, and I sampled some of it.
   B. store-bought, and I sampled some of it.
   C. not sampled until baked.
6. I clean my kitchen counters and food preparation areas with:
   A. a damp sponge that I rinse and reuse.
   B. a clean sponge or cloth and water.
   C. a clean sponge or cloth with hot water and soap.
   D. the same as above, then a bleach solution or other sanitizer.
7. When dishes are washed in my home, they are:
   A. cleaned by an automatic dishwasher and then air-dried.
   B. left to soak in the sink for several hours and then washed with soap in the same water.
   C. washed right away with hot water and soap in the sink and then air-dried.
   D. washed right away with hot water and soap in the sink and immediately towel-dried.
8. The last time I handled raw meat, poultry, or fish, I cleaned my hands afterward by:
   A. wiping them on a towel.
   B. rinsing them under hot, cold, or warm tap water.
   C. washing with soap and water.
9. Meat, poultry, and fish products are defrosted in my home by:
   A. setting them on the counter.
   B. placing them in the refrigerator.
   C. microwaving,
   D. soaking them in warm water.
10. I realize that eating raw seafood poses special problems for people with:
   A. diabetes.
   B. HIV infection.
   C. cancer.
   D. liver disease.

Answers

1. Refrigerators should stay at 40°F or less, so if you chose answer B, give yourself two points; zero for other answers.
2. Answer B is the best practice; give yourself two points if you picked it; zero for other answers.
3. If answer D best describes your household's practice, give yourself two points; if C, one point.
4. Give yourself two points if you picked answer C; zero for other answers.
5. If you answered A, you may be putting yourself at risk for infection from bacteria in raw shell eggs. Answer C—eating the baked product—will earn you two points and so will answer B. Commercial products are made with pasteurized eggs.
6. Answers C or D will earn you two points each; answer B, one point; answer A, zero.
7. Answers A and C are worth two points each; other answers, zero.
8. The only correct practice is answer C. Give yourself two points if you picked it; zero for others.
9. Give yourself two points if you picked B or C; zero for others.
10. This is a trick question: all of the answers apply. Give yourself two points for knowing one or more of the risky conditions.

Rating Your Home's Food Practices

20 points: Feel confident about the safety of foods served in your home.

12 to 19 points: Re-examine food safety practices in your home. Some key rules are being violated.

11 points or below: Take steps immediately to correct food-handling, storage, and cooking techniques used in your home. Current practices are putting you and other members of your household in danger of food-borne illness.

SOURCE: Adapted from U.S. Food and Drug Administration, Can your kitchen pass the food safety test? *FDA Consumer,* October 1998.

Food can provide ideal conditions for bacteria to thrive and produce their toxins. Disease-causing bacteria require three things: (1) warmth (40°F to 140°F), (2) moisture, and (3) nutrients. To defeat bacteria, people who prepare food should keep in mind that food-borne illness is always possible. Do these three things: keep hot food hot, keep cold food cold, and keep the kitchen clean. Keeping hot food hot includes cooking foods long enough to reach an internal temperature that will kill microbes, as described in the next section. Cooked foods must be held at 140°F or higher until served. Refrigerate foods immediately after serving a meal, and definitely before two hours have passed (one hour if room temperature approaches 90°F).

cross-contamination the contamination of a food through exposure to utensils, hands, or other surfaces that were previously in contact with a contaminated food.

mad cow disease an often-fatal illness of cattle affecting the nerves and brain. Also, called bovine spongiform encephalopathy (BSE).

Remember to do these three things:

- Keep hot food hot.
- Keep cold food cold.
- Keep the kitchen clean.

Table 14-4

Safe Refrigerator Storage Times (40°F)

1 to 2 Days
Raw ground meats, breakfast or other raw sausages, raw fish or poultry; gravies
3 to 5 Days
Raw steaks, roasts, or chops; cooked meats, vegetables, and mixed dishes; ham slices; mayonnaise salads (chicken, egg, pasta, tuna)
1 Week
Hard-cooked eggs, bacon or hot dogs (opened packages); smoked sausages.
2 to 4 Weeks
Raw eggs (in shells); bacon or hot dogs (packages unopened); dry sausages (pepperoni, hard salami); most aged and processed cheeses (Swiss, brick)
2 Months
Mayonnaise (opened jar); most dry cheeses (parmesan, romano)

Keeping cold food cold starts when you leave the grocery store. If you are running errands, shop last so that the groceries will not stay in the car too long. (If ice cream begins to melt, it has been too long.) Upon arrival home, load foods into the refrigerator or freezer immediately. Keeping foods cold applies to defrosting foods, too. Thaw meats or poultry in the refrigerator, not at room temperature. Table 14-4 lists some safe keeping times for foods stored at or below 40°F.

Keeping the kitchen clean requires using freshly washed utensils and laundered towels and washing your hands with warm water and soap for a minimum of 20 seconds before and during food handling. If you are ill or have open sores, stay away from food. Clean equipment frequently to prevent **cross-contamination** of microorganisms. Microbes love to nestle down in small, damp spaces such as the inner cells of sponges or the pores between the fibers of wooden cutting boards. You can ensure the safety of cutting boards and sponges by washing them in a dishwasher or by treating them as suggested below. Alternatively, sponges can be saved for car washing and other heavy cleaning chores, while the kitchen is cleaned with washable dishcloths that can be laundered often. Sponges with special antibacterial properties are also available for purchase, although they are expensive.

To eliminate microbes you have three choices, each with benefits and drawbacks. One is to poison the microbes on cutting boards, sponges, and other equipment with toxic chemicals such as bleach (one capful per gallon of water). The benefit is that chlorine can kill even the hardiest organism. The drawback is that chlorine that washes down household drains into the water supply forms chemicals that can harm waterways and fish.

A second option is to treat kitchen equipment with heat. Soapy water heated to 140°F kills most harmful organisms and washes most others away. This method takes effort, though, since you have to use truly scalding water heated well beyond the temperature of the tap. Thirdly, an automatic dishwasher can combine both methods: it washes in water hotter than hands can tolerate, and a chlorine-containing dishwasher detergent can be used but, of course, with the environmental disadvantage that chlorine entails. Whichever strategy you use, for a small initial investment, you can have truly safe implements to prepare your food.

KEY POINT *To prevent food-borne illness, always remember that it can happen. Keep hot foods hot, keep cold foods cold, and keep the kitchen clean.*

Are Some Foods More Likely Than Others to Make People Sick?

Some foods are more hospitable to microbial growth than others, so they pose a potential hazard. In general, foods that are high in moisture and nutrients and those that are chopped or ground are especially favorable hosts.

Meats and Poultry Meats and poultry require special handling. Raw meats and poultry bear labels to instruct consumers on meat safety (see Figure 14-1). Meats often contain all sorts of bacteria, and they provide a moist, nutritious environment that is just right for microbial growth. If you take burgers out to the grill on a plate, wash that plate in hot, soapy water before using it to hold the cooked burgers. Ground meat or poultry is handled more than meats left whole and exposes much more surface area for bacteria to land on, so experts advise cooking it well-done. For meat loaf, use a thermometer to test the internal temperature.

Though contaminated meat poses a real danger, consumers need to be aware that the media often exaggerate a story beyond its facts. Many Americans were concerned when they heard that an outbreak of **mad cow disease** in the United Kingdom had led British officials to slaughter thousands of cattle as a precautionary measure. Although the bovine disease may possibly have passed from cattle to people who consumed infected beef, no British beef or milk is imported into the United States. Thus, for U.S. consumers, mad cow disease poses almost no threat. Real threats come from other infectious agents, how-

Figure 14-1

Safe Handling Instructions for Meat and Poultry

Never allow frozen meat to defrost at room temperature or in a bath of warm water. In both cases, meat thaws from outside in, and the outside meat layer can easily warm up to temperatures that permit bacterial growth before the core defrosts.

Safe Handling Instructions

THIS PRODUCT WAS PREPARED FROM INSPECTED AND PASSED MEAT AND/OR POULTRY. SOME FOOD PRODUCTS MAY CONTAIN BACTERIA THAT CAN CAUSE ILLNESS IF THE PRODUCT IS MISHANDLED OR COOKED IMPROPERLY. FOR YOUR PROTECTION, FOLLOW THESE SAFE HANDLING INSTRUCTIONS.

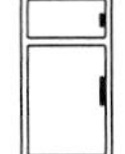

KEEP REFRIGERATED OR FROZEN. THAW IN REFRIGERATOR OR MICROWAVE.

KEEP RAW MEAT AND POULTRY SEPARATE FROM OTHER FOODS. WASH WORKING SURFACES (INCLUDING CUTTING BOARDS), UTENSILS, AND HANDS AFTER TOUCHING RAW MEAT OR POULTRY.

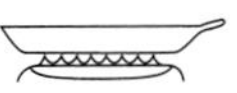

COOK THOROUGHLY.

KEEP HOT FOODS HOT. REFRIGERATE LEFTOVERS IMMEDIATELY OR DISCARD.

Microwave cooking of meats requires special care. Large, thick, dense foods such as roasts or meat loaves may register "cooked" on an internal meat thermometer, but may harbor cool spots in which dangerous microorganisms, such as the *Trichinella spiralis* parasite, sometimes present in pork, can survive. Such foods are best cooked by another method or divided into small, thin portions to be microwaved individually.

Properly cooked food hot from the oven or stove is relatively free of bacteria, but as soon as it is taken out to serve, it is reinoculated. Kitchen utensils recontaminate the food, or bacteria from the air land on its surface. Promptly after serving, even while the food is still hot, refrigerate leftovers in shallow containers for quick, even chilling. Large amounts of food refrigerated in deep containers may take hours to cool though, allowing bacteria time to multiply in the warm internal portions.

Take care when preparing meats along with foods intended to be served raw, such as chopped salads or lettuce and tomato toppers for hamburgers. A grave error is to prepare raw foods on the same board or with the same utensils as were used to prepare raw meats for cooking.

ever, in undercooked or improperly stored meats. Figure 14-2 on the next page lists safe internal temperatures of cooked meats and other important temperatures.[†]

It goes without saying that any food with an "off" appearance or odor should not be used or even tasted. You cannot rely on your senses of smell and sight alone to warn you, however, because most hazards are not detectable by odor, taste, or appearance. Also, cooking does not destroy all bacterial toxins. Even hot cooked food, if handled improperly prior to serving, can cause illness.

To protect yourself as far as you can, always keep the possibility of food poisoning in mind. Delicious looking meatballs on a buffet should be steaming hot. Food at 140°F feels hot, not just warm.

Eggs Over the past ten years, surveillance of food safety has revealed an alarming trend. Increasingly, *Salmonella* of a most virulent type is being detected in blood samples from food-poisoning victims and samples of illness-causing foods.[7] Raw, unpasteurized eggs seem especially likely to be contaminated. The Centers for Disease Control and Prevention warn that consumers should cook eggs until the whites are set firmly and the yolks begin to thicken. No longer is it safe to drop a raw egg into a food or beverage that will not be cooked before consumption. Healthy people can still safely enjoy classic foods that call for raw or undercooked eggs, such as Caesar salad dressing and hollandaise sauce, by preparing them with pasteurized egg substitutes, sold in cartons in the dairy or freezer case, instead of raw eggs. The elderly, infants and very young children, and people with compromised immune systems, however, should avoid any dishes with uncooked or undercooked eggs, including pasteurized ones.[8]

A safe hamburger is cooked well-done, appears brown (not pink) throughout, has juices that run clear, and is steaming hot. Place it on a clean plate when it's done.

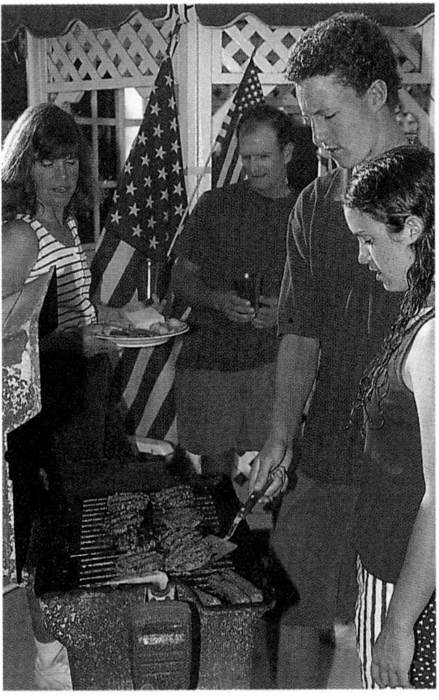

[†] The USDA's meat and poultry hotline answers questions about meat and poultry safety: 1-800-535-4555.

sushi a Japanese dish that consists of vinegar-flavored rice, seafood, and colorful vegetables, typically wrapped in seaweed. Some sushi is wrapped in raw fish; other sushi contains only cooked ingredients.

Figure 14-2

Food Safety Temperatures (Fahrenheit) and Household Thermometers

This figure depicts two kinds of thermometers: one, with a high temperature range, is used to test doneness of meats, and the other, with a much lower range that extends well below freezing, is used to test the temperatures of refrigerators and freezers. A well-equipped kitchen should have both kinds, and safety-conscious cooks should use them often.

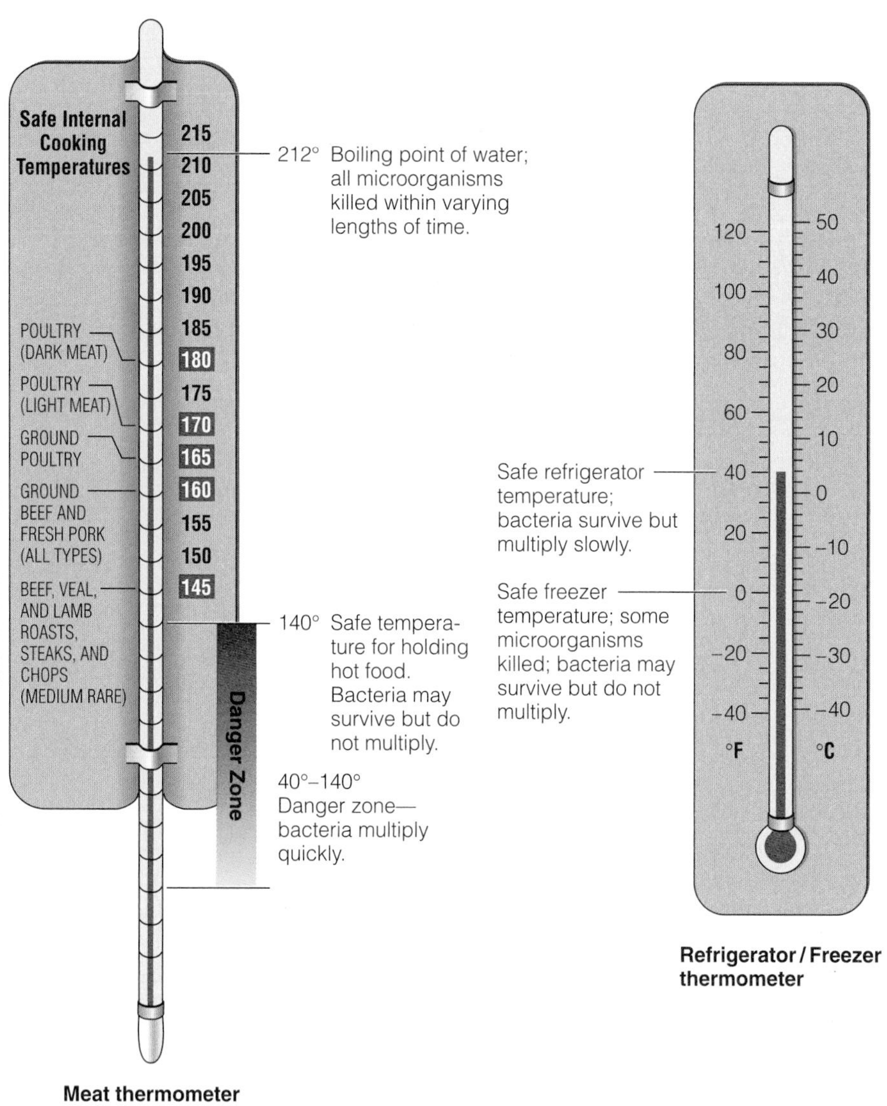

Safe Internal Cooking Temperatures
215
210
205
200
195
190
185
180
175
170
165
160
155
150
145
POULTRY (DARK MEAT)
POULTRY (LIGHT MEAT)
GROUND POULTRY
GROUND BEEF AND FRESH PORK (ALL TYPES)
BEEF, VEAL, AND LAMB ROASTS, STEAKS, AND CHOPS (MEDIUM RARE)
212° Boiling point of water; all microorganisms killed within varying lengths of time.
140° Safe temperature for holding hot food. Bacteria may survive but do not multiply.
Danger Zone
40°–140° Danger zone—bacteria multiply quickly.
Meat thermometer
Safe refrigerator temperature; bacteria survive but multiply slowly.
Safe freezer temperature; some microorganisms killed; bacteria may survive but do not multiply.
120
100
80
60
40
20
0
−20
−40
°F
50
40
30
20
10
0
−10
−20
−30
−40
°C
Refrigerator / Freezer thermometer

Seafood For adults and children alike, eating raw or lightly steamed seafood is a risky proposition even if it is prepared by a master chef. The microorganisms that lurk there are undetectable, even to an expert.[‡]

People who like **sushi** know that not all varieties are made from raw fish. Many types are made with cooked crabmeat and vegetables, avocado, or other delicacies and are perfectly safe to enjoy. Also, the rumor that freezing fish will make it safe to eat raw is only partly true. Freezing fish will kill mature parasitic worms, but only cooking can kill all worm eggs and other microorganisms that can cause illness.

As population density increases along the seashores, the offshore waters are becoming polluted, contaminating the seafood living there. Watchdog agencies monitor commercial fishing areas and try to keep harvesters out of unsafe waters. The agencies do eventually catch cheaters, but meanwhile, unwholesome food can reach the market. In

[‡]To speak with an expert on seafood safety, call the FDA's seafood hotline: 1-800-FDA-4010.

one season alone, black-market dealers may sell millions of dollars worth of clams and oysters taken illegally from closed harvesting areas.

The food-borne infections that lurk in normal-appearing seafood can be even worse than those of spoilage: viral hepatitis; worms, flukes, and other parasites; severe viral intestinal disorders; poisoning by naturally occurring toxins; and other diseases. Hepatitis infection causes prolonged illness that persists for months or years, severely damages the liver, greatly increases the risk of developing liver cancer, and, once in the body, is transmissible to others. Many types of worms depend on the blood of their host for food and reproduction; they attack digestive membranes, sometimes causing life-threatening perforations. Flukes attack the liver, damaging it.

People who have loved and eaten raw oysters and other seafood for years may be tempted to ignore these threats because they have never experienced serious illness. Some have heard that alcoholic beverages taken with raw seafood eliminate risks or that hot sauce kills the bacteria, but these assertions are not true. A study did find a correlation between taking one drink of whiskey or wine and a reduced risk of disease after eating contaminated seafood, but this evidence is no guarantee of protection.[9] Hot sauce is useless against the infectious agents that contaminate oysters.[10] Experts unanimously agree that the risks of eating raw or lightly cooked seafood today are unacceptably high due to environmental contamination.

In many states, containers of raw oysters must bear this warning: "There is a risk associated with consuming raw oysters or any raw animal protein. If you have chronic illness of the liver, stomach or blood or have immune disorders, you are at greater risk of serious illness from raw oysters and should eat oysters fully cooked. If unsure of your risk, consult a physician."

Picnics and Lunch Bags For picnics that are fun and safe and for safe packed lunches, keep these precautions in mind. Choose foods that last without refrigeration, such as fresh fruits and vegetables, breads and crackers, and canned spreads and cheeses that you can open and use on the spot. Aged cheeses, such as cheddar and Swiss, do well at environmental temperatures for an hour or two, but for longer periods, carry them in a cooler or thermal lunch bag. Mayonnaise alone is somewhat resistant to spoilage because of its acid content, but when it is mixed with chopped ingredients in pasta, meat, or vegetable salads, the mixtures spoil easily. The chopped ingredients have extensive surface areas for bacteria to invade, and foods that have been in contact with cutting boards, hands, and kitchen utensils have picked up at least a few bacteria earlier. Chill chopped salads well in shallow containers before, during, and after a picnic, and keep salad sandwiches cold until eaten. To keep lunch bag foods chilled, choose a thermal lunch bag and freeze beverages to pack in with the foods. As the beverages thaw in the hours before lunch, they keep the foods cold.

Frequently unsafe:

- Raw milk and milk products.
- Raw or undercooked seafood, meat, poultry, or eggs.

Occasionally unsafe:

- Airline food.
- Hamburgers.
- Salad bar items.
- Sandwiches.
- Soft cheeses (Mexican style, feta, brie, camembert, blue-veined).
- Sprouts.
- Unpasteurized fruit juices and ciders.
- Unwashed berries and grapes.

Rarely unsafe:

- Peeled fruit.
- Steaming-hot foods.

Honey Another danger lurks in honey. Honey can contain dormant spores of *Clostridium botulinum* that can awaken (germinate) in the human body to produce the deadly botulinum toxin mentioned earlier. Mature adults are usually protected against this threat, but infants under one year of age should never be fed honey, which can also be contaminated with environmental pollutants picked up by the bees. Honey has been implicated in several cases of sudden infant death.

KEY POINT *Some foods pose special microbial threats and so require special handling. Seafood is especially likely to be contaminated. Almost all types of food poisoning can be prevented by safe food preparation, storage, and cleanliness. Honey is unsafe for infants.*

Food Safety while Traveling

About half of the people who travel to places where cleanliness standards are lacking suffer from food-borne illnesses—commonly known as traveler's diarrhea (see Table 14-2). A bout of this illness can ruin a trip. To avoid food-borne illness while traveling:

- Before you travel, ask your physician which medicines to take with you in case you get sick.
- Wash your hands often with soap and water, especially before handling food or eating.

- Eat only cooked and canned foods. Eat raw fruits or vegetables only if you have washed them with your own clean hands in boiled water and peeled them yourself. Skip salads.
- Be aware that water, and ice made from it, may be unsafe, too. Take along disinfecting tablets or an element that boils water in a cup. Drink only treated, boiled, canned, or bottled beverages, and drink them without ice, even if they are not chilled to your liking.
- Avoid using the local water supply, even if you are just brushing your teeth, unless you boil or disinfect it first.

Here are some general rules to follow: boil it, cook it, peel it, or forget it. If you follow these recommendations, chances are excellent that you will remain well.

KEY POINT *Some special food safety concerns arise when traveling. To avoid food-borne illnesses, remember to boil it, cook it, peel it, or forget it.*

Natural Toxins in Foods

Consumers may naively think they can eliminate all poisons from their diets by eating only "natural" foods. On the contrary, nature has provided natural foods with the natural poisons they need to fend off diseases, insects, and other predators. Humans rarely suffer actual harm from such poisons, but the *potential* for harm does exist.

Although the herbs belladonna and hemlock have reputations as deadly poisons, few people know that the herb sassafras contains a cancer-causing agent and is banned from use in commercially produced foods and beverages (see Table 11-9 for more on these and other potentially harmful herbs). Equally surprising to many people is that cabbage, turnips, mustard greens, and radishes all contain small quantities of harmful goitrogens, compounds that can enlarge the thyroid gland and aggravate thyroid problems. These effects show up only under extreme conditions when people have little but cabbage to eat. Ordinarily, cabbages and their relatives are celebrated for their phytochemicals—compounds associated with low cancer rates.

Read more about the association between cancer and foods in Chapter 11.

Other natural poisons in *raw* lima and fava beans and in fruit seeds such as apricot pits are members of a group called cyanogens, precursors to the deadly poison cyanide. Many countries restrict commercially grown lima beans to those varieties with the lowest cyanogen contents. As for fruit seeds, they are seldom deliberately eaten. An occasional swallowed seed or two presents no danger, but a couple of dozen seeds could be fatal to a small child. Perhaps the most infamous cyanogen is laetrile, a compound erroneously represented as a cancer cure. True, the poison laetrile kills cancer cells, but only at doses that kill the person, too. Research over the past 100 years has proved that laetrile is an ineffective cancer treatment and dangerous to the taker.

Potatoes contain many natural poisons, including solanine, a powerful, bitter, narcotic-like substance. The small amounts of solanine normally found in potatoes are harmless, but solanine can build up to toxic levels when potatoes are exposed to light during storage. Cooking does not destroy solanine, but because most of a potato's solanine is in a green layer that develops just beneath the skin, it can be peeled off, making the potato safe to eat. If the potato tastes bitter, however, throw it out.

At certain times of the year, seafood may become contaminated with the so-called red tide toxin that occurs during algae blooms. Eating seafood contaminated with red tide causes a form of food poisoning that paralyzes the eater. The FDA monitors fishing waters for red tide algae and closes waters to fishing whenever it appears.

These examples of naturally occurring toxins should serve as a reminder of three principles. First, any substance can be toxic when consumed in excess. Practice moderation in the use of all foods. Second, poisons are poisons, whether made by people or by nature. It is not the source of a chemical that makes it hazardous, but its chemical structure. Third, by including a variety of foods in the diet, consumers ensure that toxins in foods are diluted by the volume of the other foods eaten.

Natural foods contain natural toxins that can be hazardous if consumed in excess. To avoid poisoning by toxins, eat all foods in moderation, treat chemicals from all sources with respect, and choose a variety of foods.

Environmental Contaminants

A justifiably high-ranking concern about the food supply everywhere is environmental contamination of foods. As populations increase worldwide and nations become more industrialized, the problem looms ever larger. A food **contaminant** is anything that does not belong there.

Harmfulness of Contaminants The potential harmfulness of a contaminant depends in part on the extent to which it lingers in the environment or in the human body—that is, on how **persistent** it is. Some contaminants are short-lived because microorganisms or agents such as sunlight or oxygen can break them down. Some contaminants linger in the body for only a short time because the body can rapidly excrete them or metabolize them to harmless compounds. These contaminants present little cause for concern. Some contaminants resist breakdown, however, and interact with the body's systems without being metabolized or excreted. These contaminants can pass from one species to the next and accumulate at higher concentrations in each level of the food chain, a process called **bioaccumulation.** Figure 14-3 on the next page shows how toxic chemicals accumulate in the food chain.

How much of a threat do environmental contaminants pose to the food supply? For the most part, the hazards appear to be small because the FDA monitors the presence of contaminants in foods and requires that contaminated foods be removed from the market. In the event of an accidental industrial spill or one caused by a natural event, such as a volcano, however, the hazard can suddenly become great. The following paragraphs describe how two different types of contaminants found their way into the food supply in the past. In one case, a **heavy metal** (mercury) was released into waterways by industry and accumulated in fish that people ate. In the other, an **organic halogen** (polybrominated biphenyl or PBB) was accidentally spilled into livestock feed and eaten by animals whose meat people eventually ate in turn.

Mercury A classic example of acute contamination occurred in 1953 when a number of people in Minamata, Japan, became ill with a disease no one had seen before. By 1960, 121 cases had been reported, including 23 in infants. Mortality was high; 46 died, and the survivors suffered progressive, irreversible blindness, deafness, loss of coordination, and severely impaired mental function. The cause of this misery was ultimately revealed: manufacturing plants in the region were discharging mercury into the waters of the bay, the mercury was turning to methylmercury on leaving the factories, and the fish in the bay were accumulating this poison in their bodies. Some of the people who were poisoned had been eating fish from the bay every day. The infants who contracted the disease had not eaten any fish, but their mothers had, and even though the mothers exhibited no symptoms during their pregnancies, the poison had been affecting their unborn babies.

PBB As for PBB, in 1973, half a ton of this toxic compound was accidentally mixed into some livestock feed that was distributed throughout the state of Michigan. The chemical found its way into millions of animals and then into people who ate their meat. The seriousness of the accident began to come to light when dairy farmers reported that their cows were going dry, aborting their calves, and developing abnormal growths on their hooves. More than 30,000 cattle, sheep, and swine and more than a million chickens were destroyed, but the effects on people were not prevented. An estimated 97 percent of Michigan's residents had been exposed to PBB. Some of the exposed farm residents suffered nervous system aberrations and liver disorders.

contaminant any substance occurring in food by accident; any food constituent that is not normally present.

persistent of a stubborn or enduring nature; with respect to food contaminants, the quality of remaining unaltered and unexcreted in plant foods or in the bodies of animals and human beings.

bioaccumulation the accumulation of a contaminant in the tissues of living things at higher and higher concentrations along the food chain.

heavy metal any of a number of mineral ions such as mercury and lead; so called because they are of relatively high atomic weight. Many heavy metals are poisonous.

organic halogen an organic compound containing one or more atoms of a halogen—fluorine, chlorine, iodine, or bromine.

Chemical contaminants of concern in foods:

- Heavy metals:
  - Lead.
  - Mercury.
  - Cadmium.
  - Selenium.
  - Arsenic.
- Halogens and organic halogens:
  - Chlorine.
  - Iodine.
  - Vinyl chloride.
  - Ethylene dichloride.
  - Trichloroethylene (TCE).
  - Polybrominated biphenyl (PBB).
  - Polychlorinated biphenyls (PCBs).
- Others:
  - Asbestos.
  - Dioxins.
  - Acrylonitrile.
  - Lysinoalanine.
  - Diethylstibestrol (DES).
  - Heat-induced mutagens.
  - Antibiotics (in animal feed).

Figure 14-3

Bioaccumulation of Toxins in the Food Chain

If none of the chemicals are lost along the way, one person ultimately receives all of the toxic chemicals that were present in the original several tons of producer organisms.

4. A person whose principal animal-protein source is fish may consume about 100 pounds of fish in a year.

3. Larger fish consume a few tons of plankton-eating fish in the course of their lifetimes—and the toxic chemicals from the small fish become more concentrated in the flesh of the larger species.

2. The toxic chemicals become more concentrated in the plankton-eating fish that consume several tons of producer organisms in their lifetimes.

1. Producer organisms may become contaminated with toxic chemicals.

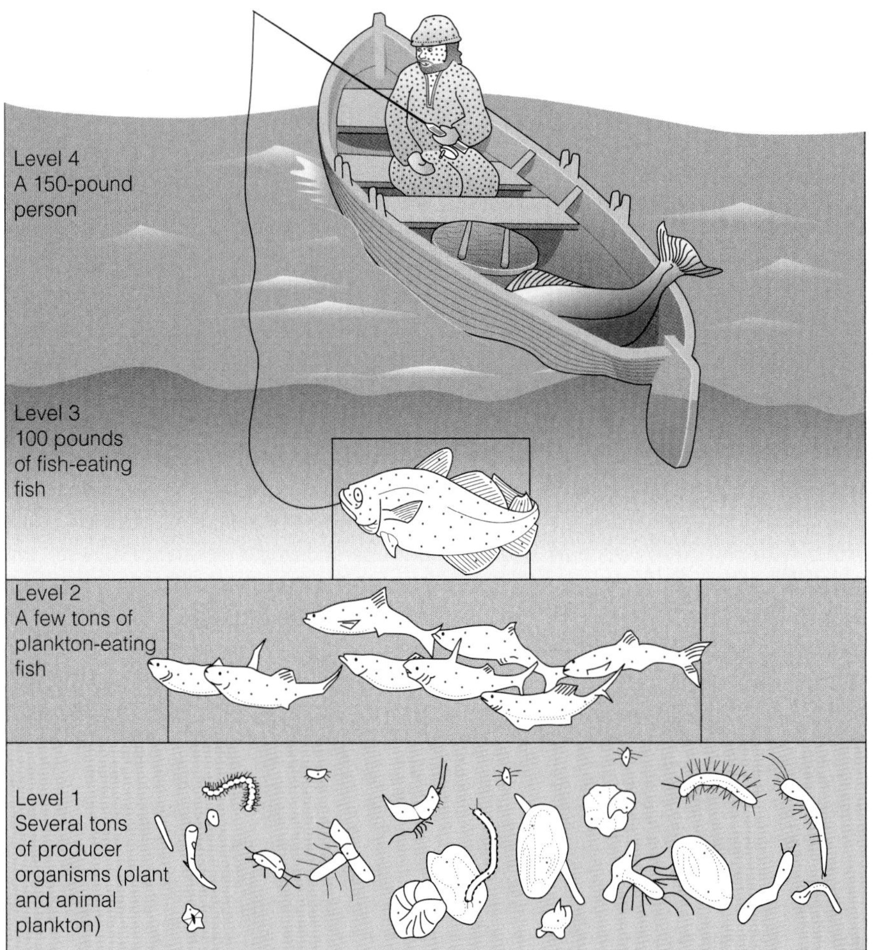

Level 4
A 150-pound person
Level 3
100 pounds of fish-eating fish
Level 2
A few tons of plankton-eating fish
Level 1
Several tons of producer organisms (plant and animal plankton)
Toxic chemicals

Mercury is a heavy metal and PBB is an organic halogen. These two classes of chemicals are among the most toxic and are still being liberated into our environment daily. Much more information is available about a host of other contaminants, but a discussion of them is far beyond the scope of this text. Table 14-5 selects a few contaminants of great concern in foods to show how pervasively a contaminant can affect the body.

KEY POINT *Persistent environmental contaminants pose a small but significant threat to the safety of food. An accidental spill can create an extreme hazard.*

Pesticides:

- Kill pests' natural predators.
- Accumulate in the food chain.
- Pollute the water, soil, and air.

pesticides chemicals used to control insects, diseases, weeds, fungi, and other pests on crops and around animals. Used broadly, the term includes *herbicides* (to kill weeds), *insecticides* (to kill insects), and *fungicides* (to kill fungi).

Pesticides

The use of **pesticides** helps to ensure the survival of some crops, but the damage pesticides do to the environment is considerable and increasing. Moreover, there is some question about whether the widespread use of pesticides has really improved the overall yield of food. Even with extensive pesticide use, the world's farmers lose large quantities of their crops to pests every year.

Do Pesticides on Foods Pose a Hazard to Consumers?

Many pesticides are broad-spectrum poisons that damage all living cells, not just those of pests. Their use, therefore, can pose hazards to the plants and animals in natural

Table 14-5

Examples of Contaminants in Foods

Name and Description	Sources	Toxic Effects	Typical Route to Food Chain
Cadmium (heavy metal)	Used in industrial processes including electroplating, plastics, batteries, alloys, pigments, smelters, and burning fuels. Present in cigarette smoke and in smoke and ash from volcanic eruptions.	No immediately detectable symptoms; slowly and irreversibly damages kidneys and liver.	Enters air in smokestack emissions, settles on ground, absorbed into food plants, consumed by farm animals, and eaten in vegetables and meat by people. Sewage sludge and fertilizers leave large amounts in soil; runoff contaminates shellfish.
Lead[a] (heavy metal)	Lead crystal decanters and glassware, painted china, old house paint, batteries, pesticides, old plumbing, and some food-processing chemicals.	Displaces calcium, iron, zinc, and other minerals from their sites of action in the nervous system, bone marrow, kidneys, and liver, causing failure to function.	Originates from industrial plants and pollutes air, water, and soil. Still present in soil from many years of leaded gasoline use.
Mercury (heavy metal)	Widely dispersed in gases from earth's crust; local high concentrations from industry, electrical equipment, paints, and agriculture.	Poisons the nervous system, especially in fetuses.	Inorganic mercury released into waterways by industry and acid rain is converted to methylmercury by bacteria and ingested by food species of fish (tuna, swordfish, and others).
Polychlorinated biphenyls (PCBs) (organic compounds)	No natural source; produced for use in electrical equipment (transformers, capacitors).	Long-lasting skin eruptions, eye irritations, growth retardation in children of exposed mothers, anorexia, fatigue, others.	Discarded electrical equipment; accidental industrial leakage, or reuse of PCB containers for food.

[a]For answers to questions concerning lead, call the National Lead Information Center at 1-800-424-LEAD.

systems, and especially to workers involved with pesticide production and transport. High doses of pesticides applied to laboratory animals cause birth defects, sterility, tumors, organ damage, and central nervous system impairment. At one time, the law stated that no traces of pesticides found to cause cancer in animals would be allowed in foods. In 1996, however, this provision was eliminated.

Ironically, pesticides also promote the survival of the very pests they are intended to wipe out. Consider a pesticide aimed at certain insects that are attacking a crop. The pesticide may kill *almost* 100 percent of them, but thanks to the genetic variability of large populations, some insects are likely to survive exposure. The resistant insects can then multiply free of competition and soon will produce many offspring—offspring that have inherited resistance to the pesticide. This new strain of insects can attack the crop with enhanced vigor. To control these resistant insects requires application of a new and more powerful pesticide—and this leads to the emergence of a population of still more resistant insects. The same effects arise from use of herbicides and fungicides. One alternative to this destructive series of events is to manage pests using a combination of natural and biological controls, as discussed in Controversy 15.

Pesticides are not produced only in laboratories; they also occur in nature. The nicotine in tobacco and psoralens in celery are examples. Natural pesticides, however, are less damaging to other living things and leave less persistent **residues** in the environment than most human-made ones.

An ideal pesticide would destroy pests in the field but vanish long before consumers ate the food; no such "perfect" pesticide yet exists, however. As Figure 14-4 demonstrates, pesticide residues on agricultural products can sometimes survive processing

residues whatever remains. In the case of pesticides, those amounts that remain on or in foods when people buy and use them.

In some small gardens, handwork can take the place of pesticides.

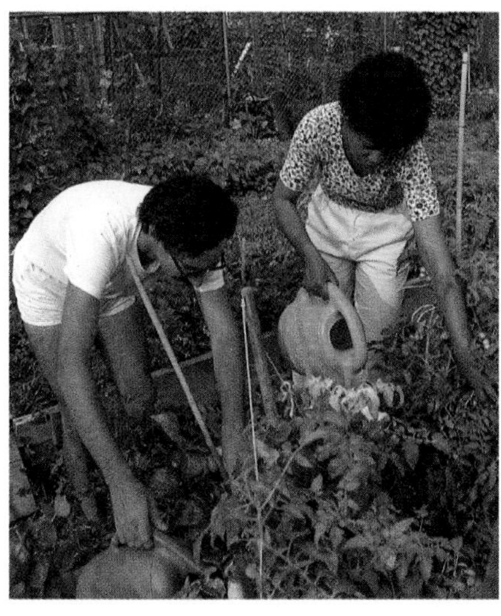

Figure 14-4

Possible Pathways of Pesticide Residues to a Fast-Food Meal

The red dots in the figure represent pesticide residues left on foods from field spraying or postharvest application. Notice that most pesticides follow fats in foods and that some processing methods, such as washing and peeling vegetables, reduce pesticide concentrations while others tend to concentrate them.

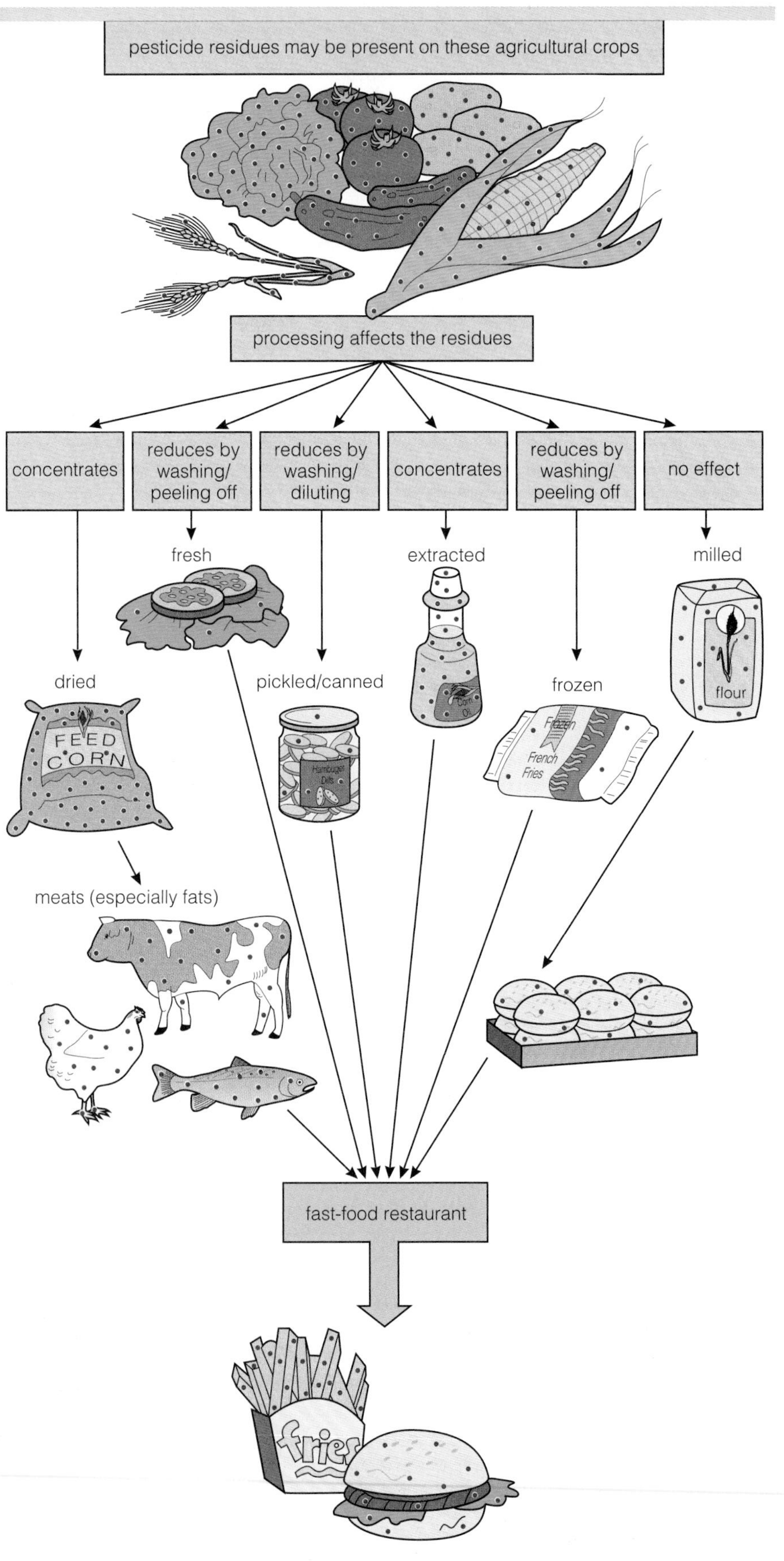
pesticide residues may be present on these agricultural crops
processing affects the residues
concentrates
reduces by washing/ peeling off
reduces by washing/ diluting
concentrates
reduces by washing/ peeling off
no effect
fresh
extracted
milled
dried
pickled/canned
frozen
flour
FEED CORN
French Fries
meats (especially fats)
fast-food restaurant
fries

and may be present in and on foods served to people. Chemical companies are working to develop safer pesticides, and government agencies vigilantly monitor the new products that appear. If a pesticide is deemed to pose a danger, its use is disallowed.

tolerance limit the maximum amount of a residue permitted in a food when a pesticide is used according to label directions.

organic foods products grown and processed without the use of synthetic chemicals such as pesticides, herbicides, fertilizers, and preservatives and without genetic engineering or irradiation.

Regulation of Pesticides The legal **tolerance limits** for pesticide residues in foods are low, generally 1/100 to 1/1,000, the level found to cause no effect in laboratory animals. Over 10,000 tolerance regulations state maximum levels for the more than 300 pesticide chemicals allowed for use on various specific crops in the United States. If a pesticide is misused, growers risk fines, lawsuits, and destruction of their crops. In 25 years of testing, the FDA has seldom found crop residues above tolerance levels, so it appears that pesticides are generally used according to regulations. This makes sense because growers are not anxious to spend extra capital on unneeded chemicals.

A loophole in federal regulations allows companies in the United States to make banned pesticides and export them to other countries. The banned pesticides then return to the United States on imported foods, a circuitous route that has been called the "circle of poison." The FDA is stepping up surveillance of imported foods and may deny entry to any imports found to contain illegal residues. The FDA collects samples of both domestic and imported foods and analyzes them using methods that can detect residues well below tolerances. If any residues exceed permitted limits, the FDA can seize the foods or order them destroyed. The overwhelming majority of the foods tested by the FDA are found to contain either no residues or residues within federally permitted limits.

A problem is that budget restraints limit the FDA's testing capacity. The FDA does not sample *all* food shipments or test for *all* pesticides. Fewer than 700 inspectors and scientists test food samples from the multitude of farms, groves, docks, airports, warehouses, and processing plants the agency oversees. The FDA cannot (nor can it be expected to) guarantee 100 percent safety in the food supply. Instead, it sets conditions so that substances do not become a hazard and acts promptly when problems or suspicions arise.

Consumer Concerns Consumers also bear some responsibility for their own health and safety with respect to pesticides. They can learn about the potential benefits and dangers of pesticide use, discuss regulations and alternatives with others, advise their government representatives about their findings, and apply pressure wherever it will help change inappropriate procedures.

Risks to health from pesticide exposure are probably small for healthy adults, but children, because of their lower body weights and immature detoxifying systems, may be more at risk for some types of pesticide poisoning.[11] Meanwhile, people can minimize their risks by following the guidelines offered in Table 14-6.[§] Consumers who want pesticide-free produce can weigh the pros and cons of organic produce. Those who choose organic shouldn't look for "perfect" fruits and vegetables, however. Pesticide-free produce may have a few minor blemishes, but these are not a hazard. The Consumer Corner on page 525 is devoted to the topic of **organic foods.**

Foods imported from other countries may contain residues of pesticides that are banned from use here.

[§]For answers to questions about any sort of pesticides, call the EPA's 24-hour national pesticide hotline: 1-800-858-PEST.

ultrahigh temperature (UHT) a process of sterilizing food by exposing it for a short time to temperatures above those normally used in processing.

modified atmosphere packaging (MAP) a preservation technique in which a perishable food is packaged in a gas-impermeable container from which air has been removed, or to which another gas mixture has been added.

canning a method of preserving food by killing all microorganisms present in the food and then sealing out air. The food, container, and lid are heated until sterile; as the food cools, the lid makes an airtight seal, preventing contamination.

freezing a method of preserving food by lowering the food's temperature to a point that halts life processes. Microorganisms do not die but remain dormant until the food is thawed.

drying a method of preserving food by removing sufficient water from the food to inhibit microbial growth.

extrusion a process by which the form of a food is changed, such as changing corn to corn chips; not a preservation measure.

Table 14-6

Ways to Reduce Pesticide Residue Intake

- Trim fat from meat and remove skin from poultry and fish; discard fats and oils in broths and pan drippings. (Pesticide residues concentrate in the animal's fat.)
- Vary meat, poultry, and fish choices from day to day and do not take fish oil capsules.
- Wash fresh produce in water, use a scrub brush, and rinse thoroughly.
- Use a knife to peel an orange or grapefruit; do not bite into the peel.
- Discard the outer leaves of leafy vegetables such as cabbage and lettuce.
- Peel waxed fruit and vegetables. (Waxes don't wash off and can seal in pesticide residues.)
- Peel vegetables such as carrots and fruits such as apples when appropriate. (Peeling removes pesticides that remain in or on the peel, but also removes fibers, vitamins, and minerals.)

People who want quick meals often rely on convenience foods and fast foods and prepare few foods at home from farm-fresh produce. The gains in convenience and speed in food preparation must be weighed against a loss of control over how, exactly, foods are processed and what, exactly, they contain. The next few sections describe the effects processing can have on foods.

KEY POINT *Pesticides can be part of a safe food protection program, but can also be hazardous when handled or used inappropriately. The FDA tests for pesticide residues in both domestic and imported foods. Consumers can take steps to minimize their ingestion of pesticide residues in foods.*

Food Processing and the Nutrients in Foods

Much of the total food eaten today, whether in restaurants or at home, has been prepared in some way by industry. People often ask what processing does to foods and to their nutritional value.

Many forms of processing aim to extend the usable life of a food—that is, to preserve it. To preserve food, a process must prevent three kinds of events: (1) microbial growth, (2) oxidative changes, and (3) enzymatic destruction of food molecules. The first two have already been discussed—microbial growth earlier in this chapter and oxidative damage in Chapter 5. Enzymatic destruction occurs as active enzymes in food cells break down their internal molecular structures and cell membranes and walls. Processes involving heat denature the enzymes, and those applying cold slow enzyme activity.

In general, food processing involves trade-offs. It makes food safer, or it gives food a longer usable lifetime, or it cuts preparation time—but at the cost of some vitamin and mineral losses. A process such as pasteurization, which makes milk safe to drink, is clearly worth that cost. Incidentally, the boxes of milk on the shelves of the grocery store that can be kept at room temperature at home have been treated with a process called **ultrahigh temperature (UHT).** The milk is exposed to temperatures above those of pasteurization for just long enough to sterilize it.

Sometimes processed foods may even gain a nutritional edge over their unprocessed counterparts, such as when fat is removed by processing from milk or other foods. The next sections describe some of the most important preservation or processing techniques—**modified atmosphere packaging (MAP), canning, freezing, drying,** and **extrusion**—and their effects on nutrients.

KEY POINT *Processing aims to protect food from microbial, oxidative, and enzymatic spoilage. Some nutrients are lost in processing.*

CONSUMER CORNER

Is Organic the Answer?

CONSUMERS WHO worry about possible health effects from pesticides on food often consider buying organic foods, but wonder whether they will be getting what they pay for. How can they be sure that the food is any better for health or even that it is really organically grown?

Farmers who want to produce and market organically grown crops employ organic farming methods as an alternative to heavy pesticide and chemical fertilizer use. For organic eggs, dairy products, and meats, farmers and ranchers raise chickens, dairy cattle, and food-producing animals without using growth hormones or other drugs to stimulate production or growth. As of this writing, national standards are being developed that will govern the production and handling of organically grown products. Labeling requirements have now been established that permit certain meat and poultry products to carry a label indicating that they are certified organic. The USDA has proposed the following standards for ranchers and farmers who produce organic meat and poultry products:

- Refrain from using growth hormones, antibiotics, or parasite-killing medications.
- Provide living conditions similar to the animals' natural habitat.
- Control the size of the herd or the amount of manure produced in one area to manage the impact on the environment.
- Use only 100% organic feed.

Fruits and vegetables, as well as other products, are permitted to carry an organic label if they meet standards set by independent certifying authorities.[1] An organic food can be defined as a product grown and processed without the use of synthetic chemicals such as pesticides, herbicides, hormones, fertilizers, or preservatives. Likewise, foods sold as organically grown have probably not been irradiated or genetically engineered.

Although it has not been determined that organic foods are more nutritious than conventional foods, organic farming may bring other benefits, especially to the environment. In addition to decreased chemical impact, crop rotation and natural fertilizers are beneficial in that they help to keep the soil nutrient-rich. Rotating crops also reduces pests and disease. Organic farming can be more expensive for the farmers, however, but consumers who believe organic foods are healthful are willing to pay more for them.

Although the definition of organic implies that these foods are safer to eat than conventional foods, this may or may not be the case. For example, the application of animal manure as organic fertilizer may expose consumers of the foods to dangerous microbial diseases, such as *E. coli* 0157:H7. Furthermore, organic foods contain no preservatives and tend to spoil faster than other foods. Those purchasing organic food are urged to buy only the amount needed, to store and cook the food properly, and to wash organic produce thoroughly before eating it.

Sophisticated farming techniques and improved distribution methods ensure that organic foods look and taste better than those of the past. Food that bears the USDA seal, shown here, on the label qualifies to be called *organic.* Both organic and conventionally grown foods arrive at the table with advantages and disadvantages. Either way, eating more fruits and vegetables brings health advantages that far outweigh any possible risks from pesticides.[2] Furthermore, research has shown that pesticide residues are well below safety standards in nearly all domestic and imported goods.[3] Wise consumers who shop for foods with an informed viewpoint stand the best chance of obtaining the benefits they desire from their choices.

MEETS USDA ORGANIC REQUIREMENTS

high-temperature–short-time (HTST) principle the rule that every 18°F (10°C) rise in processing temperature brings about an approximately tenfold increase in microbial destruction, while only doubling nutrient losses.

Modified Atmosphere Packaging

Today, in most produce departments, shoppers can choose bags of washed, trimmed, fresh, chilled salads and chopped vegetables. These products are convenient for busy cooks who have little time for preparation. True, they are more expensive to purchase than comparable loose vegetables, but, unopened, they last much longer and so save on waste. The secret to these vegetables' long shelf life is a technique called modified atmosphere packaging (MAP). The method also preserves freshness in soft pasta noodles, baked goods, prepared foods, fresh and cured meats, seafoods, dry beans and other dry products, ground and whole-bean coffee, and other foods.

Food manufacturers using MAP first package foods in plastic film or other wraps that oxygen cannot penetrate. Then, they remove the air inside the package, creating a vacuum, or they replace the air with a mixture of oxygen-free gases, such as carbon dioxide and nitrogen. By excluding oxygen, MAP:

- Slows ripening of fruits and vegetables.
- Reduces spoilage by mold and bacterial growth.
- Prevents discoloration of cut vegetables and fruits.
- Prevents spoilage of fats by rancidity.
- Slows development of "off" flavors from accelerated enzyme action that breaks down flavor and aroma molecules.
- Slows enzyme-induced breakdown of vitamins.

Chilling of all foods packaged this way is imperative to keep them fresh and safe.

MAP foods retain their vitamins much longer than the same foods exposed to the air. MAP foods also taste fresh, making them especially popular with consumers. One concern about the MAP method is that it may permit growth of the *Clostridium botulinum* bacterium in moist, low-acid foods, such as lunch meats or cooked dishes, when they are kept for long periods at too-warm temperatures. Properly stored, however, MAP foods present very little hazard and can generally be considered as safe and nutritious as fresh foods. An important exception is food that has become spoiled—such foods pose not only the threat of botulism, but other food-borne illness as well.[12]

KEY POINT
Modified atmosphere packaging makes many fresh packaged foods available to consumers. MAP foods compare well to fresh foods in terms of nutrient quality. MAP foods may pose a threat of food-borne illness if not properly stored.

Do Canned Foods Contain Any Nutrients?

Canning is one of the more effective methods of protecting food against the growth of microbes (bacteria, fungi, and yeasts) that might otherwise spoil it, but canned foods, unfortunately, do have fewer nutrients. Like other heat treatments, the canning process is based on time and temperature. Each small increase in temperature has a major killing effect on microbes and only a minor effect on nutrients. In contrast, long heating times are costly in terms of nutrient losses. Therefore industry chooses treatments that employ the **high-temperature–short-time (HTST) principle** for canning.

Which nutrients does canning affect, and how? To answer these questions, food scientists have performed many experiments. They have paid particular attention to three vulnerable water-soluble vitamins: thiamin, riboflavin, and vitamin C.

Acid stabilizes thiamin, but heat rapidly destroys it; therefore the foods that lose the most thiamin during canning are the low-acid foods such as lima beans, corn, and meat. Up to half, or even more, of the thiamin in these foods can be lost during canning. Unlike thiamin, riboflavin is stable to heat but sensitive to light, so glass-packed, not canned, foods are most likely to lose riboflavin. Vitamin C's special enemy is an enzyme (ascorbic acid oxidase) present in fruits and vegetables as well as in microorganisms. By destroying this enzyme, HTST processes such as canning actually help to preserve at least some of the product's vitamin C. Some will be destroyed by the heat of the process, though. As for the fat-soluble vitamins, they are relatively stable and are not affected much by canning.

Minerals are unaffected by heat, so they cannot be destroyed as vitamins can be. Both minerals and water-soluble vitamins can be lost, however, when they leach into canning or cooking water that the consumer then throws away. Losses are closely related to the extent to which a food's tissues have been broken, cut, or chopped and to the length of time the food is in the water.

The Food Feature later in this chapter gives tips on preserving nutrients during cooking.

Some minerals are added when foods are canned. Important in this respect is sodium chloride, table salt, which is added for flavoring. Many food companies have begun making low-salt versions of their products. Unfortunately, because the low-salt batches are smaller, these products may cost more than the higher-salt versions.

KEY POINT *Some water-soluble vitamins are destroyed by canning, but many more diffuse into the canning liquid. Fat-soluble vitamins and minerals are not affected by canning, but minerals also leach into canning liquid.*

Freezing

Freezing is an alternative to canning as a means of preserving food. People often ask how frozen foods compare with canned. In general, frozen foods' nutrient contents are similar to those of fresh foods; losses are minimal. The freezing process itself does not destroy any nutrients, but some losses may occur during the steps taken before freezing, such as the quick dunking into boiling water (blanching), washing, trimming, or grinding. Vitamin C losses are especially likely because they occur whenever tissues are broken and exposed to air (oxygen destroys vitamin C). Uncut fruits, especially if they are acidic, do not lose their vitamin C; strawberries, for example, may be kept frozen for over a year without losing any vitamin C. Mineral contents of frozen foods are much the same as for fresh.

Frozen foods may even have a nutrient advantage over fresh. Fresh foods are often shipped long distances, and to ensure that they make the trip without bruising or spoiling, they are often harvested unripe. Frozen foods are shipped frozen, so produce is allowed to ripen in the field and to develop nutrients to their fullest potential. If foods are frozen and stored under proper conditions, they will often contain more nutrients when served at the table than fresh fruits and vegetables that have stayed in the produce department of the grocery store for even a day.

Frozen foods have to be kept solidly frozen at below 32°F or 0°C, if they are to retain their nutrients. Vitamin C converts to its inactive forms rapidly at warmer temperatures. Food may seem frozen at 36°F or 2°C, but much of it is actually unfrozen, and enzyme-mediated changes can occur fast enough to completely destroy the vitamin C in only two months. If you want to maximize the nutritive value of the foods you store at home, invest in a freezer thermometer, monitor your freezer, and keep it at below freezing temperature (0°F).

KEY POINT *Foods frozen promptly and kept frozen lose few nutrients.*

Drying

Consumers wonder how dried or dehydrated foods compare with canned and frozen foods. Dried or dehydrated foods have their own special characteristics. Drying offers several advantages. It eliminates microbial spoilage (because microbes need water to grow), and it greatly reduces the weight and volume of foods (because foods are mostly water). Furthermore, commercial drying does not cause major nutrient losses. Foods dried in heated ovens at home, however, may sustain dramatic nutrient losses. Vacuum puff drying and freeze drying, which take place at cold temperatures, conserve nutrients especially well.

Sulfite additives are added during the drying of fruits such as peaches, grapes (raisins), and plums (prunes) to prevent browning. Some people suffer allergic reactions

additives substances that are added to foods, but are not normally consumed by themselves as foods.

when they consume sulfites.[13] Sulfur dioxide helps to preserve vitamin C as well, but it is highly destructive to thiamin. This is of small concern, however, because most dehydrated products with added sulfur dioxide were not major sources of thiamin before processing.

KEY POINT *Commercially dried foods retain most of their nutrients, but home-dried foods often sustain dramatic losses.*

Extrusion

Some food products, particularly cereals and snack foods, have undergone a process known as extrusion. In this process the food is heated, ground, and pushed through various kinds of screens to yield different shapes, such as breakfast "puffs," potato "tots" and snack products, the "bits" you sprinkle on salad, and so-called food novelties. Considerable nutrient losses occur during extrusion, and nutrients are usually added to compensate. But foods this far removed from the original fresh state are still lacking significant nutrients (notably, vitamin E), and consumers should not rely on them as staple foods. Enjoy them, but only as occasional snacks and as additions to enhance the appearance, taste, and variety of meals.

KEY POINT *Extrusion involves heat and destroys nutrients.*

Food Additives

People ask valid questions about **additives.** What are they, why are they there, and are they dangerous in any way? In the FDA's list of concerns presented at the start of this chapter, food additives were not a high priority. Compared with the FDA's other concerns, additives pose little danger to consumers, and the FDA has confidence in the ability of regulations already in place to control additive use in the food industry. In fact, compared with largely unregulated and untested "dietary supplements" sold directly to consumers, food additives are strictly controlled and pose little cause for concern indeed.

Manufacturers use food additives to give foods desirable characteristics: color, flavor, texture, stability, enhanced nutrient composition, or resistance to spoilage. Additives, classed by their functions, are listed with their definitions in Table 14-7, and some are discussed further in the section that follows.

Controversy 14 discusses food irradiation.

Regulations Governing Additives

The FDA has the responsibility for deciding what additives shall be in foods. The FDA's judgments on additives hinge primarily on their safety and effectiveness for the stated purpose. To obtain permission to use a new additive in food products, a manufacturer must test the additive and then satisfy the FDA that:

- It is effective (it does what it is supposed to do).
- It can be detected and measured in the final food product.

Then the manufacturer must study the effects of the additive when fed in large doses to animals under strictly controlled conditions to prove that:

- It is safe for consumption (it causes no birth defects or other injury).

Finally, the manufacturer must submit all test results to the FDA. The whole process may take many years.

The FDA then schedules a public hearing and announces the date and location in its official publication, *FDA Consumer.* Consumers are invited to participate at these hearings, where experts present testimony for and against granting permission to use

Without additives, bread would quickly mold and salad dressing would go rancid.

KRAFT
Ranch
BREAD

Table 14-7

Food Additives by Function

- **antimicrobial agents** preservatives that prevent spoilage by mold or bacterial growth. Familiar examples are acetic acid (vinegar) and sodium chloride (salt). Others are benzoic, propionic, and sorbic acids; nitrites and nitrates; and sulfur dioxide.
- **antioxidants** preservatives that prevent rancidity of fats in foods and other damage to food caused by oxygen. Examples are vitamins E and and C, BHA, BHT, proply gallate, and sulfites.
- **artificial colors** certified food colors, added to enhance appearance. (*Certified* means approved by the FDA.) Vegetable dyes are extracted from vegetables such as beta-carotene from carrots. Food colors are a mix of vegetable dyes and synthetic dyes approved by the FDA for use in food.
- **artificial flavors, flavor enhancers** chemicals that mimic natural flavors and those that enhance flavor.
- **bleaching agents** substances used to whiten foods such as flour and cheese. Peroxides are examples.
- **chelating agents** defined in Chapter 4 as molecules that bind other molecules. As additives, they prevent discoloration, flavor changes, and rancidity that might occur because of processing. Examples are citric acid, malic acid, and tartaric acid (cream of tartar).
- **nutrient additives** vitamins and minerals added to improve nutritive value.
- **preservatives** antimicrobial agents, antioxidants, chelating agents, radiation, and other additives that retard spoilage or preserve desired qualities, such as softness in baked goods.
- **thickening and stabilizing agents** ingredients that maintain emulsions, foams, or suspensions or lend a desirable thick consistency to foods. Dextrins (short chains of glucose formed as a breakdown product of starch), starch, and pectin are examples. (Gums such as carrageenan, guar, locust bean, agar, and gum arabic are others.)

the additive. Thus the consumer's rights and responsibilities are written into the provisions for deeming additives safe.

The FDA's approval of an additive does not give manufacturers free license to add it to foods with abandon. On the contrary, the FDA writes a regulation stating in what amounts, for what purposes, and in what foods the additive may be used. No additives are permanently approved; all are periodically reviewed.

The GRAS List Many substances were exempted from complying with this procedure at the time it was first instituted because they had been used for a long time and their use entailed no known hazards. Some 700 substances in all were put on the **generally recognized as safe (GRAS) list.** When substantial scientific evidence or public outcry has questioned the safety of a GRAS list additive, however, its safety has been reevaluated. All substances about which any legitimate question was raised have been removed or reclassified.

In the past, additives were held to a standard of zero cancer risk. A provision of the law known as the Delaney Clause was adopted over 40 years ago at a time when scientists' awareness of cancer causes was limited to radiation, tobacco smoke, a chemical used to make dyes, and soot. Since then researchers have identified more than three dozen human carcinogens and several hundred animal carcinogens. In addition, with advances in technology, substances once detectable only in parts per thousand can now be measured in parts per billion or even per trillion. (One part per trillion is equivalent to about one grain of sugar in an Olympic-sized swimming pool.) We cannot provide absolute protection from all carcinogens in foods, as Congressman Delaney once thought we could. In 1996, the Food Quality Protection Act eliminated the Delaney Clause from the law books.

The Margin of Safety Decisions about an additive's safety are governed by an important distinction—the distinction between **toxicity** and hazard associated with substances. Toxicity is a general property of all substances; hazard is the capacity of a substance to produce injury *under conditions of its use.* All substances can be toxic at some level of consumption, but they are called hazardous only if they are toxic in the

GRAS (generally recognized as safe) list a list, established by the FDA, of food additives long in use and believed safe.

toxicity the ability of a substance to harm living organisms. All substances are toxic if the concentration is high enough.

margin of safety in reference to food additives, a zone between the concentration normally used and that at which a hazard exists. For common table salt, for example, the margin of safety is 1/5 (five times the concentration normally used would be hazardous).

amounts ordinarily consumed. That is, an additive is not a hazard if it proves toxic only in an immense amount that people never consume. The additive is a hazard only if it is toxic as actually used.

A food additive is supposed to have a wide **margin of safety.** Most additives that involve risk are allowed in foods only at levels 100 times below those at which the risk is still known to be zero. Experiments to determine the extent of risk involve feeding test animals the substance at different concentrations throughout their lifetimes. The additive is then permitted in foods at 1/100 the level that causes no harmful effect whatever in the animals. In many foods, *naturally* occurring toxins appear at levels that bring their margins of safety close to 1/10. Even nutrients, as you have seen, involve risks at high dosage levels. The margin of safety for vitamins A and D is 1/25 to 1/40; it may be less than 1/10 in infants. For some trace elements, it is about 1/5. People consume common table salt daily in amounts only three to five times less than those that cause serious toxicity.

The margin-of-safety concept also applies to nutrients used to fortify foods. Iodine has been added to salt to prevent iodine deficiency, but it has to be added with care because it is a deadly poison in excess. Similarly, iron added to grain products has doubtless helped prevent many cases of iron-deficiency anemia in women and children, but iron in excess can cause iron overload in men. The upper limit has to be remembered.

Most additives used in foods are there because they offer benefits that outweigh their risks or that make the risks worth taking. In the case of color additives that only enhance the appearance of foods without improving their health value or safety, no amount of risk may be deemed worth taking. Only 10 of an original 80 synthetic color additives are still approved by the FDA for use in foods, and screening of these substances continues.

Manufacturers must comply with other regulations as well. Additives must not be used:

- In quantities larger than those necessary to achieve the needed effects.
- To disguise faulty or inferior products.
- To deceive the consumer.
- Where they significantly destroy nutrients.
- Where their effects can be achieved by economical, sound manufacturing processes.

The regulations in force governing the management of intentional additives are well conceived, and on the whole, they have been effective. Funding shortages limit the capabilities of watchdog agencies such as the FDA, however, and some mistakes and false reports do slip by.

The following sections focus on the food additives that receive the most publicity because people ask questions about them most often. The order is alphabetical; it is not in order of importance.

KEY POINT *The FDA regulates the use of intentional additives. Additives must be safe, effective, and measurable in the final product. Additives on the GRAS list are assumed to be safe because they have long been used. Additives used must have wide margins of safety.*

Antimicrobial Agents

Preservatives known as *antimicrobial agents* protect food from the growth of microbes that can spoil the food and cause food-borne illnesses. Three of these preservatives—salt, sugar, and nitrites—are commonly used.

Examples of common antimicrobial additives:

- Salt.
- Sugar.
- Nitrites.

Salt and Sugar The best-known, most widely used antimicrobial agents are two common substances—salt and sugar. Salt has been used since before recorded history to preserve meat and fish; sugar serves the same purpose in jams, jellies, and canned and frozen fruits. (Any jam or jelly that toots its "no preservatives" horn is exaggerating. There is no need to add extra preservatives, so most makers do not.) Both salt and sugar work by withdrawing water from the food; microbes cannot grow without water. Today, other additives such as potassium sorbate and sodium propionate are also used to extend the shelf life of baked goods, cheese, beverages, mayonnaise, margarine, and many other products.

Two long-used preservatives.

Nitrites The *nitrites*, another group of antimicrobial agents, are added to meats and meat products for three main purposes: to preserve their color (especially the pink color of hot dogs and other cured meats); to enhance their flavor by inhibiting rancidity (in cured meats); and to protect against bacterial growth. In particular, in amounts much smaller than needed to confer color, nitrites prevent the growth of the bacterium that produces the deadly botulinum toxin described earlier in the chapter.

Nitrites clearly perform important jobs, but they have been the object of controversy because they can be converted in the human body to nitrosamines, which cause cancer in animals. Some cured meats are available without nitrites. However, reducing nitrites consumed in meats would hardly make a difference in a person's overall exposure to nitrosamine-related compounds. For example, an average cigarette smoker inhales 100 times the nitrosamines that the average bacon eater ingests. Likewise, a beer drinker imbibes up to roughly five times the amount that the bacon eater receives, and cosmetics release into the skin about twice as much as is delivered from bacon. Even the air inside automobiles delivers measurable nitrites.

KEY POINT *Microbial food spoilage can be prevented by antimicrobial additives. Of these, sugar and salt have a long history of use. Nitrites added to meats have been associated with cancer in laboratory animals.*

How Do Antioxidants Protect Food?

Food can also go bad when it undergoes changes in color and flavor caused by exposure to oxygen in the air (oxidation). Often these changes involve little hazard to health, but they damage the food's appearance, taste, and nutritional quality. Antioxidants are often added to vulnerable foods to prevent the damage caused by oxidation. Familiar examples of oxidative changes are sliced apples or potatoes turning brown and oil going rancid. Antioxidant preservatives protect food from this kind of spoilage. Some 27 antioxidants, including vitamin C (ascorbate) and vitamin E (tocopherol), are approved for use in food.

Controversy 7 describes how antioxidants break the destructive chain reactions of oxidation.

Sulfites Another group of antioxidants is the sulfites. They are used to prevent oxidation in many processed foods, in alcoholic beverages (especially wine), and in drugs. They used to be popular with restaurant owners for use on salad bars because they kept raw fruits and vegetables looking fresh, but this use was banned after a few people experienced dangerous allergic reactions to the sulfites. The FDA now prohibits sulfite use on food meant to be eaten raw, with the exception of grapes, and it requires foods and drugs to list on their labels any sulfites that are present. For most people, sulfites do not pose a hazard in the amounts used in products, but they have one other drawback. As mentioned earlier, sulfites can destroy a lot of thiamin in foods. A person choosing a food that contains sulfites should not count on that food to contribute to the daily thiamin intake.

Raw grapes may be treated with sulfites. Wash them thoroughly before eating them.

The ban on sulfites has stimulated a search for alternatives, with pleasing results. Some producers now use honey to clarify browned apple juice. Agriculturists have also created a hybrid apple that does not turn brown. A combination of four GRAS additives can also substitute for sulfites.**

BHA and BHT Two other antioxidants in wide use are the well-known BHA and BHT, which prevent rancidity in baked goods and snack foods. BHT provides a refreshing change from the many tales of woe and cancer scares associated with other additives.

**The four GRAS additives are citric acid, ascorbic acid, sodium acid pyrophosphate, and calcium chloride.

Examples of common antioxidant additives:

- Vitamin C.
- Vitamin E (tocopherol).
- Sulfites.
- BHA and BHT.

Among the many tests performed on BHT were several showing that animals fed large amounts of this substance developed *less* cancer when exposed to carcinogens and lived longer than controls. BHT apparently protects against cancer through an antioxidant effect similar to that of vitamin E. To obtain this effect, though, a much larger amount of BHT must be present in the diet than the U.S. average. A caution: used experimentally at levels of intake even higher than this, the substance has *produced* cancer.

This discussion provides the opportunity to mention an important point about additives. No two additives are alike, so generalizations about them are meaningless. No single valid statement can apply to all of the 3,000-odd different substances commonly added to foods. Questions about which additives are safe and under what conditions of use must be asked and answered item by item.

KEY POINT *Antioxidants prevent oxidative changes in foods that would lead to unacceptable discoloration and texture changes in the food. Ingestion of the antioxidant sulfites can cause problems for some people; BHT may offer antioxidant effects in the body.*

Artifical Colors

As mentioned, only about ten artificial colors are still on the GRAS list; they form a highly select group that has survived considerable screening. Artificial colors are among the most intensively investigated of all additives. In fact, they are much better known than the *natural* pigments of plants, and the limits on the safety of their use can be stated with greater certainty. Examples of natural pigments commonly used by the food industry are the caramel that tints cola beverages and baked goods and the carotenoids that color margarine, cheeses, and pastas.

Foods containing tartrazine:

Orange drinks (Tang, Daybreak, Awake).
Gatorade (lime flavored).
Gelatin desserts (Jell-O, Royal).
Golden Blend Italian dressing (Kraft).
Some cake mixes and icings (Duncan Hines, Pillsbury, Cake Mate).
Imitation banana or pineapple extract (McCormick).
Seasoning salt (French's).
Macaroni and cheese dinner (Kraft).
'Cheez' curls and balls (Planter's).
Fruit chews (Skittles).
Butterscotch squares and candy corn (Brach's).

Nevertheless, the food colors, because they are dispensable, have been more heavily criticized than almost any other group of additives. Simply stated, they only make foods pretty, whereas other additives, such as preservatives, make foods safe. Hence with food colors we can afford to require that their use entail no risk, whereas with other additives we may have to compromise between the risks of using them and the risks of *not* using them.

The food color tartrazine (yellow number 5) causes an allergic reaction in susceptible people. Symptoms include hives, itching, and nasal congestion, sometimes severe enough to require medical treatment. It is not a common problem; only 1 or 2 in 10,000 individuals may experience the reaction. Still, that is over 20,000 individuals in the nation as a whole. In addition, some hyperactive children may be sensitive to tartrazine, as Chapter 13 explained. People with allergy and parents of children who react to tartrazine rightly demand to know when the dye is in foods so that they can avoid it. They cannot just avoid yellow-colored foods because tartrazine is used to confer turquoise, green, and maroon colors on foods and drugs as well. U.S. legislation is now in force requiring that tartrazine be listed on all labels of foods that contain it so that consumers can avoid it if they wish.

KEY POINT *The addition of artificial colors is tightly controlled. Some people react adversely to the colorant tartrazine.*

Color additives not only make foods attractive, but identify flavors as well. Everyone agrees that yellow jellybeans should taste lemony and black ones like licorice.

Are Artificial Flavors and the Flavor Enhancer MSG Safe to Consume?

Although only a few artificial colors are currently permitted in foods, close to 2,000 artificial flavors and flavor enhancers are approved, making them the largest single group of food additives. The safety evaluation of flavoring agents is somewhat problematic because so many flavoring agents are already in use, the flavors are strong and so are used in tiny amounts unlikely to impose risks, and they also occur naturally in a wide variety of foods.

One of the best-known members of the flavor enhancer group is monosodium glutamate, or MSG (trade name Accent), the monosodium salt of the amino acid glutamic acid. MSG is used widely in restaurants, especially Asian restaurants, as a flavor enhancer. In addition to enhancing other flavors, MSG itself may possess a basic taste independent of the well-known sweet, salty, bitter, and sour tastes.[††]

In a few sensitive individuals, MSG produces adverse reactions known as the **MSG symptom complex.** Symptoms may include burning sensations, chest and facial flushing or pain, and throbbing headaches. A probable link may lie in elevated blood levels of the MSG component glutamate, a compound known to stimulate the release of some types of pituitary hormones in experimental animals.[14] Meals containing carbohydrate seem less likely to induce adverse effects from MSG than meals of broth, so when dining on Asian-style foods, potentially sensitive people should perhaps order dishes such as soups that contain noodles and eat plenty of plain rice with main dishes to provide carbohydrate, as do Asians themselves.

MSG has been investigated extensively enough to be deemed safe for adults to use (except people who react adversely to it, of course), but it is kept out of foods for infants because very large doses have been shown to destroy brain cells in developing mice. Infants have not yet developed the capacity to fully exclude such substances from their brains and so are more sensitive to them. For other foods, the FDA requires that food label ingredient lists itemize each additive by its full name, including MSG as *monosodium glutamate.*[15]

KEY POINT *Among flavorings added to foods, the flavor enhancer MSG has been determined to cause reactions in people with sensitivities to it.*

Incidental Food Additives

Indirect or **incidental additives** are called *additives,* but they are really contaminants that find their way into food as the result of some phase of production, processing, storage, or packaging. Examples of incidental additives include tiny bits of plastic, glass, paper, tin, and the like from packages and chemicals from processing, such as the solvent used to decaffeinate some coffees.

Some microwave products are sold in "active packaging" that participates in cooking the food. Pizza, for example, may rest on a cardboard pan coated with a thin film of metal that absorbs microwave energy and may heat up to 500°F. During the intense heat, some particles of the packaging components migrate into the food. Regular plastic packages heat up less, but particles still migrate. Materials from such packaging may not be entirely safe for consumption. Until more is known, a wise choice is to use only glass or ceramic containers or those plastics labeled as safe for microwaving. Avoid reusing disposable containers, such as margarine tubs or single-use trays from frozen microwavable meals, for microwaving.

Coffee filters, milk cartons, paper plates, and frozen food packages can all be made of bleached paper and so can contaminate foods with trace amounts of compounds known as dioxins. Dioxins form during the chlorination step in making bleached paper. Dioxins can migrate into foods that come in contact with bleached paper, but the amounts entering food are infinitesimally small—one part per trillion, or the equivalent of one second in 32,000 years. Such amounts do not appear to present a health risk to people, and drinking milk from bleached cartons appears to be safe. Dioxins are persistent, however, and they leach into the environment by way of both paper mill effluent and discarded paper products in landfills. Like heavy metals and organic halogens, dioxins accumulate, becoming more and more concentrated in land, water, and animals until they build up to hazardous levels.

Incidental additives sometimes find their way into foods, but adverse effects are rare. These additives are well regulated, just as intentional additives are. All food packagers are required to perform specific tests to discover whether materials from packages are migrating into foods. If they are, their safety must be confirmed by strict procedures like those governing intentional additives.

[††] The taste produced by MSG is termed *umami.*

MSG symptom complex the acute, temporary, and self-limiting reactions experienced by sensitive people upon ingesting a large dose of MSG. The name *MSG symptom complex*, given by the FDA, replaces the former *Chinese restaurant syndrome.*

incidental additives substances that can get into food not through intentional introduction but as a result of contact with the food during growing, processing, packaging, storing, or some other stage before the food is consumed. Also called *accidental* or *indirect additives.*

MSG symptom complex:

- Breathing problems in people with asthma.
- Burning sensations on forearms, chest, and back of neck.
- Chest pain.
- Drowsiness.
- Facial pressure or tightness of skin.
- Headache.
- Nausea.
- Numbness of the neck that spreads to arms and back.
- Palpitations.
- Tingling, warmth, weakness in face, upper back, neck, and arms.
- Weakness.

SOURCE: D. J. Raiten, J. M. Talbot, and K. D. Fisher, Executive summary from the report: Analysis of adverse reactions to monosodium glutamate (MSG), *Journal of Nutrition* 125 (1995): S2892–S2906.

growth hormone a hormone (somatotropin) that promotes growth and that is produced naturally in the pituitary gland of the brain.

bovine somatotropin (BST) growth hormone of cattle, which can be produced for agricultural use by genetic engineering. Also called *bovine growth hormone (BGH)*.

human somatotropin (HST) human growth hormone.

Examples of common nutrient additives:

- Thiamin, niacin, riboflavin, folate, and iron in grain products.
- Iodine in salt.
- Vitamins A and D in milk.
- Vitamin C in fruit drinks.
- Beta-carotene in cheeses.

KEY POINT *Incidental additives are substances that get into food during processing. They are well regulated, and most present no hazard.*

Nutrient Additives

Nutrients added to improve or to maintain the nutritional value of foods make up another class of additives. Among them are the enrichment nutrients added to refined grains, the iodine added to salt, vitamins A and D added to dairy products, and the nutrients used to fortify breakfast cereals. When nutrients are added to a nutrient-poor food, it may appear from its label to be nutrient-rich. It is, but only in those nutrients chosen for addition. Nutrients are sometimes also added for other purposes. Vitamins C and E used as antioxidants and beta-carotene as a colorant are examples already mentioned.

A topic of interest to many consumers, but of small concern to the FDA, is the hormones administered to livestock that produce food. The FDA has deemed the practice safe and does not require testing of food products for traces of the drugs. The next section provides the details about the hormones that prompted the FDA's decision.

KEY POINT *Nutrients are added to foods to enrich or to fortify them. These additives do not necessarily make the foods nutritious, only rich in the vitamins and minerals that have been added.*

Growth Hormone in Meat and Milk

Some people fear the introduction of the cattle form of **growth hormone, bovine somatotropin (BST),** into meat animals or dairy herds. The hormone, produced by genetically engineered bacteria, is virtually identical to growth hormone made naturally in the pituitary gland of the animal's brain.[16]

Ranchers advocate the use of BST because it makes meat animals develop more meat and less fat. BST may also increase milk production in dairy cows by up to 25 percent, while requiring less feed.[17] To the farmers, these changes mean higher profits without the high costs of more cattle, more farmhands, or more equipment. The environment may profit as well. Smaller herds can live on smaller plots of cleared land, and less feed means the use of less resources to produce and transport it (Controversy 15 gives details).

Most consumer fears of BST stem from failure to distinguish steroid hormones, which can be taken orally because they survive digestion, from peptide hormones, which are destroyed by digestive enzymes. Estrogen, found in many oral contraceptives, is a steroid hormone; BST, however, is a peptide hormone and so is destroyed during digestion.

The amount of the hormone found in the milk of BST-treated cows is within the range that can occur naturally. About 90 percent of the BST in milk, regardless of its source, is destroyed by pasteurization. The rest is destroyed during digestion. Even if some BST were to survive to enter the bloodstream, it would have no effect on the body because the chemical structures of animal growth hormones differ widely from the structure of human growth hormone, **human somatotropin (HST).** When BST was first discovered, scientists hoped to use it to treat growth hormone–deficient children. Tests proved disappointing, for BST failed to stimulate receptors for human growth hormone and thus had no effect on the children's growth.

Cows treated with BST suffer more udder infections (mastitis) and so are given more antibiotics; these drugs then show up in the cows' milk and meat. Eating the meat could thus pose a hazard to those allergic to the drugs, but milk and meat are tested for drug residues and contaminated products are not sold. The National

Institutes of Health concludes that as BST is currently used in the United States, meat and milk from hormone-treated cows are as safe as those from untreated cows, and the FDA approves of BST use.[18]

KEY POINT *Bovine somatotropin causes cattle to produce more meat and milk on less feed than untreated cattle. The FDA has deemed the practice safe.*

To sum up the messages of this chapter, U.S. foods are safe and hazards are rare. Precautions against food poisoning are the most important measures people can take to protect themselves from illness caused by foods. For optimal nutrition, though, which lies beyond safety, people can do more. The Food Feature that follows offers pointers on the selection and cooking of foods for the healthiest possible diet.

FOOD FEATURE

Making Wise Food Choices and Cooking to Preserve Nutrients

In terms of nutrient density, canned juice is almost as nutritious as fresh, but yogurt-covered raisins are *not* as nutritious as plain raisins.

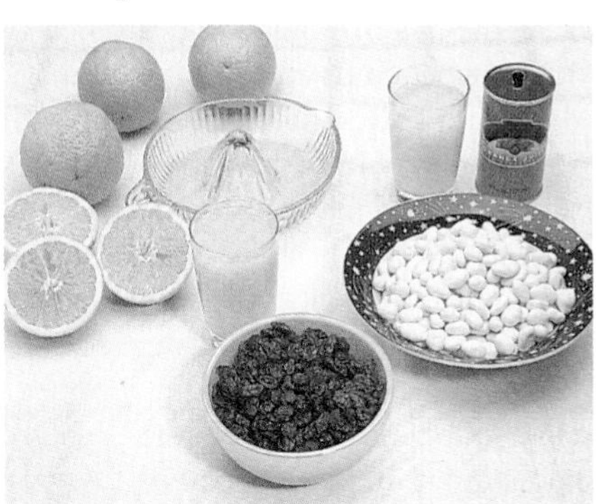

In general, the more heavily processed foods are, the less nutritious they become. Does that mean, then, that everyone should avoid all processed food? The answer is not simple: in each case it depends on the food and on the process. Consider the case of orange juice and vitamin C.

The Choice of Orange Juice

Orange juice is available in several forms, each processed a different way. Fresh juice is simply squeezed from the orange, a process that extracts the fluid juice from the fibrous structures that contain it. Each 100 calories of the fresh-squeezed juice contains 111 milligrams of vitamin C. When this juice is condensed by heat, frozen, and then reconstituted, as is the juice from the freezer case of the grocery store, 100 calories of the reconstituted juice contain just 88 milligrams of vitamin C because vitamin C is destroyed in the condensing process. Canning is even harder on vitamin C: 100 calories of canned orange juice have 82 milligrams of vitamin C.

These figures may seem to indicate that fresh juice is the superior food, and so it may be. But consider this: most people's recommended intake of vitamin C (60 milligrams) is covered by a single serving of any of the above choices. In this case, at least for vitamin C, the losses due to processing are not a problem. Besides, processing confers enormous convenience and distribution advantages. Fresh orange juice spoils. Shipping fresh juice to distant places in refrigerated trucks costs much more than shipping frozen juice (which takes up less space) or canned juice (which requires no refrigeration). The fresh product still contains active enzymes that continue to degrade its compounds (including vitamin C) and so cannot be stored indefinitely without compromising nutrient quality. The savings gained from shipping and storing canned and frozen juices are passed on to consumers. Without canned or frozen juice, people with limited incomes or those with no access to fresh juice would be deprived of this excellent food.

Processing Mischief

Some processing stories are not so rosy. In Chapter 8, for instance, you saw how processed foods are often loaded with sodium as their potassium is leached away, exactly the wrong effect for people with hypertension. A related mischief of processing is the addition of sugar and fat—palatable, high-calorie additives that reduce nutrient density. For example, consider nuts and raisins covered with "natural yogurt." This may sound like one healthy food being added to another, but a look at the ingredient panel warns that generous amounts of sugar and fat accompany the yogurt. About 75 percent of the weight of the product is sugar and fat; only 8 percent is yogurt. To pick just one nutrient for an example, here is what happens to the iron density of the raisins: 100 calories of raisins = 0.71 milligrams of iron; 100 calories of "yogurt" raisins = 0.26 milligrams of

FAT-FREE MILK
MILK
Value CUT CORN
Value GREEN BEANS

Purchase mostly fresh foods or those that processing has benefited nutritionally.

Steam vegetables or cook them in a microwave oven.

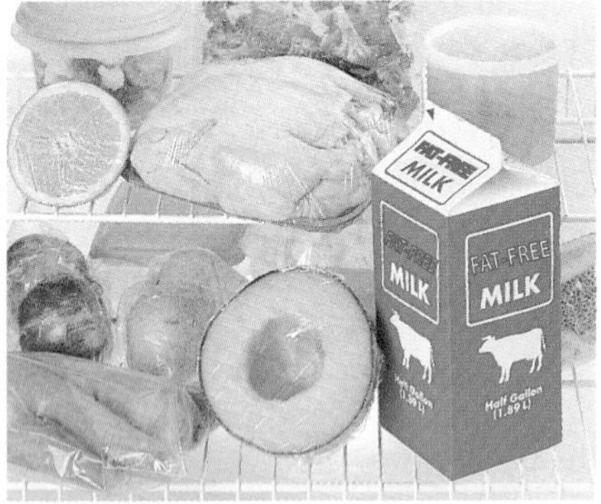
FAT-FREE MILK
MILK
FAT FREE MILK

Wrap foods tightly and refrigerate them. Space foods to allow chilled air to circulate around them.

iron. These foods taste so good that wishful thinking can easily take hold, but the reality is that sugar- and fat-coated food is candy. The word *yogurt* on the label means only that one of the ingredients of the candy coating is some small amount of yogurt.

Best Nutrient Buys

A good general rule for making food choices is to choose whole foods to the greatest extent possible and to seek out among processed foods only the ones that processing has improved nutritionally. (When processing removes fat, as in fat-free milk, it is often a benefit to the consumer.) Being realistic, few people have the time to bake all their own bread from scratch, to shop every few days for fresh meats, or to wash, peel, chop, and cook fresh fruits and vegetables at every meal. This is where food processing comes in. Commercially prepared whole-grain breads, frozen cuts of meats, bags of frozen vegetables, and canned or frozen fruit juices do little disservice to nutrition and enable the consumer to eat a wide variety of foods at great savings in time and human energy. The nutrient contents of processed foods exist on a continuum:

Whole-grain bread > refined white bread > sugared doughnuts.

Milk > fruit-flavored yogurt > canned chocolate pudding.

Corn on the cob > canned creamed corn > caramel popcorn.

Oranges > orange juice > orange-flavored drink.

Baked ham > deviled ham > fried bacon.

The nutrient continuum is paralleled by another continuum—the nutrition status of the consumer. The closer to the farm the foods you eat, the better nourished you are, but that doesn't mean you have to live in the fields.

Conserving Nutrients at Home

Wise food choices are half the story of smart nutrition self-care; skillful food preparation is the other half. In modern commercial processing, losses of vitamins seldom exceed 25 percent. In contrast, losses in food preparation at home can be close to 100 percent, and it is not unusual to see losses in the 60 to 75 percent range. The kinds of foods you buy certainly make a difference, but what you do with them in your kitchen can make an even greater difference.

Preventing Enzymatic Destruction

To develop skill in preparing foods requires some understanding of the effects of cooking and storing foods on nutrients. Vitamins are organic compounds synthesized and broken down by enzymes found in the foods that contain them. Like all enzymes, the enzymes that break down nutrients in fruits and vegetables have a temperature optimum. They work best at the temperatures at which the plants grow, normally about 70°F (25°C), which is also the room temperature in most homes. Chilling fresh produce slows down enzymatic destruction of nutrients. To protect the vitamin content, most fruits and vegetables should be vine ripened (if possible), chilled immediately after picking, and kept cold until use.

Protecting from Light and Air

Besides being vulnerable to enzyme-mediated spoilage, the vitamin riboflavin is light sensitive. It can be destroyed by the ultraviolet rays of the sun or by fluorescent light. For this reason milk is not sold (and should not be stored) in transparent glass containers. Cardboard or opaque plastic containers screen out light, protecting the riboflavin. Since grain products such as macaroni and rice are also important sources of riboflavin, cooks who store them in glass jars should stow the jars in closed cupboards.

Some vitamins are acids or antioxidants and so are most stable in an acid solution away from air. Citrus fruits, tomatoes, and many juices are acid. As long as the skin is uncut or the can is unopened, their vitamins are protected from air. If you store a cut vegetable or fruit, cover it with an airtight wrapper; close an opened carton of juice tightly and store it in the refrigerator.

Refreezing

Labels on frozen foods tell you "Do not refreeze." As food freezes, the cellular water expands into long, spiky ice crystals that puncture cell membranes and disrupt tissue structures, changing the texture of the food. There is usually no danger in eating a twice-frozen food although some nutrients are lost upon thawing and refreezing. Provided that it hasn't spoiled while it was thawed or wasn't thawed at warm temperature, the main problem with a twice-frozen food is that it may be unappealing.

Preventing Nutrient Losses in Water

Minerals and water-soluble vitamins in fresh-cut vegetables readily dissolve into the water in which they are washed, boiled, or canned. If the water is discarded, as much as half of the vitamins and minerals in foods go down the drain with it. A bit of southern folk wisdom is to serve the cooking liquid with the vegetable rather than throwing it away; this liquid is known as the "pot liquor" and may be used to moisten cornbread or to make gravies or soups. Other ways to minimize cooking losses: steam vegetables over water rather than in it, stir-fry them in small amounts of oil, or microwave them. Wash the intact food vigorously and briefly, don't soak it. Cut vegetables after washing except for those such as broccoli that you have to cut to wash adequately. For peeled vegetables, such as potatoes, add them to water that is vigorously boiling, not to cold water, to minimize the length of time the vegetables are exposed to nutrient-leaching water. Microwave ovens are excellent for conserving nutrients. They cook fast without requiring the addition of fats or excess liquid. Some special microwaving concerns appear in the margin.

During other types of cooking, minimize the destruction of vitamins by avoiding high temperatures and long cooking times. Iron destroys vitamin C by catalyzing its oxidation, but perhaps the benefit of increasing the iron content of foods by cooking in iron utensils outweighs this disadvantage. Each of these tactics is small by itself, but saving a small percentage of the vitamins in foods each day can mean saving significant amounts in a year's time.

Meanwhile, however, a law of diminishing returns operates. Most vitamin losses under reasonable conditions are not catastrophic. You need not fret over small vitamin losses that occur in your kitchen; you may waste energy or time that is valuable to you in other ways. Be assured that if you start with fresh, whole foods containing ample amounts of vitamins and are reasonably careful in their preparation, you will receive a bounty of the nutrients that they contain.

Take care when cooking in a microwave oven. Food can become extraordinarily hot or build up steam that may scald unprotected hands or face. Before cooking eggs, sausages, or any food encased in a membrane, pierce the membrane to prevent explosion of the food. Never warm baby formula or food in a microwave oven because hot spots can form that can scald the baby.

Here's a way to tell if glass or other containers are made of microwave-safe materials. Microwave the empty container for one minute and carefully touch it.

Warm = unsafe for microwave.

Lukewarm = safe for short reheating use.

Cool = safe for long microwave cooking times.

Self Check

Answers to these Self Check questions are in Appendix G.

1. Which of the following food hazards has the FDA identified as its number one concern?
   a. pesticides in food
   b. microbial food poisoning
   c. intentional food additives
   d. environmental contaminants
2. To prevent food-borne illnesses, cooked foods should be held at temperatures higher than:
   a. 85°F
   b. 100°F
   c. 140°F
   d. 212°F
3. Which of the following may be contracted from normal-appearing seafood?
   a. hepatitis
   b. worms and flukes
   c. viral intestinal disorders
   d. all of the above
4. Which of the following heavy metals is *not* among those considered to be chemical contaminants of concern?
   a. zinc
   b. cadmium
   c. lead
   d. mercury
5. Which of the following are likely sources of nitrites?
   a. cosmetics
   b. cigarette smoke
   c. bacon and hot dogs
   d. all of the above
6. Foods that smell good, look good, and taste good are always safe to eat. T F
7. Vitamins are unaffected by heat processing because they cannot be destroyed, as minerals can be. T F
8. The canning industry chooses treatments that employ the low-temperature–long-time (LTLT) principle for canning. T F
9. The artificial flavors and flavor enhancers are the largest single group of food additives. T F
10. Foods produced through biotechnology, if not substantially different from foods already in use, require no special safety testing or labeling. (Read about this in the upcoming Controversy.) T F

NUTRITION
ON
THE
NET

For further study of the topics of this chapter, access these websites and search for the phrases or words in quotation marks:

1. For the latest changes in chapter material, to find supplemental learning tools, or access links to related websites, go to the "Nutrition: Concepts and Controversies" site at: **www.wadsworth.com/nutrition/prod/allprod.html**
2. Search the Food and Drug Administration's site for more information on "food safety" at: **www.foodsafety.gov**
3. Additional information on food safety is provided through the Partnership for Food Safety Education at: **www.fightbac.org.**
4. The Centers for Disease Control and Prevention provides information about "food-borne illnesses" at: **www.cdc.gov**
5. For more information about food safety in the marketplace, contact: **www.usda.gov/fsis**
6. For topics related to food safety in the kitchen, search the U.S. Government site and look for "meat," "poultry," and "seafood" at: **www.healthfinder.gov/searchoptions/topicsaz/htm**
7. To learn more about "pesticides and food," contact the Environmental Protection Agency at: **www.epa.gov**
8. Information about organic foods can be found at: **www.ams.usda.gov/nop**
9. Additional information regarding food additives is located at: **www.fao.org/NEW/WHATSNEW.HTM**
10. Contact the Biotechnology Information Center for information about "food biotechnology" at: **www.nal.usda.gov/bic**
11. Search among thousands of current scientific and medical abstracts for any topic related to "food safety" or "food pathogens" at: **www.ncbi.nlm.nih.gov/PubMed/**

CONTROVERSY
14

FUTURE FOODS: ARE THE NEW FOOD TECHNOLOGIES SAFE?

FOOD FUTURISTS who gaze ahead into the new millennium see a world with many more people, increasing demands for foods, and shrinking farmland on which to grow food. Foods will have to be easy to grow in abundance. People in developed countries will have little or no time for cooking or sitting down to meals. Consumers will need nutritious, affordable, easy-to-prepare foods that are low in fat and taste good. And, of course, the foods must be safe to eat—free from microbial and other contamination. People of the developing world have an even greater need for foods with these qualities than do developed nations, for without changes in current conditions they face disastrous famines. Many look with hope to the new technologies to provide increased crop yields and reduced waste (more about the world's food supply in Chapter 15). Also, the production of foods must economically support food growers and producers and must have as little environmental impact as possible. These diverse needs may seem to conflict with each other, yet all are promised by advocates of new technologies. Should we believe these promises? Do the new technologies present problems of their own?

Today the world is witnessing the beginning of a revolution of applied technology in food science and agriculture. Upon surveying government, business, and university experts about technologies now under development, the FDA concluded, "the floodgates of innovation are opening." Thousands of experimental developments are feasible, and many are becoming realities.

Some people feel uneasy about such rapid change and its possible risks. They ask who will ultimately stand to benefit: manufacturers of foods or the consumers who purchase them? And who will suffer the consequences if the new technology brings unsuspected problems? This Controversy focuses on both sides of two major food technology issues that hold vast potential for changing the food supply. The first is **genetic engineering**, a form of biotechnology, and the second is **irradiation** of foods. Table C14-1 provides some definitions of terms.

The original wild corn from which today's corn was developed over centuries of selective breeding.

Genetic Engineering

For centuries farmers have been changing the genetic makeup of their plants and animals. Season after season they have selectively bred plants or animals possessing desirable traits in the hope of obtaining offspring that reliably display those traits. Today's lush, hefty, healthy agricultural crops and animals, from cabbage and squash to pigs and cattle, all demonstrate the results of those efforts.

Among the successes of selective breeding is corn. Its large ears with full, sweet kernels and high yields bear little resemblance to the original wild, native corn with its sparse two or three kernels to a stalk. Breeders have even trained corn to "stay sweet" by breeding out an enzyme that normally turns sugar to starch within days after harvest. Selective breeding, called by some the "old biotechnology," works, but slowly and imprecisely.

Table C14-1

Food Technology Terms

- **antisense gene** a gene's chemical opposite, which interferes with the native working gene and keeps it from producing proteins.
- **clone** an individual created asexually from a single ancestor, such as a plant grown from a single stem cell; a group of genetically identical individuals descended from a single common ancester, such as a colony of bacteria arising from a single bacterial cell; in genetics, a replica of a segment of DNA, such as a gene, produced by genetic engineering.
- **genetic engineering** a field within biotechnology that involves the direct, intentional manipulation of the genetic material of living things in order to obtain some desirable trait not present in the original organism; also called *recombinant DNA (rDNA) technology.*
- **irradiation** application of ionizing radiation to foods to reduce insect infestation or microbial contamination or to slow the ripening or sprouting process.
- **outcrossing** the unintended breeding of a domestic crop with a related wild species.
- **plant-pesticides** substances produced within plant tissues that kill or repel attacking organisms.
- **radiolytic products** chemicals formed during irradiation of food.
- **stem cell** an undifferentiated cell that can mature into any of a number of specific specialized cell types. A stem cell of bone marrow may mature into one of many kinds of blood cells, for example.
- **transgenic organism** an organism that grows from an embryonic, stem, or germ cell into which a new gene has been inserted. The organism carries the new gene in all of its cells.

Biotechnology represents a way of speeding up and refining the process of genetic selection. The changes in corn just mentioned required centuries, but today's genetic engineering methods could have accomplished the same things in a year or two. Genetic engineering goes beyond simple improvements in selective breeding, however. The technique wields awesome power—the power to change the most basic patterns of life in ways never before possible.[1] Its progeny reach far beyond nutrition into medicine, forestry, and even international trade policies and weapons of war.[2] Figure C14-1 compares the genetic results of selective breeding and genetic engineering.

Food Biotechnology Three areas of research in genetic engineering are most relevant to the food supply. First, new strains of agricultural crops and animals offer new desired traits, such as improved resistance to diseases or insect pests. Second, strains of microorganisms have been engineered to produce substances that occur in only small amounts or not at all in nature (such as bovine somatotropin, the cattle growth hormone mentioned in the preceding chapter). Third, agricultural crops have been developed that resist destruction by herbicides.

Plant cells make likely candidates for genetic engineering because a single plant cell can often be coaxed to reproduce an entire new plant. Cells with this talent include fertilized ova, stem or germ cells, and some embryonic cells. Each cell contains an exact replica of the genetic information contained in the original cell. If scientists have introduced any DNA fragments into that first single cell, those fragments will be faithfully reproduced in all of the cell's offspring. All of the resulting cells are **clone** cells—exact genetic replicas of the original.

For example, scientists can start with an undifferentiated cell, known as a **stem cell**, from the "eye" of a potato plant. Into that cell they can implant some DNA with genes for the protein coat (but not the infective part) of a virus that attacks potato plants. Then they can stimulate the stem cell to begin growing a whole new **transgenic** potato plant that replicates the piece of viral protein coat in each of its cells. The presence of the protein stimulates the plant to develop immunity to an attack from the real virus.

Animals, too, have stem cells and can therefore be cloned. Recently, twin calves were cloned from a stem cell of a fetal calf; a famous barnyard cousin, Dolly the sheep, was cloned from a stem cell harvested from a ewe's udder.

What Do We Stand to Gain from Biotechnology? The researchers who cloned these animals are working toward creating animals with the ability to produce needed pharmaceutical products as well as food. For example, a cow cloned with the genetic equipment to make a vaccine in its milk could provide nourishment and immunization to a whole village of people now left unprotected because they lack medical help. In the same way, researchers hope to genetically create not only animals, but also bananas, potatoes, or other foods with the ability to grow valuable pharmaceutical products in their tissues, a process whimsically called "biopharming."*

Already, products from transgenic bacteria are assisting food manufacturers. One bacterium, for example, was given the ability to make the enzyme renin, a substance necessary to produce cheese. Traditionally, renin was harvested from the stomachs of calves, an expensive process. Through recombinant DNA technology, the gene for making renin was snipped (enzymes do the snipping) from some calf DNA and transferred to a single bacterial cell, which reproduced itself into a large bacterial factory of mass produc-

*Read more about "biopharming" and biotherapy in J. Raso, The biopharm revolution, *Priorities* 10, no. 1(1998): 33–35.

Figure C14-1

*The Precision of Genetic Engineering**

Traditional Breeding

DNA is a strand of genes, much like a strand of pearls. Traditional plant breeding combines many genes at once.

donor | commercial variety | new variety (Many genes are transferred.)

desired gene + = desired gene

Genetic Engineering

Through genetic engineering, a single gene may be transferred from one strand of DNA to another.

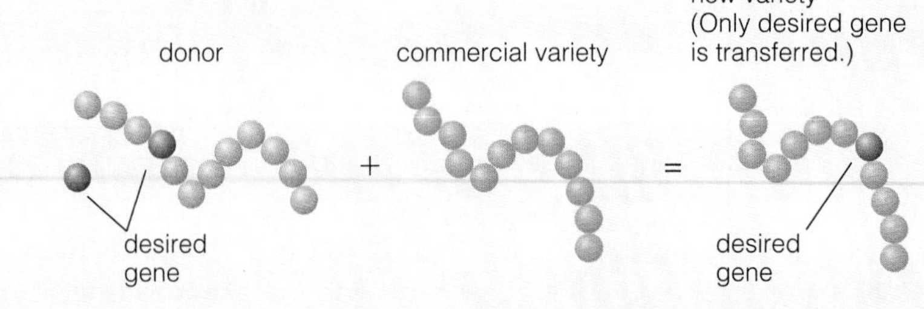
donor
commercial variety
new variety
(Only desired gene
is transferred.)
+
=
desired
gene
desired
gene

*Another name for genetic engineering is recombinant DNA (rDNA) technology.
SOURCE: © 1995 Monsanto Company.

MacGREGOR'S
84805

tion. By the same process, scientists have harvested the once-scarce human growth hormone. Today, many children with growth hormone deficiency can grow normally thanks to a reliable supply of human growth hormone produced by transgenic bacteria.

The technique just described allows an organism to make proteins native to some other living thing. Another option is to block or suppress production of unwanted cell products, as in the corn with the long-lasting sweet taste, mentioned earlier. This procedure has brought to market an especially long-lasting tomato.[3] Normally, tomatoes produce a protein that softens them after they have been picked. Scientists introduced into a tomato plant an **antisense gene**, a mirror image of the native gene that coded for the "softening" enzyme. The new antisense gene blocked the synthesis of the softening enzyme (see Figure C14-2). A vine-ripe tomato with the antisense gene can be harvested at its most flavorful and nutritious red-ripe stage and still last long enough to go to market.

Another group of crops are intended to benefit farmers by easing the task of controlling weeds. These plants are genetically engineered to withstand potent herbicides. As a result, farmers can spray whole fields with the herbicides and kill every other plant growing there, leaving only the desired crop.

Other possibilities for the near future are tomatoes and cotton that produce their own insecticides, which may render pesticide sprays unnecessary. Shrimp may soon fight diseases with genetic ammunition borrowed from sea urchins. Some plants may even be given special molecules to help them grow food in soil so polluted that

all other plants wither and die. Supporters of technology predict that these and other advances will enable farmers to reliably produce bumper crops of food every year on far fewer acres of land, with less loss of water and topsoil, and with far less use of toxic pesticides and herbicides that end up in foods and drinking water.

These projects are already in progress. Close on their heels are many more ingenious ideas. What if salt tolerance could be transplanted from a coastal marsh plant into crop plants? Could crops then be irrigated with seawater, thus conserving dwindling fresh water supplies? Would the world food supply increase if rice farmers were able to grow plants that were immune to disease? What if consumers could dictate which traits scientists insert into food plants? Would they choose to add extra cancer-fighting phytochemicals or hard-to-get nutrients? These and other ideas listed in Table C14-2 (on the following page) may sound fantastic, but many such products are already on laboratory shelves, awaiting FDA approval for use in agriculture.

Plant Pesticides and the Problem of Outcrossing

Among the newest transgenic foods are yellow squash with two viral genes that confer resistance to the most common viral diseases and a potato that produces a beetle-killing toxin. These two new plants and many others like them are currently being grown or tested in fields around the United States. The plants produce what the Environmental Protection Agency (EPA) calls **plant-pesticides**, or pesticides made by the plants themselves. The EPA has established regulations for approval of products of biotechnology such as plant-pesticides, microorganisms, or other organisms used in agriculture.[4]

Some scientists are concerned that such disease-resistant crops could lead to **outcrossing**, accidental cross-pollination with related wild weeds that would give the weeds an enormous survival advantage over other wild species and crowd them out. A possible solution to the outcrossing problem has

Figure C14-2

How Biotechnology Techniques Can Block Formation of Specific Proteins

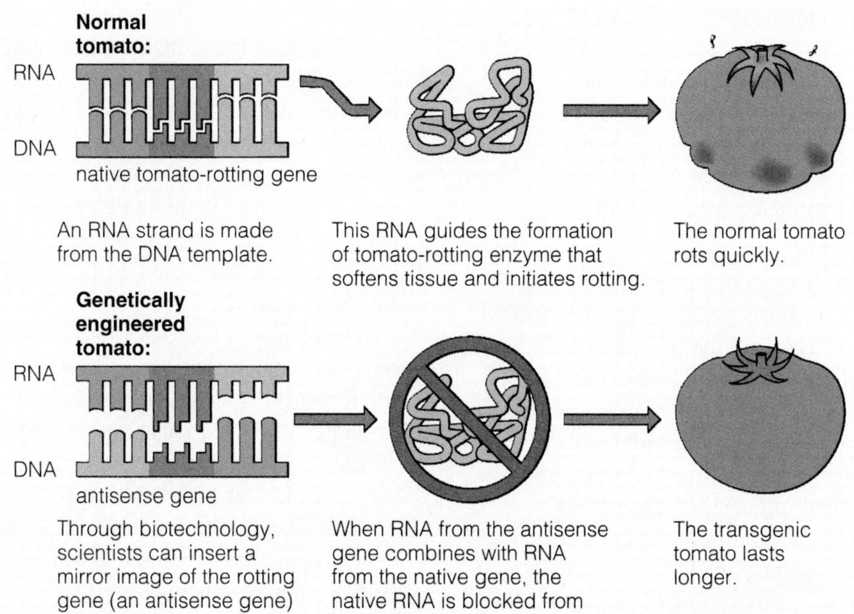

Normal tomato:
RNA
DNA
native tomato-rotting gene
An RNA strand is made from the DNA template.
This RNA guides the formation of tomato-rotting enzyme that softens tissue and initiates rotting.
The normal tomato rots quickly.
Genetically engineered tomato:
RNA
DNA
antisense gene
Through biotechnology, scientists can insert a mirror image of the rotting gene (an antisense gene) into tomatoes.
When RNA from the antisense gene combines with RNA from the native gene, the native RNA is blocked from producing the rotting enzyme.
The transgenic tomato lasts longer.

emerged in the form of the so-called *terminator technology*, a process recently patented in the United States. A terminator plant grows and bears crops normally for one season, but it destroys its own living seeds before the next growing season. This feature reduces the threat of outcrossing, but it also prevents farmers from saving seeds from one year's harvest to the next and forces them to buy expensive new seeds each year.[5] Only time will tell whether terminator plants become part of the future of farming.

Table C14-2

Food Products of Biotechnology—Present and Future

Now Available	Expected Soon	Down the Road
Disease-resistant, pest-resistant, and herbicide-resistant crops: virus-resistant papayas, potatoes, and squash; insect-protected corn, cotton, and soybeans; sugar beet and soybean plants that survive when sprayed with herbicides.	*Reduced natural toxins:* fungus-resistant foods; foods genetically modified to produce fewer toxins.	*Reduced allergens:* modified peanuts, milk, eggs, and other commonly allergenic foods.
More nutritious foods: soybeans that are lower in saturated fatty acids and higher in monounsaturated fatty acids that yield stable cooking oils.	*More nutritious foods:* peanuts, beans, and rice with improved amino acid balance; oils with less harmful fatty acids for margarine and shortening; potatoes that absorb less fat while frying; fruits and vegetables with higher concentrations of vitamins C and E; foods with higher phytochemical contents.	*Enhanced foods:* high-protein rice with complete amino acid profile; frost-resistant potatoes; other crops with resistance to drought, flood, salt, metals, heat, and cold; animals and vegetables that provide vaccines and other pharmaceutical products along with the foods they supply.
More appealing foods: peppers modified to taste sweeter; tomatoes modified to ripen on the vine without rotting before consumption.	*More appealing foods:* strawberries with improved flavor and texture; single serving–sized melons; sweeter green peas; many others.	*Almost any imaginable food.*

Biotechnology Backlash

Although both scientists and food industrialists hail biotechnology with confidence, some consumers fear what they call "Frankenfoods." They are concerned for the safety of a world where direct genetic tampering that produces effects not yet fully understood is driven solely by potential profits without moral judgment or laws to harness its effects.[6] They point out that even the scientists who developed the techniques cannot predict the ultimate outcomes of their discoveries.[†] Genetic decisions, say these critics, are best left to the powers of nature. If science and the marketplace are allowed to drive biotechnology without restraint, critics fear that these problems may result:

- *Disruption of natural ecosystems.* New, genetically unusual organisms are being released into the environment without knowledge of whether or how nature can accomodate them. Such organisms have no natural place in the food chain or in evolutionary biological systems.
- *Disease.* Newly created viruses may mutate to cause new deadly diseases that may attack plants, animals, or human beings.
- *Weapons.* Development of fatal untreatable bacterial and viral diseases for use as weapons may be progressing alongside peaceful developments.
- *Animal cruelty.* The rights and well-being of animals may be disregarded, especially those with inserted human genes that are grown for transplantable organs for human beings.
- *Human ethics.* Critics pose the question, "How many human genes does an organism have to contain before it is considered human? For instance, how many human genes would a green pepper have to contain before one would have qualms about eating it?"[7]

For many people, the most serious ethical concerns posed by the new technology center on the prospect of cloning human beings for certain traits and genetic "improvements." In the United States, federal funds are withheld from any laboratory involved in human genetic engineering experiments.

At a minimum, the critics of biotechnology have made a strong case for rigorous safety testing of new products. They contend, for example, that when a new gene has been intoduced into food, tests should make sure that other, unwanted genes have not accompanied it. If a disease-producing microorganism has donated genetic material to make the recombinant DNA, scientists should be required to prove that no dangerous characteristic from the microorganism has also entered the food. If the newly altered genetic material creates unique proteins that have never before been encountered by the human body, their effects should be studied and their presence regulated to ensure that people can eat them safely.

The concerns just described illustrate the tension that exists between the forward thrust of science and the hesitancy of consumer groups to accept all the new products scientists can create. In particular, the demands for testing and labeling of new products may benefit consumers by lending a measure of safety as this powerful technology advances.

[†]A similar example from history is scientist Alfred Nobel, founder of the Nobel Prize for Peace and inventor of dynamite. His purpose in inventing dynamite was to speed land excavation for construction. Dynamite technology was later used to make bombs for war.

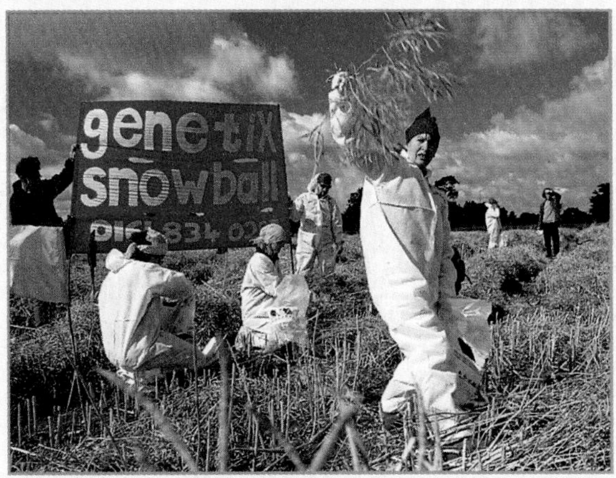
gene-tix
snowball

Protesters uprooting genetically altered plants.

The FDA's Position To help determine the safety of the food products of biotechnology intended for human consumption, the FDA has established a new National Center for Food Safety and Technology in Illinois. Studies performed at the center guide the FDA in establishing regulations for the new products.

Thus far, the FDA has taken the position that whole foods produced through biotechnology require no special safety testing or labeling if they are not substantially different from foods already in use.[8] The developer of a new food that differs significantly from traditional foods, however, is required to conduct tests and prove its safety to the FDA. Under the FDA's rules, any product with an antisense gene, such as the tomato described earlier, is assumed to be safe because the antisense gene merely prevents the snythesis of a protein and adds only a tiny fragment of genetic material. In contrast, any extra substances introduced, such as new enzymes, hormones, or resistance traits, must meet the same safety standards applied to all food additives.[9] Thus a tomato with a gene that produces an insecticide could not be marketed unless the developer proved that the insecticide "additive" was safe to eat at the levels people would normally encounter.[10]

Consumer advocacy groups question the adequacy of the FDA's labeling requirements and call for all genetically altered products to be clearly identified. In addition to the general objection that the FDA is forcing millions of consumers to be unwitting testers of the products of biotechnology, these critics raise two specific concerns. One fear is that genetic material from a source to which some people develop allergies, such as nuts, may be added to another product, such as soybeans; unless the products made from the soybeans are labeled, people who are allergic to proteins in nuts may unwittingly consume them in the altered soybeans.[11] Secondly, people with religious objections to particular foods may be unable to avoid consuming genes of prohibited organisms that have been added to permitted foods. For example, someone keeping a kosher kitchen may unknowingly purchase a food product containing genes normally found in pork. One group has recently filed a lawsuit against the FDA in the hope that the courts will ultimately rule in favor of mandated labeling of products of biotechnology.[12]

Speaking in defense of the FDA's position are the FDA itself, recognized as the nation's leading expert and advocate for food safety, and the American Dietetic Association, which represents current scientific thinking in nutrition.[13] Many other scientific orgainizations agree, contending that biotechnology can deliver on promises for an improved food supply if we give it a fair chance to do so.

A lack of scientific understanding underlies many fears of biotechnology. The same lack of understanding breeds fear in consumers who adamantly oppose another food technology—irradiation.

IRRADIATED FOODS

Today, consumers bear the lion's share of the responsibility for safety from food-borne illnesses. They must cook meats and eggs to the well-done stage to kill dangerous microorganisms lurking in the raw products. They must scrub vegetables and fruits to remove pesticide residues applied to kill molds and insects that attack food during storage. Even diligent consumers cannot be certain that foods such as grains are free of sprays. Can exposing food to ionizing irradiation relieve consumers of some of these burdens? Proponents claim that because irradiation easily kills almost all disease-producing microorganisms, its use could save the lives of millions of children, older adults, and susceptible people worldwide who die each year from food-borne illnesses.[14]

Irradiation might also replace some types of postharvest pesticides used on foods because it can kill mold spores and insect pests and their eggs. In many areas of the world, between a quarter and half of the food produced annually is lost to pests and decay after harvest. Most of this waste could be eliminated, supporters claim, if the food could be irradiated. Though many top scientists agree with these assessments, critics fear that food irradiation may also pose some serious and unnecessary threats.

The Irradiation Process

Irradiation works by exposing foods to controlled doses of gamma rays from the radioactive compound cobalt 60 or from X rays generated by machines. As radiation passes through a living cell, it disrupts the internal structures and so kills or deactivates the cell. Low doses can kill the growth cells in the "eyes" of potatoes and ends of onions, preventing them from sprouting, and they delay ripening of bananas, avocados, and other fruits. High doses can penetrate tough insect exoskeletons and mold or bacterial cell walls to destroy their life-maintaining DNA, proteins, and other molecules and thereby greatly reduce these threats to health.‡ Interestingly, irradiation can kill microbes even while food is in a frozen state, making irradiation uniquely useful for protecting such foods as whole frozen

‡Even the highest legal doses of radiation have no effect on the spores of one dangerous bacterium—*Clostridium botulinum*, the bacterium responsible for the lethal food-poisoning agent, botulinum toxin.

turkeys. Doses of radiation high enough to sterilize food completely cannot be used because they would also destroy the food. They can, however, be used to sterilize dried herbs and spices, as has been done in this country for several years.

A perspective on the doses of radiation used on foods is gained by comparing them to the lethal human dose. The lowest doses of radiation needed to delay ripening and sprouting of fragile fruits and vegetables are 10 to 20 times higher than the doses that would kill human beings. The dose required to sterilize foods is many times higher still. Needless to say, irradiation technology requires extremely cautious handling.

If Irradiation Is So Great, Why Aren't We Using It?

Many people fear irradiation. Some consumers seem willing to purchase irradiated foods, but others vigorously challenge the whole idea of food irradiation. Among registered dietitians, with their superior knowledge of food safety, opinions run toward employing higher cleanliness standards (such as greater enforcement of the HACCP prevention system, described in the preceding chapter) rather than irradiating foods to kill disease-causing organisms.[15] As sensible as this attitude may seem, recent stepped-up use of HACCP plans by U.S. food producers has not sufficiently reduced the threat of food-borne illnesses to ensure the safety of the food supply.

Among the most common reasons people give for fearing irradiation are these:

- The foods will become radioactive.
- Unique, untested chemicals will form in foods during irradiation.
- The foods will lose substantial nutrients during irradiation.
- Irradiated foods will not be safe to eat.
- The radioactive substances used to irradiate foods will endanger plant workers, the general population, and the environment.

The first of these concerns is easily put to rest. In truth, properly irradiated food does not become radioactive any more than teeth become radioactive after dental X-ray procedures. The use of radioactivity demands great care, though. Foods exposed to extremely high doses of the wrong sort of radiation can indeed become radioactive. A point of reassurance is that the foods would also be rendered inedible.

The other fears, though not so easily dismissed, have been addressed by the scientific community to the satisfaction of the World Health Organization (WHO).[16] WHO has concluded that irradiation of food is safe and is analogous to cooking in its effects on foods. The FDA has also endorsed and approved irradiation for controlling microbial contamination of fresh and frozen red meats, such as lamb, beef, and pork.[17] The FDA requires that irradiated foods either bear a label, in letters as large as those of the ingredient list, stating that the foods have been treated with radiation, or display the irradiation symbol, also called the radura logo, shown here.[18] Exceptions are permitted for spices that are mixed with processed foods and for irradiated foods served in restaurants.

Clearly, scientists seem confident that irradiation is safe and helpful. Consumers, however, are still uneasy about irradiation. Evidence about their concerns follows.

Radiolytic Products Concerns exist about chemicals called **radiolytic products**, which are produced in foods as they undergo irradiation. A very few radiolytic products are unique, appearing only in irradiated foods; most are commonly found in many foods after all sorts of processing, including cooking. Their effects on human health and nutrition are probably nil, based on the results of animal studies.

Irradiation's Effects on Nutrients Opponents of irradiation fear that foods processed by irradiation will become so nutrient poor that people eating them could develop nutrient deficiencies. Indeed, nutrients in foods are destroyed by free radicals formed during irradiation. For example, the side chains of certain individual amino acids (see Chapter 6) are open to destruction by these influences. In fact,

This radura logo is the symbol for foods treated with radiation.

this effect is one reason why irradiation is so effective at disrupting life processes. When irradiation destroys the side chains of amino acids, it also destroys the integrity and function of the protein of which they are a part.

In the same way, irradiation also destroys some vitamins such as vitamin E, beta-carotene, vitamin B_6, vitamin C, and especially thiamin. However, as pointed out by the FDA and the American Council on Science and Health (ACSH), a group vocal in its support of irradiation, the amount of nutrients lost during irradiation are generally minor. While acknowledging that doses of radiation high enough to sterilize food cause substantial vitamin losses, the ACSH describes the losses sustained during low-dose irradiation as similar to those caused by canning or other processing techniques. Though nutrient losses sustained through radiation are probably insignificant, one question above all others still remains: Are irradiated foods safe to eat?

Irradiation Safety Radiation changes foods in ways whose health effects are still not completely understood. For example, when freshly irradiated food is heated in a laboratory, it emits light detectable by instruments.[19] (People fearful of this phenomenon say, "It glows.") This light energy, called thermoluminescence, arises from overexcited electrons, stimulated by radiation.

Two decades ago, studies of safety seemed to indicate health problems in laboratory rats fed on freshly irradiated chow. The rats were reported to have developed chromosomal abnormalities, impaired fertility, and depressed immune responses.[20] In one study performed on malnourished children, in the days before research ethics would have prevented such a study, researchers reported increased chromosomal abnormalities in all but one of the children.[21] Since then, no other evidence has emerged to support the idea that irradiated food can damage the chromosomes or cause any other adverse effect on health; meanwhile much research supports its safety. In its publication *Priorities*, the ACSH attacks the validity of studies showing ill effects. It claims that arguments against food irradiation are largely "fronts for an antinuclear political agenda."

In fact, some people do oppose food irradiation on other safety grounds. The method necessitates transporting radioactive materials, exposing workers to them, and then disposing of the spent wastes, which remain radioactive for many years. These opponents claim that these processes pose an unacceptable and unnecessary risk to human health and reproduction. Birth defects are common in children who were exposed to nonlethal low doses of radiation during their fetal development and even in children whose parents were exposed to radiation before the children were conceived. These concerns are echoed by food industrialists and others who hope to gain acceptance for the process, but are unwilling to risk human health. They hope to safeguard both workers and future generations through strict operating standards and enforcement of regulations limiting radiation exposure.

Finally, some consumers worry that food manufacturers might use the technology unethically. For example, the law currently prohibits sale of old or tainted food found to have high bacterial counts, usually from insect or rodent infestation. With the availability of irradiation, unscrupulous manufacturers could sanitize the old or tainted food, causing FDA inspectors to proclaim the food wholesome and fit for consumption. An important point is that irradiation is intended to complement other traditional food safety methods, not to replace them. Even irradiation cannot protect people from food that becomes recontaminated due to poor sanitation practices either in the marketplace or at home.

In the end, some of the objectives of irradiation technology may be achievable by less expensive conventional methods of sanitation. Higher standards of cleanliness to prevent contamination, selective breeding and genetic engineering of produce to permit longer safe storage, and application of gases and pesticides that dissipate before purchase of the products could go a long way toward ensuring a safer food supply.

A Possible Alternative—High-Intensity Pulsed Light A brand new food safety technology, called *high-intensity pulsed light*, has recently been approved by the FDA. High-intensity pulsed light kills microorganisms on the surfaces of foods using an intense flash of light. It extends the shelf life of foods without affecting their nutrient quality. This new process poses none of the hazards associated with the transport and handling of radioactive materials and may prove less controversial than irradiation.

CONCLUSION

Will our impressive new technologies provide foods to meet the needs of the future? Optimists would say yes. Biotechnology holds a world of promise, and with proper safeguards and controls, it may yield products that meet the needs of consumers almost perfectly.[22] Even irradiation, with its potential for hazard, may, with proper safety controls in place, prove useful in helping to provide safe, abundant food for the world's growing population.[23] All products of technology must pass a final examination, though—the test of consumer acceptance. If the products meet people's needs, are attractive, economical, and tasty, and have been proved safe, consumers will buy them. If they fail to meet these standards, consumers will bypass them, and the technologies themselves will fade into history.

15 Hunger and the Global Environment

Jacob Lawrence (1917)
Street to M'bari, 1964

Contents

Frequently Asked Questions

Figure 15-1

U.S. Food Security and Hunger, 1995

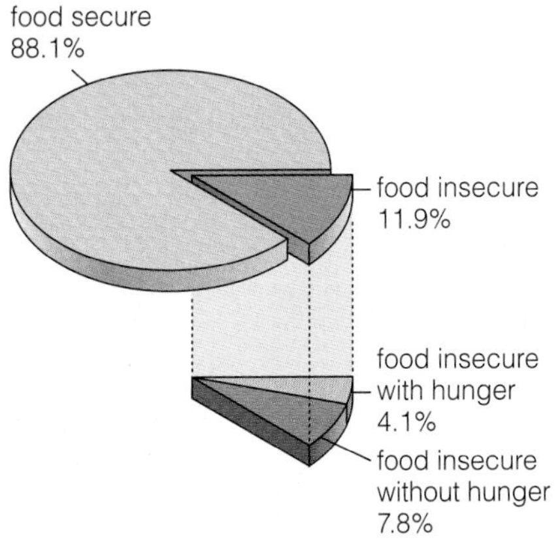
food secure
88.1%
food insecure
11.9%
food insecure
with hunger
4.1%
food insecure
without hunger
7.8%

SOURCE: USDA, Could there be hunger in America? *Nutrition Insights,* September 1998.

"Hunger, particularly childhood hunger, [is] not only a moral issue [but] a competitiveness issue. . . . Hunger compromises the ability to learn. Hungry children have higher school absence rates, and when they are in school, their powers of concentration are greatly reduced. The malnutrition that results from chronic hunger can even slow or permanently inhibit the physical development of the brain. So hunger is . . . an issue of failed beginnings for millions of American children—the future of our country."

—M. Mudd, vice president, Kraft General Foods, 1992

IN THE LATE 1990's, the United States was enjoying one of the longest economic expansions on record. Yet almost unbelievably, at the same time, more than one in every ten U.S. households had one or more members who experienced pain from hunger caused by lack of food (see Figure 15-1).[1] In these households, well over 300,000 children are hungry at least some of the time. Under the broader definition of **food insecurity** (see Table 15-1), almost three in ten U.S. households and more than 600,000 children do not know where their next meal is coming from or when it will come.

The contrast of hunger amidst plenty characterizes many of the world's nations. Over 820 million of the world's people are suffering from chronic hunger while their neighbors are secure and well fed.[2] Many are dying of starvation: tens of thousands die each day, one every two seconds.[3] As the world's population increases, more and more people will face hunger; as human beings populate and pollute the earth, they deplete its resources; yet, more and more food is needed yearly to avert widespread hunger.[4]

The tragedy described on these pages may seem at first to be beyond the influence of the ordinary person. What possible difference can one person make? Can one person's choice to recycle a bottle or to eat a vegetarian meal or to join a hunger-relief organization make a difference? In truth, such choices produce several benefits. For one, a single person's awareness and example, shared with others, may influence many people over time. For another, an action repeated becomes a habit. For still another, making choices with awareness of their impacts lends a sense of control over those impacts. That sense of personal control, in turn, helps people to take effective action in many areas.

Students, especially, can play a powerful role in bringing about change. Students everywhere are helping to change governments, human predicaments, and environmental problems for the better. Student movements persuaded 127 universities and many institutions, corporations, and government agencies to put pressure on South Africa and succeeded in ending apartheid. Student pressure opened the way for the first deaf president at a university for the deaf. Students offer major services to communities through soup kitchens, home repair programs, and child education. In fact, the young people of today are the best hope of millions for a better tomorrow.

The Need for Change

The earth's total food supply is currently more than sufficient to feed all of its people adequately, but uneven distribution of resources leaves some without enough food. In the next decade, however, we may face more global food insecurity as many forces compound to threaten world food production and distribution. At the turn of the millennium, all of the following trends are taking place:

- *Hunger, poverty, and population growth.* Millions of the world's people are starving. Fifteen children die of malnutrition every 30 seconds, but 125 children are born during that same 30 seconds. Every day, the earth gains another 220,000 new residents to feed.[5]

Table 15-1

Hunger Terms

- **famine** widespread scarcity of food in an area that causes starvation and death in a large portion of the population.
- **food insecurity** the condition of uncertain access to food of sufficient quality or quantity.
- **food poverty** hunger occurring when enough food exists in an area but some of the people cannot obtain it because they lack money, are being deprived for political reasons, live in a country at war, or suffer from other problems such as lack of transportation.
- **food shortage** hunger occurring when an area of the world lacks enough total food to feed its people.
- **hunger** lack or shortage of basic foods needed to provide the energy and nutrients that support health.

- *Loss of food-producing land.* Food-producing land is becoming saltier, eroding, and being paved over. Each year, the world's farmers try to feed some 85 million additional people with 24 billion fewer tons of topsoil. This threatens overall food security.
- *Accelerating* ***fossil fuel*** *use.* Fossil fuel use is growing, with attendant pollution of air, soil, and water; ozone depletion; and global climate changes.
- *Increasing air pollution.* As populations increase, air quality diminishes in many areas around the globe.*
- *Atmosphere and climate changes, droughts, and floods.* Climbing atmospheric levels of heat-trapping carbon dioxide are a concern. The concentration of carbon dioxide is now 26 percent higher than 200 years ago. As a result, a warming trend seems to be taking place.[6] Climate change causes both droughts and floods, which destroy crops and people's homelands.
- *Ozone loss from the outer atmosphere.* The outer atmosphere's protective ozone layer is growing thinner, permitting harmful radiation from the sun to damage crops and ecosystems and increasing the likelihood of skin cancers and cataracts in people and animals.
- *Water shortages.* The world's supplies of fresh water are dwindling and becoming polluted.
- *Deforestation and desertification.* Forests are shrinking and deserts are growing.
- *Ocean pollution.* Ocean pollution is killing fish; overfishing is depleting the numbers of those that remain.[7]
- *Extinctions of species.* More than 140 species of animals and plants are going extinct each day. Another 20 percent of all species are expected to die out in the next ten years. Thousands of animals and plants including many kinds of whales, birds, giant mammals, and colorful butterflies, will never again be seen in the universe.

These global problems are all related. The causes overlap, and so do the solutions. To think positively, this means that any initiative a person takes to help solve one problem will help solve many others. In particular, control of the earth's population is urgent, as a later section spells out. This chapter's Controversy shows how U.S. consumers and agricultural practices affect the world's resources.

fossil fuel coal, oil, and natural gas. These are nonrenewable fuels that pollute. Alternative fuels, such as solar and wind energy, are renewable and pollute less or not at all.

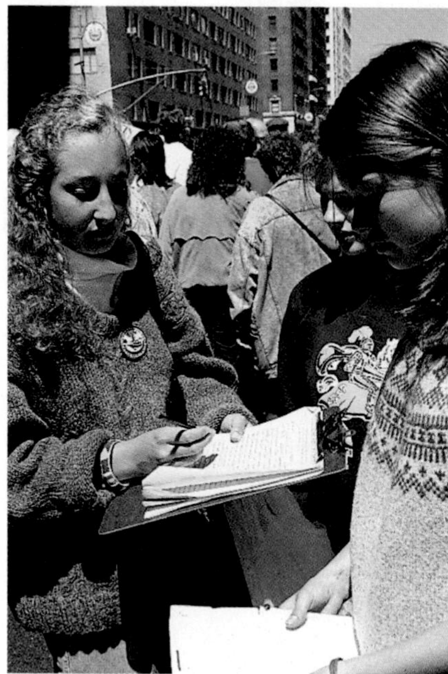

Each person's choice to get involved and be heard can help lead to needed change.

Food poverty is the prevailing form of hunger in the United States.

Hunger

The hunger of concern here is not the healthy appetite we all feel, which leads us to sit down and eat a hearty meal, but the chronic, painful **hunger** people feel when no food is available. Severe deficiencies of vitamins and minerals accompany this hunger, afflicting more than 40 percent of the world's people to some degree. An estimated two billion people, mostly women and children, suffer the effects of iron-deficiency anemia, and many more suffer milder degrees of insufficiency.[8] Iodine deficiency remains the single greatest cause of preventable brain damage and mental retardation; 750 million adults suffer from goiter. Deficiency of vitamin A, too, takes a terrible toll on the world's children: vitamin A deficiency stands out as the world's leading cause of blindness in young children and robs many millions of the ability to fight off infections. Worldwide, three-fourths of those who die each year from starvation and related illnesses are children.[9]

In developed countries, the primary cause of hunger is **food poverty.**[10] People are hungry not because there is no food nearby to purchase, but because they lack money with which to buy the food. Also contributing to food poverty are other problems such as abuse of alcohol and other drugs, mental or physical illness, depression, lack of awareness of or access to available food programs, and the reluctance of people, particularly the elderly, to accept what they perceive as "welfare" or "charity." Lack of resources remains the major cause of food poverty, and solving this problem would do much to relieve hunger.

U.S. federal food programs include:

- Special Supplemental Food Program for Women, Infants, and Children (WIC, see Chapter 12).
- WIC Farmers' Market Nutrition Program.
- Food Stamp Program.
- National School Lunch and Breakfast Programs (see Chapter 13).
- Emergency Food Assistance Program.
- Commodity Supplemental Food Program.
- Commodity Distribution to Charitable Institutions.
- Senior Nutrition Program (see Chapter 13).
- Food Distribution Program on Indian Reservations.
- Nutrition Assistance to Puerto Rico.

*Many older references have been removed to save space, but they are available in older editions of this book.

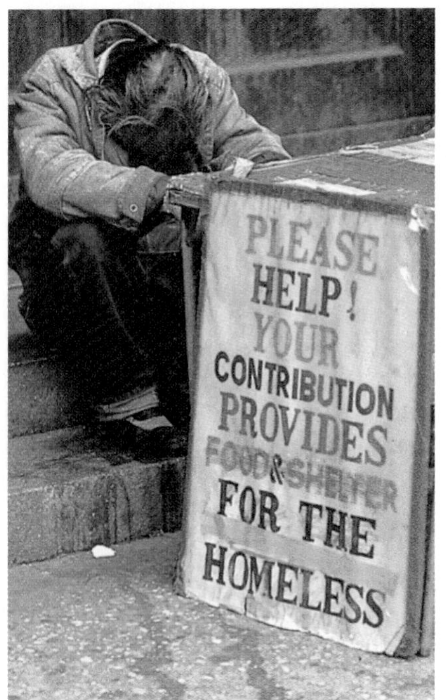
PLEASE
HELP!
YOUR
CONTRIBUTION
PROVIDES
FOOD&SHELTER
FOR THE
HOMELESS

In the United States, at least 20 million people are hungry for at least part of every month.

In the United States, food poverty reaches into many segments of society, affecting not only the chronic poor (migrant workers, ethnic minorities, the unskilled and unemployed, the homeless, and some elderly) but also the so-called new poor. Some are displaced farm families. Some are former blue-collar and white-collar workers forced out of their trades and professions into minimum-wage jobs. These people outnumber the chronic poor, and they are not on welfare; they have jobs, but the pay is too low to meet their needs. Families with incomes below a certain level are simply unable to buy sufficient amounts of nourishing foods, even if they are skilled in food shopping. Their worry about food security leads them to skip meals or cut their portions. Children in such families sometimes go hungry for an entire day until the adults find money for food.

Hunger is not always easy to recognize. Table 15-2 shows how national surveys identify it in the United States. A family that would answer "Yes" to the questions in the table is a family that suffers from the hunger caused by food poverty. The American Dietetic Association holds that "aggressive action is needed to bring an end to domestic hunger and to achieve food and nutrition security for all residents of the United States."[11] Such action is in everyone's interest because the hunger of individual families affects the nation as a whole, as the quotation at the beginning of this chapter made clear.

KEY POINT *Chronic hunger causes many deaths worldwide, especially among children. Intermittent hunger is frequently seen in U.S. children. The immediate cause of hunger is poverty.*

What U.S. Food Programs Are Directed at Stopping Domestic Hunger?

An extensive network of food assistance programs delivers life-giving food daily to millions of U.S. citizens.[12] One out of every six Americans receives food assistance of some kind, at a total cost of almost $40 billion per year.[13] Even so, the programs are not fully successful in preventing hunger, even among those who receive their benefits.[14]

Programs described in earlier chapters include children's school lunch and school breakfast programs, child care and elder care food programs, programs to supply low-income pregnant women and mothers with nourishing food (WIC), and food assistance programs for older adults such as congregate meals and Meals on Wheels.

The centerpiece of U.S. food programs for low-income people is the Food Stamp Program, administered by the U.S. Department of Agriculture (USDA). Eligible households (defined as people who live and purchase food together) receive food stamp

Table 15-2

How to Diagnose Food Insecurity in a U.S. Household

Questions like these are asked on surveys to determine the extent of food insecurity in a household. The more questions that receive a "Yes" answer, the more intense the hunger the household is experiencing.

- Do you often go hungry?
- Do you often have too little food to eat because you have no money, transportation, or kitchen appliances that work?
- Do you ever rely on nutritionally inferior foods to feed yourself or your children because you lack any of these resources?
- Do you ever eat less than you feel you should because you lack any of these resources?
- Do you ever skip meals or cut the size of meals because you lack any of these resources?
- Do you ever rely on neighbors, friends, relatives, or schools to feed any of your children because there is not enough food in the house?
- Do your children ever say they are hungry because there is not enough food in the house?
- Do you or any of your children ever go to bed hungry because there is not enough food in the house?

SOURCE: Adapted from C. A. Wehler, R. I. Scott, and J. J. Anderson, The Community Childhood Hunger Identification Project: A model of domestic hunger—Demonstration project in Seattle, Washington, *Journal of Nutrition Education* (1 supplement), January/February 1992, pp. 29S–35S; and R. R. Briefel and C. E. Woteki, Development of food sufficiency questions for the Third National Health and Nutrition Examination Survey, *Journal of Nutrition Education* (1 supplement), January/February 1992, pp. S24–S28.

These people and many others like them in the United States face food insecurity daily.

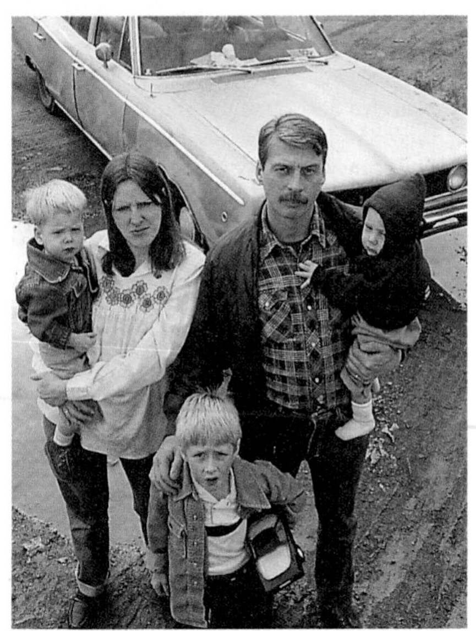

coupons or debit cards similar to credit cards through state social services or welfare agencies. Recipients may use the coupons or cards like cash to purchase food and food-bearing plants and seeds, but not to buy tobacco, cleaning items, alcohol, or other non-food items. More than 25 million people in the United States receive food stamps, and over 20 million more who have not applied for assistance are thought to be qualified to receive them.[15] Of the homeless people in the United States who are eligible for food assistance, only 15 percent of single adults and 50 percent of families receive food stamps.

In an effort to assist where federal programs fall short, concerned citizens in many communities work through local agencies and churches to help deliver food to hungry people. Food recovery, or **gleaning**, from private industry has become a national priority, and federal funds are earmarked for it as part of the Community Food Security Initiative. The initiative assists communities in grass-roots efforts to reduce hunger and improve nutrition.[16] Instead of being discarded, surplus produce from grocery stores or farms, excess food prepared for banquets or restaurant meals, and nonperishable foods from any source can be put to good use feeding those in need. Industries donating the food often qualify for tax deductions in proportion to their donations. Other initiatives to supply the hungry include community-based soup kitchens and shelters, which generally provide good-quality meals. Food pantries, food banks, and community gardens contribute groceries to families in need. Table 15-3 presents a 14-step program for developing a hunger-free community. Although these efforts provide emergency relief to hungry people, they leave unsolved the greater problems of low wages and poverty among people who lack higher education or training.

KEY POINT *Poverty and hunger are widespread in the United States, not only among the unemployed, but also among working people. Government programs to relieve poverty and hunger are not fully successful.*

Table 15-3

Fourteen Ways Communities Can Address Their Local Hunger Problems

1. Establish a community-based emergency food-delivery network.
2. Assess food-insecurity problems and evaluate community services. Create strategies for responding to unmet needs.
3. Establish a group of individuals, including low-income participants, to develop and to implement policies and programs to combat food insecurity; monitor responsiveness of existing services; and address underlying causes of hunger.
4. Participate in federally assisted nutrition programs that are easily assessible to targeted populations.
5. Integrate public and private resources, including local businesses, to relieve food insecurity.
6. Establish an education program that addresses the food needs of the community and the need for increased local citizen participation in activities to alleviate food insecurity.
7. Provide information and referral services for accessing both public and private programs and services.
8. Support programs to provide transportation and assistance in food shopping, where needed.
9. Identify high-risk populations, and target services to meet their needs.
10. Provide adequate transportation and distribution of food from all resources.
11. Coordinate food services with parks and recreation programs and other community-based outlets to which area residents have easy access.
12. Improve public transportation to human services agencies and food resources.
13. Establish nutrition education programs for low-income citizens to enhance their food purchasing and preparation skills and to make them aware of the connections between diet and health.
14. Establish a program for collecting and distributing nutritious foods, either agricultural commodities in farmers' fields or prepared foods that would have been wasted.

SOURCE: House Select Committee on Hunger, legislation introduced by Tony P. Hall, excerpted in *Seeds,* Sprouts edition, January 1992, p. 3 with permission (SEEDS Magazine; P.O. Box 6170; Waco, TX 76706). For more on developing a hunger-free community, write: Hunger Free, House Select Committee on Hunger, 505 Ford House Office Building, Washington, DC 20515.

gleaning traditionally, the practice of gathering crops left in the field after a harvest; today, refers to the recovery of excess food from various sources including restaurants, hotels, corporations, farms, wholesalers, farmers' markets, and supermarkets.

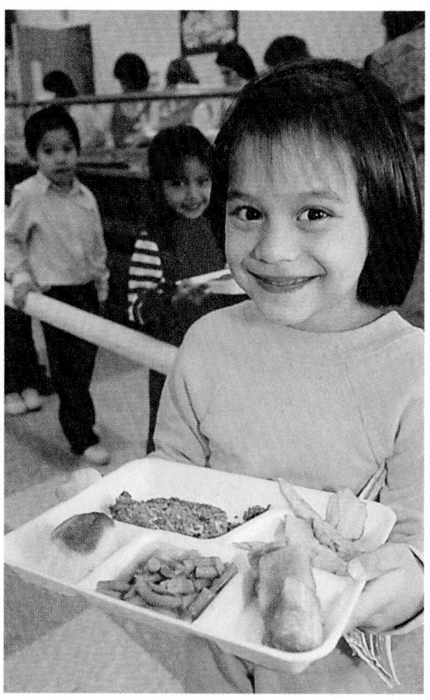

School lunches provide low-income children with nourishment at little or no cost.

For more information about gleaning, call the USDA's food gleaning hotline: 1-800-GLEAN-IT.

In 1999, Congressman Tony P. Hall conducted a survey of private U.S. food banks and reported that:

- Requests for food assistance were increasing at a rapid rate.
- People who need help are often working but not earning a living wage, are elderly, or have recently lost federal welfare or food stamp benefits.
- Fewer food banks are meeting the needs of the hungry on private food donations alone; they are turning to federal agencies for access to bulk foods, but this source is not steadily available to food banks.
- Gleaning from restaurants and other food industries is a potential partial solution, if problems of distribution can be solved.

SOURCE: T. P. Hall, Empty shelves: 1999 survey of U.S. food banks, a report available on the Internet at www.house.gov/tonyhall/pr49.htm

oral rehydration therapy (ORT) oral fluid replacement for children with severe diarrhea caused by infectious disease. ORT enables parents to mix a simple solution for their child from substances that they have at home.

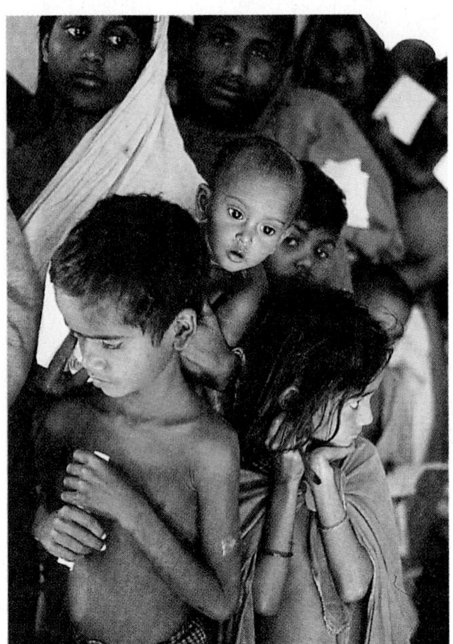

A primary cause of hunger is poverty.

The symptoms of malnutrition vary according to the nutrients lacking and the individual's stage of life. See Chapter 6 for effects of protein and energy deficiency; Chapters 7 and 8 for vitamin and mineral deficiencies; Chapter 11 for effects on immunity; Chapter 12 for effects on newborns and pregnant women; and Chapter 13 for effects on children, teens, and the elderly.

To prevent death from diarrheal disease, provide:

- Adequate sanitation.
- Safe water.
- Oral rehydration therapy (ORT). A simple recipe for ORT calls for 1 cup boiled water, 2 teaspoons sugar, and a pinch of salt.

 What Is the State of World Hunger?

In the developing world, hunger and poverty are even more intense and the causes more diverse. The primary form of hunger is still food poverty, but the poverty is more extreme. Many of the world's people face hunger every day. It can be difficult to grasp the severity of poverty in the developing world, but some statistics may help. One-fifth of the world's five billion people have no land and no possessions *at all.* They survive on less than one dollar a day each, they lack water that is safe to drink, and they cannot read or write. Many spend about 80 percent of all they earn on food, but still cannot meet their needs. The average U.S. housecat eats twice as much protein every day as one of these people, and the cost of keeping that cat is greater than that person's annual income.

Food Shortage The most visible form of hunger is **famine**, a true **food shortage** in an area that causes multitudes of people to starve and die (for definitions, see Table 15-1). The natural causes of famine—drought, flood, and pests—have, in recent years, taken second place behind the social causes. Between 1959 and 1961, for example, 15 to 30 million people died in China in the worst famine of the twentieth century. The famine was caused mainly by government policies associated with the "great leap forward," which devastated Chinese agriculture.[17] In parts of Africa, killer famines recur whenever human conflict converges with drought on a country such as Sudan that has little food in reserve even in a good year.

Since the 1990s, the violence of human conflict has become a dominant cause of famine worldwide. In all of the countries that have reported famine during this period, armed conflict has been a major cause. Farmers become warriors and agricultural fields become battlegrounds while citizens go hungry. Warring factions often repel famine relief efforts in hopes of starving their opponents before they succumb to starvation themselves. The world continues to struggle to find a middle ground between respecting the sovereignty of nations and insisting that all nations allow humanitarian assistance to reach their people.

During natural disasters without war, food aid from other countries has provided a safety net for countries whose crops fail. But food aid now does more than just offset poor harvests; it also delivers food relief to countries, such as Ethiopia, that are chronically short of food and without resources to buy it. Some people are concerned that as many nations cut their foreign aid, this food aid backup may become insufficient.

Chronic Hunger Though we usually associate world hunger with famine, the numbers affected by famine are relatively small compared with those suffering from less severe but chronic hunger. Nearly 800 million people, mostly women and children, in developing countries suffer from chronic malnutrition. The ravages to the body of nutrient deficiencies were spelled out in earlier chapters of this book.

Tens of thousands die of malnutrition every day. Many are afflicted by the diseases of poverty: parasitic and infectious diseases such as dysentery, whooping cough, measles, tuberculosis, cholera, and malaria. These interact with poor nutrition in a fatal cycle. Most children who die of malnutrition do not starve to death—they die because their health has been compromised by dehydration from infections that cause diarrhea. Currently, **oral rehydration therapy (ORT)** is saving an estimated one million lives each year by helping to stop the infection-diarrhea cycle. The ORT solution increases a body's ability to absorb fluids 25-fold. Clean or boiled drinking water is essential, however, for contaminated water will reinfect the child.

Malnourished women in poverty bear sickly infants who cannot fend off the diseases of poverty, and many succumb within the first years of life. One child in six in the world is born underweight, and ten million die by age five.[18] Breastfeeding helps prolong an infant's life, but eventually the child must be weaned to thin gruels of scant quantity made with unclean water. All too often, children sicken and die soon after weaning. Because of poverty, infection, and malnutrition, the life expectancy in some African countries averages 50 years; in Uganda it is only 42 years, little more than half of the U.S. life expectancy.

World Food Supply Most disturbingly, such misery and starvation exist side by side with ample food supplies. Today, the world's supply of grain, an index of the sufficiency of the world food supply, can feed the world for several months. The demand for food is great, but technological advances in farming have increased crop yields and prices of many foods have fallen in response.[19] Wheat and corn, for example, the staple foods of many nations, are abundantly available and are now priced at less than half of their cost of 40 years ago.

The future may not be so bright, however. At the present rate of growth, the world's population will soon outstrip the rate of food production. Technological advances have produced dramatically higher yields per acre of farmland over past decades, but progress has slowed and may not generate the greater crop yields needed to keep pace. In other words, crop yields may be increasing, but not at a pace sufficient to feed the increased numbers of people expected to arrive on the earth in coming years. Environmental degradation and dwindling water supplies may ultimately prevent further growth in the world's food output in many agricultural areas.[20] No part of the world is safely insulated against future food shortages. Developed countries may be the last to feel the effects, but they will ultimately go as the world goes.

BISCUITS
The hunger of conflict—people desperate for food in Sudan.

KEY POINT *Natural causes such as drought, flood, and pests and social causes such as armed conflicts and overpopulation all contribute to hunger and poverty of developing countries. To meet future demands for food, technology must improve food production.*

Environmental Degradation and Hunger

Hunger and poverty interact with a third force: environmental degradation. Poor people often destroy the very resources they need for survival. In desperation to obtain money for food, they sell everything they own, even the seeds that would have provided next year's crops. They cut their trees for firewood or timber to sell, then lose the soil to erosion. Without these resources, they become still poorer. Thus poverty causes environmental ruin, and the ruin leads to hunger.

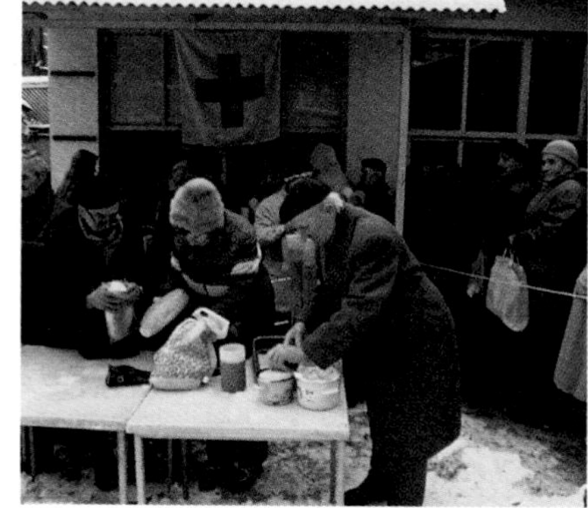
The hunger of conflict—hunger relief in Sarajevo.

Soil Erosion Soil erosion affects agriculture in every nation. Deforestation of the world's rain forests dramatically adds to land loss. In Sierra Leone, 60 percent of the land was primary rain forest in 1961, but only 6 percent was in 1994.[21] Without the forest covering to hold the soil in place, it washes off the rocks beneath, drastically reducing the land's productivity.[22]

Around the world, irrigation and fertilizer can no longer compensate for these losses by improving crop yields because all the land that can benefit from these measures is already receiving them. In fact, continuous irrigation leaves deposits of salt in the soil, and rising salt concentrations are lowering yields on close to a quarter of the world's irrigated cropland.

Grazing Lands and Fisheries Meat and fish outputs are also endangered. Grasslands for growing beef are already being fully used or overused on every continent. Despite persistent expansion of the world's fishing industry, the yield of fish from the oceans has been declining in recent years due to overfishing and pollution.[23] Big fish, such as tuna, swordfish, and shark, are being overfished. Between 1989 and 1992, Atlantic stocks of the heavily fished bluefin tuna dropped by 94 percent.[24] Cod are rapidly disappearing off the New England coast and are almost gone farther north. Inland fisheries have also suffered tremendous drops in yield as a result of environmental damage. In the early 1990s, 14,000 Canadian lakes were declared biologically dead as a result of acid rain. The Food and Agriculture Organization, an agency of the United Nations that monitors the world's food supplies, predicts that, unless something changes, the world demand for fish will outstrip the supply in about ten years' time.[25] Preventing this outcome will require commitments from the

More about overgrazing appears in Controversy 15.

carrying capacity the total number of living organisms that a given environment can support without deteriorating in quality.

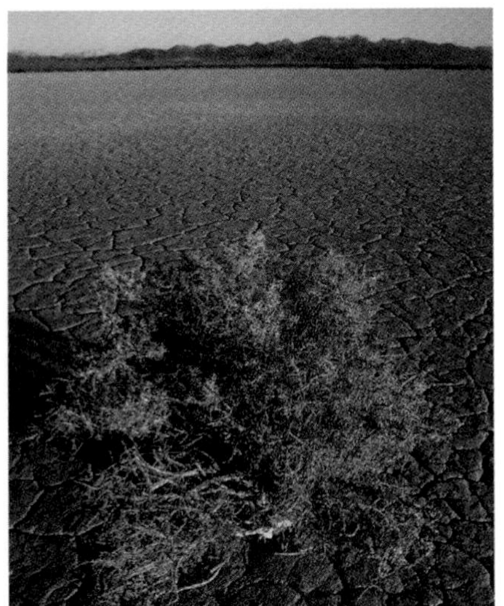

As groundwater is used up, deserts spread.

Years needed for the world's population to reach . . .

Its 1st billion	2,000,000 years
2nd billion	105 years
3rd billion	30 years
4th billion	15 years
5th billion	12 years
6th billion	11 years

Is it any wonder that food supplies may fall behind?

People wonder whether the disease AIDS will stem the growth of the world's population. The answer is no, because while millions die of this dread disease each year, many more millions are born.

world's fishing nations to refrain from overfishing and to protect the environments of ocean fisheries. Developing greater production of fish from aquaculture ("fish farms") may also help meet human demand.

Climate and Water Other forms of environmental degradation that reduce food outputs include both air pollution and climate change. According to the United Nations' Intergovernmental Panel on Climate Change, a major international collaboration involving more than 2,500 scientists from around the world, changes in climate are expected to result from a buildup of so-called greenhouse gases, such as carbon dioxide, methane, and nitrous oxide, and airborne particles.[26] Both types of pollutants are the result of human industry, agriculture, and transportation activities. Global mean surface temperature has increased by about 0.6 degree Celsius since the late nineteenth century.[27] This change is all but imperceptible in year-to-year weather patterns, but it is believed to foretell a warming trend that could threaten the health of many habitats, including human habitats, on a global scale.[28]A rise of only a degree or so in average global temperature may reduce soil moisture, impair pollination of major food crops such as rice and corn, slow growth, weaken disease resistance, and disrupt many other factors affecting crop yields.

Supplies of fresh water have shrunk to the point where they are limiting the numbers of people who can survive in some areas. In fact, water availability may limit human population growth even before food availability does.

Overpopulation The world's population is rising at an alarming rate and is expected to double by the year 2033. Before then, however, the human population will exceed the earth's estimated **carrying capacity.** Many authorities in many fields—and more every year—are calling for a reduction in the rate at which the world's population is allowed to increase. Overpopulation may well be the most serious threat that humankind faces today.

The sheer magnitude of our annual population increase is difficult to comprehend. Each month the world adds the equivalent of another New York City.[29] During six months of the terrible 1992 famine in Somalia, an estimated 300,000 people starved to death. Yet it took the world only 29 *hours* to replace their numbers!

Population stabilization has become one of the most pressing needs of our time: it appears to be the only way to enable the world's food output to keep up with demands. Without population stabilization, the world can neither support the lives of people already born nor halt environmental deterioration around the globe. And before the population problem can be resolved, it may be necessary to remedy the poverty problem. In countries around the world, economic growth has been accompanied by slowed population growth.[30] Of the many millions added to the population each year, 98 percent are born in the most poverty-stricken areas of the world.

If the present trend in population continues, the time may be approaching when the world will experience a deficit of food.[31] Skyrocketing human numbers threaten the earth's capacity to produce adequate food and imperil health in innumerable other ways. To resolve the population problem, a necessary first step is to remedy poverty problems, for reasons discussed next.

KEY POINT *Environmental degradation caused by the impacts of growing numbers of people is threatening the world's future ability to feed all of its citizens. Improvements in agriculture can no longer keep up with people's growing numbers. Human population growth is an urgent concern.*

Poverty and Overpopulation

Population growth is at the center of a giant web of factors contributing to poverty and hunger. The reverse is also true: poverty and hunger contribute to population growth (see Figure 15-2). The first of these cause-effect relationships is easy to understand. As a population grows larger, more mouths must be fed, and the worse poverty and hunger become.

How does poverty lead to overpopulation? Poverty and hunger exert an ironic effect on people, driving them to bear more children. Poverty and hunger typically go hand in hand with ignorance, including ignorance of how to control family size. Also, a family in poverty depends on its children to farm the land, haul water, and care for the adults in their old age. Poverty claims many of a family's young children, who are among the most likely to die from disease and other causes. If a family faces ongoing poverty, the parents will choose to have many children as a form of "insurance" that some will survive to adulthood. People are willing to risk having fewer children only if they are sure that their children will live. The environment also suffers, for the more people must share meager resources, the more heavily they draw on trees for fuel and on water supplies.

The world's poorest people live in the world's most damaged and inhospitable environments. There they experience, daily, tens of thousands of early deaths from malnutrition and disease. This sets up the damaging cycle shown in Figure 15-3: as more resources are needed, more hands are needed to help gather them, so more children are produced.[32]

Relieving poverty and hunger, then, may be a necessary first step in curbing population growth. When people attain better access to health care, education, and family planning, the death rate falls. After a time, the birthrate follows suit. Thus improvements in living standards help stabilize the population.[33] Wealth distribution matters, too. In countries where economic growth has occurred but only the rich have grown richer, population growth has remained high. Examples include Brazil, Mexico, the Philippines, and Thailand, where large families continue to be a major economic asset for the poor.

As a society gains economic footing, education also becomes a greater priority. A society that educates its children, both male and female, sees fertility rates decline (see Figure 15-4 on the next page). Education for girls and women yields improvements in family life, including improved nutrition, better sanitation, and elevated status for women. Educated women also take more control of family planning and can earn esteem through employment, and not solely through fertility.

Figure 15-2

Income and Birth Rate

Greater wealth means lower rates of birth. Each dot represents a country.

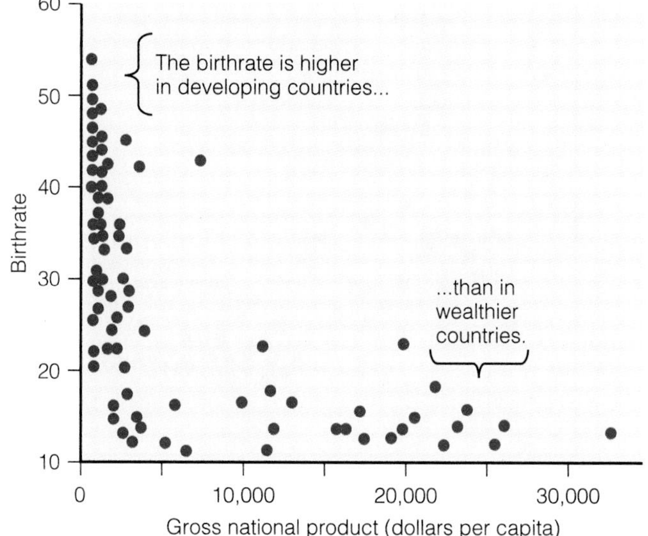

60
50
40
30
20
10
Birthrate
The birthrate is higher in developing countries...
...than in wealthier countries.
0
10,000
20,000
30,000
Gross national product (dollars per capita)

SOURCE: H. R. Pulliam and N. M. Haddad, Human population growth and the carrying capacity concept. *Bulletin of the Ecological Society of America,* September 1994, pp. 141–157.

Figure 15-3

The Worsening of Poverty, Overpopulation, and Environmental Degradation

Population growth, poverty, and environmental degradation combine to worsen each other.

poverty
environmental degradation
more children produced to gather resources
more resources needed
more mouths to feed; more poverty
more environmental degradation and povery

sustainable able to continue indefinitely. Here the term refers to the use of resources at such a rate that the earth can keep on replacing them; for example, cutting trees no faster than new ones grow and producing pollutants at a rate with which the environment and human cleanup efforts can keep pace. In a sustainable economy, resources do not become depleted, and pollution does not accumulate.

In some countries, every pair of little hands is needed to help feed the family.

KEY POINT *More people means more mouths to feed, which worsens poverty, hunger, and environmental problems. Poverty, hunger, and a degraded environment, in turn, prompt parents to have more children. Breaking this cycle requires improving the economic status of the people and providing them with health care, education, and family planning.*

Moving toward Solutions

Slowly but surely, improvements are evident in developing nations. For example, most nations have seen a rise in their gross domestic product, a key measure of economic well-being. Adult literacy rates have increased by more than 50 percent in some areas since 1970, and the proportion of children being sent to school has risen, while the proportion of chronically undernourished people has declined.[34] Today, optimism abounds, and keys to solving the world's environmental, poverty, and hunger problems are within the reach of both the poor and the rich nations, but still require efforts from them.

The poor nations need to make contraceptive technology and information more widely available, educate their citizens, assist the poor, and adopt **sustainable** development practices that slow and reverse the destruction of their forests, waterways, and soil. The rich nations need to stem their wasteful and polluting uses of resources and energy, which are contributing to global environmental degradation. They must also become willing to ease the debt burden that many poor nations face. One way is to offer to forgive the debt of the poor nations if they will take recommended steps to help their own people and conserve their resources.

What Are the Impacts of Sustainable Development Worldwide?

Many nations now agree that improvement of all nations' economies is a prerequisite to meeting the world's other urgent needs: population stabilization, arrest of environmental degradation, sustainable treatment of resources, and relief of hunger. At a summit of over

Figure 15-4

Education and Birth Rate

Higher education means lower rates of birth. Each dot represents a country.

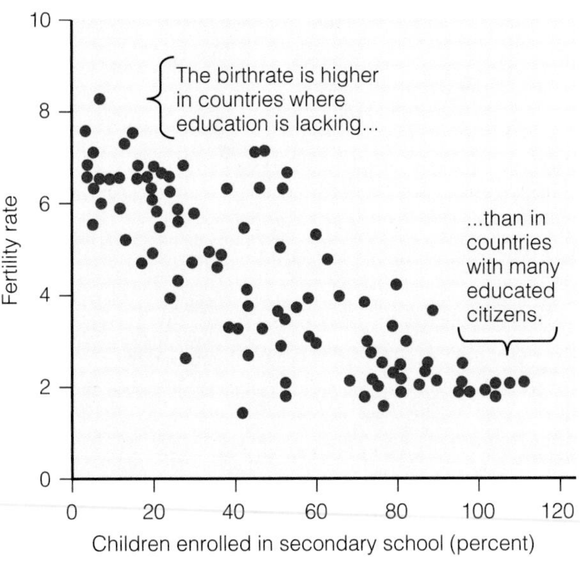
10
8
6
4
2
0
Fertility rate
The birthrate is higher in countries where education is lacking...
...than in countries with many educated citizens.
0
20
40
60
80
100
120
Children enrolled in secondary school (percent)

SOURCE: H. R. Pulliam and N. M. Haddad, Human population growth and the carrying capacity concept, *Bulletin of the Ecological Society of America,* September 1994, pp. 141–157.

100 nations, the United Nations Conference on Environment and Development, or UNCED for short, many nations agreed to a set of principles of sustainable development. The conferees defined sustainable development as development that would equitably meet both the economic and the environmental needs of present and future generations.

To rephrase a well-known adage, If you give a man a fish, he will eat for a day. If you teach him to fish, so that he can buy and maintain his own gear and bait, he will eat for a lifetime and help to feed you. Unlike food giveaways and money doles, which are only stop-gap measures, social reforms that permanently better the lot of the poor can permanently solve the hunger problem.

How Can People Engage in Activism and Simpler Lifestyles at Home?

Every segment of our society can play a role in the fight against poverty, hunger, and environmental degradation. The federal government, the states, local communities, big business and small companies, educators, and all individuals, including dietitians and foodservice managers, have many opportunities to forward the effort.

Government Action Government policies can change to promote sustainability. For example, the government can devote tax dollars and other resources to development of energy-conservation services and crop protection and to national and international education on sustainable development techniques.

Private and Community Enterprises Businesses can take initiatives to help; some already have. Several, such as AT&T, Prudential, and Kraft General Foods, are major supporters of antihunger programs. Restaurants and other food facilities can participate in the nation's gleaning effort by giving their fresh leftover foods to community distribution centers. As mentioned earlier, a new government program, the Community Food Security Initiative, aids grass-roots efforts to reduce hunger and improve nutrition within communities.

Educators Educators, including nutrition educators, have a crucial role to play. They can teach others about the underlying social and political causes of poverty, the root cause of hunger. At the college level, they can teach the relationships between hunger and population, hunger and environmental degradation, hunger and the status of women, and hunger and global economics.

Food and Nutrition Professionals Dietitians and foodservice managers have a special role to play. They are being urged by their professional organization, the American Dietetic Association (ADA), to promote the saving of resources by reuse, recycling (including composting), energy conservation, and water conservation in both their professional and their personal lives. In addition, the ADA urges its members to work for policy changes in private and government food assistance programs, to intensify education about hunger, and to be advocates on the local, state, and national levels to help end hunger in the United States.

Individuals All individuals can also become involved in these large trends. Many small decisions each day add up to large impacts on the environment. The Consumer Corner that follows sums up some of these decisions and actions.

KEY POINT *Government, business, educators, and all individuals have many opportunities to promote sustainability worldwide and wise resource use at home.*

"Never doubt that a small group of thoughtful, committed people can change the world. Indeed, it is the only thing that ever has."

—Margaret Mead

CONSUMER CORNER

Saving Money and Protecting the Environment

FIGURE 15-5 SHOWS the many ways in which consumers can "tread lightly on the earth" through their daily choices. Consider this list:

- Shop "carless" and plan to make fewer shopping trips. Motor vehicles constitute the largest single source of air pollution. Air pollution harms children and adults with lung problems, reduces crop yields, causes acid rain, and damages forests.
- Choose foods that are low on the food chain more often (see the chapter Controversy). Growing animals for their meat and dairy products by feeding them grain uses much more land and other resources than growing grain and other crops for direct use by people.
- Limit use of imported canned beef products, including stews, chili, corned beef, and pet foods. Many of these foods come at the expense of cleared rain forest land: 200 square feet of rain forest are lost *permanently* for every pound of beef produced.
- Choose chicken and small fish more often. Chickens are often grown locally and require fewer resources to produce; small fish eat low on the food chain.
- Buy more local foods grown close to home. Foods from a farmers' market require less transportation, packaging, and refrigeration than typical foods from a supermarket.
- Avoid overly packaged items; buy bulk items with minimal packaging or reusable or recyclable ones. Each can, foam tray, waxed or clay-coated cardboard container, plastic bottle, or glass jar requires land and many other resources to produce, and its disposal pollutes and costs more land.
- Use reusable pans and dishes, rather than disposable items that are used once and thrown away. Use pumps instead of spray cans—these are hard to recycle because they are made of many materials.
- Carry reusable string or cloth grocery sacks. Production of paper and plastic grocery bags represents a huge drain on resources. Paper factories use chemicals such as toxic forms of chlorine bleach, which are released into waterways in quantities so large that the chemicals can destroy whole bays and fisheries.
- Use fast cooking methods. Stir-frying, pressure cooking, and microwaving all use less energy than conventional stovetop or oven cooking methods.
- Reduce use of aluminum foil, paper towels, plastic wraps, plastic storage bags, sponges, and other disposable items. Find permanent reusable replacements for each, such as pans with lids, reusable storage containers, and washable cloths and towels.
- Use fewer electric gadgets. Mix batters, chop vegetables, and open cans by hand.
- Purchase the most efficient large appliances possible.
- Insulate the home.
- Consider using solar power, especially to heat water.
- Reduce, reuse, recycle.

The personal rewards of all these behaviors are many, from saving money to the satisfaction of knowing that you are enjoying and preserving the earth. But do they really help? They do, if enough people join in. To make the greatest impact, people can also support organizations that lobby for

This energy-saving refrigerator requires less than a twentieth of the energy used by a regular refrigerator, but chills and freezes food just as well. The motor is small, releases little heat, and is on top. In contrast, with a "regular" refrigerator's inefficient design, the large motor is below the unit, so it heats the very unit it is trying to cool.

Figure 15-5

Domestic Foodways that Respect the Environment

The refrigerator uses more energy than any other appliance in most people's homes. Consumers can take several steps to minimize the energy a refrigerator uses:

- *Set it at 37° to 40°F; set the freezer at 0°F.*
- *Clean the coils and the insulating gaskets around the doors regularly.*
- *Keep it in good repair.*

The water heater can also waste a lot of energy. Keep the water heater set at 120° to 130°F (no hotter) to save energy. For safe household dishes, sterilization is not necessary. Water of 120° to 130°F enhances the action of dishwashing detergents, making microorganisms slippery and removing them from the dishes. These measures will keep food fresh and clean while keeping energy use low.

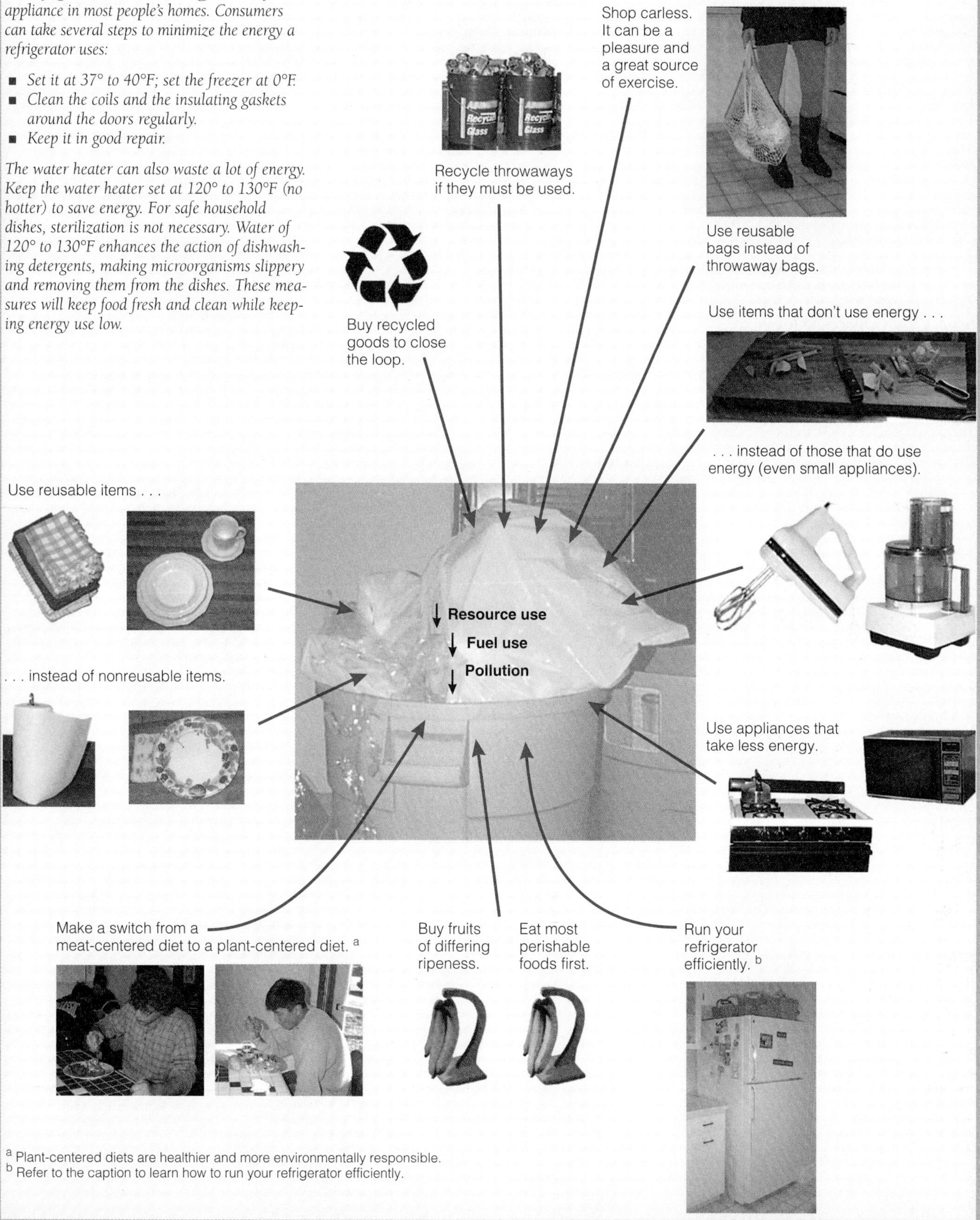

Recycle
Glass
Recycle
Glass

Recycle throwaways if they must be used.

Shop carless. It can be a pleasure and a great source of exercise.

Use reusable bags instead of throwaway bags.

Buy recycled goods to close the loop.

Use items that don't use energy . . .

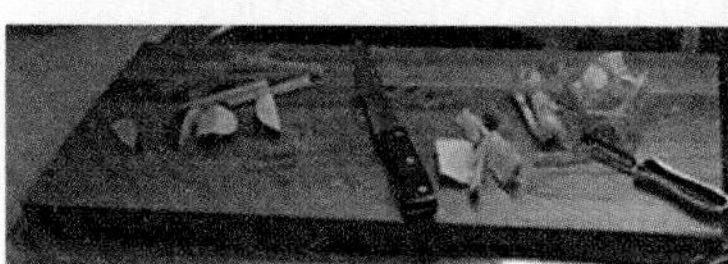

. . . instead of those that do use energy (even small appliances).

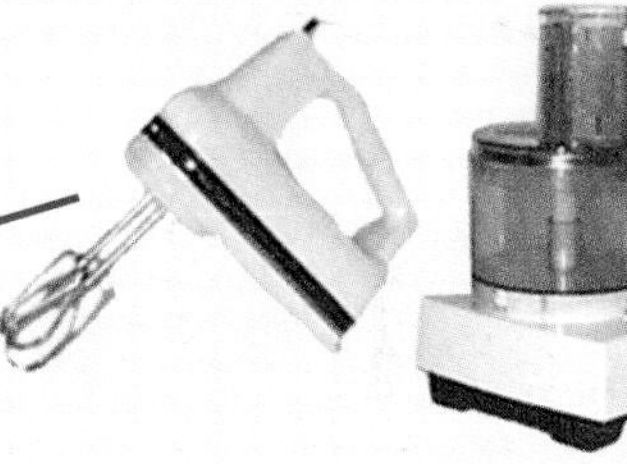

Use reusable items . . .

↓ Resource use

↓ Fuel use

↓ Pollution

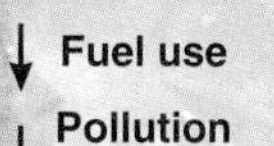

. . . instead of nonreusable items.

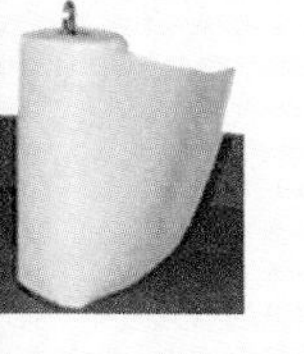

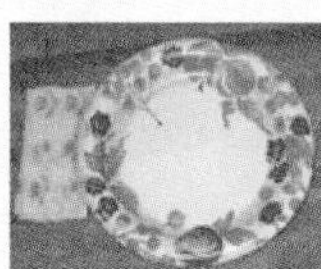

Use appliances that take less energy.

Make a switch from a meat-centered diet to a plant-centered diet. [a]

Buy fruits of differing ripeness.

Eat most perishable foods first.

Run your refrigerator efficiently. [b]

[a] Plant-centered diets are healthier and more environmentally responsible.
[b] Refer to the caption to learn how to run your refrigerator efficiently.

"We do not inherit the earth from our ancestors, we borrow it from our children." Ascribed to Chief Seattle, a nineteenth-century Native American leader.

changes in economic policies toward developing countries. According to one nutrition educator, "This [lobbying] is probably the most effective single thing that Joe or Jane Average Citizen can do [to help solve the world's hunger and environmental problems]."[1]

Another way people can assist the global community in solving its poverty and hunger problems is to join and work for international hunger-relief organizations. Table 15-4 lists some of the major ones. It's up to us, the human population, to solve these problems.

Table 15-4

Hunger-Relief Organizations People Can Join

Bread for the World Institute
1100 Wayne Ave., Suite 1000
Silver Spring, MD 20910
(301) 608-2400

International Labour Organization
1828 L St. NW, Suite 801
Washington, DC 20036
(202) 653-7652

OXFAM America
26 West St.
Boston, MA 02111-1206
(617) 482-1211

Seeds
P.O. Box 6170
Waco, TX 76706

United Nations International Children's Emergency Fund (UNICEF)
3 United Nations Plaza
New York, NY 10017-4414
(212) 326-7035

United Nations Food and Agriculture Organization (FAO)
1001 22nd St. NW, Suite 300
Washington, DC 20437
(202) 653-2400

World Health Organization (WHO)
525 23rd St. NW
Washington, DC 20037
(202) 861-3200

World Hunger Program
Brown University
Box 1831
Providence, RI 02912
(401) 863-2700

Self Check

Answers to these Self Check questions are in Appendix G.

1. Which of the following is a symptom of food insecurity?
   a. You worry about gaining weight.
   b. You sometimes rely on neighbors to feed your children because there is not enough food in the house.
   c. You shop daily to get the best prices.
   d. You buy organic foods to avoid harm from chemicals.
2. What is the primary cause of famine in the world?
   a. poverty
   b. drought
   c. social causes such as war
   d. flood
3. Which of the following is an example of environmental degradation?
   a. soil erosion
   b. damaged grazing lands
   c. air pollution
   d. all of the above
4. Which of the following activities is recommended due to the small negative impact it has on the environment?
   a. Use the oven whenever possible.
   b. Line pans with aluminum foil to reduce cleanup time.
   c. Use a pressure cooker or microwave to cook foods.
   d. Carry groceries home in paper bags rather than plastic bags.
5. Which of the following is a sustainable agricultural practice? (Read about this in the upcoming Controversy.)
   a. Plant legumes between grain crops.
   b. Use fertilizers generously.
   c. Irrigate on a large scale.
   d. Grow the same crop repeatedly on the same land.
6. At least 140 species of animals and plants are going extinct every day in the world. T F
7. Most children who die of malnutrition starve to death. T F
8. More people in the world suffer from famine than from chronic hunger. T F
9. The higher a nation's economic status, the faster its population grows over the long run. T F
10. An example of "eating lower on the food chain" would be to eat more fruits and grains and less beef. (A topic of this chapter's Controversy) T F

NUTRITION ON THE NET

For further study of the topics of this chapter, access these websites and search for the phrases or words in quotation marks:

1. For the latest changes in chapter material, to find supplemental learning tools, or access links to related websites, go to the "Nutrition: Concepts and Controversies" site at: **www.wadsworth.com/nutrition/prod/allprod.html**
2. For more information about hunger and food poverty in the United States, contact the USDA National Hunger Clearinghouse at: **www.worldhungeryear.org**
3. Visit the USDA Food Stamp Program site to learn about "food assistance programs" at: **www.fns.usda.gov/fns**
4. Additional information about "food recovery" or "gleaning" can be found at: **www.fns.usda.gov/fns**
5. To learn more about world hunger: **www.wfp.org**
6. For information about the impact of human activities on climate change, visit: **www.ipcc.ch**
7. The World Health Organization provides information about "poverty" and "overpopulation" at: **www.who.org**
8. Search the Food and Agriculture Organization site for information regarding "sustainable development" at: **www.fao.org/sd**
9. Information about food distribution in communities can be located at: **www.secondharvest.org**
10. To obtain additional information about "sustainable food production": **www.nal.usda.gov/afsic**
11. Search among thousands of current scientific and medical abstracts for any topic related to "malnutrition" at: **www.ncbi.nlm.nih.gov/PubMed**

CONTROVERSY 15

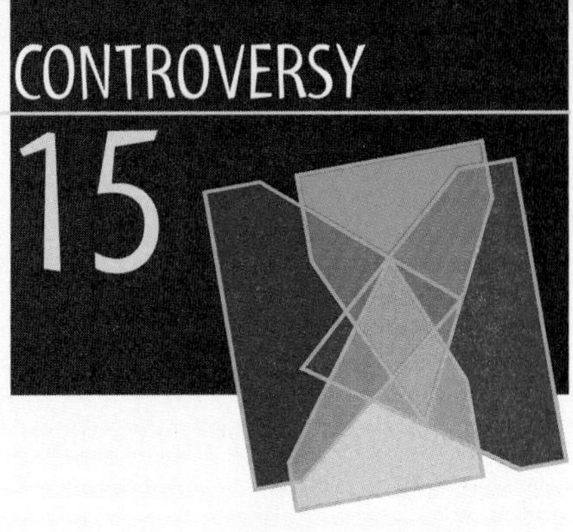

AGRIBUSINESS AND SUSTAINABLE FOOD PRODUCTION: HOW TO GO FORWARD?

WHILE SOME individuals are attempting to make their own personal lifestyles more environmentally benign, as suggested in the chapter, others are seeking ways to improve whole sectors of human enterprise, such as agriculture. To date, large agricultural enterprises have been among the world's biggest polluters and resource users. Is it possible for agriculture to become sustainable? Do our new technologies hold promise for advancing sustainability? And if so, will the change hurt farmers? These questions are addressed in this Controversy.

COSTS OF PRODUCING FOOD UNSUSTAINABLY

The environmental and social costs of agriculture and the food industry take many forms. Among them are resource waste and pollution, energy overuse, and tolls on life in farm communities. Table C15-1 offers some terms important to these concepts.

Impacts on Land and Water

Producing food has always cost the earth dearly. To grow food, we clear land—prairie, wetland, or forest—causing losses of native ecosystems and wildlife. Then we plant crops or graze animals on the land. The soil loses nutrients as each crop is taken from it, so fertilizer is applied. Some fertilizer runs off and pollutes the waterways. Some plowed soil runs off, clouds the water, and interferes with the growth of aquatic plants and animals.

Then, to protect crops against weeds and pests, we apply herbicides and pesticides. These chemicals also pollute the water and, wherever the wind carries them, the air. Most herbicides and pesticides kill not only weeds and pests, but also native plants, native insects, and animals that eat those plants and insects. Widespread use of pesticides and herbicides also causes resistant pests and weeds to evolve. Pesticide residues can even become a problem for people who eat foods produced this way, not to mention the hazard the pesticides pose to farm workers who handle and apply them.

Agricultural pesticides and herbicides, if not used conservatively, also pollute rivers, lakes, and groundwater. Pollution from "point sources," such as sewage plants or factories, is relatively easy to control, but runoff from fields and pastures enters waterways from all over broad regions and is nearly impossible to control.

Finally, we irrigate, a practice that adds salts to the soil in many areas. The water evaporates, but the salts do not. As soils become salty, plant growth fails. Irrigation can also deplete the water supply over time because water is pulled from surface waters or from underground and then evaporates or runs off. This process, carried to an extreme, can dry up whole rivers and lakes and lower the water table of entire regions. The lower the water table, the more farmers must irrigate; and the more they irrigate, the more groundwater they use up.

Soil Depletion and Losses of Species The soil can also be depleted by some agricultural practices, particularly indiscriminate land clearing (deforestation) and overuse by cattle (overgrazing). In just the past 40 years, human agricultural activities have ruined more than 10 percent of the earth's fertile land, an area the size of China and India combined. Over 20 million acres have been so damaged that they may be impossible to reclaim. With soil erosion proceeding unchecked, people in the year 2025 may see a 40 percent reduction in food-producing land per person, along with many more people to feed.[1]

Unsustainable agriculture has already destroyed many once-fertile regions, where high civilizations formerly flourished. The dry, salty deserts of North Africa were once plowed and irrigated wheat fields, the breadbasket of the Roman Empire. Mistreatment of soil and water is now causing destruction on a scale never known before.[2]

Agriculture is also weakening its own underpinnings by failing to conserve species diversity. By the year 2050, some 40,000 more plant species, existing in the 1990s, may go extinct.[3] The United Nations' Food and Agriculture Organization attributes many of the losses, which are occurring daily, to modern farming practices, as well as to population growth. Global eating habits, too, have become uniform. As people everywhere eat the same limited array of foods, demand for local, genetically diverse, native plants is insufficient to make them seem worth preserving. Yet in the future, as the climate warms and the earth changes, those may be the very plants that people will need as food sources.[4] A wild species of corn that grows in a dry climate, for example, might contain just the genetic information necessary to help make the domestic corn crop resistant to drought. Controversy 14 offered other

Table C15-1

Agricultural and Environmental Terms

- **agribusiness** agriculture practiced on a massive scale by large corporations owning vast acreages and employing intensive technological, fuel, and chemical inputs.
- **alternative (low-input,** or **sustainable) agriculture** agriculture practiced on a small scale using individualized approaches that vary with local conditions so as to minimize technological, fuel, and chemical inputs.
- **integrated pest management (IPM)** management of pests using a combination of natural and biological controls and minimal or no application of pesticides.
- **subsidies** government money, derived from taxes, used to support (subsidize) practices that otherwise would force producers to set their prices too high to compete successfully.

examples of genes that might be needed to improve food crops.

Energy Massive fossil fuel use is threatening our planet by causing ozone depletion, water pollution, ocean pollution, and other ills and by making global warming likely. In the United States, the food industry consumes about 20 percent of all the energy the nation uses. Each year we spend 1,500 liters (over 350 gallons) of oil per person to produce, process, distribute, and prepare our food. Energy is used to run farm machinery and to produce fertilizers and pesticides. Energy is also used to prepare, package, transport, refrigerate, and otherwise store, cook, and wash our foods.

The Problems of Livestock and Fishing Raising livestock also takes a toll. Like plant crops, herds of livestock occupy land that once maintained itself in a natural state. The land pays a price in losses of native plants and animals, soil erosion, water depletion, and desert formation. Alternatively, if animals are raised in concentrated areas such as cattle feedlots or giant hog "farms," a high price is paid when huge masses of animal wastes produced in overcrowded, factory-style farms leach into local soils and water supplies, polluting them.[5] In an effort to control this source of pollution, the Environmental Protection Agency offers incentives to livestock farmers who agree to clean up their wastes and allow their operations to be monitored for pollution.[6] The U.S. Senate Agricultural Committee has concluded that current regulations for handling animal wastes are inadequate and should be changed to protect against this source of pollution.[7] In addition to the waste problem, animals in such feedlots still have to be fed; grain is grown for them on other land (Figure C15-1, on page 564, compares the grain required to produce various foods). That grain may require fertilizers, herbicides, pesticides, and irrigation, too. In the United States, one-fifth of all cropland is used to produce grain for livestock—more land than is used to produce grain for people.

Vast areas under plow are exposed to erosion, and those that must be irrigated can, over time, become salty and unusable.

Other environmental costs attend fishing. Fishing easily becomes overfishing and depletes stocks of the very fish that people need to eat. Further, most nets also collect many nonfood species that are killed during harvest but returned to the sea instead of being put to use. Other kinds of aquatic animals are also vulnerable to injury and death, and populations of ecologically important nonfood animals, such as dolphins, are diminished. In short, our ways of producing foods are, for the most part, not sustainable.

FROM FAMILY FARMS TO AGRIBUSINESS

Beginning in the early 1980s and continuing today, U.S. agriculture has encountered serious economic problems: declining markets for U.S. farm products abroad as other countries increased their agricultural production and exports, reduced international economic ability to import grain, and an increase in energy and other costs of producing food domestically.[8] Many U.S. farmers, particularly those who specialize in export crops, have suffered heavy financial losses. Some have been unable to pay their debts and have had to leave farming. Between 1980 and 1994, more than 15 percent of the farms in the United States disappeared.[9] At the dawn of the year 2000 tens of thousands of farms are still struggling, especially medium-size family farms, because of competition from foreign producers and a trend toward large food-producing operations.[10]

Pure rivers represent irreplaceable water resources.

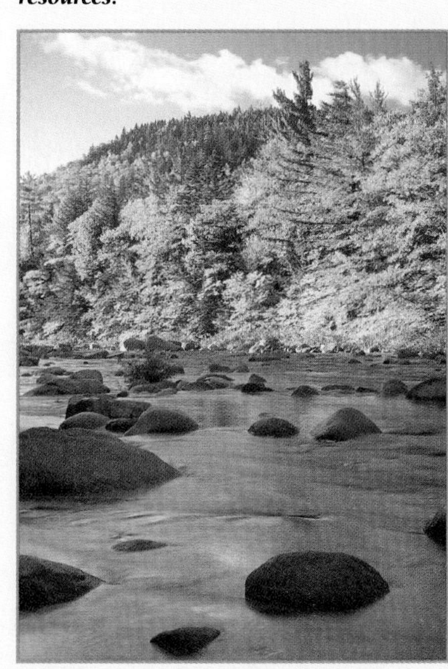

Taking the place of family farms are huge farms and ranches, many of which are being operated in Mexico or other developing nations; collectively, they are part of the massive food-producing enterprise called **agribusiness.** These huge operations in the United States tend to use little local labor, and the profits they make tend not to stay in local communities. Those in foreign countries tend to hire local laborers willing to work for much less than laborers in the United States.

Agribusinesses also tend to place a higher priority on producing abundant, inexpensive food than on protecting soil, water, and local biodiversity. When such large operations overuse fertilizers and pesticides, overuse land at the cost of soil erosion, use excessive irrigation water, and promote intensive forms of livestock production, their impacts can be enormous. Then, in an effort to compete, small farmers may be driven to adopt similar unsustainable practices.

Because of economies of scale, agribusinesses can price their products so low that consumers tend to buy more products from them than from smaller, local farms. Thus local U.S. grocers offer broccoli from Mexico, carrots from California, pineapples from Hawaii, and bananas from Central America at prices no local farmers can match, even if they could grow those products. Roadside stands offer bundles of local green vegetables and baskets of local fruits and tomatoes, but less conveniently and sometimes at higher prices than many shoppers are willing to pay.

Industrial farms generate huge masses of wastes that can contaminate local soil and water.

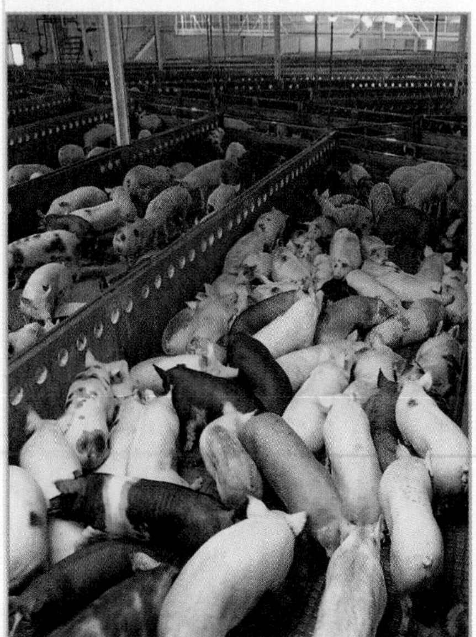

If the prices had to include a "tax" to pay for pollution cleanup, water protection, and land restoration, the prices of the products produced unsustainably would be higher. If they included a living wage, education, and benefits for the nation's silent slave labor force—the migrant farm workers—they would be higher still.

PROPOSED SOLUTIONS

For each of the problems described above, solutions are being devised, and indeed, some are being put into practice. To fully exploit these new sustainable agriculture techniques across the country will require some new learning. Sustainable agriculture is not one system but a set of practices that can be matched to particular needs in local areas. The first of these ideas, **alternative,** or **low-input agriculture**, emphasizes careful use of natural processes wherever possible, rather than chemically intensive methods.

Low-Input Agriculture

One form of low-input agriculture is **integrated pest management**. Farmers using this system employ many techniques, such as crop rotation and natural predators, to control pests rather than depending on heavy use of pesticides alone. Not all crops can grow reliably without pesticides, but many can. Table C15-2 contrasts low-input agriculture methods with unsustainable methods. Many sustainable techniques are not really new, incidentally; they would be familiar to our great-grandparents. Many farmers today are rediscovering the benefits of old techniques as they adapt and experiment with them in the search for sustainable methods.

Low-input agriculture has some apparent disadvantages, but advantages offset them. For example, as chemical use falls, yields per acre also fall somewhat, but costs per acre also fall, so the return per acre may be the same as or greater than before. More money goes to farmers and less to the fuels, fertilizers, pesticides, and irrigation. The end result of such farming is to make both farmers and consumers better off financially and environmentally.

Low-input agriculture works. As the world's population grows, and its land and water dwindle, the need to adopt sustainable agriculture and development around the globe grows urgent.[11]

Figure C15-1

Pounds of Grain Needed to Produce One Pound of Bread and One Pound of Animal Weight Gain

SOURCE: Idea and data from T. R. Reid, Feeding the planet, *National Geographic,* October 1998, pp. 58–74.

More than 30,000 U.S. farmers are successfully using sustainable techniques such as those described in Table C15-2. They see it as a food production system that can indefinitely sustain a healthy food supply, restore soil and water resources, and revitalize farming communities, while reducing reliance on fossil fuels.

Precision Agriculture An exciting development in agriculture is the application of powerful new computer technologies to food production. Through techniques collectively known as *precision agriculture*, farmers can adjust soil and crop management to meet precisely the needs of various areas of the farm. For example, a farmer growing crops in a field with hills, which tend to stay drier, and with low-lying areas, which tend to stay wetter, can adjust irrigation water to precisely meet the needs of each part of the field. Similarly, if one section of a field needs nitrogen fertilizer while another needs a different mix, the farmer can preprogram a computer to apply fertilizer of just the right type and amount for each area. Likewise, pesticide application can be programmed to prescribed applications, thus avoiding areas too close to streams or other water sources. The preprogrammed system turns off the pesticide flow when it comes to a designated safety zone.

Many of today's extraordinary technologies make precision agriculture possible. An important recent development, the *global positioning satellite (GPS)* system, is at the heart of precision farming. In the GPS system, military satellites beam accurate information about land positions and elevations of an area, such as a field, to receivers placed on farm equipment here on earth. The GPS delivers a grid map, pinpointing locations on a farm. Farmers then can use the GPS information grid to target, within a meter's accuracy, areas that need treatments. They can then program computerized farm equipment to apply chemicals or other treatments accordingly. Farmers can also use the information to adjust the depths to which they till the soil. The goal is to till deeply enough to prepare seedbeds properly and control weeds, but to avoid excessive tilling that wastes fuel and worsens erosion. Finally, at harvest, a GPS system produces an accurate accounting of crop yield, acre by acre, so that spot adjustments may be made in the next planting season.

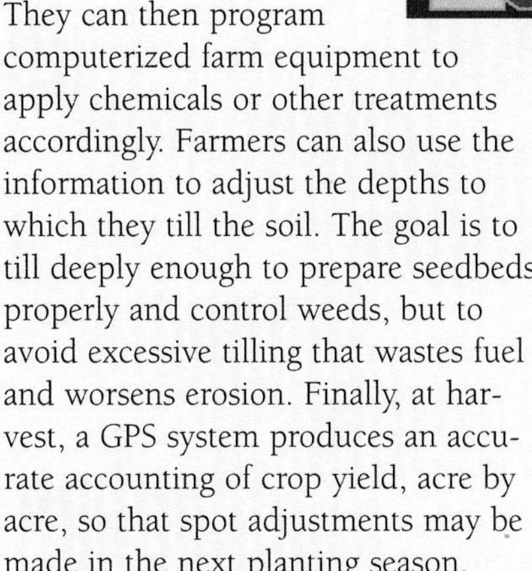

The future of precision agriculture seems bright, and the potential savings to farmers in terms of water, fertilizers, and pesticides are enormous. The accompanying reduction in polluting chemicals introduced into the environment means that everyone benefits.

Table C15-2
High-Input and Low-Input Agricultural Techniques Compared

Unsustainable Practice	Sustainable Practice
▪ Growing the same crop repeatedly on the same patch of land. This takes more and more nutrients out of the soil, makes fertilizer use necessary; favors soil erosion; and invites weeds and pests to become established, making pesticide use necessary.	▪ Rotate crops. This increases nitrogen in the soil so there is less need to buy fertilizers. If used with appropriate plowing methods, crop rotation reduces soil erosion. Crop rotation also reduces weeds and pests.
▪ Using fertilizers generously. Excess fertilizer pollutes ground and surface water and costs both farmers' household money and consumers' tax money.	▪ Reduce the use of fertilizers and use livestock manure more effectively. Store manure during the nongrowing season and apply it during the growing season. ▪ Alternate nutrient-devouring crops with nutrient-restoring crops, such as legumes. ▪ Compost on a large scale, including all plant residues not harvested. Plow the compost into the soil to improve its water-holding capacity.
▪ Feeding livestock in feedlots where their manure produces major water and soil pollutant problems. Piled in heaps, manure also releases methane, a global-warming gas.	▪ Feed livestock or buffalo on the open range where their manure will fertilize the ground on which plants grow and will release no methane. Alternatively, at least collect feedlot animals' manure and use it as fertilizer, or, at the very least, treat it before release.
▪ Spraying herbicides and pesticides over large areas to wipe out weeds and pests.	▪ Apply technology in weed and pest control. Use precision agriculture techniques if affordable or use rotary hoes twice instead of herbicides once. Spot treat weeds by hand. ▪ Rotate crops to foil pests that lay their eggs in the soil where last year's crop was grown. ▪ Use genetically resistant crops. ▪ Use biological controls such as predators that destroy the pests.
▪ Plowing the same way everywhere, allowing unsustainable water runoff and erosion.	▪ Plow in ways tailored to different areas. Conserve both soil and water by using cover crops, crop rotation, no-till planting, and contour plowing.
▪ Injecting animals with antibiotics to prevent disease in livestock.	▪ Maintain animals' health so that they can resist disease.
▪ Irrigating on a large scale.	▪ Irrigate only during dry spells and only where needed.

Agricultural Biotechnology Although not every farmer worldwide may be in a position to reap benefits from the technologies of precision agriculture, the advances of biotechnology may prove to be an essential part of a worldwide move toward sustainable agriculture. Biotechnology promises economic, environmental, and agricultural benefits by shrinking the acreage needed for crops, reducing soil losses, minimizing use of chemical insecticides, and bettering crop protection. Salt-resistant and drought-resistant crops that can grow under stressful conditions may one day make use of now unusable soils. Nitrogen-fixing cereals with legume genes may bring complete protein to populations who now lack quality protein in the diet. Likewise, transgenic livestock that produces more food while consuming less feed could help to stretch resources and reduce wastes.

Transgenic microbes also offer a wide variety of benefits to sustainable agriculture. Bioengineered microbes can contribute to continuous renewal of soil structure and fertility by fixing nitrogen and releasing other nutrients into the soil, lessening the need for chemical fertilizers and easing the environmental burden.[12] Scientific laboratories are also working toward engineering microbes that can recycle agricultural, industrial, and household wastes into fertilizers, an obvious boon to the environment. Bacterial and fungal herbicides, fungicides, and insecticides are in advanced experimental stages and promise to augment other integrated pest management systems of low-input agriculture such as crop rotation. Research is under way to ensure that transgenic microorganisms released into the environment will not turn out to be more harmful than the products they are intended to replace.

Energy Efficiency Some 6,560 calories of fuel are used to produce a can of corn, and 7,980 calories are needed to produce a package of frozen corn. Much of this energy input could be reduced, as Table C15-3 shows. The last item in the table suggests that consumers should center their diets on foods that require low energy inputs, a choice that is described next. That means choosing plants over meats most of the time.

Eating Lower on the Food Chain Studies of energy use in the U.S. food system have revealed which foods require the most and least energy to produce. The least energy is needed for grain: about one-third calorie of fuel is burned to produce each calorie of grain. Fruits and vegetables are intermediate, and most animal protein requires from 10 to 90 calories of fossil energy per calorie of usable food. An exception is livestock raised on the open range; these animals require low energy inputs as do most plant foods. So much more of our beef is grain fed, rather than range fed, however, that the average energy requirement for beef production is high. Figure C15-2 shows how much less fuel vegetarian diets require than meat diets and shows that vegan diets require the least fuel of all.

To support our meat intake, we maintain several billion livestock, about four times our own weight in animals. Livestock consume ten times as much grain each day as we do. We could use much of that grain to make grain products for ourselves and share them. The shift could free up enough grain to feed 400 million people and would necessitate burning less fuel and using less water. It could also free up much more land.

Some individuals are taking action to put into effect solutions to these problems. Some meat eaters are choosing to cut down on their meat portions or to eat range-fed beef or buffalo only. Livestock on the range eat grass, which people cannot eat.[13] "Rangeburger" buffalo also offers nutrition advantages over grain-fed beef. It is lower in fat, and the fat has more polyunsaturated fatty acids, including the omega-3 type.[14] Some people are switching to nonmeat, and even pure vegan, diets.[15] Shifting to a fish diet appears not to be a practical alternative at present, although fish farming shows promise of becoming practical in the future and could help greatly to provide nutritious meat at a price people and the environment could afford.[16]

CONCLUSION

Chapter 15 and its Controversy have suggested that, although many problems are global in scope, the actions of individual people lie at the heart of their solutions. On learning this, concerned people may take a perfectionist attitude, believing that they "should" be doing more than they realistically can, and so feel defeated. Yet, striving for perfection, even while falling short, is a way to achieve progress well worth celebrating. A positive attitude can bring about improvement, and improvement is enough to be proud of. Celebrate the changes that are possible today by making them a permanent part of your life; do the same with changes that become possible tomorrow and every day thereafter. The results may add up to more than you dared to hope for.

Table C15-3

Sustainable Energy-Saving Agricultural Techniques

- Use machinery scaled to the job at hand and operate it at efficient speeds.
- Combine operations. Harrow, plant, and fertilize in the same operation.
- Use diesel fuel. Use solar and wind energy on farms. Use methane from manure. Be open-minded to alternative energy sources.
- Use new disease- and pest-resistant plant varieties developed through genetic engineering.
- Save on technological and chemical inputs and spend some of the savings paying people to do manual jobs. Increasing labor inputs has been considered inefficient. Reverse this thinking: creating more jobs is preferable to using more machinery and fuel.
- Partially return to the techniques of using animal manure and crop rotation. This would save energy because chemical fertilizers require large energy inputs to produce.
- Choose crops that require low energy inputs (fertilizer, pesticides, irrigation).
- Educate people to cook food efficiently and to eat low on the food chain.

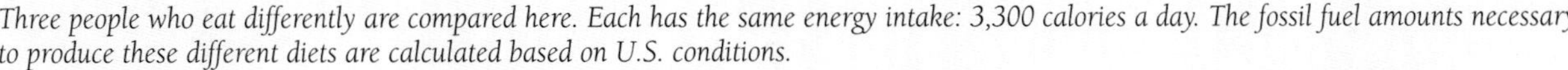

Figure C15-2

Average Amounts of Fuel Required to Feed People Eating at Different Levels on the Food Chain

Three people who eat differently are compared here. Each has the same energy intake: 3,300 calories a day. The fossil fuel amounts necessary to produce these different diets are calculated based on U.S. conditions.

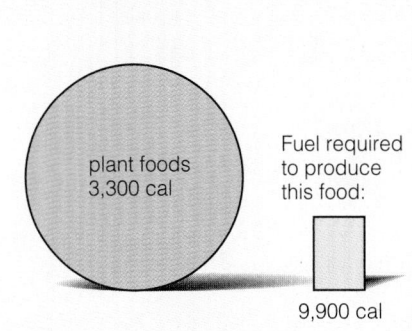

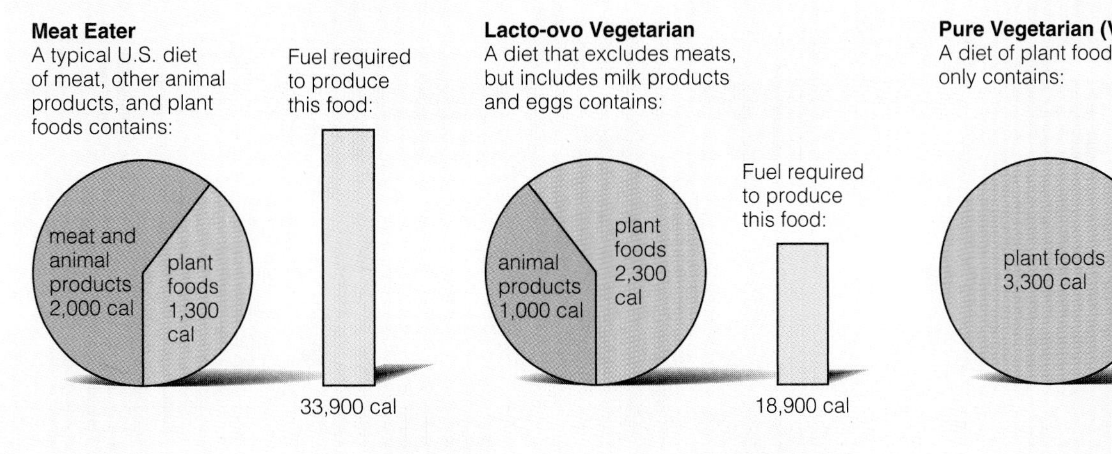
Meat Eater
A typical U.S. diet of meat, other animal products, and plant foods contains:
meat and animal products 2,000 cal
plant foods 1,300 cal
Fuel required to produce this food:
33,900 cal
Lacto-ovo Vegetarian
A diet that excludes meats, but includes milk products and eggs contains:
animal products 1,000 cal
plant foods 2,300 cal
Fuel required to produce this food:
18,900 cal
Pure Vegetarian (Vegan)
A diet of plant foods only contains:
plant foods 3,300 cal
Fuel required to produce this food:
9,900 cal

Appendixes

CONTENTS

A Table of Food Composition

This edition of the table of food composition contains more complete values for several nutrients than any comparable table.[a] These include dietary fiber; saturated, monounsaturated, and polyunsaturated fat; vitamin B_6; vitamin E; folate; magnesium; and zinc. The table includes a wide variety of foods from all food groups and is updated yearly to reflect current food patterns. For example, this edition includes many new nonfat items; several new ethnic items such as adzuki beans, tahitian taro, and gai choy chinese mustard; and a new selection of vegetarian foods.

Sources of Data

To achieve a complete and reliable listing of nutrients for all the foods, over 1,200 sources of information are researched. Government sources are the primary base for all data for most foods. In addition to USDA data (from Release 12 and surveys), provisional USDA information—both published and unpublished—is included.

Even with all the government sources available, however, some nutrient values are still missing; and as the USDA updates various data, it sometimes reports conflicting values for the same items. To fill in the missing values and resolve discrepancies, other reliable sources of information are used. These sources include journal articles, food composition tables from Canada and England, information from other nutrient data banks and publications, unpublished scientific data, and manufacturers' data.

The data for brand foods are listed as provided by the food manufacturers and the food chain restaurants. This information changes often because recipes and formulations are modified to meet consumer preferences, and the data are usually limited to those nutrients required for food labels. To provide more complete information, values for several nutrients are often estimated based on known values for major ingredients.

Accuracy

The energy and nutrients in recipes and combination foods vary widely, depending on the ingredients. The amounts of various fatty acids and cholesterol are influenced by the type of fat used (the specific type of oil, vegetable shortening, butter, margarine, etc.).

Estimates of nutrient amounts for foods and nutrients include all possible adjustments in the interest of accuracy. When multiple values are reported for a nutrient, the numbers are averaged and weighted with consideration of the original number of analyses in the separate sources. Whenever water percentages are available, estimates of nutrient amounts are adjusted for water content. When no water is given, water percentage is assumed to be that shown in the table. Whenever a reported weight appeared inconsistent, many kitchen tests were made, and the average weight of the typical product was given as tested.

When estimates of nutrient amounts in cooked foods are derived from reported amounts in raw foods, published retention factors are applied. Data for combination foods are modified to include newer data for major ingredients.

Considerable effort has been made to report the most accurate data available. The table is revised annually, and the authors welcome any suggestions or comments for future editions.

Average Values

It is important to know that many different nutrient values can be reported for foods, even by reliable sources. Many factors influence the amounts of nutrients in foods, including the mineral content of the soil, the method of processing, genetics, the diet of the animal or the fertilizer of the plant, the season of the year, methods of analysis, the difference in moisture content of the samples analyzed, the length and method of storage, and methods of cooking the food. The mineral content of water also varies according to the source.

Although each nutrient is presented as a single number, each number is actually an average of a range of data. More detailed reports from the USDA, for example, indicate the number of samples and the standard deviation of the data. One can also find different reported values for foods as older data are replaced with newer data from more recent analyses using newer analytical techniques. Therefore, nutrient data should be viewed and used only as a guide, a close approximation of nutrient content.

Dietary Fiber

There can be many different reported values for dietary fiber in foods because information depends on the type of ana-

[a]This food composition table has been prepared for West-Wadsworth Publishing Company and is copyrighted by ESHA Research in Salem, Oregon—the developer and publisher of the Food Processor®, Genesis® R&D, and the Computer Chef® nutrition software systems. The major sources for the data are from the USDA, supplemented by more than 1200 additional sources of information. Because the list of references is so extensive, it is not provided here, but is available from the publisher.

lytical technique used. The fiber data in this table are primarily from the USDA/ARS Human Nutrition Information Service in Hyattsville, Maryland; Composition of Foods by Southgate and Paul (England); and many journal articles.

Vitamin A

Vitamin A is reported in retinol equivalents (RE). The amount of vitamin A can vary by the season of the year and the maturity of the plant. Reported values in both dairy products and plants are higher in summer and early fall than in winter. The values reported here represent year-round averages. The organ meats of all animal products (liver especially) contain large amounts of vitamin A, which vary widely, depending on the background of the animal. The vitamin is also present in very small amounts in regular meat and is often reported as a trace.

Vitamin E

Vitamin E is actually a combination of various forms of this nutrient, and the measure of alpha-tocopherol equivalents (α-TE) summarizes the activity of the various types of tocopherols and tocotrienols into one measure.

Fats

Total fats, as well as the breakdown of total fats to saturated, monounsaturated, and polyunsaturated fats, are listed in the table. The fatty acids seldom add up to the total due to rounding and to other fatty acid components that are not included in these basic categories, such as *trans*-fatty acids and glycerol. *Trans*-fatty acids can comprise a large share of the total fat in margarine and shortening (hydrogenated oils) and in any foods that include them as ingredients.

Enrichment-Fortification

The mandatory enrichment values for foods are presented as appropriate, including the new values for folate enrichment folacin in grain products.

Niacin

Niacin values are for preformed niacin and do not include additional niacin that may form in the body from the conversion of tryptophan.

Using the Table

The items in this table have been organized into several categories, which are listed at the head of each right-hand page. As the key shows, each group has been color-coded to make it easier to find individual items.

In an effort to conserve space, the following abbreviations have been used in the food descriptions and nutrient breakdowns:

- diam = diameter
- ea = each
- enr = enriched
- f/ = from
- frzn = frozen
- g = grams
- liq = liquid
- pce = piece
- pkg = package
- w/ = with
- w/o = without
- t = trace
- 0 = zero (no nutrient value)
- blank space = information not available

Table A-1

Food Composition (Computer code number is for West Diet Analysis program) (For purposes of calculations, use "0" for t, <1, <.1, <.01, etc.)

Computer Code Number	Food Description	Measure	Wt (g)	H_2O (%)	Ener (cal)	Prot (g)	Carb (g)	Dietary Fiber (g)	Fat (g)	Fat Breakdown (g) Sat	Mono	Poly
	BEVERAGES											
	Alcoholic:											
	Beer:											
1	Regular (12 fl oz)	1½ c	356	92	146	1	13	1	0	0	0	0
2	Light (12 fl oz)	1½ c	354	95	99	1	5	0	0	0	0	0
1506	Nonalcoholic (12 fl oz)	1½ c	360	98	32	1	5	0	0	0	0	0
	Gin, rum, vodka, whiskey:											
3	80 proof	1½ fl oz	42	67	97	0	0	0	0	0	0	0
4	86 proof	1½ fl oz	42	64	105	0	<1	0	0	0	0	0
5	90 proof	1½ fl oz	42	62	110	0	0	0	0	0	0	0
	Liqueur:											
1359	Coffee liqueur, 53 proof	1½ fl oz	52	31	175	<1	24	0	<1	.1	t	.1
1360	Coffee & cream liqueur, 34 proof	1½ fl oz	47	46	154	1	10	0	7	4.5	2.1	.3
1361	Crème de menthe, 72 proof	1½ fl oz	50	28	186	0	21	0	<1	t	t	.1
	Wine, 4 fl oz:											
6	Dessert, sweet	½ c	118	72	181	<1	14	0	0	0	0	0
7	Red	½ c	118	88	85	<1	2	0	0	0	0	0
8	Rosé	½ c	118	89	84	<1	2	0	0	0	0	0
9	White medium	½ c	118	90	80	<1	1	0	0	0	0	0
1592	Nonalcoholic	1 c	232	98	14	1	3	0	0	0	0	0
1593	Nonalcoholic light	1 c	232	98	14	1	3	0	0	0	0	0
1409	Wine cooler, bottle (12 fl oz)	1½ c	340	90	169	<1	20	<1	<1	t	t	t
1595	Wine cooler, cup	1 c	227	90	113	<1	13	<1	<1	t	t	t
	Carbonated:											
10	Club soda (12 fl oz)	1½ c	355	100	0	0	0	0	0	0	0	0
11	Cola beverage (12 fl oz)	1½ c	372	89	153	0	39	0	0	0	0	0
12	Diet cola w/aspartame (12 fl oz)	1½ c	355	100	4	<1	<1	0	0	0	0	0
13	Diet soda pop w/saccharin (12 fl oz)	1½ c	355	100	0	0	<1	0	0	0	0	0
14	Ginger ale (12 fl oz)	1½ c	366	91	124	0	32	0	0	0	0	0
15	Grape soda (12 fl oz)	1½ c	372	89	160	0	42	0	0	0	0	0
16	Lemon-lime (12 fl oz)	1½ c	368	89	147	0	38	0	0	0	0	0
17	Orange (12 fl oz)	1½ c	372	88	179	0	46	0	0	0	0	0
18	Pepper-type soda (12 fl oz)	1½ c	368	89	151	0	38	0	<1	.1	0	0
19	Root beer (12 fl oz)	1½ c	370	89	152	0	39	0	0	0	0	0
20	Coffee, brewed	1 c	237	99	5	<1	1	0	<1	t	0	t
21	Coffee, prepared from instant	1 c	238	99	5	<1	1	0	<1	t	0	t
	Fruit drinks, noncarbonated:											
22	Fruit punch drink, canned	1 c	248	88	117	0	29	<1	<1	t	t	t
1358	Gatorade	1 c	241	93	60	0	15	0	0	0	0	0
23	Grape drink, canned	1 c	250	87	125	<1	32	<1	0	0	0	0
1304	Koolade sweetened with sugar	1 c	262	90	97	0	25	0	<1	t	t	t
1356	Koolade sweetened with nutrasweet	1 c	240	95	43	0	11	0	0	0	0	0
26	Lemonade,frzn concentrate (6-oz can)	¾ c	219	52	396	1	103	1	<1	.1	t	.1
27	Lemonade, from concentrate	1 c	248	89	99	<1	26	<1	<1	t	t	t
28	Limeade, frzn concentrate (6-oz can)	¾ c	218	50	408	<1	108	1	<1	t	t	.1
29	Limeade, from concentrate	1 c	247	89	101	0	27	<1	<1	t	t	t
24	Pineapple grapefruit, canned	1 c	250	88	118	<1	29	<1	<1	t	t	.1
25	Pineapple orange, canned	1 c	250	87	125	3	29	<1	0	0	0	0
	Fruit and vegetable juices: see Fruit and Vegetable sections											
	Ultra Slim Fast, ready to drink, can:											
30411	Chocolate Royale	1 ea	350	84	220	10	38	5	3	1	1	.5
30415	French Vanilla	1 ea	350	84	220	10	38	5	3	1	1.5	.5
30413	Strawberries n' cream	1 ea	350	83	220	10	42	5	3	1	1.5	.5
1357	Water, bottled: Perrier (6½ fl oz)	1 ea	192	100	0	0	0	0	0	0	0	0
1594	Water, bottled: Tonic water	1½ c	366	91	124	0	32	0	0	0	0	0
	Tea:											
30	Brewed, regular	1 c	237	100	2	0	1	0	<1	t	0	t
1662	Brewed, herbal	1 c	237	100	2	0	<1	0	<1	t	t	t
32	From instant, sweetened	1 c	259	91	88	<1	22	0	<1	t	t	t
31	From instant, unsweetened	1 c	237	100	2	<1	<1	0	0	0	0	0

Chol (mg)	Calc (mg)	Iron (mg)	Magn (mg)	Pota (mg)	Sodi (mg)	Zinc (mg)	VT-A (RE)	Thia (mg)	VT-E (a-TE)	Ribo (mg)	Niac (mg)	V-B6 (mg)	Fola (µg)	VT-C (mg)
0	18	.11	21	89	18	.07	0	.02	0	.09	1.61	.18	21	0
0	18	.14	18	64	11	.11	0	.03	0	.11	1.39	.12	14	0
0	25	.04	32	90	18	.04	0	.02	0	.09	1.63	.18	22	0
0	0	.02	0	1	<1	.02	0	<.01	0	<.01	<.01	0	0	0
0	0	.02	0	1	<1	.02	0	<.01	0	<.01	<.01	0	0	0
0	0	.02	0	1	<1	.02	0	<.01	0	<.01	<.01	0	0	0
0	1	.03	2	16	4	.02	0	<.01	0	.01	.07	0	0	0
7	8	.06	1	15	43	.07	20	0	.12	.03	.04	.01	0	0
0	0	.03	0	0	2	.02	0	0	0	0	<.01	0	0	0
0	9	.28	11	109	11	.08	0	.02	0	.02	.25	0	<1	0
0	9	.51	15	132	6	.11	0	.01	0	.03	.1	.04	2	0
0	9	.45	12	117	6	.07	0	<.01	0	.02	.09	.03	1	0
0	11	.38	12	94	6	.08	0	<.01	0	.01	.08	.02	<1	0
0	21	.93	23	204	16	.19	0	0	0	.02	.23	.05	2	0
0	21	.93	23	204	16	.19	0	0	0	.02	.23	.05	2	0
0	19	.92	18	152	29	.2	<1	.02	.02	.02	.15	.04	4	6
0	13	.61	12	102	19	.13	<1	.01	.02	.02	.1	.03	3	4
0	18	.04	4	7	75	.35	0	0	0	0	0	0	0	0
0	11	.11	4	4	15	.04	0	0	0	0	0	0	0	0
0	14	.11	4	0	21	.28	0	.02	0	.08	0	0	0	0
0	14	.14	4	7	57	.18	0	0	0	0	0	0	0	0
0	11	.66	4	4	26	.18	0	0	0	0	0	0	0	0
0	11	.3	4	4	56	.26	0	0	0	0	0	0	0	0
0	7	.26	4	4	40	.18	0	0	0	0	.05	0	0	0
0	19	.22	4	7	45	.37	0	0	0	0	0	0	0	0
0	11	.15	0	4	37	.15	0	0	0	0	0	0	0	0
0	18	.18	4	4	48	.26	0	0	0	0	0	0	0	0
0	5	.12	12	128	5	.05	0	0	0	0	.53	0	<1	0
0	7	.12	10	86	7	.07	0	0	0	<.01	.67	0	0	0
0	20	.52	5	62	55	.3	3	0	0	.06	.05	0	3	73
0	0	.12	2	26	96	.05	0	.01	0	0	0	0	0	0
0	7	.25	10	87	2	.07	<1	.02	0	.02	.25	.05	2	40
0	42	.13	3	3	37	.08	0	0	0	<.01	<.01	0	<1	31
0	17	.65	5	50	50	.26	2	.02	0	.05	.05	0	5	77
0	15	1.58	11	147	9	.17	22	.06	0	.21	.16	.05	22	39
0	7	.4	5	37	7	.1	5	.01	0	.05	.04	.01	5	10
0	11	.22	9	129	0	.09	0	.02	0	.02	.22	0	9	26
0	7	.07	2	32	5	.05	0	<.01	0	<.01	.05	0	2	7
0	17	.77	15	153	35	.15	9	.07	0	.04	.67	.1	26	115
0	12	.67	15	115	7	.15	13	.07	0	.05	.52	.12	27	56
5	400	2.7	140	530	220	2.24	525	.52	7	.59	7	.7	120	21
5	400	2.8	140	450	460	2.1	525	.52	7	.59	7	.7	120	21
5	400	2.7	140	450	460	2.24	525	.52	7	.59	7	.7	120	21
0	27	0	0	0	2	0	0	0	0	0	0	0	0	0
0	4	.04	0	0	15	.37	0	0	0	0	0	0	0	0
0	0	.05	7	88	7	.05	0	0	0	.03	0	0	12	0
0	5	.19	2	21	2	.09	0	.02	0	.01	0	0	1	0
0	5	.05	5	49	8	.08	0	0	0	.05	.09	<.01	10	0
0	5	.05	5	47	7	.07	0	0	0	<.01	.09	<.01	1	0

Table A–1

Food Composition (Computer code number is for West Diet Analysis program) (For purposes of calculations, use "0" for t, <1, <.1, <.01, etc.)

Computer Code Number	Food Description	Measure	Wt (g)	H_2O (%)	Ener (cal)	Prot (g)	Carb (g)	Dietary Fiber (g)	Fat (g)	Fat Breakdown (g)		
										Sat	Mono	Poly
	DAIRY											
	Butter: see Fats and Oils, #158,159,160											
	Cheese, natural:											
33	Blue	1 oz	28	42	99	6	1	0	8	5.2	2.2	.2
34	Brick	1 oz	28	41	104	6	1	0	8	5.3	2.4	.2
35	Brie	1 oz	28	48	93	6	<1	0	8	4.9	2.2	.2
36	Camembert	1 oz	28	52	84	6	<1	0	7	4.3	2	.2
37	Cheddar:	1 oz	28	37	113	7	<1	0	9	5.9	2.6	.3
38	1″ cube	1 ea	17	37	68	4	<1	0	6	3.6	1.6	.2
39	Shredded	1 c	113	37	455	28	1	0	37	24	10.6	1.1
1406	Low fat, low sodium	1 oz	28	65	48	7	1	0	2	1.2	.6	.1
	Cottage:											
984	Low sodium, low fat	1 c	225	83	162	28	6	0	2	1.4	.7	.1
40	Creamed, large curd	1 c	225	79	232	28	6	0	10	6.4	2.9	.3
41	Creamed, small curd	1 c	210	79	216	26	6	0	9	6	2.7	.3
42	With fruit	1 c	226	72	280	22	30	0	8	4.9	2.2	.2
43	Low fat 2%	1 c	226	79	203	31	8	0	4	2.8	1.2	.1
44	Low fat 1%	1 c	226	82	164	28	6	0	2	1.5	.7	.1
46	Cream	1 tbs	15	54	52	1	<1	0	5	3.3	1.5	.2
983	low fat	1 tbs	15	64	35	2	1	0	3	1.7	.9	.1
47	Edam	1 oz	28	42	100	7	<1	0	8	4.9	2.3	.2
48	Feta	1 oz	28	55	74	4	1	0	6	4.2	1.3	.2
49	Gouda	1 oz	28	41	100	7	1	0	8	4.9	2.2	.2
50	Gruyère	1 oz	28	33	116	8	<1	0	9	5.3	2.8	.5
51	Gorgonzola	1 oz	28	43	97	6	1	0	8	5		
1676	Limburger	1 oz	28	48	92	6	<1	0	8	4.7	2.4	.1
53	Monterey Jack	1 oz	28	41	104	7	<1	0	8	5.3	2.4	.3
54	Mozzarella, whole milk	1 oz	28	54	79	5	1	0	6	3.7	1.8	.2
55	Mozzarella, part-skim milk, low moisture	1 oz	28	49	78	8	1	0	5	3	1.4	.1
56	Muenster	1 oz	28	42	103	7	<1	0	8	5.3	2.4	.2
2422	Neufchatel	1 oz	28	62	73	3	1	0	7	4.1	1.9	.2
1399	Nonfat cheese (Kraft Singles)	1 oz	28	61	44	6	4	0	0	0	0	0
59	Parmesan, grated:	1 oz	28	18	128	12	1	0	8	5.5	2.4	.2
57	Cup, not pressed down	1 c	100	18	456	42	4	0	30	19.7	8.7	.7
58	Tablespoon	1 tbs	6	18	27	2	<1	0	2	1.2	.5	t
60	Provolone	1 oz	28	41	98	7	1	0	7	4.9	2.1	.2
61	Ricotta, whole milk	1 c	246	72	428	28	7	0	32	20.4	8.9	.9
62	Ricotta, part-skim milk	1 c	246	74	339	28	13	0	19	12.1	5.7	.6
63	Romano	1 oz	28	31	108	9	1	0	8	4.8	2.2	.2
64	Swiss	1 oz	28	37	105	8	1	0	8	5	2	.3
976	low fat	1 oz	28	60	50	8	1	0	1	.9	.4	t
	Pasteurized processed cheese products:											
65	American	1 oz	28	39	105	6	<1	0	9	5.5	2.5	.3
66	Swiss	1 oz	28	42	93	7	1	0	7	4.5	2	.2
67	American cheese food, jar	½ c	57	43	187	11	4	0	14	9	4.1	.4
68	American cheese spread	1 tbs	15	48	43	2	1	0	3	2	.9	.1
982	Velveeta cheese spread, low fat, low sodium, slice	1 pce	34	62	61	9	1	0	2	1.5	.7	.1
	Cream, sweet:											
69	Half & half (cream & milk)	1 c	242	81	315	7	10	0	28	17.3	8	1
70	Tablespoon	1 tbs	15	81	19	<1	1	0	2	1.1	.5	.1
71	Light, coffee or table:	1 c	240	74	468	6	9	0	46	28.8	13.4	1.7
72	Tablespoon	1 tbs	15	74	29	<1	1	0	3	1.8	.8	.1
73	Light whipping cream, liquid:	1 c	239	63	698	5	7	0	74	46.1	21.7	2.1
74	Tablespoon	1 tbs	15	63	44	<1	<1	0	5	2.9	1.4	.1
75	Heavy whipping cream, liquid:	1 c	238	58	821	5	7	0	88	54.7	25.5	3.3
76	Tablespoon	1 tbs	15	58	52	<1	<1	0	6	3.4	1.6	.2
77	Whipped cream, pressurized:	1 c	60	61	154	2	7	0	13	8.3	3.8	.5
78	Tablespoon	1 tbs	4	61	10	<1	<1	0	1	.6	.3	t
79	Cream, sour, cultured:	1 c	230	71	492	7	10	0	48	29.9	13.9	1.8
80	Tablespoon	1 tbs	14	71	30	<1	1	0	3	1.8	.8	.1

Chol (mg)	Calc (mg)	Iron (mg)	Magn (mg)	Pota (mg)	Sodi (mg)	Zinc (mg)	VT-A (RE)	Thia (mg)	VT-E (a-TE)	Ribo (mg)	Niac (mg)	V-B6 (mg)	Fola (µg)	VT-C (mg)
21	148	.09	6	72	391	.74	64	.01	.18	.11	.29	.05	10	0
26	189	.12	7	38	157	.73	85	<.01	.14	.1	.03	.02	6	0
28	51	.14	6	43	176	.67	51	.02	.18	.15	.11	.07	18	0
20	109	.09	6	52	236	.67	71	.01	.18	.14	.18	.06	17	0
29	202	.19	8	28	174	.87	85	.01	.1	.1	.02	.02	5	0
18	123	.12	5	17	106	.53	51	<.01	.06	.06	.01	.01	3	0
119	815	.77	31	111	702	3.51	342	.03	.41	.42	.09	.08	21	0
6	197	.2	8	31	6	.86	17	.01	.05	.01	.02	.02	5	0
9	137	.31	11	194	29	.85	25	.04	.25	.36	.29	.16	27	0
33	135	.31	12	190	911	.83	108	.05	.27	.37	.28	.15	27	0
31	126	.29	11	177	851	.78	101	.04	.26	.34	.26	.14	26	0
25	108	.25	9	151	915	.65	81	.04	.21	.29	.23	.12	22	0
19	155	.36	14	217	918	.95	45	.05	.13	.42	.32	.17	30	0
10	138	.32	12	193	918	.86	25	.05	.25	.37	.29	.15	28	0
16	12	.18	1	18	44	.08	66	<.01	.14	.03	.01	.01	2	0
8	17	.25	1	25	44	.11	33	<.01	.07	.04	.02	.01	3	0
25	205	.12	8	53	270	1.05	71	.01	.21	.11	.02	.02	5	0
25	138	.18	5	17	312	.81	36	.04	.01	.24	.28	.12	9	0
32	196	.07	8	34	229	1.09	49	.01	.1	.09	.02	.02	6	0
31	283	.05	10	23	94	1.09	84	.02	.1	.08	.03	.02	3	0
30	170	.18			280		43							0
25	139	.04	6	36	224	.59	88	.02	.18	.14	.04	.02	16	0
25	209	.2	8	23	150	.84	71	<.01	.09	.11	.03	.02	5	0
22	145	.05	5	19	104	.62	67	<.01	.1	.07	.02	.02	2	0
15	205	.07	7	26	148	.88	53	.01	.13	.1	.03	.02	3	0
27	201	.11	8	37	176	.79	88	<.01	.13	.09	.03	.02	3	0
21	21	.08	2	32	112	.15	74	<.01	.26	.05	.03	.01	3	0
4	221	0		81	427		126		0	.1				0
22	385	.27	14	30	521	.89	48	.01	.22	.11	.09	.03	2	0
79	1375	.95	51	107	1861	3.19	173	.04	.8	.39	.31	.1	8	0
5	82	.06	3	6	112	.19	10	<.01	.05	.02	.02	.01	<1	0
19	212	.15	8	39	245	.9	74	<.01	.1	.09	.04	.02	3	0
124	509	.93	28	258	207	2.85	330	.03	.86	.48	.26	.11	30	0
76	669	1.08	36	308	308	3.3	278	.05	.53	.45	.19	.05	32	0
29	298	.22	11	24	336	.72	39	.01	.2	.1	.02	.02	2	0
26	269	.05	10	31	73	1.09	71	.01	.14	.1	.03	.02	2	0
10	269	.05	10	31	73	1.09	18	.01	.05	.1	.02	.02	2	0
26	172	.11	6	45	400	.84	81	.01	.13	.1	.02	.02	2	0
24	216	.17	8	60	384	1.01	64	<.01	.19	.08	.01	.01	2	0
36	327	.48	17	159	678	1.7	125	.02	.4	.25	.08	.08	4	0
8	84	.05	4	36	202	.39	28	.01	.11	.06	.02	.02	1	0
12	233	.15	8	61	2	1.13	22	.01	.17	.13	.03	.03	3	0
89	254	.17	25	315	98	1.23	259	.08	.27	.36	.19	.09	6	2
6	16	.01	2	19	6	.08	16	<.01	.02	.02	.01	.01	<1	<1
159	231	.1	21	293	95	.65	437	.08	.36	.35	.14	.08	6	2
10	14	.01	1	18	6	.04	27	<.01	.02	.02	.01	<.01	<1	<1
265	166	.07	17	231	82	.6	705	.06	1.43	.3	.1	.07	9	1
17	10	<.01	1	14	5	.04	44	<.01	.09	.02	.01	<.01	1	<1
326	154	.07	17	179	89	.55	1001	.05	1.5	.26	.09	.06	9	1
21	10	<.01	1	11	6	.03	63	<.01	.09	.02	.01	<.01	1	<1
46	61	.03	6	88	78	.22	124	.02	.36	.04	.04	.02	2	0
3	4	<.01		6	5	.01	8	<.01	.02	<.01	<.01	<.01	<1	0
102	267	.14	26	331	123	.62	449	.08	1.3	.34	.15	.04	25	2
6	16	.01	2	20	7	.04	27	<.01	.08	.02	.01	<.01	2	<1

Table A-1

Food Composition (Computer code number is for West Diet Analysis program) (For purposes of calculations, use "0" for t, <1, <.1, <.01, etc.)

Computer Code Number	Food Description	Measure	Wt (g)	H_2O (%)	Ener (cal)	Prot (g)	Carb (g)	Dietary Fiber (g)	Fat (g)	Fat Breakdown (g) Sat	Mono	Poly
	DAIRY—Continued											
	Cream products—imitation and part dairy:											
81	Coffee whitener, frozen or liquid	1 tbs	15	77	20	<1	2	0	1	1.4	t	0
82	Coffee whitener, powdered	1 tsp	2	2	11	<1	1	0	1	.6	t	0
83	Dessert topping, frozen, nondairy:	1 c	75	50	239	1	17	0	19	16.4	1.2	.4
84	Tablespoon	1 tbs	5	50	16	<1	1	0	1	1.1	.1	t
85	Dessert topping, mix with whole milk:	1 c	80	67	151	3	13	0	10	8.6	.7	.2
86	Tablespoon	1 tbs	5	67	9	<1	1	0	1	.5	t	t
88	Dessert topping, pressurized	1 c	70	60	185	1	11	0	16	13.3	1.3	.2
87	Tablespoon	1 tbs	4	60	11	<1	1	0	1	.8	.1	t
91	Sour cream, imitation:	1 c	230	71	478	6	15	0	45	40.9	1.3	.1
92	Tablespoon	1 tbs	14	71	29	<1	1	0	3	2.5	.1	t
89	Sour dressing, part dairy:	1 c	235	75	418	8	11	0	39	31.3	4.6	1.1
90	Tablespoon	1 tbs	15	75	27	1	1	0	2	2	.3	.1
	Milk, fluid:											
93	Whole milk	1 c	244	88	150	8	11	0	8	5.1	2.7	.3
94	2% reduced-fat milk	1 c	244	89	121	8	12	0	5	2.9	1.3	.2
95	2% milk solids added	1 c	245	89	125	9	12	0	5	2.9	1.4	.2
96	1% lowfat milk	1 c	244	90	102	8	12	0	3	1.6	.7	.1
97	1% milk solids added	1 c	245	90	104	9	12	0	2	1.5	.7	.1
98	Nonfat milk, vitamin A added	1 c	245	91	85	8	12	0	<1	.3	.1	t
99	Nonfat milk solids added	1 c	245	90	90	9	12	0	1	.4	.2	t
100	Buttermilk, skim	1 c	245	90	99	8	12	0	2	1.3	.6	.1
	Milk, canned:											
101	Sweetened condensed	1 c	306	27	982	24	166	0	27	16.8	7.4	1
103	Evaporated, nonfat	1 c	256	79	199	19	29	0	1	.3	.2	t
	Milk, dried:											
104	Buttermilk, sweet	1 c	120	3	464	41	59	0	7	4.3	2	.3
105	Instant, nonfat, vit A added (makes 1 qt)	1 ea	91	4	326	32	47	0	1	.4	.2	t
106	Instant nonfat, vit A added	1 c	68	4	243	24	35	0	<1	.3	.1	t
107	Goat milk	1 c	244	87	168	9	11	0	10	6.5	2.7	.4
108	Kefir	1 c	233	88	149	8	11	0	8			
	Milk beverages and powdered mixes:											
	Chocolate:											
109	Whole	1 c	250	82	209	8	26	2	8	5.3	2.5	.3
110	2% fat	1 c	250	84	179	8	26	1	5	3.1	1.5	.2
111	1% fat	1 c	250	84	158	9	26	1	2	1.5	.7	.1
	Chocolate-flavored beverages:											
112	Powder containing nonfat dry milk:	1 oz	28	1	101	3	22	<1	1	.7	.4	t
113	Prepared with water	1 c	275	86	138	4	30	3	2	.9	.5	t
114	Powder without nonfat dry milk:	1 oz	28	1	98	1	25	2	1	.5	.3	t
115	Prepared with whole milk	1 c	266	81	226	9	31	1	9	5.5	2.6	.3
116	Eggnog, commercial	1 c	254	74	343	10	34	0	19	11.3	5.7	.9
974	Eggnog, 2% reduced-fat	1 c	254	85	189	12	17	0	8	3.7	2.7	.7
1027	Instant Breakfast, envelope,powder only:	1 ea	37	7	131	7	24	<1	1	.3	.1	t
1028	Prepared with whole milk	1 c	281	77	280	15	36	<1	9	5.4	2.5	.3
1029	Prepared with 2% milk	1 c	281	78	252	15	36	<1	5	3.3	1.5	.2
1283	Prepared with 1% milk	1 c	281	79	233	15	36	<1	3	1.9	.9	.1
1284	Prepared with nonfat milk	1 c	282	80	216	16	36	<1	1	.7	.3	t
117	Malted milk, chocolate, powder:	3 tsp	21	1	79	1	18	<1	1	.5	.2	.1
118	Prepared with whole milk	1 c	265	81	228	9	30	<1	9	5.5	2.6	.4
1661	Ovaltine with whole milk	1 c	265	81	225	9	29	<1	9	5.5	2.5	.4
119	Malted mix powder, natural:	3 tsp	21	2	87	2	16	<1	2	.9	.4	.3
120	Prepared with whole milk	1 c	265	81	236	10	27	<1	10	6	2.8	.6
121	Milk shakes, chocolate	1 c	166	71	211	6	34	1	6	3.8	1.8	.2
122	Milk shakes, vanilla	1 c	166	75	184	6	30	1	5	3.1	1.4	.2
	Milk desserts:											
134	Custard, baked	1 c	282	79	296	14	30	0	13	6.6	4.3	1
1548	Low-fat frozen dessert bars	1 ea	81	72	88	2	19	0	1	.2	.1	.4
	Ice cream, vanilla (about 10% fat):											
123	Hardened: ½ gallon	1 ea	1064	61	2138	37	251	0	117	72.4	33.8	4.4
124	Cup	1 c	132	61	265	5	31	0	14	9	4.2	.5
126	Soft serve	1 c	172	60	370	7	38	0	22	12.9	6	.8

Chol (mg)	Calc (mg)	Iron (mg)	Magn (mg)	Pota (mg)	Sodi (mg)	Zinc (mg)	VT-A (RE)	Thia (mg)	VT-E (a-TE)	Ribo (mg)	Niac (mg)	V-B6 (mg)	Fola (µg)	VT-C (mg)
0	1	<.01		29	12	<.01	1	0	.24	0	0	0	0	0
0		.02		16	4	.01	<1	0	<.01	<.01	0	0	0	0
0	5	.09	1	14	19	.02	64	0	.14	0	0	0	0	0
0		.01		1	1	<.01	4	0	.01	0	0	0	0	0
8	72	.03	8	121	53	.22	39	.02	.11	.09	.05	.02	3	1
<1	5	<.01		8	3	.01	2	<.01	.01	.01	<.01	<.01	<1	<1
0	4	.01	1	13	43	.01	33	0	.12	0	0	0	0	0
0		<.01		1	2	0	2	0	.01	0	0	0	0	0
0	6	.9	15	370	235	2.71	0	0	.34	0	0	0	0	0
0		.05	1	22	14	.16	0	0	.02	0	0	0	0	0
13	266	.07	23	381	113	.87	5	.09	.29	.38	.17	.04	28	2
1	17	<.01	1	24	7	.06	<1	.01	.02	.02	.01	<.01	2	<1
33	290	.12	33	371	120	.93	76	.09	.24	.39	.2	.1	12	2
18	298	.12	33	376	122	.95	139	.09	.17	.4	.21	.1	12	2
18	314	.12	35	397	128	.98	140	.1	.17	.42	.22	.11	13	2
10	300	.12	34	381	123	.95	144	.09	.1	.41	.21	.1	12	2
10	314	.12	35	397	128	.98	145	.1	.1	.42	.22	.11	13	2
4	301	.1	28	407	126	.98	149	.09	.1	.34	.22	.1	13	2
5	316	.12	35	419	130	1	149	.1	.1	.43	.22	.11	13	2
9	284	.12	27	370	257	1.03	20	.08	.15	.38	.14	.08	12	2
104	869	.58	79	1135	389	2.88	248	.27	.65	1.27	.64	.16	34	8
9	742	.74	69	850	294	2.3	300	.11	.01	.79	.44	.14	22	3
83	1420	.36	132	1910	620	4.82	65	.47	.48	1.9	1.05	.41	57	7
17	1119	.28	106	1551	500	4.01	646	.38	.02	1.58	.81	.31	45	5
12	836	.21	80	1159	373	3	483	.28	.01	1.18	.61	.23	34	4
28	327	.12	34	498	122	.73	137	.12	.22	.34	.68	.11	1	3
		.3	33	373	107									
30	280	.6	32	418	149	1.03	72	.09	.23	.4	.31	.1	12	2
17	285	.6	33	423	151	1.03	143	.09	.13	.41	.31	.1	12	2
7	288	.6	33	425	152	1.03	148	.09	.06	.41	.32	.1	12	2
1	91	.33	23	199	141	.41	1	.03	.04	.16	.16	.03	0	<1
3	129	.47	33	270	198	.6	1	.04	.06	.21	.22	.04	0	1
0	10	.88	27	165	59	.43	1	.01	.11	.04	.14	<.01	2	<1
32	301	.8	53	497	165	1.28	77	.1	.21	.43	.32	.1	12	2
149	330	.51	47	419	138	1.17	203	.09	.58	.48	.27	.13	2	4
194	269	.71	32	367	155	1.26	197	.11	1.01	.55	.21	.15	30	2
4	105	4.74	84	350	142	3.16	554	.31	5.31	.07	5.27	.42	105	28
38	396	4.86	117	719	262	4.09	630	.41	5.51	.47	5.46	.52	118	31
23	401	4.86	118	726	264	4.12	693	.41	5.41	.48	5.46	.53	118	31
14	406	4.86	118	731	266	4.12	698	.41	5.36	.48	5.47	.53	118	31
9	407	4.83	112	755	268	4.14	703	.4	5.3	.42	5.47	.52	118	31
1	13	.48	15	130	53	.17	4	.04	.08	.04	.42	.03	4	<1
34	305	.61	48	498	172	1.09	79	.13	.26	.44	.62	.13	16	3
34	384	3.76	53	620	244	1.17	901	.73	.32	1.26	10.9	1.02	32	34
4	63	.15	19	159	104	.21	18	.11	.08	.19	1.1	.09	10	1
37	355	.26	53	530	223	1.14	95	.2	.32	.59	1.31	.19	22	3
22	188	.51	28	332	161	.68	38	.1	.11	.41	.27	.08	6	1
18	203	.15	20	289	136	.6	53	.07	.1	.3	.31	.09	5	1
245	316	.85	39	431	217	1.49	169	.09	.68	.64	.24	.14	28	1
1	82	.07	10	111	47	.26	38	.03	.07	.11	.06	.03	3	1
468	1361	.96	149	2117	851	7.34	1244	.44	0	2.55	1.23	.51	53	6
58	169	.12	18	263	106	.91	154	.05	0	.32	.15	.06	7	1
157	225	.36	21	304	105	.89	265	.08	.64	.31	.16	.08	15	1

Table A-1

Food Composition (Computer code number is for West Diet Analysis program) (For purposes of calculations, use "0" for t, <1, <.1, <.01, etc.)

Computer Code Number	Food Description	Measure	Wt (g)	H_2O (%)	Ener (cal)	Prot (g)	Carb (g)	Dietary Fiber (g)	Fat (g)	Fat Breakdown (g)		
										Sat	Mono	Poly
	DAIRY—Continued											
	Ice cream, rich vanilla (16% fat):											
127	Hardened: ½ gallon	1 ea	1188	60	2554	49	264	0	154	88.9	41.5	5.5
128	Cup	1 c	148	57	357	5	33	0	24	14.8	6.9	.9
1724	Ben & Jerry's	½ c	108		230	4	21	0	17	10		
	Ice milk, vanilla (about 4% fat):											
129	Hardened: ½ gallon	1 ea	1048	68	1456	40	238	0	45	27.7	12.9	1.7
130	Cup	1 c	132	68	183	5	30	0	6	3.5	1.6	.2
131	Soft serve (about 3% fat)	1 c	176	70	222	9	38	0	5	2.9	1.3	.2
	Pudding, canned (5 oz can = .55 cup):											
135	Chocolate	1 ea	142	69	189	4	32	1	6	1	2.4	2
136	Tapioca	1 ea	142	74	169	3	27	<1	5	.9	2.3	1.9
137	Vanilla	1 ea	142	71	185	3	31	<1	5	.8	2.2	1.9
	Puddings, dry mix with whole milk:											
138	Chocolate, instant	1 c	294	74	326	9	55	3	9	5.4	2.7	.5
139	Chocolate, regular, cooked	1 c	284	74	315	9	51	3	10	5.9	2.8	.4
140	Rice, cooked	1 c	288	72	351	9	60	<1	8	5.1	2.4	.3
141	Tapioca, cooked	1 c	282	74	321	8	55	0	8	5.1	2.3	.3
142	Vanilla, instant	1 c	284	73	324	8	56	0	8	4.9	2.4	.4
143	Vanilla, regular, cooked	1 c	280	75	311	8	52	0	8	5.1	2.4	.4
	Sherbet (2% fat):											
132	½ gallon	1 ea	1542	66	2127	17	469	8	31	17.9	8.3	1.2
133	Cup	1 c	198	66	273	2	60	1	4	2.3	1.1	.2
144	Soy milk	1 c	245	93	81	7	4	3	5	.7	1	2.6
2301	Soy milk, fortified, fat free	1 c	240	88	110	6	22	1	0	0	0	0
	Yogurt, frozen, low-fat											
1584	Cup	1 c	144	65	229	6	35	0	8	4.9	2.3	.3
1512	Scoop	1 ea	79	74	78	4	15	0	<1	.1	t	t
	Yogurt, lowfat:											
1172	Fruit added with low-calorie sweetener	1 c	241	86	122	12	19	1	<1	.2	.1	t
145	Fruit added	1 c	245	74	250	11	47	<1	3	1.7	.7	.1
146	Plain	1 c	245	85	155	13	17	0	4	2.4	1	.1
147	Vanilla or coffee flavor	1 c	245	79	209	13	34	0	3	2	.8	.1
148	Yogurt, made with nonfat milk	1 c	245	85	137	15	19	0	<1	.3	.1	t
149	Yogurt, made with whole milk	1 c	245	88	150	9	11	0	8	5.1	2.2	.2
	EGGS											
	Raw, large:											
150	Whole, without shell	1 ea	50	75	74	6	1	0	5	1.5	1.9	.7
151	White	1 ea	33	88	16	3	<1	0	0	0	0	0
152	Yolk	1 ea	17	49	61	3	<1	0	5	1.6	2	.7
	Cooked:											
153	Fried in margarine	1 ea	46	69	91	6	1	0	7	1.9	2.8	1.3
154	Hard-cooked, shell removed	1 ea	50	75	77	6	1	0	5	1.6	2	.7
155	Hard-cooked, chopped	1 c	136	75	211	17	2	0	14	4.4	5.5	1.9
156	Poached, no added salt	1 ea	50	75	74	6	1	0	5	1.5	1.9	.7
157	Scrambled with milk & margarine	1 ea	61	73	101	7	1	0	7	2.2	2.9	1.3
1681	Egg substitute, liquid:	½ c	126	83	106	16	1	0	4	.8	1.1	2
1254	Egg Beaters, Fleischmann's	½ c	122		60	12	2	0	0	0	0	0
1262	Egg substitute, liquid, prepared	½ c	105	80	100	14	1	0	4	.8	1.1	1.9
	FATS and OILS											
158	Butter: Stick	½ c	114	16	817	1	<1	0	92	57.7	27.4	3.4
159	Tablespoon:	1 tbs	14	16	100	<1	<1	0	11	7.1	3.4	.4
8025	Unsalted	1 tbs	14	18	100	<1	<1	0	11	7.1	3.4	.4
160	Pat (about 1 tsp)	1 ea	5	16	36	<1	<1	0	4	2.5	1.2	.2
1682	Whipped	1 tsp	3	16	21	<1	<1	0	2	1.5	.7	.1
	Fats, cooking:											
1363	Bacon fat	1 tbs	14		125	0	0	0	14	6.3	5.9	1.1
1362	Beef fat/tallow	1 c	205	0	1849	0	0	0	205	103	87.3	8.2
1364	Chicken fat	1 c	205		1845	0	0	0	205	61.1	91.6	42.8
161	Vegetable shortening:	1 c	205	0	1812	0	0	0	205	52.1	91.2	53.5
162	Tablespoon	1 tbs	13	0	115	0	0	0	13	3.3	5.8	3.4

Chol (mg)	Calc (mg)	Iron (mg)	Magn (mg)	Pota (mg)	Sodi (mg)	Zinc (mg)	VT-A (RE)	Thia (mg)	VT-E (a-TE)	Ribo (mg)	Niac (mg)	V-B6 (mg)	Fola (µg)	VT-C (mg)
1081	1556	2.49	143	2102	725	6.18	1829	.58	4.4	2.16	1.13	.57	107	9
90	173	.07	16	235	83	.59	272	.06	0	.24	.12	.06	7	1
95	150	.36			55		214		0					0
147	1456	1.05	157	2211	891	4.61	493	.61	0	2.78	.94	.68	63	8
18	183	.13	20	279	112	.58	62	.08	0	.35	.12	.09	8	1
21	276	.11	25	389	123	.93	51	.09	0	.35	.21	.08	11	2
4	128	.72	30	256	183	.6	16	.04	.18	.22	.49	.04	4	3
1	119	.33	11	148	168	.38	0	.03	.13	.14	.44	.14	4	1
10	125	.18	11	160	192	.35	9	.03	.18	.2	.36	.02	0	0
32	300	.85	53	488	835	1.23	62	.1	.18	.41	.28	.11	12	3
34	315	1.02	43	463	293	1.28	74	.09	.17	.49	.29	.1	11	2
32	297	1.09	37	372	314	1.09	58	.22	.17	.4	1.28	.1	11	2
34	293	.17	34	372	341	.96	76	.08	.23	.4	.21	.11	11	2
31	287	.2	34	364	812	.94	71	.09	.18	.39	.21	.1	11	2
34	300	.14	36	381	448	.98	76	.08	.17	.4	.21	.09	11	2
77	833	2.16	123	1480	709	7.4	216	.39	.88	1.05	1.48	.52	77	66
10	107	.28	16	190	91	.95	28	.05	.11	.13	.19	.07	10	9
0	10	1.42	47	345	29	.56	7	.39	.02	.17	.36	.1	4	0
0	400	1.44		20	60		0	.07		.1	3			0
3	206	.43	20	304	125	.6	82	.05	.07	.32	.41	.11	9	1
1	137	.07	13	175	53	.67	1	.03	<.01	.16	.08	.04	8	1
3	369	.61	41	550	139	1.83	6	.1	.17	.45	.5	.11	32	26
10	372	.17	36	478	143	1.81	27	.09	.07	.44	.23	.1	23	2
15	448	.2	43	573	172	2.18	39	.11	.1	.52	.28	.12	27	2
12	419	.17	40	537	161	2.03	32	.1	.08	.49	.26	.11	26	2
4	488	.22	47	625	187	2.38	5	.12	.01	.57	.3	.13	30	2
31	296	.12	28	380	114	1.45	73	.07	.22	.35	.18	.08	18	1
213	24	.72	5	60	63	.55	95	.03	.52	.25	.04	.07	23	0
0	2	.01	4	47	54	<.01	0	<.01	0	.15	.03	<.01	1	0
218	23	.6	2	16	7	.53	99	.03	.54	.11	<.01	.07	25	0
211	25	.72	5	61	162	.55	114	.03	.75	.24	.03	.07	17	0
212	25	.59	5	63	62	.52	84	.03	.52	.26	.03	.06	22	0
577	68	1.62	14	171	169	1.43	228	.09	1.43	.7	.09	.16	60	0
212	24	.72	5	60	140	.55	95	.02	.52	.21	.03	.06	17	0
215	43	.73	7	84	171	.61	119	.03	.8	.27	.05	.07	18	<1
1	67	2.65	11	416	223	1.64	272	.14	.61	.38	.14	<.01	19	0
0	80	2.16		170	200		80		.59					
1	63	2.51	10	394	211	1.55	258	.11	.58	.34	.12	<.01	13	0
250	27	.18	2	30	942	.06	860	.01	1.8	.04	.05	<.01	3	0
31	3	.02		4	116	.01	106	<.01	.22	<.01	.01	0	<1	0
31	3	.02		4	2	.01	106	<.01	.22	<.01	.01	0	<1	0
11	1	.01		1	41	<.01	38	0	.08	<.01	<.01	0	<1	0
7	1	<.01		1	25	<.01	23	0	.05	<.01	<.01	0	<1	0
14		0			76	<.01	0	0	.31	0	0	0	0	0
223	0	0	0		<1	0	0	0	3.08	0	0	0	0	0
174	0	0	0	0	0	0	0	0	5.54	0	0	0	0	0
0	0	0	0	0	0	0	0	0	17	0	0	0	0	0
0	0	0	0	0	0	0	0	0	1.08	0	0	0	0	0

Table A-1

Food Composition (Computer code number is for West Diet Analysis program) (For purposes of calculations, use "0" for t, <1, <.1, <.01, etc.)

Computer Code Number	Food Description	Measure	Wt (g)	H_2O (%)	Ener (cal)	Prot (g)	Carb (g)	Dietary Fiber (g)	Fat (g)	Fat Breakdown (g) Sat	Mono	Poly
	FATS and OILS—Continued											
163	Lard:	1 c	205	0	1849	0	0	0	205	81.1	87	28.3
164	Tablespoon	1 tbs	13	0	117	0	0	0	13	5.1	5.5	1.8
	Margarine:											
165	Imitation (about 40% fat), soft:	1 c	232	58	800	1	1	0	90	17.9	36.4	32
166	Tablespoon	1 tbs	14	58	48	<1	<1	0	5	1.1	2.2	1.9
167	Regular, hard (about 80% fat):	½ c	114	16	820	1	1	0	92	18	40.8	29
168	Tablespoon	1 tbs	14	16	101	<1	<1	0	11	2.2	5	3.6
169	Pat	1 ea	5	16	36	<1	<1	0	4	.8	1.8	1.3
170	Regular, soft (about 80% fat):	1 c	227	16	1625	2	1	0	183	31.3	64.7	78.5
171	Tablespoon	1 tbs	14	16	100	<1	<1	0	11	1.9	4	4.8
2056	Saffola, unsalted	1 tbs	14	20	100	0	0	0	11	2	3	4.5
2057	Saffola, reduced fat	1 tbs	14	37	60	0	0	0	8	1.3	2.7	4.4
172	Spread (about 60% fat), hard:	1 c	227	37	1225	1	0	0	138	32	59	41.1
173	Tablespoon	1 tbs	14	37	76	<1	0	0	9	2	3.6	2.5
174	Pat	1 ea	5	37	27	<1	0	0	3	.7	1.2	1
175	Spread (about 60% fat), soft:	1 c	227	37	1225	1	0	0	138	29.3	71.5	31.3
176	Tablespoon	1 tbs	14	37	76	<1	0	0	9	1.8	4.4	1.9
2160	Touch of Butter (47% fat)	1 tbs	14		60	0	0	0	7	1.5	3.1	1.5
	Oils:											
1585	Canola:	1 c	218	0	1927	0	0	0	218	15.5	128	64.5
1586	Tablespoon	1 tbs	14	0	124	0	0	0	14	1	8.2	4.1
177	Corn:	1 c	218	0	1927	0	0	0	218	29.4	54.1	131
178	Tablespoon	1 tbs	14	0	124	0	0	0	14	1.9	3.5	8.4
179	Olive:	1 c	216	0	1909	0	0	0	216	29.4	159	21.3
180	Tablespoon	1 tbs	14	0	124	0	0	0	14	1.9	10.3	1.4
1683	Olive, extra virgin	1 tbs	14		126	0	0	0	14	2	10.8	1.3
181	Peanut:	1 c	216	0	1909	0	0	0	216	40	99.8	71.3
182	Tablespoon	1 tbs	14	0	124	0	0	0	14	2.6	6.5	4.6
183	Safflower:	1 c	218	0	1927	0	0	0	218	19.8	26.4	162
184	Tablespoon	1 tbs	14	0	124	0	0	0	14	1.3	1.7	10.4
185	Soybean:	1 c	218	0	1927	0	0	0	218	32	50.8	126
186	Tablespoon	1 tbs	14	0	124	0	0	0	14	2.1	3.3	8.1
187	Soybean/cottonseed:	1 c	218	0	1927	0	0	0	218	39.5	64.3	105
188	Tablespoon	1 tbs	14	0	124	0	0	0	14	2.5	4.1	6.7
189	Sunflower:	1 c	218	0	1927	0	0	0	218	25.3	42.5	143
190	Tablespoon	1 tbs	14	0	124	0	0	0	14	1.6	2.7	9.2
	Salad dressings/sandwich spreads:											
191	Blue cheese, regular	1 tbs	15	32	76	1	1	0	8	1.5	1.8	4.2
1040	Low calorie	1 tbs	15	79	15	1	<1	<1	1	.2	.5	.4
1684	Caesar's	1 tbs	12	36	56	1	<1	<1	5	1	3.7	.5
192	French, regular	1 tbs	16	38	69	<1	3	0	7	1.5	1.3	3.5
193	Low calorie	1 tbs	16	71	21	<1	3	0	1	.1	.2	.5
194	Italian, regular	1 tbs	15	40	70	<1	2	0	7	1	1.7	4.2
195	Low calorie	1 tbs	15	84	16	<1	1	<1	1	.2	.3	.9
	Kraft, Deliciously Right:											
2150	1000 Island	1 tbs	16		35	0	4	0	2	.5		
2153	Bacon & tomato	1 tbs	16		31	1	2	0	3	.5		
2154	Cucumber ranch	1 tbs	16		31	0	1	0	3	.5		
2151	French	1 tbs	16		25	0	3	0	1	.2		
2152	Ranch	1 tbs	16		52	0	3	0	5	.8		
199	Mayo type, regular	1 tbs	15	40	58	<1	4	0	5	.7	1.3	2.7
1030	Low calorie	1 tbs	14	54	36	<1	3	0	3	.4	.7	1.4
	Mayonnaise:											
197	Imitation, low calorie	1 tbs	15	63	35	<1	2	0	3	.5	.7	1.6
196	Regular (soybean)	1 tbs	14	17	100	<1	<1	0	11	1.7	3.2	5.8
1488	Regular, low calorie, low sodium	1 tbs	14	63	32	<1	2	0	3	.5	.6	1.4
1493	Regular, low calorie	1 tbs	15	63	35	<1	2	0	3	.5	.7	1.6
198	Ranch, regular	1 tbs	15	39	80	0	<1	0	8	1.2		
2251	Low calorie	1 tbs	14	69	30	<1	1	0	2	.5		
1685	Russian	1 tbs	15	34	74	<1	2	0	8	1.1	1.8	4.4
1502	Salad dressing, low calorie, oil free	1 tbs	15	88	4	<1	1	<1	<1	0	0	0

Chol (mg)	Calc (mg)	Iron (mg)	Magn (mg)	Pota (mg)	Sodi (mg)	Zinc (mg)	VT-A (RE)	Thia (mg)	VT-E (a-TE)	Ribo (mg)	Niac (mg)	V-B6 (mg)	Fola (μg)	VT-C (mg)
195		0			<1	.23	0	0	2.46	0	0	0	0	0
12		0			<1	.01	0	0	.16	0	0	0	0	0
0	41	0	4	59	2227	0	1853	.01	5.41	.05	.03	.01	2	<1
0	2	0		4	134	0	112	<.01	.33	<.01	<.01	<.01	<1	<1
0	34	.07	3	48	1075	0	911	.01	14.6	.04	.03	.01	1	<1
0	4	.01		6	132	0	112	<.01	1.79	<.01	<.01	<.01	<1	<1
0	1	<.01		2	47	0	40	<.01	.64	<.01	<.01	0	<1	<1
0	60	0	5	86	2447	0	1813	.02	27.2	.07	.04	.02	2	<1
0	4	0		5	151	0	112	<.01	1.68	<.01	<.01	<.01	<1	<1
	0	0			0		51							<1
	0	0			115		51							0
0	47	0	4	68	2256	0	1813	.02	11.4	.06	.04	.01	2	<1
0	3	0		4	139	0	112	<.01	.7	<.01	<.01	<.01	<1	<1
0	1	0		1	50	0	40	0	.25	<.01	<.01	0	<1	<1
0	47	0	4	68	2256	0	1813	.02	20.5	.06	.04	.01	2	<1
0	3	0		4	139	0	112	<.01	1.26	<.01	<.01	<.01	<1	<1
0	0	0		0	110		100		1.27					0
0	0	0	0	0	0	0	0	0	45.8	0	0	0	0	0
0	0	0	0	0	0	0	0	0	2.94	0	0	0	0	0
0	0	0	0	0	0	0	0	0	46	0	0	0	0	0
0	0	0	0	0	0	0	0	0	2.95	0	0	0	0	0
0		.82		0	<1	.13	0	0	26.8	0	0	0	0	0
0		.05		0	<1	.01	0	0	1.74	0	0	0	0	0
									1.74					
0		.06			<1	.02	0	0	27.9	0	0	0	0	0
0		<.01			<1	<.01	0	0	1.81	0	0	0	0	0
0	0	0	0	0	0	0	0	0	94	0	0	0	0	0
0	0	0	0	0	0	0	0	0	6.03	0	0	0	0	0
0		.04		0	0	0	0	0	39.7	0	0	0	0	0
0		<.01		0	0	0	0	0	2.55	0	0	0	0	0
0	0	0	0	0	0	0	0	0	61.5	0	0	0	0	0
0	0	0	0	0	0	0	0	0	3.95	0	0	0	0	0
0	0	0	0	0	0	0	0	0	110	0	0	0	0	0
0	0	0	0	0	0	0	0	0	7.08	0	0	0	0	0
3	12	.03	0	6	164	.04	10	<.01	1.4	.01	.01	.01	1	<1
<1	13	.07	1	1	180	.04	<1	<.01	.13	.01	.01	<.01	<1	<1
12	23	.2	3	21	207	.13	7	<.01	.72	.02	.5	.01	2	1
0	2	.06	0	13	219	.01	21	<.01	1.35	<.01	0	<.01	1	0
0	2	.06	0	13	126	.03	21	0	.19	0	0	0	0	0
0	1	.03		2	118	.02	4	<.01	1.56	<.01	0	<.01	1	0
1		.03	0	2	118	.02	0	0	.22	0	0	0	0	0
2	0	0		27	160		0		.19					0
1	0	0		21	155		0		.75					0
0	0	0		10	232		0		.73					0
0	0	0		7	130		50		.42					0
0	0	0		5	165		0		1.31					0
4	2	.03		1	107	.03	13	<.01	.6	<.01	<.01	<.01	1	0
4	2	.03		1	99	.02	9	<.01	.6	<.01	0	<.01	1	0
4		0		1	75	.02	0	0	.96	0	0	0	0	0
8	3	.07		5	79	.02	12	0	1.65	0	<.01	.08	1	0
3	0	0	0	1	15	.01	1	0	.53	<.01	0	0	<1	0
4		0		1	75	.02	0	0	.96	0	0	0	0	0
5	0	0			105		0							0
5	5	0			120		0		.7					0
3	3	.09		24	130	.06	31	.01	1.53	.01	.09	<.01	2	1
0	1	.04	2	7	256	<.01	<1	0	0	0	<.01	<.01	<1	<1

Table A-1

Food Composition (Computer code number is for West Diet Analysis program) (For purposes of calculations, use "0" for t, <1, <.1, <.01, etc.)

Computer Code Number	Food Description	Measure	Wt (g)	H_2O (%)	Ener (cal)	Prot (g)	Carb (g)	Dietary Fiber (g)	Fat (g)	Fat Breakdown (g) Sat	Mono	Poly
	FATS and OILS—Continued											
	Salad dressing, no cholesterol											
1605	Miracle Whip	1 tbs	15	57	48	0	2	0	4	1.1	1.1	2.1
203	Salad dressing, from recipe, cooked	1 tbs	16	69	25	1	2	0	2	.5	.6	.3
200	Tartar sauce, regular	1 tbs	14	34	74	<1	1	<1	8	1.5	2.6	4.1
1503	Low calorie	1 tbs	14	63	31	<1	2	<1	2	.4	.6	1.3
201	Thousand island, regular	1 tbs	16	46	60	<1	2	0	6	1	1.3	3.2
202	Low calorie	1 tbs	15	69	24	<1	2	<1	2	.2	.4	.9
204	Vinegar and oil	1 tbs	16	47	72	0	<1	0	8	1.5	2.4	3.9
	Wishbone:											
2180	Creamy Italian, lite	1 tbs	15		26	<1	2		2	.4	.9	.7
2166	Italian, lite	1 tbs	16	90	6	0	1		<1	0	.2	.1
8427	Ranch, lite	1 tbs	15	56	50	0	2	0	4	.7		
	FRUITS and FRUIT JUICES											
	Apples:											
	Fresh, raw, with peel:											
205	2¾" diam (about 3 per lb w/cores)	1 ea	138	84	81	<1	21	4	<1	.1	t	.1
206	3¼" diam (about 2 per lb w/cores)	1 ea	212	84	125	<1	32	6	1	.1	t	.2
207	Raw, peeled slices	1 c	110	84	63	<1	16	2	<1	.1	t	.1
208	Dried, sulfured	10 ea	64	32	156	1	42	6	<1	t	t	.1
209	Apple juice, bottled or canned	1 c	248	88	117	<1	29	<1	<1	t	t	.1
210	Applesauce, sweetened	1 c	255	80	194	<1	51	3	<1	.1	t	.1
211	Applesauce, unsweetened	1 c	244	88	105	<1	28	3	<1	t	t	t
	Apricots:											
212	Raw, w/o pits (about 12 per lb w/pits)	3 ea	105	86	50	1	12	3	<1	t	.2	.1
	Canned (fruit and liquid):											
213	Heavy syrup	1 c	240	78	199	1	52	4	<1	t	.1	t
214	Halves	3 ea	120	78	100	1	26	2	<1	t	t	t
215	Juice pack	1 c	244	87	117	2	30	4	<1	t	t	t
216	Halves	3 ea	108	87	52	1	13	2	<1	t	t	t
217	Dried, halves	10 ea	35	31	83	1	22	3	<1	t	.1	t
218	Dried, cooked, unsweetened, w/liquid	1 c	250	76	213	3	55	8	<1	t	.2	.1
219	Apricot nectar, canned	1 c	251	85	141	1	36	2	<1	t	.1	t
	Avocados, raw, edible part only:											
220	California	1 ea	173	73	306	4	12	8	30	4.5	19.5	3.5
221	Florida	1 ea	304	80	340	5	27	16	27	5.3	14.8	4.5
222	Mashed, fresh, average	1 c	230	74	370	5	17	11	35	5.6	22.1	4.5
	Bananas, raw, without peel:											
223	Whole, 8¾" long (175g w/peel)	1 ea	118	74	109	1	28	3	1	.2	t	.1
224	Slices	1 c	150	74	138	2	35	4	1	.3	.1	.1
1285	Bananas, dehydrated slices	½ c	50	3	173	2	44	4	1	.3	.1	.2
225	Blackberries, raw	1 c	144	86	75	1	18	8	1	t	.1	.3
	Blueberries:											
226	Fresh	1 c	145	85	81	1	20	4	1	t	.1	.2
227	Frozen, sweetened	10 oz	284	77	230	1	62	6	<1	t	.1	.2
228	Frozen, thawed	1 c	230	77	186	1	51	5	<1	t	t	.1
	Cherries:											
229	Sour, red pitted, canned water pack	1 c	244	90	88	2	22	3	<1	.1	.1	.1
230	Sweet, red pitted, raw	10 ea	68	81	49	1	11	2	1	.1	.2	.2
231	Cranberry juice cocktail,vitamin C added	1 c	253	85	144	0	36	<1	<1	t	t	.1
1411	Cranberry juice, low calorie	1 c	237	95	45	0	11	<1	0	0	0	0
232	Cranberry-apple juice,vitamin C added	1 c	245	83	164	<1	42	<1	0	0	0	0
233	Cranberry sauce, canned, strained	1 c	277	61	418	1	108	3	<1	t	.1	.2
234	Dates, whole, without pits	10 ea	83	22	228	2	61	6	<1	.2	.1	t
235	Dates, chopped	1 c	178	22	490	4	131	13	1	.3	.3	.1
236	Figs, dried	10 ea	190	28	485	6	124	18	2	.4	.5	1.1
	Fruit cocktail, canned, fruit and liq:											
237	Heavy syrup pack	1 c	248	80	181	1	47	2	<1	t	t	.1
238	Juice pack	1 c	237	87	109	1	28	2	<1	t	t	t
	Grapefruit:											
	Raw 3¾" diam (half w/rind = 241g)											
239	Pink/red, half fruit, edible part	1 ea	123	91	37	1	9	2	<1	t	t	t
240	White, half fruit, edible part	1 ea	118	90	39	1	10	1	<1	t	t	t
241	Canned sections with light syrup	1 c	254	84	152	1	39	1	<1	t	t	.1

Chol (mg)	Calc (mg)	Iron (mg)	Magn (mg)	Pota (mg)	Sodi (mg)	Zinc (mg)	VT-A (RE)	Thia (mg)	VT-E (a-TE)	Ribo (mg)	Niac (mg)	V-B6 (mg)	Fola (μg)	VT-C (mg)
0	0	<.01	0	0	102	0	2	0	.64	0	0	0	0	0
9	13	.08	1	19	117	.06	20	.01	.3	.02	.04	<.01	1	<1
7	3	.13		11	99	.02	9	<.01	2.24	<.01	0	<.01	1	<1
3	2	.09		5	83	.02	2	0	.83	<.01	.01	<.01	<1	<1
4	2	.1		18	112	.02	15	<.01	.18	<.01	<.01	<.01	1	0
2	2	.09		17	150	.02	14	<.01	.18	<.01	0	<.01	1	0
0	0	0	0	1	<1	0	0	0	1.41	0	0	0	0	0
<1	0	0			148		0	0	.56	0	0			0
0	1	0			255		2	0	.24	0	0			<1
2	0	0			120		0							0
0	10	.25	7	159	0	.05	7	.02	.44	.02	.11	.07	4	8
0	15	.38	11	244	0	.08	11	.04	.68	.03	.16	.1	6	12
0	4	.08	3	124	0	.04	4	.02	.09	.01	.1	.05	<1	4
0	9	.9	10	288	56	.13	4	0	.35	.1	.59	.08	0	2
0	17	.92	7	295	7	.07	<1	.05	.02	.04	.25	.07	<1	2
0	10	.89	8	156	8	.1	3	.03	.03	.07	.48	.07	2	4
0	7	.29	7	183	5	.07	7	.03	.02	.06	.46	.06	1	3
0	15	.57	8	311	1	.27	274	.03	.93	.04	.63	.06	9	10
0	22	.72	17	336	10	.26	295	.05	2.14	.05	.9	.13	4	7
0	11	.36	8	168	5	.13	148	.02	1.07	.03	.45	.06	2	4
0	29	.73	24	403	10	.27	412	.04	2.17	.05	.84	.13	4	12
0	13	.32	11	178	4	.12	183	.02	.96	.02	.37	.06	2	5
0	16	1.65	16	482	3	.26	253	<.01	.52	.05	1.05	.05	4	1
0	40	4.18	42	1222	7	.65	590	.01	1.25	.07	2.36	.28	0	4
0	18	.95	13	286	8	.23	331	.02	.2	.03	.65	.05	3	2
0	19	2.04	71	1096	21	.73	106	.19	2.32	.21	3.32	.48	113	14
0	33	1.61	103	1483	15	1.28	185	.33	2.37	.37	5.84	.85	162	24
0	25	2.35	90	1377	23	.97	140	.25	3.08	.28	4.42	.64	142	18
0	7	.37	34	467	1	.19	9	.05	.32	.12	.64	.68	22	11
0	9	.46	43	594	1	.24	12	.07	.4	.15	.81	.87	29	14
0	11	.57	54	746	1	.3	15	.09	0	.12	1.4	.22	7	3
0	46	.82	29	282	0	.39	23	.04	1.02	.06	.58	.08	49	30
0	9	.25	7	129	9	.16	14	.07	1.45	.07	.52	.05	9	19
0	17	1.11	6	170	3	.17	11	.06	2.02	.15	.72	.17	19	3
0	14	.9	5	138	2	.14	9	.05	1.63	.12	.58	.14	15	2
0	27	3.34	15	239	17	.17	183	.04	.32	.1	.43	.11	19	5
0	10	.26	7	152	0	.04	14	.03	.09	.04	.27	.02	3	5
0	8	.38	5	45	5	.18	1	.02	0	.02	.09	.05	<1	90
0	21	.09	5	52	7	.05	1	.02	0	.02	.08	.04	<1	76
0	17	.15	5	66	5	.1	1	.01	0	.05	.15	.05	<1	78
0	11	.61	8	72	80	.14	6	.04	.28	.06	.28	.04	3	6
0	27	.95	29	541	2	.24	4	.07	.08	.08	1.83	.16	10	0
0	57	2.05	62	1160	5	.52	9	.16	.18	.18	3.92	.34	22	0
0	274	4.24	112	1352	21	.97	25	.13	9.5	.17	1.32	.43	14	2
0	15	.72	12	218	15	.2	50	.04	.72	.05	.93	.12	6	5
0	19	.5	17	225	9	.21	73	.03	.47	.04	.95	.12	6	6
0	13	.15	10	159	0	.09	32	.04	.31	.02	.23	.05	15	47
0	14	.07	11	175	0	.08	1	.04	.29	.02	.32	.05	12	39
0	36	1.02	25	328	5	.2	0	.1	.63	.05	.62	.05	22	54

Table A-1

Food Composition (Computer code number is for West Diet Analysis program) (For purposes of calculations, use "0" for t, <1, <.1, <.01, etc.)

Computer Code Number	Food Description	Measure	Wt (g)	H_2O (%)	Ener (cal)	Prot (g)	Carb (g)	Dietary Fiber (g)	Fat (g)	Fat Breakdown (g) Sat	Mono	Poly
	FRUITS and FRUIT JUICES											
	Grapefruit juice:											
242	Fresh, white, raw	1 c	247	90	96	1	23	<1	<1	t	t	.1
243	Canned, unsweetened	1 c	247	90	94	1	22	<1	<1	t	t	.1
244	Sweetened	1 c	250	87	115	1	28	<1	<1	t	t	.1
	Frozen concentrate, unsweetened:											
245	Undiluted, 6-fl-oz can	¾ c	207	62	302	4	72	1	1	.1	.1	.2
246	Diluted with 3 cans water	1 c	247	89	101	1	24	<1	<1	t	t	.1
	Grapes, raw European (adherent skin):											
247	Thompson seedless	10 ea	50	81	35	<1	9	<1	<1	.1	t	.1
248	Tokay/Emperor, seeded types	10 ea	50	81	35	<1	9	<1	<1	.1	t	.1
	Grape juice:											
249	Bottled or canned	1 c	253	84	154	1	38	<1	<1	.1	t	.1
	Frozen concentrate, sweetened:											
250	Undiluted, 6-fl-oz can, vit C added	¾ c	216	54	387	1	96	1	1	.2	t	.2
251	Diluted with 3 cans water, vit C added	1 c	250	87	128	<1	32	<1	<1	.1	t	.1
1410	Low calorie	1 c	253	84	154	1	38	<1	<1	.1	t	.1
252	Kiwi fruit, raw, peeled (88 g with peel)	1 ea	76	83	46	1	11	3	<1	t	t	.2
253	Lemons, raw, without peel and seeds (about 4 per lb whole)	1 ea	58	89	17	1	5	2	<1	t	t	.1
	Lemon juice:											
254	Fresh:	1 c	244	91	61	1	21	1	0	0	0	0
255	Tablespoon	1 tbs	15	91	4	<1	1	<1	0	0	0	0
256	Canned or bottled, unsweetened:	1 c	244	92	51	1	16	1	1	.1	t	.2
257	Tablespoon	1 tbs	15	92	3	<1	1	<1	<1	t	t	t
258	Frozen, single strength, unsweetened:	1 c	244	92	54	1	16	1	1	.1	t	.2
2298	Tablespoon	1 tbs	15	92	3	<1	1	<1	<1	t	t	t
	Lime juice:											
260	Fresh:	1 c	246	90	66	1	22	1	<1	t	t	.1
261	Tablespoon	1 tbs	15	90	4	<1	1	<1	<1	t	t	t
262	Canned or bottled, unsweetened	1 c	246	92	52	1	16	1	1	.1	.1	.2
263	Mangos, raw, edible part (300 g w/skin & seeds)	1 ea	207	82	135	1	35	4	1	.1	.2	.1
	Melons, raw, without rind and contents:											
264	Cantaloupe, 5" diam (2⅓ lb whole with refuse), orange flesh	½ ea	276	90	97	2	23	2	1	.2	t	.3
265	Honeydew, 6½" diam (5¼ lb whole with refuse), slice = 1/10 melon	1 pce	160	90	56	1	15	1	<1	t	t	.1
266	Nectarines, raw, w/o pits, 2¼" diam	1 ea	136	86	67	1	16	2	1	.1	.2	.3
	Oranges, raw:											
267	Whole w/o peel and seeds, 2⅝" diam (180 g with peel and seeds)	1 ea	131	87	62	1	15	3	<1	t	t	t
268	Sections, without membranes	1 c	180	87	85	2	21	4	<1	t	t	t
	Orange juice:											
269	Fresh, all varieties	1 c	248	88	112	2	26	<1	<1	.1	.1	.1
270	Canned, unsweetened	1 c	249	89	105	1	24	<1	<1	t	.1	.1
271	Chilled	1 c	249	88	110	2	25	<1	1	.1	.1	.2
	Frozen concentrate:											
272	Undiluted (6-oz can)	¾ c	213	58	339	5	81	2	<1	.1	.1	.1
273	Diluted w/3 parts water by volume	1 c	249	88	112	2	27	<1	<1	t	t	t
1345	Orange juice, from dry crystals	1 c	248	88	114	0	29	0	0	0	0	0
274	Orange and grapefruit juice, canned	1 c	247	89	106	1	25	<1	<1	t	t	t
	Papayas, raw:											
275	½" slices	1 c	140	89	55	1	14	3	<1	.1	.1	t
276	Whole, 3½" diam by 5⅛" w/o seeds and skin (1 lb w/refuse)	1 ea	304	89	119	2	30	5	<1	.1	.1	.1
1031	Papaya nectar, canned	1 c	250	85	143	<1	36	1	<1	.1	.1	.1
	Peaches:											
277	Raw, whole, 2½" diam, peeled, pitted (about 4 per lb whole)	1 ea	98	88	42	1	11	2	<1	t	t	t
278	Raw, sliced	1 c	170	88	73	1	19	3	<1	t	.1	.1
	Canned, fruit and liquid:											
279	Heavy syrup pack:	1 c	262	79	194	1	52	3	<1	t	.1	.1
280	Half	1 ea	98	79	72	<1	19	1	<1	t	t	t

Chol (mg)	Calc (mg)	Iron (mg)	Magn (mg)	Pota (mg)	Sodi (mg)	Zinc (mg)	VT-A (RE)	Thia (mg)	VT-E (a-TE)	Ribo (mg)	Niac (mg)	V-B6 (mg)	Fola (μg)	VT-C (mg)
0	22	.49	30	400	2	.12	2	.1	.12	.05	.49	.11	25	94
0	17	.49	25	378	2	.22	2	.1	.12	.05	.57	.05	26	72
0	20	.9	25	405	5	.15	0	.1	.12	.06	.8	.05	26	67
0	56	1.01	79	1001	6	.37	6	.3	.37	.16	1.6	.32	26	248
0	20	.35	27	336	2	.12	2	.1	.12	.05	.54	.11	9	83
0	5	.13	3	92	1	.02	3	.05	.35	.03	.15	.05	2	5
0	5	.13	3	92	1	.02	3	.05	.35	.03	.15	.05	2	5
0	23	.61	25	334	8	.13	3	.07	0	.09	.66	.16	7	<1
0	28	.78	32	160	15	.28	6	.11	.38	.2	.93	.32	9	179
0	10	.25	10	52	5	.1	2	.04	.12	.06	.31	.1	3	60
0	23	.61	25	334	8	.13	3	.07	0	.09	.66	.16	7	<1
0	20	.31	23	252	4	.13	14	.01	.85	.04	.38	.07	29	74
0	15	.35	5	80	1	.03	2	.02	.14	.01	.06	.05	6	31
0	17	.07	15	303	2	.12	5	.07	.22	.02	.24	.12	31	112
0	1	<.01	1	19	<1	.01	<1	<.01	.01	<.01	.01	.01	2	7
0	27	.32	19	249	51	.15	5	.1	.22	.02	.48	.1	25	60
0	2	.02	1	15	3	.01	<1	.01	.01	<.01	.03	.01	2	4
0	19	.29	19	217	2	.12	2	.14	.22	.03	.33	.15	23	77
0	1	.02	1	13	<1	.01	<1	.01	.01	<.01	.02	.01	1	5
0	22	.07	15	268	2	.15	2	.05	.22	.02	.25	.11	20	72
0	1	<.01	1	16	<1	.01	<1	<.01	.01	<.01	.01	.01	1	4
0	29	.57	17	185	39	.15	5	.08	.17	.01	.4	.07	19	16
0	21	.27	19	323	4	.08	805	.12	2.32	.12	1.21	.28	29	57
0	30	.58	30	853	25	.44	889	.1	.41	.06	1.58	.32	47	116
0	10	.11	11	434	16	.11	6	.12	.24	.03	.96	.09	10	40
0	7	.2	11	288	0	.12	101	.02	1.21	.06	1.35	.03	5	7
0	52	.13	13	237	0	.09	27	.11	.31	.05	.37	.08	40	70
0	72	.18	18	326	0	.13	38	.16	.43	.07	.51	.11	54	96
0	27	.5	27	496	2	.12	50	.22	.22	.07	.99	.1	75	124
0	20	1.1	27	436	5	.17	45	.15	.22	.07	.78	.22	45	86
0	25	.42	27	473	2	.1	20	.28	.47	.05	.7	.13	45	82
0	68	.75	72	1435	6	.38	60	.6	.68	.14	1.53	.33	330	294
0	22	.25	25	473	2	.12	20	.2	.47	.04	.5	.11	109	97
0	62	.2	2	50	12	.1	551	<.01	0	.04	0	0	143	121
0	20	1.14	25	390	7	.17	30	.14	.17	.07	.83	.06	35	72
0	34	.14	14	360	4	.1	39	.04	1.57	.04	.47	.03	53	86
0	73	.3	30	781	9	.21	85	.08	3.4	.1	1.03	.06	116	188
0	25	.85	7	77	12	.37	27	.01	.05	.01	.37	.02	5	7
0	5	.11	7	193	0	.14	53	.02	.69	.04	.97	.02	3	6
0	8	.19	12	335	0	.24	92	.03	1.19	.07	1.68	.03	6	11
0	8	.71	13	241	16	.24	86	.03	2.33	.06	1.61	.05	8	7
0	3	.26	5	90	6	.09	32	.01	.87	.02	.6	.02	3	3

Table A-1

Food Composition (Computer code number is for West Diet Analysis program) (For purposes of calculations, use "0" for t, <1, <.1, <.01, etc.)

Computer Code Number	Food Description	Measure	Wt (g)	H_2O (%)	Ener (cal)	Prot (g)	Carb (g)	Dietary Fiber (g)	Fat (g)	Fat Breakdown (g) Sat	Mono	Poly
	FRUITS and FRUIT JUICES—Continued											
281	Juice pack:	1 c	248	87	109	2	29	3	<1	t	t	t
282	Half	1 ea	98	87	43	1	11	1	<1	t	t	t
283	Dried, uncooked	10 ea	130	32	311	5	80	11	1	.1	.4	.5
284	Dried, cooked, fruit and liquid	1 c	258	78	199	3	51	7	1	.1	.2	.3
	Frozen, slice, sweetened:											
285	10-oz package,vitamin C added	1 ea	284	75	267	2	68	5	<1	t	.1	.2
286	Cup, thawed measure, vitamin C added	1 c	250	75	235	2	60	4	<1	t	.1	.2
1032	Peach nectar, canned	1 c	249	86	134	1	35	1	<1	t	t	t
	Pears:											
	Fresh, with skin, cored:											
287	Bartlett, 2½" diam (about 2½ per lb)	1 ea	166	84	98	1	25	4	1	t	.1	.2
288	Bosc, 2⅕" diam (about 3 per lb)	1 ea	139	84	82	1	21	3	1	t	.1	.1
289	D'Anjou, 3" diam (about 2 per lb)	1 ea	209	84	123	1	32	5	1	t	.2	.2
	Canned, fruit and liquid:											
290	Heavy syrup pack:	1 c	266	80	197	1	51	4	<1	t	.1	.1
291	Half	1 ea	76	80	56	<1	15	1	<1	t	t	t
292	Juice pack:	1 c	248	86	124	1	32	4	<1	t	t	t
293	Half	1 ea	76	86	38	<1	10	1	<1	t	t	t
294	Dried halves	10 ea	175	27	459	3	122	13	1	.1	.2	.3
1033	Pear nectar, canned	1 c	250	84	150	<1	39	1	<1	t	t	t
	Pineapple:											
295	Fresh chunks, diced	1 c	155	86	76	1	19	2	1	t	.1	.2
	Canned, fruit and liquid:											
	Heavy syrup pack:											
296	Crushed, chunks, tidbits	½ c	127	79	99	<1	26	1	<1	t	t	.1
297	Slices	1 ea	49	79	38	<1	10	<1	<1	t	t	t
298	Juice pack, crushed, chunks, tidbits	1 c	250	83	150	1	39	2	<1	t	t	.1
299	Juice pack, slices	1 ea	47	83	28	<1	7	<1	<1	t	t	t
300	Pineapple juice, canned, unsweetened	1 c	250	85	140	1	34	<1	<1	t	t	.1
	Plantains, yellow flesh, without peel:											
301	Raw slices (whole=179 g w/o peel)	1 c	148	65	181	2	47	3	1	.2	t	.1
302	Cooked, boiled, sliced	1 c	154	67	179	1	48	4	<1	.1	t	.1
	Plums:											
303	Fresh, medium, 2⅛" diam	1 ea	66	85	36	1	9	1	<1	t	.3	.1
304	Fresh, small, 1½" diam	1 ea	28	85	15	<1	4	<1	<1	t	.1	t
	Canned, purple, with liquid:											
305	Heavy syrup pack:	1 c	258	76	230	1	60	3	<1	t	.2	.1
306	Plums	3 ea	138	76	123	<1	32	1	<1	t	.1	t
307	Juice pack:	1 c	252	84	146	1	38	3	<1	t	t	t
308	Plums	3 ea	138	84	80	1	21	1	<1	t	t	t
1698	Pomegranate, fresh	1 ea	154	81	105	1	26	1	<1	.1	.1	.1
	Prunes, dried, pitted:											
309	Uncooked (10 = 97 g w/pits, 84 g w/o pits)	10 ea	84	32	201	2	53	6	<1	t	.3	.1
310	Cooked, unsweetened, fruit & liq (250 g w/pits)	1 c	248	70	265	3	70	16	1	t	.4	.1
311	Prune juice, bottled or canned	1 c	256	81	182	2	45	3	<1	t	.1	t
	Raisins, seedless:											
312	Cup, not pressed down	1 c	145	15	435	5	115	6	1	.2	t	.2
313	One packet, ½ oz	½ oz	14	15	42	<1	11	1	<1	t	t	t
	Raspberries:											
314	Fresh	1 c	123	87	60	1	14	8	1	t	.1	.4
315	Frozen, sweetened	10 oz	284	73	293	2	74	12	<1	t	t	.3
316	Cup, thawed measure	1 c	250	73	258	2	65	11	<1	t	t	.2
317	Rhubarb, cooked, added sugar	1 c	240	68	278	1	75	5	<1	t	t	.1
	Strawberries:											
318	Fresh, whole, capped	1 c	144	92	43	1	10	3	1	t	.1	.3
	Frozen, sliced, sweetened:											
319	10-oz container	10 oz	284	73	273	2	74	5	<1	t	.1	.2
320	Cup, thawed measure	1 c	255	73	245	1	66	5	<1	t	t	.2
	Tangerines, without peel and seeds:											
321	Fresh (2⅜" whole) 116 g w/refuse	1 ea	84	88	37	1	9	2	<1	t	t	t
322	Canned, light syrup, fruit and liquid	1 c	252	83	154	1	41	2	<1	t	t	t

Chol (mg)	Calc (mg)	Iron (mg)	Magn (mg)	Pota (mg)	Sodi (mg)	Zinc (mg)	VT-A (RE)	Thia (mg)	VT-E (a-TE)	Ribo (mg)	Niac (mg)	V-B6 (mg)	Fola (µg)	VT-C (mg)
0	15	.67	17	317	10	.27	94	.02	3.72	.04	1.44	.05	8	9
0	6	.26	7	125	4	.11	37	.01	1.47	.02	.57	.02	3	4
0	36	5.28	55	1294	9	.74	281	<.01	0	.28	5.69	.09	<1	6
0	23	3.38	33	826	5	.46	52	.01	0	.05	3.92	.1	<1	10
0	9	1.05	14	369	17	.14	79	.04	2.53	.1	1.85	.05	9	268
0	7	.92	12	325	15	.12	70	.03	2.23	.09	1.63	.04	8	236
0	12	.47	10	100	17	.2	65	.01	.2	.03	.72	.02	3	13
0	18	.41	10	208	0	.2	3	.03	.83	.07	.17	.03	12	7
0	15	.35	8	174	0	.17	3	.03	.69	.06	.14	.02	10	6
0	23	.52	12	261	0	.25	4	.04	1.05	.08	.21	.04	15	8
0	13	.58	11	173	13	.21	0	.03	1.33	.06	.64	.04	3	3
0	4	.17	3	49	4	.06	0	.01	.38	.02	.18	.01	1	1
0	22	.72	17	238	10	.22	2	.03	1.24	.03	.5	.03	3	4
0	7	.22	5	73	3	.07	1	.01	.38	.01	.15	.01	1	1
0	59	3.68	58	933	10	.68	1	.01	0	.25	2.4	.13	0	12
0	12	.65	7	32	10	.17	<1	<.01	.25	.03	.32	.03	3	3
0	11	.57	22	175	2	.12	3	.14	.15	.06	.65	.13	16	24
0	18	.48	20	132	1	.15	1	.11	.13	.03	.36	.09	6	9
0	7	.19	8	51	<1	.06	<1	.04	.05	.01	.14	.04	2	4
0	35	.7	35	305	2	.25	10	.24	.25	.05	.71	.18	12	24
0	7	.13	7	57	<1	.05	2	.04	.05	.01	.13	.03	2	4
0	42	.65	32	335	2	.27	1	.14	.05	.05	.64	.24	58	27
0	4	.89	55	739	6	.21	167	.08	.4	.08	1.02	.44	33	27
0	3	.89	49	716	8	.2	140	.07	.22	.08	1.16	.37	40	17
0	3	.07	5	114	0	.07	21	.03	.4	.06	.33	.05	1	6
0	1	.03	2	48	0	.03	9	.01	.17	.03	.14	.02	1	3
0	23	2.17	13	235	49	.18	67	.04	1.81	.1	.75	.07	6	1
0	12	1.16	7	126	26	.1	36	.02	.97	.05	.4	.04	3	1
0	25	.86	20	388	3	.28	255	.06	1.76	.15	1.19	.07	7	7
0	14	.47	11	213	1	.15	139	.03	.97	.08	.65	.04	4	4
0	5	.46	5	399	5	.18	0	.05	.85	.05	.46	.16	9	9
0	43	2.08	38	626	3	.44	167	.07	1.22	.14	1.65	.22	3	3
0	57	2.75	50	828	5	.59	77	.06	<.01	.25	1.79	.54	<1	7
0	31	3.02	36	707	10	.54	1	.04	.03	.18	2.01	.56	1	10
0	71	3.02	48	1088	17	.39	1	.23	1.02	.13	1.19	.36	5	5
0	7	.29	5	105	2	.04	<1	.02	.1	.01	.11	.03	<1	<1
0	27	.7	22	187	0	.57	16	.04	.55	.11	1.11	.07	32	31
0	43	1.85	37	324	3	.51	17	.05	1.28	.13	.65	.1	74	47
0	37	1.63	32	285	2	.45	15	.05	1.13	.11	.57	.08	65	41
0	348	.5	29	230	2	.19	17	.04	.48	.05	.48	.05	13	8
0	20	.55	14	239	1	.19	4	.03	.2	.09	.33	.08	25	82
0	31	1.68	20	278	9	.17	6	.04	.4	.14	1.14	.08	42	118
0	28	1.5	18	250	8	.15	5	.04	.36	.13	1.02	.08	38	106
0	12	.08	10	132	1	.2	77	.09	.2	.02	.13	.06	17	26
0	18	.93	20	197	15	.6	212	.13	.86	.11	1.12	.11	12	50

Table A-1

Food Composition (Computer code number is for West Diet Analysis program) (For purposes of calculations, use "0" for t, <1, <.1, <.01, etc.)

Computer Code Number	Food Description	Measure	Wt (g)	H_2O (%)	Ener (cal)	Prot (g)	Carb (g)	Dietary Fiber (g)	Fat (g)	Fat Breakdown (g) Sat	Mono	Poly
	FRUITS and FRUIT JUICES—Continued											
323	Tangerine juice, canned, sweetened	1 c	249	87	125	1	30	<1	<1	t	t	.1
	Watermelon, raw, without rind and seeds:											
324	Piece, 1/16 wedge	1 pce	286	91	91	2	20	1	1	.1	.3	.4
325	Diced	1 c	152	91	49	1	11	1	1	.1	.2	.2
	BAKED GOODS: BREADS, CAKES, COOKIES, CRACKERS, PIES											
326	Bagels, plain, enriched, 3½" diam.	1 ea	71	33	195	7	38	2	1	.2	.1	.5
1663	Bagel, oat bran	1 ea	71	33	181	8	38	3	1	.1	.2	.3
	Biscuits:											
327	From home recipe	1 ea	60	29	212	4	27	1	10	2.6	4.2	2.5
328	From mix	1 ea	57	29	191	4	28	1	7	1.6	2.4	2.5
329	From refrigerated dough	1 ea	74	27	276	4	34	1	13	8.7	3.4	.5
330	Bread crumbs, dry, grated (see # 364, 365 for soft crumbs)	1 c	108	6	427	13	78	3	6	1.4	2.3	1.7
2087	Bread sticks, brown & serve	1 ea	57	34	150	7	28	1	1	.5	.5	.5
	Breads:											
331	Boston brown, canned, 3¼" slice	1 pce	45	47	88	2	19	2	1	.1	.1	.3
332	Cracked wheat (¼ cracked-wheat & ¾ enr wheat flour): 1-lb loaf	1 ea	454	36	1180	39	225	25	18	4.2	8.6	3.1
333	Slice (18 per loaf)	1 pce	25	36	65	2	12	1	1	.2	.5	.2
334	Slice, toasted	1 pce	23	30	65	2	12	1	1	.2	.5	.2
335	French/Vienna, enriched: 1-lb loaf	1 ea	454	34	1243	40	236	14	14	2.9	5.5	3.1
337	Slice, 4¾ x 4 x ½"	1 pce	25	34	68	2	13	1	1	.2	.3	.2
336	French, slice, 5 x 2½"	1 pce	25	34	68	2	13	1	1	.2	.3	.2
	French toast: see Mixed Dishes, and Fast Foods, #691											
2083	Honey wheatberry	1 pce	38	38	100	3	18	2	1	0	.5	
338	Italian, enriched: 1-lb loaf	1 ea	454	36	1230	40	227	12	16	3.9	3.7	6.3
339	Slice, 4½ x 3¼ x ¾"	1 pce	30	36	81	3	15	1	1	.3	.2	.4
340	Mixed grain, enriched: 1-lb loaf	1 ea	454	38	1135	45	211	29	17	3.7	6.9	4.2
341	Slice (18 per loaf)	1 pce	26	38	65	3	12	2	1	.2	.4	.2
342	Slice, toasted	1 pce	24	32	65	3	12	2	1	.2	.4	.2
343	Oatmeal, enriched: 1-lb loaf	1 ea	454	37	1221	38	220	18	20	3.2	7.2	7.7
344	Slice (18 per loaf)	1 pce	27	37	73	2	13	1	1	.2	.4	.5
345	Slice, toasted	1 pce	25	31	73	2	13	1	1	.2	.4	.5
346	Pita pocket bread, enr, 6½" round	1 ea	60	32	165	5	33	1	1	.1	.1	.3
	Pumpernickel (⅔ rye & ⅓ enr wheat flr):											
347	1-lb loaf	1 ea	454	38	1135	40	216	29	14	2	4.2	5.6
348	Slice, 5 x 4 x ⅜"	1 pce	26	38	65	2	12	2	1	.1	.2	.3
349	Slice, toasted	1 pce	29	32	80	3	15	2	1	.1	.3	.4
350	Raisin, enriched: 1-lb loaf	1 ea	454	34	1243	36	237	19	20	4.9	10.5	3.1
351	Slice (18 per loaf)	1 pce	26	34	71	2	14	1	1	.3	.6	.2
352	Slice, toasted	1 pce	24	28	71	2	14	1	1	.3	.6	.2
353	Rye, light (⅓ rye & ⅔ enr wheat flr): 1-lb loaf	1 ea	454	37	1175	39	219	26	15	2.9	6	3.6
354	Slice, 4¾ x 3¾ x 7/16"	1 pce	32	37	83	3	15	2	1	.2	.4	.3
355	Slice, toasted	1 pce	24	31	68	2	13	2	1	.2	.3	.2
356	Wheat (enr wheat & whole-wheat flour): 1-lb loaf	1 ea	454	37	1160	43	213	25	19	3.9	7.3	4.5
357	Slice (18 per loaf)	1 pce	25	37	65	2	12	1	1	.2	.4	.2
358	Slice, toasted	1 pce	23	32	65	2	12	1	1	.2	.4	.2
359	White, enriched: 1-lb loaf	1 ea	454	35	1293	36	225	9	26	5.4	5.9	12.6
360	Slice	1 pce	42	35	120	3	21	1	2	.5	.5	1.2
361	Slice, toasted	1 pce	38	29	119	3	21	1	2	.5	.5	1.2
366	Whole-wheat: 1-lb loaf	1 ea	454	38	1116	44	209	31	19	4.2	7.6	4.5
367	Slice (16 per loaf)	1 pce	28	38	69	3	13	2	1	.3	.5	.3
368	Slice, toasted	1 pce	25	30	69	3	13	2	1	.3	.5	.3
	Bread stuffing, prepared from mix:											
369	Dry type	1 c	200	65	356	6	43	6	17	3.5	7.6	5.2
370	Moist type, with egg and margarine	1 c	232	65	390	9	51	5	17	3.4	7.4	4.9

Chol (mg)	Calc (mg)	Iron (mg)	Magn (mg)	Pota (mg)	Sodi (mg)	Zinc (mg)	VT-A (RE)	Thia (mg)	VT-E (a-TE)	Ribo (mg)	Niac (mg)	V-B6 (mg)	Fola (μg)	VT-C (mg)
0	45	.5	20	443	2	.07	105	.15	.22	.05	.25	.08	11	55
0	23	.49	31	332	6	.2	106	.23	.43	.06	.57	.41	6	27
0	12	.26	17	176	3	.11	56	.12	.23	.03	.3	.22	3	15
0	52	2.53	21	72	379	.62	0	.38	.02	.22	3.24	.04	62	0
0	9	2.19	40	145	360	1.48	<1	.23	.17	.24	2.1	.14	57	<1
2	141	1.74	11	73	348	.32	14	.21	1.45	.19	1.77	.02	37	<1
2	105	1.17	14	107	544	.35	15	.2	.23	.2	1.72	.04	3	<1
5	89	1.64	9	87	584	.29	24	.27	.44	.18	1.63	.03	6	0
0	245	6.61	50	239	931	1.32	<1	.83	.95	.47	7.4	.11	118	0
0	60	2.7			290		0	.22		.1	1.6			0
<1	31	.94	28	143	284	.22	5	.01	.13	.05	.5	.04	5	0
														0
0	195	12.8	236	804	2442	5.63	0	1.63	2.56	1.09	16.7	1.38	277	
0	11	.7	13	44	135	.31	0	.09	.14	.06	.92	.08	15	0
0	11	.7	13	44	135	.31	0	.07	.14	.05	.83	.07	7	0
0	341	11.5	123	513	2764	3.95	0	2.36	1.07	1.49	21.6	.19	431	0
0	19	.63	7	28	152	.22	0	.13	.06	.08	1.19	.01	24	0
0	19	.63	7	28	152	.22	0	.13	.06	.08	1.19	.01	24	0
0	20	.72			200		0	.12	.24	.07	.8			0
0	354	13.3	123	499	2651	3.9	0	2.15	1.26	1.33	19.9	.22	431	0
0	23	.88	8	33	175	.26	0	.14	.08	.09	1.31	.01	28	0
0	413	15.8	241	926	2210	5.77	0	1.85	2.79	1.55	19.8	1.51	363	1
0	24	.9	14	53	127	.33	0	.11	.16	.09	1.14	.09	21	<1
0	24	.9	14	53	127	.33	0	.08	.16	.08	1.02	.08	16	<1
0	300	12.3	168	645	2719	4.63	9	1.81	1.56	1.09	14.3	.31	281	2
0	18	.73	10	38	162	.27	1	.11	.09	.06	.85	.02	17	<1
0	18	.73	10	38	163	.28	<1	.09	.09	.06	.77	.02	13	<1
0	52	1.57	16	72	322	.5	0	.36	.02	.2	2.78	.02	57	0
0	309	13	245	944	3046	6.72	0	1.48	2.3	1.38	14	.57	363	0
0	18	.75	14	54	174	.38	0	.08	.13	.08	.8	.03	21	0
0	21	.91	17	66	214	.47	0	.08	.17	.09	.89	.04	20	0
0	300	13.2	118	1030	1770	3.27	1	1.54	3.44	1.81	15.8	.31	395	2
0	17	.75	7	59	101	.19	<1	.09	.2	.1	.9	.02	23	<1
0	17	.76	7	59	102	.19	<1	.07	.2	.09	.81	.02	18	<1
0	331	12.8	182	754	2996	5.18	2	1.97	2.51	1.52	17.3	.34	390	1
0	23	.91	13	53	211	.36	<1	.14	.18	.11	1.22	.02	27	<1
0	19	.74	10	44	174	.3	<1	.09	.15	.08	.9	.02	17	<1
0	572	15.8	209	627	2447	4.77	0	2.09	3	1.45	20.5	.49	204	0
0	26	.83	11	50	133	.26	0	.1	.14	.07	1.03	.02	19	0
0	26	.83	11	50	132	.26	0	.08	.14	.06	.93	.02	15	0
14	259	13.5	86	663	1629	2.91	100	1.84	4.95	1.74	16.3	.23	413	1
1	24	1.25	8	61	151	.27	9	.17	.46	.16	1.51	.02	38	<1
1	24	1.24	8	61	150	.27	8	.13	.46	.14	1.35	.02	12	<1
0	327	15	390	1144	2392	8.81	0	1.59	4.72	.93	17.4	.81	227	0
0	20	.92	24	71	148	.54	0	.1	.29	.06	1.08	.05	14	0
0	20	.93	24	71	148	.54	0	.08	.23	.05	.97	.04	9	0
0	64	2.18	24	148	1086	.56	162	.27	2.8	.21	2.96	.08	202	0
0	148	3.8	35	304	1069	.74	160	.39	2.78	.33	3.69	.12	39	4

Table A-1

Food Composition (Computer code number is for West Diet Analysis program) (For purposes of calculations, use "0" for t, <1, <.1, <.01, etc.)

Computer Code Number	Food Description	Measure	Wt (g)	H_2O (%)	Ener (cal)	Prot (g)	Carb (g)	Dietary Fiber (g)	Fat (g)	Fat Breakdown (g) Sat	Mono	Poly
	BAKED GOODS: BREADS, CAKES, COOKIES, CRACKERS, PIES—Continued											
	Cakes, prepared from mixes using enriched flour and veg shortening, w/frostings made from margarine:											
	Angel food:											
371	Whole cake, 9 ¾" diam tube	1 ea	340	33	877	20	197	5	3	.4	.2	1.2
372	Piece, 1/12 of cake	1 pce	28	33	72	2	16	<1	<1	t	t	.1
373	Boston cream pie, ⅛ of cake	1 pce	123	45	310	3	53	2	10	3.1	5.4	1.2
	Coffee cake:											
374	Whole cake, 7¾ x 5⅝ x 1¼"	1 ea	336	30	1068	18	177	4	32	6.3	13	10.7
375	Piece, ⅙ of cake	1 pce	56	30	178	3	30	1	5	1	2.2	1.8
	Devil's food, chocolate frosting:											
376	Whole cake, 2 layer, 8 or 9" diam	1 ea	1021	23	3747	42	557	29	167	47.9	91.9	19.5
377	Piece, 1/16 of cake	1 pce	64	23	235	3	35	2	10	3	5.8	1.2
378	Cupcake, 2½" diam	1 ea	42	23	154	2	23	1	7	2	3.8	.8
	Gingerbread:											
379	Whole cake, 8" square	1 ea	603	33	1863	24	306	7	61	15.8	34	8.1
380	Piece, 1/9 of cake	1 pce	67	33	207	3	34	1	7	1.8	3.8	.9
	Yellow, chocolate frosting, 2 layer:											
381	Whole cake, 8 or 9" in diam	1 ea	1024	22	3880	39	567	18	178	49	99	21.4
382	Piece, 1/16 of cake	1 pce	64	22	243	2	35	1	11	3.1	6.2	1.3
	Cakes from home recipes w/enr flour:											
	Carrot cake, made with veg oil, cream cheese frosting:											
383	Whole, 9 x 13" cake	1 ea	1776	21	7743	82	838	21	469	86.8	116	242
384	Piece, 1/16 of cake, 2¼ x 3¼" slice	1 pce	111	21	484	5	52	1	29	5.4	7.2	15.1
	Fruitcake, dark:											
385	Whole cake, 7½"diam tube, 2¼"high	1 ea	1376	25	4458	40	848	51	125	15.4	57.4	44.6
386	Piece, 1/32 of cake, ⅔" arc	1 pce	43	25	139	1	26	2	4	.5	1.8	1.4
	Sheet, plain, made w/veg shortening, no frosting:											
387	Whole cake, 9" square	1 ea	774	24	2817	35	433	3	108	29.9	51.5	25.5
388	Piece, 1/9 of cake	1 pce	86	24	313	4	48	<1	12	3.3	5.7	2.8
	Sheet, plain, made w/margarine, uncooked white frosting:											
389	Whole cake, 9" square	1 ea	576	22	2148	20	339	2	83	13.8	35.5	29.5
390	Piece, 1/9 of cake	1 pce	64	22	239	2	38	<1	9	1.5	3.9	3.3
	Cakes, commerical:											
	Cheesecake:											
401	Whole cake, 9" diam	1 ea	960	46	3081	53	245	4	216	111	74.4	13.2
402	Piece, 1/12 of cake	1 pce	80	46	257	4	20	<1	18	9.2	6.2	1.1
	Pound cake:											
393	Loaf, 8½ x 3½ x 3"	1 ea	340	25	1319	19	166	2	68	38.1	19	3.7
394	Slice, 1/17 of loaf, 2" slice	1 pce	28	25	109	2	14	<1	6	3.1	1.6	.3
	Snack cakes:											
395	Chocolate w/creme filling,Ding Dong	1 ea	50	20	188	2	30	<1	7	1.6	2.7	2.1
396	Sponge cake w/creme filling,Twinkie	1 ea	43	20	157	1	27	<1	5	1.2	1.9	1.5
1677	Sponge cake, 1/12 of 12" cake	1 pce	38	30	110	2	23	<1	1	.3	.4	.2
	White, white frosting, 2 layer:											
397	Whole cake, 8 or 9" diam	1 ea	1136	20	4260	37	716	11	153	68.2	60.1	15.4
398	Piece, 1/16 of cake	1 pce	71	20	266	2	45	1	10	4.3	3.8	1
	Yellow, chocolate frosting, 2 layer:											
399	Whole cake, 8 or 9" in diam	1 ea	1024	22	3880	39	567	18	178	49	99	21.4
400	Piece, 1/16 of cake	1 pce	64	22	243	2	35	1	11	3.1	6.2	1.3
1332	Bagel chips	5 pce	70	3	298	6	52	6	7	1.2	2	3.4
2225	Bagel chips, onion garlic, toasted	1 oz	28		193	5	31	3	8	1.7	5.2	0
1035	Cheese puffs/Cheetos	1 c	20	1	111	2	11	<1	7	1.3	4.1	1
	Cookies made with enriched flour:											
	Brownies with nuts:											
403	Commercial w/frosting, 1½ x 1¾ x ⅞"	1 ea	61	14	247	3	39	1	10	2.6	5.1	1.6
1902	Fat free fudge, Entenmann's	1 pce	40	24	110	2	27	1	0	0	0	0

Chol (mg)	Calc (mg)	Iron (mg)	Magn (mg)	Pota (mg)	Sodi (mg)	Zinc (mg)	VT-A (RE)	Thia (mg)	VT-E (a-TE)	Ribo (mg)	Niac (mg)	V-B6 (mg)	Fola (µg)	VT-C (mg)
0	476	1.77	41	316	2546	.24	0	.35	.34	1.67	3	.1	119	0
0	39	.15	3	26	210	.02	0	.03	.03	.14	.25	.01	10	0
45	28	.47	7	48	177	.2	28	.5	1.3	.33	.23	.03	18	<1
165	457	4.8	60	376	1414	1.51	134	.56	5.58	.59	5.11	.17	228	1
27	76	.8	10	63	236	.25	22	.09	.93	.1	.85	.03	38	<1
470	439	22.5	347	2042	3410	7.04	286	.28	17.3	1.36	5.89	.32	174	1
29	27	1.41	22	128	214	.44	18	.02	1.08	.08	.37	.02	11	<1
19	18	.92	14	84	140	.29	12	.01	.71	.06	.24	.01	7	<1
211	416	20	96	1453	2761	2.47	96	1.14	8.26	1.12	9.41	.23	60	1
23	46	2.22	11	161	307	.27	11	.13	.92	.12	1.05	.02	7	<1
563	379	21.3	307	1822	3450	6.35	276	1.23	27.6	1.61	12.8	.3	225	1
35	24	1.33	19	114	216	.4	17	.08	1.73	.1	.8	.02	14	<1
959	444	22.2	320	1989	4368	8.7	6819	2.42	74.9	2.77	17.9	1.35	213	19
60	28	1.39	20	124	273	.54	426	.15	4.68	.17	1.12	.08	13	1
69	454	28.5	220	2105	3715	3.72	261	.69	42.9	1.36	10.9	.63	261	5
2	14	.89	7	66	116	.12	8	.02	1.34	.04	.34	.02	8	<1
503	495	11.7	108	611	2322	2.74	372	1.24	11	1.39	10.1	.26	54	2
56	55	1.3	12	68	258	.3	41	.14	1.22	.15	1.12	.03	6	<1
323	357	6.16	35	305	1981	1.44	109	.58	10.9	.4	2.88	.2	156	1
36	40	.68	4	34	220	.16	12	.06	1.22	.04	.32	.02	17	<1
528	490	6.05	106	864	1987	4.9	1545	.27	10.1	1.85	1.87	.5	173	6
44	41	.5	9	72	166	.41	129	.02	.84	.15	.16	.04	14	<1
751	119	4.69	37	405	1353	1.56	530	.47	2.24	.78	4.45	.12	139	<1
62	10	.39	3	33	111	.13	44	.04	.18	.06	.37	.01	11	<1
8	36	1.68	20	61	213	.28	2	.11	1.01	.15	1.22	.01	14	<1
7	19	.55	3	39	157	.13	2	.07	.83	.06	.52	.01	12	<1
39	27	1.03	4	38	93	.19	17	.09	.17	.1	.73	.02	15	0
91	545	9.09	60	659	2658	1.76	368	1.14	20.4	1.48	10.2	.16	64	1
6	34	.57	4	41	166	.11	23	.07	1.28	.09	.64	.01	4	<1
563	379	21.3	307	1822	3450	6.35	276	1.23	27.6	1.61	12.8	.3	225	1
35	24	1.33	19	114	216	.4	17	.08	1.73	.1	.8	.02	14	<1
0	9	1.38	41	167	419	.9	0	.13	.46	.12	1.57	.19	58	0
0	0	2.52			490		0	.39	<.01	.24	3.5			0
1	12	.47	4	33	210	.08	7	.05	1.02	.07	.65	.03	24	<1
10	18	1.37	19	91	190	.44	12	.16	1.3	.13	1.05	.02	13	<1
0	0	1.08		90	140		0		.01					0

Table A-1

Food Composition (Computer code number is for West Diet Analysis program) (For purposes of calculations, use "0" for t, <1, <.1, <.01, etc.)

Computer Code Number	Food Description	Measure	Wt (g)	H_2O (%)	Ener (cal)	Prot (g)	Carb (g)	Dietary Fiber (g)	Fat (g)	Fat Breakdown (g) Sat	Mono	Poly
	BAKED GOODS: BREADS, CAKES, COOKIES, CRACKERS, PIES—Continued											
	Chocolate chip cookies:											
405	Commercial, 2¼" diam	4 ea	60	12	275	2	35	2	15	4.5	7.8	1.6
406	Home recipe, 2¼" diam	4 ea	64	6	312	4	37	2	18	5.2	6.7	5.4
407	From refrigerated dough, 2¼" diam	4 ea	64	13	284	3	39	1	13	4.5	6.5	1.3
408	Fig bars	4 ea	64	16	223	2	45	3	5	.9	2.6	.8
2052	Fruit bar, no fat	1 ea	28		90	2	21	0	0	0	0	0
2162	Fudge, fat free, Snackwell	1 ea	16	14	53	1	12	<1	<1	.1	.1	t
409	Oatmeal raisin, 2⅝" diam	4 ea	60	6	261	4	41	2	10	1.9	4.1	3
410	Peanut butter, home recipe, 2⅝"diam	4 ea	80	6	380	7	47	2	19	3.5	8.7	5.8
411	Sandwich-type, all	4 ea	40	2	189	2	28	1	8	1.7	4.7	1.1
412	Shortbread, commercial, small	4 ea	32	4	161	2	21	1	8	2	4.3	1
413	Shortbread, home recipe, large	2 ea	22	3	120	1	12	<1	7	4.5	2.1	.3
414	Sugar from refrigerated dough, 2" diam	4 ea	48	5	232	2	31	<1	11	2.8	6.2	1.4
1874	Vanilla sandwich, Snackwell's	2 ea	26	4	109	1	21	1	2	.5	.8	.2
415	Vanilla wafers	10 ea	40	5	176	2	29	1	6	1.4	2.4	1.5
416	Corn chips	1 c	26	1	140	2	15	1	9	1.2	2.5	4.3
	Crackers (enriched):											
417	Cheese	10 ea	10	3	50	1	6	<1	3	.9	.9	.5
418	Cheese with peanut butter	4 ea	28	4	135	4	16	1	6	1.4	3.4	1.2
	Fat free, enriched:											
2161	Cracked pepper, Snackwell	1 ea	15	2	60	2	12	<1	<1	.1	t	.1
2159	Wheat, Snackwell	7 ea	15	1	60	2	12	1	<1	.1	.1	.1
2075	Whole wheat, herb seasoned	5 ea	14	5	50	2	11	2	0	0	0	0
2077	Whole wheat, onion	5 ea	14	4	50	2	11	2	0	0	0	0
419	Graham, enriched	2 ea	14	4	59	1	11	<1	1	.4	.7	.2
420	Melba toast, plain, enriched	1 pce	5	5	19	1	4	<1	<1	t	t	.1
1514	Rice cakes, unsalted, enriched	2 ea	18	6	70	1	15	1	<1	.1	.2	.2
421	Rye wafer, whole grain	2 ea	22	5	73	2	18	5	<1	t	t	.1
422	Saltine-enriched	4 ea	12	4	52	1	9	<1	1	.3	.8	.2
1971	Saltine, unsalted tops, enriched	2 ea	6		25	1	4	0	<1	0	0	0
423	Snack-type, round like Ritz, enriched	3 ea	9	3	45	1	5	<1	2	.4	1	.7
424	Wheat, thin, enriched	4 ea	8	3	38	1	5	<1	2	.7	.8	.2
425	Whole-wheat wafers	2 ea	8	3	35	1	5	1	1	.2	.8	.2
426	Croissants, 4½ x 4 x 1¾"	1 ea	57	23	231	5	26	1	12	6.7	3.2	.7
1699	Croutons, seasoned	½ c	20	4	93	2	13	1	4	1	1.9	.5
	Danish pastry:											
427	Packaged ring, plain, 12 oz	1 ea	340	21	1349	19	181	1	65	13.5	40.9	6.4
428	Round piece, plain, 4¼" diam, 1" high	1 ea	88	21	349	5	47	<1	17	3.5	10.6	1.6
429	Ounce, plain	1 oz	28	21	111	2	15	<1	5	1.1	3.4	.5
430	Round piece with fruit	1 ea	94	29	335	5	45		16	3.3	10.1	1.6
	Desserts, 3 x 3" piece:											
1348	Apple crisp	1 pce	78	61	127	1	25	1	3	.6	1.2	.9
1353	Apple cobbler	1 pce	104	57	199	2	35	2	6	1.2	2.8	2
1349	Cherry crisp	1 pce	138	77	146	2	24	1	5	.9	2.5	1.8
1352	Cherry cobbler	1 pce	129	66	198	2	34	1	6	1.2	2.8	1.9
1350	Peach crisp	1 pce	139	75	155	2	27	2	5	.9	2.5	1.7
1351	Peach cobbler	1 pce	130	64	204	2	36	2	6	1.2	2.8	1.9
	Doughnuts:											
431	Cake type, plain, 3¼" diam	1 ea	47	21	198	2	23	1	11	1.8	4.5	3.8
432	Yeast-leavened, glazed, 3¾" diam	1 ea	60	25	242	4	27	1	14	3.5	7.8	1.7
	English muffins:											
433	Plain, enriched	1 ea	57	42	134	4	26	2	1	.2	.2	.5
434	Toasted	1 ea	52	37	133	4	26	2	1	.1	.2	.5
1504	Whole wheat	1 ea	66	46	134	6	27	4	1	.2	.3	.6
1414	Granola bar, soft	1 ea	28	6	124	2	19	1	5	2	1.1	1.5
1415	Granola bar, hard	1 ea	25	4	118	3	16	1	5	.6	1.1	3
1985	Granola bar, fat free, all flavors	1 ea	42	10	140	2	35	3	0	0	0	0
	Muffins, 2½" diam, 1½" high:											
	From home recipe:											
435	Blueberry	1 ea	57	39	165	4	23	1	6	1.4	1.6	3.1
436	Bran, wheat	1 ea	57	35	164	4	24	4	7	1.5	1.8	3.6
437	Cornmeal	1 ea	57	32	183	4	25	2	7	1.6	1.8	3.5

Chol (mg)	Calc (mg)	Iron (mg)	Magn (mg)	Pota (mg)	Sodi (mg)	Zinc (mg)	VT-A (RE)	Thia (mg)	VT-E (a-TE)	Ribo (mg)	Niac (mg)	V-B6 (mg)	Fola (μg)	VT-C (mg)
0	9	1.45	21	56	196	.28	1	.07	1.74	.12	.97	.1	23	0
20	25	1.57	35	143	231	.59	105	.12	1.86	.11	.87	.05	21	<1
15	16	1.44	15	115	134	.32	11	.12	1.31	.12	1.27	.03	36	0
0	41	1.86	17	132	224	.25	3	.1	.45	.14	1.2	.05	17	<1
0	0	.36			95		0		.01					0
0	3	.29	5	26	71	.08	<1	.02	<.01	.02	.26	<.01		0
20	60	1.59	25	143	323	.52	98	.15	1.5	.1	.76	.04	18	<1
25	31	1.78	31	185	414	.66	125	.18	3.04	.17	2.81	.07	44	<1
0	10	1.55	18	70	242	.32	<1	.03	1.21	.07	.83	.01	17	0
6	11	.88	5	32	146	.17	4	.11	.98	.1	1.07	.01	19	0
20	4	.58	3	15	102	.09	67	.08	.18	.06	.64	<.01	2	0
15	43	.88	4	78	225	.13	5	.09	1.54	.06	1.16	.01	25	0
<1	17	.61	5	28	95	.16	<1	.05		.07	.69	.01		0
23	19	.95	6	39	125	.14	7	.11	.54	.13	1.24	.03	20	0
0	33	.34	20	37	164	.33	2	.01	.35	.04	.31	.06	5	0
1	15	.48	4	14	99	.11	3	.06	.1	.04	.47	.05	8	0
1	22	.82	16	69	278	.3	10	.11	1.24	.1	1.83	.42	25	0
<1	26	.73	4	19	148	.14	<1	.05		.06	.78	.01		<1
<1	28	.58	7	43	169	.21	<1	.04		.07	.73	.02		0
0	0				80		100							2
0	0	0			80		100							2
0	3	.52	4	19	85	.11	0	.03	.27	.04	.58	.01	8	0
0	5	.18	3	10	41	.1	0	.02	.01	.01	.21	<.01	6	0
0	2	.27	24	52	5	.54	1	.01	.02	.03	1.41	.03	4	0
0	9	1.31	27	109	175	.62	<1	.09	.44	.06	.35	.06	3	<1
0	14	.65	3	15	156	.09	0	.07	.2	.05	.63	<.01	15	0
0		.36		5	50				.1					
0	11	.32	2	12	76	.06	0	.04	.4	.03	.36	<.01	7	0
2	3	.28	5	16	70	.13	<1	.04	.02	.03	.34	.01	1	0
0	4	.25	8	24	53	.17	0	.02	.31	.01	.36	.01	3	0
43	21	1.16	9	67	424	.43	78	.22	.24	.14	1.25	.03	35	<1
1	19	.56	8	36	248	.19	1	.1	.32	.08	.93	.02	18	0
105	143	6.94	54	371	1261	1.87	20	.99	3.06	.75	8.5	.2	211	10
27	37	1.8	14	96	326	.48	5	.25	.79	.19	2.2	.05	55	3
9	12	.57	4	30	104	.15	2	.08	.25	.06	.7	.02	17	1
19	22	1.4	14	110	333	.48	24	.29	.85	.21	1.8	.06	31	2
0	22	.58	5	76	142	.12	24	.07		.06	.6	.03	4	2
1	21	.79	6	106	288	.16	76	.1	1.11	.09	.74	.04	3	<1
0	26	2.14	11	154	74	.15	150	.06	.93	.08	.6	.06	11	3
1	28	1.81	9	133	294	.2	135	.1	1.01	.11	.85	.05	9	2
0	20	.89	12	189	70	.19	108	.06	1.13	.05	1.06	.03	6	5
1	24	.91	10	159	291	.23	105	.09	1.16	.09	1.19	.03	6	3
17	21	.92	9	60	257	.26	8	.1	1.63	.11	.87	.03	22	<1
4	26	1.22	13	65	205	.46	6	.22	1.75	.13	1.71	.03	26	0
0	99	1.43	12	75	264	.4	0	.25	.07	.16	2.21	.02	46	<1
0	98	1.41	11	74	262	.39	0	.2	.07	.14	1.98	.02	38	<1
0	175	1.62	47	139	420	1.06	0	.2	.46	.09	2.25	.11	28	0
<1	29	.72	21	91	78	.42	0	.08	.34	.05	.14	.03	7	0
0	15	.74	24	84	73	.51	4	.07	.33	.03	.39	.02	6	<1
0	0	3.6			5		100							0
22	107	1.29	9	69	251	.31	16	.15	1.03	.16	1.26	.02	7	1
20	106	2.39	44	181	335	1.57	136	.19	1.31	.25	2.29	.18	30	4
26	147	1.49	13	82	333	.35	23	.17	1.08	.18	1.36	.05	10	<1

Table A-1

Food Composition (Computer code number is for West Diet Analysis program) (For purposes of calculations, use "0" for t, <1, <.1, <.01, etc.)

Computer Code Number	Food Description	Measure	Wt (g)	H_2O (%)	Ener (cal)	Prot (g)	Carb (g)	Dietary Fiber (g)	Fat (g)	Fat Breakdown (g) Sat	Mono	Poly
	BAKED GOODS: BREADS, CAKES, COOKIES, CRACKERS, PIES—Continued											
	From commercial mix:											
438	Blueberry	1 ea	50	36	150	3	24	1	4	.7	1.8	1.5
439	Bran, wheat	1 ea	50	35	138	3	23	2	5	1.2	2.3	.7
440	Cornmeal	1 ea	50	30	161	4	25	1	5	1.4	2.6	.6
1864	Nabisco Newtons, fat free, all flavors	1 ea	23		69	1	16		0	0	0	0
	Pancakes, 4" diam:											
441	Buckwheat, from mix w/ egg and milk	1 ea	30	54	62	2	8	1	2	.6	.6	.8
442	Plain, from home recipe	1 ea	38	53	86	2	11	1	4	.8	.9	1.7
443	Plain, from mix; egg, milk, oil added	1 ea	38	53	74	2	14	<1	1	.2	.3	.3
1468	Pan dulce, sweet roll w/topping	1 ea	79	21	291	5	48	1	9	2	3.9	2.7
	Piecrust,with enriched flour, vegetable shortening, baked:											
444	Home recipe, 9" shell	1 ea	180	10	949	12	85	3	62	15.5	27.4	16.4
	From mix:											
445	Piecrust for 2-crust pie	1 ea	320	10	1686	21	152	5	111	27.6	48.6	29.2
446	1 pie shell	1 ea	160	11	802	11	81	3	49	12.3	27.7	6.2
	Pies, 9" diam; pie crust made with vegetable shortening, enriched flour:											
447	Apple: Whole pie	1 ea	1000	52	2370	19	340	16	110	21.1	59.4	20.9
448	Piece, ⅙ of pie	1 pce	167	52	396	3	57	3	18	3.5	9.9	3.5
449	Banana cream: Whole pie	1 ea	1152	48	3098	51	379	8	157	43.3	65.9	38
450	Piece, ⅙ of pie	1 pce	192	48	516	8	63	1	26	7.2	11	6.3
451	Blueberry: Whole pie	1 ea	1176	51	2881	32	394	16	140	34.3	60.2	36.2
452	Piece, ⅙ of pie	1 pce	196	51	480	5	66	3	23	5.7	10	6
453	Cherry: Whole pie	1 ea	1140	46	3078	32	439	17	139	34.1	60.5	37.1
454	Piece, ⅙ of pie	1 pce	240	46	648	7	92	4	29	7.2	12.7	7.8
455	Chocolate cream: Whole pie	1 ea	1194	63	2150	49	281	12	97	35.5	38.2	18.6
456	Piece, ⅙ of pie	1 pce	199	63	358	8	47	2	16	5.9	6.4	3.1
457	Custard: Whole pie	1 ea	630	61	1323	35	131	10	73	17.5	36.3	12.1
458	Piece, ⅙ of pie	1 pce	105	61	221	6	22	2	12	2.9	6	2
459	Lemon meringue: Whole pie	1 ea	678	42	1817	10	320	8	59	10.6	24.6	19.6
460	Piece, ⅙ of pie	1 pce	113	42	303	2	53	1	10	1.8	4.1	3.3
461	Peach: Whole pie	1 ea	1111	45	2994	26	443	16	130	31.1	55.7	37.4
462	Piece, ⅙ of pie	1 pce	139	45	375	3	55	2	16	3.9	7	4.7
463	Pecan: Whole pie	1 ea	678	19	2712	27	388	24	125	25.5	73.2	20.1
464	Piece, ⅙ of pie	1 pce	113	19	452	5	65	4	21	4.2	12.2	3.4
465	Pumpkin: Whole pie	1 ea	654	58	1373	25	179	18	62	13.2	32.8	10.5
466	Piece, ⅙ of pie	1 pce	109	58	229	4	30	3	10	2.2	5.5	1.7
467	Pies, fried, commercial: Apple	1 ea	85	40	266	2	33	1	14	6.5	5.8	1.2
468	Pies, fried, commercial: Cherry	1 ea	128	38	404	4	54	3	21	3.1	9.5	6.9
	Pretzels, made with enriched flour:											
469	Thin sticks, 2¼" long	1 oz	28	3	107	3	22	1	1	.2	.4	.3
470	Dutch twists	10 pce	60	3	229	5	47	2	2	.4	.8	.7
471	Thin twists, 3¼ x 2¼ x ¼"	10 pce	60	3	229	5	47	2	2	.4	.8	.7
	Rolls & buns, enriched, commercial:											
472	Cloverleaf rolls, 2½" diam, 2" high	1 ea	28	32	84	2	14	1	2	.5	1	.3
473	Hot dog buns	1 ea	43	34	123	4	22	1	2	.5	1.1	.4
474	Hamburger buns	1 ea	43	34	123	4	22	1	2	.5	1.1	.4
475	Hard roll, white, 3¾" diam, 2" high	1 ea	57	31	167	6	30	1	2	.3	.6	1
476	Submarine rolls/hoagies, 11¼ x 3 x 2½"	1 ea	135	31	392	12	75	4	4	.9	1.3	1.4
	Rolls & buns, enriched, home recipe:											
477	Dinner rolls 2½" diam, 2" high	1 ea	35	29	112	3	19	1	3	.7	1.1	.7
	Sports/fitness bar:											
2043	Forza energy bar	1 ea	70	18	231	10	45	4	1			
2042	Power bar	1 ea	65		230	10	45	3	2			
2041	Tiger sports bar	1 ea	65	17	229	11	40	4	2			
478	Toaster pastries, fortified (Poptarts)	1 ea	52	12	204	2	37	1	5	.8	2.1	2
2132	Toaster strudel pastry—cream cheese	1 ea	54	32	188	3	24	<1	9	2.7		
2134	Toaster strudel pastry—french toast	1 ea	54	32	188	3	24	<1	9	2.9		

Chol (mg)	Calc (mg)	Iron (mg)	Magn (mg)	Pota (mg)	Sodi (mg)	Zinc (mg)	VT-A (RE)	Thia (mg)	VT-E (a-TE)	Ribo (mg)	Niac (mg)	V-B6 (mg)	Fola (μg)	VT-C (mg)
23	12	.56	5	39	219	.19	11	.07	.7	.16	1.12	.04	5	<1
34	16	1.27	28	73	234	.57	15	.1	.75	.12	1.44	.09	8	0
31	37	.97	10	65	398	.32	22	.12	.75	.14	1.05	.05	5	<1
					77									
20	77	.56	17	70	160	.35	20	.05	.62	.08	.4	.04	5	<1
22	83	.68	6	50	167	.21	20	.08	.36	.11	.6	.02	14	<1
5	48	.59	8	66	239	.15	3	.08	.32	.08	.65	.03	3	<1
26	13	1.82	10	57	140	.35	87	.23	1.35	.21	1.98	.04	22	<1
0	18	5.2	25	121	976	.79	0	.7	9.94	.5	5.96	.04	121	0
0	32	9.25	45	214	1734	1.41	0	1.25	17.7	.89	10.6	.08	214	0
0	96	3.44	24	99	1166	.62	0	.48	8.83	.3	3.79	.09	19	0
0	110	4.5	70	650	2660	1.6	300	.28	16.5	.27	2.63	.38	220	32
0	18	.75	12	109	444	.27	50	.05	2.76	.04	.44	.06	37	5
588	864	12	184	1900	2764	5.53	806	1.6	16.9	2.38	12.1	1.53	311	18
98	144	2	31	317	461	.92	134	.27	2.82	.4	2.02	.25	52	3
0	82	14.5	94	588	2175	2.35	47	1.8	24.7	1.55	14	.4	270	8
0	14	2.41	16	98	363	.39	8	.3	4.12	.26	2.33	.07	45	1
0	114	21.1	103	878	2177	2.28	547	1.69	21.7	1.43	14.6	.39	308	11
0	24	4.44	22	185	458	.48	115	.35	4.56	.3	3.07	.08	65	2
109	1028	8.84	170	1705	2085	4.93	235	1.03	11.4	2.43	7.34	.37	59	6
18	171	1.47	28	284	348	.82	39	.17	1.9	.41	1.22	.06	10	1
208	504	3.65	69	668	1512	3.28	315	.25	7.5	1.31	1.84	.3	126	2
35	84	.61	12	111	252	.55	52	.04	1.25	.22	.31	.05	21	<1
305	380	4.14	102	603	990	3.32	353	.42	9.7	1.42	4.4	.2	88	22
51	63	.69	17	101	165	.55	59	.07	1.62	.24	.73	.03	15	4
0	59	12.1	79	1047	2025	1.88	386	1.41	26.2	1.19	15.3	.2	58	589
0	7	1.52	10	131	253	.23	48	.18	3.28	.15	1.91	.02	7	74
217	115	7.05	122	502	2874	3.86	319	.62	17.2	.83	1.69	.14	183	7
36	19	1.18	20	84	479	.64	53	.1	2.86	.14	.28	.02	30	1
131	392	5.17	98	1007	1844	2.94	3139	.36	10.5	1	1.22	.37	131	10
22	65	.86	16	168	307	.49	523	.06	1.75	.17	.2	.06	22	2
13	13	.88	8	51	325	.17	33	.1	.37	.08	.98	.03	4	1
0	28	1.56	13	83	479	.29	22	.18	.55	.14	1.83	.04	23	2
0	10	1.21	10	41	480	.24	0	.13	.06	.17	1.47	.03	48	0
0	22	2.59	21	88	1029	.51	0	.28	.13	.37	3.15	.07	103	0
0	22	2.59	21	88	1029	.51	0	.28	.13	.37	3.15	.07	103	0
<1	33	.88	6	37	146	.22	0	.14	.22	.09	1.13	.01	27	<1
0	60	1.36	9	61	241	.27	0	.21	.2	.13	1.69	.02	41	0
0	60	1.36	9	61	241	.27	0	.21	.2	.13	1.69	.02	41	0
0	54	1.87	15	62	310	.54	0	.27	.1	.19	2.42	.03	54	0
0	122	3.78	27	122	783	.85	0	.54	.1	.33	4.47	.05	40	0
13	21	1.04	7	53	145	.24	28	.14	.35	.14	1.21	.02	15	<1
0	300	6.3	160	220	65	5.25		1.5	20	1.7	20	2	400	60
0	300	5.4	140	150	110	5.25		1.5		1.7	20	2	400	60
	349	4.49	140	279	100		50	1.5	19.9	1.69	19.9	1.99	399	60
0	13	1.81	9	58	218	.34	149	.15	.97	.19	2.05	.2	34	<1
12	12	.97			217		17		1					0
12	12	.97			217		17		1					0

Table A-1

Food Composition (Computer code number is for West Diet Analysis program) (For purposes of calculations, use "0" for t, <1, <.1, <.01, etc.)

Computer Code Number	Food Description	Measure	Wt (g)	H_2O (%)	Ener (cal)	Prot (g)	Carb (g)	Dietary Fiber (g)	Fat (g)	Fat Breakdown (g) Sat	Mono	Poly
	BAKED GOODS: BREADS, CAKES, COOKIES, CRACKERS, PIES—Continued											
	Tortilla chips:											
1271	Plain	10 pce	18	2	90	1	11	1	5	.9	2.8	.7
1036	Nacho flavor	1 c	26	2	129	2	16	1	7	1.3	3.9	.9
1037	Taco flavor	1 pce	18	2	86	1	11	1	4	.8	2.6	.6
	Tortillas:											
479	Corn, enriched, 6″ diam	1 ea	26	44	58	2	12	1	1	.1	.2	.3
480	Flour, 8″ diam	1 ea	49	27	159	4	27	2	3	.6	1.4	1.4
1301	Flour, 10″ diam	1 ea	72	27	234	6	40	2	5	.8	2.1	2
481	Taco shells	1 ea	14	4	63	1	9	1	3	.4	1.5	.6
	Waffles, 7″ diam:											
482	From home recipe	1 ea	75	42	218	6	25	1	11	2.2	2.6	5.1
483	From mix, egg/milk added	1 ea	75	42	218	5	26	1	10	1.7	2.7	5.2
1510	Whole grain, prepared from frozen	1 ea	39	43	107	4	13	1	5	1.6	1.9	.9
	GRAIN PRODUCTS: CEREAL, FLOUR, GRAIN, PASTA and NOODLES, POPCORN											
484	Barley, pearled, dry, uncooked	1 c	200	10	704	20	155	31	2	.5	.3	1.1
485	Barley, pearled, cooked	1 c	157	69	193	4	44	6	1	.1	.1	.3
2009	Breakfast bars, fat free, all flavors	1 ea	38	25	110	2	26	3	0	0	0	0
	Breakfast bar, Snackwell:											
2165	Apple-cinnamon	1 ea	37	16	119	1	29	1	<1	.1	t	.1
2164	Blueberry	1 ea	37	16	121	1	29	1	<1	t	t	.1
2163	Strawberry	1 ea	37	16	120	1	29	1	<1	t	t	.1
	Breakfast cereals, hot, cooked w/o salt added:											
	Corn grits (hominy) enriched:											
486	Regular/quick prep w/o salt, yellow:	1 c	242	85	145	3	31	<1	<1	.1	.1	.2
487	Instant, prepared from packet, white	1 ea	137	82	89	2	21	1	<1	t	t	.1
	Cream of wheat:											
488	Regular, quick, instant	1 c	239	87	129	4	27	1	<1	.1	.1	.3
489	Mix and eat, plain, packet	1 ea	142	82	102	3	21	<1	<1	t	t	.2
1664	Farina cereal, cooked w/o salt	1 c	233	88	117	3	25	3	<1	t	t	.1
490	Malt-O-Meal, cooked w/o salt	1 c	240	88	122	4	26	1	<1	.1	.1	t
494	Maypo	1 c	216	83	153	5	29	5	2	.4	.7	.8
	Oatmeal or rolled oats:											
491	Regular, quick, instant,nonfortified cooked w/o salt	1 c	234	85	145	6	25	4	2	.4	.7	.9
	Instant, fortified:											
492	Plain, from packet	½ c	118	85	70	4	12	2	1	.2	.4	.4
493	Flavored, from packet	½ c	109	76	106	3	21	2	1	.2	.5	.5
	Breakfast cereals, ready to eat:											
495	All-Bran	1 c	62	3	160	8	46	20	2	.4	.4	1.3
1306	Alpha Bits	1 c	28	1	110	2	24	1	1	.1	.2	.2
1307	Apple Jacks	1 c	33	3	120	2	30	1	<1	.1	.1	.2
1308	Bran Buds	1 c	90	3	240	8	72	36	2	.4	.4	1.4
1305	Bran Chex	1 c	49	2	156	5	39	8	1	.2	.3	.7
1309	Honey BucWheat Crisp	1 c	38	5	147	4	31	3	1	.2	.3	.6
1310	C.W. Post, plain	1 c	97	2	421	9	73	7	13	1.7	6	4.7
1311	C.W. Post, with raisins	1 c	103	4	446	9	74	14	15	11	1.7	1.4
496	Cap'n Crunch	1 c	37	2	147	2	32	1	2	.5	.4	.3
1312	Cap'n Crunchberries	1 c	35	2	140	2	30	1	2	.5	.3	.3
1313	Cap'n Crunch, peanut butter	1 c	35	2	146	3	28	1	3	.7	1.1	.7
497	Cheerios	1 c	23	3	84	2	17	2	1	.3	.5	.2
1314	Cocoa Krispies	1 c	41	2	159	3	36	1	1	.7	.2	.2
1316	Cocoa Pebbles	1 c	32	2	131	1	27	1	2	1.1	.4	.1
1315	Corn Bran	1 c	36	3	120	2	30	6	1	.3	.3	.4
1317	Corn Chex	1 c	28	2	110	2	25	<1	<1	t	t	t
498	Corn Flakes, Kellogg's	1 c	28	3	100	2	24	1	<1	.1	t	.1
499	Corn Flakes, Post Toasties	1 c	24	3	93	2	21	1	<1	t	t	t
1340	Corn Pops	1 c	31	3	120	1	28	<1	<1	.1	.1	t
1318	Cracklin' Oat Bran	1 c	65	4	252	5	48	8	8	3.4	3.8	.9
1038	Crispy Wheat 'N Raisins	1 c	43	7	150	3	35	3	1	.1	.1	.2

Chol (mg)	Calc (mg)	Iron (mg)	Magn (mg)	Pota (mg)	Sodi (mg)	Zinc (mg)	VT-A (RE)	Thia (mg)	VT-E (a-TE)	Ribo (mg)	Niac (mg)	V-B6 (mg)	Fola (µg)	VT-C (mg)
0	28	.27	16	35	95	.27	4	.01	.24	.03	.23	.05	2	0
1	38	.37	21	56	184	.31	11	.03	.35	.05	.37	.07	4	<1
1	28	.36	16	39	142	.23	16	.04	.24	.04	.36	.05	4	<1
0	45	.36	17	40	42	.24	6	.03	.04	.02	.39	.06	30	0
0	61	1.62	13	64	234	.35	0	.26	.62	.14	1.75	.02	60	0
0	90	2.38	19	94	344	.51	0	.38	.91	.21	2.57	.04	89	0
0	35	.36	15	34	25	.19	6	.04	.59	.02	.24	.04	4	0
52	191	1.73	14	119	383	.51	49	.2	1.73	.26	1.55	.04	34	<1
38	93	1.22	15	134	458	.35	19	.15	1.5	.19	1.23	.07	9	<1
39	84	.69	15	91	150	.45	25	.08	.53	.13	.75	.04	7	<1
0	58	5	158	560	18	4.26	4	.38	.26	.23	9.2	.52	46	0
0	17	2.09	34	146	5	1.29	2	.13	.08	.1	3.23	.18	25	0
0	20	.72			25		20							1
<1	17	5	6	68	103	3.88	260	.39		.44	5.2	.52		<1
<1	14	4.83	5	43	107	3.85	260	.39		.44	5.2	.52		<1
<1	14	4.82	6	47	102	3.83	260	.39		.44	5.2	.52		2
0	0	1.55	10	53	0	.17	14	.24	.12	.14	1.96	.06	75	0
0	8	8.19	11	38	289	.21	0	.15	.03	.08	1.38	.05	47	0
0	50	10.3	12	45	139	.33	0	.24	.03	0	1.43	.03	108	0
0	20	8.09	7	38	241	.24	125	.43	.02	.28	4.97	.57	101	0
0	5	1.17	5	30	0	.16	0	.19	.03	.12	1.28	.02	54	0
0	5	9.6	5	31	2	.17	0	.48	.03	.24	5.76	.02	5	0
0	112	7.56	45	190	233	1.34	633	.65	1.51	.65	8.42	.86	9	26
0	19	1.59	56	131	2	1.15	5	.26	.23	.05	.3	.05	9	0
0	109	4.2	28	66	190	.58	302	.35	.14	.19	3.65	.49	100	0
0	112	4.45	34	91	169	.66	306	.35	.14	.25	3.92	.51	100	<1
0	200	9	280	620	560	7.5	450	.75	1.14	.85	10	1	186	30
0	8	2.66	16	54	178	1.48	371	.36	.02	.42	4.93	.5	99	0
0	0	4.5	8	35	150	3.75	225	.38	.05	.43	5	.5	116	15
0	60	13.5	240	809	599	11.3	676	1.17	1.42	1.26	15	1.53	270	45
0	29	14	69	216	345	6.47	11	.64	.56	.26	8.62	.88	173	26
0	54	10.9	43	142	361	.68	913	.9	8.99	1.03	12.1	1.88	11	36
<1	47	15.4	67	198	167	1.64	1284	1.26	.68	1.46	17.1	1.75	342	0
<1	50	16.4	74	261	161	1.64	1363	1.34	.72	1.55	18.1	1.85	364	0
0	7	6.18	13	47	286	5.14	5	.51	.18	.58	6.85	.68	137	0
0	9	6.06	13	49	256	5.39	6	.5	.25	.57	6.72	.67	135	<1
0	3	5.85	24	80	264	4.87	5	.49	.19	.55	6.48	.65	130	0
0	42	6.21	25	68	218	2.88	288	.29	.16	.33	3.84	.38	77	11
0	0	2.38	11	79	278	1.97	298	.49	.19	.57	6.6	.66	123	20
0	5	2.02	13	53	180	1.7	424	.42	.04	.48	5.63	.58	113	0
0	27	10.1	19	75	338	5	5	.1	.19	.56	6.66	.67	134	0
0	3	8.01	4	23	306	.1	14	.36	.07	.07	4.93	.5	99	15
0	0	8.68	3	25	300	.17	225	.36	.03	.43	5	.5	99	15
0	1	.63	4	28	252	.07	318	.31	.06	.36	4.22	.43	85	0
0	0	1.86	2	25	120	1.55	225	.4	.03	.43	5.18	.5	109	15
0	26	2.41	79	305	226	1.95	299	.5	.43	.56	6.63	.66	181	20
0	54	3.52	33	180	223	.85	293	.29	.45	.33	3.91	.39	78	0

Table A-1

Food Composition (Computer code number is for West Diet Analysis program) (For purposes of calculations, use "0" for t, <1, <.1, <.01, etc.)

Computer Code Number	Food Description	Measure	Wt (g)	H_2O (%)	Ener (cal)	Prot (g)	Carb (g)	Dietary Fiber (g)	Fat (g)	Fat Breakdown (g) Sat	Mono	Poly
	GRAIN PRODUCTS: CEREAL, FLOUR, GRAIN, PASTA and NOODLES, POPCORN—Continued											
1319	Fortified Oat Flakes	1 c	48	3	180	8	36	1	1	.2	.3	.4
500	40% Bran Flakes, Kellogg's	1 c	39	4	121	4	32	7	1	.2	.2	.5
501	40% Bran Flakes, Post	1 c	47	3	152	5	37	9	1	.1	.1	.4
502	Froot Loops	1 c	32	2	120	2	28	1	1	.4	.2	.3
518	Frosted Flakes	1 c	41	3	159	2	37	1	<1	.1	t	.1
1320	Frosted Mini-Wheats	1 c	51	5	170	5	41	5	1	.2	.1	.6
1321	Frosted Rice Krispies	1 c	35	2	135	2	32	<1	<1	.1	.1	.1
1324	Fruit & Fibre w/dates	1 c	57	9	193	5	43	8	3	.4	1.3	1
1322	Fruity Pebbles	1 c	32	3	130	1	28	<1	2	1.4	.1	.1
503	Golden Grahams	1 c	39	3	150	2	33	1	1	.2	.4	.2
504	Granola, homemade	½ c	61	5	285	9	32	6	15	2.9	4.8	6.5
505	Granola, low fat	½ c	47	3	181	5	38	3	3	0		
1670	Granola, low fat, commercial	½ c	45	5	165	4	35	2	2	.6	.7	.9
505	Grape Nuts	½ c	55	3	196	7	45	5	<1	t	t	.1
1326	Grape Nuts Flakes	1 c	39	3	144	4	32	4	1	.6	.1	.2
1665	Heartland Natural with raisins	1 c	110	5	468	11	76	6	16	4	4.2	6.2
1327	Honey & Nut Corn Flakes	1 c	37	2	148	3	31	1	2	.3	.7	.6
506	Honey Nut Cheerios	1 c	33	2	126	3	27	2	1	.3	.5	.2
1328	HoneyBran	1 c	35	2	119	3	29	4	1	.3	.1	.3
1329	HoneyComb	1 c	22	1	86	1	20	1	<1	.2	.1	.1
1330	King Vitaman	1 c	21	2	81	2	18	1	1	.2	.3	.2
1039	Kix	1 c	19	2	72	1	16	1	<1	.1	.1	t
1331	Life	1 c	44	4	167	4	35	3	2	.3	.6	.8
507	Lucky Charms	1 c	32	2	124	2	27	1	1	.2	.4	.2
1323	Mueslix Five Grain	1 c	82	5	279	7	63	7	3	.5	1	1.2
508	Nature Valley Granola	1 c	113	4	510	12	74	7	20	2.6	13.3	3.8
1666	Nutri Grain Almond Raisin	1 c	40	6	147	3	31	3	2	.1	1	1.2
1336	100% Bran	1 c	66	3	178	8	48	19	3	.6	.6	1.9
509	100% Natural cereal, plain	1 c	104	2	462	11	71	8	17	7.5	7.5	2.3
1337	100% Natural with apples & cinnamon	1 c	104	2	477	11	70	7	20	15.5	1.8	1.3
1338	100% Natural with raisins & dates	1 c	110	3	496	12	72	7	20	13.6	3.7	1.7
510	Product 19	1 c	33	4	110	2	28	1	<1	t	.2	.2
1339	Quisp	1 c	30	3	121	2	25	1	2	.5	.4	.2
511	Raisin Bran, Kellogg's	1 c	61	9	200	6	47	8	1	.1	.1	.4
512	Raisin Bran, Post	1 c	56	9	172	5	42	8	1	.2	.1	.5
1667	Raisin Squares	1 c	71	9	241	6	55	7	2	.2	.2	.6
1041	Rice Chex	1 c	33	3	130	2	29	1	<1	t	t	t
513	Rice Krispies, Kellogg's	1 c	28	2	111	2	25	<1	<1	t	t	t
514	Rice, puffed	1 c	14	4	54	1	12	<1	<1	t	t	t
515	Shredded Wheat	1 c	43	5	154	5	35	4	1	.1	.1	.4
516	Special K	1 c	31	3	110	6	22	1	<1	t	t	.2
517	Super Golden Crisp	1 c	33	1	123	2	30	<1	<1	.1	.1	.1
519	Honey Smacks	1 c	36	3	133	3	32	1	1	.4	.1	.3
1341	Tasteeos	1 c	24	2	94	3	19	3	1	.2	.2	.2
1342	Team	1 c	42	4	164	3	36	1	1	.1	.2	.3
520	Total, wheat, with added calcium	1 c	40	3	140	4	32	4	1	.2	.2	.1
521	Trix	1 c	28	2	114	1	24	1	2	.4	.9	.3
1344	Wheat Chex	1 c	46	2	169	5	38	4	1	.2	.1	.5
1043	Wheat cereal, puffed, fortified	1 c	12	4	44	2	9	1	<1	t	t	.1
522	Wheaties	1 c	29	3	106	3	23	2	1	.2	.2	.1
523	Buckwheat flour, dark	1 c	120	11	402	15	85	12	4	.8	1.1	1.1
525	Buckwheat, whole grain, dry	1 c	170	10	583	23	122	17	6	1.3	1.8	1.8
526	Bulgar, dry, uncooked	1 c	140	9	479	17	106	26	2	.3	.2	.8
527	Bulgar, cooked	1 c	182	78	151	6	34	8	<1	.1	.1	.2
	Cornmeal:											
528	Whole-ground, unbolted, dry	1 c	122	10	442	10	94	9	4	.6	1.2	2
530	Degermed, enriched, dry	1 c	138	12	505	12	107	10	2	.3	.6	1
38041	Degermed, enriched, baked	1 c	138	12	505	12	107	10	2	.3	.6	1
	Macaroni, cooked:											
532	Enriched	1 c	140	66	197	7	40	2	1	.1	.1	.4
533	Whole wheat	1 c	140	67	174	7	37	4	1	.1	.1	.3

Chol (mg)	Calc (mg)	Iron (mg)	Magn (mg)	Pota (mg)	Sodi (mg)	Zinc (mg)	VT-A (RE)	Thia (mg)	VT-E (a-TE)	Ribo (mg)	Niac (mg)	V-B6 (mg)	Fola (μg)	VT-C (mg)
0	68	13.7	58	228	220	2.54	636	.62	.34	.72	8.45	.86	169	0
0	0	10.9	81	229	309	5.03	505	.51	7.22	.58	6.71	.66	138	20
0	21	13.4	102	251	431	2.49	622	.61	.54	.7	8.27	.85	166	0
0	0	4.51	8	35	150	3.75	225	.37	.12	.43	5	.5	96	15
0	0	6.15	4	26	264	.2	298	.5	.05	.56	6.61	.66	123	20
0	0	15	60	170	0	1.5	0	.37	.46	.42	4.64	.5	102	0
0	0	2.42	8	27	256	.42	303	.49	.03	.56	6.72	.66	140	20
0	30	10.1	81	335	270	3.02	725	.75	1.32	.85	10.1	1	201	0
0	4	2.02	9	24	178	1.7	424	.42	.03	.48	5.63	.58	113	0
0	19	5.85	12	69	357	4.88	293	.49	.29	.55	6.51	.65	130	19
0	49	2.56	109	328	15	2.48	2	.45	7.87	.17	1.25	.19	52	1
0		2.71	36	143	90	5.64	226	.56	7.57	.64	7.52	.75	151	
0	15	1.35	30	127	101	2.84	169	.27	4.03	.31	3.74	.36	90	0
0	5	15.7	37	184	382	1.21	728	.71	.14	.82	9.68	.99	194	0
0	16	11.2	43	136	220	.78	516	.51	.1	.58	6.86	.7	138	0
0	66	4.02	141	415	226	2.83	7	.32	.77	.14	1.54	.2	44	1
0	0	3.03	3	40	249	.2	152	.26	.09	.3	3.37	.33	74	10
0	22	4.95	32	94	285	4.13	248	.41	.34	.47	5.51	.55	110	16
0	16	5.57	46	151	202	.9	463	.45	.81	.52	6.16	.63	23	19
0	4	2.09	7	25	124	1.17	291	.29	.09	.33	3.87	.4	78	0
0	3	5.92	18	58	176	2.65	212	.26	1.42	.3	3.53	.35	71	8
0	28	5.13	6	26	167	2.38	238	.24	.05	.27	3.17	.32	63	9
0	134	12.3	43	109	240	5.5	2	.55	.22	.62	7.35	.73	147	0
0	35	4.8	21	58	217	4	240	.4	.14	.45	5.34	.53	107	16
0	38	8.94	82	369	107	7.46	747	.75	8.94	.84	9.84	.99	197	1
0	85	3.53	107	375	183	2.27	0	.35	7.97	.12	1.25	.16	17	0
0	122	1.14	13	147	139	3.06	0	.32	4.38	.35	4.08	.41	80	0
0	46	8.12	312	652	457	5.74	0	1.58	1.53	1.78	20.9	2.11	47	63
1	100	3.11	109	457	28	2.5	1	.36	1.19	.17	1.84	.19	26	<1
0	157	2.89	72	514	52	2	6	.33	.73	.57	1.87	.11	17	1
0	160	3.12	124	538	47	2.11	7	.31	.77	.65	2.09	.16	45	0
0	0	19.8	18	55	308	16.5	248	1.65	24.4	1.88	22	2.21	429	66
0	6	5.1	15	40	216	4.26	4	.42	.15	.48	5.67	.56	113	0
0	40	4.5	80	350	390	3.75	225	.37	.56	.43	5	.5	122	0
0	26	8.9	95	345	365	2.97	741	.73	1.3	.84	9.86	1.01	198	0
0	0	21.7	54	335	4	1.99	0	.5	.38	.57	6.67	.64	142	0
0	5	9.44	8	38	276	.45	2	.43	.04	.01	5.81	.59	116	17
0	5	.7	12	27	206	.46	371	.52	.03	.59	6.92	.69	138	15
0	1	.41	4	16	1	.15	0	.06	.01	.01	.87	0	1	0
0	16	1.81	57	155	4	1.42	0	.11	.23	.12	2.26	.11	21	0
0	0	8.4	16	55	250	3.75	225	.53	.08	.59	7.01	.71	93	15
0	7	2.08	20	48	51	1.75	437	.43	.12	.49	5.81	.59	116	0
0	0	2.4	21	53	67	.4	300	.5	.18	.58	6.66	.68	133	20
0	11	6.86	26	71	183	.69	318	.31	.17	.36	4.22	.43	85	13
0	6	12	12	71	260	.58	556	.55	.1	.63	7.39	.76	7	22
0	344	24	43	129	265	20	500	2	31.3	2.27	26.8	2.67	533	80
0	30	4.2	3	16	184	3.5	210	.35	.56	.4	4.68	.47	93	14
0	18	13.2	58	173	308	1.23	0	.6	.17	.17	8.1	.83	162	24
0	3	.56	16	44	1	.37	<1	.05	.08	.03	1.43	.02	4	0
0	53	7.83	31	101	215	.68	218	.36	.36	.41	4.84	.48	97	14
0	49	4.87	301	692	13	3.74	0	.5	1.24	.23	7.38	.7	65	0
0	31	3.74	393	782	2	4.08	0	.17	1.75	.72	11.9	.36	51	0
0	49	3.44	230	574	24	2.7	0	.32	.22	.16	7.15	.48	38	0
0	18	1.75	58	124	9	1.04	0	.1	.05	.05	1.82	.15	33	0
0	7	4.21	155	350	43	2.22	57	.47	.82	.24	4.43	.37	31	0
0	7	5.7	55	224	4	.99	57	.99	.45	.56	6.94	.35	258	0
0	7	5.7	55	224	4	.99	57	.79	.5	.5	6.25	.32	181	0
0	10	1.96	25	43	1	.74	0	.29	.04	.14	2.34	.05	98	0
0	21	1.48	42	62	4	1.13	0	.15	.14	.06	.99	.11	7	0

Table A-1

Food Composition (Computer code number is for West Diet Analysis program) (For purposes of calculations, use "0" for t, <1, <.1, <.01, etc.)

Computer Code Number	Food Description	Measure	Wt (g)	H_2O (%)	Ener (cal)	Prot (g)	Carb (g)	Dietary Fiber (g)	Fat (g)	Fat Breakdown (g) Sat	Mono	Poly
	GRAIN PRODUCTS: CEREAL, FLOUR, GRAIN, PASTA and NOODLES, POPCORN—Continued											
534	Vegetable, enriched	1 c	134	68	172	6	36	2	<1	t	t	.1
535	Millet, cooked	1 c	240	71	286	8	57	3	2	.4	.4	1.2
	Noodles (see also Pasta and Spaghetti):											
1507	Cellophane noodles, cooked	1 c	190	79	160	<1	39	<1	<1	t	t	t
1995	Cellophane noodles, dry	1 c	140	13	491	<1	121	1	<1	t	t	t
537	Chow mein, dry	1 c	45	1	237	4	26	2	14	2	3.5	7.8
536	Egg noodles, cooked, enriched	1 c	160	69	213	8	40	2	2	.5	.7	.7
538	Spinach noodles, dry	3½ oz	100	8	372	13	75	11	2	.2	.2	.6
1343	Oat bran, dry	¼ c	24	7	59	4	16	4	2	.3	.6	.7
	Pasta, cooked:											
1418	Fresh	2 oz	57	69	75	3	14	1	1	.1	.1	.2
1417	Linguini/Rotini	1 c	140	66	197	7	40	4	1	.1	.1	.4
	Popcorn:											
539	Air popped, plain	1 c	8	4	31	1	6	1	<1	t	.1	.2
1042	Microwaved, low fat, low sodium	1 c	6	3	25	1	4	1	1	.1	.2	.3
540	Popped in vegetable oil/salted	1 c	11	3	55	1	6	1	3	.5	.9	1.5
541	Sugar-syrup coated	1 c	35	3	151	1	28	2	4	1.3	1	1.6
	Rice:											
542	Brown rice, cooked	1 c	195	73	216	5	45	4	2	.4	.6	.6
2215	Mexican rice, cooked	1 c	226		820	16	180	6	30	4	1	1
2216	Spanish rice, cooked	1 c	246	85	130	3	28	2	1			
	White, enriched, all types:											
543	Regular/long grain, dry	1 c	185	12	675	13	148	2	1	.3	.4	.3
544	Regular/long grain, cooked	1 c	158	68	205	4	45	1	<1	.1	.1	.1
545	Instant, prepared without salt	1 c	165	76	162	3	35	1	<1	.1	.1	.1
	Parboiled/converted rice:											
546	Raw, dry	1 c	185	10	686	13	151	3	1	.3	.3	.3
547	Cooked	1 c	175	72	200	4	43	1	<1	.1	.1	.1
1486	Sticky rice (glutinous), cooked	1 c	174	77	169	4	37	2	<1	.1	.1	.1
548	Wild rice, cooked	1 c	164	74	166	7	35	3	1	.1	.1	.4
1700	Rice and pasta (Rice-a-Roni), cooked	1 c	202	72	246	5	43	1	6	1.1	2.3	1.9
549	Rye flour, medium	1 c	102	10	361	10	79	15	2	.2	.2	.8
1044	Soy flour, low-fat	1 c	88	3	325	45	30	9	6	.9	1.3	3.3
	Spaghetti pasta:											
550	Without salt, enriched	1 c	140	66	197	7	40	4	1	.1	.1	.4
551	With salt, enriched	1 c	140	66	197	7	40	2	1	.1	.1	.4
552	Whole-wheat spaghetti, cooked	1 c	140	67	174	7	37	6	1	.1	.1	.3
1302	Tapioca-pearl, dry	1 c	152	11	544	<1	135	1	<1	t	t	t
553	Wheat bran, crude	1 c	58	10	125	9	37	25	2	.4	.4	1.3
554	Wheat germ, raw	1 c	115	11	414	27	60	15	11	1.9	1.6	6.9
555	Wheat germ, toasted	1 c	113	6	432	33	56	15	12	2.1	1.7	7.5
1669	Wheat germ, with brown sugar & honey	1 c	113	3	420	30	66	11	9	1.5	1.2	5.5
556	Rolled wheat, cooked	1 c	240	84	149	5	33	4	1	.1	.1	.5
557	Whole-grain wheat, cooked	1 c	150	86	84	4	20	3	<1	.1	.1	.2
	Wheat flour (unbleached):											
	All-purpose white flour, enriched:											
558	Sifted	1 c	115	12	419	12	88	3	1	.2	.1	.5
559	Unsifted	1 c	125	12	455	13	95	3	1	.2	.1	.5
560	Cake or pastry, enriched, sifted	1 c	96	12	348	8	75	2	1	.1	.1	.4
561	Self-rising, enriched, unsifted	1 c	125	11	443	12	93	3	1	.2	.1	.5
562	Whole wheat, from hard wheats	1 c	120	10	407	16	87	15	2	.4	.3	.9
	MEATS: FISH and SHELLFISH											
1045	Bass, baked or broiled	4 oz	113	69	165	27	0	0	5	1.4	1.5	2.3
1046	Bluefish, baked or broiled	4 oz	113	63	180	29	0	0	6	1.4	2	2.7
1686	Catfish, breaded/flour fried	4 oz	113	49	325	21	14	1	20	5	9	4.7
	Clams:											
563	Raw meat only	1 ea	145	82	107	19	4	0	1	.3	.4	.7
564	Canned, drained	1 c	160	64	237	41	8	0	3	.7	.9	1.5
1290	Steamed, meat only	10 ea	95	64	141	24	5	0	2	.4	.5	.9

Chol (mg)	Calc (mg)	Iron (mg)	Magn (mg)	Pota (mg)	Sodi (mg)	Zinc (mg)	VT-A (RE)	Thia (mg)	VT-E (a-TE)	Ribo (mg)	Niac (mg)	V-B6 (mg)	Fola (μg)	VT-C (mg)
0	15	.66	25	41	8	.59	7	.15	.05	.08	1.43	.03	87	0
0	7	1.51	106	149	5	2.18	0	.25	.43	.2	3.19	.26	46	0
0	14	1	3	5	9	.23	0	.07	.06	0	.09	.02	1	0
0	35	3.04	4	14	14	.57	0	.21	.18	0	.28	.07	3	0
0	9	2.13	23	54	198	.63	4	.26	.07	.19	2.68	.05	40	0
53	19	2.54	30	45	11	.99	10	.3	.08	.13	2.38	.06	102	0
0	58	2.13	174	376	36	2.76	46	.37	.04	.2	4.55	.32	48	0
0	14	1.3	56	136	1	.75	0	.28	.41	.05	.22	.04	12	0
19	3	.65	10	14	3	.32	3	.12	.09	.09	.56	.02	36	0
0	10	1.96	25	43	1	.74	0	.29	.08	.14	2.34	.05	98	0
0	1	.21	10	24	<1	.27	2	.02	.01	.02	.15	.02	2	0
0	1	.14	9	14	29	.23	1	.02	.06	.01	.12	.01	1	0
0	1	.31	12	25	97	.29	2	.01	.03	.01	.17	.02	2	<1
2	15	.61	12	38	72	.2	3	.02	.42	.02	.77	.01	1	0
0	19	.82	84	84	10	1.23	0	.19	.53	.05	2.98	.28	8	0
0	300	9			2700		120							96
0		.72			1340									
0	52	7.97	46	213	9	2.02	0	1.07	.24	.09	7.75	.3	427	0
0	16	1.9	19	55	2	.77	0	.26	.08	.02	2.34	.15	92	0
0	13	1.04	8	7	5	.4	0	.12	.08	.08	1.45	.02	68	0
0	111	6.59	57	222	9	1.78	0	1.1	.24	.13	6.72	.65	427	0
0	33	1.98	21	65	5	.54	0	.44	.09	.03	2.45	.03	87	0
0	3	.24	9	17	9	.71	0	.03	.07	.02	.5	.04	2	0
0	5	.98	52	166	5	2.2	0	.08	.38	.14	2.12	.22	43	0
2	16	1.9	24	85	1147	.57	0	.25	.27	.16	3.6	.2	89	<1
0	24	2.16	76	347	3	2.03	0	.29	1.36	.12	1.76	.27	19	0
0	165	5.27	202	2261	16	1.04	4	.33	.17	.25	1.9	.46	361	0
0	10	1.96	25	43	1	.74	0	.29	.08	.14	2.34	.05	98	0
0	10	1.96	25	43	140	.74	0	.29	.38	.14	2.34	.05	98	0
0	21	1.48	42	62	4	1.13	0	.15	.07	.06	.99	.11	7	0
0	30	2.4	2	17	2	.18	0	.01	0	0	0	.01	6	0
0	42	6.15	354	686	1	4.22	0	.3	1.35	.33	7.89	.75	46	0
0	45	7.2	275	1025	14	14.1	0	2.16	20.7	.57	7.83	1.5	323	0
0	51	10.3	362	1070	5	18.9	0	1.89	20.5	.93	6.32	1.11	398	7
0	56	9.1	307	1089	12	15.7	11	1.51	24.9	.78	5.34	.56	376	0
0	17	1.49	53	170	0	1.15	0	.17	.48	.12	2.14	.17	26	0
0	9	.88	35	99	1	.73	0	.12	.3	.03	1.5	.08	12	0
0	17	5.34	25	123	2	.8	0	.9	.07	.57	6.79	.05	177	0
0	19	5.8	27	134	2	.87	0	.98	.07	.62	7.38	.05	193	0
0	13	7.03	15	101	2	.59	0	.86	.06	.41	6.52	.03	148	0
0	423	5.84	24	155	1587	.77	0	.84	.07	.52	7.29	.06	193	0
0	41	4.66	166	486	6	3.52	0	.54	1.48	.26	7.64	.41	53	0
98	116	2.16	43	515	102	.94	40	.1	.84	.1	1.72	.16	19	2
86	10	.7	47	539	87	1.18	156	.08	.71	.11	8.19	.52	2	0
92	41	1.44	34	376	598	1.05	33	.4	2.48	.18	3.37	.21	19	1
49	67	20.3	13	455	81	1.99	131	.12	1.45	.31	2.57	.09	23	19
107	147	44.8	29	1004	179	4.37	274	.24	3.04	.68	5.36	.18	46	35
64	87	26.6	17	597	106	2.59	162	.14	1.86	.4	3.18	.1	27	21

Table A-1

Food Composition (Computer code number is for West Diet Analysis program) (For purposes of calculations, use "0" for t, <1, <.1, <.01, etc.)

Computer Code Number	Food Description	Measure	Wt (g)	H_2O (%)	Ener (cal)	Prot (g)	Carb (g)	Dietary Fiber (g)	Fat (g)	Fat Breakdown (g) Sat	Mono	Poly
	MEATS: FISH and SHELLFISH—Continued											
	Cod:											
565	Baked	4 oz	113	76	119	26	0	0	1	.2	.1	.5
566	Batter fried	4 oz	113	67	196	20	8	<1	9	2.2	3.6	2.6
567	Poached, no added fat	4 oz	113	77	116	25	0	0	1	.2	.1	.3
	Crab, meat only:											
1048	Blue crab, cooked	1 c	118	77	120	24	0	0	2	.3	.3	.8
1049	Dungeness crab, cooked	1 c	118	73	130	26	1	0	1	.2	.3	.5
568	Blue crab, canned	1 c	135	76	134	28	0	0	2	.4	.3	.6
1587	Crab, imitation, from surimi	4 oz	113	74	115	14	11	0	1	.3	.2	.8
569	Fish sticks, breaded pollock	2 ea	56	46	152	9	13	<1	7	1.8	2.8	1.8
572	Flounder/sole, baked	4 oz	113	73	132	27	0	0	2	.5	.4	.9
1599	Grouper, baked or broiled	4 oz	113	73	133	28	0	0	1	.4	.4	.6
573	Haddock, breaded, fried	4 oz	113	55	264	22	14	1	13	3.2	5.4	3.3
1050	Haddock, smoked	4 oz	113	71	131	28	0	0	1	.3	.3	.5
	Halibut:											
17291	Baked	4 oz	113	72	158	30	0	0	3	.7	1	1.4
1051	Smoked	4 oz	113	64	203	34			4	.6	1.2	1.5
1054	Raw	4 oz	113	78	124	23	0	0	3	.7	.8	1.1
575	Herring, pickled	4 oz	113	55	296	16	11	0	20	4.4	11	4.8
1052	Lobster meat, cooked w/moist heat	1 c	145	76	142	30	2	0	1	.2	.2	.5
1687	Ocean perch, baked/broiled	4 oz	113	73	137	27	0	0	2	.4	1	.7
576	Ocean perch, breaded/fried	4 oz	113	59	249	22	9	<1	13	3.2	5.7	3.4
1056	Octopus, raw	4 oz	113	80	93	17	2	0	1	.3	.2	.3
	Oysters:											
577	Raw, Eastern	1 c	248	85	169	17	10	0	6	2	.9	2.8
578	Raw, Pacific	1 c	248	82	201	23	12	0	6	1.3	.9	2.2
	Cooked:											
579	Eastern, breaded, fried, medium	5 ea	73	65	144	6	8	<1	9	2.7	1.7	4.6
580	Western, simmered	5 ea	125	64	204	24	12	0	6	1.9	.9	2.7
581	Pollock, baked, broiled, or poached	4 oz	113	74	128	27	0	0	1	.3	.2	.6
	Salmon:											
582	Canned pink, solids and liquid	4 oz	113	69	157	22	0	0	7	1.7	2.1	2.3
583	Broiled or baked	4 oz	113	62	244	31	0	0	12	2.2	6	2.7
584	Smoked	4 oz	113	72	132	21	0	0	5	1.2	2.3	1.1
585	Atlantic sardines, canned, drained, 2 = 24 g	4 oz	113	60	235	28	0	0	13	1.9	4.4	6.4
586	Scallops, breaded, cooked from frozen	6 ea	93	58	200	17	9	<1	10	2	2.5	5.3
1588	Scallops, imitation, from surimi	4 oz	113	74	112	14	12	0	1	.1	.1	.3
1688	Scallops, steamed/boiled	½ c	60	76	64	10	1	0	2	.3	.7	.6
	Shrimp:											
587	Cooked, boiled, 2 large = 11g	16 ea	88	77	87	18	0	0	1	.2	.2	.5
588	Canned, drained	½ c	64	73	77	15	1	0	1	.3	.3	.7
589	Fried, 2 large = 15 g, breaded	12 ea	90	53	218	19	10	<1	11	2	3	5.9
1057	Raw, large, about 7g each	14 ea	98	76	104	20	1	0	2	.3	.3	1
1589	Shrimp, imitation, from surimi	4 oz	113	75	114	14	10	0	2	.3	.2	1
1053	Snapper, baked or broiled	4 oz	113	70	145	30	0	0	2	.4	.4	.7
1060	Squid, fried in flour	4 oz	113	64	198	20	9	0	8	2.1	3.1	2.4
1590	Surimi	4 oz	113	76	112	17	8	0	1	.2	.2	.6
1058	Swordfish, raw	4 oz	113	76	137	22	0	0	5	1.3	1.7	1
1059	Swordfish, baked or broiled	4 oz	113	69	175	29	0	0	6	1.7	2.2	1.3
590	Trout, baked or broiled	4 oz	113	70	170	26	0	0	7	1.8	2	2.2
	Tuna, light, canned, drained solids:											
591	Oil pack	1 c	145	60	287	42	0	0	12	2.2	4.3	4.2
592	Water pack	1 c	154	74	179	39	0	0	1	.4	.2	.5
1061	Bluefin tuna, fresh	4 oz	113	68	163	26	0	0	6	1.4	1.8	1.6
	MEATS: BEEF, LAMB, PORK and others											
	BEEF, cooked, trimmed to ½" outer fat:											
	Braised, simmered, pot roasted:											
	Relatively fat, choice chuck blade:											
593	Lean and fat, piece 2½ x 2½ x ¾"	4 oz	113	47	393	30	0	0	29	13	14.8	1.2
594	Lean only	4 oz	113	55	297	35	0	0	16	7.3	8.2	.7

Chol (mg)	Calc (mg)	Iron (mg)	Magn (mg)	Pota (mg)	Sodi (mg)	Zinc (mg)	VT-A (RE)	Thia (mg)	VT-E (a-TE)	Ribo (mg)	Niac (mg)	V-B6 (mg)	Fola (µg)	VT-C (mg)
62	16	.55	47	276	88	.65	16	.1	.39	.09	2.84	.32	9	1
64	43	.92	36	443	124	.62	17	.13	.92	.13	2.58	.23	10	1
61	23	.54	41	496	69	.63	14	.09	.32	.08	2.48	.28	8	1
118	123	1.07	39	382	329	4.98	2	.12	1.18	.06	3.89	.21	60	4
90	70	.51	68	481	446	6.45	37	.07	1.33	.24	4.27	.2	50	4
120	136	1.13	53	505	450	5.43	2	.11	1.35	.11	1.85	.2	57	4
23	15	.44	49	102	950	.37	23	.04	.12	.03	.2	.03	2	0
63	11	.41	14	146	326	.37	17	.07	.77	.1	1.19	.03	10	0
77	20	.38	65	389	119	.71	12	.09	2.6	.13	2.46	.27	10	0
53	24	1.29	42	537	60	.58	56	.09	.71	.01	.43	.4	11	0
96	63	1.92	46	345	523	.59	33	.08	1.56	.14	4.49	.28	19	<1
87	55	1.58	61	469	862	.56	25	.05	.56	.05	5.73	.45	17	0
46	68	1.21	121	651	78	.6	61	.08	1.23	.1	8.05	.45	16	0
59	87	1.56	154	833	2260	.78	86	.11	1.11	.14	10.8	.64	22	0
36	53	.95	94	509	61	.47	53	.07	.96	.08	6.61	.39	14	0
15	87	1.38	9	78	983	.6	292	.04	1.81	.16	3.73	.19	3	0
104	88	.57	51	510	551	4.23	38	.01	2.1	.1	1.55	.11	16	0
61	155	1.33	44	396	108	.69	16	.15	1.84	.15	2.76	.3	12	1
71	136	1.57	38	323	431	.67	23	.14	2.41	.18	2.68	.24	15	1
54	60	5.99	34	396	260	1.9	51	.03	1.36	.04	2.37	.41	18	6
131	112	16.5	117	387	523	225	74	.25	1.98	.24	3.42	.15	25	9
124	20	12.7	55	417	263	41.2	201	.17	2.11	.58	4.98	.12	25	20
59	45	5.07	42	178	304	63.6	66	.11	1.66	.15	1.2	.05	23	3
125	20	11.5	55	378	265	41.5	183	.16	2.21	.55	4.53	.11	19	16
108	7	.32	82	437	131	.68	26	.08	.32	.09	1.86	.08	4	0
62	241	.95	38	368	626	1.04	19	.03	1.53	.21	7.39	.34	17	0
98	8	.62	35	424	75	.58	71	.24	1.42	.19	7.54	.25	6	0
26	12	.96	20	198	886	.35	29	.03	1.53	.11	5.33	.31	2	0
160	432	3.3	44	449	571	1.48	76	.09	.34	.26	5.93	.19	13	0
57	39	.76	55	310	432	.99	20	.04	1.77	.1	1.4	.13	34	2
25	9	.35	49	116	898	.37	23	.01	.12	.02	.35	.03	2	0
19	15	.15	32	168	246	.55	27	.01	.81	.04	.6	.08	7	1
172	34	2.72	30	160	197	1.37	58	.03	.66	.03	2.28	.11	3	2
111	38	1.75	26	134	108	.81	11	.02	.59	.02	1.77	.07	1	1
159	60	1.13	36	203	310	1.24	50	.12	1.35	.12	2.76	.09	7	1
149	51	2.36	36	181	145	1.09	53	.03	.8	.03	2.5	.1	3	2
41	21	.68	49	101	797	.37	23	.03	.12	.04	.19	.03	2	0
53	45	.27	42	590	64	.5	40	.06	.71	<.01	.39	.52	7	2
294	44	1.14	43	315	346	1.97	12	.06	2.09	.52	2.94	.07	16	5
34	10	.29	49	127	162	.37	23	.02	.28	.02	.25	.03	2	0
44	5	.91	30	325	102	1.3	41	.04	.56	.11	10.9	.37	2	1
56	7	1.18	38	417	130	1.66	46	.05	.71	.13	13.3	.43	3	1
78	97	.43	35	506	63	.58	17	.17	.57	.11	6.52	.39	21	2
26	19	2.02	45	300	513	1.31	33	.05	1.74	.17	18	.16	8	0
46	17	2.36	42	365	521	1.19	26	.05	.82	.11	20.5	.54	6	0
43	9	1.15	56	285	44	.68	740	.27	1.13	.28	9.77	.51	2	0
112	11	3.45	21	275	67	7.57	0	.08	.26	.27	3.54	.32	10	0
120	15	4.16	26	297	80	11.6	0	.09	.16	.32	3.02	.33	7	0

Table A-1

Food Composition (Computer code number is for West Diet Analysis program) (For purposes of calculations, use "0" for t, <1, <.1, <.01, etc.)

Computer Code Number	Food Description	Measure	Wt (g)	H_2O (%)	Ener (cal)	Prot (g)	Carb (g)	Dietary Fiber (g)	Fat (g)	Fat Breakdown (g) Sat	Mono	Poly
	MEATS: BEEF, LAMB, PORK and others—Continued											
	Relatively lean, like choice round:											
595	Lean and fat, pce 4⅛ x 2½ x ¾"	4 oz	113	52	311	32	0	0	19	8.5	9.7	.8
596	Lean only	4 oz	113	57	249	36	0	0	11	4.8	5.4	.5
	Ground beef, broiled, patty 3 x ⅝":											
597	Extra lean, about 16% fat	4 oz	113	54	299	32	0	0	18	8	9	.8
598	Lean, 21% fat	4 oz	113	53	316	32	0	0	20	8.9	10.1	.8
	Roasts, oven cooked, no added liquid:											
	Relatively fat, prime rib:											
601	Lean and fat, piece 4⅛ x 2¼ x ½"	4 oz	113	46	425	25	0	0	35	15.8	17.9	1.5
602	Lean only	4 oz	113	58	271	31	0	0	16	7	7.9	.7
	Relatively lean, choice round:											
603	Lean and fat, piece 2½ x 2½ x ¾"	4 oz	113	59	272	30	0	0	16	7.1	8.1	.7
604	Lean only	4 oz	113	65	198	33	0	0	6	2.9	3.3	.3
1701	Steak, rib, broiled, lean	4 oz	113	58	250	32	0	0	13	5.7	6.4	.5
	Steak, broiled, relatively lean,											
606	choice sirloin, lean only	4 oz	113	62	228	34	0	0	9	4	4.6	.4
	Steak, broiled, relatively fat,											
	choice T-bone:											
1063	Lean and fat	4 oz	113	52	349	26	0	0	26	11.8	13.3	1.1
1064	Lean only	4 oz	113	61	232	30	0	0	11	5.1	5.8	.5
	Variety meats:											
1086	Brains, panfried	4 oz	113	71	221	14	0	0	18	6.7	7	3.9
599	Heart, simmered	4 oz	113	64	198	32	<1	0	6	2.9	1.5	1.5
600	Liver, fried	4 oz	113	56	245	30	9	0	9	3.1	1.8	1.9
1062	Tongue, cooked	4 oz	113	56	320	25	<1	0	23	10.1	10.7	.9
607	Beef, canned, corned	4 oz	113	58	283	31	0	0	17	7.5	8.5	.7
608	Beef, dried, cured	1 oz	28	56	46	8	<1	0	1	.5	.5	.1
	LAMB, domestic, cooked:											
	Chop, arm, braised (5.6 oz raw w/bone):											
609	Lean and fat	1 ea	70	44	242	21	0	0	17	7.8	7.3	1.4
610	Lean only	1 ea	55	49	153	19	0	0	8	3.6	3.4	.6
	Chop, loin, broiled (4.2 oz raw w/bone):											
611	Lean and fat	1 ea	64	52	202	16	0	0	15	6.8	6.4	1.2
612	Lean only	1 ea	46	61	99	14	0	0	4	2.1	1.9	.4
1067	Cutlet, avg of lean cuts, cooked	4 oz	113	54	330	28	0	0	23	10.9	10.2	1.9
	Leg, roasted, 3 oz = 4⅛ x 2¼ x ½":											
613	Lean and fat	4 oz	113	57	292	29	0	0	19	8.7	8.1	1.6
614	Lean only	4 oz	113	64	216	32	0	0	9	4.1	3.8	.7
615	Rib, roasted, lean and fat	4 oz	113	48	406	24	0	0	34	15.7	14.7	2.8
616	Rib, roasted, lean only	4 oz	113	60	262	30	0	0	15	7	6.6	1.2
1065	Shoulder, roasted, lean and fat	4 oz	113	56	312	25	0	0	23	10.5	9.8	1.9
1066	Shoulder, roasted, lean only	4 oz	113	63	231	28	0	0	12	5.7	5.3	1.1
	Variety meats:											
1069	Brains, panfried	4 oz	113	76	164	14	0	0	11	4.4	3.7	1.9
1068	Heart, braised	4 oz	113	64	209	28	2	0	9	3.9	2.7	1.1
1070	Sweetbreads, cooked	4 oz	113	60	264	26	0	0	17	8.1	6.5	1.4
1071	Tongue, cooked	4 oz	113	58	311	24	0	0	23	8.8	11.6	1.4
	PORK, cured, cooked (see also Sausages and Lunch Meats)											
617	Bacon, medium slices	3 pce	19	13	109	6	<1	0	9	3.3	4.5	1.1
1087	Breakfast strips, cooked	2 pce	23	27	106	7	<1	0	8	2.9	3.8	1.3
618	Canadian-style bacon	2 pce	47	62	87	11	1	0	4	1.3	1.9	.4
	Ham, roasted:											
619	Lean and fat, 2 pces 4⅛ x 2¼ x ¼"	4 oz	113	64	201	25	0	0	10	3.4	5	1.7
620	Lean only	4 oz	113	68	164	24	2	0	6	2.1	3	.6
621	Ham, canned, roasted, 8% fat	4 oz	113	69	154	24	1	0	6	1.8	2.8	.5
	PORK, fresh, cooked:											
	Chops, loin (cut 3 per lb with bone):											
1291	Braised, lean and fat	1 ea	89	58	213	24	0	0	12	4.5	5.4	1
1292	Lean only	1 ea	80	61	163	23	0	0	7	2.7	3.3	.6
622	Broiled, lean and fat	1 ea	82	58	197	23	0	0	11	3.9	4.8	.8
623	Broiled, lean only	1 ea	74	61	149	22	0	0	6	2.2	2.7	.4

Chol (mg)	Calc (mg)	Iron (mg)	Magn (mg)	Pota (mg)	Sodi (mg)	Zinc (mg)	VT-A (RE)	Thia (mg)	VT-E (a-TE)	Ribo (mg)	Niac (mg)	V-B6 (mg)	Fola (μg)	VT-C (mg)
108	7	3.53	25	319	56	5.55	0	.08	.21	.27	4.21	.37	11	0
108	6	3.91	28	348	58	6.19	0	.08	.2	.29	4.61	.41	12	0
112	10	3.13	28	417	93	7.27	0	.08	.2	.36	6.61	.36	12	0
114	14	2.77	27	394	101	7.01	0	.07	.23	.27	6.75	.34	12	0
96	12	2.61	21	334	71	5.92	0	.08	.27	.19	3.8	.26	8	0
91	11	2.95	28	425	84	7.84	0	.09	.14	.24	4.64	.34	9	0
81	7	2.07	27	406	67	4.87	0	.09	.23	.18	3.92	.4	7	0
78	6	2.2	30	446	70	5.36	0	.1	.12	.19	4.24	.43	8	0
90	15	2.9	30	445	78	7.9	0	.11	.16	.25	5.42	.45	9	0
101	12	3.8	36	455	75	7.37	0	.15	.16	.33	4.84	.51	11	0
76	9	3.06	26	363	72	5.03	0	.1	.24	.24	4.46	.37	8	0
67	7	3.58	32	427	80	6	0	.12	.16	.28	5.23	.44	9	0
2254	10	2.51	17	400	179	1.53	0	.15	2.37	.29	4.27	.44	7	4
218	7	8.49	28	263	71	3.54	0	.16	.81	1.74	4.6	.24	2	2
545	12	7.1	26	411	120	6.16	12123	.24	.72	4.68	16.3	1.62	249	26
121	8	3.83	19	203	68	5.42	0	.03	.4	.4	2.43	.18	6	1
97	14	2.35	16	154	1136	4.03	0	.02	.17	.17	2.75	.15	10	0
12	2	1.26	9	124	972	1.47	0	.02	.04	.06	1.53	.1	3	0
84	17	1.67	18	214	50	4.26	0	.05	.1	.17	4.66	.08	13	0
67	14	1.49	16	186	42	4.02	0	.04	.1	.15	3.48	.07	12	0
64	13	1.16	15	209	49	2.23	0	.06	.08	.16	4.54	.08	11	0
44	9	.92	13	173	39	1.9	0	.05	.07	.13	3.15	.07	11	0
110	12	2.26	25	340	77	4.67	0	.12	.15	.32	7.48	.16	19	0
105	12	2.24	27	354	75	4.97	0	.11	.17	.3	7.45	.17	23	0
101	9	2.4	29	382	77	5.58	0	.12	.2	.33	7.16	.19	26	0
110	25	1.81	23	306	82	3.94	0	.1	.11	.24	7.63	.12	17	0
99	24	2	26	356	91	5.05	0	.1	.17	.26	6.96	.17	25	0
104	23	2.23	26	284	75	5.91	0	.1	.16	.27	6.95	.15	24	0
98	21	2.41	28	299	77	6.83	0	.1	.2	.29	6.51	.17	28	0
2308	14	1.9	16	232	151	1.54	0	.12	1.73	.27	2.79	.12	6	14
281	16	6.24	27	212	71	4.16	0	.19	.79	1.34	4.93	.34	2	8
452	14	2.4	21	329	59	3.03	0	.02	.78	.24	2.89	.06	15	23
214	11	2.97	18	179	76	3.38	0	.09	.36	.47	4.17	.19	3	8
16	2	.31	5	92	303	.62	0	.13	.1	.05	1.39	.05	1	0
24	3	.45	6	107	483	.85	0	.17	.08	.08	1.75	.08	1	0
27	5	.38	10	183	727	.8	0	.39	.15	.09	3.25	.21	2	0
67	9	1.51	25	462	1695	2.79	0	.82	.45	.37	6.95	.35	3	0
60	9	1.67	16	324	1359	3.25	0	.85	.29	.23	4.54	.45	3	0
34	7	1.04	24	393	1282	2.52	0	1.18	.29	.28	5.53	.51	6	0
71	19	.95	17	333	43	2.12	2	.56	.3	.23	3.93	.33	3	1
63	14	.9	16	310	40	1.98	2	.53	.3	.21	3.67	.31	3	<1
67	27	.66	20	294	48	1.85	2	.88	.27	.24	4.3	.35	5	<1
61	23	.63	20	278	44	1.76	2	.85	.31	.23	4.1	.35	4	<1

Table A-1

Food Composition (Computer code number is for West Diet Analysis program) (For purposes of calculations, use "0" for t, <1, <.1, <.01, etc.)

Computer Code Number	Food Description	Measure	Wt (g)	H_2O (%)	Ener (cal)	Prot (g)	Carb (g)	Dietary Fiber (g)	Fat (g)	Fat Breakdown (g) Sat	Mono	Poly
	MEATS: BEEF, LAMB, PORK and others—Continued											
624	Panfried, lean and fat	1 ea	78	53	216	23	0	0	13	4.7	5.5	1.5
625	Panfried, lean only	1 ea	63	59	152	16	0	0	10	3.2	3.9	1.2
626	Leg, roasted, lean and fat	4 oz	113	55	308	30	0	0	20	7.3	8.9	1.9
627	Leg, roasted, lean only	4 oz	113	61	233	35	0	0	9	3.2	4.3	.9
628	Rib, roasted, lean and fat	4 oz	113	56	288	31	0	0	17	6.7	7.9	1.4
629	Rib, roasted, lean only	4 oz	113	59	252	32	0	0	13	4.9	5.9	1
630	Shoulder, braised, lean and fat	4 oz	113	48	372	32	0	0	26	9.6	11.8	2.6
631	Shoulder, braised, lean only	4 oz	113	54	280	36	0	0	14	4.7	6.5	1.3
1088	Spareribs, cooked, yield from 1 lb raw with bone	4 oz	113	40	449	33	0	0	34	12.5	15.3	3.1
1095	Rabbit, roasted (1 cup meat = 140 g)	4 oz	113	61	223	33	0	0	9	4	2	3
	VEAL, cooked:											
632	Cutlet, braised or broiled, 4⅛ x 2¼ x ½"	4 oz	113	52	321	34	0	0	19	7.6	7.6	1.3
633	Rib roasted, lean, 2 pieces 4⅛ x 2¼ x ¼"	4 oz	113	60	258	27	0	0	16	6.1	6.1	1.1
634	Liver, panfried	4 oz	113	67	186	24	3	0	8	3.3	1.9	2.2
1096	Venison (deer meat), roasted	4 oz	113	65	179	34	0	0	4	1.4	1	.7
	MEATS: POULTRY and POULTRY PRODUCTS											
	CHICKEN, cooked:											
	Fried, batter dipped:											
635	Breast	1 ea	280	52	728	69	25	1	37	9.9	15.3	8.6
636	Drumstick	1 ea	72	53	193	16	6	<1	11	3	4.7	2.7
637	Thigh	1 ea	86	51	238	19	8	<1	14	3.8	5.9	3.3
638	Wing	1 ea	49	46	159	10	5	<1	11	2.9	4.5	2.5
	Fried, flour coated:											
639	Breast	1 ea	196	57	435	62	3	<1	17	4.9	7.1	3.8
1212	Breast, without skin	1 ea	86	60	161	29	<1	<1	4	1.1	1.5	.9
640	Drumstick	1 ea	49	57	120	13	1	<1	7	1.8	2.7	1.6
641	Thigh	1 ea	62	54	162	17	2	<1	9	2.5	3.7	2.1
1099	Thigh, without skin	1 ea	52	59	113	15	1	<1	5	1.4	2	1.3
642	Wing	1 ea	32	49	103	8	1	<1	7	1.9	2.9	1.6
	Roasted:											
643	All types of meat	1 c	140	64	266	40	0	0	10	2.9	3.8	2.4
644	Dark meat	1 c	140	63	287	38	0	0	14	3.7	5.2	3.2
645	Light meat	1 c	140	65	242	43	0	0	6	1.8	2.2	1.4
646	Breast, without skin	1 ea	172	65	284	53	0	0	6	1.8	2.2	1.3
647	Drumstick, without skin	1 ea	44	67	76	12	0	0	2	.7	.8	.6
1703	Leg, without skin	1 ea	95	65	181	26	0	0	8	2.2	2.9	1.9
648	Thigh	1 ea	62	59	153	16	0	0	10	2.7	3.9	2.1
1100	Thigh, without skin	1 ea	52	63	109	13	0	0	6	1.6	2.2	1.3
649	Stewed, all types	1 c	140	67	248	38	0	0	9	2.6	3.5	2.2
656	Canned, boneless chicken	4 oz	113	69	186	25	0	0	9	2.5	3.6	2
1102	Gizzards, simmered	1 c	145	67	222	39	2	0	5	1.5	1.3	1.5
1101	Hearts, simmered	1 c	145	65	268	38	<1	0	11	3.3	2.9	3.3
2300	Liver, simmered: Ounce	3 oz	85	68	133	21	1	0	5	1.6	1.1	.8
1098	Liver, simmered: Piece = 20 g	6 ea	120	68	188	29	1	0	7	2.2	1.6	1.1
	DUCK, roasted:											
1293	Meat with skin, about 2.7 cups	½ ea	382	52	1287	73	0	0	108	36.9	49.3	13.9
651	Meat only, about 1.5 cups	½ ea	221	64	444	52	0	0	25	9.2	8.2	3.2
	GOOSE, domesticated, roasted:											
1294	Meat only, about 4.2 cups	½ ea	591	57	1406	173	0	0	75	23.6	40.2	10.9
1295	Meat with skin, about 5.5 cups	½ ea	774	52	2360	195	0	0	170	53.2	80.5	24.8
	TURKEY:											
	Roasted, meat only:											
652	Dark meat	4 oz	113	63	211	33	0	0	8	2.7	1.8	2.5
653	Light meat	4 oz	113	66	177	34	0	0	4	1.2	.6	1
654	All types, chopped or diced	1 c	140	65	238	42	0	0	7	2.3	1.5	2
1103	Ground, cooked	4 oz	113	59	266	31	0	0	15	4.1	5.5	3.6
1106	Gizzard, cooked	2 ea	134	65	218	39	1	0	5	1.5	1	1.5
1107	Heart, cooked	4 ea	64	64	113	17	1	0	4	1.1	.8	1.1
1108	Liver, cooked	1 ea	75	66	127	18	3	0	4	1.4	1.1	.8

Chol (mg)	Calc (mg)	Iron (mg)	Magn (mg)	Pota (mg)	Sodi (mg)	Zinc (mg)	VT-A (RE)	Thia (mg)	VT-E (a-TE)	Ribo (mg)	Niac (mg)	V-B6 (mg)	Fola (μg)	VT-C (mg)
72	21	.71	23	332	62	1.8	2	.89	.32	.24	4.37	.37	5	1
52	14	.67	16	230	49	2.44	1	.46	.3	.23	2.8	.26	3	<1
106	16	1.14	25	398	68	3.34	3	.72	.34	.35	5.16	.45	11	<1
108	8	1.29	33	442	73	3.4	3	.91	.46	.4	5.56	.38	3	<1
82	32	1.06	24	476	52	2.33	2	.82	.41	.34	6.92	.37	3	<1
80	29	1.11	25	494	53	2.41	2	.86	.55	.36	7.25	.38	3	<1
123	20	1.82	21	417	99	4.72	3	.61	.5	.35	5.89	.4	5	<1
129	9	2.2	25	458	115	5.62	3	.68	.58	.41	6.71	.46	6	<1
137	53	2.09	27	362	105	5.2	3	.46	.52	.43	6.19	.4	5	0
93	21	2.57	24	433	53	2.57	0	.1	.96	.24	9.53	.53	12	0
133	32	1.23	27	316	90	4.1	0	.04	.45	.34	10.2	.29	16	0
124	12	1.1	25	333	104	4.62	0	.06	.4	.3	7.89	.28	15	0
634	8	2.96	21	232	60	10.8	9095	.15	.42	2.19	9.58	.55	858	35
127	8	5.05	27	379	61	3.11	0	.2	.28	.68	7.58	.42	5	0
238	56	3.5	67	563	770	2.66	56	.32	2.97	.41	29.4	1.2	42	0
62	12	.97	14	134	194	1.68	19	.08	.88	.15	3.67	.19	13	0
80	15	1.25	18	165	248	1.75	25	.1	1.05	.19	4.92	.22	16	0
39	10	.63	8	68	157	.68	17	.05	.52	.07	2.58	.15	9	0
174	31	2.33	59	508	149	2.16	29	.16	1.12	.26	26.9	1.14	12	0
78	14	.98	27	237	68	.93	6	.07	.36	.11	12.7	.55	3	0
44	6	.66	11	112	44	1.42	12	.04	.41	.11	2.96	.17	5	0
60	9	.92	15	147	55	1.56	18	.06	.52	.15	4.31	.2	7	0
53	7	.76	13	135	49	1.45	11	.05	.3	.13	3.7	.2	5	0
26	5	.4	6	57	25	.56	12	.02	.18	.04	2.14	.13	2	0
125	21	1.69	35	340	120	2.94	22	.1	.58	.25	12.8	.66	8	0
130	21	1.86	32	336	130	3.92	31	.1	.81	.32	9.17	.5	11	0
119	21	1.48	38	346	108	1.72	13	.09	.37	.16	17.4	.84	6	0
146	26	1.79	50	440	127	1.72	10	.12	.66	.2	23.6	1.03	7	0
41	5	.57	11	108	42	1.4	8	.03	.25	.1	2.68	.17	4	0
89	11	1.24	23	230	86	2.72	18	.07	.55	.22	6	.35	8	0
58	7	.83	14	138	52	1.46	30	.04	.35	.13	3.95	.19	4	0
49	6	.68	12	124	46	1.34	10	.04	.3	.12	3.4	.18	4	0
116	20	1.64	29	252	98	2.79	21	.07	.42	.23	8.57	.36	8	0
70	16	1.79	14	156	568	1.59	38	.02	.24	.15	7.15	.4	5	2
281	14	6.02	29	260	97	6.35	81	.04	2.29	.35	5.77	.17	77	2
351	28	13.1	29	191	70	10.6	13	.1	2.32	1.07	4.06	.46	116	3
536	12	7.2	18	119	43	3.69	4176	.13	1.45	1.49	3.78	.49	655	13
757	17	10.2	25	168	61	5.21	5895	.18	2.04	2.1	5.34	.7	924	19
321	42	10.3	61	779	225	7.11	241	.66	2.5	1.03	18.5	.69	23	0
197	26	5.97	44	557	144	5.75	51	.57	1.55	1.04	11.3	.55	22	0
567	83	17	148	2293	449	18.7	71	.54	9.16	2.3	24.1	2.78	71	0
704	101	21.9	170	2546	542	20.3	163	.6	13.5	2.5	32.3	2.86	15	0
96	36	2.63	27	328	89	5.04	0	.07	.94	.28	4.12	.41	10	0
78	21	1.53	32	345	72	2.31	0	.07	.12	.15	7.73	.61	7	0
106	35	2.49	36	417	98	4.34	0	.09	.59	.25	7.62	.64	10	0
115	28	2.18	27	305	121	3.23	0	.06	.45	.19	5.45	.44	8	0
311	20	7.29	25	283	72	5.57	74	.04	.27	.44	4.11	.16	70	2
145	8	4.41	14	117	35	3.37	5	.04	.13	.56	2.08	.2	51	1
470	8	5.85	11	146	48	2.32	2805	.04	2.41	1.07	4.46	.39	500	1

Table A-1

Food Composition (Computer code number is for West Diet Analysis program) (For purposes of calculations, use "0" for t, <1, <.1, <.01, etc.)

Computer Code Number	Food Description	Measure	Wt (g)	H_2O (%)	Ener (cal)	Prot (g)	Carb (g)	Dietary Fiber (g)	Fat (g)	Fat Breakdown (g) Sat	Mono	Poly
	MEATS: POULTRY and POULTRY PRODUCTS—Continued											
	POULTRY FOOD PRODUCTS (see also items in Sausages & Lunchmeats section):											
1567	Chicken patty, breaded, cooked	1 ea	75	49	213	12	11	<1	13	4.1	6.4	1.6
659	Turkey and gravy, frozen package	3 oz	85	85	57	5	4	<1	2	.8	.8	.4
	Turkey breast, Louis Rich:											
1104	Barbecued	2 oz	56	72	58	12	2	0	<1	.2	.2	.1
1943	Hickory smoked	1 pce	80		80	16	2	0	1	0		
1947	Honey roasted	1 pce	80		80	16	3	0	1	.5		
1945	Oven roasted	1 pce	80		70	16		0	1	0		
661	Turkey patty, breaded, fried	2 oz	57	50	161	8	9	<1	10	2.7	4.3	2.7
662	Turkey, frozen, roasted, seasoned	4 oz	113	68	175	24	3	0	7	2.1	1.4	1.9
1704	Turkey roll, light meat	1 pce	28	72	41	5	<1	0	2	.6	.7	.5
	MEATS: SAUSAGES and LUNCHMEATS (see also Poultry Food Products)											
1072	Beerwurst/beer salami, beef	1 oz	28	53	92	3	<1	0	8	3.6	3.9	.3
1074	Beerwurst/beer salami, pork	1 oz	28	61	67	4	1	0	5	1.8	2.5	.7
1075	Berliner sausage	1 oz	28	61	64	4	1	0	5	1.7	2.2	.4
	Bologna:											
1297	Beef	1 pce	23	55	72	3	<1	0	7	2.8	3.2	.3
2115	Beef, light, Oscar Mayer	1 pce	28	65	56	3	2	0	4	1.6	2	.1
663	Beef & pork	1 pce	28	54	88	3	1	0	8	3	3.7	.7
2155	Healthy Favorites	1 pce	23		22	3	1	0	<1	0		
1298	Pork	1 pce	23	61	57	4	<1	0	5	1.6	2.2	.5
2114	Regular, light, Oscar Mayer	1 pce	28	65	56	3	2	0	4	1.6	2	.4
664	Turkey	1 pce	28	65	56	4	<1	0	4	1.4	1.3	1.2
1970	Turkey, Louis Rich	1 pce	28	67	57	3	<1	0	5	1.5	1.8	1.3
665	Braunschweiger sausage	2 pce	57	48	205	8	2	0	18	6.2	8.5	2.1
1073	Bratwurst, link	1 ea	70	51	226	10	2	0	19	6.9	9.3	2
666	Brown & serve sausage links, cooked	2 ea	26	45	102	4	1	0	10	3.4	4.4	1
1089	Cheesefurter/cheese smokie	2 ea	86	52	281	12	1	0	25	9	11.8	2.6
2157	Chicken breast, Healthy Favorites	4 pce	52		40	9	1	0	0	0	0	0
1556	Chorizo, pork & beef	1 ea	60	32	273	15	1	0	23	8.6	11	2.1
1090	Corned beef loaf, jellied	1 pce	28	69	43	6	0	0	2	.7	.7	.1
	Frankfurters:											
1077	Beef, large link, 8/package	1 ea	57	55	180	7	1	0	16	6.9	7.9	.8
1078	Beef and pork, large link, 8/package	1 ea	57	54	182	6	1	0	17	6.2	8	1.6
667	Beef and pork, small link, 10/pkg	1 ea	45	54	144	5	1	0	13	4.9	6.3	1.2
668	Turkey frankfurter, 10/package	1 ea	45	63	102	6	1	0	8	2.7	2.5	2.2
1968	Turkey/chicken frank 8/pkg	1 ea	43		80	6	1	0	6	2		
	Ham:											
669	Ham lunchmeat, canned, 3 x 2 x ½"	1 pce	21	52	70	3	<1	0	6	2.3	3	.7
670	Chopped ham, packaged	2 pce	42	64	96	7	0	0	7	2.4	3.4	.9
2156	Honey ham, Healthy Favorites	4 pce	52	73	55	9	2	0	1	.4	.8	.1
2113	Oscar Mayer lower sodium ham	1 pce	21	73	23	3	1	0	1	.3	.4	.1
673	Turkey ham lunchmeat	2 pce	57	71	73	11	<1	0	3	1	.7	.9
1091	Kielbasa sausage	1 pce	26	54	81	3	1	0	7	2.6	3.4	.8
1092	Knockwurst sausage, link	1 ea	68	55	209	8	1	0	19	6.9	8.7	2
1093	Mortadella lunchmeat	2 pce	30	52	93	5	1	0	8	2.8	3.4	.9
1097	Olive loaf lunchmeat	2 pce	57	58	134	7	5	<1	9	3.3	4.5	1.1
1952	Turkey breast, fat free	1 pce	28	77	22	4	1	0	<1	.1	.1	t
1080	Turkey pastrami	2 pce	57	71	80	10	1	0	4	1	1.2	.9
1969	Turkey salami	1 pce	28	72	41	4	<1	0	3	.9	1	.8
1081	Pepperoni sausage	2 pce	11	27	55	2	<1	0	5	1.8	2.3	.5
1094	Pickle & pimento loaf	2 pce	57	57	149	7	3	<1	12	4.5	5.5	1.5
1082	Polish sausage	1 oz	28	53	91	4	<1	0	8	2.9	3.8	.9
674	Pork sausage, cooked, link, small	2 ea	26	45	96	5	<1	0	8	2.8	4.1	.8
1079	Pork sausage, cooked, patty	4 oz	113	45	417	22	1	0	35	12.1	17.7	3.3
675	Salami, pork and beef	2 pce	57	60	143	8	1	0	11	4.6	5.2	1.1
677	Salami, pork and beef, dry	3 pce	30	35	125	7	1	0	10	3.7	5.1	1
676	Salami, turkey	2 pce	57	66	112	9	<1	0	8	2.3	2.6	2
	Sandwich spreads:											
1300	Ham salad spread	2 tbs	30	63	65	3	3	0	5	1.5	2.2	.8
678	Pork and beef	2 tbs	30	60	70	2	4	<1	5	1.8	2.3	.8
1296	Chicken/turkey	2 tbs	26	66	52	3	2	0	4	.9	.8	1.6

Chol (mg)	Calc (mg)	Iron (mg)	Magn (mg)	Pota (mg)	Sodi (mg)	Zinc (mg)	VT-A (RE)	Thia (mg)	VT-E (a-TE)	Ribo (mg)	Niac (mg)	V-B6 (mg)	Fola (µg)	VT-C (mg)
45	12	.94	15	185	399	.78	22	.07	1.46	.1	5.04	.23	8	<1
15	12	.79	7	52	471	.59	11	.02	.3	.11	1.53	.08	3	0
25	14	.62	16	175	599	.59	0	.02		.06	5.35	.22	2	0
35	0	.72			1060		0							0
35	0	.72			940		0							0
35	0				910		0							0
35	8	1.25	9	157	456	.82	6	.06	1.36	.11	1.31	.11	16	0
60	6	1.84	25	337	768	2.87	0	.05	.43	.18	7.09	.3	6	0
12	11	.36	4	70	137	.44	0	.02	.04	.06	1.96	.09	1	0
17	3	.42	3	49	288	.68	0	.02	.05	.03	.95	.05	1	0
16	2	.21	4	71	347	.48	0	.15	.06	.05	.91	.1	1	0
13	3	.32	4	79	363	.69	0	.11	.06	.06	.87	.06	1	0
13	3	.38	3	36	226	.5	0	.01	.04	.02	.55	.03	1	0
13	4	.34	4	44	314	.53	0							0
15	3	.42	3	50	285	.54	0	.05	.06	.04	.72	.05	1	0
7		.18			255									
14	3	.18	3	65	272	.47	0	.12	.06	.04	.9	.06	1	0
15	14	.39	5	46	312	.45	0							0
28	23	.43	4	56	246	.49	0	.01	.15	.05	.99	.06	2	0
22	34	.45	5	51	242	.57	0	.01		.05	1.08	.05		0
89	5	5.34	6	113	652	1.6	2405	.14	.2	.87	4.77	.19	25	0
44	34	.72	11	197	778	1.47	0	.17	.19	.16	2.31	.09	3	0
16	2	.62	4	70	248	.3	0	.21	.06	.09	.96	.06	1	0
58	50	.93	11	177	931	1.94	33	.21	.27	.14	2.49	.11	3	0
25		.72			620									
53	5	.95	11	239	741	2.05	0	.38	.13	.18	3.08	.32	1	0
13	3	.57	3	28	267	1.15	0	0	.05	.03	.49	.03	2	0
35	11	.81	2	95	585	1.24	0	.03	.11	.06	1.38	.07	2	0
28	6	.66	6	95	638	1.05	0	.11	.14	.07	1.5	.07	2	0
22	5	.52	4	75	504	.83	0	.09	.11	.05	1.18	.06	2	0
48	48	.83	6	81	642	1.4	0	.02	.28	.08	1.86	.1	4	0
40	60	1.08			480		0							0
13	1	.15	2	45	271	.31	0	.08	.05	.04	.66	.04	1	<1
21	3	.35	7	134	576	.81	0	.26	.11	.09	1.63	.15	<1	0
24	6	.7	18	144	635	1.02	0							0
9	1	.3	5	197	174	.42	0							0
32	6	1.57	9	185	568	1.68	0	.03	.36	.14	2.01	.14	3	0
17	11	.38	4	70	280	.52	0	.06	.06	.06	.75	.05	1	0
39	7	.62	7	135	687	1.13	0	.23	.39	.09	1.86	.12	1	0
17	5	.42	3	49	374	.63	0	.04	.07	.05	.8	.04	1	0
22	62	.31	11	169	846	.79	34	.17	.14	.15	1.05	.13	1	0
9	3	.34	8	59	387	.24	0							0
31	5	.95	8	148	596	1.23	0	.03	.12	.14	2.01	.15	3	0
21	11	.35	6	61	281	.65	0							0
9	1	.15	2	38	224	.27	0	.03	.02	.03	.55	.03	<1	0
21	54	.58	10	194	792	.8	4	.17	.14	.14	1.17	.11	3	0
20	3	.4	4	66	245	.54	0	.14	.06	.04	.96	.05	1	<1
22	8	.33	4	94	336	.65	0	.19	.07	.07	1.18	.09	1	<1
94	36	1.42	19	408	1462	2.84	0	.84	.29	.29	5.11	.37	2	2
37	7	1.52	9	113	607	1.22	0	.14	.12	.21	2.02	.12	1	0
24	2	.45	5	113	558	.97	0	.18	.08	.09	1.46	.15	1	0
47	11	.92	9	139	572	1.03	0	.04	.32	.1	2.01	.14	2	0
11	2	.18	3	45	274	.33	0	.13	.52	.04	.63	.04	<1	0
11	4	.24	2	33	304	.31	3	.05	.52	.04	.52	.04	1	0
8	3	.16	3	48	98	.27	11	.01	.57	.02	.43	.03	1	<1

Table A-1

Food Composition (Computer code number is for West Diet Analysis program) (For purposes of calculations, use "0" for t, <1, <.1, <.01, etc.)

Computer Code Number	Food Description	Measure	Wt (g)	H_2O (%)	Ener (cal)	Prot (g)	Carb (g)	Dietary Fiber (g)	Fat (g)	Fat Breakdown (g) Sat	Mono	Poly
	MEATS: SAUSAGES and LUNCHMEATS—Continued											
1084	Smoked link sausage, beef and pork	1 ea	68	52	228	9	1	0	21	7.2	9.7	2.2
1083	Smoked link sausage, pork	1 ea	68	39	265	15	1	0	22	7.7	9.9	2.6
1085	Summer sausage	2 pce	46	51	154	7	<1	0	14	5.5	6	.6
1076	Turkey breakfast sausage	1 pce	28	60	64	6	0	0	5	1.6	1.8	1.2
679	Vienna sausage, canned	2 ea	32	60	89	3	1	0	8	3	4	.5
	MIXED DISHES and FAST FOODS											
	MIXED DISHES:											
1445	Almond Chicken	1 c	242	77	275	20	18	4	14	2	5.3	5.8
1981	Baked beans, fat free, honey	½ c	120	73	110	7	24	7	0	0	0	0
1454	Bean cake	1 ea	32	23	130	2	16	1	7	1	2.9	2.6
680	Beef stew w/ vegetables, homemade	1 c	245	82	218	16	15	2	10	4.9	4.5	.5
1109	Beef stew w/ vegetables, canned	1 c	245	82	194	14	17	2	8	2.4	3.1	.3
1116	Beef, macaroni, tomato sauce casserole	1 c	226	76	255	16	26	2	10	3.8	4.1	.5
2295	Beef fajita	1 ea	223	63	409	17	46	4	17	5.1	7.6	3.9
1265	Beef flauta	1 ea	113	49	360	16	13	2	27	4.9	11.6	9.1
681	Beef pot pie, homemade	1 pce	210	55	517	21	39	3	30	8.4	14.7	7.3
1898	Broccoli, batter fried	1 c	85	74	123	3	9	2	9	1.3	2.2	4.9
1462	Buffalo wings/spicy chicken wings	2 pce	32	53	98	8	<1	<1	7	1.8	2.8	1.6
1675	Carrot raisin salad	½ c	88	58	204	1	21	2	14	2	3.9	7.3
2248	Cheeseburger deluxe	1 ea	219	52	563	28	38		33	15	12.6	2
682	Chicken à la king, homemade	1 c	245	68	468	27	12	1	34	12.7	14.3	6.2
683	Chicken & noodles, homemade	1 c	240	71	367	22	26	2	18	5.9	7.1	3.5
684	Chicken chow mein, canned	1 c	250	89	95	6	18	2	1	0	.1	.8
685	Chicken chow mein, homemade	1 c	250	78	255	31	10	1	10	2.4	4.3	3.1
1266	Chicken fajita	1 ea	223	61	405	22	50	4	13	2.4	6	3.5
1264	Chicken flauta	1 ea	113	52	343	14	13	2	27	4.3	11.1	9.6
686	Chicken pot pie, homemade (⅓)	1 pce	232	57	545	23	42	3	31	10.9	14.5	5.8
1672	Chili con carne	½ c	127	77	128	12	11	2	4	1.7	1.7	.3
1112	Chicken salad with celery	½ c	78	53	268	11	1	<1	25	3.1	4.5	15.8
1382	Chicken teriyaki, breast	1 ea	128	67	176	26	7	<1	4	.9	1	.9
687	Chili with beans, canned	1 c	256	75	287	15	30	11	14	6	6	.9
1479	Chinese pastry	1 oz	28	46	67	1	13	<1	1	.2	.4	.8
688	Chop suey with beef & pork	1 c	220	63	425	22	31	3	24	5	8.6	9.3
690	Coleslaw	1 c	132	74	195	2	17	2	15	2.1	3.2	8.5
689	Corn pudding	1 c	250	76	273	11	32	4	13	6.4	4.3	1.8
1110	Corned beef hash, canned	1 c	220	67	398	19	23	1	25	11.9	10.9	.9
1255	Deviled egg (½ egg + filling)	1 ea	31	69	62	4	<1	0	5	1.2	1.7	1.5
	Egg foo yung patty:											
1467	Meatless	1 ea	86	78	113	6	3	1	8	1.9	3.3	2.1
1458	With beef	1 ea	86	74	129	9	3	<1	9	2.2	3.2	2.4
1465	With chicken	1 ea	86	74	130	9	4	<1	9	2.1	3.1	2.5
1602	Egg roll, meatless	1 ea	64	70	102	3	10	1	6	1.2	2.5	1.6
1550	Egg roll, with meat	1 ea	64	66	114	5	9	1	6	1.5	2.7	1.5
1113	Egg salad	1 c	183	57	586	17	3	0	56	10.6	17.4	24.2
691	French toast w/wheat bread, homemade	1 pce	65	54	151	5	16	<1	7	2	3	1.7
1355	Green pepper, stuffed	1 ea	172	75	229	11	20	2	11	5	4.9	.5
1487	Hot & sour soup (Chinese)	1 c	244	88	133	12	5	<1	6	2	2.5	1
2242	Hamburger deluxe	1 ea	110	49	279	13	27		13	4.1	5.3	2.6
1997	Hummous/hummus	¼ c	62	65	106	3	12	3	5	.8	2.2	2
	Lasagna:											
1346	With meat, homemade	1 pce	245	67	382	22	39	3	15	7.7	5	.9
1111	Without meat, homemade	1 pce	218	69	298	15	39	3	9	5.4	2.4	.6
1117	Frozen entree	1 ea	340	75	390	24	42	4	14	6.7	5.5	.8
1606	Lo mein, meatless	1 c	200	82	134	6	27	3	1	.1	.1	.3
1607	Lo mein, with meat	1 c	200	70	285	17	31	2	10	1.9	2.9	4.5
692	Macaroni & cheese, canned	1 c	240	80	228	9	26	1	10	4.2	3.1	1.4
693	Macaroni & cheese, homemade	1 c	200	58	430	17	40	1	22	8.9	8.8	3.6
1115	Macaroni salad, no cheese	1 c	177	60	461	5	28	2	37	4	6	25.5
1120	Meat loaf, beef	1 pce	87	63	182	16	4	<1	11	4.4	4.7	.5
1119	Meat loaf, beef and pork (⅓)	1 pce	87	60	205	15	4	<1	14	5.2	6.3	.9
1303	Moussaka (lamb & eggplant)	1 c	250	82	237	16	13	4	13	4.6	5.4	1.9

Chol (mg)	Calc (mg)	Iron (mg)	Magn (mg)	Pota (mg)	Sodi (mg)	Zinc (mg)	VT-A (RE)	Thia (mg)	VT-E (a-TE)	Ribo (mg)	Niac (mg)	V-B6 (mg)	Fola (µg)	VT-C (mg)
48	7	.99	8	129	643	1.43	0	.18	.15	.12	2.2	.12	1	0
46	20	.79	13	228	1020	1.92	0	.48	.17	.17	3.08	.24	3	1
34	6	1.17	6	125	571	1.18	0	.07	.1	.15	1.98	.12	1	0
23	5	.51	6	75	188	.96	0	.03	.14	.08	1.4	.08	1	0
17	3	.28	2	32	305	.51	0	.03	.07	.03	.51	.04	1	0
35	81	2	59	551	615	1.54	75	.08	2.64	.19	8.59	.42	31	10
0	40	2.7			135		450							12
0	3	.67	6	57	55	.16	0	.07	1.14	.05	.55	.02	9	0
64	29	2.94	40	613	292	5.29	568	.15	.49	.17	4.66	.28	37	17
34	29	2.21	39	426	1006	4.24	262	.07	.34	.12	2.45	.2	31	7
39	26	2.7	40	522	862	3.14	97	.22	.57	.22	4.31	.29	20	14
26	76	3.69	38	427	850	2.38	52	.46	2.08	.3	4.73	.32	25	29
45	50	2.15	29	292	187	4.18	15	.07	4	.15	2.13	.25	10	14
44	29	3.78	6	334	596	3.17	519	.29	3.78	.29	4.83	.24	29	6
16	67	.94	20	242	62	.38	102	.08	2.1	.13	.75	.11	43	53
26	5	.4	6	59	61	.56	17	.01	.23	.04	2.06	.13	1	<1
10	26	.75	14	317	118	.19	1452	.08	5.03	.05	.64	.22	9	5
88	206	4.66	44	445	1108	4.6	129	.39	1.18	.46	7.38	.28	81	8
186	127	2.45	20	404	760	1.8	272	.1	.98	.42	5.39	.23	11	12
96	26	2.16	26	149	600	1.53	10	.05		.17	4.32	.19	10	0
7	45	1.25	14	418	725	1.3	28	.05	.05	.1	1	.09	12	12
77	57	2.5	28	473	718	2.12	50	.07	.75	.22	4.25	.41	19	10
41	83	3.7	51	532	439	1.77	55	.48	2.04	.37	6.64	.35	41	22
37	52	.97	27	243	189	1.18	21	.05	4.06	.1	3.21	.22	8	14
72	70	3.02	25	343	594	2	735	.32	3.25	.32	4.87	.46	29	5
67	34	2.6	23	347	505	1.79	84	.06	.81	.57	1.24	.16	23	1
48	16	.62	11	138	201	.79	31	.03	6.27	.07	3.28	.34	8	1
80	27	1.75	36	309	1866	1.94	16	.08	.35	.2	8.69	.46	13	3
43	120	8.78	115	934	1336	5.12	87	.12	1.88	.27	.92	.34	58	4
0	8	.51	6	28	3	.18	<1	.05	.25	<.01	.41	.02	1	0
46	39	4.16	54	515	818	3.52	134	.36	1.82	.37	5.63	.44	44	20
7	45	.96	12	236	356	.26	66	.05	5.28	.04	.11	.14	51	11
250	100	1.4	37	403	138	1.25	90	1.03	.52	.32	2.47	.29	63	7
73	29	4.4	36	440	1188	3.3	0	.02	.48	.2	4.62	.43	20	0
121	15	.35	3	36	94	.3	49	.02	.86	.14	.02	.05	13	0
184	31	1.04	12	118	310	.7	86	.04	1.57	.25	.44	.09	30	5
180	26	1.11	11	145	184	1.16	92	.05	1.79	.24	.74	.15	22	3
182	27	.86	12	144	187	.81	95	.05	1.87	.25	.96	.13	22	3
30	12	.76	9	98	306	.25	15	.08	.81	.1	.81	.05	12	3
37	13	.78	10	124	304	.46	14	.16	.78	.13	1.31	.1	9	2
574	74	1.8	13	180	665	1.42	260	.08	8.87	.66	.09	.46	62	0
76	64	1.09	11	86	311	.44	81	.13	.31	.21	1.06	.05	15	<1
34	16	1.77	20	233	201	2.28	44	.15	.75	.1	2.74	.3	17	55
23	29	1.83	27	351	1562	1.17	2	.19	.12	.22	4.58	.15	12	1
26	63	2.63	22	227	504	2.06	9	.23	.82	.2	3.69	.12	52	2
0	31	.97	18	108	151	.68	1	.06	.62	.03	.25	.25	37	5
56	258	3.22	50	461	745	3.25	158	.23	1.15	.33	3.97	.21	19	15
31	252	2.5	44	375	714	1.77	156	.22	1.07	.27	2.49	.17	17	15
55	263	3.44	64	752	823	3.7	248	.29	3.45	.39	5.07	.32	28	41
0	47	2.06	33	389	623	.92	130	.23	.35	.24	2.83	.19	49	12
30	25	2.11	39	246	276	1.63	6	.37	1.51	.24	4.25	.28	41	8
24	199	.96	31	139	730	1.2	73	.12	.14	.24	.96	.02	8	<1
42	362	1.8	37	240	1086	1.2	234	.2	.12	.4	1.8	.05	10	1
27	31	1.56	20	170	352	.53	44	.18	10.3	.1	1.43	.33	20	4
84	29	1.61	14	187	145	3.23	23	.05	.31	.22	2.61	.15	11	1
84	33	1.42	14	213	381	2.68	23	.19	.32	.22	2.68	.17	10	1
97	68	1.79	40	557	432	2.56	105	.15	.81	.31	4.14	.23	45	6

Table A-1

Food Composition (Computer code number is for West Diet Analysis program) (For purposes of calculations, use "0" for t, <1, <.1, <.01, etc.)

Computer Code Number	Food Description	Measure	Wt (g)	H_2O (%)	Ener (cal)	Prot (g)	Carb (g)	Dietary Fiber (g)	Fat (g)	Fat Breakdown (g) Sat	Mono	Poly
	MIXED DISHES and FAST FOODS—Continued											
1899	Mushrooms, batter fried	5 ea	70	66	148	2	8	1	12	2.1	3	6.4
715	Potato salad with mayonnaise and eggs	½ c	125	76	179	3	14	2	10	1.8	3.1	4.7
1674	Pizza, combination, 1/12 of 12" round	1 pce	79	48	184	13	21		5	1.5	2.5	.9
1673	Pizza, pepperoni, 1/12 of 12" round	1 pce	71	46	181	10	20		7	2.2	3.1	1.2
694	Quiche Lorraine ⅛ of 8" quiche	1 pce	176	54	508	20	20	1	39	17.6	13.8	4.9
1449	Ramen noodles, cooked	1 c	227	82	156	6	29	3	2	.4	.5	.5
1671	Ravioli, meat	½ c	125	68	194	10	18	1	9	2.9	3.6	1
1597	Fried rice (meatless)	1 c	166	68	264	5	34	1	12	1.7	3	6.3
2142	Roast beef hash	½ c	117	66	230	9	11	1	16	7	5.8	3.2
	Spaghetti (enriched) in tomato sauce											
	With cheese:											
695	Canned	1 c	250	80	190	5	38	2	1	0	.4	.5
696	Home recipe	1 c	250	77	260	9	37	2	9	2	5.4	1.2
	With meatballs:											
697	Canned	1 c	250	78	258	12	28	6	10	2.1	3.9	3.9
698	Home recipe	1 c	248	70	332	19	39	8	12	3.3	6.3	2.2
716	Spinach soufflé	1 c	136	74	219	11	3	3	19	9.5	5.7	2.2
1553	Sweet & sour pork	1 c	226	77	231	15	25	1	8	2.2	2.9	2.4
1263	Sweet & sour chicken breast	1 ea	131	79	117	8	15	1	3	.6	.8	1.5
1515	Three bean salad	1 c	150	82	139	4	13	3	8	1.2	1.9	4.9
717	Tuna salad	1 c	205	63	383	33	19	0	19	3.2	5.9	8.4
1121	Tuna noodle casserole, homemade	1 c	202	75	237	17	25	1	7	1.9	1.5	3.2
1270	Waldorf salad	1 c	137	58	408	4	13	2	40	4.1	7.3	27
	FAST FOODS and SANDWICHES (see end of this appendix for additional Fast Foods)											
699	Burrito, beef & bean	1 ea	116	52	255	11	33	3	9	4.2	3.5	.6
700	Burrito, bean	1 ea	109	52	225	7	36	4	7	3.5	2.4	.6
2106	Burrito, chicken con queso	1 ea	306	76	280	12	53	5	6	1.5		
701	Cheeseburger with bun, regular	1 ea	154	55	359	18	28		20	9.2	7.2	1.5
702	Cheeseburger with bun, 4-oz patty	1 ea	166	51	417	21	35		21	8.7	7.8	2.7
703	Chicken patty sandwich	1 ea	182	47	515	24	39	1	29	8.5	10.4	8.4
704	Corndog	1 ea	175	47	460	17	56		19	5.2	9.1	3.5
1922	Corndog, chicken	1 ea	113	52	271	13	26		13			
705	Enchilada	1 ea	163	63	319	10	28		19	10.6	6.3	.8
706	English muffin with egg, cheese, bacon	1 ea	146	49	383	20	31	1	20	9	6.8	2.1
	Fish sandwich:											
707	Regular, with cheese	1 ea	183	45	523	21	48	<1	28	8.1	8.9	9.4
708	Large, no cheese	1 ea	158	47	431	17	41	<1	23	5.2	7.7	8.2
709	Hamburger with bun, regular	1 ea	107	45	275	14	33	1	10	3.5	3.7	1.8
710	Hamburger with bun, 4-oz patty	1 ea	215	50	576	32	39		32	12	14.1	2.8
711	Hotdog/frankfurter with bun	1 ea	98	54	242	10	18		14	5.1	6.8	1.7
	Lunchables:											
2129	Bologna & American cheese	1 ea	128		450	18	19	0	34	15		
2130	Ham & cheese	1 ea	128		320	22	19	0	17	8		
2117	Honey ham & Amer. w/choc pudding	1 ea	176		390	18	34	<1	20	9		
2118	Honey turkey & cheddar w/Jello	1 ea	163		320	17	27	<1	16	9		
2131	Pepperoni & American cheese	1 ea	128		480	20	19	0	36	17		
2125	Salami & American cheese	1 ea	128		430	18	18	0	32	15		
2127	Turkey & cheddar cheese	1 ea	128		360	20	20	1	22	11		
712	Pizza, cheese, ⅛ of 15" round	1 pce	63	48	140	8	20	1	3	1.5	1	.5
	SANDWICHES:											
	Avocado, chesse, tomato & lettuce:											
1276	On white bread, firm	1 ea	210	58	478	15	41	5	29	8.8	11.3	7.2
1278	On part whole wheat	1 ea	201	59	444	14	34	7	29	8.6	11.4	7.3
1277	On whole wheat	1 ea	214	58	468	16	40	8	30	8.7	11.6	7.5
	Bacon, lettuce & tomato sandwich:											
1137	On white bread, soft	1 ea	124	53	308	10	28	2	18	4.5	6.1	6.1
1139	On part whole wheat	1 ea	124	54	303	10	26	3	17	4.3	6.2	6.1
1138	On whole wheat	1 ea	137	53	328	12	32	4	18	4.4	6.4	6.3

Chol (mg)	Calc (mg)	Iron (mg)	Magn (mg)	Pota (mg)	Sodi (mg)	Zinc (mg)	VT-A (RE)	Thia (mg)	VT-E (a-TE)	Ribo (mg)	Niac (mg)	V-B6 (mg)	Fola (µg)	VT-C (mg)
14	54	.76	8	180	121	.42	10	.07	.92	.22	1.65	.05	8	1
85	24	.81	19	318	661	.39	41	.1	2.33	.07	1.11	.18	8	12
20	101	1.53	18	179	382	1.11	101	.21		.17	1.96	.09	32	2
14	65	.94	9	153	267	.52	55	.13		.23	3.05	.06	37	2
205	201	1.9	27	271	549	1.66	243	.23	1.91	.44	4.71	.19	17	3
38	20	1.89	24	51	1349	.76	204	.22	.09	.1	1.75	.06	9	<1
84	32	2.03	20	259	619	1.67	94	.15	1.52	.22	2.95	.14	14	11
42	30	1.84	24	134	286	.89	21	.21	2.46	.11	2.25	.15	22	4
40	10	.9	22	362	695	2.99	0	.09		.12	2.33	.3	12	0
7	40	2.75	21	303	955	1.12	120	.35	2.13	.27	4.5	.13	6	10
7	80	2.25	26	408	955	1.3	140	.25	2.75	.17	2.25	.2	8	12
22	52	3.25	20	245	1220	2.39	100	.15	1.5	.17	2.25	.12	5	5
74	124	3.72	40	665	1009	2.45	159	.25	1.64	.3	3.97	.2	10	22
184	230	1.35	38	201	763	1.29	675	.09	1.22	.3	.48	.12	80	3
38	28	1.36	34	390	1219	1.46	28	.55	.62	.21	3.6	.41	10	20
23	16	.79	21	187	732	.66	20	.06	.39	.08	3.06	.18	6	12
0	35	1.42	25	224	514	.54	23	.07	1.96	.09	.4	.04	53	4
27	35	2.05	39	365	824	1.15	55	.06	1.95	.14	13.7	.17	16	5
41	34	2.3	30	182	772	1.2	13	.18	1.18	.15	7.78	.2	10	1
21	43	.88	39	270	236	.63	39	.1	8.67	.05	.36	.36	27	6
24	53	2.46	42	329	670	1.93	32	.27	.7	.42	2.71	.19	58	1
2	57	2.27	44	328	495	.76	16	.32	.87	.3	2.04	.15	44	1
10	40	.72			600		40							15
52	182	2.65	26	229	976	2.62	71	.32	1.34	.23	6.38	.15	65	2
60	171	3.42	30	335	1050	3.49	65	.35		.28	8.05	.18	61	2
60	60	4.68	35	353	957	1.87	31	.33	.55	.24	6.81	.2	100	9
79	102	6.18	17	263	973	1.31	37	.28	.7	.7	4.17	.09	103	0
64					668									
44	324	1.32	50	240	784	2.51	186	.08	1.47	.42	1.91	.39	65	1
234	207	3.29	34	213	784	1.81	158	.48	.6	.53	3.93	.16	47	1
68	185	3.5	37	353	939	1.17	97	.46	1.83	.42	4.23	.11	91	3
55	84	2.61	33	340	615	.99	30	.33	.87	.22	3.4	.11	85	3
43	51	2.46	22	215	564	2.05	13	.26	.43	.32	4.7	.13	52	3
103	92	5.55	45	527	742	5.81	4	.34	1.61	.41	6.73	.37	84	1
44	23	2.31	13	143	670	1.98	0	.23	.27	.27	3.65	.05	48	<1
85	300	2.7			1620		60							0
60	300	1.8			1770		80							
55	250	2.7			1540		40							
50	20	6			1360		80							
95	250	2.7			1840		60							
80	250	2.7			1740		60							
70	300	1.8			1650		60							
9	117	.58	16	110	336	.81	74	.18		.16	2.48	.04	35	1
34	294	3.06	54	581	550	1.71	140	.37	4.55	.39	3.77	.32	80	11
31	291	3.1	67	617	525	1.91	140	.35	4.55	.39	3.94	.35	80	11
31	281	3.53	102	679	593	2.68	140	.36	4.17	.37	4.29	.42	92	11
21	52	1.99	20	233	590	.96	31	.35	2.34	.19	3.2	.14	34	12
20	63	2.27	33	283	604	1.19	31	.36	2.7	.21	3.62	.17	37	12
20	55	2.68	64	342	670	1.9	31	.37	2.36	.2	3.97	.24	48	12

Table A-1

Food Composition (Computer code number is for West Diet Analysis program) (For purposes of calculations, use "0" for t, <1, <.1, <.01, etc.)

Computer Code Number	Food Description	Measure	Wt (g)	H_2O (%)	Ener (cal)	Prot (g)	Carb (g)	Dietary Fiber (g)	Fat (g)	Fat Breakdown (g) Sat	Mono	Poly
	MIXED DISHES and FAST FOODS—Continued											
	Cheese, grilled:											
1140	On white bread, soft	1 ea	119	37	400	18	30	1	24	13.2	7.5	2
1142	On part whole wheat	1 ea	119	37	396	18	28	3	24	13	7.6	2.1
1141	On whole wheat	1 ea	132	38	420	20	33	4	24	13.1	7.8	2.2
1596	Chicken fillet	1 ea	182	47	515	24	39	1	29	8.5	10.4	8.4
	Chicken salad:											
1143	On white bread, soft	1 ea	110	40	369	11	31	1	23	3.7	6.1	12.1
1145	On part whole wheat	1 ea	110	41	364	11	29	4	23	3.5	6.1	12.1
1144	On whole wheat	1 ea	123	41	387	13	34	5	23	3.6	6.3	12.2
1146	Corned beef & swiss on rye	1 ea	156	49	420	28	22	<1	26	9.4	7.4	6.3
	Egg salad:											
1147	On white bread, soft	1 ea	117	43	380	10	31	1	25	4.4	6.8	12
1149	On part whole wheat	1 ea	116	43	374	10	29	3	25	4.1	6.8	12
1148	On whole wheat	1 ea	130	43	400	12	35	5	25	4.3	7.1	12.2
	Ham:											
1279	On rye bread	1 ea	150	60	283	22	21	<1	13	2.4	3.8	6
1151	On white bread, soft	1 ea	157	55	334	22	30	1	14	3	4.4	5.9
1153	On part whole wheat	1 ea	156	55	328	22	28	3	14	2.7	4.4	6
1152	On whole wheat	1 ea	169	54	352	24	34	4	15	2.9	4.7	6.1
	Ham & cheese:											
1280	On white bread, soft	1 ea	157	49	403	23	30	1	22	8.4	5.9	6.1
1282	On part whole wheat	1 ea	156	50	397	23	28	3	21	8.1	5.9	6.2
1281	On whole wheat	1 ea	170	49	424	24	34	4	22	8.3	6.2	6.4
1150	Ham & swiss on rye	1 ea	150	54	339	22	22	<1	19	6.5	5.1	6
	Ham salad:											
1154	On white bread, soft	1 ea	131	47	362	11	37	1	20	4.8	6.7	7.4
1156	On part whole wheat	1 ea	131	48	357	11	35	3	20	4.5	6.8	7.5
1155	On whole wheat	1 ea	144	47	380	12	40	4	20	4.7	7	7.6
1157	Patty melt: Ground beef & cheese on rye	1 ea	182	46	561	37	22	3	37	13.2	11.7	8.4
	Peanut butter & jelly:											
1158	On white bread, soft	1 ea	101	26	351	12	47	3	15	3.1	6.7	3.9
1160	On part whole wheat	1 ea	101	27	346	12	45	5	15	2.9	6.7	4
1159	On whole wheat	1 ea	114	28	370	13	50	6	15	3	7	4.1
1161	Reuben, grilled: Corned beef, swiss cheese, sauerkraut on rye	1 ea	239	64	462	28	25	2	29	9.9	9.5	7.1
	Roast beef:											
713	On a bun	1 ea	139	49	346	21	33		14	3.6	6.8	1.7
1162	On white bread, soft	1 ea	157	46	404	29	34	1	17	3.4	4.2	8.2
1164	On part whole wheat	1 ea	156	47	398	29	32	3	17	3.2	4.3	8.3
1163	On whole wheat	1 ea	169	46	422	31	38	4	17	3.3	4.5	8.4
	Tuna salad:											
1165	On white bread, soft	1 ea	122	46	327	14	35	2	15	2.5	3.8	7.9
1167	On part whole wheat	1 ea	122	47	322	14	33	4	15	2.2	3.8	8
1166	On whole wheat	1 ea	135	46	346	16	39	5	15	2.3	4.1	8.1
	Turkey:											
1168	On white bread, soft	1 ea	156	54	346	24	29	1	15	2.4	3.2	8.3
1170	On part whole wheat	1 ea	155	54	338	24	27	3	14	2.1	3.2	8.3
1169	On whole wheat	1 ea	169	53	365	26	33	4	15	2.3	3.5	8.5
	Turkey ham:											
1272	On rye bread	1 ea	150	60	280	21	20	<1	14	2.5	2.8	6.9
1273	On white bread, soft	1 ea	156	55	331	21	29	1	14	3	3.4	6.8
1275	On part whole wheat	1 ea	156	56	326	21	28	3	14	3	4.2	5.6
1274	On whole wheat	1 ea	169	55	350	23	33	4	15	2.9	3.7	7
714	Taco	1 ea	171	58	369	21	27		20	11.4	6.6	1
	Tostada:											
1114	With refried beans	1 ea	144	66	223	10	26	7	10	5.4	3	.7
1118	With beans & beef	1 ea	225	70	333	16	30	4	17	11.5	3.5	.6
1354	With beans & chicken	1 ea	156	68	248	19	18	3	11	5.3	3.9	1.6
	NUTS, SEEDS, and PRODUCTS											
	Almonds:											
1365	Dry roasted, salted	1 c	138	3	810	22	33	19	71	6.7	46.2	14.9

Chol (mg)	Calc (mg)	Iron (mg)	Magn (mg)	Pota (mg)	Sodi (mg)	Zinc (mg)	VT-A (RE)	Thia (mg)	VT-E (a-TE)	Ribo (mg)	Niac (mg)	V-B6 (mg)	Fola (µg)	VT-C (mg)
55	399	1.81	25	154	1143	2.05	212	.24	1.13	.34	1.91	.06	24	<1
54	412	2.12	39	209	1160	2.31	212	.25	1.53	.36	2.39	.1	28	<1
54	402	2.54	73	271	1226	3.08	212	.26	1.14	.35	2.74	.17	40	<1
60	60	4.68	35	353	957	1.87	31	.33	.55	.24	6.81	.2	100	9
32	60	2.04	18	139	460	.8	24	.25	6.16	.18	3.68	.25	26	1
31	73	2.35	33	195	475	1.06	24	.26	6.57	.2	4.16	.29	30	1
30	63	2.79	68	259	543	1.85	24	.27	6.14	.19	4.5	.36	42	1
82	268	3.12	28	225	1392	3.65	81	.19	2.59	.33	2.72	.17	19	1
157	71	2.18	16	113	526	.74	76	.26	4.52	.31	1.98	.2	37	0
155	83	2.47	31	169	539	1	75	.27	4.91	.33	2.45	.23	41	0
155	73	2.92	66	234	611	1.8	76	.28	4.53	.32	2.82	.3	53	0
47	48	2.3	26	364	1566	2.11	8	.99	2.36	.31	5.45	.47	15	23
47	60	2.39	29	368	1619	2.04	8	1.02	2.34	.33	5.99	.47	24	22
45	71	2.68	43	421	1630	2.29	8	1.03	2.73	.35	6.45	.5	27	22
45	62	3.1	77	483	1696	3.05	8	1.04	2.34	.33	6.78	.57	39	22
61	232	2.28	30	315	1620	2.34	90	.76	2.53	.36	4.64	.36	25	15
59	244	2.58	45	368	1630	2.59	90	.77	2.93	.39	5.09	.39	28	15
59	236	3.02	79	432	1707	3.37	90	.78	2.56	.37	5.47	.46	40	15
57	258	2.25	29	344	1602	2.59	79	.72	2.52	.36	4.06	.35	16	15
30	56	2.07	19	159	921	1.06	8	.51	3.29	.22	3.25	.17	22	4
29	69	2.38	33	216	936	1.32	8	.52	3.69	.24	3.74	.21	26	4
29	59	2.81	69	279	1001	2.11	8	.53	3.29	.23	4.09	.28	38	4
113	222	4.19	36	391	701	7.11	123	.25	3.5	.46	6.14	.35	25	<1
2	60	2.25	56	245	293	1.07	<1	.27	.12	.17	5.33	.13	40	<1
0	72	2.55	70	299	308	1.32	<1	.28	.51	.19	5.8	.17	44	<1
0	63	2.97	104	361	375	2.09	<1	.29	.14	.17	6.14	.24	56	<1
80	288	4.24	38	361	1949	3.73	130	.21	4.46	.34	2.79	.27	38	13
51	54	4.23	31	316	792	3.39	21	.37	.19	.31	5.87	.26	57	2
45	60	3.98	28	432	1595	3.78	12	.29	3.3	.3	6.39	.39	30	12
43	72	4.27	42	485	1607	4.02	12	.31	3.69	.32	6.84	.42	34	12
43	62	4.7	77	547	1672	4.79	12	.31	3.31	.31	7.18	.49	45	12
14	60	2.24	22	161	567	.67	22	.25	2.72	.18	5.53	.11	25	1
13	73	2.55	37	217	582	.93	22	.26	3.12	.2	6	.16	29	1
12	63	2.98	72	280	649	1.72	22	.27	2.71	.19	6.34	.23	41	1
45	56	2	29	302	1585	1.33	12	.26	3.46	.23	8.97	.4	24	0
43	68	2.28	43	354	1589	1.57	11	.27	3.82	.25	9.38	.44	28	0
43	59	2.73	77	418	1665	2.35	12	.28	3.47	.23	9.77	.51	39	0
55	51	4.06	25	342	1185	3	8	.22	2.8	.33	4.3	.29	17	<1
55	62	4.09	28	346	1248	2.9	8	.27	2.75	.35	4.87	.28	25	0
53	74	4.37	42	400	1262	3.15	8	.28	3.57	.37	5.34	.32	29	0
53	65	4.81	76	462	1329	3.91	8	.29	2.76	.35	5.68	.39	41	0
56	221	2.41	70	474	802	3.93	147	.15	1.88	.44	3.21	.24	68	2
30	210	1.89	59	403	543	1.9	85	.1	1.15	.33	1.32	.16	43	1
74	189	2.45	67	491	871	3.17	173	.09	1.8	.49	2.86	.25	85	4
53	168	1.79	47	365	433	2.28	86	.11	1.87	.2	4.52	.32	53	3
0	389	5.24	420	1062	1076	6.76	0	.18	7.66	.83	3.89	.1	88	1

Table A-1

Food Composition (Computer code number is for West Diet Analysis program) (For purposes of calculations, use "0" for t, <1, <.1, <.01, etc.)

Computer Code Number	Food Description	Measure	Wt (g)	H_2O (%)	Ener (cal)	Prot (g)	Carb (g)	Dietary Fiber (g)	Fat (g)	Fat Breakdown (g) Sat	Mono	Poly
	NUTS, SEEDS, and PRODUCTS—Continued											
	Almonds:											
718	Slivered, packed, unsalted	1 c	108	4	636	22	22	12	56	5.3	36.6	11.9
719	Whole, dried, unsalted	1 c	142	4	836	28	29	15	74	7	48.1	15.6
720	Ounce	1 oz	28	4	165	6	6	3	15	1.4	9.5	3.1
721	Almond butter:	1 tbs	16	1	101	2	3	1	9	.9	6.1	2
4572	Salted	1 tbs	16	1	101	2	3	1	9	.9	6.1	2
722	Brazil nuts, dry (about 7)	1 c	140	3	918	20	18	8	93	22.7	32.2	33.7
	Cashew nuts, dry roasted:											
723	Salted:	1 c	137	2	786	21	45	4	64	12.8	37.4	10.7
724	Ounce	1 oz	28	2	161	4	9	1	13	2.6	7.6	2.2
4621	Unsalted:	1 c	137	2	786	21	45	4	64	12.8	37.4	10.7
4621	Ounce	1 oz	28	2	161	4	9	1	13	2.6	7.6	2.2
725	Oil roasted:	1 c	130	4	749	23	37	5	63	12.6	36.9	10.6
726	Ounce	1 oz	28	4	161	5	8	1	13	2.7	7.9	2.3
4622	Unsalted:	1 c	130	4	749	21	37	5	63	12.6	36.9	10.6
4622	Ounce	1 oz	28	4	161	5	8	1	13	2.7	7.9	2.3
727	Cashew butter, unsalted	1 tbs	16	3	94	3	4	<1	8	1.6	4.7	1.3
4662	Cashew butter, salted	1 tbs	16	3	94	3	4	<1	8	1.6	4.7	1.3
728	Chestnuts, European, roasted (1 cup = approx 17 kernels)	1 c	143	40	350	5	76	7	3	.6	1.1	1.2
	Coconut, raw:											
729	Piece 2 x 2 x ½"	1 pce	45	47	159	2	7	4	15	13.5	.6	.2
730	Shredded/grated, unpacked	½ c	40	47	142	1	6	4	13	12	.6	.1
	Coconut, dried, shredded/grated:											
731	Unsweetened	1 c	78	3	515	6	19	13	50	45.1	2.1	.6
732	Sweetened	1 c	93	13	466	3	44	4	33	29.6	1.4	.4
733	Filberts/hazelnuts, chopped:	1 c	135	5	853	18	21	8	84	6.2	66.3	8.1
734	Ounce	1 oz	28	5	177	4	4	2	17	1.3	13.7	1.7
735	Macadamias, oil roasted, salted:	1 c	134	2	962	10	17	12	103	15.4	80.9	1.8
736	Ounce	1 oz	28	2	201	2	4	3	21	3.2	16.9	.4
1368	Macadamias, oil roasted, unsalted	1 c	134	2	962	10	17	12	103	15.4	80.9	1.8
	Mixed nuts:											
737	Dry roasted, salted	1 c	137	2	814	24	35	12	71	9.4	43	14.8
738	Oil roasted, salted	1 c	142	2	876	24	30	13	80	12.4	45	18.9
1369	Oil roasted, unsalted	1 c	142	2	876	27	30	14	80	12.4	45	18.9
	Peanuts:											
739	Oil roasted, salted	1 c	144	2	837	38	27	13	71	9.8	35.3	22.5
740	Ounce	1 oz	28	2	163	7	5	3	14	1.9	6.9	4.4
1370	Oil roasted, unsalted	1 c	144	2	837	38	27	10	71	9.8	35.3	22.5
741	Dried, salted	1 c	146	2	854	35	31	12	73	10.1	36.1	22.9
742	Ounce	1 oz	28	2	164	7	6	2	14	1.9	6.9	4.4
743	Peanut butter:	½ c	128	1	759	33	25	8	65	14.3	31.1	17.7
1371	Tablespoon	2 tbs	32	1	190	8	6	2	16	3.6	7.8	4.4
744	Pecan halves, dried, unsalted:	1 c	108	5	720	9	20	8	73	5.8	45.6	18
745	Ounce	1 oz	28	5	187	2	5	2	19	1.5	11.8	4.7
1372	Pecan halves, dry roasted, salted	¼ c	28	1	185	2	6	3	18	1.4	11.3	4.5
746	Pine nuts/piñons, dried	1 oz	28	6	176	3	5	3	17	2.6	6.4	7.2
747	Pistachios, dried, shelled	1 oz	28	4	162	6	7	3	14	1.8	9.2	2
1373	Pistachios,dry roasted,salted,shelled	1 c	128	2	776	19	35	14	68	8.8	45.7	10.2
748	Pumpkin kernels, dried, unsalted	1 oz	28	7	151	7	5	1	13	2.4	4	5.8
1374	Pumpkin kernels, roasted, salted	1 c	227	7	1184	75	30	9	96	18.1	29.7	43.6
749	Sesame seeds, hulled, dried	¼ c	38	5	223	10	4	3	21	2.9	7.9	9.1
	Sunflower seed kernels:											
750	Dry	¼ c	36	5	205	8	7	4	18	1.9	3.4	11.8
751	Oil roasted	¼ c	34	3	209	7	5	2	20	2	3.7	12.9
752	Tahini (sesame butter)	1 tbs	15	3	91	3	3	1	8	1.2	3.2	3.7
1334	Trail mix w/chocolate chips	1 c	146	7	707	21	66	8	47	9.3	19.8	16.5
753	Black walnuts, chopped:	1 c	125	4	759	31	15	6	71	4.8	15.9	46.9
754	Ounce	1 oz	28	4	170	7	3	1	16	1.1	3.6	10.5
755	English walnuts, chopped:	1 c	120	4	770	17	22	6	74	7.2	17	46.9
756	Ounce	1 oz	28	4	180	4	5	1	17	1.7	4	10.9

Chol (mg)	Calc (mg)	Iron (mg)	Magn (mg)	Pota (mg)	Sodi (mg)	Zinc (mg)	VT-A (RE)	Thia (mg)	VT-E (a-TE)	Ribo (mg)	Niac (mg)	V-B6 (mg)	Fola (µg)	VT-C (mg)
0	287	3.95	320	791	12	3.15	0	.23	25.9	.84	3.63	.12	63	1
0	378	5.2	420	1039	16	4.15	0	.3	34.1	1.11	4.77	.16	83	1
0	74	1.02	83	205	3	.82	0	.06	6.72	.22	.94	.03	16	<1
0	43	.59	48	121	2	.49	0	.02	3.25	.1	.46	.01	10	<1
0	43	.59	48	121	72	.49	0	.02	3.25	.1	.46	.01	10	<1
0	246	4.76	315	840	3	6.43	0	1.4	10.6	.17	2.27	.35	6	1
0	62	8.22	356	774	877	7.67	0	.27	.78	.27	1.92	.35	95	0
0	13	1.68	73	158	179	1.57	0	.06	.16	.06	.39	.07	19	0
0	62	8.22	356	774	22	7.67	0	.27	.78	.27	1.92	.35	95	0
0	13	1.68	73	158	4	1.57	0	.06	.16	.06	.39	.07	19	0
0	53	5.33	332	689	814	6.18	0	.55	2.03	.23	2.34	.32	88	0
0	11	1.15	71	148	175	1.33	0	.12	.44	.05	.5	.07	19	0
0	53	5.33	332	689	22	6.18	0	.55	2.03	.23	2.34	.32	88	0
0	11	1.15	71	148	5	1.33	0	.12	.44	.05	.5	.07	19	0
0	7	.8	41	87	2	.83	0	.05	.25	.03	.26	.04	11	0
0	7	.8	41	87	98	.83	0	.05	.25	.03	.26	.04	11	0
0	41	1.3	47	847	3	.81	3	.35	1.72	.25	1.92	.71	100	37
0	6	1.09	14	160	9	.49	0	.03	.33	.01	.24	.02	12	1
0	6	.97	13	142	8	.44	0	.03	.29	.01	.22	.02	11	1
0	20	2.59	70	424	29	1.57	0	.05	1.05	.08	.47	.23	7	1
0	14	1.79	46	313	244	1.69	0	.03	1.26	.02	.44	.25	8	1
0	254	4.41	385	601	4	3.24	9	.67	32.3	.15	1.54	.83	97	1
0	53	.92	80	125	1	.67	2	.14	6.69	.03	.32	.17	20	<1
0	60	2.41	157	441	348	1.47	1	.28	.55	.15	2.71	.26	21	0
0	13	.5	33	92	73	.31	<1	.06	.11	.03	.57	.05	4	0
0	60	2.41	157	441	9	1.47	1	.28	.55	.15	2.71	.26	21	0
0	96	5.07	308	818	917	5.21	1	.27	8.22	.27	6.44	.41	69	1
0	153	4.56	334	825	926	7.21	3	.71	8.52	.31	7.19	.34	118	1
0	153	4.56	334	825	16	7.21	3	.71	8.52	.31	7.19	.34	118	1
0	127	2.64	266	982	624	9.55	0	.36	10.7	.16	20.6	.37	181	0
0	25	.51	52	191	121	1.86	0	.07	2.07	.03	4	.07	35	0
0	127	2.64	266	982	9	9.55	0	.36	10.7	.16	20.6	.37	181	0
0	79	3.3	257	961	1186	4.83	0	.64	10.8	.14	19.7	.37	212	0
0	15	.63	49	184	228	.93	0	.12	2.07	.03	3.78	.07	41	0
0	49	2.36	204	856	598	3.74	0	.11	12.8	.13	17.2	.58	95	0
0	12	.59	51	214	149	.93	0	.03	3.2	.03	4.29	.14	24	0
0	39	2.3	138	423	1	5.91	14	.92	3.35	.14	.96	.2	42	2
0	10	.6	36	110	<1	1.53	4	.24	.87	.04	.25	.05	11	1
0	10	.61	37	104	218	1.59	4	.09	.84	.03	.26	.05	11	1
0	2	.86	65	176	20	1.2	1	.35	.98	.06	1.22	.03	16	1
0	38	1.9	44	306	2	.37	6	.23	1.46	.05	.3	.07	16	2
0	90	4.06	166	1241	998	1.74	31	.54	8.26	.31	1.8	.33	76	9
0	12	4.2	150	226	5	2.09	11	.06	.28	.09	.49	.06	16	1
0	98	33.8	1212	1829	1305	16.9	86	.48	2.27	.72	3.95	.2	130	4
0	50	2.96	132	155	15	3.91	3	.27	.86	.03	1.78	.05	36	0
0	42	2.44	127	248	1	1.82	2	.82	18.1	.09	1.62	.28	82	<1
0	19	2.28	43	164	1	1.77	2	.11	17.1	.09	1.4	.27	80	<1
0	21	.95	53	69	<1	1.58	1	.24	.34	.02	.85	.02	15	0
6	159	4.95	235	946	177	4.58	7	.6	15.6	.33	6.44	.38	95	2
0	72	3.84	253	655	1	4.28	37	.27	3.28	.14	.86	.69	82	4
0	16	.86	57	147	<1	.96	8	.06	.73	.03	.19	.15	18	1
0	113	2.93	203	602	12	3.28	14	.46	3.14	.18	1.25	.67	79	4
0	26	.68	47	141	3	.76	3	.11	.73	.04	.29	.16	18	1

A

Table A-1

Food Composition (Computer code number is for West Diet Analysis program) (For purposes of calculations, use "0" for t, <1, <.1, <.01, etc.)

Computer Code Number	Food Description	Measure	Wt (g)	H_2O (%)	Ener (cal)	Prot (g)	Carb (g)	Dietary Fiber (g)	Fat (g)	Fat Breakdown (g) Sat	Mono	Poly
	SWEETENERS and SWEETS (see also Dairy [milk desserts] and Baked Goods)											
757	Apple butter	2 tbs	36	52	66	<1	17	<1	<1	t	t	t
1124	Butterscotch topping	2 tbs	41	32	103	1	27	<1	<1	t	t	0
1125	Caramel topping	2 tbs	41	32	103	1	27	<1	<1	t	t	0
	Cake frosting, creamy vanilla:											
1127	Canned	2 tbs	39	13	163	<1	27	<1	7	1.9	3.4	.9
1123	From mix	2 tbs	39	12	165	<1	28	<1	6	1.3	2.6	2.2
	Cake frosting, lite:											
2061	Milk chocolate	1 tbs	16	18	58	<1	11	<1	1	.4		
2062	Vanilla	1 tbs	16	15	60	0	12	<1	1	.4		
	Candy:											
1128	Almond Joy candy bar	1 oz	28	10	131	1	16	1	7	4.8	1.8	.4
2069	Butterscotch morsels	¼ c	43	1	246	0	31	0	12	12.5	0	0
758	Caramel, plain or chocolate	1 pce	10	8	38	<1	8	<1	1	.7	.1	t
1961	Chewing gum, sugarless	1 pce	3		6	0	2		0	0	0	0
	Chocolate (see also #784, 785, 971):											
	Milk chocolate:											
759	Plain	1 oz	28	1	144	2	17	1	9	5.2	2.8	.3
760	With almonds	1 oz	28	1	147	3	15	2	10	4.8	3.8	.6
761	With peanuts	1 oz	28	1	155	5	11	2	11	3.4	5.1	2.5
762	With rice cereal	1 oz	28	2	139	2	18	1	7	4.4	2.4	.2
763	Semisweet chocolate chips	1 c	168	1	805	7	106	10	50	29.9	16.8	1.6
764	Sweet dark chocolate (candy bar)	1 ea	41	1	226	2	25	2	13	8.3	4.6	.4
765	Fondant candy, uncoated (mints, candy corn, other)	1 pce	16	7	57	0	15	0	0	0	0	0
1697	Fruit Roll-Up (small)	1 ea	14	11	49	<1	12	<1	<1	.1	.2	.1
766	Fudge, chocolate	1 pce	17	10	65	<1	13	<1	1	.9	.4	.1
767	Gumdrops	1 c	182	1	703	0	180	0	0	0	0	0
768	Hard candy, all flavors	1 pce	6	1	22	0	6	0	<1	0	0	0
769	Jellybeans	10 pce	11	6	40	0	10	0	<1	t	t	t
1134	M&M's plain chocolate candy	10 pce	7	2	34	<1	5	<1	1	.9	.5	t
1135	M&M's peanut chocolate candy	10 pce	20	2	103	2	12	1	5	2.1	2.2	.8
1130	Mars almond bar	1 ea	50	4	234	4	31	1	11	2.7	5.5	2.8
1129	Milky Way candy bar	1 ea	60	6	254	3	43	1	10	4.7	3.6	.4
1708	Milk chocolate-coated peanuts	1 c	149	2	773	19	74	7	50	21.8	19.4	6.4
1709	Peanut brittle, recipe	1 c	147	2	666	11	102	3	28	7.4	12.5	6.9
1132	Reese's peanut butter cup	2 ea	50	3	271	5	27	2	16	5.5	6.5	2.8
1133	Skor English toffee candy bar	1 ea	39	3	217	2	22	1	13	8.5	4.3	.5
1131	Snickers candy bar (2.2oz)	1 ea	62	5	297	5	37	2	15	5.6	6.5	3
1482	Fruit juice bar (2.5 fl oz)	1 ea	77	78	63	1	16	0	<1	t	0	t
771	Gelatin dessert/Jello, prepared	½ c	135	85	80	2	19	0	0	0	0	0
1702	SugarFree	½ c	117	98	8	1	1	0	0	0	0	0
772	Honey:	1 c	339	17	1030	1	279	1	0	0	0	0
773	Tablespoon	1 tbs	21	17	64	<1	17	<1	0	0	0	0
774	Jams or preserves:	1 tbs	20	29	54	<1	14	<1	<1	0	t	t
775	Packet	1 ea	14	34	34	<1	9	<1	<1	t	t	0
776	Jellies:	1 tbs	19	28	51	<1	13	<1	<1	t	t	t
777	Packet	1 ea	14	28	38	<1	10	<1	<1	t	t	t
1136	Marmalade	1 tbs	20	33	49	<1	13	<1	0	0	0	0
770	Marshmallows	1 ea	7	16	22	<1	6	<1	<1	t	t	t
1126	Marshmallow creme topping	2 tbs	38	18	118	1	30	<1	<1	t	t	t
778	Popsicle/ice pops	1 ea	128	80	92	0	24	0	0	0	0	0
	Sugars:											
779	Brown sugar	1 c	220	2	827	0	214	0	0	0	0	0
780	White sugar, granulated:	1 c	200		774	0	200	0	0	0	0	0
781	Tablespoon	1 tbs	12		46	0	12	0	0	0	0	0
782	Packet	1 ea	6		23	0	6	0	0	0	0	0
783	White sugar, powdered, sifted	1 c	100		389	0	99	0	<1	t	t	t
	Sweeteners:											
1711	Equal, packet	1 ea	1	12	4	<1	1	0	<1	0	t	t
1712	Sweet 'N Low, packet	1 ea	1		0	0	1	0	0	0	0	0

Chol (mg)	Calc (mg)	Iron (mg)	Magn (mg)	Pota (mg)	Sodi (mg)	Zinc (mg)	VT-A (RE)	Thia (mg)	VT-E (a-TE)	Ribo (mg)	Niac (mg)	V-B6 (mg)	Fola (μg)	VT-C (mg)
0	2	.05	1	33	0	.02	0	<.01	.01	<.01	.03	.01	<1	1
<1	22	.08	3	34	143	.08	11	<.01	0	.04	.02	.01	1	<1
<1	22	.08	3	34	143	.08	11	<.01	0	.04	.02	.01	1	<1
0	1	.04		14	35	0	88	0	.79	<.01	<.01	0	0	0
0	4	.09	1	9	87	.04	42	.01	.79	.01	.13	<.01	0	0
0	1	.24			40		<1							0
0		.02			29		0							0
1	17	.39	18	69	41	.22	1	.01	.63	.04	.13	.02		<1
0	0	0		80	46		0	.03		.04	.03			0
1	14	.01	2	21	24	.04	1	<.01	.05	.02	.02	<.01	<1	<1
				0	0									
6	53	.39	17	108	23	.39	15	.02	.35	.08	.09	.01	2	<1
5	63	.46	25	124	21	.37	4	.02	.53	.12	.21	.01	3	<1
3	32	.52	34	150	11	.68	6	.08	1.3	.05	2.12	.04	23	0
5	48	.21	14	96	41	.31	3	.02	.35	.08	.13	.02	3	<1
0	54	5.26	193	613	18	2.72	3	.09	2	.15	.72	.06	5	0
<1	11	.98	47	139	3	.61	1	.01	.41	.1	.27	.02	1	0
0		.01		3	6	.01	<1	0	0	<.01	0	0	0	0
0	4	.14	3	41	9	.03	2	.01	.04	<.01	.01	.04	1	1
2	7	.08	4	17	10	.07	8	<.01	.02	.01	.02	<.01	<1	<1
0	5	.73	2	9	80	0	0	0	0	<.01	<.01	0	0	0
0		.02			2	<.01	0	0	0	0	0	0	0	0
0		.12		4	3	.01	0	0	0	0	0	0	0	0
1	7	.08	3	19	4	.07	4	<.01	.06	.01	.02	<.01	<1	<1
2	20	.23	12	69	10	.27	5	.03	.43	.04	.41	.02	7	<1
4	84	.55	36	163	85	.55	22	.02	.3	.16	.47	.03	9	<1
8	78	.46	20	145	144	.43	34	.02	.39	.13	.21	.03	6	1
13	155	1.95	134	748	61	2.8	0	.17	3.8	.26	6.33	.31	12	0
19	44	2.03	73	306	664	1.43	69	.28	2.41	.08	5.15	.15	103	0
2	39	.6	42	176	159	.7	9	.02	.66	.1	1.99	.04	27	<1
20	51	.02	13	93	108	.3	27	.01	.53	.13	.03	.01		<1
8	58	.47	42	209	165	.88	24	.13	.95	.1	2.26	.07	25	<1
0	4	.15	3	41	3	.04	2	.01	0	.01	.12	.02	5	7
0	3	.04	1	1	57	.04	0	0	0	<.01	<.01	<.01	0	0
0	2	.01	1	0	56	.03	0	0	0	<.01	<.01	<.01	0	0
0	20	1.42	7	176	14	.75	0	0	0	.13	.41	.08	7	2
0	1	.09		11	1	.05	0	0	0	.01	.02	<.01	<1	<1
0	4	.2	1	18	2	.01	<1	<.01	.02	.01	.04	<.01	2	<1
0	3	.07	1	11	6	.01	<1	0	0	<.01	<.01	<.01	5	1
0	2	.04	1	12	7	.01	<1	0	0	<.01	.01	<.01	<1	<1
0	1	.03	1	9	5	.01	<1	0	0	<.01	<.01	<.01	<1	<1
0	8	.03		7	11	.01	1	<.01	0	<.01	.01	<.01	7	1
0		.02			3	<.01	<1	0	0	0	<.01	0	<1	0
0	1	.08	1	2	17	.01	<1	0	0	0	.03	<.01	<1	0
0	0	0	1	5	15	.03	0	0	0	0	0	0	0	0
0	187	4.2	64	761	86	.4	0	.02	0	.01	.18	.06	2	0
0	2	.12	0	4	2	.06	0	0	0	.04	0	0	0	0
0		.01	0		<1	<.01	0	0	0	<.01	0	0	0	0
0		<.01	0		<1	<.01	0	0	0	<.01	0	0	0	0
0	1	.06	0	2	1	.03	0	0	0	0	0	0	0	0
0		<.01			<1	0	0	0	0	0	0	0	0	0
0	0	0			0		0							0

A

Table A-1

Food Composition (Computer code number is for West Diet Analysis program) (For purposes of calculations, use "0" for t, <1, <.1, <.01, etc.)

Computer Code Number	Food Description	Measure	Wt (g)	H_2O (%)	Ener (cal)	Prot (g)	Carb (g)	Dietary Fiber (g)	Fat (g)	Fat Breakdown (g) Sat	Mono	Poly
	SWEETENERS and SWEETS—Continued											
	Syrups, chocolate:											
785	Hot fudge type	2 tbs	43	22	149	2	27	1	4	2.4	1.6	1.4
784	Thin type	2 tbs	38	29	93	1	25	1	<1	.3	.2	t
786	Molasses, blackstrap	2 tbs	41	29	96	0	25	0	0	0	0	0
1710	Light cane syrup	2 tbs	41	24	103	0	27	0	0	0	0	0
787	Pancake table syrup (corn and maple)	2 tbs	40	24	115	0	30	0	0	0	0	0
	VEGETABLES and LEGUMES											
788	Alfalfa sprouts	1 c	33	91	10	1	1	1	<1	t	.1	t
1815	Amaranth leaves, raw, chopped	1 c	28	92	7	1	1	<1	<1	t	t	t
1816	Amaranth leaves, raw, each	1 ea	14	92	4	<1	1	<1	<1	t	t	t
1817	Amaranth leaves, cooked	1 c	132	91	28	3	5	2	<1	.1	.1	.1
1987	Arugula, raw, chopped	½ c	10	92	2	<1	<1	<1	<1	t	t	t
789	Artichokes, cooked globe (300 g with refuse)	1 ea	120	84	60	4	13	6	<1	t	t	.1
1177	Artichoke hearts, cooked from frozen	1 c	168	86	76	5	15	8	1	.2	t	.4
1176	Artichoke hearts, marinated	1 c	130	81	128	3	10	6	10	1.5	2.3	5.9
2021	Artichoke hearts, in water	½ c	100	91	37	2	6	0	0	0	0	0
	Asparagus, green, cooked:											
	From fresh:											
790	Cuts and tips	½ c	90	92	22	2	4	1	<1	.1	t	.1
791	Spears, ½" diam at base	4 ea	60	92	14	2	3	1	<1	t	t	.1
	From frozen:											
792	Cuts and tips	½ c	90	91	25	3	4	1	<1	.1	t	.2
793	Spears, ½" diam at base	4 ea	60	91	17	2	3	1	<1	.1	t	.1
794	Canned, spears, ½" diam at base	4 ea	72	94	14	2	2	1	<1	.1	t	.2
795	Bamboo shoots, canned, drained slices	1 c	131	94	25	2	4	2	1	.1	t	.2
1795	Bamboo shoots, raw slices	1 c	151	91	41	4	8	3	<1	.1	t	.2
1798	Bamboo shoots, cooked slices	1 c	120	96	14	2	2	1	<1	.1	t	.1
	Beans (see also alphabetical listing this section):											
1990	Adzuki beans, cooked	½ c	115	66	147	9	28	1	<1	t	t	t
796	Black beans, cooked	½ c	86	66	114	8	20	7	<1	.1	t	.2
	Canned beans (white/navy):											
803	With pork and tomato sauce	½ c	127	73	124	7	25	6	1	.5	.6	.2
804	With sweet sauce	½ c	130	71	144	7	27	7	2	.7	.8	.2
805	With frankfurters	½ c	130	69	185	9	20	9	9	3.1	3.7	1.1
	Lima beans:											
797	Thick seeded (Fordhooks), cooked from frozen	½ c	85	73	85	5	16	5	<1	.1	t	.1
798	Thin seeded (Baby), cooked from frozen	½ c	90	72	94	6	18	5	<1	.1	t	.1
799	Cooked from dry, drained	½ c	94	70	108	7	20	7	<1	.1	t	.2
1998	Red Mexican, cooked f/dry	½ c	112	70	126	8	23	9	<1	.1	.1	.2
	Snap bean/green string beans cuts and french style:											
800	Cooked from fresh	½ c	63	89	22	1	5	2	<1	t	t	.1
801	Cooked from frozen	½ c	68	91	19	1	4	2	<1	t	t	.1
802	Canned, drained	½ c	68	93	14	1	3	1	<1	t	t	t
1713	Snap bean, yellow, cooked f/fresh	½ c	63	89	22	1	5	2	<1	t	t	.1
	Bean sprouts (mung):											
806	Raw	½ c	52	90	16	2	3	1	<1	t	t	t
807	Cooked, stir fried	½ c	62	84	31	3	7	1	<1	t	t	t
808	Cooked, boiled, drained	½ c	62	93	13	1	3	<1	<1	t	t	t
1788	Canned, drained	½ c	63	96	8	1	1	<1	<1	t	t	t
	Beets, cooked from fresh:											
809	Sliced or diced	½ c	85	87	37	1	8	2	<1	t	t	.1
810	Whole beets, 2" diam	2 ea	100	87	44	2	10	2	<1	t	t	.1
	Beets, canned:											
811	Sliced or diced	½ c	79	91	24	1	6	1	<1	t	t	t
812	Pickled slices	½ c	114	82	74	1	19	2	<1	t	t	t
813	Beet greens, cooked, drained	½ c	72	89	19	2	4	2	<1	t	t	.1

Chol (mg)	Calc (mg)	Iron (mg)	Magn (mg)	Pota (mg)	Sodi (mg)	Zinc (mg)	VT-A (RE)	Thia (mg)	VT-E (a-TE)	Ribo (mg)	Niac (mg)	V-B6 (mg)	Fola (μg)	VT-C (mg)
5	43	.52	21	92	56	.34	9	.01	0	.09	.09	.01	2	<1
0	5	5.17	25	183	58	.28	494	<.01	.01	.31	12.8	.01	2	<1
0	353	7.18	88	1021	23	.41	0	.01	0	.02	.44	.29	<1	0
0	68	1.76	100	376	6	.12	0	.03	0	.02	.08	.27	0	0
0		.04	1	1	33	.02	0	<.01	0	<.01	.01	0	0	0
0	11	.32	9	26	2	.3	5	.02	—	.04	.16	.01	12	3
0	60	.65	15	171	6	.25	82	.01	.22	.04	.18	.05	24	12
0	30	.32	8	85	3	.13	41	<.01	.11	.02	.09	.03	12	6
0	276	2.98	73	846	28	1.16	366	.03	.66	.18	.74	.23	75	54
0	16	.15	5	37	3	.05	24	<.01	.04	.01	.03	.01	10	1
0	54	1.55	72	425	114	.59	22	.08	.23	.08	1.2	.13	61	12
0	35	.94	52	444	89	.6	27	.1	.32	.26	1.54	.15	200	8
0	30	1.24	37	335	688	.41	21	.05	1.43	.13	1.06	.11	114	40
0	0	1.35		0	250		12							4
0	18	.66	9	144	10	.38	49	.11	.34	.11	.97	.11	131	10
0	12	.44	6	96	7	.25	32	.07	.23	.08	.65	.07	88	6
0	21	.58	12	196	4	.5	74	.06	1.13	.09	.94	.02	122	22
0	14	.38	8	131	2	.34	49	.04	.75	.06	.62	.01	81	15
0	11	1.32	7	124	207	.29	38	.04	.31	.07	.69	.08	69	13
0	10	.42	5	105	9	.85	1	.03	.5	.03	.18	.18	4	1
0	20	.75	5	805	6	1.66	3	.23	1.51	.11	.91	.36	11	6
0	14	.29	4	640	5	.56	0	.02	.8	.06	.36	.12	3	1
0	32	2.3	60	612	9	2.04	1	.13	.11	.07	.82	.11	139	0
0	23	1.81	60	305	1	.96	1	.21	.07	.05	.43	.06	128	0
9	71	4.17	44	381	559	7.44	15	.07	.69	.06	.63	.09	29	4
9	79	2.16	44	346	437	1.95	14	.06	.7	.08	.46	.11	49	4
8	62	2.25	36	306	559	2.43	19	.07	.61	.07	1.17	.06	39	3
0	19	1.16	29	347	45	.37	16	.06	.25	.05	.91	.1	18	11
0	25	1.76	50	370	26	.49	15	.06	.58	.05	.69	.1	14	5
0	16	2.25	40	478	2	.89	0	.15	.17	.05	.4	.15	78	0
0	42	1.86	48	369	240	.87	<1	.13	.08	.07	.37	.11	94	2
0	29	.81	16	188	2	.23	42	.05	.09	.06	.39	.03	21	6
0	33	.6	16	86	6	.33	27	.02	.09	.06	.26	.04	16	3
0	18	.61	9	74	178	.2	24	.01	.09	.04	.14	.02	22	3
0	29	.81	16	188	2	.23	5	.05	.18	.06	.39	.03	21	6
0	7	.47	11	77	3	.21	1	.04	.02	.06	.39	.05	32	7
0	8	1.18	20	136	6	.56	2	.09	.01	.11	.74	.08	43	10
0	7	.4	9	63	6	.29	1	.03	.01	.06	.51	.03	18	7
0	9	.27	6	17	88	.18	1	.02	.01	.04	.14	.02	6	<1
0	14	.67	20	259	65	.3	3	.02	.25	.03	.28	.06	68	3
0	16	.79	23	305	77	.35	4	.03	.3	.04	.33	.07	80	4
0	12	1.44	13	117	153	.17	1	.01	.24	.03	.12	.04	24	3
0	12	.47	17	169	301	.3	1	.01	.15	.05	.29	.06	30	3
0	82	1.37	49	654	174	.36	367	.08	.22	.21	.36	.09	10	18

Table A-1

Food Composition (Computer code number is for West Diet Analysis program) (For purposes of calculations, use "0" for t, <1, <.1, <.01, etc.)

Computer Code Number	Food Description	Measure	Wt (g)	H_2O (%)	Ener (cal)	Prot (g)	Carb (g)	Dietary Fiber (g)	Fat (g)	Fat Breakdown (g) Sat	Mono	Poly
	VEGETABLES and LEGUMES—Continued											
	Broccoli, raw:											
817	Chopped	½ c	44	91	12	1	2	1	<1	t	t	.1
818	Spears	1 ea	31	91	9	1	2	1	<1	t	t	.1
	Broccoli, cooked from fresh:											
819	Spears	1 ea	180	91	50	5	9	5	1	.1	t	.3
820	Chopped	½ c	78	91	22	2	4	2	<1	t	t	.1
	Broccoli, cooked from frozen:											
821	Spear, small piece	½ c	92	91	26	3	5	3	<1	t	t	.1
822	Chopped	½ c	92	91	26	3	5	3	<1	t	t	.1
1603	Broccoflower, steamed	½ c	78	90	25	2	5	2	<1	t	t	.1
823	Brussels sprouts, cooked from fresh	½ c	78	87	30	2	7	2	<1	.1	t	.2
824	Brussels sprouts, cooked from frozen	½ c	78	87	33	3	6	3	<1	.1	t	.2
	Cabbage, common varieties:											
825	Raw, shredded or chopped	1 c	70	92	17	1	4	2	<1	t	t	.1
826	Cooked, drained	1 c	150	94	33	2	7	3	1	.1	t	.3
	Cabbage, Chinese:											
1178	Bok choy, raw, shredded	1 c	70	95	9	1	2	1	<1	t	t	.1
827	Bok choy, cooked, drained	1 c	170	96	20	3	3	3	<1	t	t	.1
1937	Kim chee style	1 c	150	92	31	2	6	2	<1	t	t	.2
828	Pe Tsai, raw, chopped	1 c	76	94	12	1	2	2	<1	t	t	.1
1796	Pe Tsai, cooked	1 c	119	95	17	2	3	3	<1	t	t	.1
	Cabbage, red, coarsely chopped:											
829	Raw	1 c	89	92	24	1	5	2	<1	t	t	.1
830	Cooked, drained	1 c	150	94	31	2	7	3	<1	t	t	.1
831	Cabbage, savoy, coarsely chopped, raw	1 c	70	91	19	1	4	2	<1	t	t	t
1785	Cabbage, savoy, cooked	1 c	145	92	35	3	8	4	<1	t	t	.1
1896	Capers	1 ea	5	86	5	<1	<1	<1	<1			
	Carrots, raw:											
832	Whole, 7½ x 1⅛"	1 ea	72	88	31	1	7	2	<1	t	t	.1
833	Grated	½ c	55	88	24	1	6	2	<1	t	t	t
	Carrots, cooked, sliced, drained:											
834	From raw	½ c	78	87	35	1	8	3	<1	t	t	.1
835	From frozen	½ c	73	90	26	1	6	3	<1	t	t	t
836	Carrots, canned, sliced, drained	½ c	73	93	17	<1	4	1	<1	t	t	.1
837	Carrot juice, canned	1 c	236	89	94	2	22	2	<1	.1	t	.2
	Cauliflower, flowerets:											
838	Raw	½ c	50	92	12	1	3	1	<1	t	t	t
839	Cooked from fresh, drained	½ c	62	93	14	1	3	2	<1	t	t	.1
840	Cooked, from frozen, drained	½ c	90	94	17	1	3	2	<1	t	t	.1
	Celery, pascal type, raw:											
841	Large outer stalk, 8 x 1½"(root end)	1 ea	40	95	6	<1	1	1	<1	t	t	t
842	Diced	1 c	120	95	19	1	4	2	<1	t	t	.1
1789	Celeriac/celery root, cooked	1 c	155	92	39	1	9	2	<1	.1	.1	.2
1179	Chard, swiss, raw, chopped	1 c	36	93	7	1	1	1	<1	t	t	t
1180	Chard, swiss, cooked	1 c	175	93	35	3	7	4	<1	t	t	t
1855	Chayote fruit, raw	1 ea	203	94	39	2	9	3	<1	.1	t	.1
1856	Chayote fruit, cooked	1 c	160	93	38	1	8	4	1	.2	.1	.3
	Chickpeas (see Garbanzo Beans #854)											
	Collards, cooked, drained:											
843	From raw	½ c	95	92	25	2	5	3	<1	t	t	.1
844	From frozen	½ c	85	88	31	3	6	3	<1	.1	t	.2
	Corn, yellow, cooked, drained:											
845	From raw, on cob, 5" long	1 ea	77	73	72	2	17	2	1	.1	.2	.3
846	From frozen, on cob, 3½" long	1 ea	63	73	59	2	14	2	<1	.1	.1	.2
847	Kernels, cooked from frozen	½ c	82	77	66	2	16	2	<1	.1	.1	.2
	Corn, canned:											
848	Cream style	½ c	128	79	92	2	23	2	1	.1	.2	.3
849	Whole kernel, vacuum pack	½ c	105	77	83	3	20	2	1	.1	.2	.2
	Cowpeas (see Black-eyed peas #814-816)											
850	Cucumber slices with peel	7 pce	28	96	4	<1	1	<1	<1	t	t	t
1948	Cucumber, kim chee style	1 c	150	91	31	2	7	2	<1	t	0	.1

Chol (mg)	Calc (mg)	Iron (mg)	Magn (mg)	Pota (mg)	Sodi (mg)	Zinc (mg)	VT-A (RE)	Thia (mg)	VT-E (a-TE)	Ribo (mg)	Niac (mg)	V-B6 (mg)	Fola (μg)	VT-C (mg)
0	21	.39	11	143	12	.18	68	.03	.73	.05	.28	.07	31	41
0	15	.27	8	101	8	.12	48	.02	.51	.04	.2	.05	22	29
0	83	1.51	43	526	47	.68	250	.1	3.04	.2	1.03	.26	90	134
0	36	.65	19	228	20	.3	108	.04	1.32	.09	.45	.11	39	58
0	47	.56	18	166	22	.28	174	.05	.95	.07	.42	.12	28	37
0	47	.56	18	166	22	.28	174	.05	1.52	.07	.42	.12	52	37
0	25	.55	16	251	18	.39	5	.06	.23	.07	.59	.14	38	49
0	28	.94	16	247	16	.26	56	.08	.66	.06	.47	.14	47	48
0	19	.58	19	254	18	.28	46	.08	.45	.09	.42	.22	79	36
0	33	.41	10	172	13	.13	9	.03	.07	.03	.21	.07	30	22
0	46	.25	12	146	12	.13	19	.09	.16	.08	.42	.17	30	30
0	73	.56	13	176	45	.13	210	.03	.08	.05	.35	.14	46	31
0	158	1.77	19	631	58	.29	437	.05	.2	.11	.73	.28	69	44
0	145	1.28	27	375	995	.35	426	.07	.24	.1	.75	.34	88	80
0	58	.24	10	181	7	.17	91	.03	.09	.04	.3	.18	60	20
0	38	.36	12	268	11	.21	115	.05	.14	.05	.59	.21	63	19
0	45	.44	13	183	10	.19	4	.04	.09	.03	.27	.19	18	51
0	55	.52	16	210	12	.22	4	.05	.18	.03	.3	.21	19	52
0	24	.28	20	161	20	.19	70	.05	.07	.02	.21	.13	56	22
0	43	.55	35	267	35	.33	129	.07	.15	.03	.03	.22	67	25
0	2	.05			105		1							0
0	19	.36	11	233	25	.14	2025	.07	.33	.04	.67	.11	10	7
0	15	.27	8	178	19	.11	1547	.05	.25	.03	.51	.08	8	5
0	24	.48	10	177	51	.23	1914	.03	.33	.04	.39	.19	11	2
0	20	.34	7	115	43	.17	1292	.02	.31	.03	.32	.09	8	2
0	18	.47	6	131	177	.19	1005	.01	.31	.02	.4	.08	7	2
0	57	1.09	33	689	68	.42	2584	.22	.02	.13	.91	.51	9	20
0	11	.22	7	152	15	.14	1	.03	.02	.03	.26	.11	28	23
0	10	.2	6	88	9	.11	1	.03	.02	.03	.25	.11	27	27
0	15	.37	8	125	16	.12	2	.03	.04	.05	.28	.08	37	28
0	16	.16	4	115	35	.05	5	.02	.14	.02	.13	.03	11	3
0	48	.48	13	344	104	.16	16	.05	.43	.05	.39	.1	34	8
0	40	.67	19	268	95	.31	0	.04	.31	.06	.66	.16	5	6
0	18	.65	29	136	77	.13	119	.01	.68	.03	.14	.04	5	11
0	102	3.96	151	961	313	.58	550	.06	3.31	.15	.63	.15	15	31
0	34	.69	24	254	4	1.5	12	.05	.24	.06	.95	.15	189	16
0	21	.35	19	277	2	.5	8	.04	.19	.06	.67	.19	29	13
0	113	.44	16	247	9	.4	297	.04	.84	.1	.55	.12	88	17
0	179	.95	25	213	42	.23	508	.04	.42	.1	.54	.1	65	22
0	2	.47	22	193	3	.48	16	.13	.07	.05	1.17	.17	23	4
0	2	.38	18	158	3	.4	13	.11	.06	.04	.96	.14	19	3
0	3	.29	16	121	4	.33	18	.07	.07	.06	1.07	.11	25	3
0	4	.49	22	172	365	.68	13	.03	.11	.07	1.23	.08	57	6
0	5	.44	24	195	286	.48	25	.04	.09	.08	1.23	.06	52	9
0	4	.07	3	40	1	.06	6	.01	.02	.01	.06	.01	4	1
0	13	7.23	12	176	1531	.76	49	.04	.24	.04	.69	.16	34	5

Table A-1

Food Composition (Computer code number is for West Diet Analysis program) (For purposes of calculations, use "0" for t, <1, <.1, <.01, etc.)

Computer Code Number	Food Description	Measure	Wt (g)	H_2O (%)	Ener (cal)	Prot (g)	Carb (g)	Dietary Fiber (g)	Fat (g)	Fat Breakdown (g) Sat	Mono	Poly
	VEGETABLES and LEGUMES—Continued											
	Dandelion Greens:											
851	Raw	1 c	55	86	25	1	5	2	<1	.1	t	.2
852	Chopped, cooked, drained	1 c	105	90	35	2	7	3	1	.2	t	.3
853	Eggplant, cooked	1 c	99	92	28	1	7	2	<1	t	t	.1
1714	Endive, fresh, chopped	1 c	50	94	8	1	2	2	<1	t	t	t
856	Escarole/curly endive, chopped	1 c	50	94	8	1	2	2	<1	t	t	t
854	Garbanzo beans (chickpeas), cooked	1 c	164	60	269	14	45	12	4	.4	1	1.9
1939	Grape leaf, raw:	1 ea	3	73	3	<1	1	<1	<1	t	t	t
7914	Cup	1 c	14	73	13	1	2	2	<1	t	t	.1
855	Great northern beans, cooked	1 c	177	69	209	15	37	12	1	.2	t	.3
857	Jerusalem artichoke, raw slices	1 c	150	78	114	3	26	2	<1	0	t	t
1794	Jicama	1 c	120	90	46	1	11	6	<1	t	t	t
	Kale, cooked, drained:											
858	From raw	1 c	130	91	36	2	7	3	1	.1	t	.3
859	From frozen	1 c	130	90	39	4	7	3	1	.1	t	.3
860	Kidney beans, canned	1 c	256	77	218	13	40	16	1	.1	.1	.5
1181	Kohlrabi, raw slices	1 c	135	91	36	2	8	5	<1	t	t	.1
861	Kohlrabi, cooked	1 c	165	90	48	3	11	2	<1	t	t	.1
1183	Leeks, raw, chopped	1 c	89	83	54	1	13	2	<1	t	t	.1
1182	Leeks, cooked, chopped	1 c	104	91	32	1	8	1	<1	t	t	.1
862	Lentils, cooked from dry	1 c	198	70	230	18	40	16	1	.1	.1	.3
1288	Lentils, sprouted, stir-fried	1 c	124	69	125	11	26	5	1	.1	.1	.2
1289	Lentils, sprouted, raw	1 c	77	67	82	7	17	3	<1	t	.1	.2
	Lettuce:											
	Butterhead/Boston types:											
863	Head, 5" diameter	¼ ea	41	96	5	1	1	<1	<1	t	t	t
864	Leaves, inner or outer	4 ea	30	96	4	<1	1	<1	<1	t	t	t
	Iceberg/crisphead:											
867	Chopped or shredded	1 c	55	96	7	1	1	1	<1	t	t	.1
865	Head, 6" diameter	1 ea	539	96	65	5	11	8	1	.1	t	.5
866	Wedge, ¼ head	1 ea	135	96	16	1	3	2	<1	t	t	.1
868	Looseleaf, chopped	½ c	28	94	5	<1	1	1	<1	t	t	t
869	Romaine, chopped	½ c	28	95	4	<1	1	<1	<1	t	t	t
870	Romaine, inner leaf	3 pce	30	95	4	<1	1	1	<1	t	t	t
1930	Luffa, cooked (Chinese okra)	1 c	178	89	57	3	13	6	<1	.1	t	.1
	Mushrooms:											
871	Raw, sliced	½ c	35	92	9	1	2	<1	<1	t	t	.1
872	Cooked from fresh, pieces	½ c	78	91	21	2	4	2	<1	t	t	.1
1962	Stir fried, shitake slices	½ c	73	83	40	1	10	2	<1	t	t	t
873	Canned, drained	½ c	78	91	19	1	4	2	<1	t	t	.1
1951	Mushroom caps, pickled	8 ea	47	92	11	1	2	1	<1	t	t	.1
	Mustard greens:											
874	Cooked from raw	½ c	70	94	10	2	1	1	<1	t	.1	t
875	Cooked from frozen	½ c	75	94	14	2	2	2	<1	t	.1	t
876	Navy beans, cooked from dry	1 c	182	63	258	16	48	12	1	.3	.1	.4
	Okra, cooked:											
877	From fresh pods	8 ea	85	90	27	2	6	2	<1	t	t	t
878	From frozen slices	1 c	184	91	51	4	11	5	1	.1	.1	.1
1236	Batter fried from fresh	1 c	92	69	175	3	11	2	13	2.1	3.4	7.1
1930	Chinese, (Luffa), cooked	1 c	178	89	57	3	13	6	<1	.1	t	.1
	Onions:											
879	Raw, chopped	½ c	80	90	30	1	7	1	<1	t	t	t
880	Raw, sliced	½ c	58	90	22	1	5	1	<1	t	t	t
881	Cooked, drained, chopped	½ c	105	88	46	1	11	1	<1	t	t	.1
882	Dehydrated flakes	¼ c	14	4	49	1	12	1	<1	t	t	t
1934	Onions, pearl, cooked	½ c	93	87	41	1	9	1	<1	t	t	.1
883	Spring/green onions, bulb and top, chopped	½ c	50	90	16	1	4	1	<1	t	t	t
884	Onion rings, breaded, heated f/frozen	2 ea	20	28	81	1	8	<1	5	1.7	2.2	1
1917	Palm hearts, cooked slices	1 c	146	69	150	4	39	2	<1	.1	.1	t
885	Parsley, raw, chopped	½ c	30	88	11	1	2	1	<1	t	.1	t
888	Parsnips, sliced, cooked	½ c	78	78	63	1	15	3	<1	t	.1	t

PAGE KEY: A–4 = Beverages A–6 = Dairy A–10 = Eggs A–10 = Fat/Oil A–14 = Fruit A–20 = Bakery A–28 = Grain A–32 = Fish A–34 = Meats A–38 = Poultry A–40 = Sausage A–42 = Mixed/Fast A–46 = Nuts/Seeds A–50 = Sweets A–52 = Vegetables/Legumes A–62 = Vegetarian Foods A–64 = Misc A–66 = Soups/Sauces A–68 = Fast A–84 = Convenience A–88 = Baby foods

Chol (mg)	Calc (mg)	Iron (mg)	Magn (mg)	Pota (mg)	Sodi (mg)	Zinc (mg)	VT-A (RE)	Thia (mg)	VT-E (a-TE)	Ribo (mg)	Niac (mg)	V-B6 (mg)	Fola (µg)	VT-C (mg)
0	103	1.71	20	218	42	.23	770	.1	1.38	.14	.44	.14	15	19
0	147	1.89	25	244	46	.29	1228	.14	2.63	.18	.54	.17	13	19
0	6	.35	13	246	3	.15	6	.07	.03	.02	.59	.08	14	1
0	26	.41	7	157	11	.39	103	.04	.22	.04	.2	.01	71	3
0	26	.41	7	157	11	.39	103	.04	.22	.04	.2	.01	71	3
0	80	4.74	79	477	11	2.51	5	.19	.57	.1	.86	.23	282	2
0	11	.08	3	8	<1	.02	81	<.01	.06	.01	.07	.01	2	<1
0	51	.37	13	38	1	.09	378	.01	.28	.05	.33	.06	12	2
0	120	3.77	88	692	4	1.56	<1	.28	.53	.1	1.21	.21	181	2
0	21	5.1	25	644	6	.18	3	.3	.28	.09	1.95	.12	20	6
0	14	.72	14	180	5	.19	2	.02	5.48	.03	.24	.05	14	24
0	94	1.17	23	296	30	.31	962	.07	1.11	.09	.65	.18	17	53
0	179	1.22	23	417	19	.23	826	.06	.23	.15	.87	.11	19	33
0	61	3.23	72	658	873	1.41	0	.27	.13	.22	1.17	.06	130	3
0	32	.54	26	473	27	.04	5	.07	.65	.03	.54	.2	22	84
0	41	.66	31	561	35	.51	7	.07	2.76	.03	.64	.25	20	89
0	52	1.87	25	160	18	.11	9	.05	.82	.03	.36	.21	57	11
0	31	1.14	15	90	10	.06	5	.03	.63	.02	.21	.12	25	4
0	38	6.59	71	731	4	2.51	2	.33	.22	.14	2.1	.35	358	3
0	17	3.84	43	352	12	1.98	5	.27	.11	.11	1.49	.2	83	16
0	19	2.47	28	248	8	1.16	4	.18	.07	.1	.87	.15	77	13
0	13	.12	5	105	2	.07	40	.02	.18	.02	.12	.02	30	3
0	10	.09	4	77	1	.05	29	.02	.13	.02	.09	.01	22	2
0	10	.27	5	87	5	.12	18	.02	.15	.02	.1	.02	31	2
0	102	2.7	48	852	48	1.19	178	.25	1.51	.16	1.01	.22	302	21
0	26	.67	12	213	12	.3	45	.06	.38	.04	.25	.05	76	5
0	19	.39	3	74	3	.08	53	.01	.12	.02	.11	.01	14	5
0	10	.31	2	81	2	.07	73	.03	.12	.03	.14	.01	38	7
0	11	.33	2	87	2	.07	78	.03	.13	.03	.15	.01	41	7
0	112	.8	101	570	420	.97	103	.23	1.22	.1	1.54	.33	81	29
0	2	.43	3	130	1	.26	0	.04	.04	.16	1.44	.03	7	1
0	5	1.36	9	278	2	.68	0	.06	.09	.23	3.48	.07	14	3
0	2	.32	10	85	3	.97	0	.03	.09	.12	1.1	.12	15	<1
0	9	.62	12	101	332	.56	0	.07	.09	.02	1.24	.05	10	0
0	2	.5	5	139	95	.28	0	.03	.05	.16	1.42	.03	6	1
0	52	.49	10	141	11	.08	212	.03	1.41	.04	.3	.07	51	18
0	76	.84	10	104	19	.15	335	.03	1.31	.04	.19	.08	52	10
0	127	4.51	107	670	2	1.93	<1	.37	.73	.11	.97	.3	255	2
0	54	.38	48	274	4	.47	49	.11	.59	.05	.74	.16	39	14
0	177	1.23	94	431	6	1.14	94	.18	1.27	.23	1.44	.09	269	22
15	104	.77	37	214	137	.5	43	.13	3.08	.1	.75	.13	37	10
0	112	.8	101	570	420	.97	103	.23	1.22	.1	1.54	.33	81	29
0	16	.18	8	126	2	.15	0	.03	.1	.02	.12	.09	15	5
0	12	.13	6	91	2	.11	0	.02	.07	.01	.09	.07	11	4
0	23	.25	12	174	3	.22	0	.04	.14	.02	.17	.13	16	5
0	36	.22	13	227	3	.26	0	.07	.19	.01	.14	.22	23	10
0	21	.22	10	154	218	.19	0	.04	.12	.02	.15	.12	14	5
0	36	.74	10	138	8	.19	19	.03	.06	.04	.26	.03	32	9
0	6	.34	4	26	75	.08	5	.06	.14	.03	.72	.01	13	<1
0	26	2.47	15	2636	20	5.45	10	.07	.73	.25	1.25	1.06	30	10
0	41	1.86	15	166	17	.32	156	.03	.54	.03	.39	.03	46	40
0	29	.45	23	286	8	.2	0	.06	.78	.04	.56	.07	45	10

Table A-1

Food Composition (Computer code number is for West Diet Analysis program) (For purposes of calculations, use "0" for t, <1, <.1, <.01, etc.)

Computer Code Number	Food Description	Measure	Wt (g)	H_2O (%)	Ener (cal)	Prot (g)	Carb (g)	Dietary Fiber (g)	Fat (g)	Fat Breakdown (g) Sat	Mono	Poly
	VEGETABLES and LEGUMES—Continued											
	Peas:											
	Black-eyed, cooked:											
814	From dry, drained	½ c	86	70	100	7	18	6	<1	.1	t	.2
815	From fresh, drained	½ c	82	75	79	3	17	4	<1	.1	t	.1
816	From frozen, drained	½ c	85	66	112	7	20	5	1	.1	.1	.2
889	Edible pod peas, cooked	½ c	80	89	34	3	6	2	<1	t	t	.1
890	Green, canned, drained:	½ c	85	82	59	4	11	3	<1	.1	t	.1
5267	Unsalted	½ c	124	86	66	4	12	4	<1	.1	t	.2
891	Green, cooked from frozen	½ c	80	79	62	4	11	4	<1	t	t	.1
1786	Snow peas, raw	½ c	49	89	21	1	4	1	<1	t	t	t
1787	Snow peas, raw	10 ea	34	89	14	1	3	1	<1	t	t	t
892	Split, green, cooked from dry	½ c	98	69	116	8	21	8	<1	.1	.1	.2
1187	Peas & carrots, cooked from frozen	½ c	80	86	38	2	8	2	<1	.1	t	.2
1186	Peas & carrots, canned w/liquid	½ c	128	88	49	3	11	3	<1	.1	t	.2
	Peppers, hot:											
893	Hot green chili, canned	½ c	68	92	14	1	3	1	<1	t	t	t
894	Hot green chili, raw	1 ea	45	88	18	1	4	1	<1	t	t	t
1715	Hot red chili, raw, diced	1 tbs	9	88	4	<1	1	<1	<1	t	t	t
1988	Jalapeno, raw	1 ea	45	90	11	<1	2		<1			
895	Jalapeno, chopped, canned	½ c	68	89	18	1	3	2	1	.1	t	.3
1918	Jalapeno wheels, in brine (Ortega)	2 tbs	29		10	0	2		0	0	0	0
	Peppers, sweet, green:											
896	Whole pod (90 g with refuse), raw	1 ea	74	92	20	1	5	1	<1	t	t	.1
897	Cooked, chopped (1 pod cooked = 73g)	½ c	68	92	19	1	5	1	<1	t	t	.1
	Peppers, sweet, red:											
1286	Raw, chopped	½ c	75	92	20	1	5	1	<1	t	t	.1
1807	Raw, each	1 ea	74	92	20	1	5	1	<1	t	t	.1
1287	Cooked, chopped	½ c	68	92	19	1	5	1	<1	t	t	.1
	Peppers, sweet, yellow:											
1872	Raw, large	1 ea	186	92	50	2	12	2	<1	.1	t	.2
1873	Strips	10 pce	52	92	14	1	3	<1	<1	t	t	.1
898	Pinto beans, cooked from dry	½ c	85	64	116	7	22	7	<1	.1	.1	.2
1191	Poi, two finger	½ c	120	72	134	<1	33	<1	<1	t	t	.1
	Potatoes:											
	Baked in oven, 4¾"x2⅓" diam											
899	With skin	1 ea	202	71	220	5	51	5	<1	.1	t	.1
900	Flesh only	1 ea	156	75	145	3	34	2	<1	t	t	.1
901	Skin only	1 ea	58	47	115	2	27	5	<1	t	t	t
	Baked in microwave, 4¾"x 2⅓"dm:											
902	With skin	1 ea	202	72	212	5	49	5	<1	.1	t	.1
903	Flesh only	1 ea	156	74	156	3	36	2	<1	t	t	.1
904	Skin only	1 ea	58	63	77	3	17	3	<1	t	t	t
	Boiled, about 2½" diam:											
905	Peeled after boiling	1 ea	136	77	118	3	27	2	<1	t	t	.1
906	Peeled before boiling	1 ea	135	77	116	2	27	2	<1	t	t	.1
	French fried, strips 2–3½" long:											
907	Oven heated	10 ea	50	35	167	2	20	2	9	3	5.7	.7
908	Fried in vegetable oil	10 ea	50	38	158	2	20	2	8	1.9	4.7	.7
1188	Fried in veg and animal oil	10 ea	50	38	158	2	20	2	8	1.9	4.7	.7
909	Hashed browns from frozen	1 c	156	56	340	5	44	3	18	7	8	2.1
	Mashed:											
910	Home recipe with whole milk	½ c	105	78	81	2	18	2	1	.4	.2	.1
911	Home recipe with milk and marg	½ c	105	76	111	2	17	2	4	1.1	1.9	1.3
912	Prepared from flakes; water, milk, margarine, salt added	½ c	110	76	124	2	16	3	6	1.6	2.5	1.7
	Potato products, prepared:											
	Au gratin:											
913	From dry mix	½ c	123	79	114	3	16	1	5	3.5	1.5	.2
914	From home recipe, using butter	½ c	122	74	161	7	14	2	9	4.8	3.2	1.3
	Scalloped:											
915	From dry mix	½ c	122	79	113	3	16	1	5	3.2	1.5	.2
916	From home recipe, using butter	½ c	123	81	106	4	13	2	5	1.7	1.7	.9

Chol (mg)	Calc (mg)	Iron (mg)	Magn (mg)	Pota (mg)	Sodi (mg)	Zinc (mg)	VT-A (RE)	Thia (mg)	VT-E (a-TE)	Ribo (mg)	Niac (mg)	V-B6 (mg)	Fola (µg)	VT-C (mg)
0	21	2.16	46	239	3	1.11	2	.17	.24	.05	.43	.09	179	<1
0	105	.92	43	343	3	.84	65	.08	.18	.12	1.15	.05	104	2
0	20	1.8	42	319	4	1.21	7	.22	.33	.05	.62	.08	120	2
0	34	1.58	21	192	3	.3	10	.1	.31	.06	.43	.11	23	38
0	17	.81	14	147	214	.6	65	.1	.32	.07	.62	.05	38	8
0	22	1.26	21	124	11	.87	47	.14	.47	.09	1.04	.08	35	12
0	19	1.26	23	134	70	.75	54	.23	.14	.08	1.18	.09	47	8
0	21	1.02	12	98	2	.13	7	.07	.19	.04	.29	.08	20	29
0	15	.71	8	68	1	.09	5	.05	.13	.03	.2	.05	14	20
0	14	1.26	35	355	2	.98	1	.19	.38	.05	.87	.05	64	<1
0	18	.75	13	126	54	.36	621	.18	.26	.05	.92	.07	21	6
0	29	.96	18	128	333	.74	739	.09	.24	.07	.74	.11	23	8
0	5	.34	10	127	798	.12	41	.01	.47	.03	.54	.1	7	46
0	8	.54	11	153	3	.13	35	.04	.31	.04	.43	.12	10	109
0	2	.11	2	31	1	.03	97	.01	.06	.01	.09	.02	2	22
				2	2		30		.37					53
0	16	1.28	10	131	1136	.23	116	.03	.47	.03	.27	.13	10	7
0				55	390		10		.2					21
0	7	.34	7	131	1	.09	47	.05	.51	.02	.38	.18	16	66
0	6	.31	7	113	1	.08	40	.04	.47	.02	.32	.16	11	51
0	7	.34	7	133	1	.09	428	.05	.52	.02	.38	.19	16	143
0	7	.34	7	131	1	.09	422	.05	.51	.02	.38	.18	16	141
0	6	.31	7	113	1	.08	256	.04	.47	.02	.32	.16	11	116
0	20	.86	22	394	4	.32	45	.05	1.28	.05	1.66	.31	48	342
0	6	.24	6	110	1	.09	12	.01	.36	.01	.46	.09	13	96
0	41	2.22	47	398	2	.92	<1	.16	.8	.08	.34	.13	146	2
0	19	1.06	29	220	14	.26	2	.16	.22	.05	1.32	.33	26	5
0	20	2.75	54	844	16	.65	0	.22	.1	.07	3.33	.7	22	26
0	8	.55	39	610	8	.45	0	.16	.06	.03	2.18	.47	14	20
0	20	4.08	25	332	12	.28	0	.07	.02	.06	1.78	.36	12	8
0	22	2.5	54	903	16	.73	0	.24	.1	.06	3.45	.69	24	30
0	8	.64	39	641	11	.51	0	.2	.06	.04	2.54	.5	19	24
0	27	3.45	21	377	9	.3	0	.04	.02	.04	1.29	.28	10	9
0	7	.42	30	515	5	.41	0	.14	.07	.03	1.96	.41	14	18
0	11	.42	27	443	7	.36	0	.13	.07	.03	1.77	.36	12	10
0	6	.83	11	270	307	.2	0	.04	.25	.02	1.34	.11	11	3
0	9	.38	17	366	108	.19	0	.09	.25	.01	1.63	.12	14	5
6	9	.38	17	366	108	.19	0	.09	.25	.01	1.63	.12	14	5
0	23	2.36	26	680	53	.5	0	.17	.3	.03	3.78	.2	10	10
2	27	.28	19	314	318	.3	6	.09	.05	.04	1.18	.24	9	7
2	27	.27	19	303	310	.28	21	.09	.31	.04	1.13	.23	8	6
4	54	.24	20	256	365	.2	23	.12	.77	.05	.74	.01	8	11
18	102	.39	18	269	540	.29	38	.02	1.48	.1	1.15	.05	8	4
18	145	.78	24	483	528	.84	46	.08	.64	.14	1.21	.21	13	12
13	44	.46	17	248	416	.3	26	.02	.18	.07	1.26	.05	12	4
7	70	.7	23	465	412	.49	23	.08	.4	.11	1.29	.22	13	13

Table A-1

Food Composition (Computer code number is for West Diet Analysis program) (For purposes of calculations, use "0" for t, <1, <.1, <.01, etc.)

Computer Code Number	Food Description	Measure	Wt (g)	H_2O (%)	Ener (cal)	Prot (g)	Carb (g)	Dietary Fiber (g)	Fat (g)	Fat Breakdown (g) Sat	Mono	Poly
	VEGETABLES and LEGUMES—Continued											
	Potato Salad (see Mixed Dishes #715)											
1192	Potato puffs, cooked from frozen	½ c	64	53	142	2	19	2	7	3.3	2.8	.5
918	Pumpkin, cooked from fresh, mashed	½ c	123	94	25	1	6	1	<1	t	t	t
919	Pumpkin, canned	½ c	123	90	42	1	10	4	<1	.2	t	t
1891	Radicchio, raw, shredded	½ c	20	93	5	<1	1	<1	<1	t	t	t
1894	Radicchio, raw, leaf	10 ea	80	93	18	1	4	1	<1	t	t	.1
920	Red radishes	10 ea	45	95	8	<1	2	1	<1	t	t	t
1793	Daikon radishes (Chinese) raw	½ c	44	95	8	<1	2	1	<1	t	t	t
921	Refried beans, canned	½ c	126	76	118	7	19	7	2	.6	.7	.2
1375	Rutabaga, cooked cubes	½ c	85	89	33	1	7	2	<1	t	t	.1
922	Sauerkraut, canned with liquid	½ c	118	92	22	1	5	3	<1	t	t	.1
923	Seaweed, kelp, raw	½ c	40	82	17	1	4	1	<1	.1	t	t
924	Seaweed, spirulina, dried	½ c	8	5	23	5	2	<1	1	.2	.1	.2
1866	Shallots, raw, chopped	1 tbs	10	80	7	<1	2	<1	<1	t	t	t
1557	Snow peas, stir-fried	½ c	83	89	35	2	6	2	<1	t	t	.1
925	Soybeans, cooked from dry	½ c	86	63	149	15	9	5	8	1.1	1.7	4.4
1996	Soybeans, dry roasted	½ c	86	1	387	34	28	7	19	2.7	4.1	10.6
	Soybean products:											
926	Miso	½ c	138	41	284	17	39	7	8	1.2	1.9	4.7
	Soy milk (see #144 and #2301 under Dairy)											
	Tofu (soybean curd):											
7540	Extra firm, silken	½ c	126	88	69	9	3	<1	2	.4	.4	1.3
7542	Firm, silken	½ c	126	87	77	9	3	<1	3	.5	.7	1.9
927	Regular	½ c	124	87	76	8	2	<1	5	.7	1	2.6
7541	Soft, silken	½ c	124	89	68	6	4	<1	3	.4	.6	1.9
	Spinach:											
928	Raw, chopped	½ c	28	92	6	1	1	1	<1	t	t	t
929	Cooked, from fresh, drained	½ c	90	91	21	3	3	2	<1	t	t	.1
930	Cooked from frozen (leaf)	½ c	95	90	27	3	5	3	<1	t	t	.1
931	Canned, drained solids:	½ c	107	92	25	3	4	3	1	.1	t	.2
5149	Unsalted	½ c	107	92	25	3	4	3	1	.1	t	.2
	Spinach soufflé (see Mixed Dishes)											
	Squash, summer varieties,cooked w/skin:											
932	Varieties averaged	½ c	90	94	18	1	4	1	<1	.1	t	.1
933	Crookneck	½ c	90	94	18	1	4	1	<1	.1	t	.1
934	Zucchini	½ c	90	95	14	1	4	1	<1	t	t	t
	Squash, winter varieties, cooked:											
	Average of all varieties, baked:											
935	Mashed	1 c	245	89	96	2	21	7	2	.3	.1	.6
936	Cubes	1 c	205	89	80	2	18	6	1	.3	.1	.5
937	Acorn, baked, mashed	½ c	123	83	69	1	18	5	<1	t	t	.1
1218	Acorn, boiled, mashed	½ c	122	90	41	1	11	3	<1	t	t	t
	Butternut squash:											
938	Baked cubes	½ c	103	88	41	1	11	3	<1	t	t	t
1219	Baked, mashed	½ c	103	88	41	1	11	3	<1	t	t	t
1193	Cooked from frozen	½ c	120	88	47	1	12	3	<1	t	t	t
1194	Hubbard, baked, mashed	½ c	120	85	60	3	13	3	1	.2	.1	.3
1195	Hubbard, boiled, mashed	½ c	118	91	35	2	8	3	<1	.1	t	.2
1196	Spaghetti, baked or boiled	½ c	77	92	22	<1	5	1	<1	t	t	.1
1189	Succotash, cooked from frozen	½ c	85	74	79	4	17	3	1	.1	.1	.4
	Sweet potatoes:											
939	Baked in skin, peeled, 5 x 2″ diam	1 ea	114	73	117	2	28	3	<1	t	t	.1
940	Boiled without skin, 5 x 2″ diam	1 ea	151	73	159	2	37	3	<1	.1	t	.2
941	Candied, 2½ x 2″	1 pce	105	67	144	1	29	3	3	1.4	.7	.2
	Canned:											
942	Solid pack	½ c	128	74	129	3	30	2	<1	.1	t	.1
943	Vacuum pack, mashed	½ c	127	76	116	2	27	2	<1	.1	t	.1
944	Vacuum pack, 3¾ x 1″	2 pce	80	76	73	1	17	1	<1	t	t	.1
1940	Taro shoots, cooked slices	1 c	140	95	20	1	4	1	<1	t	t	t
1941	Taro, tahitian, cooked slices	1 c	137	86	60	6	9	1	1	.2	.1	.4
	Tomatillos:											
1877	Raw, each	1 ea	34	92	11	<1	2	1	<1	t	.1	.1
1875	Raw, chopped	1 c	132	92	42	1	8	3	1	.2	.2	.6

Chol (mg)	Calc (mg)	Iron (mg)	Magn (mg)	Pota (mg)	Sodi (mg)	Zinc (mg)	VT-A (RE)	Thia (mg)	VT-E (a-TE)	Ribo (mg)	Niac (mg)	V-B6 (mg)	Fola (μg)	VT-C (mg)
0	19	1	12	243	477	.19	1	.12	.03	.05	1.38	.15	11	4
0	18	.7	11	283	1	.28	1330	.04	1.3	.1	.51	.05	10	6
0	32	1.71	28	253	6	.21	2713	.03	1.3	.07	.45	.07	15	5
0	4	.11	3	60	4	.12	1	<.01	.45	.01	.05	.01	12	2
0	15	.45	10	242	18	.5	2	.01	1.81	.02	.2	.05	48	6
0	9	.13	4	104	11	.13	<1	<.01	0	.02	.13	.03	12	10
0	12	.18	7	100	9	.07	0	.01	0	.01	.09	.02	12	10
10	44	2.09	42	336	377	1.47	0	.03	.39	.02	.4	.18	14	8
0	41	.45	20	277	17	.3	48	.07	.13	.03	.61	.09	13	16
0	35	1.73	15	201	780	.22	2	.02	.12	.03	.17	.15	28	17
0	67	1.14	48	36	93	.49	5	.02	.35	.06	.19	<.01	72	1
0	10	2.28	16	109	84	.16	5	.19	.4	.29	1.02	.03	8	1
0	4	.12	2	33	1	.04	125	.01	.01	<.01	.02	.03	3	1
0	36	1.73	20	166	3	.22	11	.11	.32	.06	.47	.13	28	42
0	88	4.42	74	443	1	.99	1	.13	1.68	.24	.34	.2	46	1
0	120	3.4	196	1173	2	4.1	2	.37	3.96	.65	.91	.19	176	4
0	91	3.78	58	226	5032	4.58	12	.13	.01	.34	1.19	.3	45	0
0	39	1.5	34	195	80	.76	0	.1	.18	.04	.3	.01		0
0	41	1.3	34	244	45	.77	0	.13	.24	.05	.31	.01		0
0	138	1.38	33	149	10	.79	1	.06	.01	.05	.66	.06	55	<1
0	38	1.02	36	223	6	.64	0	.12	.25	.05	.37	.01		0
0	28	.76	22	156	22	.15	188	.02	.53	.05	.2	.05	54	8
0	122	3.21	78	419	63	.68	737	.09	.86	.21	.44	.22	131	9
0	139	1.44	66	283	82	.66	739	.06	.91	.16	.4	.14	103	12
0	136	2.46	81	370	29	.49	939	.02	1.39	.15	.41	.11	105	15
0	136	2.46	81	370	29	.49	939	.02	1.39	.15	.41	.11	105	15
0	24	.32	22	173	1	.35	26	.04	.11	.04	.46	.06	18	5
0	24	.32	22	173	1	.35	26	.04	.11	.04	.46	.08	18	5
0	12	.31	20	228	3	.16	22	.04	.11	.04	.38	.07	15	4
0	34	.81	20	1070	2	.64	872	.21	.29	.06	1.72	.18	69	23
0	29	.68	16	896	2	.53	730	.17	.25	.05	1.44	.15	57	20
0	54	1.14	53	538	5	.21	53	.2	.15	.02	1.08	.24	23	13
0	32	.68	32	321	4	.13	32	.12	.15	.01	.65	.14	14	8
0	42	.62	30	293	4	.13	721	.07	.17	.02	1	.13	20	16
0	42	.62	30	293	4	.13	721	.07	.17	.02	1	.13	20	16
0	23	.7	11	160	2	.14	401	.06	.16	.05	.56	.08	20	4
0	20	.56	26	430	10	.18	725	.09	.14	.06	.67	.21	19	11
0	12	.33	15	253	6	.12	473	.05	.14	.03	.39	.12	11	8
0	16	.26	8	90	14	.15	8	.03	.09	.02	.62	.08	6	3
0	13	.76	20	225	38	.38	20	.06	.31	.06	1.11	.08	28	5
0	32	.51	23	397	11	.33	2487	.08	.32	.14	.69	.27	26	28
0	32	.85	15	278	20	.41	2574	.08	.42	.21	.97	.37	17	26
8	27	1.19	12	198	73	.16	440	.02	3.99	.04	.41	.04	12	7
0	38	1.7	31	269	96	.27	1936	.03	.35	.11	1.22	.3	14	7
0	28	1.13	28	396	67	.23	1013	.05	.32	.07	.94	.24	21	33
0	18	.71	18	250	42	.14	638	.03	.2	.05	.59	.15	13	21
0	20	.57	11	482	3	.76	7	.05	1.4	.07	1.13	.16	4	26
0	204	2.14	70	854	74	.14	241	.06	3.7	.27	.66	.16	10	52
0	2	.21	7	91	<1	.07	4	.01	.13	.01	.63	.02	2	4
0	9	.82	26	354	1	.29	14	.06	.5	.05	2.44	.07	9	15

Table A-1

Food Composition (Computer code number is for West Diet Analysis program) (For purposes of calculations, use "0" for t, <1, <.1, <.01, etc.)

Computer Code Number	Food Description	Measure	Wt (g)	H_2O (%)	Ener (cal)	Prot (g)	Carb (g)	Dietary Fiber (g)	Fat (g)	Fat Breakdown (g) Sat	Mono	Poly
	VEGETABLES and LEGUMES—Continued											
	Tomatoes:											
945	Raw, whole, 2 ⅗" diam	1 ea	123	94	26	1	6	1	<1	.1	.1	.2
946	Raw, chopped	1 c	180	94	38	2	8	2	1	.1	.1	.2
947	Cooked from raw	1 c	240	92	65	3	14	2	1	.1	.2	.4
948	Canned, solids and liquid:	1 c	240	94	46	2	10	2	<1	t	t	.1
5741	Unsalted	1 c	240	94	46	2	10	2	<1	t	t	.1
1879	Tomatoes, sundried:	1 c	54	15	139	8	30	7	2	.2	.3	.6
1881	Pieces	10 pce	20	15	52	3	11	2	1	.1	.1	.2
1885	Oil pack, drained	10 pce	30	54	64	2	7	2	4	.6	2.6	.6
2020	Tomato, raw	1 ea	123	94	26	1	6	1	<1	.1	.1	.2
949	Tomato juice, canned:	1 c	243	94	41	2	10	1	<1	t	t	.1
5397	Unsalted	1 c	243	94	41	2	10	2	<1	t	t	.1
	Tomato products, canned:											
950	Paste, no added salt	1 c	262	74	215	10	51	11	1	.2	.2	.6
951	Puree, no added salt	1 c	250	87	100	4	24	5	<1	.1	.1	.2
952	Sauce, with salt	1 c	245	89	73	3	18	3	<1	.1	.1	.2
953	Turnips, cubes, cooked from fresh	1 c	156	94	33	1	8	3	<1	t	t	.1
	Turnip greens, cooked:											
954	From fresh, leaves and stems	1 c	144	93	29	2	6	5	<1	.1	t	.1
955	From frozen, chopped	1 c	164	90	49	6	8	6	1	.2	t	.3
956	Vegetable juice cocktail, canned	1 c	242	93	46	2	11	2	<1	t	t	.1
	Vegetables, mixed:											
957	Canned, drained	½ c	81	87	38	2	7	2	<1	t	t	.1
958	Frozen, cooked, drained	½ c	91	83	54	3	12	4	<1	t	t	.1
1818	Water chestnuts, Chinese, raw	½ c	62	73	60	1	15	2	<1	t	t	t
	Water chestnuts, canned:											
959	Slices	½ c	70	86	35	1	9	2	<1	t	t	t
960	Whole	4 ea	28	86	14	<1	3	1	<1	t	t	t
1190	Watercress, fresh, chopped	½ c	17	95	2	<1	<1	<1	<1	t	t	t
	VEGETARIAN FOODS:											
7509	Bacon strips, meatless	3 ea	15	49	46	2	1	<1	4	.7	1.1	2.3
1511	Baked beans, canned	½ c	127	73	118	6	26	6	1	.1	t	.2
7526	Bakon crumbles	¼ c	7	16	28	2	1	<1	2			
7548	Chicken, breaded, fried, meatless	1 pce	57	70	97	6	3	3	7	1	2.9	2.5
7547	Chicken slices, meatless	2 ea	60	59	132	10	4	3	8	1.3	2	4.4
7557	Chili w/meat substitute	½ c	107	64	141	19	15	4	2	.3	.6	.9
7549	Fish stick, meatless	2 ea	57	45	165	13	5	3	10	1.6	2.5	5.4
7550	Frankfurter, meatless	1 ea	51	58	102	10	4	2	5	.8	1.2	2.7
7504	GardenBurger, patty	1 ea	45	53	87	5	13	3	1	.4	.3	.7
7505	GardenSausage, patty	1 ea	35	15	117	4	22	5	1	.7	.3	t
7551	Luncheon slice, meatless	1 sl	67	46	188	17	6	3	11	1.7	2.6	5.6
7560	Meatloaf, meatless	1 ea	71	58	142	15	6	3	6	1	1.5	3.3
1171	Nuteena	1 ea	55	58	162	6	6	2	13	5.1	5.8	1.7
7556	Pot pie, meatless	1 ea	227	59	524	15	41	5	34	9.5	12.6	9.8
7554	Soyburger, patty	1 ea	71	58	142	15	6	3	6	1	1.5	3.3
7562	Soyburger w/cheese, patty	1 ea	135	50	316	21	29	4	13	4.2	3.9	3.7
7564	Tempeh	1 c	166	55	330	31	28	9	13	1.9	2.9	7.2
7670	Vegan burger, patty	1 ea	78	71	75	11	6	4	<1	.1	.3	.2
	Vegetarian foods, Green Giant:											
7677	Breakfast links	3 ea	68	65	114	12	5	4	5	.7	1.2	3.1
7676	Breakfast patties	2 ea	57	65	95	10	5	3	4	.6	1	2.6
	Burger, harvest, patty:											
7673	Italian	1 ea	90	65	139	17	8	5	4	1.4	.3	.4
7674	Original	1 ea	90	65	137	18	8	5	4	1.3	.1	.4
7675	Southwestern	1 ea	90	65	135	16	9	5	4	1.4	.2	.4
	Vegetarian foods, Loma Linda											
7727	Chik nuggets, frozen	5 pce	85	47	245	12	13	5	16	2.5	4	8.8
7753	Chik-fried, frozen	1 pce	57	51	178	11	1	1	15	1.9	3.7	8.7
7744	Franks, big, canned	1 ea	51	59	110	10	2	2	7	1.1	1.7	3.8
7747	Linketts, canned	1 ea	35	60	72	7	1		4	.7	1.2	2.5
1173	Redi-burger, patty	1 ea	85	59	172	16	5		10	1.5	2.4	5.8
7755	Swiss stake w/gravy, canned	1 pce	92	71	120	9	8	4	6	.8	1.5	3.3

Chol (mg)	Calc (mg)	Iron (mg)	Magn (mg)	Pota (mg)	Sodi (mg)	Zinc (mg)	VT-A (RE)	Thia (mg)	VT-E (a-TE)	Ribo (mg)	Niac (mg)	V-B6 (mg)	Fola (µg)	VT-C (mg)
0	6	.55	13	273	11	.11	76	.07	.47	.06	.77	.1	18	23
0	9	.81	20	400	16	.16	112	.11	.68	.09	1.13	.14	27	34
0	14	1.34	34	670	26	.26	178	.17	.91	.14	1.8	.23	31	55
0	72	1.32	29	530	355	.38	144	.11	.77	.07	1.76	.22	19	34
0	72	1.32	29	545	24	.38	144	.11	.91	.07	1.76	.22	19	34
0	59	4.91	105	1850	1131	1.07	47	.28	<.01	.26	4.89	.18	37	21
0	22	1.82	39	685	419	.4	17	.11	<.01	.1	1.81	.07	14	8
0	14	.8	24	470	80	.23	39	.06	.16	.11	1.09	.1	7	31
0	6	.55	13	273	11	.11	76	.07	.47	.06	.77	.1	18	23
0	22	1.41	27	535	877	.34	136	.11	2.21	.07	1.64	.27	48	44
0	22	1.41	27	535	24	.34	136	.11	2.21	.07	1.64	.27	48	44
0	92	5.08	134	2454	231	2.1	639	.41	11.3	.5	8.44	1	59	111
0	42	3.1	60	1065	85	.55	320	.18	6.3	.13	4.3	.38	27	26
0	34	1.89	47	909	1482	.61	240	.16	3.43	.14	2.82	.38	23	32
0	34	.34	12	211	78	.31	0	.04	.05	.04	.47	.1	14	18
0	197	1.15	32	292	42	.2	792	.06	2.48	.1	.59	.26	170	39
0	249	3.18	43	367	25	.67	1308	.09	4.79	.12	.77	.11	65	36
0	27	1.02	27	467	653	.48	283	.1	.77	.07	1.76	.34	51	67
0	22	.85	13	236	121	.33	944	.04	.49	.04	.47	.06	19	4
0	23	.75	20	154	32	.45	389	.06	.33	.11	.77	.07	17	3
0	7	.04	14	362	9	.31	0	.09	.74	.12	.62	.2	10	2
0	3	.61	3	83	6	.27	0	.01	.35	.02	.25	.11	4	1
0	1	.24	1	33	2	.11	0	<.01	.14	.01	.1	.04	2	<1
0	20	.03	4	56	7	.02	80	.01	.17	.02	.03	.02	2	7
0	3	.36	3	25	220	.06	1	.66	1.04	.07	1.13	.07	6	0
0	63	.37	41	376	504	1.78	22	.19	.67	.08	.54	.17	30	4
	8	.44	11	120	172	.25	0	.06		.02	.12	.02	7	0
0	13	.97	7	171	228	.37	0	.4	1.11	.27	2.68	.28	32	0
0	21	.78	10	198	474	.42	0	.66	1.61	.24	3.18	.42	46	0
0	53	4.24	36	362	527	1.26	78	.12	1.25	.07	1.21	.15	82	16
0	54	1.14	13	342	279	.8	0	.63	2.25	.51	6.84	.85	58	0
0	17	.92	9	76	219	.61	0	.56	.98	.61	8.16	.5	40	0
0	36	1.35		129	112		18	.05		.09		.06		<1
0	181	.33		307	78		3	.08		.2		.13		<1
0	27	1.54	15	188	576	1.07	0	.64	2.01	.37	7.37	.74	67	0
0	21	1.49	13	128	391	1.28	0	.64	1.23	.43	7.1	.85	55	0
0	9	.27	33	166	119	.46	0	.1		.35	1.04	.45	49	0
20	66	2.9	31	331	538	1.05	729	.65	4	.4	4.47	.31	40	10
0	21	1.49	13	128	391	1.28	0	.64	1.23	.43	7.1	.85	55	0
13	146	2.71	26	211	931	1.97	45	.77	1.43	.55	8.14	.86	69	1
0	154	3.75	116	609	10	3	115	.22	.03	.18	7.69	.5	86	0
0	79	2.66	15	398	351	.69	0	.23	.01	.51	3.78	.18	225	0
0	65	1.84			340	4.56	0	.18		.09	.27	.18		0
0	54	2			285	3.82	0	.15		.07	2.28	.15		0
0	74	2.61			374	6.93	3	.28		.14	4.05	.28		0
0	76	2.7			378	7.2	0	.29		.14	4.32	.29		0
0	71	2.52			371	6.66	3	.27		.13	4.05	.27		0
2	40	1.4		153	709	.43	0	.67		.3	2.89	.45		0
4	2	.63		76	503	.2	0	.98		.46	2.1	.35		0
2	8	.77		51	243	.89	0	.26		.46	1.98	.14		0
1	4	.39		29	160	.46	0	.13		.22	.64	.29		0
1	12	1.06	16	121	455	1.11	0	.14		.3	1.9	.51	21	0
2	24	.31		225	433	.41	0	1.25		.65	5.41	1		0

Table A-1

Food Composition (Computer code number is for West Diet Analysis program) (For purposes of calculations, use "0" for t, <1, <.1, <.01, etc.)

Computer Code Number	Food Description	Measure	Wt (g)	H_2O (%)	Ener (cal)	Prot (g)	Carb (g)	Dietary Fiber (g)	Fat (g)	Fat Breakdown (g) Sat	Mono	Poly
	VEGETARIAN FOODS:—Continued											
1174	Vege-Burger, patty	1 ea	55	71	66	10	2	2	2	.4	.6	.5
	Vegetarian foods, Morningstar Farms:											
7672	Better-n-burgers, svg	1 ea	78	71	75	11	6	4	<1	.1	.3	.2
7766	Better-n-eggs	¼ c	57	88	23	5	<1	0	<1	.1	.1	.1
57436	Breakfast links	2 pce	45	60	63	8	2	2	2	.5	.7	1.3
7752	Breakfast strips	2 pce	16	43	56	2	2	<1	4	.7	1.1	2.6
7725	Burger crumbles, svg	1 ea	55	60	116	11	3	3	6	1.6	2.3	2.5
7726	Burger, spicy black bean	1 ea	78	60	113	11	15	5	1	.2	.3	.4
7665	Chik pattie	1 ea	71	51	177	7	15	2	10	1.3	2.6	5.9
7724	Frank, deli	1 ea	45	52	109	10	3	3	7	1	2.1	3.5
7722	Garden vege pattie	1 ea	67	60	104	11	9	4	4	.5	1.1	2.2
7746	Grillers	1 ea	64	55	139	14	5	3	7	1.7	2.2	3
7664	Prime pattie	1 ea	64	64	94	16	4	3	2	.2	.4	.6
	Vegetarian foods, Worthington:											
7634	Beef style, meatless, frzn	3 pce	55	58	113	9	4	3	7	1.2	2.7	2.6
7732	Burger, meatless, patty	¼ c	55	71	60	9	2	1	2	.3	.5	1.1
1846	Chik slices, canned	2 pce	60	78	62	6	1	1	4	.6	.9	2.3
1833	Chili, canned	½ c	106	73	136	9	10	4	7	1.1	1.7	4.1
1835	Choplets, slices, canned	2 pce	92	72	93	17	3	2	2	.9	.3	.3
7608	Corned beef style, meatless, frzn	4 pce	57	55	138	10	5	2	9	1.9	4.1	3.1
1831	Country stew, canned	1 c	240	81	208	13	20	5	9	1.6	2.3	4.8
7632	Egg roll, meatless, frzn	1 ea	85	53	181	6	20	2	8	1.7	4.5	2.3
1838	Numete, slices, canned	1 pce	55	58	132	6	5	3	10	2.4	4.4	2.7
1839	Prime stakes, slices, canned	1 pce	92	71	136	9	4	4	9	1.4	2.9	4.9
1840	Protose, slices, canned	1 pce	55	53	131	13	5	3	7	1	3	2.4
7606	Roast, dinner, meatless, frzn	1 ea	85	63	180	12	5	3	12	2.2	5	5.2
1842	Saucette links, canned	1 pce	38	62	86	6	1		6	1.1	1.6	3.8
1844	Savory slices, canned	1 pce	28	66	48	3	2	1	3	1.2	1.3	.6
7735	Stakelets, frzn	1 pce	71	58	145	12	6	2	8	1.4	2.7	3.9
1847	Turkee slices, canned	1 pce	33	64	68	5	1	1	5	.8	1.9	2.1
	MISCELLANEOUS											
	Baking powders for home use:											
	Sodium aluminum sulfate:											
962	With monocalcium phosphate monohydrate	1 tsp	5	2	6	<1	2	0	0	0	0	0
963	With monocalcium phosphate monohydrate, calcium sulfate	1 tsp	5	5	3	0	1	<1	0	0	0	0
964	Straight phosphate	1 tsp	5	4	3	<1	1	<1	0	0	0	0
965	Low sodium	1 tsp	5	6	5	<1	2	<1	<1	t	0	t
1204	Baking soda	1 tsp	5		0	0	0	0	0	0	0	0
966	Basil, dried	1 tbs	5	6	13	1	3	2	<1	t	t	.1
2068	Cajun seasoning	1 tsp	3	5	6	<1	1	<1	<1			
961	Carob flour	1 c	103	4	185	5	92	41	1	.1	.2	.2
967	Catsup:	¼ c	61	67	63	1	17	1	<1	t	t	.1
968	Tablespoon	1 tbs	15	67	16	<1	4	<1	<1	t	t	t
1200	Cayenne/red pepper	1 tbs	5	8	16	1	3	1	1	.2	.1	.4
969	Celery seed	1 tsp	2	6	8	<1	1	<1	<1	t	.3	.1
1203	Chili powder:	1 tbs	8	8	25	1	4	3	1	.2	.3	.6
970	Teaspoon	1 tsp	3	8	9	<1	2	1	<1	.1	.1	.2
	Chocolate:											
971	Baking, unsweetened, square	1 oz	28	1	146	3	8	4	15	9.1	5.2	.5
	For other chocolate items, see Sweeteners & Sweets											
972	Cilantro/coriander, fresh	1 tbs	1	93		<1	<1	<1	<1	0	t	0
2287	Cinnamon	1 tsp	2	10	5	<1	2	1	<1	t	t	t
1197	Cornstarch	1 tbs	8	8	30	<1	7	<1	<1	t	t	t
2239	Curry powder	1 tsp	2	10	6	<1	1	1	<1	t	.1	.1
1202	Dill weed, dried	1 tbs	3	7	8	1	2	<1	<1	t	.1	t
975	Garlic cloves	1 ea	3	59	4	<1	1	<1	<1	t	0	t
2238	Garlic powder	1 tsp	3	6	10	<1	2	<1	<1	t	t	t
977	Gelatin, dry, unsweetened: Envelope	1 ea	7	13	23	6	0	0	<1	t	t	t
978	Ginger root, slices, raw	2 pce	5	82	3	<1	1	<1	<1	t	t	t

Chol (mg)	Calc (mg)	Iron (mg)	Magn (mg)	Pota (mg)	Sodi (mg)	Zinc (mg)	VT-A (RE)	Thia (mg)	VT-E (a-TE)	Ribo (mg)	Niac (mg)	V-B6 (mg)	Fola (µg)	VT-C (mg)
0	8	.5	12	30	114	.58	0	.2		.25	.78	.31	15	0
0	79	2.66	15	398	351	.69	0	.23	.01	.51	3.78	.18	225	0
2	7	.63		68	90	.51	64	.01		.26	0	.11		0
1	15	2.14	16	59	338	.36	0	6.95		.22	5.19	.33	12	0
<1	7	.27		15	220	.05	0	.75		.04	.6	.07		0
0	40	3.2	1	89	238	.82	0	4.96	.34	.18	1.49	.27		0
1	56	1.84	44	269	499	.93	14	8.03	.36	.14	0	.21		0
1	11	1.02		163	536	.31	0	2.15		.16	1.51	.14		0
2	16	.26	4	50	524	.38	0	.14	1.26	.02	0	.01		0
1	34	.72	29	180	382	.59	20	6.47	.98	.1	0	0	29	0
2	43	1.16		127	256	.49	0	11.7		.24	2.99	.37		0
1	46	2.14		142	247	.74	0	.51		.25	.92	.41		2
0	4	2.63		44	624	.22	0	.89		.34	6.46	.56		0
0	4	1.73		25	269	.38	0	.13		.1	1.96	.24		0
1	9	.73		111	257	.26	0	.06		.05	.37	.08		0
0	20	1.49		195	523	.57	0	.02		.03	1.04	.31		0
0	6	.37		40	500	.65	0	.05		.05	0	.05		0
1	6	1.17		58	524	.26	0	10.6		.07	1.36	.3		0
2	51	5.09		270	826	1.03	216	1.85		.29	4.22	.86		0
1	15	.57		96	384	.31	0	1.22		.19	0	.03		0
0	10	1.12		155	272	.56	0	.08		.06	.54	.2		0
2	12	.38		82	445	.38	0	.12		.13	1.98	.38		0
<1	1	1.84		50	283	.7	0	.18		.13	1.34	.24		0
2	36	2.87		38	566	.64	0	2.13		.25	6.02	.6		0
1	9	1.15		25	205	.26	0	.59		.08	.09	.13		0
<1		.47		14	179	.08	0	.08		.06	.48	.1		0
2	49	.99		95	484	.5	0	1.51		.12	3.1	.26		0
1	3	.47		16	203	.11	0	1.13		.05	.39	.09		0
0	97	0		7	547	0	0	0	0	0	0	0	0	0
0	294	.55	1	1	530	<.01	0	0	0	0	0	0	0	0
0	368	.56	2		395	<.01	0	0	0	0	0	0	0	0
0	217	.41	1	505	4	.04	0	0	<.01	0	0	0	0	0
0	0	0	0	0	1368	0	0	0	0	0	0	0	0	0
0	106	2.1	21	172	2	.29	47	.01	.08	.02	.35	.06	14	3
				29	474									
0	358	3.03	56	852	36	.95	1	.05	.65	.47	1.96	.38	30	<1
0	12	.43	13	293	723	.14	62	.05	.9	.04	.84	.11	9	9
0	3	.1	3	72	178	.03	15	.01	.22	.01	.21	.03	2	2
0	7	.39	8	101	1	.12	208	.02	.24	.05	.43	.1	5	4
0	35	.9	9	28	3	.14	<1	.01	.02	.01	.06	.01	<1	<1
0	22	1.14	14	153	81	.22	279	.03	.08	.06	.63	.15	8	5
0	8	.43	5	57	30	.08	105	.01	.03	.02	.24	.06	3	2
0	21	1.77	87	233	4	1.12	3	.02	.34	.05	.31	.03	2	0
0	1	.02		5	<1	<.01	3	<.01	.02	<.01	.01	<.01	<1	<1
0	25	.76	1	10	1	.04	1	<.01	0	<.01	.03	<.01	1	1
0		.04			1	<.01	0	0	0	0	0	0	0	0
0	10	.59	5	31	1	.08	2	<.01	.01	.01	.07	.01	3	<1
0	53	1.46	13	99	6	.1	18	.01		.01	.08	.04		1
0	5	.05	1	12	1	.03	0	.01	0	<.01	.02	.04	<1	1
0	2	.08	2	33	1	.08	0	.01	0	<.01	.02	.08	<1	1
0	4	.08	2	1	14	.01	0	<.01	0	.02	.01	0	2	0
0	1	.02	2	21	1	.02	0	<.01	.01	<.01	.03	.01	1	<1

Table A-1

Food Composition (Computer code number is for West Diet Analysis program) (For purposes of calculations, use "0" for t, <1, <.1, <.01, etc.)

Computer Code Number	Food Description	Measure	Wt (g)	H_2O (%)	Ener (cal)	Prot (g)	Carb (g)	Dietary Fiber (g)	Fat (g)	Fat Breakdown (g) Sat	Mono	Poly
	MISCELLANEOUS—Continued											
1198	Horseradish, prepared	1 tbs	15	85	7	<1	2	<1	<1	t	t	.1
1997	Hummous/hummus	1 c	246	65	421	12	50	12	21	3.1	8.8	7.8
1909	Mustard, country dijon	1 tsp	5		5	<1	<1	0	0	0	0	0
2019	Mustard, gai choy Chinese	1 tbs	16	94	3	<1	1		<1			
979	Mustard, prepared (1 packet = 1 tsp)	1 tsp	5	80	4	<1	<1	<1	<1	t	.2	t
	Miso (see #926 under Vegetables and Legumes, Soybean products)											
980	Olives, green	5 ea	20	78	23	<1	<1	<1	3	.3	1.9	.2
981	Olives, ripe, pitted	5 ea	22	80	25	<1	1	1	2	.3	1.7	.2
26008	Onion powder	1 tsp	2	5	7	<1	2	<1	<1	t	t	t
2237	Oregano, ground	1 tsp	2	7	6	<1	1	1	<1	.1	t	.1
2236	Paprika	1 tsp	2	10	6	<1	1	<1	<1	t	t	.2
887	Parsley, freeze dried	¼ c	1	2	3	<1	<1	<1	<1	t	t	t
	Parsley, fresh (see #885 and #886)											
985	Pepper, black	1 tsp	2	10	5	<1	1	1	<1	t	t	t
	Pickles:											
986	Dill, medium, 3¾ x 1¼" diam	1 ea	65	92	12	<1	3	1	<1	t	t	t
987	Fresh pack, slices, 1½" diam x ¼"	2 pce	15	79	11	<1	3	<1	<1	0	0	t
988	Sweet, medium	1 ea	35	65	41	<1	11	<1	<1	t	t	t
989	Pickle relish, sweet	1 tbs	15	63	21	<1	5	<1	<1	t	t	t
	Popcorn (see Grain Products #539-541)											
917	Potato chips:	10 pce	20	2	107	1	11	1	7	2.2	2	2.4
44076	Unsalted	1 oz	28	2	150	2	15	1	10	3.1	2.8	3.4
1201	Sage, ground	1 tsp	1	8	3	<1	1	<1	<1	.1	t	t
1347	Salsa, from recipe	1 tbs	15	93	3	<1	1	<1	<1	t	t	t
2218	Salsa, pico de gallo, medium	1 tbs	15	92	2	0	1	<1	0	0	0	0
990	Salt	1 tsp	6		0	0	0	0	0	0	0	0
	Salt Substitutes:											
1205	Morton, salt substitute	1 tsp	6			0	<1		0	0	0	0
1207	Morton, light salt	1 tsp	6			0	<1		0	0	0	0
2067	Seasoned salt, no MSG	1 tsp	5	5	4	<1	1	<1	<1			
991	Vinegar, cider	½ c	120	94	17	0	7	0	0	0	0	0
2172	Balsamic	1 tbs	15	64	21	0	4	0	0	0	0	0
2176	Malt	1 tbs	15	90	5	0	<1	0	0	0	0	0
2182	Tarragon	1 tbs	15	95	3	0	<1	0	0	0	0	0
2181	White wine	1 tbs	15	89	5	0	<1	0	0	0	0	0
	Yeast:											
992	Baker's, dry, active, package	1 ea	7	8	21	3	3	1	<1	t	.2	t
993	Brewer's, dry	1 tbs	8	5	23	3	3	3	<1	t	t	0
	SOUPS, SAUCES, and GRAVIES											
	SOUPS, canned, condensed:											
	Unprepared, condensed:											
1210	Cream of celery	1 c	251	85	181	3	18	2	11	2.8	2.6	5
1215	Cream of chicken	1 c	251	82	233	7	18	<1	15	4.2	6.5	3
1216	Cream of mushroom	1 c	251	81	259	4	19	1	19	5.1	3.6	8.9
1220	Onion	1 c	246	86	113	8	16	2	3	.5	1.5	1.3
	Prepared w/equal volume of whole milk:											
994	Clam chowder, New England	1 c	248	85	164	9	17	1	7	2.9	2.3	1.1
1209	Cream of celery	1 c	248	86	164	6	14	1	10	3.9	2.5	2.6
995	Cream of chicken	1 c	248	85	191	8	15	<1	11	4.6	4.5	1.6
996	Cream of mushroom	1 c	248	85	203	6	15	<1	14	5.1	3	4.6
1214	Cream of potato	1 c	248	87	149	6	17	<1	6	3.8	1.7	.6
1213	Oyster stew	1 c	245	89	135	6	10	0	8	5	2.1	.3
997	Tomato	1 c	248	85	161	6	22	3	6	2.9	1.6	1.1
	Prepared with equal volume of water:											
998	Bean with bacon	1 c	253	84	172	8	23	9	6	1.5	2.2	1.8
999	Beef broth/bouillon/consommé	1 c	240	98	17	3	<1	0	1	.3	.2	t
1000	Beef noodle	1 c	244	92	83	5	9	1	3	1.1	1.2	.5
1001	Chicken noodle	1 c	241	92	75	4	9	1	2	.7	1.1	.6
1002	Chicken rice	1 c	241	94	60	4	7	1	2	.5	.9	.4
1208	Chili beef	1 c	250	85	170	7	21	9	7	3.3	2.8	.3
1003	Clam chowder, Manhattan	1 c	244	92	78	2	12	1	2	.4	.4	1.3

PAGE KEY: A–4 = Beverages A–6 = Dairy A–10 = Eggs A–10 = Fat/Oil A–14 = Fruit A–20 = Bakery A–28 = Grain A–32 = Fish A–34 = Meats A–38 = Poultry A–40 = Sausage A–42 = Mixed/Fast A–46 = Nuts/Seeds A–50 = Sweets A–52 = Vegetables/Legumes A–62 = Vegetarian Foods A–64 = Misc A–66 = Soups/Sauces A–68 = Fast A–84 = Convenience A–88 = Baby foods

Chol (mg)	Calc (mg)	Iron (mg)	Magn (mg)	Pota (mg)	Sodi (mg)	Zinc (mg)	VT-A (RE)	Thia (mg)	VT-E (a-TE)	Ribo (mg)	Niac (mg)	V-B6 (mg)	Fola (μg)	VT-C (mg)
0	8	.06	4	37	47	.12	<1	<.01	<.01	<.01	.06	.01	9	4
0	123	3.86	71	428	600	2.71	5	.23	2.46	.13	1.01	.98	146	19
0				10	120									
0	4	.1	2	6	63	.03	0	0	.09	0	0	<.01	0	0
0	12	.32	4	11	480	.01	6	0	.6	0	0	<.01	<1	0
0	19	.73	1	2	192	.05	9	<.01	.66	0	.01	<.01	0	<1
0	7	.05	2	19	1	.05	0	.01	<.01	<.01	.01	.03	3	<1
0	31	.88	5	33	<1	.09	14	.01	.03	.01	.12	.02	5	1
0	4	.47	4	47	1	.08	121	.01	.01	.03	.31	.04	2	1
0	2	.54	4	63	4	.06	63	.01	.06	.02	.1	.01	15	1
0	9	.58	4	25	1	.03	<1	<.01	.02	<.01	.02	.01	<1	<1
0	6	.34	7	75	833	.09	21	.01	.1	.02	.04	.01	1	1
0	5	.27	1	30	101	0	2	0	.02	<.01	0	<.01	0	1
0	1	.21	1	11	329	.03	5	<.01	.06	.01	.06	<.01	<1	<1
0	3	.12	1	30	107	.01	1	0	.02	<.01	0	0	0	1
0	5	.33	13	255	119	.22	0	.03	.98	.04	.77	.13	9	6
0	7	.46	19	357	2	.3	0	.05	1.37	.05	1.07	.18	13	9
0	16	.28	4	11	<1	.05	6	.01	.02	<.01	.06	.01	3	<1
0	1	.06	1	24	58	.02	22	.01	.04	<.01	.06	.01	2	5
0					130									
0	1	.02			2325	.01	0	0	0	0	0	0	0	0
	33			3018	<1									
	2		4	1560	1170									
				15	1542									
0	7	.72	26	120	1	0	0	0	0	0	0	0	0	0
	2	.07		10	3		<1	.07		.07	.07			<1
	2	.07		13	4		<1	.07		.07	.07			1
		.07		2	1		<1	.07		.07	.07			<1
	1	.07		12	1		<1	.07		.07	.07			<1
0	4	1.16	7	140	3	.45	<1	.16	.01	.38	2.79	.11	164	<1
0	17	1.38	18	151	10	.63	0	1.25		.34	3.03	.4	313	0
28	80	1.26	13	246	1900	.3	60	.06	.38	.1	.66	.02	5	<1
20	68	1.2	5	176	1972	1.26	113	.06	.33	.12	1.64	.03	3	<1
3	65	1.05	10	168	1736	1.19	0	.06	2.61	.17	1.62	.02	8	2
0	54	1.35	5	138	2115	1.23	0	.07	.57	.05	1.21	.1	30	2
22	186	1.49	22	300	992	.8	40	.07	.15	.24	1.03	.13	10	3
32	186	.69	22	310	1009	.2	67	.07	.97	.25	.44	.06	8	1
27	181	.67	17	273	1046	.67	94	.07	.24	.26	.92	.07	8	1
20	179	.59	20	270	918	.64	37	.08	1.34	.28	.91	.06	10	2
22	166	.55	17	322	1061	.67	67	.08	.1	.24	.64	.09	9	1
32	167	1.05	20	235	1041	10.3	44	.07	.49	.23	.34	.06	10	4
17	159	1.81	22	449	744	.29	109	.13	2.6	.25	1.52	.16	21	68
3	81	2.05	45	402	951	1.03	89	.09	.08	.03	.57	.04	32	2
0	14	.41	5	130	782	0	0	<.01	0	.05	1.87	.02	5	0
5	15	1.1	5	100	952	1.54	63	.07	<.01	.06	1.07	.04	19	<1
7	17	.77	5	55	1106	.39	72	.05	.07	.06	1.39	.03	22	<1
7	17	.75	0	101	815	.26	65	.02	.05	.02	1.13	.02	1	<1
12	42	2.13	30	525	1035	1.4	150	.06	.17	.07	1.07	.16	17	4
2	27	1.63	12	188	578	.98	98	.03	.73	.04	.82	.1	10	4

Table A-1

Food Composition (Computer code number is for West Diet Analysis program) (For purposes of calculations, use "0" for t, <1, <.1, <.01, etc.)

Computer Code Number	Food Description	Measure	Wt (g)	H_2O (%)	Ener (cal)	Prot (g)	Carb (g)	Dietary Fiber (g)	Fat (g)	Fat Breakdown (g) Sat	Mono	Poly
	SOUPS, SAUCES, and GRAVIES—Continued											
1004	Cream of chicken	1 c	244	91	117	3	9	<1	7	2.1	3.3	1.5
1005	Cream of mushroom	1 c	244	90	129	2	9	<1	9	2.4	1.7	4.2
1006	Minestrone	1 c	241	91	82	4	11	1	3	.6	.7	1.1
1211	Onion	1 c	241	93	58	4	8	1	2	.3	.7	.7
1007	Split pea & ham	1 c	253	82	190	10	28	2	4	1.8	1.8	.6
1008	Tomato	1 c	244	90	85	2	17	<1	2	.4	.4	1
1009	Vegetable beef	1 c	244	92	78	6	10	<1	2	.9	.8	.1
1010	Vegetarian vegetable	1 c	241	92	72	2	12	<1	2	.3	.8	.7
	Ready to serve:											
1707	Chunky chicken soup	1 c	251	84	178	13	17	2	7	2	3	1.4
	SOUPS, dehydrated:											
	Prepared with water:											
1299	Beef broth/bouillon	1 c	244	97	19	1	2	0	1	.3	.3	t
1376	Chicken broth	1 c	244	97	22	1	1	0	1	.3	.4	.4
1013	Chicken noodle	1 c	252	94	53	3	7	1	1	.3	.5	.4
1122	Cream of chicken	1 c	261	91	107	2	13	<1	5	3.4	1.2	.4
1014	Onion	1 c	246	96	27	1	5	1	1	.1	.3	.1
1217	Split pea	1 c	255	87	125	7	21	3	1	.4	.7	.3
1015	Tomato vegetable	1 c	253	93	56	2	10	<1	1	.4	.3	.1
	Unprepared, dry products:											
1011	Beef bouillon, packet	1 ea	6	3	14	1	1	0	1	.3	.2	t
1012	Onion soup, packet	1 ea	39	4	115	5	21	4	2	.5	1.4	.3
	SAUCES											
	From dry mixes, prepared with milk:											
1016	Cheese sauce	1 c	279	77	307	17	23	1	17	9.3	5.3	1.6
1017	Hollandaise	1 c	259	84	240	5	14	<1	20	11.6	5.9	.9
1018	White sauce	1 c	264	81	240	10	21	<1	13	6.4	4.7	1.7
	From home recipe:											
1206	Lowfat cheese sauce	¼ c	61	73	85	6	4	<1	5	2.1	1.9	.9
1019	White sauce, medium	¼ c	72	77	102	2	6	<1	8	2.3	3.2	2
	Ready to serve:											
2202	Alfredo sauce, reduced fat	¼ c	69		170	5	16	0	10	6		
1020	Barbeque sauce	1 tbs	16	81	12	<1	2	<1	<1	t	.1	.1
1706	Chili sauce, tomato base	1 tbs	17	68	18	<1	4	<1	<1	t	t	t
2126	Creole sauce	¼ c	62	89	25	1	4	1	1	.1	.2	.3
2124	Hoisin sauce	1 tbs	17	47	35	<1	7	0	1	0		
2199	Pesto sauce	2 tbs	16		83	2	1	0	8	1.8	5.4	.7
1021	Soy sauce	1 tbs	16	71	8	1	1	<1	<1	t	t	t
2123	Szechuan sauce	1 tbs	16	71	21	<1	3	<1	1	.1	.3	.4
1380	Teriyaki sauce	1 tbs	18	68	15	1	3	<1	0	0	0	0
	Spaghetti sauce, canned:											
1377	Plain	1 c	249	75	271	5	40	8	12	1.7	6.1	3.3
1378	With meat	1 c	250	74	300	8	37	8	14	2.7	7	3.2
1379	With mushrooms	½ c	123	84	108	2	13	1	3	.4	1.5	.8
	GRAVIES											
	Canned:											
1022	Beef	1 c	233	87	123	9	11	1	5	2.7	2.2	.2
1023	Chicken	1 c	238	85	188	5	13	1	14	3.4	6.1	3.6
1024	Mushroom	1 c	238	89	119	3	13	1	6	1	2.8	2.4
1025	From dry mix, brown	1 c	258	92	75	2	13	<1	2	.8	.7	.1
1026	From dry mix, chicken	1 c	260	91	83	3	14	<1	2	.5	.9	.4
	FAST FOOD RESTAURANTS											
	ARBY'S											
1402	Bac'n cheddar deluxe	1 ea	231	59	512	21	39	<1	31	8.7	12.7	10.1
	Roast beef sandwiches:											
1403	Regular	1 ea	155	47	383	22	35	1	18	7	8	3.5
1404	Junior	1 ea	89	48	233	11	23	<1	11	4.1	5.2	2.5
1405	Super	1 ea	254	58	552	24	54	1	28	7.6	12.2	8.4
1407	Beef 'n cheddar	1 ea	194	50	508	25	43		26	7.7	12	6.8
1408	Chicken breast sandwich	1 ea	204	52	445	22	52	1	22	3	9.7	10.1
1412	Ham'n cheese sandwich	1 ea	169	54	355	25	34	<1	14	5.1	5.8	3.8
1726	Italian sub sandwich	1 ea	297	58	671	34	47		39	12.8	15.7	8.5

Chol (mg)	Calc (mg)	Iron (mg)	Magn (mg)	Pota (mg)	Sodi (mg)	Zinc (mg)	VT-A (RE)	Thia (mg)	VT-E (a-TE)	Ribo (mg)	Niac (mg)	V-B6 (mg)	Fola (µg)	VT-C (mg)
10	34	.61	2	88	986	.63	56	.03	.2	.06	.82	.02	2	<1
2	46	.51	5	100	881	.59	0	.05	1.24	.09	.72	.01	5	1
2	34	.92	7	313	911	.73	234	.05	.07	.04	.94	.1	36	1
0	26	.67	2	67	1053	.61	0	.03	.29	.02	.6	.05	15	1
8	23	2.28	48	400	1006	1.32	45	.15	.15	.08	1.47	.07	3	2
0	12	1.76	7	264	695	.24	68	.09	2.49	.05	1.42	.11	15	66
5	17	1.12	5	173	791	1.54	190	.04	.32	.05	1.03	.08	10	2
0	22	1.08	7	210	822	.46	301	.05	.79	.05	.92	.05	11	1
30	25	1.73	8	176	889	1	131	.08	.18	.17	4.42	.05	5	1
0	10	.02	7	37	1361	.07	<1	<.01	.02	.02	.36	0	0	0
0	15	.07	5	24	1483	.01	12	.01	.02	.03	.19	0	2	0
3	33	.5	8	30	1282	.2	5	.07	.02	.06	.88	.01	18	<1
3	76	.26	5	214	1184	1.57	123	.1	.15	.2	2.61	.05	5	1
0	12	.15	5	64	849	.06	<1	.03	.1	.06	.48	0	1	<1
3	20	.94	43	224	1147	.56	5	.21	.13	.14	1.26	.05	39	0
0	8	.63	20	104	1146	.17	20	.06	.81	.05	.79	.05	10	6
1	4	.06	3	27	1018	0	<1	<.01	.01	.01	.27	.01	2	0
2	55	.58	25	260	3493	.23	1	.11	.42	.24	1.99	.04	6	1
53	569	.28	47	552	1565	.97	117	.15	.33	.56	.32	.14	13	2
52	124	.9	8	124	1564	.7	220	.04	.26	.18	.06	.5	22	<1
34	425	.26	264	444	797	.55	92	.08	1.58	.45	.53	.07	16	3
11	166	.25	10	100	389	.73	58	.03	.55	.14	.16	.03	4	<1
8	75	.21	9	100	82	.26	89	.05	.98	.12	.28	.03	4	1
30	150	0	8	80	600		80	0		.1	0			0
0	3	.14	3	28	130	.03	14	<.01	.18	<.01	.14	.01	1	1
0	3	.14	2	63	227	.05	24	.01	.05	.01	.27	.02	1	3
0	35	.31	9	187	339	.1	24	.03	.61	.02	.53	.07	9	0
0	0	0			250		0							0
4	64	.09	6	15	129	.29	39	.01		.03	0	.02	<1	0
0	3	.32	5	29	914	.06	0	.01	0	.02	.54	.03	2	0
0	2	.12	2	13	218	.02	10	<.01	.07	<.01	.1	.01	1	<1
0	4	.31	11	40	690	.02	0	<.01	0	.01	.23	.02	4	0
0	70	1.62	60	956	1235	.52	306	.14	4.98	.15	3.76	.88	54	28
15	68	1.94	60	952	1179	1.37	577	.13	5.91	.17	4.51	.87	52	26
0	15	1	15	332	494	.34	241	.08	1.35	.08	.93	.16	12	9
7	14	1.63	5	189	1304	2.33	0	.07	.15	.08	1.54	.02	5	0
5	48	1.12	5	259	1373	1.9	264	.04	.37	.1	1.05	.02	5	0
0	17	1.57	5	252	1356	1.67	0	.08	.19	.15	1.6	.05	29	0
3	67	.23	10	57	1075	.31	0	.04	.05	.08	.81	0	0	0
3	39	.26	10	62	1133	.32	0	.05	.05	.15	.78	.03	3	3
38	110	4.32		491	1094	3	40	.34		.46	9.6			11
43	60	4.86	16	422	936	3.75	0	.28		.48	11	.2	14	1
22	40	2.7	8	201	519	1.5		.18		.25	6.6	.1	7	
43	90	6.48	25	533	1174	3.75	30	.39		.58	12.4	.3	21	9
52	150	6.12		321	1166	3		.42		.63	9.8			1
45	60	2.88	30	330	1019	.15		.22		.54	9	.38	18	5
55	170	2.7	31	382	1400	.9	40	.82		.37	7.8	.31	26	24
69	410	4.32		565	2062		100	.91		.49	8.2			11

Table A-1

Food Composition (Computer code number is for West Diet Analysis program) (For purposes of calculations, use "0" for t, <1, <.1, <.01, etc.)

Computer Code Number	Food Description	Measure	Wt (g)	H_2O (%)	Ener (cal)	Prot (g)	Carb (g)	Dietary Fiber (g)	Fat (g)	Fat Breakdown (g) Sat	Mono	Poly
	FAST FOOD RESTAURANTS											
	ARBY'S—Continued											
1413	Turkey sandwich, deluxe	1 ea	195	69	260	20	33	<1	6	1.6	2.3	2.4
1680	Turkey sub sandwich	1 ea	277	62	486	33	46		19	5.3	6	7
	Milkshakes:											
1419	Chocolate	1 ea	340	74	451	10	76	<1	12	2.8	7	1.7
1420	Jamocha	1 ea	326	75	368	9	59	0	10	2.5	6.4	1.6
1421	Vanilla	1 ea	312	77	330	10	46	0	11	3.9	5.3	2.3
1728	Salad, roast chicken	1 ea	400	88	204	24	12		7	3.3	.9	.9
1729	Sports drink, Upper Ten	1 ea	358	88	169	0	42		0	0	0	0
	Source: Arby's											
	BURGER KING											
1423	Croissant sandwich, egg, sausage & cheese	1 ea	176	46	600	22	25	1	46	16		
	Whopper sandwiches:											
1425	Whopper	1 ea	270	58	640	27	45	3	39	11		
1426	Whopper with cheese	1 ea	294	57	730	33	46	3	46	16		
	Sandwiches:											
1629	BK broiler chicken sandwich	1 ea	248	59	550	30	41	2	29	6		
1432	Cheeseburger	1 ea	138	48	380	23	28	1	19	9		
1434	Chicken sandwich	1 ea	229	45	710	26	54	2	43	9		
1427	Double beef	1 ea	351	57	870	46	45	3	56	19		
1428	Double beef & cheese	1 ea	375	56	960	52	46	3	63	24		
1433	Double cheeseburger with bacon	1 ea	218	48	640	44	28	1	39	18		
1431	Hamburger	1 ea	126	48	330	20	28	1	15	6		
1437	Ocean catch fish fillet	1 ea	255	51	700	26	56	3	41	6		
1435	Chicken tenders	1 ea	88	50	230	16	14	2	12	3		
1439	French fries (salted)	1 svg	116	40	370	5	43	3	20	5		
1630	French toast sticks	1 svg	141	33	500	4	60	1	27	7		
1440	Onion rings	1 svg	124	51	310	4	41	6	14	2	8	4
1441	Milk shakes, chocolate	1 ea	284	75	320	9	54	3	7	4		
1442	Milk shakes, vanilla	1 ea	284	75	300	9	53	1	6	4		
1443	Fried apple pie	1 ea	113	47	300	3	39	2	15	3		
	Source: Burger King Corporation											
	CHICK-FIL-A											
	Sandwiches:											
69153	Chargrilled chicken	1 ea	150	54	280	27	36	1	3	1		
69152	Chicken	1 ea	167	61	290	24	29	1	9	2		
69155	Chicken salad	1 ea	167	55	320	25	42	1	5	2		
69154	Chicken salad club	1 ea	232	62	390	33	38	2	12	5		
	Salads:											
52139	Carrot and raisin	1 ea	76	53	150	5	28	2	2	0		
52136	Chicken plate	1 ea	468	85	290	21	40	6	5	0		
52134	Chicken garden, charbroiled	1 ea	397	89	170	26	10	5	3	1		
52135	Chick-n-strips	1 ea	451	86	290	32	21	5	9	2		
52138	Cole slaw	1 ea	79	70	130	6	11	1	6	1		
52137	Tossed salad	1 ea	130	85	70	5	13	1	0	0	0	0
15263	Chicken nuggets, svg	1 ea	110	51	290	28	12	60	12	3		
15262	Chicken-n-strips, svg	1 ea	119	59	230	29	10	0	8	2		
50885	Hearty breast of chicken soup, svg	1 ea	215	86	110	16	10	1	1	0		
7973	Waffle potato fries, svg	1 ea	85	28	290	1	49	0	10	4		
46489	Cheesecake, svg	1 ea	88	52	270	13	7	0	21	9		
49134	Fudge nut brownie, svg	1 ea	88	8	416	12	49	0	19	3.6		
20601	Icedream, svg	1 ea	127	74	140	11	16	0	4	1		
48214	Lemon pie, svg	1 ea	99	56	280	1	19	0	22	6		
	Source: Chick-Fil-A											

PAGE KEY: A–4 = Beverages A–6 = Dairy A–10 = Eggs A–10 = Fat/Oil A–14 = Fruit A–20 = Bakery A–28 = Grain A–32 = Fish A–34 = Meats A–38 = Poultry A–40 = Sausage A–42 = Mixed/Fast A–46 = Nuts/Seeds A–50 = Sweets A–52 = Vegetables/Legumes A–62 = Vegetarian Foods A–64 = Misc A–66 = Soups/Sauces A–68 = Fast A–84 = Convenience A–88 = Baby foods

Chol (mg)	Calc (mg)	Iron (mg)	Magn (mg)	Pota (mg)	Sodi (mg)	Zinc (mg)	VT-A (RE)	Thia (mg)	VT-E (a-TE)	Ribo (mg)	Niac (mg)	V-B6 (mg)	Fola (μg)	VT-C (mg)
33	130	3.42	30	353	1262	1.5	40	.08		.41	15.4	.52	20	12
51	400	4.68		500	2033		20	13.2		.54	18.8			
36	250	.72	48	410	341	1.5	60	.12		.68	.8	.14	14	5
35	250	2.7	36	525	262	1.5	60	.12		.68	.8	.14	14	2
32	300	2.7	36	686	281	1.5	60	.12		.68	4	.14	37	2
43	170	1.98		877	508		485	.33		.54	5.6			51
0				0	40									
260	150	3.6			1140		80							0
90	80	4.5			870		100	.33		.41	7	.35		9
115	250	4.5			1350		150	.34		.48	7	.33		9
80	60	5.4			480		60							6
65	100	2.7			770		60							0
60	100	3.6			1400		0							0
170	80	7.2			940		100	.34		.56	10			9
195	250	7.2			1420		150	.35		.63	10			9
145	200	4.5			1240		80	.31		.42	6			0
55	40	1.8			530		20	.28		.31	4.89			0
90	60	2.7			980		20							1
35	0	.72			530		0							0
0	0	1.08			240		0							4
0	60	2.7			490		0							0
0	100	1.44			810		0							0
20	200	1.8			230		60	.13		.55	.13			0
20	300	0			230		60	.11		.57	.13			4
0	0	1.44			230		0							6
40					640									
50					870									
10					810									
70					980									
6					650									
35					570									
25					650									
20					430									
15					430									
0					0									
14					770									
20					380									
45					760									
5					960									
10					510									
36					773									
40					240									
5					550									

Table A-1

Food Composition (Computer code number is for West Diet Analysis program) (For purposes of calculations, use "0" for t, <1, <.1, <.01, etc.)

Computer Code Number	Food Description	Measure	Wt (g)	H_2O (%)	Ener (cal)	Prot (g)	Carb (g)	Dietary Fiber (g)	Fat (g)	Fat Breakdown (g) Sat	Mono	Poly
	FAST FOOD RESTAURANTS—Continued											
	DAIRY QUEEN											
	Ice cream cones:											
1446	Small vanilla	1 ea	142	63	230	6	38	0	7	4.5		
1447	Regular vanilla	1 ea	213	64	350	8	57	0	10	7		
1448	Large vanilla	1 ea	253	65	410	10	65	0	12	8		
1450	Chocolate dipped	1 ea	234	59	510	9	63	1	25	13		
1453	Chocolate sundae	1 ea	241	62	410	8	73	0	10	6		
1455	Banana split	1 ea	369	67	510	8	96	3	12	8		
1456	Peanut buster parfait	1 ea	305	51	730	16	99	2	31	17		
1457	Hot fudge brownie delight	1 ea	305	52	710	11	102	1	29	14	12	2
1459	Buster bar	1 ea	149	45	450	10	41	2	28	12		
1645	Breeze, strawberry, regular	1 ea	383	70	460	13	99	1	1	1	0	0
1460	Dilly bar	1 ea	85	55	210	3	21	0	13	7	3	3
1461	DQ ice cream sandwich	1 ea	61	46	150	3	24	1	5	2		
1463	Milk shakes, regular	1 ea	397	71	520	12	88	<1	14	8	2	2
1464	Milk shakes, large	1 ea	461	71	600	13	101	<1	16	10	2	2
1466	Milk shakes, malted	1 ea	418	68	610	13	106	<1	14	8	2	2
1470	Misty slush, small	1 ea	454	88	220	0	56	0	0	0	0	0
2250	Starkiss	1 ea	85	75	80	0	21	0	0	0	0	0
	Yogurt:											
1641	Yogurt cone, regular	1 ea	213	66	280	9	59	0	1	.5		
1643	Yogurt sundae, strawberry	1 ea	255	69	300	9	66	1	<1	.5	0	0
	Sandwiches:											
1481	Cheeseburger, double	1 ea	219	55	540	35	30	2	31	16		
1480	Cheeseburger, single	1 ea	152	55	340	20	29	2	17	8		
1474	Chicken	1 ea	191	56	430	24	37	2	20	4		
1647	Chicken fillet, grilled	1 ea	184	64	310	24	30	3	10	2.5		
1475	Fish fillet sandwich	1 ea	170	57	370	16	39	2	16	3.5		
1476	Fish fillet with cheese	1 ea	184	56	420	19	40	2	21	6	7	8
1477	Hamburger, single	1 ea	128	56	269	16	27	2	11	4.6	5.6	.9
1478	Hamburger, double	1 ea	212	62	440	30	29	2	22	10		
	Hotdog:											
1483	Regular	1 ea	99	57	240	9	19	1	14	5		
1484	With cheese	1 ea	113	55	290	12	20	1	18	8	8	2
1485	With chili	1 ea	128	61	280	12	21	2	16	6		
1489	French fries, small	1 ea	71	39	210	3	29	3	10	2	5	3
1490	French fries, large	1 ea	128	40	390	5	52	6	18	4	8	6
1491	Onion rings	1 ea	85	46	240	4	29	2	12	2.5		
	Source: International Dairy Queen											
	HARDEES											
1734	Frisco burger hamburger	1 ea	242		760	36	43		50	18		
1736	Frisco grilled chicken salad	1 ea	278		120	18	2		4	1		
1737	Peach shake	1 ea	345		390	10	77		4	3		
	Source: Hardees											
	JACK IN THE BOX											
	Breakfast items:											
1492	Breakfast Jack sandwich	1 ea	121	49	300	18	30	0	12	5	5	2.5
1494	Sausage crescent	1 ea	156	39	580	22	28	0	43	16		
1495	Supreme crescent	1 ea	153	40	530	23	34	0	33	10	18.9	7.8
1496	Pancake platter	1 ea	231	45	610	15	87	0	22	9	7.6	3.5
1497	Scrambled egg platter	1 ea	213	52	560	18	50	0	32	9	16.6	4.4
	Sandwiches:											
1654	Bacon cheeseburger	1 ea	242	49	710	35	41	0	45	15	15.7	8.7
1499	Cheeseburger	1 ea	110	41	330	16	32	0	15	6	5.9	2.3
1739	Chicken caesar pita sandwich	1 ea	237	59	520	27	44	4	26	6		
1655	Chicken sandwich	1 ea	160	52	400	20	38	0	18	4		
1656	Chicken sandwich, sourdough ranch	1 ea	225		490	29	45	1	21	6		
1505	Chicken supreme	1 ea	245	55	620	25	48	0	36	11	14.8	11.4
1583	Double cheeseburger	1 ea	152	44	450	24	35	0	24	12	11.6	3.1

PAGE KEY: A–4 = Beverages A–6 = Dairy A–10 = Eggs A–10 = Fat/Oil A–14 = Fruit A–20 = Bakery A–28 = Grain A–32 = Fish A–34 = Meats A–38 = Poultry A–40 = Sausage A–42 = Mixed/Fast A–46 = Nuts/Seeds A–50 = Sweets A–52 = Vegetables/Legumes A–62 = Vegetarian Foods A–64 = Misc A–66 = Soups/Sauces A–68 = Fast A–84 = Convenience A–88 = Baby foods

Chol (mg)	Calc (mg)	Iron (mg)	Magn (mg)	Pota (mg)	Sodi (mg)	Zinc (mg)	VT-A (RE)	Thia (mg)	VT-E (a-TE)	Ribo (mg)	Niac (mg)	V-B6 (mg)	Fola (µg)	VT-C (mg)
20	200	1.08		250	115		122	.05		.28				
30	300	1.8		390	170		150	.09		.38	.16	.13		2
40	350	1.8		451	200		200	.11		.4	.2			2
30	300	1.8		435	200		150	.09		.38	.16	.13		2
30	250	1.44		394	210		150	.08		.35	.4	.19		0
30	250	1.8		860	180		200	.15		.25	.4	.2		15
35	300	1.8		660	400		150	.15		.51	3	.22		1
35	300	5.4		510	340		80	.15		.68	.3	.18		1
15	150	1.08		400	280		80	.09		.17	3	.08		0
10	450	2.7		530	270		0	.13		.73				9
10	100	.36		170	75		60	.03		.14		.06		0
5	60	.72		105	115		40	.03		.25	.4	.05		0
45	400	1.44		570	230		80	.12		.59	.8	.19		<1
50	450	1.44		660	260		200	.15		.68	.8			<1
45	400	1.44		570	230		80	.12		.59	.8	.19		<1
0	0	0			20		0							0
0	0	0			10		0							0
5	300	1.8		285	170		0	.09		.38				2
5	300	1.8		352	180		0	.09		.49				6
115	250	4.5		426	1130		150	.29		.49	6.78			4
55	150	3.6		263	850		60	.29		.33	3.89			4
55	40	1.8		350	760		0	.37		.34	11			0
50	200	2.7		330	1040		0	.3		1.02	12			0
45	40	1.8		280	630		0	.3		.22	3			0
60	100	1.8		290	850		80	.3		.25	5			0
42	56	2.5		234	584		37	.27		.23	3.6			3
90	60	4.5		444	680		60	.32		.45	7.49			6
25	60	1.8		170	730		20	.22		.14	2			4
40	150	1.8		180	950		60	.22		.17	2			4
35	60	1.8		262	870		80	.23		.14	3			4
0	0	.72		430	115		0	.09		.03	2			5
0	0	1.44		780	200		0	.15		.07	3			9
0	0	1.08		90	135		0	.09		.05	.4			0
70					1280									
60					520									
25					290									
185	200	2.7		220	890		80	.47		.41	3			9
185	150	2.7		260	1010		100	.6		.51	4.6			0
210	150	3.6		270	930		150	.65		.54	4.2			12
100	100	1.8		310	890		80	.03		.85	7			6
380	150	4.5		450	1060		150			.66	5			9
110	250	5.4		540	1240		80	.24		.48	8.8	.39		9
35	200	2.7		200	510		60	.23		.23	3			1
55	250	2.7		490	1050		80							2
45	150	1.8		180	1290		40							0
65	150	1.8		340	1060									0
75	200	2.7		190	1520		100	.39		.32	11			2
75	250	3.6		320	900		100	.15		.34	6			0

Table A-1

Food Composition (Computer code number is for West Diet Analysis program) (For purposes of calculations, use "0" for t, <1, <.1, <.01, etc.)

Computer Code Number	Food Description	Measure	Wt (g)	H_2O (%)	Ener (cal)	Prot (g)	Carb (g)	Dietary Fiber (g)	Fat (g)	Fat Breakdown (g)		
										Sat	Mono	Poly
	FAST FOOD RESTAURANTS—Continued											
	JACK IN THE BOX—Continued											
1651	Grilled sourdough burger	1 ea	223	48	670	32	39	0	43	16	17.8	7.9
1498	Hamburger	1 ea	97	42	280	13	31	0	11	4	4.9	2
1500	Jumbo Jack burger	1 ea	229	55	560	26	41	0	32	10	13	8
1501	Jumbo Jack burger with cheese	1 ea	242	55	610	29	41	0	36	12	15	9
1740	Monterey roast beef sandwich	1 ea	238	57	540	30	40	3	30	9		
1508	Tacos, regular	1 ea	78	57	190	7	15	2	11	4		
1509	Tacos, super	1 ea	126	59	280	12	22	3	17	6		
	Teriyaki bowl:											
1679	Beef	1 ea	440	62	640	28	124	7	3	1		
1668	Chicken	1 ea	440	62	580	28	115	6	1			
1516	French fries	1 ea	109	38	350	4	45	4	17	4		
1517	Hash browns	1 ea	57	53	160	1	14	1	11	2.5	6.8	.3
1518	Onion rings	1 ea	103	34	380	5	38	0	23	6	15.2	.9
	Milkshakes:											
1519	Chocolate	1 ea	322	72	390	9	74	0	6	3.5	2.1	
1520	Strawberry	1 ea	298	74	330	9	60	0	7	4	2	
1521	Vanilla	1 ea	304	73	350	9	62	0	7	4	1.8	
1522	Apple turnover	1 ea	110	34	350	3	48	0	19	4	10.6	1.5
	Source: Jack in the Box Restaurant, Inc											
	KENTUCKY FRIED CHICKEN											
	Rotisserie gold:											
1472	Dark qtr, no skin	1 ea	117	66	217	27	0	0	12	3.5		
1473	Dark qtr, w/skin	1 ea	146	62	333	30	1		24	6.6		
1513	White qtr with wing, w/skin	1 ea	176	65	335	40	1		19	5.4		
1525	White qtr with wing, no skin	1 ea	117	63	199	37	0	0	6	1.7		
	Original Recipe:											
1253	Center breast	1 ea	103	52	260	25	9	<1	14	3.8	7.8	2
1251	Side breast	1 ea	83	47	245	18	9	<1	15	4.2	8.8	2.2
1250	Drumstick	1 ea	57	54	152	14	3	<1	8	2.2	4.1	1.3
1252	Thigh	1 ea	95	49	287	18	8	<1	21	5.3	9.4	3.1
1249	Wing	1 ea	53	45	172	12	5	<1	12	3	6	1.8
	Hot & spicy:											
1451	Center breast	1 ea	125	48	360	28	13		22	5		
1452	Side breast	1 ea	120	43	400	22	16		28	6		
1430	Thigh	1 ea	119	47	370	24	10		27	6		
1471	Wing	1 ea	61	38	220	14	5		16	4		
	Extra crispy recipe:											
1261	Center breast	1 ea	118	48	330	26	14	<1	20	4.8	10.8	2.1
1259	Side breast	1 ea	116	40	400	21	19	<1	27	5.5	12.9	2.3
1258	Drumstick	1 ea	65	49	190	14	6	<1	12	3.4	7.7	1.7
1260	Thigh	1 ea	109	43	380	23	7	<1	30	7.7	16	4.2
1257	Wing	1 ea	59	34	240	13	8	<1	17	4.4	10.7	2.5
1390	Baked beans	½ c	167	70	200	8	36	6	3	1.5	.7	.4
1526	Breadstick	1 ea	33	30	110	3	17	0	3	0		
1388	Buttermilk biscuit	1 ea	65	28	234	5	28	<1	13	3.4	6.2	2.3
1391	Chicken little sandwich	1 ea	47	35	169	6	14	<1	10	2	4.7	3.4
1269	Coleslaw	1 svg	90	75	114	1	13	<1	6	1	1.7	3.4
1527	Cornbread	1 ea	56	26	228	3	25	1	13	2		
1268	Corn-on-the-cob	1 ea	151	70	222	4	27	8	12	2	1	1.5
1429	Chicken, hot wings	1 svg	135	38	471	27	18		33			
1386	Kentucky fries	1 svg	77	42	228	3	26	3	12	3.2		
1381	Kentucky nuggets	6 ea	95	48	284	16	15	<1	18	4		
1534	Macaroni & cheese	1 svg	114	71	162	7	15	0	8	3		
1387	Mashed potatoes & gravy	1 svg	120	80	103	1	16	<1	5	.4	.5	.2
1530	Pasta salad	1 svg	108	78	135	2	14	1	8	1		
1389	Potato salad	½ c	188	74	271	5	27	3	16	3	4.2	7.3
1383	Potato wedges	1 svg	92	59	192	3	25	3	9	3		
1535	Red beans & rice	1 svg	112	76	114	4	18	3	3	1		
1529	Vegetable medley salad	1 ea	114	77	126	1	21	3	4	1		
	Source: Kentucky Fried Chicken Corp											

Chol (mg)	Calc (mg)	Iron (mg)	Magn (mg)	Pota (mg)	Sodi (mg)	Zinc (mg)	VT-A (RE)	Thia (mg)	VT-E (a-TE)	Ribo (mg)	Niac (mg)	V-B6 (mg)	Fola (μg)	VT-C (mg)
110	200	4.5		510	1140		150	.65		.48	8	.33		6
25	100	2.7		190	430		20	.15		.26	2			1
65	100	4.5		450	700		40	.36		.29	1.8			6
80	200	5.4		460	780		60	.36		.44	1.6			6
75	300	3.6		500	1270		80							5
20	100	1.08	35	240	410	1.2	0	.07		.17	1	.13		0
30	150	1.8	45	370	720	1.8	0	.12		.08	1.4	.18		2
25	150	4.5		430	930		1000							6
30	100	1.8		380	1220		1100							9
0	0	1.08		690	190		0	.18		.03	3.8			24
0	0	.36		190	310		0	.05			1			6
0	20	1.8		130	450		0	.29		.17	2.6			2
25	300	.72		680	210		0	.15		.6	.4			0
30	300	0		550	180		0	.15		.43	.4			0
30	300	0		570	180		0	.15		.34	.4			0
0	0	1.8		80	460		0	.2		.12	1.8			9
128	10	.18			772		15							1
163	10	.18			980		15							1
157	10	.18			1104		15							1
97	10	.18			667		15							1
92	30	.72			609		15	.09		.17	11.5			
78	68	1.2			604		15	.06		.13	6.9			
75	21	1.1			269		15	.05		.12	3.2			
112	40	1.08			591		31	.08		.3	5.5			
59	30	.54			383		15	.03		.08	3.7			
80	20	.72			750		15							6
80	40	1.08			850		15							6
100	20	1.08			670		15							6
65	20	.72			440		30							
75	33	.8			740		15	.11		.13	13.1			
75	20	.72			710		15	.09		.1	8.5			
65	20	.36			310		30	.06		.12	3.7			
90	49	1.2			520		30	.1		.21	6.5			
65	20	.36			320		30			.04	.06	3.3		
5	61	2.17	44	348	812	1.96	38	.09		.06	.76	.11	49	3
0	30	.18			15		0							0
3	43	1.92			565		28	.26		.2	2.77			
18	23	1.7			331		5	.16		.12	2.2			
5	30	.36			177		32	.03		.03	.2			27
42	60	.72			194		10							
0	0	.36			76		20	.14		.11	1.8			2
150	40	3.24			1230		15							6
4	11	.98			535		0							0
66	2	.1			865		15	.02		.02	1	.05		<1
16	120	.72			531		190							0
<1	20	.4			388		15			.04	1.2			
1	20	1.08			663		110							7
16	16	3.25	23	385	636	.44	120	.1		.03	.9	.29	11	
3					428		0							
4	10	.72			315									
0	20	.36			240		375							5

Table A-1

Food Composition (Computer code number is for West Diet Analysis program) (For purposes of calculations, use "0" for t, <1, <.1, <.01, etc.)

Computer Code Number	Food Description	Measure	Wt (g)	H_2O (%)	Ener (cal)	Prot (g)	Carb (g)	Dietary Fiber (g)	Fat (g)	Fat Breakdown (g) Sat	Mono	Poly
	FAST FOOD RESTAURANTS—Continued											
	LONG JOHN SILVER'S											
1528	Chicken plank dinner, 3 piece	1 ea	399	56	890	32	101		44	9.5	24.8	9.4
1531	Clam chowder	1 ea	198	86	140	11	10	1	6	1.8	2.5	1.7
1532	Clam dinner	1 ea	361	46	990	24	114		52	10.9	31.4	9.9
	Fish, batter fried:											
1523	Fish & fryes (fries), 3 piece	1 ea	384	54	980	31	92		50	11.3	28.4	9.7
1524	Fish & fryes, 2 piece	1 ea	261	54	610	27	52		37	7.9	23.5	5.3
2240	Fish and lemon crumb dinner, 3 piece	1 ea	493	71	610	39	86		13	2.2	3.9	5.3
2241	Fish and lemon crumb dinner, 2 piece	1 ea	334	77	330	24	46		5	.9	1.6	1.2
1533	Fish & chicken dinner	1 ea	431	55	950	36	102		49	10.6	28.8	9.5
1537	Shrimp dinner, batter fried	1 ea	331	54	840	18	88		47	9.7	27.2	9.1
	Salads:											
1541	Cole slaw	1 ea	98	70	140	1	20	1	6	1	1.5	3.5
1539	Ocean chef salad	1 ea	234	89	110	12	13	2	1	.4	.4	.2
1540	Seafood salad	1 ea	278	79	380	15	12	2	31	5.1	8.2	17.5
1542	Fryes (fries) serving	1 ea	85	43	250	3	28	1	15	2.5	7.4	5.1
1543	Hush puppies	1 ea	24	40	70	2	10	<1	2	.4	1.3	.2
	Source: Long John Silver's, Lexington KY											
	McDONALD'S											
	Sandwiches:											
1221	Big mac	1 ea	216	53	510	25	46	3	26	9.3	7.5	4.1
1226	Cheeseburger	1 ea	122	46	318	15	36	2	13	5.6	3.8	1.1
1224	Filet-o-fish	1 ea	145	49	364	14	41	1	16	3.7	3.8	5.6
1225	Hamburger	1 ea	108	49	266	12	35	2	9	3.2	2.8	.9
1444	McChicken	1 ea	189	52	491	17	42	2	29	5.4	8.5	10.2
1591	McLean deluxe	1 ea	214	64	345	23	37	2	12	4.4	3.6	1.2
1438	McLean deluxe with cheese	1 ea	228	63	398	26	38	2	16	6.8	4.6	1.3
1222	Quarter-pounder	1 ea	171	52	415	23	36	2	20	7.8	6.7	1.3
1223	Quarter-pounder with cheese	1 ea	199	50	520	28	37	2	29	12.6	8.7	1.6
1227	French fries, small serving	1 ea	68	40	207	3	26	2	10	1.7	3.1	2.5
1228	Chicken McNuggets	4 pce	73	51	198	12	10	0	12	2.5	3.7	2.4
	Sauces (packet):											
1229	Hot mustard	1 ea	30	60	63	1	7	1	4	.5	1.1	2
1230	Barbecue	1 ea	32	58	53	<1	12	<1	<1	.1	.1	.2
1231	Sweet & sour	1 ea	32	57	55	<1	12	<1	<1	.1	.1	.3
	Low-fat (frozen yogurt) milk shakes:											
1232	Chocolate	1 ea	295		348	13	62	1	6	3.5	.1	.7
1233	Strawberry	1 ea	294		343	12	63	<1	5	3.4	.1	.6
1234	Vanilla	1 ea	293		308	12	54	<1	5	3.3	.1	.6
	Low-fat (frozen yogurt) sundaes:											
1237	Hot caramel	1 ea	182	56	307	7	62	1	3	2	.3	1
1235	Hot fudge	1 ea	179	60	293	8	53	2	5	4.7	.1	.4
1267	Strawberry	1 ea	178	65	239	6	51	1	1	.7	.1	.2
1238	Vanilla	1 ea	90	68	118	4	24	<1	1	.5	.2	t
1241	Cookies, McDonaldland	1 ea	56	3	258	4	41	1	9	1.7	6.4	.8
1242	Cookies, chocolaty chip	1 ea	56	3	282	3	36	1	14	3.9	4.4	1
1240	Muffin, apple bran, fat-free	1 ea	75	39	182	4	40	1	1	.2	.1	.3
1239	Pie, apple	1 ea	84	35	289	3	37	1	14	3.7	4.5	.3
	Breakfast items:											
1243	English muffin with spread	1 ea	63	33	189	5	30	2	6	2.4	1.5	1.3
1244	Egg McMuffin	1 ea	137	57	289	17	27	1	13	.7	4.5	1.6
1245	Hotcakes with marg & syrup	1 ea	222	44	557	8	100	2	14	2.4	4.6	5.8
1246	Scrambled eggs	1 ea	102	73	170	13	1	0	12	3.6	5.3	1.7
1247	Pork sausage	1 ea	43	45	173	6	<1	0	16	5.5	6.4	2.1
1248	Hashbrown potatoes	1 ea	53	55	130	1	13	1	8	1.3	2.3	1.9
1392	Sausage McMuffin	1 ea	112	42	361	13	26	1	23	8.3	8.2	2.8
1393	Sausage McMuffin with egg	1 ea	163	52	443	19	27	1	29	10	10.7	3.6
1394	Biscuit with biscuit spread	1 ea	76	32	260	4	32	1	13	3.8	3.7	.8

PAGE KEY: A–4 = Beverages A–6 = Dairy A–10 = Eggs A–10 = Fat/Oil A–14 = Fruit A–20 = Bakery A–28 = Grain A–32 = Fish A–34 = Meats A–38 = Poultry A–40 = Sausage A–42 = Mixed/Fast A–46 = Nuts/Seeds A–50 = Sweets A–52 = Vegetables/Legumes A–62 = Vegetarian Foods A–64 = Misc A–66 = Soups/Sauces A–68 = Fast A–84 = Convenience A–88 = Baby foods

Chol (mg)	Calc (mg)	Iron (mg)	Magn (mg)	Pota (mg)	Sodi (mg)	Zinc (mg)	VT-A (RE)	Thia (mg)	VT-E (a-TE)	Ribo (mg)	Niac (mg)	V-B6 (mg)	Fola (μg)	VT-C (mg)
55	200	4.5		1170	2000	3	40	.52		.51	16			9
20	200	1.8		380	590	.6	150	.09		.25	2			
75	200	4.5		910	1830	3	40	.75		.42	12			12
70	200	4.5		1120	1530	3	40	.45		.42	8			15
60	40	1.8		900	1480	1.2		.37		.34	8			9
125	200	5.4		990	1420	2.25	700	.75		.59	24			6
75	80	1.8		440	640	.9	1000	.3		.25	14			18
75	200	4.5		1280	2090	3	40	.6		.59	14			9
100	200	3.6		840	1630	3	40	.45		.42	9			9
15	60	.72		190	260	.6	40	.06		.07	2			
40	100	3.6		95	730	.3	500	.12		.14	3			21
55	150	4.5		130	980	.9	200	.15		.25	3			21
0	200	.72		370	500	.3	0	.09			1.6			6
	40	.72		65	25	.3		.06		.03	.8			
76	202	4.32	46	456	932	4.81	66	.49	1.01	.44	6.08	.25	49	3
42	134	2.73	27	281	766	2.62	64	.33	.46	.31	3.81	.15	24	2
37	123	1.85	32	266	708	.7	21	.32	1.52	.23	2.58	.07	30	0
28	126	2.73	24	260	531	2.25	22	.33	.23	.26	3.81	.14	21	2
52	128	2.5	32	319	797	1.06	29	.91	6.16	.24	7.74	.38	37	1
59	131	4.29	40	537	811	4.9	74	.42	.63	.34	7.16	.29	44	8
73	139	4.29	43	559	1046	5.26	115	.42	.85	.39	7.16	.3	47	8
70	127	4.33	33	405	692	4.66	33	.39	.36	.32	6.78	.24	27	3
97	143	4.5			1160		115	.39	.81	.43	6.78	.26	33	3
0	9	.53	26	469	135	.32	0	.05	.83	0	1.94	.24	26	8
42	9	.65	17	210	353	.69	0	.08	.96	.11	5.15	.21		0
3	7	.78		29	85		4	.01		.01	.15			0
0	4	0		51	277		0	.01		.01	.17			4
0	2	.16		7	158		74	0		.01	.08			0
24	372	1.04		543	241		46	.12		.51	.4	.1		3
24	366	.29		542	170		46	.12		.51	.4	.11		3
24	360	.29		533	193		45	.12		.51	.31			3
7	246	.15		344	197		18	.09		.34	.27			2
5	258	.59		441	190		7	.09		.34	.29			2
5	221	.25		325	115		6	.06		.34	.25			2
3	132	.23		175	84		4	<.01		.01	.23			1
0	10	1.73	11	62	267	.38	0	.24	.99	.16	2.01	.03		0
3	28	1.78	24	142	229	.4	0	.14	.92	.16	1.48			0
0	34	1.29	13	77	215	.33	0	.14	0	.14	1.32	.03	5	1
0	17	1.23	7	69	221	.23		.19	1.5	.12	1.55	.04	9	27
13	103	1.59	13	69	386	.42	33	.25	.13	.31	2.61	.04	57	1
234	151	2.44	24	199	730	1.56	100	.49	.85	.45	3.33	.15	33	2
11	108	1.98	27	285	746	.53	119	.24	1.2	.26	1.86	.09	<1	<1
424	50	1.19	10	126	143	1.06	168	.07	.92	.51	.06	.12	44	0
33	7	.5	7	102	292	.78	0	.18	.26	.06	1.7	.09		0
0	7	.3	11	213	332	.15	0	.08	.58	.02	.9	.08	8	3
46	132	2.07	22	191	751	1.51	48	.56	.66	.27	3.76	.14	16	0
257	156	2.8	26	251	821	2.07	117	.59	1.11	.49	3.79	.19	33	0
0	68	1.85	9	105	836	.3	2	.29	.81	.23	2.23	.03	5	0

Table A-1

Food Composition (Computer code number is for West Diet Analysis program) (For purposes of calculations, use "0" for t, <1, <.1, <.01, etc.)

Computer Code Number	Food Description	Measure	Wt (g)	H_2O (%)	Ener (cal)	Prot (g)	Carb (g)	Dietary Fiber (g)	Fat (g)	Fat Breakdown (g) Sat	Mono	Poly
	FAST FOOD RESTAURANTS—Continued											
	McDONALD'S—Continued											
1395	Biscuit with sausage	1 ea	119	37	433	10	32	1	29	8.6	10.1	2.8
1396	Biscuit with sausage & egg	1 ea	170	48	518	16	33	1	35	10.5	12.7	3.7
1397	Biscuit with bacon, egg, cheese	1 ea	152	46	450	17	33	1	27	8.7	8.9	2.3
	Salads:											
1398	Chef salad	1 ea	313	86	206	19	9	3	11	4.2	3	1.2
1400	Garden salad	1 ea	234	92	84	6	7	3	4	1.1	1.4	.7
1401	Chunky chicken salad	1 ea	296	87	164	23	8	3	5	1.3	1.6	1
	Source: McDonald's Corporation											
	PIZZA HUT											
	Pan pizza:											
1657	Cheese	2 pce	216	51	522	24	56	4	22	10	6.8	3.4
1658	Pepperoni	2 pce	208	49	531	22	56	4	24	8	9.9	3.7
1659	Supreme	2 pce	273	56	622	30	56	6	30	12	12	4.2
1660	Super supreme	2 pce	286	56	645	30	56	6	34	12		
	Thin 'n crispy pizza:											
1649	Cheese	2 pce	174	52	411	22	42	4	16	8	4.4	2.3
1623	Pepperoni	2 pce	168	48	431	22	42	2	20	8		
1622	Supreme	2 pce	232	57	514	28	42	4	26	10		
1620	Super supreme	2 pce	247	57	541	28	44	4	28	12		
	Hand tossed pizza:											
1619	Cheese	2 pce	216	53	470	26	58	4	14	7.9		
1618	Pepperoni	2 pce	208	51	477	24	58	4	16	8		
1648	Supreme	2 pce	273	56	568	32	60	6	24	10		
1617	Super supreme	2 pce	286	57	591	32	60	6	26	10		
	Personal pan pizza:											
1610	Pepperoni	1 ea	255	50	637	27	69	5	28	10	11.8	4.5
1609	Supreme	1 ea	327	57	721	33	70	6	34	12	14.7	5.6
	Source: Pizza Hut											
	SUBWAY											
	Deli style sandwich:											
69104	Bologna	1 ea	171	64	292	10	38	2	12	4		
69102	Ham	1 ea	171	69	234	11	37	2	4	1		
69103	Roast beef	1 ea	180	69	245	13	38	2	4	1		
69105	Seafood and crab:	1 ea	178	66	298	12	37	2	11	2		
69106	With light mayo	1 ea	178	68	256	12	37	2	7	2		
69108	Tuna:	1 ea	178		354	11	37	2	18	3		
69107	With light mayo	1 ea	178	67	279	11	38	2	9	2		
69101	Turkey	1 ea	180	69	235	12	38	2	4	1		
	Sandwiches, 6 inch:											
	B.L.T.:											
69135	On white bread	1 ea	191	67	311	14	38	3	10	3		
69136	On wheat bread	1 ea	198	65	327	14	44	3	10	3		
	Chicken taco sub:											
69131	On white bread	1 ea	286	70	421	24	43	3	16	5		
69132	On wheat bread	1 ea	293	69	436	25	49	4	16	5		
	Club :											
69117	On white bread	1 ea	246	73	297	21	40	3	5	1		
69118	On wheat bread	1 ea	253	71	312	21	46	3	5	1		
	Cold cut trio:											
69113	On white bread	1 ea	246	71	362	19	39	3	13	4		
69114	On wheat bread	1 ea	253	68	378	20	46	3	13	4		
	Ham:											
69115	On white bread	1 ea	232	73	287	18	39	3	5	1		
69115	On wheat bread	1 ea	239	71	302	19	45	3	5	1		

Chol (mg)	Calc (mg)	Iron (mg)	Magn (mg)	Pota (mg)	Sodi (mg)	Zinc (mg)	VT-A (RE)	Thia (mg)	VT-E (a-TE)	Ribo (mg)	Niac (mg)	V-B6 (mg)	Fola (μg)	VT-C (mg)
33	75	2.35	15	207	1128	1.08	2	.48	1.07	.29	3.93	.12	5	0
245	100	2.95	20	271	1199	1.61	59	.51	1.53	.55	3.96	.18	27	0
238	103	2.6	20	245	1315	1.64	99	.39	1.49	.57	3.32	.13	30	0
179	157	1.81	40	605	727	2.16	1179	.33	1.45	.37	4.32	.36	100	22
139	52	1.34	24	407	61	.73	1114	.12	.95	.24	.65	.16	96	22
76	54	1.62	44	673	318	1.52	1973	.51	1.28	.21	8.46	.52	83	30
50	288	3	63	337	1002	4.32	211	.6		.64	5.48	.18		7
48	206	3.21	55	399	1140	4.14	190	.62		.48	5.31	.16	0	8
60	234	4.6	81	620	1529	6	195	.86		.84	6.4	.33		11
68	236	4.39	80	592	1649	5.99	201	.83		.73	7.13			12
50	291	2.06	56	307	1070	4.23	217	.46		.46	5.65	.18		6
50	208	2.2	51	330	1255	4.02	199	.48		.49	5.97			7
62	238	3.6	79	631	1591	5.4	197	.7		.57	6.27			12
70	238	3.41	73	563	1762	5.47	208	.72		.53	6.57			9
50	284	3	71	388	1242	4.6	198	.48		.48	5.3			10
48	202	3.21	84	610	1380	6.01	187	.72		.56	7.59			13
60	232	4.6	87	589	1769	5.48	192	.82		.66	8.45			14
68	232	4.39	89	607	1889	5.65	198	.84		.68	8.71			14
55	250	4	60	406	1338	3.8	233	.56		.66	8.16	.2		10
66	276	5.19	74	603	1757	4.69	240	.73		.82	9.91	.4		14
20	39	3			744		113							14
14	24	3			773		113							14
13	23	3			638		113							14
17	24	3			544		113							14
16	24	3			556		118							14
18	26	3			557		116							14
16	26	3			583		126							14
12	26	3			944		113							14
16	27	3			945		120							15
16	33	3			957		120							15
52	118	4			1264		209							18
52	124	4			1275		209							18
26	29	4			1341		120							15
26	35	4			1352		120							15
64	49	4			1401		130							16
64	55	4			1412		130							16
28	28	3			1308		120							15
28	35	3			1319		120							15

Table A–1

Food Composition (Computer code number is for West Diet Analysis program) (For purposes of calculations, use "0" for t, <1, <.1, <.01, etc.)

Computer Code Number	Food Description	Measure	Wt (g)	H_2O (%)	Ener (cal)	Prot (g)	Carb (g)	Dietary Fiber (g)	Fat (g)	Fat Breakdown (g) Sat	Mono	Poly
	FAST FOOD RESTAURANTS—Continued											
	SUBWAY—Continued											
	Italian B.M.T.											
69139	On white bread	1 ea	246	66	445	21	39	3	21	8		
69140	On wheat bread	1 ea	253	64	460	21	45	3	22	7		
	Meatball:											
69129	On white bread	1 ea	260	70	404	18	44	3	16	6		
69130	On wheat bread	1 ea	267	67	419	19	51	3	16	6		
	Melt with turkey, ham, bacon, cheese:											
69127	On white bread	1 ea	251	70	366	22	40	3	12	5		
69128	On wheat bread	1 ea	258	68	382	23	46	3	12	5		
	Pizza sub:											
69133	On white bread	1 ea	250	66	448	19	41	3	22	9		
69134	On wheat bread	1 ea	257	65	464	19	48	3	22	9		
	Roast beef:											
69121	On white bread	1 ea	232	72	288	19	39	3	5	1		
69122	On wheat bread	1 ea	239	70	303	20	45	3	5	1		
	Roasted chicken breast:											
69125	On white bread	1 ea	246	70	332	26	41	3	6	1		
69126	On wheat bread	1 ea	253	68	348	27	47	3	6	1		
	Seafood and crab:											
69145	On white bread:	1 ea	246	69	415	19	38	3	19	3		
69147	With light mayo	1 ea	246	72	332	19	39	3	10	2		
69146	On wheat bread:	1 ea	253	67	430	20	44	3	19	3		
69148	With light mayo	1 ea	253	70	347	20	45	3	10	2		
	Spicy italian:											
69123	On white bread	1 ea	232	64	467	20	38	3	24	9		
69124	On wheat bread	1 ea	239	62	482	21	44	3	25	9		
	Steak and cheese:											
69119	On white bread	1 ea	257	68	383	29	41	3	10	6		
69120	On wheat bread	1 ea	264	67	398	30	47	3	10	6		
	Tuna:											
69141	On white bread:	1 ea	246	62	527	18	38	3	32	5		
69143	With light mayo	1 ea	246	70	376	18	39	3	15	2		
69142	On wheat bread:	1 ea	253	62	542	19	44	3	32	5		
69144	With light mayo	1 ea	253	68	391	19	46	3	15	2		
	Turkey:											
69111	On white bread	1 ea	232	73	273	17	40	3	4	1		
69112	On wheat bread	1 ea	239	71	289	18	46	3	4	1		
	Turkey breast and ham:											
69137	On white bread	1 ea	232	73	280	18	39	3	5	1		
69138	On wheat bread	1 ea	239	71	295	18	46	3	5	1		
	Veggie delite:											
69109	On white bread	1 ea	175	71	222	9	38	3	3	0		
69110	On wheat bread	1 ea	182	69	237	9	44	3	3	0		
	Salads:											
52128	B.L.T.	1 ea	276	91	140	7	10	2	8	3		
52124	B.M.T., classic Italian	1 ea	331	86	274	14	11	1	20	7		
52127	Chicken taco	1 ea	370	87	250	18	15	2	14	5		
52115	Club	1 ea	331	91	126	14	12	1	3	1		
52120	Cold cut trio	1 ea	330	89	191	13	11	1	11	3		
52123	Ham	1 ea	316	91	116	12	11	1	3	1		
52129	Meatball	1 ea	345	88	233	12	16	2	14	5		
52131	Melt	1 ea	336	88	195	16	12	1	10	4		
52121	Pizza	1 ea	335	86	277	12	13	2	20	8		
52126	Roast beef	1 ea	316	92	117	12	11	1	3	1		
52119	Roasted chicken breast	1 ea	331	89	162	20	13	1	4	1		
52117	Seafood and crab:	1 5	331	88	244	13	10	2	17	3		
52116	With light mayo	1 5	331	90	161	13	11	2	8	1		
52130	Steak and cheese	1 ea	342	87	212	22	13	1	8	5		
52122	Tuna:	1 ea	331	84	356	12	10	1	30	5		
52118	With light mayo	1 ea	331	89	205	12	11	1	13	2		
52114	Turkey breast	1 ea	316	92	102	11	12	1	2	1		
52125	With ham	1 ea	316	92	109	11	11	1	3	1		

Chol (mg)	Calc (mg)	Iron (mg)	Magn (mg)	Pota (mg)	Sodi (mg)	Zinc (mg)	VT-A (RE)	Thia (mg)	VT-E (a-TE)	Ribo (mg)	Niac (mg)	V-B6 (mg)	Fola (μg)	VT-C (mg)
56	44	4			1652		151							15
56	50	4			1664		151							15
33	32	4			1035		142							16
33	39	4			1046		142							16
42	93	4			1735		155							15
42	100	3			1746		156							15
50	103	4			1609		238							16
50	110	3			1621		238							16
20	25	4			928		120							15
20	32	3			939		120							15
48	35	3			967		123							15
48	42	3			978		123							15
34	28	3			849		121							15
32	28	3			873		131							15
34	34	3			860		121							15
32	34	3			884		131							15
57	40	4			1592		169							15
57	47	4			1604		169							15
70	88	5			1106		175							18
70	95	5			1117		176							18
36	32	3			875		125							15
32	32	3			928		146							15
36	38	3			886		126							15
32	38	3			940		146							15
19	30	4			1391		120							15
19	37	3			1403		120							15
24	29	3			1350		120							15
24	36	3			1361		120							15
0	25	3			582		120							15
0	32	3			593		120							15
16	24	1			672		273							32
56	41	2			1379		303							32
52	115	3			990		361							35
26	26	2			1067		273							32
64	46	2			1127		282							33
28	25	2			1034		273							32
33	30	2			761		295							33
42	90	2			1461		308							32
50	100	2			1336		390							33
20	23	2			654		273							32
48	32	2			693		276							32
34	25	2			575		273							32
32	25	2			599		284							32
70	86	3			832		328							35
36	29	2			601		278							32
32	29	2			654		298							32
19	28	2			1117		273							32
24	27	2			1076		273							32

Table A-1

Food Composition (Computer code number is for West Diet Analysis program) (For purposes of calculations, use "0" for t, <1, <.1, <.01, etc.)

Computer Code Number	Food Description	Measure	Wt (g)	H_2O (%)	Ener (cal)	Prot (g)	Carb (g)	Dietary Fiber (g)	Fat (g)	Fat Breakdown (g) Sat	Mono	Poly
	FAST FOOD RESTAURANTS—Continued											
	SUBWAY—Continued											
52113	Veggie delite	1 ea	260	94	51	2	10	1	1	0		
	Cookies:											
47662	Brazil nut and chocolate chip	1 ea	48	12	229	3	27	1	12	3.5		
47655	Chocolate chip:	1 ea	48	14	209	2	29	1	10	3.5		
47658	With M&M's	1 ea	48	14	209	2	29	1	10	3		
47659	Chocolate chunk	1 ea	48	14	209	2	29	1	10	3.5		
47656	Oatmeal raisin	1 ea	48	15	199	3	29	1	8	2		
47657	Peanut butter	1 ea	48	13	219	3	26	1	12	2.5		
47660	Sugar	1 ea	48	11	229	2	28	0	12	3		
47661	White chip macademia	1 ea	48	12	229	2	28	1	12	2.5		
	Source: Subway International											
	TACO BELL											
	Breakfast burrito:											
1601	Bacon breakfast burrito	1 ea	99	48	291	11	23		17	4		
1627	Country breakfast burrito	1 ea	113	55	220	8	26	2	14	5		
1626	Fiesta breakfast burrito	1 ea	92	44	280	9	25	2	16	6		
1625	Grande breakfast burrito	1 ea	177	56	420	13	43	3	22	7		
1604	Sausage breakfast burrito	1 ea	106	49	303	11	23		19	6		
	Burritos:											
1544	Bean with red sauce	1 ea	198	58	380	13	55	13	12	4		
1545	Beef with red sauce	1 ea	198	57	432	22	42	4	19	8	6.7	.7
1546	Beef & bean with red sauce	1 ea	198	57	412	17	50	5	16	6	6.1	2.1
1569	Big beef supreme	1 ea	298	64	520	24	54	11	23	10		
1552	Chicken burrito	1 ea	171	58	345	17	41		13	5		
1547	Supreme with red sauce	1 ea	248	64	428	16	50	10	18	7.8		
1571	7 layer burrito	1 ea	234	61	438	13	55	11	19	5.8		
1538	Chilito	1 ea	156	49	391	17	41		18	9		
1549	Chilito, steak	1 ea	257	62	496	26	47		23	10		
	Tacos:											
1551	Taco	1 ea	78	58	180	9	12	3	10	4		
1554	Soft taco	1 ea	99	63	242	10	13	3	11	4.4		
1536	Soft taco supreme	1 ea	128	64	234	11	21	3	13	6.3		
1568	Soft taco, chicken	1 ea	128	63	212	15	22	2	7	2.6		
1572	Soft taco, steak	1 ea	100	63	180	12	16	2	8	1.9		
1555	Tostada with red sauce	1 ea	156	67	264	9	27	11	13	4.4		
1558	Mexican pizza	1 ea	223	53	578	21	43	8	35	10.1		
1559	Taco salad with salsa	1 ea	585	71	923	33	70	17	56	16.3		
1560	Nachos, regular	1 ea	106	40	343	5	36	3	19	4.3		
1561	Nachos, bellgrande	1 ea	287	51	708	19	77	16	36	10.1		
1562	Pintos & cheese with red sauce	1 ea	128	68	203	10	19	11	10	4.3		
1563	Taco sauce, packet	1 ea	9	94	2	<1	<1	<1	<1	0	0	0
1564	Salsa	1 ea	10	28	27	1	6		<1	0	0	0
1565	Cinnamon twists	1 ea	35	6	175	1	24	0	7	0		
1628	Caramel roll	1 ea	85	19	353	6	46		16	4		
	Source: Taco Bell Corporation											
	WENDY'S											
	Hamburgers:											
1566	Single on white bun, no toppings	1 ea	133	44	360	24	31	2	16	6		
1570	Cheeseburger, bacon	1 ea	166	55	380	20	34	2	19	7	10.3	1.4
1730	Chicken sandwich, grilled	1 ea	189	62	310	27	35	2	8	1.5		
	Baked potatoes:											
1573	Plain	1 ea	284	71	310	7	71	7	0	0	0	0
1574	With bacon & cheese	1 ea	380	69	530	17	78	7	18	4	10.7	3.3
1575	With broccoli & cheese	1 ea	411	74	470	9	80	9	14	2.5	8	2.5
1576	With cheese	1 ea	383	68	570	14	78	7	23	8	9.2	4.8

Chol (mg)	Calc (mg)	Iron (mg)	Magn (mg)	Pota (mg)	Sodi (mg)	Zinc (mg)	VT-A (RE)	Thia (mg)	VT-E (a-TE)	Ribo (mg)	Niac (mg)	V-B6 (mg)	Fola (µg)	VT-C (mg)
0	23	1			308		136							32
10	32	1.99			115		0							0
10	16	1.99			139		0							0
15	16	1			139		0							0
10	16	1			139		0							0
15	32	1			159		0							0
0	16	1			179		0							0
20	0	.72			179		0							0
10	16	1			139		0							0
181	80	1.8			652		310							
195	80	1.08			690		250							0
25	80	.72			580		150							0
205	100	1.8			1050		500							0
183	80	1.8			661		320							
10	150	2.7		495	1100		450	.04		2.02	1.98	.31		0
57	160	3.96		380	1303		530	.4		2.14	3.44	.32		1
32	170	3.78	50	442	1221	2.67	450	.49		.41	3.09	.59	38	1
55	150	2.7			1520		600							5
57	140	2.52			854		440							1
34	146	8.75	48	410	1196		486	.39		2.04	2.81	.34		5
21	165	2.98			1058		248							5
47	300	3.06			980		950							
78	200	2.7			1313		970							2
25	80	1.08		159	330		100	.05		.14	1.2	.12		0
27	88	1.19		211	363		110	.42		.24	2.95	1.08		0
31	90	1.62			532		135							3
37	85	1.52			571		63							1
19	62	1.13			797		31							0
13	132	1.59		401	573		441	.05		.17	.63	.26		1
46	253	3.65	80	408	1054	5.37	405	.32		.33	2.96	1.12	60	5
65	326	6.84		1048	1931	1.67	1736	.51		.76	4.8	.56	10	26
5	107	.77		160	610	1.68	64	.17		.16	.68	.19	10	0
32	184	3.31		674	1205		138	.1		.34	2.17			3
16	160	1.92	110	384	693	2.17	267	.05		.15	.43	.21	68	0
0	0	.07		9	75		30	0			.02			<1
0	50	.6		376	709		168	.02		.14	0			10
0	0	.45		27	238		50	.1		.04	.71	.04		0
15	60	1.44			312		330							4
65	110	4.14		296	580		0	.43		.38	6.71			0
60	170	3.42	38	375	850	5.9	80	.3		.31	6.43	.26	28	6
65	100	2.7			790		40							6
0	30	3.78	75	1187	25	.74	0	.31	.14	.12	4.3	.8	31	36
20	180	4.32	87	1498	1390	2.75	100	.24		.19	5.04	.94	36	36
5	210	4.5	93	1745	470	.97	350	.34		.29	4.5	.97	74	72
30	380	4.14	85	1510	640	.67	200	.25		.28	3.6	.88	36	36

Table A-1

Food Composition (Computer code number is for West Diet Analysis program) (For purposes of calculations, use "0" for t, <1, <.1, <.01, etc.)

Computer Code Number	Food Description	Measure	Wt (g)	H_2O (%)	Ener (cal)	Prot (g)	Carb (g)	Dietary Fiber (g)	Fat (g)	Fat Breakdown (g) Sat	Mono	Poly
	FAST FOOD RESTAURANTS—Continued											
	WENDY'S—Continued											
1577	With chili & cheese	1 ea	439	69	630	20	83	9	24	9		
1578	With sour cream & chives	1 ea	314	71	380	8	74	8	6	4		
1579	Chili	1 ea	227	81	210	15	21	5	7	2.5		
1582	Chocolate chip cookies	1 ea	57	6	270	3	36	1	13	6		
1580	French fries	1 ea	130	41	390	5	50	5	19	3	11.9	2.4
1581	Frosty dairy dessert	1 ea	227	68	330	8	56	0	8	5		
	Source: Wendy's International											
	CONVENIENCE FOODS and MEALS											
	BUDGET GOURMET											
1695	Chicken cacciatore	1 ea	312	80	300	20	27		13			
1692	Linguini & shrimp	1 ea	284	77	330	15	33		15			
1691	Scallops & shrimp	1 ea	326	79	320	16	43		9			
2245	Seafood newburg	1 ea	284	74	350	17	43		12			
1693	Sirloin tips with country gravy	1 ea	284	80	310	16	21		18			
1694	Sweet & sour chicken with rice	1 ea	284	72	350	18	53		7			
1689	Teriyaki chicken	1 ea	340	77	360	20	44		12			
1690	Veal parmigiana	1 ea	340	75	440	26	39		20			
1696	Yankee pot roast	1 ea	312	77	380	27	22		21			
	Source: The All American Gourmet Co.											
	HAAGEN DAZS											
1755	Ice cream bar, vanilla almond	1 ea	107		371	6	26		27	14.1	10	3
	Sorbet:											
1758	Lemon	½ c	113		140	0	35		0	0	0	0
1760	Orange	½ c	113		140	0	36		0	0	0	0
1759	Raspberry	½ c	113		110	0	27		0	0	0	0
	Yogurt, frozen:											
1753	Chocolate	½ c	98		171	8	26		4	2	2	0
1754	Strawberry	½ c	98		171	6	27		4	2	2	0
	Yogurt extra, frozen:											
1752	Brownie nut	½ c	101		220	8	29		9	4	4	1
1751	Raspberry rendezvous	½ c	101		132	4	26		2	1	1	0
	Source: Pillsbury											
	HEALTHY CHOICE											
	Entrees:											
2112	Fish, lemon pepper	1 ea	303	78	290	14	47	7	5	1		
1624	Lasagna	1 ea	383	76	390	26	60	9	5	2		
2111	Meatloaf, traditional	1 ea	340	79	320	16	46	7	8	4		
2104	Zucchini lasagna	1 ea	396	80	329	20	58	11	1	1		
2110	Dinner, pasta shells marinara	1 ea	340	74	360	25	59	5	3	1.5		
	Low-fat ice cream:											
973	Brownie	½ c	71	60	120	3	22	2	2	1	.3	.7
259	Butter pecan	½ c	71	60	120	3	22	1	2	1	.3	.7
650	Chocolate chip	½ c	71	62	120	3	21	<1	2	1	1	0
1608	Cookie & cream	½ c	71	62	120	3	21	<1	2	1.5	.5	0
650	Chocolate chip	½ c	71	62	120	3	21	<1	2	1	1	0
45	Rocky road	½ c	71	53	140	3	28	2	1	1	.5	0
1621	Vanilla	½ c	71	66	100	3	18	1	2	.5	1.5	0
391	Vanilla fudge	½ c	71	62	120	3	21	1	2	1.5		
	Source: ConAgra Frozen Foods, Omaha, NE											

Chol (mg)	Calc (mg)	Iron (mg)	Magn (mg)	Pota (mg)	Sodi (mg)	Zinc (mg)	VT-A (RE)	Thia (mg)	VT-E (a-TE)	Ribo (mg)	Niac (mg)	V-B6 (mg)	Fola (μg)	VT-C (mg)
40	330	5.04	122	1745	770	4.15	200	.33		.29	4.5	.99	55	36
15	80	4.32	71	1438	40	.91	300	.23		.14	3.04	.8	32	48
30	80	2.9		501	800		80	.11		.15	2.66			4
30	10	1.8	13	89	120	.41	0	.05		.06	.36	.03	5	0
0	20	1.08	55	845	120	.62	0	.18		.04	3.6	.33	40	6
35	310	1.08	46	544	200	.97	150	.11		.47	.32	.13	17	0
60	150	1.8			810		40	.23		.51	5			21
75	10	3.6			1250		1000	.3		.17	3			2
70	150	.72			690		150			.26	3			12
70	100	.72			660		40	.23		.26	2			
40	60	.36			570		150	.15		.17	4	.28		2
40	60	.72			640		80	.12		.34	3			2
55	80	1.4			610		300	.15		.34	6			12
165	30	4.5			1160		1000	.45		.6	6			6
70	150	1.8			690		600	.15		.43	7			6
90	161	.38		221	85		161			.18				
0				30	20									7
0				80	20									20
0				60	15									7
40	147	.71		241	45		20			.17				
50	147			141	45		20	.03		.17				5
55	152	.73		250	60		20			.14				
20	81			97	25		0			.1				5
25	20	1.08			360		100							30
15	150	3.6		500	550		100	.3		.26	2			6
35	40	1.8			460		150							54
10	199	2.69			309		249							0
25	400	1.8			390		100							4
2	80	0		268	55		40							0
2	100	0		211	60		40							0
2	100	0		240	50		40							0
2	100			254	90		60	.03		.15				2
2	100	0		240	50		40							0
2	100	0		168	60		40	.03		.15				0
5	100			254	50		60	.05		.22				2
2	100	0		296	50		40							0

Table A-1

Food Composition (Computer code number is for West Diet Analysis program) (For purposes of calculations, use "0" for t, <1, <.1, <.01, etc.)

Computer Code Number	Food Description	Measure	Wt (g)	H_2O (%)	Ener (cal)	Prot (g)	Carb (g)	Dietary Fiber (g)	Fat (g)	Fat Breakdown (g) Sat	Mono	Poly
	CONVENIENCE FOODS and MEALS—Continued											
	HEALTH VALLEY											
	Soups, fat-free:											
2001	Beef broth, no salt added	1 c	240	98	18	5	0	0	0	0	0	0
2073	Beef broth, w/salt	1 c	240	98	30	5	2	0	0	0	0	0
2016	Black bean & vegetable	1 c	240	85	110	11	24	12	0	0	0	0
2017	Chicken broth	1 c	240	97	30	6	0	0	0	0	0	0
2018	14 garden vegetable	1 c	240	90	80	6	17	4	0	0	0	0
2015	Lentil & carrot	1 c	240	85	90	10	25	14	0	0	0	0
2014	Split pea & carrot	1 c	240	89	110	8	17	4	0	0	0	0
2013	Tomato vegetable	1 c	240	90	80	6	17	5	0	0	0	0
	Source: Health Valley											
	LA CHOY											
2100	Egg rolls, mini, chicken	1 svg	106	53	220	8	35	3	6	1.5		
2099	Egg rolls, mini, shrimp	1 svg	106	56	210	7	35	3	4	1		
	Source: Beatrice/Hunt Wesson											
	LEAN CUISINE											
	Dinners:											
1639	Baked cheese ravioli	1 ea	241	77	250	12	32	4	8	3	2	1
1632	Chicken chow mein	1 ea	255	81	210	13	28	2	5	1	2	1
1633	Lasagna	1 ea	291	79	270	19	34	5	6	2.5	1.5	.5
1634	Macaroni & cheese	1 ea	255	78	270	13	39	2	7	3.5	1.5	.5
1631	Spaghetti w/meatballs	1 ea	269	74	290	17	40	4	7	2	3	1.5
	Pizza:											
1636	French bread sausage pizza	1 ea	170	53	420	19	41	4	20	5	13.9	1.1
	Source: Stouffer's Foods Corp, Solon, OH											
	TASTE ADVENTURE SOUPS											
1905	Black bean	1 c	242		139	6	28	6	1			
1904	Curry lentil	1 c	241		138	6	30	5	1			
1906	Lentil chili	1 c	242		181	11	33	6	1			
1903	Split pea	1 c	244		140	5	27	5	1			
	Source: Taste Adventure Soups											
	WEIGHT WATCHERS											
	Cheese, fat-free slices:											
1978	Cheddar, sharp	2 pce	21	65	30	5	2	0	0	0	0	0
1980	Swiss	2 pce	21	65	30	5	2	0	0	0	0	0
1977	White	2 pce	21	65	30	5	2	0	0	0	0	0
1979	Yellow	2 pce	21	65	30	5	2	0	0	0	0	0
	Dinners:											
2029	Chicken chow mein	1 ea	255	81	200	12	34	3	2	.5		
1646	Oven fried fish	1 ea	218	78	230	15	25	2	8	2.5	5	2
1972	Margarine, reduced fat	1 tbs	14	49	59	0	0	0	7	1.5		
	Pizza:											
1653	Cheese	1 ea	163	48	390	23	49	6	12	4	3	1
1650	Deluxe combination pizza	1 ea	186	56	380	23	47	6	11	3.5	5	2
1652	Pepperoni pizza	1 ea	158	48	390	23	46	4	12	4	5	2
	Desserts:											
1644	Chocolate brownie	1 ea	182	75	190	6	35	4	4	1	2	1
2024	Chocolate eclair	1 ea	60	45	151	3	24	2	5	1.5		
2247	Chocolate mousse	1 ea	78	44	190	6	33	3	4	1.5		
1642	Strawberry cheesecake	1 ea	111	62	180	7	28	2	5	2	1	2

Chol (mg)	Calc (mg)	Iron (mg)	Magn (mg)	Pota (mg)	Sodi (mg)	Zinc (mg)	VT-A (RE)	Thia (mg)	VT-E (a-TE)	Ribo (mg)	Niac (mg)	V-B6 (mg)	Fola (μg)	VT-C (mg)
0				196	74						.98			
0	0	0		196	160		0				.98			5
0	40	3.6		676	280		2000	.34		.11	1.35	.22	135	9
0	20	1.8		147	170		0			.03	2.45			1
0	40	1.8		406	250		2000	.26		.08	2.25	.18	27	15
0	60	5.4		439	220		2000	.1		.16	5.63	.45	27	2
0	40	5.4		439	230		2000	.1		.16	5.63	.45		9
0	40	5.4		609	240		2000	.1		.08	2.25	.13	<1	9
5	20	1.44			460		20							0
5	20	1.44			510		20							0
55	200	1.08	42	400	500	1.5	150	.06		.25	1.2	.2	48	6
35	20	.36	30	300	510	1.1	20	.15		.17	5			6
25	150	1.8	44	620	560	2.9	100	.15		.25	3	.32		12
20	250	.72		170	550		20	.12		.25	1.2			0
30	100	2.7	47	480	520	2.5	80	.15		.25	3	.2		4
35	250	2.7	39	340	900	2.2	80	.45		.51	5	.07		6
				650	565									
				467	584									
				650	448									
				484	591									
0	99	0		64	306		56							0
0	99	0		74	276		56							0
0	99	0		64	306		56							0
0	99	0		64	306		56							0
25	40	.72		360	430		300							36
25	20	1.44		370	450		40	.09		.14	1.6			0
0	0	0		5	128		49							0
35	700	1.8		290	590		80	.3		.51	3	.06		6
40	500	3.6		370	550		150	.3		.51	3	.2		5
45	450	1.8		320	650		80	.23		.51	3			5
5	80	1.08		230	160		0	.06		.03	.2	.03		0
0	40	0		65	151		0							0
5	60	1.8		320	150		0							0
15	80	.36		115	230		40	.06		.07	1.6			2

Table A–1

Food Composition (Computer code number is for West Diet Analysis program) (For purposes of calculations, use "0" for t, <1, <.1, <.01, etc.)

Computer Code Number	Food Description	Measure	Wt (g)	H_2O (%)	Ener (cal)	Prot (g)	Carb (g)	Dietary Fiber (g)	Fat (g)	Fat Breakdown (g) Sat	Mono	Poly
	CONVENIENCE FOODS and MEALS—Continued											
	WEIGHT WATCHERS—CONTINUED											
2027	Triple chocolate cheesecake	1 ea	89	52	199	7	32	1	5	2.5		
	Source: Weight Watchers											
	SWEET SUCCESS:											
	Drinks, prepared:											
1776	Chocolate chip	1 c	265	81	180	15	30	6	3	1.6		
1777	Chocolate fudge	1 c	265	81	180	15	30	6	2			
1774	Chocolate mocha	1 c	265	81	180	15	30	6	1	1		
1778	Milk chocolate	1 c	265	81	180	15	30	6	2	1		
1775	Vanilla	1 c	265	81	180	15	33	6	1	.6		
	Drinks, ready to drink:											
2147	Chocolate mint	1 c	297	82	187	11	36	6	3	0		
2148	Strawberry	1 c	265	82	167	10	32	5	3	0		
	Shakes:											
1771	Chocolate almond	1 c	250	82	158	9	30	5	2	0	2.1	.2
1773	Chocolate fudge	1 c	250	82	158	9	30	5	2	0	2.1	.2
1768	Chocolate mocha	1 c	250	82	158	9	30	5	2	0	.6	1.8
1769	Chocolate raspberry truffle	1 c	250	82	158	9	30	5	2	0	2.2	.2
1770	Vanilla creme	1 c	250	82	158	9	30	5	2	0	2.1	.3
	Snack bars:											
1767	Chocolate brownie	1 ea	33	9	120	2	23	3	4	2	.5	.6
1766	Chocolate chip	1 ea	33	9	120	2	23	3	4	2	.4	.5
1921	Oatmeal raisin	1 ea	33	9	120	2	23	3	4	2		
1765	Peanut butter	1 ea	33	9	120	2	23	3	4	2	.6	.6
	Source: Foodway National Inc, Boise, ID											
	BABY FOODS											
1720	Apple juice	½ c	125	88	59	0	15	<1	<1	t	t	t
1721	Applesauce, strained	1 tbs	16	89	7	<1	2	<1	<1	t	t	t
1716	Carrots, strained	1 tbs	14	92	4	<1	1	<1	<1	t	t	t
1718	Cereal, mixed, milk added	1 tbs	15	75	17	1	2	<1	1	.3		
1719	Cereal, rice, milk added	1 tbs	15	75	17	<1	3	<1	1	.3		
1723	Chicken and noodles, strained	1 tbs	16	88	8	<1	1	<1	<1	.1	.1	t
1722	Peas, strained	1 tbs	15	87	6	1	1	<1	<1	t	t	t
1717	Teething biscuits	1 ea	11	6	43	1	8	<1	<1	.2	.2	.1

PAGE KEY: A–4 = Beverages A–6 = Dairy A–10 = Eggs A–10 = Fat/Oil A–14 = Fruit A–20 = Bakery A–28 = Grain A–32 = Fish A–34 = Meats A–38 = Poultry A–40 = Sausage A–42 = Mixed/Fast A–46 = Nuts/Seeds A–50 = Sweets A–52 = Vegetables/Legumes A–62 = Vegetarian Foods A–64 = Misc A–66 = Soups/Sauces A–68 = Fast A–84 = Convenience A–88 = Baby foods

Chol (mg)	Calc (mg)	Iron (mg)	Magn (mg)	Pota (mg)	Sodi (mg)	Zinc (mg)	VT-A (RE)	Thia (mg)	VT-E (a-TE)	Ribo (mg)	Niac (mg)	V-B6 (mg)	Fola (μg)	VT-C (mg)
10	80	1.08		169	199		0							0
6	500	6.3	140	600	288	5.25	350	.52	7.05	.59	7	.7	140	21
6	500	6.3	140	750	336	5.25	350	.52	7.05	.59	7	.7	140	21
6	500	6.3	140	800	336	5.25	350	.52	7.05	.59	7	.7	140	21
6	500	6.3	140	750	336	5.25	350	.52	7.05	.59	7	.7	140	21
6	500	6.3	140	830	312	5.25	250	.52	7.05	.59	7	.7	140	21
6	470	5.94	131	526	226	5.05	329	.5	6.56	.56	6.53	.65	131	20
5	419	5.3	117	310	175	4.51	294	.45	5.86	.5	5.83	.58	117	17
5	396	5	110	443	190	4.25	277	.42	5.53	.47	5.5	.55	110	16
5	396	5	110	443	175	4.25	277	.42	5.53	.47	5.5	.55	110	16
5	396	5	110	403	175	4.25	277	.42	5.53	.47	5.5	.55	110	16
5	383	5	110	443	175	4.25	277	.42	5.53	.47	5.5	.55	110	16
5	396	5	110	293	175	4.25	277	.42	5.53	.47	5.5	.55	110	16
3	150	2.71	60	140	45	.59	150	.22	3.01	.25	3	.3	60	9
3	150	2.71	60	110	40	.59	150	.22	3.01	.25	3	.3	60	9
3	150	2.71	60		30	.59	150	.22	3.01	.25	3	.3	60	9
3	150	2.71	60	125	35	.59	150	.22	3.01	.25	3	.3	60	9
0	5	.71	4	114	4	.04	2	.01	.75	.02	.1	.04	<1	72
0	1	.03		11	<1	<.01	<1	<.01	.1	<.01	.01	<.01	<1	6
0	3	.05	1	27	5	.02	160	<.01	.07	.01	.06	.01	2	1
2	33	1.56	4	30	7	.11	4	.06		.09	.87	.01	2	<1
2	36	1.83	7	28	7	.1	4	.07		.07	.78	.02	1	<1
3	4	.07	1	6	3	.05	18	<.01	.04	.01	.07	<.01	2	<1
0	3	.14	2	17	1	.05	8	.01	.08	.01	.15	.01	4	1
0	29	.39	4	35	40	.1	1	.03	.05	.06	.48	.01	5	1

Canadiana: Recommendations, Choice System, and Labels

CONTENTS

B

Chapter 2 introduced the 1991 Recommended Nutrient Intakes (RNI), food guides, and food labels. This appendix presents details for Canadians. The U.S. Exchange System is found in Appendix D, and its Canadian equivalent, the Cana-

Table B-1
Recommended Nutrient Intakes for Canadians, 1990

				Vitamins					
Age	**Sex**	**Weight (kg)**	**Protein (g/day)[a]**	**Vitamin A (RE/day)[b]**	**Vitamin E (mg/day)[c]**	**Vitamin C (mg/day)[d]**	**Iron (mg/day)**	**Iodine (μg/day)**	**Zinc (mg/day)**
Infants (months)									
0–4	Both	6	12[e]	400	3	20	0.3[f]	30	2[g]
5–12	Both	9	12	400	3	20	7	40	3
Children (years)									
1	Both	11	13	400	3	20	6	55	4
2–3	Both	14	16	400	4	20	6	65	4
4–6	Both	18	19	500	5	25	8	85	5
7–9	M	25	26	700	7	25	8	110	7
	F	25	26	700	6	25	8	95	7
10–12	M	34	34	800	8	25	8	125	9
	F	36	36	800	7	25	8	110	9
13–15	M	50	49	900	9	30	10	160	12
	F	48	46	800	7	30	13	160	9
16–18	M	62	58	1000	10	40	10	160	12
	F	53	47	800	7	30	12	160	9
Adults (years)									
19–24	M	71	61	1000	10	40	9	160	12
	F	58	50	800	7	30	13	160	9
25–49	M	74	64	1000	9	40	9	160	12
	F	59	51	800	6	30	13[h]	160	9
50–74	M	73	63	1000	7	40	9	160	12
	F	63	54	800	6	30	8	160	9
75+	M	69	59	1000	6	40	9	160	12
	F	64	55	800	5	30	8	160	9
Pregnancy (additional amount needed)									
1st trimester			5	0	2	0	0	25	6
2nd trimester			20	0	2	10	5	25	6
3rd trimester			24	0	2	10	10	25	6
Lactation (additional amount needed)			20	400	3	25	0	50	6

NOTE: Recommended intakes of energy and of certain nutrients are not listed in this table because of the nature of the variables upon which they are based. The figures for energy are estimates of average requirements for expected patterns of activity (see Table B-2). For nutrients not shown, the following amounts are recommended based on at least 2000 kcalories per day and body weights as given: thiamin, 0.4 milligram per 1000 kcalories (0.48 milligram/5000 kilojoules); riboflavin, 0.5 milligram per 1000 kcalories (0.6 milligram/5000 kilojoules); niacin, 7.2 niacin equivalents per 1000 kcalories (8.6 niacin equivalents/5000 kilojoules); vitamin B_6, 15 micrograms, as pyridoxine, per gram of protein. Recommended intakes during periods of growth are taken as appropriate for individuals representative of the midpoint in each age group. All recommended intakes are designed to cover individual variations in essentially all of a healthy population subsisting upon a variety of common foods available in Canada.
Source: Health and Welfare Canada, *Nutrition Recommendations: The Report of the Scientific Review Committee* (Ottawa: Canadian Government Publishing Centre, 1990), Table 20, p. 204.

[a]The primary units are expressed per kilogram of body weight. The figures shown here are examples.
[b]One retinol equivalent (RE) corresponds to the biological activity of 1 microgram of retinol, 6 micrograms of beta-carotene, or 12 micrograms of other carotenes.
[c]Expressed as δ-α-tocopherol equivalents, relative to which β- and γ-tocopherol and α-tocotrienol have activities of 0.5, 0.1, and 0.3, respectively.
[d]Cigarette smokers should increase intake by 50 percent.
[e]The assumption is made that the protein is from breast milk or has the same biological value as breast milk and that, between 3 and 9 months, adjustment for the quality of the protein is made.
[f]Based on the assumption that breast milk is the source of iron.
[g]Based on the assumption that breast milk is the source of zinc.
[h]After menopause, the recommended intake is 8 milligrams per day.

dian Choice System, is presented here. Appendix E includes addresses of Canadian governmental agencies and professional organizations that may provide additional information.

RNI

As Chapter 2 mentioned, a major revision of the nutrient recommendations is underway in both the United States and Canada. The Dietary Reference Intakes (DRI) reports are replacing the 1989 RDA in the United States and the 1991 RNI in Canada. Recommendations from the DRI reports are presented on the inside front cover. For nutrients that do not yet have new values, the RNI will continue to serve health professionals in Canada (Tables B-1 and B-2).

Table B-2
Average Energy Requirements for Canadians

Age	Sex	Average Height (cm)	Average Weight (kg)	Requirements[a]					
				(kcal/kg)[b]	(MJ/kg)[b]	(kcal/day)	(MJ/day)	(kcal/cm)	(MJ/cm)
Infants (months)									
0–2	Both	55	4.5	120–100	0.50–0.42	500	2.0	9	0.04
3–5	Both	63	7.0	100–95	0.42–0.40	700	2.8	11	0.05
6–8	Both	69	8.5	95–97	0.40–0.41	800	3.4	11.5	0.05
9–11	Both	73	9.5	97–99	0.41	950	3.8	12.5	0.05
Children and Adults (years)									
1	Both	82	11	101	0.42	1100	4.8	13.5	0.06
2–3	Both	95	14	94	0.39	1300	5.6	13.5	0.06
4–6	Both	107	18	100	0.42	1800	7.6	17	0.07
7–9	M	126	25	88	0.37	2200	9.2	17.5	0.07
	F	125	25	76	0.32	1900	8.0	15	0.06
10–12	M	141	34	73	0.30	2500	10.4	17.5	0.07
	F	143	36	61	0.25	2200	9.2	15.5	0.06
13–15	M	159	50	57	0.24	2800	12.0	17.5	0.07
	F	157	48	46	0.19	2200	9.2	14	0.06
16–18	M	172	62	51	0.21	3200	13.2	18.5	0.08
	F	160	53	40	0.17	2100	8.8	13	0.05
19–24	M	175	71	42	0.18	3000	12.6		
	F	160	58	36	0.15	2100	8.8		
25–49	M	172	74	36	0.15	2700	11.3		
	F	160	59	32	0.13	1900	8.0		
50–74	M	170	73	31	0.13	2300	9.7		
	F	158	63	29	0.12	1800	7.6		
75+	M	168	69	29	0.12	2000	8.4		
	F	155	64	23	0.10	1500	6.3		

[a]Requirements can be expected to vary within a range of ±30 percent.
[b]First and last figures are averages at the beginning and end of the three-month period.
Source: Health and Welfare Canada, *Nutrition Recommendations: The Report of the Scientific Review Committee* (Ottawa: Canadian Government Publishing Centre, 1990), Tables 5 and 6, pp. 25, 27.

Choice System for Meal Planning

The *Good Health Eating Guide* is the Canadian choice system of meal planning.[a] It is similar to the U.S. exchange system in the following ways:

[a]The tables for the Canadian choice system are adapted from the *Good Health Eating Guide Resource,* copyright 1994, with permission of the Canadian Diabetes Association.

- Foods are divided into lists according to carbohydrate, protein, and fat content.
- Foods are interchangeable within a group.
- Most foods are eaten in measured amounts.
- An energy value is given for each food group.

Tables B-3 through B-10 present the Canadian choice system.

Table B-3
Canadian Choice System: Starch Foods

1 starch choice = 15 g carbohydrate (starch), 2 g protein, 290 kJ (68 kcal)

Food	Measure	Mass (Weight)
Breads		
Bagels	½	30 g
Bread crumbs	50 mL (¼ c)	30 g
Bread cubes	250 mL (1 c)	30 g
Bread sticks	2	20 g
Brewis, cooked	50 mL (¼ c)	45 g
Chapati	1	20 g
Cookies, plain	2	20 g
English muffins, crumpets	½	30 g
Flour	40 mL (2½ tbs)	20 g
Hamburger buns	½	30 g
Hot dog buns	½	30 g
Kaiser rolls	½	30 g
Matzo, 15 cm	1	20 g
Melba toast, rectangular	4	15 g
Melba toast, rounds	7	15 g
Pita, 20 cm (8") diameter	¼	30 g
Pita, 15 cm (6") diameter	½	30 g
Plain rolls	1 small	30 g
Pretzels	7	20 g
Raisin bread	1 slice	30 g
Rice cakes	2	30 g
Roti	1	20 g
Rusks	2	20 g
Rye, coarse or pumpernickel	½ slice	30 g
Soda crackers	6	20 g
Tortilla, corn (taco shell)	1	10 g
Tortilla, flour (9" diameter)	1	30 g
White (French and Italian)	1 slice	25 g
Whole-wheat, cracked-wheat, rye, white enriched	1 slice	30 g
Cereals		
Bran flakes, 100% bran	125 mL (½ c)	30 g
Cooked cereals, cooked	125 mL (½ c)	125 g
Dry	30 mL (2 tbs)	20 g
Cornmeal, cooked	125 mL (½ c)	125 g
Dry	30 mL (2 tbs)	20 g
Ready-to-eat unsweetened cereals	125 mL (½ c)	20 g
Shredded wheat biscuits, rectangular or round	1	20 g
Shredded wheat, bite size	125 mL (½ c)	20 g
Wheat germ	75 mL (⅓ c)	30 g
Cornflakes	300 mL (1¼ c)	30 g
Rice Krispies	250 mL (1 c)	30 g

B

Table B-3 (continued)

Canadian Choice System: Starch Foods

1 starch choice = 15 g carbohydrate (starch), 2 g protein, 290 kJ (68 kcal)

Food	Measure	Mass (Weight)
Cheerios	200 mL (¾ c)	20 g
Muffets	1	20 g
Puffed rice	300 mL (1¼ c)	15 g
Puffed wheat	425 mL (1⅔ c)	20 g
Grains		
Barley, cooked	125 mL (½ c)	120 g
Dry	30 mL (2 tbs)	20 g
Bulgur, kasha, cooked, moist	125 mL (½ c)	70 g
Cooked, crumbly	75 mL (⅓ c)	40 g
Dry	30 mL (2 tbs)	20 g
Rice, cooked, brown and white (short and long grain)	75 mL (⅓ c)	60 g
Rice, cooked, wild	75 mL (⅓ c)	60 g
Tapioca, pearl and granulated, quick cooking, dry	30 mL (2 tbs)	15 g
Couscous, cooked moist	125 mL (½ c)	70 g
Dry	30 mL (2 tbs)	20 g
Quinoa, cooked moist	125 mL (½ c)	70 g
Dry	30 mL (2 tbs)	20 g
Pastas		
Macaroni, cooked	125 mL (½ c)	70 g
Noodles, cooked	125 mL (½ c)	80 g
Spaghetti, cooked	125 mL (½ c)	70 g
Starchy Vegetables		
Beans and peas, dried, cooked	125 mL (½ c)	98 g
Breadfruit	1 slice	75 g
Corn, canned, whole kernel	125 mL (½ c)	85 g
Corn on the cob	½ medium cob	140 g
Cornstarch	30 mL (2 tbs)	15 g
Plantains	⅓ small	50 g
Popcorn, air-popped, unbuttered	750 mL (3 c)	20 g
Potatoes, whole (with or without skin)	½ medium	95 g
Yams, sweet potatoes, (with or without skin)	½	75 g

Food	Choices per Serving	Measure	Mass (Weight)
NOTE: Food items found in this category provide more than 1 starch choice:			
Bran flakes	1 starch + ½ sugar	150 mL (⅔ c)	24 g
Croissant, small	1 starch + 1½ fats	1 small	35 g
Large	1 starch + 1½ fats	½ large	30 g
Corn, canned creamed	1 starch + ½ sugar	12 mL (½ c)	113 g
Potato chips	1 starch + 2 fats	15 chips	30 g
Tortilla chips (nachos)	1 starch + 1½ fats	13 chips	30 g
Corn chips	1 starch + 2 fats	30 chips	30 g
Cheese twists	1 starch + 1½ fats	30 chips	30 g
Cheese puffs	1 starch + 2 fats	27 chips	30 g
Tea biscuit	1 starch + 2 fats	1	30 g
Pancakes, homemade using 50 mL (¼ c) batter (6″ diameter)	1½ starches + 1 fat	1 medium	50 g
Potatoes, french fried (homemade or frozen)	1 starch + 1 fat	10 regular size	35 g
Soup, canned* (prepared with equal volume of water)	1 starch	250 mL (1 c)	260 g
Waffles, packaged	1 starch + 1 fat	1	35 g

*Soup can vary according to brand and type. Check the label for Food Choice Values and Symbols or the core nutrient listing.

Table B-4
Canadian Choice System: Fruits and Vegetables

1 fruits and vegetables choice = 10 g carbohydrate, 1 g protein, 190 kJ (44 kcal)

Food	Measure	Mass (Weight)
Fruits (fresh, frozen, without sugar, canned in water)		
Apples, raw (with or without skin)	½ medium	75 g
Sauce unsweetened	125 mL (½ c)	120 g
Sweetened	see *Combined Food Choices*	
Apple butter	20 mL (4 tsp)	20 g
Apricots, raw	2 medium	115 g
Canned, in water	4 halves, plus 30 mL (2 tbs) liquid	110 g
Bake-apples (cloudberries), raw	125 mL (½ c)	120 g
Bananas, with peel	½ small	75 g
Peeled	½ small	50 g
Berries (blackberries, boysenberries, raspberries) (blueberries, loganberries, huckleberries)	250 mL (1 c) 125 mL (½ c)	150 g 70 g
Raw	125 mL (½ c)	70 g
Cantaloupe, wedge with rind	¼	160 g
Cubed or diced	175 mL (⅔ c)	135 g
Cherries, raw, with pits	10	75 g
Raw, without pits	10	70 g
Canned, in water, with pits	75 mL (⅓ c), plus 30 mL (2 tbs) liquid	90 g
Canned, in water, without pits	75 mL (⅓ c), plus 30 mL (2 tbs) liquid	85 g
Crabapples, raw	1 small	55 g
Cranberries, raw	250 mL (1 c)	100 g
Figs, raw	1 medium	50 g
Canned, in water	3 medium, plus 30 mL (2 tbs) liquid	100 g
Foxberries, raw	250 mL (1 c)	100 g
Fruit cocktail, canned, in water	125 mL (½ c), plus 30 mL (2 tbs) liquid	120 g
Fruit, mixed, cut-up	125 mL (½ c)	120 g
Gooseberries, raw	250 mL (1 c)	150 g
Canned, in water	250 mL (1 c), plus 30 mL (2 tbs) liquid	230 g
Grapefruit, raw, with rind	½ small	185 g
Raw, sectioned	125 mL (½ c)	100 g
Canned, in water	125 mL (½ c), plus 30 mL (2 tbs) liquid	120 g
Grapes, raw, slip skin	125 mL (½ c)	75 g
Raw, seedless	125 mL (½ c)	75 g
Canned, in water	75 mL (⅓ c), plus 30 mL (2 tbs) liquid	115 g
Guavas, raw	½	50 g
Honeydew melon, raw, with rind	⅒ wedge	130 g
Cubed or diced	175 mL (⅔ c)	115 g
Kiwis, raw, with skin	1 medium	76 g
Kumquats, raw	3	60 g
Loquats, raw	8	130 g
Lychee fruit, raw	8	120 g
Mandarin oranges, raw, with rind	1	135 g
Raw, sectioned	125 mL (½ c)	100 g
Canned, in water	125 mL (½ c), plus 30 mL (2 tbs) liquid	100 g
Mangoes, raw, without skin and seed,		
Diced	75 mL (⅓ c)	65 g
Nectarines	½ medium	75 g
Oranges, raw, with rind	1 small	130 g
Raw, sectioned	125 mL (½ c)	95 g
Papayas, raw, with skin and seeds	¼ medium	150 g

Table B-4 (continued)

Canadian Choice System: Fruits and Vegetables

1 fruits and vegetables choice = 10 g carbohydrate, 1 g protein, 190 kJ (44 kcal)

Food	Measure	Mass (Weight)
Papayas (continued)		
Raw, without skin and seeds	¼ medium	100 g
Cubed or diced	125 mL (½ c)	100 g
Peaches, raw, with seed and skin	1 large	100 g
Raw, sliced or diced	125 mL (½ c)	85 g
Canned in water, halves or slices	125 mL (½ c), plus 30 mL (2 tbs) liquid	120 g
Pears, raw, with skin and core	½	90 g
Raw, without skin and core	½	85 g
Canned, in water, halves	1 half plus 30 mL (2 tbs) liquid	90 g
Persimmons, raw, native	1	30 g
Raw, Japanese	¼	50 g
Pineapple, raw	1 slice	75 g
Raw, diced	125 mL (½ c)	75 g
Canned, in juice, diced	75 mL (⅓ c), plus 15 mL (1 tbs) liquid	55 g
Canned, in juice, sliced	1 slice, plus 15 mL (1 tbs) liquid	55 g
Canned, in water, diced	125 mL (½ c), plus 30 mL (2 tbs) liquid	100 g
Canned, in water, sliced	2 slices, plus 15 mL (1 tbs) liquid	100 g
Plums, raw	2 small	60 g
Damson	6	65 g
Japanese	1	70 g
Canned, in apple juice	2, plus 30 mL (2 tbs) liquid	70 g
Canned, in water	3, plus 30 mL (2 tbs) liquid	100 g
Pomegranates, raw	½	140 g
Strawberries, raw	250 mL (1 c)	150 g
Frozen/canned, in water	250 mL (1 c), plus 30 mL (2 tbs) liquid	240 g
Rhubarb	250 mL (1 c)	150 g
Tangelos, raw	1	205 g
Tangerines, raw	1 medium	115 g
Raw, sectioned	125 mL (½ c)	100 g
Watermelon, raw, with rind	1 wedge	310 g
Cubed or diced	250 mL (1 c)	160 g
Dried Fruit		
Apples	5 pieces	15 g
Apricots	4 halves	15 g
Banana flakes	30 mL (2 tbs)	15 g
Currants	30 mL (2 tbs)	15 g
Dates, without pits	2	15 g
Peaches	½	15 g
Pears	½	15 g
Prunes, raw, with pits	2	15 g
Raw, without pits	2	10 g
Stewed, no liquid	2	20 g
Stewed, with liquid	2, plus 15 mL (1 tbs) liquid	35 g
Raisins	30 mL (2 tbs)	15 g
Juices (no sugar added or unsweetened)		
Apricot, grape, guava, mango, prune	50 mL (¼ c)	55 g
Apple, carrot, papaya, pear, pineapple, pomegranate	75 mL (⅓ c)	80 g
Cranberry (see Sugars Section)		
Clamato (see Sugars Section)		

(continued on the next page)

Table B-4 (continued)
Canadian Choice System: Fruits and Vegetables

1 fruits and vegetables choice = 10 g carbohydrate, 1 g protein, 190 kJ (44 kcal)

Food	Measure	Mass (Weight)
Grapefruit, loganberry, orange, raspberry, tangelo, tangerine	125 mL (½ c)	130 g
Tomato, tomato-based mixed vegetables	250 mL (1 c)	255 g
Vegetables (fresh, frozen, or canned)		
Artichokes, French, globe	2 small	50 g
Beets, diced or sliced	125 mL (½ c)	85 g
Carrots, diced, cooked or uncooked	125 mL (½ c)	75 g
Chestnuts, fresh	5	20 g
Parsnips, mashed	125 mL (½ c)	80 g
Peas, fresh or frozen	125 mL (½ c)	80 g
Canned	75 mL (⅓ c)	55 g
Pumpkin, mashed	125 mL (½ c)	45 g
Rutabagas, mashed	125 mL (½ c)	85 g
Sauerkraut	250 mL (1 c)	235 g
Snow peas	250 mL (1 c)	135 g
Squash, yellow or winter, mashed	125 mL (½ c)	115 g
Succotash	75 mL (⅓ c)	55 g
Tomatoes, canned	250 mL (1 c)	240 g
Tomato paste	50 mL (¼ c)	55 g
Tomato sauce*	75 mL (⅓ c)	100 g
Turnips, mashed	125 mL (½ c)	115 g
Vegetables, mixed	125 mL (½ c)	90 g
Water chestnuts	8 medium	50 g

*Tomato sauce varies according to brand name. Check the label or discuss with your dietitian.

Table B-5
Canadian Choice System: Milk

Type of Milk	Carbohydrate (g)	Protein (g)	Fat (g)	Energy
Nonfat (0%)	6	4	0	170 kJ (40 kcal)
1%	6	4	1	206 kJ (49 kcal)
2%	6	4	2	244 kJ (58 kcal)
Whole (4%)	6	4	4	319 kJ (76 kcal)

Food	Measure	Mass (Weight)
Buttermilk (higher in salt)	125 mL (½ c)	125 g
Evaporated milk	50 mL (¼ c)	50 g
Milk	125 mL (½ c)	125 g
Powdered milk, regular	30 mL (2 tbs)	15 g
Instant	50 mL (¼ c)	15 g
Plain yogurt	125 mL (½ c)	125 g

Food	Choices per Serving	Measure	Mass (Weight)
NOTE: Food items found in this category provide more than 1 milk choice:			
Milkshake	1 milk + 3 sugars + ½ protein	250 mL (1 c)	300 g
Chocolate milk, 2%	2 milks 2% + 1 sugar	250 mL (1 c)	300 g
Frozen yogurt	1 milk + 1 sugar	125 mL (½ c)	125 g

Table B-6
Canadian Choice System: Sugars

1 sugar choice = 10 g carbohydrate (sugar), 167 kJ (40 kcal)

Food	Measure	Mass (Weight)
Beverages		
Condensed milk	15 mL (1 tbs)	
Flavoured fruit crystals*	75 mL (⅓ c)	
Iced tea mixes*	75 mL (⅓ c)	
Regular soft drinks	125 mL (½ c)	
Sweet drink mixes*	75 mL (⅓ c)	
Tonic water	125 mL (½ c)	
*These beverages have been made with water.		
Miscellaneous		
Bubble gum (large square)	1 piece	5 g
Cranberry cocktail	75 mL (⅓ c)	80 g
Cranberry cocktail, light	350 mL (1⅓ c)	260 g
Cranberry sauce	30 mL (2 tbs)	
Hard candy mints	2	5 g
Honey, molasses, corn and cane syrup	10 mL (2 tsp)	15 g
Jelly bean	4	10 g
Licorice	1 short stick	10 g
Marshmallows	2 large	15 g
Popsicle	1 stick (½ popsicle)	
Powdered gelatin mix (Jello®) (reconstituted)	50 mL (¼ c)	
Regular jam, jelly, marmalade	15 mL (1 tbs)	
Sugar, white, brown, icing, maple	10 mL (2 tsp)	10 g
Sweet pickles	2 small	100 g
Sweet relish	30 mL (2 tbs)	

Food	Choices per Serving	Measures	Mass (Weight)
The following food items provide more than 1 sugar choice:			
Brownie	1 sugar + 1 fat	1	20 g
Clamato juice	1½ sugars	175 mL (⅔ c)	
Fruit salad, light syrup	1 sugar + 1 fruits & vegetables	125 mL (½ c)	130 g
Aero® bar	2½ sugars + 2½ fats	1 bar	43 g
Smarties®	4½ sugars + 2 fats	1 box	60 g
Sherbet	3 sugars + ½ fat	125 mL (½ c)	95 g

Table B-7
Canadian Choice System: Protein Foods

1 protein choice = 7 g protein, 3 g fat, 230 kJ (55 kcal)

Food	Measure	Mass (Weight)
Cheese		
Low-fat cheese, about 7% milk fat	1 slice	30 g
Cottage cheese, 2% milkfat or less	50 mL (¼ c)	55 g
Ricotta, about 7% milkfat	50 mL (¼ c)	60 g
Fish		
Anchovies (see *Extras,* Table B-9)		
Canned, drained (e.g., mackerel, salmon, tuna packed in water)	50 mL (¼ c)	30 g
Cod tongues, cheeks	75 mL (⅓ c)	50 g
Fillet or steak (e.g., Boston blue, cod, flounder, haddock, halibut, mackerel, orange roughy, perch, pickerel, pike, salmon, shad, snapper, sole, swordfish, trout, tuna, whitefish)	1 piece	30 g
Herring	⅓ fish	30 g
Sardines, smelts	2 medium or 3 small	30 g
Squid, octopus	50 mL (¼ c)	40 g
Shellfish		
Clams, mussels, oysters, scallops, snails	3 medium	30 g
Crab, lobster, flaked	50 mL (¼ c)	30 g
Shrimp, fresh	5 large	30 g
Frozen	10 medium	30 g
Canned	18 small	30 g
Dry pack	50 mL (¼ c)	30 g
Meat and Poultry (e.g., beef, chicken, goat, ham, lamb, pork, turkey, veal, wild game)		
Back, peameal bacon	3 slices, thin	30 g
Chop	½ chop, with bone	40 g
Minced or ground, lean or extra-lean	30 mL (2 tbs)	30 g
Sliced, lean	1 slice	30 g
Steak, lean	1 piece	30 g
Organ Meats		
Hearts, liver	1 slice	30 g
Kidneys, sweetbreads, chopped	50 mL (¼ c)	30 g
Tongue	1 slice	30 g
Tripe	5 pieces	60 g
Soyabean		
Bean curd or tofu	½ block	70 g
Eggs		
In shell, raw or cooked	1 medium	50 g
Without shell, cooked or poached in water	1 medium	45 g
Scrambled	50 mL (¼ c)	55 g

Food	Choices per Serving	Measures	Mass (Weight)
Note: The following choices provide more than 1 protein exchange:			
Cheese			
Cheeses	1 protein + 1 fat	1 piece	25 g
Cheese, coarsely grated (e.g., cheddar)	1 protein + 1 fat	50 mL (¼ c)	25 g
Cheese, dry, finely grated (e.g., parmesan)	1 protein + 1 fat	45 mL	15 g
Cheese, ricotta, high fat	1 protein + 1 fat	50 mL (¼ c)	55 g
Fish			
Eel	1 protein + 1 fat	1 slice	50 g

Table B-7 (continued)
Canadian Choice System: Protein Foods

1 protein choice = 7 g protein, 3 g fat, 230 kJ (55 kcal)

Food	Choices per Serving	Measures	Mass (Weight)
Meat			
Bologna	1 protein + 1 fat	1 slice	20 g
Canned lunch meats	1 protein + 1 fat	1 slice	20 g
Corned beef, canned	1 protein + 1 fat	1 slice	25 g
Corned beef, fresh	1 protein + 1 fat	1 slice	25 g
Ground beef, medium-fat	1 protein + 1 fat	30 mL (2 tbs)	25 g
Meat spreads, canned	1 protein + 1 fat	45 mL	35 g
Mutton chop	1 protein + 1 fat	½ chop, with bone	35 g
Paté (see *Fats and Oils* group, Table B-8)			
Sausages, garlic, Polish or knockwurst	1 protein + 1 fat	1 slice	50 g
Sausages, pork, links	1 protein + 1 fat	1 link	25 g
Spareribs or shortribs, with bone	1 protein + 1 fat	1 large	65 g
Stewing beef	1 protein + 1 fat	1 cube	25 g
Summer sausage or salami	1 protein + 1 fat	1 slice	40 g
Weiners, hot dog	1 protein + 1 fat	½ medium	25 g
Miscellaneous			
Blood pudding	1 protein + 1 fat	1 slice	25 g
Peanut butter	1 protein + 1 fat	15 mL (1 tbs)	15 g

Table B-8
Canadian Choice System: Fats and Oils

1 fat choice = 5 g fat, 190 kJ (45 kcal)

Food	Measure	Mass (Weight)	Food	Measure	Mass (Weight)
Avocado*	⅛	30 g	Nuts (continued):		
Bacon, side, crisp*	1 slice	5 g	Sesame seeds	15 mL (1 tbs)	10 g
Butter*	5 mL (1 tsp)	5 g	Sunflower seeds		
Cheese spread	15 mL (1 tbs)	15 g	Shelled	15 mL (1 tbs)	10 g
Coconut, fresh*	45 mL (3 tbs)	15 g	In shell	45 mL (3 tbs)	15 g
Coconut, dried*	15 mL (1 tbs)	10 g	Walnuts	4 halves	10 g
Cream, Half and half			Oil, cooking and salad	5 mL (1 tsp)	5 g
(cereal), 10%*	30 mL (2 tbs)	30 g	Olives, green	10	45 g
Light (coffee), 20%*	15 mL (1 tbs)	15 g	Ripe black	7	57 g
Whipping, 32 to 37%*	15 mL (1 tbs)	15 g	Pâté, liverwurst, meat spreads	15 mL (1 tbs)	15 g
Cream cheese*	15 mL (1 tbs)	15 g			
Gravy*	30 mL (2 tbs)	30 g	Salad dressing: blue, French, Italian, mayonnaise,	10 mL (2 tsp)	10 g
Lard*	5 mL (1 tsp)	5 g			
Margarine	5 mL (1 tsp)	5 g			
Nuts, shelled:			Thousand Island	5 mL (1 tsp)	5 g
Almonds	8	5 g	Salad dressing, low-calorie	30 mL (2 tbs)	30 g
Brazil nuts	2	10 g			
Cashews	5	10 g	Salt pork, raw or cooked*	5 mL (1 tsp)	5 g
Filberts, hazelnuts	5	10 g			
Macadamia	3	5 g	Sesame oil	5 mL (1 tsp)	5 g
Peanuts	10	10g	Sour cream		
Pecans	5 halves	5 g	12% milkfat	30 mL (2 tbs)	30 g
Pignolias, pine nuts	25 mL (5 tsp)	10 g	7% milkfat	60 mL (4 tbs)	60 g
Pistachios, shelled	20	10 g	Shortening*	5 mL (1 tsp)	
Pistachios, in shell	20	20 g			
Pumpkin and squash seeds	20 mL (4 tsp)	10 g			

*These items contain higher amounts of saturated fat.

B

Table B-9
Canadian Choice System: Extras

Extras have no more than 2.5 g carbohydrate, 60 kJ (14 kcal)

Vegetables 125 mL (½ c)

Artichokes
Asparagus
Bamboo shoots
Bean sprouts, mung or soya
Beans, string, green, or yellow
Bitter melon (balsam pear)
Bok choy
Broccoli
Brussels sprouts
Cabbage
Cauliflower
Celery
Chard
Cucumbers
Eggplant
Endive
Fiddleheads
Greens: beet, collard, dandelion, mustard, turnip, etc.
Kale
Kohlrabi
Leeks
Lettuce
Mushrooms
Okra
Onions, green or mature
Parsley
Peppers, green, yellow or red
Radishes
Rapini
Rhubarb
Sauerkraut
Shallots
Spinach
Sprouts: alfalfa, radish, etc.
Tomato wedges
Watercress
Zucchini

Free Foods (may be used without measuring)

Artificial sweetener, such as cyclamate or aspartame	Lime juice or lime wedges Marjoram, cinnamon, etc.
Baking powder, baking soda	Mineral water
Bouillon from cube, powder, or liquid	Mustard Parsley
Bouillon or clear broth	Pimentos
Chowchow, unsweetened	Salt, pepper, thyme
Coffee, clear	Soda water, club soda
Consommé	Soya sauce
Dulse	Sugar-free Crystal Drink
Flavorings and extracts	Sugar-free Jelly Powder
Garlic	Sugar-free soft drinks
Gelatin, unsweetened	Tea, clear
Ginger root	Vinegar
Herbal teas, unsweetened	Water
Horseradish, uncreamed	Worcestershire sauce
Lemon juice or lemon wedges	

Condiments

Food	**Measure**
Anchovies	2 fillets
Barbecue sauce	15 mL (1 tbs)
Bran, natural	30 mL (2 tbs)
Brewer's yeast	5 mL (1 tsp)
Carob powder	5 mL (1 tsp)
Catsup	5 mL (1 tsp)
Chili sauce	5 mL (1 tsp)
Cocoa powder	5 mL (1 tsp)
Cranberry sauce, unsweetened	15 mL (1 tbs)
Dietetic fruit spreads	5 mL (1 tsp)
Maraschino cherries	1
Nondairy coffee whitener	5 mL (1 tsp)
Nuts, chopped pieces	5 mL (1 tsp)
Pickles	
Unsweetened dill	2
Sour mixed	11
Sugar substitutes, granular	5 mL (1 tsp)
Whipped toppings	15 mL (1 tbs)

Table B-10
Canadian Choice System: Combined Food Choices

Food	Choices per Serving	Measure	Mass (Weight)
Angel food cake	½ starch + 2½ sugars	1/12 cake	50 g
Apple crisp	½ starch + 1½ fruits & vegetables + 1 sugar + 1–2 fats	125 mL (½ c)	
Applesauce, sweetened	1 fruits & vegetables + 1 sugar	125 mL (½ c)	
Beans and pork in tomato sauce	1 starch + ½ fruits & vegetables + ½ sugar + 1 protein	125 mL (½ c)	135 g
Beef burrito	2 starches + 3 proteins + 3 fats		110 g
Brownie	1 sugar + 1 fat	1	20 g
Cabbage rolls*	1 starch + 2 proteins	3	310 g
Caesar salad	2–4 fats	20 mL dressing (4 tsp)	
Cheesecake	½ starch + 2 sugars + ½ protein + 5 fats	1 piece	80 g
Chicken fingers	1 starch + 2 proteins + 2 fats	6 small	100 g
Chicken and snow pea Oriental	2 starches + ½ fruits & vegetables + 3 proteins + 1 fat	500 mL (2 c)	
Chili	1½ starches + ½ fruits & vegetables + 3½ protein	300 mL (1¼ c)	325 g
Chips			
Potato chips	1 starch + 2 fats	15 chips	30 g
Corn chips	1 starch + 2 fats	30 chips	30 g
Tortilla chips	1 starch + 1½ fats	13 chips	
Cheese twist	1 starch + 1½ fats	30 chips	30 g
Chocolate bar			
Aero®	2½ sugars + 2½ fats	bar	43 g
Smarties®	4½ sugars + 2 fats	package	60 g
Chocolate cake (without icing)	1 starch + 2 sugars + 3 fats	1/10 of a 8″ pan	
Chocolate devil's food cake (without icing)	2 starches + 2 sugars + 3 fats	1/12 of a 9″ pan	
Chocolate milk	2 milks 2% + 1 sugar	250 mL (1 c)	300 g
Clubhouse (triple-decker) sandwich	3 starches + 3 proteins + 4 fats		
Cookies			
Chocolate chip	½ starch + ½ sugar + 1½ fats	2	22 g
Oatmeal	1 starch + 1 sugar + 1 fat	2	40 g
Donut (chocolate glazed)	1 starch + 1½ sugars + 2 fats	1	65 g
Egg roll	1 starch + ½ protein + 1 fat		75 g
Four bean salad	1 starch + ½ protein + 1 fat	125 mL (½ c)	
French toast	1 starch + ½ protein + 2 fats	1 slice	65 g
Fruit in heavy syrup	1 fruits & vegetables + 1½ sugars	125 mL (½ c)	
Granola bar	½ starch + 1 sugar + 1–2 fats		30 g
Granola cereal	1 starch + 1 sugar + 2 fats	125 mL (½ c)	45 g
Hamburger	2 starches + 3 proteins + 2 fats	junior size	
Ice cream and cone, plain flavour			
Ice cream	½ milk + 2–3 sugars + 1–2 fats		100 g
Cone	½ sugar		4 g
Lasagna			
Regular cheese	1 starch + 1 fruits & vegetables + 3 proteins + 2 fats	3″ x 4″ piece	
Low-fat cheese	1 starch + 1 fruits & vegetables + 3 proteins	3″ x 4″ piece	
Legumes			
Dried beans (kidney, navy, pinto, fava, chick peas)	2 starches + 2 protein	250 mL (1 c)	175 g
Dried peas	2 starches + 2 protein	250 mL (1 c)	196 g
Lentils	2 starches + 2 protein	250 mL (1 c)	196 g

* If eaten with sauce, add ½ fruits & vegetables exchange.

B

Table B-10 (continued)

Canadian Choice System: Combined Food Choices

Food	Choices per serving	Measure	Mass (Weight)
Macaroni and cheese	2 starches + 2 proteins + 2 fats	250 mL (1 c)	210 g
Minestrone soup	1½ starches + ½ fruits & vegetables + ½ fat	250 mL (1 c)	
Muffin	1 starch + ½ sugar + 1 fat	1 small	45 g
Nuts (dry or roasted without any oil added).			
Almonds, dried sliced	½ protein + 2 fats	50 mL (¼ c)	22 g
Brazil nuts, dried unblanched	½ protein + 2½ fats	5 large	23 g
Cashew nuts, dry roasted	½ starch + ½ protein + 2 fats	50 mL (¼ c)	28 g
Filbert hazelnut, dry	½ protein + 3½ fats	50 mL (¼ c)	30 g
Macadamia nuts, dried	½ protein + 4 fats	50 mL (¼ c)	28 g
Peanuts, raw	1 protein + 2 fats	50 mL (¼ c)	30 g
Pecans, dry roasted	½ fruits & vegetables + 3 fats	50 mL (¼ c)	22 g
Pine nuts, pignolia dried	1 protein + 3 fats	50 mL (¼ c)	34 g
Pistachio nuts, dried	½ fruits & vegetables + ½ protein + 2½ fats	50 mL (¼ c)	27 g
Pumpkin seeds, roasted	2 proteins + 2½ fats	50 mL (¼ c)	47 g
Sesame seeds, whole dried	½ fruits & vegetables + ½ protein + 2½ fats	50 mL (¼ c)	30 g
Sunflower kernel, dried	½ protein + 1½ fats	50 mL (¼ c)	17 g
Walnuts, dried chopped	½ protein + 3 fats	50 mL (¼ c)	26 g
Perogies	2 starches + 1 protein + 1 fat	3	
Pie, fruit	1 starch + 1 fruits & vegetables + 2 sugars + 3 fats	1 piece	120 g
Pizza, cheese	1 starch + 1 protein + 1 fat	1 slice (⅛ of a 12")	50 g
Pork stir fry	½ to 1 fruits & vegetables + 3 proteins	200 mL (¾ c)	
Potato salad	1 starch + 1 fat	125 mL (½ c)	130 g
Potatoes, scalloped	2 starches + 1 milk + 1–2 fats	200 mL (¾ c)	210 g
Pudding, bread or rice	1 starch + 1 sugar + 1 fat	125 mL (½ c)	
Pudding, vanilla	1 milk + 2 sugars	125 mL (½ c)	
Raisin bran cereal	1 starch + ½ fruits & vegetables + ½ sugar	175 mL (⅔ c)	40 g
Rice krispie squares	½ starch + 1½ sugars + ½ fat	1 square	30 g
Shepherd's pie	2 starches + 1 fruits & vegetables + 3 proteins	325 mL (1⅓ c)	
Sherbet, orange	3 sugars + ½ fat	125 mL (½ c)	
Spaghetti and meat sauce	2 starches + 1 fruits & vegetables + 2 proteins + 3 fats	250 mL (1 c)	
Stew	2 starches + 2 fruits & vegetables + 3 proteins + ½ fat	200 mL (¾ c)	
Sundae	4 sugars + 3 fats	125 mL (½ c)	
Tuna casserole	1 starch + 2 proteins + ½ fat	125 mL (½ c)	
Yogurt, fruit bottom	1 fruits & vegetables + 1 milk + 1 sugar	125 mL (½ c)	125 g
Yogurt, frozen	1 milk + 1 sugar	125 mL (½ c)	125 g

Food Labels

Consumers can gather a lot of information from a nutrition label. Figure B-1 demonstrates the reading of a food label and Table B-11 defines terms.

Figure B-1

Example of a Food Label

Kellogg's
RaisinBran*

OUR COMMITMENT TO QUALITY

Kellogg's is committed to providing foods of outstanding quality and freshness. If this product in any way falls below the high standards you've come to expect from Kellogg's, please send your comments and both top flaps to:

Consumer Affairs
KELLOGG CANADA INC.
Etobicoke, Ontario M9W 5P2

IF IT DOESN'T SAY Kellogg's* ON THE BOX, IT'S NOT Kellogg's* IN THE BOX.
SI LE NOM Kellogg's* N'EST PAS SUR LA BOÎTE, CE N'EST PAS Kellogg's* DANS LA BOÎTE.

- **HIGH IN FIBRE**
- **LOW IN FAT**
- **PRESERVATIVE FREE**
- **SOURCE ÉLEVÉE DE FIBRES**
- **FAIBLE EN MATIÈRES GRASSES**
- **SANS AGENT DE CONSERVATION**

NUTRITION INFORMATION
APPORT NUTRITIONNEL

	Per 40 g serving cereal (175 mL, ¾ cup) / Par ration de 40 g de céréale (175 mL, ¾ tasse)	Per 40 g serving cereal with 125 mL Partly Skimmed Milk (2%) / Par ration de 40 g de céréale avec 125 mL de lait partiellement écrémé (2,0 %)	
ENERGY	130Cal 540kJ	195Cal 810kJ	ÉNERGIE
PROTEIN	3.0g	7.3g	PROTÉINES
FAT	0.4g	2.9g	MATIÈRES GRASSES
CARBOHYDRATE	32g	38g	GLUCIDES
SUGARS*	11g	18g	*SUCRES
STARCH	16g	16g	AMIDON
DIETARY FIBRE	4.6g	4.6g	FIBRES ALIMENTAIRES
SODIUM	235mg	300mg	SODIUM
POTASSIUM	240mg	440mg	POTASSIUM

% of Recommended Daily Intake
% de l'apport quotidien conseillé

VITAMIN A	0%	7%	VITAMINE A
VITAMIN D	0%	23%	VITAMINE D
VITAMIN B1	62%	66%	VITAMINE B1
VITAMIN B2	3%	16%	VITAMINE B2
NIACIN	13%	18%	NIACINE
VITAMIN B6	13%	16%	VITAMINE B6
FOLACIN	11%	14%	FOLACINE
VITAMIN B12	0%	25%	VITAMINE B12
PANTOTHENATE	9%	15%	PANTOTHÉNATE
CALCIUM	1%	15%	CALCIUM
PHOSPHORUS	12%	23%	PHOSPHORE
MAGNESIUM	20%	27%	MAGNÉSIUM
IRON	38%	39%	FER
ZINC	16%	22%	ZINC

*Approximately half of the sugars occur naturally in the raisins.
Environ la moitié des sucres se retrouvent à l'état naturel dans les fruits.

Canadian Diabetes Association Food Choice Values: 40 g (175 mL, ¾ cup) cereal. Système des choix d'aliments de l'Association canadienne du diabète : 40 g (175 mL, ¾ tasse) céréale = 1 [■] + ½ [⬮] + ½ [✱] choices/choix

INGREDIENTS / INGRÉDIENTS

WHOLE WHEAT, RAISINS (COATED WITH SUGAR, HYDROGENATED VEGETABLE OIL), WHEAT BRAN, SUGAR/GLUCOSE-FRUCTOSE, SALT, MALT (CORN FLOUR, MALTED BARLEY), VITAMINS (THIAMIN HYDROCHLORIDE, PYRIDOXINE HYDROCHLORIDE, FOLIC ACID, d-CALCIUM PANTOTHENATE), MINERALS (IRON, ZINC OXIDE).

BLÉ ENTIER, RAISINS SECS (ENROBÉS DE SUCRE, D'HUILE VÉGÉTALE HYDROGÉNÉE), SON DE BLÉ, SUCRE/GLUCOSE-FRUCTOSE, SEL, MALT (FARINE DE MAÏS, ORGE MALTÉ), VITAMINES (CHLORHYDRATE DE THIAMINE, CHLORHYDRATE DE PYRIDOXINE, ACIDE FOLIQUE, PANTOTHÉNATE DE d-CALCIUM), MINÉRAUX (FER, OXYDE DE ZINC).

Made by / Produit par
KELLOGG CANADA INC.
ETOBICOKE, ONTARIO
CANADA M9W 5P2
*Registered trademark of / *Marque déposée de
KELLOGG CANADA INC. © 1994
00094

WHAT YOU WILL FIND ON A LABEL:

Nutrition Claims

- in Canada, it is optional for a company to decide to use claims,
- when claims appear on a label, they must follow government laws

Nutrition Information

- gives detailed nutrition facts about the product, including serving size and core list
- does not have to appear by law on food products in Canada
- refers to the food as packaged, so if you add milk, eggs or other food, the nutritional content of the food you eat can be very different

Serving Size

- the amount of food for which the information is given
- check the serving size: the serving size on the label may not be the same as the serving size you would actually eat (for example, the serving size of cereal may be ¾ cup, much smaller than your regular serving

Core List

- the energy (in Calories and kilojoules), grams of protein, fat and carbohydrate for each serving
- some products break down fat into monounsaturates, polyunsaturates, saturates, and cholesterol (to find out what these mean, look at the Fats & Oils section)
- carbohydrates may include the amount of sugars, starch and fibre, or may list these items separately

Sodium and Potassium (in milligrams)

Vitamins and Minerals (as percent of your recommended daily intake)

Canadian Diabetes Association Food Choice Values and Symbols

- the Values and Symbols are tools to help you fit the food into your meal plan, they are not an endorsement by CDA
- it is up to the food company to decide if it wants its foods analyzed and assigned symbols
- when they are on a label, they have been assigned by a dietitian working for CDA, so you can be sure the information is correct

Ingredients

- must be found on all food labels by law
- ingredients are listed in decreasing order by weight, so what you see first is what you get the most of

B

Table B-11
Terms on Food Labels

Term	Definition
Energy	
kcalorie reduced	50% or fewer kcalories than the regular version.
light	term may be used to describe anything (for example, light in colour, texture, flavour, taste, or kcalories); read the label to find out what is "light" about the product.
low kcalorie	kcalorie-reduced and no more than 15 kcalories per serving.
Fat and Cholesterol	
low cholesterol	no more than 3 mg of cholesterol per 100 g of the food and low in saturated fat; *does not* always mean low in total fat.
low fat	no more than 3 g of fat per serving; *does not* always mean low in kcalories.
lower fat	at least 25% less fat than the comparison food; be aware that 80% fat-free still means the food is 20% fat.
Carbohydrates: Fibre and Sugar	
carbohydrate reduced	not more than 50% of the carbohydrate found in the regular version; *does not* always mean the product is lower in kcalories because other ingredients such as fat may have increased.
source of dietary fibre	a product that provides 2–4 g of fibre.
high source of dietary fibre	a product that provides 4–6 g of fibre.
very high source of fibre	a product that provides 6 g (or more) of fibre.
sugar free	low in carbohydrates and kcalories; can be used as an extra food in the exchange system.
unsweetened or no sugar added	no sugar was added to the product; sugar may be found naturally in the food (for example, fruit canned in its own juice).

Canada's *Food Guide*

Canada's *Food Guide to Healthy Eating,* shown in Figure B-2, gives detailed information for selecting foods to meet the nutritional needs of all Canadians four years of age and older. Like the U.S. Daily Food Guide, Canada's *Food Guide* also takes a total diet approach, rather than emphasizing a single food, meal, or day's meals and snacks.

Figure B-2
Canada's Food Guide to Healthy Eating

Healthy Canada

Health and Welfare Canada | Santé et Bien-être social Canada

CANADA'S
Food Guide
TO HEALTHY EATING

Enjoy a variety of foods from each group every day.

Choose lower-fat foods more often.

EVAP MILK
POWDER
SKIM MILK CHEESE
YOGOURT 2.2%
0.1%
2%
SKIM
TOFU

Grain Products
Choose whole grain and enriched products more often.

Vegetables & Fruit
Choose dark green and orange vegetables and orange fruit more often.

Milk Products
Choose lower-fat milk products more often.

Meat & Alternatives
Choose leaner meats, poultry and fish, as well as dried peas, beans and lentils more often.

CANADA'S
Food Guide
TO HEALTHY EATING
FOR PEOPLE FOUR YEARS AND OVER

Different People Need Different Amounts of Food

The amount of food you need every day from the 4 food groups and other foods depends on your age, body size, activity level, whether you are male or female and if you are pregnant or breast-feeding. That's why the Food Guide gives a lower and higher number of servings for each food group. For example, young children can choose the lower number of servings, while male teenagers can go to the higher number. Most other people can choose servings somewhere in between.

Grain Products
5–12
SERVINGS PER DAY

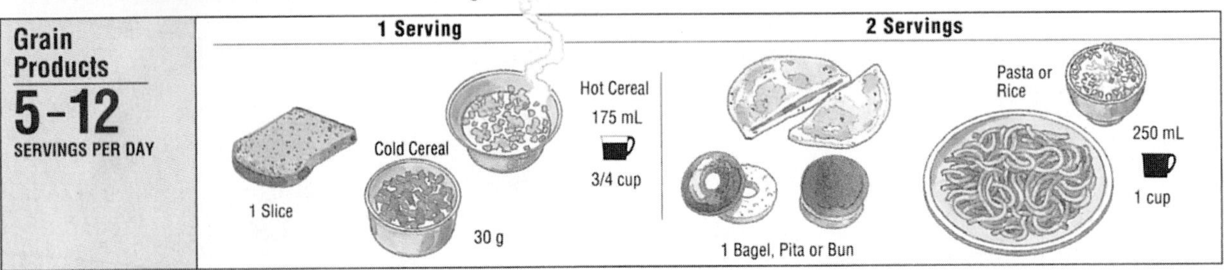
1 Serving
2 Servings
1 Slice
Cold Cereal
30 g
Hot Cereal
175 mL
3/4 cup
1 Bagel, Pita or Bun
Pasta or Rice
250 mL
1 cup

Vegetables & Fruit
5–10
SERVINGS PER DAY

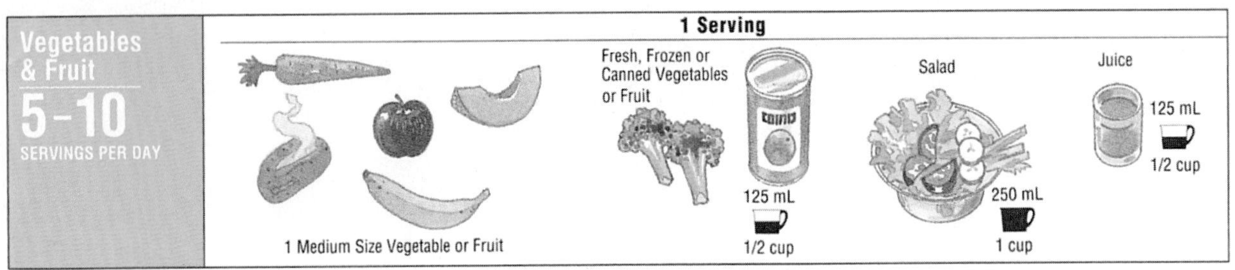
1 Serving
1 Medium Size Vegetable or Fruit
Fresh, Frozen or Canned Vegetables or Fruit
125 mL
1/2 cup
Salad
250 mL
1 cup
Juice
125 mL
1/2 cup

Milk Products
SERVINGS PER DAY
Children 4–9 years: 2–3
Youth 10–16 years: 3–4
Adults: 2–4
Pregnant & Breast-feeding Women: 3–4

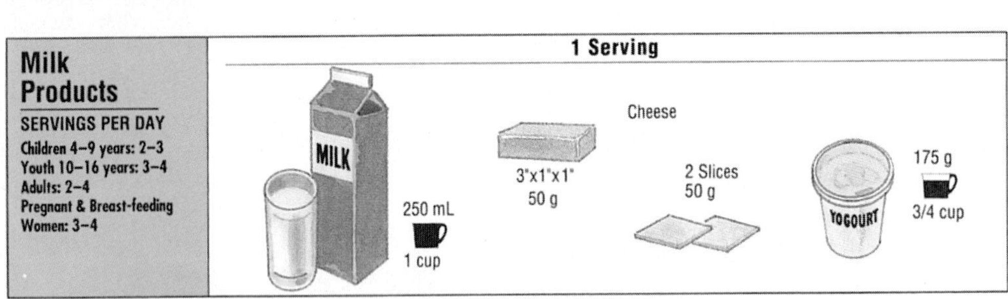
1 Serving
MILK
250 mL
1 cup
Cheese
3"x1"x1"
50 g
2 Slices
50 g
YOGOURT
175 g
3/4 cup

Meat & Alternatives
2–3
SERVINGS PER DAY

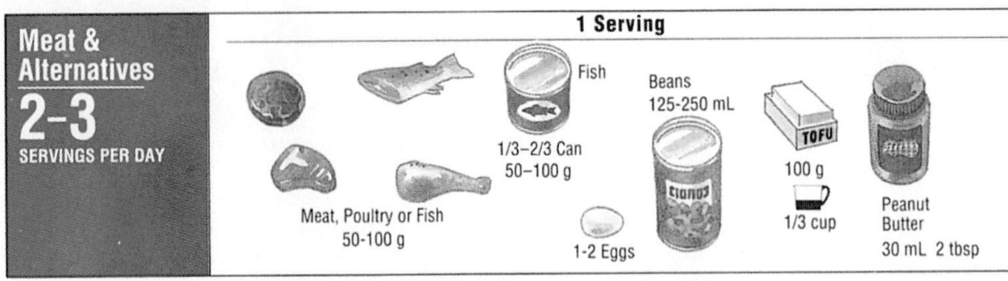
1 Serving
Meat, Poultry or Fish
50-100 g
Fish
1/3–2/3 Can
50–100 g
1-2 Eggs
Beans
125-250 mL
TOFU
100 g
1/3 cup
Peanut Butter
30 mL 2 tbsp

Other Foods

Taste and enjoyment can also come from other foods and beverages that are not part of the 4 food groups. Some of these foods are higher in fat or Calories, so use these foods in moderation.

Enjoy eating well, being active and feeling good about yourself. That's VITALIT*E*

© Minister of Supply and Services Canada 1992 Cat. No. H39-252/1992E No changes permitted. Reprint permission not required.

Aids to Calculation

CONTENTS

C

Mathematical problems have been worked out for you as examples at appropriate places in the text. This appendix aims to help with the use of the metric system and with those problems not fully explained elsewhere.

Conversion Factors

Conversion factors are useful mathematical tools in everyday calculations, like the ones encountered in the study of nutrition. A conversion factor is a fraction in which the numerator (top) and the denominator (bottom) express the same quantity in different units. For example, 2.2 pounds (lb) and 1 kilogram (kg) are equivalent; they express the same weight. The conversion factor used to change pounds to kilograms or vice versa is:

$$\frac{2.2 \text{ lb}}{1 \text{ kg}} \quad \text{or} \quad \frac{1 \text{ kg}}{2.2 \text{ lb}}$$

Because both factors equal 1, measurements can be multiplied by the factor without changing the value of the measurement. Thus the units can be changed.

The correct factor to use in a problem is the one with the unit you are seeking in the numerator (top) of the fraction. Following are two examples of problems commonly encountered in nutrition study; they illustrate the usefulness of conversion factors.

Example 1

Convert the weight of 130 pounds to kilograms.

1. Choose the conversion factor in which the unit you are seeking is on top:

$$\frac{1 \text{ kg}}{2.2 \text{ lb}}$$

2. Multiply 130 pounds by the factor:

$$130 \text{ lb} \times \frac{1 \text{ kg}}{2.2 \text{ lb}} = \frac{130 \text{ kg}}{2.2}$$

= 59 kg (rounded off to the nearest whole number)

Example 2

How many grams (g) of saturated fat are contained in a 3-ounce (oz) hamburger?

1. Appendix A shows that a 4-ounce hamburger contains 7 grams of saturated fat. You are seeking grams of saturated fat; therefore, the conversion factor is:

$$\frac{7 \text{ g saturated fat}}{4 \text{ oz hamburger}}$$

2. Multiply 3 ounces of hamburger by the conversion factor:

$$3 \text{ oz hamburger} \times \frac{7 \text{ g saturated fat}}{4 \text{ oz hamburger}} = \frac{3 \times 7}{4} = \frac{21}{4}$$

= 5 g saturated fat (rounded off to the nearest whole number

Energy Units

1 calorie[a] (cal) = 4.2 kilojoules
1 millijoule (MJ) = 240 cal
1 kilojoule (kJ) = 0.24 cal
1 gram (g) carbohydrate = 4 cal = 17 kJ
1 g fat = 9 cal = 37 kJ
1 g protein = 4 cal = 17 kJ
1 g alcohol = 7 cal = 29 kJ

Percentages

A percentage is a comparison between a number of items (perhaps your intake of energy) and a standard number (perhaps the number of calories recommended for your age and sex—your energy RDA). The standard number is the number you divide by. The answer you get after the division must be multiplied by 100 to be stated as a percentage (*percent* means "per 100").

Example 3

What percentage of the 1989 RDA for energy is your energy intake?

1. Find your energy RDA (see 1989 RDA, inside front cover). We'll use 2,100 calories to demonstrate.
2. Total your energy intake for a day—for example, 1,200 calories.
3. Divide your calorie intake by the RDA calories:

1,200 cal (your intake) ÷ 2,100 cal (RDA) = 0.571

4. Multiply your answer by 100 to state it as a percentage:

0.571 × 100 = 57.1 = 57% (rounded off to the nearest whole number)

In some problems in nutrition, the percentage may be more than 100. For example, suppose your daily intake of vitamin A is 3,200 RE and your RDA (male) is 1,000 RE. Your intake as a percentage of the RDA is more than 100 percent (that is, you consume more than 100 percent of your vitamin A RDA). The following calculations show your vitamin A intake as a percentage of the RDA:

3,200 ÷ 1,000 = 3.2
3.2 × 100 = 320% of RDA

[a]Note: Throughout this book and in the appendixes, the term *calorie* is used to mean kilocalorie. Thus, when converting calories to kilojoules, do not enlarge the calorie values—they are kilocalorie values.

C

Example 4

Food labels express nutrients and energy contents of foods as percentages of the Daily Values. If a serving of a food contains 200 milligrams of calcium, for example, what percentage of the calcium Daily Value does the food provide?

1. Find the calcium Daily Value on the inside front cover, page c.
2. Divide the milligrams of calcium in the food by the Daily Value standard:

$$\frac{200}{1{,}000} = 0.2$$

3. Multiply by 100:

$$0.2 \times 100 = 20\% \text{ of the Daily Value}$$

Example 5

This example demonstrates how to calculate the percentage of fat in a day's meals.

1. Recall the general formula for finding percentages of calories from a nutrient:

(one nutrient's calories ÷ total calories) × 100 = the percentage of calories from that nutrient

2. Say a day's meals provide 1,754 calories and 54 grams of fat. First, convert fat grams to fat calories:

54 g × 9 cal per g = 486 cal from fat

3. Then apply the general formula for finding percentage of calories from fat:

(fat calories ÷ total calories) × 100 = percentage of calories from fat

(486 ÷ 1,754) × 100 = 27. 7 (28%, rounded)

Ratios

A ratio is a comparison of two or three values in which one of the values is reduced to 1. A ratio compares identical units and so is expressed without units. For example, Table 8-6 in Chapter 8 compares the milligrams of potassium to the milligrams of sodium in selected foods.

Example 6

Find the potassium-to-sodium ratio of your diet.

1. Using Appendix A and a record of your food intake, find how many milligrams of potassium and sodium you consumed. For this exercise, we assume 3,000 milligrams potassium and 2,500 milligrams sodium.
2. Divide the potassium milligrams by the sodium milligrams:

3,000 mg potassium ÷ 2,500 mg sodium = 1.2

3. The potassium-to-sodium ratio is usually expressed as correct to one decimal point: 1.2.

The potassium-to-sodium ratio of this diet is 1.2:1 (read as "one point two to one" or simply "one point two"). A ratio greater than 1 means that the first value (in this case, milligrams of potassium) is greater than the second (sodium). When the second value is larger, the ratio is less than 1.

Weights and Measures

Length

1 inch (in) = 2.54 centimeters (cm)
1 foot (ft) = 30.48 cm
1 meter (m) = 39.37 in

Temperature

Steam = 100° Celsius[a] (C) or 212° Fahrenheit (F)
Body temperature = 37°C or 98.6°F
Ice = 0°C or 32°F

To convert Fahrenheit (t_F) to Celsius:

$$t_C = \frac{5}{9}(t_F - 32)$$

To convert Celsius (t_C) to Fahrenheit:

$$t_F = \frac{9}{5}(t_C + 32)$$

Volume

Used to measure fluids or pourable dry substances such as cereal.
1 milliliter (ml) = 1/5 teaspoon or 0.034 fluid ounce or 1/1000 liter
1 teaspoon (tsp or t) = 5 ml or about 5 grams (weight) salt
1 tablespoon (tbs or T) = 3 tsp or 15 ml
1 ounce, fluid (fl oz) = 2 tbs or 30 ml
1 cup (c) = 8 fl oz or 16 tbs or 250 ml
1 quart (qt) = 32 fl oz or 4 c or 0.95 liter
1 liter (1) = 1.06 qt or 1,000 ml
1 gallon (gal) = 16 c or 4 qt or 128 fl oz or 3.79 l

Weight

1 microgram (µg or mcg) = 1/1000 milligram
1 milligram (mg) = 1,000 µg or 1/1,000 gram
1 gram (g) = 1,000 mg or 1/1,000 kilogram
1 ounce, weight (oz) = about 28 g or 1/16 pound
1 pound (lb) = 16 oz (wt) or about 454 g
1 kilogram (kg) = 1,000 g or 2.2 lb

International Units (IU)

To convert IU to:

- RE:[b] from animal sources, divide by 3.33; and from vegetables and fruits, divide by 10.
- µg vitamin D: divide by 40 or multiply by 0.025.
- mg a-TE:[c] divide by 1.5.

Sodium

To convert milligrams of sodium to grams of salt:

mg sodium ÷ 400 = g of salt

The reverse is also true:

g salt × 400 = mg sodium

Folate

To convert micrograms of synthetic folate in supplements and enriched foods to Dietary Folate Equivalents (µg DFE):

µg synthetic folate × 1.7 = µg DFE

For naturally occurring folate, assign each microgram folate a value of 1 µg DFE:

µg folate = µg DFE

---

[a]Also known as centigrade.
[b]Retinol equivalents (vitamin A).
[c]Alpha-tocopherol equivalents (vitamin E).

U.S. Food Exchange System

CONTENTS

The U.S. food exchange system is intended to help people with diabetes control the levels of glucose and lipids in the blood by controlling the grams of carbohydrate and fat they consume. Other diet planners have found the system invaluable for achieving calorie control and moderation.

Planning a Diet

Unlike the Daily Food Guide of Chapter 2, which sorts foods primarily by their protein, vitamin, and mineral contents, the exchange system sorts foods into three main groups by their proportions of carbohydrate, fat, and protein. These three groups—the carbohydrate group, the fat group, and the meat and meat substitute group (protein)—are each subdivided into several exchange lists of foods (Table D-1, see p. D-2).

Portion Sizes

All of the food portions in a given list provide approximately the same amounts of energy nutrients (carbohydrate, fat, and protein) and the same number of calories. Portion sizes are strictly defined so that every item on a given list provides roughly the same amount of energy. Any food on a list can then be exchanged, or traded, for any other food on that same list without affecting a plan's balance or total calories.

To apply the system successfully, users must become familiar with portion sizes. A convenient way to remember the portion sizes and energy values is to keep in mind a typical item from each list. Table D-1 below includes some representative portion sizes; Figure D-1 shows the foods on each of the exchange lists and their accurate portion sizes.

The Foods on the Lists

Foods are not always on the exchange list where you might first expect them to be because they are grouped according to their energy-nutrient contents rather than by their source (such as milks), their outward appearance, or their vitamin and mineral contents. For example, cheeses are grouped with meats in the exchange system because, like meats, cheeses contribute energy from protein and fat but provide negligible carbohydrate. (In the food group plans presented earlier, cheeses are classed with milk because they are milk products with a similar calcium content.)

For similar reasons, starchy vegetables such as corn, green peas, and potatoes are listed on the starch list in the exchange system, rather than with

(text continued on page D–4)

TABLE D-1

Exchange Groups and Lists

List	Portion Size	Carbohydrate (g)	Protein (g)	Fat (g)	Energy (cal)
Carbohydrate Group					
Starch	1 slice; ½ c	15	3	1 or less	80
Fruit	varies	15	—	—	60
Milk	1 c				
Nonfat[a]		12	8	0–3	90
Low-fat		12	8	5	120
Whole		12	8	8	150
Other carbohydrates	varies	15	varies	varies	varies
Vegetable	½ c	5	2	—	25
Meat and Meat Substitute Group	1 oz				
Very lean		—	7	0–1	35
Lean		—	7	3	55
Medium-fat		—	7	5	75
High-fat		—	7	8	100
Fat Group	1 tsp pure fat	—	—	5	45

[a]Nonfat is the same as fat-free or skim.

FIGURE D-1

THE EXCHANGE SYSTEM: EXAMPLE FOODS, PORTION SIZES, AND ENERGY-NUTRIENT CONTRIBUTIONS

THE CARBOHYDRATE GROUP

Starch
1 starch exchange is like:
1 slice bread
¾ c ready-to-eat cereal
½ c cooked pasta
⅓ c cooked rice
½ c cooked beans[a]
½ c corn, peas, or yams
1 small (3 oz) potato
½ bagel, English muffin, or bun
1 tortilla, waffle, or roll
(1 starch = 15 g carbohydrate, 3 g protein, 0–1 g fat, and 80 cal)

Vegetables
1 vegetable exchange is like:
½ c cooked carrots, greens, green beans, brussels sprouts, beets, broccoli, cauliflower, or spinach
1 c raw carrots, radishes, or salad greens
1 lg tomato
(1 vegetable = 5 g carbohydrate, 2 g protein, and 25 cal)

Fruits
1 fruit exchange is like:
1 small banana, nectarine, apple, or orange
½ large grapefruit or pear
½ c orange, apple, or grapefruit juice
17 small grapes
⅓ cantaloupe (or 1 c cubes)
2 tbs raisins
(1 fruit = 15 g carbohydrate and 60 cal)

[a] ½ c cooked beans = 1 very lean meat exchange *plus* 1 starch exchange.

THE MEAT AND MEAT SUBSTITUTES GROUP (PROTEIN)

NONFAT CHEESE
EGG SUBSTITUTE
TUNA
Packed in Water

Meat and substitutes (very lean)
1 very lean meat exchange is like:
1 oz chicken (white meat, no skin)
1 oz cod, flounder, or trout
1 oz tuna (canned in water)
1 oz clams, crab, lobster, scallops, shrimp, or imitation seafood
1 oz fat-free cheese
½ c cooked beans, peas, or lentils
¼ c nonfat or low-fat cream cheese
2 egg whites (or ¼ c egg substitute)
(1 very lean meat = 7 g protein, 0–1 g fat, and 35 cal)

LOW FAT CHEESE
TUNA
Packed in Oil

Meats and substitutes (lean)
1 lean meat exchange is like:
1 oz beef or pork tenderloin
1 oz chicken (dark meat, no skin)
1 oz herring or salmon
1 oz tuna (canned in oil, drained)
1 oz low-fat cheese or luncheon meats
(1 lean meat = 7 g protein, 3 g fat, and 55 cal)

RICOTTA CHEESE
TOFU
RICOTTA CHEESE
TOFU

Meats and substitutes (medium-fat)
1 medium-fat meat exchange is like:
1 oz ground beef
1 oz pork chop
1 egg
¼ c ricotta
4 oz tofu
(1 medium-fat meat = 7 g protein, 5 g fat, and 75 cal)

D

Other carbohydrates
1 other carbohydrates exchange is like:
2 small cookies
1 small brownie or cake
5 vanilla wafers
1 granola bar
½ c ice cream
(1 other carbohydrate = 15 g carbohydrate and may be exchanged for 1 starch, 1 fruit, or 1 milk. Because many items on this list contain added sugar and fat, their fat and calorie values vary and their portion sizes are small.)

chunky
Peanut
Butter
CHEDDAR
CHEESE

Meats and substitutes (high-fat)
1 high-fat meat exchange is like:
1 oz pork sausage
1 oz luncheon meat (such as bologna)
1 oz regular cheese (such as cheddar or Swiss)
1 small hot dog (turkey or chicken)[b]
2 tbs peanut butter[c]
(1 high-fat meat = 7 g protein, 8 g fat, and 100 cal)

[b]A beef or pork hot dog counts as 1 high-fat meat exchange *plus* 1 fat exchange.
[c]Peanut butter counts as 1 high-fat meat exchange *plus* 1 fat exchange.

Milks (nonfat and very-low-fat)
1 nonfat milk exchange is like:
1 c nonfat milk
¾ c nonfat yogurt, plain
1 c nonfat or lowfat buttermilk
½ c evaporated nonfat milk
⅓ c dry nonfat milk
(1 nonfat milk = 12 g carbohydrate, 8 g protein, 0–3 g fat, and 90 cal)

DRY
NONFAT
MILK
LOW-FAT
BUTTERMILK
NONFAT
MILK
EVAPORATED
NONFAT
MILK
PLAIN
NONFAT
YOGURT

Milks (low-fat)
1 low-fat milk exchange is like:
1 c 2% milk
¾ c low-fat yogurt, plain
(1 low-fat milk = 12 g carbohydrate, 8 g protein, 5 g fat, and 120 cal)

2%
MILK
PLAIN
LOW-FAT
YOGURT

Milks (whole)
1 whole milk exchange is like:
1 c whole milk
½ c evaporated whole milk
(1 whole milk = 12 g carbohydrate, 8 g protein, 8 g fat, and 150 cal)

WHOLE
MILK
EVAPORATED
WHOLE
MILK

THE FAT GROUP

Fats
1 fat exchange is like:
1 tsp butter
1 tsp margarine or mayonnaise (1 tbs reduced fat)
1 tsp any oil
1 tbs salad dressing (2 tbs reduced fat)
8 large black olives
10 large peanuts
⅛ medium avocado
1 slice bacon
2 tbs shredded coconut
1 tbs cream cheese (2 tbs reduced fat)
(1 fat = 5 g fat and 45 cal)

CORN
OIL
Salad
Dressing
Tasty
Mayonnaise
MARGARINE
CREAM
CHEESE
BUTTER

D

the vegetables. Likewise, olives are not classed as a "fruit" as a botanist would claim; they are classified as a "fat" because their fat content makes them more similar to butter than to berries. Bacon is also on the fat list to remind users of its high fat content. These groupings permit you to see the characteristics of foods that are significant to energy intake.

Users of the exchange lists learn to view mixtures of foods, such as casseroles and soups, as combinations of foods from different exchange lists. They also learn to interpret food labels with the exchange system in mind. Knowing that foods on the starch list provide 15 grams of carbohydrate and those on the vegetable list provide 5, you can interpret the label of a lasagna dinner that lists 37 grams of carbohydrate as "2 starches" (mostly noodles) and "1 vegetable" (the sauce).

Controlling Energy, Fat, and Sodium

The exchange system helps people control their energy intakes by paying close attention to portion sizes. A portion of any food on a given list provides roughly the same amount of energy nutrients and total calories. The portion sizes have been adjusted so that all portions have the same energy value. For example, 17 grapes count as one fruit portion, as does 1⁄2 grapefruit. A whole grapefruit counts as two portions.

A *portion* in the exchange system is not the same as a *serving* in the Daily Food Guide, especially when it comes to meats. The exchange system lists meats and most cheeses in single ounces; that is, 1 *portion* (or *exchange*) of meat is 1 ounce, whereas one *serving* is 2 to 3 ounces. Calculating meat by the ounce encourages the planner to keep close track of the exact amounts eaten. This, in turn, helps control energy and fat intakes.

By allocating items like bacon and avocados to the fat list, the exchange system alerts consumers to foods that are unexpectedly high in fat. Even the starch list specifies which grain products contain added fat (such as biscuits, muffins, and waffles). In addition, the exchange system encourages users to think of nonfat milk as milk and of whole milk as milk with added fat, and to think of very lean meats as meats and of lean, medium-fat, and high-fat meats as meats with added fat. To that end, foods on the milk and meat lists are separated into categories based on their fat contents.

Control of food energy and fat intake can be highly successful with the exchange system. Exchange plans do not, however, guarantee adequate intakes of vitamins and minerals. Food group plans work better from that standpoint because the food groupings are based on similarities in vitamin-mineral content. In the exchange system, for example, meats are grouped with cheeses, yet the meats are iron-rich and calcium-poor, whereas the cheeses are iron-poor and calcium-rich. To take advantage of the strengths of both food group plans and exchange patterns, and to compensate for their weaknesses, diet planners often combine these two diet planning tools, as the following section shows.

People wishing to control the sodium in their diets can begin by eliminating any foods bearing this symbol [ ] found on any exchange list. The symbol identifies each food that, in one exchange, provides 400 milligrams or more of sodium. Other foods may also contribute substantially to sodium, however (consult Chapter 8 for details).

Combining Food Group Plans and Exchange Lists

A diet planner may find that using a food group plan together with the exchange lists eases the task of choosing foods that will provide all the nutrients. The food group plan ensures that all classes of nutritious foods are included, thus promoting adequacy, balance, and variety. The exchange system classifies the food selections by their energy-yielding nutrients, thus controlling energy and fat intakes.

Table D-2 shows how to use the Daily Food Guide plan together with the exchange lists to plan a diet. The Daily Food Guide ensures that a certain number of servings is chosen from each of the five food groups (see the first column of the table). The second column translates the number of servings (using the midpoint) into exchanges. With the addition of a small amount of fat, this sample diet

TABLE D-2

Diet Planning with the Exchange System Using the Daily Food Guide Pattern

Pattern from Daily Food Guide Plan	Selections Made Using the Exchange System	Energy Cost (cal)
Grains (breads and cereals)—6 to 11 servings	Starch list—select 9 exchanges	720
Vegetables—3 to 5 servings	Vegetable list—select 4 exchanges	100
Fruits—2 to 4 servings	Fruit list—select 3 exchanges	180
Meat—2 to 3 servings[a]	Meat list—select 6 lean exchanges	330
Milk—2 servings	Milk list—select 2 nonfat exchanges	180
	Fat list—select 5 exchanges	225
Total		1,735

[a]In the food group plan, 1 serving is 2 to 3 ounces; in the exchange system, 1 exchange is 1 ounce. The Daily Food Guide suggests that amounts should total 5 to 7 ounces of meat daily.

plan provides about 1,750 calories. Most people can meet their needs for all the nutrients within this reasonable energy allowance. (Table D-3 shows patterns for other energy intakes.) The next step in diet planning is to assign the exchanges to meals and snacks. The final plan might look like the one in Table D-4. To aid you in the development of your own diet plan, Tables D-5 through D-13 present the U.S. exchange system in detail.

Next, a person could begin to fill in the plan with real foods to create a menu. For example, the breakfast plan calls for 2 starches, 1 fruit, and 1 nonfat milk. A person might select a bowl of shredded wheat with banana slices and milk (1 cup shredded wheat = 2 starches, 1 small banana = 1 fruit, and 1 cup nonfat milk = 1 milk); or a bagel and a bowl of cantaloupe pieces topped with yogurt (1 bagel = 2 starches, ⅓ cantaloupe melon = 1 fruit, and ¾ cup nonfat plain yogurt = 1 milk). A person who wanted butter on the bagel could move a fat exchange or two from dinner to breakfast. If willing to use two fat exchanges at breakfast, the person could have pancakes with strawberries and milk (4 small pancakes = 2 starches plus 2 fats, 1¼ cup strawberries = 1 fruit, and a cup of nonfat milk = 1 milk). Then the person could move on to complete the menu for lunch, dinner and snacks.

U.S. Exchange Lists for Meal Planning[a]

[a]SOURCE: *Exchange Lists for Meal Planning* (Alexandria, Va.: American Diabetes Association and American Dietetic Association). For a copy of the 33-page booklet, call (800) 232-3472 or (800) 366-1655.

TABLE D-3

Diet Patterns for Different Energy Intakes

	ENERGY LEVEL (cal)						
EXCHANGE	1,200	1,500	1,800	2,000	2,200	2,600	3,000
Starch	6	7	8	9	11	13	15
Meat (lean)	4	5	6	6	6	7	8
Vegetable	3	4	5	5	5	6	6
Fruit	2	3	4	4	4	5	6
Milk (nonfat)	2	2	2	3	3	3	3
Fat	3	5	6	7	8	10	12

NOTE: These patterns follow the Daily Food Guide plan and supply less than 30 percent of calories as fat.

TABLE D-4

A Sample Diet Plan

Exchange	Breakfast	Lunch	Snack	Dinner	Evening Snack
9 starch	2	2	1	3	1
4 vegetable				4	
3 fruit	1	1	1		
6 lean meat		2		4	
2 nonfat milk	1	1			1
5 fat		1		4	

NOTE: This diet plan is one of many possibilities. It follows the number of servings suggested by the Daily Food Guide and meets dietary recommendations to provide 55 to 60 percent of its calories from carbohydrate, 15 to 20 percent from protein, and less than 30 percent from fat.

TABLE D-5

U.S. Exchange System: Starch List

1 starch exchange = 15 g carbohydrate, 3 g protein, 0–1 g fat, and 80 cal
Note: In general, a starch serving is ½ c cereal, grain, pasta, or starchy vegetable; 1 oz of bread; ¾ to 1 oz snack food.

Serving Size	Food
Bread	
½ (1 oz)	Bagels
2 slices (1½ oz)	Bread, reduced-calorie
1 slice (1 oz)	Bread, white (including French and Italian), whole-wheat, pumpernickel, rye
2 (⅔ oz)	Bread sticks, crisp, 4″ × ½″
½	English muffins
½ (1 oz)	Hot dog or hamburger buns
½	Pita, 6″ across
1 (1 oz)	Plain rolls, small
1 slice (1 oz)	Raisin bread, unfrosted
1	Tortillas, corn, 6″ across
1	Tortillas, flour, 7–8″ across
1	Waffles, 4½″ square, reduced-fat
Cereals and Grains	
½ c	Bran cereals
½ c	Bulgur, cooked
½ c	Cereals, cooked
¾ c	Cereals, unsweetened, ready-to-eat
3 tbs	Cornmeal (dry)
⅓ c	Couscous
3 tbs	Flour (dry)
¼ c	Granola, low-fat
¼ c	Grape nuts
½ c	Grits, cooked
½ c	Kasha
¼ c	Millet
¼ c	Muesli
½ c	Oats
½ c	Pasta, cooked
1½ c	Puffed cereals
½ c	Rice milk
⅓ c	Rice, white or brown, cooked
½ c	Shredded wheat
½ c	Sugar-frosted cereal
3 tbs	Wheat germ
Starchy Vegetables	
⅓ c	Baked beans
½ c	Corn
1 (5 oz)	Corn on cob, medium
1 c	Mixed vegetables with corn, peas, or pasta
½ c	Peas, green
½ c	Plantains
1 small (3 oz)	Potatoes, baked or boiled
½ c	Potatoes, mashed
1 c	Squash, winter (acorn, butternut)
½ c	Yams, sweet potatoes, plain
Crackers and Snacks	
8	Animal crackers
3	Graham crackers, 2½″ square
¾ oz	Matzoh
4 slices	Melba toast
24	Oyster crackers
3 c	Popcorn (popped, no fat added or low-fat microwave)
¾ oz	Pretzels
2	Rice cakes, 4″ across
6	Saltine-type crackers
15–20 (¾ oz)	Snack chips, fat-free (tortilla, potato)
2–5 (¾ oz)	Whole-wheat crackers, no fat added
Dried Beans, Peas, and Lentils	
½ c	Beans and peas, cooked (garbanzo, lentils, pinto, kidney, white, split, black-eyed)
⅔ c	Lima beans
3 tbs	Miso ✎
Starchy Foods Prepared with Fat **Count as 1 starch + 1 fat exchange.**	
1	Biscuit, 2½″ across
½ c	Chow mein noodles
1 (2 oz)	Corn bread, 2″ cube
6	Crackers, round butter type
1 c	Croutons
16–25 (3 oz)	French-fried potatoes
¼ c	Granola
1 (1½ oz)	Muffin, small
2	Pancake, 4″ across
3 c	Popcorn, microwave
3	Sandwich crackers, cheese or peanut butter filling
⅓ c	Stuffing, bread (prepared)
2	Taco shell, 6″ across
1	Waffle, 4½″ square
4–6 (1 oz)	Whole-wheat crackers, fat added

✎ = 400 mg or more of sodium per serving.

D

TABLE D-6

U.S. Exchange System: Fruit List

1 fruit exchange = 15 g carbohydrate and 60 cal

Note: In general, a fruit serving is 1 small to medium fresh fruit; ½ c canned or fresh fruit or fruit juice; ¼ c dried fruit. The weights given include skin, core, seeds, and rind.

Serving Size	Food	Serving Size	Food
1 (4 oz)	Apples, unpeeled, small	½ (8 oz) or 1 c cubes	Papayas
½ c	Applesauce, unsweetened	1 (6 oz)	Peaches, medium, fresh
4 rings	Apples, dried	½ c	Peaches, canned
4 whole (5½ oz)	Apricots, fresh	½ (4 oz)	Pears, large, fresh
8 halves	Apricots, dried	½ c	Pears, canned
½ c	Apricots, canned	¾ c	Pineapple, fresh
1 (4 oz)	Bananas, small	½ c	Pineapple, canned
¾ c	Blackberries	2 (5 oz)	Plums, small
¾ c	Blueberries	½ c	Plums, canned
⅓ melon (11 oz) or 1 c cubes	Cantaloupe, small	3	Prunes, dried
		2 tbs	Raisins
12 (3 oz)	Cherries, sweet, fresh	1 c	Raspberries
½ c	Cherries, sweet, canned	1¼ c whole berries	Strawberries
3	Dates	2 (8 oz)	Tangerines, small
1½ large or 2 medium (3½ oz)	Figs, fresh	1 slice (13½ oz) or 1¼ c cubes	Watermelon
1½	Figs, dried	**Fruit Juice**	
½ c	Fruit cocktail	½ c	Apple juice/cider
½ (11 oz)	Grapefruit, large	⅓ c	Cranberry juice cocktail
¾ c	Grapefruit sections, canned	1 c	Cranberry juice cocktail, reduced-calorie
17 (3 oz)	Grapes, small		
1 slice (10 oz) or 1 c cubes	Honeydew melon	⅓ c	Fruit juice blends, 100% juice
		⅓ c	Grape juice
1 (3½ oz)	Kiwi	½ c	Grapefruit juice
¾ c	Mandarin oranges, canned	½ c	Orange juice
½ (5½ oz) or ½ c	Mangoes, small	½ c	Pineapple juice
1 (5 oz)	Nectarines, small	⅓ c	Prune juice
1 (6½ oz)	Oranges, small		

TABLE D-7

U.S. Exchange System: Milk List

Serving Size	Food	Serving Size	Food
Nonfat and Very-Low-Fat Milk		**Low-Fat Milk**	
1 nonfat/low-fat milk exchange = 12 g carbohydrate, 8 g protein, 0–3 g fat, 90 cal		1 low-fat milk exchange = 12 g carbohydrate, 8 g protein, 5 g fat, 120 cal	
1 c	Nonfat milk	1 c	2% milk
1 c	½% milk	¾ c	Plain low-fat yogurt
1 c	1% milk	1 c	Sweet acidophilus milk
1 c	Nonfat or low-fat buttermilk	**Whole Milk**	
½ c	Evaporated nonfat milk	1 whole milk exchange = 12 g carbohydrate, 8 g protein, 8 g fat, 150 cal	
⅓ c dry	Dry nonfat milk		
¾ c	Plain nonfat yogurt	1 c	Whole milk
1 c	Nonfat or low-fat fruit-flavored yogurt sweetened with aspartame or with a nonnutritive sweetener	½ c	Evaporated whole milk
		1 c	Goat's milk
		1 c	Kefir

D

TABLE D-8

U.S. Exchange System: Other Carbohydrates List

1 other carbohydrate exchange = 15 g carbohydrate, or 1 starch, or 1 fruit, or 1 milk exchange

Food	Serving Size	Exchanges per Serving
Angel food cake, unfrosted	1/12 cake	2 carbohydrates
Brownies, small, unfrosted	2″ square	1 carbohydrate, 1 fat
Cake, unfrosted	2″ square	1 carbohydrate, 1 fat
Cake, frosted	2″ square	2 carbohydrates, 1 fat
Cookie, fat-free	2 small	1 carbohydrate
Cookies or sandwich cookies	2 small	1 carbohydrate, 1 fat
Cupcakes, frosted	1 small	2 carbohydrates, 1 fat
Cranberry sauce, jellied	¼ c	2 carbohydrates
Doughnuts, plain cake	1 medium, (1½ oz)	1½ carbohydrates, 2 fats
Doughnuts, glazed	3¾″ across (2 oz)	2 carbohydrates, 2 fats
Fruit juice bars, frozen, 100% juice	1 bar (3 oz)	1 carbohydrate
Fruit snacks, chewy (pureed fruit concentrate)	1 roll (¾ oz)	1 carbohydrate
Fruit spreads, 100% fruit	1 tbs	1 carbohydrate
Gelatin, regular	½ c	1 carbohydrate
Gingersnaps	3	1 carbohydrate
Granola bars	1 bar	1 carbohydrate, 1 fat
Granola bars, fat-free	1 bar	2 carbohydrates
Hummus	⅓ c	1 carbohydrate, 1 fat
Ice cream	½ c	1 carbohydrate, 2 fats
Ice cream, light	½ c	1 carbohydrate, 1 fat
Ice cream, fat-free, no sugar added	½ c	1 carbohydrate
Jam or jelly, regular	1 tbs	1 carbohydrate
Milk, chocolate, whole	1 c	2 carbohydrates, 1 fat
Pie, fruit, 2 crusts	⅙ pie	3 carbohydrates, 2 fats
Pie, pumpkin or custard	⅛ pie	1 carbohydrate, 2 fats
Potato chips	12–18 (1 oz)	1 carbohydrate, 2 fats
Pudding, regular (made with low-fat milk)	½ c	2 carbohydrates
Pudding, sugar-free (made with low-fat milk)	½ c	1 carbohydrate
Salad dressing, fat-free 🧂	¼ c	1 carbohydrate
Sherbet, sorbet	½ c	2 carbohydrates
Spaghetti or pasta sauce, canned 🧂	½ c	1 carbohydrate, 1 fat
Sweet roll or danish	1 (2½ oz)	2½ carbohydrates, 2 fats
Syrup, light	2 tbs	1 carbohydrate
Syrup, regular	1 tbs	1 carbohydrate
Syrup, regular	¼ c	4 carbohydrates
Tortilla chips	6–12 (1 oz)	1 carbohydrate, 2 fats
Vanilla wafers	5	1 carbohydrate, 1 fat
Yogurt, frozen, low-fat, fat-free	⅓ c	1 carbohydrate, 0–1 fat
Yogurt, frozen, fat-free, no sugar added	½ c	1 carbohydrate
Yogurt, low-fat with fruit	1 c	3 carbohydrates, 0–1 fat

🧂 = 400 mg or more sodium per exchange.

TABLE D-9

U.S. Exchange System: Vegetable List

1 vegetable exchange = 5 g carbohydrate, 2 g protein, 25 cal

Note: In general, a vegetable serving is ½ c cooked vegetables or vegetable juice; 1 c raw vegetables. Starchy vegetables such as corn, peas, and potatoes are on the starch list.

Artichokes
Artichoke hearts
Asparagus
Beans (green, wax, Italian)
Bean sprouts
Beets
Broccoli
Brussels sprouts
Cabbage
Carrots
Cauliflower
Celery
Cucumbers
Eggplant
Green onions or scallions
Greens (collard, kale, mustard, turnip)
Kohlrabi
Leeks
Mixed vegetables (without corn, peas, or pasta)
Mushrooms
Okra
Onions
Pea pods
Peppers (all varieties)
Radishes
Salad greens (endive, escarole, lettuce, romaine, spinach)
Sauerkraut
Spinach
Summer squash (crookneck)
Tomatoes
Tomaties, canned
Tomato sauce
Tomato/vegetable juice
Trunips
Water chestnuts
Watercress
Zucchini

= 400 mg or more sodium per exchange.

D

TABLE D-10

U.S. Exchange System: Meat and Meat Substitutes List

Note: In general, a meat serving is 1 oz meat, poultry, or cheese; ½ c dried beans (weigh meat and poultry and measure beans after cooking).

Serving Size	Food
Very Lean Meat and Substitutes	
1 very lean meat exchange = 7 g protein, 0–1 g fat, 35 cal	
1 oz	Poultry: Chicken or turkey (white meat, no skin), Cornish hen (no skin)
1 oz	Fish: Fresh or frozen cod, flounder, haddock, halibut, trout; tuna, fresh or canned in water
1 oz	Shellfish: Clams, crab, lobster, scallops, shrimp, imitation shellfish
1 oz	Game: Duck or pheasant (no skin), venison, buffalo, ostrich
	Cheese with ≤ 1 g fat/oz:
¼ c	Nonfat or low-fat cottage cheese
1 oz	Fat-free cheese
	Other:
1 oz	Processed sandwich meats with ≤ 1 g fat/oz (such as deli thin, shaved meats, chipped beef 🧂, turkey ham)
2	Egg whites
¼ c	Egg substitutes, plain
1 oz	Hot dogs with ≤ 1 g fat/oz
1 oz	Kidney (high in cholesterol)
1 oz	Sausage with ≤ 1 g fat/oz 🧂
Count as one very lean meat and one starch exchange:	
½c	Dried beans, peas, lentils (cooked)
Lean Meat and Substitutes	
1 lean meat exchange = 7 g protein, 3 g fat, 55 cal	
1 oz	Beef: USDA Select or Choice grades of lean beef trimmed of fat (round, sirloin, and flank steak); tenderloin; roast (rib, chuck, rump); steak (T-bone, porterhouse, cubed), ground round
1 oz	Pork: Lean pork (fresh ham); canned, cured, or boiled ham; Canadian bacon 🧂; tenderloin, center loin chop
1 oz	Lamb: Roast, chop, leg
1 oz	Veal: Lean chop, roast
1 oz	Poultry: Chicken, turkey (dark meat, no skin), chicken white meat (with skin), domestic duck or goose (well drained of fat, no skin)
	Fish:
1 oz	Herring (uncreamed or smoked)
6 medium	Oysters
1 oz	Salmon (fresh or canned), catfish
2 medium	Sardines (canned)
1 oz	Tuna (canned in oil, drained)
1 oz	Game: Goose (no skin), rabbit
	Cheese:
¼ c	4.5%-fat cottage cheese
2 tbs	Grated Parmesan
1 oz	Cheeses with ≤ 3 g fat/oz
	Other:
1½ oz	Hot dogs with ≤ 3 g fat/oz 🧂
1 oz	Processed sandwich meat with ≤ 3 g fat/oz (turkey pastrami or kielbasa)
1 oz	Liver, heart (high in cholesterol)
Medium-Fat Meat and Substitutes	
1 medium-fat meat exchange = 7 g protein, 5 g fat, and 75 cal	
1 oz	Beef: Most beef products (ground beef, meat loaf, corned beef, short ribs, Prime grades of meat trimmed of fat, such as prime rib)
1 oz	Pork: Top loin, chop, Boston butt, cutlet
1 oz	Lamb: Rib roast, ground
1 oz	Veal: Cutlet (ground or cubed, unbreaded)
1 oz	Poultry: Chicken dark meat (with skin), ground turkey or ground chicken, fried chicken (with skin)
1 oz	Fish: Any fried fish product
	Cheese with ≤ 5 g fat/oz:
1 oz	Feta
1 oz	Mozzarella
¼ c (2 oz)	Ricotta
	Other:
1	Egg (high in cholesterol, limit to 3/week)
1 oz	Sausage with ≤ 5 g fat/oz
1 c	Soy milk
¼ c	Tempeh
4 oz or ½ c	Tofu
High-Fat Meat and Substitutes	
1 high-fat meat exchange = 7 g protein, 8 g fat, 100 cal	
1 oz	Pork: Spareribs, ground pork, pork sausage
1 oz	Cheese: All regular cheeses (American 🧂, cheddar, Monterey Jack, Swiss)
	Other:
1 oz	Processed sandwich meats with ≤ 8 g fat/oz (bologna, pimento loaf, salami)
1 oz	Sausage (bratwurst, Italian, knockwurst, Polish, smoked)
1 (10/lb)	Hot dog (turkey or chicken) 🧂
3 slices (20 slices/lb)	Bacon
Count as one high-fat meat plus one fat exchange:	
1 (10/lb)	Hot dog (beef, pork, or combination) 🧂
2 tbs	Peanut butter (contains unsaturated fat)

🧂 = 400 mg or more sodium per exchange.

TABLE D-11

U.S. Exchange System: Fat List

1 fat exchange = 5 g fat, 45 cal
Note: In general, a fat serving is 1 tsp regular butter, margarine, or vegetable oil; 1 tbs regular salad dressing. Many fat-free and reduced fat foods are on the Free Foods List.

Serving Size	Food
Monounsaturated Fats	
⅛ medium (1 oz)	Avocados
1 tsp	Oil (canola, olive, peanut)
8 large	Olives, ripe (black)
10 large	Olives, green, stuffed 🖉
6 nuts	Almonds, cashews
6 nuts	Mixed nuts (50% peanuts)
10 nuts	Peanuts
4 halves	Pecans
2 tsp	Peanut butter, smooth or crunchy
1 tbs	Sesame seeds
2 tsp	Tahini paste
Polyunsaturated Fats	
1 tsp	Margarine, stick, tub, or squeeze
1 tbs	Margarine, lower-fat (30% to 50% vegetable oil)
1 tsp	Mayonnaise, regular
1 tbs	Mayonnaise, reduced-fat
4 halves	Nuts, walnuts, English
1 tsp	Oil (corn, safflower, soybean)
1 tbs	Salad dressing, regular
2 tbs	Salad dressing, reduced-fat
2 tsp	Mayonnaise-type salad dressing, regular 🖉
1 tbs	Mayonnaise-type salad dressing, reduced-fat
1 tbs	Seeds: pumpkin, sunflower
Saturated Fats[a]	
1 slice (20 slices/lb)	Bacon, cooked
1 tsp	Bacon, grease
1 tsp	Butter, stick
2 tsp	Butter, whipped
1 tbs	Butter, reduced-fat
2 tbs (½ oz)	Chitterlings, boiled
2 tbs	Coconut, sweetened, shredded
2 tbs	Cream, half and half
1 tbs (½ oz)	Cream cheese, regular
2 tbs (1 oz)	Cream cheese, reduced-fat
	Fatback or salt pork[b]
1 tsp	Shortening or lard
2 tbs	Sour cream, regular
3 tbs	Sour cream, reduced-fat

🖉 = 400 mg or more sodium per exchange
[a]Saturated fats can raise blood cholesterol levels.
[b]Use a piece 1″ × 1″ × ¼″ if you plan to eat the fatback cooked with vegetables. Use a piece 2″ × 1″ × ½″ when eating only the vegetables with the fatback removed.

TABLE D-12

U.S. Exchange System: Free Foods List

Note: A serving of free food contains fewer than 20 calories; those with serving sizes should be limited to three servings a day whereas those without serving sizes can be eaten freely.

Serving Size	Food
Fat-Free or Reduced-Fat Foods	
1 tbs	Cream cheese, fat-free
1 tbs	Creamers, nondairy, liquid
2 tsp	Creamers, nondairy, powdered
1 tbs	Mayonnaise, fat-free
1 tsp	Mayonnaise, reduced-fat
4 tbs	Margarine, fat-free
1 tsp	Margarine, reduced-fat
1 tbs	Mayonnaise type salad dressing, nonfat
1 tsp	Mayonnaise type salad dressing, reduced-fat
	Nonstick cooking spray
1 tbs	Salad dressing, fat-free
2 tbs	Salad dressing, fat-free, Italian
¼ c	Salsa
1 tbs	Sour cream, fat-free, reduced-fat
2 tbs	Whipped topping, regular or light
Sugar-Free or Low-Sugar Foods	
1 piece	Candy, hard, sugar-free
	Gelatin dessert, sugar-free
	Gelatin, unflavored
	Gum, sugar-free
2 tsp	Jam or jelly, low-sugar or light
	Sugar substitutes
2 tbs	Syrup, sugar-free
Drinks	
	Bouillon, broth, consommé 🖉
	Bouillon or broth, low-sodium
	Carbonated or mineral water
1 tbs	Cocoa powder, unsweetened
	Coffee
	Club soda
	Diet soft drinks, sugar-free
	Drink mixes, sugar-free
	Tea
	Tonic water, sugar-free
Condiments	
1 tbs	Catsup
	Horseradish
	Lemon juice
	Lime juice
	Mustard
1½ large	Pickles, dill 🖉
	Soy sauce, regular or light 🖉
1 tbs	Taco sauce
	Vinegar
Seasonings	
Flavoring extracts	Spices
Garlic	Hot pepper sauces
Herbs, fresh or dried	Wine, used in cooking
Pimento	Worcestershire sauce

🖉 = 400 mg or more sodium per exchange.

TABLE D-13

U.S. Exchange System: Combination Foods List

Food	Serving Size	Exchanges per Serving
Entrées		
Tuna noodle casserole, lasagna, spaghetti with meatballs, chili with beans, macaroni and cheese 🧂	1 c (8 oz)	2 carbohydrates, 2 medium-fat meats
Chow mein (without noodles or rice)	2 c (16 oz)	1 carbohydrate, 2 lean meats
Pizza, cheese, thin crust 🧂 (5 oz)	¼ of 10″	2 carbohydrates, 2 medium-fat meats, 1 fat
Pizza, meat topping, thin crust 🧂 (5 oz)	¼ of 10″	2 carbohydrates, 2 medium-fat meats, 2 fats
Potpie 🧂	1 (7 oz)	2 carbohydrates, 1 medium-fat meat, 4 fats
Frozen Entrées		
Salisbury steak with gravy, mashed potato	1 (11 oz)	2 carbohydrates, 3 medium-fat meats, 3–4 fats
Turkey with gravy, mashed potato, dressing 🧂	1 (11 oz)	2 carbohydrates, 2 medium-fat meats, 2 fats
Entrée with less than 300 calories 🧂	1 (8 oz)	2 carbohydrates, 3 lean meats
Soups		
Bean 🧂	1 c	1 carbohydrate, 1 very lean meat
Cream (made with water) 🧂	1 c (8 oz)	1 carbohydrate, 1 fat
Split pea (made with water) 🧂	½ c (4 oz)	1 carbohydrate
Tomato (made with water) 🧂	1 c (8 oz)	1 carbohydrate
Vegetable beef, chicken noodle, or other broth-type 🧂	1 c (8 oz)	1 carbohydrate
Fast Foods		
Burritos with beef 🧂	2	4 carbohydrates, 2 medium-fat meats, 2 fats
Chicken nuggets 🧂	6	1 carbohydrate, 2 medium-fat meats, 1 fat
Chicken breast and wing, breaded and fried 🧂	1	1 carbohydrate, 4 medium-fat meats, 2 fats
Fish sandwich/tartar sauce 🧂	1	3 carbohydrates, 1 medium-fat meat, 3 fats
French fries, thin	20–25	2 carbohydrates, 2 fats
Hamburger, regular	1	2 carbohydrates, 2 medium-fat meats
Hamburger, large 🧂	1	2 carbohydrates, 3 medium-fat meats, 1 fat
Hot dog with bun 🧂	1	1 carbohydrate, 1 high-fat meat, 1 fat
Individual pan pizza 🧂	1	5 carbohydrates, 3 medium-fat meats, 3 fats
Soft serve cone	1 medium	2 carbohydrates, 1 fat
Submarine sandwich 🧂	1 (6″)	3 carbohydrates, 1 vegetable, 2 medium-fat meats, 1 fat
Taco, hard shell 🧂	1 (6 oz)	2 carbohydrates, 2 medium-fat meats, 2 fats
Taco, soft shell 🧂	1 (3 oz)	1 carbohydrate, 1 medium-fat meat, 1 fat

🧂 = 400 mg or more sodium per exchange.

CONTENTS

E

Nutrition Resources

People interested in nutrition often want to know where they can find reliable nutrition information. Wherever you live, there are several sources you can turn to:

- The Department of Health may have a nutrition expert.
- The local extension agent is often an expert.
- The food editor of your local paper may be well informed.
- The dietitian at the local hospital had to fulfill a set of qualifications before he or she became an RD (see Controversy 1).
- There may be knowledgeable professors of nutrition or biochemistry at a nearby college or university.

In addition, you may be interested in building a nutrition library of your own. Books you can buy, journals you can subscribe to, and addresses you can contact for general information are given below.

Books

For students seeking to establish a personal library of nutrition references, the authors of this text recommend the following books:

- *Present Knowledge in Nutrition,* 7th ed. (Washington, D.C.: International Life Sciences Institute—Nutrition Foundation, 1996).

This 646-page paperback has a chapter on each of 64 topics, including energy, obesity, each of the nutrients, several diseases, malnutrition, growth and its assessment, immunity, alcohol, fiber, exercise, drugs, and toxins. Watch for an update; new editions come out every few years.

- M. E. Shils, J. A. Olson, and M. Shike, eds., *Modern Nutrition in Health and Disease,* 8th ed. (Philadelphia: Lea & Febiger, 1994).

This two-volume set is a major technical reference book on nutrition topics. It contains encyclopedic articles on the nutrients, foods, the diet, metabolism, malnutrition, age-related needs, and nutrition in disease.

- Committee on Dietary Reference Intakes, *Dietary Reference Intakes for Calcium, Phosphorus, Magnesium, Vitamin D, and Fluoride* (Washington, D.C.: National Academy Press, 1997).
- Committee on Dietary Reference Intakes, *Dietary Reference Intakes for Thiamin, Riboflavin, Niacin, Vitamin* B_6*, Folate, Vitamin* B_{12}*, Pantothenic Acid, Biotin, and Choline* (Washington, D.C.: National Academy Press, 1998).

These two reports review the function of each nutrient, dietary sources, and deficiency and toxicity symptoms as well as provide recommendations for intakes. Watch for additional reports on the Dietary Reference Intakes for the remaining nutrients. Until they are published, you may need the following:

- Committee on Dietary Allowances, *Recommended Dietary Allowances,* 10th ed. (Washington, D.C.: National Academy Press, 1989).

The Canadian equivalent is *Nutrition Recommendations,* available by mail from the Canadian Government Publishing Centre, Supply and Services Canada, Ottawa, Ontario K1A OS9, Canada.

- Committee on Diet and Health, *Diet and Health Implications for Reducing Chronic Disease Risk* (Washington, D.C.: National Academy Press, 1989).

This 749-page book presents the integral relationship between diet and chronic disease prevention. Its nutrient chapters provide evidence on how diet influences disease development, and its disease chapters review the dietary patterns implicated in each chronic disease.

- E. M. N. Hamilton and S. A. S. Gropper, *The Biochemistry of Human Nutrition: A Desk Reference* (St. Paul, Minn.: West, 1987).

This 324-page paperback presents the biochemical concepts necessary for an understanding of nutrition. It is a handy reference book for those who have forgotten the basics of biochemistry or for those who are learning biochemistry for the first time.

We also recommend three of our own books that explore current topics in nutrition, health, and the life span:

- S. R. Rolfes, L. K. DeBruyne, and E. N. Whitney, *Life Span Nutrition: Conception through Life* (Belmont, Calif.: West/Wadsworth, 1998).
- E. N. Whitney and S. R. Rolfes, *Understanding Nutrition* (Belmont, Calif.: West/Wadsworth, 1999).
- E. N. Whitney, C. B. Cataldo, L. K. DeBruyne, and S. R. Rolfes, *Nutrition for Health and Health Care* (St. Paul, Minn.: West, 1995).

Journals

Nutrition Today is an excellent magazine for the interested layperson. It makes a point of raising controversial issues and providing a forum for conflicting opinions. Six issues per year are published. Order from Williams and Wilkins, 351 West Camden Street, Baltimore, MD 21201-2436.

The *Journal of the American Dietetic Association,* the official publication of the ADA, contains articles of interest to dietitians and nutritionists, news of legislative action on food and nutrition, and a very useful section of abstracts of articles from many other journals of nutrition and related areas. There are 12 issues per year, available from the American Dietetic Association (see "Addresses," later).

Nutrition Reviews, a publication of the International Life Sciences Institute, does much of the work for the library researcher, compiling recent evidence on current topics and presenting extensive bibliographies. Twelve issues per year are available from Nutrition Reviews, P.O. Box 1897, Lawrence, KS 66044-8897.

Nutrition and the M.D. is a monthly newsletter that provides up-to-date, easy-to-read, practical information on nutrition for health care providers. It is available from Lippincott-Raven Publishers, 12107 Insurance Way, Hagerstown, MD 21740.

Other journals that deserve mention here are *Food Technology, Journal of Nutrition, American Journal of Clinical Nutrition, Nutrition Research,* and *Journal of Nutrition Education. FDA Consumer,* a government publication with many articles of interest to the consumer, is available from the Food and Drug Administration (see "Addresses," below). Many other journals of value are referred to throughout this book.

Addresses

Many of the organizations listed below will provide publication lists free on request. Government and international agencies and professional nutrition organizations are listed first, followed by organizations in the following areas: aging, alcohol and drug abuse, consumer organizations, fitness, food safety, health and disease, infancy and childhood, pregnancy and lactation, trade and industry organizations, weight control and eating disorders, and world hunger.

U.S. Government

- Federal Trade Commission (FTC)
  Public Reference Branch
  (202) 326-2222
  www.ftc.gov
- Food and Drug Administration (FDA)
  Office of Consumer Affairs, HFE 1
  Room 16-85
  5600 Fishers Lane
  Rockville, MD 20857
  (301) 443-1544
  www.fda.gov
- FDA Consumer Information Line
  (301) 827-4420
- FDA Office of Food Labeling, HFS 150
  200 C Street SW
  Washington, DC 20204
  (202) 205-4561; fax (202) 205-4564
  www.cfsan.fda.gov
- FDA Office of Plant and Dairy Foods and Beverages
  HFS 300
  200 C Street SW
  Washington, DC 20204
  (202) 205-4064; fax (202) 205-4422
- FDA Office of Special Nutritionals, HFS 450
  200 C Street SW
  Washington, DC 20204
  (202) 205-4168; fax (202) 205-5295
- Food and Nutrition Information Center
  National Agricultural Library, Room 304
  10301 Baltimore Avenue
  Beltsville, MD 20705-2351
  (301) 504-5719; fax (301) 504-6409
  www.nal.usda.gov/fnic
- Food Research Action Center (FRAC)
  1875 Connecticut Avenue NW, Suite 540
  Washington, DC 20009
  (202) 986-2200; fax (202) 986-2525
- Superintendent of Documents
  U.S. Government Printing Office
  Washington, DC 20402
  (202) 512-1071
  www.access.gpo.gov/su_docs
- U.S. Department of Agriculture (USDA)
  14th Street SW and Independence Avenue
  Washington, DC 20250
  (202) 720-2791
  www.usda.gov/fcs
- USDA Center for Nutrition Policy and Promotion
  1120 20th Street NW, Suite 200
  North Lobby
  Washington, DC 20036
  (202) 208-2417
  www.usda.gov/fcs/cnpp.htm
- USDA Food Safety and Inspection Service
  Food Safety Education Office,
  Room 1180-S
  Washington, DC 20250
  (202) 690-0351
  www.usda.gov/fsis
- U.S. Department of Education (DOE)
  Accreditation Agency Evaluation Branch
  7th and D Street SW
  ROB 3, Room 3915
  Washington, DC 20202-5244
  (202) 708-7417
- U.S. Department of Health and Human Services
  200 Independence Avenue SW
  Washington, DC 20201
  (202) 619-0257
  www.os.dhhs.gov
- U.S. Environmental Protection Agency (EPA)
  401 Main Street SW
  Washington, DC 20460
  (202) 260-2090
  www.epa.gov
- U.S. Public Health Service
  Assistant Secretary of Health
  Humphrey Building, Room 725-H
  200 Independence Avenue SW
  Washington, DC 20201
  (202) 690-7694

Canadian Government

Federal

- Bureau of Nutritional Sciences
  Food Directorate
  Health Protection Branch
  3-West
  Sir Frederick Banting Research Centre, 2203A
  Tunney's Pasture
  Ottawa, Ontario K1A 0L2
  www.hc-sc.gc.ca
- Canadian Food Inspection Agency
  Agriculture and Agri-Food Canada
  59 Camelot Drive
  Nepean, Ontario K1A 0Y9
  (613) 225-CFIA or (613) 225-2342
  www.agr.ca
- Nutrition & Healthy Eating Unit
  Strategies and Systems for Health Directorate, 1917C
  17th Floor—Jeanne Mance Bldg.
  Tunney's Pasture
  Ottawa, Ontario K1A 1B4
  www.hc-sc.gc.ca
- Nutrition Specialist
  Health Support Services
  Indian and Northern Health Services Directorate
  Medical Services Branch
  20th Floor—Jeanne Mance Bldg., 1920B
  Tunney's Pasture
  Ottawa, Ontario K1A 0L3
  www.hc-sc.gc.ca

Provincial and Territorial

- Population Health Strategies Branch
  Alberta Health
  23rd Floor, TELUS Plaza, North Tower
  10025 Jasper Avenue
  Edmonton, AB T5J 2N3
  www.health.gov.ab.ca
- Nutritionist
  Preventive Services Branch
  Ministry of Health
  1520 Blanshard Street
  Victoria BC V8W 3C8
  www.hlth.gov.bc.ca

- Executive Director
Health Programs
2nd Floor 800 Portage Avenue
Winnipeg, MB R3G 0P4
www.gov.mb.ca/health
- Project Manager
Public Health Management Services
Health and Community Services
P.O. Box 5100
520 King Street
Fredericton, NB E3B 5G8
www.gov.nb.ca/hcs-ssc
- Director, Health Promotion
Department of Health
Government of Newfoundland and Labrador
P.O. Box 8700
Confederation Building, West Block
St. John's, NF A1B 4J6
www.gov.nf.ca/health
- Consultant, Nutrition
Health & Wellness Promotion
Population Health
Department of Health and Social Services
Government of the Northwest Territories
Centre Square Tower, 6th Floor
P.O. Box 1320
Yellowknife, NT X1A 2L9
www.hlthss.gov.nt.ca
- Public Health Nutritionist
Central Health Region
201 Brownlow Avenue, Unit 4
Dartmouth, NS B3B 1W2
www.gov.ns.ca/health
- Senior Consultant, Nutrition
Public Health Branch
Ministry of Health, 8th Floor
5700 Yonge St.
North York, ON M2M 4K5
www.gov.on.ca/health
- Coordinator, Health Information Resource Centre
Department of Health and Social Services
1 Rochford Street, Box 2000
Charlottetown, PEI C1A 7N8
www.gov/pe.ca/infopei/health
- Responsables de la santé cardio-vasculaire et de la nutrition
Ministère de la Santé et des Services sociaux, Service de la Prévention en Santé
3e étage
1075, chemin Sainte-Foy
Québec (Québec) G1S 2M1
www.msss.gouv.qc.ca (French only)
- Health Promotion Unit
Population Health Branch
Saskatchewan Health
3475 Albert Street
Regina, SK S4S 6X6
www.gov.sk.ca/govt/health
- Director, Nutrition Services
Yukon Hospital Corporation
#5 Hospital Road
Whitehorse, YT Y1A 3H7
www.hss.gov.yk.ca

International Agencies

- Food and Agriculture Organization of the United Nations (FAO)
Liaison Office for North America
2175 K Street, Suite 300
Washington, DC 20437
(202) 653-2400
www.fao.org
- International Food Information Council Foundation
1100 Connecticut Avenue NW, Suite 430
Washington, DC 20036
(202) 296-6540
ificinfo.health.org
- UNICEF
3 United Nations Plaza
New York, NY 10017
(212) 326-7000
www.unicef.com
- World Health Organization (WHO)
Regional Office
525 23rd Street NW
Washington, DC 20037
(202) 974-3000
www.who.org

Professional Nutrition Organizations

- American Academy of Nutritional Sciences
9650 Rockville Pike
Bethesda, MD 20814
(303) 530-7050; fax (301) 571-1892
www.nutrition.org
- American Dietetic Association (ADA)
216 West Jackson Boulevard, Suite 800
Chicago, IL 60606-6995
(800) 877-1600; (312) 899-0040
www.eatright.org
- ADA, The Nutrition Hotline
(800) 366-1655
- American Society for Clinical Nutrition
9650 Rockville Pike
Bethesda, MD 20814-3998
(301) 530-7110; fax (301) 571-1863
www.faseb.org/ascn
- Dietitians of Canada
480 University Avenue, Suite 604
Toronto, Ontario M5G 1V2, Canada
(416) 596-0857; fax (416) 596-0603
www.dietitians.ca
- Human Nutrition Institute (INACG)
1126 Sixteenth Street NW
Washington, DC 20036
(202) 659-0789
www.ilsi.org
- National Academy of Sciences/National Research Council (NAS/NRC)
2101 Constitution Avenue, NW
Washington, DC 20418
(202) 334-2000
www.nas.edu
- National Institute of Nutrition
265 Carling Avenue, Suite 302
Ottawa, Ontario K1S 2E1
(613) 235-3355; fax (613) 235-7032
www.nin.ca
- Society for Nutrition Education
7101 Wisconsin Avenue, Suite 901
Bethesda, MD 20814-4805
(301) 656-4938

Aging

- Administration on Aging
330 Independence Avenue SW
Washington, DC 20201
(202) 619-0724
www.aoa.dhhs.gov
- American Association of Retired Persons (AARP)
601 E Street NW
Washington, DC 20049
(202) 434-2277
www.aarp.org
- Canadian Association of Gerantology
www.cagacg.ca
- National Aging Information Center
330 Independence Avenue SW
Washington, DC 20201
(202) 619-7501
www.aoa.dhhs.gov/naic
- National Institute on Aging
Public Information Office
31 Center Drive, MSC 2292
Bethesda, MD 20892
(301) 496-1752
www.nih.gov/nia

Alcohol and Drug Abuse

- Al-Anon Family Group Headquarters, Inc.
1600 Corporate Landing Parkway
Virginia Beach, VA 23454-5617
(800) 356-9996
www.al-anon.alateen.org

- Alateen
  1600 Corporate Landing Parkway
  Virginia Beach, VA 23454-5617
  (800) 356-9996
  www.al-anon.alateen.org
- Alcohol & Drug Abuse Information Line
  Adcare Hospital
  (800) 252-6465
- Alcoholics Anonymous (AA)
  General Service Office
  475 Riverside Drive
  New York, NY 10115
  (212) 870-3400
  www.aa.org
- Narcotics Anonymous (NA)
  P.O. Box 9999
  Van Nuys, CA 91409
  (818) 773-9999; fax (818) 700-0700
  www.wsoinc.com
- National Clearinghouse for Alcohol and Drug Information (NCADI)
  P.O. Box 2345
  Rockville, MD 20847-2345
  (800) 729-6686
  www.health.org
- National Council on Alcoholism and Drug Dependence (NCADD)
  12 West 21st Street
  New York, NY 10010
  (800) NCA-CALL or (800) 622-2255
  (212) 206-6770; fax (212) 645-1690
  www.ncadd.org
- U.S. Center for Substance Abuse Prevention
  1010 Wayne Avenue, Suite 850
  Silver Spring, MD 20910
  (301) 459-1591 ext. 244; fax (301) 495-2919
  www.covesoft.com/csap.html

Consumer Organizations

- Center for Science in the Public Interest (CSPI)
  1875 Connecticut Avenue NW,
  Suite 300
  Washington, DC 20009-5728
  (202) 332-9110; fax (202) 265-4954
  www.cspinet.org
- Choice in Dying, Inc.
  1035 30th Street NW
  Washington, DC 20007
  (202) 338-9790; fax (202) 338-0242
  www.choices.org
- Consumer Information Center
  Pueblo, CO 81009
  (888) 8 PUEBLO or (888) 878-3256
  www.pueblo.gsa.gov
- Consumers Union of US Inc.
  101 Truman Avenue
  Yonkers, NY 10703-1057
  (914) 378-2000
  www.consunion.org
- National Council Against Health Fraud, Inc. (NCAHF)
  P.O. Box 1276
  Loma Linda, CA 92354
  (909) 824-4690
  www.ncahf.org

Fitness

- American College of Sports Medicine
  P.O. Box 1440
  Indianapolis, IN 46206-1440
  (317) 637-9200
  www.acsm.org/sportsmed
- American Council on Exercise (ACE)
  5820 Oberlin Drive, Suite 102
  San Diego, CA 92121
  (800) 529-8227
  www.acefitness.org
- President's Council on Physical Fitness and Sports
  Humphrey Building, Room 738
  200 Independence Avenue SW
  Washington, DC 20201
  (202) 690-9000; fax (202) 690-5211
  www.indiana.edu/~preschal
- Shape Up America!
  6707 Democracy Boulevard, Suite 306
  Bethesda, MD 20817
  (301) 493-5368
  www.shapeup.org

Food Safety

- Alliance for Food & Fiber
  Food Safety Hotline
  (800) 266-0200
- Canadian Food Inspection Agency
  Agriculture and Agri-Food Canada
  59 Camelot Drive
  Nepean, Ontario K1A 0Y9
  (613) 225-CFIA or (613) 225-2342
  www.agr.ca
- FDA Center for Food Safety and Applied Nutrition
  200 C Street SW
  Washington, DC 20204
  (800) FDA-4010 or (800) 332-4010
  vm.cfsan.fda.gov
- National Lead Information Center
  (800) LEAD-FYI or (800) 532-3394
  (800) 424-LEAD or (800) 424-5323
- National Pesticide Telecommunications Network (NPTN)
  Oregon State University
  333 Weniger Hall
  Corvallis, OR 97331-6502
  (541) 737-6091
  www.ace.orst.edu/info/nptn
- USDA Meat and Poultry Hotline
  (800) 535-4555
- U.S. EPA Safe Drinking Water Hotline
  (800) 426-4791

Health and Disease

- Alzheimer's Disease Education and Referral Center
  P. O. Box 8250
  Silver Spring, MD 20907-8250
  (800) 438-4380
  www.alzheimers.org
- Alzheimer's Disease Information and Referral Service
  919 North Michigan Avenue, Suite 1000
  Chicago, IL 60611
  (800) 272-3900
  www.alz.org
- Alzheimer Society of Canada
  www.alzheimer.ca
- American Academy of Allergy, Asthma, and Immunology
  611 East Wells Street
  Milwaukee, WI 53202
  (414) 272-6071; fax (414) 276-3349
  www.aaaai.org
- American Cancer Society
  National Home Office
  1599 Clifton Road NE
  Atlanta, GA 30329-4251
  (800) ACS-2345 or (800) 227-2345
  www.cancer.org
- American Council on Science and Health
  1995 Broadway, 2nd Floor
  New York, NY 10023-5860
  (212) 362-7044; fax (212) 362-4919
  www.acsh.org
- American Dental Association
  211 East Chicago Avenue
  Chicago, IL 60611
  (312) 440-2800
  www.ada.org
- American Diabetes Association
  1660 Duke Street
  Alexandria, VA 22314
  (800) 232-3472 or (703) 549-1500
  www.diabetes.org

- American Heart Association
  Box BHG, National Center
  7320 Greenville Avenue
  Dallas, TX 75231
  (800) 275-0448 or (214) 373-6300
  www.amhrt.org
- American Institute for Cancer Research
  1759 R Street NW
  Washington, DC 20009
  (800) 843-8114 or (202) 328-7744; fax (202) 328-7226
  www.aicr.org
- American Medical Association
  515 North State Street
  Chicago, IL 60610
  (312) 464-5000
  www.ama-assn.org
- American Public Health Association (APHA)
  1015 Fifteenth Street NW, Suite 300
  Washington, DC 20005
  (202) 789-5600
  www.apha.org
- American Red Cross
  National Headquarters
  8111 Gatehouse Road
  Falls Church, VA 22042
  (703) 206-7180
  www.redcross.org
- Canadian Cancer Society
  www.cancer.ca
- Canadian Diabetes Association
  15 Toronto Street, Suite 800
  Toronto, ON M5C 2E3
  (800) BANTING or (800) 226-8464
  (416) 363-3373
  www.diabetes.ca
- Canadian Heart and Stroke Foundation
  www.hsf.ca
- Canadian Public Health Association
  400-1565 Carling Avenue
  Ottawa, Ontario K1Z 8R1
  (613) 725-3769; fax (613) 725-9826
  www.cpha.ca
- Centers for Disease Control and Prevention (CDC)
  1600 Clifton Road NE
  Atlanta, GA 30333
  (404) 639-3311
  www.cdc.gov
- The Food Allergy Network
  10400 Eaton Place, Suite 107
  Fairfax, VA 22030-2208
  (800) 929-4040 or (703) 691-3179
  www.foodallergy.org
- Internet Health Resources
  www.ihr.com
- National AIDS Hotline (CDC)
  (800) 342-AIDS (English)
  (800) 344-SIDA (Spanish)
  (800) 2437-TTY (Deaf)
  (900) 820-2437
- National Cancer Institute
  Office of Cancer Communications
  Building 31, Room 10824
  Bethesda, MD 20892
  (800) 4-CANCER or (800) 422-6237
  www.nci.nih.gov
- National Diabetes Information Clearinghouse
  1 Information Way
  Bethesda, MD 20892-3560
  (301) 654-3327
  www.niddk.nih.gov
- National Digestive Disease Information Clearinghouse (NDDIC)
  2 Information Way
  Bethesda, MD 20892-3570
  (301) 654-3810
  www.niddk.nih.gov
- National Health Information Center (NHIC)
  Office of Disease Prevention and Health Promotion
  (800) 336-4797
  nhic-nt.health.org
- National Heart, Lung, and Blood Institute Information Center
  P.O. Box 30105
  Bethesda, MD 20824-0105
  (301) 251-1222
  www.nhlbi.nih.gov/nhlbi/nhlbi.htm
- National Institute of Allergy and Infectious Diseases
  Office of Communications
  Building 31, Room 7A50
  31 Center Drive, MSC2520
  Bethesda, MD 20892-2520
  (301) 496-5717
  www.niaid.nih.gov
- National Institute of Dental Research (NIDR)
  National Institute of Health
  Bethesda, MD 20892-2190
  (301) 496-4261
  www.nidr.nih.gov
- National Institutes of Health (NIH)
  9000 Rockville Pike
  Bethesda, MD 20892
  (301) 496-2433
  www.nih.gov
- National Osteoporosis Foundation
  1150 17th Street NW, Suite 500
  Washington, DC 20036
  (202) 223-2226
  www.nof.org
- Office of Disease Prevention and Health Promotion
  odphp.osophs.dhhs.gov
- Office on Smoking and Health (OSH)
  www.americanheart.org/heart.org/Heart_and_stroke_A_Z_Guide/osh.html

Infancy and Childhood

- American Academy of Pediatrics
  141 Northwest Point Boulevard
  Elk Grove Village, IL 60007-1098
  (847) 228-5005
  www.aap.org
- Association of Birth Defect Children, Inc.
  930 Woodcock Road, Suite 225
  Orlando, FL 32803
  (407) 245-7035
  www.birthdefects.org
- Canadian Paediatric Society
  100-2204 Walkley Road
  Ottawa, ON K1G 4G8
  (613) 526-9397; fax (613) 526-3332
  www.cps.ca
- National Center for Education in Maternal & Child Health
  2000 15th Street North, Suite 701
  Arlington, VA 22201-2617
  (703) 524-7802
  www.ncemch.org

Pregnancy and Lactation

- American College of Obstetricians and Gynecologists Resource Center
  409 12th Street SW
  Washington, DC 20024-2188
  (202) 638-5577
  www.acog.org
- La Leche International, Inc.
  1400 N. Meacham Road
  Schaumburg, IL 60173
  (847) 519-7730
  www.lalecheleague.org
- March of Dimes Birth Defects Foundation
  1275 Mamaroneck Avenue
  White Plains, NY 10605
  (914) 428-7100
  www.modimes.org

Trade and Industry Organizations

- Beech-Nut Nutrition Corporation
  P.O. 618
  St. Louis, MO 63188-0618
  (800) 523-6633
  www.beechnut.com

- Borden Inc.
180 East Broad Street
Columbus, OH 43215
(800) 426-7336
- Campbell Soup Company
Consumer Response Center
Campbell Place, Box 26B
Camden, NJ 08103-1701
(800) 257-8443
www.campbellssoup.com
- Elan Pharma/Hi Chem Diagnostics
2 Thurber Boulevard
Smithfield, RI 02917
(401) 233-3526
www.hi-chem.com
- General Mills, Inc.
Number One General Mills Boulevard
Minneapolis, MN 55426
(800) 328-6787
www.generalmills.com
- Hoffmann-LaRoche, Inc.
340 Kingsland Street
Nutley, NJ 07110
(973) 235-5000
- Kellogg Company
P.O. Box 3599
Battle Creek, MI 49016-3599
(616) 961-2000
www.kelloggs.com
- Kraft Foods
Consumer Response and Information Center
One Kraft Court
Glenview, IL 60025
(800) 323-0768
www.kraftfoods.com
- Mead Johnson Nutritionals
2400 West Lloyd Expressway
Evansville, IN 47721
(800) 247-7893
www.meadjohnson.com
- Nabisco Consumer Affairs
100 DeForest Avenue
East Hanover, NJ 07936
(800) NABISCO or (800) 932-7800
www.nabisco.com
- National Dairy Council
10255 West Higgins Road, Suite 900
Rosemond, IL 60018-5616
(847) 803-2000
www.dairyinfo.com
- NutraSweet/KELCO
P.O. Box 2986
Chicago, IL 60654-0986
www.equal.com
- Pillsbury Company
Consumer Relations
P.O. Box 550
Minneapolis, MN 55440
(800) 767-4466
www.pillsbury.com
- Procter and Gamble Company
One Procter and Gamble Plaza
Cincinnati, OH 45202
(513) 983-1100
www.pg.com/info
- Ross Laboratories, Abbot Laboratory
625 Cleveland Avenue
Columbus, OH 43215
(800) 227-5767
www.abbot.com
- Sherwood Medical
1915 Olive Street
St. Louis, MO 63103
(800) 428-4400
- Sunkist Growers
Consumer Affairs
Fresh Fruit Division
14130 Riverside Drive
Sherman Oaks, CA 91423
(800) CITRUS-5 or (800) 248-7875
www.sunkist.com
- United Fresh Fruit and Vegetable Association
727 North Washington Street
Alexandria, VA 22314
(703) 836-3410
- USA Rice Federation
4301 North Fairfax Drive, Suite 305
Arlington, VA 22203
Phone: (703) 351-8161
www.usarice.com
- Weight Watchers International, Inc.
Consumer Affairs Department/IN
175 Crossways Park West
Woodbury, NY 11797
(516) 390-1400; fax (516) 390-1632.
www.weightwatchers.com

Weight Control and Eating Disorders

- American Anorexia & Bulimia Association, Inc.
165 West 46th Street #1108
New York, NY 10036
(212) 575-6200
members.aol.com/amanbu
- Anorexia Nervosa and Related Eating Disorders (ANRED)
P.O. Box 5102
Eugene, OR 97405
(541) 344-1144
www.anred.com
- National Association of Anorexia Nervosa and Associated Disorders, Inc. (ANAD)
P.O. Box 7
Highland Park, IL 60035
(847) 831-3438
members.aol.com/anad20/index.html
- National Eating Disorder Information Centre
200 Elizabeth Street, College Wing 1-304
Toronto, Ontario M5G 2C4
(519) 253-7421; fax (519) 253-7545
- Overeaters Anonymous (OA)
World Service Office
6075 Zenith Court NE
Rio Rancho, NM 87124
(505) 891-2664; fax (505) 891-4320
www.overeatersanonymous.org
- TOPS (Take Off Pounds Sensibly)
4575 South Fifth Street
P.O. Box 07360
Milwaukee, WI 53207-0360
(800) 932-8677 or (414) 482-4620
www.tops.org

World Hunger

- Bread for the World
1100 Wayne Avenue, Suite 1000
Silver Spring, MD 20910
(301) 608-2400
www.bread.org
- Center on Hunger, Poverty and Nutrition Policy
Tufts University School of Nutrition
11 Curtis Avenue
Medford, MA 02155
(617) 627-3956
- Freedom from Hunger
P.O. Box 2000
1644 DaVinci Court
Davis, CA 95617
(530) 758-6200
www.freefromhunger.org
- Oxfam America
26 West Street
Boston, MA 02111
(617) 482-1211
www.oxfamamerica.org
- SEEDS Magazine
P.O. Box 6170
Waco, TX 76706
(254) 755-7745
www.helwys.com/seedhome.htm
- Worldwatch Institute
1776 Massachusetts Avenue NW, Suite 800
Washington, DC 20036
(202) 452-1999
www.worldwatch.org

Notes

F

Chapter 1

1. What everyone should know about media: Five core concepts of media literacy, 1998, a teaching resource available from Center for Media Literacy, 4727 Wilshire Blvd, #403, Los Angeles, CA 90010 *(www.medialit.org)*.

2. National Center for Health Statistics, *Healthy People 2000 Review: Health United States 1992* (Hyattsville, Md.: Public Health Service, 1993), p. 252.

3. N. S. Scrimshaw, Nutrition and health from womb to tomb, *Nutrition Today,* March/April 1996, pp. 55–67.

4. Y. Okada and coauthors, Small volumes of enteral feedings normalise immune function in infants receiving parenteral nutrition, *Journal of Pediatric Surgery* 33 (1998): 16–19.

5. A. S. Levine and C. J. Billington, Why do we eat? A neural systems approach, *Annual Review of Nutrition* 17 (1997): 597–619; R. D. Mattes, Physiologic responses to sensory stimulation by food: Nutritional implications, *Journal of the American Dietetic Association* 97 (1997): 406–410, 413.

6. W. W. Souba, Nutritional support, *New England Journal of Medicine* 336 (1997): 41–48.

7. V. Woolf, *A Room of One's Own* (New York: Harcourt Brace, 1929), p. 30.

8. J. Avorn and coauthors, Reduction of bacteriuria and pyuria after ingestion of cranberry juice, *Journal of the American Medical Association* 271 (1994): 751–754.

9. D. Woznicki and H. Nguyen, We are what we read: Shaping the nation's nutrition IQ through the media, *Priorities* 7 (1995): 20–23.

10. J. P. Goldberg, Nutrition and health communication: The message and the media over half a century, *Nutrition Reviews* 50 (1992): 71–77.

11. *Teaching Tolerance,* Southern Poverty Law Center, 400 Washington Avenue, Montgomery, AL 36104.

12. M. Nestle and coauthors, Behavioral and social influences on food choices, *Nutrition Reviews* 56 (1998): 550–574.

13. U.S. Department of Agriculture, *Dietary Guidelines for Americans,* Garden Bulletin No. 232 (Washington, D.C.: Government Printing Office, 1995), p. 8.

14. J. Cousminer and G. Hartman, Understanding America's regional taste preferences, *Food Technology* 50 (1996): 73–77.

15. A. Drewnowski, Taste preferences and food intake, *Annual Review of Nutrition* 17 (1997): 237–253.

16. S. Haberman and D. Luffey, Weighing in college students' diet and exercise behaviors, *Journal of American College Health* 46 (1998): 189–191.

17. *Healthy People 2000 Review, 1998–1999* available from www.cdc.gov/nchswww/releases/99news/99news/99hp2000.htm.

18. Belsville Human Nutrition Research Center, Pyramid servings data: Results from USDA's 1994 Continuing Survey of Food Intakes by Individuals, as cited in Research and evaluation activities in USDA, *Family Economics and Nutrition Review* 10 (1997): 66–67.

19. F. M. Clydesdale, A proposal for the establishment of scientific criteria for health claims for functional foods, *Nutrition Reviews* 55 (1997): 413–422.

20. C. W. Enns, J. D. Goldman, and A. Cook, Trends in food and nutrient intakes by adults: NFCS 1977–78, CSFII 1989–91, and CSFII 1994–95, *Family Economics and Nutrition Review* 10 (1997): 2–19.

21. S. M. Krebs-Smith and coauthors, Characterizing food intake patterns of American adults, *American Journal of Clinical Nutrition* 65 (1997): 1264S–1268S.

22. Convenient food and the new household economics, *Journal of the American Dietetic Association* 92 (1992): 981.

Controversy 1

1. P. Kurtzweil, Phony doctor sentenced, *FDA Consumer,* March 1995, p. 32.

2. Top health frauds, *FDA Backgrounder,* November 6, 1996 (available from *www.fda.gov/opacom/backgrounders*).

3. W. M. Silberg, G. D. Lundberg, and R. A. Musacchio, Assessing, controlling, and assuring the quality of medical information on the Internet, *Journal of the American Medical Association* 277 (1997): 1244–1245.

4. E. A. Young, National Dairy Council award for excellence in medical/dental nutrition education lecture, 1994: Nutrition education in medical schools—the prospect before us, *American Journal of Clinical Nutrition* 60 (1994): 631–638.

5. National Nutrition Monitoring and Related Research Act of 1990, Public Law 101-445, as quoted in C. H. Halstead, Toward standardized training of physicians in clinical nutrition, *American Journal of Clinical Nutrition* 56 (1992): 1–3.

6. Position of the American Dietetic Association: Nutrition education of health professionals. *Journal of the American Dietetic Association* 98 (1998): 343–346.

7. C. Gopalan, Dietetics and nutrition: Impact of scientific advances and development. *Journal of the American Dietetic Association* 97 (1997): 737–741.

8. B. Haughton and J. Shaw, Functional roles of today's public health nutritionist, *Journal of the American Dietetic Association* 92 (1992): 1218–1222.

Chapter 2

1. USDA Center for Nutrition Policy and Promotion, *The Healthy Eating Index,* 1994–1996 (Washington, D.C.: Government Printing Office, 1998), or available from www.usda.gov/cnpp

2. Standing Committee on the Scientific Evaluation of Dietary Reference Intakes, Food and Nutrition Board, Institute of Medicine, *Dietary Reference Intakes for Calcium, Phosphorus, Magnesium, Vitamin D, and Fluoride* (Washington, D.C.: National Academy Press, 1997), p. S-5.

3. Standing Committee on the Scientific Evaluation of Dietary Reference Intakes, 1997, p. S-3.

4. Dietary Reference Intakes, *Nutrition Reviews* 55 (1997): 319–326.

5. A. P. Simopoulos, Diet and gene interactions. *Food Technology* 51 (1997): 66–69.

6. M. Ward, Dietary guidelines: Does one size fit all? *School Foodservice and Nutrition,* June/July 1998, pp. 34–40.

7. M. F. Picciano, L. D. McBean, and V. A. Stallings, How to grow a healthy child: A conference report, *Nutrition Today* 34 (1999): 6–14.

8. Simopoulos, 1997.

9. J. L. Dodd, Incorporating genetics into dietary guidance, *Food Technology* 51 (1997): 80–82.

10. J. Dollahite, D. Franklin, and R. McNew. Problems encountered in meeting the Recommended Dietary Allowances for menus designed according to the Dietary Guidelines for Americans, *Journal of the American Dietetic Association* 95 (1995): 341–345.

11. E. N. Siguel, Role of essential fatty acids: Dangers in the U.S. Department of Agriculture dietary recommendations and in low-fat diets. *American Journal of Clinical Nutrition* (1994): 973–974.

12. L. R. Young and M. Nestle, Portion sizes in dietary assessment: Issues and policy implications, *Nutrition Reviews* 53 (1995): 149–158.

13. L. R. Young and M. Nestle, Variation in perceptions of a "medium" food portion: Implications for dietary guidance, *Journal of the American Dietetic Association* 98 (1998): 458–459.

Controversy 2

1. M. de Lorgeril and coauthors, Mediterranean dietary pattern in a randomized trial: Prolonged survival and possible reduced cancer rate, *Archives of Internal Medicine* 158 (1998): 1181–1187. For a debate on issues surrounding Mediterranean diets, see articles by P. Crothy and K. D. Gifford in *Nutrition Today* 33 (1998).

2. A. Trichopoulou, Correspondence, *New England Journal of Medicine* 327 (1992): 53.

3. A. Trichopoulou and P. Lagiou, Healthy traditional Mediterranean diet: An expression of culture, history, and lifestyle, *Nutrition Reviews* 55 (1997): 383–389.

4. M. de Lorgeril and coauthors, Mediterranean diet, traditional risk factors, and the rate of cardiovascular complications after myocardial infarction: Final report of the Lyon Diet Heart Study, *Circulation,* 99 (1999): 779–785; S. Renaud and coauthors, Cretan Mediterranean diet for prevention of coronary heart disease, *American Journal of Clinical Nutrition* 61 (1995): S1360–S1367; B. Haber, The Mediterranean diet: A view from history, *American Journal of Clinical Nutrition* 66 (1997): S1053–S1057.

5. A. Drewnowski and B. M. Popkin, The nutrition transition: New trends in the global diet, *Nutrition Reviews* 55 (1997): 31–43.

6. Drewnowski and Popkin, 1997; M. Nestle, Mediterranean diets: Historical and research overview, *American Journal of Clinical Nutrition* 61 (1995): S1313–S1320; A. P. Simopoulos, The Mediterranean Food Guide, *Nutrition Today,* March/April 1995, pp. 54–61.

7. L. Serra-Majem and coauthors, Nutrition policies in Mediterranean Europe, *Nutrition Reviews* 55 (1997): S42–S57.

8. A. Ghiselli, A. D'Amieis, and A. Giacosa, The antioxidant potential of the Mediterranean diet, *European Journal of Cancer Prevention* 6 (1997): S15–S19.

9. J. W. Anderson, B. M. Smith, and J. J. Gustafson, Health benefits and practical aspects of high-fiber diets, *American Journal of Clinical Nutrition* 59 (1994): S1242–S1247.

10. P. B. Geil and J. W. Anderson, Nutrition and health implications of dry beans: A review, *Journal of the American College of Nutrition* 13 (1994): 549–558; American Dietetic Association, Nutrition recommendations and principles for people with diabetes mellitus, *Diabetes Care* 17 (1994): 519.

11. S. Mackay and J. J. Ball, Do beans and oat bran add to the effectiveness of a low-fat diet? *European Journal of Clinical Nutrition* 46 (1992): 641–648.

12. J. W. Anderson and coauthors, Meta-analysis of the effects of soy protein intake on serum lipids, *New England Journal of Medicine* 333 (1995): 276–282.

13. L. Schweigerer and J. Fotis, Genistein, a dietary-derived inhibitor of in vitro angiogenesis, *Proceedings of the National Academy of Sciences,* 90 (1993): 2690–2694.

14. P. Leatherwood and P. Pollet, as cited by Geil and Anderson, 1994.

15. T. W. A. de Bruin and coauthors, Different postprandial metabolism of olive oil and soybean oil: A possible mechanism of the high-density lipoprotein conserving effect of olive oil, *American Journal of Clinical Nutrition* 58 (1993): 477–483.

16. M. B. Katan, P. L. Zock, and R. P. Mensink, Effects of fats and fatty acids on blood lipids in humans: An overview, *American Journal of Clinical Nutrition* 60 (1994): S1017–S1022.

17. F. Visioli, G. Bellomo, and C. Galli, Free radical-scavenging properties of olive oil polyphenols, *Biochemical and Biophysical Research Communications* 247 (1998): 60–64.

18. B. Halliwell and S. Chirico, Lipid peroxidation: Its mechanism, measurement, and significance, *American Journal of Clinical Nutrition* 57 (1993): S715–S725.

19. D. M. Hegsted, Dietary fat and serum lipids: An evaluation of the experimental data, *American Journal of Clinical Nutrition* 57 (1993): 875–883.

20. Y. Delneste, A. Dounet-Hughes, and E. J. Schiffin, Functional Foods: Mechanisms of action on immunocompetent cells, *Nutrition Reviews* 56 (1998): 593–598.

21. H. Link-Amster and coauthors, Modulation of a specific humoral immune response and changes in intestinal flora mediated through fermented milk intake, *FEMS Immunology and Medical Microbiology* 10 (1994): 55–63.

22. P. Michetti and coauthors, Effect of whey-based culture supernatant of Lactobacillus acidophilus (johnsonii) La1 on helicobacter pylori infection in humans, *Digestion* 60 (1999): 203–209; M. H. Coconnier and coauthors. Antagonistic activity against Helicobacter infection in vitro and in vivo by the human Lactobacillus acidophilus strain LB, *Applied and Environmental Microbiology* 64 (1998): 4573–4580.

23. M. M. Velraeds and coauthors, Interference in initial adhesion of uropathogenic bacteria and yeasts to silicone rubber by *Lactobacillus* acidophilus biosurfactant, *Journal of Medical Microbiology* 47 (1998): 1081–1085; K. Gupta and coauthors, Inverse association of H202-producing lactobacilli and vaginal *Escherichia* coli colonization in women with recurrent urinary tract infections, *Journal of Infectious Diseases* 178 (1998): 446–450; E. Hilton and coauthors, Ingestion of yogurt containing *Lactobacillus acidophilus* as prophylaxis for candidal vaginitis, *Annals of Internal Medicine* 116 (1992): 353–357.

24. A. Tjønneland and coauthors, Wine intake and diet in a random sample of 48,763 Danish men and women, *American Journal of Clinical Nutrition* 69 (1999): 49–54.

25. J. M. Gaziano and coauthors, Moderate alcohol intake, increased levels of high-density lipoprotein and its subfractions, and decreased risk of myocardial infarction, *New England Journal of Medicine* 329 (1993): 1829–1834; P. R. Ridker and coauthors, Association of moderate alcohol consumption and plasma concentration of endogenous tissue-type plasminogen activator, *Journal of the American Medical Association* 272 (1994): 929–933.

26. C. S. Fuchs and coauthors, Alcohol consumption and mortality among women, *New England Journal of Medicine* 332 (1995): 1245–1250.

27. T. A. Pearson and P. Terry, What to advise patients about drinking alcohol—The clinician's conundrum, *Journal of the American Medical Association* 272 (1994): 967–968.

28. K. Meister, The not-so-great Mediterranean diet pyramid, *Priorities* 7 (1995): 14–18.

Consumer Corner 2

1. J. E. Foulke, Cooking up the new food label, *FDA Consumer,* May 1993, pp. 33–38.

2. USDA, Away-From-Home Foods Increasingly Important to Quality of American Diet, 1999, available from www.econ.ag.gov

3. Foods in menu claims must meet FDA rule, *FDA Consumer,* October 1996, p. 5.

Fitness for Life 2

1. S. N. Blair and coauthors, Changes in physical fitness and all-cause mortality, *Journal of the American Medical Association* 263 (1995): 1093–1098.

Chapter 3

1. A. P. Simopoulos and coauthors, Impact of diet and genetic interactions on chronic disease risk, A symposium in *Food Technology* 51 (1997): 65–82.

2. D. P. Huston, The biology of the immune system, *Journal of the American Medical Association* 278 (1997): 1804–1814.

3. A Drewnowski, Why do we like fat? *Journal of the American Dietetic Association* 97 (1997): S58–S62.

4. G. A. Falciglia and P. A. Norton, Evidence for a genetic influence on preference for some foods, *Journal of the American Dietetic Association* 94 (1994): 154–158.

5. J. L. Jeraci, B. A. Lewis, and P. J. Van Soest, Interaction between human gut bacteria and fiberous structures, in G. A. Spiller, ed., *Dietary Fiber in Human Nutrition* (Boca Raton, Fla.: CRC Press, 1993), pp. 371–376.

6. M. Robinson and coauthors, Heartburn requiring frequent antacid use may indicate significant illness, *Archives of Internal Medicine* 158 (1998): 2373–2376.

7. L-E. Hansson and coauthors, The risk of stomach cancer in patients with gastric or duodenal ulcer disease, *New England Journal of Medicine* 335 (1996): 242–249; Knowledge about causes of peptic ulcer disease—United States, March-April 1997, *Morbidity and Mortality Weekly Report* (1997): 985–987.

Controversy 3

1. B. Bower, Banquets in the ruins, *Science News* 153 (1998): 331–333.

2. S. M. Garn, From the Miocene to olestra: A historical perspective on fat consumption, *Journal of the American Dietetic Association* 97 (1997): S54–S57.

3. H. V. Kohnlein and O. Receveur, Dietary change and traditional food systems of indigenous peoples, *Annual Review of Nutrition* 16 (1996): 417–442.

4. L. Seachrist, Gene ups obesity, accelerates diabetes, *Science News* 148 (1995): 103.

5. W. Hoy, A. Light, and D. Megill, Cardiovascular disease in Navajo Indians with type 2 diabetes, *Public Health Reports* 100 (1995): 87–94.

6. N. J. Murphy and coauthors, Dietary change and obesity associated with glucose intolerance in Alaskan natives, *Journal of the American Dietetic Association* 95 (1995): 676–682.

7. R. Fabsitz, Administrator of the Strong Heart Study, as quoted by K. A. Fackelmann, *Science News* 142 (1992): 168–170.

8. M. Segal, Native food preparation fosters botulism, *FDA Consumer*, January/February 1992, pp. 23–26.

9. Kohnlein and Receveur, 1996.

Chapter 4

1. J. H. Cummings and coauthors, A new look at dietary carbohydrate: Chemistry, physiology and health (Paris Carbohydrate Group), *European Journal of Clinical Nutrition* 51 (1997): 417–423.

2. J. L. Jeraci, B. A. Lewis, and P. J. Van Soest, Interaction between human gut bacteria and fibrous substrates, in G. A. Spiller, ed., *Dietary Fiber in Human Nutrition* (Boca Raton, Fla.: CRC Press, 1993), pp. 371–376.

3. J. W. Anderson and coauthors, Postprandial serum glucose, insulin, and lipoprotein responses to high- and low-fiber diets, *Metabolism: Clinical and Experimental* 44 (1995): 848–854; D. J. Jenkins and coauthors, Effect of dietary fiber on plasma lipoproteins, in G. A. Spiller, ed., *Handbook of Lipids in Human Nutrition* (Boca Raton, Fla.: CRC Press, 1996), pp. 173–198.

4. J. Salmerón and coauthors, Dietary fiber, glycemic load, and risk of non-insulin-dependent diabetes mellitus in women, *Journal of the American Medical Association* 277 (1997): 472–477; J. Salmerón and coauthors, Dietary fiber, glycemic load, and risk of NIDDM in men, *Diabetes Care* 20 (1997): 545–550; T. M. S. Wolever and coauthors, Low dietary fiber and high protein intakes associated with newly diagnosed diabetes in a remote aboriginal community, *American Journal of Clinical Nutrition* 66 (1997): 1470–1474.

5. B. S. Reddy, Role of dietary fiber in colon cancer: An overview, *American Journal of Medicine* 106 (1999): S16–S19; D. Kritchevsky, Protective role of wheat bran fiber: Preclinical data, *American Journal of Medicine* 106 (1999): S28–S31; E. Negri and coauthors, Fiber intake and risk of colorectal cancer, *Cancer Epidemiology, Biomarkers, and Prevention* 7 (1998): 667–671; the epidemiological study showing no effect of fiber on cancer of the colon was C. S. Fuchs and coauthors, Dietary fiber and the risk of colorectal cancer and adenoma in women, *New England Journal of Medicine* 340 (1999): 169–176; ongoing studies are described in J. Faivre and A. Giacosa, Primary prevention of colorectal cancer through fibre supplementation, *European Journal of Cancer Prevention* 7 (1998): S29–S32.

6. L. Brown and coauthors, Cholesterol-lowering effects of dietary fiber: A meta-analysis, *American Journal of Clinical Nutrition* 69 (1999): 30–42; E. B. Rinim and coauthors, Vegetable, fruit, and cereal fiber intake and risk of coronary heart disease among men, *Journal of the American Medical Association* 275 (1996): 447–451; S. R. Glore and coauthors, Soluble fiber and serum lipids: A literature review, *Journal of the American Dietetic Association* 94 (1994): 425–436.

7. Jenkins and coauthors, 1996.

8. Dietary fiber: Importance of function as well as amount, *Lancet* 340 (1992): 1133–1134.

9. L. Cara and coauthors, Effects of oat bran, rice bran, wheat fiber, and wheat germ on postprandial lipemia in healthy adults, *American Journal of Clinical Nutrition* 55 (1992): 81–88.

10. K. Alaimo and coauthors, Dietary intake of vitamins, minerals, and fiber of persons ages 2 months and over in the United States: Third National Health and Nutrition Examination Survey, Phase 1, 1988–1991, as cited in Position of the American Dietetic Association: Health implications of dietary fiber, *Journal of the American Dietetic Association* 97 (1997): 1157–1159.

11. D. D. Gallaher and B. O. Schneeman, Dietary fiber, in E. E. Ziegler and L. J. Filer, Jr., eds., *Present Knowledge in Nutrition*, 7th ed. (Washington, D.C.: International Life Sciences Institute Press, 1996), pp. 87–97.

12. WHO Study Group on Diet, Nutrition, and Prevention of Noncommunicable Diseases, Diet, nutrition, and the prevention of chronic diseases, *Nutrition Reviews* 49 (1991): 291–301.

13. D. A. T. Southgate, Digestion and metabolism of sugars, *American Journal of Clinical Nutrition* 62 (1995): S203–S211.

14. K. N. Englyst and coauthors, Rapidly available glucose in foods: An in vitro measurement that reflects the glycemic response, *American Journal of Clinical Nutrition* 69 (1999): 448–454.

15. N.-G. Asp, J. M. M. van Amelsvoort, and J. G. A. J. Hautvast, Nutritional implications of resistant starch, *Nutrition Research Reviews* 9 (1996): 1–31; K. R. Silvester, H. N. Englyst, and J. H. Cummings, Ileal recovery of starch from whole diets containing resistant starch measured in vitro and fermentation of ileal efferent, *American Journal of Clinical Nutrition* 62 (1995): 403–411.

16. M. A. Eastwood, The physiological effects of dietary fiber: An update, *Annual Review of Nutrition* 12 (1992): 19–35.

17. G. Annison and D. L. Topping, Nutritional role of resistant starch: Chemical structures vs physiological function, *Annual Review of Nutrition* 14 (1994): 297–320.

18. Asp, van Amelsvoort, and Hautvast, 1996.

19. F. L. Suarez and D. A. Savaiano, Diet, genetics, and lactose intolerance, *Food Technology* 51 (1997): 74–76; B. Levine, Most frequently asked questions about lactose intolerance, *Nutrition Today* 31 (1996): 78–79.

20. J. L. Rosado, Lactose digestion and maldigestion: Implications for dietary habits in developing countries, *Nutrition Research Reviews* (1997): 137–149.

21. F. L. Suarez and coauthors, Tolerance to the daily ingestion of two cups of milk by individuals claiming lactose intolerance, *American Journal of Clinical Nutrition* 65 (1997): 1502–1506; F. L. Suarez, D. A. Savaiano, and M. D. Levitt, A comparison of symptoms after the consumption of milk or lactose-hydrolyzed milk by people with self-reported severe lactose intolerance, *New England Journal of Medicine* 333 (1995): 1–4.

22. F. Suarez and M. D. Levitt, Abdominal symptoms and lactose: The discrepancy between patients' claims and the results of blinded trials, *American Journal of Clinical Nutrition* 64 (1996): 251–252.

23. F. L. Suarez and coauthors, Lactose maldigestion is not an impediment to the intake of 1500 mg calcium daily as dairy products, *American Journal of Clinical Nutrition* 68 (1998): 1118–1122.

24. J. C. B. Miller, Importance of glycemic index in diabetes, *American Journal of Clinical Nutrition* (supplement) 59 (1994): S747–S752.

25. K. Indar-Brown, C. Noreberg, and Z. Madar, Glycemic and insulinemic responses after ingestion of ethnic foods by NIDDM and healthy subjects, *American Journal of Clinical Nutrition* 55 (1992): 89–95.

26. Salmerón and coauthors, Dietary fiber, glycemic load, and risk of NIDDM in men, 1997.

27. E. Jequier, Carbohydrates as a source of energy, *American Journal of Clinical Nutrition* (supplement) 59 (1994): S682–S685.

28. National diabetes month—November 1996, *Mortality and Morbidity Weekly Report* 45 (1996): 937; Blindness caused by diabetes—Massachusetts, 1987–1994, *Mortality and Morbidity Weekly Report* 45 (1996): 937–941.

29. Report of the Expert Committee on the diagnosis and classification of diabetes mellitus, *Clinical Diabetes,* July/August 1997, pp. 158–174; C. R. Scott and coauthors, Characteristics of youth-onset noninsulin-dependent diabetes mellitus and insulin-dependent diabetes mellitus at diagnosis, *Pediatrics* 100 (1997): 84–91.

30. A. A. Skolnick, First Type 1 diabetes prevention trials, *Journal of the American Medical Association* 277 (1997): 1101–1102.

31. J. R. Baker, Autoimmune endocrine disease, *Journal of the American Medical Association* 278 (1997): 1931–1937.

32. Evidence supporting the link between cow's milk formula and diabetes is found in American Academy of Pediatrics Work Group on cow's milk protein and diabetes: Infant feeding practices and their possible relationship to the etiology of diabetes mellitus, *Pediatrics* 94 (1994): 752–754; evidence against the link is in J. M. Norris and coauthors, Lack of association between early exposure to cow's milk protein and ß-cell autoimmunity, *Journal of the American Medical Association* 276 (1996): 609–614; see also Letters, *Journal of the American Medical Association* 276 (1996): 1799–1801.

33. F. Holleman and J. B. L. Hoekstra, Insulin lispro, *New England Journal of Medicine* 337 (1997): 176–183.

34. A. Rubenstein, as quoted by Skolnick, 1997.

35. New Type II diabetes drug may reduce insulin needs, *FDA Consumer,* April 1997, p. 4.

36. G. A. Colditz and coauthors, Weight gain as a risk factor for clinical diabetes mellitus in women, *Annals of Internal Medicine* 122 (1995): 481–486.

37. M. J. Busby, Glucose tolerance in women: The effects of age, body composition, and sex hormones, *Journal of the American Geriatric Society* 40 (1992): 497–502.

38. J. Randal, Insulin key to diabetes but not a full cure, *FDA Consumer,* May 1992, pp. 15–19.

39. D. M. Nathan, Long-term complications of diabetes mellitus, *New England Journal of Medicine* 328 (1993): 1676–1685.

40. A. Ceriello and coauthors, Antioxidant defences are reduced during the oral glucose tolerance test in normal and non-insulin-dependent diabetic subjects, *European Journal of Clinical Investigation* 28 (1998): 329–333.

41. E. Lipkin, New strategies for the treatment of type 2 diabetes, *Journal of the American Dietetic Association* 99 (1999) 329–334; A. Monk and coauthors, Practice guidelines for medical nutrition therapy provided by dietitians for persons with non-insulin-dependent diabetes mellitus, *Journal of the American Dietetic Association* 95 (1995): 999–1006.

42. B. Vessby, Dietary carbohydrates in diabetes, *American Journal of Clinical Nutrition* (supplement) 59 (1994): S742–S746.

43. Lipkin, 1999; E. J. Mayer-Davis and coauthors, Dietary fat and insulin sensitivity in a triethnic population: The role of obesity. The Insulin Resistance Atherosclerosis Study (IRAS), *American Journal of Clinical Nutrition* 65 (1997): 70–87.

44. Position statement: Nutrition recommendations and principles for people with diabetes mellitus, *Diabetes Care* 17 (1994): 519–522.

45. M. Wei and coauthors, The association between cardiorespiratory fitness and impaired fasting glucose and type 2 diabetes mellitus in men, *Annals of Internal Medicine* 130 (1999): 89–96; Joint position statement of the American College of Sports Medicine and the American Diabetes Association, Diabetes and exercise, *Medicine and Science in Sports and Exercise* 29 (1997): i–vi.

46. G. Perseghin and coauthors, Increased glucose transport-phosphorylation and muscle glycogen synthesis after exercise training in insulin-resistant subjects, *New England Journal of Medicine* 335 (1996): 1357–1362; A. Kriska, Physical activity and the prevention of Type II (non-insulin dependent) diabetes, *President's Council on Physical Fitness and Sports Research Digest,* June 1997, available from the President's Council on Physical Fitness and Sports, HHH Building, Room 738H, 200 Independence Avenue, W.W., Washington, DC 20201.

47. C. V. Ford, Somatization and fashionable diagnoses: Illness as a way of life, *Scandinavian Journal of Work Environment and Health* (Supplement 3) 23 (1997): 7–16.

48. K. S. Polonsky, A practical approach to fasting hypoglycemia, *New England Journal of Medicine* 326 (1992): 1020–1021.

49. A. S. Kabadi, Pancreatic alpha-cell function in idiopathic reactive hypoglycemia, *Metabolism* 46 (1997): 639–643; F. Leonetti and coauthors, Increased nonoxidative glucose metabolism in idiopathic reactive hypoglycemia, *Metabolism: Clinical and Experimental* 45 (1996): 606–610.

50. D. J. A. Jenkins and coauthors, Low glycemic index: Lente carbohydrates and physiological effects of altered food frequency, *American Journal of Clinical Nutrition* (supplement) 59 (1994): S706–S709.

51. D. Flanagan and coauthors, Gin and tonic and reactive hypoglycemia: What is important—the gin, the tonic, or both?, *Journal of Clinical Endocrinology and Metabolism* 83 (1998): 796–800.

Controversy 4

1. A. Drewnowski, Taste preferences and food intake, *Annual Review of Nutrition* 17 (1997): 237–253.

2. USDA, *A Dietary Assessment of the U.S. Food Supply: Comparing Per Capita Food Consumption with the Food Guide Pyramid Serving Recommendations* (Washington, D.C.: Government Printing Office, 1998), AER number 772.

3. Executive summary, from *Third Report on Nutrition Monitoring in the United States* (Washington, D.C.: Government Printing Office, 1996).

4. U.S. Department of Agriculture, *Sugar and Sweetener Situation and Outlook Report,* March 1996, pp. 3–4.

5. Executive summary, 1996.

6. C. J. Lewis and coauthors, Nutrient intakes and body weights of persons consuming high and moderate levels of added sugars, *Journal of the American Dietetic Association* 92 (1992): 708–713.

7. Drewnowski, 1997.

8. J. Salmerón and coauthors, Dietary fiber, glycemic load, and risk of non-insulin-dependent diabetes mellitus in women, *Journal of the American Medical Association* 277 (1997): 472–477; J. Salmerón and coauthors, Dietary fiber, glycemic load, and risk of NIDDM in men, *Diabetes Care* 20 (1997): 545–550.

9. M. K. Lockwood and C. D. Eckhert, Sucrose-induced lipid, glucose, and insulin elevations, microvascular injury, and selenium, *American Journal of Physiology* 262 (1992): R144–R149.

10. L. C. Hudgins and coauthors, Human fatty acid synthesis is reduced after the substitution of dietary starch for sugar, *American Journal of Clinical Nutrition* 67 (1998): 631–639.

11. K. N. Frayn and S. M. Kingman, Dietary sugars and lipid metabolism in humans, *American Journal of Clinical Nutrition* 62 (1995): S250–S263.

12. Frayn and Kingman, 1995.

13. American Heart Association Nutrition Committee, Dietary guidelines for healthy American adults, *Circulation* 94 (1996): 1795–1800.

14. J. W. White and M. Wolraich, Effect of sugar on behavior and mental performance, *American Journal of Clinical Nutrition* 62 (1995): S242–S249; M. L. Wolraich, D. B. Wilson, and J. W. White, The effect of sugar on behavior or cognition of children: A meta-analysis, *Journal of the American Medical Association* 274 (1995): 1617–1621.

15. S. Gibson and S. Williams, Dental caries in pre-school children: Associations with social class, toothbrushing habit and consumption of sugars and sugar-containing foods. Further analysis of data from the National Diet and Nutrition Survey of children aged 1.5–4.5 years, *Caries Research* 32 (1999): 101–113; K. G. König and J. M. Navia, Nutritional role of sugars in oral health, *American Journal of Clinical Nutrition* 62 (1995): S275–S283.

16. B. Szepesi, Carbohydrates, in E. E. Ziegler and L. J. Filer, Jr., eds., *Present Knowledge in Nutrition,* 6th ed. (Washington, D.C.: International Life Sciences Institute Press, 1996), pp. 33–43.

17. F. R. J. Bonet, Undigestible sugars in food products, *American Journal of Clinical Nutrition* 59 (1994): S763–S769.

18. K. McNutt, Why some consumers don't believe some nutrition claims, *Nutrition Today* 32 (1997): 252–256.

19. Diarrhea from sugarless candy, *Health Gazette,* April 1998, p. 2.

20. K. McNutt and A. Sentko, Isomalt: The sugar that isn't, a 1997 fact sheet available from Consumer Choices, Inc., 28W176 Belleau Drive, Winfield, IL 60190.

21. S. S. Natah and coauthors, Metabolic response to lactitol and xylitol in healthy men, *American Journal of Clinical Nutrition* 65 (1997): 947–950.

22. L. Corcoran and M. Jacobson, Saccharin: Bittersweet, *Nutrition Action Healthletter,* April 1998, pp. 11–13.

23. Position of the American Dietetic Association: Use of nutritive and nonnutritive sweeteners, *Journal of the American Dietetic Association* 98 (1998): 581–587.

24. S. Cohen and coauthors, Saccharin and urothelial proliferation: A threshold phenomenon, *FASEB Journal* 6 (1992): A1594.

25. Quantitative Regulation (180.37), D. Plumb, FDA Center for Food and Nutrition Safety, personal communication, July 6, 1992.

26. A bibliography of 167 research articles on aspartame can be found in J. Van de Kamp, Adverse effects of aspartame, *Current Bibliographies in Medicine* (Washington, D.C.: Government Printing Office, 1991).

27. Notebook, *FDA Consumer,* September 1996, p. 29.

28. MediaScams, *Priorities,* Number 4, 1996, p. 8.

29. P. A. Spiers and coauthors, Aspartame: Neuropsychologic and neurophysiologic evaluation of acute and chronic effects, *American Journal of Clinical Nutrition* 68 (1998): 531–537; J. G. Gurney and coauthors, Aspartame consumption in relation to childhood brain tumor risk: Results from a case-control study, *Journal of the National Cancer Institute* 89 (1997): 1072–1074.

30. A. Drewnowski, Intense sweeteners and control of appetite, *Nutrition Reviews* 53 (1995): 1–7.

31. A. Drewnowski and coauthors, Comparing the effects of aspartame and sucrose on motivational ratings, taste preferences, and energy intake in humans, *American Journal of Clinical Nutrition* 59 (1994): 338–345.

32. G. H. Anderson, Sugars, sweetness, and food intake, *American Journal of Clinical Nutrition* 62 (1995): S195–S202.

33. Position of the American Dietetic Association, 1998.

34. Position of the American Dietetic Association, 1998.

Chapter 5

1. A. H. Lichenstein and coauthors, Dietary fat consumption and health, *Nutrition Reviews* 56 (1998): S3–S19.

2. R. L. Leibel, Fat as fuel and metabolic signal, *Nutrition Reviews* 50 (1992): II12–II16.

3. A. Drewnowski, Why do we like fat? *Journal of the American Dietetic Association* 97 (1997): S58–S62.

4. A. Drewnowski and B. M. Popkin, The nutrition transition: New trends in the global diet, *Nutrition Reviews* 55 (1997): 31–43.

5. J. E. Blundell and coauthors, Control of human appetite: Implications for the intake of dietary fat, *Annual Review of Nutrition* 16 (1996): 285–319.

6. S. M. Grundy, Dietary fat, in E. E. Ziegler and L. J. Filer, Jr., eds., *Present Knowledge in Nutrition,* 7th ed. (Washington, D.C.: International Life Sciences Institute Press, 1996), pp. 44–57.

7. S. A. Morgan, K. O'Dea, and A. J. Sinclair, A low-fat diet supplemented with monounsaturated fat results in less HDL-C lowering than a very-low-fat diet, *Journal of the American Dietetic Association* 97 (1997): 151–156; M. B. Katan, P. L. Zock, and R. P. Mensink, Effects of fats and fatty acids on blood lipids in humans: An overview, *American Journal of Clinical Nutrition* 60 (1994): S1017–S1022.

8. N. R. Simonsen and coauthors, Tissue stores of individuals' monounsaturated fatty acids and breast cancer: The EURAMIC study, European Community Multicenter Study on Antioxidants, Myocardial Infarction, and Breast Cancer, *American Journal of Clinical Nutrition* 68 (1998): 134–141; R. McPherson and G. A. Spiller, Effects of dietary fatty acids and cholesterol on cardiovascular disease risk factors in man, in G. A. Spiller, ed., *Handbook of Lipids in Human Nutrition* (Boca Raton, Fla.: CRC Press, 1996), pp. 41–49.

9. P. M. Kris-Etherton and S. Yu, Individual fatty acid effects on plasma lipids and lipoproteins: Human studies, *American Journal of Clinical Nutrition* 65 (1997): S1628–S1644.

10. E. Sarkkinen and coauthors, Effect of apolipoprotein E polymorphism on serum lipid resonse to the separate modification of dietary fat and dietary cholesterol, *American Journal of Clinical Nutrition* 68 (1998): 1215–1222; McPherson and Spiller, 1996.

11. NIH Consensus Development Panel, Triglyceride, high-density lipoprotein, and coronary heart disease, *Journal of the American Medical Association* 269 (1993): 505–510.

12. F. B. Hu and coauthors, Dietary fat intakes and the risk of coronary heart disease in women, *New England Journal of Medicine* 337 (1997): 1491–1499.

13. B. Halliwell, Oxidation of low-density lipoproteins: Questions of initiation, propagation, and the effect of antioxidants, *American Journal of Clinical Nutrition* 61 (1995): S670–S677.

14. A. P. Simopoulos, Omega-3 fatty acids, part II: Epidemiological aspects of omega-3 fatty acids in disease states, in G. A. Spiller, ed., *Handbook of Lipids in Human Nutrition* (Boca Raton, Fla.: CRC Press, 1996), pp. 75–89.

15. J. P. Middaugh, Cardiovascular deaths among Alaskan Natives, 1980–1986, *American Journal of Public Health* 80 (1990): 282–285; J. Dyerberg, Linolenate-derived polyunsaturated fatty acids and prevention of atherosclerosis, *Nutrition Reviews* 44 (1986): 125–134.

16. T. Tashiro and coauthors, n-3 versus n-6 polyunsaturated fatty acids in critical illness, *Nutrition* 14 (1998): 551–553.

17. A. P. Simopoulos, Omega-3 fatty acids: Part I: Metabolic effects of omega-3 fatty acids and essentiality, in G. A. Spiller, ed., *Handbook of Lipids in Human Nutrition* (Boca Raton, Fla.: CRC Press, 1996), pp. 51–73.

18. S. L. Connor and W. E. Connor, Are fish oils beneficial in the prevention and treatment of coronary artery disease? *American Journal of Clinical Nutrition* 66 (1997): S1020–S1031.

19. N. F. Sheard, Fish consumption and risks of sudden cardiac death, *Nutrition Reviews* 56 (1998): 177–179; C. M. Albert and coauthors, Fish consumption and risk of sudden cardiac death, *Journal of the American Medical Association* 279 (1998): 23–28; R. F. Gillum, M. E. Mussolino, and J. H. Madans, The relationship between fish consumption and stroke incidence: The NHANES I Epidemiologic Follow-up Study, *Archives of Internal Medicine* 156 (1996): 537–542; D. S. Siscovick and coauthors, Dietary intake and cell membrane levels of long-chain n-3 polyunsaturated fatty acids and the risk of primary cardiac arrest, *Journal of the American Medical Association* 274 (1995): 1363–1367.

20. M. B. Katan, Fish and heart disease: What is the real story? *Nutrition Reviews* 53 (1995): 228–230.

21. P. C. Calder and coauthors, Dietary fish oil suppresses human colon tumor growth in athymic mice, *Clinical Science* 94 (1998): 303–311.

22. K. Ando and coauthors, Effect of n-3 fatty acid supplementation on lipid peroxidation and protein aggregation in rat erythrocyte membranes, *Lipids* 33 (1998): 505–512.

23. P. Griffini and coauthors, Dietary omega-3 polyunsaturated fatty acids promote colon carcinoma metastasis in rat liver, *Cancer Research* 58 (1998): 3312–3319.

24. L. Hansen and M. S. Rose, Sensory acceptability is inversely related to development of fat rancidity in bread made from stored flour, *Journal of the American Dietetic Association* 96 (1996): 792–793.

25. E. A. Emken, Trans-fatty acids and coronary heart disease: Physiochemical properties, intake, and metabolism, *American Journal of Clinical Nutrition* 62 (1995): S655–S708.

26. ASCN/AIN Task Force on *trans*-fatty acids, Position paper on *trans*-fatty acids, *American Journal of Clinical Nutrition* 63 (1996): 663–670;

F

A. Aro and coauthors, Stearic acid, *trans* fatty acids, and dairy fat: Effects on serum and lipoprotein lipids, apolipoproteins, lipoprotein (a), and lipid transfer proteins in healthy subjects, *American Journal of Clinical Nutrition* 65 (1977): 1419–1426; a dissenting view is found in G. J. Nelson, Dietary fat, *trans*-fatty acids, and risk of coronary heart disease, *Nutrition Reviews* 56 (1998): 250–252; S. Shapiro, Do *trans* fatty acids increase the risk of coronary artery disease? A critique of the epidemiologic evidence, *American Journal of Clinical Nutrition* 66 (1997): S1011–S1017.

27. M. Noakes and P. M. Clifton, Oil blends containing partially hydrogenated or interesterified fats: Differential effects on plasma lipids, *American Journal of Clinical Nutrition* 68 (1998): 242–247; J. T. Judd and coauthors, Effects of margarine compared with those of butter on blood lipid profiles related to cardiovascular disease risk factors in normolipemic adults fed controlled diets, *American Journal of Clinical Nutrition* 68 (1998): 768–777; M. B. Katan, P. L. Zock, and R. P. Mensink, Trans-fatty acids and their effects on lipoproteins in humans, *Annual Review of Nutrition* 15 (1995): 473–493; M. B. Katan, Commentary on the supplement trans-fatty acids and coronary heart disease risk, *American Journal of Clinical Nutrition* 62 (1995): 518–519.

28. M. W. Gillman and coauthors, Margarine intake and subsequent coronary heart disease in men, *Epidemiology* 8 (1997): 144–149.

29. L. Kohlmeier and coauthors, Adipose tissue *trans*-fatty acids and breast cancer in the European Community Multicenter Study on Antioxidants, Myocardial Infarction, and Breast Cancer, *Cancer, Epidemiology, Biomarkers and Prevention* 6 (1997): 705–710; C. Ip, Review of the effects of *trans*-fatty acids, oleic acid, n-3 polyunsaturated fatty acids, and conjugated linoleic acid on mammary carcinogenesis in animals, *American Journal of Clinical Nutrition* 66 (1997): S1523–S1529; M. L. Gurr, Dietary fatty acids with *trans* unsaturation, *Nutrition Research Reviews* 9 (1996): 259–279.

30. M. A. Hallikainen and M. I. Uusitupa, Effects of 2 low-fat stanol ester-containing margarines on serum cholesterol concentrations as part of a low-fat diet in hypercholesterolemic subjects, *American Journal of Clinical Nutrition* 69 (1999): 403–410.

31. D. B. Allison and coauthors, Estimated intakes of *trans* fatty and other fatty acids in the US population, *Journal of the American Dietetic Association* 99 (1999): 166–174; A. P. Simopoulos, Trans fatty acids, in G. A. Spiller, ed., *Handbook of Lipids in Human Nutrition* (Boca Raton, Fla.: CRC Press, 1996), pp. 91–99.

32. W. C. Willett and A. Ascherio, Response to the International Life Sciences Institute report on trans-fatty acids, *American Journal of Clinical Nutrition* 62 (1995): 524–526; W. C. Willett and A. Ascherio, Trans-fatty acids: Are the effects only marginal? *American Journal of Public Health* 84 (1994): 722–724; M. B. Katan, European researcher calls for reconsideration of trans-fatty acid, *Journal of the American Dietetic Association* 94 (1994): 1097–1098.

33. WHO and FAL Joint Consultation: Fats and oils in human nutrition, *Nutrition Reviews* 53 (1995): 202–205.

34. Is total fat consumption really decreasing? *Nutrition Insights,* April 1998.

35. B. Hardin, Genetic testing helps single out leanness gene, *Agricultural Research Service News,* June 1999, available from www.ars.usda.gov/is/pr/1999/990622.htm

Consumer Corner 5

1. R. S. Sandler and coauthors, Gastrointestinal symptoms in 3181 volunteers ingesting snack foods containing olestra or triglycerides: A 6-week randomized, placebo-controlled trial, *Annals of Internal Medicine* 130 (1999): 253–261.

2. J. A. Westrate and K. H. van het Hof, Sucrose polyester and plasma carotenoid concentrations in healthy subjects, *American Journal of Clinical Nutrition* 62 (1995): 591–597.

3. J. R. Cotton, J. A. Westrate, and J. E. Blundell, Replacement of dietary fat with sucrose polyester: Effects on energy intake and appetite control in nonobese males, *American Journal of Clinical Nutrition* 63 (1996): 891–896.

4. Position of the American Dietetic Association: Fat replacements, *Journal of the American Dietetic Association* 98 (1998): 463–468.

Controversy 5

1. Substance Abuse and Mental Health Services Administration (SAMHSA), Prevalence of past-month alcohol, cigarette, and other drug use, U.S. population, age 12 and older, 1996, *Substance Abuse and Mental Health Statistics Source Book* (Washington, D.C.: Government Printing Office, 1998), pp. 20–21.

2. K. A. Douglas and coauthors, Results from the 1995 National College Health Risk Behavior Survey, *Journal of College Health,* September 1997, pp. 55–66.

3. M. W. Westerterp-Plantenga and C. R. Verwegen, The appetizing effect of an aperitif in overweight and normal-weight humans, *American Journal of Clinical Nutrition* 69 (1999): 205–212.

4. P. M. Suter, E. Häsler, and W. Vetter, Effects of alcohol on energy metabolism and body weight regulation: Is alcohol a risk factor for obesity? *Nutrition Reviews* 55 (1997): 157–171.

5. Suter, Häsler, and Vetter, 1997.

6. B. B. Duncan and coauthors, Association of the waist-to-hip ratio is different with wine than with beer or hard liquor consumption, *American Journal of Epidemiology* 142 (1995): 1034–1038.

7. J. P. Flatt, Body weight, fat storage, and alcohol metabolism, *Nutrition Reviews* 50 (1992): 267–270.

8. M. J. Thun and coauthors, Alcohol consumption and mortality among middle-aged and elderly U.S. adults, *New England Journal of Medicine* 337 (1997): 1705–1714; E. B. Rimm and coauthors, Review of moderate alcohol consumption and reduced risk of coronary heart disease: The effect due to beer, wine, or spirits? *British Medical Journal* 321 (1996): 731–736.

9. C. L. Hart and coauthors, Alcohol consumption and mortality from all causes, coronary heart disease, and stroke: results from a prospective cohort study of Scottish men with 21 years of follow up, *British Medical Journal* 318 (1999): 1725–1729.

10. J. M. Gaziano and J. E. Buring, Alcohol intake, lipids and risks of myocardial infarction, *Novartis Foundation Symposium* 216 (1998): 86–95.

11. E. V. Leino and coauthors, Alcohol consumption and mortality: II Studies of male populations, *Addiction* 93 (1998): 205–218.

12. S. A. Smith-Warner and coauthors, Alcohol and breast cancer in women, *Journal of the American Medical Association* 279 (1998): 535–540.

13. R. Brouillard, F. George, and A. Fougerousse, Polyphenols produced during red wine aging, *Biofactors* 6 (1997): 403–410.

14. S. I. Oh and coauthors, Chronic ethanol consumption affects glutathione status in rat liver, *Journal of Nutrition* 128 (1998): 758-763; I. D. Norton and coauthors, Chronic ethanol administration causes oxidative stress in the rat pancreas, *Journal of Laboratory and Clinical Medicine* 131 (1998): 442–446.

15. M. Serafini, G. Maiani, and A. Ferro-Luzzi, Alcohol-free red wine enhances plasma antioxidant capacity in humans, *Journal of Nutrition* 128 (1998): 1003–1007.

16. H. Wechsler and coauthors, Health and behavioral consequences of binge drinking in college: A national survey of students at 140 campuses, *Journal of the American Medical Association* 272 (1994): 1672–1677.

17. K. L. Graves, Risky sexual behavior and alcohol use among young adults: Results from a national survey, *American Journal of Health Promotion* 10 (1995): 27–36.

18. K. A. Bradley and coauthors, Medical risks for women who drink alcohol, *Journal of General Internal Medicine* 13 (1998): 627–639.

19. M. A. Emanuele and coauthors, Reversal of ethanol-induced testosterone suppression in prepubertal male rats by opiate blockage, *Alcoholism: Clinical and Experimental Research* 22 (1998): 1199–1204.

20. Smith-Warner and coauthors, 1998.

21. J. E. Foulke, Urethane in alcoholic beverages under investigation, *FDA Consumer,* January/February 1993, pp. 19–23.

22. J. D. Piette, P. G. Barnett, and R. H. Moos, First-time admissions with alcohol-related medical problems: A 10-year follow-up of a national sample of alcoholic patients, *Journal of Studies on Alcohol* 59 (1998): 89–96; A. L. Klatsky, M. A. Armstrong, and G. D. Friedman, Alcohol and mortality, *Annals of Internal Medicine* 117 (1992): 646–654.

23. SAMHSA, 1998, pp. 170–717.

24. X. Wang, Chronic alcohol intake interferes with retinoid metabolism and signalling, *Nutrition Review* 57 (1999): 51–59.

25. Committee on Substance Abuse, Alcohol use and abuse: A pediatric concern, *Pediatrics* 95 (1995): 439–442.

Chapter 6

1. C. Wu, Understanding how proteins fold, *Science News* 152 (1997): 270; E. Strauss, How proteins take shape: Guardians give a new twist to protein folding, Science News 142 (1997): 155.

2. M. C. Crim and H. N. Munro, Proteins and amino acids, in M. E. Shils, J. A. Olson, and M. Shike, eds., *Modern Nutrition in Health and Disease* (Philadelphia: Lea & Febiger, 1994), p. 31.

3. Position of the American Dietetic Association: Vegetarian diets, *Journal of the American Dietetic Association* 97 (1997): 1317–1321; V. R. Young and P. L. Pellett, Plant proteins in relation to human protein and amino acid nutrition, *American Journal of Clinical Nutrition* 59 (1994): S1203–S1212.

4. *Protein Quality Evaluation: Report of a Joint FAO/WHO Expert Consultation,* Food and Nutrition paper no. 51 (Rome, Italy: FAO and WHO, 1990).

5. A thorough review of methodology in whole-body protein turnover is provided by J. C. Waterlow, Whole-body protein turnover in humans—past, present, and future, *Annual Review of Nutrition* 15 (1995): 57–92.

6. S. M. Smith and coauthors, Nutrition in space, *Nutrition Today* 32 (1997): 6–12.

7. A. Miyamoto and coauthors, Medical baseline data collection on bone and muscle change with space flight, *Bone* 22 (1998): S79–S82.

8. D. G. Schroeder and R. Martorell, Enhancing child survival by preventing malnutrition, *American Journal of Clinical Nutrition* 65 (1997): 1080–1081.

9. A full discussion of PEM appears in B. Torun and F. Chew, Protein-energy malnutrition, in M. E. Shils, J. A. Olson, and M. Shike, eds., *Modern Nutrition in Health and Disease* (Philadelphia: Lea & Febiger, 1994), pp. 950–976.

10. B. Woodward, Protein, calories, and immune defenses, *Nutrition Reviews* 56 (1998): S84–S92.

11. J. C. Waterlow, Childhood malnutrition in developing nations: Looking back and forward, *Annual Review of Nutrition* 14 (1994): 1–19.

12. Woodward, 1998; Protein deficiency abets tuberculosis, *Science News* 150 (1996): 374.

13. Nutritional status of poor children in the United States, *Journal of the American Dietetic Association* 95 (1995): 248–250; M. Mowe, T. Bohmer, and E. Kindt, Reduced nutritional status in an elderly population (>70 yrs) is probable before disease and possibly contributes to the development of disease, *American Journal of Clinical Nutrition* 59 (1994): 317–324; J. Dye, Malnourished children in the United States: Caught in the cycle of poverty, *Journal of the American Dietetic Association* 94 (1994): 218; M. L. Taylor and S. A. Koblinsky, Dietary intake and growth status of young homeless children, *Journal of the American Dietetic Association* 93 (1993): 464–466.

14. Distinguishing malnutrition from the effects of stress and disease, *Nutrition and the M.D.,* April 1998, p. 6.

15. J. W. Anderson and coauthors, Meta-analysis of the effects of soy protein intake on serum lipids, *New England Journal of Medicine* 333 (1995): 276–282.

16. E. Brändle, H. G. Sieberth, and R. E. Hautmann, Effect of chronic dietary protein intake on the renal function in healthy subjects, *European Journal of Clinical Nutrition* 50 (1996): 734–740.

17. M. T. Pedrini and coauthors, The effect of dietary protein restriction on the progression of diabetic and nondiabetic renal diseases: A meta-analysis, *Annals of Internal Medicine* 124 (1996): 627–632.

18. Two points of view concerning protein and bone loss are found in U. S. Barzel and L. K. Massey, Excess dietary protein can adversely affect bone, *Journal of Nutrition* 128 (1998): 1051–1053; and R. P. Heaney, Excess dietary protein may not adversely affect bone, *Journal of Nutrition* 128 (1998): 1054–1057.

19. L. K. Massey, Does excess dietary protein adversely affect bone? Symposium overview, *Journal of Nutrition* 128 (1998): 1048–1050.

20. B. J. Abelow, T. R. Holford, and K. L. Insogna, Cross-cultural associations between dietary animal protein and hip fracture: A hypothesis, *Calcified Tissue International* 50 (1996): 14–18. D. Feskanich and coauthors, Protein consumption and bone fractures in women, *American Journal of Epidemiology* 143 (1996): 472–479.

21. R. G. Munger, J. R. Cerhan, and B. C-H. Chiu, Prospective study of dietary protein intake and risk of hip fracture in postmenopausal women, *American Journal of Clinical Nutrition* 69 (1999): 147–152; J. P. Bonjour, M. A. Schurch, and R. Rizzoli, Nutritional aspects of hip fractures, *Bone* 18 1996): S139–S144.

22. R. P. Heaney, Protein intake and the calcium economy, *Journal of the American Dietetic Association* 93 (1993): 1259–1260.

23. Heaney, 1998.

24. J. A. Metz, J. J. B. Anderson, and P. N. Gallagher, Intakes of calcium, phosphorus, and protein, and physical activity levels are related to radial bone mass in young adult women, *American Journal of Clinical Nutrition* 58 (1993): 537–542; Heaney, 1993.

25. J. A. Eisman, Genetics, calcium intake and osteoporosis, *Proceedings of the Nutrition Society* 57 (1998): 187–193; A. S. Ryan and coauthors, Resistive training maintains bone mineral density in postmenopausal women, *Calcified Tissue International* 62 (1998): 295–299.

Fitness for Life 6

1. Position of the American Dietetic Association, *Journal of the American Dietetic Association* 97 (1997): 1317–1321.

2. P. W. R. Lemon, Is increased dietary protein necessary or beneficial for individuals with a physically active lifestyle? *Nutrition Reviews* 54 (1996): S169–S175.

Consumer Corner 6

1. N. W. Flodin, The metabolic roles, pharmacology and toxicology of lysine, *Journal of the American College of Nutrition* 16 (1997): 7–21.

2. FDA, Impurities confirmed in dietary supplement 5-hydroxy-L-tryptophan, FDA Talk Paper, August 1998, available from Food and Drug Administration, U.S. Department of Health and Human Services, Public Health Service, 5600 Fishers Lane, Rockville, MD 20857.

3. H. N. Christensen, Amino acid nutrition: A two-step absorptive process, *Nutrition Reviews* 51 (1993): 95–100.

4. S. A. Anderson and D. J. Raiten, eds., *Safety of Amino Acids Used as Supplements* (Bethesda, Md.: Federation of American Societies for Experimental Biology, 1992).

Controversy 6

1. L. B. Szabo, The health risks of new-wave vegetarianism, *Canadian Medical Association Journal* 156 (1997): 1454–1455.

2. U.S. Department of Agriculture, *Dietary Guidelines for Americans,* Home and Garden Bulletin no. 232 (Washington, D.C.: Government Printing Office, 1995), p. 8.

3. J. Raloff, Soya-nara, heart disease: The United States' top-selling legume gains heartfelt respect, *Science News* 153 (1998): 348–349; Position of the American Dietetic Association: Vegetarian diets, *Journal of the American Dietetic Association* 97 (1997): 1317–1321; P. Walter, Effects of vegetarian diets on aging and longevity, *Nutrition Reviews* 55 (1997): S61–S68.

4. M. Thorogood, The epidemiology of vegetarianism and health, *Nutrition Research Reviews* 8 (1995): 179–192.

5. P. N. Appleby and coauthors, Low body mass index in non-meat eaters: The possible roles of animal fat, dietary fiber and alcohol, *International Journal of Obesity and Related Metabolic Disorders* 22 (1998): 454–460; T. Key, Prevalence of obesity is low in people who do not eat meat, *British Medical Journal* 313 (1996): 816–817.

6. T. J. Parsons and coauthors, Reduced bone mass in Dutch adolescents fed a macrobiotic diet in early life, *Journal of Bone Mineral Research* 12 (1997): 1486–1494.

7. A. Golay and E. Bobbioni, The role of dietary fat in obesity, *International Journal of Obesity and Related Metabolic Disorders* 21 (1997): S2–S11.

F

8. M. J. Toth and E. F. Poehlman, Sympathetic nervous system activity and resting metabolic rate in vegetarians, *Metabolism: Clinical and Experimental* 43 (1994): 621–625.

9. R. R. Wolfe, Metabolic interactions between glucose and fatty acids in humans, *American Journal of Clinical Nutrition* 67 (1998): S519–S526.

10. Thorogood, 1995.

11. L. J. Beilin, Vegetarian and other complex diets, fats, fiber, and hypertension, *American Journal of Clinical Nutrition* 59 (1994): S1130–S1135.

12. J. I. Mann and coauthors, Dietary determinants of ischaemic heart disease in health conscious individuals, *Heart* 78 (1997): 450–455; M. Krajcovicova-Kudlackova and coauthors, Plasma fatty acid profile and alternative nutrition, *Annals of Nutrition and Metabolism* 41 (1997): 365–370.

13. G. E. Fraser, Diet and coronary heart disease: Beyond dietary fats and low-density lipoprotein cholesterol, *American Journal of Clinical Nutrition* 59 (1994): S1117–S1123.

14. S. K. Harman and W. R. Parnell, The nutritional health of New Zealand vegetarian and nonvegetarian Seventh-Day Adventists: Selected vitamin, mineral, and lipid levels, *New Zealand Medical Journal* 27 (1998): 91–94; D. C. Knight and J. A. Eden, A review of the clinical effects of phytoestrogens, *Obstetrics and Gynecology* 87 (1996): 897–904.

15. T. J. A. Key and coauthors, Dietary habits and mortality in 11,000 vegetarians and health conscious people: Results of a 17 year follow up, *British Medical Journal* 313 (1996): 775–779.

16. Raloff, 1998.

17. J. W. Anderson, B. M. Johnstone, and M. E. Cook-Newell, Meta-analysis of the effects of soy protein intake on serum lipids, *New England Journal of Medicine* 333 (1995): 276–282.

18. J. A. Conquer and B. J. Holub, Supplementation with an algae source of docosahexaenoic acid increases (n-3) fatty acid status and alters selected risk factors for heart disease in vegetarian subjects, *Journal of Nutrition* 126 (1996): 3032–3039.

19. Harman and Parnell, 1998.

20. R. Frentzel-Beyme and J. Chang-Claude, Vegetarian diets and colon cancer: The German experience, *American Journal of Clinical Nutrition* 59 (1994): S1143–S1152.

21. B.C.-H. Chiu and coauthors, Diet and risk of non-Hodgkins lymphoma in older women, *Journal of the American Medical Association* 275 (1996): 1315–1321.

22. J. G. Holmen and coauthors, Long-term effects of a change from a mixed diet to a lacto-vegetarian diet on human urinary and fecal mutagenic activity, *Mutagenesis* 13 (1998): 167–171; J. G. Holmen and coauthors, Dietary influence on some proposed risk factors for colon cancer: Fecal and urinary mutagenic activity and the activity of some intestinal bacterial enzymes, *Cancer Detection and Prevention* 21 (1997): 258–266.

23. B. C. Pence and coauthors, Feeding of a well-cooked beef diet containing a high heterocyclic amine content enhances colon and stomach carcinogenesis in 1,2-dimethylhydrazine-treated rats, *Nutrition and Cancer* 30 (1998): 220–226.

24. A. Hackett, I. Nathan, and L. Burgess, Is a vegetarian diet adequate for children, *Nutrition and Health* 12 (1998): 189–195.

25. Hackett, Nathan, and Burgess,1998.

26. L. H. Allen and coauthors, Interactive effects of dietary quality on the growth and attained size of young Mexican children, *American Journal of Clinical Nutrition* 56 (1992): 329–333.

27. T. A. B. Sanders and S. Reddy, Vegetarian diets and children, *American Journal of Clinical Nutrition* 59 (1994): S1176–S1181.

28. L. H. Allen, The nutrition CRSP: What is marginal malnutrition, and does it affect human function? *Nutrition Reviews* 51 (1993): 255–267.

29. I. Nathan, A. F. Hackett, and S. Kirby, A longitudinal study of the growth of matched pairs of vegetarian and omnivorous children, aged 7–11 years, in the north-west of England, *European Journal of Clinical Nutrition* 51 (1997): 20–25.

30. P. J. Grattan-Smith and coauthors, The neurological syndrome of infantile cobalamin deficiency: Developmental regression and involuntary movements, *Movement Disorders* 12 (1997): 39–46.

31. K. Lovblad and coauthors, Retardation of myelination due to dietary vitamin B12 deficiency: Cranial MRI findings, *Pediatric Radiology* 27 (1997): 155–158.

32. U. von Schenck, C. Bender-Gotze, and B. Kolezko, Persistence of neurological damage induced by dietary vitamin B-12 deficiency in infancy, *Archives of Diseases in Childhood* 77 (1997): 137–139.

33. R. R. Hunt, L. A. Matthys, and L. K. Johnson, Zinc absorption, mineral balance, and blood lipids in women consuming controlled lacto-ovo vegetarian and omnivorous diets for 8 wk, *American Journal of Clinical Nutrition* 67 (1998): 421–430.

34. C. Lamgerg-Allardt and coauthors, Low serum 25-hydroxyvitamin D concentrations and secondary hyperparathyroidism in middle-aged white strict vegetarians, *American Journal of Clinical Nutrition* 58 (1993): 684–689.

35. Position of the American Dietetic Association, 1997; V. R. Young and P. L. Pellett, Plant proteins in relation to human protein and amino acid nutrition, *American Journal of Clinical Nutrition* 59 (1994): S1203–S1212.

Chapter 7

1. W. Mertz, A balanced approach to nutrition for health: The need for biologically essential minerals and vitamins, *Journal of the American Dietetic Association* 94 (1994): 1259–1262.

2. C. D. Berdanier, Vitamin A, in *Advanced Nutrition: Micronutrients* (Boca Raton, Fla.: CRC Press, 1998), pp. 22–37.

3. M. H. Zile, Vitamin A and embryonic development: An overview, *Journal of Nutrition* 128 (1998): S455–S458; J. A. Olson, Vitamin A, in E. E. Ziegler and L. J. Filer, Jr., eds., *Present Knowledge in Nutrition*, 7th ed. (Washington, D.C.: International Life Sciences Institute Press, 1996), pp. 109–119.

4. P. Christian and coauthors, Working after the sun goes down: Exploring how night blindness impairs women's work activities in rural Nepal, *European Journal of Clinical Nutrition* 52 (1998): 519–524.

5. A. Sommer, Xerophthalmia and vitamin A status, *Progress in Retinal and Eye Research* 17 (1998): 9–31.

6. L. M. DeLuca, Vitamin A in epithelial differentiation and skin carcinogenesis, *Nutrition Reviews* (supplement) 52 (1994): S45–S52.

7. R. D. Semba, The role of vitamin A and related retinoids in immune function, *Nutrition Reviews* 56 (1998): S38–S48.

8. J. Sakar, Vitamin A is required for regulation of polymeric immunoglobulin receptor (pIgR) expression by interleukin-4 and interferon-gamma in a human intestinal epithelial cell line, *Journal of Nutrition* 128 (1998): 1063–1069; M. Cippitelli and coauthors, Retinoic acid–induced transcriptional modulation of the human interferon-gamma promoter, *Journal of Biological Chemistry* 43 (1996): 26783–26793.

9. W. W. Fawzi and coauthors, Dietary vitamin A intake in relation to child growth, *Epidemiology* 8 (1997): 402–407; C. B. Stephensen and coauthors, Vitamin A is excreted in the urine during acute infection, *American Journal of Clinical Nutrition* 60 (1994): 388–392.

10. S. L. Teitelbaum and coauthors, Cellular and molecular mechanisms of bone resorption, *Mineral and Electrolyte Metabolism* 21 (1995): 193–196.

11. W. W. Fawzi and coauthors, The effect of vitamin A supplementation on the growth of preschool children in the Sudan, *American Journal of Public Health* 87 (1997): 1359–1362.

12. World-wide vitamin A deficiency targeted as a public health problem, *Nutrition Today* 30 (1995): 53.

13. W. W. Fawzi and coauthors, Dietary vitamin A intake and risk of mortality among children, *American Journal of Clinical Nutrition* 59 (1994): 401–408; N. M. P. Daulaire and coauthors, Childhood mortality after a high dose of vitamin A in high risk populations, *British Medical Journal* 304 (1992): 207–210.

14. K. J. Rothman and coauthors, Teratogenicity of high vitamin intake, *New England Journal of Medicine* 333 (1995): 1369–1373.

15. J. W. Coates and coauthors, Gastric ulceration and suspected vitamin A toxicosis in grower pigs fed fish silage, *Canadian Veterinary Journal* 39 (1998): 167–170.

16. Berdanier, 1998.

17. E. R. Greenberg and coauthors, Mortality associated with low plasma concentration of beta carotene and effect of oral supplementation, *Journal of the American Medical Association* 275 (1996): 699–703.

18. Standing Committee on the Scientific Evaluation of Dietary Reference Intakes, *Dietary Reference Intakes Proposed Definition and Plan for Review of the Dietary Antioxidants and Related Compounds* (Washington, D.C.: National Academy Press), pp. 7–8.

19. I. Nemere and M. C. Farach-Carson, Membrane receptors for steroid hormones: A case for specific cell surface binding sites for vitamin D metabolites and estrogens, *Biochemical and Biophysical Research Communications* 248 (1998): 443–449; C. Carlberg and P. Polly, Gene regulation by vitamin D3, *Critical Reviews in Eukaryotic Gene Expression* 8 (1998): 19–42.

20. Standing Committee on the Scientific Evaluation of Dietary Reference Intakes, Food and Nutrition Board, Institute of Medicine, *Dietary Reference Intakes for Calcium, Phosphorus, Magnesium, Vitamin D, and Fluoride* (Washington, D.C.: National Academy Press, 1997), pp. 7-1–7-30.

21. J. B. Randlov and coauthors, Acute cardiovascular effect of 1,25-dihydroxycholecalciferol in essential hypertension, *American Journal of Hypertension* 11 (1998): 659–666; N. Niederhoffer and coauthors, Calcification of medial elastic fibers and aortic elasticity, *Hypertension* 29 (1997): 999–1006.

22. S. Blank and coauthors, An outbreak of hypervitaminosis D associated with the overfortification of milk from a home-delivery dairy, *American Journal of Public Health* 85 (1995): 656–659.

23. J. L. Giunta, Dental changes in hypervitaminosis D, *Oral Surgery, Oral Medicine, Oral Pathology, Oral Radiology and Endodontics* 85 (1998): 410–413.

24. M. F. Holick, Vitamin D, in M. E. Shils, J. A. Olson, and M. Shike, eds., *Modern Nutrition in Health and Disease* (Philadelphia: Lea & Febiger, 1994), p. 313.

25. S. S. Harris and B. Dawson-Hughes, Seasonal changes in plasma 25-hydroxyvitamin D concentrations of young American black and white women, *American Journal of Clinical Nutrition* 67 (1998): 1232–1236.

26. S. N. Meydani and A. A. Beharka, Recent developments in vitamin E and immune response, *Nutrition Reviews* 56 (1998): S49–S58.

27. M. J. Fryer, The possible role of nitric oxide and impaired mitochondrial function in ataxia due to severe vitamin E deficiency, *Medical Hypotheses* 50 (1998): 353–354.

28. M. A. Beck, The influence of antioxidant nutrients on viral infection, *Nutrition Reviews* 56 (1998): S140–S146.

29. S. N. Meydani and coauthors, Assessment of the safety of supplementation with different amounts of vitamin E in healthy older adults, *American Journal of Clinical Nutrition* 68 (1998): 311–318; S. N. Meydani and coauthors, Assessment of the safety of high-dose, short-term supplementation with vitamin E in healthy older adults, *American Journal of Clinical Nutrition* 60 (1994): 704–709.

30. R. J. Sokol, Vitamin E, in E. E. Ziegler and L. J. Filer, Jr., eds., *Present Knowledge in Nutrition,* 7th ed. (Washington, D.C.: International Life Sciences Institute Press, 1996), pp. 130–136.

31. G. Ferland, The vitamin K–dependent proteins: An update, *Nutrition Reviews* 56 (1998): 223–230.

32. D. Feskanich and coauthors, Vitamin K and hip fractures in women: A prospective study, *American Journal of Clinical Nutrition* 69 (1999): 77–79.

33. C. G. Harper and coauthors, Prevalence of Wernicke-Korsakoff syndrome in Australia: Has thiamin fortification made a difference? *Medical Journal of Australia* 168 (1998): 542–545.

34. J. B. Hack and R. S. Hoffman, Thiamine before glucose to prevent Wernicke encephalopathy: Examining the conventional wisdom (letter), *Journal of the American Medical Association* 279 (1998): 583–584.

35. R. S. Rivlin and P. Dutta, Vitamin B_2 (riboflavin), *Nutrition Today* 30 (1995): 62–67.

36. P. W. Jungnickel and coauthors, Effect of two aspirin pretreatment regimens on niacin-induced cutaneous reactions, *Journal of General Internal Medicine* 12 (1997): 591–596.

37. American Society of Health-System Pharmacists, ASHP therapeutic position statement on the safe use of niacin in the management of dyslipidemias, *American Journal of Health-System Pharmacy* 54 (1997): 2815–2819.

38. M. T. Behme, Nicotinamide and diabetes prevention, *Nutrition Reviews* 53 (1995): 137–139.

39. D. Callanan, B. A. Blodi, and D. F. Martin, Macular edema associated with nicotinic acid, *Journal of the American Medical Association* 279 (1998): 1702; S. P. Lawrence, Transient focal hepatic defects related to sustained-release niacin, *Journal of Clinical Gastroenterology* 16 (1993): 234–236.

40. American Society of Health-System Pharmacists, 1997.

41. A. Kwasniewska, A. Tukendorf, and M. Semczuk, Folate deficiency and cervical intraepithelial neoplasia, *European Journal of Gynaecological Oncology* 18 (1997): 526–530.

42. Standing Committee On the Scientific Evaluation of Dietary Reference Intakes, Food and Nutrition Board, Institutes of Medicine, *Dietary Reference Intakes for Thiamin, Riboflavin, Niacin, Vitamin B_6, Folate, Vitamin B_{12}, Pantothenic Acid, Biotin, and Choline* (Washington, D.C.: National Academy Press, 1998), p. 8-1.

43. R. J. Hine, What practitioners need to know about folic acid, *Journal of the American Dietetic Association* 96 (1996): 451–452. A full discussion of folate and neural tube defects is found in C. E. Butterworth and A. Bendich, Folic acid and the prevention of birth defects, *Annual Review of Nutrition* 16 (1996): 73–97.

44. I. A. Brouwer and coauthors, Low-dose folic acid supplementation decreases plasma homocysteine concentrations: A randomized trial, *American Journal of Clinical Nutrition* 69 (1999): 99–104; T. K. A. B. Eskes, Open or closed? A world of difference: A history of homocysteine research, *Nutrition Reviews* 56 (1998): 236–244.

45. P. Kelly and coauthors, Unmetabolized folic acid in serum: Acute studies in subjects consuming fortified food and supplements, *American Journal of Clinical Nutrition* 65 (1997): 1790–1795.

46. L. B. Bailey, Dietary Reference Intakes for folate: The debut of Dietary Folate Equivalents, *Nutrition Reviews* 56 (1998): 294–299.

47. S. P. Rothenberg, Increasing the dietary intake of folate: Pros and cons, *Seminars in Hematology* 36 (1999): 65–74.

48. D. S. Inagaki and coauthors, Effects of vitamin B_6 deficiency on cytokine levels and lymphocytes in mice, *Bioscience, Biotechnology and Biochemistry* 62 (1998): 1008–1010; L. C. Rall and S. N. Meydani, Vitamin B_6 and immune competence, *Nutrition Reviews* 51 (1993): 217–225.

49. T. R. Guilarte, Vitamin B_6 and cognitive development: Recent research findings from human and animal studies, *Nutrition Reviews* 51 (1993): 193–198.

50. J. E. Leklem, Vitamin B-6, in E. E. Ziegler and L. J. Filer, Jr., eds., *Present Knowledge in Nutrition,* 7th ed. (Washington, D.C.: International Life Sciences Institute Press, 1996), pp. 174–183.

51. E. B. Rimm and coauthors, Folate and vitamin B_6 from diet and supplements in relation to risk of coronary heart disease among women, *Journal of the American Medical Association* 279 (1998): 359–364.

52. Y. C. Huang and coauthors, Vitamin B-6 requirement and status assessment of young women fed a high-protein diet with various levels of vitamin B-6, *American Journal of Clinical Nutrition* 67 (1998): 208–220; C. M. Hansen, J. E. Leklem, and L. T. Miller, Changes in vitamin B-6 status indicators of women fed a constant protein diet with varying levels of vitamin B-6, *American Journal of Clinical Nutrition* 66 (1997): 1379–1387.

53. Standing Committee on the Scientific Evaluation of Dietary Reference Intakes, 1998, pp. 7-1–7-27.

54. C. J. Schorah and coauthors, The responsiveness of plasma homocysteine to small increases in dietary folic acid: A primary care study, *European Journal of Clinical Nutrition* 52 (1998): 407–411; J. B. Ubbink, P. J. Becker, and W. J. H. Vermaak, Will an increased dietary folate intake reduce the incidence of cardiovascular disease? *Nutrition Reviews* 54 (1996): 213–216.

55. Rimm and coauthors, 1998; A. R. Folsom and coauthors, Prospective study of coronary heart disease incidence in relation to fasting total homocysteine, related genetic polymorphisms, and B vitamins: The Atherosclerosis Risk in Communities (ARIC) study, *Circulation* 98 (1998): 204–210; N. Pancharuniti and coauthors, Plasma homocyst(e)ine, folate, and vitamin B_{12} concentrations and risk for early-onset coronary artery disease, *American Journal of Clinical Nutrition* 59 (1994): 940–948.

56. Rimm and coauthors, 1998.

57. J. V. Woodside and coauthors, Effect of B-group vitamins and antioxidant vitamins on hyperhomocysteinemia: A double-blind, randomized, factorial design, controlled trial, *American Journal of Clinical Nutrition* 67 (1998): 858–866.

58. P. F. Jacques and coauthors, The effect of folic acid fortification on plasma folate and total homocysteine concentrations, *New England*

Journal of Medicine 340 (1999): 1449–1454.

59 G. P. Oakely, Eat right *and* take a multivitamin, *New England Journal of Medicine* 338 (1998): 1060–1061.

60 J. N. Hathcock, Vitamins and minerals: Efficacy and safety, *American Journal of Clinical Nutrition* 66 (1997): 427–437.

61. Woodside and coauthors, 1998.

62. J. Selhum and A. D'Angelo, Relationship between homocysteine and thrombotic disease, *American Journal of the Medical Sciences* 316 (1998): 129–141; J. B. Ubbink and coauthors, Effective homocysteine metabolism may protect South African blacks against coronary heart disease, *American Journal of Clinical Nutrition* 62 (1995): 802–808.

63. S. Mendiratta, Z. C. Qu, and J. M. May, Erythrocyte ascorbate recycling: Antioxidant effects in blood, *Free Radical Biology and Medicine* 24 (1998): 789–797; J. M. May, Ascorbate function and metabolism in the human erythrocyte, *Frontiers in Bioscience* 3 (1998): D1–D10.

64. D. L. Tibble, L. J. Giuliano, and S. P. Fortmann, Reduced plasma ascorbic acid concentrations in nonsmokers regularly exposed to environmental tobacco smoke, *American Journal of Clinical Nutrition* 58 (1993): 886–890.

65. Are older Americans making better food choices to meet diet and health recommendations? *Nutrition Reviews* 51 (1993): 20–22; B. J. Rolls, Aging and appetite, *Nutrition Reviews* 50 (1992): 422–426.

66. C. S. Johnson and M. F. Yen, Megadose of vitamin C delays insulin response to a glucose challenge in normoglycemic adults, *American Journal of Clinical Nutrition* 60 (1994): 735–738.

67. M. Levine and coauthors, Criteria and recommendations for vitamin C intake, *Journal of the American Medical Association* 281 (1999): 1415–1428.

68. R. A. Jacob, Vitamin C, in M. E. Shils, J. A. Olson, and M. Shike, eds., *Modern Nutrition in Health and Disease* (Philadelphia: Lea & Febiger, 1994), pp. 432–448.

69. V. Herbert, Vitamin C supplements are dangerous for iron-overloaded persons, *Journal of the American Dietetic Association* 93 (1993): 526–527.

70. M. Levine and coauthors, 1999; C. S. Johnston, Biomarkers for establishing a Tolerable Upper Intake Level for vitamin C, *Nutrition Reviews* 57 (1999): 71–77.

71. L. H. Allen, The Nutrition CRSP: What is marginal malnutrition, and does it affect human function? *Nutrition Reviews* 51 (1993): 255–267.

72. R. Oren and Y. Ilan, Reversible hepatic injury induced by long-term vitamin A ingestion, *American Journal of Medicine* 93 (1992): 703–704.

73. T. E. Kowalski and coauthors, Vitamin A hepatotoxicity: A cautionary note regarding 25,000 IU supplements, *American Journal of Medicine* 97 (1994): 523–528.

74. G. J. Handelman, High-dose vitamin supplements for cigarette smokers: Caution is indicated, *Nutrition Reviews* 55 (1997): 369–370.

75. I. H. Rosenburg and coauthors, Dietary supplements: Recent chronology and legislation, *Nutrition Reviews* 53 (1995): 31–36.

76. M. C. Nesheim, Regulation of dietary supplements, *Nutrition Today* 33 (1998): 62–68.

77. U.S. Department of Health and Human Services, Food and Drug Administration, HHS News (electronic bulletin board), June 2, 1997; Adverse events associated with ephedrine-containing products—Texas, December 1993–September 1995, *Morbidity and Mortality Weekly Report* 45 (1996): 689–693.

78. C. S. Johnston and B. Luo, Comparison of the absorption and excretion of three commercially available sources of vitamin C, *Journal of the American Dietetic Association* 94 (1994): 779–781.

Consumer Corner 7

1. S. B. Mossad, Treatment of the common cold, *British Medical Journal* 317 (1998): 33–36.

2. L. Pauling, *Vitamin C and the Common Cold* (San Francisco: Freeman, 1970); T. C. Chalmers, Effects of ascorbic acid on the common cold: An evaluation of the evidence, *American Journal of Medicine* 58 (1975): 532–536, as cited by H. Hemiliä and Z. S. Herman, Vitamin C and the common cold: A retrospective analysis of Chalmers' review, *Journal of the American College of Nutrition* 14 (1995): 116–123.

3. H. Hemiliä, Does vitamin C alleviate the symptoms of the common cold? A review of the current evidence, *Scandinavian Journal of Infectious Diseases* 26 (1994): 1–6.

4. Hemiliä and Herman, 1995.

5. H. Hemiliä, Vitamin C supplementation and common cold symptoms: factors affecting the magnitude of the benefit, *Medical Hypotheses* 52 (1999): 171–178.

6. M. Del Rio and coauthors, Improvement by several antioxidants of macrophage function in vito, *Life Sciences* 63 (1998): 871–881.

7. C. S. Johnston, L. J. Martin, and X. Cai, Antihistamine effect of supplemental ascorbic acid and neutrophil chemotaxis, *Journal of the American College of Nutrition* 11 (1992): 172–176.

8. M. Levine and coauthors, Criteria and recommendations for vitamin C intake, *Journal of the American Medical Association* 281 (1999): 1415–1423.

Controversy 7

1. B. Halliwell, Antioxidants in human health and disease, *Annual Review of Nutrition* 16 (1996): 33–50; J. D. Morrow and coauthors, Increase in circulating products of lipid peroxidation (F_2-isoprostanes) in smokers—Smoking as a cause of oxidative damage, *New England Journal of Medicine* 332 (1995): 1198–1203.

2. M. Meydani, Isopostanes as oxidant stress markers in coronary reperfusion, *Nutrition Reviews* 55 (1997): 404–407.

3. C. L. Rock, R. A. Jacob, and P. E. Bowen, Update on the biological characteristics of the antioxidant micronutrient: Vitamin C, vitamin E, and the carotenoids, *Journal of the American Dietetic Association* 96 (1996): 693–702.

4. B. Caballero, Vitamin E improves the action of insulin, *Nutrition Reviews* 51 (1993): 339–340; G. Paolisso and coauthors, Pharmacologic doses of vitamin E improve insulin action in healthy subjects and non-insulin-dependent diabetic subjects, *American Journal of Clinical Nutrition* 57 (1993): 650–656.

5. G. Cao and coauthors, Increases in human plasma antioxidant capacity after consumption of controlled diets high in fruit and vegetables, *American Journal of Clinical Nutrition* 68 (1998) 1081–1087.

6. M. P. Longnecker and coauthors, Intake of carrots, spinach, and supplements containing vitamin A in relation to breast cancer, *Cancer, Epidemiology, Biomarkers and Prevention* 6 (1997): 887–892.

7. P. Walter, Effects of vegetarian diets on aging and longevity, *Nutrition Reviews* 55 (1997): S61–S68.

8. Y. M. Peng and coauthors, Concentrations of carotenoids, tocopherols, and retinol in paired plasma and cervical tissue of patients with cervical cancer, precancer, and noncancerous diseases, *Cancer, Epidemiology, Biomarkers and Prevention* 7 (1998): 347-350; Y. Kumagai and coauthors, Serum antioxidant vitamins and risk of lung and stomach cancers in Shenyang, China, *Cancer Letters* 129 (1998): 145–149; A. R. Giuliano and coauthors, Antioxidant nutrients: Associations with persistent human papillomavirus infection, *Cancer, Epidemiology, Biomarkers and Prevention* 6 (1997): 917–923.

9. S. T. Mayne, Beta-carotene, carotenoids, and disease prevention in humans, *FASEB Journal* 10 (1996): 690–701.

10. E. R. Greenberg and coauthors, A clinical trial of antioxidant vitamins to prevent colorectal adenoma, *New England Journal of Medicine* 33 (1994): 141–147.

11. O. P. Heinonen, J. K. Huttunen, and D. Albanes (and other participants in the alpha-tocopherol, beta-carotene cancer prevention study group), The effect of vitamin E and beta carotene on the incidence of lung cancer and other cancers in male smokers, *New England Journal of Medicine* 330 (1994): 1029–1035.

12. K. Smigel, Beta-carotene fails to prevent cancer in two major studies; CARET intervention stopped, *Journal of the National Cancer Institute* 88 (1996): 145; G. S. Omenn and coauthors, Effects of a combination of beta-carotene and vitamin A on lung cancer and cardiovascular disease, *New England Journal of Medicine* 334 (1996): 1150–1155.

13. Smigel, 1996; Omenn and coauthors, 1996.

14. P. James, K. Norum, and I. Rosenberg, Meeting summary, *Nutrition Reviews* (supplement) 52 (1994): S87–S90.

15. M. A. Wagstaff and coauthors, Malignant melanoma: Diet, alcohol, and obesity, *Journal of the American Dietetic Association* 94 (1994): 1210.

16. Greenberg and coauthors, 1994; Heinonen, Huttunen, and Albanes, 1994.

17. J. C. Fleet, Dietary selenium repletion may reduce cancer incidence in people at high risk who live in areas with low soil selenium, *Nutrition Reviews* 55 (1997): 277–286.

18. L. C. Clark and coauthors, Effects of selenium supplementation for cancer prevention in patients with carcinoma of the skin, *Journal of the American Medical Association* 276 (1996): 1957–1963.

19. V. Herbert, Selenium supplementation and cancer rates, *Journal of the American Medical Association* 277 (1997): 880.

20. K. Yoshizawa and coauthors, Study of prediagnostic selenium level in toenails and the risk of advanced prostate cancer, *Journal of the National Cancer Institute* 90 (1998): 1219–1224.

21. Fleet, 1997.

22. C. D. Berdanier, Trace minerals, in *Advanced Nutrition: Micronutrients* (Boca Raton, Fla.: CRC Press, 1998), pp. 183–208.

23. L. A. Levin, Opthalmology, *Journal of the American Medical Association* 273 (1995): 1703–1705.

24. R. C. Rose, S. P. Richer, and A. M. Bode, Ocular oxidants and antioxidant protection, *Proceedings of the Society for Experimental Biology and Medicine* 217 (1998): 397–407.

25. J. M. Teikari and coauthors, Retinal vascular changes following supplementation with alpha-tocopherol or beta-carotene, *Acta Ophthalmologica Scandinavica* 76 (1998): 68–73.

26. J. E. Buring and C. H. Hennekens, Antioxidant vitamins and cardiovascular disease, *Nutrition Reviews* 55 (1997): S53–S60.

27. B. Halliwell, Oxidation of low-density lipoproteins: Questions of initiation, propagation, and the effect of antioxidants, *American Journal of Clinical Nutrition* 61 (1995): S670–S677.

28. E. K. Parkkala-Sarataho and coauthors, A randomized, single-blind, placebo-controlled trial of the effects of 200 mg α-tocopherol on the oxidation resistance of atherogenic lipoproteins, *American Journal of Clinical Nutrition* 68 (1998): 1034–1041.

29. M. J. Stampfer and coauthors, Vitamin E consumption and the risk of coronary disease in women, *New England Journal of Medicine* 328 (1993): 1444–1449; E. B. Rimm and coauthors, Vitamin E consumption and the risk of coronary disease in men, *New England Journal of Medicine* 328 (1993): 1450–1456.

30. H. N. Hodis and coauthors, Serial coronary angiographic evidence that antioxidant vitamin intake reduces progression of coronary artery atherosclerosis, *Journal of the American Medical Association* 273 (1995): 1849–1854.

31. L. H. Kushi and coauthors, Dietary antioxidant vitamins and death from coronary heart disease in postmenopausal women, *New England Journal of Medicine* 334 (1996): 1156–1162.

32. N. G. Stephens and coauthors, Randomized controlled trial of vitamin E in patients with coronary disease: Cambridge Heart Antioxidant Study (CHAOS), *Lancet* 347 (1996): 781–786.

33. G. D. Plotnick, M. C. Corretti, and R. A. Vogel, Effect of antioxidant vitamins on the transient impairment of endothelium-dependent brachial artery vasoactivity following a single high-fat meal, *Journal of the American Medical Association* 278 (1997): 1682–1686.

34. D. Harats, Citrus fruit supplementation reduces lipoprotein oxidation in young men ingesting a diet high in saturated fat: Presumptive evidence for an interaction between vitamins C and E in vivo, *American Journal of Clinical Nutrition* 67 (1998): 240–245.

35. D. Kritchevsky, Antioxidant vitamins in the prevention of cardiovascular disease, *Nutrition Today,* January/February 1992, pp. 30–33.

36. M. M. Mahfouz, H. Kawano, and F. A. Kummerow, Effect of cholesterol-rich diets with and without added vitamins E and C on the severity of atherosclerosis in rabbits, *American Journal of Clinical Nutrition* 66 (1997): 1240–1249.

37. J. P. Moran and coauthors, Plasma ascorbic acid concentrations relate inversely to blood pressure in human subjects, *American Journal of Clinical Nutrition* 57 (1993): 213–217.

38. K. Nyyssönen and coauthors, Vitamin C deficiency and risk of myocardial infarction: Prospective population study of men from eastern Finland, *British Medical Journal* 314 (1997): 634–638.

39. M. Abbey, M. Noakes, and P. J. Nestel, Dietary supplementation with orange and carrot juice in cigarette smokers lowers oxidation products in copper-oxidized low-density lipoproteins, *Journal of the American Dietetic Association* 95 (1995): 671–675.

40. Heinonen, Huttunen, and Albanes, 1994.

F

Chapter 8

1. S. M. Kleiner, Water: An essential but overlooked nutrient, *Journal of the American Dietetic Association* 99 (1999): 200–206.

2. J. Chmielnicka and B. Sowa, Cadmium interaction with essential metals (Zn, Cu, Fe), metabolism metallothionein, and ceruloplasmin in pregnant rats and fetuses, *Ecotoxicology and Environmental Safety* 36 (1996): 277–281.

3. Standing Committee on the Scientific Evaluation of Dietary Reference Intakes, Food and Nutrition Board, Institute of Medicine, *Dietary Reference Intakes for Calcium, Phosphorus, Magnesium, Vitamin D, and Fluoride* (Washington, D.C.: National Academy Press, 1997), pp. 4-1–4-57.

4. C. M. Weaver and coauthors, Differences in calcium metabolism between adolescent and adult females, *American Journal of Clinical Nutrition* 61 (1995): 577–581.

5. Standing Committee on the Scientific Evaluation of Dietary Reference Intakes, 1997, pp. 5-6–5-7.

6. Standing Committee on the Scientific Evaluation of Dietary Reference Intakes, 1997, pp. 6-1–6-45.

7. J. L. Groff, S. S. Gropper, and S. M. Hunt, Macrominerals, in *Advanced Nutrition and Human Metabolism,* 2nd ed. (St. Paul, Minn.: West Publishing Company, 1995), pp. 325–351.

8. Standing Committee on the Scientific Evaluation of Dietary Reference Intakes, 1997, pp. 6-40–6-41; B. C. Coleman, Too many Maalox moments can kill, *Tallahassee Democrat,* 30 August 1995.

9. Scientists' statement regarding data on the sodium-hypertension relationship and sodium health claims on food labeling, *Nutrition Reviews* 55 (1997): 172–175.

10. N. M. Karanja and coauthors, Descriptive characteristics of the dietary patterns used in the Dietary Approaches to Stop Hypertension Trial, *Journal of the American Dietetic Association* 99 (1999): S19–S27; L. P. Svetkey and coauthors, Effects of dietary patterns on blood pressure: Subgroup analysis of the Dietary Approaches to Stop Hypertension (DASH) randomized clinical trial, *Archives of Internal Medicine* 159 (1999): 285–293.

11. M. H. Alderman, H. Cohen, and S. Madhavan, Dietary sodium intake and mortality: The National Health and Nutrition Examination Survey, *Lancet* 351 (1998): 781–785.

12. N. R. Poulter, Dietary sodium intake and mortality: NHANES, *The Lancet* 352 (1998): 987–988.

13. J. Stamler, The INTERSALT Study: Background, methods, findings, and implications, *American Journal of Clinical Nutrition* 65 (1997): S626–S642.

14. E. Saltos and S. Bowman, Dietary guidance on sodium: Should we take it with a grain of salt? *Family Economics and Nutrition Review, USDA* 11 (1998): 49–51.

15. N. A. Graudal and coauthors, Effects of sodium restriction on blood pressure, renin, aldosterone, catecholamines, cholesterol, and triglycerides, *Journal of the American Medical Association* 279 (1998): 1383–1391.

16. M. E. Reusser and D. A. McCarron, Micronutrient effects on blood pressure regulation, *Nutrition Reviews* 52 (1994): 367–375.

17. F. P. Cappuccio and coauthors, Double-blind randomized trial of modest salt restriction in older people, *Lancet* 350 (1997): 850–854.

18. T. F. Antonios and G. A. MacGregor, Deleterious effects of salt intake other than effects on blood pressure, *Clinical and Experimental Pharmacology and Physiology* 22 (1995): 180–184.

19. H. Hwang, J. Dwyer, and R. M. Russell, Diet, Heliobacter pylori infection, food preservation and gastric cancer risk: Are there new roles for preventative factors? *Nutrition Reviews* 52 (1994): 75–83.

20. P. K. Whelton and coauthors, Effects of oral potassium on blood pressure, *Journal of the American Medical Association* 277 (1997): 1624–1632.

21. N. Bleichrodt and coauthors, The benefits of adequate iodine intake, *Nutrition Reviews* 54 (1996): S72–S78; C. Xue-Yi and coauthors, Timing of vulnerability of the brain to iodine deficiency in endemic cretinism, *New England Journal of Medicine* 331 (1994): 1739–1744.

22. S. C. Das, U. P. Isichei, and P. O. Obekpa, Iodine deficiency disorders in pre-adolescent and adolescent children in Nigeria, West Africa, *West African Journal of Medicine* 17 (1998): 113–120; G. R. DeLong, Effects of nutrition on brain development in humans, *American Journal of Clinical Nutrition* 57 (1993): S286–S290.

23. B. Elnagar and coauthors, Control of iodine deficiency using iodination of water in a goiter endemic area, *International Journal of Food Sciences and Nutrition* 48 (1997): 119–127; B. S. Hetzel, Iodine deficiency and fetal brain damage, *New England Journal of Medicine* 331 (1994): 1770–1771.

24. D. C. Rocky and J. P. Cello, Evaluation of the gastrointestinal tract in patients with iron-deficiency anemia, *New England Journal of Medicine* 329 (1993): 1691–1695.

25. Centers for Disease Control and Prevention, Recommendations to prevent and control iron deficiency in the United States, *Morbidity and Mortality Weekly Report* (April supplement) 47 (1998), 3; R. Yip and P. R. Dallman, Iron, in E. E. Ziegler and L. J. Filer, Jr., eds., *Present Knowledge in Nutrition,* 7th ed. (Washington, D.C.: International Life Sciences Institute Press, 1996), pp. 277–292.

26. Iron deficiency, *Nutrition and the M.D.,* August 1994, p. 3.

27. J. M. McCord, Effects of positive iron status at a cellular level, *Nutrition Reviews* 54 (1996): 85–88.

28. D. P. Mascotti, D. Rup, and R. E. Thach, Regulation of iron metabolism, *Annual Review of Nutrition* 15 (1995): 239–261.

29. J. T. Salonen and coauthors, High stored iron levels are associated with excess risk of myocardial infarction in Eastern Finnish men, *Circulation* 86 (1992): 803–811.

30. H. van Jaarsveld, G. F. Pool, and H. C. Barnard, Dietary iron concentration alters LDL oxidatively: The effect of antioxidants, *Research Communications in Molecular Pathology and Pharmacology* 99 (1998): 69–80.

31. C. T. Sempos, A. C. Looker, and R. F. Gillum, Iron and heart disease: The epidemiological data, *Nutrition Reviews* 54 (1996): 73–84.

32. R. L. Nelson and coauthors, Body iron stores and risk of colonic neoplasia, *Journal of the National Cancer Institute* 86 (1994): 455–460.

33. P. Idjradinata, W. E. Watkins, and E. Pollit, Adverse effect of iron supplementation on weight gain of iron-replete young children, *Lancet* 343 (1994): 1252–1254.

34. V. Herbert, S. Shaw, and E. Jayatilleke, Vitamin C supplements are harmful to lethal for the over 10% of Americans with high iron stores, *FASEB Journal* 8 (1994): A678.

35. FDA wants warnings on labels for iron supplements, *Journal of the American Dietetic Association* 95 (1995): 7.

36. R. J. Cousins, Zinc, in E. E. Ziegler and L. J. Filer, Jr., eds., *Present Knowledge in Nutrition,* 7th ed. (Washington, D.C.: International Life Sciences Institute Press, 1996), pp. 293–306.

37. A. H. Shankar and A. S. Prasad, Zinc and immune function: The biological basis of altered resistance to infection, *American Journal of Clinical Nutrition* 69 (1998): S447–S463.

38. N. W. Solomons, Mild human zinc deficiency produces an imbalance between cell-mediated and humoral immunity, *Nutrition Reviews* 56 (1998): 27–28.

39. M. S. Golub and coauthors, Activity and attention in zinc-deprived adolescent monkeys, *American Journal of Clinical Nutrition* 64 (1996): 908–915.

40. H. H. Sandstead, Requirements and toxicity of essential trace elements, illustrated by zinc and copper, *American Journal of Clinical Nutrition* 61 (1995): S621–S624.

41. Cousins, 1996.

42. M. L. Macknin and coauthors, Zinc gluconate lozenges for treating the common cold in children: A randomized controlled trial, *Journal of the American Medical Association* 279 (1998): 1962, 1967; S. B. Mossad and coauthors, Zinc gluconate lozenges for treating the common cold: A randomized, double-blind, placebo-controlled study, *Annals of Internal Medicine* 125 (1996): 81–88.

43. Cousins, 1996.

44. D. H. Holben and A. M. Smith, The diverse role of selenium within selenoproteins: A review, *Journal of the American Dietetic Association* 99 (1999): 836–843.

45. K. Yoshizawa and coauthors, Study of prediagnostic selenium level in toenails and the risk of advanced prostate cancer, *Journal of the National Cancer Institute* 90 (1998): 1219–1224; O. A. Levander and R. F. Burk, Selenium, in E. E. Ziegler and L. J. Filer, Jr., eds., *Present Knowledge in Nutrition,* 7th ed. (Washington, D.C.: International Life Sciences Institute Press, 1996), pp. 320–328.

46. P. R. Larsen and M. J. Berry, Nutritional and hormonal regulation of thyroid hormone deiodases, *Annual Review of Nutrition* 15 (1995): 323–352.

47. Standing Committee on the Scientific Evaluation of Dietary Reference Intakes, 1977, pp. 8-1–8-20.

48. Fluorides and fluorosis, *Nutrition and the M.D.,* February 1997, pp. 4–5.

49. American Academy of Pediatrics, Fluoride supplementation for children: Interim policy recommendations, *Pediatrics* 95 (1995): 777.

50. Position of the American Dietetic Association: The impact of fluoride on dental health, *Journal of the American Dietetic Association* 94 (1994): 1428–1431.

51. B. D. Gessner and coauthors, Acute fluoride poisoning from a public water system, *New England Journal of Medicine* 330 (1994): 95–99.

52. D. Littlefield, Chromium decreases blood glucose in a patient with diabetes (letter), *Journal of the American Dietetic Association* 94 (1994): 1368.

53. R. A. Anderson, Effects of chromium on body composition and weight loss, *Nutrition Reviews* 56 (1998): 266–270; W. J. Pasman, M. S. Weterterp-Plantenga, and W. H. Saris, The effectiveness of long-term supplementation of carbohydrate, chromium, fiber and caffeine on weight maintenance, *International Journal of Obesity Related Metabolic Disorders* 21 (1997): 1143–1151; H. C. Lukaski and coauthors, Chromium supplementation and resistance training: Effects on body composition, strength, and trace element status of men, *American Journal of Clinical Nutrition* 63 (1996): 954–965.

54. V. A. Dubrovskaya and K. E. Wetterhahn, Effects of Cr(IV) on the expression of the oxidative stress genes in human lung cells, *Carcinogenesis* (1998): 1401–1407.

55. R. A. Anderson, Chromium supplements safe, *U.S.D.A. Food and Nutrition Research Brief,* October 1997, pp. 5–6, available from www.nal.usda.gov/fnic/usda/fnrb/fnrb1097.html.

56. W. Mertz, Interaction of chromium with insulin: A progress report, *Nutrition Reviews* 56 (1998): 174–177.

57. R. Uauy, M. Olivares, and M. Gonzalez, Essentiality of copper in humans, *American Journal of Clinical Nutrition* 67 (1998): S952–S959.

58. A. Cordano, Clinical manifestations of nutritional copper deficiency in infants and children, *American Journal of Clinical Nutrition* 67 (1998): S1012–S1016.

59. D. S. Kelly and coauthors, Effects of low-copper diets on human immune response, *American Journal of Clinical Nutrition* 62 (1995): 412–416; Decreased dietary copper impairs vascular function, *Nutrition Reviews* 51 (1993): 188–189.

60. Sandstead, 1995.

61. D. J. Fitzgerald, Safety guidelines for copper in water, *American Journal of Clinical Nutrition* 67 (1998): S1098–S1102.

62. National Dairy Council, Calcium Summit, *Breaking News* available on the Internet at www.nationaldairycouncil.org/nwbdbrek.html

Consumer Corner 8

1. Another example is found in Outbreaks of cyclosporiasis—United States, 1997, *Morbidity and Mortality Weekly Report* 46 (1997): 451–452; Position of the American Dietetic Association: Food and water safety, *Journal of the American Dietetic Association* 97 (1997): 184–189.

2. T. J. Doyle and coauthors, The association of drinking water source and chlorination by-products with cancer incidence among postmenopausal women in Iowa: A prospective cohort study, *American Journal of Public Health* 87 (1997): 1168–1176.

3. R. Lipin, Electron beam cleans dirty water, *Science News* 148 (1995): 171.

4. National Sanitation Foundation, Consumer guide to bottled water, *Information for Consumers,* March 1998; V. Lambert, Bottled water: New trends, new rules, *FDA Consumer,* June 1993, pp. 9–11.

5. Food and Drug Administration, Quality standards for foods with no identity standards: Bottled waters, *Federal Register* 59 (1994): 61529–61538.

6. A. M. Weissman, Bottled water use in an immigrant community: A public health issue? (letter), *American Journal of Public Health* 87 (1997): 1379–1380.

7. P. Garzon and M. J. Eisenberg, Variation in the mineral content of commercially available bottled waters: Implications for health and disease, *American Journal of Medicine* 105 (1998): 125–130.

Controversy 8

1. J. D. Zuckerman, Hip fracture, *New England Journal of Medicine* 334 (1996): 1519–1525.

2. F. H. Anderson, Osteoporosis in men, *International Journal of Clinical Practice* 52 (1998): 176–180.

3. National Center for Injury Prevention and Control, Falls and hip fractures among the elderly, *Unintentional Injury Fact Sheet,* 1998 (available from www.cdc.gov); Standing Committee on the Scientific Evaluation of Dietary Reference Intakes, Food and Nutrition Board, Institute of Medicine, *Dietary Reference Intakes for Calcium, Phosphorus, Magnesium, Vitamin D, and Fluoride* (Washington, D.C.: National Academy Press, 1997), pp. 4-1–4-57.

4. Zuckerman,1996.

5. T. D. Galsworthy and P. L. Wilson, Osteoporosis: It steals more than bone, *American Journal of Nursing* 96 (1996): 27–32.

6. Centers for Disease Control and Prevention, Osteoporosis among estrogen-deficient women—United States, 1988–1994, *Morbidity and Mortality Weekly Report* 47 (1998): 969–973.

7. M. K. Jeffcoat, Osteoporosis: A possible modifying factor in oral bone loss, *Annals of Periodontology* 3 (1998): 312–321.

8. E. Siris and coauthors, Design of NORA, the National Osteoporosis Risk Assessment Program: A longitudinal US registry of postmenopausal women, *Osteoporosis International* 8 (1998): S62–S69.

9. C. J. Rosen and L. R. Donahue, Insulin-like growth factor and bone: The osteoporosis connection revisited, *Proceedings of the Society for Experimental Biological Medicine* 219 (1998): 1–7.

10. I. R. Reid, The roles of calcium and vitamin D in the prevention of osteoporosis, *Endocrinology and Metabolism Clinics of North America* 27 (1998): 389–398; K. O. O'Brien, Combined calcium and vitamin D supplementation reduces bone loss and fracture incidence in older men and women, *Nutrition Reviews* 56 (1998): 148–150.

11. R. P. Heaney, Bone mass, nutrition, and other lifestyle factors, *Nutrition Reviews* 54 (1996): S3–S10.

12. A. E. Andersen and J. Holman, Diagnosing and treating bone mineral deficiency, *Eating Disorders Review,* January/February 1997, pp. 1–4.

13. E. Barrett-Connor, Hormone replacement therapy, *British Medical Journal* 317 (1998): 457–461; A. Volpe and coauthors, Oral contraceptives and bone metabolism, *European Journal of Contraception and Reproductive Health Care* 2 (1997): 225–228.

14. E. Velazquez and G. A. Bellabarba, Testosterone replacement therapy, *Archives of Andrology* 41 (1998): 79–90.

15. C. J. Strange, Boning up on osteoporosis, *FDA Consumer,* September 1996, pp. 15–20.

16. S. H. Ralston, Science, medicine, and the future: Osteoporosis, *British Medical Journal* 315 (1997): 469–472; S. R. Cummings and coauthors, Risk factors for hip fractures in white women, *New England Journal of Medicine* 332 (1995): 767–773.

17. R. J. Wood and J. C. Fleet, The genetics of osteoporosis: Vitamin D receptor polymorphisms, *Annual Review of Nutrition* 18 (1998): 233–258.

18. W. C. Annie and coauthors, Age-related osteoporosis in Chinese: An evaluation of the response of intestinal calcium absorption and calcitropic hormones to dietary calcium deprivation, *American Journal of Clinical Nutrition* 68 (1998): 1291–1297.

19. T. V. Nguyen, P. N. Sambrook, and J. A. Eisman, Bone loss, physical activity, and weight change in elderly women: The Dubbo Osteoporosis Epidemiology Study, *Journal of Bone Mineral Research* 13 (1998): 1458–1467.

20. S. Suleiman and coauthors, Effect of calcium intake and physical activity level protect bone mass and turnover in healthy white postmenopausal women, *American Journal of Clinical Nutrition* 66 (1997): 937–943.

21. N. K. Henderson, C. P. White, and J. A. Eisman, The roles of exercise and fall risk reduction in the prevention of osteoporosis, *Endocrinology and Metabolism Clinics of North America* 27 (1998): 369–387.

22. R. D. Lewis and C. M. Modlesky, Nutrition, physical activity, and bone health in women, *International Journal of Sports Nutrition* 8 (1998): 250–284.

23. E. Ernst, Exercise for female osteoporosis: A systematic review of randomized clinical trials, *Sports Medicine* 25 (1998): 359–368.

24. J. F. Aloia and coauthors, To what extent is bone mass determined by fat-free or fat mass? *American Journal of Clinical Nutrition* 61 (1995): 1110–1114; Cummings and coauthors, 1995; S. L. Edelstein and E. Barrett-Connor, Relation between body size and bone mineral density in elderly men and women, *American Journal of Epidemiology* 138 (1993): 160–169; I. R. Reid and coauthors, Determinants of total body and regional bone mineral density in normal post-menopausal women—A key role for fat mass, *Journal of Clinical Endocrinology and Metabolism* 75 (1992): 45–51.

25. J. L. Hopper and E. Seeman, The bone density of female twins discordant for tobacco use, *New England Journal of Medicine* 330 (1994): 387–392.

26. M. N. Hadley and S. V. Reddy, Smoking and the human vertebral column: A review of the impact of cigarette use on vertebral bone metabolism and spinal fusion, *Neurosurgery* 41 (1997): 116–124.

27. C. W. Slemenda, Cigarettes and the skeleton, *New England Journal of Medicine* 330 (1994): 430–431.

28. S. W. Sampson, Alcohol, osteoporosis, and bone regulating hormones, *Alcoholism: Clinical and Experimental Research* 21 (1997): 400–403; R. F. Klein, Alcohol-induced bone disease: Impact of ethanol on osteoblast proliferation, *Alcoholism: Clinical and Experimental Research* 21 (1997): 392–399.

29. H. W. Sampson and D. Shipley, Moderate alcohol consumption does not augment bone density in ovariectomized rats, *Alcohol: Clinical and Experimental Research* 21 (1997): 1165–1168; Klein, 1997.

30. T. Lloyd and coauthors, Dietary caffeine intake and bone status of postmenopausal women, *American Journal of Clinical Nutrition* 65 (1997): 1826–1830.

31. S. S. Harris and B. Dawson-Hughes, Caffeine and bone loss in healthy postmenopausal women, *American Journal of Clinical Nutrition* 60 (1994): 573–578.

32. U. S. Barzel and L. K. Massey, Excess dietary protein can adversely affect bone, *Journal of Nutrition* 128 (1998): 1051–1053.

33. D. Feskanich and coauthors, Protein consumption and bone fractures in women, *American Journal of Epidemiology* 43 (1996): 472–479.

34. R. P. Heaney, Excess dietary protein may not adversely affect bone, *Journal of Nutrition* 128 (1998): 1054–1057.

35. R. Itoh, N. Nishiyama, and Y. Suyama, Dietary protein intake and urinary excretion of calcium: A cross-sectional study in a healthy Japanese population, *American Journal of Clinical Nutrition* 67 (1998): 438–444.

36. A. D. Maravilla nd coauthors, Acute effects of soft drink intake on calcium and phosphate metabolism in immature and adult rats (abstract) *Revista de Investigacion Clinica* 50 (1998): 185–189; K. Lau and coauthors, Differing effects of acid versus neutral phosphate therapy of hypercalcuria, *Kidney International* (1979) as cited by Barzel and Massey, 1998.

37. E. Petridou and coauthors, The role of dairy products and non alcoholic beverages in bone fractures among schoolage children, *Scandinavian Journal of Social Medicine* 25 (1997): 119–125; G. Wyshak and

R. E. Frisch, Carbonated beverages, dietary calcium, the dietary calcium/phosphorus ratio, and bone fractures in girls and boys, *Journal of Adolescent Health* 15 (1994): 210–215.

38. L. Harnack, J. Stang, and M. Story, Soft drink consumption among U.S. children and adolescents: Nutritional consequences, *Journal of the American Dietetic Association* 99 (1999): 436–441; Federal Interagency Forum on Child and Family Statistics, *America's Children: Key National Indicators of Well-Being, 1999,* available on the Internet at http://childstats.gov or from the National Maternal and Child Health Clearinghouse, 2070 Chain Bridge Road, Suite 450, Vienna, VA 22182; USDA, Third Report on Nutrition Monitoring in the United States, Executive summary, *Journal of Nutrition* 126 (1996): S1907–S1936.

39. Standing Committee on the Scientific Evaluation of Dietary Reference Intakes, 1997.

40. A. Devine and coauthors, A longitudinal study of the effect of sodium and calcium intakes on regional bone density in postmenopausal women, *American Journal of Clinical Nutrition* 62 (1995): 740–745.

41. F. Ginty, A. Flynn, and K. D. Cashman, The effect of dietary sodium intake on biochemical markers of bone metabolism in young women, *British Journal of Nutrition* 79 (1998): 343–350.

42. P. T. Packard and R. P. Heaney, Medical nutrition therapy for patients with osteoporosis, *Journal of the American Dietetic Association* 97 (1997): 414–417.

43. J-P. Bonjour, M-A. Schurch, and R. Rizzoli, Nutritional aspects of hip fractures, *Bone* 18 (1996): S139–S144.

44. G. Ferland, The vitamin K–dependent proteins: An update, *Nutrition Reviews* 56 (1998): 223–230.

45. A. M. Craciun and coauthors, Improved bone metabolism in female elite athletes after vitamin K supplementation, *International Journal of Sports Medicine* 19 (1998): 479–484.

46. D. Feskanich and coauthors, Vitamin K intake and hip fractures in women: A prospective study, *American Journal of Clinical Nutrition* 69 (1999): 74–79.

47. I. R. Reid and coauthors, Effect of calcium supplementation on bone loss in postmenopausal women, *New England Journal of Medicine* 328 (1993): 460–464.

48. S. J. Whiting and R. J. Wood, Adverse effects of high-calcium diets in humans, *Nutrition Reviews* 55 (1997): 1–9.

49. Whiting and Wood, 1997.

50. D. I. Levenson and R. S. Bockman, A review of calcium preparations, *Nutrition Reviews* 52 (1994): 221–232.

51. Committee on Diet and Health, *Diet and Health: Implications for Reducing Chronic Disease Risk* (Washington, D.C.: National Academy Press, 1989), p. 17.

Chapter 9

1. J. P. Flatt, How NOT to approach the obesity problem (comment), *Obesity Research* 5 (1997): 632–633.

2. J. Calles-Escandon and E. T. Poehlman, Aging, fat oxidation, and exercise, *Aging* 9 (1997): 57–63.

3. Y. Schutz and E. Jéquier, Energy needs: Assessment and requirements, in M. E. Shils, J. A. Olson, and M. Shike, eds., *Modern Nutrition in Health and Disease* (Philadelphia: Lea & Febiger, 1994), pp. 101–111.

4. L. O. Schulz and D. A. Schoeller, A compilation of total daily energy expenditures and body weights in healthy adults, *American Journal of Clinical Nutrition* 60 (1994): 676–681.

5. T. J. Horton and C. A. Geissler, Effect of habitual exercise on daily energy expenditure and metabolic rate during standardized activity, *American Journal of Clinical Nutrition* 59 (1994): 13–19.

6. P. N. Singh and K. D. Lindsted, Body mass and 26-year risk of mortality from specific diseases among women who never smoked, *Epidemiology* 9 (1998): 246–254.

7. World Health Organization, as quoted in J. M. Rippe, S. Crossley, and R. Ringer, Obesity as a chronic disease: Modern medical and lifestyle management, *Journal of the American Dietetic Association* 98 (1998): S9–S15.

8. National Heart, Lung, and Blood Institute Expert Panel, National Institutes of Health, *Clinical Guidelines on the Identification, Evaluation, and Treatment of Overweight and Obesity in Adults* (Washington, D.C.: Government Printing Office, 1998).

9. Update: Prevalence of overweight among children, adolescents, and adults—United States, 1988–1994, *Morbidity and Mortality Weekly Report* 46 (1997): 199–202.

10. S. E. Gariballa and coauthors, Nutritional status of hospitalized acute stroke patients, *British Journal of Nutrition* 79 (1998): 481–487.

11. R. H. Eckel and R. M. Krauss, American Heart Association call to action: Obesity as a major risk factor for coronary heart disease, *Circulation* 97 (1998): 2099–2100.

12. S. W. Farrell and coauthors, Influences of cardiorespiratory fitness levels and other predictors on cardiovascular disease mortality in men, *Medicine and Science in Sports and Exercise* 30 (1998): 899–904.

13. National Heart, Lung, and Blood Institute Expert Panel, 1998.

14. 1997 Heart and Stroke Statistical Update, Dallas, Texas, as cited in Rippe, Crossley, and Ringer, 1998.

15. D. A. McCarron and M. E. Reusser, Body weight and blood pressure regulation, *American Journal of Clinical Nutrition* 63 (1996): S423–S425.

16. Rippe, Crossley, and Ringer, 1998.

17. G. A. Colditz and coauthors, Weight gain as a risk factor for clinical diabetes mellitus in women, *Annals of Internal Medicine* 122 (1995): 481–486.

18. J. E. Manson and coauthors, Body weight and mortality among women, *New England Journal of Medicine* 333 (1995): 677–685.

19. T. K. Young and D. E. Gelskey, Is noncentral obesity metabolically benign? Implications for prevention from a population survey, *Journal of the American Medical Association* 274 (1995): 1939–1941.

20. K. N. Frayn, Regulation of fatty acid delivery in vivo, *Advances in Experimental Medicine and Biology* 441 (1998): 171–179.

21. M. D. Jensen, Lipolysis: Contribution from regional fat, *Annual Review of Nutrition* 17 (1997): 127–139; P. R. Jones and D. A. Edwards, Areas of fat loss in overweight young females following an 8-week period of energy intake reduction, *Annals of Human Biology* 26 (1999): 151–162.

22. K. M. Flegal and coauthors, The influence of smoking cessation on the prevalence of overweight in the United States, *New England Journal of Medicine* 333 (1995): 1165–1170.

23. S. L. Gortmaker and coauthors, Social and economic consequences of overweight in adolescence and young adulthood, *New England Journal of Medicine* 329 (1993): 1008–1012.

24. J. C. Seidell, Societal and personal costs of obesity, *Experimental and Clinical Endocrinology and Diabetes* 106 (1998): S7–S9.

25. National Heart, Lung, and Blood Institute Expert Panel, 1998.

26. National Heart, Lung, and Blood Institute Expert Panel, 1998.

27. W. H. Dietz and M. C. Bellizzi, Introduction: The use of body mass index to assess obesity in children, *The American Journal of Clinical Nutrition* 70 (1999): S123–S125; M. C. Bellizzi and W. H. Dietz, Workshop on childhood obesity: Summary of the discussion, *The American Journal of Clinical Nutrition* 70 (1999): S173–S175.

28. M. E. J. Lean, H. S. Han, and C. E. Morrison, Waist circumference as a measure for indicating need for weight management, *British Medical Journal* 311 (1995): 158–161.

29. C. Orphanidou and coauthors, Accuracy of subcutaneous fat measurement: Comparison of skinfold calipers, ultrasound, and computed tomography, *Journal of the American Dietetic Association* 94 (1994): 855–858.

30. National Heart, Lung, and Blood Institute Expert Panel, 1998.

31. G. A. Bray, An approach to the classification and evaluation of obesity, in P. Bjorntorp and B. N. Brodoff, eds., *Obesity* (Philadelphia: J. B. Lippincott, 1992), pp. 294–308.

32. J. E. Blundell and A. Tremblay, Appetite control and energy (fuel) balance, *Nutrition Research Reviews* 8 (1995): 225–242.

33. A. S. Livine and C. J. Billington, Why do we eat? A neural systems approach, *Annual Review of Nutrition* 17 (1997): 597–619.

34. A. Geliebter and coauthors, Reduced stomach capacity in obese subjects after dieting, *American Journal of Clinical Nutrition* 63 (1996): 170–173.

35. T. A. Spiegel and coauthors, Contribution of gastric and postgastric feedback to satiation and satiety in women, *Physiology and Behavior* 62 (1997): 1125–1136.

36. A. L. Hirschberg, Hormonal regulation of appetite and food intake, *Annals of Medicine* 30 (1998): 7–20.

37. S. H. A. Holt and coauthors, A satiety index of common foods, *European Journal of Clinical Nutrition* 49 (1995): 675–690.

38. L. Pérusse and C. Bouchard, Genotype-environment interaction in human obesity, *Nutrition Reviews* 57 (1999): S31–S38.

39. J. P. Foreyt and W. S. C. Poston II, Diet, genetics, and obesity, *Food Technology* 51 (1997): 70–73; C. Bouchard, Human variation in body mass: Evidence for a role of the genes, *Nutrition Reviews* 55 (1997): S21–S30.

40. C. Bouchard and A. Tremblay, Genetic influences on the response of body fat and fat distribution to positive and negative energy balances in human identical twins, *Journal of Nutrition* 127 (1997): S943–S947.

41. Y. Zhang and coauthors, Positional cloning of the mouse obese gene and its human homologue, *Nature* 372 (1994): 425–431.

42. Fifth obesity gene found in mice, *Science News* 149 (1996): 257–272.

43. H. Qian and coauthors, Brain administration of leptin causes deletion of adipocytes by apoptosis, *Endocrinology* 139 (1998): 791–794.

44. C. S. Mantzoros, The role of leptin in human obesity and disease: A review of current evidence, *Annals of Internal Medicine* 130 (1999): 671–680; F. Rohner-Jeanrenaud and B. Jeanrenaud, Obesity, leptin and the brain, *New England Journal of Medicine* 334 (1996): 324–325.

45. J. M. Friedman, Leptin, leptin receptors, and the control of body weight, *Nutrition Reviews* 56 (1998): S38–S46; R. V. Considine and coauthors, Serum immunoreactive–leptin concentrations in normal-weight and obese humans, *New England Journal of Medicine* 334 (1996): 292–295.

46. C. T. Montague and coauthors, Congenital leptin deficiency is associated with severe early-onset obesity in humans, *Nature* 387 (1997): 903–908.

47. R. B. Ceddia, Pivotal role of leptin in insulin effects, *Brazilian Journal of Medical and Biological Research* 31 (1998): 715–722; J. Girard, Is leptin the link between obesity and insulin resistance? *Diabetes and Metabolism* 23 (1997): 16–24.

48. Nir Barzilai and coauthors, Leptin selectively decreases visceral adiposity and enhances insulin action, *Journal of Clinical Investigation* 100 (1997): 3105–3110.

49. T. Thomas and coauthors, Leptin acts on human marrow stromal cells to enhance differentiation to osteoblasts and to inhibit differentiation to adipocytes, *Endocrinology* 140 (1999): 1630–1638; P. Trayhurn and coauthors, Hormonal and neuroendocrine regulation of energy balance—The role of leptin, *Arch Tierernahr* 51 (1998): 177–185; G. Frunbeck, S. A. Jebb, and A. M. Prentice, Leptin: Physiology and pathophysiology, *Clinical Physiology* 18 (1998): 399–419; M. R. Sierra-Honigmann and coauthors, Biological action of leptin as an angiogenic factor, *Science* 281 (1998): 1683–1686.

50. A. J. Stunkard, Body weight regulation, an address presented at the conference, *Obesity Update: Pathophysiology, Clinical Consequences and Therapeutic Options*, in Atlanta, Georgia, August 31–September 2, 1992.

51. R. L. Leibel, M. Rosenbaum, and J. Hirsch, Changes in energy expenditure resulting from altered body weight, *New England Journal of Medicine* 332 (1995): 621–628; E. Saltzman and S. B. Roberts, The role of energy expenditure in energy regulation: Findings from a decade of research, *Nutrition Reviews* 53 (1995): 209–220.

52. G. Wolf, A new uncoupling protein: A potential component of the human body weight regulation system, *Nutrition Reviews* 55 (1997): 178–179.

53. A. G. Dulloo and J. Jacquet, Adaptive reduction in basal metabolic rate in response to food deprivation in humans: A role for feedback signals from fat stores, *American Journal of Clinical Nutrition* 68 (1998): 599–606.

54. M. A. McCrory and coauthors, Dietary variety within food groups: Association with energy intake and body fatness in men and women, *American Journal of Clinical Nutrition* 69 (1999): 440–447.

55. J. E. Blundell and J. I. Macdiarmid, Fat as a risk factor for overconsumption: Satiation, satiety, and patterns of eating, *Journal of the American Dietetic Association* 97 (1997): S63–S69; L. H. Nelson and L. A. Tucker, Diet composition related to body fat in a multivariate study of 203 men, *Journal of the American Dietetic Association* 96 (1996): 771–777.

56. Prevalence of physical inactivity during leisure time among overweight persons—1994, *Morbidity and Mortality Weekly Report* 45 (1996): 183–188; R. Rising and coauthors, Determinants of total daily energy expenditure: Variability in physical activity, *American Journal of Clinical Nutrition* 59 (1994): 800–804.

57. J. M. Rippe and S. Hess, The role of physical activity in the prevention and management of obesity, *Journal of the American Dietetic Association* 98 (1998): S31–S38.

58. R. C. Klesges, M. L. Shelton, and L. M. Klesges, Effects of television on metabolic rate: Potential implications for childhood obesity, *Pediatrics* 91 (1993): 281–286.

59. National Heart, Lung, and Blood Institute Expert Panel, 1998.

60. J. Polivy, Psychological consequences of food restriction, *Journal of the American Dietetic Association* 96 (1996): 589–592.

61. M. E. Sweeny and coauthors, Severe vs moderate energy restriction with and without exercise in the treatment of obesity: Efficiency of weight loss, *American Journal of Clinical Nutrition* 57 (1993): 127–134.

62. S. M. Shick and coauthors, Persons successful at long-term weight loss and maintenance continue to consume a low-energy, low-fat diet, *Journal of the American Dietetic Association* 98 (1998): 408–413.

63. P. M. Suter, E. Häsler, and W. Vetter, Effects of alcohol on energy metabolism and body weight regulation: Is alcohol a risk factor for obesity? *Nutrition Reviews* 55 (1997): 157–171; Y. Sakurai and coauthors, Relation of total and beverage-specific alcohol intake to body mass index and waist-to-hip ratio: A study of self-defense officials in Japan, *European Journal of Epidemiology* 13 (1997): 893–898.

64. National Heart, Lung, and Blood Institute Expert Panel, 1998.

65. D. Festi and coauthors, Gallbladder motility and gallstone formation in obese patients following very low calorie diets: Use it (fat) to lose it (well), *International Journal of Obesity and Related Metabolic Disorders* 22 (1998): 592–600.

66. National Heart, Lung, and Blood Institute Expert Panel, 1998, p. 75.

67. National Heart, Lung, and Blood Institute Expert Panel, 1998, p. 75.

68. A. Golay and E. Bobbioni, The role of dietary fat in obesity, *International Journal of Obesity and Related Metabolic Disorders* 21 (1997): S2–S11.

69. W. J. McCarthy, Strategies for achieving long-term weight maintenance (letter), *Journal of the American Dietetic Association* 98 (1998): 1273.

70. D. G. Schlundt and coauthors, The role of breakfast in the treatment of obesity: A randomized clinical trial, *American Journal of Clinical Nutrition* 55 (1992): 645–651.

71. M. T. McGuire and coauthors, Long-term maintenance of weight loss: Do people who lose weight through various weight loss methods use different behaviors to maintain their weight? *International Journal of Obesity and Related Metabolic Disorders* 22 (1998): 572–577.

72. W. J. Pasman, M. S. Westerterp-Plantenga, and W. H. Saris, The effectiveness of long-term supplementation of carbohydrate, chromium, fiber and caffeine on weight maintenance, *International Journal of Obesity and Related Metabolic Disorders* 21 (1997): 1143–1151.

73. McCarthy, 1998.

74. P. D. Wood, Clinical applications of diet and physical activity in weight loss, *Nutrition Reviews* 54 (1998): S131–S135; M. L. Klem and coauthors, A descriptive study of individuals successful at long-term maintenance of substantial weight loss, *American Journal of Clinical Nutrition* 66 (1997): 239–246.

75. K. P. G. Kempen, W. H. M. Saris, and K. R. Westerterp, Energy balance during an 8-wk energy-restricted diet with and without exercise in obese women, *American Journal of Clinical Nutrition* 62 (1995): 722–729.

76. G. Haus and coauthors, Key modifiable factors in weight maintenance: Fat intake, exercise, and weight cycling, *Journal of the American Dietetic Association* 94 (1994): 409–413.

77. M. L. Klem and coauthors, A descriptive study of individuals successful at long-term maintenance of substantial weight loss, *American Journal of Clinical Nutrition* 66 (1997): 239–246.

Consumer Corner 9

1. P. S. Powers and coauthors, Outcome of gastric restriction procedures: Weight, psychiatric diagnoses, and satisfaction, *Obesity Surgery* 7 (1997): 471–477.

2. National Heart, Lung, and Blood Institute Expert Panel, National Institutes of Health, *Clinical Guidelines on the Identification, Evaluation, and Treatment of Overweight and Obesity in Adults* (Washington, D.C.: Government Printing Office, 1998).

3. Powers and coauthors, 1997.

4. R. Weiner, D. Wagner, and H. Bockhorn, Laparoscopic gastric banding for morbid obesity, *Journal of Laparoendoscopic and Advanced Surgical Techniques* 9 (1999): 23–30.

5. National Heart, Lung, and Blood Institute Expert Panel, 1998; J. D. Halverson, Metabolic risk of obesity surgery and long-term follow-up, *American Journal of Clinical Nutrition* 55 (1992): S602–S605.

6. National Task Force on the Prevention and Treatment of Obesity, Long-term pharmacotherapy in the management of obesity, *Journal of the American Medical Association* 276 (1996): 1907–1915.

7. National Heart, Lung, and Blood Institute Expert Panel, 1998; R. L. Atkinson, Use of drugs in the treatment of obesity, *Annual Review of Nutrition* 17 (1997): 383–403.

8. M. A. Khan and coauthors, The prevalence of cardiac valvular insufficiency assessed by transthoracic echocardiography in obese patients treated with appetite-suppressant drugs, *New England Journal of Medicine* 339 (1998): 713–718; H. M. Connolly and coauthors, Valvular heart disease associated with fenfluramine-phentermine, *New England Journal of Medicine* 337 (1997): 581–588.

9. L. J. Aronne, Modern medical management of obesity: The role of pharmaceutical intervention, *Journal of the American Dietetic Association* 98 (1998): S23–S26.

10. M. H. Davidson and coauthors, Weight control and risk factor reduction in obese subjects treated for 2 years with orlistat, *Journal of the American Medical Association* 281 (1999): 235–242.

11. National Heart, Lung, and Blood Institute Expert Panel, 1998.

12. P. A. Rushing and coauthors, Acute administration of phenylpropanolamine fails to affect resting energy expenditure in men of normal weight, *Obesity Research* 5 (1997): 470–473.

13. J. Tapia, Cerebral hemorrhage associated with the use of phenylpropanolamine, *Revista Medica Chile* 124 (1996): 1499–1503.

14. FDA warns against drug promotion or "herbal fen-phen," *FDA Talk Paper,* 1997, available from the Food and Drug Administration, 5600 Fishers Lane, Rockville, MD 20857.

15. U.S. Department of Health and Human Services, Food and Drug Administration, HHS News (electronic bulletin board), June 2, 1997; Adverse events associated with ephedrine-containing products—Texas, December 1993–September 1995, *Morbidity and Mortality Weekly Report* 45 (1996): 689–693.

16. P. Kurtzweil, Dieter's brews make tea time a dangerous affair, *FDA Consumer*, July/August 1997, pp. 6–11.

Fitness for Life 9

1. National Heart, Lung, and Blood Institute Expert Panel, National Institutes of Health, *Clinical Guidelines on the Identification, Evaluation, and Treatment of Overweight and Obesity in Adults* (Washington, D.C.: Government Printing Office, 1998), p. 78.

2. C. J. Zelasko, Exercise for weight loss: What are the facts? *Journal of the American Dietetic Association* 95 (1995): 1414–1417.

Controversy 9

1. National Heart, Lung, and Blood Institute Expert Panel, National Institutes of Health, *Clinical Guidelines on the Identification, Evaluation, and Treatment of Overweight and Obesity in Adults* (Washington, D.C.: Government Printing Office, 1998).

2. A. M. Wolf and G. A. Colditz, Social and economic effects of body weight in the United States, *American Journal of Clinical Nutrition* 63 (1996): S466–S469.

3. J. E. Manson and coauthors, Body weight and mortality among women, *New England Journal of Medicine* 333 (1995): 677–685.

4. K. D. Lindsted and P. N. Singh, Body mass and 26-year risk of mortality among women who never smoked: Findings from the Aventist mortality study, *American Journal of Epidemiology* 146 (1997): 1–11; R. P. Troiano, The relationship between body weight and mortality: A quantitative analysis of combined information from existing studies, *International Journal of Obesity* 20 (1996): 63–75.

5. S. W. Farrell and coauthors, Influences of cardiorespiratory fitness levels and other predictors on cardiovascular disease mortality in men, *Medicine and Science in Sports and Exercise* 30 (1998): 899–904.

6. L. Pérusse and C. Bouchard, Genotype-environment interaction in human obesity, *Nutrition Reviews* 57 (1999): S31–S38.

7. *The Third Report on Nutrition Monitoring in the United States: Executive Summary* (Washington, D.C: Government Printing Office, 1996); M. Stern, Epidemiology of obesity and its link to heart disease, *Metabolism: Clinical and Experimental* 44 (1995): 1–3.

8. J. G. Meisler and S. St. Jeor, Summary and recommendations from the American Health Foundation's Expert Panel on Healthy Weight, *American Journal of Clinical Nutrition* 63 (1996): S474–S477.

9. J. Stevens and coauthors, The effect of age on the association between body-mass index and mortality, *New England Journal of Medicine* 338 (1998): 1–7.

10. I-M. Lee, Is weight loss hazardous? *Nutrition Reviews* 54 (1996): S116–S124.

11. R. E. Andersen, T. A. Wadden, and R. J. Herzog, Changes in bone mineral content in obese dieting women, *Metabolism: Clinical and Experimental* 46 (1997): 857–861.

12. M. J. Kretsch and coauthors, Cognitive effects of a long-term weight reducing diet, *International Journal of Obesity and Related Metabolic Disorders* 21 (1997): 14–21.

13. M. J. Kreych and coauthors, Cognitive function, iron status, and hemoglobin concentration in obese dieting women, *European Journal of Clinical Nutrition* 52 (1998): 512–518.

14. J. Plivy, Psychological consequences of food restriction, *Journal of the American Dietetic Association* 96 (1996): 589–592.

15. A. G. Dulloo, J. Jacquet, and L. Girardier, Poststarvation hyperphasia and body fat overshooting in humans: A role for feedback signals from lean and fat tissues, *American Journal of Clinical Nutrition* 65 (1997): 717–723.

16. G. C. Patton and coauthors, Onset of adolescent eating disorders: Population based cohort study over 3 years, *British Journal of Medicine* 318 (1999): 765–768.

17. Lee, 1996.

18. A. Frank, Accreditation and certification: The rationalization of obesity-management services, *American Journal of Clinical Nutrition* 62 (1995): 439–440.

19. G. L. Blackburn, Treatment of obesity is imperative: Physicians can make a difference, an address presented at the conference, *Obesity Update*, 1992.

20. Farrell and coauthors, 1998; L. DiPietro and coauthors, Improvements in cardiorespiratory fitness attenuate age-related weight gain in healthy men and women: The Aerobics Center Longitudinal Study, *International Journal of Obesity and Related Metabolic Disorders* 22 (1998): 55–62.

Chapter 10

1. A. L. Dunn and coauthors, Comparison of lifestyle and structured interventions to increase physical activity and cardiorespiratory fitness, *Journal of the American Medical Association* 281 (1999): 327–334.

2. U. M. Kujala and coauthors, Relationship of leisure-time physical activity and mortality, *Journal of the American Medical Association* 279 (1998): 440–444; S. N. Blair and coauthors, Changes in physical fitness and all-cause mortality, *Journal of the American Medical Association* 263 (1995): 1093–1098; R. S. Paffenbarger and coauthors, The association of changes in physical-activity level and other lifestyle characteristics with mortality among men, *New England Journal of Medicine* 328 (1993): 538–545.

3. S. A. Oliveria and P. J. Christos, The epidemiology of physical activity and cancer, *Annals of the New York Academy of Sciences* 833 (1997): 79–90; G. Perseghin and coauthors, Increased glucose transport-phosphorylation and muscle glycogen synthesis after exercise training in insulin-resistant subjects, *New England Journal of Medicine* 335 (1996): 1357–1362; S. N. Blair, Physical inactivity and cardiovascular disease risk in women, *Medicine and Science in Sports and Exercise* 28 (1996): 9–10; NIH Consensus Development Panel on Physical Activity and Cardiovascular Health, Physical activity and cardiovascular health, *Journal of the American Medical Association* 276 (1996): 241–246; R. R. Pate and coauthors, Physical activity and public health: A recommendation from the Centers for Disease Control and Prevention and the American College of Sports Medicine, *Journal of the American Medical Association* 273 (1995): 402–407; A. M. Bovens and coauthors, Physical activity, fitness, and selected risk factors for CHD in active men and women, *Medicine and Science in Sports and Exercise* 25 (1993): 572–576.

4. American College of Sports Medicine, Position stand: The recommended quantity and quality of exercise for developing and maintaining cardiorespiratory and muscular fitness, and flexibility in healthy adults, *Medicine and Science in Sports and Exercise* 30 (1998): 975–991.

5. D. A. Leaf, D. L. Parker, and D. Schaad, Changes in VO_2 max, physical activity, and body fat with chronic exercise: Effects on plasma lipids, *Medicine and Science in Sports and Exercise* 29 (1997): 1152–1159; P. T. Williams, Relationship of distance run per week to coronary heart disease risk factors in 8283 male runners, *Archives of Internal Medicine* 157 (1997): 191–198.

6. P. A. Ades and coauthors, Weight training improves walking endurance in healthy elderly persons, *Annals of Internal Medicine* 124 (1996): 568–572.

7. J. M. Guralnik and coauthors, Lower extremity function in persons over the age of 70 years as a predictor of subsequent disability, *New England Journal of Medicine* 332 (1995): 556–561.

8. J. E. Layne and M. E. Nelson, The effects of progressive resistance training on bone density: A review, *Medicine and Science in Sports and Exercise* 31 (1999): 25–30; American College of Sports Medicine, Position stand: Osteoporosis and exercise, *Medicine and Science in Sports and Exercise* 27 (1995): i–vii.

9. C. M. L. Snow-Harter, Effects of resistance and endurance exercise on bone mineral status of young women: A randomized exercise intervention trial, *Journal of Bone Mineral Research* 7 (1992): 761–769.

10. P. T. Williams, High-density lipoprotein cholesterol and other risk factors for coronary heart disease in female runners, *New England Journal of Medicine* 334 (1996): 1298–1303.

11. E. F. Coyle, Substrate utilization during exercise in active people, *American Journal of Clinical Nutrition* 61 (1995): S968–S974.

12. C. M. Donovan and K. D. Sumida, Training enhanced hepatic gluconeogenesis: The importance for glucose homeostasis during exercise, *Medicine and Science in Sports and Exercise* 29 (1997): 628–634.

13. C. Williams and C. Chryssanthopoulos, Pre-exercise food intake and performance, in A. P. Simopoulos and K. N. Pavlou, eds., *Nutrition and Fitness: Metabolic and Behavioral Aspects in Health and Disease* (New York: Karger, 1997), pp. 33–45; Coyle, 1995.

14. Coyle, 1995.

15. Williams and Chryssanthopoulos, 1997.

16. J. A. M. Parkin and coauthors, Muscle glycogen storage following prolonged exercise: Effect of timing of ingestion of high glycemic index food, *Medicine and Science in Sports and Exercise* 29 (1997): 220–224; Coyle, 1995.

17. A. R. Coggan, Plasma glucose metabolism during exercise: Effect of endurance training in humans, *Medicine and Science in Sports and Exercise* 29 (1997): 620–627.

18. Diabetes in the elite athlete, *Sports Medicine Digest* 18 (1998): 109–111; C. S. Moy and coauthors, Insulin dependent diabetes mellitus, physical activity, and death, *American Journal of Epidemiology* 137 (1993): 74–81.

19. D. M. Muolo and coauthors, Effect of dietary fat on metabolic adjustments to maximal VO_2 and endurance in runners, *Medicine and Science in Sports and Exercise* 26 (1994): 81–88.

20. J. W. Helge, B. Wulff, and B. Kiens, Impact of a fat-rich diet on endurance in man: Role of the dietary period, *Medicine and Science in Sports and Exercise* 30 (1998): 456–461.

21. W. M. Sherman and N. Leenders, Fat loading: The next magic bullet? *International Journal of Sports Nutrition* 5 (1995): S1–S12.

22 P. W. R. Lemon, Is increased dietary protein necessary or beneficial for individuals with a physically active lifestyle? *Nutrition Reviews* 54 (1996): S169–S175.

23. Lemon, 1996.

24. Position of the American Dietetic Association and the Canadian Dietetic Association: Nutrition for physical fitness and athletic performance for adults, *Journal of the American Dietetic Association* 93 (1993): 691–695.

25. Lemon, 1996.

26. M. J. Webster, Physiological and performance responses to supplementation with thiamin and pantothenic acid derivatives, *European Journal of Applied Physiology* 77 (1998): 486–491; A. Singh, F. M. Moses, and P. A. Deuster, Chronic multivitamin-mineral supplementation does not enhance physical performance, *Medicine and Science in Sports and Exercise* 24 (1992): 726–732.

27. R. Murray and coauthors, Physiological and performance responses to nicotinic-acid ingestion during exercise, *Medicine and Science in Sports and Exercise* 27 (1995): 1057–1062.

28. G. M. Fogelholm, Dietary and biochemical indices of nutritional status in male athletes, *Journal of the American College of Nutrition* 11 (1992): 181–191.

29. R. S. Virk and coauthors, Effect of vitamin B-6 supplementation on fuels, catecholamines, and amino acids during exercise in men, *Medicine and Science in Sports and Exercise* 31 (1999): 400–408.

30. J. M. McBride and coauthors, Effect of resistance exercise on free radical production, *Medicine and Science in Sports and Exercise* 30 (1998): 67–72; D. A. Leaf and coauthors, The effect of exercise intensity on lipid peroxidation, *Medicine and Science in Sports and Exercise* 29 (1997): 1036–1039; R. A. Fielding and M. Meydani, Exercise, free radical generation, and aging, *Aging: Clinical and Experimental Research* 9 (1997): 12–18; M. Kanter, Free radicals and exercise: Effects of nutritional antioxidant supplementation, *Exercise and Sports Science Review* 23 (1995): 375–397.

31. A. H. Goldfarb, Antioxidants: Role of supplementation to prevent exercise-induced oxidative stress, *Medicine and Science in Sports and Exercise* 25 (1993): 232–236.

32. L. L. Ji, Oxidative stress during exercise: Implication of antioxidant nutrients, *Free Radical Biology and Medicine* 18 (1995): 1079–1086.

33. McBride and coauthors, 1998; L. Grievink and coauthors, Acute effects of ozone on pulmonary function of cyclists receiving antioxidant supplements, *Occupational and Environmental Medicine* 55 (1998): 13–17; K. V. Reddy and coauthors, Pulmonary lipid peroxidation and antioxidant defenses during exhaustive physical exercise: The role of vitamin E and selenium, *Nutrition* 14 (1998): 448–451; M. Kanter, Free radicals, exercise and antioxidant supplementation, *Proceedings of the Nutrition Society* 57 (1998): 9–13; M. Meydani and coauthors, Protective effect of vitamin E on exercise-induced oxidative damage in young and older adults, *American Journal of Physiology* 264 (1993): R992–R998.

34. L. Packer, Oxidants, antioxidant nutrients, and the athlete, *Journal of Sports Science* 15 (1997): 353–363.

35. L. E. Armstrong and C. M. Maresh, Vitamin and mineral supplements as nutritional aids to exercise performance and health, *Nutrition Reviews* 54 (1996): S149–S158.

36. American College of Sports Medicine, 1995.

37. P. M. Clarkson and E. M. Haymes, Exercise and mineral status of athletes: Calcium, magnesium, phosphorus, and iron, *Medicine and Science in Sports and Exercise* 27 (1995): 831–843.

38. R. V. West, The female athlete. The triad of disordered eating, amenorrhea, and osteoporosis, *Sports Medicine* 26 (1998): 63–71.

39. M. F. Waller and E. M. Haymes, The effects of heat and exercise on sweat iron loss, *Medicine and Science in Sports and Exercise* 28 (1996): 197–203.

40. P. M. Clarkson, Minerals: Exercise performance and supplementation in athletes, in C. Williams and J. T. Devlin, eds., *Foods, Nutrition and Sports Performance: An Internatinal Scientific Consensus* (London: E & FN Spon, 1992), pp. 113–146.

41. E. Coleman, Nutritional concerns of vegetarian athletes, *Sports Medicine Digest,* 20 (1998): 22–23.

F

42 Z. Y. Haas, Iron depletion without anemia and physical performance in young women, *American Journal of Clinical Nutrition* 66 (1997): 334–341; Clarkson and Haymes, 1995.

43. Clarkson, 1992.

44. Clarkson and Haymes, 1995.

45. C. V. Gisolfi, Fluid balance for optimal performance, *Nutrition Reviews* 54 (1996): S159–S168.

46. J. E. Greenleaf, Problems: Thirst, drinking behavior, and involuntary dehydration, *Medicine and Science in Sports and Exercise* 24 (1992): 645–656.

47. American College of Sports Medicine, Position stand: Heat and cold illness during distance running, *Medicine and Science in Sports and Exercise* 28 (1996): i–x.

48. R. J. Maughan, Fluid and electrolyte loss and replacement in exercise, in C. Williams and J. T. Devlin, eds., *Foods, Nutrition and Sports Performance: An International Scientific Consensus* (London: E & FN Spon, 1992), pp. 147–178.

49. J. H. Wilmore and coauthors, Role of taste preference on fluid intake during and after 90 minutes of running at 60% of VO_2 max in the heat, *Medicine and Science in Sports and Exercise* 30 (1998): 587–595.

50. X. Shi and C. V. Gisolfi, Fluid and carbohydrate replacement during intermittent exercise, *Sports Medicine* 25 (1998): 157–172.

51. American College of Sports Medicine, 1996.

52. M. St. Louis and coauthors, The emergence of grade A eggs as a major source of Salmonella enteritidis infections, *Journal of the American Medical Association* 259 (1988): 2103–2107.

Consumer Corner 10

1. R. M. Philen and coauthors, Survey of advertising for nutritional supplements in health and bodybuilding magazines, *Journal of the American Medical Association* 268 (1992): 1008–1011.

2. D. H. Catlin and T. H. Murray, Performance-enhancing drugs, fair competition and Olympic sport, *Journal of the American Medical Association* 276 (1996): 231–237.

3. E. Coleman, Ergogenic aids for athletes, *Nutrition and the M.D.*, July 1998, pp. 1–4.

4. G. Lombardi and coauthors, Is growth hormone bad for your heart? Cardiovascular impact of GH deficiency and of acromegaly, *Journal of Endocrinology* 155 (1997): S33–S37.

5. K. D. Mittleman, M. R. Ricci, and S. P. Bailey, Branched-chain amino acids prolong exercise during heat stress in men and women, *Medicine and Science in Sports and Exercise* 30 (1998): 83–91; P. Calders and coauthors, Pre-exercise branched-chain amino acid administration increases endurance performance in rats, *Medicine and Science in Sports and Exercise* 29 (1997): 1182–1186; M. D. Vukovich and coauthors, Effects of a low-dose amino acid supplement on adaptations to cycling training in untrained individuals, *International Journal of Sports Nutrition* 7 (1997): 298–309; A. J. M. Wagenmakers, J. H. Coakley, and R. H. T. Edwards, Metabolism of branched-chain amino acids and ammonia during exercise: Clues from McArdle's disease, *International Journal of Sports Nutrition* 11 (1990): S101–S113.

6. Mittleman, Ricci, and Bailey, 1998; E. Coleman, Branched-amino acids and fatigue, *Sports Medicine Digest* 18 (1996):44.

7. G. W. Evans, The effect of chromium picolinate on insulin-controlled parameters in humans, *Journal of Biosocial Medicine Research* 11 (1998): 163–180.

8. H. C. Lukaski and coauthors, Chromium supplementation and resistance training: Effects on body composition, strength, and trace element status of men, *American Journal of Clinical Nutrition* 63 (1996): 954–965; M. A. Hallmark and coauthors, Effects of chromium and resistance training on muscle strength and body composition, *Medicine and Science in Sports and Exercise* 28 (1996): 139–144.

9. W. R. Martin and R. E. Fuller, Suspected chromium picolinate-induced rhabdomyolysis, *Pharmacotherapy* 18 (1998): 860–862.

10. E. A. Applegate and L. E. Grivetti, Search for the competitive edge: A history of dietary fads and supplements, *Journal of Nutrition* 127 (1997): S869–S873.

11. R. B. Kreider and coauthors, Effects of creatine supplementation on body composition, strength, and sprint performance, *Medicine and Science in Sports and Exercise* 30 (1998): 73–82; J. S. Volek and coauthors, Creatine supplementation enhances muscular performance during high-intensity resistance exercise, *Journal of the American Dietetic Association* 97 (1997): 765–770; S. M. Tolar, Creatine is an ergogen for anaerobic exercise, *Nutrition Reviews* 55 (1997): 21–23; C. P. Earnest and coauthors, The effect of creatine monohydrate ingestion on anaerobic power indices, muscular strength, and body composition, *Acta Physiologica Scandinavica* 153 (1995): 207–209.

12. J. D. Gilliam, C. Hohzom, and A. D. Martin, Effect of oral creatine supplementation on isokinetic force production, *Medicine and Science in Sports and Exercise* 30 (1998): S140; L. M. Odland and coauthors, Effect of oral creatine supplementation on muscle [PCr] and short-term maximum power output, *Medicine and Science in Sports and Exercise* 29 (1997): 216–219.

13. C. Nelson, How effective is creatine? *Sports Medicine Digest* 20 (1998): 73, 78–81.

14. W. Y. Ensign and coauthors, Effects of creatine supplementation on short-term anaerobic exercise performance of U.S. Navy SEALS, *Medicine and Science in Sports and Exercise* 30 (1998): S265.

15. T. Ziegenfuss and coauthors, Performance benefits following a five-day creatine loading procedure persist for at least four weeks, *Medicine and Science in Sports and Exercise* 30 (1998): S265.

16. E. B. Feldman, Creatine: A dietary supplement and ergogenic aid, *Nutrition Reviews* 57 (1999): 45–50.

17. T. Noakes, as quoted in M. Gaie, Olympic athletes face heat, other health hurdles, *Journal of the American Medical Association* 276 (1996): 231–237.

Controversy 10

1. L. Kann and coauthors, Youth risk behavior surveillance—United States 1993, *Morbidity and Mortality Weekly Report* 44 (1995): 1–56.

2. S. Haberman and D. Luffey, Weighing in college students' diet and exercise behaviors, *Journal of American College Health* 46 (1998): 189–191.

3. J. C. Rosen, Assessment and treatment of body image disturbance, in K. D. Brownell and C. G. Fairburn. eds., *Eating Disorders and Obesity* (New York: Guilford Press, 1995), pp. 369–373.

4. G. C. Patton and coauthors, Onset of adolescent eating disorders: Population based cohort study over 3 years, *British Journal of Medicine* 318 (1999): 765–768. D. Neumark-Sztainer, R. Butler, and H. Palti, Dieting and binge eating: Which dieters are at risk? *Journal of the American Dietetic Association* 95 (1995): 586–588.

5. D. Neumark-Sztainer, Excessive weight preoccupation, *Nutrition Today*, March/April 1995, pp. 68–74.

6. Girls in the 90's: Working to undo stereotypes, *Eating Disorders Review*, September/October 1997, pp. 6–7.

7. R. V. West, The female athlete. The triad of disordered eating, amenorrhea and osteoporosis, *Sports Medicine* 26 (1998): 63–71.

8. N. Bettle and coauthors, Adolescent ballet school students: Their quest for body weight change, *Psychopathology* 31 (1998): 153–159; J. H. Wilson, Nutrition, physical activity and bone health in women, *Nutrition Research Reviews* 7 (1994): 67–91.

9. K. K. Yeager and coauthors, The female athlete triad: Disordered eating, amenorrhea, osteoporosis, *Medicine and Science in Sports and Exercise* 25 (1993): 775–777.

10. Yeager and coauthors, 1993.

11. N. A. Armsey, Stress injury to bone in the female athlete, *Clinical Sports Medicine* 16 (1997): 197–224.

12. M. Hotta and coauthors, The importance of body weight history in the occurrence and recovery of osteoporosis in patients with anorexia nervosa: Evaluation by dual X-ray absorptiometry and bone metabolic markers, *European Journal of Endocrinology* 139 (1998): 276–283.

13. E. R. Brooks, B. W. Ogden, and D. S. Cavalier, Compromised bone density 11.4 years after diagnosis of anorexia nervosa, *Journal of Women's Health* 7 (1998): 567–574.

14. J. Sundgot-Borgen and coauthors, Normal bone mass in bulimic women, *Journal of Clinical Endocrinology and Metabolism* 83 (1998): 3144–3149.

15. Hyperthermia and dehydration-related deaths associated with intentional rapid weight loss in three collegiate wrestlers—North Carolina, Wisconsin, and Michigan, November-December 1977, *Morbidity and Mortality Weekly Report* 47 (1998): 105–108.

16. American College of Sports Medicine, Position stand: Weight loss in wrestlers, *Medicine and Science in Sports and Exercise* 28 (1996): ix–xii.

17. A. E. Andersen, Eating disorders in males, in K. D. Brownell and C. G. Fairburn, eds., *Eating Disorders and Obesity* (New York: Guilford Press, 1995), pp. 177–182.

18. A. Gila and coauthors, Subjective body-image dimensions in normal and anorexic adolescents, *British Journal of Medical Psychology* 71 (1998): 175–184.

19. G. Waller, Perceived control in eating disorders: Relationship with reported sexual abuse, *International Journal of Eating Disorders* 23 (1998): 213–216.

20. T. Pryor and M. W. Weiderman, Personality features and expressed concerns of adolescents with eating disorders, *Adolescence* 33 (1998): 291–300.

21. P. Santonastaso, A. Sala, and A. Favaro, Water intoxication in anorexia nervosa: A case report, *International Journal of Eating Disorders* 24 (1998): 439–442.

22. D. M. McLoughlin and coauthors, Structural and functional changes in skeletal muscle in anorexia nervosa, *Acta Neuropathologica* 95 (1998): 632–640.

23. G. Addolorato and coauthors, A case of marked cerebellar atrophy in a woman with anorexia nervosa and cerebral atrophy and a review of the literature, *International Journal of Eating Disorders* 24 (1998): 443–447; L. M. Allende and coauthors, Immunodeficiency associated with anorexia nervosa is secondary and improves after refeeding, *Immunology* 94 (1998): 543–551; V. W. Swayze, Brain imaging and eating disorders, *Eating Disorders Review,* May/June 1997, pp. 1–4.

24. G. F. Russell, J. Treasure, and I. Eisler, Mothers with anorexia nervosa who underfeed their children: Their recognition and management, *Psychological Medicine* 28 (1998): 93–108.

25. Position of the American Dietetic Association: Nutrition intervention in the treatment of anorexia nervosa, bulimia nervosa, and binge eating, *Journal of the American Dietetic Association* 94 (1994): 902–907.

26. A. E. Becker and coauthors, Eating disorders, *New England Journal of Medicine* 340 (1999): 1092–1098.

27. M. M. Fichter, Inpatient treatment of anorexia nervosa, in K. D. Brownell and C. G. Fairburn, eds., *Eating Disorders and Obesity* (New York: Guilford Press, 1995), pp. 336–343.

28. J. T. Dwyer, Adolescence, in E. E. Ziegler and L. J. Filer, Jr., eds., *Present Knowledge in Nutrition,* 7th ed. (Washington, D.C.: International Life Sciences Institute Press, 1996), pp. 404–413.

29. M. Moukaddem and coauthors, Increase in diet-induced thermogenesis at the start of refeeding in severely malnourished anorexia nervosa patients, *American Journal of Clinical Nutrition* 66 (1997): 133–140.

30. E. D. Eckert and coauthors, Leptin in anorexia nervosa, *Journal of Clinical Endocrinology and Metabolism* 83 (1998): 791–795.

31. American Psychiatric Association Workgroup on Eating Disorders, Practice guidelines for eating disorders, I. Disease definition, epidemiology, and natural history, *American Journal of Psychiatry* 150 (1993): 212–228.

32. K. M. Pike, Long-term course of anorexia nervosa: Response, relapse, remission, and recovery, *Clinical Psychology Reviews* 18 (1998): 447–475.

33. H. W. Hoek, The distribution of eating disorders, in K. D. Brownell and C. G. Fairburn, eds., *Eating Disorders and Obesity* (New York: Guilford Press, 1995), pp. 207–211.

34. N. A. Troop, A. Holbrey, and J. L. Treasure, Stress, coping, and crisis support in eating disorders, *International Journal of Eating Disorders* 24 (1998): 157–166.

35. C. G. Fairburn and coauthors, Risk factors for bulimia nervosa: A community-based case-control study, *Archives of General Psychiatry* 54 (1997): 509–517.

36. Waller, 1998.

37. K. A. Gendall and coauthors, The nutrient intake of women with bulimia nervosa, *International Journal of Eating Disorders* 21 (1997): 115–127.

38. G. T. Wilson, Eating disorders and addictive disorders, in K. D. Brownell and C. G. Fairburn, eds., *Eating Disorders and Obesity* (New York: Guilford Press, 1995), pp. 165–170.

39. P. K. Keel and J. E. Mitchell, Outcome in bulimia nervosa, *American Journal of Psychiatry* 154 (1997): 313–321.

40. Federal Interagency Forum on Child and Family Statistics, *America's Children: Key National Indicators of Well-Being, 1999,* a report from the National Institutes of Health, available from National Maternal and Child Health Clearinghouse, 2070 Chain Bridge Road, Suite 450, Vienna, VA 22182 or on the Internet at http://childstats.gov.

41. W. Kaye, K. Gendall, and M. Strober, Serotonin neuronal function and selective serotonin reuptake inhibitor treatment in anorexia and bulimia nervosa, *Biological Psychiatry* 44 (1998): 825–838; B. Baranowska and coauthors, Neuropeptide Y, galanin, and leptin release in obese women and in women with anorexia nervosa, *Metabolism* 46 (1997): 1384–1389; B. E. Wolfe, E. Metzger, and D. C. Jimerson, Re-search update on serotonin function in bulimia nervosa and anorexia nervosa, *Psychopharmacology Bulletin* 33 (1997): 345–354.

42. E. LeShan, *Winning the Losing Game: Why I Will Never Be Fat Again* (New York: Crowell, 1979).

Chapter 11

1. A. E. Platt, Confronting infectious diseases, in L. R. Brown, ed., *State of the World 1996: A Worldwatch Institute Report on Progress toward a Sustainable Society* (New York: Norton, 1996), pp. 114–132.

2. R. K. Chandra, Nutrition and the immune system: An introduction, *American Journal of Clinical Nutrition* 66 (1997): S460–S463.

3. N. S. Scrimshaw and J. P. SanGiovanni, Synergism of nutrition, infection, and immunity: An overview, *American Journal of Clinical Nutrition* 66 (1997): S464–S477.

4. A. Nimmagadda and coauthors, The significance of vitamin A and carotenoid status in persons infected by the human immunodeficiency virus, *Clinical Infectious Diseases* 26 (1998): 711–718; M. K. Baum and coauthors, High risk of HIV-related mortality is associated with selenium deficiency, *Journal of Acquired Immune Deficiency Syndromes and Human Retrovirology* 15 (1997): 370–374.

5. J. S. Young, HIV and medical nutrition therapy, *Journal of the American Dietetic Association* 97 (1997): S161–S166.

6. W. J. Evans, R. Roubenoff, and A. Shevitz, Exercise and the treatment of wasting: Aging and human immunodeficiency virus infection, *Seminars in Oncology* 25 (1998): 112–122.

7. S. N. Meydani and A. A. Beharka, Recent developments in vitamin E and immune response, *Nutrition Reviews* 56 (1998): S49–S58.

8. Scrimshaw and SanGiovanni, 1997.

9. Position of the American Dietetic Association: The role of nutrition in health promotion and disease prevention programs, *Journal of the American Dietetic Association* 98 (1998): 205–208.

10. W. R. Fair, N. E. Fleshner, and W. Heston, Cancer of the prostate: A nutritional disease? *Urology* 50 (1997): 840–848.

11. A. P. Simopoulos, Diet and gene interactions, *Food Technology* 51 (1997): 66–69.

12. R. H. Eckel and R. M. Krauss, American Heart Association call to action: Obesity as a major risk factor for coronary heart disease, *Circulation* 97 (1998): 2099–2100.

13. Age-adjusted death rates for 1997, percentage of change in age-adjusted death rates for the 15 leading causes of death, 1996–1997 and 1979–1997, and ratio of age-adjusted death rates, by sex and race of decedent, 1997—United States, *Morbidity and Mortality Weekly Report* 48 (1999): 664.

14. P. D. Reaven and J. L. Witztum, Oxidized low density lipoproteins in atherogenesis: Role of dietary modification, *Annual Review of Nutrition* 16 (1996): 51–71.

15. K. K. Griendling and W. Alexander, Oxidative stress and cardiovascular disease (editorial), *Circulation* 96 (1997): 3264–3265; H. Sies,

Oxidative stress: Oxidants and antioxidants, *Experimental Physiology* 82 (1997): 291–295.

16. L. Cominacini and coauthors, Modified LDL in the pathogenesis of atherosclerosis: Role of nutrition, in G. A. Spiller, ed., *Handbook of Lipids in Human Nutrition* (Boca Raton, Fla.: CRC Press, 1996), pp. 155–162.

17. D. S. Siscovick and coauthors, Dietary intake and cell membrane levels of long-chain n-3 polyunsaturated fatty acids and the risk of primary cardiac arrest, *Journal of the American Medical Association* 274 (1995): 1363–1367.

18. L. Kuller and coauthors, Prevalence of subclinical atherosclerosis and cardiovascular disease and association with risk factors in the Cardiovascular Health Study, *American Journal of Epidemiology* 139 (1994): 1164–1179.

19. J. U. Opara and J. H. Levine, The deadly quartet—the insulin resistance syndrome, *Southern Medical Journal* 90 (1997): 1162–1168.

20. M. E. Daly and coauthors, Dietary carbohydrates and insulin sensitivity: A review of the evidence and clinical implications, *American Journal of Clinical Nutrition* 66 (1997): 1072–1085.

21. J. P. Despres and coauthors, Hyperinsulinemia as an independent risk factor for ischemic heart disease, *New England Journal of Medicine* 334 (1996): 952–957.

22. B. A. Griffin and A. Zampelas, Influence of dietary fatty acids on the atherogenic lipoprotein phenotype, *Nutrition Research Reviews* 8 (1995): 1–26.

23. NIH Consensus Conference, Triglyceride, high-density lipoprotein, and coronary heart disease, *Journal of the American Medical Association* 269 (1993): 505–510.

24. J. Jeppesen and coauthors, Triglyceride concentration and ischemic heart disease: An eight-year follow-up in the Copenhagen Male Study, *Circulation* 97 (1998): 1029–1036.

25. P. O. Kwiterovich, The effect of dietary fat, antioxidants, and pro-oxidants on blood lipids, lipoproteins, and atherosclerosis, *Journal of the American Dietetic Association* 97 (1997): S31–S41.

26. A. Ascherio and coauthors, *Trans*-fatty acids and coronary heart disease, *New England Journal of Medicine* 340 (1999): 1994–1998.

27. A dietary intervention trial for nutritional management of cardiovascular risk factors, *Nutrition Reviews* 55 (1997): 54–60.

28. F. B. Hu and coauthors, Dietary fat intake and the risk of coronary heart disease in women, *New England Journal of Medicine* 337 (1997): 1491–1499.

29. American Heart Association, *Trans*-fatty acids, a monograph available at www.americanheart.org on the Internet.

30. M. de Longeril and coauthors, Mediterranean diet, traditional risk factors and the rate of cardiovascular complications after myocardial infarction: Final report of the Lyon Diet Heart Study, *Circulation* 99 (1999): 779–785.

31. E. B. Rimm and coauthors, Vegetable, fruit, and cereal fiber intake and risk of coronary heart disease among men, *Journal of the American Medical Association* 275 (1996): 447–451.

32. R. E. Andersen and coauthors, Relation of weight loss to changes in serum lipids and lipoproteins in obese women, *American Journal of Clinical Nutrition* 62 (1995): 350–357.

33. K. Robinson and coauthors, Low circulating folate and vitamin B_6 concentrations: Risk factors for stroke, peripheral vascular disease, and coronary artery disease, *Circulation* 97 (1998): 437–443; E. B. Rimm and coauthors, Folate and vitamin B_6 from diet and supplements in relation to risk of coronary heart disease among women, *Journal of the American Medical Association* 279 (1998): 359–364.

34. F. Nappo and coauthors, Impairment of endothelial functions by acute hyperhomocysteinemia and reversal by antioxidant nutrients, *Journal of the American Medical Association* 281 (1999): 2113–2118; J. S. Stamler and A. Slivka, Biological chemistry of thiols in vascular-related disease, *Nutrition Reviews* 54 (1996): 1–30; J. B. Ubbink, Homocysteine—An atherogenic and thrombogenic factor? *Nutrition Reviews* 53 (1995): 323–332.

35. C. La Vecchia, A. Decarli, and R. Pagano, Vegetable consumption and risk of chronic disease, *Epidemiology* 9 (1998): 208–210.

36. S. W. Farrell and coauthors, Influences of cardiorespiratory fitness levels and other predictors on cardiovascular disease mortality in men, *Medicine and Science in Sports and Exercise* 30 (1998): 899–905.

37. S. N. Blair and coauthors, Influences of cardiorespiratory fitness and other precursors on cardiovascular disease and all-cause mortality in men and women, *Journal of the American Medical Association* 276 (1996): 205–210; A. L. Macnair, Physical activity, not diet, should be the focus of measures for the primary prevention of cardiovascular disease, *Nutrition Research Reviews* 7 (1994): 43–65.

38. G. A. Bray, Obesity, in E. E. Ziegler and L. J. Filer, Jr., eds., *Present Knowledge in Nutrition*, 7th ed. (Washington, D.C.: International Life Sciences Institute Press, 1996), pp. 19–32; C. Bouchard, G. A. Bray, and V. S. Hubbard, Basic and clinical aspects of regional fat distribution, *American Journal of Clinical Nutrition* 52 (1990): 946–950.

39. D. Ornish, Avoiding revascularization with lifestyle changes: The Multicenter Lifestyle Demonstration Project, *American Journal of Cardiology* 82 (1998): T72–T76.

40. C. T. Valmadrid and coauthors, Alcohol intake and the risk of coronary heart disease mortality in persons with older-onset diabetes mellitus, *Journal of the American Medical Association* 282 (1999): 239–246; M. J. Thun and coauthors, Alcohol consumption and mortality among middle-aged and elderly U.S. adults, *New England Journal of Medicine* 337 (1997): 1705–1714; E. B. Rimm and coauthors, Review of moderate alcohol consumption and reduced risk of coronary heart disease: The effect due to beer, wine, or spirits? *British Medical Journal* 321 (1996): 731–736.

41. R. Locher, P. M. Suter, and W. Vetter, Ethanol suppresses smooth muscle cell proliferation in the postprandial state: A new antiatherosclerotic mechanism of ethanol? *American Journal of Clinical Nutrition* 67 (1998): 338–341; P. R. Ridker and coauthors, Association of moderate alcohol consumption and plasma concentration of endogenous tissue-type plasminogen activator, *Journal of the American Medical Association* 272 (1994): 929–933; J. M. Gaziano and coauthors, Moderate alcohol intake, increased levels of high-density lipoprotein and its subfractions, and decreased risk of myocardial infarction, *New England Journal of Medicine* 329 (1993): 1829–1834.

42. C. L. Hart and coauthors, Alcohol consumption and mortality from all causes, coronary heart disease, and stroke: results from a prospective cohort study of Scottish men with 21 years of follow up, *British Medical Journal* 318 (1999): 1725–1729.

43. M. J. Williams, N. J. Restieaux, and C. J. Low, Myocardial infarction in young people with normal coronary arteries, *Heart* 79 (1998): 191–194.

44. T. A. Pearson and P. Terry, What to advise patients about drinking alcohol—The clinician's conundrum, *Journal of the American Medical Association* 272 (1994): 967–968.

45. The Expert Panel, Summary of the second report of the National Cholesterol Education Program (NCEP) Expert Panel on Detection, Evaluation, and Treatment of High Blood Cholesterol in Adults (Adult Treatment Panel II), *Journal of the American Medical Association* 269 (1993): 3015–3023.

46. P. Bjorntorp, Stress and cardiovascular disease, *Acta Physiologica Scandinavica Supplementum* 640 (1997): 144–148.

47. U.S. Department of Agriculture, Third Report on Nutrition Monitoring in the United States: Executive summary, *Journal of Nutrition* 126 (1996): S1907–S1936.

48. R. F. Gillum, M. E. Mussolino, and J. H. Madans, Body fat distribution and hypertension incidence in women and men: The NHANES I Epidemiologic Follow-up Study, *International Journal of Obesity and Related Metabolic Disorders* 22 (1998): 127–134.

49. M. R. Sierra-Honigmann and coauthors, Biological action of leptin as an angiogenic factor, *Science* 281 (1998): 1683–1686.

50. W. B. Kannel, Blood pressure as a cardiovascular risk factor, *Journal of the American Medical Association* 275 (1996): 1571–1575.

51. M. L. Nurminen, R. Korpela, and H. Vapaatalo, Dietary factors in the pathogenesis and treatment of hypertension, *Annals of Medicine* 30 (1998): 143–150; D. A. McCarron and coauthors, Comprehensive nutrition plan improves cardiovascular risk factors in essential hypertension, *American Journal of Hypertension* 11 (1998): 31–40.

52. A. W. Cowley, Genetic and nongenetic determinants of salt sensitivity and blood pressure, *American Journal of Clinical Nutrition* 65 (1997): 587S–593S; D. L. Ely, Overview of dietary sodium effects on and interactions with cardiovascular and neuroendocrine functions, *American Journal of Clinical Nutrition* 65 (1997): S594–S605.

53. J. Stamler, The INTERSALT Study: Background, methods, findings, and implications, *American Journal of Clinical Nutrition* 65 (1997): S626–S642.

54. J. A. Staessen and coauthors, Salt and blood pressure in community-based intervention trials, *American Journal of Clinical Nutrition* 65 (1997): S661–S670.

55. Stamler, 1997; J. A. Cutler, D. Follmann, and P. S. Allender, Randomized trials of sodium reduction: An overview, *American Journal of Clinical Nutrition* 65 (1997): S643–S651.

56. E. Saltos and S. Bowman, Dietary guidance on sodium: Should we take it with a grain of salt? *USDA Family Economics and Nutrition Review* vol. 11, no. 4, 1998, pp. 49–51.

57. D. A. McCarron and M. E. Reusser, Body weight and blood pressure regulation, *American Journal of Clinical Nutrition* 63 (1996): S423–S425.

58. P. F. Kokkinos and coauthors, Effects of regular exercise on blood pressure and left ventricular hypertrophy in African-American men with severe hypertension, *New England Journal of Medicine* 333 (1995): 1462–1467.

59. M. Hillbom, S. Juvela, and V. Karttunen, Mechanisms of alcohol-related strokes, in *Alcohol and Cardiovascular Diseases: Novartis Foundation Symposium 216* (New York: John Wiley & Sons, 1998).

60. U.S. Department of Agriculture and U.S. Department of Health and Human Services, *Nutrition and Your Health: Dietary Guidelines for Americans,* Home and Garden Bulletin no. 232 (Washington, D.C.: Government Printing Office, 1995); R. G. Victor and J. Hansen, Alcohol and blood pressure—A drink a day . . . , *New England Journal of Medicine* 332 (1995): 1782–1783.

61. M. E. Reusser and D. A. McCarron, Micronutrient effects on blood pressure regulation, *Nutrition Reviews* 52 (1994): 367–375.

62. D. A. McCarron, Dietary calcium and lower blood pressure: We can all benefit, *Journal of the American Medical Association* 275 (1996): 1128–1129.

63. D. A. McCarron, Role of adequate dietary calcium intake in the prevention and management of salt-sensitive hypertension, *American Journal of Clinical Nutrition* 65 (1997): S712–S716; C. G. Osborne and coauthors, Evidence for the relationship of calcium to blood pressure, *Nutrition Reviews* 54 (1996): 365–381; McCarron, 1996.

64. P. K. Whelton and coauthors, Effects of oral potassium on blood pressure: Meta-analysis of randomized controlled clinical trials, *Journal of the American Medical Association* 277 (1997): 1624–1632.

65. C. J. Bates and coauthors, Does vitamin C reduce blood pressure? Results of a large study of people aged 65 or older, *Journal of Hypertension* 16 (1998): 925–932.

66. L. J. Appel and coauthors, A clinical trial of the effects of dietary patterns on blood pressure, *New England Journal of Medicine* 336 (1997): 1117–1124.

67. P. T. Strickland and J. D. Groopman, Biomarkers for assessing environmental exposure to carcinogens in the diet, *American Journal of Clinical Nutrition* 61 (1995): S710–S720.

68. S. A. Oliveria and P. J. Christos, The epidemiology of physical activity and cancer, *Annals of the New York Academy of Sciences* 833 (1997): 79–90.

69. Committee on Comparative Toxicity of Naturally Occurring Carcinogens, *Carcinogens and Anticarcinogens in the Human Diet* (Washington, D.C.: National Academy Press, 1996), pp. 1–18.

70. T. Sugimura, Cancer prevention: Past, present, future, *Mutation Research* 402 (1998): 7–14; M. J. Hill, Nutrition and human cancer, *Annals of the New York Academy of Sciences* 883 (1997): 68–78.

71. D. M. DeMarini, Dietary interventions of human carcinogenesis, *Mutation Research* 400 (1998): 457–465.

72. L. F. Macrae, Wheat bran fiber and development of adenomatous polyps: Evidence from randomized, controlled, clinical trials, *American Journal of Medicine* 106 (1999): S38–S42; Chatenoud and coauthors, Whole grain food intake and cancer risk, *International Journal of Cancer* 77 (1998): 24–28.

73. B. S. Reddy, Dietary fat and colon cancer: Animal model studies, *Lipids* 27 (1992): 807–813.

74. J. R. Hecht, Dietary fat and colon cancer, *Advances in Experimental Medicine and Biology* 399 (1996): 157–163; H. S. Black, Effect of a low-fat diet on the incidence of actinic keratosis, *New England Journal of Medicine* 330 (1994): 1272–1275.

75. E. Giovannucci and B. Goldin, The role of fat, fatty acids, and total energy intake in the etiology of human colon cancer, *American Journal of Clinical Nutrition* 66 (1997): S1564–S1571.

76. D. D. Hensrud and D. C. Heimburger, Diet, nutrients, and gastrointestinal cancer, *Gastroenterology Clinics of North America* 27 (1998): 325–346; B. C.-H. Chiu and coauthors, Diet and risk of non-Hodgkin lymphoma in older women, *Journal of the American Medical Association* 275 (1996): 1315–1321.

77. M. D. Holmes and coauthors, Association of dietary intake of fat and fatty acids with risk of breast cancer, *Journal of the American Medical Association* 281 (1999): 914–920; D. J. Hunter and coauthors, Cohort studies of fat intake and the risk of breast cancer—A pooled analysis, *New England Journal of Medicine* 334 (1996): 356–361.

78. C. M. Yang and coauthors, Thermally oxidized dietary fat and colon carcinogenesis in rodents, *Nutrition and Cancer* 30 (1998): 69–73; J. G. Ernhardt and coauthors, A diet rich in fat and poor in dietary fiber increases the in vitro formation of reactive oxygen species in human feces, *Journal of Nutrition* 127 (1997): 706–709.

79. M. A. Belury, Conjugated dienoic linoleate: A polyunsaturated fatty acid with unique chemoprotective properties, *Nutrition Reviews* 53 (1995): 83–89.

80. D. Y Kim, K. H. Chung, and J. H. Lee, Stimulatory effects of high-fat diets on colon cell proliferation depend on the type of dietary fat and site of the colon, *Nutrition and Cancer* 30 (1998): 118–123.

81. P. C. Calder and coauthors, Dietary fish oil suppresses human colon tumor growth in athymic mice, *Clinical Science* 94 (1998): 303–311.

82. B. C. Pence and coauthors, Feeding of a well-cooked beef diet containing a high heterocyclic amine content enhances colon and stomach carcinogenesis in 1, 2-dimethylhydrazine-treated rats, *Nutrition and Cancer* 30 (1998): 220–226; X. M. Zhang and coauthors, Initiation and promotion of colonic aberrant crypt foci in rats by 5-hydroxymethyl-2-furaldehyde in thermolyzed sucrose, *Carcinogenesis* 14 (1993): 773–775.

83. Giovannucci and Goldin, 1997.

84. D. Kritchevsky, Caloric restriction and experimental mammary carcinogenesis, *Breast Cancer Research and Treatment* 46 (1997): 161–167; Committee on Comparative Toxicity of Naturally Occurring Carcinogens, 1996, pp. 35–126.

85. Giovannucci and Goldin, 1997.

86. B. J. Caan and coauthors, Body size and the risk of colon cancer in a large case-control study, *International Journal of Obesity and Related Metabolic Disorders* 22 (1998): 178–184; C. La Vecchia and coauthors, Diabetes mellitus and colorectal cancer risk, *Cancer, Epidemiology, Biomarkers, and Prevention* 6 (1997): 1007–1010; N. Koohestani and coauthors, Insulin resistance and promotion of aberrant crypt foci in the colons of rats on a high-fat diet, *Nutrition and Cancer* 29 (1997): 69–76; E. Giovannucci, Insulin and colon cancer, *Cancer Causes Control* 6 (1995): 164–179.

87. B. S. Reddy, Role of dietary fiber in colon cancer: An overview, *American Journal of Medicine* 106 (1999): S16–S19; D. Kritchevsky, Protective role of wheat bran fiber: Preclinical data, *American Journal of Medicine* 106 (1999): S28–S31; M. C. Jansen and coauthors, Dietary fiber and plant foods in relation to colorectal cancer mortality: The Seven Countries Study, *International Journal of Cancer* 81 (1999): 174–179; M. J. Hill, Cereals, cereal fibre and colorectal cancer risk: A review of the epidemiological literature, *European Journal of Cancer Prevention* 6 (1997): 219–225; L. Le Marchand and coauthors, Dietary fiber and colorectal cancer risk, *Epidemiology* 8 (1997): 658–665.

88. The study finding no effect was C. S. Fuchs and coauthors, Dietary fiber and the risk of colorectal cancer and adenoma in women, *New England Journal of Medicine* 340 (1999): 169–176; some additional work supporting an effect includes J. Faivre and A. Giacosa, Primary prevention of colorectal cancer through fibre supplementation, *European Journal of Cancer Prevention* 7 (1998): S29–S32, and E. Negri and coauthors, Fiber intake and risk of colorectal cancer, *Cancer Epidemiology, Biomarkers, and Prevention* 7 (1998): 667–671.

89. I. Zusman and coauthors, The immune response of rat spleen to dietary fibers and to low doses of carcinogen: Morphometric and immunohistochemical studies, *Oncology Reports* 5 (1998): 1577–1581.

90. D. S. Michaud and coauthors, Fluid intake and the risk of bladder cancer in men, *New England Journal of Medicine* 340 (1999): 1390–1397.

91. K. A. Steinmetz and J. D. Potter, Vegetables, fruit, and cancer prevention: A review, *Journal of the American Dietetic Association* 96 (1996): 1027–1039.

92. National Academy of Sciences, quoted in First International Conference on East-West Perspectives on Functional Foods, *Nutrition Today,* March/April 1996, pp. 70–73.

93. K. McNutt, Medicinals in food, Part I: Is science coming full circle? *Nutrition Today,* September/October 1995, pp. 218–222.

94. *The Surgeon General's Report on Nutrition and Health, Summary and Recommendations* (Washington, D.C.: Department of Health and Human Services—Public Health Service publication no. 88-50211, 1988).

95. A. K. Kant, A. Shatzkin, and R. G. Ziegler, Dietary diversity and subsequent cause-specific mortality in the NHANES I Epidemiologic Follow-up Study, *Journal of the American College of Nutrition* 14 (1995): 233–238.

Consumer Corner 11

1. M. Larkin, NIH's Office of Alternative Medicine: A wise use of tax dollars? *Priorities* 6 (1994): 32–36.

2. M. M. Lipman, The power of placebos, *Consumer Reports on Health,* February 1996, p. 23.

3. P. Lipkin, An ancient salve dampens pain, *Science News* 149 (1996): 20.

4. N. Ahmad and coauthors, Green tea constituent epigallocatechin-3-gallate and induction of apoptosis and cell cycle arrest in human carcinoma cells, *Journal of the National Cancer Institute* 89 (1997): 1881–1886.

5. A photo finish for total Taxol synthesis, *Science News* 145 (1994): 223.

6. J. A. Bakerlink and coauthors, Multiple organ failure after ingestion of pennyroyal oil from herbal tea in two infants, *Pediatrics* 98 (1996): 944–947.

7. S. B. Markowitz and coauthors, Lead poisoning due to *Hai Ge Fen:* The porphyrin content of individual erythrocytes, *Journal of the American Medical Association* 271 (1994): 932–934.

8. V. E. Tyler, What pharmacists should know about herbal remedies, *Journal of the American Pharmaceutical Association* 36 (1996): 944–947.

Controversy 11

1. A. King and G. Young, Characteristics and occurrence of phenolic phytochemicals, *Journal of the American Dietetic Association* 99 (1999): 213–218.

2. F. M. Clydesdale, A proposal for the establishment of scientific criteria for health claims for functional foods, *Nutrition Reviews* 55 (1997): 413–422.

3. Standing Committee on the Scientific Evaluation of Dietary Reference Intakes and Its Panel on Dietary Antioxidants and Related Compounds, *Dietary Reference Intakes: Proposed Definition and Plan for Review of Dietary Antioxidants and Related Compounds* (Washington, D.C.: National Academy Press, 1998), p. 8.

4. H. Mo and C. E. Elson, Apoptosis and cell-cycle arrest in human and murine tumor cells are initiated by isoprenoids, *Journal of Nutrition* 129 (1999) 804–813.

5. W. R. Bidlack, Interrelationships of food, nutrition, diet, and health: The National Association of State Unviersities and Land Grant Colleges white paper, *Journal of the American College of Nutrition* 15 (1996): 422–433.

6. J. A. Milner, Garlic: Its anticarcinogenic and antitumorigenic properties, *Nutrition Reviews* 54 (1996): S82–S86.

7. S. V. Singh and coauthors, Differential induction of NAD(P)H: quinone oxidoreductase by anti-carcinogenic organosulfides from garlic, *Biochemical and Biophysical Research Communications* 27 (1998): 917–920; S. Fukushima and coauthors, Cancer prevention by organosulfur compounds from garlic and onion, *Journal of Cell Biochemistry* 27 (1997): 100–105.

8. J. L. Isaacsohn and coauthors, Garlic powder and plasma lipids and lipoproteins: A multicenter, randomized placebo-controlled trial, *Archives of Internal Medicine* 158 (1998): 1189–1194; H. K. Berthold, T. Sudhop, and K. von Bergmann, Effect of garlic oil preparation on serum lipoproteins and cholesterol metabolism: A randomized controlled trial, *Journal of the American Medical Association* 279 (1998): 1900–1902; A. J. Adler and B. J. Holub, Effect of garlic and fish oil supplementation on serum lipid and lipoprotein concentrations in hypercholesterolemic men, *American Journal of Clinical Nutrition* 65 (1997): 445–450.

9. J. T. Pinto and coauthors, Effects of garlic thioallyl derivatives on growth, glutathione concentration, and polyamine formation of human prostate carcinoma cells in culture, *American Journal of Clinical Nutrition* 66 (1997): 398–405.

10. L. O. Dragsted, M. Strube, and T. Leth, Dietary levels of plant phenols and other non-nutritive components: Could they prevent cancer? *European Journal of Cancer Prevention* 6 (1997): 522–528.

11. D. Ingram and coauthors, Case-control study of phyto-estrogens and breast cancer, *Lancet* 350 (1997): 990–994; H. Adlercreutz and W. Mazur, Phyto-estrogens and western diets, *Annals of Medicine* 29 (1997): 95–120.

12. S. M. Potter, Soy protein and cardiovascular disease: The impact of bioactive components in soy, *Nutrition Reviews* 56 (1998): 231–235.

13. M. S. Anthony, CHD protection by soy and its phytoestrogens: Beyond plasma lipid concentrations, *Soy Connection*, Summer 1995, pp. 1, 4.

14. M. S. Morton and coauthors, Lignans and isoflavonoids in plasma and prostatic fluid in men: Samples from Portugal, Hong Kong, and the United Kingdom, *Prostate* 32 (1997): 122–128.

15. A. Cassidy, S. Bingham, and K. D. R. Setchell, Biological effects of a diet of soy protein rich in isoflavones on the menstrual cycle of premenopausal women, *American Journal of Clinical Nutrition* 60 (1994): 333–340.

16. S. K. Clinton, Lycopene: Chemistry, biology, and implications for human health and disease, *Nutrition Reviews* 56 (1998): 35–51.

17. Fat-soluble vitamins, in C. D. Berdanier, *Advanced Nutrition: Micronutrients* (Boca Raton, Fla.: CRC Press, 1998), pp. 21–72; H. Nishino, Cancer prevention by natural carotenoids, *Journal of Cellular Biochemistry* (supplement) 37 (1997); 86–91.

18. G. R. Beecher, Nutrient contents of tomatoes and tomato products, *Proceedings of the Society for Experimental and Biological Medicine* 218 (1998): 98–100.

19. H. Gerster, The potential role of lycopene for human health, *Journal of the American College of Nutrition* 16 (1997): 109–126.

20. P. A. Kantesky and coauthors, Dietary intake and blood levels of lycopene: Association with cervical dysplasia among non-Hispanic black women, *Nutrition and Cancer* 31 (1998): 31–40.

21. J. D. Ribaya-Mercado and coauthors, Skin lycopene is destroyed preferentially over β-carotene during ultravolet irradiation in humans, *Journal of Nutrition* 125 (1995): 1854–1859, as cited in Clinton, 1998.

22. D. A. Cooper, A. L. Eldridge, and J. C. Peters, Dietary carotenoids and lung cancer: A review of recent research, *Nutrition Reviews* 57 (1999): 133–145.

23. M. A. Wagstaff and coauthors, Oregano flavonoids as lipid antioxidants, *Journal of the American Dietetic Association* 93 (1993): 1217.

24. M. G. Hertog and coauthors, Flavonoid intake and long-term risk of coronary heart disease and cancer in the seven countries study, *Archives of Internal Medicine* 155 (1995): 381–386.

25. J. V. Formica and W. Regelson, Review of the biology of quercetin and related bioflavonoids, *Food Chemistry and Toxicology* 33 (1995): 1061–1080.

26. S. V. Nigdikar and coauthors, Consumption of red wine polyphenols reduces the susceptibility of low-density lipoproteins to oxidation in vivo, *American Journal of Chemical Nutrition* 68 (1998): 258–265.

27. J. D. Folts, Fruits, vegetables, and stroke risk, *Journal of the American Medical Association* 274 (1995): 1197.

28. Formica and Regelson, 1995.

29. A. Drewnowski, S. A. Henderson, and A. B. Shore, Taste responses to naringin, a flavonoid, and the acceptance of grapefruit juice are related to genetic sensitivity to 6-n-propylthiouracil, *American Journal of Clinical Nutrition* 66 (1997): 391–397.

30. P. C. Hollman and M. B. Katan, Absorption, metabolism, and health effects of dietary flavonoids in man, *Biomedicine Pharmacother-*

apy 51 (1997): 305–310.

31. For a review of studies relating to phytochemicals and cancer see K. Steinmetz and J. D. Potter, Vegetables, fruit, and cancer prevention: A review, *Journal of the American Dietetic Association* 96 (1996): 1027–1039. For an entire journal issue on many topics relating to phytochemicals and functional foods, see *Nutrition Reviews* 54 (1996), number 11, part II.

32. Steinmetz and Potter, 1996.

33. W. J. Craig, Phytochemicals: Guardians of our health, *Journal of the American Dietetic Association* 97 (1997): S199–S204.

34. E. Giovannucci and coauthors, Intake of carotenoids and retinol in relation to risk of prostate cancer, *Journal of the National Cancer Institute* 87 (1995): 1767–1776.

35. N. Ahmad and H. Mukhtar, Green tea polyphenols and cancer: Biologic mechanisms and practical implications, *Nutrition Reviews* 57 (1999): 78–83; K. Imai, K. Suga, and K. Nakachi, Cancer preventive effects of drinking green tea among a Japanese population, *Preventive Medicine* 26 (1997): 769–775.

36. King and Young, 1999.

37. R. L. Prior and G. Cao, Antioxidant capacity and polyphenolic components of teas: Implications for altering in vivo antioxidant status, *Proceedings of the Society for Experimental Biology and Medicine* 220 (1999): 255–261.

38. Dragsted, Strube, and Leth, 1997.

39. M. Nestle, Broccoli sprouts in cancer prevention, *Nutrition Reviews* 56 (1998): 127–130.

40. B. Liebman and D. Schardt, Wild claims, weak evidence, *Nutrition Action Healthletter,* July/August 1998, pp. 8–9.

41. J. Foote and B. Cohen, Medicinal herb use and the renal patient, *Journal of Renal Nutrition* 8 (1998): 40–42.

42. D. R. Farr, Functional foods, *Cancer Letters* 114 (1997): 59–63.

43. R. Kava, FYI: When is a vegetable not a vegetable? When it's a pill? *Priorities* 4 (1995): pp. 35–36.

44. A. F. Subar and coauthors, Fruit and vegetable intake in the United States: The baseline survey of the Five A Day for Better Health program, *American Journal of Health Promotion* 9 (1995): 352–360.

Chapter 12

1. B. Guyer and coauthors, Annual summary of vital statistics—1997, *Pediatrics* 102 (1998): 1333–1349.

2. M. M. Werler and coauthors, Prepregnant weight in relation to risk of neural tube defects, *Journal of the American Medical Association* 275 (1996): 1089–1092; G. M. Shaw, E. M. Velie, and D. Schaffer, Risk of neural tube defect—Affected pregnancies among obese women, *Journal of the American Medical Association* 275 (1996): 1093–1096.

3. R. L. Goldenberg and T. Tamura, Prepregnancy weight and pregnancy outcome, *Journal of the American Medical Association* 275 (1996): 1127–1128.

4. W. W. Hay and coauthors, Workshop summary: Fetal growth: Its regulation and disorders, *Pediatrics* 99 (1997): 585–591.

5. D. J. Barker and P. M. Clark, Fetal undernutrition and disease in later life, *Reviews of Reproduction* 2 (1997): 105–112.

6. J. Newnham, Consequences of fetal growth restriction, *Current Opinion in Obstetrics and Gynecology* 10 (1998): 145–149; W. P. T. James, Long-term fetal programming of body composition and longevity, *Nutrition Reviews* 55 (1997): S31–S43.

7. R. Uauy and coauthors, Role of essential fatty acids in the function of the developing nervous system, *Lipids* 31 (1996): S167–S176.

8. B. Burke and coauthors, *Preventing Neural Tube Birth Defects: A Prevention Model and Resource Guide* (Atlanta, Ga.: Centers for Disease Control and Prevention, 1998).

9. G. J. Locksmith and P. Duff, Preventing neural tube defects: The importance of periconceptional folic acid supplements, *Obstetrics and Gynecology* 91 (1998): 1027–1034.

10. Burke and coauthors, 1998.

11. J. E. Brown and coauthors, Predictors of red cell folate level in women attempting pregnancy, *Journal of the American Medical Association* 277 (1997): 548–552.

12. Standing Committee on the Scientific Evaluation of Dietary Reference Intakes, Food and Nutrition Board, Institute of Medicine, *Dietary Reference Intakes for Thiamin, Riboflavin, Niacin, Vitamin B_6, Folate, Vitamin B_{12}, Pantothenic Acid, Biotin, and Choline* (Washington, D.C.: National Academy Press, 1998), pp. 8-44–8-45.

13. Standing Committee on the Scientific Evaluation of Dietary Reference Intakes, Food and Nutrition Board, Institute of Medicine, *Dietary Reference Intakes for Calcium, Phosphorus, Magnesium, Vitamin D, and Fluoride* (Washington, D.C: National Academy Press, 1997), p. 4-38.

14. L. H. Allen, Pregnancy and iron deficiency: Unresolved issues, *Nutrition Reviews* 55 (1997): 91–101; Committee on Nutritional Status during Pregnancy and Lactation, Food and Nutrition Board, *Nutrition during Pregnancy* (Washington, D.C.: National Academy Press, 1990), pp. 272–298.

15. T. O. Scholl and coauthors, Use of multivitamin /mineral prenatal supplements: Influence on the outcome of pregnancy, *American Journal of Epidemiology* 146 (1997): 134–141.

16. A. L. Owen and G. M. Owen, Twenty years of WIC: A review of some effects of the program, *Journal of the American Dietetic Association* 97 (1997): 777–782.

17. Committee on Adolescence, American Academy of Pediatrics, Adolescent pregnancy—Current trends and issues: 1998, *Pediatrics* 103 (1999): 516–520.

18. Position of the American Dietetic Association: Nutrition care for pregnant adolescents, *Journal of the American Dietetic Association* 94 (1994): 449–450.

19. M. Erick, Battling morning (noon and night) sickness: New approaches for treating an age-old problem, *Journal of the American Dietetic Association* 94 (1994): 147–148.

20. C. D. Drews and coauthors, The relationship between idiopathic mental retardation and maternal smoking during pregnancy, *Pediatrics* 97 (1996): 547–553; D. L. Olds, C. R. Henderson, Jr., and R. Tatelbaum, Intellectual impairment in children of women who smoke cigarettes during pregnancy, *Pediatrics* 93 (1994): 221–227; D. M. Fergusson, L. J. Horwood, and M. T. Lynskey, Maternal smoking before and after pregnancy: Effects on behavioral outcomes in middle childhood *Pediatrics* 92 (1993): 815–822.

21. Guyer and coauthors, 1998.

22. Environmental tobacco smoke affects birth weight, *Journal of the American Medical Association* 279 (1998): 739.

23. E. Cutz and coauthors, Maternal smoking and pulmonary neuroendocrine cells in sudden infant death syndrome, *Pediatrics* 98 (1996): 668–672; H. S. Klonoff-Cohen and coauthors, The effect of passive smoking and tobacco exposure through breast milk on sudden infant death syndrome, *Journal of the American Medical Association* 273 (1995): 795–798.

24. G. Koren, A. Pastuszak, and S. Ito, Drugs in pregnancy, *New England Journal of Medicine* 338 (1998): 1128–1137.

25. F. D. Eyler and coauthors, Birth outcome from a prospective, matched study of prenatal crack/cocaine use: II. Interactive and dose effects on neurobehavioral assessment, *Pediatrics* 101 (1998): 237–241; T. A. King and coauthors, Neurologic manifestations of in utero cocaine exposure in near-term infants, *Pediatrics* 96 (1995): 259–264.

26. K. J. Rothman and coauthors, Teratogenicity of high vitamin intake, *New England Journal of Medicine* 333 (1995): 1369–1373.

27. T. S. Hinds and coauthors, The effect of caffeine on pregnancy variables, *Nutrition Reviews* 54 (1996): 203–207.

28. J. L. Mills and coauthors, Moderate caffeine use and the risk of spontaneous abortion and intrauterine growth retardation, *Journal of the American Medical Association* 269 (1993): 593–597.

29. J. O. Beattie, Alcohol exposure and the fetus, *European Journal of Clinical Nutrition* 46 (1992): S7–S17.

30. K. Strömand and A. Hellström, Fetal alcohol syndrome: An opthalmological and socioeducational prospective study, *Pediatrics* 97 (1996): 845–850; Committee on Substance Abuse and Committee on Children with Disabilities, American Academy of Pediatrics, Fetal alcohol syndrome and fetal alcohol effects, *Pediatrics* 91 (1993): 1004–1006.

31. Update: Trends in fetal alcohol syndrome—United States, 1979–1993, *Morbidity and Mortality Weekly Reports* 44 (1995): 249–251.

32. S. N. Mattson and coauthors, Heavy prenatal alcohol exposure with or without physical features of fetal alcohol syndrome leads to IQ deficits, *Journal of Pediatrics* 131 (1997): 718–721.

33. Committee on Substance Abuse and Committee on Children with Disabilities, 1993.

34. D. B. Carr and S. Gabbe, Gestational diabetes: Detection, management, and implications, *Clinical Diabetes* 16 (1998): 4–11.

35. C. D. Naylor and coauthors, Cesarean delivery in relation to birth weight and gestational glucose tolerance: Pathophysiology or practice style? *Journal of the American Medical Association* 275 (1996): 1165–1170.

36. The Expert Committee on the Diagnosis and Classification of Diabetes Mellitus, Report of the Expert Committee on the diagnosis and classification of diabetes mellitus, *Diabetes Care* (supplement 1) 21 (1998): 5–19.

37. Position of the American Dietetic Association: Promotion of breast feeding, *Journal of the American Dietetic Association* 97 (1997): 662–666.

38. American Academy of Pediatrics, Work Group on Breastfeeding, Breastfeeding and the use of human milk, *Pediatrics* 100 (1997): 1035–1039.

39. A. C. Goedhart and J. G. Bindels, The composition of human milk as a model for the design of infant formulas: Recent findings and possible applications, *Nutrition Research Reviews* 7 (1994): 1–23.

40. A. L. Wright and coauthors, Increasing breastfeeding rates to reduce infant illness at the community level, *Pediatrics* 101 (1998): 837–844.

41. J. Raisler and coauthors, Breast-feeding and infant illness: A dose-response relationship? *American Journal of Public Health* 89 (1999): 25–30.

42. D. S. Newburg and coauthors, Role of human-milk lactadherin in protection against symptomatic rotavirus infection, *Lancet* 351 (1998): 1160–1164.

43. A. H. Cushing and coauthors, Breastfeeding reduces risk of respiratory illness in infants, *American Journal of Epidemiology* 147 (1998): 863–870; D. S. Newburg and J. M. Street, Bioactive materials in human milk, *Nutrition Today* 32 (1997): 191–201.

44. M. A. Atkinson and T. M. Ellis, Infants diets and insulin-dependent diabetes: Evaluating the "cows' milk hypothesis" and a role for anti-bovine serum albumin immunity, *Journal of the American College of Nutrition* 16 (1997): 334–340.

45. Newburg and Street, 1997.

46. Breast milk and subsequent intelligence quotient in children born preterm, *Nutrition Reviews* 50 (1992): 334–335.

47. K. G. Dewey, Energy and protein requirements during lactation, *Annual Review of Nutrition* 17 (1997): 19–36.

48. Position of the American Dietetic Association, 1997.

49. F. M. Kramer and coauthors, Breast feeding reduces maternal lower-body fat, *Journal of the American Dietetic Association* 93 (1993): 429–433.

50. M. J. Heinig and K. G. Dewey, Health effects of breast feeding for mothers: A critical review, *Nutrition Research Reviews* 10 (1997): 35–56.

51. K. G. Dewey and coauthors, A randomized study of the effects of aerobic exercise by lactating women on breast-milk volume and composition, *New England Journal of Medicine* 330 (1994): 449–453; A. Prentice, Should lactating women exercise? *Nutrition Reviews* 52 (1994): 358–360.

52. J. Liston, Breastfeeding and the use of recreational drugs—Alcohol, caffeine, nicotine, and marijuana, *Breastfeeding Reviews* 2 (1998): 27–30.

53. Committee on Drugs, American Academy of Pediatrics, The transfer of drugs and other chemicals into human milk, *Pediatrics* 93 (1994): 137–150.

54. R. F. Black, Transmission of HIV-1 in the breast-feeding process, *Journal of the American Dietetic Association* 96 (1996): 267–274.

55. Black, 1996.

56. Black, 1996.

57. American Academy of Pediatrics, Committee on Nutrition, *Pediatric Nutrition Handbook*, 4th ed., ed. R. E. Kleinman (Elk Grove Village, Ill.: American Academy of Pediatrics, 1998), pp. 277–278.

Fitness for Life 12

1. K. G. Dewey and M. A. McCrory, Effects of dieting and physical activity on pregnancy and lactation, *American Journal of Clinical Nutrition* (supplement) 59 (1994): S446–S453.

2. B. Sternfeld and coauthors, Exercise during pregnancy and pregnancy outcome, *Medicine and Science in Sports and Exercise* 27 (1995): 634–640.

3. J. M. Pivarnik, Potential effects of maternal physical activity on birth weight: Brief review, *Medicine and Science in Sports and Exercise* 30 (1998): 400–406.

Consumer Corner 12

1. C. R. Howard and coauthors, Antenatal formula advertising: Another potential threat to breast-feeding, *Pediatrics* 94 (1994): 102–104.

2. J. B. Schwartz and coauthors, Does WIC participation improve breastfeeding practices? *American Journal of Public Health* 85 (1995): 729–731.

3. Position of the American Dietetic Association: Promotion of breast-feeding, *Journal of the American Dietetic Association* 97 (1997): 662-666.

4. I. B. Stehlin, Infant formula: Second best but good enough, *FDA Consumer*, June 1996, pp. 17–20.

5. A. N. J. Malik and W. A. M. Cutting, Breast feeding: The baby friendly initiative, *British Medical Journal* 316 (1998): 1548–1549.

Controversy 12

1. E. B. Feldman, How grapefruit juice potentiates drug bioavailability, *Nutrition Reviews* 55 (1997): 398–400.

2. K. Sasaki and coauthors, Bilobalide, a constituent of Ginkgo biloba L., potentiates drug metabolizing enzyme activities in mice: Possible mechanism for anticonvulsant activity against 4-O-methylpyridoxine induced convulsions, *Research Communications in Molecular Pathology and Pharmacology* 96 (1997): 45–56.

3. T. J. Green and coauthors, Oral contraceptives did not affect biochemical folate indexes and homocysteine concentrations in adolescent females, *Journal of the American Dietetic Association* 98 (1998): 49–55.

4. G. Berg, L. Kohlmeier, and H. Brenner, Use of oral contraceptives and serum beta-carotene, *European Journal of Clinical Nutrition* 51 (1997): 181–187.

5. S. S. Harris and B. Dawson-Hughes, The association of oral contraceptive use with plasma 25-hydroxyvitamin D levels, *Journal of the American College of Nutrition* 17 (1998): 282–284.

6. S. M. Vaziri and coauthors, The impact of female hormone usage on the lipid profile: The Framingham Offspring Study, *Archives of Internal Medicine* 153 (1993): 2200–2206.

7. R. A. Hatcher and coauthors, *Contraceptive Technology* (New York: Irvington, 1992), pp. 240–241.

8. J. R. Huges and coauthors, Endorsement of DSM-IV dependence criteria among caffeine users, *Drug Alcohol Depend* 52 (1998): 99–107.

9. J. D. Lane and B. G. Phillips-Bute, Caffeine deprivation affects vigilance performance and mood, *Physiology and Behavior* 65 (1998): 171–175; M. Robelin and P. J. Rogers, Mood and psychomotor performance effects of the first, but not subsequent, cup-of-coffee equivalent doses of caffeine consumed after overnight caffeine abstinence, *Behavioral Pharmacology* 9 (1998): 611–618.

10. P. J. Durlach, The effects of a low dose of caffeine on cognitive performance, *Psychopharmacology* 140 (1998): 116–119.

11. W. C. Willett and coauthors, Coffee consumption and coronary heart disease in women: A ten-year follow-up, *Journal of the American Medical Asosciation* 275 (1996): 458–462.

12. M. F. Leitzmann and coauthors, A prospective study of coffee consumption and the risk of symptomatic gallstone disease in men, *Journal of the American Medical Association* 281 (1999): 2106–2112.

13. A. Rodgers, Effect of cola consumption on urinary biochemical and physiochemical risk factors associated with calcium oxalate urolithianis, *Urological Research* 27 (1999): 77–81.

F

Chapter 13

1. J. Skinner and coauthors, Toddlers' food preferences: Concordance with family members' preferences, *Journal of Nutrition Education* 30 (1998): 17–22.

2. U.S. data from E. Kennedy and J. Goldberg, What are American children eating and implications for public policy, *Nutrition Reviews* 53 (1995): 111–126.

3. A. C. Looker and coauthors, Prevalence of iron deficiency in the United States, *Journal of the American Medical Association* 277 (1997): 973–976.

4. Recommendations to prevent and control iron deficiency in the United States, *Morbidity and Mortality Weekly Report,* April 3, 1998, pp. 1–36.

5. E. Pollitt, Iron deficiency and educational deficiency, *Nutrition Reviews* 55 (1997): 133–141.

6. E. K. Hurtado, A. H. Claussen, and K. G. Scott, Early childhood anemia and mild or moderate mental retardation, *American Journal of Clinical Nutrition* 69 (1999): 115–119.

7. T. D. Matte, Reducing blood lead levels: Benefits and strategies, *Journal of the American Medical Association* 281 (1999): 2340–2342. P. Mushak and A. F. Crocetti, Lead and nutrition, *Nutrition Today,* February 1996, pp. 12–17.

8. H. L. Needleman and coauthors, Bone lead levels and delinquent behavior, *Journal of the American Medical Association* 275 (1996): 363–369.

9. Federal Interagency Forum on Child and Family Statistics, *America's Children: Key National Indicators of Well-Being, 1999* (Washington, D.C.: Government Printing Office, 1999). Yearly updates available from http://www.childstats.gov.

10. D. Farley, Dangers of lead still linger, *FDA Consumer,* January/February 1998, pp. 16–21.

11. H. A. Sampson and A. W. Burks, Mechanisms of food allergy, *Annual Review of Nutrition* 16 (1996): 161–177.

12. K. Deibel and coauthors, A comprehensive approach to reducing the risk of allergens in foods, *Journal of Food Protection* 60 (1997): 436–441.

13. B. Wuthrich, Food-induced cutaneous adverse reactions, *Allergy* 53 (1998): 131–135.

14. G. Iacono, Intolerance of cow's milk and chronic constipation in children, *New England Journal of Medicine* 339 (1998): 1100–1104.

15. S. L. Taylor, S. L. Hefle, and A. Munoz-Furlong, Food allergies and avoidance diets, *Nutrition Today* 34 (1999): 15–22; A. W. Burks and coauthors, Atopic dermatitis and food hypersensitivity reactions, *Journal of Pediatrics* 132 (1998): 132–136.

16. T. Spencer, J. Biederman, and T. Wilens, Growth deficits in children with attention deficit hyperactivity disorder, *Pediatrics* 102 (1998): 501–506.

17. J. Breakey, The role of diet and behaviour in childhood, *Journal of Paediatrics and Child Health* 33 (1997): 190–194.

18. K. S. Rowe and K. J. Rowe, Synthetic food coloring and behavior: A dose response effect in a double-blind, placebo-controlled, repeated-measures study, *Journal of Pediatrics* 124 (1994): 691–698.

19. J. M. Murphy and coauthors, Relationship between hunger and psychosocial functioning in low-income American children, *Journal of the American Academy of Child and Adolescent Psychiatry* 37 (1998): 163–170.

20. R. E. Andersen and coauthors, Relationship of physical activity and television watching with body weight and level of fatness among children, *Journal of the American Medical Association* 279 (1998): 938–942.

21. R. C. Klesges, M. L. Shelton, and L. M. Klesges, Effects of television on metabolic rate: Potential implications for childhood obesity, *Pediatrics* 91 (1993): 281–286.

22. N. D. Wong and coauthors, Television viewing and pediatric hypercholesterolemia, *Pediatrics* 90 (1992): 75–79.

23. Update: Prevalence of overweight among children, adolescents, and adults—United States, 1988–1994, *Morbidity and Mortality Weekly Report* 46 (1997): 199–202.

24. J. O. Fisher and L. L. Birch, Fat preferences and fat consumption of 3- to 5-year-old children are related to parental obesity, *Journal of the American Dietetic Association* 95 (1995): 759–764.

25. M. I. Goran, Energy expenditure, body composition, and disease risk in children and adolescents, *Proceedings of the Nutrition Society* 56 (1997): 195–209.

26. O. T. Raitakari and coauthors, Effects of persistent physical activity on coronary risk factors in children and young adults: The Cardiovascular Risk in Young Finns Study, *American Journal of Epidemiology* 140 (1994): 195–205.

27. D. S. Hardin and coauthors, Treatment of childhood syndrome X (abstract), *Pediatrics* 100 (1997): E5.

28. American Heart Association, Dietary guidelines for healthy children, 1996, available on the Internet at www.americanheart.org/Heart_and_Stroke_A_Z_Guide/dietgk.html or from the American Heart Association, (800) 275-0448.

29. D. Neumark-Sztainer and coauthors, Adolescents engaging in unhealthy weight control behaviors: Are they at risk for other health-compromising behaviors? *American Journal of Public Health* 88 (1998): 952–955.

30. A. F. Subar and coauthors, Dietary sources of nutrients among US children, 1989–1991, *Pediatrics* 102 (1998): 913–923; C. H. Ruxton and T. R. Kirk, Breakfast: A review of associations with measures of dietary intake, physiology, and biochemistry, *British Journal of Nutrition* 78 (1997): 199–214.

31. Murphy and coauthors, 1998.

32. B. A. Bidgood and C. Gameron, Meal/snack missing and dietary adequacy of primary school children, *Journal of the Canadian Dietetic Association* 53 (1992): 164–168.

33. Position of the American Dietetic Association: Dietary guidance for healthy children aged 2 to 11 years, *Journal of the American Dietetic Association* 99 (1999): 93–101.

34. J. Anding and coauthors, Blood lipids, cardiovascular fitness, obesity, and blood pressure: The presence of potential coronary heart disease risk factors in adolescents, *Journal of the American Dietetic Association* 96 (1996): 238–242.

35. K. Schuster, Feds put schools on a lowfat diet, *Food Management,* August 1994, pp. 78–84; Evidence that children grow well on a low-fat diet was provided by L. Van Horn, the principal investigator of the Dietary Intervention Study in Children (DISC), which follows children up to age 18, as cited in Kids grow well on low-fat diet, *Nutrition and the M.D.,* January 1996, pp. 6–7.

36. The Writing Group for the DISC Collaborative Research Group, Efficacy and safety of lowering dietary intake of fat and cholesterol in children with elevated low-density lipoprotein cholesterol: The Dietary Intervention Study in Children (DISC), *Journal of the American Medical Association* 273 (1995): 1429–1435.

37. M. Story, M. Hayes, and B. Kalina, Availability of foods in high schools: Is there cause for concern? *Journal of the American Dietetic Association* 96 (1996): 123–126.

38. V. C. Lysen and R. Walker, Osteoporosis risk factors in eighth grade students, *Journal of School Health* 67 (1997): 317–321.

39. L. Harnack, J. Stang, and M. Story, Soft drink consumption among U.S. children and adolescents: Nutritional consequences, *Journal of the American Dietetic Association* 99 (1999): 436–441.

40. J. Cadogan and coauthors, Milk intake and bone mineral acquisition in adolescent girls: Randomised, controlled intervention trial, *British Medical Journal* 315 (1997): 1255–1260.

41. A. M. Siega-Riz, T. Carson, and B. Popkin, Three squares or mostly snacks—What do teens really eat? A sociodemographic study of meal patterns, *Journal of Adolescent Health* 22 (1998): 29–36.

42. Siega-Riz, Carson, and Popkin, 1998.

43. E. J. van Hoogdalem, I. J. Terpstra, and A. L. Baven, Evaluation of the effect of zinc acetate on the stratum corneum penetration kinetics of erythromycin in healthy male volunteers, *Skin Pharmacology* 9 (1996): 104–110.

44. H. Kerschner and J. M. Pegues, Productive aging: A quality of life agenda, *Journal of the American Dietetic Association* 98 (1998): 1445–1448.

45. Centers for Disease Control and Prevention, National Center for Health Statistics, *Monthly Vital Statistics Report,* October 1996, p. 4.

46. K. G. Manton and J. W. Vaupel, Survival after the age of 80 in the United States, Sweden, France, England, and Japan, *New England Journal of Medicine* 333 (1995): 1232–1235.

47. K. G. Manton and E. Stallard, Longevity in the United States: Age and sex-specific evidence on life span limits from mortality patterns 1960–1990, *Journal of Gerontology* 51A (1996): B362–B375.

48. D. A. Banks and M. Fossel, Telomeres, cancer, and aging: Altering the human life span, *Journal of the American Medical Association* 278 (1997): 1345–1348.

49. D. B. Allison and coauthors, Body mass index and all-cause mortality among people age 70 and over: The Longitudinal Study of Aging, *International Journal of Obesity and Related Metabolic Disorders* 21 (1997): 424–431.

50. L. Di Pietro, The epidemiology of physical activity and physical function in older people, *Medicine and Science in Sports and Exercise* 28 (1996): 596–660.

51. K. Buzina-Suboticanec and coauthors, Aging, nutritional status and immune response, *International Journal of Vitamin and Nutrition Research* 68 (1998): 133–141.

52. A. M. Egbert, The dwindles: Failure to thrive in older patients, *Nutrition Reviews* 54 (1996): S25–S30.

53. Targeting arthritis: The nation's leading cause of disability, *At-A-Glance,* 1998, available from Centers for Disease Control and Prevention, Mail Stop K-13, 4770 Buford Highway NE, Atlanta, GA 30341-3724.

54. T. McAlindon and D. T. Felson, Nutrition: Risk factors for osteoarthritis, *Annals of the Rheumatic Diseases* 56 (1997): 397–400.

55. McAlindon and Felson, 1997; S. E. Edmonds and coauthors, Putative analgesic activity of repeated oral doses of vitamin E in the treatment of rheumatoid arthritis: Results of a prospective placebo controlled double blind trial, *Annals of the Rheumatic Diseases* 56 (1997): 649–655.

56. J. J. Shrander and coauthors, Does food intolerance play a role in juvenile chronic arthritis? *British Journal of Rheumatology* 36 (1997): 905–908; J. A. Shapiro and coauthors, Diet and rheumatoid arthritis in women: A possible protective effect of fish consumption, *Epidemiology* 7 (1996): 256–263.

57. Standing Committee on the Scientific Evaluation of Dietary Reference Intakes, Food and Nutrition Board, Institute of Medicine, *Dietary Reference Intakes for Calcium, Phosphorus, Magnesium, Vitamin D, and Fluoride* (Washington, D.C.: National Academy Press, 1997).

58. C. Ho and coauthors, Practitioners' guide to meeting the vitamin B_{12} Recommended Dietary Allowance for people aged 51 years and older, *Journal of the American Dietetic Association* 99 (1999): 725–727; Standing Committee On the Scientific Evaluation of Dietary Reference Intakes, *Dietary Reference Intakes for Thiamin, Riboflavin, Niacin, Vitamin B_6, Folate, Vitamin B_{12}, Pantothenic Acid, Biotin, and Choline* (Washington, D.C.: National Academy Press, 1998), pp. 7-1–7-27.

59. B. R. Hammond and coauthors, Dietary modification of human macular pigment density, *Investigative Opthalmology and Visual Science* 38 (1997): 1795–1801; J. M. Seddon and coauthors, Dietary carotenoids, vitamins A, C, and E, and advanced age-related macular degeneration: Eye Disease Case-Control Study Group, *Journal of the American Medical Association* 272 (1994): 1413–1420.

60. S. T. Mayne, Beta-carotene, carotenoids, and disease prevention in humans, *FASEB Journal* 10 (1996): 690–701.

61. M. C. Leske and coauthors, Antioxidant vitamins and nuclear opacities: The longitudinal study of cataract, *Opthalmology* 105 (1998): 831–836; P. F. Jacques and coauthors, Long-term vitamin C supplement use and prevalence of early age-related lens opacities, *American Journal of Clinical Nutrition* 66 (1997): 911–916.

62. J. C. Chidester and A. A. Spangler, Fluid intake in the institutionalized elderly, *Journal of the American Dietetic Association* 97 (1997): 23–28; S. A. Gilmore and coauthors, Clinical indicators associated with unintentional weight loss and pressure ulcers in elderly residents of nursing facilities, *Journal of the American Dietetic Association* 95 (1995): 984–992.

63. R. J. Wood, P. M. Suter, and R. M. Russell, Mineral requirements of elderly people, *American Journal of Clinical Nutrition* 62 (1995): 493–505.

64. F. Girodon and coauthors, Impact of trace elements and vitamin supplementation on immunity and infections in institutionalized elderly patients: A randomized controlled trial, *Archives of Internal Medicine* 159 (1999): 748–754; C. Fortes and coauthors, The effect of zinc and vitamin A supplementation on immune response in an older population, *Journal of the American Geriatric Society* 46 (1998): 19–26; M. A. Johnson and K. H. Porter, Micronutrient supplementation and infection in institutionalized elders, *Nutrition Reviews* 55 (1997): 400–404.

65. S. N. Meydani and coauthors, Vitamin E supplementation and in vivo immune response in healthy elderly subjects, *Journal of the American Medical Association* 277 (1997): 1380–1386.

66. R. D. Chandra, Graying of the immune system: Can nutrient supplements improve immunity in the elderly? *Journal of the American Medical Association* 277 (1997): 1398–1399.

67. Johnson and Porter, 1997.

68. N. B. Belloc and L. Breslow, Relationship of physical health status and health practices, *Preventive Medicine* 1 (1972): 409–421.

69. R. Weindruch, Caloric restriction and aging, *New England Journal of Medicine* 337 (1997): 986–994.

70. P. R. Johnson and coauthors, Longevity in obese and lean male and female rats of the Zucker strain: Prevention of hyperphagia, *American Journal of Clinical Nutrition* 66 (1997): 890–903.

71. E. J. M. Velthuis-te Wierik and coauthors, Energy restriction, a useful intervention to retard human aging? Results of a feasibility study, *European Journal of Clinical Nutrition* 48 (1994): 138–148.

72. L. E. Rikans and K. R. Hornbrook, Lipid peroxidation, antioxidant protection and aging, *Biochimica et Biophysica Acta* 1362 (1997): 116–127.

73. G. Paolisso and coauthors, Oxidative stress and advancing age: Results in healthy centenarians, *Journal of the American Geriatric Society* 46 (1998): 833–838.

74. Health report, *Time,* August 29, 1994, p. 21.

75. H. C. Hendrie, Epidemiology of dementia and Alzheimer's disease, *American Journal of Psychiatry* 6 (1998): S3–S18.

76. M. A. Lovely and coauthors, Copper, iron and zinc in Alzheimer's disease senile plaques, *Journal of the Neurological Sciences* 158 (1998): 47–52; C. R. Cornett, W. R. Markesbery, and W. D. Ehmann, Imbalances of trace elements related to oxidative damage in Alzheimer's disease brain, *Neurotoxicology* 19 (1998): 339–345.

77. M. A. Lovely, C. Xie, and W. R. Markesbery, Protections against amyloid beta peptide toxicity by zinc, *Brain Research* 823 (1999): 88–95; C. R. Cornett, W. R. Markesbery, and W. D. Ehmann, Imbalances of trace elements related to oxidative damage in Alzheimer's disease brain, *Neurotoxicology* (1998): 339–345; M. A. Lovely and coauthors, Copper, iron and zinc in Alzheimer's disease senile plaques, *Journal of the Neurological Sciences* 158 (1998): 47–52; M. P. Cuajungco and G. J. Lees, Zinc metabolism in the brain, *Neurobiology of Disease* 4 (1997): 137–169; F. C. Potocnik and coauthors, Zinc and platelet membrane microviscosity in Alzheimer's disease: The in vivo effect of zinc on platelet membranes and cognition, *South African Medical Journal* 87 (1997): 1116–1119.

78. E. M. Reiman and coauthors, Preclinical evidence of Alzheimer's disease in persons homozygous for the e4 allele for apolipoprotein E, *New England Journal of Medicine* 334 (1996): 752–758; National Institute of Aging, *Progress Report on Alzheimer's Disease,* NIH publication no. 94-3885 (Washington, D.C.: Government Printing Office, 1994).

79. M. Sano and coauthors, A controlled trial of selegiline, alpha-tocopherol, or both as treatment for Alzheimer's disease, *New England Journal of Medicine* 336 (1997): 1216–1222.

80. Pl. L. LeBars and coauthors, A placebo-controlled, double-blind, randomized trial of an extract of Ginkgo biloba for dementia. North American EGb study group, *Journal of the American Medical Association* 278 (1997): 1327–1332.

81. G. W. Small, Treatment of Alzheimer's disease: Current approaches and promising developments, *American Journal of Medicine* 27 (1998): S32–S38.

82. D. Schenk and coauthors, Immunization with amyloid-beta attenuates Alzheimer-disease-like pathology in the PDAPP mouse, *Nature* 400 (1999): 173–177.

83. R. M. Ortega and coauthors, Dietary intake and cognitive function in a group of elderly people, *American Journal of Clinical Nutrition* 66

(1997): 803–809; M. Riggs and coauthors, Relations of vitamin B_{12}, vitamin B_6, folate, and homocysteine to cognitive performance in the Normative Aging Study, *American Journal of Clinical Nutrition* 63 (1996): 306–314.

84. K. J. Joshipura, W. C. Willett, and C. W. Douglass, The impact of edentulousness on food and nutrient intake, *Journal of the American Dental Association* 127 (1996): 459–467.

85. An excellent review of the many problems associated with alcoholism in the elderly is found in AMA Council on Scientific Affairs, Alcoholism in the elderly, *Journal of the American Medical Association* 275 (1996): 797–801.

86. Federal program nourishes poor elderly, *Journal of the American Medical Association* 278 (1997): 1301.

87. J. Weinberg, Psychologic implications of the nutritional needs of the elderly, *Journal of the American Dietetic Association* 60 (1972): 293–296.

88. P. Kurtzweil, Growing older, eating better, *FDA Consumer,* March 1996, pp. 12–16.

89. The takeout-food trend: Who carries out their meals and why, *Journal of the American Dietetic Association* 98 (1998): 820.

90. Economic Research Service of the United States Department of Agriculture, USDA report encourages Americans need to remember nutritional needs when eating out, USDA News Release No. 0060.99 (1999) (available on the Internet at www.econ.ag.gov/whatsnew/news/diet.htm).

91. L. H. Clemens, D. L. Slawson, and R. C. Klesges, The effect of eating out on quality of diet in premenopausal women, *Journal of the American Dietetic Association* 99 (1999): 442–444.

Consumer Corner 13

1. G. R. Kraemer and R. R. Kraemer, Premenstrual syndrome: Diagnosis and treatment experiences, *Journal of Women's Health* 7 (1998): 893–907.

2. P. J. Schmidt and coauthors, Differential behavioral effects of gonadal steroids in women with and in those without premenstrual syndrome, *New England Journal of Medicine* 338 (1998): 209–216.

3. D. R. Rubinow, P. J. Schmidt, and C. A. Roca, Estrogen-serotonin interactions: Implications for affective regulation, *Biological Psychiatry* 44 (1998): 839–850.

4. G. A. L. Meijer and coauthors, Sleeping metabolic rate in relation to body composition and the menstrual cycle, *American Journal of Clinical Nutrition* 55 (1992): 637–641; M. Tai and coauthors, Resting metabolic rate during four phases of the menstrual cycle (abstract), *American Journal of Clinical Nutrition* 56 (1992): 101.

5. A. F. Walker and coauthors, Magnesium supplementation alleviates premenstrual symptoms of fluid retention, *Journal of Women's Health* 7 (1998): 1157–1165.

6. S. Thys-Jacobs and coauthors, Calcium carbonate and the premenstrual syndrome: Effects on premenstrual and menstrual symptoms, *American Journal of Obstetrics and Gynecology* 179 (1998): 444–452.

7. S. S. Girdler and coauthors, Dysregulation of cardiovascular and neuroendocrine responses to stress in premenstrual dysphoric disorder, *Psychiatry Research* 81 (1998): 163–178.

Fitness for Life 13

1. American College of Sports Medicine, Position stand: Exercise and physical activity for older adults, *Medicine and Science in Sports and Exercise* 30 (1998): 992–1008; S. G. Wannamethee, A. G. Shaper, and M. Walker, Changes in physical activity, mortality, and incidence of coronary heart disease in older men, *Lancet* 351 (1998): 1603–1608; S. Shinkai, M. Konishi, and R. J. Shepard, Aging, exercise, training, and the immune system, *Exercise Immunology Review* 3 (1997): 68–95.

2. I. H. Rosenberg, Nutrition in the elderly, *Nutrition Reviews* 50 (1992): 349–350.

3. W. J. Evans, Exercise training guidelines for the elderly, *Medicine and Science in Sports and Exercise* 31 (1999): 12–17.

4. R. Marcus and A. R. Hoffman, Growth hormone as therapy for older men and women, *Annual Review of Pharmacology and Toxicology* 38 (1998): 45–61; S. A. Lieberman and A. R. Hoffman, The somatopause: Should growth hormone deficiency in older people be treated? *Clinical Geriatric Medicine* 13 (1997): 671–684.

Controversy 13

1. F. Bellisle and coauthors, Functional food science and behaviour and psychological functions, *British Journal of Nutrition* 80 (1998): S173–S193.

2. P. Norton, G. Falciglia, and D. Gist, Physiologic control of food intake by neural and chemical mechanisms, *Journal of the American Dietetic Association* 93 (1993): 450–454.

3. J. D. Fernstrom, Dietary amino acids and brain function, *Journal of the American Dietetic Association* 94 (1994): 71–77.

4. P. J. Rogers, Food, mood, and appetite, *Nutrition Research Reviews* 8 (1995): 243–269.

5. P. J. Rogers and H. M. Lloyd, Nutrition and mental performance, *Proceedings of the Nutrition Society* 53 (1994): 443–456.

6. Rogers and Lloyd, 1994.

7. I. Blum and coauthors, Food preferences, body weight, and platelet-poor plasma serotonin and catecholamines, *American Journal of Clinical Nutrition* 57 (1993): 486–489.

8. M. W. Green and P. J. Rogers, Impaired cognitive functioning during spontaneous dieting, *Psychological Medicine* 25 (1995): 1003–1010.

9. Brain neurochemistry and macronutrient selection: A role for serotonin feedback? *Nutrition Reviews* 50 (1992): 21–22.

10. C. Benkelfat and coauthors, Mood-lowering effect of tryptophan depletion, *Archives of General Psychiatry* 51 (1994): 687–697.

11. Blum and coauthors, 1993.

12. Benkelfat and coauthors, 1994.

13. A. B. Bruner and coauthors, Randomised study of cognitive effects of iron supplementation in non-anaemic iron-deficient adolescent girls, *Lancet* 348 (1996): 992–996.

14. J. R. Kaplan and coauthors, Demonstration of an association among dietary cholesterol, central serotonergic activity, and social behavior in monkeys, *Psychosomatic Medicine* 56 (1994): 479–484.

15. Kaplan and coauthors, 1994.

16. T. Hamazaki and coauthors, The effect of docosa hexaenoic acid on aggression in young adults: A placebo-controlled double-blind study, *Journal of Clinical Investigation* 97 (1996): 1129–1133.

17. P. Hollingsworth, Beverages: Redefining new age, *Food Technology* 51 (1997): 44–51.

18. Dr. James McGaugh, director of the center for neurobiology and memory at the University of California at Irvine, as quoted in A. Purvis, Ultra think fast, *Time,* June 8, 1992, p. 80.

19. V. Lambert, Using "smart" drugs and drinks may not be smart, *FDA Consumer*, April 1993, pp. 24–26.

20. P. L. Le Bars an coauthors, A placebo-controlled, double-blind, randomized trial of an extract of ginkgo biloba for dementia, *Journal of the American Medical Association* 278 (1997): 1327–1332; S. Stoll and coauthors, Ginkgo biloba extract (EGB761) independently improves changes in passive-avoidance learning and brain membrane fluidity in the aging mouse, *Pharmacopsychiatry* 29 (1996): 144–149; P. F. Smith, C. L. Darlington, and K. Maclennan, Gingko biloba: An ancient Chinese tree with neuroprotective properties, *Asia Pacific Journal of Pharmacology* 10 (1995), and supplements 1, 3–4.

21. Rogers, 1995.

22. E. diTomaso, M. Beltramo, and D. Piomelli, Brain cannabinoids in chocolate (scientific correspondence), *Nature* 382 (1996): 677–678.

23. Bellisle and coauthors, 1998; Rogers, 1995.

24. Rogers and Lloyd, 1994, 447.

25. Bellisle and coauthors, 1998.

26. C. R. Markus and coauthors, Does carbohydrate-rich, protein-poor food prevent a deterioration of mood and cognitive performance of stress-prone subjects when subjected to a stressful task? *Appetite* 31 (1998): 49–65.

27. H. M. Lloyd and P. J. Rogers, Acute effects of breakfasts of differing fat and carbohydrate content on morning mood and cognitive performance, *Proceedings of the Nutrition Society* 53 (1994): 239A.

28. Rogers and Lloyd, 1994, 446.

Chapter 14

1. Science-based, unified approach needed to safeguard the nation's food supply, *Public Health Reports* 113 (1998): 482–483.

2. Other features of the botulinum toxin are in L. Vangelova, Botulinum toxin: A poison that can heal, *FDA Consumer,* December 1995, pp. 16–19.

3. P. Kurtzweil, Questions keep sprouting about sprouts, *FDA Consumer,* January/February 1999, pp. 18–22.

4. National Advisory Committee on Microbiological Criteria for Foods, Hazard analysis and critical control point principles and application guidelines, *Journal of Food Protection* 61 (1998): 762–775.

5. USDA, Second progress report on *Salmonella* testing for raw meat and poultry products, *FSIS Backgrounder,* January 21, 1999.

6. Centers for Disease Control and Prevention, Salmonella and campylobacter illnesses on the decline, CDC Press Release, March 11, 1999 (available from www.hhs.gov/news/press/1999pres/990311b.html).

7. Outbreaks of *Salmonella* serotype enteritidis infection associated with consumption of raw shell eggs—United States, 1994–1995, *Morbidity and Mortality Weekly Report* 45 (1996): 737–742.

8. Even pasteurized eggs may contain harmful bacteria, *Tufts University Health and Nutrition Letter,* July 1998, p. 6.

9. J. A. Desenclos and coauthors, The protective effect of alcohol on the occurrence of epidemic oyster-borne hepatitis A, *Epidemiology* 3 (1992): 371–374.

10. Y. Sun and J. D. Oliver, Hot sauce: No elimination of *Vibrio vulnificus, Journal of Food Protection* 58 (1995): 441–442.

11. 'Provocative' report issued on use of pesticides, *Journal of the American Medical Association* 275 (1996): 899.

12. A. E. Larson and coauthors, Evaluation of the botulism hazard from vegetables in modified atmosphere packaging, *Journal of Food Protection* 60 (1997): 1208–1214.

13. R. Papazian, Sulfites: Safe for most, dangerous for some, *FDA Consumer,* December 1996, pp. 11–14.

14. D. J. Raiten, J. M. Talbot, and K. D. Fisher, eds., Executive summary from the report: Analysis of adverse reactions to monosodium glutamate (MSG), *Journal of Nutrition* 125 (1995): S2892–S2906.

15. Monosodium glutamate, *FDA Medical Bulletin* 26 (1996).

16. T. D. Etherton, P. M. Kris-Etherton, and E. W. Mills, Recombinant bovine and porcine somatotropin: Safety and benefits of these biotechnologies, *Journal of the American Dietetic Association* 93 (1993): 177–180.

17. Position of the American Dietetic Association: Biotechnology and the future of food, *Journal of the American Dietetic Association* 95 (1995): 1429–1432.

18. K. L. Ropp, New animal drug increases milk production, *FDA Consumer,* May 1994, pp. 24–27.

Consumer Corner 14

1. United States Department of Agriculture, Organic labeling claim allowed on meat and poultry products, USDA Press Release, January 14, 1999 (available from www.usda.gov/news/releases/1999/01/0015).

2. *FDA Consumer,* March/April 1998, p. 36.

3. Pesticides not pesty, *Priorities* (1996): 6.

Controversy 14

1. J. Henkel, Genetic engineering: Fast forwarding to future foods, *FDA Consumer,* April 1995, pp. 6–11.

2. R. A. Dixon and coauthors, Metabolic engineering: Prospects for crop improvement through genetic manipulation of phenylpropanoid biosynthesis and defense responses—A review, *Gene* 179 (1996): 61–71.

3. First biotech tomato marketed, *FDA Consumer,* September 1994, pp. 3–4.

4. EPA Office of Pollution Prevention and Toxics, TSCA biotechnology program, 1998, accessible on the Internet at www.epa.gov; EPA proposes regulation of gene altered pesticidal plants, *EDF Letter,* March 1995, p. 2; Bioengineering regulations go public, *Science News,* December 3, 1994, p. 383.

5. H. Shand and P. Mooney, Terminator seeds threaten an end to farming, *Earth Island Journal,* Fall 1998, pp. 30–31.

6. J. Walsh, Brave new farm, *Time,* January 11, 1999, pp. 86–88.

7. R. Epstein, Redesigning the world: Ethical questions about genetic engineering, an essay available on the Internet at online.sfsu.edu/~rone/GE%20Essays/Redesigning.htm

8. Food and Drug Administration, Statement of policy: Foods derived from new plant varieties, *Federal Register,* May 29, 1992.

9. International Food Information Council, Food biotechnology, *Backgrounder,* 1998 (available from http://www.starpass.net/winific.html); Biotechnology of food: Background information from the FDA, *Nutrition Today,* July/August 1994, pp. 19–20.

10. S. L. Huttner, Getting the products of biotechnology to market, *Priorities,* March 1995, pp. 11–14.

11. International Food Information Council, 1998; J. A. Nordlee and coauthors, Identification of a brazil-nut allergen in transgenic soybeans, *New England Journal of Medicine* 334 (1996): 688–692.

12. J. Puzzanghera, Coalition sues over genetically altered foods, *Tallahassee Democrat,* May 28, 1998, p. 3A.

13. Position of the American Dietetic Association: Biotechnology and the future of food, *Journal of the American Dietetic Association* 95 (1995): 1429–1432.

14. J. Farkas, Irradiation as a method for decontaminating food: A review, *International Journal of Food Microbiology* 44 (1998): 189–204.

15. J. N. Giamalva, M. Redfern, and W. C. Bailey, Dietitians employed by health care facilities preferred a HACCP system over irradiation or chemical rinses for reducing risk of foodborne disease, *Journal of the American Dietetic Association* 98 (1998): 885–888.

16. World Health Organization, *Weekly Epidemiological Record,* January 16, 1998, pp. 9–11.

17. S. L. Nightingale, Irradiation of meat approved for pathogen control, *Journal of the American Medical Association* 279 (1998): 9.

18. FDA clarifies radiation labeling, *Journal of the American Dietetic Association* 98 (1998): 1403.

19. L. Heide and K. W. Bogl, Detection methods for irradiated foods—Luminescence and viscosity measures, *International Journal of Radiation Biology* 57 (1990): 201–219; T. Autio and S. Pinnioja, Identification of irradiated foods by the thermoluminescence of mineral contamination, *Zeitschrift für Lebensmittel-Untersuchung und-Forschung* 191 (1990): 177–180; A. M. Sjoberg and coauthors, Methods for detection of irradiation of spices, *Zeitschrift für Lebensmittel-Untersuchung und-Forschung* 190 (1990): 99–103.

20. Vijayalaxmi and S. G. Srikantia, A preview of the studies on the wholesomeness of irradiated wheat, conducted at the National Institute of Nutrition, India, *Radiation, Physics, and Chemistry* 34 (1989): 941–952.

21. Vijayalaxmi and Srikantia, 1989.

22. T. D. Etherton, The impact of biotechnology on animal agriculture and the consumer, *Nutrition Today,* July/August 1994, pp. 12–18.

23. Position of the American Dietetic Association: Food irradiation, *Journal of the American Dietetic Association* 96 (1996): 69–72.

Chapter 15

1. USDA Center for Nutrition Policy and Promotion, Could there be hunger in America? *Nutrition Insight,* September 1998.

2. United Nations Food and Agriculture Organization (FAO), *The State of Food and Agriculture,* 1998 (available on the Internet at www.fao.org.

3. L. N. Burby, *World Hunger* (San Diego, Calif.: Lucent Books, 1995), pp. 13–16.

4. L. R. Brown, Facing food scarcity, *World Watch,* November/December 1995, pp. 10–20.

5. World Health Organization, Executive summary, World Health Report 1998: Life in the 21st century—A vision for all (available from www.ch/whr/1998/exsum98e.htm).

6. C. Flavin and O. Tunali, Climate of hope: New strategies for stabilizing the world's atmosphere, *World Watch Paper* 130, June 1996, a 68-page monograph with several hundred references.

7. United Nations Food and Agriculture Organization (FAO), 1998.

8. The sweeping toll of malnutrition, *Public Health Reports* 112 (1997): 186–187.

9. Burby, 1995, p. 13.

10. P. Uvin, The state of world hunger, *Nutrition Reviews* 52 (1994): 151–161.

11. Position of the American Dietetic Association: Domestic food and nutrition security, *Journal of the American Dietetic Association* 98 (1999): 337–342.

12. C. S. Kramer-LeBlanc and K. McMurry, Discussion paper on domestic food security, *Family Economics and Nutrition Review* 11 (1998): 49–78.

13. P. P. Basiotis, C. S. Kramer-LeBlanc, and E. T. Kennedy, Maintaining nutrition security and diet quality: The role of the Food Stamp Program and WIC, *Family Economics and Nutrition Review* 11 (1998): 4–16.

14. USDA Center for Nutrition Policy and Promotion, 1998.

15. USDA Center for Nutrition Policy and Promotion, 1998; U.S. Department of Commerce, *Statistical Abstract of the United States, 1994* (Washington, D.C.: Bureau of the Census, 1994).

16. L. Trivers and J. Borland, Glickman announces new community anti-hunger initiative, USDA press release, February 11, 1999 (available from www.usda.gov/news/releases/); Kramer-LeBlanc and McMurry, 1998.

17. R. W. Kates, Ending deaths from famine: The opportunity in Somalia, *New England Journal of Medicine* 328 (1993): 1055–1057.

18. World Health Organization, 1998.

19. World Health Organization, 1998.

20. J. Raloff, Can grain yields keep pace? *Science News* 152 (1997): 104–105.

21. R. D. Kaplan, The coming anarchy, *Atlantic Monthly,* February 1994, pp. 44–76.

22. J. W. Clay and coauthors, *The Spoils of Famine: Ethiopian Famine Policy and Peasant Agriculture* (Cambridge, Mass.: Cultural Survival, 1988), as cited in L. R. Brown and H. Kane, *Full House* (New York: W. W. Norton, 1994), pp. 146–157.

23. United Nations Food and Agriculture Organization (FAO), Elements for a plan of action on fishing capacity, Preparatory meeting for the FAO consultation on the management of fishing capacity, shark fisheries, and incidental catch of seabirds in longline fisheries, July 22–24, 1998 (available from www.nmfs.gov).

24. United Nations Food and Agriculture Organization (FAO), as cited in World Resources Institute (WRI), *World Resources 1992–93* (New York: Oxford University Press, 1992); bluefin tuna figure from D. Meadows and coauthors, *Beyond the Limits* (Post Mills, Vt.: Chelsea Green Publishing Company, 1992), as cited in Brown and Kane, 1994, pp. 75–88.

25. United Nations Food and Agriculture Organization (FAO), 1998.

26. The Intergovernmental Panel on Climate Change (IPCC), *Climate Change 1995: IPCC Second Assessment Report,* available on the Internet at www.ipcc.ch/cc95/synt.htm

27. R. N. Harris and D. S. Chapman, Borehole temperatures and a baseline for 20th-century global warming estimates, *Science* 275 (1997): 1618–1621.

28. W. J. Martens, R. Slooff, and E. K. Jackson, Climate change, human health, and sustainable development, *Bulletin of the World Health Organization* 75 (1997): 583–588.

29. Population Reference Bureau (PRB), *1993 World Population Data Sheet* (Washington, D.C.: Population Reference Bureau, 1993), as cited in Brown and Kane, 1994, pp. 49–61.

30. P. S. Dasgupta, Population, poverty and the local environment, *Scientific American,* February 1995, pp. 40–45.

31. L. R. Brown, *Who Will Feed China? Wake-up Call for a Small Planet* (New York: W. W. Norton, 1995).

32. Dasgupta, 1995.

33. Dasgupta, 1995.

34. World Health Organization, 1998.

Consumer Corner 15

1. S. Smith, professor of nutrition, University of New Hampshire, Durham, N.H., personal communication, summer 1993.

Controversy 15

1. World Resources Institute, *The 1992 Information Please Environmental Almanac* (Boston: Houghton Mifflin, 1992), p. 13.

2. C. B. Heiser, Jr., *Seeds to Civilization: The Story of Food* (Cambridge, Mass.: Harvard University Press, 1990), p. 13.

3. J. Dixon, Agency warns of threats posed by plant extinction, *Tallahassee Democrat,* March 24, 1992.

4. Dixon, 1992.

5. P. Smith and J. Warrick, Boss hog: North Carolina's pork revolution, *Amicus Journal,* Spring 1996, pp. 36–42.

6. EPA and hog farmers agree, *Progressive Farmer,* January 1999, pp. 6–7.

7. A. Beers, Animal waste report gives traction to reform bill, *Food Chemical News* 39 (1998): 24–26.

8. USDA Interagency Agricultural Projections Committee, *USDA Agricultural Baseline Projections to 2008,* 1999 (available on the Internet at usda.mannlib.cornell.edu).

9. U.S. Department of Commerce, *Statistical Abstract of the United States, 1994* (Washington, D.C.: Bureau of the Census, 1994), p. 668.

10. USDA National Agricultural Statistics Service, *Farms and Land in Farms, 1998* (Washington, D.C.: USDA Interagency Agricultural Projections Committee, 1999).

11. H. Carsalade, FAO, Sustainable Development Department, Sustainable food security, March 20, 1998 (available from www.fao.org).

12. R. P. Tengerdy and G. Szakacs, Perspectives in agrobiotechnology, *Journal of Biotechnology* 66 (1998): 91–99.

13. A. T. Durning and H. B. Brough, Reforming the livestock economy, in L. R. Brown and coauthors, *State of the World 1992* (New York: W. W. Norton, 1992), pp. 66–82.

14. S. Smith, professor of nutrition, University of New Hampshire, Durham, N.H., personal communication, August 1993.

15. A. Zorc, From family farm to agribusiness: The spoilage of America's meat industry, *Co-op America Quarterly,* Spring 1992, pp. 10–13, 19; Eating green, *Nutrition Action Health Letter,* January/February 1992, pp. 1, 5–7.

16. C. Flavin and J. E. Young, Shaping the next industrial revolution, in L. R. Brown, *State of the World 1993* (New York: W. W. Norton, 1993), pp. 180–199.

Answers to the Self Check Questions

G

Chapter 1

1. d (p. 10)
2. a (p. 6)
3. a (p. 5)
4. c (p. 17)
5. b (p. 18)
6. False. Heart disease and cancer are influenced by many factors with genetics and diet among them. (p. 3)
7. True (p. 6)
8. False. Only when a finding has been repeatedly confirmed by science is it wise to change your diet accordingly. (p. 11)
9. True (p. 25)

Chapter 2

1. b (p. 30)
2. d (p. 29)
3. b (p. 41)
4. d (p. 34)
5. d (p. 35)
6. True (p. 32)
7. False. The DRI are estimates of the needs of healthy persons only. Medical problems alter nutrient needs. (p. 31)
8. True (p. 39)
9. False. By law, food labels must state as a percentage of the Daily Values the amounts of vitamins A and C present in a food. (p. 48)
10. True (p. 50)
11. False. The diets of the Mediterranean are generally low in animal protein and high in carbohydrates and fiber. (p. 60)

Chapter 3

1. c (pp. 69, 81)
2. d (p. 71)
3. a (pp. 78, 80)
4. c (p. 78)
5. d (pp. 76, 80, 81)
6. True (p. 67)
7. False. Vitamin and other deficiencies easily damage the linings of the body cavities. (p. 73)
8. False. The process of digestion occurs mainly in the small intestine. (p. 78)
9. False. The digestive tract works efficiently to digest all foods simultaneously, regardless of composition. (p. 78)
10. True (p. 78)
11. False. Absorption of the majority of nutrients takes place across the specialized cells of the small intestine. (p. 81)
12. False. Stone Age people achieved their abundant nutrient intakes using only the meats, fruits, and vegetables food groups. (p. 90)

Chapter 4

1. b (p. 95)
2. d (p. 96)
3. a (p. 96)
4. c (p. 102)
5. a (p. 108)
6. b (p. 109)
7. True (p. 109)
8. True (p. 110)
9. True (p. 112)
10. False. Type 1 diabetes is most often controlled by insulin injections. (p. 112)
11. True (p. 107)
12. False. Whole-grain bread remains more nutritious despite the enrichment of white flour. (p. 104)
13. True (p. 103)
14. False. Using artificial sweeteners has not been proved to help people lose weight. (p. 130)

Chapter 5

1. c (p. 135)
2. a (p. 138)
3. c (p. 138)
4. b (p. 141)
5. d (p. 158)
6. d (p. 167)
7. True (p. 142)
8. True (p. 143)
9. False. Consuming large amounts of *trans*-fatty acids elevates serum LDL cholesterol and thus raises the risk of heart disease and heart attack. (p. 151)
10. False. When olestra is present in the digestive tract, fat-soluble vitamins, including vitamin E, become unavailable for absorption. (p. 154)

Chapter 6

1. b (p. 176)
2. b (p. 181)
3. c (p. 184)
4. a (p. 194)
5. d (pp. 204, 205)
6. True (p. 186)
7. False. Excess protein in the diet may have adverse effects such as obesity, enlarged liver or kidneys, worsened kidney disease, and accelerated adult bone loss. (p. 197)
8. False. Impoverished people living on Indian reservations, in inner cities, and in rural areas of the United States, as well as some elderly, homeless, and ill people in hospitals, are often diagnosed with PEM. (pp. 196, 197)
9. True (p. 191)
10. True (p. 194)

G

Chapter 7

1. d (p. 211)
2. c (pp. 212, 243)
3. d (pp. 215, 243)
4. a (p. 216)
5. d (pp. 214, 230)
6. d (p. 211)
7. True (p. 240)
8. False. With rare exceptions, nutrients are absorbed best from foods where they are dispersed among other ingredients that facilitate their absorption. (p. 239)
9. True (p. 242)
10. True (p. 241)
11. True (p. 257)

Chapter 8

1. b (p. 267)
2. c (p. 268)
3. b (p. 273)
4. d (p. 283)
5. a (p. 286)
6. False. The FDA requires yearly tests of bottled water to ensure that it meets the same standards as those set for the purity and sanitation of U.S. tap water. (p. 270)
7. False. You can survive only a few days without water. (p. 266)
8. True (p. 280)
9. True (p. 272)
10. False. Butter, cream, and cream cheese are almost pure fat and contain negligible calcium, whereas large servings of vegetables such as broccoli, are good sources of available calcium. (p. 295)
11. False. Actions to prevent osteoporosis are best begun in childhood and adolescence when the bones are growing most rapidly. (pp. 304, 305)

Chapter 9

1. b (p. 316)
2. d (p. 318)
3. c (p. 321)
4. a (p. 326)
5. d (p. 337)
6. a (p. 333)
7. False. The thermic effect of food is believed to have negligible effects on total energy expenditure. (p. 315)
8. True (p. 317)
9. False. The BMI are unsuitable for use with athletes and adults over age 65. (p. 320)
10. False. No special protein supplements can help speed weight gain beyond the effects of ordinary foods. (p. 342)

Chapter 10

1. c (p. 359)
2. b (p. 360)
3. a (p. 366)
4. d (p. 373)
5. d (p. 377)
6. False. The guidelines for developing physical fitness and more rigorous than those for obtaining health benefits. (p. 356)
7. False. The average resting pulse rate for adults is around 70 beats per minute, but the rate is lower for active people. (p. 358)
8. True (p. 369)
9. True (p. 375)
10. False. Anorexia nervosa occurs most often in women, but men account for about 1 in 20 eating disorder cases in the general population. (p. 384)

Chapter 11

1. b (p. 397)
2. c (p. 400)
3. d (p. 407)
4. d (p. 411)
5. c (p. 424)
6. False. The best way to plan a diet to support the immune system is to meet the recommended intake for each nutrient while not ingesting a dose from a supplement that would cause harm. (p. 426)
7. True (p. 399)
8. True (p. 405)
9. False. Hypertension is more severe and occurs earlier in life among African Americans than those of European or Asian descent. (p. 406)
10. False. Vegetarians have lower mortality rates from cancer than the rest of the population, even when cancers linked to smoking and alcohol are taken out of the picture. (p. 410)

Chapter 12

1. c (p. 435)
2. b (p. 437)
3. d (pp. 444, 445)
4. a (p. 450)
5. d (p. 453)
6. True (p. 433)
7. True (p. 447)
8. False. In general, the effect of nutritional deprivation of the mother is to reduce the quantity, not the quality, of her milk. (p. 450)
9. False. There is no proof for the theory that "stuffing the baby" at bedtime will promote sleeping through the night. (p. 458)
10. True (p. 466)

Chapter 13

1. d (p. 472)
2. c (p. 474)
3. c (p. 476)
4. d (p. 485)
5. b (p. 486)
6. False. Research to date does not support the idea that food allergies or intolerances cause hyperactivity in children, but studies continue. (p. 477)
7. True (p. 478)
8. True (p. 490)
9. False. Vitamin A absorption appears to increase with aging. (p. 491)
10. True (p. 504)

Chapter 14

1. b (p. 508)
2. c (p. 513)
3. d (p. 517)
4. a (p. 519)
5. d (p. 531)
6. False. Most hazardous contamination is not detectable by a food's odor, appearance, or taste. (p. 515)

7. False. Minerals are unaffected by heat processing because they cannot be destroyed, as vitamins can be. (p. 527)
8. False. The canning industry chooses treatments that employ the high-temperature–short-time [HTST] principle for canning. (p. 526)
9. True (p. 532)
10. True (p. 543)

Chapter 15

1. b (p. 550)
2. c (p. 552)
3. d (pp. 553, 554)
4. c (p. 558)
5. a (p. 565)
6. True (p. 549)
7. False. Most children who die of malnutrition do not starve to death—they die because their health has been compromised by dehydration from infections that cause diarrhea. (p. 552)
8. False. The number of people affected by famine is relatively small compared with the number suffering from less severe but chronic hunger. (p. 552)
9. False. The link between improved economic status and slowed population growth has been demonstrated in country after country. (p. 555)
10. True (p. 566)

—A—

absorb to take in, as nutrients are taken into the intestinal cells after digestion; the main function of the digestive tract with respect to nutrients.

acid reducers and **acid controllers** drugs that reduce the acid output of the stomach. They are most suitable for treating severe, persistent forms of heartburn, but are useless for neutralizing acid already present in the stomach.

acid-base balance equilibrium between acid and base concentrations in the body fluids.

acidosis (acid-DOH-sis) blood acidity above normal, indicating excess acid (*osis* means "too much in the blood").

acids compounds that release hydrogens in a watery solution.

acne chronic inflammation of the skin's follicles and oil-producing glands, which leads to an accumulation of oils inside the ducts that surround hairs; usually associated with the maturation of young adults.

adaptive thermogenesis adjustments in energy expenditure related to changes in environment such as cold and to physiological events such as underfeeding or trauma.

added sugars sugars added to a food for any purpose, such as to add sweetness or bulk or to aid in browning (baked goods).

additives substances that are added to foods, but are not normally consumed by themselves as foods.

adequacy the dietary characteristic of providing all of the essential nutrients, fiber, and energy in amounts sufficient to maintain health and body weight.

adipose tissue the body's fat tissue, consisting of massses of fat-storing cells and blood vessels to nourish them.

aerobic (air-ROE-bic) requiring oxygen.

AIDS acquired immune deficiency syndrome; caused by infection with HIV, a virus that is transmitted primarily by sexual contact, contact with infected blood, needles shared among drug users, or fluids transferred from an infected mother to her fetus or infant.

alkalosis (al-kah-LOH-sis) blood alkalinity above normal (*alka* means "base"; *osis* means "too much in the blood").

allergy an immune reaction to a foreign substance, such as a component of food. Also called *hypersensitivity* by researchers.

alpha-lactalbumin (lact-AL-byoo-min) the chief protein in human breast milk. The chief protein in cow's milk is *casein* (CAY-seen).

amenorrhea the absence or cessation of menstruation.

amine (a-MEEN) **group** the nitrogen-containing portion of an amino acid.

amino (a-MEEN-o) **acids** the building blocks of protein. Each has an amine group at one end, an acid group at the other, and a distinctive side chain.

amino acid pools amino acids dissolved in cellular fluid that provide cells with ready raw materials from which to build new proteins or other molecules.

amniotic (am-nee-OTT-ic) **sac** the "bag of waters" in the uterus in which the fetus floats.

anaerobic (AN-air-ROE-bic) not requiring oxygen.

anaphylactic (an-AFF-ill-LAC-tic) **shock** a life-threatening whole-body allergic reaction to an offending substance.

anemia the condition of inadequate or impaired red blood cells; a reduced number or volume of red blood cells along with too little hemoglobin in the blood. Anemia is not a disease, but a symptom of another problem; its name literally means "too little blood."

anencephaly (an-en-SEFF-ah-lee) a severe neural tube defect in which the brain fails to form; anencephaly leads to death soon after birth.

aneurysm (AN-you-rism) the ballooning out of an artery wall at a point that is weakened by deterioration.

antacids medications that react directly and immediately with the acid of the stomach, neutralizing it. Antacids are most suitable for treating occasional heartburn.

antibodies (AN-te-bod-ees) large proteins of the blood, produced by the immune system in response to an invasion of the body by foreign substances (antigens). Antibodies combine with and inactivate the antigens. Also defined in Chapter 3.

antibodies proteins, made by cells of the immune system, that are expressly designed to combine with and to inactivate specific antigens.

antigen a substance foreign to the body that elicits the formation of antibodies or an inflammation reaction from immune system cells.

antioxidant (anti-OX-ih-dant) a compound that protects other compounds from oxygen by itself reacting with oxygen (*anti* means "against"; *oxy* means "oxygen").

antipromoters compounds in foods that act in several ways to oppose the formation of cancer.

aorta (ay-OR-tuh) the large, primary artery that conducts blood from the heart to the body's smaller arteries.

Apgar score a system of scoring an infant's physical condition right after birth. Heart rate, respiration, muscle tone, response to stimuli, and color are ranked 0, 1, or 2. A low score indicates that medical attention is required to facilitate survival.

appendicitis inflammation and/or infection of the appendix, a sac protruding from the intestine.

appetite the psychological desire to eat; a learned motivation and a positive sensation that accompanies the sight, smell, or thought of appealing foods.

arousal heightened activity of certain brain centers associated with attention, excitement, and anxiety.

arteries blood vessels that carry blood containing fresh oxygen supplies from the heart to the tissues.

arthritis a usually painful inflammation of joints caused by many conditions, including infections, metabolic disturbances, or injury; usually results in altered joint structure and loss of function.

ascorbic acid one of the active forms of vitamin C (the other is *dehydroascorbic* acid); an antioxidant nutrient.

atherosclerosis (ath-er-oh-scler-OH-sis) the most common form of cardiovascular disease; characterized by plaques along the inner walls of the arteries (*scleros* means "hard"; *osis* means "too much").

atrophy (AT-tro-fee) a decrease in size (for example, of a muscle) because of disuse.

—B—

balance the dietary characteristic of providing foods of a number of types in proportion to each other, such that foods rich in some nutrients do not crowd out of the diet foods that are rich in other nutrients. Also called *proportionality*.

balance study a laboratory study in which a person is fed a controlled diet and the intake and excretion of a nutrient are measured.

basal metabolic rate (BMR) the rate at which the body uses energy to support its basal metabolism.

basal metabolism the sum total of all the involuntary activities that are necessary to sustain life, including circulation, respiration, temperature maintenance, hormone secretion, nerve activity, and new tissue synthesis, but excluding digestion and voluntary activities. Basal metabolism is the largest component of the average person's daily energy expenditure.

bases compounds that accept hydrogens from solutions.

B-cells lymphocytes that produce antibodies. *B* stands for bursa, an organ in the chicken in which B-cells were first identified.

behavior modification alteration of behavior using methods based on the theory that actions can be controlled by manipulating the environmental factors that cue, or trigger, the actions.

beriberi the thiamin-deficiency disease; characterized by loss of sensation in the hands and feet, muscular weakness, advancing paralysis, and abnormal heart action.

beta-carotene an orange pigment with antioxidant activity; a vitamin A precursor made by plants and stored in human fat tissue.

bicarbonate a common alkaline chemical; a secretion of the pancreas; also, the active ingredient of baking soda.

bile an emulsifier made by the liver from cholesterol and stored in the gallbladder. Bile does not digest fat as enzymes do but emulsifies it so that enzymes in the watery fluids may contact it and split the fatty acids from their glycerol for absorption.

bioaccumulation the accumulation of a contaminant in the tissues of living things at higher and higher concentrations along the food chain.

bioavailability absorbability; the individual differences in the proportion of a nutrient that is available for absorption from various sources.

bioelectrical impedance a technique to measure body fatness by measuring the body's electrical conductivity.

biosensor a genetically altered microbe that provides a rapid, low-cost, and accurate test for toxic products of microbial agents in foods.

biotechnology the science that manipulates biological systems or organisms to modify their products or components or create new products.

biotin (BY-o-tin) a B vitamin; a coenzyme necessary for fat synthesis and other metabolic reactions.

bladder the sac that holds urine until time for elimination.

blood the fluid of the cardiovascular system; composed of water, red and white blood cells, other formed particles, nutrients, oxygen, and other constituents.

body composition the proportions of muscle, bone, fat, and other tissue that make up a person's total body weight.

body mass index (BMI) an indicator of obesity, calculated by dividing the weight of a person by the square of the person's height.

body system a group of related organs that work together to perform a function. Examples are the circulatory system, respiratory system, and nervous system.

bottled water drinking water sold in bottles.

botulism an often-fatal food poisoning caused by botulinum toxin, a toxin produced by the *Clostridium botulinum* bacterium that grows without oxygen in nonacidic canned foods.

bovine somatotropin (BST) growth hormone of cattle, which can be produced for agricultural use by genetic engineering. Also called *bovine growth hormone (BGH)*.

branched-chain amino acids amino acids that, unlike the others, can provide energy directly to muscle tissue: leucine, isoleucine, and valine.

brown fat adipose tissue abundant in hibernating animals and human infants. Brown fat cells are packed with pigmented, energy-burning enzymes that release heat rather than accomplishing other tasks. These enzymes give the cells a darkened appearance under a microscope.

buffers molecules that can help to keep the pH of a solution from changing by gathering or releasing H ions.

—C—

caloric effect the drop in cancer incidence seen whenever intake of food energy (calories) is restricted.

calorie control control of energy intake; a feature of a sound diet plan.

calories units of energy. Strictly speaking, the unit used to measure the energy in foods is a kilocalorie (*kcalorie*, or *Calorie*): it is the amount of heat energy necessary to raise the temperature of a kilogram (a liter) of water 1 degree Celsius. This book follows the common practice of using the lowercase term *calorie* (abbreviated *cal*) to mean the same thing.

cancer a disease in which cells multiply out of control and disrupt normal functioning of one or more organs.

canning a method of preserving food by killing all microorganisms present in the food and then sealing out air. The food, container, and lid are heated until sterile; as the food cools, the lid makes an airtight seal, preventing contamination.

capillaries minute, weblike blood vessels that connect arteries to veins and permit transfer of materials between blood and tissues.

carbohydrate loading a regimen of exhausting exercise, followed by eating a high-carbohydrate diet, that enables muscles to temporarily store glycogen beyond their normal capacity; also called *glycogen loading* or *glycogen supercompensation*.

carbohydrate sweeteners ingredients composed of carbohydrates that contain sugars used for sweetening food products, including glucose, fructose, corn syrup, concentrated grape juice, and other sweet carbohydrates.

carbohydrates compounds composed of single or multiple sugars. The name means "carbon and water," and a chemical shorthand for carbohydrate is CHO, signifying carbon (C), hydrogen (H), and oxygen (O).

carcinogen (car-SIN-oh-jen) a cancer-causing substance (*carcin* means "cancer"; *gen* means "gives rise to").

carcinogenesis the origination or beginning of cancer.

cardiac output the volume of blood discharged by the heart each minute.

cardiorespiratory endurance the ability to perform large-muscle dynamic exercise of moderate-to-high intensity for prolonged periods.

cardiovascular disease (CVD) disease of the heart and blood vessels; also called *coronary heart disease (CHD)*. The two most common forms of CVD are atherosclerosis and hypertension.

carnitine a nonessential nutrient that functions in cellular activities.

carrying capacity the total number of living organisms that a given environment can support without deteriorating in quality.

cataracts (CAT-uh-racts) thickening of the lens of the eye that can lead to blindness. Cataracts can be caused by injury, viral infection, toxic substances, genetic disorders, and, possibly, some nutrient deficiencies or imbalances.

cell differentiation (dih-fer-en-she-AY-shun) the process by which immature cells are stimulated to mature and gain the ability to perform functions characteristic of their cell type.

cells the smallest units in which independent life can exist. All living things are single cells or organisms made of cells.

central obesity excess fat in the abdomen and around the trunk.

certified lactation consultant a health-care provider, often a registered nurse, with specialized training in breast and infant anatomy and physiology who teaches the mechanics of breastfeeding to new mothers. Certification is granted after passing a standardized post-training examination.

cesarean (see-ZAIR-ee-un) **section** surgical childbirth, in which the infant is taken through an incision in the woman's abdomen.

chelating (KEE-late-ing) **agents** molecules that surround other molecules and are therefore useful in either preventing or promoting movement of substances from place to place.

chlorophyll the green pigment of plants that captures energy from sunlight for use in photosynthesis.

cholesterol (koh-LESS-ter-all) a member of the group of lipids known as sterols; a soft, waxy substance made in the body for a variety of purposes and also found in animal-derived foods.

choline (KOH-leen) a nonessential nutrient used to make the phospholipid lecithin and other molecules.

chronic diseases long-duration degenerative diseases characterized by deterioration of the body organs; examples include heart disease, cancer, and diabetes.

chylomicrons (KYE-low-MY-krons) clusters formed when lipids from a meal are combined with carrier proteins in the intestinal lining. Chylomicrons transport food fats through the watery body fluids to the liver and other tissues.

chyme (KIME) the fluid resulting from the actions of the stomach upon a meal.

coenzyme (co-EN-zime) a small molecule that works with an enzyme to promote the enzyme's activity. Many coenzymes have B vitamins as part of their structure (*co* means "with").

collagen (COLL-a-jen) the chief protein of most connective tissues, including scars, ligaments, and tendons, and the underlying matrix on which bones and teeth are built.

colon the large intestine.

colostrum (co-LAHS-trum) a milklike secretion from the breasts during the first day or so after delivery before milk appears; rich in protective factors.

complementary proteins two or more proteins whose amino acid assortments complement each other in such a way that the essential amino acids missing from one are supplied by the other.

complete proteins proteins containing all the essential amino acids in the right balance to meet human needs.

complex carbohydrates long chains of sugar units arranged to form starch or fiber; also called *polysaccharides*.

conditionally essential amino acid an amino acid that is normally nonessential, but must be supplied by the diet in special circumstances when the need for it exceeds the body's ability to produce it.

constipation infrequent, difficult bowel movements often caused by diet, inactivity, dehydration, or medication.

contaminant any substance occurring in food by accident; any food constituent that is not normally present.

cornea (KOR-nee-uh) the hard, transparent membrane covering the outside of the eye.

cortex the outermost layer of something. The brain's cortex is the part of the brain where conscious thought takes place.

cretinism (CREE-tin-ism) severe mental and physical retardation of an infant caused by the mother's iodine deficiency during pregnancy.

critical period a finite period during development in which certain events may occur that will have irreversible effects on later developmental stages. A critical period is usually a period of cell division in a body organ.

cross-contamination the contamination of a food through exposure to utensils, hands, or other surfaces that were previously in contact with a contaminated food.

cruciferous vegetables vegetables with cross-shaped blossoms. Their intake is associated with low cancer rates in human populations. Examples include broccoli, brussels sprouts, cabbage, cauliflower, rutabagas, and turnips.

cuisine a style of cooking.

—D—

degenerative diseases chronic, irreversible diseases characterized by degeneration of body organs due in part to such personal lifestyle elements as poor food choices, smoking, alcohol use, and lack of physical activity. Also called *lifestyle diseases, chronic diseases,* or the *diseases of old age*.

dehydration loss of water. The symptoms progress rapidly, from thirst to weakness to exhaustion and delirium, and end in death.

denaturation the change in a protein's shape brought about by heat, acids, bases, alcohol, salts of heavy metals, or other agents.

diabetes (dye-uh-BEET-eez) a disease (technically termed *diabetes mellitus*) characterized by elevated blood glucose and inadequate or ineffective insulin, which renders a person unable to regulate blood glucose normally.

diarrhea frequent, watery bowel movements usually caused by diet, stress, or irritation of the colon. Severe, prolonged diarrhea robs the body of fluid and certain minerals, causing dehydration and imbalances that can be dangerous if left untreated.

diastolic (dye-as-TOL-ik) **pressure** the second figure in a blood pressure reading (the "lub" of the heartbeat), which reflects the arterial pressure when the heart is between beats.

diet the foods (including beverages) a person usually eats and drinks.

Dietary Folate Equivalent (DFE) a unit of measure that mathematically equalizes the difference in absorption between less absorbable food folate and highly absorbable synthetic folate added to enriched foods and found in supplements.

Dietary Reference Intakes a set of four lists of values for the dietary nutrient intakes of healthy people in the United States and Canada. The values include: Estimated Average Requirement (EAR), Recommended Dietary Allowances (RDA), Adequate Intakes (AI), and Tolerable Upper Intake Levels (UL).

dietary supplement as defined by DSHEA a product, other than tobacco, that is added to the diet and contains one of the following ingredients: a vitamin, mineral, herb, botanical (plant extract), amino acid, metabolite, constituent, or extract, or a combination of any of these ingredients.

digest to break molecules into smaller molecules; a main function of the digestive tract with respect to food.

digestive system the body system composed of organs that break down complex food particles into smaller, absorbable products. The *digestive tract* and *alimentary canal* are names for the tubular organs that extend from the mouth to the anus. The whole system, including the pancreas, liver, and gallbladder, is sometimes called the *gastrointestinal*, or *GI*, system.

dipeptides (dye-PEP-tides) protein fragments that are two amino acids long. A peptide is a strand of amino acids (*di* means "two").

disaccharides pairs of single sugars linked together (*di* means "two").

diuretics (dye-you-RET-ics) compounds, usually medications, causing increased urinary water excretion; "water pills."

diverticulosis (dye-ver-tic-you-LOH-sis) outpocketing or ballooning out of areas of the intestinal wall, caused by weakening of the muscle layers that encase the intestine.

drying a method of preserving food by removing sufficient water from the food to inhibit microbial growth.

dysentery (DISS-en-terry) an infection of the digestive tract that causes diarrhea.

—E—

edema (eh-DEEM-uh) swelling of body tissue caused by leakage of fluid from the blood vessels, seen in protein deficiency (among other conditions).

electrolytes compounds that partly dissociate in water to form ions, such as the potassium ion (K^+) and the chloride ion (Cl^-).

elemental diets diets composed of purified ingredients of known chemical composition; intended to supply all essential nutrients to people who cannot eat foods.

embolism an embolus that causes sudden closure of a blood vessel.

embolus (EM-boh-luss) a thrombus that breaks loose (*embol* means "to insert").

embryo (EM-bree-oh) the stage of human gestation from the third to the eighth week after conception.

emulsification the process of mixing lipid with water by adding an emulsifier.

emulsifier (ee-MULL-sih-fire) a compound with both water-soluble and fat-soluble portions that can attract fats and oils into water to form an emulsion.

endorphins, endogenous opiates compounds of the brain whose actions mimic those of opiate drugs (morphine, heroin) in reducing pain and producing pleasure. In appetite control, endorphins are released on seeing, smelling, or tasting delicious food and are believed to enhance the drive to eat or continue eating.

energy the capacity to do work. The energy in food is chemical energy; it can be converted to mechanical, electrical, heat, or other forms of energy in the body. Food energy is measured in calories.

energy-yielding nutrients the nutrients the body can use for energy. They may also supply building blocks for body structures.

enterotoxins poisons that act upon mucous membranes, such as those of the digestive tract.

environmental tobacco smoke (ETS) the combination of exhaled smoke (mainstream smoke) and smoke from lighted cigarettes, pipes, or cigars (sidestream smoke) that enters the air and may be inhaled by other people.

enzyme (EN-zime) a protein catalyst. A catalyst is a compound that facilitates a chemical reaction without itself being altered in the process.

EPA, DHA eicosapentaenoic (EYE-cossa-PENTA-ee-NO-ick) acid, docosahexaenoic (DOE-cossa-HEXA-ee-NO-ick) acid; omega-3 fatty acids made from linolenic acid in the tissues of fish.

epinephrine the major hormone that elicits the stress response.

epiphyseal (eh-PIFF-ih-seal) **plate** a thick, cartilage-like layer that forms new cells that are eventually calcified, lengthening the bone (*epiphysis* means "growing" in Greek).

epithelial (ep-ith-THEE-lee-ull) **tissue** the layers of the body that serve as selective barriers to environmental factors. Examples are the cornea, the skin, the respiratory tract lining, and the lining of the digestive tract.

erythrocyte (eh-REETH-ro-sight) **hemolysis** (HE-moh-LIE-sis, he-MOLL-ih-sis) rupture of the red blood cells, caused by vitamin E deficiency (*erythro* means "red"; *cyte* means "cell"; *hemo* means "blood"; *lysis* means "breaking").

essential amino acids amino acids that either cannot be synthesized at all by the body or cannot be synthesized in amounts sufficient to meet physiological need. Also called *indispensable amino acids*.

essential fatty acids fatty acids that the body needs but cannot make in amounts sufficient to meet physiological needs.

essential nutrients the nutrients the body cannot make for itself (or cannot make fast enough) from other raw materials; nutrients that must be obtained from food to prevent deficiencies.

ethnic foods foods associated with particular cultural subgroups within a population.

exchange system a diet planning tool that organizes foods with respect to their nutrient contents and calorie amounts. Foods on any single exchange list can be used interchangeably. See the U.S. Exchange System, Appendix D, for details.

extracellular fluid fluid residing outside the cells.

extrusion a process by which the form of a food is changed, such as changing corn to corn chips; not a preservation measure.

—F—

fasting hypoglycemia hypoglycemia that occurs after 8 to 14 hours of fasting.

fat cells cells that specialize in the storage of fat and that form the fat tissue.

fatfold test measurement of the thickness of a fold of skin on the back of the arm (over the triceps muscle), below the shoulder blade (subscapular), or in other places, using a caliper. Also called *skinfold test.*

fats lipids that are solid at room temperature (70° F or 25° C).

fatty acids organic acids composed of carbon chains of various lengths. Each fatty acid has an acid end and hydrogens attached to all of the carbon atoms of the chain.

feces waste material remaining after digestion and absorption are complete, eventually discharged from the body.

fetal alcohol effect (FAE) partial abnormalities from prenatal alcohol exposure, not sufficient for diagnosis with FAS, but impairing to the child. Also called *alcohol-related birth defects (ARBD)* or *subclinical FAS.*

fetal alcohol syndrome (FAS) the cluster of symptoms seen in an infant or child whose mother consumed excess alcohol during her pregnancy. FAS includes, but is not limited to, brain damage, growth retardation, mental retardation, and facial abnormalities.

fetus (FEET-us) the stage of human gestation from eight weeks after conception until the birth of an infant.

fibers the indigestible polysaccharides in food, consisting mostly of cellulose, hemicellulose, and pectin. Also called *nonstarch polysaccharides.*

fight-or-flight reaction the body's instinctive hormone- and nerve-mediated reaction to danger. Also known as the *stress response.*

flexibility the capacity of the joints to move through a full range of motion; the ability to bend and recover without injury.

fluid and electrolyte balance maintenance of the proper amounts and kinds of fluids and minerals in each compartment of the body.

fluid and electrolyte imbalance failure to maintain the proper amount and kind of fluid in every body compartment; a medical emergency.

fluorapatite (floor-APP-uh-tight) a crystal of bones and teeth, formed when fluoride displaces the hydroxy portion of hydroxyapatite. Fluorapatite resists being dissolved back into body fluid.

fluorosis (floor-OH-sis) discoloration of the teeth due to ingestion of too much fluoride during tooth development.

folate (FOH-late) a B vitamin that acts as part of a coenzyme important in the manufacture of new cells. Other names for folate are *folacin* and *folic acid.*

food medically, any substance that the body can take in and assimilate that will enable it to stay alive and to grow; the carrier of nourishment; socially, a more limited number of such substances defined as acceptable by each culture.

food aversion an intense dislike of a food, possibly biological in nature, resulting from an illness or other negative experience associated with that food.

food group plans diet planning tools that sort foods into groups based on origin and nutrient content and then specify that people should eat certain minimum numbers of servings of foods from each group.

food intolerance an adverse effect of a food or food additive not involving the immune response.

food-borne illness illness transmitted to human beings through food and water; caused by a poisonous substance *(food intoxication)* or an infectious agent *(food-borne infection)*. Also called *food poisoning.*

foodways the sum of a culture's habits, customs, beliefs, and preferences concerning food.

fossil fuel coal, oil, and natural gas. These are nonrenewable fuels that pollute. Alternative fuels, such as solar and wind energy, are renewable and pollute less or not at all.

freezing a method of preserving food by lowering the food's temperature to a point that halts life processes. Microorganisms do not die but remain dormant until the food is thawed.

fructose (FROOK-tose) a monosaccharide; sometimes known as fruit sugar (*fruct* means "fruit"; *ose* means "sugar").

—G—

galactose (ga-LACK-tose) a monosaccharide; part of the disaccharide lactose (milk sugar).

gastric juice the digestive secretion of the stomach.

gastro-esophageal reflux disease (GERD) a severe and chronic splashing of stomach acid and enzymes into the esophagus, throat, mouth, or airway that causes inflammation and injury to those organs. Untreated GERD may increase the risk of esophageal cancer; treatment may require surgery or management with medication.

gatekeeper with respect to nutrition, a key person who controls other people's access to foods and thereby affects their nutrition profoundly. Examples are the spouse who buys and cooks the food, the parent who feeds the children, and the caretaker in a day-care center.

genes units of a cell's inheritance, made of the chemical DNA (deoxyribonucleic acid). Each gene directs the making of a protein to do the body's work.

gestation the period of about 40 weeks (three trimesters) from conception to birth; the term of a pregnancy.

gestational diabetes abnormal glucose tolerance appearing during pregnancy, with subsequent return to normal after the end of pregnancy.

gleaning traditionally, the practice of gathering crops left in the field after a harvest; today, refers to the recovery of excess food from various sources including restaurants, hotels, corporations, farms, wholesalers, farmers' markets, and supermarkets.

glucagon a hormone secreted by the pancreas that stimulates the liver to release glucose into the blood when blood glucose concentration dips.

glucose (GLOO-cose) a single sugar used in both plant and animal tissues for quick energy; sometimes known as blood sugar or *dextrose*.

glucose tolerance the ability of the body to respond to dietary carbohydrate by promptly regulating its blood glucose concentration to a normal level.

glycemic (gligh-SEEM-ic) **effect** a measure of the extent to which a food raises the blood glucose concentration and elicits an insulin response as compared with pure glucose.

glycerol (GLISS-er-all) an organic compound, three carbons long, of interest here because it serves as the backbone for triglycerides.

glycogen (GLY-co-gen) a polysaccharide composed of glucose that is made and stored by liver and muscle tissues of human beings and animals as a storage form of glucose.

goiter (GOY-ter) enlargement of the thyroid gland due to iodine deficiency is *simple goiter;* enlargement due to an excess is *toxic goiter.*

gout a painful form of arthritis resulting from a metabolic abnormality in which excessive amounts of the waste product uric acid collect in the blood and uric acid salt is deposited as crystals in the joints.

grams units of weight. A gram (g) is the weight of a cubic centimeter (cc) or milliliter (ml) of water under defined conditions of temperature and pressure. About 28 grams equal an ounce.

granules small grains. Starch granules are packages of starch molecules. Various plant species make starch granules of varying shapes.

GRAS (generally recognized as safe) list a list, established by the FDA, of food additives long in use and believed safe.

growth hormone a hormone (somatotropin) that promotes growth and that is produced naturally in the pituitary gland of the brain.

—H—

hard water water with high calcium and magnesium concentrations.

hazard a state of danger; used to refer to any circumstance in which harm is possible under normal conditions of use.

Hazard Analysis Critical Control Point (HACCP) a systematic plan to identify and correct potential microbial hazards in the manufacturing, distribution, and commercial use of food products.

heart attack the event in which the vessels that feed the heart muscle become closed off by an embolism, thrombus, or other cause with resulting sudden tissue death. A heart attack is also called a *myocardial infarction* (*myo* means "muscle"; *cardial* means "of the heart"; *infarct* means "tissue death").

heartburn a burning sensation in the chest (in the area of the heart) area caused by backflow of stomach acid into the esophagus.

heat stroke an acute and life-threatening reaction to heat buildup in the body.

heavy metal any of a number of mineral ions such as mercury and lead; so called because they are of relatively high atomic weight. Many heavy metals are poisonous.

heme (HEEM): the iron-containing portion of the hemoglobin and myoglobin molecules.

hemoglobin (HEEM-oh-globe-in) the oxygen-carrying protein of the blood; found in the red blood cells (*hemo* means "blood"; *globin* means "spherical protein").

hemorrhoids (HEM-or-oids) swollen, hardened (varicose) veins in the rectum, usually caused by the pressure resulting from constipation.

hernia a protrusion of an organ or part of an organ through the wall of the body chamber that normally contains the organ. An example is a *hiatal* (high-AY-tal) *hernia,* in which part of the stomach protrudes up through the diaphragm into the chest cavity, which contains the esophagus, heart, and lungs.

hiccups spasms of both the vocal cords and the diaphragm, causing periodic, audible, short, inhaled coughs. Can be caused by irritation of the diaphragm, indigestion, or other causes.

high-density lipoproteins (HDL) lipoproteins that return cholesterol from storage places to the liver for dismantling and disposal; contain a large proportion of protein.

high-temperature–short-time (HTST) principle the rule that every 18°F (10°C) rise in processing temperature brings about an approximately tenfold increase in microbial destruction, while only doubling nutrient losses.

histamine a substance that participates in causing inflammation; produced by cells of the immune system as part of a local immune reaction to an antigen.

hormones chemicals that are secreted by glands into the blood in response to conditions in the body that require regulation. These chemicals serve as messengers, acting on other organs to maintain constant conditions.

human somatotropin (HST) human growth hormone.

hunger the physiological craving for food; the progressive discomfort, illness, and pain resulting from the lack of food.

hydrogenation (high-dro-gen-AY-shun) the process of adding hydrogen to unsaturated fatty acids to make fat more solid and resistant to the chemical change of oxidation.

hydroxyapatite (hi-DROX-ee-APP-uh-tight) the chief crystal of bone, formed from calcium and phosphorus.

hyperactivity (in children) a syndrome characterized by inattention, impulsiveness, and excess motor activity; usually diagnosed before age seven, lasts six months or more, and usually does not entail mental illness or mental retardation. Also *called attention-deficit/hyperactivity disorder (ADHD)* or *hyperkinesis* and may be associated with minimal brain damage.

hyperglycemia (HIGH-per-gligh-SEEM-ee-uh) an abnormally high blood glucose concentration (hyper means "too much"; *glyce* means "glucose"; *emia* means "in the blood").

hypertension high blood pressure.

hypertrophy (high-PURR-tro-fee) an increase in size (for example, of a muscle) in response to use.

hypoglycemia a blood glucose concentration below normal, a symptom that may indicate any of several diseases, including impending diabetes.

hypothalamus (high-poh-THAL-uh-mus) a part of the brain that senses a variety of conditions in the blood, such as temperature, glucose content, salt content, and others. It signals other parts of the brain or body to adjust those conditions when necessary.

hypothermia a below-normal body temperature.

—I—

immune system a system of tissues and organs that defend the body against antigens, foreign materials that have penetrated the skin or body linings.

immunity specific disease resistance, derived from the immune system's memory of prior exposure to specific disease agents and its ability to mount a swift defense against them.

implantation the stage of development, during the first two weeks after conception, in which the fertilized egg (fertilized ovum or zygote) embeds itself in the wall of the uterus and begins to develop.

incidental additives substances that can get into food not through intentional introduction but as a result of contact with the food during growing, processing, packaging, storing, or some other stage before the food is consumed. Also called *accidental* or *indirect additives*.

incomplete proteins proteins lacking, or low in, one or more of the essential amino acids.

infectious diseases diseases caused by bacteria, viruses, parasites, and other microbes, which can be transmitted from one person to another through air, water, or food; by contact; or through vector organisms such as mosquitoes or fleas.

initiation an event, probably occurring in a cell's genetic material, caused by radiation or by a chemical carcinogen that can give rise to cancer.

inositol (in-OSS-ih-tall) a nonessential nutrient found in cell membranes.

insoluble fibers the tough, fibrous structures of fruits, vegetables, and grains; indigestible food components that do not dissolve in water.

insulin a hormone secreted by the pancreas in response to a high blood glucose concentration. It assists cells in drawing glucose from the blood.

insulin resistance syndrome a combination of four risk factors—diabetes, obesity, hypertension, and high blood cholesterol—that greatly increase a person's risk of developing CVD. Also called *syndrome X*.

intestine the body's long, tubular organ of digestion and the site of nutrient absorption.

intrinsic factor a factor found inside a system. The intrinsic factor necessary to prevent pernicious anemia is now known to be a compound that helps in the absorption of vitamin B_{12}.

ions (EYE-ons) electrically charged particles, such as sodium (positively charged) or chloride (negatively charged).

iron deficiency the condition of having depleted iron stores, which, at the extreme, causes iron-deficiency anemia.

iron-deficiency anemia a form of anemia caused by iron deficiency and characterized by red blood cell shrinkage and color loss. Accompanying symptoms are weakness, apathy, headaches, pallor, intolerance to cold, and inability to pay attention. (For other anemias, see the index.)

iron overload the state of having more iron in the body than it needs or can handle. Too much iron is toxic and can damage the liver.

irritable bowel syndrome intermittent disturbance of bowel function, especially diarrhea or alternating diarrhea and constipation; associated with diet, lack of physical activity, or psychological stress.

IU (international unit) a measure of fat-soluble vitamin activity.

—K—

kefir a yogurt-based beverage.

keratin (KERR-uh-tin) the normal protein of hair and nails.

keratinization accumulation of keratin in a tissue; a sign of vitamin A deficiency.

ketone (kee-tone) **bodies** acidic, fat-related compounds that can arise from the incomplete breakdown of fat when carbohydrate is not available.

ketosis (kee-TOE-sis) an undesirable high concentration of ketone bodies, such as acetone, in the blood or urine.

kidneys a pair of organs that filter wastes from the blood, make urine, and release it to the bladder for excretion from the body.

kwashiorkor (kwash-ee-OR-core, kwashee-or-CORE) a disease related to protein malnutrition, with a set of recognizable symptoms, such as edema.

—L—

lactase the intestinal enzyme that splits the disaccharide lactose to monosaccharides during digestion.

lactation production and secretion of breast milk for the purpose of nourishing an infant.

lactic acid a product of the incomplete breakdown of glucose during anaerobic metabolism. When oxygen becomes available, lactic acid can be completely broken down for energy or converted back to glucose.

lactoferrin (lack-toe-FERR-in) a factor in breast milk that binds iron and keeps it from supporting the growth of the infant's intestinal bacteria.

lactose a disaccharide composed of glucose and galactose; sometimes known as milk sugar (*lact* means "milk"; *ose* means "sugar").

lactose intolerance inability to digest lactose due to a lack of the enzyme lactase.

large intestine the portion of the intestine that completes the absorption process.

learning disabilities a group of conditions resulting in an altered ability to learn basic cognitive skills such as reading, writing, and mathematics.

leavened (LEV-end) literally, "lightened" by yeast cells, which digest some carbohydrate components of the dough and leave behind bubbles of gas that make the bread rise.

lecithin (LESS-ih-thin) a phospholipid manufactured by the liver and also found in many foods; a major constituent of cell membranes.

legumes (leg-GOOMS, LEG-yooms) plants of the bean and pea family that have roots with nodules containing special bacteria. These bacteria can trap nitrogen from the air in the soil and make it into compounds that become part of the plant's seeds. The seeds are rich in protein compared with those of most other plant foods.

leptin an appetite-suppressing hormone produced in the fat cells that conveys information about body fatness to the brain; believed to be involved in the maintenance of body composition (*leptos* means "slender").

life expectancy the average number of years lived by people in a given society.

life span the maximum number of years of life attainable by a member of a species.

limiting amino acid an essential amino acid that is present in dietary protein in an insufficient amount, thereby limiting the body's ability to build protein.

linoleic (lin-oh-LAY-ic) **acid** and **linolenic** (lin-oh-LEN-ic) **acid** polyunsaturated fatty acids that are essential nutrients for human beings.

lipid (LIP-id) a family of compounds soluble in organic solvents but not in water. Lipids include triglycerides (fats and oils), phospholipids, and sterols.

lipoic (lip-OH-ic) acid a nonessential nutrient.

lipoproteins (LYE-poh-PRO-teens, LIH-poh-PRO-teens) clusters of lipids associated with protein, which serve as transport vehicles for lipids in blood and lymph. Major lipoprotein classes are the chylomicrons, the VLDL, the LDL, and the HDL.

liver a large, lobed organ that lies just under the ribs. It filters the blood, removes and processes nutrients, manufactures materials for export to other parts of the body, and destroys toxins or stores them to keep them out of the circulation.

longevity long duration of life.

low birthweight a birthweight of less than 5½ pounds (2,500 grams); used as a predictor of probable health problems in the newborn and as a probable indicator of poor nutrition status of the mother before and/or during pregnancy.

low-density lipoproteins (LDL) lipoproteins that transport lipids from the liver to other tissues such as muscle and fat; contain a large proportion of cholesterol.

LPL (lipoprotein lipase) an enzyme mounted on the surfaces of fat cells that splits triglycerides in the blood into fatty acids and glycerol to be absorbed into the cells for reassembly and storage.

lungs the body's organs of gas exchange. Blood circulating through the lungs releases its carbon dioxide and picks up fresh oxygen to carry to the tissues.

lymph (LIMF) the fluid that moves from the bloodstream into tissue spaces and then travels in its own vessels, which eventually drain back into the bloodstream.

lymphocytes (LIM-foe-sites) white blood cells that participate in the immune response; B-cells and T-cells.

—M—

macular degeneration a common, progressive loss of function of the part of the retina that is most crucial to focused vision. This degeneration often leads to blindness.

mad cow disease an often-fatal virus of cattle affecting the nerves and brain. Formerly called bovine spongiform encephalopathy (BSE).

major minerals essential mineral nutrients found in the human body in amounts larger than 5 grams.

malnutrition any condition caused by excess or deficient food energy or nutrient intake or by an imbalance of nutrients. Nutrient or energy deficiencies are classed as forms of undernutrition; nutrient or energy excesses are classed as forms of overnutrition.

maltose a disaccharide composed of two glucose units; sometimes known as malt sugar.

marasmus (ma-RAZ-mus) the calorie-deficiency disease; starvation.

margin of safety in reference to food additives, a zone between the concentration normally used and that at which a hazard exists. For common table salt, for example, the margin of safety is ⅕ (five times the concentration normally used would be hazardous).

metabolism the sum of all physical and chemical changes taking place in living cells; including all reactions by which the body obtains and spends the energy from food.

metastasis (meh-TASS-ta-sis) movement of cancer cells from one body part to another, usually by way of the body fluids.

MFP factor a factor (identity unknown) present in meat, fish, and poultry that enhances the absorption of nonheme iron present in the same foods or in other foods eaten at the same time.

microbes bacteria, viruses, or other organisms invisible to the naked eye, some of which cause diseases. Also called *microorganisms*.

microvilli (MY-croh-VILL-ee, MY-croh-VILL-eye) tiny, hairlike projections on each cell of every villus that can trap nutrient particles and transport them into the cells (singular: *microvillus*).

milk anemia iron-deficiency anemia caused by drinking so much milk that iron-rich foods are displaced from the diet.

minerals naturally occurring, inorganic, homogeneous substances; chemical elements.

moderation the dietary characteristic of providing constituents within set limits, not to excess.

modified atmosphere packaging (MAP) a preservation technique in which a perishable food is packaged in a gas-impermeable container from which air has been removed, or to which another gas mixture has been added.

monoglycerides (mon-oh-GLISS-er-ides) products of the digestion of lipids; consist of glycerol molecules with one fatty acid attached (*mono* means "one"; *glyceride* means "a compound of glycerol").

monosaccharides single sugar units (*mono* means "one"; *saccharide* means "sugar unit").

monounsaturated fats triglycerides in which most of the fatty acids have one point of unsaturation (are monounsaturated).

monounsaturated fatty acid a fatty acid containing one point of unsaturation.

MSG symptom complex the acute, temporary, and self-limiting reactions experienced by sensitive people upon ingesting a large dose of MSG. The name *MSG symptom complex,* given by the FDA, replaces the former *Chinese restaurant syndrome.*

mucus (MYOO-cus) a slippery coating of the digestive tract lining (and other body linings) that protects the cells from exposure to digestive juices (and other destructive agents). The adjective form is *mucous* (same pronunciation). The digestive tract lining is a *mucous membrane.*

muscle endurance the ability of a muscle to contract repeatedly within a given time without becoming exhausted.

muscle strength the ability of muscles to work against resistance.

mutual supplementation the strategy of combining two incomplete protein sources so that the amino acids in one food make up for those lacking in the other food. Such protein combinations are sometimes called *complementary proteins.*

myoglobin (MYE-oh-globe-in) the oxygen-holding protein of the muscles (*myo* means "muscle").

—N—

naturally occurring sugars sugars that are not added to a food but are present as its original constituents, such as the sugars of fruit or milk.

nephrons the working units in the kidneys, consisting of intermeshed blood vessels and tubules.

neural tube the embryonic tissue that later forms the brain and spinal cord.

neural tube defects a group of nervous system abnormalities caused by interruption of the normal early development of the neural tube.

neuropeptide Y (NPY) a neurotransmitter whose functions in the brain's hypothalamus include the stimulation of appetite, especially appetite for carbohydrate-rich foods.

neurotoxins poisons that act upon the cells of the nervous system.

neurotransmitters chemicals that are released at the end of a nerve cell when a nerve impulse arrives there; they diffuse across the gap to the next cell and alter the membrane of that second cell to either inhibit or excite it.

niacin a B vitamin needed in energy metabolism. Niacin can be eaten preformed or can be made in the body from tryptophan, one of the amino acids. Other forms of niacin are *nicotinic acid, niacinamide,* and *nicotinamide.*

niacin equivalents the amount of niacin present in food, including the niacin that can theoretically be made from its precursor tryptophan that is present in the food.

night blindness slow recovery of vision after exposure to flashes of bright light at night; an early symptom of vitamin A deficiency.

nitrogen balance the amount of nitrogen consumed compared with the amount excreted in a given time period.

nonnutrients a term used in this book to mean compounds other than the six nutrients that are present in foods.

norepinephrine a compound related to epinephrine that helps to elicit the stress response.

nori a type of seaweed popular in Asian, particularly Japanese, cooking.

nutrient density a measure of nutrients provided per calorie of food.

nutrients components of food that are indispensable to the body's functioning. They provide energy, serve as building material, help maintain or repair body parts, and support growth. The nutrients include water, carbohydrate, fat, protein, vitamins, and minerals.

nutrition the study of the nutrients in foods and in the body; sometimes also the study of human behaviors related to food.

—O—

obesity overfatness with adverse health effects, as determined by reliable measures and interpreted with good medical judgment.

oils lipids that are liquid at room temperature (70°F or 25°C).

omega-3 fatty acid a polyunsaturated fatty acid with its endmost double bond three carbons from the end of the carbon chain. Also called n-3 fatty acid. Linolenic acid is an example.

omega-6 fatty acid a polyunsaturated fatty acid with its endmost double bond six carbons from the end of the carbon chain. Also called n-6 fatty acid. Linoleic acid is an example.

omnivores people who eat foods of both plant and animal origin, including animal flesh.

oral rehydration therapy (ORT) oral fluid replacement for children with severe diarrhea caused by infectious disease. ORT enables parents to mix a simple solution for their child from substances that they have at home.

organic carbon containing. Four of the six classes of nutrients are organic: carbohydrate, fat, protein, and vitamins. Strictly speaking, organic compounds include only those made by living things and do not include carbon dioxide and a few carbon salts.

organic foods products grown and processed without the use of synthetic chemicals such as pesticides, herbicides, fertilizers, and preservatives and without genetic engineering or irradiation.

organic halogen an organic compound containing one or more atoms of a halogen—fluorine, chlorine, iodine, or bromine.

organs discrete structural units made of tissues that perform specific jobs. Examples are the heart, liver, and brain.

osteomalacia (OS-tee-o-mal-AY-shuh) the vitamin D–deficiency disease in adults (*osteo* means "bone"; *mal* means "bad"). Symptoms include bending of the spine and bowing of the legs.

osteoporosis (OSS-tee-oh-pore-OH-sis) a reduction of the bone mass of older persons in which the bones become porous and fragile (*osteo* means "bones"; *poros* means "porous"); also known as **adult bone loss.**

overload an extra physical demand placed on the body; an increase in the frequency, duration, or intensity of an activity. A principle of training is that for a body system to improve, it must be worked at frequencies, durations, or intensities that increase by increments.

ovum the egg, produced by the mother, that unites with a sperm from the father to produce a new individual.

oxidation interaction of a compound with oxygen; in this case, a damaging effect by a chemically reactive form of oxygen.

—P—

pancreas an organ with two main functions. One is an endocrine function—the making of hormones such as insulin, which it releases directly into the blood (*endo* means "into" the blood). The other is an exocrine function—the making of digestive enzymes, which it releases through a duct into the small intestine to assist in digestion (*exo* means "out" into a body cavity or onto the skin surface).

pancreatic juice fluid secreted by the pancreas that contains both enzymes to digest carbohydrate, fat, and protein and sodium bicarbonate, a neutralizing agent.

pantothenic (PAN-to-THEN-ic) **acid** a B vitamin.

pasteurization the treatment of milk with heat sufficient to kill certain pathogens (disease-causing microbes) commonly transmitted through milk; not a sterilization process. Pasteurized milk retains bacteria that cause milk spoilage. Raw milk, even if labeled "certified," transmits many food-borne diseases to people each year and should be avoided.

peak bone mass the highest attainable bone density for an individual; developed during the first three decades of life.

pellagra (pell-AY-gra) the niacin-deficiency disease (*pellis* means "skin"; *agra* means "rough"). Symptoms include the "4 Ds": diarrhea, dermatitis, dementia, and, ultimately, death.

peptide bond a bond that connects one amino acid with another, forming a link in a protein chain.

peristalsis (perri-STALL-sis) the wavelike muscular squeezing of the esophagus, stomach, and small intestine that pushes their contents along.

pernicious (per-NISH-us) **anemia** a vitamin B_{12}-deficiency disease, caused by lack of intrinsic factor and characterized by large, immature red blood cells and damage to the nervous system (*pernicious* means "highly injurious or destructive").

persistent of a stubborn or enduring nature; with respect to food contaminants, the quality of remaining unaltered and unexcreted in plant foods or in the bodies of animals and human beings.

pesticides chemicals used to control insects, diseases, weeds, fungi, and other pests on crops and around animals. Used broadly, the term includes *herbicides* (to kill weeds), *insecticides* (to kill insects), and *fungicides* (to kill fungi).

pH a measure of acidity on a point scale. A solution with a pH of 1 is a strong acid; a solution with a pH of 7 is neutral; a solution with a pH of 14 is a strong base.

phagocytes (FAG-oh-sites) white blood cells that can ingest and destroy antigens. The process by which phagocytes engulf materials is called *phagocytosis*. The Latin word *phagein* means "to eat."

phospholipids (FOSS-foh-LIP-ids) one of the three main classes of dietary lipids. These lipids are similar to triglycerides, but each has a phosphorus-containing acid in place of one of the fatty acids. Phospholipids are present in all cell membranes.

photosynthesis the process by which green plants make carbohydrates from carbon dioxide and water using the green pigment chlorophyll to capture the sun's energy (*photo* means "light"; *synthesis* means "making").

phytates compounds present in plant foods (particularly whole grains) that bind iron and prevent its absorption.

phytochemicals nonnutrient compounds in plant-derived foods that have biological activity in the body.

pica (PIE-ka) a craving for nonfood substances. Also known as *geophagia* (gee-oh-FAY-gee-uh) when referring to clay eating, and *pagophagia* (pag-oh-FAY-gee-uh) when referring to ice craving (*geo* means "earth"; *pago* means "frost"; *phagia* means "to eat").

placenta (pla-SEN-tuh) the organ that develops inside the uterus in early pregnancy in which maternal and fetal blood circulate in close proximity and exchange materials.

plaques (PLACKS) mounds of lipid material mixed with smooth muscle cells and calcium that develop in the artery walls in atherosclerosis (*placken* means "patch"). The same word is also used to describe the accumulation of a different kind of deposits on teeth, which promote dental caries.

plasma the cell-free fluid part of blood and lymph.

platelets tiny cell-like fragments in the blood, important in blood clot formation (*platelet* means "little plate").

point of unsaturation a site in a molecule where the bonding is such that additional hydrogen atoms can easily be attached.

polypeptides protein fragments of many (more than ten) amino acids bonded together (*poly* means "many"). A chain of between four and ten amino acids is called an oligopeptide.

polysaccharides another term for complex carbohydrates; compounds composed of long strands of glucose units linked together (*poly* means "many"). Also called *complex carbohydrates*.

polyunsaturated fats triglycerides in which most of the fatty acids have two or more points of unsaturation (are polyunsaturated).

polyunsaturated fatty acid (PUFA) a fatty acid with two or more points of unsaturation.

postprandial hypoglycemia a drop in blood glucose that follows a meal and is accompanied by symptoms of the stress response; also called *reactive hypoglycemia*.

precursors, provitamins compounds that can be converted into active vitamins.

preeclampsia a potentially dangerous condition during pregnancy characterized by edema, hypertension, and protein in the urine.

pregame meal a meal eaten three to four hours before athletic competition.

premenstrual syndrome (PMS) a cluster of symptoms that some women experience prior to and during menstruation. They include, among others, abdominal cramps, back pain, swelling, headache, painful breasts, and mood changes.

promoters factors that do not initiate cancer but speed up its development once initiation has taken place.

prooxidant a compound that triggers reactions involving oxygen.

protein digestibility–corrected amino acid score (PDCAAS) a measuring tool used to determine protein quality. The PDCAAS reflects a protein's digestibility as well as the proportions of amino acids that it provides.

protein efficiency ratio (PER) a measure of protein quality assessed by determining how well a given protein supports weight gain in growing rats. The PER is used to judge the quality of protein in infant formulas and baby foods.

protein-energy malnutrition (PEM) the world's most widespread malnutrition problem, including both marasmus and kwashiorkor and states in which they overlap; also called *protein-calorie malnutrition (PCM)*.

proteins compounds composed of carbon, hydrogen, oxygen, and nitrogen and arranged as strands of amino acids. Some amino acids also contain the element sulfur.

protein-sparing action the action of carbohydrate and fat in providing energy that allows protein to be used for purposes it alone can serve.

puberty the period in life when a person develops sexual maturity and the ability to reproduce.

pyloric (pye-LORE-ick) **valve** the circular muscle of the lower stomach that regulates the flow of partly digested food into the small intestine. Also called *pyloric sphincter*.

—R—

RE (retinol equivalent) a measure of vitamin A activity; the amount of retinol that the body will derive from a food containing vitamin A (preformed retinol) or its precursor beta-carotene.

Recommended Dietary Allowances (RDA) formerly, the name of the nutrient intake standards of the United States. Currently, the RDA constitute a part of the Dietary Reference Intakes (DRI). RDA are the average daily amounts of nutrients considered adequate to meet the known nutrient needs of practically all healthy people.

requirement the amount of a nutrient that will just prevent the development of specific deficiency signs; distinguished from the DRI recommended intake value, which is a generous allowance with a margin of safety.

residues whatever remains. In the case of pesticides, those amounts that remain on or in foods when people buy and use them.

resistant starch the fraction of starch in a food that is slowly digested, or not digested, by human enzymes.

retina (RET-in-uh) the layer of light-sensitive nerve cells lining the back of the inside of the eye.

retinol one of the active forms of vitamin A made from beta-carotene in animal and human bodies; an antioxidant nutrient. Other active forms are *retinal* and *retinoic acid*.

rhodopsin (roh-DOP-sin) the light-sensitive pigment of the cells in the retina; it contains vitamin A (*rod* refers to the rod-shaped cells; *opsin* means "visual protein").

riboflavin (RIBE-o-flay-vin) a B vitamin active in the body's energy-releasing mechanisms.

rickets the vitamin D–deficiency disease in children; characterized by abnormal growth of bone and manifested in bowed legs or knock-knees, outward-bowed chest, and knobs on the ribs.

risk factors factors known to be related to (or correlated with) diseases but not proved to be causal.

roughage (RUFF-idge) the rough parts of food; an imprecise term that has largely been replaced by the term *fiber*.

—S—

safety the practical certainty that injury will not result from the use of a substance.

salts compounds composed of charged particles (ions). An example is potassium chloride (K^+Cl^-).

satiation (SAY-she-AY-shun) the perception of fullness that builds throughout a meal, eventually reaching the degree of fullness and satisfaction that halts eating. Satiation generally determines how much food is consumed at one sitting.

satiety (sah-TIE-eh-tee) the perception of fullness that lingers in the hours after a meal and inhibits eating until the next mealtime. Satiety generally determines the length of time between meals.

saturated fats triglycerides in which most of the fatty acids are saturated.

saturated fatty acid a fatty acid carrying the maximum possible number of hydrogen atoms (having no points of unsaturation). A saturated fat is a triglyceride that contains three saturated fatty acids.

scurvy the vitamin C-deficiency disease.

self-efficacy a person's belief in his or her ability to succeed in an undertaking.

senile dementia the loss of brain function beyond the normal loss of physical adeptness and memory that occurs with aging.

set-point theory the theory that the body tends to maintain a certain weight by means of its own internal controls.

side chain the unique chemical structure attached to the backbone of each amino acid that differentiates one amino acid from another.

simple carbohydrates sugars, including both single sugar units and linked pairs of sugar units. The basic sugar unit is a molecule containing six carbon atoms, together with oxygen and hydrogen atoms.

small intestine the 20-foot length of small-diameter intestine, below the stomach and above the large intestine, that is the major site of digestion of food and absorption of nutrients.

smoking point the temperature at which fat gives off an acrid blue gas.

soft water water with a high sodium concentration.

soluble fibers food components that readily dissolve in water and often impart gummy or gel-like characteristics to foods. An example is pectin from fruit, which is used to thicken jellies. Soluble fibers are indigestible by human enzymes but may be broken down to absorbable products by bacteria in the digestive tract.

Special Supplemental Food Program for Women, Infants, and Children (WIC) a USDA program to provide nutrition support to low-income women who are pregnant or who have infants or preschool children. WIC offers coupons redeemable for specific foods to supply the nutrients deemed most needed for growth and development.

sphincter (SFINK-ter) a circular muscle surrounding, and able to close, a body opening.

spina bifida (SPY-na BIFF-ih-duh) one of the most common types of neural tube defects; the infant is born with gaps in the bones of the spine, leaving the spinal cord protected only by a sheath of skin in those spots, or with no protection at all. The spinal cord may bulge and protrude through the gaps in the vertebral column.

stanol esters compounds belonging to the sterol family of lipids, derived from plants, that have been shown experimentally to reduce blood cholesterol when consumed daily in addition to a low-fat diet.

starch a plant polysaccharide composed of glucose. After cooking, starch is highly digestible by human beings; raw starch often resists digestion.

sterols (STEER-alls) one of the three main classes of dietary lipids. Sterols have a structure similar to that of cholesterol.

stomach a muscular, elastic, pouchlike organ of the digestive tract that grinds and churns swallowed food and mixes it with acid and enzymes, forming chyme.

stone-ground flour flour made by grinding kernels of grain between heavy wheels made of limestone, a kind of rock derived from the shells and bones of marine animals. As the stones scrape together, bits of the limestone mix with the flour, enriching it with calcium.

stress fracture a bone injury or break caused by the stress of exercise on the bone surface.

stroke the sudden shutting off of the blood flow to the brain by a thrombus, embolism, or the bursting of a vessel (hemorrhage).

stroke volume the amount of oxygenated blood ejected from the heart toward body tissues at each beat.

subclinical, or **marginal, deficiency** a nutrient deficiency that has no detectable (clinical) symptoms. The term is often used to scare consumers into buying unneeded nutrient supplements.

subcutaneous fat fat stored directly under the skin (sub means "beneath"; *cutaneous* refers to the skin).

sucrose (SOO-crose) a disaccharide composed of glucose and fructose; sometimes known as table, beet, or cane sugar.

sugars simple carbohydrates, that is, molecules of either single sugar units or pairs of those sugar units bonded together.

supplements pills, liquids, or powders that contain purified nutrients or other ingredients.

sushi a Japanese dish that consists of vinegar-flavored rice, seafood, and colorful vegetables, typically wrapped in seaweed. Some sushi is wrapped in raw fish; other sushi contains only cooked ingredients.

sustainable able to continue indefinitely. Here the term refers to the use of resources at such a rate that the earth can keep on replacing them; for example, cutting trees no faster than new ones grow and producing pollutants at a rate with which the environment and human cleanup efforts can keep pace. In a sustainable economy, resources do not become depleted, and pollution does not accumulate.

systolic (sis-TOL-ik) **pressure** the first figure in a blood pressure reading (the "dub" of the heartbeat), which reflects arterial pressure caused by the contraction of the heart's left ventricle.

—T—

tannins compounds in tea (especially black tea) and coffee that bind iron. Tannins also denature proteins.

T-cells lymphocytes that attack antigens. *T* stands for the thymus gland of the neck, where the T-cells are stored and matured.

textured vegetable protein processed soybean protein used in products formulated to look and taste like meat, fish, or poultry.

thermic effect of food (TEF) the body's speeded-up metabolism in response to having eaten a meal. Also called *diet-induced thermogenesis.*

thermogenesis the generation and release of body heat associated with the breakdown of body fuels.

thiamin (THIGH-uh-min) a B vitamin involved in the body's use of fuels.

thrombosis a thrombus that has grown enough to close off a blood vessel. A *coronary thrombosis* is the closing off of a vessel that feeds the heart muscle. A *cerebral thrombosis* is the closing off of a vessel that feeds the brain (*coronary* means "crowning" [the heart]; *thrombo* means "clot"; the *cerebrum* is part of the brain).

thrombus a stationary clot.

tissues systems of cells working together to perform specialized tasks. Examples are muscles, nerves, blood, and bone.

tocopherol (tuh-KOFF-er-all): a kind of alcohol. The active form of vitamin E is alpha-tocopherol.

tofu (TOE-foo) a curd made from soybeans that is rich in protein, often rich in calcium, and variable in fat content; used in many Asian and vegetarian dishes in place of meat.

tolerance limit the maximum amount of a residue permitted in a food when a pesticide is used according to label directions.

toxicity the ability of a substance to harm living organisms. All substances are toxic if the concentration is high enough.

trace minerals essential mineral nutrients found in the human body in amounts less than 5 grams.

training regular practice of an activity, which leads to physical adaptations of the body with improvement in flexibility, strength, or endurance.

***trans*-fatty acids** fatty acids with unusual shapes that can arise when polyunsaturated oils are hydrogenated.

triglycerides (try-GLISS-er-ides) one of the three main classes of dietary lipids and the chief form of fat in foods. A triglyceride is made up of three units of fatty acids and one unit of glycerol (fatty acids and glycerol are defined later). Triglycerides are also called *triacylglycerols.*

trimester one-third of gestation, about 13 to 14 weeks.

tripeptides (try-PEP-tides) protein fragments that are three amino acids long (*tri* means "three").

type 1 diabetes the type of diabetes in which the person produces no or very little insulin; also known as *juvenile-onset* or *insulin-dependent diabetes,* although some cases arise in adulthood.

type 2 diabetes the type of diabetes in which the person makes plenty of insulin, but the body cells resist insulin's action; also called *adult-onset* or *noninsulin-dependent diabetes.*

—U—

ulcer erosion in the topmost, and sometimes underlying, layers of cells that form a lining. Ulcers of the digestive tract commonly form in the esophagus, stomach, or upper small intestine.

ultrahigh temperature (UHT) a process of sterilizing food by exposing it for a short time to temperatures above those normally used in processing.

underwater weighing a measure of density and volume used to determine body fat content.

unsaturated fatty acid a fatty acid that lacks some hydrogen atoms and has one or more points of unsaturation. An unsaturated fat is a triglyceride that contains one or more unsaturated fatty acids.

urea (yoo-REE-uh) the principal nitrogen-excretion product of metabolism; generated mostly by removal of amine groups from unneeded amino acids or from amino acids being sacrificed to a need for energy.

uterus (YOO-ter-us) the womb, the muscular organ within which the infant develops before birth.

—V—

variety the dietary characteristic of providing a wide selection of foods—the opposite of monotony.

vegetarians people who exclude from their diets animal flesh and possibly other animal products such as milk, cheese, and eggs.

veins blood vessels that carry blood, with the carbon dioxide it has collected, from the tissues back to the heart.

very-low-density lipoproteins (VLDL) lipoproteins that transport triglycerides and other lipids from the liver to various tissues in the body.

villi (VILL-ee, VILL-eye) fingerlike projections of the sheets of cells that line the intestinal tract. The villi make the surface area much greater than it would otherwise be (singular: *villus*).

visceral fat fat stored within the abdominal cavity in association with the internal abdominal organs. Also called *intra-abdominal fat*.

vitamin B_6 a B vitamin needed in protein metabolism. Its three active forms are *pyridoxine, pyridoxal,* and *pyridoxamine*.

vitamin B_{12} a B vitamin that enables folate to get into cells and also helps maintain the sheath around nerve cells. Vitamin B_{12}'s scientific name, not often used, is *cyanocobalamin*.

vitamins organic compounds that are vital to life and indispensable to body functions, but are needed only in minute amounts; noncaloric essential nutrients.

VO_2 max the maximum rate of oxygen consumption by an individual (measured at sea level).

voluntary activities intentional activities (such as walking, sitting, or running) conducted by voluntary muscles.

—W—

wasting the progressive, relentless loss of the body's tissues that accompanies certain diseases and shortens survival time.

water balance the balance between water intake and water excretion, which keeps the body's water content constant.

water intoxication the rare condition in which body water content is too high. Symptoms are headache, muscular weakness, lack of concentration, poor memory, and loss of appetite.

weight cycling repeated rounds of weight loss and subsequent regain, with reduced ability to lose weight with each attempt. Also called *yo-yo dieting*.

weight training (also called **resistance training**) the use of free weights or weight machines to provide resistance for developing muscle strength and endurance. A person's own body weight may also be used to provide resistance as when a person does push-ups, pull-ups, or sit-ups.

Wernicke-Korsakoff (VER-nih-kee KORE-sah-kof) **syndrome** a form of thiamin deficiency affecting the brain tissues. The syndrome is associated with alcohol abuse and is characterized by mental confusion and disorientation, loss of memory, jerky eye movements, and staggering gait.

—X—

xerophthalmia (ZEER-ahf-THALL-me-uh) hardening of the cornea of the eye in advanced vitamin A deficiency that can lead to blindness (*xero* means "dry"; *ophthalm* means "eye").

xerosis (zeer-OH-sis) drying of the cornea; a symptom of vitamin A deficiency.

—Z—

zygote (ZYE-goat) the term that describes the product of the union of ovum and sperm during the first two weeks after fertilization.

A

the inside front cover. The page letters Y, and Z refer to the tables ending on the last pages of the book. The page numbers preceded by A through G are appendix page numbers. The boldfaced page numbers are the pages on which the margin definitions appear. These terms are also defined in the glossary. Page numbers followed by *n* indicate footnotes. Page numbers followed by *t* indicate tables. Italicized page numbers indicate figures.

•••C•••

•••D•••

•••F•••

•••G•••

•••H•••

•••L•••

•••O•••

•••P•••

•••S•••

T

U

•••V•••

•••X•••

•••Y•••

•••Z•••

CHAPTER OPENER ART CREDITS

CO 1: Collection of Frances O. Stem Babinsky
CO 2: Hermitage Museum, St. Petersburg, Russia. Photo © Bridgeman Art Library/SuperStock
CO 3: Oil on canvas, 72.7 × 44.5 cm., The Baltimore Museum of Art: The Cone Collection, formed by Dr. Claribel Cone and Miss Etta Cone of Baltimore, Maryland. BMA 1950.213
CO 4: Gouache on paper, 20 × 33 cm., Victoria & Albert Museum, London. Photo: Art Resource, NY
CO 5: Oil pastel, 30 × 22 in. Creative Growth Art Center, Oakland, CA
CO 6: Collection of Siri von Reis, New York City
CO 7: Oil on canvas, 17 3/8 × 23 in. High Museum of Art, Atlanta, Georgia: Purchase, 75.37
CO 8: Teresa Fasolino/The Newborn Group
CO 9: Oil on canvas, 33 1/2 × 23 in. John Berggruen Gallery, San Francisco, CA
CO 10: Oil on canvas. Musée National Fernand Leger, Biot, France. Photo: Erich Lessing/Art Resource, NY. © 2000 Artists Rights Society (ARS), NY/ADAGP, Paris
CO 11: Oil on canvas, 51 × 68 in. Acquired 1923, The Phillips Collection, Washington, DC
CO 12: Mixed media. Private Collection. Photo: Art Resource, NY. © 2000 Michael Escoffery/Artists Rights Society (ARS), NY
CO 13: Secretaría de Educacción Pública Murals, Mexico City. Courtesy: Banco de Mexico, Fiduciario en el Fideicomiso Relativo a los Museos Diego Rivera y Frida Kahlo; y Instituto Nacional de Bellas Artes y Literatura, Mexico. Photo: ET Archive, London/SuperStock
CO 14: Haitian Private Collection. Photo: © Van Hoorick/SuperStock
CO 15: Gouache with graphite on paper, .565 × .784 mm. (22 1/4 × 30 7/8 in.). Gift of Mr. and Mrs. James T. Dyke, © 1999 Board of Trustees, National Gallery of Art, Washington, DC. and Courtesy of Jacob Lawrence

PHOTO CREDITS

Page **2,** © 1999 PhotoDisc, Inc.; **7,** © Bob Daemmrich/Stock, Boston; **8,** CORBIS/Richard Fukuhara ©; **13,** © Michael Newman/PhotoEdit; **16** (left), © David R. Frazier/Photo Library; **16** (right), © Polara Studios Inc.; **18,** CORBIS/Japack Company ©; **19,** © 1999 PhotoDisc, Inc.; **22,** © Howard Grey/Tony Stone Worldwide; **26,** Courtesy of Marilyn Herbert; **30,** © 1999 PhotoDisc, Inc.; **35,** © Lori Adamski Peak/Tony Stone Imgaes; **36,** © 1999 Daphne Hougard; **40,** © Felicia Martinez/PhotoEdit; **41,** © Felicia Martinez/PhotoEdit; **42** (top), © Tony Freeman/PhotoEdit; **42** (bottom), © Felicia Martinez/PhotoEdit; **43** (top), © Michael Newman/PhotoEdit; **43** (bottom), © Bonnie Kamin/PhotoEdit; **46,** © Tony Freeman/PhotoEdit; **47,** © David Young-Wolff/PhotoEdit; **52** (left), CORBIS/Japack Company ©; **52** (center first three images), © Polara Studios Inc.; **52** (bottom two images), © Thomas Harm and Tom Peterson/Quest Photographic, Inc.; **52** (right), © Thomas Harm and Tom Peterson/Quest Photographic, Inc.; **60,** © Robert Frerck/Tony Stone Worldwide; **62** (left), CORBIS; **62** (right), © 1999 PhotoDisc, Inc.; **71,** © Alan Oddie/PhotoEdit; **81,** © 1999 PhotoDisc, Inc.; **82,** From D. W. Fawcet, *The Cell,* 2d ed. (Philadelphia: Saunders, 1981), color by Kidd & Company; **98,** © Mary Kate Denny/PhotoEdit; **103,** © Steven Rothfeld/Tony Stone Worldwide; **111,** CORBIS/R. W. Jones ©; **115,** © Polara Studios Inc.; **116,** CORBIS/Japack Company ©; **118,** © Thomas Harm and Tom Peterson/Quest Photographic, Inc.; **124,** © Polara Studios Inc.; **137** (top), © David R. Frazier/Photo Library; **137** (bottom), © Thomas Harm and Tom Peterson/Quest Photographic, Inc.; **143,** © Peter Correz/Tony Stone Images; **146** (top left and center, bottom left and center) © Polara Studios Inc.; **146** (top and bottom right), © Thomas Harm and Tom Peterson/Quest Photographic, Inc.; **152,** © Felicia Martinez/PhotoEdit; 155, © 1999 PhotoDisc, Inc.; **156,** © Thomas Harm and Tom Peterson/Quest Photographic, Inc.; **158,** © Polara Studios Inc.; **159,** © Polara Studios Inc.; **160,** CORBIS/Japack Company ©; **167,** CORBIS/Charles O'Rear ©; **173,** © A. Glauberman/Photo Researchers; **176,** © Mary Conner/PhotoEdit; **188,** CORBIS/Japack Company ©; **192** (top and bottom), © Michael Newman/PhotoEdit; **192** (center), Felicia Martinez/PhotoEdit; **195** (top), © Alan Oddie/PhotoEdit; **195** (bottom), United Nations Food and Agriculture Organization/Photo Library; **198,** CORBIS/Japack Company ©; **200,** © Michael Newman/PhotoEdit; **201,** © Michael Newman/PhotoEdit; **203,** © Williams & Edwards/The Image Bank; **205,** © Thomas Harm and Tom Peterson/Quest Photographic, Inc.; **206,** © Thomas Harm and Tom Peterson/Quest Photographic, Inc.; **212** (top), ©Tony Freeman/PhotoEdit; **213** (top), © David Farr/Image Smythe; **213** (bottom), *Nutrition Today,* H. Stanstead, J. Carter, and W. Darby, Nutritional Deficiencies, Nutrition Today Aid #5 (Nutrition Today: Annapolis, MD) 1975; **217,** © Biophoto Association/Science Source/Photo Reserchers; **218,** CORBIS/Fotografia ©; **226,** CORBIS/L. V. Bergman & Associates ©; **227,** *Nutrition Today,* C. Butterworth & G. Blackburn, Hospital Nutrition and How to Assesss the Nutritional Status of a Patient. Nutrition Today Teaching Aid #18 (Nutrition Today: Annapolis, MD) 1975; **230,** © Martin M. Rotker; **231,** *Nutrition Today,* C. Butterworth & G. Blackburn, Hospital Nutrition and How to Assesss the Nutritional Status of a Patient. Nutrition Today Teaching Aid #18 (Nutrition Today: Annapolis, MD) 1975; **235,** CORBIS/L. V. Bergman & Associates ©; **236,** CORBIS/L. V. Bergman & Assosiates ©; **248,** CORBIS/Japack Company ©; **256,** © J. Share/Tony Stone Images; **261,** © Thomas Braise/Tony Stone Worldwide; **264,** © Lawrence Migdale/Stock, Boston; **265,** © Camera M.D. Studios, Inc.; **266,** © Felicia Martinez/PhotoEdit; **270,** © 1999 PhotoDisc, Inc; **281,** © Felicia Martinez/PhotoEdit; **284,** CORBIS/L. V. Bergman & Associates ©; **285,** CORBIS/L. V. Bergman & Associates ©; **287,** © Michael Newman/PhotoEdit; **288,** © Ray Stanyard; **290,** Reproduced with permission of *Nutrition Today* Magazine, P. O. Box 1829, Annapolis, MD 21404, March 1968; **292,** © Dr. P. Marazzi/Science Photo Library/Photo Researchers; **295,** CORBIS/Japack Company ©; **297,** © 1999 PhotoDisc, Inc.; **304,** Courtesy of Gjon Mills; **305,** With permission from Dempster et al., J. bone Min. Res I, 15-21, 1986; **307,** CORBIS/Kevin R. Morris ©; **315,** © Jurgen Reisch/Tony Stone Images; **321,** © Rich Schaff; **322,** © David Young-Wolff/PhotoEdit; **325,** © R. Benali/Liaison Gamma; **329,** © David Madison/Tony Stone Images; **333,** © 1999 PhotoDisc, Inc.; **335,** CORBIS; **343,** CORBIS; **344,** CORBIS/Japack Company ©; **346** (top), © Michael Newman/PhotoEdit; **346** (center and bottom), © Michael Bridwell/PhotoEdit; **350,** © Cleo Photography/PhotoEdit; **356,** © 1999 PhotoDisc, Inc.; **363,** CORBIS/Kevin R. Morris ©; **366,** © David Madison/Tony Stone Images; **369,** © Polara Studios Inc.; **370,** CORBIS/Bill Ross ©; **376,** © 1999 PhotoDisc, Inc.; **378,** CORBIS/Japack Company ©; **379,** © Polara Studios Inc.; **384,** © Tony Freeman/PhotoEdit; **388,** © Tony Freeman/PhotoEdit; **389,** © Michael Newman/PhotoEdit; **398,** Courtesy of Zeneca Pharmaceutical Division, Cheshire, England; **399,** © Science Photo Library/Photo Researchers; **401,** © Kathy Ferguson/PhotoEdit; **404,** CORBIS/Jenny Woodcock, Photolibrary ©; **410,** © Elizabeth Simpson/FPG International LLC; **412,** CORBIS/Paul A. Souders ©; **413,** © Mary Kate Denny/PhotoEdit; **417,** CORBIS/Japack Company ©; **418,** © Tom McCarthy/PhotoEdit; **420,** © Felicia Martinez/PhotoEdit; **430,** © Felicia Martinez/PhotoEdit; **432,** © The Stock Market/Jose I. Pelaez/R, 1999; **434** (photos 1, 2, and 3), Petit Format/Nestle/ Photo Researchers; **434** (photo 4), © Anthony M. Vanelli; **440,** © Bill Bachmann/Stock, Boston; **446,** © 1995 George Steinmetz; **448,** © Myrleen Ferguson Cate/PhotoEdit; **455,** © C. J. Allen/Stock, Boston; **456,** © 1999 PhotoDisc, Inc.; **457,** © Michael Bridwell/PhotoEdit; **458,** © Polara Studios Inc.; **459,** © Jonathan Nourok/PhotoEdit; **460,** Courtesy of Pamela R. Erickson; **461,** CORBIS/Japack Company ©; **463,** CORBIS; **466,** © 1999 PhotoDisc, Inc.; **470,** © Anthony Vanelli; **472,** © Lawrence Migdale/Stock, Boston; **475,** © Tony Freeman/PhotoEdit; **476,** © Thomas Harm and Tom Peterson/Quest Photographic, Inc.; **482,** © Henley and Savage/Tony Stone Worldwide; **483,** © D. M. Phillips/Visuals Unlimited; **489,** Courtesy of Dr. William Evans, from *BioMarkers: The 10 Keys to Prolonging Vitality;* **491,** © Brian Bailey/Tony Stone Images; **497** (top), © D & I MacDonald/PhotoEdit; **497** (bottom), CORBIS/Japack Company ©; **498,** © Susan Van Etten/ PhotoEdit; **499,** © Jeffrey Mark Dunn/Stock, Boston; **501,** © Michael Newman/PhotoEdit; **508,** © Jeff Greenberg/PhotoEdit; **509,** © New England Stock Photo; **515,** © David Young-Wolff/PhotoEdit; **521,** © Joseph Schuyler/ Stock, Boston; **523,** © George Loun/Visual Unlimited; **528,** © Polara Studios Inc.; **531** (top), © Polara Studios Inc.; **531** (bottom), © Joe McDonald/ Visuals Unlimited; **532,** © 1999 PhotoDisc, Inc.; **535** (top), CORBIS/Japack Company ©; **535** (bottom), © Michael Newman/PhotoEdit; **536,** © Thomas Harm and Tom Peterson/Quest Photographic, Inc.; **539,** Smithsonian photo by Antonio Montaner; **541,** © Tony Freeman/PhotoEdit; **543,** © Adrian Arbib/Still Pictures; **549,** © Elsa Peterson/Stock, Boston; **550** (top), © Robert Breener/PhotoEdit; **550** (bottom), © Rick Brown/Stock, Boston; **551,** © Bob Daemmrich/Stock, Boston; **552,** © David Austin/Stock, Boston; **553** (top), © L. Gilbert/Sygma Photos; **553** (bottom), © 1999 PhotoDisc, Inc.; **554,** CORBIS/Craig Aurness ©; **556,** © David Austin/Stock, Boston; **558,** Courtesy of Sun Frost; **560,** Courtesy of NASA; **562,** © Paul Stover/Tony Stone Images; **563** (left), Tony Stone Worldwide; **563** (right), © Larry Ulrich/Tony Stone Images; **564,** Oklahoma Publishing Co., May 1997.